TARGET ORGAN TOXICOLOGY SERIES

Series Editors

A. Wallace Hayes, John A. Thomas, and Donald E. Gardner

TOXICOLOGY OF THE KIDNEY, THIRD EDITION
Joan B. Tarloff and Lawrence H. Lash, editors, 1200 pp., 2004

OVARIAN TOXICOLOGY
Patricia B. Hoyer, editor, 248 pp., 2004

CARDIOVASCULAR TOXICOLOGY, THIRD EDITION
Daniel Acosta, Jr., editor, 616 pp., 2001

NUTRITIONAL TOXICOLOGY, SECOND EDITION
Frank N. Kotsonis and Maureen A. Mackey, editors, 480 pp., 2001

TOXICOLOGY OF SKIN
Howard I. Maibach, editor, 558 pp., 2000

TOXICOLOGY OF THE LUNG, THIRD EDITION
Donald E. Gardner, James D. Crapo, and Roger O. McClellan, editors, 668 pp., 1999

NEUROTOXICOLOGY, SECOND EDITION
Hugh A. Tilson and G. Jean Harry, editors, 386 pp., 1999

TOXICANT–RECEPTOR INTERACTIONS: MODULATION OF SIGNAL TRANSDUCTIONS AND GENE EXPRESSION
Michael S. Denison and William G. Helferich, editors, 256 pp., 1998

TOXICOLOGY OF THE LIVER, SECOND EDITION
Gabriel L. Plaa and William R. Hewitt, editors, 444 pp., 1997

FREE RADICAL TOXICOLOGY
Kendall B. Wallace, editor, 454 pp., 1997

ENDOCRINE TOXICOLOGY, SECOND EDITION
Raphael J. Witorsch, editor, 336 pp., 1995

CARCINOGENESIS
Michael P. Waalkes and Jerrold M. Ward, editors, 496 pp., 1994

(Continued)

DEVELOPMENTAL TOXICOLOGY, SECOND EDITION
Carole A. Kimmel and Judy Buelke-Sam, editors, 496 pp., 1994

IMMUNOTOXICOLOGY AND IMMUNOPHARMACOLOGY, SECOND EDITION
Jack H. Dean, Michael I. Luster, Albert E. Munson, and Ian Kimber, editors, 784 pp., 1994

NUTRITIONAL TOXICOLOGY
Frank N. Kotsonis, Maureen A. Mackey, and Jerry J. Hjelle, editors, 336 pp., 1994

OPHTHALMIC TOXICOLOGY
George C. Y. Chiou, editor, 352 pp., 1992

TOXICOLOGY OF THE BLOOD AND BONE MARROW
Richard D. Irons, editor, 192 pp., 1985

TOXICOLOGY OF THE EYE, EAR, AND OTHER SPECIAL SENSES
A. Wallace Hayes, editor, 264 pp., 1985

CUTANEOUS TOXICITY
Victor A. Drill and Paul Lazar, editors, 288 pp., 1984

Target Organ Toxicology Series

Toxicology of the Kidney

Third Edition

Edited by
Joan B. Tarloff
Lawrence H. Lash

CRC PRESS

Boca Raton London New York Washington, D.C.

Library of Congress Cataloging-in-Publication Data

Catalog record is available from the Library of Congress

Direct all inquiries to CRC Press, 2000 N.W. Corporate Blvd., Boca Raton, Florida 33431.

Visit the CRC Press Web site at www.crcpress.com

International Standard Book Number 0-415-24864-7
Printed in the United States of America 1 2 3 4 5 6 7 8 9 0
Printed on acid-free paper

CONTENTS

PREFACE

In the last 10 years, there has been an explosion of research at the cellular and molecular levels. This generation of new knowledge has been fueled at least in part by breakthrough methodologies in the fields of molecular biology, proteomics and genomics. It is fitting that the second edition of *Toxicology of the Kidney*, published in 1993, undergoes a facelift.

The new edition provides an update on many of the key elements of the previous editions including *in vitro* models for studying renal function and toxicity and classic nephrotoxicants. In addition, we now incorporate cutting edge information on newly emerging areas of renal research. Among these are a consideration of mechanisms of cell injury, signaling pathways, and the interface between basic renal science and clinical outcomes, and biomarkers of renal disease.

The book takes a more mechanistic approach to the study of nephrotoxicity and includes topics such as apoptosis vs. necrosis, endocytosis, cell repair and growth, oxidative processes and cellular energetics. We are particularly pleased to include chapters dealing with clinical aspects of nephrotoxicity and information concerning risk assessment.

This book is intended for basic scientists interested in renal function and disease processes as well as clinicians who desire a deeper biochemical and molecular understanding of renal toxicity. Because of the developing and emerging status of many of these topics, reviews in the literature may not be readily available. Hence, this volume should provide a concise compendium of many of the key topics that will continue to play a central role in the understanding and treatment of nephrotoxicity in the 21st century.

We thank Robin Goldstein and Peter Bach for setting up the framework and the colleagues who contributed outstanding chapters for this updated edition. We enjoyed working on this book and are delighted in the outcome.

CONTRIBUTORS

George L. Bakris, MD
Rush Presbyterian/St. Luke's Medical Center, Department of Preventive Medicine, Chicago, Illinois

Alexei G. Basnakian, MD, PhD
Department of Internal Medicine, UAMS College of Medicine, University of Arkansas for Medical Sciences, Central Arkansas Veterans Healthcare System, Little Rock, Arkansas

Ramon G.B. Bonegio, MD
Renal Section, Boston Medical Center and Boston University School of Medicine, Boston, Massachusetts

Douglas Charney, MD
Department of Pathology, St. Lukes-Roosevelt Hospital, New York, New York

Marc E. De Broe, MD, PhD
University of Antwerp, Department of Nephrology-Hypertension, Antwerpen, Belgium

Marjo de Graauw, MSc
Division of Toxicology, Leiden/Amsterdam Center for Drug Research, Leiden University, Leiden, The Netherlands

Wolfgang Dekant, PhD
Department of Toxicology, University of Würzburg, Würzburg, Germany

Gary L. Diamond, PhD
Syracuse Research Corporation, Syracuse, New York

Daniel R. Dietrich, PhD
Environmental Toxicology, University of Konstanz, Konstanz, Germany

Sue M. Ford, PhD
St. John's University, College of Pharmacy & Allied Health Professions, Jamaica, New York

A. Jay Gandolfi, PhD
Department of Pharmacology and Toxicology, College of Pharmacy, University of Arizona, Tucson, Arizona

Jay F. Harriman, PhD
WIL Research Laboratories, Inc., 1407 George Road, Ashland, Ohio

Susan G. Emeigh Hart, VMD, PhD
AstraZeneca Pharmaceuticals, Wilmington, Delaware

Raoef Imamdi, MSc
Division of Toxicology, Leiden/Amsterdam Center for Drug Research, Leiden University, Leiden, The Netherlands

Nicholas Kaperonis, MD
Department of Preventive Medicine, Rush Hypertension/Clinical Research Center, Chicago, Illinois

Gur P. Kaushal, PhD
Department of Internal Medicine, UAMS College of Medicine, University of Arkansas for Medical Sciences, Central Arkansas Veterans Healthcare System, Little Rock, Arkansas

Kanwar Nasir M. Khan, DVM, PhD
Pfizer R & D, Ann Arbor, Michigan

Lewis B. Kinter, PhD
AstraZeneca Pharmaceuticals, Wilmington, Delaware

Donald S. Kirkpatrick, PhD
Department of Pharmacology and Toxicology, College of Pharmacy, University of Arizona, Tucson, Arizona

Mark Krause, MD
Department of Medicine, University of Illinois Medical Center at Chicago, Chicago, Illinois

Lawrence H. Lash, PhD
Department of Pharmacology, Wayne State University School of Medicine, Detroit, Michigan

Serrine S. Lau, PhD
Department of Pharmacology and Toxicology, University of Arizona College of Pharmacy, Tucson, Arizona

Wilfred Lieberthal, MD
Amgen, Thousand Oaks, California

Margaret R.E. McCredie
Department of Preventive and Social Medicine, University of Otago, Dunedin, New Zealand

Marie-Paule Mingeot-Leclercq
Cellular and Molecular Pharmacology Unit, Catholic University of Louvain, Brussels, Belgium

Bruce A. Molitoris, MD
Roudebush Veterans Affairs Medical Center, Indianapolis, Indiana

Terrence J. Monks, PhD
Department of Pharmacology and Toxicology, University of Arizona, College of Pharmacy, Tucson, Arizona

Marina Noris, Chem.Pharm.D.
Laboratory of Immunology and Genetics, of Rare Diseases and Organ Transplantation, Clinical Research Center for Rare Diseases, Mario Negri Institute for Pharmacological Research, Bergamo, Italy

Evelyn O'Brien, PhD
Environmental Toxicology, University of Konstanz, Konstanz, Germany

Alan R. Parrish, PhD
Department of Medical Pharmacology & Toxicology, College of Medicine, Texas A&M University System Health Science Center, College Station, Texas

Norberto Perico, MD
Laboratory of Drug Development, Clinical Research Center for Rare Diseases, Mario Negri Institute for Pharmacological Research, Bergamo, Italy

Didier Portilla, MD
Department of Internal Medicine, UAMS College of Medicine, University of Arkansas for Medical Sciences, Central Arkansas Veterans Healthcare System, Little Rock, Arizona

Lorraine C. Racusen, MD
Department of Pathology, The Johns Hopkins Medical Institutions, Baltimore, Maryland

Gary O. Rankin, PhD
Department of Pharmacology, Marshall University School of Medicine, Huntington, West Virginia

Giuseppe Remuzzi, MD
Negri Bergamo Laboratories, Mario Negri Institute for Pharmacological Research, Bergamo, Italy

Jeff M. Sands, MD
Emory University School of Medicine, Atlanta, Georgia

Rick G. Schnellmann, PhD
Department of Pharmaceutical Sciences, Medical University of South Carolina, Charleston, South Carolina

Rani S. Sellers, DVM, PhD
Purdue Pharma, Ardsley, New York

Hélène Servais
Cellular and Molecular Pharmacology Unit, Catholic University of Louvain, Brussels, Belgium

Sudhir V. Shah, MD
Department of Internal Medicine, UAMS College of Medicine, University of Arkansas for Medical Sciences, Central Arkansas Veterans Healthcare System, Little Rock, Arkansas

John H. Stewart
Department of Renal Medicine, Westmead Hospital, Sydney, Australia

David P. Sundin, PhD
Department of Medicine/Nephrology Division, Indiana University School of Medicine, Indianapolis, Indiana

Joan B. Tarloff, PhD
Department of Pharmaceutical Sciences, Philadelphia College of Pharmacy, University of the Sciences in Philadelphia, Philadelphia, Pennsylvania

Paul M. Tulkens
Cellular and Molecular Pharmacology Unit, Catholic University of Louvain, Brussels, Belgium

Norishi Ueda
Tokoname City Hospital, Aichi, Japan

Monica A. Valentovic, PhD
Department of Pharmacology, Marshall University School of Medicine, Huntington, West Virginia

Bob van de Water, PhD
Division of Toxicology, Leiden/Amsterdam Center for Drug Research, Leiden University, Leiden, The Netherlands

Jill W. Verlander, DVM
University of Florida College of Medicine, Gainesville, Florida

Rudolfs K. Zalups, PhD
Division of Basic Medical Sciences, Mercer University School of Medicine, Macon, Georgia

THE EDITORS

Lawrence H. Lash, PhD, is professor of pharmacology at Wayne State University in Detroit, Michigan.

Dr. Lash received his B.A. in biology from Case Western Reserve University in Cleveland, Ohio in 1980 and his Ph.D. in biochemistry from Emory University in Atlanta, Georgia in 1985. After a postdoctoral fellowship in pharmacology and toxicology at the University of Rochester in Rochester, New York in the laboratory of M.W. Anders, Dr. Lash joined the faculty at Wayne State University in 1988 as an assistant professor.

Dr. Lash has served on the Alcohol and Toxicology-4 study section of the National Institutes of Health and as an ad hoc reviewer for grants from the National Institute of Diabetes and Digestive and Kidney Diseases, the National Institute of Environmental Health Sciences, the National Science Foundation, Veterans Administration, and Medical Research Council of Canada. He served as a consultant for the National Research Council subcommittee on urinary toxicology, helping to prepare a book on biomarkers in urinary toxicology, and to the U.S. Environmental Protection Agency in their human health risk assessments for trichloroethylene and perchloroethylene. Dr. Lash has been the chair of the Toxicology Division of ASPET, serves on the ASPET Scientific Council, and has been president of the Michigan Regional Chapter of the Society of Toxicology. He is currently an associate editor for *Toxicology and Applied Pharmacology* and *The Journal of Pharmacology and Experimental Therapeutics* and is on the editorial board of *Drug Metabolism and Disposition.*

Dr. Lash's primary research interests are in the areas of development of *in vitro* models for the study of mechanisms of nephrotoxicity, trichloroethylene metabolism and mechanisms of renal cellular injury, and glutathione metabolism and transport. He has most recently developed the use of human proximal tubular cells for the study of trichloroethylene-induced renal cellular injury and has defined the carriers involved in and studied the molecular biology of transport of glutathione into renal mitochondria. He has published more than 120 peer-reviewed papers, reviews, and book chapters, has edited four books, and has presented invited lectures at 15 international meetings. Dr. Lash's research has been funded by the National Institutes of Health, the U.S. Environmental Protection Agency, and the pharmaceutical industry.

Dr. Lash is a member of the American Society of Biochemistry and Molecular Biology, the American Society of Pharmacology and Experimental Therapeutics, the Society of Toxicology, the American Society of Nephrology, the International Society for the Study of Xenobiotics, and the American Association for the Advancement of Science.

Joan B. Tarloff, Ph.D., is professor of Pharmacology and Toxicology at the Philadelphia College of Pharmacy, University of the Sciences in Philadelphia, Pennsylvania.

Dr. Tarloff received a B.S. in pharmacy from the University of Toledo in Toledo, Ohio in 1972 and a Ph.D. in Pharmacology from the Medical College of Ohio in Toledo, Ohio, in 1982. She completed postdoctoral fellowships in physiology at the Medical College of Ohio and in toxicology at SmithKline Beckman in Philadelphia, Pennsylvania; in the laboratory of Jerry Hook. Dr. Tarloff joined the faculty at the Philadelphia College of Pharmacy & Science in 1988 as an assistant professor and director of the undergraduate toxicology program.

Dr. Tarloff serves as an ad hoc reviewer for small business grants through special emphasis panels formed by the National Institutes of Health and is a member of the editorial advisory board for *Toxicology and Applied Pharmacology*. She is a councilor for the *in vitro* specialty section of the Society of Toxicology and secretary/treasurer for the Toxicology Division of ASPET. In addition, she is a past-president of the Mid-Atlantic Society of Toxicology.

Dr. Tarloff is heavily involved in undergraduate and graduate toxicology education. She serves on a Society of Toxicology subcommittee examining the impact of undergraduate toxicology education and has served as a consultant for program evaluations at St. John's University and at Long Island University.

Dr. Tarloff's primary research interests are defining the mechanisms by which acetaminophen and its nephrotoxic metabolite, para-aminophenol, cause cell death. Her studies have been funded by Hoffman-LaRoche and the National Institute of General Medical Sciences.

Dr. Tarloff is a member of the American Society of Pharmacology and Experimental Therapeutics, the Society of Toxicology, the International Society for the Study of Xenobiotics, the American Association of Colleges of Pharmacy, and the American Association for the Advancement of Science.

I

Basic Principles

1

Anatomy and Physiology of the Kidneys

Jeff M. Sands and Jill W. Verlander

INTRODUCTION

Toxins are generally cleared from the body either by the kidney or by the liver and gastrointestinal tract. The kidney is required to maintain total body salt, water, potassium, and acid–base balance, while excreting waste products and toxins. The kidney is therefore responsible for preserving the body's internal environment. The kidney uses three general mechanisms to accomplish these goals: glomerular filtration, tubular reabsorption, and tubular secretion. A basic understanding of these processes is important because they can affect the excretion or retention of various toxins. The movement of water and solute across the glomerular capillary wall to form an ultrafiltrate of plasma is called glomerular filtration; the rate at which this occurs is called the glomerular filtration rate (GFR). The movement of a substance from tubular fluid back into the plasma is called tubular reabsorption and results in retention of that substance within the body. The movement of a substance from plasma into the tubular fluid is called tubular secretion and results in excretion of that substance into the urine.

Several highly specialized nephron segments composed of distinct epithelial cell types and with specific transport properties are found in the kidney (Figure 1.1). Importantly, renal function results not only from

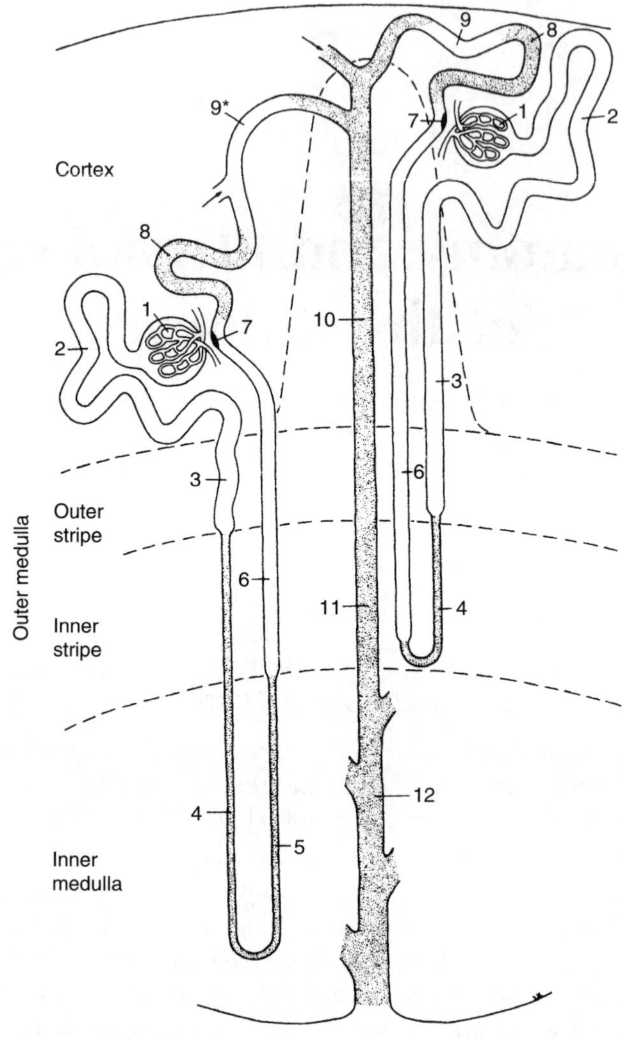

Figure 1.1 A long-looped (left) and a short-looped (right) nephron together with the collecting system (not drawn to scale). Within the cortex, the medullary ray is delineated by a dashed line. (1) Renal corpuscle including Bowman's capsule and the glomerulus (glomerular tuft), (2) proximal convoluted tubule, (3) proximal straight tubule, (4) thin descending limb, (5) thin ascending limb, (6) medullary thick ascending limb, (7) macula densa located within the final portion of the cortical thick ascending limb, (8) distal convoluted tubule, (9) connecting tubule, (10) cortical collecting duct, (11) outer medullary collecting duct, (12) inner medullary collecting duct. (Reproduced with permission of the American Physiological Society from *Am J Physiol* 254:F1–F8 (1988).)

the flow of tubular fluid from one nephron segment to the next, but also from the three-dimensional organization of the kidney. This is particularly important in the cortex where the macula densa abuts other elements of the juxtaglomerular apparatus (JGA) in the glomerulus, and in the medulla where both the loops of Henle and the vasa recta are arranged in hair-pin or U-shaped configurations and are adjacent to collecting ducts.

Due to the enormous quantity of blood filtered by the kidney, about 180 l/day, the kidney is very susceptible to toxic injury. Approximately 25% of a person's cardiac output goes to the kidneys, thereby exposing the kidney to toxins in excess of the exposure of other tissues. In addition, toxins can be concentrated in renal tissue by the processes of tubular reabsorption and the counter-current arrangement of structures in the medulla. This frequently results in renal plasma or tubular fluid toxin concentrations that exceed those measured in systemic plasma, which increases the potential for renal toxicity.

The kidney also plays a pivotal role in the production and regulation of several important hormones, including 1,25-$(OH)_2$-cholecalciferol (vitamin D), erythropoietin, and renin. The kidney performs the 1-hydroxylation of 25-OH-cholecalciferol to create its active form, 1,25-$(OH)_2$-cholecalciferol. This hormone is critically important to body calcium and phosphorus homeostasis. Patients with chronic renal failure are deficient in 1,25-$(OH)_2$-cholecalciferol and develop secondary hyperparathyroidism unless given replacement therapy. Chronic renal failure also results in anemia due to deficient erythropoietin production, necessitating replacement therapy with erythropoietin. Renin is produced by cells in the JGA in response to decreases in intrarenal perfusion. Renin is released into the systemic circulation, after which it is converted by angiotensin converting enzyme in several tissues to angiotensin II. Angiotensin II stimulates aldosterone production by the adrenal gland. It is also a potent vasoconstrictor that helps to restore renal perfusion by increasing blood pressure, and enhances sodium uptake in the proximal tubule. Aldosterone increases intravascular volume and blood pressure by stimulating sodium reabsorption by the collecting duct. However, inappropriate stimulation of the renin–angiotensin II–aldosterone system can lead to hypertension or exacerbate congestive heart failure.

In summary, the kidneys are involved in several critical bodily functions, including maintenance of blood pressure via regulation of sodium excretion, maintenance of plasma osmolality via regulation of water excretion, regulation of acid–base balance, regulation of plasma potassium, calcium, and phosphorus, and elaboration of hormones such as 1,25-$(OH)_2$-cholecalciferol, erythropoietin, and renin. This chapter will provide an overview of renal anatomy and physiology, with an emphasis on the major ultrastructural and transport features of each nephron segment. This chapter can not, however, provide an in-depth review of every aspect of renal anatomy or physiology, and the reader is referred to the major nephrology textbooks for a more-detailed discussion of specific aspects of renal function.

MAMMALIAN KIDNEY: OVERALL STRUCTURAL ORGANIZATION

Mammals have two kidneys, located retroperitoneally and posteriorly, near the lower ribs. In humans, the kidney may be biopsied percutaneously by inserting a biopsy needle through the back and into the kidney under ultrasound guidance. Each kidney normally receives its blood supply from a single renal artery, which originates from the aorta and drains into a single renal vein that connects to the inferior vena cava. Normally, urine empties from each kidney into a single ureter and then into the bladder. However, there can be anatomic anomalies in which a person will have more than one ureter draining a single kidney.

The kidney can be subdivided into three regions: cortex, outer medulla, and inner medulla (Figure 1.1 and Figure 1.2). The inner medulla can be further subdivided into the base and the tip (or papilla). Most small mammals, such as rat, mouse, and rabbit, have a single papilla in each kidney, and so are termed unipapillate. The papilla is shaped like an inverted pyramid. It descends into the renal pelvis, which is an upward-expanded extension of the ureter. Larger mammals, including humans, have more than one papilla per kidney, and so are termed multipapillate. Each papilla descends into a renal fornix and these fornices merge to form the renal pelvis. Urine exits from the tip of the papilla(e) via the ducts of Bellini into the renal pelvis in all mammals, regardless of whether they are unipapillate or multipapillate.

Cortex

The renal cortex can be divided into the cortical labyrinth and the medullary rays (Figure 1.1 and Figure 1.2). Glomeruli, proximal

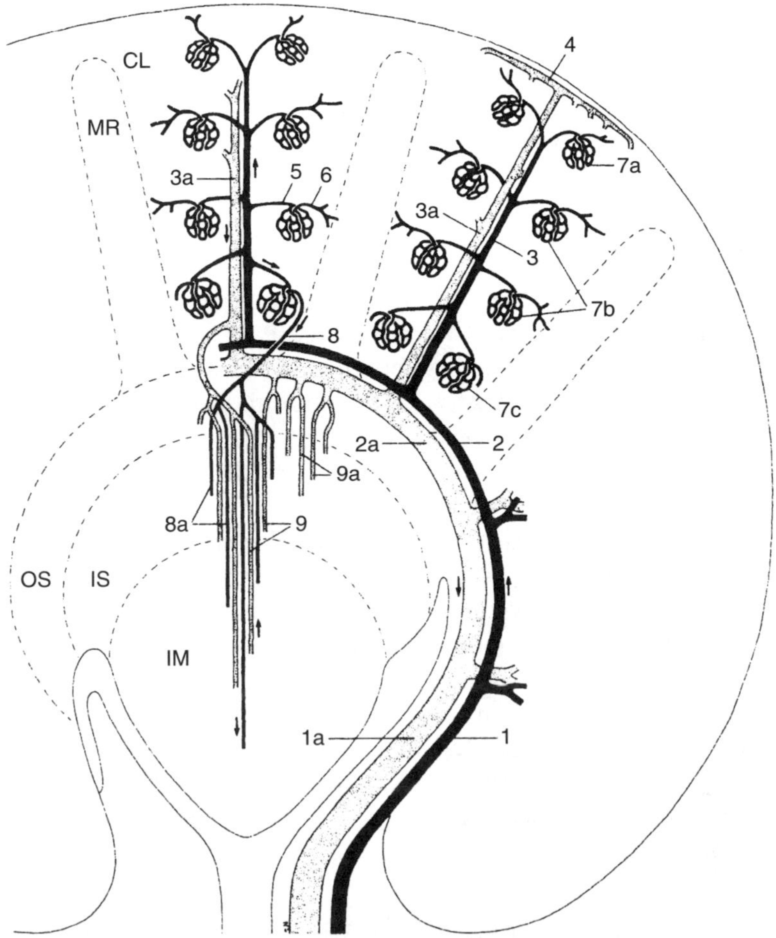

Figure 1.2 Course and distribution of intrarenal blood vessels (peritubular capillaries are not shown, diagram is not drawn to scale). Within the cortex, the medullary rays (MR) are delineated from the cortical labyrinth (CL) by a dashed line. (OS) Outer stripe of the outer medulla, (IS) inner stripe of the outer medulla, (IM) inner medulla, (P) renal pelvis (which is lined by the papillary surface epithelium). (1,1a) interlobar artery and vein, (2,2a) arcuate artery and vein, (3,3a) cortical radial artery and vein, (4) stellate vein, (5) afferent arteriole, (6) efferent arteriole, (7a,7b,7c) superficial, midcortical, and juxtamedullary glomerulus, (8,8a) juxtamedullary efferent arteriole, descending vasa recta, (9,9a) ascending vasa recta within a vascular bundle, ascending vasa recta independent of a vascular bundle. (Reproduced with permission of the American Physiological Society from *Am J Physiol* 254:F1–F8 (1988).)

convoluted tubules, macula densa, distal convoluted tubules, connecting tubules, initial collecting ducts, interlobular arteries, and afferent and efferent capillary networks are located in the cortical labyrinth. Proximal straight tubules, cortical thick ascending limbs, and cortical collecting ducts are located in the medullary rays.

Outer Medulla

The outer medulla is divided into two subregions: the outer stripe and the inner stripe (Figure 1.1 and Figure 1.2). The outer stripe is located just inside the cortex and contains proximal straight tubules (also called pars recta or thick descending limbs), medullary thick ascending limbs, and outer medullary collecting ducts. The inner stripe is located inside the outer stripe and contains thin descending limbs, thick ascending limbs, and outer medullary collecting ducts. The outer and inner stripes are thus distinguished by the presence of proximal tubules in the outer stripe but not in the inner stripe, and by thin descending limbs in the inner stripe but not in the outer stripe.

The blood supply to both outer medullary subregions is provided by the descending and ascending vasa recta (Figure 1.2). However, there is some difference in how the vessels are organized within the outer vs. inner stripes: vascular bundles, which are cone-shaped aggregations of descending and ascending vasa recta, are formed in the outer stripe; these vascular bundles become quite prominent in the inner stripe (Lemley and Kriz, 1987).

Inner Medulla

The inner medulla can be divided into two subregions: the base (which is adjacent to the outer medulla), and the tip (or true papilla). Both subregions contain thin descending limbs, thin ascending limbs, inner medullary collecting ducts, and descending and ascending vasa recta (Figure 1.1 and Figure 1.2). As the collecting ducts descend through the inner medulla, they merge, ultimately forming the ducts of Bellini, which discharge urine from the papillary tip into the pelvis. The deepest portion of the inner medulla is called the tip or papilla, which is found in the renal pelvis and is covered by the papillary surface epithelium (Figure 1.2).

RENAL VASCULATURE

Each kidney typically has a single renal artery, which originates from the aorta and divides into several interlobar arteries as it enters the renal sinus (Figure 1.2). The interlobar arteries enter the kidney tissue at the corticomedullary border and give rise to the arcuate arteries, which in turn give rise to the interlobular arteries. The interlobular arteries ascend into the cortex, where they divide into afferent arterioles that supply blood to the glomerular capillaries. After exiting the renal corpuscle, the glomerular capillaries reform into a second arteriolar network, the efferent arterioles.

Efferent arterioles give rise to different structures, depending on the location of the glomerulus (Figure 1.2). Efferent arterioles from superficial and midcortical nephrons form peritubular capillary networks that surround proximal and distal convoluted tubules located in the cortical labyrinth. However, an efferent arteriole originating from a specific glomerulus does not necessarily provide blood to the peritubular capillary surrounding the proximal tubule that originated from that glomerulus. The precise organization of the postglomerular vasculature is complex and poorly understood.

Efferent arterioles from the juxtamedullary nephrons give rise to descending vasa recta (Figure 1.2). Descending vasa recta give rise to capillary networks at various levels within the medulla and supply it with oxygenated blood. As vasa recta descend deeper into the medulla, blood flow slows and so reduces oxygen delivery – the inner medulla is essentially anaerobic. This is clinically important because the low pO_2, along with the high osmolality, produce a strong stimulus for SS (sickle cell anemia), or even SA (sickle trait), causing erythrocytes to become sickled. The sickled erythrocytes further reduce medullary blood flow, cause microinfarcts, and eventually result in papillary necrosis, both in patients with sickle cell anemia and in those with sickle trait.

Venous drainage within the medulla is provided by the ascending vasa recta (Figure 1.2). Countercurrent exchange between descending and ascending vasa recta allows for rapid osmotic equilibration between arterial and venous capillary blood. This is important because any increase in the osmolality of the blood exiting from the ascending vasa recta will dissipate the effectiveness of countercurrent exchange. A decrease in countercurrent exchange will result in solute being carried out of the medulla and back into the systemic circulation. This loss of solute from the medulla means that some of the work performed to generate a hypertonic medulla is wasted, and will result in a reduction in urine concentrating ability. At the corticomedullary junction, ascending vasa recta merge with the venules draining the cortex to form interlobular veins. The interlobular veins merge to form arcuate veins, then interlobar veins, and ultimately the renal vein within the renal sinus. The renal vein drains directly into the inferior vena cava.

NEPHRON HETEROGENEITY

Nephrons are divided into superficial, midcortical, and juxtamedullary, based on the location of their glomerulus (Figure 1.1 and Figure 1.2). Glomeruli located near the surface of the kidney give rise to superficial nephrons, which generally descend only into the outer medulla. These nephrons are generally short-looped nephrons that lack thin ascending limbs and do not penetrate into the inner medulla; their loops of Henle bend at the inner–outer medullary border and their thin descending limbs connect directly to medullary thick ascending limbs. Superficial nephrons have proximal straight tubules that are located centrally within the medullary rays. Animals that produce highly concentrated urine, such as the rat, have thin descending limbs that descend within the vascular bundles, allowing countercurrent exchange between ascending vasa recta and thin descending limbs from short-looped nephrons, especially within the inner stripe (Kriz and Bankir, 1983; Lemley and Kriz, 1987).

Glomeruli located deep within the cortex, near the corticomedullary border, give rise to juxtamedullary nephrons. These nephrons are generally long-looped nephrons that enter the inner medulla and have thin ascending limbs (Figure 1.1 and Figure 1.2). In general, mammals that produce highly concentrated urine have longer papillae and a higher percentage of long-looped nephrons.

Glomeruli located between the superficial and juxtamedullary glomeruli give rise to midcortical nephrons (Figure 1.2). These nephrons may be either long-looped or short-looped. In general, the deeper the location of the glomerulus, the more likely it is that its nephron will be long-looped, i.e., descend into the inner medulla.

GLOMERULUS

Anatomy

The filtering unit of the kidney, the glomerulus, contains several capillary loops (Figure 1.3). The glomerulus plus Bowman's capsule constitute the renal corpuscle. Blood enters the glomerulus through the afferent arteriole and exits via the efferent arteriole, with glomerular capillaries located between these two arterioles. Glomerular capillaries are lined by large flat endothelial cells with rounded fenestra that are spanned by thin diaphragms. Glomerular endothelial cells are covered by negatively charged polyanionic surface glycoproteins (Tisher and Brenner, 1989). The glomerular basement membrane is located between the endothelial cells and the glomerular epithelial cells (Figure 1.4). The glomerular filtration barrier therefore consists of

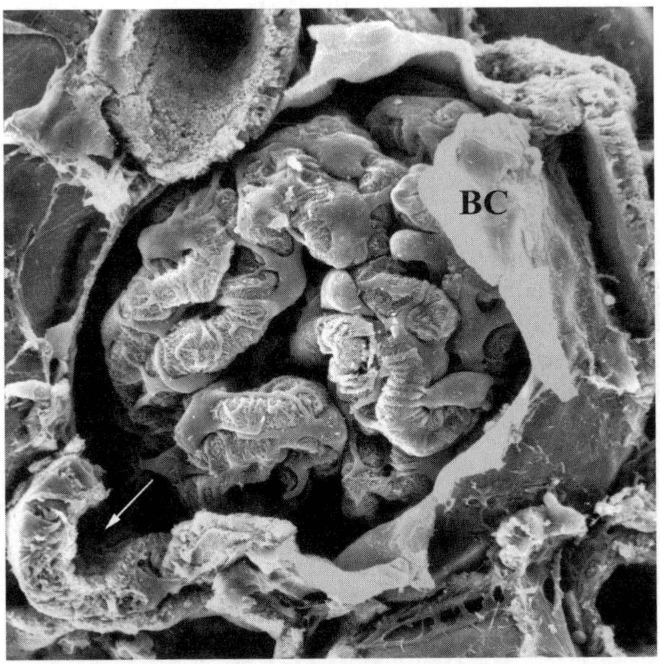

Figure 1.3 Scanning electron micrograph of a rat glomerulus. The Bowman's capsule (BC) is cut away, revealing the capillary tuft. The proximal tubule (arrowed) emerges from the urinary pole of the Bowman's capsule. (Original magnification ×1,600.)

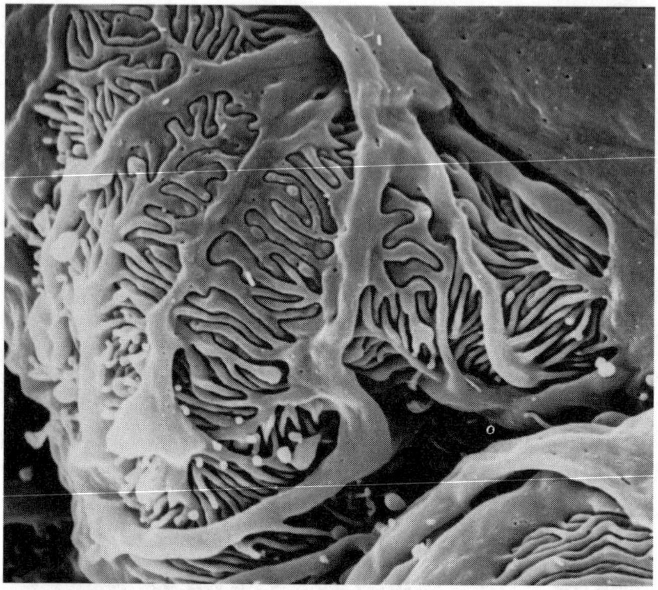

Figure 1.4 Scanning electron micrograph of a visceral epithelial cell. The complex interdigitations of the podocyte foot processes are clearly shown. (Original magnification ×14,000.)

three structures: endothelial cells, glomerular basement membrane, and visceral epithelial cells (Figure 1.5). Bowman's capsule is a cup-shaped extension of the proximal tubule. An ultrafiltrate of plasma is formed in Bowman's space, located between the visceral epithelial cells (which are also called podocytes) and the parietal epithelial cells (simple squamous epithelial cells), by filtering the capillary blood across this filtration barrier (Figure 1.6). The parietal epithelial cell layer gives rise to the proximal convoluted tubule (Figure 1.7).

The hilum of the glomerulus contains extraglomerular mesangial cells and mesangial matrix that surround the afferent and efferent arterioles as they enter and exit the glomerulus (Figure 1.8). Mesangial cells are similar to vascular smooth muscle cells. Glomerular mesangial cells in cell culture contract or relax in response to several vasoactive compounds when tested *in vitro*. Therefore, glomerular mesangial cells are thought to be involved in regulating glomerular hemodynamics, although this has not been established *in vivo*. Glomerular mesangial cells are thought to proliferate during

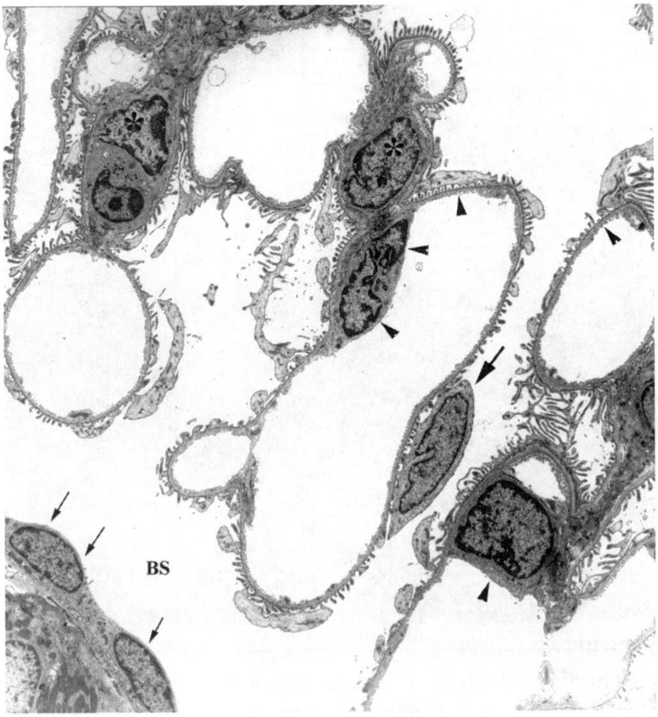

Figure 1.5 Transmission electron micrograph of a rat glomerulus. Several glomerular capillary loops are shown. The lumens are lined with fenestrated endothelial cells (arrowheads). The capillary loops are covered with visceral epithelial cells, including the cell body (large arrow) and cytoplasmic extensions, known as foot processes. Mesangial cells (asterisks) are interposed between capillary loops. The capillary tuft is covered by Bowman's capsule, which is lined with parietal epithelial cells (small arrows); the ultrafiltrate forms in Bowman's space (BS). (Original magnification ×3,500.)

glomerular injury and hence play a role in the formation of proliferative glomerular lesions. This hypothesis also lacks *in vivo* confirmation.

Physiology

The glomerular ultrafiltrate normally contains water and solute, and other small molecules and electrolytes normally found in plasma. The ultrafiltrate from a healthy glomerulus will not contain plasma proteins

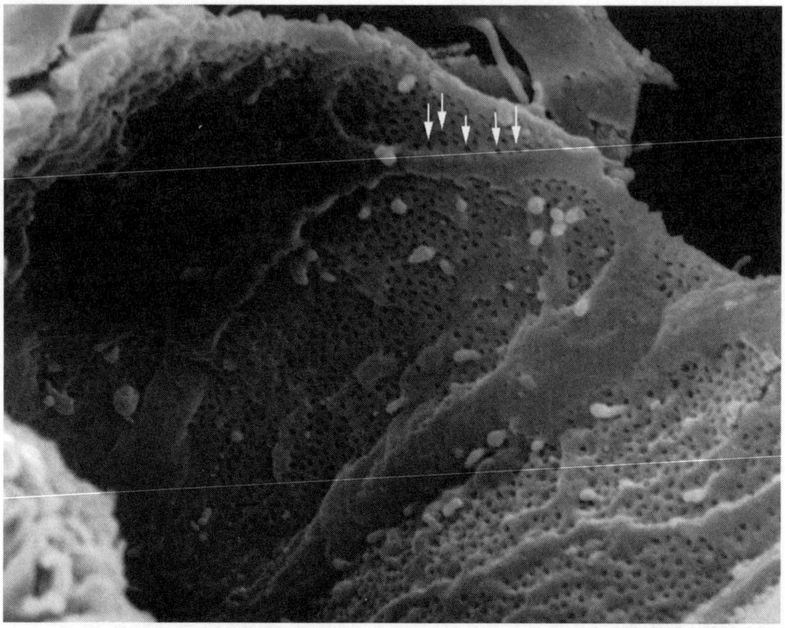

Figure 1.6 Scanning electron micrograph of the lumen of a rat glomerular capillary. Most of the endothelial surface is perforated by numerous small fenestrae (arrows). (Original magnification ×19,000.)

because the glomerular capillary wall forms both a size- and a charge-selective barrier that prevents passage of plasma proteins. In disease states, both the size and charge barrier properties of the glomerulus are compromised, resulting in the appearance of proteins and cells in the ultrafiltrate and urine (Maddox and Brenner, 1991; Valtin, 1973). These abnormal urinary constituents can be detected by urinalysis.

Filtration across the glomerular capillary bed is driven by the same Starling forces that drive fluid movement across the systemic capillary bed. However, glomerular capillaries connect or lie between two arterioles, hence the hydrostatic pressure in glomerular capillaries is affected by both the afferent and efferent arteriolar tone. Hydrostatic pressure within glomerular capillaries normally exceeds that in Bowman's space, favoring the production of an ultrafiltrate. However, oncotic pressure within glomerular capillaries will increase as ultrafiltration proceeds. In some species, such as the rat, the oncotic

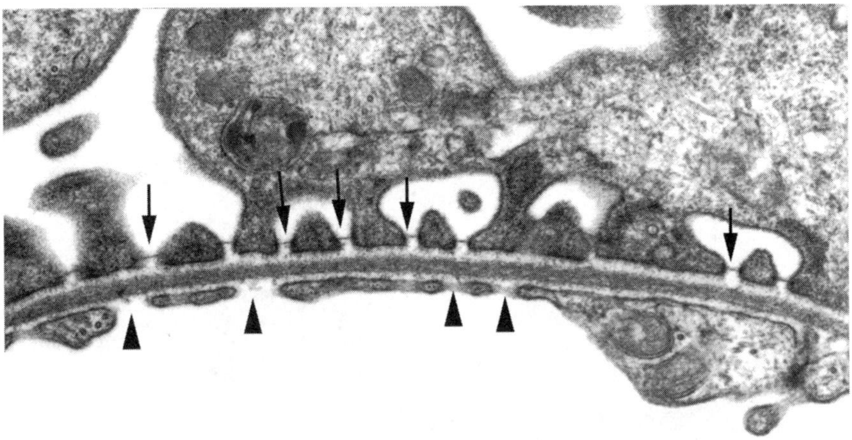

Figure 1.7 Transmission electron micrograph of a rat glomerular capillary wall. The foot processes of the visceral epithelial cells are connected by slit diaphragms (arrows). The capillary endothelial cells are perforated with numerous fenestrae (arrowheads). The basement membrane is interposed between the visceral epithelial cells and the endothelial cells, and is made up of three layers. The central layer, the lamina densa, is visible as the dark band between the cell layers; the lamina rara externa and lamina rara interna are the light layers next to the epithelial cells and the endothelial cells, respectively. (Original magnification ×23,800.)

pressure can increase until it equals the hydrostatic pressure, at which point ultrafiltration ceases; this condition is called filtration equilibrium. It is not known whether filtration equilibrium occurs in humans (Valtin, 1973).

GFR can be calculated as:

$$\begin{aligned} \text{GFR} &= K_f S(\Delta\text{hydraulic pressure} - \text{oncotic pressure}) \\ &= K_f S[(P_{gc} - P_{bs}) - \sigma(\pi_p - \pi_{bs})] \end{aligned}$$

where K_f is the ultrafiltration coefficient (or unit permeability) of the glomerular capillary wall, S is the surface area available for filtration, P is the hydraulic pressure, π is the oncotic pressure, gc is glomerular capillary, bs is Bowman's space, p is plasma; and σ is the reflection coefficient of proteins across the capillary wall.

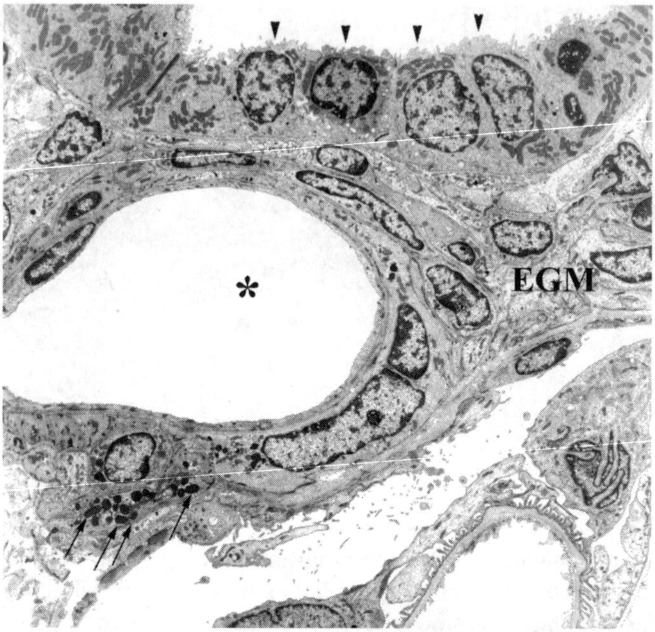

Figure 1.8 Transmission electron micrograph of the juxtaglomerular apparatus in rat kidney. The macula densa cells (arrowheads) are taller and narrower than the adjacent cortical thick ascending limb cells, and little cytoplasm separates the macula densa cell nuclei. The extraglomerular mesangium (EGM) is located between the afferent (asterisk in lumen) and efferent arterioles, the macula densa, and the capillary tuft (bottom). In the wall of the afferent arteriole are juxtaglomerular cells, which contain electron-dense renin granules (arrows). (Original magnification ×4,000.)

To measure GFR, either in humans or in animals, one uses a substance that is filtered by the glomerulus but is neither reabsorbed nor secreted by the tubules, e.g., inulin or iothalamate. These substances are infused intravenously to achieve a stable plasma level and their clearance is then measured. GFR is calculated as:

$$\text{Clearance} = \frac{(U_i V)}{P_i}$$

where i is the infused substance (inulin or iothalamate), U is the urine concentration, V is the urine flow rate, and P is the plasma concentration.

It is important to measure GFR, both for evaluating renal function and for its use in determining drug dosage. In clinical practice endogenous creatinine clearance is generally used, rather than infusing an exogenous substance such as inulin or iothalamate. Creatinine has the advantage of being safe, inexpensive, and an endogenous molecule. However, it undergoes tubular secretion, making it an imperfect substance for measuring GFR. If one has not measured GFR, one can estimate GFR using the Cockcroft-Gault formula (Cockcroft and Gault, 1976) or the following formulas (which assume a stable rate of creatinine production):

$$\text{For men: GFR} = \frac{[28 - (\text{age in years}/6)] \times \text{body weight (in kg)}}{P_{\text{creatinine}}}$$

$$\text{For women: GFR} = \frac{[22 - (\text{age in years}/9)] \times \text{body weight (in kg)}}{P_{\text{creatinine}}}$$

Elderly individuals will frequently have a reduced GFR, even if they have a normal serum creatinine, which makes it imperative that GFR be measured, or at least estimated using one of the preceding formulas, to avoid overdosing a geriatric patient (Johnson et al., 2002).

The kidney maintains glomerular filtration and renal blood flow over a wide range of systemic blood pressures due to autoregulatory mechanisms. Although the precise mechanisms are not known, angiotensin II and prostaglandins are important components of renal autoregulation. Toxins or drugs that interfere with either angiotensin II (such as angiotensin II type-1 receptor antagonists or angiotensin converting enzyme inhibitors) or prostaglandin production (such as nonsteroidal anti-inflammatory drugs) therefore prevent appropriate autoregulation.

In addition to autoregulation of renal blood flow, two other mechanisms are present to preserve glomerular filtration: glomerulotubular balance, and tubuloglomerular feedback. Glomerulotubular balance is a mechanism that allows proximal tubule reabsorption to vary with changes in glomerular filtration. An increase in glomerular filtration results in an increase in the oncotic pressure within the efferent arterioles, which increases reabsorption from the proximal convoluted tubules. Conversely, if glomerular filtration decreases, oncotic pressure decreases, and proximal tubular reabsorption decreases. Therefore, glomerulotubular balance tends to stabilize the delivery of tubular fluid to the end of the proximal convoluted tubule.

Tubuloglomerular feedback is a regulatory mechanism that is controlled by the macula densa cells of the JGA. The JGA is composed of both vascular and tubular elements, including the afferent and efferent arterioles, the extraglomerular mesangium, and the macula densa. The macula densa cells form a specialized portion of the cortical thick ascending limb that abuts extraglomerular mesangium at the hilum of the glomerulus between the efferent and afferent arterioles. Located within the walls of the afferent arterioles are specialized cells, called juxtaglomerular cells, that contain renin granules. Juxtaglomerular cells release renin in response to decreases in renal perfusion pressure. Renin is converted to angiotensin II, which causes vasoconstriction, increases sodium reabsorption in the proximal tubule and thick ascending limb, and stimulates aldosterone production. Both vasoconstriction and the increase in sodium reabsorption mediated by angiotensin II and aldosterone increase blood pressure, thereby increasing renal perfusion. The increase in renal perfusion will increase hydrostatic pressure within the glomerular capillaries and restore glomerular filtration. Therefore, tubuloglomerular feedback is a mechanism that allows the kidney to sense decreases in intravascular volume and activate the renin–angiotensin system to restore glomerular filtration. The exact signal that the macula densa cells sense is not known, but it is currently thought to involve some combination of tubular fluid Na^+, Cl^-, or NaCl concentration or osmolality, or a combination of both.

Since the glomeruli filter approximately 180 liters of plasma per day, they are often damaged as "innocent bystanders." One example is the trapping of immune complexes within the glomerular membrane; the subsequent activation of complement results in damage to the glomerulus. Another example is toxicity from aminoglycoside antibiotics. Aminoglycosides neutralize the anionic charges in the glomerular membrane, thus eliminating the charge-barrier. This can result in ultrastructural changes, reduced glomerular filtration, and acute renal failure (Hook and Hewitt, 1986). One of the earliest clinical manifestations of aminoglycoside-induced acute renal failure is proteinuria, presumably due to destruction of the charge-barrier that normally prevents loss of negatively charged plasma proteins.

PROXIMAL TUBULE

Proximal Convoluted Tubule

Anatomy

The proximal tubule originates from the parietal epithelial cell layer of the glomerulus (Figure 1.9) and is commonly divided into two portions, the proximal convoluted tubule (PCT) and the proximal straight tubule (PST). It is actually composed of three distinct subsegments: S_1, S_2, and S_3 (or P_1, P_2, and P_3) (Maunsbach, 1966, 1973). The initial portion of the proximal tubule is the S_1 subsegment (Figure 1.10), which is located entirely within the PCT. Cells in the S_1 subsegment have a tall brush border (apical plasma membrane), an extensively interdigitated basolateral plasma membrane, and a well-developed vacuolar lysosomal system. As is typical of cells performing extensive active transport, S_1 cells have numerous mitochondria that are predominantly located near the basolateral plasma membrane.

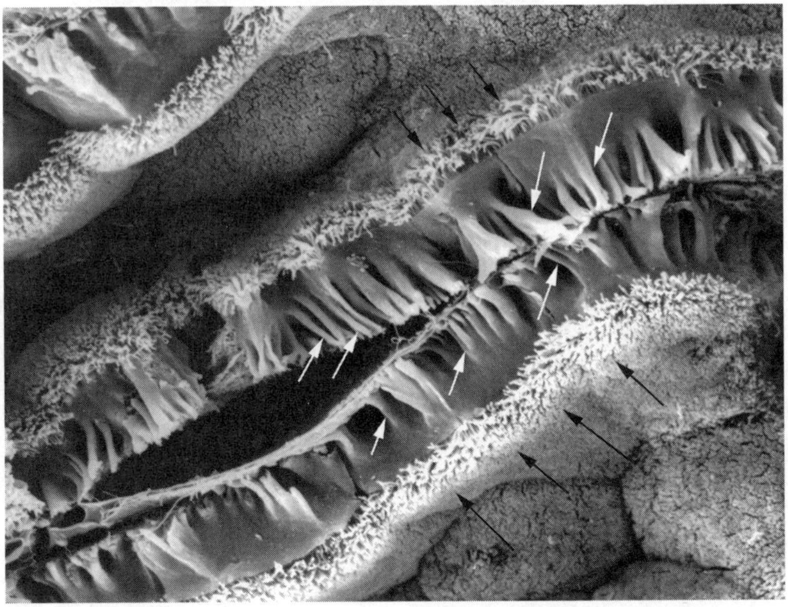

Figure 1.9 Scanning electron micrograph of rat proximal tubules illustrating the lush brush border on the luminal surface of the epithelium (black arrows) and the basolateral plasma membrane infoldings (white arrows). (Original magnification ×3,400.)

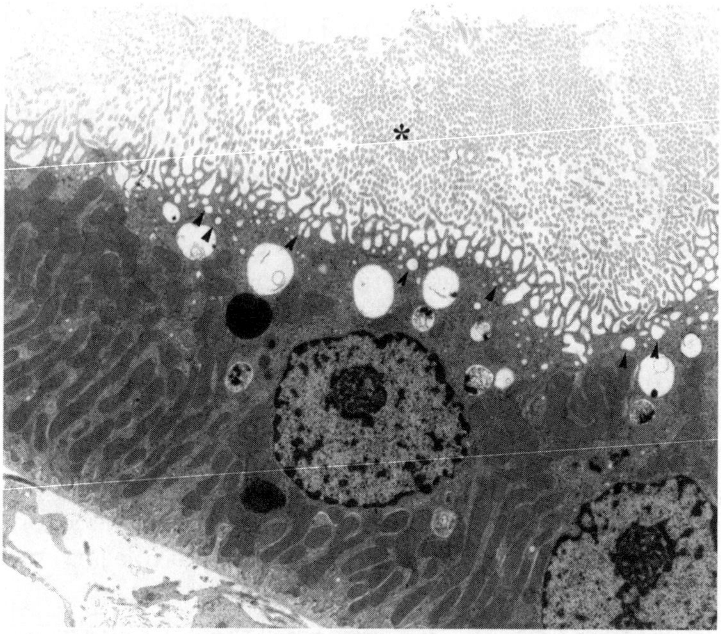

Figure 1.10 Transmission electron micrograph of the S_1 segment of the rat proximal tubule. The S_1 segment is a tall epithelium with a long, apical brush border (star), numerous apical endocytic vesicles (arrowheads), and numerous vertically oriented mitochondria interposed between extensive basolateral plasma membrane infoldings. (Original magnification ×7,200.)

S_1 cells have a high metabolic rate, which makes them particularly susceptible to ischemic or toxic damage. S_1 cells also metabolize many small proteins that appear in glomerular ultrafiltrate by reabsorbing them from the tubular fluid and into their intracellular compartment. This intracellular uptake makes proximal tubule cells especially vulnerable to injury by toxins, such as heavy metals and aminoglycoside antibiotics.

Both the PCT and PST contain S_2 cells (Figure 1.11). The transition from S_1 to S_2 is gradual, beginning around 1 mm from the glomerulus in the rat. S_2 cells are not as tall as S_1 cells and have fewer mitochondria and a shorter brush border. In S_2 cells, the basolateral plasma membrane infoldings are not as deep as in S_1, and along the basement membrane there are many shallow plasma membrane infoldings that create small cytoplasmic extensions called micropedici.

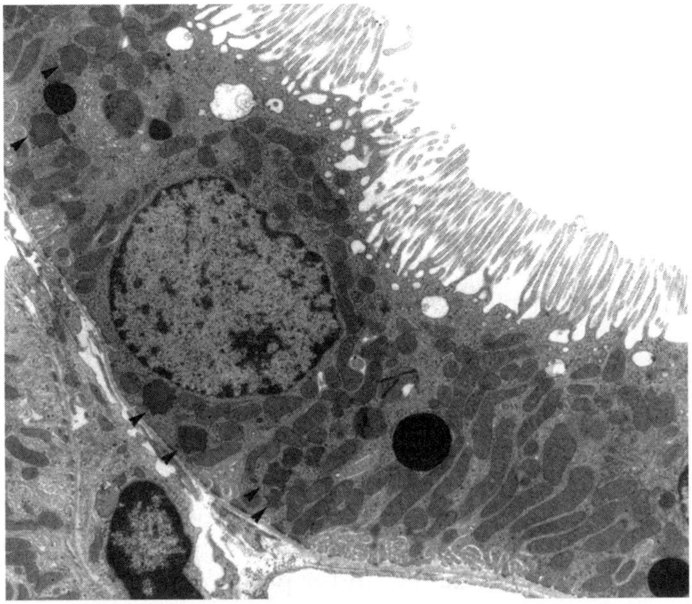

Figure 1.11 Transmission electron micrograph of the S_2 segment of the rat proximal tubule. Compared with S_1 and S_3, the S_2 segment has a lower epithelium and a shorter brush border. The mitochondria and basolateral plasma membrane infoldings are less prominent than in S_1 and basal micropedici are frequently present. There are numerous peroxisomes (arrowheads) in this cell, which suggests that it is from the proximal straight tubule (pars recta). (Original magnification ×9,100.)

The relative sizes of lysosomal vacuoles in S_1 and S_2 cells can vary with gender and species. For example, in male rats the lysosomes in S_2 cells are larger than in S_1 cells (Maunsbach, 1966, 1973).

Physiology

Over half of the glomerular ultrafiltrate is reabsorbed by the PCT, making it the nephron segment responsible for high volume reabsorption. The PCT reabsorbs solute and water isosmotically. Water is reabsorbed through aquaporin-1 (AQP1) water channels located in both the apical and basolateral plasma membranes (Nielsen et al., 1993; Preston et al., 1992); reabsorption of NaCl and $NaHCO_3$ provide the osmotic driving force for water reabsorption. A variety of transcellular and paracellular mechanisms mediate NaCl and $NaHCO_3$ reabsorption, including: Na^+–H^+ antiporters (NHE), Na–HCO_3 cotransporters (NBC), and Cl^--formate exchangers. Almost 100% of the filtered bicarbonate is reabsorbed under normal conditions.

The PCT reabsorbs most of the amino acids, glucose, and small peptides that are filtered by the glomerulus (Berry and Rector Jr., 1991). Many of the transport proteins responsible for the secondary active, Na^+-coupled transport of amino acids, glucose, and small peptides have been cloned and several isoforms of each transporter have been identified. These Na^+-coupled cotransporters all take advantage of the low intracellular Na^+ concentration generated by the Na^+-pump to drive active transport of the desired substrate, i.e., amino acids, glucose, or small peptides. The need to actively reabsorb large amounts of these substrates places a high metabolic demand on the PCT cells and makes them very sensitive to anoxic or toxic injury. Damage to the PCT cells presents clinically as Fanconi's syndrome with aminoaciduria, glucosuria, and a proximal (type II) renal tubular acidosis.

The PCT also has a multidrug resistance (MDR) transporter and both secretes and reabsorbs numerous organic anions and cations (Kartner and Ling, 1989). Secretion of organic anions, such as *p*-aminohippurate (PAH), occurs by Na^+-dependent secondary active transport across the basolateral plasma membrane from peritubular capillaries into PCT cells. One example of basolateral plasma membrane organic anion transport is the Na^+-α-ketoglutarate cotransporter coupled to an α-ketoglutarate-organic anion countertransporter. This combination of transporters results in net Na^+-organic anion uptake across the basolateral plasma membrane (Pritchard, 1995); transport into the PCT lumen across the apical plasma membrane usually does not require active transport. A facilitated transporter for organic anions and an OH^-/urate$^-$ exchanger are examples of apical plasma membrane secretory transport pathways (Pritchard, 1995).

The major source of cytochrome P450 in kidney is the PCT (Badr, 1995). Organic anions, such as PAH, increase cytochrome P450-mediated arachidonic acid metabolism (Badr, 1995). However, a causal link has not been established between organic anion secretion and cytochrome P450-mediated arachidonic acid metabolism (Badr, 1995).

Sodium-dependent secondary active transporters also mediate organic cation secretion. Organic cations generally enter the cell by a facilitated transport pathway located in the basolateral plasma membrane and are secreted across the apical plasma membrane into the tubule lumen by an amiloride-inhibitable Na^+–H^+ antiporter coupled to a proton-organic cation exchanger. This combination of transport processes results in net Na^+-organic cation exchange. A second apical plasma membrane secretory mechanism is an organic cation ATPase (Pritchard, 1995). Because many pharmaceuticals or toxins are organic cations or anions, they are able to be transported on these transporters, making the PCT very sensitive to toxic injury due to accumulation within the cell.

Peptides and proteins that are smaller than albumin are normally filtered across the glomerulus (Johnson and Maack, 1995). The peptides can be reabsorbed by Na^+-coupled secondary active transport pathways, while the small proteins can be reabsorbed by nonspecific endocytic reabsorption pathways. PCT cells take up many small peptides, including insulin, glucagon, and other small peptide hormones. Thus, whenever glomerular filtration is reduced, as occurs in renal insufficiency, the filtration and subsequent degradation of these peptide hormones by the PCT is reduced, resulting in an increase in their plasma half-life. These small peptides are degraded within lysosomal vacuoles present in PCT cells. Two lysosomal enzymes, cathepsin B and cathepsin L, are very abundant in the PCT, but are also found in the PST (Olbricht et al., 1986).

Finally, the PCT reabsorbs divalent cations, such as calcium and magnesium. Calcium is reabsorbed in parallel with sodium. Magnesium is also reabsorbed passively in the PCT, but it is less permeable than calcium and therefore needs to achieve a tubular fluid to plasma ratio of 1.5 before it is reabsorbed (Quamme and Dirks, 1980).

Proximal Straight Tubule

Anatomy

The length of any individual PST depends on the location of its glomerulus within the cortex (Figure 1.1) because all PSTs undergo an abrupt transition to thin descending limbs at the junction of the outer and inner stripes of the outer medulla (Woodhall et al., 1978). A PST originating from a superficial or midcortical nephron contains two cell types, S_2 and S_3, whereas a PST from a juxtamedullary nephron has predominantly S_3 cells (Figure 1.12). As a PST descends from the cortex into the outer stripe of the outer medulla, its cell type changes

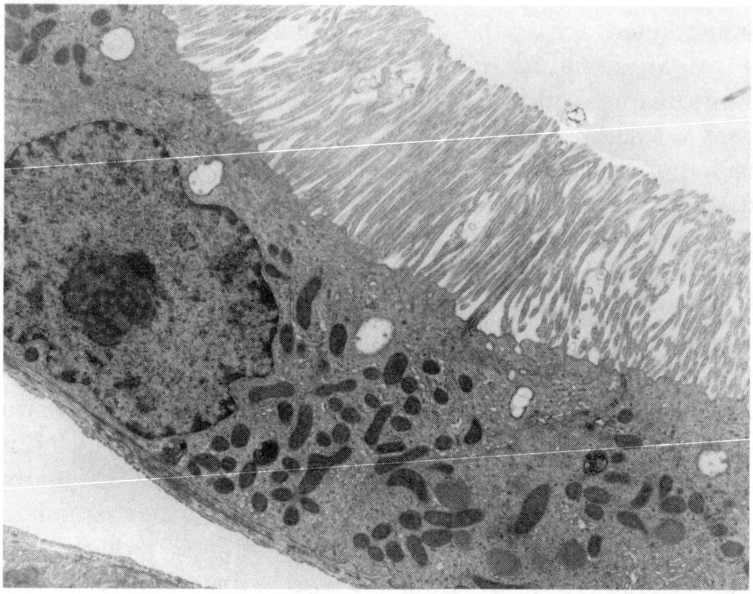

Figure 1.12 Transmission electron micrograph of the S_3 segment of the rat proximal tubule. The S_3 segment is a tall epithelium with a very long brush border and relatively few endocytic vesicles and mitochondria compared with S_1 and S_3. (Original magnification ×7,900.)

from S_2 to S_3; only S_3 proximal tubule cells are found in the outer stripe of the outer medulla (Woodhall et al., 1978). Compared with S_1 or S_2 cells, S_3 cells are cuboidal and have the longest brush border, fewest mitochondria, fewest basolateral plasma membrane invaginations, and least developed endocytic apparatus (Maunsbach, 1966). Numerous peroxisomes are found in both S_2 and S_3 cells in the PST (Maunsbach, 1966).

Physiology

The reabsorption rates for almost all solutes are lower in PSTs than PCTs. The PST has secondary active, Na^+-coupled transporters that mediate the reabsorption of amino acids, glucose, ketone bodies, pyruvate, lactate, and short chain fatty acids (Berry and Rector Jr., 1991). The PST has an extremely high osmotic water permeability that is mediated by AQP1 water channels (Nielsen et al., 1993; Preston et al.,

1992). The rabbit PST has a low passive urea permeability (Knepper, 1983a, 1983b). In addition, one study found active urea secretion in the PST (Kawamura and Kokko, 1976), although a second study did not (Knepper, 1983a, 1983b).

Potassium transport is heterogeneous in the PST, both when comparing PSTs from superficial vs. juxtamedullary nephrons, and along the length of a PST from superficial nephrons (Work et al., 1982). The highest passive K^+ permeability is found in PSTs from juxtamedullary nephrons, which would favor significant K^+ secretion if the medullary interstitial K^+ concentration exceeds that of tubular fluid (Work et al., 1982). Net K^+ secretion is present in both the S_2 and S_3 subsegments of PSTs from superficial nephrons and in PSTs from juxtamedullary nephrons (which are almost exclusively made up of S_3 cells) (Work et al., 1982).

The PST reabsorbs bicarbonate at only 20 to 25% of the rate seen in the PCT but bicarbonate reabsorption is greater in PSTs from juxtamedullary than superficial nephrons (Warnock and Burg, 1977). Because the HCO_3^- concentration in fluid leaving the PCT has already been reduced to a low level, the PST does not normally need to contribute to net HCO_3^- reabsorption. A spontaneous luminal disequilibrium pH, which results from the absence of functional luminal carbonic anhydrase, is present in rabbit S_3, but not in rabbit S_2 or rat PSTs (Garvin and Knepper, 1987; Kurtz et al., 1986); a disequilibrium pH favors NH_3 diffusion into the PST lumen (Kurtz et al., 1986).

LOOP OF HENLE

Thin Descending Limb

Anatomy

Thin descending limbs (tDLs) are present in both short-looped and long-looped nephrons (Figure 1.1). They are conventionally divided into three subtypes in rabbit, rat, and hamster: types I, II, and III (Kriz, 1988), but the chinchilla has an additional papillary subsegment (type III distal) located in the deepest 20% of the inner medulla of the longest tDL (Chou et al., 1993).

Only a single type of tDL cell, called the type I tDL cell, is found in short-looped nephrons. The type I cell forms a squamous epithelium with little cellular interdigitation and deep tight junctions (Schwartz and

Venkatachalam, 1974; Schwartz et al., 1979). In long-looped nephrons, the tDL contains two cell types. The proximal portion contains cells with well-developed lateral interdigitation and shallow tight junctions, called type II tDL cells. The cell type changes to type III tDL cells as the tDL of long-looped nephrons enters the inner medulla (Figure 1.13). This classification holds true for the species that have been studied in detail, such as rat and hamster, but may not be true in other species.

Until recently, it was thought that tDLs and thin ascending limbs (tALs) were sequentially arranged. However, studies have shown the existence of mixed-type thin limbs (Pannabecker et al., 2000). These thin limbs have alternating portions of descending and ascending limb morphology, and immunohistochemistry shows that the transport proteins present in a portion of a mixed-type thin limb correlate with the morphological appearance (Pannabecker et al., 2000). It therefore seems that if a portion of a mixed-type thin limb has descending limb morphology, it will also have descending limb function.

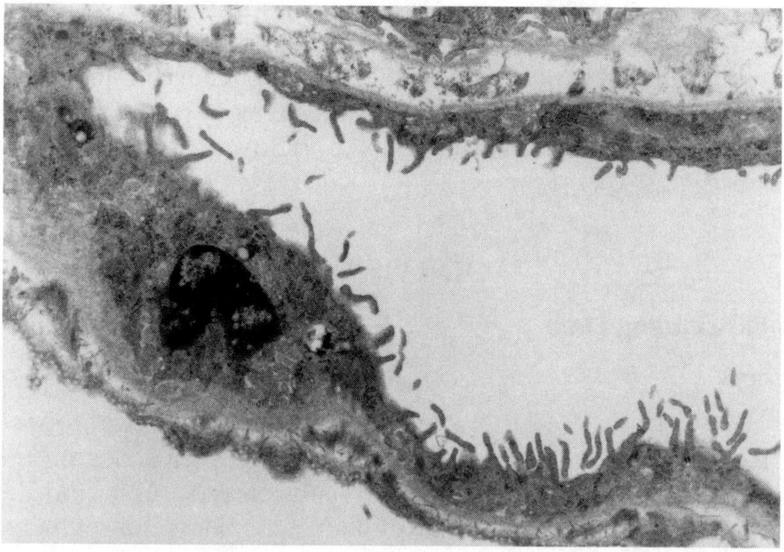

Figure 1.13 Transmission electron micrograph of a thin descending limb in the inner medulla of a rat. A type III epithelium is illustrated, which exhibits prominent apical plasma membrane microprojections. (Original magnification ×12,200.)

Physiology

The osmotic water permeability of types I, II, and III tDL cells is extremely high in rat, rabbit, hamster, and chinchilla and is mediated by AQP1 water channels in both the apical and basolateral plasma membranes (Chou et al., 1993; Nielsen et al., 1993, 1995; Preston et al., 1992; Sabolic et al., 1992). There is enough AQP1 protein in tDL cells to explain the measured rates of transepithelial water transport (Maeda et al., 1995). The chinchilla type III distal tDL is the only tDL subsegment that has a low osmotic water permeability (Chou and Knepper, 1992; Chou et al., 1993). This pre-bend segment could be an important site of solute reabsorption in the inner medulla (Layton and Davies, 1993; Layton et al., 1996).

Urea permeability is heterogeneous in different tDL subsegments. Urea permeability is low in rabbit types I and II, chinchilla type II, and hamster types I and II tDL (Chou and Knepper, 1993; Imai et al., 1984, 1988; Kokko, 1972; Stoner and Roch-Ramel, 1979). In rat types I and II tDL, urea permeability is higher than in the other species but is still relatively low (Morgan and Berliner, 1968). Urea permeability is higher in hamster and chinchilla type III tDL (Chou and Knepper, 1993; Imai et al., 1988a, 1988b; Layton et al., 1996), and is quite high in chinchilla type III distal tDL (Chou and Knepper, 1993). The urea reflection coefficient is close to unity in both hamster and rabbit tDLs, indicating that solvent drag of urea does not occur (Imai et al., 1984; Kokko, 1970).

Sodium permeability is low in hamster types I and III tDL (Imai et al., 1984, 1988b) and rabbit types I and II tDL (Abramow and Orci, 1980; Kokko, 1970; Rocha and Kokko, 1973a, 1973b). However, hamster type II and chinchilla types III and III distal tDL have high Na^+ permeabilities (Chou and Knepper, 1993; Imai et al., 1984, 1988b). Na^+-pump activity is very low in all tDL segments that have been measured (Katz, 1986; Terada and Knepper, 1989). The NaCl reflection coefficient is close to unity in rabbit types I and II and hamster type III tDL (Kokko, 1970), but 0.83 in hamster type II tDL (Imai et al., 1984; Kokko, 1970; Tabei and Imai, 1987).

Thin Ascending Limb

Anatomy

Only long-looped nephrons have a tAL, which begins slightly before the bend of Henle's loop in most species. It abruptly becomes the medullary

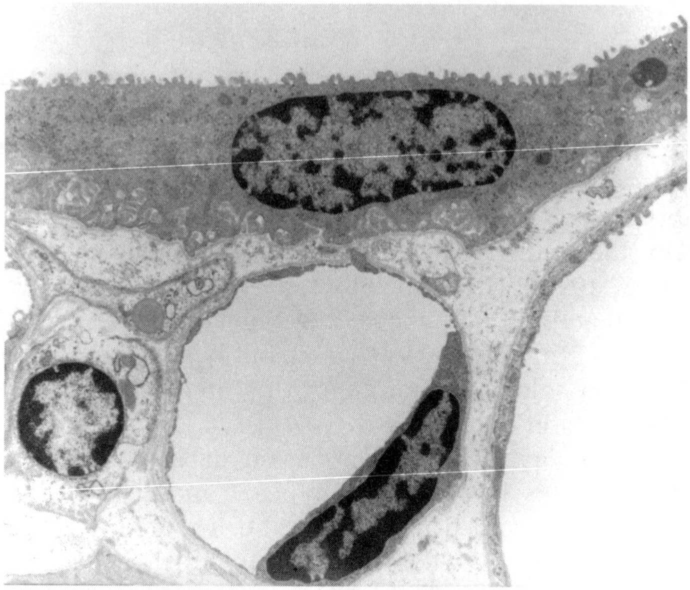

Figure 1.14 Transmission electron micrograph of structures in the rat initial inner medulla, including a principal cell in the initial inner medullary collecting duct ($IMCD_1$)(top), an interstitial cell (below left), a vasa recta (below center), and a thin ascending limb of Henle's loop with the typical morphology of type IV epithelial cells (below right). The vasa recta capillary endothelial cell is fenestrated, indicating that it is from the ascending portion of the vasa recta. (Original magnification ×10,300.)

thick ascending limb at the inner–outer medullary border. The tAL contains flat, moderately interdigitated cells called type IV cells (Figure 1.14) which are connected by shallow tight junctions (Schwartz et al., 1979). More recent studies have shown the existence of mixed-type thin limbs, and the portions of a mixed-type thin limb that have ascending limb morphology have ascending limb function (Pannabecker et al., 2000).

Physiology

The tAL is water impermeable in all species studied (Chou and Knepper, 1992; Chou et al., 1993; Imai, 1977; Imai and Kokko, 1974) and does not express any aquaporin water channels (Ecelbarger et al., 1995; Frigeri et al., 1995; Fushimi et al., 1993; Maeda et al., 1995;

Nielsen et al., 1995; Sabolic et al., 1992; Terris et al., 1995). Although the tAL has a urea permeability that is lower than its NaCl permeability (Chou and Knepper, 1993; Imai, 1977; Kondo and Imai, 1987a, 1987b; Morgan and Berliner, 1968), it is higher than the value required for operation of the passive mechanism for concentrating urine, especially in the chinchilla tAL (Chou and Knepper, 1993; Layton et al., 1996).

The tAL has a very low level of Na^+-pump activity (Katz, 1986; Terada and Knepper, 1989) that is unlikely to support active Na^+ transport (Kondo et al., 1993). However, some *in vivo* studies have reported active Na^+ transport in rat and hamster tALs (Marsh and Azen, 1975; Pennell et al., 1974). The tAL does have a high passive NaCl permeability in all species studied (Chou and Knepper, 1993; Imai, 1977; Imai and Kokko, 1974; Kondo and Imai, 1987a, 1987b; Morgan and Berliner, 1968), which is thought to occur paracellularly since no apical plasma membrane transport pathway has been demonstrated (Koyama et al., 1991; Takahashi et al., 1995a). Chloride transport occurs transcellularly in the tAL, mediated by the ClC-K1 Cl^- channel, which is present in both the apical and basolateral plasma membranes (Liu et al., 2002; Uchida et al., 1995). Chloride transport is stimulated by vasopressin (Takahashi et al., 1995b).

Thick Ascending Limb

Anatomy

The thick ascending limb (TAL) is divided into two subsegments: the medullary thick ascending limb (mTAL) located in the outer medulla, and the cortical thick ascending limb (cTAL) located in the medullary rays within the cortex (Figure 1.15 and Figure 1.16). The mTAL originates at the inner–outer medullary border, either from the tAL of long-looped nephrons or from the tDL of short-looped nephrons. By definition, it becomes the cTAL at the junction of the outer medulla and cortex, although there is no abrupt transition in the cell type (Woodhall et al., 1978). Long-looped (juxtamedullary) nephrons have a longer mTAL than short-looped (superficial or midcortical) nephrons since the latter complete their hairpin turn within the outer medulla (Figure 1.1). Adjacent mTAL cells are extensively interdigitated and joined to one another by shallow tight junctions (Woodhall et al., 1978). They contain a large number of elongated rod-shaped mitochondria and one or two cilia on their apical surface

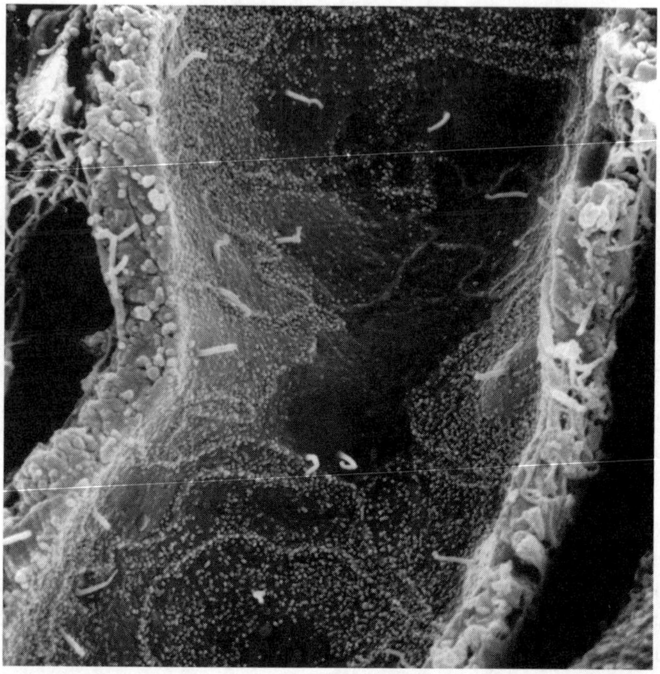

Figure 1.15 Scanning electron micrograph of rat cortical thick ascending limb of the loop of Henle. Both rough and smooth type thick ascending limb cells are present. In both cell types, the cell borders exhibit an undulating path; in this example the undulations of the cell borders are few and shallow, typical of the thick ascending limb as it nears the distal convoluted tubule. In deeper thick ascending limb segments, the frequency and amplitude of the undulations are much greater. The apical surface contains a cilium and short microprojections, which are most prevalent near the cell margins. The rough cell type has short microprojections over the entire apical surface of the cell. (Original magnification ×6,500.)

(Woodhall et al., 1978). Cells in the mTAL can be divided into two types, based on whether their luminal surfaces are smooth or rough; the smooth type cell predominates in the mTAL (Woodhall et al., 1978).

The cTAL originates at the junction of outer medulla and cortex. Although this transition is not abrupt, the cTAL epithelium does become much thinner than the mTAL and individual cTAL cells contain fewer mitochondria than mTAL cells (Woodhall et al., 1978). The cTAL contains a mixture of rough and smooth type cells

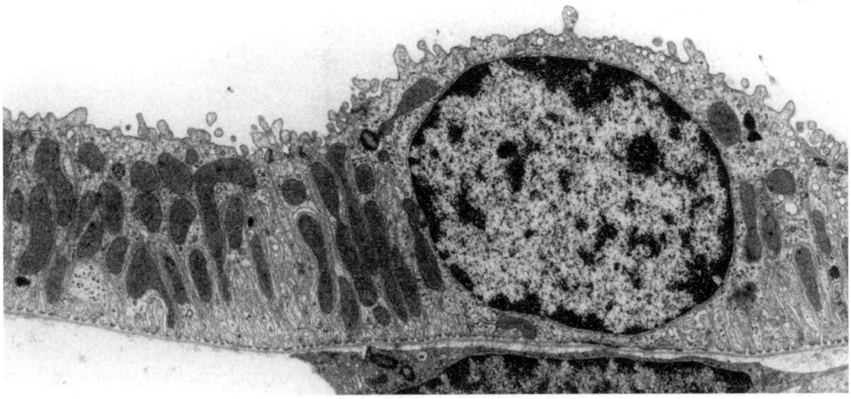

Figure 1.16 Transmission electron micrograph of rat cortical thick ascending limb of the loop of Henle. The epithelial cells contain prominent, vertically arranged mitochondria and extensive infoldings of the basolateral plasma membrane. The apical surface is covered with variable numbers of short plasma membrane microprojections and subapical cytoplasmic vesicles are present. (Original magnification ×9,100.)

(Figure 1.15), although the rough-type cell predominates (Woodhall et al., 1978). The cTAL extends a variable distance beyond the macula densa, depending on the species.

Physiology

Both the mTAL and cTAL contain a cytochrome P450 enzyme that metabolizes arachidonic acid to a variety of biologically active products (Badr, 1995; Wang et al., 1996). Inhibition of cytochrome P450 with stannous chloride reduces blood pressure to normal values in spontaneously hypertensive rats (Badr, 1995). Rat mTALs and cTALs actively reabsorb total ammonia and HCO_3^-; this reabsorption can be inhibited by loop diuretics (Garvin et al., 1988; Good, 1988). Active NH_4^+ reabsorption occurs because NH_4^+ can substitute for K^+ on both the Na^+–K^+–$2Cl^-$ cotransporter and the Na^+/K^+-ATPase.

Both the mTAL and cTAL are water impermeable (Hall and Varney, 1980; Rocha and Kokko, 1973a, 1973b) and do not express any aquaporin water channels (Ecelbarger et al., 1995; Frigeri et al., 1995; Fushimi et al., 1993; Maeda et al., 1995; Nielsen et al., 1995;

Sabolic et al., 1992; Terris et al., 1995). Active solute reabsorption is therefore the primary mechanism for diluting the luminal fluid in the TAL. Urea permeability is lower in the mTAL than the cTAL in both rabbit and rat (Knepper, 1983, 1983b; Rocha and Kokko, 1974). In rat, however, the transition to higher urea permeability occurs between the inner and outer stripes of the mTAL, rather than between the outer medulla and cortex (Knepper, 1983a, 1983b). This higher urea permeability could result in dilution of luminal fluid by passive urea reabsorption from the TAL from the inner–outer stripe to the corticomedullary border.

The major solute reabsorbed across the mTAL and the cTAL is NaCl. It is actively reabsorbed by the bumetanide-sensitive Na^+–K^+–$2Cl^-$ cotransporter (NKCC2, BSC1) in the apical plasma membrane and the Na^+-pump in the basolateral plasma membrane (Hebert, 1998; Knepper et al., 1999; Mount et al., 1999). The mTAL from short-looped nephrons can lower NaCl concentration within the luminal fluid at the bend of the loop from ≈ 300 to $\approx 117-140$ mM at the corticomedullary border (Burg and Green, 1973; Horster, 1978), while the cTAL can lower the NaCl concentration to ≈ 32 mM (Rocha and Kokko, 1973a, 1973b). However, the mTAL can reabsorb more NaCl than the cTAL, as evidenced by the higher Na^+ pump activity in the mTAL in all species studied (Garg et al., 1982; Katz, 1986). The mTAL therefore has a higher capacity to reabsorb NaCl but can generate only modest transepithelial NaCl gradients, while the cTAL can generate significantly higher transepithelial NaCl gradients.

Both the mTAL and cTAL are major sites for regulated reabsorption of the divalent cations Ca^{2+} and Mg^{2+}. In these nephron segments, 20% of the filtered Ca^{2+}, and 50 to 60% of the filtered Mg^{2+} are reabsorbed, primarily by paracellular transport driven by the lumen-positive voltage (Quamme and Dirks, 1980; Suki, 1979). Loop-diuretics inhibit Ca^{2+} and Mg^{2+} reabsorption since they inhibit the generation of the lumen positive voltage necessary to drive their reabsorption.

DISTAL CONVOLUTED TUBULE, CONNECTING SEGMENT, AND INITIAL COLLECTING TUBULE

Anatomy

"Distal tubule" refers to the nephron segment between the macula densa and the first junction of two tubules, but it is actually composed of several structurally and functionally distinct nephron segments. The most proximal portion of the distal tubule begins approximately 100 to 150 μm beyond the macula densa and contains cTAL cells (Crayen and Thoenes, 1975). The cTAL cells transition to distal convoluted tubule (DCT) cells, which are found in the second portion of the distal tubule or true DCT. The DCT appears bright under the dissection microscope and is rather short. The DCT cell (Figure 1.17) has deep tight junctions, short blunt apical microvilli, deep basolateral invaginations, and many long mitochondria that are oriented vertically, i.e., virtually perpendicular to the basolateral cell membrane, and interposed among the basolateral plasma membrane infoldings (Crayen and Thoenes, 1975).

The third portion of the distal tubule is also named the connecting segment or connecting tubule (CNT) (Figure 1.18). It appears granular under the light microscope and is wider than the DCT.

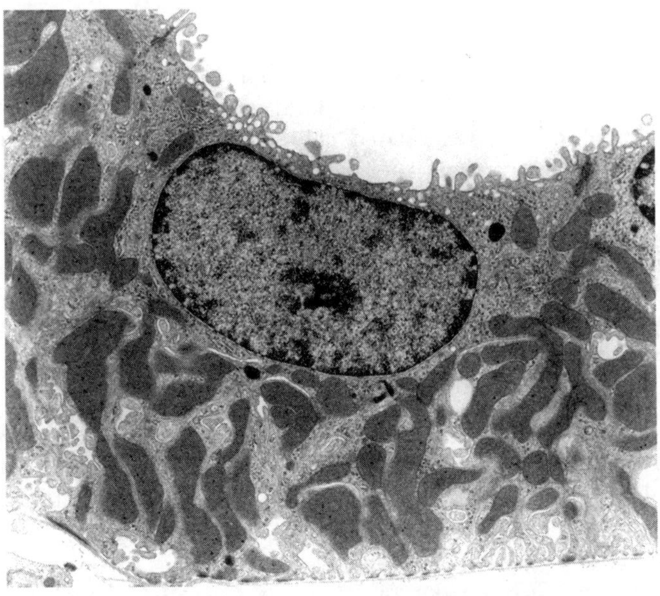

Figure 1.17 Transmission electron micrograph of a rat distal convoluted tubule cell (early or bright portion of the distal tubule). Distal convoluted tubule cells are tall and cuboidal with prominent mitochondria that are largely vertically oriented and interposed between extensive basolateral plasma membrane infoldings. (Original magnification ×12,500.)

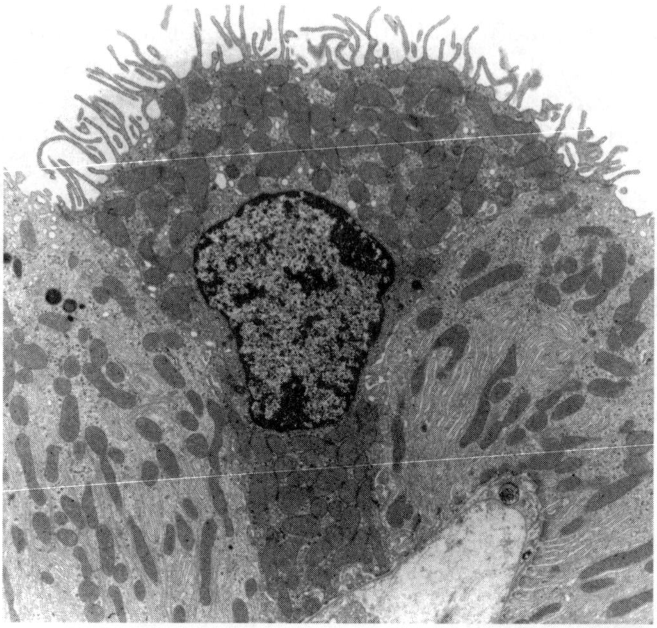

Figure 1.18 Transmission electron micrograph of a non-A, non-B intercalated cell in the rat connecting segment (late or granular portion of the distal convoluted tubule). This intercalated cell subtype is most frequently found in the connecting segment and is characterized by a very high mitochondrial density and prominent apical plasma membrane microprojections. The intercalated cell is interposed between two connecting segment cells, which have extensive basolateral plasma membrane infoldings, vertically oriented mitochondria, and short, sparse apical plasma membrane microprojections. (Original magnification ×11,000.)

This segment is occasionally called the granular distal tubule and contains two major cell types: intercalated or dark cells, and connecting tubule cells (Crayen and Thoenes, 1975; Kaissling, 1978; Kaissling and Kriz, 1979; Kaissling et al., 1977). In the CNT, at least three intercalated cell subtypes are recognized, the type A and type B intercalated cell subtypes (discussed below), and a third type, the so-called non-A, non-B intercalated cell (Figure 1.18). Under basal conditions, the ultrastructural characteristics of the non-A, non-B intercalated cell include a very high mitochondrial density, relatively sparse apical cytoplasmic vesicles, strikingly prominent apical plasma membrane microprojections, and, frequently, bulging of the apical surface into the tubule lumen. The non-A, non-B cell contains the vacuolar proton pump as well as the anion exchanger, pendrin, in the apical plasma membrane and apical cytoplasmic vesicles, but is negative for the anion exchanger, AE1. Although the non-A, non-B intercalated cell has been observed in the initial collecting duct (ICT) and the cortical collecting duct (CCD), it is most prevalent in the CNT. The ICT is the fourth and final portion of the distal tubule and contains type A, type B, and non-A, non-B intercalated cells, and principal or light cells.

Physiology

Both the DCT and CNT exhibit high rates of Na^+ reabsorption (Costanzo, 1985), resulting in a lumen-negative voltage. NaCl reabsorption is mediated by the thiazide-inhibitable Na^+–Cl^- cotransporter (NCC1, TSC) in rat DCT (Costanzo, 1985; Ellison et al., 1987; Kim et al., 1998; Verlander et al., 1998) and rabbit CNT (Shimizu et al., 1988). Potassium is transported by the K^+–Cl^- cotransporter (KCC1) in the apical plasma membrane of rat DCT and ICT (Velazquez et al., 1987) and a conductive K^+ channel in rabbit DCT (Taniguchi et al., 1989). Calcium is reabsorbed in the DCT, but at a lower rate than in the cTAL (Costanzo, 1985). The DCT is water impermeable (Imai, 1979). However, the vasopressin-regulated water channel, aquaporin-2 (AQP2), is present in rat CNT, indicating that this segment should be permeable to water (Kishore et al., 1996).

COLLECTING DUCT

Cortical Collecting Duct

Anatomy

The cortical collecting duct (CCD) originates at the convergence of two initial collecting tubules and extends to the corticomedullary border (Figure 1.1). The distal end of the CCD is defined by location and not by a distinct or abrupt change in its appearance or cell type. The CCD contains two cell types: principal (or light) cells, and intercalated (or dark) cells (Figure 1.19), at a ratio of approximately 2 : 1 (Stokes et al., 1975). Compared with intercalated cells, principal cells have fewer cellular organelles, including mitochondria, and less electron-dense cytoplasm. Their apical plasma membrane has short,

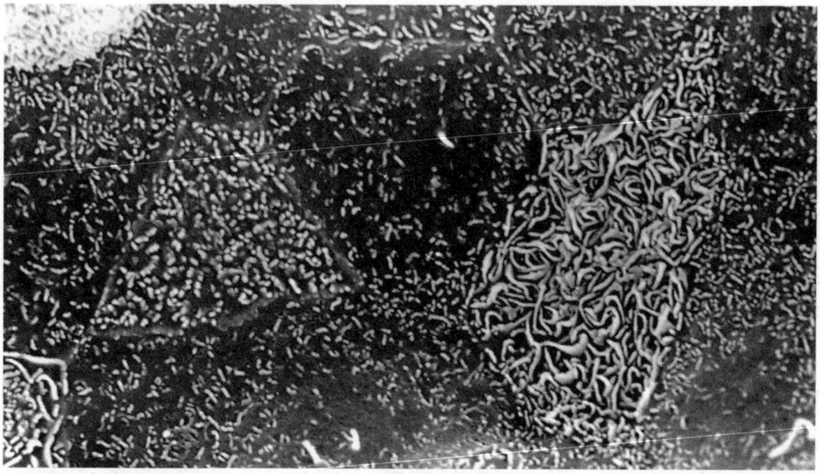

Figure 1.19 Scanning electron micrograph of the luminal surface of rat cortical collecting duct. Three distinct cell types are illustrated: principal cell (center), with a central cilium and short, sparse apical microprojections, Type B intercalated cell (left), with stubby microvilli covering the apical surface, and Type A intercalated cell (right), with extensive microplicae (folds) covering the apical surface. (Original magnification ×8,200.)

sparse microprojections, whereas the apical plasma membrane of many intercalated cells have more densely packed microvilli or microplicae. Intercalated cells also contain abundant cytoplasmic vesicles, whereas principal cells have fewer vesicles (Kaissling and Kriz, 1979). Carbonic anhydrase is abundant in intercalated cells, but is either absent (rat, rabbit) or present at relatively low amounts (mouse) in principal cells in the CCD (Brown et al., 1982; Schwartz et al., 2000).

Recent evidence indicates that there are at least two populations of intercalated cells in the CCD: type A (or H^+-secreting) cells (Figure 1.20), type B (or HCO_3-secreting) cells (Figure 1.21), and occasionally the non-A, non-B intercalated cell discussed previously.

The type A and type B intercalated cells are structurally distinct at the electron microscopic level. Under basal conditions, type A intercalated cells are characterized by a prominent apical tubulo-vesicular compartment, centrally located nucleus, evenly distributed mitochondria, and moderate apical plasma membrane microprojections. By contrast, under basal conditions the type B intercalated cell

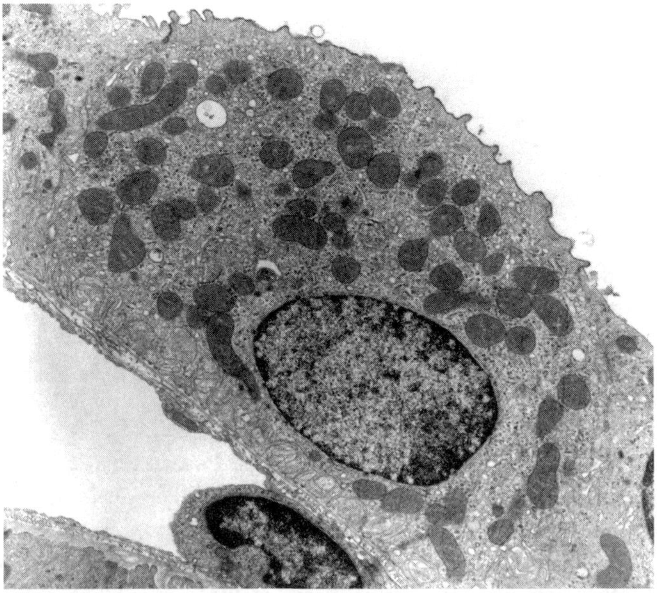

Figure 1.20 Transmission electron micrograph of a type A intercalated cell in rat cortical collecting duct. Under basal conditions, the type A intercalated cell typically has prominent apical tubulovesicles, a central nucleus, evenly distributed mitochondria, and moderate apical plasma membrane microprojections. (Original magnification $\times$10,200.)

typically has small cytoplasmic vesicles throughout the cell except for a subapical, vesicle-free band of cytoplasm, an eccentric nucleus, clusters of mitochondria, few apical plasma membrane microprojections, and a relatively dark cytoplasm. The outer medullary collecting duct (OMCD) contains only acid-secreting intercalated cells similar to type A cells (Verlander et al., 1987), thus there is marked axial heterogeneity of intercalated cells along the collecting duct. All OMCD intercalated cells stain for the band-3 like anion exchanger, AE1, on their basolateral plasma membrane, whereas only type A intercalated cells in the CCD stain for AE1 (Alper et al., 1989; Verlander et al., 1988). Type A cells express the H^+-ATPase in their apical plasma membrane and apical cytoplasmic vesicles, whereas type B cells express the H^+-ATPase in their basolateral plasma membrane and in cytoplasmic vesicles throughout the cell (Alper et al., 1989). In addition, studies have demonstrated that type B intercalated cells

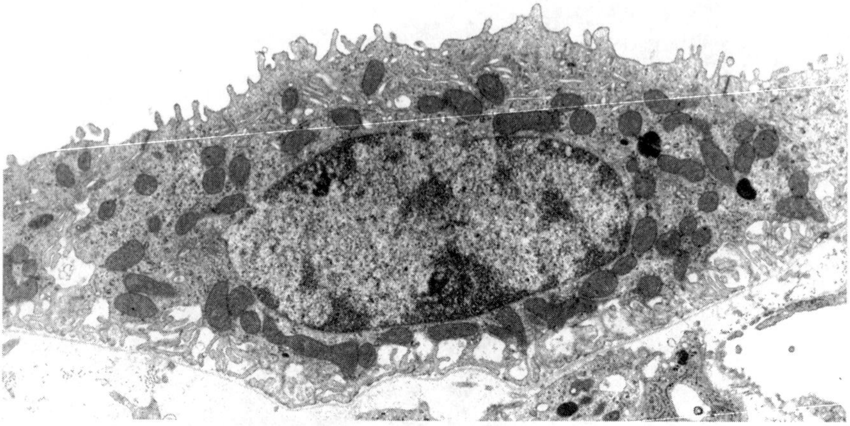

Figure 1.21 Transmission electron micrograph of a Type B intercalated cell in a cortical collecting duct from rat kidney. The type B intercalated cell under basal conditions is characterized by a smooth apical surface, dark cytoplasm, an eccentric nucleus, small cytoplasmic vesicles throughout the cell, a subapical band of cytoplasm that is relatively free of vesicles, and numerous, clustered mitochondria. (Original magnification ×12,500.)

also express the anion exchanger pendrin in the apical plasma membrane and apical cytoplasmic vesicles, whereas the type A intercalated cell is negative for pendrin (Kim et al., 2002; Royaux et al., 2001; Wall et al., 2003).

Physiology

The collecting duct is responsible for regulating the final composition of urine. In contrast to the proximal tubule, the collecting duct has a low capacity for reabsorption. However, the collecting duct regulates the reabsorption of water, solute, electrolytes, bicarbonate, and protons.

In the absence of vasopressin, the CCD has an extremely low osmotic water permeability (Grantham and Burg, 1966; Reif et al., 1984; Sands et al., 1987). Vasopressin increases osmotic water permeability by a factor of 10 to 100 in both rabbit and rat (Grantham and Burg, 1966; Reif et al., 1984; Sands et al., 1987). Vasopressin-stimulated osmotic water permeability is inhibited by a postcyclic AMP mechanism by arachidonic acid metabolites that are produced by cytochrome P450 (Badr, 1995). The CCD has a low urea permeability in both rat and rabbit that is unaffected by vasopressin (Knepper, 1983a, 1983b; Sands and Knepper, 1987). Vasopressin-induced water reabsorption therefore increases the urea concentration within the lumen of the CCD.

Aldosterone-mediated Na^+ reabsorption and K^+ secretion occur primarily in the CCD (Grantham et al., 1970). Active Na^+ reabsorption is mediated by the amiloride-sensitive Na^+ channel (ENaC) located in the apical plasma membrane of principal cells in the CCD and is responsible for the generation of a lumen-negative voltage (Grantham et al., 1970; Palmer and Frindt, 1986). Vasopressin can also stimulate Na^+ reabsorption in the CCD, while bradykinin can inhibit it (Tomita et al., 1985). Sodium exits the principal cell via the Na^+-pump located in the basolateral plasma membrane. The CCD has the highest Na^+-pump activity of any collecting duct segment, but less than that of the ICT or CNT (Garg et al., 1981, 1982; Katz, 1986; Katz et al., 1979; O'Neil and Dubinsky, 1984; Schmidt and Horster, 1977; Terada and Knepper, 1989).

Both paracellular and transcellular pathways mediate Cl^- transport across the CCD. Chloride reabsorption is primarily by a passive mechanism in the rabbit CCD, although some evidence for Cl^- reabsorption against an electrochemical gradient does exist (Hanley and Kokko, 1978). Active Cl^- reabsorption occurs in the rat CCD; it is stimulated by vasopressin and inhibited by bradykinin (Tomita et al., 1986).

Both active and passive K^+ transport pathways are present in the CCD. The apical and basolateral plasma membranes exhibit large K^+ conductances, with the apical exceeding the basolateral and thus partially accounting for active K^+ secretion (Koeppen et al., 1983). The apical membrane K^+ conductance is mediated by a high conductance Ca^{2+}-activated (maxi) K^+ channel and a low conductance K^+ channel (Frindt and Palmer, 1989). The apical membrane also contains the K^+–Cl^- cotransporter (KCC1) (Wingo, 1989a, 1989b). Aldosterone increases the K^+ permeability of the rabbit CCD (Stokes, 1985). Vasopressin stimulates K^+ secretion in the rat CCD (Tomita et al., 1985).

Acid–base transport in the CCD has been extensively studied. The CCD from normal rabbits does not transport HCO_3^-, while the CCD from normal rats reabsorbs HCO_3^- (Atkins and Burg, 1985; Tomita et al., 1986). Interestingly, vasopressin converts this HCO_3^- secretion to reabsorption (Tomita et al., 1986). The CCD from NH_4Cl-treated rats and rabbits reabsorbs HCO_3, while the CCD from $NaHCO_3$-treated and deoxycorticosterone acetate-treated rats and rabbits secretes HCO_3^- (Atkins and Burg, 1985; Lombard et al., 1983). Ammonia secretion is dependent on the presence of an acidic luminal disequilibrium pH in CCD generated by the lack of luminal carbonic anhydrase (Star et al., 1987). Interestingly, ammonia may also be acting as a paracrine factor that regulates bicarbonate transport (Frank et al., 2000, 2002). Angiotensin II inhibits H^+-ATPase activity via type 1 angiotensin II (AT_1) receptors in the rat CCD (Tojo et al., 1994) and stimulates luminal alkalinization by stimulating B cells via basolateral plasma membrane AT_1 receptors in the rabbit outer CCD (Weiner et al., 1995). Pendrin has been shown to be a Cl^-/HCO_3^- exchanger (Royaux et al., 2001), and ammonium transporter proteins RhBG and RhCG have been identified (Verlander et al., 2003) in the mouse CCD.

Outer Medullary Collecting Duct

Anatomy

The outer medullary collecting duct (OMCD) is defined as starting at the corticomedullary border and ending at the inner–outer medullary border (Figure 1.1). The OMCD from the outer stripe of the outer medulla resembles the CCD, both morphologically and functionally; except that neither type B intercalated cells nor HCO_3-secretion are present in the OMCD (Figure 1.22) (Stokes et al., 1981). The number of intercalated cells progressively decreases as the OMCD descends from the outer stripe to the inner medulla.

Physiology

The primary function of the OMCD is urinary acidification. Both rat and rabbit OMCD reabsorb HCO_3^- (Atkins and Burg, 1985; Lombard et al., 1983). Aldosterone stimulates HCO_3^- reabsorption by a Na^+-independent mechanism in both outer and inner stripe OMCD (Star et al., 1987). A disequilibrium pH is found in the outer stripe OMCD but not in inner stripe OMCD, indicating the presence of functional luminal carbonic anhydrase only in inner stripe OMCD (Brown and Kumpulainen, 1985; Star et al., 1987). Consistent with this finding, ammonia secretion, mediated predominantly by NH_3 diffusion, occurs three times faster in the outer than inner stripe OMCD (Star et al., 1987). The apical plasma membrane of the inner stripe OMCD possesses both an H^+/K^+-ATPase in intercalated and principal cells and an H^+-ATPase in intercalated cells (Wingo, 1989a, 1989b; Zeidel et al., 1986). It also has a nonselective cation channel (Xia et al., 2001). The basolateral plasma

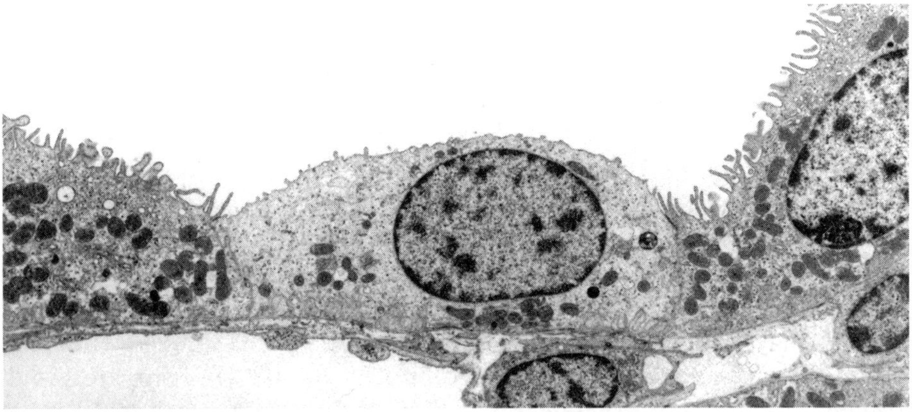

Figure 1.22 Transmission electron micrograph of rat outer medullary collecting duct from the outer stripe ($OMCD_o$). In the $OMCD_o$, two cell types are present: the principal cell and the intercalated cell. Compared with intercalated cells, the principal cell (center) has a pale cytoplasm, few mitochondria, a smooth apical surface. Intercalated cells (right and left of the principal cell) have darker cytoplasm, numerous mitochondria, and prominent apical plasma membrane microprojections. In the OMCD, only one intercalated cell subtype is recognized, an acid-secreting cell similar to the type A intercalated cell in the CCD. (Original magnification ×5,000.)

membrane of intercalated cells contains a Cl^-/HCO_3^- exchanger (AE1) and a Na^+–H^+ exchanger (Breyer and Jacobson, 1989). NKCC1 is present in the basolateral plasma membrane of OMCD cells but does not appear to play a significant functional role in acid–base transport (Wall et al., 2001; Wall and Fischer, 2002).

The rabbit OMCD has a low osmotic water permeability in the absence of vasopressin which is increased 20- to 30-fold by vasopressin (Horster and Zink, 1982; Rocha and Kokko, 1974). The OMCD also has a low urea permeability in both rat and rabbit (Knepper, 1983a, 1983b; Rocha and Kokko, 1974). Active Na^+ transport does not occur in the OMCD (Stokes et al., 1981).

Inner Medullary Collecting Duct

Anatomy

The inner medullary collecting duct (IMCD) consists of three subsegments: $IMCD_1$, $IMCD_2$, and $IMCD_3$ (Clapp et al., 1987; Clapp et al., 1989).

The $IMCD_1$ (or initial IMCD) is located in the first-third of the inner medulla closest to the inner–outer medullary border (Figure 1.23) and in the rat contains 90% principal cells plus 10% intercalated cells (Clapp et al., 1987). These cells are morphologically similar to those found in the inner stripe OMCD (Clapp et al., 1987). The $IMCD_2$ and $IMCD_3$ are located in the middle and deepest thirds of the inner medulla, respectively (Figure 1.24). The $IMCD_2$ and $IMCD_3$ are also called the terminal IMCD or papillary collecting duct and they contain a unique cell type, the IMCD cell (Figure 1.25) (Clapp et al., 1989). The $IMCD_2$ contains both principal cells and IMCD cells, while the $IMCD_3$ contains only IMCD cells (Clapp et al., 1989). Compared with principal cells, IMCD cells are taller, more columnar, with more prominent microvilli, have fewer basal infoldings, and do not have a

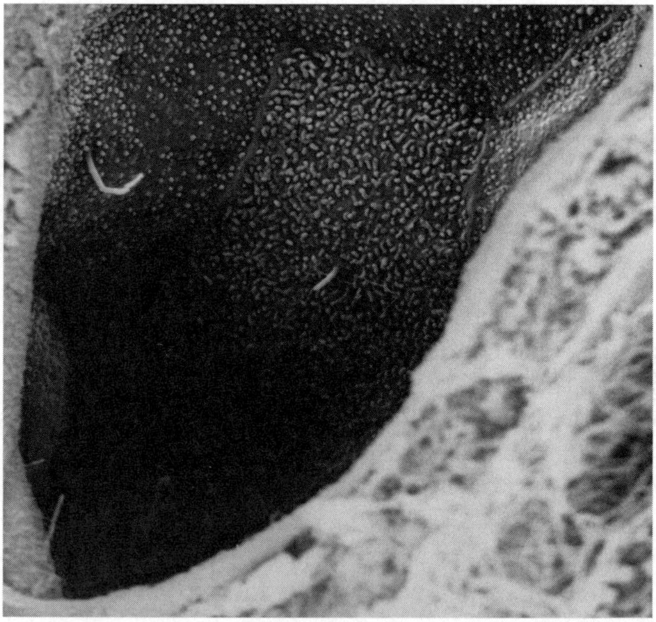

Figure 1.23 Scanning electron micrograph of a rabbit initial inner medullary collecting duct (IMCD). This segment of the collecting duct contains both principal cells and intercalated cells, although the intercalated cells account for only about 10% of the cells in the rat. In the rabbit, the frequency of intercalated cells in the initial IMCD varies between individuals, and intercalated cells are often absent, as in this example. The principal cells have relatively few short apical microprojections and a single central cilium. (Original magnification ×8,600.)

Figure 1.24 Scanning electron micrograph of the terminal inner medullary collecting duct (IMCD) in rat kidney. The epithelial cells in this segment are almost exclusively IMCD cells, which are covered with densely packed short microprojections. The lateral cell margins are prominent (arrows) and there are no cilia. (Original magnification ×5,700.)

central cilium (Clapp et al., 1989). The IMCD cell has a lighter staining cytoplasm and fewer organelles, including mitochondria, but abundant free ribosomes, small coated vesicles, and lysosomes (Clapp et al., 1989). There are only rare intercalated cells in the $IMCD_2$ and none in the $IMCD_3$ (Clapp et al., 1989).

Physiology

The $IMCD_1$ has a low osmotic water permeability in the absence of vasopressin, which is increased 10- to 30-fold by vasopressin (Imai et al., 1988a, 1988b; Sands et al., 1987, 1996). Facilitated urea permeability is low in the $IMCD_1$ and is unaffected by vasopressin in mammals consuming a standard diet and water *ad libitum* (Imai et al., 1988a, 1988b; Isozaki et al., 1993; Kato et al., 1998; Sands and Knepper, 1987;

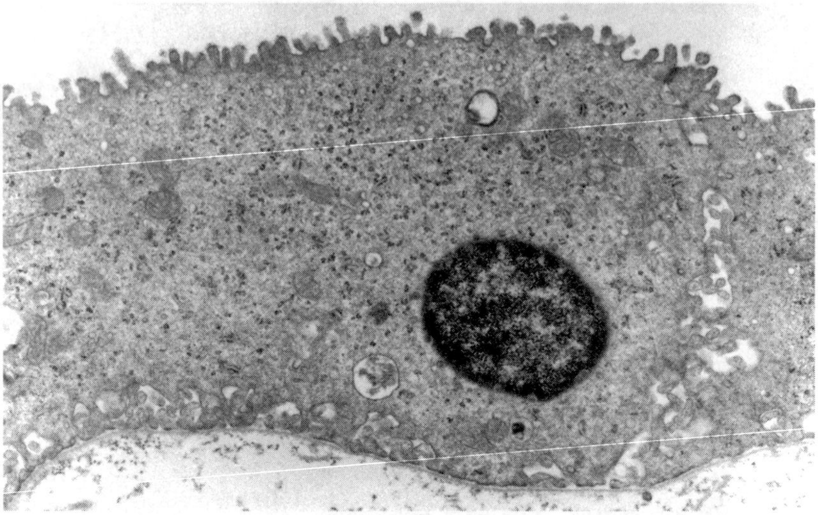

Figure 1.25 Transmission electron micrograph of an $IMCD_2$ cell in the rat inner medullary collecting duct (IMCD). The $IMCD_2$ cells are cuboidal and contain relatively few mitochondria and cytoplasmic vesicles. The apical surface is covered with short microprojections and there is no cilium. The IMCD epithelium becomes progressively taller as the collecting duct nears the papillary tip. (Original magnification ×12,200.)

Sands et al., 1987). The $IMCD_1$ from normal rats does not show any active urea transport (Gillin and Sands, 1992; Isozaki et al., 1993; Kato and Sands, 1998a, 1998b). However, sodium-dependent active urea reabsorption is present in the $IMCD_1$ from hypercalcemic (Kato and Sands, 1999), low-protein fed (Isozaki et al., 1993, 1994a, 1994b), or furosemide-treated rats (Kato and Sands, 1998a, 1998b).

The $IMCD_2$ and $IMCD_3$ have higher basal (no vasopressin) osmotic water permeabilities than other portions of the collecting duct (Imai et al., 1988a, 1988b; Sands et al., 1987, 1996); vasopressin increases osmotic water permeability by a factor of 10 (Sands et al., 1987, 1996). Osmotic water permeability is also affected by the *in vivo* condition of the rat. Terminal IMCD subsegments from water diuretic rats have lower basal and vasopressin-stimulated osmotic water permeabilities than terminal IMCD subsegments from dehydrated rats (Kato et al., 1998; Lankford et al., 1991). Although there is no difference in basal osmotic water permeability in terminal IMCD subsegments from hypercalcemic and low-protein fed rats, vasopressin-stimulated osmotic water permeability is markedly reduced in these animals compared with controls (Sands et al., 1996, 1998). Both cisplatinum and lithium treatment markedly reduce the abundance of the AQP2 water channel in the inner medulla (Ecelbarger et al., 2001; Kishore et al., 2000; Klein et al., 2002; Kwon et al., 2000; Marples et al., 1995).

The terminal IMCD subsegments also have higher basal facilitated urea permeabilities than more proximal portions of the collecting duct (Imai et al., 1988a, 1988b; Kato et al., 1998; Kondo and Imai, 1987a, 1987b; Sands and Knepper, 1987; Sands et al., 1987). Both vasopressin and hypertonicity can stimulate facilitated urea permeability by a factor of four to six, and together they increase facilitated urea permeability by a factor of 10 (Gillin and Sands, 1992; Imai et al., 1988a, 1988b; Kondo and Imai, 1987a, 1987b; Sands et al., 1987; Sands and Schrader, 1991). Angiotensin II can also increase urea permeability, but only in the presence of vasopressin (Kato et al., 2000). Both vasopressin and angiotensin II increase urea permeability by phosphorylating the UT-A1 urea transporter (Kato et al., 2000; Zhang et al., 2002).

The pattern of urea and water permeabilities in IMCD subsegments from normal rats results in urea being progressively concentrated within the collecting duct lumen from cortex through the base of the inner medulla by water reabsorption. Thus, a tubular fluid that has a high urea concentration is delivered to the highly urea-permeable terminal IMCD subsegments where it can be reabsorbed (Chandhoke and Saidel, 1981; Sands et al., 1987).

Facilitated urea permeability in terminal IMCD subsegments is affected by the *in vivo* condition of the rat from which the tubule is harvested. Both basal facilitated urea permeability and the abundance of the vasopressin-regulated urea transporter protein (UT-A1) are increased in the $IMCD_3$ from rats with reduced urine concentrating ability, regardless of etiology (Kato et al., 1998; Kato and Sands, 1999; Naruse et al., 1997; Sands et al., 1997, 1998; Terris et al., 1998), with the exception of lithium-treated rats (Klein et al., 2002). In lithium-treated rats, the abundances of the UT-A1 and UT-B urea transporters is markedly reduced, and the ability of vasopressin to phosphorylate UT-A1 is blunted (Klein et al., 2002).

No active urea transport occurs in the $IMCD_2$ from normal rats (Isozaki et al., 1993; Kato and Sands, 1998a, 1998b). However,

Na^+-dependent active urea secretion is present in the apical plasma membrane of the $IMCD_3$ from normal rats (Kato and Sands, 1998a, 1998b). Water diuresis increases active urea secretion by 500% in the $IMCD_3$ (Kato and Sands, 1998a, 1998b) and induces its expression in the $IMCD_2$ (Kato and Sands, 1999).

Ammonia secretion and HCO_3^- reabsorption occur in the rat $IMCD_2$; HCO_3^- reabsorption is stimulated by NH_4Cl-loading, mineralocorticoids, and vasopressin (Wall et al., 1990). Ammonium is transported on the Na^+-pump and its rate of transport is increased during hypokalemia (Wall, 2000; Wall et al., 2002). The rat $IMCD_2$ is able to lower luminal pH, whereas the rabbit $IMCD_2$ cannot (Wall et al., 1990). An acidic disequilibrium pH is present in the $IMCD_2$, consistent with the absence of luminal carbonic anhydrase (Wall et al., 1990). Although the Na^+ and Cl^- permeabilities are low in the terminal IMCD subsegments (Imai et al., 1988a, 1988b; Rocha and Kudo, 1982; Sands et al., 1988), they are the highest Na^+ or Cl^- permeabilities found in the entire collecting duct (Sands et al., 1992).

PAPILLARY SURFACE EPITHELIUM

Anatomy

The surface of the true papilla is covered by the papillary surface epithelium (PSE) (Figure 1.2). The PSE is bathed by urine on its apical plasma membrane and by inner medullary interstitial fluid on its basolateral plasma membrane. Hamster PSE cells are morphologically similar to IMCD cells (Lacy and Schmidt-Nielsen, 1979). However, the PSE is a simple cuboidal epithelium, which is distinct from IMCD cells in rat and rabbit, except at the papillary tip (Khorshid and Moffat, 1974; Sands et al., 1986; Silverblatt, 1974). The PSE becomes a columnar epithelium, similar to IMCD cells, near the openings of the ducts of Bellini (Sands et al., 1986; Silverblatt, 1974). The PSE cell has an apical plasma membrane with scattered short microvilli surrounded by an electron-dense fuzz and a cytoplasm containing relatively few mitochondria (Khorshid and Moffat, 1974); its lateral cell membrane is tortuous and contains some dilatations of the intercellular space which increases when interstitial osmolality increases (Bonventre et al., 1978; Sands et al., 1986). The PSE also contains some intercalated cells, based on the expression of the α2c isoform of the H^+/K^+-ATPase in some cells (Campbell-Thompson et al., 1995).

Physiology

The osmotic water permeability (Packer et al., 1989) and the urea permeability (Sands and Knepper, 1987) of the PSE are low and unaffected by vasopressin. The PSE has a higher Cl^- permeability than the $IMCD_2$ (Packer et al., 1989). The PSE expresses a Na^+–K^+–Cl^- cotransporter in its apical plasma membrane that is stimulated by vasopressin and inhibited by bumetanide (Sands et al., 1986). The PSE is able to acidify its apical surface (Chandhoke et al., 1990). The PSE expresses a K^+ conductive pathway in its basolateral plasma membrane (Reeves, 1994; Sands et al., 1985).

ACKNOWLEDGMENTS

This work was supported by National Institutes of Health grants R01-DK41707, R01-DK63657, and P01-DK50268.

REFERENCES

Abramow M, and Orci L (1980) On the "tightness" of the rabbit descending limb of the loop of Henle-physiological and morphological evidence. *Int J Biochem* 12:23–27.

Alper SL, Natale J, Gluck S, Lodish HF, and Brown D (1989) Subtypes of intercalated cells in rat kidney collecting duct defined by antibodies against erythroid band 3 and renal vacuolar H^+-ATPase. *Proc Natl Acad Sci USA* 86:5429–5433.

Atkins JL, and Burg MB (1985) Bicarbonate transport by isolated perfused rat collecting ducts. *Am J Physiol* 249:F485–F489.

Badr KF (1995) Kidney and endocrine system. Part 1: Eicosanoids, in *Textbook of Nephrology*, 3rd ed., vol. 1, Massry SG and Glassock RJ, Eds., Williams and Wilkins Co., Baltimore, pp. 182–191.

Berry CA, and Rector FC Jr (1991) Renal transport of glucose, amino acids, sodium, chloride, and water, in *The Kidney*, 4th ed. Brenner BM, and Rector FC Jr, Eds., W.B. Saunders, Philadelphia, pp. 245–282.

Bonventre JV, Karnovsky MJ, and Lechene CP (1978) Renal papillary epithelial morphology in antidiuresis and water diuresis. *Am J Physiol* 235:F69–F76.

Breyer MD, and Jacobson HR (1989) Regulation of Rabbit Medullary Collecting Duct Cell pH by Basolateral Na^+/H^+ and Cl^-/Base Exchange. *J Clin Invest* 84:996–1004.

Brown D, and Kumpulainen T (1985) Immunocytochemical localization of carbonic anhydrase on ultrathin frozen sections with protein A-gold. *Histochemistry* 83:153–158.

Brown D, Roth J, Kumpulainen T, and Orci L (1982) Ultrastructural immunocytochemical localization of carbonic anhydrase. *Histochemistry* 75:209–213.

Burg MB, and Green N (1973) Function of the thick ascending limb of Henle's loop. *Am J Physiol* 224:659–668.

Campbell-Thompson ML, Verlander JW, Curran KA, Campbell WG, Cain BD, Wingo CS, and McGuigan JE (1995) In situ hybridization of H^+-K^+-ATPase β-subunit mRNA in rat and rabbit kidney. *Am J Physiol* 269:F345–F354.

Chandhoke PS, and Saidel GM (1981) Mathematical model of mass transport throughout the kidney. Effects of nephron heterogeneity and tubular-vascular organization. *Ann Biomed Eng* 9:263–301.

Chandhoke PS, Packer RK, and Knepper MA (1990) Apical acidification by rabbit papillary surface epithelium. *Am J Physiol* 258:F893–F899.

Chou C-L, and Knepper MA (1992) In vitro perfusion of chinchilla thin limb segments: Segmentation and osmotic water permeability. *Am J Physiol* 263:F417–F426.

Chou C-L, and Knepper MA (1993) In vitro perfusion of chinchilla thin limb segments: Urea and NaCl permeabilities. *Am J Physiol* 264:F337–F343.

Chou C-L, Nielsen S, and Knepper MA (1993) Structural-functional correlation in chinchilla long loop of Henle thin limbs: A novel papillary subsegment. *Am J Physiol* 265:F863–F874.

Clapp WL, Madsen KM, Verlander JW, and Tisher CC (1987) Intercalated cells of the rat inner medullary collecting duct. *Kidney Int* 31:1080–1087.

Clapp WL, Madsen KM, Verlander JW, and Tisher CC (1989) Morphologic heterogeneity along the rat inner medullary collecting duct. *Lab Invest* 60:219–230.

Cockcroft D, and Gault M (1976) Prediction of creatinine clearance from serum creatinine. *Nephron* 16:31–41.

Costanzo LS (1985) Localization of diuretic action in microperfused rat distal tubules: Ca and Na transport. *Am J Physiol* 248:F527–F535.

Crayen M, and Thoenes W (1975) Architekus and cytologische Characterisiering des distalen Tubulus der Rattenniere. *Fortsschr Zool* 23:270–288.

Ecelbarger CA, Sands JM, Doran JJ, Cacini W, and Kishore BK (2001) Expression of salt and urea transporters in rat kidney during cisplatin-induced polyuria. *Kidney Int* 60:2274–2282.

Ecelbarger CA, Terris J, Frindt G, Echevarria M, Marples D, Nielsen S, and Knepper MA. (1995) Aquaporin-3 water channel localization and regulation in rat kidney. *Am J Physiol* 269:F663–F672.

Ellison DH, Velazquez H, and Wright FS (1987) Thiazide-sensitive sodium chloride cotransport in early distal tubule. *Pflügers Arch* 409:182–187.

Frank AE, Wingo CS, Andrews PM, Ageloff S, Knepper MA, and Weiner ID (2002) Mechanisms through which ammonia regulates cortical collecting duct net proton secretion. *Am J Physiol* 282:F1120–F1128.

Frank AE, Wingo CS, and Weiner ID (2000) Effects of ammonia on bicarbonate transport in the cortical collecting duct. *Am J Physiol* 278:F219–F226.

Frigeri A, Gropper MA, Turck CW, and Verkman AS (1995) Immunolocalization of the mercurial-insensitive water channel and glycerol intrinsic protein in epithelial cell plasma membranes. *Proc Natl Acad Sci USA* 92:4328–4331.

Frindt G, and Palmer LG (1989) Low-conductance K channels in apical membrane of rat cortical collecting tubule. *Am J Physiol* 256:F143–F151.

Fushimi K, Uchida S, Hara Y, Hirata Y, Marumo F, and Sasaki S (1993) Cloning and expression of apical membrane water channel of rat kidney collecting tubule. *Nature* 361:549–552.

Garg LC, Knepper MA, and Burg MB (1981) Mineralocorticoid effects on Na-K-ATPase in individual nephron segments. *Am J Physiol* 240:F536–F544.

Garg LC, Mackie S, and Tisher CC (1982) Effect of low potassium-diet on Na-K-ATPase in rat nephron segments. *Pflügers Arch* 394:113–117.

Garvin JL, and Knepper MA (1987) Bicarbonate and ammonia transport in isolated perfused rat proximal straight tubules. *Am J Physiol* 253:F277–F281.

Garvin JL, Burg MB, and Knepper MA (1988) Active NH_4^+ absorption by the thick ascending limb. *Am J Physiol* 255:F57–F65.

Gillin AG, and Sands JM (1992) Characteristics of osmolarity-stimulated urea transport in the rat IMCD. *Am J Physiol* 262:F1061–F1067.

Good DW (1988) Active absorption of NH_4^+ by rat medullary thick ascending limb: Inhibition by potassium. *Am J Physiol* 255:F78–F87.

Grantham JJ, and Burg MB (1966) Effect of vasopressin and cyclic AMP on permeability of isolated collecting tubules. *Am J Physiol* 211:255–259.

Grantham JJ, Burg MB, and Orloff J (1970) The nature of transtubular Na and K transport in isolated rabbit renal collecting tubules. *J Clin Invest* 49:1815–1826.

Hall DA, and Varney DM (1980) Effect of vasopressin in electrical potential differences and chloride transport in mouse medullary thick ascending limb of Henle. *J Clin Invest* 66:792–802.

Hanley MJ, and Kokko JP (1978) Study of chloride transport across the rabbit cortical collecting tubule. *J Clin Invest* 62:39–44.

Hebert SC (1998) Roles of Na-K-2Cl and Na-Cl cotransporters and ROMK potassium channels in urinary concentrating mechanism. *Am J Physiol* 275:F325–F327.

Hook JB, and Hewitt WR (1986) Toxic responses of the kidney, in *Casarett and Doull's Toxicology: The Basic Science of Poisons*, 3rd ed., Klaassen CD, Amdur MO, and Doul J, Eds., Macmillan, New York, pp. 310–329.

Horster M (1978) Loop of Henle functional differentiation: In vitro perfusion of the isolated thick ascending segment. *Pflügers Arch* 378:15–24.

Horster MF, and Zink H (1982) Functional differentiation of the medullary collecting tubule: Influence of vasopressin. *Kidney Int* 22:360–365.

Imai M (1977) Function of the thin ascending limbs of Henle of rats and hamsters perfused *in vitro*. *Am J Physiol* 232:F201–F209.

Imai M (1979) The connecting tubule: A functional subdivision of the rabbit distal nephron segments. *Kidney Int* 15:346–356.

Imai M, and Kokko JP (1974) Sodium, chloride, urea, and water transport in the thin ascending limb of Henle. *J Clin Invest* 53:393–402.

Imai M, Hayashi M, and Araki M (1984) Functional heterogeneity of the descending limbs of Henle's loop. I. Internephron heterogeneity in the hamster kidney. *Pflügers Arch* 402:385–392.

Imai M, Taniguchi J, and Yoshitomi K (1988a) Osmotic work across inner medullary collecting duct accomplished by difference in reflection coefficients for urea and NaCl. *Pflügers Arch* 412:557–567.

Imai M, Taniguchi J, and Yoshitomi K (1988b) Transition of permeability properties along the descending limb of long-loop nephron. *Am J Physiol* 254:F323–F328.

Isozaki T, Gillin AG, Swanson CE, and Sands JM (1994a) Protein restriction sequentially induces new urea transport processes in rat initial IMCDs. *Am J Physiol* 266:F756–F761.

Isozaki T, Lea JP, Tumlin JA, and Sands JM (1994b) Sodium-dependent net urea transport in rat initial IMCDs. *J Clin Invest* 94:1513–1517.

Isozaki T, Verlander JW, and Sands JM (1993) Low protein diet alters urea transport and cell structure in rat initial inner medullary collecting duct. *J Clin Invest* 92:2448–2457.

Johnson TMI, Sands JM, and Ouslander JG (2002) A prospective evaluation of the glomerular filtration rate in older adults with frequent nighttime urination. *J Urol* 167:146–150.

Johnson V, and Maack T (1995) Renal handling of organic compounds. Part 5: Renal tubular handling of proteins and peptides, in *Textbook of Nephrology*, 3rd ed., Vol. 1, Massry SG, and Glassock RJ, Eds., Williams and Wilkins Co., Baltimore, pp. 101–107.

Kaissling B (1978) Ultrastructural characterization of the connecting tubule and the different segments of the collecting duct system in the rabbit kidney, in *Biochemical Nephrology*, Guder WG, and Schmidt V, Eds., Huber, Bern, pp. 435–445.

Kaissling B, and Kriz W (1979) Structural analysis of the rabbit kidney. In Advances in Anatomy: Embryology and Cell Biology, Vol. 56A. Brodal et al., Eds., Springer Verlag, Berlin, pp. 1–123.

Kaissling B, Peter S, and Kriz W (1977) The transition of the thick ascending limb of Henle's loop into the distal convoluted tubule in the nephron of the rat kidney. *Cell Tissue Res* 182:111–118.

Kartner N, and Ling V (1989) Multidrug resistance in cancer. *Scientific Am* 260:44–51.

Kato A, and Sands JM (1998a) Active sodium-urea counter-transport is inducible in the basolateral membrane of rat renal initial inner medullary collecting ducts. *J Clin Invest* 102:1008–1015.

Kato A, and Sands JM (1998b) Evidence for sodium-dependent active urea secretion in the deepest subsegment of the rat inner medullary collecting duct. *J Clin Invest* 101:423–428.

Kato A, and Sands JM (1999) Urea transport processes are induced in rat IMCD subsegments when urine concentrating ability is reduced. *Am J Physiol* 276:F62–F71.

Kato A, Klein JD, Zhang C, and Sands JM (2000) Angiotensin II increases vasopressin-stimulated facilitated urea permeability in rat terminal IMCDs. *Am J Physiol* 279:F835–F840.

Kato A, Naruse M, Knepper MA, and Sands JM (1998) Long-term regulation of inner medullary collecting duct urea transport in rat. *J Am Soc Nephrol* 9:737–745.

Katz AI (1986) Distribution and function of classes of ATPases along the nephron. *Kidney Int* 29:21–31.

Katz AI, Doucet A, and Morel F (1979) Na-K-ATPase activity along the rabbit, rat and mouse nephron. *Am J Physiol* 237:F114–F120.

Kawamura S, and Kokko JP (1976) Urea secretion by the straight segment of the proximal tubule. *J Clin Invest* 58:604–612.

Khorshid MR, and Moffat DB (1974) The epithelia lining the renal pelvis in the rat. *J Anat* 118:561–569.

Kim GH, Masilamani S, Turner R, Mitchell C, Wade JB, and Knepper MA (1998) The thiazide-sensitive Na-Cl cotransporter is an aldosterone-induced protein. *Proc Natl Acad Sci USA* 95:14552–14557.

Kim YH, Kwon TH, Frische S, Kim J, Tisher CC, Madsen KM, and Nielsen S (2002) Immunocytochemical localization of pendrin in intercalated cell subtypes in rat and mouse kidney. *Am J Physiol* 283:F744–F754.

Kishore BK, Krane CM, Di Iulio D, Menon AG, and Cacini W (2000) Expression of renal aquaporins 1, 2, and 3 in a rat model of cisplatin-induced polyuria. *Kidney Int* 58:701–711.

Kishore BK, Mandon B, Oza NB, DiGiovanni SR, Coleman RA, Ostrowski NL, Wade JB, and Knepper MA (1996) Rat renal arcade segment expresses vasopressin-regulated water channel and vasopressin V_2 receptor. *J Clin Invest* 97:2763–2771.

Klein JD, Gunn RB, Roberts BR, and Sands JM (2002) Down-regulation of urea transporters in the renal inner medulla of lithium-fed rats. *Kidney Int* 61:995–1002.

Knepper MA (1983a) Urea transport in nephron segments from medullary rays of rabbits. *Am J Physiol* 244:F622–F627.

Knepper MA (1983b) Urea transport in isolated thick ascending limbs and collecting ducts from rats. *Am J Physiol* 245:F634–F639.

Knepper MA, Kim GH, Fernández-Llama P, and Ecelbarger CA (1999) Regulation of thick ascending limb transport by vasopressin. *J Am Soc Nephrol* 10:628–634.

Koeppen BM, Biagi BA, and Giebisch G (1983) Intracellular microelectrode characterization of the rabbit cortical collecting duct. *Am J Physiol* 244:F35–F47.

Kokko JP (1970) Sodium chloride and water transport in the descending limb of Henle. *J Clin Invest* 49:1838–1846.

Kokko JP (1972) Urea transport in the proximal tubule and the descending limb of Henle. *J Clin Invest* 51:1999–2008.

Kondo Y, and Imai M (1987a) Effects of glutaraldehyde fixation on renal tubular function. I. Preservation of vasopressin-stimulated water and urea pathways in rat papillary collecting duct. *Pflügers Arch* 408:479–483.

Kondo Y, and Imai M (1987b) Effect of glutaraldehyde on renal tubular function. II. Selective inhibition of Cl^- transport in the hamster thin ascending limb of Henle's loop. *Pflügers Arch* 408:484–490.

Kondo Y, Abe K, Igarashi Y, Kudo K, Tada K, and Yoshinaga K (1993) Direct evidence for the absence of active Na^+ reabsorption in hamster ascending thin limb of Henle's loop. *J Clin Invest* 91:5–11.

Koyama S, Yoshitomi K, and Imai M (1991) Effect of protamine on ion conductance of ascending thin limb of Henle's loop from hamsters. *Am J Physiol* 261:F593–F599.

Kriz W (1988) A standard nomenclature for structures of the kidney. *Am J Physiol* 254:F1–F8.

Kriz W, and Bankir L (1983) Structural organization of the renal medullary counterflow system. *Fed Proc* 42:2379–2385.

Kurtz I, Star R, Balaban RS, Garvin JL, and Knepper MA (1986) Spontaneous Luminal Disequilibrium pH in S_3 Proximal Tubules. Role in ammonia and bicarbonate transport. *J Clin Invest* 78:989–996.

Kwon TH, Laursen UH, Marples D, Maunsbach AB, Knepper MA, Frokiaer J, and Nielsen S (2000) Altered expression of renal AQPs and Na^+ transporters in rats with lithium-induced NDI. *Am J Physiol* 279:F552–F564.

Lacy ER, and Schmidt-Nielsen B (1979) Anatomy of the renal pelvis in the hamster. *Am J Anat* 154:291–320.

Lankford SP, Chou C-L, Terada Y, Wall SM, Wade JB, and Knepper MA (1991) Regulation of collecting duct water permeability independent of cAMP-mediated AVP response. *Am J Physiol* 261:F554–F566.

Layton HE, and Davies JM (1993) Distributed solute and water reabsorption in a central core model of the renal medulla. *Math Biosci* 116:169–196.

Layton HE, Knepper MA, and Chou C-L (1996) Permeability criteria for effective function of passive countercurrent multiplier. *Am J Physiol* 270:F9–F20.

Lemley KV, and Kriz W (1987) Cycles and separations: The histotopography of the urinary concentrating process. *Kidney Int* 31:538–548.

Liu W, Moritomo T, Kondo Y, Iinuma K, Uchida S, Sasaki S, Marumo F, and Imai M (2002) Analysis of NaCl transport in the ascending thin limb of Henle's loop in CLC-K1 null mice. *Am J Physiol* 282:F451–F457.

Lombard WE, Kokko JP, and Jacobson HR (1983) Bicarbonate transport in cortical and outer medullary collecting tubules. *Am J Physiol* 244:F289–F296.

Maddox DA, and Brenner BM (1991) Glomerular ultrafiltration, in *The Kidney*, 4th ed., Brenner BM, and Rector FC Jr, Eds., W.B. Saunders, Philadelphia, pp. 205–244.

Maeda Y, Smith BL, Agre P, and Knepper MA (1995) Quantification of Aquaporin-CHIP water channel protein in microdissected renal tubules by fluorescence-based ELISA. *J Clin Invest* 95:422–428.

Marples D, Christensen S, Christensen EI, Ottosen PD, and Nielsen S (1995) Lithium-induced downregulation of Aquaporin-2 water channel expression in rat kidney medulla. *J Clin Invest* 95:1838–1845.

Marsh DJ, and Azen SP (1975) Mechanism of NaCl reabsorption by hamster thin ascending limb of Henle's loop. *Am J Physiol* 228:71–79.

Maunsbach AB (1966) Observations on the segmentation of the proximal tubule in the rat kidney. Comparison of results from phase contrast, fluorescence, and electron microscopy. *J Ultrastruct Res* 16:239–258.

Maunsbach AB (1973) Ultrastructure of the proximal tubule, in *Handbook of Physiology*, Section 8: Renal Physiology, Orloff J, Ed., American Physiological Society, Washington, D.C., pp. 31–81.

Morgan T, and Berliner RW (1968) Permeability of the loop of Henle, vasa recta, and collecting duct to water, urea, and sodium. *Am J Physiol* 215:108–115.

Mount DB, Baekgaard A, Hall AE, Plata C, Xu J, Beier DR, Gamba G, and Hebert SC (1999) Isoforms of the Na-K-2Cl cotransporter in murine TAL. I. Molecular characterization and intrarenal localization. *Am J Physiol* 276:F347–F358.

Naruse M, Klein JD, Ashkar ZM, Jacobs JD, and Sands JM (1997) Glucocorticoids downregulate the rat vasopressin-regulated urea transporter in rat terminal inner medullary collecting ducts. *J Am Soc Nephrol* 8:517–523.

Nielsen S, Pallone T, Smith BL, Christensen EI, Agre P, and Maunsbach AB (1995) Aquaporin-1 water channels in short and long loop descending thin limbs and in descending vasa recta in rat kidney. *Am J Physiol* 268:F1023–F1037.

Nielsen S, Smith BL, Christensen EI, Knepper MA, and Agre P (1993) CHIP28 water channels are localized in constitutively water-permeable segments of the nephron. *J Cell Biol* 120:371–383.

O'Neil RG, and Dubinsky WP (1984) Micromethodology for measuring ATPase activity in renal tubules: mineralocorticoid influence. *Am J Physiol* 247:C314–C320.

Olbricht CJ, Cannon JK, Garg LC, and Tisher CC (1986) Activities of cathepsins B and L in isolated nephron segments from proteinuric and nonproteinuric rats. *Am J Physiol* 250:F1055–F1063.

Packer RK, Sands JM, and Knepper MA (1989) Chloride and osmotic water permeabilities of isolated rabbit renal papillary surface epithelium. *Am J Physiol* 257:F218–F224.

Palmer LG, and Frindt G (1986) Amiloride-sensitive Na channels from the apical membrane of the rat cortical collecting tubule. *Proc Natl Acad Sci USA* 83:2767–2770.

Pannabecker TL, Dahlmann A, Brokl OH, and Dantzler WH (2000) Mixed descending- and ascending-type thin limbs of Henle's loop in mammalian renal inner medulla. *Am J Physiol* 278:F202–F208.

Pennell JP, Lacy FB, and Jamison RL (1974) An *in vivo* study of the concentrating process in the descending limb of Henle's loop. *Kidney Int* 5:337–347.

Preston GM, Carroll TP, Guggino WB, and Agre P (1992) Appearance of water channels in Xenopus oocytes expressing red cell CHIP28 protein. *Science* 256:385–387.

Pritchard JB (1995) Renal handling of organic compounds. Part 4: Renal handling of organic acids and bases, in *Textbook of Nephrology*, 3rd ed., Vol. 1, Massry SG, and Glassock RJ, Eds., Williams and Wilkins Co., Baltimore, pp. 96–101.

Quamme GA, and Dirks JH (1980) Magnesium transport in the nephron. *Am J Physiol* 239:F393–F401.

Reeves WB (1994) Conductive properties of papillary surface epithelium. *Am J Physiol* 266:F259–F265.

Reif MC, Troutman SL, and Schafer JA (1984) Sustained response to vasopressin in isolated rat cortical collecting tubule. *Kidney Int* 26:725–732.

Rocha AS, and Kokko JP (1973a) Membrane characteristics regulating potassium transport out of the isolated perfused descending limb of Henle. *Kidney Int* 4:326–330.

Rocha AS, and Kokko JP (1973b) Sodium chloride and water transport in the medullary thick ascending limb of Henle. Evidence for active chloride transport. *J Clin Invest* 52:612–623.

Rocha AS, and Kokko JP (1974) Permeability of medullary nephron segments to urea and water: Effect of vasopressin. *Kidney Int* 6:379–387.

Rocha AS, and Kudo LH (1982) Water, urea, sodium, chloride, and potassium transport in the *in vitro* perfused papillary collecting duct. *Kidney Int* 22:485–491.

Royaux IE, Wall SM, Karniski LP, Everett LA, Suzuki K, Knepper MA, and Green ED (2001) Pendrin, encoded by the Pendred syndrome gene, resides in the apical region of renal intercalated cells and mediates bicarbonate secretion. *Proc Natl Acad Sci USA* 98:4221–4226.

Sabolic I, Valenti G, Verbavatz J-M, Van Hoek AN, Verkman AS, Ausiello DA, and Brown D (1992) Localization of the CHIP28 water channel in rat kidney. *Am J Physiol* 263:C1225–C1233.

Sands JM, and Knepper MA (1987) Urea permeability of mammalian inner medullary collecting duct system and papillary surface epithelium. *J Clin Invest* 79:138–147.

Sands JM, and Schrader DC (1991) An independent effect of osmolality on urea transport in rat terminal IMCDs. *J Clin Invest* 88:137–142.

Sands JM, Flores FX, Kato A, Baum, Brown EM, Ward DT, Hebert SC, and Harris HW (1998) Vasopressin-elicited water and urea permeabilities are altered in the inner medullary collecting duct in hypercalcemic rats. *Am J Physiol* 274:F978–F985.

Sands JM, Ivy EJ, and Beeuwkes III R (1985) Transmembrane potential difference of renal papillary epithelial cells. Effect of urea and DDAVP. *Am J Physiol* 248:F762–F766.

Sands JM, Jacobson HR, and Kokko JP (1992) Intrarenal Heterogeneity. Vascular and Tubular, in *The Kidney: Physiology and Pathophysiology*, 2nd ed., Seldin DW, and Giebisch G, Eds., Raven Press, New York City, pp. 1087–1155.

Sands JM, Knepper MA, and Spring KR (1986) Na-K-Cl cotransport in apical membrane of rabbit renal papillary surface epithelium. Am J Physiol 251:F475–F484.

Sands JM, Naruse M, Jacobs JD, Wilcox JN, and Klein JD (1996) Changes in aquaporin-2 protein contribute to the urine concentrating defect in rats fed a low-protein diet. *J Clin Invest* 97:2807–2814.

Sands JM, Nonoguchi H, and Knepper MA (1987) Vasopressin effects on urea and H_20 transport in inner medullary collecting duct subsegments. *Am J Physiol* 253:F823–F832.

Sands JM, Nonoguchi H, and Knepper MA (1988) Hormone effects on NaCl permeability of rat inner medullary collecting duct subsegments. *Am J Physiol* 255:F421–F428.

Sands JM, Timmer RT, and Gunn RB (1997) Urea transporters in kidney and erythrocytes. *Am J Physiol* 273:F321–F339.

Schmidt U, and Horster M (1977) Na-K-activated ATPase: Activity maturation in rabbit nephron segments dissected *in vitro*. *Am J Physiol* 233:F55–F60.

Schwartz GJ, Kittelberger AM, Barnhart DA, and Vijayakumar S (2000) Carbonic anhydrase IV is expressed in H^+-secreting cells of rabbit kidney. *Am J Physiol* 278:F894–F904.

Schwartz MM, and Venkatachalam MA (1974) Structural differences in thin limbs of Henle: Physiological implications. *Kidney Int* 6:193–208.

Schwartz MM, Karnovsky MJ, and Venkatachalam MA (1979) Regional membrane specialization in the thin limb of Henle's loops as seen by freeze-fracture electron microscopy. *Kidney Int* 16:577–589.

Shimizu T, Yoshitomi K, Nakamura M, and Imai M (1988) Site and mechanism of action of trichlormethiazide in rabbit distal nephron segments perfused *in vitro*. *J Clin Invest* 82:721–730.

Silverblatt FJ (1974) Ultrastructure of the renal pelvic epithelium of the rat. *Kidney Int* 5:214–220.

Star RA, Burg MB, and Knepper MA (1987) Luminal disequilibrium pH and ammonia transport in outer medullary collecting duct. *Am J Physiol* 252:F1148–F1157.

Stokes JB (1985) Mineralocorticoid effect of K^+ permeability of the rabbit cortical collecting tubule. *Kidney Int* 28:640–645.

Stokes JB, Ingram MJ, Williams AD, Ingram D (1981) Heterogeneity of the rabbit collecting tubule: Localization of mineralocorticoid hormone action to the cortical portion. *Kidney Int* 20:340–347.

Stokes JB, Tisher CC, and Kokko JP (1975) Structural-functional heterogeneity along the rabbit collecting tubule. *Kidney Int* 14:585–593.

Stoner LC, and Roch-Ramel F (1979) The effects of pressure on the water permeability of the descending limb of Henle's loops of rabbits. *Pflügers Arch* 382:7–15.

Suki WN (1979) Calcium transport in the nephron. *Am J Physiol* 237:F1–F6.

Tabei K, and Imai M (1987) K transport in upper portion of descending limbs of long-loop nephron from hamster. *Am J Physiol* 252:F387–F392.

Takahashi N, Kondo Y, Fujiwara I, Ito O, Igarashi Y, and Abe K (1995a) Characterization of Na^+ transport across the cell membranes of the ascending thin limb of Henle's loop. *Kidney Int* 47:789–794.

Takahashi N, Kondo Y, Ito O, Igarashi Y, Omata K, and Abe K (1995b) Vasopressin stimulates Cl^- transport in ascending thin limb of Henle's loop in hamster. *J Clin Invest* 95:1623–1627.

Taniguchi J, Yoshitomi K, and Imai M (1989) K^+ channel currents in basolateral membrane of distal convoluted tubule of rabbit kidney. *Am J Physiol* 256:F246–F254.

Terada Y, and Knepper MA (1989) Na^+-K^+-ATPase activities in renal tubule segments of rat inner medulla. *Am J Physiol* 256:F218–F223.

Terris J, Ecelbarger CA, Marples D, Knepper MA, and Nielsen S (1995) Distribution of aquaporin-4 water channel expression within rat kidney. *Am J Physiol* 269:F775–F785.

Terris J, Ecelbarger CA, Sands JM, and Knepper MA (1998) Long-term regulation of collecting duct urea transporter proteins in rat. *J Am Soc Nephrol* 9:729–736.

Tisher CC, and Brenner BM (1989) Structure and function of the glomerulus, in *Renal pathology with clinical and functional correlations*, Vol. 1, Tisher CC, and Brenner BM, Eds., J.B. Lippincott, Philadelphia, pp. 92–110.

Tojo A, Tisher CC, and Madsen KM (1994) Angiotensin II regulates H^+-ATPase activity in rat cortical collecting duct. *Am J Physiol* 267:F1045–F1051.

Tomita K, Pisano JJ, Burg MB, and Knepper MA (1986) Effects of Vasopressin and Bradykinin on Anion Transport by the Rat Cortical Collecting Duct. Evidence for an electroneutral sodium chloride transport pathway. *J Clin Invest* 77:136–141.

Tomita K, Pisano JJ, and Knepper MA (1985) Control of sodium and potassium transport in the cortical collecting duct of the rat. Effects of bradykinin, vasopressin, and deoxycorticosterone. *J Clin Invest* 76:132–136.

Uchida S, Sasaki S, Nitta K, Uchida K, Horita S, Nihei H, and Marumo F (1995) Localization and functional characterization of rat kidney-specific chloride channel, ClC-K1. *J Clin Invest* 95:104–113.

Valtin H (1973) *Renal Function: Mechanisms Preserving Fluid and Solute Balance in Health*, Little, Brown, and Company, Boston.

Velazquez H, Ellison DH, and Wright FS (1987) Chloride-dependent potassium secretion in early and late renal distal tubules. *Am J Physiol* 253:F555–F562.

Verlander JW, Madsen KM, Low PS, Allen DP, and Tisher CC (1988) Immunocytochemical localization of band 3 protein in the rat collecting duct. *Am J Physiol* 255:F115–F125.

Verlander JW, Madsen KM, and Tisher CC (1987) Effect of acute respiratory acidosis on two populations of intercalated cells in rat cortical collecting duct. *Am J Physiol* 253:F1142–F1156.

Verlander JW, Miller RT, Frank AE, Royaux IE, Kim YH, and Weiner ID (2003) Localization of the ammonium transporter proteins RhBG and RhCG in mouse kidney. *Am J Physiol* 284:F323–F337.

Verlander JW, Tran TM, Zhang L, Kaplan MR, and Hebert SC (1998) Estradiol enhances thiazide-sensitive NaCl cotransporter density in the apical plasma membrane of the distal convoluted tubule in ovariectomized rats. *J Clin Invest* 101:1661–1669.

Wall SM (2000) Impact of K^+ homeostasis on net acid secretion in rat terminal inner medullary collecting duct: Role of the Na,K-ATPase. *Am J Kidney Dis* 36:1079–1088.

Wall SM, and Fischer MP (2002) Contribution of the Na^+-K^+-$2Cl^-$ cotransporter (NKCC1) to transepithelial transport of H^+, NH_4^+, K^+, and Na^+ in rat outer medullary collecting duct. *J Am Soc Nephrol* 13:827–835.

Wall SM, Fischer MP, Kim GH, Nguyen BM, and Hassell KA (2002) In rat inner medullary collecting duct, NH_4^+ uptake by the Na,K-ATPase is increased during hypokalemia. *Am J Physiol* 282:F91–F102.

Wall SM, Fischer MP, Mehta P, Hassell KA, and Park SJ (2001) Contribution of the Na^+-K^+-$2Cl^-$ cotransporter NKCC1 to Cl^- secretion in rat OMCD. *Am J Physiol* 280:F913–F921.

Wall SM, Hassell KA, Royaux IE, Green ED, Chang JY, Shipley GL, and Verlander JW (2003) Localization of pendrin in mouse kidney. *Am J Physiol* 284:F229–F241.

Wall SM, Sands JM, Flessner MF, Nonoguchi H, Spring KR, and Knepper MA (1990) Net acid transport by isolated perfused inner medullary collecting ducts. *Am J Physiol* 258:F75–F84.

Wang WH, Lu M, and Hebert SC (1996) Cytochrome P-450 metabolites mediate extracellular Ca^{2+}- induced inhibition of apical K^+ channels in the TAL. *Am J Physiol* 271:C103–C111.

Warnock DG, and Burg MB (1977) Urinary acidification: CO_2 transport by the rabbit proximal straight tubule. *Am J Physiol* 232:F20–F25.

Weiner ID, New AR, Milton AE, and Tisher CC (1995) Regulation of luminal alkalinization and acidification in the cortical collecting duct by angiotensin II. *Am J Physiol* 269:F730–F738.

Wingo CS (1989a) Active Proton Secretion and Potassium Absorption in the Rabbit Outer Medullary Collecting Duct. Functional evidence for proton-potassium-adenosine triphosphatase. *J Clin Invest* 84:361–365.

Wingo CS (1989b) Reversible chloride-dependent potassium flux across the rabbit cortical collecting tubule. *Am J Physiol* 256:F697–F704.

Woodhall PB, Tisher CC, Simonton CA, and Robinson RR (1978) Relationship between para-aminohippurate secretion and cellular morphology in rabbit proximal tubules. *J Clin Invest* 61:1320–1329.

Work J, Troutman SL, and Schafer JA (1982) Transport of potassium in the rabbit pars recta. *Am J Physiol* 242:F226–F237.

Xia SL, Noh SH, Verlander JW, Gelband CH, and Wingo CS (2001) Apical membrane of native $OMCD_i$ cells has nonselective cation channels. *Am J Physiol* 281:F48–F55.

Zeidel ML, Silva P, and Seifter JL (1986) Intracellular pH Regulation and Proton Transport by Rabbit Renal Medullary Collecting Duct Cells. Role of plasma membrane proton adenosine triphosphatase. *J Clin Invest* 77:113–120.

Zhang C, Sands JM, and Klein JD (2002) Vasopressin rapidly increases the phosphorylation of the UT-A1 urea transporter activity in rat IMCDs through PKA. *Am J Physiol* 282:F85–F90.

2

Molecular and Cell Biology of Normal and Diseased or Intoxicated Kidney

Lawrence H. Lash

INTRODUCTION

The use of sensitive methods to detect changes in expression of renal cell proteins or secretion of proteins into urine has provided information about the molecular steps involved in the progression from chemical exposure to cellular dysfunction and toxicity. Such information can also be used for several other purposes. For example, changes in gene expression are among the earliest cellular responses to toxicity (Stevens et al., 2000).

A series of biochemical effects, such as protein thiol oxidation or alterations in calcium ion homeostasis, act as signals that lead to a series of changes in gene expression. As part of the very early stages of the response of renal proximal tubular (PT) cells to chemical toxicants, these changes may include altered expression of heat shock proteins or other proteins that serve as chaperones. Because these changes occur very early after exposure to the toxicant, they may be used as molecular markers of exposure. Changes in expression of other proteins may serve as additional indicators of exposure and may also be indicators of the type of chemical that is causing the renal dysfunction. In other words, by observing the nature of the gene expression change, one can make conclusions about what type of chemical (e.g., oxidant, reactive electrophile, mutagen, etc.) is involved in the toxicologic process.

Besides using gene expression changes as indicators of exposure or as measures of nephrotoxicity, certain changes can be used as indicators of specific patterns of exposure, such as differences in nephron cell type (e.g., glomerular versus PT cell, or PT cell versus distal tubular cell). In some cases, noninvasive methods can be used to provide an indication or biomarker for exposure, such as excretion into the urine of specific proteins or enzymes. These types of measurements range in sensitivity from highly sensitive to rather insensitive.

Other effects may serve as markers for renal cellular repair and regeneration. When renal tissue undergoes repair and regeneration, differentiation status is altered as new cells exhibit many properties that are not characteristic of mature epithelial cells. Various cell surface markers differ in these newly synthesized cells. Similarly, renal tissue that has undergone neoplastic transformation expresses many properties that are different from those of the mature epithelial cells.

This chapter focuses on some of the molecular changes that are elicited by toxicant exposure or that are observed in diseased states. The ability of these molecular changes to serve as biomarkers of exposure or of mechanisms of toxicity are discussed. Initial discussion concerns how measurements of urinary enzymes or proteins can be used as markers for exposure or effect. Several cellular proteins whose level of expression indicate various aspects of renal cellular function and can provide significant mechanistic clues about the toxicant of interest are then discussed. Finally, the last two sections briefly consider selected markers that are characteristic of renal tissue undergoing repair and regeneration and renal tissue that has undergone neoplastic transformation. Various aspects of these highlighted molecular changes are considered in greater detail in other chapters of this volume, and the reader is referred to these for more in depth material. For example, in Chapter 8, the function of signaling pathways and stress response mechanisms in the renal cellular response to nephrotoxicants are described, and mechanisms of chemically induced nephrocarcinogenesis are discussed in Chapter 9.

URINARY ENZYMES AND OTHER PROTEINS

Perhaps one of the most widely used, standard approaches to studying nephrotoxicity involves detection of changes in expression or levels of selected proteins that are secreted into urine as a consequence of exposure to nephrotoxic chemicals. Advantages of this approach include that this is a noninvasive method that can be used in both experimental animals and in humans in the clinical setting and that this method can be highly sensitive for selected proteins that are normally not found in urine. Significant disadvantages also exist, and include a lack of sensitivity or specificity for some markers. A major goal in the field of toxicology is therefore to identify biomarkers, particularly ones that can be measured by noninvasive means (i.e., by collection of body fluids), that are both highly sensitive and are specific to responses or effects in a given nephron segment or cell type.

Much of the rationale behind these types of assays and approaches is described in a publication from the U.S. National Academy of Science, "Biomarkers in Urinary Toxicology" (National Research Council, 1995). This report focuses on biomarkers of susceptibility, exposure, and effect. For the first case, these are mostly genetic markers (e.g., genetic polymorphisms) that are indicators of how an individual will respond to a toxicant, and are discussed in this chapter. The second and third types of biomarkers include chemicals or proteins that are either indicators that the tissue has been exposed to a chemical of interest or that some change in the tissue has occurred that indicates a response to a chemical of interest.

Biomarkers of exposure can include metabolites of a particular chemical, or molecules from the tissue that are secreted after exposure to the chemical. Such a biomarker is usually a very early indicator that should be very sensitive and should only be an indicator that the chemical of interest is present and should not produce any modulation in cellular or tissue function. In contrast, a biomarker of effect is more readily understood to be a chemical or metabolite that is produced by the tissue in response to exposure and is usually a consequence of some alteration in cellular function. If the biomarker of exposure is good, then it can also provide mechanistic information, such as the specific cell type that is affected by the chemical exposure or the type of chemical that is causing the tissue dysfunction. This section focuses briefly on biomarkers of effects, as they are more easily defined and identified.

One important function of a biomarker of effect in which the kidneys are the target tissue is the ability to distinguish between toxicants that affect the glomerulus versus those that affect the tubular epithelial cells. Moreover, some markers can be used to distinguish further between effects on different segments of the nephron. Table 2.1 lists a few selected biomarkers that are excreted into the urine and can be used to

Table 2.1 Selected Urinary Biomarkers for Assessment of Nephrotoxicity

Marker	*Comments*	*Specificity*
Proteinuria	< 20 kDa molecular weight (e.g., extrarenal proteins such as β2-microglobulin)	PT cells
Proteinuria	> 20 kDa molecular weight (e.g., extrarenal proteins such as albumin)	Glomerulus
N-acetylglucosaminidase (NAG)	Secreted from lysosomes of damaged cells	PT cells
Brush-border membrane proteins	E.g., alkaline phosphatase, γ-glutamyltransferase	PT cells
Tamm-Horsfall protein (THP)	Called urinary mucoprotein	Thick ascending limb, distal tubule

distinguish the region of the kidney that is affected by a toxicant. It should be apparent from the listing that the largest variety of sensitive biomarkers of effect are available for the PT region of the tubular epithelium. Of these markers, γ-glutamyltransferase (GGT), *N*-acetylglucosaminidase (NAG), and β2-μglobulin (β2MG) are excellent indicators of effects on the PT cells, whereas Tamm-Horsfall protein (THP) is a selective and sensitive marker of the distal nephron (thick ascending limb cells in particular).

MOLECULAR MARKERS OF RENAL CELLULAR FUNCTION AND CYTOTOXICITY

Many of the biomarkers of effect that also provide information about how renal cellular function is altered by exposure to a nephrotoxicant include small to large proteins that fall into the category of stress-response proteins. These stress-response proteins play critical roles in the ability of cells to adapt to changes in the environment, and have been portrayed as a cellular analogy to the "fight-or-flight" reaction that is critical to the survival of the organism (Goligorsky, 2001). At the cellular level, there are a myriad of signals that can induce stress-response proteins, such as oxidant stress, alkylation of critical functional groups on proteins, lipids, or nucleic acids, osmotic stress, oxygen deprivation, or

Table 2.2 Selected Biomarkers Linked to Renal Cellular Injury

Classification	*Example(s)*
Immunologic factors: humoral	Antibodies, antibody fragments, components of complement cascade, coagulation factors
Immunologic factors: cellular	Lymphocytes, phagocytes, eosinophils, basophils, etc.
Growth factors and cytokines	Platelet-derived growth factor (PDGF), transforming growth factor (TGF), tumor necrosis factor (TNF), interleukins, etc.
Lipid mediators	Prostaglandins, thromboxanes, leukotrienes
Extracellular matrix components	Collagens, procollagen, laminin, fibronectin
Adhesion molecules	Cadherins, catenins, actin, integrins
Transcription factors and protooncogenes	c-myc, c-fos, c-jun, c-Ha-ras, Egr-1, etc.
Tubular proteins (antigens)	Tamm-Horsfall protein, brush-border membrane proteins
Heat shock proteins	Low molecular weight (HSP25/27); high molecular weight (HSP70, HSP72, HSP90)

Source: Classification is based on Discussion in "Biomarkers in Urinary Toxicology" (National Research Council, 1995).

elevated temperature (i.e., heat). Depending on the mechanism by which the nephrotoxicity has been produced, a large array of markers of effect (both protein and nonprotein) can be detected; some are listed in Table 2.2.

Heat Shock Proteins

One of the earliest described stresses that leads to induction of specific proteins is heat, hence the name heat shock proteins, or HSPs. Many other disturbances in cellular function and homeostasis, such as those listed above, were subsequently demonstrated to act analogously to heat in causing increased expression of selected proteins. As a general principle, therefore, almost any type of perturbation in cellular homeostasis leads to a coordinated response that is genetically controlled and ultimately serves a protective and adaptive function.

The HSPs are thought to act primarily as molecular chaperones (Beck et al., 2000). The term molecular chaperone is used to designate proteins that function transiently in the folding or assembly (or both) of other proteins or oligomeric complexes. These proteins can also function to regulate the translocation of proteins to specific subcellular compartments. Although many of these HSPs are expressed constitutively, expression of many HSPs is also markedly induced by various physical, chemical, or pathological stresses. Furthermore, HSPs are divided into small and large, based on their molecular weight. Each class of HSP seems to play specific roles in cellular function during various stresses and may play central roles in recovery of renal cellular function after sublethal stress. While these highly conserved proteins are expressed throughout the tissues of the body, in the kidneys they exhibit specific distributions along the nephron, and function in such diverse processes as regulation of folding, assembly, subcellular distribution, and degradation of proteins, and preservation and restructuring of the cytoskeleton (Beck et al., 2000).

The small HSPs (HSP25/27 and the crystallins) appear to largely function in the regulation of actin dynamics. These small stress proteins (designated HSP25 in rodents and HSP27 in humans) are single-copy gene products. Several studies have demonstrated a close correlation between subcellular distribution and protein and mRNA expression levels of these small HSPs and conditions that lead to specific changes in actin cytoskeletal structure and distribution (e.g., Aufricht et al., 1998; Shelden et al., 2002; Van Why et al., 2003). For example, ATP depletion by either renal ischemia *in vivo* or the use of metabolic inhibitors *in vitro* leads to marked changes in cytoskeletal organization and changes in expression and subcellular distribution of HSP25/27. HSP27 mRNA expression is also increased after exposure of cultured human PT cells to other types of agents that are known to interact with cytoskeletal proteins, such as $CdCl_2$ (Somji et al., 1999a) and the nephrotoxic cysteine conjugate of trichloroethylene *S*-(1,2-dichlorovinyl)-L-cysteine (DCVC).

Among the many effects associated with exposure of the kidneys to DCVC are alterations in cytoskeletal structure and function (van de Water et al., 1996, 1999a). Figure 2.1 demonstrates the concentration-dependent increase in expression of HSP27 protein in primary cultures

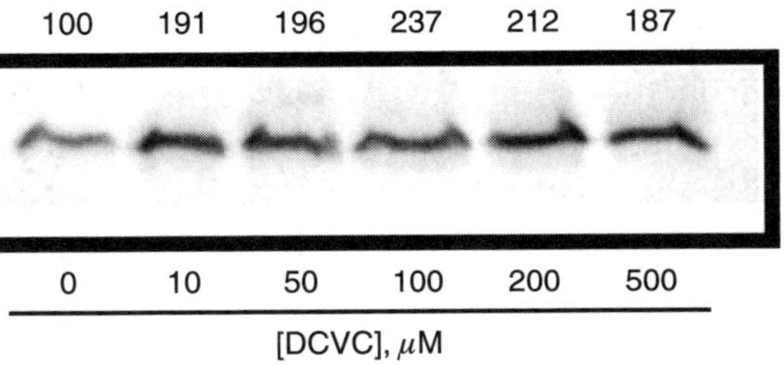

Figure 2.1 Effect of *S*-(1,2-dichlorovinyl)-l-cysteine (DCVC) on expression of HSP27 in human proximal tubular (hPT) cells. Confluent primary cultures of hPT cells were incubated for 2 hr with DCVC (0, 10, 50, 100, 200, or 500 μM). Samples were processed and analyzed for expression of HSP27 by immunoblot analysis, using specific polyclonal antibody (purchased from a commercial vendor) that recognize the human protein. Numbers under the bands represent the concentration of DCVC with which the hPT cells were incubated. Numbers above the bands represent the relative density of protein bands, as quantified with NIH Image software v. 1.6.2.

of human PT cells exposed to DCVC (Lash LH, Putt DA, and Hueni SE, unpublished data).

As can be readily observed, even as low a concentration of 10 μM DCVC increased expression of HSP27 protein by approximately twofold after a 2-hour incubation. This dose–response relationship suggests that HSP27 can be used as a very sensitive marker for DCVC exposure because at this concentration of DCVC and incubation time, little cytotoxicity occurs in human PT cells (Lash et al., 2001).

Expression of the larger HSPs is also altered in response to some of the same stimuli that alter expression of the small HSPs. Therefore, for example, a loss of HSP70 mRNA expression occurs in the inner medulla after induction of a nonoliguric, acute renal failure (Cowley and Gudapaty, 1995). Sens and colleagues demonstrated increased expression of both HSP70 mRNA and protein in human PT cells under some, but not all, conditions of exposure to toxicants (Somjo et al., 1999b). The authors concluded that induction of expression of HSP70 mRNA does not result from a direct effect on the transcription of a single gene but, rather, results from a complex interplay of multiple genes.

An important aspect of how cells respond to external stimuli involves an understanding of the impact of prior stress on subsequent exposure to stress. Prior heat stress can make renal cells relatively resistant to toxicity from additional toxicant exposure. This adaptive or protective response appears to include both small HSPs (HSP25/27) as well as many forms of the larger HSPs, including HSP72. For example, He and Lemasters (2003) demonstrated a selective role for HSP25, as opposed to the larger HSPs, in suppression of the mitochondrial membrane permeability transition due to various toxicants, including $HgCl_2$ and calcium ion overload. Borkan and colleagues, however, demonstrated protective roles involving a prominent member of the larger HSPs (Wang et al., 1999). They showed that HSP72 co-immunoprecipitated with Bcl-2 and prevented apoptosis of opossum kidney proximal tubular cells induced by ATP depletion. Prior heat stress increased the interaction between HSP72 and Bcl-2, thereby further preventing apoptosis. Along similar lines, Borkan and colleagues in another study (Ruchalski et al., 2003) showed that HSP72 inhibited release of apoptosis-inducing factor in ATP-depleted renal epithelial cells (cells treated with cyanide and incubated in medium without glucose), and in an additional study (Li et al., 2002) showed that prior heat stress prevented mitochondrial injury in ATP-depleted renal epithelial cells by a large increase in HSP72 expression.

Heme Oxygenase-1

Elimination of excess free heme in cells is accomplished by metabolism by heme oxygenases (HOs), which catalyze the rate-limiting step in heme catabolism, involving oxidative cleavage of the porphyrin ring to generate biliverdin IXα, heme iron, and carbon monoxide. Of the two functional HO isoforms, heme oxygenase-1 (HO-1) is considered the more important as its expression is highly inducible (Maines and Panahian, 2001). HO-1, also known as HSP32, is considered an oxidative stress responsive gene that rapidly responds to various stresses. It is present in many tissues, but appears to be particularly important in the kidneys because activity levels of biliverdin reductase, which produces the biologically active metabolite biliverdin, are highest in the kidneys of any tissue, including the liver (McCoubrey et al., 1995). As with the other HSPs, HO-1 induction by a prior exposure to a nephrotoxicant can protect the kidneys against injury from subsequent exposures to nephrotoxicants or pathological conditions.

Integrins and Cytoskeletal Proteins

As alluded to above, structure and function of the cytoskeleton in renal PT cells can be markedly and rapidly altered by exposure to various toxicants or in pathological states, such as ischemia. Specific patterns of expression of selected cytoskeletal proteins are observed during kidney development and in disease or intoxicated states.

One prominent family of cytoskeletal proteins whose altered state can be a molecular biomarker for exposure to nephrotoxicants is the integrins (see Kreidberg and Symons, 2000, for a recent review). Integrins are heterodimers composed of a single α-subunit and a single β-subunit. While there are more than 20 different integrin heterodimers expressed in mammalian kidney, those containing the $\beta1$-subunit are the most common. $\beta1$-Integrins serve as receptors for extracellular matrix proteins, including fibronectin, collagen, vitronectin, thrombospondin, and certain forms of laminin.

Alterations in cytoskeletal structure and cell surface blebbing during the early stages of exposure to toxic chemicals or in pathological states are common observations (e.g., see Chen and Wagner, 2001, and references cited therein). Disruption of integrin localization and expression by such chemical exposures is likely to be a proximate mechanism by which the characteristic blebbing occurs. Several studies illustrate this point, in particular the work of Goligorski and colleagues. For example, in one early study (Gailit et al., 1993) using a primate kidney epithelial cell line (BS-C-1 cells), oxidative stress produced by a brief, nonlethal, exposure to H_2O_2 caused a disruption of focal contacts, disappearance of the cytoskeletal protein talin from the basal cell surface, and a redistribution of integrin $\alpha3$-subunits from a predominantly basal localization to a predominantly apical localization. This oxidative stress also reduced cell adhesion to various extracellular matrix proteins, including type IV collagen, laminin, fibronectin, and vitronectin. Expression and membrane localization of integrins may be used as an early marker for cell injury.

Repair of renal structure after acute injury requires repopulation of the epithelium and is, therefore, dependent on proliferation. Additionally, appropriate differentiation of the newly produced cells is needed, and occurs by a process involving sequential expression of various stress proteins, cell-cycle specific genes, and then other organ-specific genes. The actin cytoskeleton is critical to these repopulation and differentiation processes. Integrins are among the group of actin-associated proteins whose proper localization and expression are essential to these repair processes (Paller, 1997). One of the events that follows binding of integrins to the extracellular matrix is activation of focal adhesion kinase (FAK), which is a 125 kDa nonreceptor tyrosine kinase that is localized at focal adhesions. After integrin binding to the extracellular matrix, FAK is autophosphorylated on tyrosine-397, which then serves as a docking site for the SH2 family of Src family kinases as well as PI-3 kinase. FAK phosphorylation status may therefore serve as an additional biomarker of renal cellular function.

As noted above, van de Water and colleagues demonstrated that acute renal cellular injury induced by the nephrotoxicant DCVC is associated with F-actin disorganization and cleavage of FAK (van de Water et al., 1996, 1999a). They also demonstrated that enhancement of FAK function by stable transfection of LLC-PK_1 cells with a green fluorescent protein–FAK construct delayed the onset of DCVC-induced apoptosis (van de Water et al., 2001). This provides additional evidence for the critical role of FAK and integrins in renal cellular function and differentiation.

Kidney Injury Molecule-1

Kidney injury molecule-1 (KIM-1) is a type I membrane glycoprotein with extracellular immunoglobulin and mucin domains, which has been proposed as a novel marker for renal injury and repair (Bailly et al., 2002; Ichimura et al., 1998, 2003). Although KIM-1 mRNA and protein are expressed at very low levels in normal rodent kidney, expression increases dramatically after injury from either exposure to nephrotoxicants (e.g., *S*-(1,1,2,2-tetrafluoroethyl)-L-cysteine, folic acid, or cisplatin) or after ischemia. Rat and human cDNA clones for KIM-1 were isolated and characterized, and the complete cDNA sequence for rat KIM-1 is listed in GenBank (accession number AF035963). Subsequent to renal cellular injury, KIM-1 expression is localized to regenerating PT epithelial cells (Ichimura et al., 1998). Localization was established by immunohistochemistry and RNA *in situ* hybridization, and was restricted to regenerating PT cells and not to adjacent, undamaged cells. This was demonstrated by selective detection in cells that were bromodeoxyuridine- and vimentin-positive (see below for further discussion of vimentin). Moreover, the protein was primarily localized to the apical membrane of regenerating S_3 PT cells, although cytoplasmic expression was also detected.

Bailly et al. (2002) showed that the soluble form of KIM-1 is constitutively shed into the culture medium of several human cell lines (including the PT cell line HK-2 cells) expressing either endogenous or recombinant human KIM-1b (a splice variant). Release of KIM-1 into the medium was enhanced by treatment with a phorbol ester and was blocked by two metalloproteinase inhibitors. The authors proposed, therefore, that shedding of KIM-1 in regenerating kidney constitutes an active mechanism that allows dedifferentiated regenerating cells to repopulate denuded patches of the basement membrane to reconstitute a continuous epithelial layer.

Ichimura et al. (2003) extended these investigations to *in vivo* treatment of rats with nephrotoxicants. Their studies demonstrated detection of KIM-1 protein in the urine of rats treated with either *S*-(1,1,2,2-tetrafluoroethyl)-L-cysteine, folic acid, or cisplatin. The increase in urinary KIM-1 occurred despite no significant increase in serum creatinine. The authors concluded that both up-regulation of KIM-1 expression and its presence in urine in response to diverse nephrotoxicants suggest that KIM-1 may serve as a sensitive and early biomarker for tubular injury and repair.

α2u-Globulin

Although there is a general consensus that renal cellular accumulation of α2u-globulin in response to various halogenated organic solvents is a male rat-specific response that has no relevance for humans (National Research Council, 1995), a very brief discussion is included here for completeness (see Lash et al., 2000, for a thorough discussion of the relevance of the α2u-globulin response for human health risk assessment for trichloroethylene, perchloroethylene, and other hydrocarbons). α2u-Globulin is the unique major component of the urinary protein load in male rats, although homologous proteins exist in other species, including humans. The generally accepted mechanism of chemically induced accumulation of α2u-globulin in male rats is that the chemical interferes with reabsorption of proteins from the glomerular filtrate by the proximal tubules, leading to excessive accumulation of α2u-globulin in lysosomes of renal PT cells and the so-called "hyaline (protein) droplet nephropathy." While most of the chemicals that have been shown to induce this response are halogenated organic solvents, the mycotoxin ochratoxin A also causes this response; this is believed to be central to the markedly higher susceptibility of male rats to renal toxicity and tumors (Rásonyi et al., 1999).

As noted above, homologous proteins are found in female rats and in other species, including humans. Nonetheless, hyaline droplet nephropathy is only observed in male rats; the reasons for this sex and species specificity remain unclear. Therefore, while urinary excretion of α2u-globulin is an excellent biomarker for exposure to, and nephrotoxicity from, numerous halogenated solvents in male rats, it cannot be used as an indicator of responses that might occur in female rats or in other species of interest.

Endoplasmic Reticulum Stress Response and DNA Damage Genes

Thiol oxidants and alkylating agents and, under the appropriate conditions, thiol reductants are cytotoxic to eukaryotic cells, including renal PT cells. Stevens and colleagues (Chen et al., 1992; Halleck et al., 1997; van de Water et al., 1999b) demonstrated that certain types of toxicant exposures activate members of the glucose-regulated protein (Grp) family. The Grp proteins, exemplified by Grp78 and Grp94, are found in the endoplasmic reticulum (ER) and function as chaperones that mediate protein folding and maturation. Because the glutathione pool is largely oxidized within the ER, which is thought to be important for protein disulfide bond formation and proper protein folding, thiols including dithiothreitol disrupt these processes and lead to increased transcription of Grp78 and Grp94.

In addition to the ER stress response proteins, Stevens and colleagues (Chen et al., 1992; Haleck et al., 1997) have found that, concomitant with these inductions, increase in expression of the growth arrest and DNA damage 153 (*gadd153*) gene are observed. GADD153 is a member of the C/EBP family of leucine zipper proteins, but is unique in how it interacts with DNA. Because of certain structural features, GADD153 can act as a negative regulator of CCAAT/enhancer-binding-protein (C/EBP)-mediated gene regulation and is thus believed to be important in regulating growth arrest as a response to some toxicants and DNA-damaging agents. The list of chemicals or conditions that lead to disruption of ER function and induction of both ER stress genes and *gadd153* include: hypoxia, calcium ionophores, glucose deprivation, agents that block glycosylation of proteins, thapsigargin (Ca^{2+}-ATPase inhibitor), thiol alkylating agents such as iodoacetamide, nephrotoxic cysteine conjugates such as DCVC, and ultraviolet irradiation. Hence, like the list of agents that induce HSPs, a diverse group of chemicals and pathological conditions can induce the ER stress response. Interestingly, Halleck et al. (1997) suggest that while oxidative stress primarily induces HSPs and related proteins, reductive stress primarily induces ER stress proteins and related proteins. This observation should not, however, be construed as meaning that ER stress proteins and GADD153 are not induced by oxidants. Guyton et al. (1996) showed, for example, that H_2O_2 induces GADD153 in HeLa cells.

Moreover, like the HSPs, a so-called preconditioning whereby exposure to a nonlethal dose of some chemical or other agent that induces the ER stress response also provides protection from oxidative injury (Hung et al., 2003). The protection mechanism is mediated through the mitogen-activated protein kinase (MAPK) signaling pathway, and involves enhanced activation of extracellular signal-regulated kinase 1/2 (ERK1/2) and decreased activation of c-Jun amino-terminal kinase (JNK). The authors concluded that the prior ER stress response modulates the balance between ERK and JNK activation, thereby preventing or diminishing cell death after subsequent oxidant exposure.

MAPK PATHWAY

The mitogen-activated protein kinase (MAPK) signaling pathway has emerged over the past decade as a central component in the regulation of many cellular processes, including apoptosis, necrosis, cell proliferation, and in various disease processes. More importantly, and particularly for this chapter, the MAPK pathway has been shown to play important roles in renal repair and regeneration after chemical- or ischemia-induced acute renal failure. As described by Tian et al. (2000) in their review, the basic MAPK signaling pathway can be divided into three modules, consisting of: a particular MAPK, its upstream activator (MAPK kinase, MKK), and a further upstream activator (MAPK kinase kinase, MKKK) (Figure 2.2).

As shown in Figure 2.2, MAPKs are divided into three families on the basis of sequence similarities, upstream activators, and substrate specificity. It should be apparent from the duplicate listings in several cases for upstream activators and effectors that there is much cross talk between the three families. The three MAPK families (extracellular signal-regulated kinases (ERKs), c-Jun amino-terminal kinases (JNKs), and p38 kinases) exhibit some specificity in their ultimate effector, with ERK1,2 being involved in responses to mitogenic (i.e., growth factor related) signals, whereas JNKs and p38 MAPKs generally respond to signals of stress and inflammation.

MAPK activation patterns have been suggested to be essential for renal regeneration and repair processes after toxicant exposure or the

	Induced by mitogens	Induced by stress and mitogens	Induced by stress	
MKKK	Raf-1, Mos MEKK1-3	MEKK1-4 TAK, ASK1, MLK3	TAK,ASK1, MLK3 PAK	
MKK	MEK1,2	MKK4,7	MKK3,6	MKK5
MAPK	ERK1,2	JNK1-3	p38α-δ	ERK5
Effector	c-Myc, Elk-1	Elk-1, c-Jun, ATF-2	ATF-2 CHOP (= GaDD153)	MEF2c

Figure 2.2 Scheme showing putative activators and effectors of the major MAPK families. Selected MAPK effectors (transcription factor substrates) are shown. Abbreviations: ERK, extracellular signal-regulated kinase; MAPK, mitogen-activated protein kinase; MEK, MAPK/ERK kinase; MEKK, MAPK/ERK kinase; MKK, MAPK kinase; MKKK, MAPK kinase kinase; PAK, p21-activated kinase; JNK, c-Jun N-terminal kinase; CHOP, CCAAT/enhancer binding protein (also known as GADD153: growth arrest and DNA damage 153); MEF, monocyte enhancer factor. (Based on a figure from Tian et al., 2000.)

occurrence of pathological conditions such as ischemia. Di Mari et al. (1999) demonstrated the requirement for ERK activation after ischemic injury. Interestingly, these authors observed a nephron segment specific pattern of MAPK activation after ischemic injury; whereas JNK was activated in both cortex and inner stripe of the outer medulla, the ERK pathway was only activated in the inner stripe. Moreover, *in vitro* studies conducted with mouse PT cells and thick ascending limb cells in culture showed that oxidative stress caused JNK activation in both cell types but caused ERK activation only in thick ascending limb cells. Similarly, Monks and colleagues (Ramachandiran et al., 2002) found that reactive oxygen species generated from 2,3,5-tris-(glutathione-*S*-yl)hydroquinone, which is a well-established nephrotoxicant that produces oxidative stress, are associated with activation of all three major families of MAPKs in LLC-PK_1 cells. Therefore, in contrast with other well-documented functions in promotion of cell survival and repair, activation of ERKs and p38 MAPK are coupled with cell death.

A study by Chaturvedi et al. (2002) provided results that were more consistent with those showing the standard protective roles for ERK activation. In this study, the investigators examined oxalate-induced cytotoxicity in LLC-PK_1 cells. Oxalate is a metabolic end product and is important in the pathogenesis of renal stone disease. While exposure of renal epithelial cells to oxalate was known to be associated with unscheduled DNA synthesis, altered gene expression, and apoptosis, the signaling pathways that coordinate these responses were unknown. The authors of the study demonstrated that exposure of LLC-PK_1 cells to 1 m*M* oxalate causes a robust phosphorylation and activation of p38 MAPK but only a modest activation of JNK and no effect on activation of ERKs. Therefore, as in the study by Di Mari et al. (1999), a higher ratio of ERK activation to JNK activation is associated with proliferative responses that lead to cell survival.

Other Growth Factors and Cell Cycle Regulators

In addition to the various proteins noted above, altered expression of other growth factors or growth factor receptors is also associated with renal injury and these factors and receptors can potentially be used as biomarkers for injury and regeneration. For example, folate-induced acute renal failure has been associated with a decreased mRNA expression for epidermal growth factor (EGF), which is expressed predominantly in the cortex and outer medulla (comprised predominantly of PT cells) (Cowley and Gudapaty, 1995). This finding is consistent with the observation that the apoptosis that occurs in primary cultures of mouse PT cells after growth factor deprivation was prevented by addition of EGF (Lieberthal et al., 1998).

Megyesi et al. (2002) demonstrated the critical importance of cell-cycle regulators in the progression of repair after acute renal injury. Although regulation of the cell cycle involves a complex interaction between both negative and positive factors controlling transcription and translation, protein activity and stability, and compartmentalization of substrates and enzymes, the studies focused on p21 and 14-3-3σ. p21 is a 21 kDa protein that is expressed most prominently in terminally differentiated or senescent cells, is induced by p53, and is required for p53-dependent cell-cycle arrest after DNA damage. Induction of p21 is associated with cell-cycle arrest and the protein acts, at least in part, by inhibiting one or more cyclin-cdk kinase activities. Proteins of the 14-3-3 family were originally discovered as abundant acidic proteins from brain. They bind to several proteins involved in cell-cycle regulation and signal transduction and act primarily to inhibit the G_2 to M transition by binding to phosphorylated Cdc25 proteins, thereby inhibiting their activity. Induction of p21 protected mice from acute renal failure caused by treatment with cisplatin or ischemia (Megyesi et al., 2002). 14-3-3σ was induced after acute renal failure. The authors suggest that it is likely that this protein functions to help coordinate the cell cycle to maximize recovery of renal epithelial cells from injury and to reduce the extent of injury itself.

Along the same lines, Javelaud and Besançon (2002) showed that inactivation of p21 sensitized HT116 human colon cancer cells to apoptosis. Their study suggested that p21 normally exerts an antagonistic effect on the mitochondrial pathway of apoptosis. Nowak et al. (2003) found that p21 knockout mice were more sensitive to apoptosis induced by cisplatin. However, it actually appears that besides the overall sensitivity to cisplatin-induced apoptosis, the mechanism of renal cellular apoptosis induced by cisplatin differs in mice with or without p21. In the presence of p21, cisplatin activates caspase-3 through a caspase-8 and caspase-9 independent pathway, whereas in the absence of p21, caspase-9 activation is required for activation of caspase-3.

MOLECULAR MARKERS OF RENAL CELLULAR REPAIR AND REGENERATION

It has already been noted that vimentin (which is not expressed in differentiated epithelial cells but is a marker for endothelial cells) is expressed when kidneys undergo repair and regeneration after injury, because the newly synthesized cells are dedifferentiated. Wallin et al. (1992) showed that both *in vivo* exposure of rats and *in vitro* exposure of rat PT cell primary cultures to DCVC leads to renal PT cell proliferation and vimentin expression during the recovery phase. The authors concluded that there are five phases of cell injury and nephrogenic repair:

- Resting cells exposed to toxicant or pathological condition
- Necrosis and sloughing of dead cells
- Change in cell shape, migration and/or cell division
- Repopulation of basement membrane and reestablishment of cell–cell contact
- Differentiation

Distinct biomarkers should be present to enable identification of each phase of this progression, as discussed above.

An illustration of vimentin expression after exposure to a nephrotoxicant is given in Figure 2.3. Primary cultures of rat PT cells were

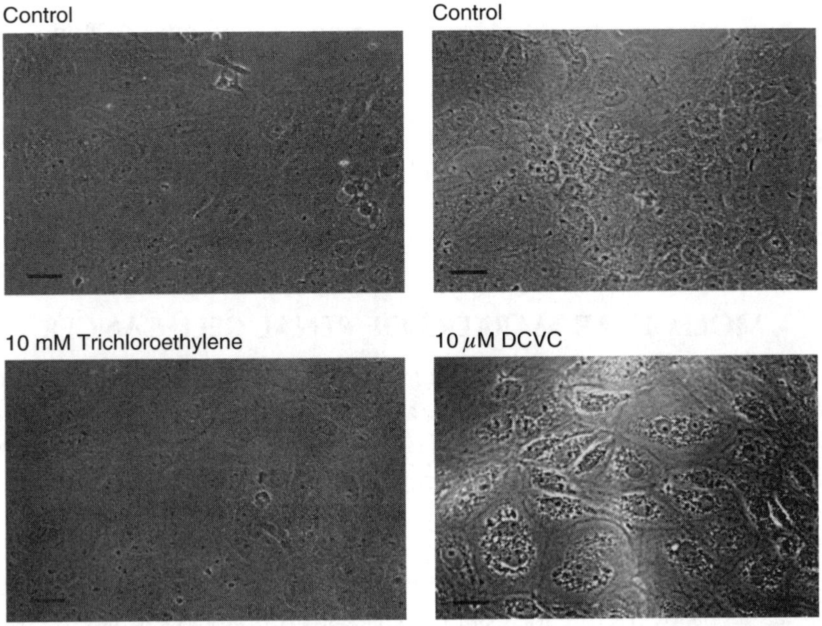

Figure 2.3 Effect of trichloroethylene and DCVC on vimentin expression in primary cultures of rat renal PT cells. Freshly isolated PT cells were seeded at a density of 0.5 to 1.0×10^6 cells/ml and allowed to grow for 24 hr prior to treatment with medium (control), trichloroethylene (10 m*M*), or DCVC (10 μ*M*). After treatment for 72 hr, cells were washed twice with sterile PBS and allowed to recover in appropriate media for 3 hr. Vimentin expression was visualized using a monoclonal Texas Red-conjugated anti-mouse vimentin antibody. Photomicrographs were taken at ×100 magnification on a Zeiss Confocal Laser Microscope. Bar = 5 μm.

incubated with either 10 m*M* trichloroethylene or 10 μ*M* DCVC for 72 hr. Control cells do not express vimentin, however cells incubated with either trichloroethylene or its metabolite DCVC clearly expressed the protein. These results indicate that the cells were undergoing repair and regeneration.

Protein kinase C (PKC) is another signaling molecule whose status plays important roles in renal repair and whose state of expression and phosphorylation have been associated with maintenance of renal cellular function during repair and regeneration. Stevens and colleagues (Dong et al., 1993, 1994) showed that the various isozymes of PKC are differentially expressed or phosphorylated during different stages of development, toxicity, or regeneration and proliferation. Hence, PKC status may be used as a biomarker for renal cellular function and regeneration, much like vimentin. PKC appears to be particularly important for the maintenance of mitochondrial function and active Na^+ transport in regenerating rat kidney after exposure to either DCVC or tert-butyl hydroperoxide. (Nowak, 2003; Nowak et al., 2004) (See Chapter 11 for additional discussion of this point.)

MOLECULAR MARKERS OF RENAL CELL CANCER

Although the subject of renal carcinogenesis is presented in Chapter 9, a few points regarding molecular markers are appropriate here. As discussed above, vimentin is not expressed in mature, quiescent renal epithelial cells but is expressed in regenerating cells. Similarly, a vimentin metaplasia was described by Ward et al. (1992) in renal cortical tubules of preneoplastic, neoplastic, aging, and regenerative lesions of both rats and humans. This finding is consistent with the preneoplastic or neoplastic tissue being in a state of dedifferentiation similar to that of regenerating tissue.

Yoon et al. (2002) demonstrated that loss of tuberin expression was required in chemically induced nephrocarcinogenicity. Initial toxicant exposure is associated with increased tuberin expression, suggesting that the tuberous sclerosis-2 (*Tsc2*) gene is an acute-phase response gene. High ERK activity, however, was associated with reexpression of the *Tsc2* locus in tuberin-negative cells, consistent with tuberin being a tumor suppressor and having growth inhibitory effects. The reader is referred to Chapter 9 for a more detailed discussion of this molecule and its role in a model for renal cancer.

von Hippel-Lindau (VHL) disease is a relatively rare hereditary disorder that involves multiple tumor target sites, including the kidneys (see Kaelin, 2003, for a brief review and references cited therein). Inactivation of the *VHL* tumor suppressor gene appears to be an early and causal event in the development of clear cell renal cell carcinoma and hemangioblastomas. Absence of the protein product, pVHL, may thus be used as a marker of dedifferentiated or neoplastic renal cells.

SUMMARY AND CONCLUSIONS

This brief discussion of molecular markers in normal and diseased or intoxicated kidney should make it apparent that much progress has been achieved over the past several years. There have been steps toward identifying mechanisms of injury and repair as well as specific molecular markers that can be used to identify either a toxicant exposure, pathological state, or renal cells that are undergoing repair and regeneration. Although this chapter has tried to categorize the various markers for organizational reasons, there is in reality a high degree of interaction between the molecular markers and pathways discussed. This interaction is illustrated in Figure 2.4, which summarizes

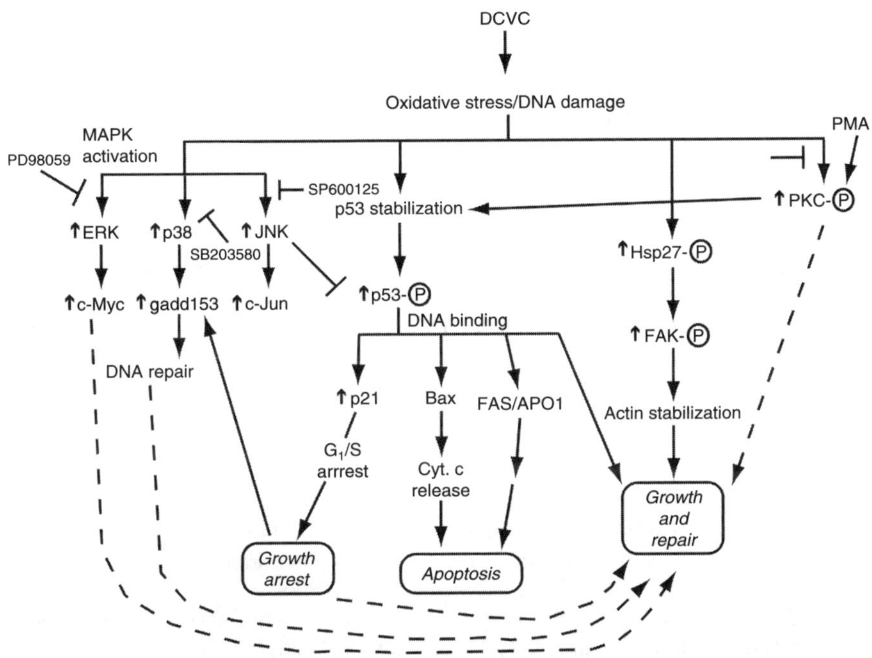

Figure 2.4 Scheme showing predicted effects of DCVC on cellular signaling pathways in renal proximal tubule. Arrows indicate predicted, stimulatory effects. Lines with cross bars indicate inhibitory effects. Dashed lines indicate processes leading to one of the three ultimate cellular responses. The Ps within circles (Ⓟ) indicate that the protein is phosphorylated. It is apparent that there is much cross talk between effects, and therefore multiple effects are expected. The specific changes that occur and the ultimate fate of the kidney cell is determined by the toxicant dose and exposure time. Abbreviations: DCVC, S-(1,2-dichlorovinyl)-l-cysteine; PKC, protein kinase C; PMA, phorbol myristic acid; HSP, heat shock protein; FAK, focal adhesion kinase; MAPK, mitogen-activated protein kinase; ERK, extracellular signal-regulated protein kinase; JNK, c-Jun N-terminal kinase; *gadd153*, growth arrest and DNA damage 153 gene.

the current knowledge of how exposure of renal PT cells to DCVC can cause growth arrest, apoptosis, or growth and repair.

REFERENCES

Aufricht C, Ardito T, Thulin G, Kashgarianb M, Siegel NJ, and Van Why SK (1998) Heat-shock protein 25 induction and redistribution during actin reorganization after renal ischemia. Am J Physiol 274:F215–F222.

Bailly V, Zhang Z, Meier W, Cate R, Sanicola M, and Bonventre JV (2002) Shedding of kidney injury molecule-1, a putative adhesion protein involved in renal regeneration. J Biol Chem 277:39739–39748.

Beck, F-X, Neuhofer W, and Müller E (2000) Molecular chaperones in the kidney: Distribution, putative roles, and regulation. Am J Physiol 279:F203–F215.

Chaturvedi LS, Koul S, Sekhon A, Bhandari A, Menon M, and Koul HK (2002) J Biol Chem 277:13321–13330.

Chen J, and Wagner MC (2001) Altered membrane-cytoskeleton linkage and membrane blebbing in energy-depleted renal proximal tubular cells. Am J Physiol 280:F619–F627.

Chen Q, Yu K, Holbrook NJ, and Stevens JL (1992) Activation of the growth arrest and DNA damage-inducible gene *gadd153* by nephrotoxic cysteine conjugates and dithiothreitol. J Biol Chem 267:8207–8212.

Cowley BD Jr, and Gudapaty S (1995) Temporal alterations in regional gene expression after nephrotoxic renal injury. J Lab Clin Med 125:187–199.

Di Mari JF, Davis R, and Safirstein RL (1999) MAPK activation determines renal epithelial cell survival during oxidative injury. Am J Physiol 277:F195–F203.

Dong L, Stevens JL, Fabbro D, and Jaken S (1993) Regulation of protein kinase C isozymes in kidney regeneration. Cancer Res 53:4542–4549.

Dong L, Stevens JL, Fabbro D, and Jaken S (1994) Protein kinase C isozyme expression and down-modulation in growing, quiescent, and transformed renal proximal tubule epithelial cells. Cell Growth Differentiation 5:881–890.

Gailit J, Colflesh D, Rabiner I, Simone J, and Goligorsky MS (1993) Redistribution and dysfunction of integrins in cultured renal epithelial cells exposed to oxidative stress. Am J Physiol 264:F149–F157.

Goligorsky MS (2001) The concept of cellular "fight-or-flight" reaction to stress. Am J Physiol 280:F551–F561.

Guyton KZ, Xu Q, and Holbrook NJ (1996) Induction of the mammalian stress response gene GADD153 by oxidative stress: role of AP-1 element. Biochem J 314:547–554.

Halleck MM, Holbrook NJ, Skinner J, Liu H, and Stevens JL (1997) The molecular response to reductive stress in LLC-PK_1 renal epithelial cells: coordinate transcriptional regulation by *gadd153* and *grp78* genes by thiols. Cell Stress & Chaperones 2:31–40.

He L, and Lemasters JJ (2003) Heat shock suppresses the permeability transition in rat liver mitochondria. J Biol Chem 278:16755–16760.

Hung C-C, Ichimura T, Stevens JL, and Bonventre JV (2003) Protection of renal epithelial cells against oxidative injury by endoplasmic reticulum stress preconditioning is mediated by ERK1/2 activation. J Biol Chem 278:29317–29326.

Ichimura T, Bonventre JV, Bailly V, Wei H, Hession CA, Cate RL, and Sanicola M (1998) Kidney injury molecule-1 (KIM-1), a putative epithelial cell adhesion molecule containing a novel immunoglobulin domain, is up-regulated in renal cells after injury. J Biol Chem 273:4135–4142.

Ichimura T, Hung CC, Yang SA, Stevens JL, and Bonventre JV (2003) Kidney injury molecule-1 (Kim-1): A tissue and urinary biomarker for nephrotoxicant-induced renal injury. Am J Physiol, in press (10.1152/ajprenal.00285.2002).

Javelaud D, and Besançon F (2002) Inactivation of $p21^{WAF1}$ sensitizes cells to apoptosis via an increase of both $p14^{ARF}$ and p53 levels and an alteration of the Bax/Bcl-2 ratio. J Biol Chem 277:37949–37954.

Kaelin WG Jr (2003) The von Hippel-Lindau gene, kidney cancer, and oxygen sensing. J Am Soc Nephrol 14:2703–2711.

Kreidberg JA, and Symons JM (2000) Integrins in kidney development, function and disease. Am J Physiol 279:F233–F242.

Lash LH, Hueni SE, and Putt DA (2001) Apoptosis, necrosis and cell proliferation induced by *S*-(1,2-dichlorovinyl)- L-cysteine in primary cultures of human proximal tubular cells. Toxicol Appl Pharmacol 177:1–16.

Lash LH, Parker JC, and Scott CS (2000) Modes of action of trichloroethylene for kidney tumorigenesis. Environ Health Perspec 108 (Suppl 2):225–240.

Li F, Mao HP, Ruchalski KL, Wang YH, Choy W, Schwartz JH, and Borkan SC (2002) Heat stress prevents mitochondrial injury in ATP-depleted renal epithelial cells. Am J Physiol 283:C917–C926.

Lieberthal W, Triaca V, Koh JS, Pagano PJ, and Levine JS (1998) Role of superoxide in apoptosis induced by growth factor withdrawal. Am J Physiol 275:F691–F702.

Maines MD, and Panahian N (2001) The heme oxygenase system and cellular defense mechanisms. Do HO-1 and HO-2 have different functions? Adv Exp Med Biol 502:249–272.

McCoubrey WK Jr, Cooklis MA, and Maines MD (1995) The structure, organization and differential expression of the rat gene encoding biliverdin reductase. Gene (Amst) 160:235–240.

Megyesi J, Andrade L, Viera JM Jr, Safirstein RL, and Price PM (2002) Coordination of the cell cycle is an important determinant of the syndrome of acute renal failure. Am J Physiol 283:F810–F816.

National Research Council (1995) Biomarkers in Urinary Toxicology. National Academies Press, Washington, D.C.

Nowak G (2003) Protein kinase C mediates repair of mitochondrial and transport functions after toxicant-induced injury in renal cells. J Pharmacol Exp Ther 306:157–165.

Nowak G, Price PM, and Schnellmann RG (2003) Lack of a functional $p21^{WAF1/CIP1}$ gene accelerates caspase-independent apoptosis induced by cisplatin in renal cells. Am J Physiol 285:F440–F450.

Nowak G, Bakajsova D, and Clifton GL (2004) Protein kinase C-ε modulates mitochondrial function and active Na^+ transport following oxidant injury in renal cells. Am J Physiol, 286:F307–F316.

Paller MS (1997) Integrins and repair after acute renal injury. Kidney Int 52 (Suppl 61):S52–S55.

Ramachandiran S, Huang Q, Dong J, Lau SS, and Monks TJ (2002) Mitogen-activated protein kinases contribute to reactive oxygen species-induced cell death in renal proximal tubule epithelial cells. Chem Res Toxicol 15:1635–1642.

Rásonyi T, Schlatter J, and Dietrich DR (1999) The role of α2u-globulin in ochratoxin A induced renal toxicity and tumors in F344 rats. Toxicol Lett 104:83–92.

Ruchalski K, Mao H, Singh SK, Wang Y, Mosser DD, Li F, Schwartz JH, and Borkan SC (2003) HSP72 inhibits apoptosis-inducing factor release in ATP-depleted renal epithelial cells. Am J Physiol 285:C1483–C1493.

Shelden EA, Borrelli MJ, Pollock FM, and Bonham R (2002) Heat shock protein 27 associated with basolateral cell boundaries in heat-shocked and ATP-depleted epithelial cells. J Am Soc Nephrol 13:332–341.

Somji S, Sens DA, Garrett SH, Sens MA, and Todd JH (1999a) Heat shock protein 27 expression in human proximal tubule cells exposed to lethal and sublethal concentrations of $CdCl_2$. Environ Health Perspec 107:545–552.

Somji S, Todd JH, Sens MA, Garrett SH, and Sens DA (1999b) Expression of the constitutive and inducible forms of heat shock protein 70 in human proximal tubule cells exposed to heat, sodium arsenite, and $CdCl_2$. Environ Health Perspec 107:887–893.

Stevens JL, Liu H, Halleck M, Bowes RC III, Chen QM, and van de Water B (2000) Toxicol Lett 112-113:479–486.

Tian W, Zhang Z, and Cohen DM (2000) MAPK signaling and the kidney. Am J Physiol 279:F593–F604.

van de Water B, Kruidering M, and Nagelkerke JF (1996) F-actin disorganization in apoptotic cell death of cultured rat renal proximal tubular cells. Am J Physiol 270:F593–F603.

van de Water B, Nagelkerke JF, and Stevens JL (1999a) Dephosphorylation of focal adhesion kinase (FAK) and loss of focal contacts precede caspase-mediated cleavage of FAK during apoptosis in renal epithelial cells. J Biol Chem 274:13328–13337.

van de Water B, Wang Y, Asmellash S, Liu H, Zhan Y, Miller E, and Stevens JL (1999b) Distinct endoplasmic reticulum signaling pathways regulate apoptotic and necrotic cell death following iodoacetamide treatment. Chem Res Toxicol 12:943–951.

van de Water B, Houtepen F, Huigsloot M, and Tijdens IB (2001) Suppression of chemically induced apoptosis but not necrosis of renal proximal tubular epithelial (LLC-PK_1) cells by focal adhesion kinase (FAK): Role of FAK in maintaining focal adhesion organization after acute renal cell injury. J Biol Chem 276:36183–36193.

Van Why SK, Mann AS, Ardito T, Thulin G, Ferris S, Macleod MA, Kashgarian M, and Siegelo NJ (2003) Hsp27 associates with actin and limits injury in energy depleted renal epithelia. J Am Soc Nephrol 14:98–106.

Wallin A, Zhang G, Jones, TW, Jaken S, and Stevens SL (1992) Mechanism of the nephrogenic repair response: Studies on proliferation and vimentin expression after ^{35}S-1,2-dichlorovinyl-L-cysteine nephrotoxicity *in vivo* and in cultured proximal tubule epithelial cells. Lab Invest 66:474–484.

Wang Y, Knowlton AA, Christensen TG, Shih T, and Borkan SC (1999) Prior heat stress inhibits apoptosis in adenosine triphosphate-depleted renal tubular cells. Kidney Int 55:2224–2235.

Ward JM, Stevens JL, Konishi N, Kurata Y, Uno H, Diwan BA, and Ohmori T (1992) Vimentin metaplasia in renal cortical tubules of preneoplastic, neoplastic, aging, and regenerative lesions of rats and human. Am J Pathol 141:955–964.

Yoon H-S, Monks TJ, Everitt JI, Walker CL, and Lau SS (2002) Cell proliferation is insufficient, but loss of tuberin is necessary, for chemically induced nephrocarcinogenicity. Am J Physiol 283:F262–F270.

3

Assessing Renal Effects of Toxicants *In Vivo*

Susan Emeigh Hart and Lewis B. Kinter

INTRODUCTION

The susceptibility of the mammalian kidney to the adverse effects of toxicants can be attributed to anatomical, physiological, and biochemical aspects of renal function. The kidneys receive a disproportionately large blood flow to support glomerular filtration and to sustain renal metabolism; high blood flow results in greater exposure to blood-borne chemicals than occurs in less-well perfused organs. Solute and water reabsorption along the nephron enriches the tubular fluid in those chemicals that remain in the tubular lumen, exposing the cells of the nephron to higher concentrations of toxicants than occur in the general circulation; chemicals with low aqueous solubilities may precipitate or crystalize, causing obstruction of the nephron or obliteration of renal tissue. Chemicals that are substrates for organic solute transport systems gain entry and accumulate within proximal tubule cells in amounts that are not reflected in cells outside of the kidney, leading to nephrotoxicity. Intrarenal metabolism of nontoxic chemicals can produce toxic metabolites that subsequently damage renal structures and functions.

Finally, the very processes that permit the kidney to regulate body fluid composition by producing urine more concentrated than plasma implicate a unique (and inhospitable) microenvironment within the renal medulla, outside the systemic homeostatic envelope, and another unique target for some toxicants. For example, erythrocytes are observed to alternatively crenate (shrink) and swell (in antidiuresis or diuresis, respectively) while passing through the osmotic gradient within the renal medulla, surviving only by the flexibility of their structure and the rapidity with which they equilibrate their internal volume in response to the changing osmotic pressure gradients. Papillary necrosis associated with sickle-cell erythrocytes reflects reduced ability of these cells to survive exposure to rapidly changing osmotic gradients, resulting in lysis and stasis within the vasa recta structures; toxicants that bind to red-cell proteins may simulate the sickle-cell effect. The significance of these observations extends beyond their implications for unique susceptibilities to toxicants. *In vitro* techniques that remove renal cells or tissue from their (as yet poorly) defined *in vivo* microenvironment and replace with standard extracellular media are at risk of introducing essentially arbitrary conditions more closely aligned with investigator convenience than reality. In studies of peptide antagonists of adenylate cyclase coupled vasopressin receptors (V2 receptors), evaluations of V2 receptor functions in renal tissue homogenates, isolated perfused cortical collecting tubules, and semi-purified renal medullary membranes (including human tissues) failed to detect V2 agonist activity of these compounds that was immediately apparent in phase I clinical trials (Kinter et al., 1988, 1993).

The anatomical and functional complexities and unique microenvironments of the kidney dictate that its functions be evaluated in conjunction with the other physiological systems upon which those functions depend (e.g., in intact animals). The investigation of suspected renal toxicants using intact animal preparations relies upon detecting and monitoring indicators (biomarkers) of renal injury and decrements in renal function based upon analyses of blood, urine, and tubular fluid by utilizing techniques developed over the centuries of clinical practice and refined over past approximately 100 years of renal research (Figure 3.1). Renal biomarkers and functional assessments can be combined or studied independently. A typical tiered strategy for detection of renal toxicants would consist of an assessment of renal biomarkers and histopathology within traditional general toxicology studies, followed by specific functional assessments, as required, and mechanism of action studies where appropriate (Figure 3.2). This chapter surveys the

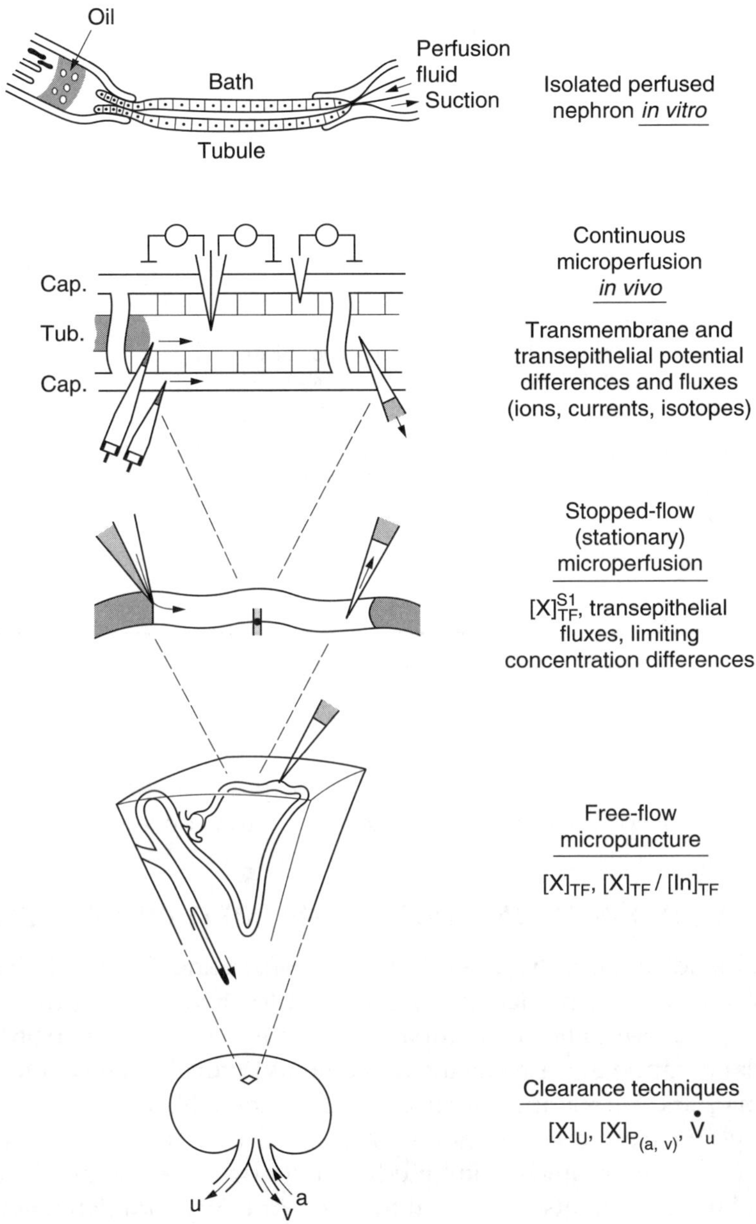

Figure 3.1 Schematic of methodologies used to assess renal function – from clearance technologies to isolated tubules and cells. (Source: K. Heirholzer, unpublished. Courtesy of G. Giebisch.)

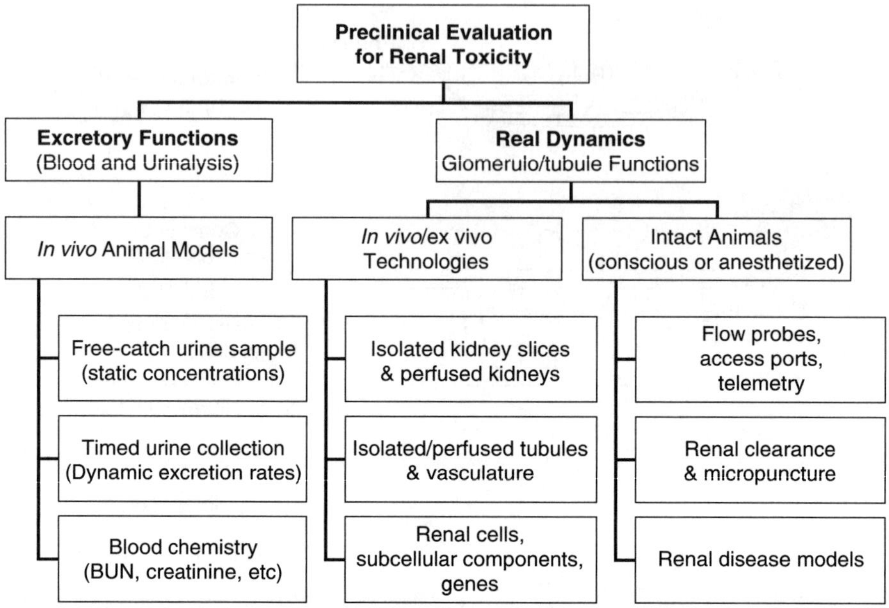

Figure 3.2 Hierarchy of renal function testing.

biomarkers (excluding histopathology) and functional assessment techniques used to evaluate renal actions of toxicants in intact animals.

OVERVIEW OF *IN VIVO* KIDNEY METHODOLOGIES

Animal studies are indispensable for demonstrating that the kidney is a site of injury for a particular toxicant and for investigating the specific mechanism. Toxicants can impact renal tubular (and interstitial) and vascular components, producing decrements in renal homeostatic, hemodynamic, and endocrine functions. Fundamentally, *in vivo* techniques employ analyses of components of blood (or plasma and serum) and urine (or tubular fluid) combined with knowledge of association of particular components with renal function or injury, or relationships that can be derived through using combinations of those components (e.g., clearance techniques) to identify decrements in renal functions or the presence of tissue injury. Both endogenous and exogenous substances are recognized. The following sections present and discuss biomarkers and tracers from blood and urine accepted for renal assessments in animals, and briefly discuss the techniques for their collection and analysis. Electronic and imaging techniques not dependent upon collections of blood and urine are also briefly presented. Finally, animal studies can yield tissue for morphological analyses, crucial for identification or confirmation of specific renal structures impacted by toxicants. The methodologies for preparation, evaluation, and interpretation of renal histology/histopathology are beyond the scope of this chapter and those interested in this topic are referred to Khan and Alden (2001), Greaves (2000), and Marcussen and Lundemose (1996).

MARKERS OF RENAL FUNCTION IN BLOOD

General Considerations and Limitations

The kidney functions to rid the body of metabolic end products (particularly organic acids and nitrogenous wastes) while maintaining water and electrolyte homeostasis in the face of variable intake and metabolic need. If a metabolic end product or protein is small enough to be filtered at the glomerulus, and is neither extensively reabsorbed nor secreted, the excretion rate is primarily determined by the glomerular filtration rate (GFR) (Price and Finney, 2000), which is considered a sensitive index of functional nephron mass (Newman and Price, 1999). Therefore, assessment of the plasma level of such a constituent can serve as a biomarker of renal function, and several have been used extensively for this purpose.

There are limitations to the utility of plasma markers of renal function. Because these tests serve primarily as measures of GFR, they are highly sensitive to modification by any factor that alters renal perfusion. Both prerenal factors (dehydration, blood loss, altered vasomotor tone, age-related decreases in renal blood flow [in rats]) and postrenal factors (obstruction or extravasation of urine to the peritoneal cavity) may cause elevations of the commonly measured analytes that do not reflect primary kidney injury. When these parameters are elevated as a consequence of primary renal injury they cannot be used to determine the location of that injury (glomerulus vs. tubule or tubule segment affected) (Baum et al., 1975; Corman and Michel, 1987; Finco, 1997; Newman and Price, 1999).

Additionally, these tests are limited by a lack of sensitivity to extremely low-level alterations in functional nephron mass, as assessed by GFR. This is due to the contributions of renal secretion or reabsorption of some analytes to their overall excretion, which can compensate for their decreased filtration, to wide variations in baseline levels of some analytes and to the inherent imprecision in the assays used (Finn and Porter, 1998; Price, 2002; Shemesh et al., 1985; Starr et al., 2002). A final contributing factor is the large renal functional reserve (e.g., capacity); for most common analytes, 25 to 50% of the nephron mass must be lost before the plasma level is elevated sufficiently to detect the functional decrement (Baum et al., 1975; Price, 2002; Swan, 1997). However, these tests can be performed on easily obtained samples using validated methods for which published reference ranges are available for most species; thus these tests can provide useful biomarkers for renal injury.

The appropriate comparator for a plasma analyte value in most toxicology studies is the mean value of the concurrent control group because it will reflect the effects of gender, diet, age, handling, test vehicle administration, and slight strain differences. When the changes in analyte values are substantial and the group sizes in the study are adequate for statistical analysis, this comparison may be sufficient. For smaller group sizes (where statistical comparisons are of limited value) or low-level changes in analyte values, comparison with a reference interval may help in determining the biological significance of the changes seen. There are numerous sources of published reference intervals for the more common plasma analytes (good sources for age, sex, and strain-specific data are the various suppliers of laboratory animals), but if possible, especially for the less-common analytes, reference data should be generated in the actual laboratory that will be analyzing the study samples. This is to control for laboratory variations resulting from differences in instrumentation or methods used. Additionally, reference intervals should be established separately for each species by gender and by age group; the latter is particularly important in establishing reference intervals for renal analytes as many of these have been shown to change as a function of age (Corman and Michel, 1987; Loeb, 1998; Loeb and Quimby, 1999; Slaughter and Everds).

Sample Collection Considerations

An important factor in the success of any study that relies on plasma analytes as a toxicologic endpoint is the ability of the investigator to consistently collect high-quality blood samples from the test species. For small laboratory rodents, test methods used and the frequency of sampling may be limited by the small volumes of blood obtainable and the need for anesthesia to collect them. For larger species, diurnal and annual variations in the levels of kidney-related analytes must be considered.

Table 3.1 lists recommended sample collection routes and maximum volumes obtainable in commonly used laboratory animal species.

Table 3.1 Recommended Blood Sample Volumes and Routes of Blood Collection for Commonly Used Toxicology Species

Species	*Blood volume (ml)*	*Maximum blood sample volume (ml)*[a]	*Collection site*	*Anesthesia needed?*	*Maximum sample size from site (ml)*
Mouse	1.8–2.5	0.3–0.5	Orbital sinus	Yes	0.1–0.4
			Hindleg vein[b,c]	No	0.1–0.15
			Tail vein[c]	No	0.1–0.15
			Tail tip amputation	Yes	0.1–0.2
			Jugular vein[c]	No	0.1–0.4
			Cardiac puncture[d]	Yes	0.5–1
Hamster	8	0.5–2	Orbital sinus	Yes	0.1–0.5
			Hindleg vein[b,c]	No	0.5–2
			Jugular vein[c]	No	0.5–2
			Cardiac puncture[d]	Yes	1–5
Rat	16–17	2.0–2.4	Orbital sinus	Yes	0.1–2
			Hindleg vein[b,c]	No	0.1–2
			Tail vein[c]	No	0.1–2
			Tail tip amputation	Yes	0.1–0.2
			Jugular vein[c]	No	0.1–3
			Sublingual vein[c]	Yes	0.2–1
			Caudal vena cava[d]	Yes	1–10
			Cardiac puncture[d]	Yes	1–10
Marmoset	25	3.5	Hindleg vein[b,c]	No	0.1–2
			Tail vein[c]	No	0.1–0.5
Guinea Pig	35	5.0	Orbital sinus	Yes	0.1–2
			Hindleg vein[b,c]	No	1–2
			Ear vein[c]	Topical	0.1–0.5
			Caudal vena cava[d]	Yes	1–5
			Cardiac puncture[d]	Yes	1–5

Continued

Table 3.1 Continued

Species	*Blood volume (ml)*	*Maximum blood sample volume (ml)*[a]	*Collection site*	*Anesthesia needed?*	*Maximum sample size from site (ml)*
Ferret	105	5–10	Orbital sinus	Yes	1–3
			Jugular vein[c]	No	1–10
			Tail vein[c]	No	1–3
			Cephalic vein[c]	No	1–5
			Cardiac puncture[d]	Yes	1–10
Rabbit	210–224	15–30	Ear vein[c]	Topical	1–5
			Ear artery[c]	Topical	1–30
			Jugular vein[c]	No	1–20
			Cardiac puncture[d]	Yes	10–60
Macaque	280–325	20–50	Hindleg vein[b,c]	No	1–20
			Cephalic vein[c]	No	1–20
Dog	850–1600	100–500	Cephalic vein[c]	No	1–10
			Jugular vein[c]	No	1–30
			Hindleg vein[b,c]	No	1–10
Minipig	975	140–200	Anterior vena cava[c]	No	1–100
			Ear vein	No	1–10

[a]For a single blood sample. For repeated sampling, the recommendation is not to exceed 20% of total blood volume in a 24 hr period.
[b]Either saphenous or femoral veins may be used.
[c]Recomended route(s) for repeated sampling.
[d]Reserved for terminal bleeding procedures.
Information from: Baumans et al., 2001; Diehl et al., 2001; Flecknell, 1995; Fox, 1988; Loeb, 1997; Loeb and Quimby, 1999; Nanhas and Provost, 2002.

The choice of method should be dictated by the sample size and number of samples needed, the ability to use anesthesia when necessary, and the training and experience of the phlebotomist. The sample site and anesthetic technique used must be kept the same within a study, and between studies whose results are to be compared. Plasma analyte concentrations can vary considerably within a species, depending on site of blood collection and the anesthetic technique used (Katein et al., 2002; Loeb and Quimby, 1999; Nanhas and Provost, 2002; Schnell et al., 2002). Decapitation is not recommended for the collection of blood samples from rodents because of the contamination and dilution of the blood by tissue fluids (Loeb and Quimby, 1999). If repeated sampling over a period of time is needed (such as for time course assessments), rodents can be implanted with venous catheters. These allow blood samples to be withdrawn without the need for either anesthesia or restraint and, because lost blood volume can be replaced with saline through the catheter, larger total quantities of blood can be withdrawn daily without affecting normal volume control processes (Loeb, 1997; Thrivikraman et al., 2002).

For collection of blood from the retroorbital sinus, heparinized glass microhematocrit tubes may be used. The blood may be collected in these tubes, dripped from the tube into an appropriately sized container, or collected directly into a capillary-action microcontainer (Becton-Dickinson, Rutherford, NJ). Venipuncture of the small peripheral veins (saphenous, ear, tail, and sublingual) in rabbits and rodents may be performed using 25 to 27 gauge needles; the blood is either collected using a tuberculin syringe or directly into capillary-action microcontainers. Warming the tail (using a warm-water bath or other gentle heat source) before venipuncture will increase the yield, especially in mice. If syringe collection is used, care must be taken to avoid rapid sample withdrawal as this can cause hemolysis in the sample or collapse the vessel from which the sample is being withdrawn. Larger vessels in rabbits and rodents (ear arteries, jugular, and femoral veins) can be sampled using slightly larger needles and syringes (23 to 25 gauge needles, 1 to 3 ml syringes) and the blood is subsequently transferred to the appropriate container. In dogs and larger species, blood may be withdrawn using needles and syringes or directly into appropriately sized vacuum containers (Becton-Dickinson, Rutherford, NJ). In all cases where blood is transferred from a syringe to another container, the needle should be removed from the hub and the blood expelled slowly into the container to avoid hemolysis (this is a particular concern with dog blood, where the erythrocytes are more fragile than in other species) (Diehl et al., 2001; Loeb, 1997; Loeb and Quimby, 1999).

Either plasma or serum is appropriate for renal functional analyte assessment. If serum is to be used, blood should be collected into serum-separator tubes. These contain a small plug of a polymer gel material that is designed to become displaced between the cells and the serum during centrifugation, which increases the ease of decanting the serum and helps prevent contamination or alteration of the serum composition by either hemolysis or prolonged contact with the erythrocytes. The blood must be allowed to clot for at least 60 minutes before centrifugation, and centrifugation must be at least 1100 g for the serum separator to function effectively.

If plasma is used, an anticoagulant must be added to the blood sample at the time of collection. Any of the commonly used anticoagulants may be used, although ethylenediaminetetraacetate (EDTA) and sodium fluoride should be avoided as they may interfere with plasma urea assessment. Heparin is preferred because it is least likely to interfere with most biochemical analyses. If electrolyte assessments are to be made on the plasma samples, lithium heparin is recommended; especially in small blood samples where the amount of potassium in the heparin may falsely elevate measured plasma levels (Young and Bermes, 1999). The amount of calcium-chelating anticoagulants (citrate, oxalate, and EDTA) used may need to be increased for rabbit blood because of the higher serum calcium levels in this species (Loeb and Quimby, 1997).

Endogenous Small Molecules

The ideal endogenous marker of GFR (and, indirectly, functional nephron mass) will be primarily excreted through the kidney, freely filtered at the glomerulus, and neither secreted nor reabsorbed by the tubule. In addition, it should be delivered in a steady fashion to the plasma and be readily detectable using available technology. This requires that production of the substrate, and so its delivery to the plasma, occurs at a stable rate and is not influenced by other disease or physiologic processes, and that the substrate is either not or minimally protein-bound (Price and Finney, 2000). Both urea and creatinine for the most part fit these criteria and so have been used extensively as plasma indices of renal function; however, each falls short of the ideal in several important ways, which must be taken into consideration in interpreting point-in-time assessments of these analytes as indicators of diminished renal function.

Urea is the primary end product of protein catabolism in most species. It is synthesized exclusively in the liver from ammonia and carbon dioxide by the urea-cycle enzymes and is released to the plasma, where its subsequent excretion is predominantly (>90%) via the kidney (with lesser amounts excreted through sweat and the gastrointestinal tract) (Finco, 1997; Newman and Price, 1999). Urea is freely filtered at the glomerulus, but is equally freely diffusable out of the tubule and a highly variable but significant amount (40 to 70%) reenters the extracellular fluid (and ultimately the plasma). The extent of urea reabsorption is in general in inverse proportion to tubular flow rates (i.e., at low flow rates, a greater proportion of urea is reabsorbed and returns to the plasma) (Bankir et al., 1996; Loeb, 1998; Newman and Price, 1999).

Tubular flow rates are primarily determined by the overall level of renal perfusion, and urea reabsorption in the medulla also increases in response to changes in vasoactive hormones (primarily vasopressin) and to tubular flow rate (Bankir et al., 1996; Bankir and Trinh-Trang-Tan, 2002; Conte et al., 1987). Therefore, plasma urea levels will invariably underestimate GFR and are more sensitive to extrarenal influences than those of other analytes (Baum et al., 1975; Kaplan and Kohn, 1992; Newman and Price, 1999).

The delivery of urea to the peripheral circulation is also variable because the rate of synthesis is not constant, and is dependent on both dietary protein intake and the functional integrity of the liver. Baseline plasma urea levels will be elevated as the consequence of increased urea synthesis in circumstances of increased protein catabolism, including: increased dietary protein intake (particularly pronounced in dogs), gastrointestinal hemorrhage, fever, severe burns, corticosteroid administration, sustained exercise or muscle wasting. Conversely, urea synthesis and baseline plasma levels decrease with low-protein diets, modest food restriction in rodents, hepatic insufficiency, hyperglycemia, and decreased circulating plasma amino-acid levels (Finco 1997; Hamberg, 1997; Newman and Price, 1999; Pickering and Pickering, 1984; Tauson and Wamberg, 1998). In extreme malnutrition, however, the effect of diminished protein intake is offset by increased urea synthesis resulting from muscle catabolism, as well as concomitant decreases in renal plasma flow and dehydration, with the net result that baseline plasma urea levels are usually elevated (Benabe and Martinez-Maldonado, 1998; Levin et al., 1993).

Creatinine, a muscle-derived byproduct of creatine metabolism, is another small molecular weight compound whose excretion is almost exclusively renal as the consequence of glomerular filtration. It has several significant advantages over urea as a plasma marker of GFR. It is synthesized and delivered to the plasma at a fairly consistent rate; moreover, the rate of synthesis decreases as plasma levels increase (presumably due to a feedback control over synthesis). Therefore, the day-to-day plasma level in a given individual does not fluctuate widely, although consumption of high-meat diets may elevate plasma creatinine and baseline levels will be elevated in individuals with higher muscle mass or following sustained exercise or acute muscle damage.

Conversely, serum creatinine will be lower in individuals who have undergone loss of muscle mass (Finco, 1997; Newman and Price, 1999). Diurnal variations in serum creatinine levels have also been documented for humans (Finco, 1997) and dogs (Loeb and Quimby, 1999), with plasma levels in general being slightly higher in the afternoon in both species.

Additionally, creatinine also undergoes tubular secretion by both the organic anion and cation pathways in the proximal tubule in most species. Although the degree of secretion is quantitatively small, secretion contributes proportionately more to the level of excretion in circumstances of decreased GFR. Creatinine secretion has also been shown to increase in some disease states that do not directly affect GFR or renal blood flow (Rocco et al., 2002; Sandsoe et al., 2002). The result is that plasma creatinine will consistently overestimate GFR and will equally be consistently insensitive to low-level decreases in GFR (Andreev et al., 1999; Newman and Price, 1999; Shemesh et al., 1985; Starr et al., 2002; Swan, 1997). This insensitivity is augmented by the feedback inhibition of serum creatinine on its synthesis and release from muscle; it has been shown (Watson et al., 2002) that the daily endogenous input of creatinine is lower in individuals with functional renal impairment. Additionally, drugs that compete with creatinine for renal transport pathways (notable examples are cimetidine, trimethoprim, and salicylates) will decrease the secretion component and elevate baseline plasma levels, even in the absence of a reduction in GFR (Andreev et al., 1999; Sandsoe et al., 2002; Shemesh et al., 1985).

Despite these potential shortcomings, serum creatinine assessment has been considered the "gold standard" for assessment of GFR in humans, and is widely accepted as an index of GFR in most animal species (Finco, 1997; Starr et al., 2002). In humans, GFR can be reliably estimated from single point-in-time creatinine measurements using one of several algorithms that correct for the effects of age and gender on muscle mass (Cockcroft and Gault, 1976; Finn and Porter, 1998; Newman and Price, 1999). A similar algorithm, which also uses the inverse of the serum creatinine concentration, has been calculated in one study for dogs. It demonstrated no influence of diet, gender, or age on the predictive value of plasma creatinine levels in this species (Finco et al., 1995).

Measurement of both urea and creatinine and calculation of the urea : creatinine ratio can provide some useful information about the etiology of the underlying process. If the ratio is elevated as the consequence of elevation of both analytes, the cause may be primary renal disease; however, extremely high urea:creatinine, when both are elevated (where urea is elevated markedly out of proportion to creatinine), more likely indicates decreased renal blood flow (because decreased tubular flow enhances creatinine excretion while increasing urea reabsorption), urinary tract obstruction (for the same reason), or extravasation of urine into the peritoneal cavity (because urea is more readily reabsorbed than creatinine from the peritoneum). Elevation in the urea:creatinine ratio as a consequence of pure urea elevation can be seen with gastrointestinal hemorrhage, high-protein diet (as a transient effect), increased protein catabolism, or loss of muscle mass (in this circumstance, plasma creatinine may be reduced). A decreased ratio can indicate early acute tubular necrosis, but more likely pinpoints primary liver dysfunction (due to decreased urea synthesis), decreased protein intake, ingestion of high-quality protein diets used in the management of renal failure, extremely muscular individuals, or circumstances of tissue anabolism (Baum et al., 1975; Newman and Price, 1999).

Chemical and enzymatic assay methods are available for both of these analytes. Urea is most commonly assayed by combined urease methods, in which the urea is first converted to two ammonium ions. These assays are quite specific for urea, but the values can be falsely elevated with increased circulating ammonia (such as occurs in aged plasma samples, metabolic disorders, and portocaval shunting). If the assay results are expressed as urea nitrogen (mg/dl), a correction factor of 2.14 is used to convert to the equivalent mass units (i.e., also mg/dl) of urea. However, in conversion of mass units of urea nitrogen to SI units for urea (the way these data are normally expressed for toxicology studies and clinically), a factor of 0.357 converts mg/dl of urea nitrogen to mmol/ of urea (Newman and Price, 1999). Creatinine is most commonly measured by chemical methods, most of which are variations of the Jaffe reaction of creatinine with picrate to generate an orange chromogen. Numerous endogenous and exogenous substrates present in plasma can interfere with the Jaffe reaction, adding to the lack of sensitivity of this method at low creatinine levels (this is of particular concern in the dog compared with other species) (Schwendenwein and Gabler, 2001). The effect of these can be minimized by using appropriate substrate extraction or by the use of kinetic assessments. Several enzymatic assays (based on reactions with creatinase or creatinine deaminase) have also been developed, but are much more costly and are no more sensitive at low concentrations (Finco et al., 1995; Finco, 1997;

Newman and Price, 1999). Approximately 50% of renal function must be lost before serum creatinine or BUN concentrations rise.

Low Molecular Weight Proteins

Some small proteins (less than approximately 66 kDa) fulfill the criteria outlined above as ideal markers of GFR; they are extensively filtered through the glomerulus and are neither reabsorbed nor secreted in their intact forms. Several of these have been proposed as alternatives to plasma urea or creatinine as sensitive indices of GFR.

The most extensively characterized of these is cystatin C (g-trace), a 120 amino acid nonglycosylated basic cysteine protease inhibitor with a molecular weight of ≈13 kDa. It is constitutively expressed by all nucleated cells and is produced and released to the plasma at a steady rate. Its production and plasma levels are not affected significantly by age, gender, coexisting disease, or frequently-used concurrent therapies, and it is freely filtered at the glomerulus, with a sieving coefficient of 0.7 (Abrahamson et al., 1990; Bökenkamp et al., 2002; Laterza et al., 2002; Newman and Price, 1999; Price and Finney, 2000; Simonsen et al., 1985; Tenstad et al., 1996), all of which should make it an ideal marker for GFR.

Use of cystatin C has been extensively compared with serum creatinine measurements in human medicine, where it has been shown to be either identical or superior in sensitivity to serum creatinine as an index to GFR in all age groups and renal disease states, especially in circumstances of low-level impairment of renal function. It is also not affected by nonrenal diseases that augment creatinine secretion (Baigent et al., 2001; Buehrig et al., 2001; Donadio et al., 2001a, 2001b, 2001c; Kazama et al., 2002; Laztera et al., 2002; Newman et al., 1994; Olivieri et al., 2002; Price and Finney, 2000; Rocco et al., 2002; Woitas et al., 2001). Cystatin C has been shown to be an equally sensitive marker of renal injury in the dog (Almy et al., 2002; Braun et al., 2002) and the rat (Bökenkamp et al., 2001), but not the cat (Martin et al., 2002), but its utility has not been explored in other species.

Several particle-enhanced turbidometric immunoassays are commercially available; a sandwich enzyme immunoassay is also available (Finney et al., 1997; Jensen et al., 2001; Pergande and Jung, 1993). Although the antibodies used in these tests are specifically directed against human cystatin C, there is significant homology across species (Esnard et al., 1988; Poulik et al., 1981). These tests have been used successfully to detect cystatin C from dogs, rats, mice, and cats (Bökenkamp et al., 2001; Braun et al., 2002; Hakansson et al., 1996; Martin et al., 2002).

Other low molecular weight proteins in the serum that have been used as indices of GFR in humans include the CNS-derived glycoprotein prostaglandin D synthase (b-trace protein), the plasma proteins α1-microglobulin and β2-microglobulin, and the retinoid transport protein retinol binding protein (RBP). In general, these proteins are better predictors of low-level decrements of GFR than is serum creatinine, but they are not superior to serum cystatin C in this respect. Additionally, it has been shown that their serum levels are significantly influenced by age, inflammation or febrile illness, liver disease, and corticosteroid administration (Bökenkamp et al., 2002; Donadio et al., 2001a, 2001b; Donaldson et al., 1990; Filler et al., 2002; Jung et al., 1987; Melegos et al., 1999; Priem et al., 1999, 2001; Woitas et al., 2001). The majority of these are detected by immunoassays with reagents specific for the human proteins; the cross-reactivity of these reagents with other species and the usefulness of these markers in animal models have not been well established (Loeb, 1998).

MARKERS OF RENAL FUNCTION IN URINE

General Considerations and Limitations

While plasma biomarkers can provide information about the level of renal perfusion and functional nephron mass, assessment of kidney function and identification of the site of injury within the nephron is best performed through examination of its end product – urine. The levels of proteins and small molecules normally filtered, excluded from filtration, secreted, or reabsorbed by the tubules can be used as indicators of the functional status of certain nephron segments. Injury can be assessed by examination of cellular enzymes that are preferentially leaked into the urine (Loeb, 1998), or proteins that are either upregulated or leaked in circumstances of cellular injury and subsequently appear in the urine. Most of the commonly used proteins and enzymes are of high enough molecular weight that the contribution to the urine from extrarenal sources is negligible (as they are effectively excluded by the glomerular filter). However, this may not be the case where there is also glomerular injury, or a potential contribution from another site in the urogenital tract (such as secretion from the accessory sex glands in the male), or from fecal contamination of a urine specimen.

Urinalysis may be conducted on spontaneously voided urine specimens from test animals. Quantitative urine collection methods are used to document pharmacodynamic changes in renal excretory functions more precisely and to define renal clearances. Quantitative urine collection for renal excretory function differs from urinalysis of spontaneously voided urine only in that quantities of urine are collected over specific intervals. Good-quality urine samples can be more difficult to collect from commonly used laboratory animals, and the samples collected can be highly variable in terms of quantity and concentration, which necessitates normalizing the parameter being measured to allow for intraanimal variation. This can be done by collection of all of the urine excreted during a timed interval (usually 24 hours, although shorter intervals are standard for routine toxicology studies) and expressing the parameter of interest as the total excreted per unit of time. (This method works best in species where the bladder can be thoroughly emptied by catheterization at the beginning and at the end of the collection period to eliminate the variability inherent to incomplete urine collection over the time period.) Where accurate timed urine samples cannot be collected (as is frequently the case with laboratory animals), the urine creatinine concentration can be assessed and used to normalize the quantity of the analyte of interest. The daily excretion of creatinine is fairly consistent in all species and therefore its quantity in a spot urine sample serves as an accurate index of the 24 hr urine output. This adjustment has been shown to accurately correct for incomplete timed urine sample collection in rodents (Haas et al., 1997), and works well for other species.

From a practical perspective, it is useful to apply a tiered strategy in analyses of urine to gain different levels of perspective on kidney function. First, urine osmolality, urine volume, and urinary pH provide basic insights into overall fluid homeostatic and acid–base state. These parameters are highly dependent upon intakes and metabolic state, and treatment-related changes are most easily detected by comparison with untreated controls or appropriate historical controls. Urine volume reflects the excess between ingested water and water produced as a byproduct of metabolism and insensible water losses. Urine osmolality is the ratio of excreted osmotically active solute (mOsmoles) to excreted water (in kg). This parameter reflects the urine concentrating ability, and is susceptible to changes in either water or solute excretion. Urine pH is an indicator of relative total body acid (or base) load. Animals maintained (or studied) in a controlled environment will exhibit urine volume, osmolality, and pH values within established ranges, unless a disturbance has taken place.

Second, sodium, potassium, chloride, and urea account for most of the total urinary solutes. The sum of sodium and potassium concentrations (mEq/l) or excretions (mEq/time) should almost equal chloride (mEq/l or mEq/time):

$$\text{sodium} + \text{potassium} = \text{chloride} \tag{3.1}$$

Additionally, the sum of the electrolytes plus urea should approximately equal urine osmolality:

$$\text{sodium} + \text{potassium} + \text{chloride} + \text{urea} = \text{urine osmolality} \tag{3.2}$$

Deviations represent electrolyte and nonelectrolyte "gaps" and suggest the presence of unusual solutes in urine. Endogenous creatinine can be used to calculate the fractional excretion of other solutes with respect to glomerular and tubular components; expressing excretion of urinary solutes per mg (or mmol) creatinine accounts for changes related to increases or decreases in filtered loads of individual solutes.

Third, and finally, renal excretion of divalent cations (Mg^{2+}, PO_4^{2-}, Ca^{2+}), other electrolytes and nonelectrolytes, proteins, hormones, vitamins, lipids, enzymes, and other chemicals may be determined. These analyses are conducted to test specific hypotheses when discrepancies have been established, based upon the prior tiers of testing or when nephrotoxicity is suspected (see Table 3.2).

Sample Collection Considerations

As for plasma analytes, collection of a good-quality sample from the test species is paramount for obtaining high-quality data from the urinalysis. Samples must be collected in clean containers and must be kept free of contamination from food, drinking water, feces, blood, and bacteria. For best results, samples should be analyzed promptly (ideally within one hour) after collection, but where analysis must be delayed, the sample must be protected from chemical degradation and evaporation; best accomplished by using collection containers that are appropriately sized for the species in question, tightly sealing the containers when possible, and keeping the specimen cold (4°C) until analysis. If chilled or frozen, samples must be allowed to slowly equilibrate to room temperature before analysis.

Table 3.2 Acute Effects of Gadolinium Chelate of Dicyclohexenetriaminepentaacetic Acid [Gd(DCTPA)2-] and Furosemide on Water and Divalent Cation Excretion in Male Rats

	0–4 Hours												
	PH (units)	Uvol (mL)	Uosm (mOsm)	Protein (mg/ml)	Ca (ug)	Mg (ug)	PO4 (ug)	NH4 (ug)	Urea (mmol)	Na (mmol)	K (mmol)	Cl (mmol)	Cr (mmol)
Treatment: Control (0.1 M sodium acetate, 1.0 mL/kg i.v.)													
Mean	7.75	2.40	768	1.57	0.09	0.42	0.90	0.24	0.84	0.14	0.38	0.25	0.01
SEM	0.05	0.33	96	0.24	0.03	0.08	0.24	0.05	1.00	0.02	0.01	0.02	0.00
Treatment: Gd-DCTPA (100 umol/kg i.v.)													
Mean	7.32	1.48	1207	2.86	0.18	0.47	2.15	0.37	0.97	0.14	0.37	0.20	0.01
SEM	0.32	0.44	345	0.67	0.11	0.14	1.42	0.10	0.28	0.04	0.07	0.02	0.00
Treatment: Furosemide (1.0 mg/kg i.v.)													
Mean	7.63	***4.95***	***356***	***0.27***	0.28	0.36	0.60	0.26	***0.29***	***0.47***	0.25	0.64	***0.00***
SEM	0.16	***0.46***	***19***	***0.07***	0.03	0.05	0.16	0.03	***0.05***	***0.05***	0.04	0.07	***0.00***
		Uvol/kg (ml/kg)	Osm/kg (MOsm/kg)	Protein/kg (mg/kg)	Ca/kg (ug/kg)	Mg/kg	PO4/kg	NH4/kg	Urea/kg (mmol/kg)	Na/kg (mmol/kg)	K/kg	Cl/kg	Cr/kg
Control													
Mean		8.22	5.76	11.55	0.31	1.43	3.04	0.83	2.84	0.47	1.28	0.85	0.04
SEM		1.22	0.36	0.83	0.09	0.29	0.80	0.17	0.13	0.07	0.07	0.09	0.00
Gd-DCTPA													
Mean		4.98	5.48	8.43	0.60	1.59	7.06	1.26	3.21	0.45	1.23	0.68	0.04
SEM		1.48	1.56	3.49	0.37	0.47	4.56	0.32	0.88	0.11	0.23	0.07	0.01

						Furosemide							
Mean	***17.12***	6.13	4.69	0.96	1.24	2.06	0.90	***1.00***	1.61	0.86	2.20	***0.01***	
SEM	***1.73***	0.71	1.22	0.13	0.16	0.56	0.09	***0.16***	0.19	0.14	0.28	***0.00***	
		Uvol/Cr (ml/mmol)	Osm/Cr (mOsm/ mmol)	Protein/Cr (mg/mmol)	Ca/Cr (ug/mmol)	Mg/Cr	PO4/CR	NH4/Cr	Urea/Cr (mmol /mmol)	Na/Cr (units)	K/Cr	Cl/Cr	Na/K
						Control							
Mean		234.45	163.74	325.22	8.43	38.88	91.75	23.31	80.75	13.45	36.42	24.27	0.36
SEM		35.23	9.4	9.72	2.45	6.57	26.37	4.82	3.13	2.06	2.14	2.48	0.04
						Gd-DCTPA							
Mean		250.7	195.37	394.85	32.77	66.76	242.16	50.63	88.22	13.54	36.01	21.51	0.37
SEM		87.53	8.22	153.03	16.52	18.75	87.31	6.54	3.93	3.73	2.31	3.6	0.08
						Furosemide							
Mean		***1429.51***	***500.02***	345.75	***76.30***	***97.24***	***144.42***	***75.73***	74.99	***135.3***	***66.34***	***182.55***	***1.97***
SEM		***240.71***	***77.01***	111.49	***9.38***	***11.91***	***26.08***	***13.65***	5.59	***24.45***	***7.05***	***31.85***	***0.19***

Notes. Values are means (w/standard error, SEN for groups ($n = 6$) treated with Gd(DCTPA) (0.1 mg/Gd/kg), furosemide (positive control agent), and control (acetate vehicle) Italicized, bolded values are significantly different ($p < 0.05$) compared with acetate-treated controls. A single exposure to Gd(DCTPA) was associated with a consistent, high incidence of struvite bladder stone formation within 30 days. Note elevated 4-hr phosphate excretion in Gd(DCTPA)-treated rats, a possible explanation for struvite bladder stone formation.

Point-in-time samples are more easily collected from larger species, either by free-catch, urethral catheterization, manual compression of the bladder, or cystocentesis (withdrawal of urine directly from the bladder using a needle and syringe). Catheterization may require sedation or anesthesia, especially in females of all species and in male pigs, whose urethral recess makes this process difficult (Van Metre and Angelos, 1999). Point-in-time samples from laboratory rodents may be obtained by taking advantage of the fact that these animals frequently urinate when they are handled or shortly after being removed from their home cage. A skilled, quick, and prepared operator, who is ready with a small container or plain microcapillary tube, may be able to obtain a small sample (Loeb and Quimby, 1999). Alternatively, the animal can be placed in a small confined space on plastic food wrap and observed carefully for urination, as it has been shown that most rodents will urinate within 20 minutes after removal from their home cages; the sample thus generated can be collected by micropipette (Kurien and Scofield, 1999). Manual compression of the bladder can also be performed on rodents, and cystocentesis can be performed by a skilled operator using a small (25 gauge) needle and syringe (Loeb and Quimby, 1999). For repeated point-in-time urine samples from rats over short periods of time (one to two weeks), the urethra or the ureter can be cannulated. Cannulated rats can also be used to collect accurate and complete timed urine samples (Horst et al., 1988; Mandavilli et al., 1991).

Timed urine collections in all species (including large animals) can also be obtained by the use of specially designed metabolism cages. A metabolism cage consists of an animal chamber mounted above an excrement collection system. The animal chamber must be equipped with feeder and waterer units if an animal is to be housed in the metabolism cage for more than a few hours. These need to be sized appropriately to the test species of interest (especially regarding the collection container, which must minimize surface area available for evaporation) and designed to eliminate contamination of the urine by food or drinking water.

In selecting the size of the collection container, the anticipated urine volume should be considered: over 24 hours a mouse will produce 0.25 to 1 ml, a rat, 10 ml, a hamster, 5 to 8 ml, a rabbit, 600 ml, and a dog or minipig, 500 ml or more (Loeb, 1998; McClure, 1999; Van Metre and Angelos, 1999). The collection system usually consists of a funnel and a urine/feces separator. Many design variations of metabolism cages are available, and one example is shown in Figure 3.3.

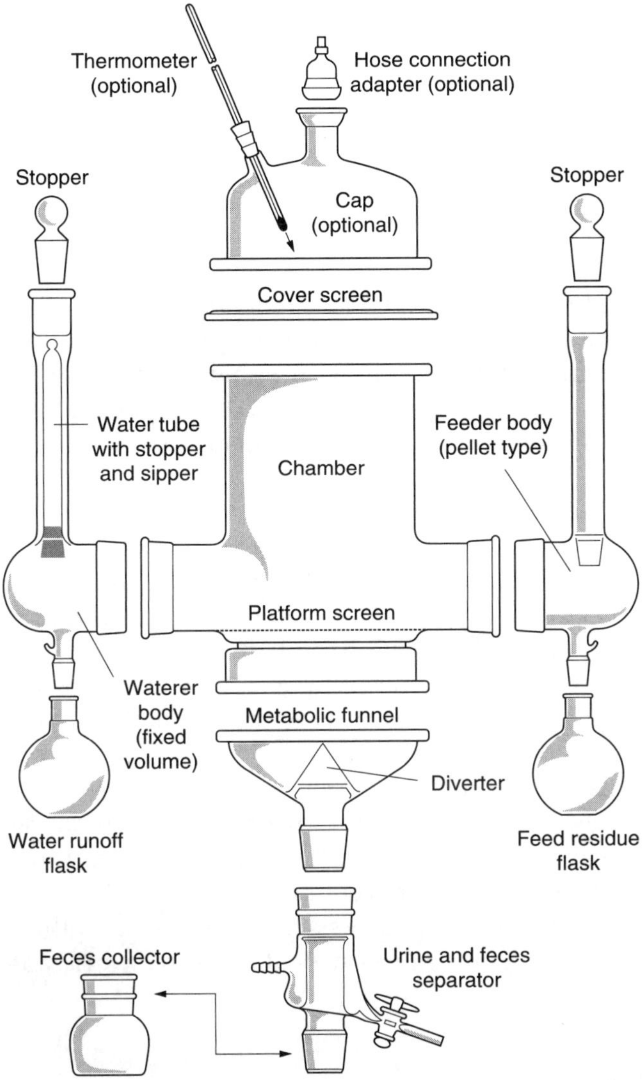

Figure 3.3 Schematic of a metabolism cage for rodents. This particular design is suitable for simultaneous collection of respiratory gases. (Courtesy of Braintree Scientific Inc., Braintree, MA.)

Whatever type of metabolism cage is used, the following general precautions are offered.

First, for studies of >24 hours duration, test animals should be acclimated to the metabolism cage for several days prior to study initiation.

During this period, animals should be monitored frequently to ensure that they learn to use the feeder and waterer systems properly. Test animals should be maintaining or gaining weight prior to study initiation. Feeder and waterer systems should provide ample food and water to meet the animals' needs and all separator systems should function properly. For chronic studies (>5 to 7 days duration) it is useful to have a complete exchange of feeder, waterer, and urine/feces collector and separator systems so that soiled units may be rapidly exchanged with clean, dry and filled units at regular intervals.

Second, cages should be decontaminated, cleaned, rinsed with distilled or deionized water, and thoroughly dried prior to use. Surfaces used to collect urine may be siliconized or sprayed with a suitable hydrophobic material (e.g., PAM®, General Foods) to facilitate urine collection.

Third, all surfaces contacting urine should be rinsed with distilled or deionized water or appropriate solvents to collect any residuals at appropriate intervals. To preserve the quality of the specimens during prolonged collection times, the opening of the collection vial needs to be small to prevent evaporation and the vial should be surrounded with either wet ice or frozen cold packs to chill the sample promptly once it is deposited (Loeb, 1998; Loeb and Quimby, 1999). Urine may also be collected under mineral oil to prevent evaporative losses. For very small or antidiuretic animals (e.g., hamsters and gerbils), placing the cage over a shallow pan of oil and skimming feces from the surface may be necessary, while urine is collected from under the oil with a pipette. To maximize the volume of the sample collected from rodents in metabolism cages, food should be withheld (this also reduces the risk of contamination of the sample) and water provided (Lee et al., 1998).

Routine Urinalysis

Routine urinalysis consists of a core battery of tests that serves as an assessment of both the functional health of the kidney and numerous other body systems. For standard toxicology studies using laboratory animals, visual assessment (color, clarity), volume, specific gravity or osmolality, pH, and quantitative or semiquantitative determination of total protein and glucose are recommended (Weingand et al., 1996). Other constituents are routinely measured in human clinical medicine, thus commercially available "dipstick" test strips also contain reagents for the semiquantitative determination of ketones, bilirubin, urobilinogen, hemoglobin, nitrite, and leukocyte esterases (Newman and Price, 1999). In addition, microscopic examination of the sediment from a centrifuged urine sample also may be performed.

In the majority of these tests, abnormal results can result from functional decrements in several organ systems and therefore they are not selective for renal injury or malfunction. Additionally, significant contributions to the urine composition in specimens obtained in metabolism cages or in free-catch or catheter-collected samples may come from the lower urinary tract or genital systems (although abnormal components in urine derived by cystocentesis are, in general, renal in origin, contributions from the urinary bladder are still possible in such samples) (Loeb and Quimby, 1999). The dipstick tests are inexpensive and easy to perform, and the results of some selected analyses may indicate altered renal function if the results are interpreted with knowledge of both the function of other body systems and the limitations of the methods involved. In using dipsticks in toxicology studies, care must be taken to ensure that test article or metabolites excreted in the urine do not generate false-positive reactions.

Measurement of the urine volume is critical if subsequent quantitative assessment of an analyte is to be performed as this allows determination of the total quantity of both the analyte and creatinine in the sample. In addition, excessively large volume samples (see section above for expected volumes by species) collected in metabolism cages may have substantial contamination with drinking water, especially if they are also very dilute (occasionally, normal animals may excrete large volume samples of normal osmolality) (Loeb, 1998; Loeb and Quimby, 1999). Urine volume, combined with assessment of urine concentration (specific gravity or osmolality), can also serve as an index to renal function; with severe acute loss of functional nephron mass, urine output is decreased (oliguria) or absent (anuria), while loss of the ability of the kidney to adequately concentrate urine results in the excretion of large volumes of dilute urine. However, because of the role of the kidney in water and electrolyte homeostasis, these results must be interpreted with knowledge of such factors as hydration state, water consumption, diet, and the presence of other disorders that alter renal fluid or electrolyte handling (Finco, 1997; Loeb, 1998; Loeb and Quimby, 1999; Newman and Price, 1999; Stonard, 1987).

In all species (with the exception of the gerbil, where dipstick positive glucosuria is normal) (McClure, 1999), over 99% of filtered glucose is reabsorbed by the proximal tubule so that only trace amounts appear in the urine. This process is saturable, so that when plasma glucose (and therefore the filtered glucose load) is increased the filtrate will contain more than the tubules can reabsorb and glucose will appear in the urine; significant glucosuria therefore is usually indicative of an increased filtered load. If plasma glucose is normal, however, the appearance of glucose in the urine may indicate a functional deficit in the proximal tubule, and glucosuria may precede the appearance of actual tubular necrosis or injury (Aleo et al., 2002; Finco, 1997; Loeb and Quimby, 1999; Newman and Price, 1999; Stonard et al., 1987). The commercially available dipstick reagents are not sensitive enough to detect the low levels of glucose found in normal urine and so can be used to detect significant glucosuria in most species. However, many of these strips use a glucose oxidase method that can show a false-positive result in species with high urinary ascorbate levels, such as the dog and mouse, or in urine contaminated with hypochlorite (used as a disinfectant) (Finco, 1997; Loeb and Quimby, 1999). Furthermore, test article formulations used in toxicology studies may contain glucose or other metabolizable sugars in quantities that upon administration may transiently overwhelm tubule reabsorption mechanisms and generate positive test results.

Microscopic examination of urine sediment is a useful adjunct to urine dipstick analysis, especially in the case of a positive reaction for blood, as the method is equally sensitive to hemoglobin or myoglobin. The presence of more than one or two erythrocytes or neutrophils per high-power field confirms a positive result of dipstick analysis for blood or leukocyte esterase. Casts (clear cylindrical structures composed primarily of coagulated Tamm-Horsfall mucoprotein, a secreted product of the distal tubule) in small numbers are a normal component of urine in most species; increased numbers of these, or the presence of granules (cells or cellular debris) within casts, may be indicative of tubular injury. The presence and numbers of casts are highly influenced by the pH and osmolality of the urine; Tamm-Horsfall mucoprotein coagulates more readily in acid and high-electrolyte concentration, but dissolves in alkaline urine. Casts are therefore less likely to be detected at high urine pH. The presence of renal tubular epithelial cells in urine sediment strongly indicates tubular injury, but differentiating these cells from epithelial cells originating from the bladder or lower urinary tract can be difficult (Finco, 1997; Finn and Porter, 1998; Hofmann et al., 1994; Newman and Price, 1999; Stonard, 1987). Normal transitional cells are positive for cytokeratin 20, and immunostaining of the sediment may therefore be useful in ascertaining the origin of the epithelial cells present in a urine sample (Deshpande and McKee, 2002).

The origin of bleeding or inflammation in the kidney can occasionally be determined by sediment examination. The presence of erythrocytes, leukocytes, or epithelial cells within casts is strong presumptive evidence of their renal origin, but additional assessment is needed with free cells. In humans, the source of bleeding can be determined either by phase-contrast microscopic examination of the sediment and semiquantitative assessment of the numbers of "dysmorpic" erythrocytes (Pillsworth et al., 1987) or by taking advantage of the fact that erythrocytes of renal origin become coated with Tamm-Horsfall mucoprotein as they pass into the urine. Several authors (Abrass and Laird, 1987; Janssens et al., 1992) have developed immunoassays for Tamm-Horsfall mucoprotein-coated erythrocytes or leukocytes and a high proportion (80% or greater) of coated erythrocytes is considered evidence of a renal origin for hemorrhage or inflammation.

High-resolution ^{1}H-NMR spectroscopy is used to analyze low molecular weight compounds present in urine (Nicholson et al., 1984). Both quantitative and qualitative analyses are possible using NMR spectroscopy. Bales and colleagues compared concentrations of urinary creatinine determined by NMR and the traditional Jaffe reaction. There was excellent agreement between the two methods in samples of human urine (creatinine concentration of 9.98 ± 0.36 mmol/l determined by NMR vs. 10.2 ± 0.1 mmol/l determined colorimetrically), suggesting that ^{1}H-NMR spectroscopy can be used for quantitative urinalysis (Bales et al., 1984). Gartland and coworkers used qualitative ^{1}H-NMR spectroscopy to assess urinary excretion of low molecular weight compounds following administration of chemicals that selectively injure the proximal tubule or collecting duct (Gartland et al., 1989). Sodium chromate, mercuric chloride, hexachlorobutadiene, and cisplatin injure the proximal tubule (Evan and Dail, 1974; Haagsma et al., 1979; Leonard, et al., 1971; Lock and Ishmael, 1979). These compounds increased urinary excretion of glucose, acetate, amino acids (such as alanine, lysine, and valine) lactic acid, and acetate (Gartland et al., 1989). In contrast, compounds that selectively injure the renal papilla, such as propylene imine and 2-bromoethanamine hydrobromide, increased urinary excretion of trimethylamine N-oxide, dimethylamine acetate, succinate, lactate, and alanine (Gartland et al., 1989; Halman and Price, 1984; Murray et al., 1972).

An emerging strategy is to attempt to associate urine NMR spectroscopic profiles with particular classes of toxicants without identifying the compounds responsible for the individual spectroscopic peaks (metabonomics). NMR urinalysis has been coupled with metabonomics technology to evaluate onset and reversal of activity of 2-bromoethylamine and 4-aminophenol (nephrotoxicants) and furosemide (control); the results suggest that the metabonomics approach may support rapid throughput detection of site-specific nephron injury using NMR spectroscopic urinalysis of urine (Robertson et al., 2000).

Urine Protein

A small amount of protein in the urine is normal for most species, resulting from filtration through the glomerular membrane (and incomplete reabsorption by the proximal tubule), secretion by the tubule into the urine, and from the normal turnover of tubular epithelial cells. In most species, albumin (filtered by the glomerulus) and Tamm-Horsfall mucoprotein (secreted from the distal tubule) are the most abundant proteins present in the urine. In male rats and mice, high levels of α2-microglobulin and major urinary protein (MUP), respectively, are also excreted into the urine and account for the normally high proteinuria in these species (Loeb and Quimby, 1999). Changes in the levels and types of proteins present in the urine can therefore reflect changes in the integrity of the glomerular filtration membrane, the endocytic capability of the proximal tubules, alterations in proteins synthesized and excreted in response to physiologic stimuli, or any combination thereof.

Part of the routine urinalysis consists of assessment of urinary protein excretion, usually based on semiquantitative analysis by dipstick reagent test. While in human samples this test performs well as an index to increased protein excretion, it is insufficiently sensitive or specific to assist in the identification of the nature of the underlying deficit. The dipstick test is based on a bromphenol blue method, which is most sensitive for albumin (Newman and Price, 1999), therefore proteinuria that does not result primarily from an increased albumin excretion may not be detected by this method. Furthermore, since these tests are designed for human urine, false positives are frequent in species, like the dog, whose normal urine protein levels are just above the lower limits for humans (Finco, 1997). Dipstick tests are invariably positive in male rats and mice, which have normal high proteinuria (Loeb and Quimby, 1999). A positive result with a urine dipstick test must therefore be followed by more detailed quantitative and qualitative assessments of the increase in protein excretion to determine the site and nature of the renal injury present.

The severity of proteinuria (determined by quantitation of the level of protein excreted in a 24 hour period) can be used as a first indication of the site of tubular malfunction. A number of different methods have been developed, including turbidometric, colorimetric (biuret and Lowry assays as examples) and dye binding assays. All have their advantages and limitations; in general, biuret assays tend to detect all types of proteins with equal sensitivity but require large sample volumes, turbidometric assays can suffer from lack of precision with variations in urine ionic strength, and dye-based methods can suffer from interference with exogenous and endogenous urine substances. The Folin phenol (Lowry), Coomassie brilliant blue, and Ponceau S methods have been recommended as being particularly precise for urine samples (Dilena et al., 1983; Finco, 1997; Newman and Price, 1999; Peterson et al., 1969).

The sensitivity of any of these tests to the proteins of interest depends in large measure on which protein is used to generate the standard curve. Albumin is most commonly used because it is the most abundant protein in urine; however, while this is adequate for most methods (Dilena et al., 1983), the standard curve will generally underestimate the abundance of many other proteins of interest in urine (Guder and Hofmann, 1992). As a general rule, excretion of markedly elevated levels of protein is indicative of glomerular disease, whereas low-level proteinuria indicates tubular damage or very early or low-grade glomerular injury (Peterson et al., 1969; Finco, 1997).

Further determination of the severity and site of nephron injury requires qualitative identification and quantitation of the individual proteins present in the urine. Table 3.3 lists the urinary proteins that have traditionally been used for functional assessments, as well as some site-specific markers of nephron injury. Selection of an appropriate battery of these tests will allow fairly accurate characterization of the nature of the functional deficit present.

Albumin is the most abundant protein in the urine of most species, and the most easily measured by a variety of methods. It is also the protein most likely to be found at increased levels early with a variety of renal disorders, particularly with glomerular injury, although it is not a specific marker for any one site (Finn and Porter, 1998; Guder and Hofmann, 1992; Price et al., 1996). While high-level increases of albumin in urine (greater than 3.5 g/24 hr, or an albumin : total protein ratio of greater than 0.4 in humans) are invariably the result of glomerular malfunction, low-level increases in albumin levels (less than 200 mg/24 hr in humans), especially in circumstances where total urine protein excretion is not elevated (microalbuminuria), can result either from increased glomerular filtration or decreased tubular reabsorption. Albumin values in this range must be interpreted in comparison to the excretion of other protein biomarkers (see Table 3.3 for examples) (Finn and Porter, 1998; Guder et al., 1998). Concomitant elevation of albumin and one or more of the low molecular weight proteins (which are freely filtered by the normal glomerulus) indicates that the proteinuria has resulted from decreased tubular reabsorption of proteins, while albumin elevation alone or concurrently with a high molecular weight protein (normally excluded from the filtrate by the glomerulus) indicates primary glomerular injury (Finn and Porter, 1998; Guder et al., 1998; Petersen et al., 1969; Umbreit and Wiedemann, 2000). Additionally, examination of the relative clearance of albumin and other, less-permeable, high molecular weight proteins (immunoglobulins or α2-microglobulin can used in this calculation) has been shown to be a useful indicator of the degree of glomerular injury (the higher the ratio, the more severe the damage) (Tencer et al., 1998, 2000).

Elevation of any of the filtered low molecular weight proteins in urine is an indication of a primary defect in tubular uptake, either as a consequence of decreased nephron mass or competition for the endocytic pathway by a competing substrate (Aleo et al., 2002, 2003; Finn and Porter, 1998). The most commonly used of these in human urine are α1- and β2-microglobulins, despite their relative instability in acid or contaminated urine. In particular, α1-microglobulin is present in fairly high abundance in normal urine, shows robust elevation with tubular disease, and is slightly more stable to degradation than β2-microglobulin, therefore making it the preferred marker of tubular malfunction in human bioassays (Donaldson et al., 1989; Holdt-Lehmann et al., 2000; Price, 2000, 2002; Price et al., 1996, 1997). These tests have been used less extensively in animal models because the majority of them are based on immunometric methods (gel immunodiffusion, nephelometry, or enzyme-linked immunosorbent assay [ELISA]) and the antibodies do not cross-react with homologous proteins in animal urine (the exception is cystatin C, where there is good cross-reactivity across a variety of species) (Aleo, M., personal communication; Esnard et al., 1988; Loeb, 1998; MacNeil et al., 1991; Poulik et al., 1981; Twyman et al., 2000; Viau et al., 1986). Newer methods based on proteomics

Table 3.3 Urinary Proteins Commonly Used as Biomarkers of Renal Injury

Protein	*Nephron segment*	*Species*	*Comments*	*References*
Filtered proteins:				
High molecular weight:				
Albumin	Glomerulus Proximal tubule	All	Specific for glomerular injury when other low molecular weight proteins in urine at normal levels Ratio of albumin: β_2-microglobulin can be used to distinguish glomerular from tubular proteinuria Microalbuminuria may also result from systemic vasculopathy or inflammation	Gosling, 1995 Peterson et al., 1969 Price, 2000, 2002 Price et al., 1997
IgG	Glomerulus	Human	Reliable index of glomerular injury, and indicates more severe glomerular injury	Finn and Porter, 1998 Price, 2000, 2002 Price et al., 1997
Transferrin	Glomerulus	Human Rat	Slightly higher molecular radius than albumin, index to low level increase in glomerular pore size (selective proteinuria) May be a more sensitive indicator of low level glomerular injury than microalbuminuria	Finn and Porter, 1998 Price, 2000 Price et al., 2000 Suzuki et al., 1996 Umbreit and Weidemann, 2000

Continued

Table 3.3 Continued

Protein	Nephron segment	Species	Comments	References
Low molecular weight:				
Retinol binding protein	Proximal tubule	Human Rat	Megalin-mediated endocytosis; appearance in urine precedes evidence of lysosomal dysfunction or necrosis Human antibody does not cross-react with rat RBP (can use urinary retinol as an indirect index of RBP excretion)	Aleo M., personal communication Aleo et al., 2002, 2003 Price, 2000, 2002 Price et al., 1997
$\beta_2$microglobulin	Proximal tubule	Human Rat	More sensitive index of tubular function but less stable to acid pH and/or bacterial proteases than α_2-microglobulin Human antibody does not cross-react with other species (rat specific antibody exists, not commercially available)	Donaldson et al., 1989 Finn and Porter, 1998 Loeb, 1998 Price, 2000, 2002 Price et al., 1997 Viau et al., 1986
$\alpha_2$microglobulin	Proximal tubule	Human	Unstable in acid urine Sensitive index to severity of renal injury following a variety of insults, even in the absence of overt necrosis	Bruning et al., 1999 Donaldson et al., 1989 Finn and Porter, 1998 Herget-Rosenthal et al., 2001 Price, 2000, 2002 Price et al., 1997
Cystatin C	Proximal tubule	Human Rat	Very stable in urine Normalized to creatinine, provides sensitive index to both GFR and tubular protein reabsorption	Finn and Porter, 1998 Herget-Rosenthal et al., 2001

			Sensitive index to severity of renal injury following a variety of insults Substantial cross-species homology and cross-reactivity with commercially available antibodies	Uchida and Gotoh, 2002
Released from damaged tissue:				
Fibronectin	Glomerulus (Proximal tubule?)	Human Rat	Precedes microalbuminuria in diabetes Plasma contribution may be a factor with severe glomerular injury Decreased tubular proteolytic degradation may contribute to increased urinary levels With chronic glomerular injury, upregulation contributes to increased urine levels	Adhikary et al., 1999 Gwinner et al., 1993 Kanauchi et al., 1995 Manabe et al., 2001 Price et al., 1997
Collagen IV	Glomerulus	Human Mouse Rat	Basement membrane specific Precedes microalbuminuria in diabetes With chronic glomerular injury, upregulation contributes to increased urine levels	Adhikary et al., 1999 Cohen et al., 2001 Makino et al., 1995 Manabe et al., 2001 Okonogi et al., 2001
α-Glutathione-*S*-transferase	Proximal tubule	Human Rat	Antibody appears to have good cross-species reactivity Highest urinary levels seen with selective straight (S_3) segment toxicants in rats Sensitive index of damage (brush border loss) in the absence of overt necrosis May be measured with either enzymatic or immunoassay	Bomhard et al., 1990 Bruning et al., 1999 Dixit R., personal communication Sundberg et al., 1994a, 1994c Usuda et al.1998 Van Kreel et al., 2002

Continued

Table 3.3 Continued

Protein	*Nephron segment*	*Species*	*Comments*	*References*
π-Gutathione-*S*-transferase	Distal tubule	Human	Questionable sensitivity (excretion in urine only seen after azotemia is apparent) Antibodies are very isoform-selective	Branten et al., 2000 Sundberg et al., 1994a, 1994b, 1994c
μ-Glutathione-*S*-transferase	Distal tubule	Rat	Similar distribution to π-GST in humans Questionable sensitivity (excretion in urine only seen when azotemia is apparent)	Dixit R., personal communication
Pap X 5C10 antigen (Pap A1)	Papillary collecting ducts	Rat	Of several papilla-specific antigens identified, this one most consistently released to urine after papillary toxicant administration. Present in urine as complexes with Tamm-Horsfall mucoprotein Monoclonal antibody not commercially available	Falkenberg et al., 1996 Hildebrand et al., 1999
Upregulated and secreted with injury:				
Clusterin	Proximal tubule	Rat Monkey	Elevated rapidly (1 hour) in both acute and chronic injury; stays elevated as long as the stimulus is present Upregulation/secretion specifically indicates injury; not seen with nontoxic homologues that do not cause necrosis Some upregulation seen in medulla with oxidative stress	Aulitzky et al., 1992 Davis et al., 2003 Eti et al., 1993 Hidaka et al., 2001 Huang et al., 2001 Nath et al., 1994

Kidney injury molecule-1	Proximal tubule	Rat Mouse Human	Biomarker for acute tubular necrosis/ischemia Upregulation demonstrated in rat kidney with injury, use as a urine biomarker not yet established in rodents	Han et al., 2002
Liver fatty acid binding protein (L-FABP)	"Glomerulus"	Human	Source is proximal tubule; upregulated and secreted in response to glomerular injury/proteinuria Small contribution (up to 3%) from serum-derived L-FABP with glomerular injury or concurrent liver injury Homologues (kidney and heart FABPs) exist in both human and rat kidney; these differ in distribution and regulation but may cross-react with antibodies to L-FABP. H-FABP in rat urine may reflect increased serum levels secondary to myocardial damage.	Kamijo et al., 2001 Kimura et al., 1989 Lam et al., 1988 Maatman et al., 1991 Volders et al., 1993
Cysteine-rich protein 61 (CYR61)	Proximal straight tubule	Mouse Rat	Rapid upregulation and secretion into urine (3 hours post-injury) with renal ischemia High heparin affinity allows concentration of the protein (using heparin-Sepharose) for detection of low levels	Muramatsu et al., 2002

technology (concentration of proteins by acetone precipitation or ultracentrifugation, separation by 2-d gel electrophoresis, or chromatographic techniques with subsequent identification and quantitation by mass spectrometry) are being developed for urine. These methods should overcome this difficulty, as well as potentially identify novel and specific biomarkers for renal function and injury (Bandara and Kennedy, 2002; Chapman, 2002; Thongboonkerd et al., 2002a, 2002b).

Urine Lipids

Excretion of small quantities of neutral lipid (visible as droplets in the sediment), phospholipid, and lipoprotein is considered normal in dogs, cats, mice, and humans (Finco, 1997; Gross et al., 1991; Streather et al., 1993). Increased excretion of all of these, occasionally resulting in visibly increased urine turbidity, has been described as a chronic change following both experimental and naturally-occurring glomerular injury in humans (Martin and Small, 1984; Mimura et al., 1984; Neverov and Nikitina, 1992; Sigitova et al., 2000; Streather et al., 1993) and rodents (de Mendoza et al., 1976; Gherardi and Calandra, 1982). Increased urinary phospholipid excretion has also been described as an antecedent finding to microscopic evidence of papillary injury in experimental renal papillary necrosis in rodents (Thanh et al., 2001).

Unlike the profile seen with glomerular injury in rats (de Mendoza et al., 1976; Gherardi and Calandra, 1982), concomitantly increased levels of lipoprotein have not been described following papillotoxins. Conversely, increased urinary sphingomyelin is detected in renal papillary necrosis (Thanh et al., 2001), but elevated levels are not detected in the urine of humans with nephrotic syndrome (Mimura et al., 1984). The level and nature of the excreted lipid can be determined by thin-layer chromatography and comparison to known standards (Mimura et al., 1984; Thanh et al., 2001).

Urinary Enzymes

Unlike urine proteins, urinary enzyme activity provides a means to determine the presence and location of renal tubular injury, as opposed to an index to the functional status of the nephron. The reason for this is the fact that the enzymes most commonly used in these determinations are quite large (greater than 80 kDa) and thus their appearance in the urine results from leakage from damaged tubular cells, not from increased filtration or decreased uptake (exceptions being lysozyme and alanine aminopeptidase, as summarized in Table 3.4) (Plummer et al., 1986). The advantages urine enzymes provide over urine protein for assessment of tubular injury are: increased sensitivity (enzyme levels in urine are frequently elevated in advance of overt evidence of renal malfunction and can be used to predict its onset) (Price, 1982; Westhuyzen et al., 2003), dose–response (the amount of enzyme activity in the urine accurately reflects the degree of tubular injury present), ease of analysis (most methods are fairly simple chromatographic assays, which have been well validated for the same enzymes present in plasma and can be performed using automated equipment), and utility for a variety of species (activity, as opposed to antigen mass is measured, so cross-reactivity of antibodies or probes across species is not a consideration). Additionally, repeated enzyme measurements over time can be used to determine the reversibility or progression of renal lesions (Clemo, 1998; Plummer et al., 1986; Price, 1982; Stonard et al., 1987; Vanderlinde, 1981).

An additional advantage to the use of urinary enzymes is the ability to localize the injury to a specific nephron site. The subcellular location of the enzyme can also serve as an index to the severity of the underlying injury (as a general rule, brush border enzymes indicate less severe damage than cytosolic, mitochondrial, or microsomal enzymes) (Dubach et al., 1988; Price, 1982). Table 3.4 lists the most commonly used and well-validated enzyme biomarkers for nephron injury, including their nephron segment and subcellular location. For best results in determining the presence and site of nephron damage, a battery of enzymes, as opposed to a single biomarker, should be examined, especially since the most-sensitive assay is highly variable, depending on the toxicant and species (Price, 1982).

There are some special considerations for collection and handling of samples that are critical for accurate assessment of urine enzyme activity. Contamination of collected specimens must be carefully avoided as enzymes present in feces, food, or bacteria can contribute significantly to the activity present in the urine. Enzyme activity can also result from the presence of increased numbers of erythrocytes, leukocytes, or epithelial cells in the urine. Centrifugation of urine to remove contaminating cells will help to mitigate this source of error, and examination of the resulting sediment will allow the investigator to discard the results from heavily contaminated samples. In addition, normal urine frequently contains a variety of low molecular weight substances (urea is a notable example) that can act as inhibitors of the enzymes of interest (as can a toxicant of

Table 3.4 Enzymes Used as Biomarkers of Nephron Injury

Enzyme	*Nephron segment*	*Cellular location*	*Comments*	*References*
N-acetyl-β-D-glucosaminidase (NAG)	Proximal tubule Papilla Glomerulus	Lysosomal	Used as a screening biomarker for chronic exposure to nephrotoxicants in humans Validated as a sensitive biomarker for renal injury in rat, mouse, dog, and Cynomolgus monkey Excretion rate is consistent in a 24 hour period, assessment in spot samples normalized to creatine accurately reflects 24 hour excretion Increased activity usually indicates proximal tubular injury but is not specific (increases also seen with glomerular disease, obstructive nephropathy and papillary injury) Ratio of NAG activity to that of a brush border enzyme may allow distinction between papillary and tubular injury (only NAG is elevated with pure papillary injury) Increased urinary activity seen with pregnancy Basal excretion increases significantly in rats with age Basal excretion higher in intact male dogs (due to contribution from seminal fluids)	Casadevall et al., 1995 Clemo, 1998 Finn and Porter, 1998 Higashiyama et al., 1983 Loeb, 1998 Nakamura et al., 1983 Price, 1982, 2002 Price et al., 1996, 1997 Stonard et al., 1987 Taylor et al., 1997

Lysozyme (muramidase)	Proximal tubule	Lysosomal	Two renal sources of urine activity; decreased uptake of the filtered load (proximal tubular functional deficit) and release from damaged renal proximal tubule cells (proximal tubular injury) Increased plasma levels from non-renal diseases (leukemia, inflammation or other neoplasia) may result in increased urine activity (secondary to saturation of renal uptake) Extremely variable secretion rate over a 24 hour period limits usefulness with timed or spot urine samples Immunochemical as well as enzyme assays exist (poor cross-reactivity of antibodies across species)	Clemo, 1998 Davey et al., 1984 Houser and Milner, 1991 Hysing and Tolleshaug, 1986 Lofti and Djalali, 2000 Montagne et al., 1998 Nishimura, 1987 Ohata et al., 1998 Price, 1982, 2002 Taylor et al., 1992 Vanderlinde, 1981
β-galactosidase	Proximal tubule	Lysosomal	Elevated following some glomerular toxicants before proteinuria evident Pronounced diurnal variation in basal excretion rate in humans Increased urinary activity following testosterone treatment in mice	Clemo, 1998 Koenig et al., 1978 Price, 1982
β-glucuronidase	Proximal tubule	Lysosomal	Contribution to urine activity from preputial gland in male rat Increased urinary activity following testosterone treatment in mice Pronounced diurnal variation in basal excretion rate in humans	Clemo, 1998 Coonrod and Paterson, 1969 Koenig et al., 1978 Price, 1982

Continued

Table 3.4 Continued

Enzyme	*Nephron segment*	*Cellular location*	*Comments*	*References*
β-glucosidase	Proximal tubule	Lysosomal	Pronounced diurnal variation in basal excretion rate in humans	Clemo, 1998
Alanine aminopeptidase (AAP)	Proximal tubule	Brush border Microsomes	Appears in urine earliest with acute tubular necrosis Renal and serum isoforms exist, but serum isoform is smaller (90 kDa vs. 230 kDa), not normally filtered. May contribute to activity with concomitant glomerular disease Some increase in urine activity may result from non-renal diseases Basal excretion increases significantly in rats with age Basal excretion rate higher in male rats and humans, but not in dogs	Casadevall et al., 1995 Clemo, 1998 Grötsch et al., 1985 Holdt-Lehmann et al., 2000 Nakamura et al., 1983 Price, 1982, 2002 Vanderline, 1981
Leucine aminopeptidase	Proximal tubule	Brush border	Elevations also seen in early glomerular diseases, preceding microalbuminuria	Bedir et al., 1996 Guder and Ross, 1984
Neutral brush border endopeptidase EC	Proximal tubule	Brush border	Assay substrate is also cleaved by leukocyte elastase; concomitant pyuria many falsely elevate activity	Vlaskou et al., 2000
γ-glutamyl tranferase (GGT, γ-glutamyl transpeptidase)	Proximal tubule (particularly S_3 segment in rodents)	Brush border	Somewhat unstable in urine relative to other enzymes (even with added stabilizers), should be assayed quickly after collection	Clemo, 1998 Gossett et al., 1987 Grötsch et al., 1985

			Within day variation of excretion rate noted in dogs, may limit accuracy of spot sample assessment even with creatinine correction	Guder and Ross, 1984; Jung and Grutzmann, 1988
			Basal excretion rate higher in male rats	Loeb, 1998
			Increased urinary activity seen with pregnancy	Mueller et al., 1986; Price, 1982, 2002
Alkaline phosphatase (intestinal var.)	Proximal tubule (S_3 segment "specific")	Brush border	Increased urine activity seen with pregnancy	Clemo, 1998
			Use of this enzyme as a selective marker of S_3 involvement requires isoenzyme character-ization (by ELISA) as well as determination of activity	Finn and Porter, 1998; Price, 1982, 2002; Verpooten et al., 1989
			More resistant to shedding than other brush border enzymes, indicates more severe damage	
α-Glutathione-*S*-transferase (GST) (ligandin)	Proximal tubule	Cytosol	Usually assessed by ELISA as quantity of protein (Table 3.3) but rapid assay for enzyme activity is also available	Van Kreel et al., 2002
			Assay is not specific for the α isoform, but it represents the majority of the activity	
π or μ GST	Distal tubule Collecting duct	Cytosol	Rapid assay for enzyme activity is avaiable, but does not distinguish this from the α isoform. Usually measured quantitatively by ELISA (Table 3.3)	Price, 2002; Van Kreel et al., 2002

Continued

Table 3.4 Continued

Enzyme	*Nephron segment*	*Cellular location*	*Comments*	*References*
Lactate dehydrogenase (LDH)	Entire nephron	Cytosol	Most commonly used as an indicator of distal tubule damage Increased cell sloughing from any nephron segment or urothelium will elevate urine activity	Clemo, 1998 Price, 1982
Aspartate aminotransferase (AST)	Entire nephron	Mitochondria (Cytosol)	Enzyme levels are highest in proximal tubule (particularly convoluted segment) and thick ascending limb, tracking mitochondrial density	Clemo, 1998 Guder and Ross, 1984

interest). These substances can usually be removed by dialysis, dilution, Sephadex filtration, ultrafiltration, or gel filtration, but these methodologies have not been well validated for all enzymes in all species and it may require some effort on the part of the investigator to determine the need and optimal method for sample preparation. The effects of dilution and pH on enzyme activity must also be considered as these factors are highly variable in urine samples. Most enzymes are very labile at acid pH and the activity of many can be decreased substantially with concentration (Clemo, 1998; Loeb, 1998; Mueller et al., 1989; Plummer et al., 1986; Price, 1982; Vanderlinde, 1981).

Finally, enzymes are generally less stable in urine than they are in high-protein matrices, such as serum or plasma. Urinary enzyme assessment therefore needs to be performed promptly (within a few hours of sample collection) or a stabilizing substrate needs to be added to the sample if storage is anticipated. This is a particular problem with γ-glutamyl transferase in most species, even with added stabilizers. The addition of albumin, ethylene glycol, glycerol, or erythritol will satisfactorily preserve the activity of most commonly assessed enzymes when the samples are stored at $-20°C$ (Jung and Grutzmann, 1988; Loeb et al., 1997; Mueller et al., 1986).

The appropriate sample and expression of activity is somewhat controversial. Because of pronounced diurnal variation in excretion rate of some enzymes (Gossett et al., 1987; Maruhn et al., 1977; Price, 1982), either 24 hour urine collection or timed urine samples collected at the same time each day is recommended, with the activity expressed per unit of time (Price, 1982; Plummer et al., 1986). For spot urine samples, or those where accurate timed collection is not possible, normalization of activity per unit of creatinine can be done, and this has been shown to be reasonably well correlated to 24 hour enzyme activity (Grauer et al., 1995; Vanderlinde, 1981). For this method to be accurate, however, diet and age-matched controls must be included (because both of these profoundly influence creatinine excretion, especially in humans). Furthermore, with severe acute renal disease, chronic renal disease (with pronounced loss of nephron mass), or prerenal reduction of GFR, creatinine excretion will also be reduced, resulting in lowering of urine creatinine and falsely elevated enzyme activity (Casadevall et al., 1995; Plummer et al., 1986; Price, 1982). Normalization of enzyme activity per unit of volume or osmolality is not recommended because of the high degree of variability of these parameters (Plummer et al., 1986; Price, 1982).

ANIMAL TECHNIQUES

Accessibility to the inflow (systemic blood) and outflow (urine) of the kidney has permitted development of sophisticated *in vivo* techniques using intact animals. These techniques utilize the analytes described above, and additional exogenous tracers, to gain more-specific information about the sites and mechanisms of toxicant effects.

Renal Clearance

Clearance methods may be useful in determining the effects of toxicants on glomerular filtration rate and renal blood flow. In addition, clearance methods are used to determine the mechanisms involved in the renal excretion of a toxicant. The renal clearance of a compound is defined as the volume of plasma from which that compound is completed removed by the kidneys per unit time (Pitts, 1974). The material removed from plasma is generally excreted in the urine.

Clearance procedures can be conducted in all laboratory animal species; anesthetized or conscious animal models may be used. Tracers are administered intravenously to near steady-state concentrations; a priming dose loads the plasma and extracellular compartments and an infusion replaces renal losses. Measurement of arterial pressure is advisable, especially in anesthetized preparations, to ensure that renal perfusion pressure remains within the autoregulatory range (usually 80 to 120 mmHg); vascular access ports can be helpful to provide continuous arterial access for pressure measurements (Mann et al., 1987). Urine can be collected directly from the ureters (anesthetized preparations) or via cannulation of the urinary bladder (conscious preparations); with the latter, inflation of the bladder with distilled water followed by air is used to wash out residual urine volume. Once steady-state plasma tracer levels are approached, a series of timed urine collections (clearance periods) are performed, with blood samples collected at either the midpoint or beginning and end of the clearance periods. The urine and blood (plasma or serum) samples are then analyzed for tracers. Renal clearance (Cl) of any compound (X) can be determined by comparing the urinary excretion rate of compound X to the plasma concentration of compound X. The urinary excretion rate is calculated as:

$$\text{Urinary excretion rate (mg/min)} = U_X\text{(ml/min)} \times V\text{(ml/min)} \qquad (3.3)$$

where U_X represents the concentration of substance X in urine (in mg/ml) and V represents the volume of urine collected per unit time (in ml/min). Thus, the clearance equation may be constructed:

$$\mathrm{Cl}_X(\text{ml/minute}) = \frac{U_X(\text{mg/ml}) \times V(\text{ml/min})}{P_X(\text{mg/ml})} \tag{3.4}$$

where P_X is the concentration of compound X in plasma (in mg/ml). More comprehensive reviews of the theories underlying clearance methods, and assumptions involved in calculating renal clearances, may be found in Vander (1985) and Valtin (1983).

The preferred indicator for the estimation of glomerular filtration rate is the fructose polysaccharide, inulin (mol wt ~5200). Since inulin is freely filtered and then is neither secreted nor reabsorbed, all of the inulin that appears in urine must have originated from plasma through the process of glomerular filtration. Measurement of inulin clearance serves as the "gold standard" method for estimating GFR (Finco et al., 1993). An example of the application of inulin clearance for characterizing the effect of a toxicant on GFR is shown in Figure 3.4. Other indicators suitable for measuring GFR include creatinine, isotopes of vitamin B_{12}, sodium iodothalamate, and metal chelates of ethylenediaminetetraacetate (EDTA) and diethylaminotriaminepentaacetate (DPTA) (Sarkar et al., 1991). Of these, only creatinine requires no infusion of an exogenous substance, and is particularly useful when it is necessary to monitor GFR over a protracted period of time (days to decades) in the same individual; the limitations of creatinine clearance are the same as previously described.

The indicator most commonly used for estimation of renal plasma flow (RPF) is *p*-aminohippuratic acid (PAH), which is both freely filtered by the glomeruli and actively secreted by the organic acid transport pathway of the proximal tubule (Figure 3.5). PAH extraction by the kidneys varies from about 70 to 90% in rats, dogs, and humans (Brenner et al., 1976). One reason why PAH extraction is incomplete is that blood flow to the inner cortical glomeruli is directed to the medulla and does not perfuse proximal tubule segments. Heterogeneity in the distributions of organic acid transporters in the cortex may also contribute to the less than complete extraction of PAH. Therefore, RPF estimated using PAH clearance is often termed effective renal plasma flow (ERPF).

As renal venous plasma is difficult to obtain, PAH extraction is often assumed to be complete. Using these assumptions, RPF is slightly

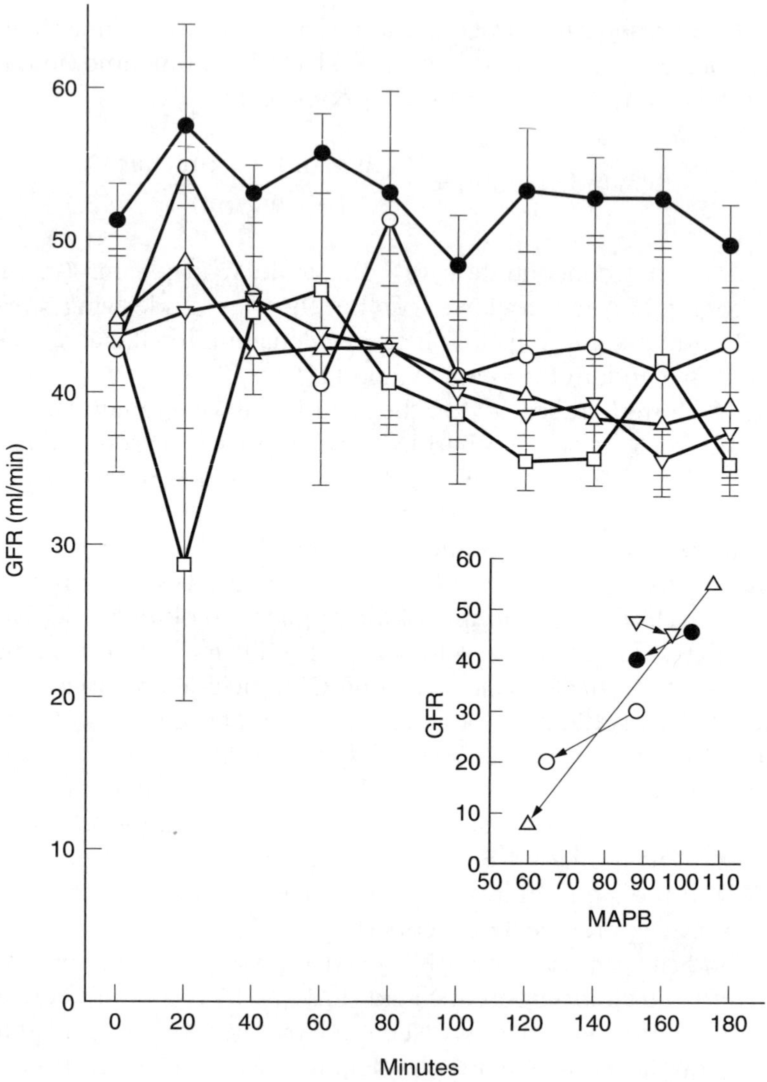

Figure 3.4 Effect of toxicant on glomerular filtration rate (GFR) in conscious dogs. Toxicant or vehicle given as prime plus infusion at 0 min. Solid circles represent vehicle; open circles, triangle, inverted triangle, and square represent increasing doses of toxicant. The insert gives individual data for the highest dose, showing the relationship between the decreases in GFR and simultaneously measured mean arterial pressure (MABP). (From Mann et al., 1991.)

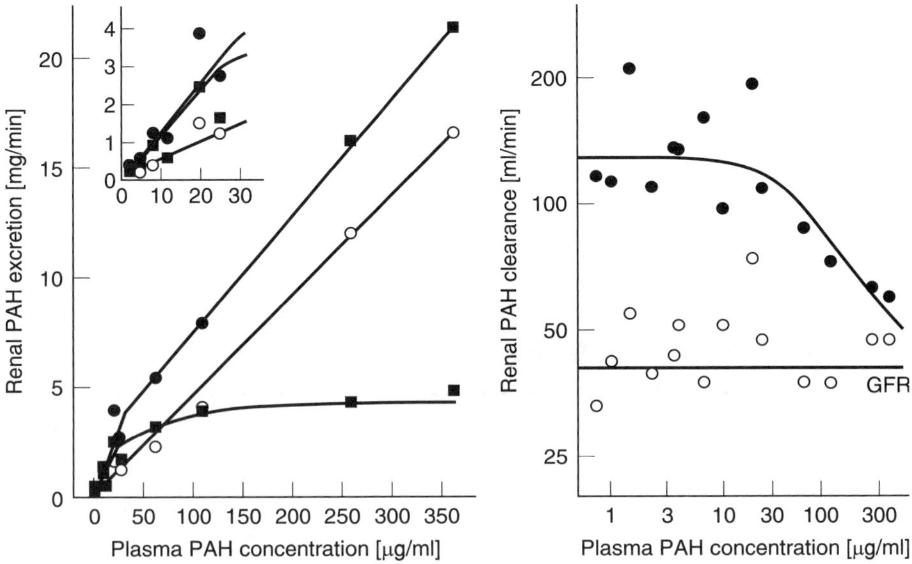

Figure 3.5 Relationships between plasma concentration of p-aminohippurate (PAH), renal PAH excretion, and renal PAH clearance (closed symbols). Simultaneously measured inulin clearance (GFR, open circles) is shown. (From Mann and Kinter, 1993a.)

underestimated. Renal plasma flow is converted to renal blood flow (RBF) by dividing RPF by the plasma fraction of whole blood, as estimated from the hematocrit (Hct):

$$\text{RBF} = \frac{\text{RBF}}{(1 - \text{Hct})} \tag{3.5}$$

The clearances of other compounds can be compared with inulin clearances to determine how the kidney functions with the elimination of the test compound. Typically, the renal clearances of a compound of interest (*X*) and inulin are determined. A clearance ratio is constructed from these data by dividing the renal clearance of the test compound (*X*) by the renal clearance of inulin:

$$\text{Clearance ratio} = \frac{\text{renal clearance of test compund } (\text{Cl}_X, (\text{ml/min}))}{\text{renal clearance of inulin } (\text{Cl}_{\text{inulin}}, (\text{ml/min}))} \tag{3.6}$$

For example, if a substance has a clearance less than that of simultaneously administered inulin, then some test substance must be reabsorbed

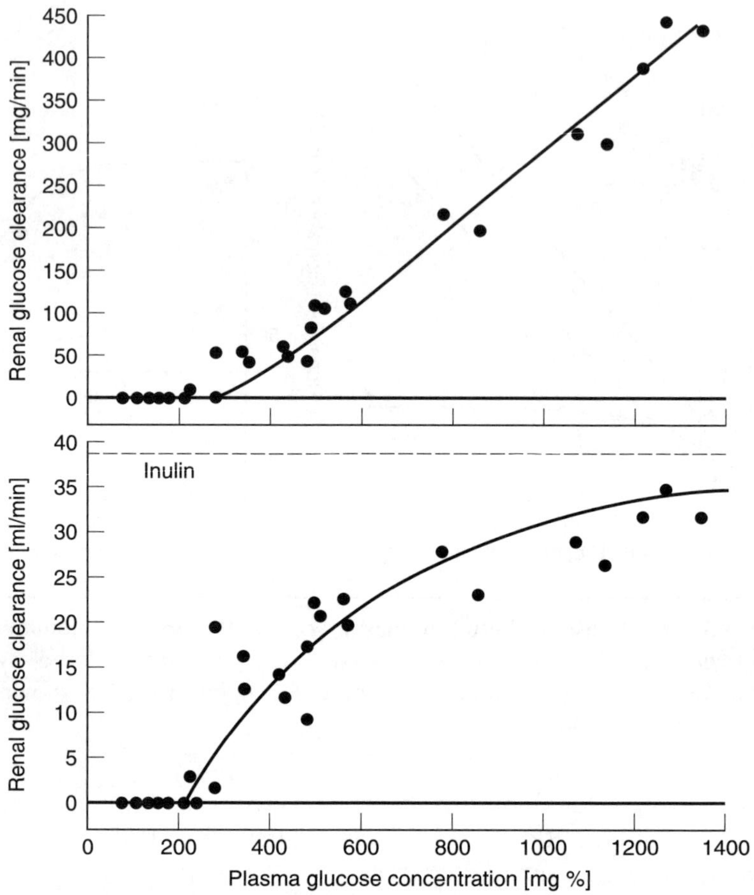

Figure 3.6 Relationships between plasma glucose concentration, renal glucose excretion, and renal glucose clearance (closed symbols). (From Mann and Kinter, 1993b.)

following filtration to account from the lower clearance relative to inulin (Figure 3.6). (Glucose is an example of a compound that is reabsorbed, and so the clearance ratio for glucose [$Cl_{glucose}/Cl_{inulin}$] is always less than unity.) Alternately, if a test substance has a clearance greater than simultaneously administered inulin, then, to account for the higher clearance relative to inulin, the test compound must be filtered and secreted by the renal tubules (Figure 3.7). PAH is an example of a compound that is secreted by the renal tubules and the clearance ratio for PAH (Cl_{PAH}/Cl_{inulin}) is always greater than unity (Figure 3.5).

Renal Clearance Variants

Renal clearance principles have led to development of several related procedures that can be applied in certain circumstances to further elucidate renal effects of toxicants.

Plasma Clearance

GFR and RPF can be measured without collection of urine by measuring the disappearance of the appropriate tracers from the blood over time. In this technique, precise amounts of tracers (usually radiolabeled) are injected intravenously as a bolus and serial samples are obtained at precise intervals. Disappearance of the tracer from the blood is used to calculate the renal clearance (Bailey et al., 1970).

Stop-Flow Clearance

This procedure can be used to approximate the site of renal injury. Following collection of baseline clearances in an anesthetized preparation, in which the ureters are catheterized, the ureters are clamped until the ureteral pressure plateaus (indicating cessation of GFR). The clamps are released and urine is collected in multiple timed aliquots (Figure 3.7). The distribution of urinary markers as a function of serial volume can be used to estimate the site of injury in the nephron tree (see Pitts, 1968; Figure 3.8).

In Vivo Micropuncture

Although micropuncture is used infrequently in toxicology, it is an important method for determining the effects of drugs or chemicals on single-nephron function in the intact kidney and can provide information concerning tubular handling of a drug or chemical. (Felley-Bosco and Diezi, 1989) Micropuncture is a highly specialized technique that requires considerable equipment and experience to be utilized properly. Studies may be undertaken in small rodents and dogs, but studies in various strains of rats are most common. Munich-Wistar rats are frequently used for micropuncture studies because the glomeruli are visible and accessible at the surface of the kidney (Ramsey and Knox, 1996). The basic experimental design in micropuncture is to identify the specific nephron segment of interest and insert a micropipette into the lumen of that segment (see Figure 3.1). A column of castor oil is initially injected into the tubular lumen to prevent collection of fluid downstream

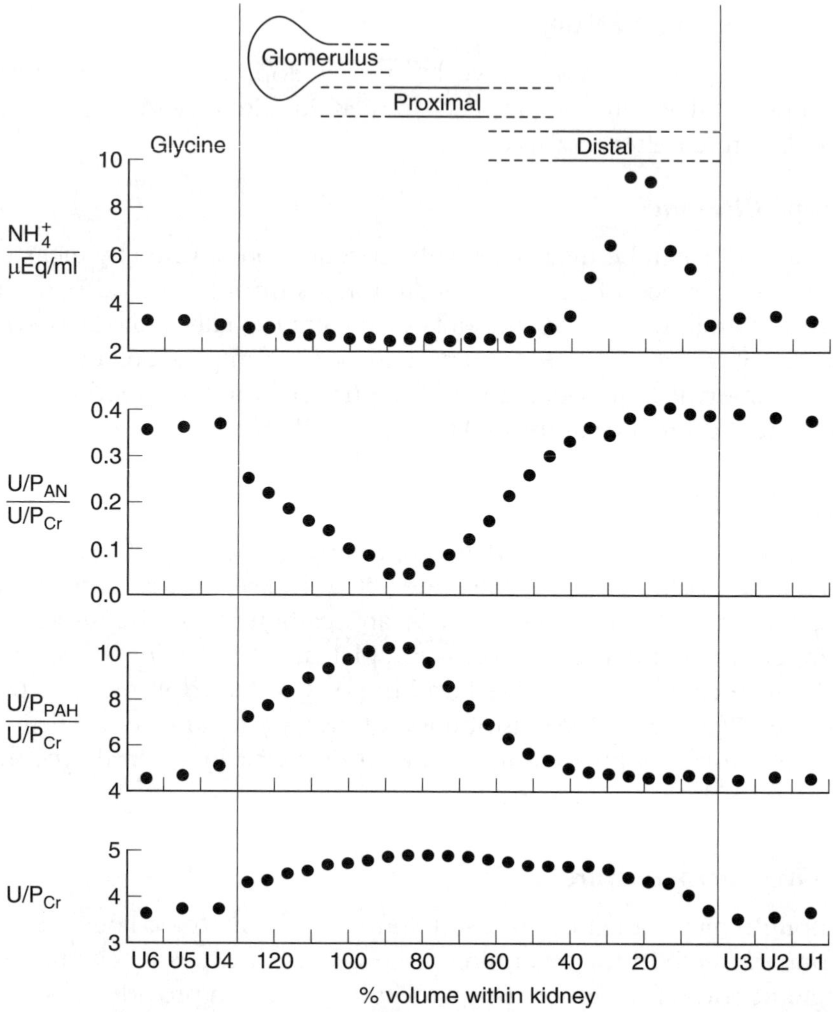

Figure 3.7 Example of a stop-flow clearance study (from Pitts, 1968).

from the micropuncture site and to allow complete collection of tubular fluid that reaches the micropipette (Ramsey and Knox, 1996). Nephron segments are identified by using a dye such as lissamine green (Ramsey and Knox, 1996). Experiments are performed under microscopic control to visualize the micropipette and tubule segment of interest.

A particularly useful aspect of micropuncture relates to the ability to collect glomerular filtrate from a single nephron by positioning the micropipette in a glomerulus rather than in a tubular lumen. Using glomerular micropuncture, the ultrafiltration coefficient of a substance may be calculated based on comparison of the tubular fluid to plasma concentration ratio of the substance in question with the tubular fluid to plasma concentration ratio of a standard substance, such as inulin. For a substance that is freely filtered, this ratio should approximate unity. Substances that are protein-bound, or are in some other manner restricted from filtration, will have a ratio less than unity.

In some cases, obtaining tubular fluid samples from thin loops of Henle or collecting ducts is possible. For these experiments, tubule segments in the papilla are subjected to micropuncture. Micropuncture of papillary structures requires the use of animals in which the papilla can be exposed, such as golden hamsters and young Wistar rats (Ramsey and Knox, 1996). Alternately, loop function can be inferred by comparing tubular fluid obtained from the later proximal convoluted tubule with that from the distal tubule. Using this method, Senekjian and coworkers determined that in male Sprague-Dawley rats the fractional delivery rate of gentamicin to the late portion of the proximal convoluted tubule was $65.5 \pm 5.3\%$, compared with a fractional delivery rate of $30.7 \pm 3.7\%$ to the distal tubule (Senekjian et al., 1981).

In Vivo Microperfusion

Microperfusion is a variation of micropuncture that can be used to isolate the effect of a toxicant on a specific segment of a nephron. The segment is isolated from the glomerular filtrate between a wax and oil block, allowing the segment to be perfused with a solution of controlled composition (see Figure 3.1). The solution is then quantitatively collected from the same nephron segment following perfusion. Essentially, microperfusion is analogous to *in vitro* perfusion of isolated nephron segments, except that a single nephron is perfusion *in situ*. With microperfusion, it is possible to perfuse the loop of Henle by inserting the perfusion pipette into the last accessible loop of the proximal tubule and the collection pipette into the first accessible loop of the distal tubule. Similarly, the distal tubule may be perfused as long as multiple loops are visible at the surface of the kidney (Ramsey and Knox, 1996). Microperfusion was used to define the renal handling of cadmium. In these experiments, radiolabeled cadmium (1 m*M* as $^{109}CdCl_2$) was microinjected into the lumens of proximal or distal tubules and the amount of cadmium excreted in the ureteral or pelvic urine was determined (Felley-Bosco and Diezi, 1987).

Renal Blood Flow

Toxicants may alter RBF either directly, by causing vasoconstriction of the renal artery or arterioles, or indirectly, by releasing one or more of the many regulators of RBF. Prostaglandins, such as PGE_2, are vasodilators, while thromboxanes are vasoconstrictors; inhibitors of prostaglandin synthesis, such as nonsteroidal anti-inflammatory agents, have the potential to decrease RBF by removing one vasodilatory influence. Prolonged reductions in RBF may have deleterious effects on the kidneys by causing hypoxia and injury to the proximal tubule and thick ascending limb of the loop of Henle. Several methods are available with which to assess effects of toxicants on RBF. These methods are more invasive than is determination of PAH clearance, however they have the advantage of allowing unambiguous interpretation of RBF in the presence of a toxicant that might reduce organic anion or cation secretion.

Flow Probes

Probes utilizing electromagnetic or Doppler technology may be positioned around a renal artery to allow direct measurement of RBF (Haywood et al., 1981; Yagil, 1990). Detection of blood flow using Doppler systems is based on changes in the emitted ultrasonic frequency, a Doppler shift, caused by reflection of the signal of moving blood cells. The Doppler shift is proportional to the velocity of blood flow. Examples of use of flow probes to study renal vasodilatory effects of drugs are included in Kinter et al. (1990, 1994).

Microspheres

Another method involves injection of radiolabeled microspheres with subsequent determination of the amount of radioactivity present in kidney tissue (Knox at al., 1984). The theory used with microspheres is that the size of the sphere determines where along the vascular tree the sphere will impact. Spheres are milled to a diameter that allows impaction in the glomerular capillaries, thereby enabling measurement of effective RBF; this is the amount of blood that actually reaches the glomeruli and is, therefore, available for filtration (Knox at al., 1984).

When RBF was measured in dogs by using either electromagnetic flow probes or radioactive microspheres, excellent correlation ($r = .96$) was obtained between the two methods (Daugaard, 1990). Similarly, RBF determined in dogs by electromagnetic flow meter was highly correlated with flow determined by ^{51}Cr-EDTA clearance over a wide range of RBFs (Daugaard, 1990). In contrast to the excellent correlation observed with electromagnetic flow probes, radioactive microspheres, and ^{51}Cr-EDTA clearance, RBF measured by PAH clearance was not as well correlated. For example, in control dogs the difference between RBF determined by flowmeter and PAH clearance averaged 19%. Following cisplatin administration, the difference between the two methods was even greater (57%) (Daugaard, 1990). Following cisplatin administration, PAH clearance in control dogs decreased from 112 ± 4 ml/min to 37 ± 5 ml/min within 72 hours, while the extraction of PAH fell from 0.73 ± 0.03 to 0.42 ± 0.04 (Daugaard, 1990). Since cisplatin damages the proximal tubule, changes in PAH clearance in these experiments might be due to cisplatin-induced alterations in PAH secretion rather than changes in RBF.

Capillary Microperfusion

Capillary microperfusion is a form of microperfusion experiment in which the peritubular capillaries are perfused with solutions of known composition, allowing examination of basolateral events in the intact kidney. This technique has been used extensively to investigate interactions among organic anions that undergo tubular secretion, as well as to identify substrates that interact with both the organic anion and cation transport systems (Ullrich et al., 1991a, 1991b, 1994).

Renal Imaging Modalities

Static imaging modalities (x-ray) have been used to provide information on effects of toxicants on renal structures, whereas dynamic modalities have increasingly been used to provide information on effects of toxicants on renal function (GFR and RBF).

Scintigraphic Imaging

Washout of radioactive inert gases (e.g., krypton or xenon) or radioactive indicators (e.g., ^{99c} Tc-DPTA, ^{113m}In-DPTA, or ^{99c}Tc-mercaptoacetyltriglycine) may be used to measure RBF, RPF, and GFR (Reba et al., 1968). These techniques require intraarterial or intravenous administration of the tracer, followed by monitoring of the amount of tracer in the kidneys by use of an external detector (Fommei and Volterrani, 1995). These techniques do not yield absolute flow, but rather flow per unit volume or tissue mass. Uptake of radioactive microspheres will yield absolute blood flows but requires removal of the kidney at the end of the experiment. With scintigraphic imaging, RBF and glomerular filtration are estimated using the Fick principle, taking advantage of the characteristics of the respective renal clearance tracers described previously.

Radionuclides used for the measurement and imaging of GFR, RBF, and the renal parenchyma are listed in Table 3.5. ^{90c}Tc represents an almost ideal radionuclide for scintigraphic imaging, the gamma-photon of 140 keV energy is well within the range of optimally imaged energies and there are no particle emissions. The half-life is six hours and the nuclide is produced from the longer-lived ^{99}Mo parent (67 hour half-life). These characteristics have led to the development of many ^{99m}Tc-labeled radiopharmaceuticals, several of which have replaced older, less-desirable, agents. Dominance of ^{99m}Tc has resulted in modifications of the gamma-camera that optimize imaging with this radionuclide, even to the detriment of other, higher energy, radionuclides.

Ultrasonic Imaging

Ultrasonic imaging offers substantial increases in spatial resolution and image quality above that provided by scintigraphy. The quality and utility of ultrasonic imaging has improved dramatically in recent years, and further improvements are to be anticipated. Ultrasonic imaging relies upon processing and display of ultrasonic radiation reflected from body tissues. Animals are usually tranquilized or anesthetized to reduce body movements. A probe is positioned manually on the abdomen or flank to collect images of the kidney. New pulsed Doppler and phase shift signal processing techniques can be applied to collect additional information on regional blood flow in the kidney, and may be used to detect renal tumors using differences in regional blood flows. Ultrasonic imaging is relatively inexpensive (compared with other imaging modalities) and easy to perform. However, use of ultrasonic techniques to quantitate renal functions is limited by manual (e.g., variable) use of the probe, and the current lack of contrast medica that interact with renal functions in a predictable fashion.

Table 3.5 Radiopharmaceutical Products Available for Determining Glomerular Filtration Rate (GFR) and Renal Blood Flow (RBF), and for Imaging Renal Parenchyma

Radiopharmaceuticals for estimating GFR	
^{3}H-inulin	^{125}I-diatrizoate
^{14}C-inulin	^{125,131}I-iothalamate (Conray – 60)
^{14}C-carboxy inulin	^{131}I-diatrizoate (Hypaque, Renografin)
^{14}C-hydroxy-methyl inulin	^{51}Cr-,^{99m}Tc-,^{111,113m}In-,^{140}La-,^{169}Yb-EDTA
^{131}I-chloroiodopropyl inulin	^{51}Cr-,^{99m}Tc-,^{111,113m}In-,^{140}La-,^{169}Yb-DTPA
^{131}I-propargyl inulin	^{57,58}Co-hydroxycobalamin
	^{57,58}Co-cyanocobalamin
Radiopharmaceuticals for estimating RBF	
^{3}H-para-aminohippuric acid (PAH)	^{125,131}I-iodopyracet (Diodrast)
^{14}C-PAH	^{131}I-orthoiodohippurate (Hippuran, OIH)
^{99m}Tc-hippuran analogs	^{67,68}Ga-*N*-succinyl desferioxamine
^{99m}Tc-iminodiacetic PAH (PAHIDA)	^{97}Ru-ruthenocenyl-glycine (Ruppuran)
^{99m}Tc-mercaptoacetyltriglycine (^{99m}Tc-Mag3)	^{99m}Tc-thiodiglycolic acid
^{99m}Tc-mercaptosuccinyltriglycine (^{99m}Tc-MSG3)	
^{99m}Tc-N,N′bis(mercaptoacetyl)-2,3-diaminopropanoate (CO2-DADS-A)	
Radiopharmaceuticals for imaging renal parenchyma	
^{197,203}Hg-chlormerodrin	^{197}Hg-bichloride
^{99m}Tc-penicilliamine (TPEN)	^{99m}Tc-penicillamine-acetazolamide (TPAC)
^{99m}Tc-2,3-dimercaptosuccinic acid (^{99m}Tc-DMSA)	^{99m}Tc-RGD (a ligand for integrins)
^{99m}Tc-gluconate and ^{99m}Tc-glucoheptonate	
^{203}Hg-3-chloromercuri-2-methoxypropyl urea (neohydrin)	

Computer-Assisted Tomography

Computer-assisted tomography (CT) uses x-ray imaging technology and available contrast media to create high-quality dynamic images, primarily of renal vasculature (Saggar et al., 2003).

Magnetic Resonance Imaging

Magnetic resonance imaging (MRI) produces the highest spatial resolution and image quality currently available. Images are based upon radio frequency information on proton spin shifts induced by changes in an imposed magnetic field. Spatial resolution is a function of differences in proton microenvironments and magnetic field strength. To achieve the highest quality images, animals are usually anesthetized prior to being placed within the imaging magnet. Serial images of the area of interest may be collected, permitting a three-dimensional reconstruction of the kidney. MRI will detect much smaller abnormalities in renal structure than will ultrasonic imaging. For example, MRI imaging at 1.5 T will detect swelling of the renal cortex and dissipation of corticomedullary intensity gradients associated with treatment with a loop diuretic (Sarkar et al., 1988). Using a combination of MRI at 4.7 T and histopathology, Bosch and colleagues observed marked cortical hypertrophy and increased glomerular size in kidneys of obese Zucker rats in comparison with kidneys from lean littermates (Bosch et al., 1993). In addiction, paramagnetic-based MRI contrast agents can be used to image GFR and regional RBF (Sarkar et al., 1991).

ACKNOWLEDGEMENTS

The authors acknowledge and thank Ms. Kuen-Ru Lin (Smith College) and Mr. Jeffrey Long (University of Delaware) for their assistance in preparation of the manuscript.

REFERENCES

Abrahamson K, Olaffson I, Palsdottir A, Ulvsback M, Lundwall A, Hensson O and Grubb A (1990) Structure and expression of the human cystatin C gene. *Biochem J* 268:287–294.

Abrass CK and Laird CW (1987) Tamm-Horsfall protein coating of free cells in urine. *Am J Kidney Dis* 9:44–50.

Adhikary LP, Yamamoto T, Isome M, Nakano Y, Kawasaki K, Yaoita E, and Kihara I (1999) Expression profile of extracelllar matrix and its regulatory proteins during the process of interstitial fibrosis after anti-glomerular basement membrane antibody-induced glomerular sclerosis in Sprague-Dawley rats. *Pathol Int* 49:716–725.

Aleo MD, Navetta KA, Emeigh Hart SG, Harrell JM, Whitman-Sherman JL, Krull DK, Wilhelms MB, Boucher GG and Jakowski AB (2002) Mechanism-based urinary biomarkers of aminoglycoside-induced phospholipidosis. *Comparative Clinical Pathology* 11:193–194.

Aleo MD, Navetta KA, Emeigh Hart SG, Harrell JM, Whitman-Sherman JL, Krull DK, Wilhelms MB, Boucher GG and Jakowski AB (2003) Mechanism-based urinary biomarkers of renal phospholipidosis and injury. *Toxicol Sci* 72 (Supplement):243.

Almy FS, Christopher MM, King DP and Brown SA (2002) Evaluation of cystatin C as an endogenous marker of glomerular filtration rate in dogs. *J Vet Intern Med* 16:45–51.

Andreev E, Koopman M and Arisz L (1999) A rise in plasma creatinine that is not a sign of renal failure: which drugs can be responsible? *J Intern Med* 246:247–252.

Aulitzky WK, Schlegel PN, Wu D, Cheng CY, Chen CC, Li PS, Goldstein M, Reidenberg M and Bardin CW (1992) Measurement of urinary clusterin as an index of nephrotoxicity. *Proc Soc Exp Biol Med* 199: 93–96.

Baigent C, Landray MJ, Newman D, Lip GYH, Townend JN and Wheeler DC (2001) Increased cystatin C is associated with markers of chronic inflammation and of endothelial dysfunction in normocreatinemic individuals. *J Am Soc Nephrol* 12:190A.

Bailey RR, Rogers TGH and Tait JJ (1970) Measurement of glomerular filtration rate using a single injection of 51Cr-Edetic acid. *Australas Ann Med* 19:255–258.

Bales JR, Higham DP, Howe I, Nicholson JK and Sadler PJ (1984) Use of high-resolution proton nuclear magnetic resonance spectroscopy for rapid multi-component analysis of urine. *Clin Chem* 30:426–432

Bandara LR and Kennedy S (2002) Toxicoproteomics – a new preclinical tool. *Drug Discov Today* 7:411–418.

Bankir L, Bouby N, Trinh-Trang-Tam MM, Ahloulay M and Promeneur D (1996) Direct and indirect cost of urea excretion. *Kidney Int* 49:1598–1607.

Bankir LT and Trinh-Trang-Tan MM (2002) Renal urea transporters. Direct and indirect regulation by vasopressin. *Exp Physiol* 85:243S–252S.

Baum N, Dichoso CC and Carlton CE (1975) Blood urea nitrogen and serum creatinine. Physiology and interpretations. *Urology* 5:583–588.

Baumans V, Remie R, Hackbarth HJ and Timmerman (2001) Experimental Procedures, in *Principals of Laboratory Animal Science* (van Zutphen LFM, Baumans V and Beynen AC, Eds), Elsevier, Amsterdam.

Bedir A, Özener IC and Emerk K (1996) Urinary leucine aminopeptidase is a more sensitive indicator of early renal damage in non-insulin dependent diabetes than microalbuminuria. *Nephron* 74:110–113.

Benabe JE and Martinez-Maldonado M (1998) The impact of malnutrition on kidney function. *Miner Electrolyte Metab* 4:20–26.

Bökenkamp A, Giuliano C and Christian D (2001) Cystatin C in a rat model of end-stage renal failure. *Renal Failure* 23:431–438.

Bökenkamp A, van Wijk JAE, Lentze MJ and Stoffel-Wagner B (2002) Effect of corticosteroid therapy on serum cystatin C and β_2-microglobulin concentrations. *Clin Chem* 48:1123–1126.

Bomhard E, Maruhn D, Vogel O and Mager H (1990) Determination of urinary glutathione-*S*-transferase and lactate dehydrogenase for differentiation between proximal and distal nephron damage. *Arch Toxicol* 64:269–278.

Bosch CS, Ackerman JJH, Tilton RG and Shalwitz RA (1993) *In vivo* NMR imaging and spectroscopic investigation of renal pathology in lean and obese rat kidneys. *Magn Reson Med* 29:335–344.

Branten, AJ, Mulder TP, Peters WH, Assmann KJ and Wetzels JF (2000) Urinary excretion of glutathione S transferases alpha and pi in patients with proteinuria: Reflection of the site of tubular injury. *Nephron* 85:120–126.

Braun J-P, Perxachs A, Péchereau D and de la Farge F (2002) Plasma cystatin C in the dog: reference values and variations with renal failure. *Comp Clin Pathol* 11:44–49.

Brenner BM, Zatz R and Ichikawa I (1976) in *The Kidney*, 3rd edition (Brenner BM and Rector FC, Eds) pp. 93–123, WB Saunders, Philadelphia.

Bruning T, Sundberg AG, Birner G, Lammert M, Bolt HM, Appelkvist EL, Nilsson R and Dallner G (1999) Glutathione transferase alpha as a marker for tubular damage after trichloroethylene exposure. *Arch Toxicol* 73:246–254.

Buehrig CK, Larson TS, Bergert JH, Pond GR and Bergstrahl EJ (2001) Cystatin C is superior to creatinine for the assessment of renal function. *J Am Soc Nephrol* 12:194A.

Casadevall G, Piera C, Setoain J and Queralt J (1995) Age-dependent enzymuria, proteinuria and changes in renal blood flow and glomerular filtration rate in rats. *Mech Ageing Dev* 82:51–60.

Chapman K (2002) The ProteinChip® biomarker system from Ciphergen Biosystems: A novel proteomics platform for rapid biomarker discovery and validation. *Biochem Soc Trans* 30:82–87.

Clemo FAS (1998) Urinary enzyme evaluation of nephrotoxicity in the dog. *Toxicol Pathol* 26:29–32.

Cockcroft DW and Gault MH (1976) Prediction of creatinine clearance from serum creatinine. *Nephron* 16:31–41.

Cohen MP, Lautenslager GT and Shearman CW (2001) Increased urinary type IV collagen marks the development of glomerular pathology in diabetic *d/db* mice. *Metabolism* 50:1435–1440.

Coonrod D and Paterson PY (1969) Urine β-glucuronidase in renal injury. I. Enzyme assay conditions and response to mercuric chloride in rats. *J Lab Clin Med* 73:6–16.

Conte G, Dal Canton A, Terribile M, Cianciaruso B, Di Minno G, Pannain M, Russo D and Andreucci VE (1987) Renal handling of urea in subjects with persistent azotemia and normal renal function. *Kidney Int* 32:721–727.

Corman B and Michel JB (1987) Glomerular filtration, renal blood flow, and solute excretion in conscious aging rats. *Am J Physiol* 22:R555–R560.

Daugaard G (1990) Cisplatin nephrotoxicity: Experimental and clinical studies. *Dan Med Bull* 37:1–12.

Davey PG, Cowley DM, Geddes AM and Terry J (1984) Clinical evaluation of β_2-microglobulin, muramidase, and alanine aminopeptidase as markers of gentamicin nephrotoxicity. *Contrib Nephrol* 42:100–106.

Davis JW, Goodsaid FM, Bral CM, Mandakas G, Obert LA, Garner CE, Smith RJ and Rosenblum IY (2003) Genomic markers of nephrotoxicity in female Cynomolgus monkeys. *Toxicol Sci* 72 (Supplement):61.

de Mendoza SG, Kashyap ML, Chen CY and Lutmer RF (1976) High density lipoproteinuria in nephrotic syndrome. *Metabolism* 25:1143–1149.

Deshpande V and McKee GT (2002) Cytokeratin 20 (CK20): A useful adjunct to urine cytology. *Lab Invest* 82:70A–71A.

Diehl KH, Hull R, Morton D, Pfister R, Rabemempianina Y, Smith D, Vidal JM and van de Vorstenbosch C (2001) A good practice guide to the administration of substances and removal of blood, including routes and volumes. *J Appl Toxicol* 21:15–23.

Dilena BA, Penberthy LA and Fraser CG (1983) Six methods for determining urinary protein compared. *Clin Chem* 29:553–557.

Donadio C, Lucchesi A, Ardini M and Giordani R (2001a) Cystatin C, β_2-microglobulin and retinol-binding protein as indicators of glomerular filtration rate: Comparison with creatinine. *J Pharm Biomed Anal* 24:835–842.

Donadio C, Lucchesi A, Ardini M and Giordani R (2001b) Serum levels of β-trace protein (prostaglandin D synthetase) and glomerular filtration rate. *J Am Soc Nephrol* 12: 727A.

Donadio C, Lucchesi A, Ardini M and Giordani R (2001c) Appraisal of cystatin C as indicator of glomerular filtration rate. *J Am Soc Nephrol* 12:727A.

Donaldson MD, Chambers RE, Woolridge MW and Whicher JT (1989) Stability of alpha-1-microglobulin, β-2-microglobulin, and retinol binding protein in urine. *Clin Chim Acta* 179:73–77.

Donaldson MD, Chambers RE, Woolridge MW and Whicher JT (1990) α_1-Microglobulin, β_2-microglobulin and retinol binding protein in childhood febrile illness and renal disease. *Pediatr Nephrol* 4:314–318.

Dubach UC, Le Hir M and Ghandi R (1988) Use of urinary enzymes as markers of nephrotoxicity. *Toxicol Lett* 46:193–196.

Esnard A, Esnard F and Gauthier F (1988) Purification of the cystatin C – like inhibitors from urine of nephropathic rats. *Biol Chem Hoppe Seyler* 369 (Suppl):219–222.

Eti S, Cheng CY, Marshall A and Reidenberg MM (1993) Urinary clusterin in chronic nephrotoxicity in the rat. *Proc Soc Exp Biol Med* 202:487–490.

Evan AP and Dail WG (1974) The effects of sodium chromate on the proximal tubules of the rat kidney. Fine structural damage and lysozymuria. *Lab Invest* 30:704–715.

Falkenberg FW, Hildebrand H, Lutte L, Schwengberg S, Henke B, Greshake D, Schmidt B, Friederich A, Rinke M, Schlüter G and Bomhard E (1996) Urinary antigens as markers of papillary toxicity. I. Identification and characterization of rat kidney papillary antigens with monoclonal antibodies. *Arch Toxicol* 71:80–92.

Felley-Bosco E and Diezi J (1987) Fate of cadmium in rat renal tubules: A microinjection study. *Toxicol Appl Pharmacol* 91:204–211.

Filler G, Priem F, Lepage N, Sinha P, Vollmer I, Clark H, Keeley E, Matzinger M, Akbari A, Althaus H and Jung K (2002) β-trace protein, cystatin C, β_2-microglobulin and creatinine compared for detecting impaired glomerular filtration rates in children. *Clin Chem* 48:729–736.

Finco DR (1997) Kidney function, in *Clinical Biochemistry of Domestic Animals* (Kaneko JJ, Harvey JW and Bruss ML, Eds), Academic Press, San Diego.

Finco DR, Tabaru H, Brown SA and Barsanti JA (1993) Endogenous creatinine clearance measurement of glomerular filtration rate in dogs. *Am. J. Vet. Res.* 54:1575–1578.

Finco DR, Brown SA, Vaden SL and Ferguson DC (1995) Relationship between plasma creatinine concentration and glomerular filtration rate in dogs. *J Vet Pharmacol Ther* 18: 418–421.

Finn WF and Porter GA (1998) Urinary biomarkers and nephrotoxicity, in *Clinical Nephrotoxins* (DeBroe ME., Porter GA, Bennett WM and Verpooten GA, Eds) Kluwer Academic Publishers, Dordrecht (Netherlands).

Finney H, Newman DJ, Gruber W, Merle P and Price CP (1997) Initial evaluation of cystatin C measurement by particle-enhanced immunonephelometry on the Behring nephelemeter systems (BNA, BN II). *Clin Chem* 43:1016–1022.

Flecknell PA (1995) Non-surgical experimental procedures, in *Laboratory Animals – An Introduction for Experimenters* (Tuffey AA Ed.), John Wiley and Sons, Ltd., Chichester.

Fommei E and Volterrani D (1995) Renal nuclear medicine. *Sem Nucl Med* 25:183–194.

Fox JG (1988) *Biology and Diseases of the Ferret*, Lea and Febiger, Philadelphia.

Gartland KPR, Bonner FW and Nicholson JK (1989) Investigations into the biochemical effects of region-specific nephrotoxins. *Molec Pharmacol* 35:242–250.

Gherardi E and Calandra S, (1982) Plasma and urinary lipids and lipoproteins during the development of nephrotic syndrome induced in the rat by puromycin amino-nucleoside. *Biochim Bioph Acta* 710:188–196.

Gosling P (1995) Microalbuminuria: a sensitive indicator of non-renal disease? *Ann Clin Biochem* 32:439–441.

Gossett KA, Turnwald GH, Kearney MT, Greco DS and Cleghorn B (1987) Evaluation of gamma-glutamyl transpeptidase-to-creatinine ratio from spot samples of urine supernatant, as an indicator of urinary enzyme excretion in dogs. *Am J Vet Res* 48:455–457.

Grauer GF, Greco DS, Behrend EN, Mani I, Fettman MJ and Allen TA (1995) Estimation of quantitative enzymuria in dogs with gentamicin-induced nephrotoxicosis using urine enzyme/creatinine ratios from spot urine samples. *J Vet Intern Med* 9:324–327.

Greaves PC (2000) *Histopathology of Preclinical Toxicology Studies: Interpretation and Relevance in Drug Safety Evaluation.* 2nd Edition, Elsevier Science, Amsterdam.

Gross SK, Daniel PF, Evans JE and McClure RH (1991) Lipid composition of lysosomal multilamellar bodies of male mouse urine. *J Lipid Res* 32:157–164.

Grötsch H, Hropot M, Klaus E and Malerczyk V (1985) Enzymuria of the rat: Biorhythms and sex differences. *J Clin Chem Clin Biochem* 23:343–347.

Guder, WG, and Hofmann, W (1992) Markers for the diagnosis and monitoring of renal lesions, *Clin Neph* 38:S3–S7.

Guder WG and Ross BD (1984) Enzyme distribution along the nephron. *Kidney Int* 26: 101–111.

Guder WG, Ivandic M and Hofmann W (1998) Physiopathology of proteinuria and laboratory diagnostic strategy based on single protein analysis. *Clin Chem Lab Med* 36: 935–939.

Gwinner W, Jackle-Meyer I, and Stolte H (1993) Origin of urinary fibronectin. *Lab Invest* 69:250–255.

Haagsma BH and Pound AW (1979) Mercuric chloride-induced renal tubular necrosis in the rat. *Br J Exp Pathol* 60:341–352.

Haas M, Kluppel AC, Moolenaar F, Meijer DK, de Jong PE and de Zeeuw D (1997) Urine collection in the freely-moving rat: Reliability for measurement of short-term renal effects. *J Pharmacol Toxicol Methods* 38:47–51.

Hakansson K, Changgoo H, Anders G, Stefan K and Magnus A (1996) Mouse and rat cystatin C: *Escherichia coli* production, characterization, and tissue distribution. *Comp Biochem Physiol – B: Comp Biochem* 114:303–311.

Halman J and Price RG (1984) in *Selected Topics in Clinical Enzymology*, Vol. 2, (Goldberg DM and Werner M, Eds) pp. 435–444, Walter de Gruyter and Co., Berlin.

Hamberg O (1997) Regulation of urea synthesis by diet protein and carbohydrate in normal man and in patients with cirrhosis. Relationship to glucagon and insulin. *Dan Med Bull* 44:225–241.

Han WK, Bailly V, Abichandani R, Thadhani R and Bonventure JV (2002) Kidney injury molecule-1 (KIM-1): A novel biomarker for human renal proximal tubule injury. *Kidney Int* 62:237–244.

Haywood JR, Shaffer RA, Fastenow C, Fink GD and Brody MJ (1981) Regional blood flow measurement with pulsed Doppler flowmeter in conscious rat. *Am J Physiol* 241:H273–H278.

Herget-Rosenthal S, Poppen D, Pietruck F, Marggraf G, Phillip T and Kribben A (2001) Prediction of severity of acute tubular necrosis by tubular proteinuria. *J Am Soc Nephrol* 12:170A.

Hidaka S, Kranzlin B, Gretz N and Witzgall R (2001) Urinary clusterin helps to differentiate between tubular and glomerualr injuries and serves as a marker for disease severity in polycystic kidney disease. *J Am Soc Nephrol* 12:782A.

Higashiyama N, Nishiyama S, Itoh T, and Nakamura M (1983) Effect of castration on urinary *N*-acetyl-β-D-glucosaminidase levels in male beagles. *Renal Physiol* 6:226–231.

Hildebrand H, Rinke M, Schlüter G, Bomhard E and Falkenberg FW (1999) Urinary antigens as markers of papillary toxicity. II. Applications of monoclonal antibodies for the determination of papillary antigens in urine. *Arch Toxicol* 73:233–245.

Hofmann W, Regenboom C, Edel H and Guder WG (1994) Diagnostic strategies in urinalysis. *Kidney Int* 46 (Suppl 47):S111–S114.

Holdt-Lehmann B, Lehmann A, Korten G, Nagel H-R, Nizze H and Schuff-Werner P (2000) Diagnostic value of urinary alanine aminopeptidase and *N*-acetyl-β-D-glucosaminidase in comparison to α_1-microglobulin as a marker in evaluating tubular dysfunction in glomerulonephritis patients. *Clin Chim Acta* 297:93–102.

Horst PJ, Bauer M, Veelken R and Unger T (1988) A new method for collecting urine directly from the ureter in conscious unrestrained rats. *Renal Physiol Biochem* 11:325–331.

Houser MT and Milner LS (1991) Renal tubular protein handling in experimental renal disease. *Nephron* 58:461–465.

Huang Q, Dunn RT, Jayadev S, DiSorbo O, Pack FD, Farr SB, Stoll RE and Blanchard KT (2001) Assessment of cisplatin-induced nephrotoxicity by microarray technology. *Toxicol Sci* 63:196–207.

Hysing J and Tolleshaug H (1986) Quantitative aspects of the uptake and degradation of lysozyme in the rat kidney *in vivo*. *Biochim Biophy Acta* 887:42–50.

Janssens PMW, Kornaat N, Tieleman R, Monnens LAH and Willems JL (1992) Localizing the site of hematuria by immunocytochemical staining of erythrocytes in urine. *Clin Chem* 38:216–222.

Jensen AL, Bomholt M and Moe L (2001) Preliminary evaluation of a particle-enhanced turbidimetric immunoassay (PETIA) for the determination of cystatin C-like immunoreactivity in dogs. *Vet Clin Pathol* 30:86–90.

Jung K, Schulze BD, Sydow K, Pergande M, Precht K and Schrieber G (1987) Diagnostic value of low-molecular mass proteins in serum for detection of reduced glomerular filtration rate. *J Clin Chem Clin Biochem* 25:499–503.

Jung K and Grutzmann KD (1988) Quality control material for activity determinations of urinary enzymes. *Clin Biochem* 21:53–57.

Kamijo A, Kimura K, Sugaya T, Hikawa A, Yamanouti M, Hirata Y, Goto A, Fujita T and Omata M (2001) Urinary excretion of liver type fatty acid binding protein (L-FABP) is a new clinical marker for progression of chronic glomerular disease. *J Am Soc Nephrol* 12:74A.

Kanauchi M, Nishioka H and Dohi K (1995) Diagnostic significance of urinary fibronectin in diabetic nephropathy. *Nippon Jinzo Gakkai Shi* 37:127–133.

Kaplan AA and Kohn OF (1992) Fractional excretion of urea as a guide to renal dysfunction. *Am J Nephrol* 12:49–54.

Katein AM, O'Bryan SM and Bounous DI (2002) Specimen collection comparison for clinical pathology analysis. *Vet Pathol* 39:614.

Kazama JJ, Keiko K, Ken A, Hiroki M and Fumitake G (2002) Serum cystatin C reliably detects renal dysfunction in patients with various renal diseases. *Nephron* 91:13–20.

Khan KN and Alden CA (2001) Kidney, in *Handbook of Toxicologic Pathology* (Haschek-Hock WA, Wallig MA and Rousseaux CG, Eds), Academic Press, New York.

Kimura H, Odani S, Suzuki J, Arakawa M and Ono T (1989) Kidney fatty-acid binding protein: Identification as alpha-2U-globulin. *FEBS Lett* 246:101–104.

Kinter LB, Horner E, Mann WA, Weinstock J and Ruffolo RR (1990) Characterization of the hemodynamic activities of fenoldopam and its enantiomers in the dog. *Chirality* 2: 219–225.

Kinter LB, Mann WA, Weinstock J and Ruffolo RR (1994) Effects of catechol ring flourination on cardiovascular and renal activities of fenoldopam enantiomers. *Chirality* 6:446–455.

Kinter LB, Huffman WF and Stassen FL (1988) Antagonists of the antidiuretic activity of vasopressin. *Am J Physiol* 254:F165–177.

Kinter LB, Caltabiano S and Huffman WF (1993) On the antidiuretic agonist activity of antidiuretic hormone receptor antagonists. *Biochem Pharmacol* 45:1731–1737.

Knox FG, Ritman EL and Romero JC (1984) Intrarenal distribution of blood flow: Evolution of a new approach to measuremen. *Kidney Int* 25:473–479.

Koenig H, Goldstone A and Hughes C (1978) Lysosomal enzymuria in the testosterone-treated mouse. A manifestation of cell defecation of residual bodies. *Lab Invest* 39:329–341.

Kurien BT and Scofield RH (1999) Mouse urine collection using clear plastic wrap. *Lab Animals* 33:83–86.

Lam KT, Borkan S, Claffey KP, Schwartz JH, Chobanian AV and Brecher B (1988) Properties and differential regulation of two fatty acids binding proteins in the rat kidney, *J Biol Chem* 263:15762–15768.

Laterza OF, Price CP and Scott MG (2002) Cystatin C: An improved estimator of glomerular filtration rate? *Clin Chem* 48:699–707.

Lee KM, Reed LL, Bove DL and Dill JA (1998) Effects of water dilution, housing and food on rat urine collected from the metabolism cage. *Lab Animal Sci* 48:520–525.

Leonard BJ, Eccleston E, Jones D, Todd P and Walpole A (1971) Antileukaemic and nephrotoxic properties of platinum compounds. *Nature (London)* 234:43–45.

Levin S, Semler D and Rubin Z (1993) Effects of two week feed restriction on some common toxicologic parameters in Sprague-Dawley rats. *Toxicol Pathol* 21:1–14.

Lock EA and Ishmael J (1979) The acute toxic effects of hexachloro-1 : 3-butadiene on the rat kidney. *Arch Toxicol* 43:47–57.

Loeb WF (1997) Clinical biochemistry of laboratory rodents and rabbits, in *Clinical Biochemistry of Domestic Animals* (Kaneko JJ, Harvey JW and Bruss ML, Eds), Academic Press, San Diego.

Loeb WF (1998) The measurement of renal injury. *Toxicol Pathol* 26:26–28.

Loeb WF and Quimby FW (1999) *The Clinical Chemistry of Laboratory Animals,* Taylor and Francis, Philadelphia.

Loeb WF, Das SR and Trout JR (1997) The effect of erythritol on the stability of γ-glutamyl transpeptidase and *N*-acetyl glocosaminidase in human urine. *Toxicol Pathol* 25:264–267.

Lofti AS and Djalali M (2000) Possible correlation between urinary muramidase (E.C.3.2.1.27) and oesophageal cancer. *Cancer Lett* 158:113–117.

Maatman RG, Van Kuppevelt TH and Verkamp JH (1991) Two types of fatty acid-binding protein in human kidney. Isolation, characterization and localization. *Biochem J* 273:759–766.

MacNeil ML, Mueller PW, Caudill SP and Steinberg KK (1991) Considerations when measuring urinary albumin: Precision, substances that may interfere, and conditions for sample storage. *Clin Chem* 37:2120–2123.

Makino H, Hayashi Y, Shikata K, Hirata K, Akiyama K, Ogura T, Obata K, and Ota Z (1995) Urinary detection of type IV collagen and its increase in glomerulonephritis. *Res Commun Molec Pathol Pharmacol* 88:215–223.

Manabe N, Kinoshita A, Yamaguchi M, Furuya Y, Nagano N, Yamada-Uchio K, Akashi N, Miyamoto-Kuramitsu K and Miyamoto H (2001) Changes in quantitative profile of extracellular matrix components in the kidneys of rats with adriamycin-induced nephropathy. *J Vet Med Sci* 63:125–133.

Mandavilli U, Schmidt J, Rattner DW, Watson WT and Warshaw AL (1991). Continuous complete collection of uncontaminated urine in conscious rodents. *Lab Animal Sci* 41: 258–261.

Mann WA and Kinter LB (1993a) Characterization of the renal handling of p-aminohippurate (PAH) in the beagle dog (*Canis familiaris*). *Gen Pharmac* 24:367–372.

Mann WA and Kinter LB (1993b) Characterization of maximal intravenous dose volumes in the beagle dog (*Canis familiaris*). *Gen Pharmac* 24:357–366.

Mann WA, Landi MS, Horner E, Woodward P, Campbell S and Kinter LB (1987) A simple procedure for direct blood pressure measurements in conscious dogs. *Lab Animal Sci* 37:105–108.

Mann WA, Welzel GE, Goldstein RS, Sozio RS, Cyronak MJ, Kao J and Kinter LB (1991) Characterization of the renal effects and renal elimination of sulotroban in dogs. *J Pharmacol Exp Ther.* 259:1231–1240.

Marcussen N and Lundemose JB (1996) Histopathologic methods in renal toxicology, in *Methods in Renal Toxicology* (Zalups RZ and Lash LH, Eds), CRC Press, New York.

Martin C, Pechereau D, de la Farge F and Braun JP (2002) Plasma cystatin C in the cat: Current techniques do not allow to use it for the diagnosis of renal failure. *Rev Med Vet (Toulouse)* 153:305–310.

Martin RS and Small DM (1984) Physicochemical characterization of the urinary lipid from humans with nephrotic syndrome. *J Lab Clin Med* 103:798-810.

Maruhn D, Strozyk K, Gielow L and Bock KD (1977) Diurnal variations of urinary enzyme excretion. *Clin Chim Acta* 75:427–433.

McClure DE (1999) Clinical pathology and sample collection in the laboratory rodent. *Vet Clin North Am* 2:565–590.

Melegos DN, Grass L, Pierratos A and Diamandis EP (1999) Highly elevated levels of prostaglandin D synthase in the serum of patients with renal failure. *Urology* 53:32–37.

Mimura K, Yukawa S, Maeda T, Kinoshita M, Yamada Y, Saika Y, Miyai T, Nagae M, Mune M and Nomoto H (1984) Selective urinary excretion of phosphatidyl ethanolamine in patients with chronic glomerular diseases. *Metabolism* 6:882–890.

Montagne P, Culliere ML, Mole C, Bene MC and Faure G (1998) Microparticle-enhanced nephelometric immunoassay of lysozyme in milk and other human body fluids. *Clin Chem* 44:1610–1615.

Mueller PW, MacNeil ML and Steinberg KK (1986) Stabilization of alanine aminopeptidase, gamma glutamyltranspeptidase, and *N*-acetyl-β-D-glucosaminidase activity in normal urines. *Arch Environ Contam Toxicol* 15:343–347.

Mueller PW, MacNeil ML and Steinberg KK (1989) *N*-acetyl-β-D-glucosaminidase assay in urine: urea inhibition. *J Anal Toxicol* 13:188–190.

Muramatsu Y, Tsujie M, Kohda Y, Pham B, Perantoni AO, Zhao H, Jo S-K, Yuen PST, Craig L, Hu X and Star RA (2002) Early detection of cysteine-rich protein 61 (CYR61, CCN1) in urine following renal ischemic reperfusion injury. *Kidney Int* 62:1601–1610.

Murray G, Wyllie G, Hill GS, Ramsden PW and Hepinstall RH (1972) Experimental papillary necrosis of the kidney. I. Morphologic and functional data. *Am J Pathol* 67:285–302.

Nakamura M, Itoh T, Miyata K, Higashiyama N, Takesue H and Nishiyama S (1983) Differences in urinary *N*-acetyl-β-D-glucosaminidase activity between male and female Beagle dogs. *Renal Physiol* 6:130–133.

Nanhas K and Provost J-P (2002) Blood sampling in the rat: current practices and limitations. *Comp Clin Pathol* 11:14–37.

Nath KA, Dvergsten J, Correa-Rotter R, Hostetter TH, Manivel JC and Rosenberg ME (1994) Induction of clusterin in acute and chronic oxidative renal disease in the rat and its dissociation from cell injury. *Lab Invest* 71:209–218.

Neverov NI and Nikitina EA (1992) Lipiduria in the nephrotic syndrome. *Terapevticheskii Arkhiv* 64:16–18.

Newman DJ and Price CP (1999) Renal function and nitrogen metabolites in *Tietz Textbook of Clinical Chemistry* (Burtis CA and Ashwood ER, Eds), W. B. Saunders Company, Philadelphia.

Newman DJ, Thakkar H, Edwards RG, Wilkie M, White T, Grubb AO and Price CP (1994) Serum cystatin C: A replacement for creatinine as a biochemical marker of GFR. *Kidney Int* 47 (Suppl):S20–21.

Nicholson JK, O'Flynn MP, Sadler PJ, MacLeod AF, Juul SM and Sonksen PH (1984) Proton-nuclear-magnetic-resonance studies of serum, plasma and urine from fasting normal and diabetic subject. *Biochem J* 217:365–375.

Nishimura N (1987) The mechanism of cadmium-induced lysozyme enhancement in rabbit kidney. *Arch Toxicol* 61:105–115.

Ohata H, Hashimoto T, Monose K, Takahashi A and Terao T (1988) Urinalysis for detection of chemically induced renal damage (3) – Establishment and application of radioimmunoassay for lysozyme of rat urine. *Arch Toxicol* 62:60–65.

Okonogi H, Nishimura M, Utsunomiya Y, Hamaguchi K, Tsuchida H, Miura Y, Suzuki S, Kawamura T, Hosoya T and Yamada K (2001) Urinary type IV collagen excretion reflects renal morphological alterations and type IV collagen expression in patients with type 2 diabetes mellitus. *Clin Nephrol* 55:357–364.

Oliveri O, Bassi A, Pizzolo F, Tinazzi E and Corrocher R (2002) Cystatin C versus creatinine in renovascular disease. *Am J Hypertens* 15:174A.

Pergande M and Jung K (1993) Sandwich enzyme immunoassay of cystatin C in serum with commercially available antibodies. *Clin Chem* 39:1885–1890.

Peterson PA, Evrin P-E and Berggård I (1969) Differentiation of glomerular, tubular and normal proteinuria: Determinations of urinary excretion of β_2-microglobulin, albumin, and total protein. *J Clin Invest* 48:1189–1198.

Pickering RG and Pickering CE (1984) The effects of reduced dietary intake upon the body and organ weights, and some clinical chemistry and haematologic variates of the young Wistar rat. *Toxicol Lett* 21:271–277.

Pillsworth TJ, Haver VM, Abrass CK and Delaney CJ (1987) Differentiation of renal from non-renal hematuria by microscopic examination of erythrocytes in urine. *Clin Chem* 33:1791–1795.

Pitts RF (1968) *Physiology of the Kidney and Body Fluids*, 3rd ed., Year Book Medical Publishers, Chicago.

Plummer DT, Noorazar S, Obatomi DK and Haslam JD (1986) Assessment of renal injury by urinary enzymes. *Uremia Invest* 9:97–102.

Poulik MD, Shinnick CS and Smithies O (1981) Partial amino acid sequences of human and dog post-gamma globulins. *Molec Immunol* 18:569–572.

Price RG (1982) Urinary enzymes, nephrotoxicity and renal disease. *Toxicol* 23:99–134.

Price RG, Taylor SA, Chivers I, Arce-Thomas M, Crutcher E, Franchini I, Alinovi R, Cavazzini S, Bergmaschi E, Mutti A, Vettori MV, Lauwerys R, Bernard A, Kabanda A, Roels H, Thielemans N, Hotz PH, De Broe ME, Elseviers MM, Nuyts GD, Gelpi E, Hotter G, Rosello J, Ramis I, Stolte H, Fels LM and Eisenberger U (1996) Development and validation of new screening tests for nephrotoxic effects. *Hum Exp Toxicol* 15:S10–S19.

Price RG, Berndt WO, Finn WF, Aresini G, Manley SE, Fels LM, Shaikh ZA and Mutti A (1997) Urinary biomarkers to detect significant effects of environmental and occupational exposure to nephrotoxins. III. Minimal battery of tests to assess subclinical nephrotoxicity for epidemiological studies based on current knowledge. *Renal Failure* 19: 535–552.

Price RG (2000) Urinalysis to exclude and monitor nephrotoxicity. *Clin Chim Acta* 297:173–182.

Price RG (2002) Early markers of nephrotoxicity. *Comp Clin Pathol* 11:2–7.

Price CP and Finney H (2000) Developments in the assessment of glomerular filtration rate. *Clin Chim Acta*, 297:55–66.

Priem F, Althaus H, Birnbaum M, Sinha P, Conradt HS and Jung K (1999) β-trace protein in serum: A new marker of glomerular filtration rate in the creatinine-blind range. *Clin Chem* 45:567–568.

Priem F, Althaus H, Jung K and Sinha P (2001) β-trace protein is not better than cystatin C as an indicator of reduced glomerular filtration rate. *Clin Chem* 47:2181.

Ramsey CR and Knox FG (1996), in *Renal Methods in Toxicology* (Lash LH and Zalups RK, Eds), CRC Press, Boca Raton.

Reba RC, Hosain F and Wagner NH (1968) Indium-113m diethylenetriaminepenta-acetic acid (DTPA): A new radiopharmaceutical for study of the kidneys. *Radiol* 90:147–149.

Robertson DG, Reiley MD, Sigler RE, Wells DF, Paterson DA and Braden TK (2000) Metabonomics: Ecvaluation of nuclear magnetic resonance (NMR) and pattern recognition technology for rapid *in vivo* screening of liver and kidney toxicants. *Toxicol Sci* 57:326–337.

Rocco O, Michele M, Plebiani M, Piccoli P, De Martin S, Floreani P, Padrini R and Pietro P (2002) Diagnostic value of plasma cystatin C as a glomerular filtration marker in decompensated liver cirrhosis. *Clin Chem* 48:850–858.

Saggar K, Sandhu P, Sandhu JS and Aulakh BS (2003) Role of computed tomographic angiography (CTA) in evaluation of living renal donors. *Dial Transplant* 23:316–323.

Sandsoe G, Alberto F, Nives CC, Lorenzo B, Erica V and Federico M (2002) Cimetidine administration and tubular creatinine secretion in patients with compensated cirrhosis. *Clin Sci* 102:91–98.

Sarkar SK, Holland GA, Lenkinski R, Mattingly MA and Kinter LB (1988) Renal imaging studies at 1.5 and 9.4 T: Effects of diuretics. *Magn Reson. Med* 7:117–24.

Sarkar SK, Rycyna RE, Lenkinski RE, Solleveld HA and Kinter LB (1991) Yb-DTPA, a novel contrast agent in magnetic resonance imaging: Application to rat kidney. *Magn Reson Med* 17:328–334.

Schnell MA, Hardy C, Hawley M, Propert KJ and Wilson JM (2002) Effect of blood collection technique in mice on clinical pathology parameters. *Human Gene Therapy* 13: 155–162.

Schwendenwein I and Gabler C (2001) Evaluation of an enzymatic creatinine assay. *Vet Clin Pathol* 30:163–164.

Senekjian HO, Knight TF and Weinman EJ (1981) Micropuncture study of the handling of gentamicin by the rat kidney. *Kidney Int* 19:416–423.

Shao R and Tarloff JB (1996) Lack of correlation between para-aminophenol toxicity *in vivo* and *in vitro* in female Sprague-Dawley rats. *Fundam Appl Toxicol* 31:268–278.

Shemesh O, Golbetz H, Kriss JP and Myers BD (1985) Limitations of creatinine as a filtration marker in glomerulonephropathic patients. *Kidney Int* 28:830–838.

Sigitova ON, Madsudova AN and Myasoutova LI (2000) Tests for cell membrane destabilization in assessing activity/progression of chronic glomerulonephritis. *Terapevticheskii Arkhiv* 72:26–30.

Simonsen O, Grubb A and Thysell H (1985) The blood serum concentration of cystatin C (γ-trace) as a measure of glomerular filtration rate. *Scand J Clin Lab Invest* 45:97–101.

Slaughter HS and Everds NE, personal communication.

Starr R, Hostetter T and Hortin GL (2002) New markers for kidney disease. *Clin Chem* 48:1375–1376.

Stonard MD (1987), in *Nephrotoxicity in the Experimental and Clinical Situation* (Bach PH and Lock EA, Eds) pp. 563–592, Martinus Nijhoff, Boston.

Stonard MD, Gore CW, Oliver GJA and Smith IK (1987) Urinary enzymes and protein patterns as indicators of injury to different regions of the kidney. *Fundam Appl Toxicol* 9: 339–351.

Streather CP, Varghese Z, Moorhead JF and Scoble JE (1993) Lipiduria in renal disease. *Am J Hypertens* 11:353S–357S.

Sundberg AG, Appelkvist EL, Backman L and Dallner G (1994a) Quantitation of glutathione transferase-pi in the urine by radioimmunoassay. *Nephron* 66:162–169.

Sundberg AG, Appelkvist EL, Backman L and Dallner G (1994b) Urinary pi-class glutathione transferase as an indicator of tubular damage in the human kidney. *Nephron* 67:308–316.

Sundberg A, Appelkvist EL, Dallner G and Nilsson R (1994c) Glutathione transferases in the urine: sensitive methods for detection of kidney damage induced by nephrotoxic agents in humans. *Environ. Health Perspec.* 102 (Supp.3):293–296.

Suzuki C, Bondy G, Gurofsky S, Hierlihy L and Curran I (1996) Evaluation of glutathione-*S*-transferase α (GSTα) and transferrin as urinary markers of fumonisin B_1-induced nephrotoxicity in the rat. *Cell Molec Biol* 42:S67.

Swan SK (1997) The search continues – an ideal marker of GFR. *Clin Chem* 43:913-914.

Tauson A-H and Wamberg S (1998) Effects of protein supply on plasma urea and creatinine concentrations in female mink (*Mustela vison*). *J Nutr* 128:2584S–2586S.

Taylor DC, Cripps AW and Clancy RL (1992) Measurement of lysozyme by an enzyme-linked immunosorbent assay. *J Immunol Methods* 146:55–61.

Taylor SA, Chivers ID, Price RG, Arce-Thomas M, Milligan P, Francini I, Alinovi R, Cavazzini S, Bergamaschi E, Vittori M, Mutti A, Lauwerys RR, Bernard AM, Roels HA, De Broe ME, Nuyts GD, Elseviers MM, Hotter G, Ramis I, Rosello J, Gelpi E, Stolte H, Eisenberger U and Fels LM (1997) The assessment of biomarkers to detect nephrotoxicity using an integrated database. *Environ Res* 75:23–33.

Tencer, J, Torffvit, O, Thysell, H, Rippe, B, and Grubb, A (1998) Proteinuria selectivity index based on α_2-macroglobulin or IgM is superior to the IgG based index in differentiating glomerular diseases, *Kidney Int* 54:2098–2105.

Tencer J, Bakoush O and Torffvit O (2000) Diagnostic and prognostic significance of proteinuria selectivity index in glomerular diseases. *Clin Chim Acta* 297:73–83.

Tenstad O, Roals AB, Grubb A and Aukland K (1996) Renal handling of radiolabeled cystatin C in the rat. *Scand J Clin Lab Invest* 56:409–414.

Thanh NTK, Stevenson G, Obatomi DK, Aicher B, Baumeister M and Bach PH (2001) Urinary lipid change during the development of chemically induced renal papillary necrosis. A study using mefenamic acid and *N*-phenylantranilic acid. *Biomarkers* 6:417–427.

Thongboonkerd V, Klein JB and McLeish KR (2002a) Proteomic identification of biomarkers of glomerular diseases. *J Am Soc Nephrol* 13:120A.

Thongboonkerd V, McLeish KR, Arthur JM and Klein JB (2002b) Proteomic analysis of normal human urinary proteins isolated by acetone precipitation or ultracentrifugation. *Kidney Int* 62:1461–1469.

Thrivikraman KV, Hout RL and Plotsky PM (2002) Jugular vein catheterization for repeated blood sampling in the unrestrained conscious rat. *Brain Res Protoc* 10:84–94.

Twyman SJ, Overton J and Rowe DJ (2000) Measurement of urinary retinol binding protein by immunonephelometry. *Clin Chim Acta* 297:155–161.

Uchida K and Gotoh A (2002) Measurement of cystatin-C and creatinine in urine. *Clin Chim Acta* 323:121–128.

Ullrich KJ, Fritzsch G, Rumrich G and David C (1994) Polysubstrates: substances that interact with renal contraluminal PAH, sulfate, and NMeN transport: Sulfamoyl-, sulfonylurea-, thiazide- and benzeneamino-carboxylate (nicotinate) compounds. *J Pharmacol Exp Ther* 269:684–92.

Ullrich KJ, Rumrich G, Papavassiliou F, Kloss S and Fritzsch G (1991a) Contraluminal p-aminohippurate transport in the proximal tubule of the rat kidney. VII. Specificity: Cyclic nucleotides, eicosanoids. *Pflugers Arch* 418:360–370.

Ullrich KJ, Rumrich G, Papavassiliou F and Hierholzer K (1991b) Contraluminal p-aminohippurate transport in the proximal tubule of the rat kidney. VIII. Transport of corticosteroids. *Pflugers Arch* 418:371–382.

Umbreit A and Wiedemann G (2000) Determination of urinary protein fractions. A comparison with different electrophoretic methods and quantitatively determined protein concentrations. *Clin Chim Acta* 297:163–172.

Usuda K, Kono K, Dote T, Nishiura K, Miyata K, Nishiura H, Shimahara M and Sugimoto K (1998) Urinary biomarkers monitoring for experimental fluoride nephrotoxicity. *Arch Toxicol* 72:104–109.

Valtin H (1983) *Renal Function*, 2nd ed, Little Brown, Boston.

Vander AJ (1985) *Renal function*, 3rd ed, McGraw-Hill, New York.

Vanderlinde RE (1981) Urinary enzyme measurements in the diagnosis of renal disorders. *Ann Clin Lab Sci* 11:189–201.

Van Kreel BK, Janssen MA and Koostra G (2002) Functional relationship of alpha-glutathione *S*-transferase and glutathione *S*-transferase activity in machine-preserved non-heart-beating donor kidneys. *Transplant Int* 15:546–549.

Verpooten GF, Nouwen EJ, Hoylaerts MF, Hendrix PG and de Broe ME (1989) Segment-specific localization of intestinal-type alkaline phosphatase in human kidney. *Kidney Int* 36:617–625.

Vlaskou D, Hofmann W, Guder WG, Siskos PA and Dionyssiou-Asteriou A (2000) Human neutral brush border endopeptidase EC 3.4.24.11 in urine, its isolation, characterization and activity in renal diseases. *Clin Chim Acta* 297:103–121.

Volders PG, Vork MM, Glatz JF, and Smits JF (1993) Fatty acid-binding proteinuria diagnoses myocardial infarction in the rat. *Molec Cell Biochem* 123:185–190.

Van Metre DC and Angelos SM (1999) Miniature pigs. *Vet Clin North Am* 2:519–537.

Viau C, Bernard A and Lauwerys R (1986) Determination of rat β_2-microglobulin in urine and serum. I. Development of an immunoassay based on latex particles agglutination. *J Appl Toxicol* 6:185–189.

Watson ADJ, Lefebvre HP, Concordet D, Laroute V, Ferré J-P, Braun J-P, Conchou F and Toutain P-L (2002) Plasma exogenous creatinine clearance test in dogs: Comparison with other methods and proposed limited sampling strategy. *J Vet Intern Med* 16:22–33.

Weingand K, Brown G, Hall R, Davies D, Gossett K, Neptun D, Waner T, Matsuzawa T, Salemik P, Froelke W, Provost J-P, Negro GD, Batchelor J, Nomura M, Groetsch H, Boink A, Kimball J, Woodman D, York M, Fabianson-Johnson E, Lupart M and Melloni E (1996) Harmonization of animal clinical pathology testing in toxicity and safety studies. *Fundam Appl Toxicol* 29:198–201.

Westhuyzen J, Endre ZH, Reece G, Reith DM, Saltissi D and Morgan TJ (2003) Measurement of tubular enzymuria facilitates the early detection of acute renal impairment in the intensive care unit. *Nephrol Dial Transp* 18:543–551.

Woitas RP, Stoffel-Wagner B, Poege U, Schiedermaier P, Spengler U and Sauerbruch T (2001) Low-molecular weight proteins as markers for glomerular filtration rate. *Clin Chem* 47:2179–2180.

Yagil Y (1990) Acute effect of cyclosporin on inner medullary blood flow in normal and postischemic rat kidney. *Am J Physiol* 258:F1139–1144.

Young DS and Bermes EW (1999) Specimen collection and processing: sources of biologic variation, in *Tietz Textbook of Clinical Chemistry* (Burtis CA and Ashwood ER, Eds), W. B. Saunders Company, Philadelphia.

4

In vitro Techniques in Screening and Mechanistic Studies: Organ Perfusion, Slices, and Nephron Components

Donald S. Kirkpatrick and A. Jay Gandolfi

INTRODUCTION

In vitro methods for studying the kidney were in development as early as the late 19th century as a means to explore the physiology of the kidney and its subunits. In modern toxicology, these methods have led to the elucidation of specific mechanisms of action for numerous classes of renal toxicants. *In vitro* methods have helped to build an understanding of how metals, chemotherapeutic agents, antimicrobial agents, cysteine conjugates, oxidants, and halogenated hydrocarbons cause kidney injury. The increased demand for *in vitro* systems has been driven by a number of scientific needs. Good *in vitro* models make mechanistic studies possible, provide alternatives to animal testing, act as high-throughput models for screening xenobiotics, and allow for the translation of toxicological data to human populations.

In vitro systems are, by definition, simplifications of the whole organs from which they come. By simplifying the complex and heterogeneous organ, it becomes possible for researchers to manipulate experimental conditions in order to test specific physiological and toxicological phenomena that might otherwise be immeasurable. The increased acceptance of *in vitro* techniques, and of the data generated from them, has allowed a reduction in the need to perform certain types of experiments in living animals. In addition, many of these *in vitro* techniques are satisfying the need for cost-efficient, high-throughput screening of the vast combinatorial libraries of toxicants and potential drugs. Finally, because many of these methods can be and have been translated for use with human tissue, they remain a very feasible way of translating *in vivo* and *in vitro* animal data to relevant cases of human exposure.

This review provides a thorough discussion of *in vitro* models that retain all or most of the structural heterogeneity of the functional kidney. Included in this group of *in vitro* techniques are whole-kidney perfusion, renal slices, and isolated nephron components. Structurally intact models of the kidney continue to be appropriate for the study of renal metabolism, cellular and subcellular disposition of toxicants, general renal toxicology, and the biomolecular effects of toxicants within the cells of the kidney.

Selecting an appropriate *in vitro* model for the study of renal toxicology remains complicated; no single model can suffice under all circumstances. The functional complexity of the kidney, involving coordination of cellular, subunit, and whole-organ function, must often be thoroughly dissected in order to address issues of human renal toxicology. Cell culture and subcellular models, ones in which the physical structure of the kidney has been eliminated, exhibit both significant advantages and disadvantages compared with the more morphologically intact models described here. (In Chapter 5, the discussion of *in vitro* models will be concluded by examining renal cell culture and cell-free kidney models.)

When deciding on the best *in vitro* technique for studying renal toxicology, the following questions should be addressed:

- Are cells that undergo a relevant *in vivo* exposure to the toxicant present?
- Do the exposed cells react to the toxicant in a similar fashion to that seen *in vivo*?
- Do the endpoints being measured reflect the toxic effect seen *in vivo*?

ORGAN PERFUSION

The isolated perfused kidney (IPK) is the least disrupted *in vitro* model for study of renal toxicology. During whole-organ perfusion experiments, the kidney is isolated from an animal and perfused with oxygenated, nutrient-rich solutions. During the procedure, the kidney remains completely intact in order to preserve the physiological functions of the whole organ. Thus, there is minimal disruption of cellular and functional heterogeneity within the IPK model prior to experimentation. The topic of IPK has been thoroughly reviewed on a number of occasions (Bekersky, 1983; Diamond, 1996; Maack, 1980; Newton and Hook, 1981; Tarloff and Goldstein, 1994).

History

The first reported use of whole-organ kidney perfusion occurred in 1876 when researchers described the production of hippurate using benzoic acid and glycine (Diamond, 1996). Organ perfusion techniques advanced during the 1950s through the work of Weiss et al. (1959). Since then, whole-organ perfusion techniques have been expanded to include their use in studies of disposition, metabolism, clearance, and transport of both essential and xenobiotic substrates. In toxicology, it has been established that this procedure will work for the study of glomerular and proximal tubular specific toxicants. Experience has shown this *in vitro* model to be the most representative of physiological events seen *in vivo*.

General Concept

In whole-organ perfusion experiments, kidneys are removed from a single animal and perfused with physiological solutions. Both the perfusate and toxicants are administered via the renal artery. In these experiments, unfiltered perfusate leaving the kidney via the renal vein can be recirculated through the system or collected for analysis. Urine produced by the kidney and excreted via the ureter can also be collected for analysis. A schematic diagram of an IPK is shown in Figure 4.1. The major advantage of the IPK is that it is the only model in which both glomerular filtration and tubular function remain intact. The perfusion process mimics renal blood flow and maintains a level of forward pressure in the kidney necessary to drive the filtration process. At the same time, complications from other organ systems on the kidney are

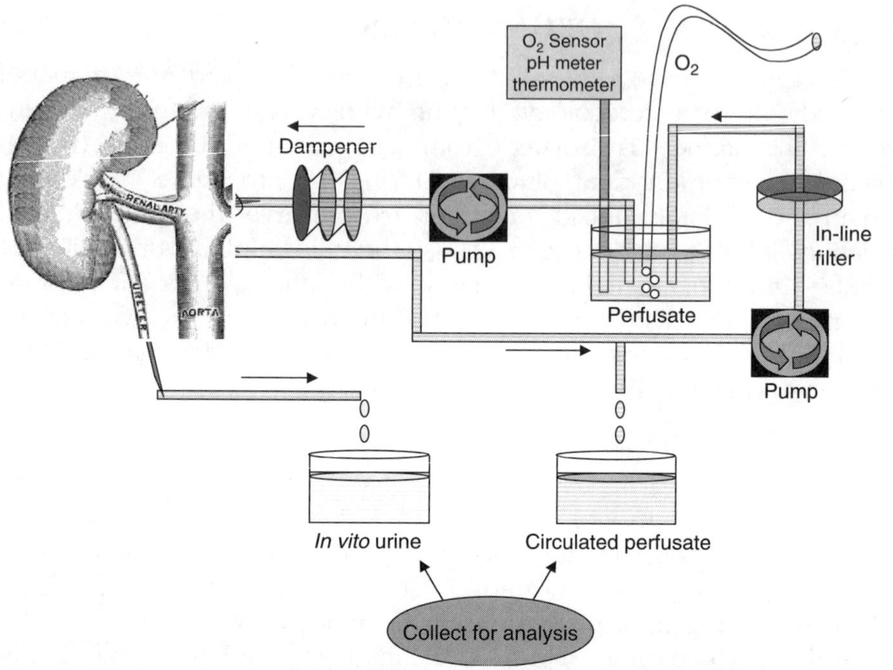

Figure 4.1 Hypothetical IPK perfusion apparatus. Perfusate flows through the circuit in the clockwise direction. Perfusate leaving the isolated kidney through the ureter or renal vein can be collected for analysis.

eliminated. For this reason, IPK is the most appropriate *in vitro* model to use when investigating the direct effects of a toxicant on overall renal function.

In toxicology, IPK is primarily utilized for the study of xenobiotic-induced effects on renal function. The model is also used to study kidney-specific metabolism, the kinetics of urinary excretion, and the process of chemical accumulation within renal cells. Unfortunately, the IPK has thus far not been shown to be compatible with molecular analyses such as gene expression profiling or the study of signal transduction cascades.

Significant limitations in the technique restrict its use in the study of many renal toxicology issues. While glomerular filtration and proximal tubular function are nicely retained in IPK preparations, significant abnormalities in distal tubular and collecting duct function exist. Because of these changes, the IPK model does not maintain the hemodynamic characteristics of a normal kidney. Abnormalities in the collecting duct significantly alter the IPK's ability to both concentrate and dilute urine as it would *in vivo* (Maack, 1980). Typically, an IPK preparation is useful for short studies, in the range of one to four hours.

Recent Advances

Technical limitations of the IPK have restricted its coupling to many of the advances in molecular toxicology. While molecular data from this model are difficult to collect and interpret, the IPK has been put to use to evaluate the effects of many biomolecules on renal function.

Vascular endothelial growth factor (VEGF), a protein endogenous to the kidney but also a known mitogen, has been shown to elicit effects on renal vascular perfusion using the IPK model (Klanke et al., 1998). Other biomolecules whose effects on renal function have also recently been described using this model include platelet-activating factor (Cailleaux et al., 1999, Monteiro et al., 1999), vascular kinins B1 and B2 (Bagate et al., 1999), and parathyroid hormone (Mokuda and Sakamoto, 1999).

With the advent of transgenic and knockout animals, the IPK will again provide an important research model for studying the interactions between toxicants and specific genes or proteins involved in maintaining renal homeostasis. In fact, IPKs from transgenic rats with a gene conferring constitutive hypertension were used by Lo et al. (1993) to evaluate the effect of the transgene on functional parameters such as sodium excretion and excretion of catecholamines.

Current Applications and Limits

As with the other *in vitro* systems discussed in this chapter, use of the IPK model is greatly limited by the viability of the isolated kidney. The technical difficulty and expensive equipment required to obtain consistent results have thus far limited the use of the IPK model beyond its current scope. Since an animal must be used in every individual IPK experiment, this technique is an unacceptable solution for those researchers who seek either high-throughput methods or alternatives to the use of animals. The issue of tissue viability will probably turn out to be the most formidable obstacle preventing the coupling of this technique with many modern molecular methods.

Methods for Kidney Perfusion

Surgery and Organ Preparation

Animals from which kidneys used in these procedures are obtained include rats, dogs, rabbits, and pigs. The donor animal is put under anesthesia during removal of the kidney. Special care must be taken to minimize stress on the animal and physical damage to the kidney, both have been shown to significantly affect the functional capabilities of the isolated organ.

Following removal of the kidney from the body, handling of the tissue should also be minimized because mechanical stimulation is believed to increase resistance of the renal vasculature. In order to maintain constant oxygen delivery, the renal artery is usually cannulated via the mesenteric artery prior to surgical isolation of the tissue (Diamond, 1996). Because the functional lifespan of the IPK is typically limited to between one and four hours, rapid isolation of the tissue is essential for obtaining reproducible data.

Cannulation is the technique used to introduce and collect perfusate from the IPK. Usually, media is introduced to the kidney through the renal artery via the mesenteric artery to mimic physiologic delivery of blood to the organ (Maack, 1980). In this manner, perfusate passes through the renal microvasculature and into the glomerulus via the afferent arterioles. Portions of the perfusate are filtered through the glomerulus and into the tubule, while the remainder passes into the efferent arteriole and leaves the kidney via the renal vein.

In order to cannulate the ureter, renal artery, or renal vein, the beveled edge of a plastic cannula is inserted and fixed in place with a suture. Selection of the appropriate diameter cannula is critical for maintaining the necessary positive pressure into the renal artery. Selecting the appropriate diameter cannula is even more important when the ureter or renal vein is cannulated. If a cannula with too small a diameter is used back-pressure can build up, retarding perfusion and leading to organ damage (Diamond, 1996).

Perfusion System

Perfusion of the kidney is normally controlled by the gentle circulation of a peristaltic pump. The pump circulates volumes of oxygenated media through the kidney. To prevent physiologic responses to pulsations in flow caused by the peristaltic pump, a dampener can be added between the influx pump and the renal artery. A normal perfusion apparatus is equipped with a system for gas exchange into the perfusate and includes monitors for pressure, pH, and O_2 (Diamond, 1996). In-line filters are often used to prevent blockage of perfusion by debris in the perfusate (Maack, 1980).

The important variables to consider when deciding upon perfusion conditions are the contents of the perfusate, the physical variables of the perfusate, and the pressure of the solutions as they enter the isolated kidney. Solutions used to perfuse the kidney are normally bicarbonate buffers, prepared as modified forms of Krebs-Ringer or Krebs-Henseleit solutions. Experimental evidence has reinforced the critical nature of perfusate additives such as glucose, oxygen, amino acids, and colloids (both protein and nonprotein) (Bekersky, 1983). Bovine serum albumin (BSA) and other colloids have been utilized for their ability to maintain osmolarity of the introduced solutions, and likewise glomerular filtration rate (GFR) (Bekersky, 1983; Diamond, 1996; Maack, 1980). Increased performance of IPKs has been obtained by the supplementation of perfusate solutions with erythrocytes or noncellular oxygen carriers that can enhance the oxygen delivering capacity of the solution (Diamond, 1996).

Special attention must be paid to the level of pressure used to perfuse the kidney. Excessive perfusion pressure often results in mechanical damage to cells of the isolated kidney. Unfortunately, in order to maintain normal GFR, perfusion flow rates must often be maintained at above normal physiological levels during experimental periods (Maack, 1980). Selection of perfusion flow rate and pressure is a delicate balance between maintaining the necessary GFR and avoiding pressure induced injury to the nephrons. The researcher must determine these criteria empirically.

While IPK preparations can be maintained for upto four hours, optimal performance for the IPK model exists occurs during the first one to two hours (Maack, 1980). The achievement of this extended viability has directly resulted from the implementation of improved perfusate formulations and filtration or dialysis of perfusate during the pumping process. With continued advances in this area, there is the potential to extend the current functional lifespan of the IPK even further.

Viability Evaluations

The main measure for retained viability of an IPK is the maintenance of GFR. GFR is often measured using inulin clearance. Inulin is introduced into the IPK preparation via the cannulated renal artery. Continued performance of the organ is determined by measuring the amount of inulin in the perfusate leaving the kidney via the ureter. The status of an IPK preparation can also be assessed by monitoring the rate of urine production, the relative excretion of sodium compared with potassium, and the continued reabsorption of glucose.

Application of the Isolated Perfused Kidney to Toxicity Studies

Current use of the IPK model has changed very little over the course of the past decade. IPK experiments still provide data about renal metabolism of xenobiotics and the effects of toxicants on the kidney's functional parameters. Much of the recent progress in imaging and molecular biology has yet to be coupled with IPK techniques.

To date, models of the IPK have been utilized to study renal biotransformation and clearance of chemicals, including enalapril (Sirianni and Pang, 1999), menadione (Redegeld et al., 1991b), bacitracin (Drapeau et al., 1992), and 1-naphthol (Redegeld et al., 1991a). Because the IPK is an intact system with near normal physiology, the model has also been used to describe transport of the renal toxicant cisplatin (Miura et al., 1987), therapeutic agents such as morphine (Van Crugten et al., 1991), and the chelating agent DMPS (Klotzbach and Diamond, 1988). Finally, many researchers have utilized IPK models in the mechanistic study of xenobiotics that block normal kidney physiology. Some of these include cyclosporin A (Benigni et al., 1988), Russell's pit viper venom (Willinger et al., 1995b), aminoglycosides (Cojocel et al., 1983a), cisplatin (Miura et al., 1987), opiates (Ellis and Adam, 1991), fluoride (Willinger et al., 1995a), and nonsteroidal anti-inflammatory drugs (NSAIDs), e.g., naproxin (Cox et al., 1990), indomethacin (Cox et al., 1991a), salicylate (Cox et al., 1991b), and ibuprofen (Cox et al., 1991c).

The IPK can be analyzed using a significant number of functional and biochemical endpoints. The IPK model has been previously analyzed using all of the following methods:

- Organic anion transport and accumulation (tetraethylammonium [TEA] and *p*-aminohippurate [PAH])
- Enzyme leakage (lactate dehydrogenase [LDH], alkaline phosphatase [ALP], γ-glutamyl transpeptidase [GGT])
- Toxicant accumulation
- Lipid peroxidation (malondialdehyde formation)

- Covalent binding (precipitatable radioactivity)
- Metabolism (high performance liquid chromatography [HPLC] analysis of metabolites)
- Functional endpoints
  - Urine production
  - GFR (inulin clearance)
  - Sodium retention and elimination
  - Potassium retention and elimination

Currently, IPK models are mainly used to study two areas of renal toxicology: the renal disposition and biotransformation of toxicants, and the effects of toxicants on renal function. An example is the use of an IPK to study the disposition of pentamidine in the kidney (Poola et al., 2002).

There has been a trend for the IPK model to be utilized by multiple groups to study agents that prevent or reverse nephrotoxicity. IPK has been used to study the protective effects of sulphonylurea drugs in reducing hypoxic damage (Engbersen et al., 2000), melatonin in preventing cyclosporine-induced nephrotoxicity (Longoni et al., 2002), and pirfenidone and spironolactone in reversing streptozotocin-induced renal fibrosis (Miric et al., 2001). Havt et al. (2001) used the IPK model to show that WEB 2086, a triazolobenzodiazepine substance, is capable of reversing nephrotoxicity of the Bothrops jararacussu snake venom. Other mechanistic studies have found success using the IPK model. In 1999, Monteiro and colleagues used IPK preparations to discern the effects of cholera toxin on glomerular function (Monteiro et al., 1999). Data obtained using the IPK model implicated the platelet-activating factor pathway in this glomerular dysfunction. Such a discovery could not be made in other *in vitro* models because the IPK model is the only one that maintains glomerular function.

RENAL SLICES

Cultured tissue slices have been used as experimental models for numerous organs, including kidney, liver, lung, heart (Parrish et al., 1995), prostate (Parrish et al., 2002), and brain (Schurr and Rigor, 1995). Tissue slice models allow for analysis of cellular, biochemical, and molecular endpoints using a simplified *in vitro* model without sacrificing the extracellular architecture of the organ. Kidney slices simplify study of the kidney by eliminating variables such as extrarenal influences and renal physiology.

History

Forster first described kidney slices in 1948. During the decades that followed, slices cut by hand were utilized to study the numerous transport functions of the mammalian kidney. These studies led to a greater understanding of renal urate efflux (Berndt, 1965), PAH and nicotinamide transport (Ross and Farrah, 1966; Ross et al., 1968), and renal control of cellular water and electrolytes (Macknight, 1968a, 1968b). Usefulness of renal slices, as with other slice models, was for a long time limited by the inability to effectively produce slices of equal thickness. These issues were addressed by Krumdieck et al. in 1980 and Smith et al. in 1985 with the advent of mechanical slicers that could produce slices with less than 10% variation in thickness. These slicers included the added advantage of bathing the tissue in an oxygenated media during the slicing process. Ruegg et al. (1987a) were among the first to use these techniques to produce precision-cut slices for the study of renal toxicology. At the current time, protocols are in place and studies have been published utilizing kidney slices from numerous species, including rabbit, rat, hamster, dog, mouse, human, and nonhuman primate.

Uses and Acceptance

Kidney slices have been used as an experimental model to investigate the nephrotoxic mechanisms of toxicants, including metals, halogenated hydrocarbons, antibiotics, and chemotherapeutic agents (McGuinness et al., 1993). Slices allow the assessment of toxicity among the variety of cell types present in the kidney while eliminating the physiological complications that go along with more-intact *in vitro* models. In a slice, toxicants can interact directly with the target cells. Studies utilizing slices are particularly useful for identifying the susceptibility of particular cell types to a toxicant, investigating the metabolic capabilities of cells within a particular region, and assessing cellular endpoints of toxicity. Because of the cellular heterogeneity that a kidney slice provides, it is possible to investigate the effects of a toxicant on cell–cell interactions and intercellular communications. Finally, when nonrenal toxicity is dose-limiting and prevents *in vivo* study, mechanistic studies can still be performed using this *in vitro* model.

Limitations

The greatest limitation facing the use of renal slices is the finite lifespan of the tissue in culture. The kidney slice model is typically known to be reliable at timepoints out to 48 hours with standard roller culture systems, although longer incubations are occasionally performed. When tissue slices are cultured using a bubble culture apparatus, slices have been incubated up to five days (Gandolfi, unpublished observations). Besides limited life span, mechanical damage to tissue inflicted during the preparation of slices does stimulate physiological and morphological changes in the kidney cells. While many of the morphological changes have been documented, most of the molecular and cellular effects caused by slicing remain unknown. The most prominent morphologic change is the closure of the tubules approximately 90 minutes following slice preparation. The implications of mechanical damage on gene expression, cellular signaling, and protein processing are not fully understood.

Recent Advances

Recent studies using kidney slices have demonstrated that this model is useful for the study of molecular effects caused by renal toxicants. Parrish et al. presented evidence for altered gene expression and transcription factor binding caused by benzo(a)pyrene in rat renal slices (Parrish et al., 1998) and inorganic arsenic in rabbit renal slices (Parrish et al., 1999). Slices have also been used to show alterations in the mitogen-activated protein kinase (MAPK) signal transduction cascade following inorganic mercury exposure in rabbit renal-cortical slices (Turney et al., 1999). Kirkpatrick et al. (2003) described effects of arsenite on the ubiquitin-proteasome pathway using the same rabbit renal-cortical slice model.

Advances in proton induced x-ray emission (PIXE) have allowed for thorough disposition studies of many metals in the kidney. Because cells in a tissue slice can accumulate metals within cells in a manner similar to that *in vivo*, and because a tissue slice provides a thin matrix of tissue for analysis, the model is well suited to PIXE analysis. Using slices analyzed by PIXE, renal disposition studies of metals, including mercury, cadmium, chromium, and arsenite, have been performed with detection limits in the low ppm range (Keith et al., 1995; Lowe et al., 1993).

Current Applications and Limits

Renal slice techniques are currently an accepted method for addressing a wide range of toxicological endpoints, including xenobiotic disposition and metabolism, extent of gross renal toxicity, and mechanism of action.

While kidney slices have proved to be reliable in number of areas, some limitations do exist. Slice techniques are not reliable tools for evaluating the carcinogenic potential of xenobiotics. The limited functional viability is often too short to allow for the identification of the molecular biomarkers that precede carcinogenesis. While DNA and protein damage are identifiable, elucidating the mechanisms for oncogenic activation has been difficult. In kidney slices, it has proved to be hard to dissect away low abundance cellular signaling because of the heterogeneity that a tissue slice maintains. Furthermore, slices have received limited use in dissecting signal transduction networks because the model has not been effectively coupled with gene transfection techniques. As technology for producing inducible transgenic and knockout animals moves forward, slices from transgenic animals may hold promise for high-throughput mechanistic study of signal transduction pathways.

Methods for Renal Slice Preparation

The original method for production of renal slices involved freehand cutting of the kidney tissue, using a razor blade. With practice, reasonably consistent slices with thicknesses between 200 and 500 μm were readily prepared (McGuinness et al., 1993). During this time, numerous studies were performed with "thick" slices (>400 μm) and with slices whose thickness varied across the slice.

The advantages of early kidney slice techniques lay in their simplicity. Freehand slicing of the kidney required no specific equipment besides a razor blade and a solid glass stage on which to perform the cutting (Berndt, 1965; Forster, 1948). Slices could be rapidly prepared, and skilled technicians could produce slices with a minimal amount of variability. Unfortunately, the technique required that tissue be handled extensively during slice preparation, often resulting in physiological alterations and damage to the tissue. Because thin tissue slices were difficult to reproduce effectively, oxygen and nutrient diffusion was limited and cellular viability suffered.

Evolution of Precision-Cut Renal Slices

The progression from handmade renal slices to precision-cut tissue slices occurred in response to the need for a consistent, and high-throughput

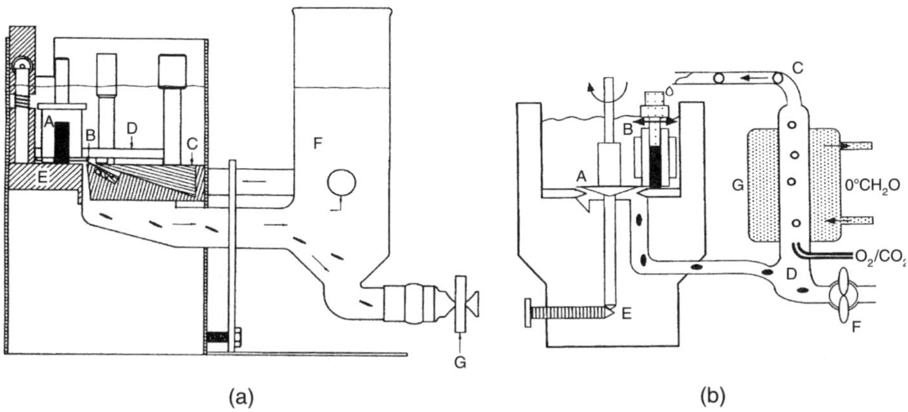

Figure 4.2 Mechanical tissue slicers. (a,b) Krumdieck/Alabama Research and Development slicer. Tissue cores are held in a cylinder (A) and slices are produced by drawing it across an oscillating blade (B). A solid wedge protects the blade (C) as the pivot arm (D) draws the tissue across it. Tissue position is gently maintained on the slicing stage (E) by weights positioned atop the tissue cylinder. Completed slices collect in the glass trap (F) and are collected through a valve held shut by a pinch clamp (G). (c) Brendel-Vitron slicer. Slices are produced as tissue is drawn across a rotating circular blade (A). Tissue cores are held in place by weights positioned atop the tissue holding cylinder (B). Buffer is recirculated into the slicer through the top of a glass trap (C) as slices are collected in the bottom of the trap (D). The thickness of the slices is controlled by a micrometer-type screw (E) that positions the height of the blade. Slices are collected through a valve positioned below the glass trap. Solutions in the slicer and the trap are kept cold by an ice-water condenser jacketing the glass trap (G). (Obtained from Zalups and Lash, 1996.)

in vitro model to use for studying the kidney (Parrish et al., 1995). The most important breakthrough toward this goal was the invention of the mechanical slicer. The mechanical slicers shown in Figure 4.2 (Krumdieck® and Brendel-Vitron®) allow the production of thin, reproducible slices. Using a mechanical slicer, kidney slices can be cut thin enough to permit sufficient oxygen and nutrient diffusion. Early work showed that the diffusion path of oxygen into a slice was approximately 150 μm on each surface, thus the optimal thickness of renal slices for use in toxicology is 250 to 300 μm. At this thickness, oxygen diffusion is sufficient across both surfaces of the tissue to allow adequate exchange of gases into the cells.

Positional Slices (Rabbit, Dog, Human, Nonhuman Primate)

The size of kidneys isolated from larger mammals, such as rabbits, dogs, and primates, provides a unique opportunity for a researcher to perform renal toxicology studies on cells from a specific physiologic region of the kidney. This opportunity has been utilized to examine the toxic effects of chemicals on the proximal tubular cells found in slices prepared from the cortical region of the kidney.

The preparation and use of positional slices in toxicology have been previously reviewed (Ruegg et al., 1987b, 1989). Positional slices are prepared from cores cut perpendicular to the corticopapillary axis. When positional slices are collected in sequence, it is possible to identify a regular distribution of cell types corresponding to the depth of the slice within the tissue (Figure 4.3). The outer two slices from a core lack cells from the straight proximal tubule (S_3 region). The next eight sequential slices represent the cortex. When continuing deeper into the core it is possible to make slices from the outer and inner stripes as well as from the inner medulla itself.

The cortical region is easily identifiable as the darkest outer stripe of the mammalian kidney. A method developed for use in proximal tubule isolation, involving direct injection of an iron oxide mixture, can be used to define more clearly the boundary between cortical and medullary sections of the kidney (Groves and Schnellmann, 1996). Iron oxide injected through the renal artery becomes trapped within the glomeruli and accumulates. Because glomeruli are positioned in the cortex and not the medulla, the cortical region is differentially stained gray. The stained cortical area can be isolated by discarding the unstained tissue from the medullary regions of the kidney. Iron oxide is then removed using a deferroxamine solution and a magnetic stir bar.

Oriented Slices (Rat, Mouse)

The small size of kidneys isolated from most rodent species makes the production of positional slices impractical. Conversely, because of the diminutive nature of the organs, rodent kidneys lend themselves well to oriented slicing. Oriented slices are heterogeneous, and contain each physiological region of the kidney in a normal anatomical arrangement. Most commonly, the entire rodent kidney is placed vertically into the cylindrical tissue receptacle. Normal mouse kidneys usually fit snugly into a 5 mm receptacle, while rat kidneys fit well into either 8 mm

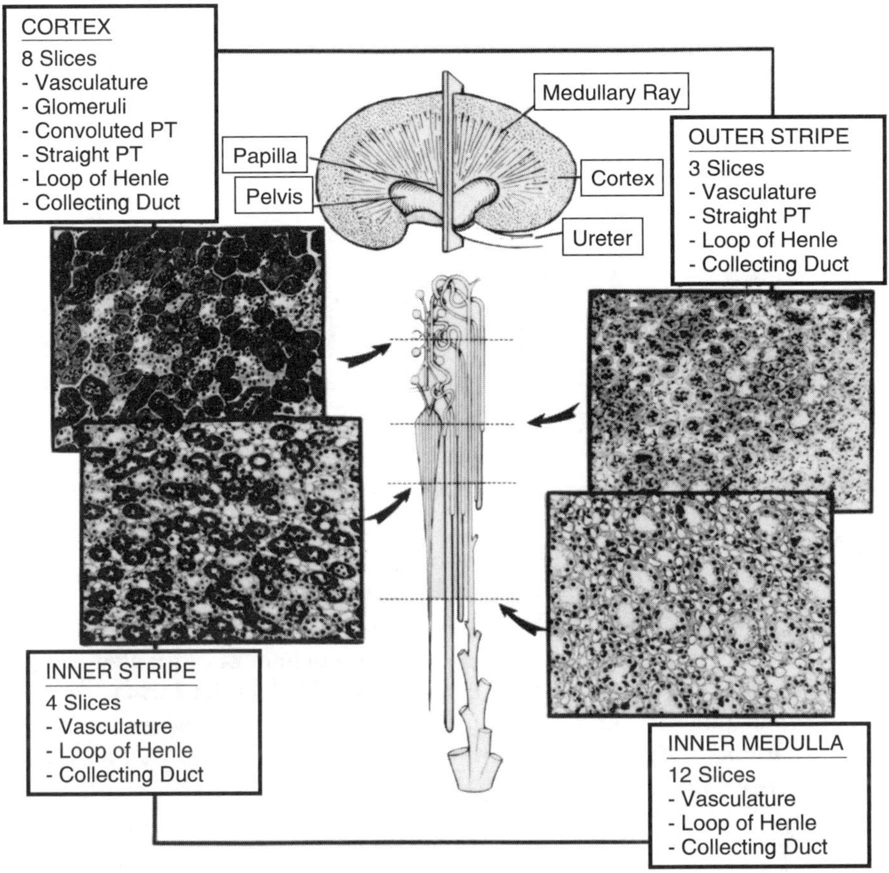

Figure 4.3 Positional slicing. Cores of kidney tissue are produced through the cortico-papillary axis. Slices from the distinct regions of the kidney can be collected according to their stratification within the core. Cell types present in the cortex, outer stripe, inner stripe and inner medulla are noted.

or 10 mm tissue holders. A small region of the kidney is regularly trimmed to attain an acceptable fit for the rodent kidney into the tissue slicer. Usually, the tissue is trimmed along a longitudinal axis, removing portions of the renal pelvis (Figure 4.4a). Removal of this region of the kidney ensures the maximal number of histologically similar slices containing comparable distributions of both cortical and medullary cells (Figures 4b to d).

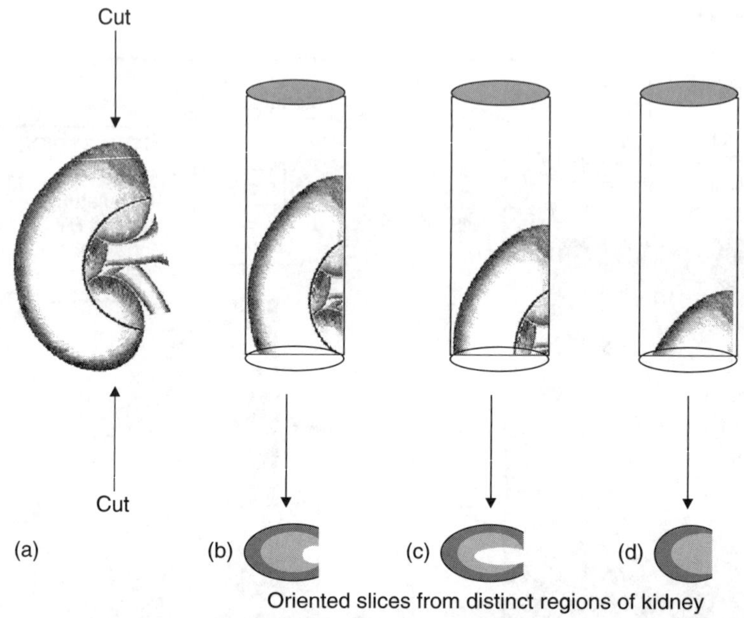

Figure 4.4 Processing of the kidney and production of "oriented slices." (a) Excision of the renal pelvis prior to insertion of kidney into tissue cylinder. (b to d) Production of oriented slices from lower, mid, and upper sections of the kidney. (Portions were adapted from "Gray's Anatomy of the Human Body," 1918, obtained through http://www.bartleby.com/107/.)

Before coring can begin, the kidney must be removed from the donor animal quickly and carefully. Practice has shown that the best slices are obtained when the kidney is rapidly cooled in cold (4°C) Krebs bicarbonate buffer immediately upon isolation. It seems that at physiological temperature, the high energy demand imposed by cellular metabolism can only be maintained for a short period before energy depletion irreversibly damages the tissue.

Aside from the efficient isolation and cooling of the tissue, the most important step in the slicing procedure is preparation of cores from the kidney tissue. Both the Krumdieck® and Brendel-Vitron® slicers produce precision-cut tissue slices from cylindrical cores of tissue prepared from the intact organ. Techniques used for coring kidney tissue are highly dependent upon the species from which tissue has been isolated. While mouse tissue requires the use of small diameter (2 to 5 mm) coring tools, animal species with larger kidney sizes (rabbit, dog, etc.) allow for the production of more easily manageable slices (8 to 10 mm). In larger animals, such as rabbits, it is often preferable to produce tissue cores prior to trimming away tissue from unwanted regions. When choosing core size, the goal remains to gain maximum utility (largest number of cores and slices) from the kidneys of the sacrificed animal while maintaining the utmost level of quality (larger cores tend to produce higher quality slices).

Tissue Slicing Systems

Precision-cut slices are currently prepared using one of the two predominant commercially available slicers. Similarities and differences between the Krumdieck and Brendel-Vitron mechanical slicers have been sufficiently reviewed previously (Gandolfi et al., 1996). The two systems share some important characteristics. In both, slices are cut from cores of renal tissue that are submersed in oxygenated buffer (Figure 4.2). The most important advance provided by the mechanical slicers is the consistent production of tissue slices that are thin enough to permit ample oxygen and nutrient diffusion (250 to 300 μm). It is possible to produce slices varying in thickness by only 25 μm in a single experiment.

Incubation of Slices

A number of methods are in place for incubation and maintenance of tissue slices. Currently, the most effective system for maintaining functional slices from all organs, including the kidney, is dynamic roller culture (Figure 4.5a). In dynamic roller culture, tissue slices rest on screens made of titanium. As a nonreactive metal, titanium does not leach material into the culture media, nor does it become contaminated with chemicals, biochemicals, or radiochemicals added to the incubation. A tissue-holding screen sits within a cylindrical titanium roller, which fits conveniently into a conventional 25 ml liquid scintillation vial. A small volume of tissue culture media, sufficient to partially submerse the slice, is aliquoted into the scintillation vial to nourish the tissue (usually 1.7 ml with a conventional set-up). The incubator (37°C, 95% O_2, 5% CO_2) that holds the scintillation vial (containing the titanium roller and tissue slice) rotates slowly (3.5 rpm) allowing the slice to pass delicately through the nutrient-rich media and then back into the air to permit maximum oxygen diffusion into the cells.

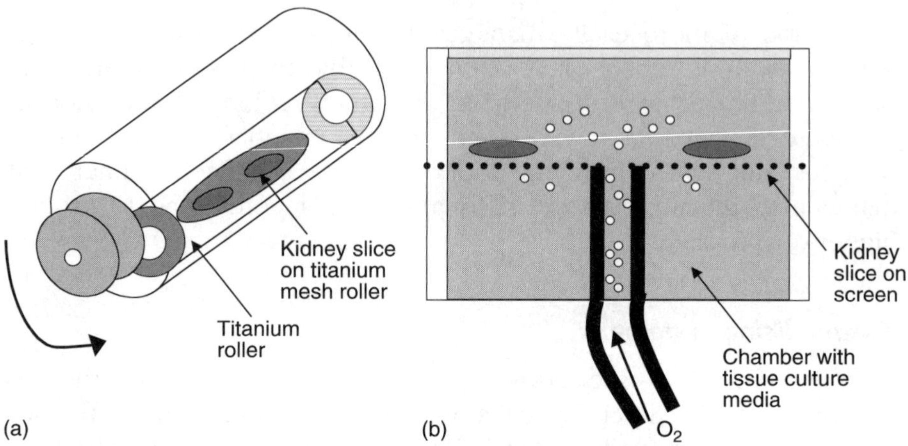

Figure 4.5 Dynamic-roller culture and bubble culture set-ups. (a) 25 ml liquid scintillation vial containing a titanium roller with a fitted titanium mesh screen. Two 8 mm renal slices are placed on the titanium mesh roller, bathing in 1.7 ml of tissue culture media. An aperture in the cap allows for oxygen to enter the vial as it rotates in a 37°C rolling incubator. (b) Tissue slices rest on screens and are oxygenated from below by a bubbling stream of gas. Up to 20 slices can rest in a bubble-culture apparatus.

For experiments where longer-term incubations, or where more highly oxygenated tissue is required, bubble culture is an alternative (Ruegg et al., 1987b). This culture apparatus is a chamber into which oxygen gas is constantly bubbled, allowing the tissue to stay highly oxygenated (Figure 4.5b). All the while, slices remain submerged in tissue culture media. Up to 20 slices may be maintained on the stainless steel screens of a single bubble-culture apparatus. Despite prolonging the viability of tissue slices, experience has shown that experiments using a bubble-culture apparatus are more complicated to perform than those with roller culture. Significantly more preparation is required for a bubble-culture experiment. Furthermore, bubble culture is an open system that is not conducive to the testing of volatile compounds or compounds that cause foam within the bubble apparatus.

Optimization (Media, System, Supplements)

Optimization of renal slice conditions involves balancing the energy needs of the tissue against the mechanical and physical strains of the culture conditions. The most critical variables that need to be considered are the nutrient requirements, buffer systems, and oxygen demand. The most common culture media are serum-free formulations of Waymouth's and Dulbecco's Modified Eagle/Hams F12 Mixture (1 : 1) (Gandolfi et al., 1996; Parrish et al., 1999). In dynamic organ culture, optimal oxygen delivery has been empirically determined to occur when slices are maintained in an environment of 95% O_2 : 5% CO_2. Lower levels of oxygen are not sufficient to meet the energy demands of renal slices, while 100% oxygen presence unnecessarily inflicts oxidative stress on the tissue.

For purposes of toxicology studies, a preincubation period is essential. Preincubation of slices for one hour allows the cells to acclimate from the chilled Krebs bicarbonate buffer to the culture conditions. This period is also important for allowing the sloughing of cells that have been physically destroyed or mechanically upset during the slicing procedure.

Viability Evaluation

Viability of slices can be monitored using nearly all of the indices used for cell culture systems. Slice viability measurements are most commonly obtained by measuring intracellular potassium or enzyme leakage. Because slices are intact tissue, histological staining is also used to evaluate slices. Other methods have been, and continue to be, put to use for renal toxicology in slices. Many of these are listed in Table 4.1.

Cortical/Medullary Slices

As previously described, both cortical and medullary specific slices can be produced when experimenting with kidneys from larger mammals. For toxicological investigations, slices from the cortical region have been the most appropriate. Studies using rabbit renal cortical slices have played a significant role in understanding the toxic effects of metals in the cells of the convoluted proximal tubule (Ruegg et al., 1987a). The mechanism of action for cysteine conjugates of trichloroethylene has been thoroughly delineated in renal cortical slices (Wolfgang et al., 1989a, 1989b, 1990). In addition, the disposition and tubule-specific nephrotoxicity of cisplatin in the kidney has been elucidated using rat (Leibbrandt and Wolfgang, 1995; Safirstein et al., 1984) and dog (Toutain et al., 1996) renal cortical slice models.

Table 4.1 Toxicants Studied Using Renal Slices

Toxicant	*Endpoint*	*Reference*
Mercury chloride	Histology, intracellular K^+, molecular signalling, toxicant uptake and accumulation	Ruegg et al., 1987a, 1989 Turney et al.,1999
Potassium dichromate	Histology, intracellular K^+	Ruegg et al., 1987a
Dichlorovinyl-cysteine (DCVC)	Histology, intracellular K^+, ATP content, toxicant accumulation, toxicant covalent binding	Wolfgang et al., 1989a, 1989b
Cisplatin	Histology, intracellular K^+, enzyme leakage, ATP content, toxicant accumulation, protein synthesis, organic anion accumulation	Liebbrandt and Wolfgang, 1995, Phelps et al., 1987
Atractyloside	Membrane dye exclusion, enzyme leakage, mitochondrial function, ATP content, lipid peroxidation, GSH content, organic anion accumulation, maintenance of gluconeogenesis	Obatomi et al., 1998a, 1998b
Cyclosporin	Metabolism	Vickers et al., 1992, 1996
1,2 Dichloropropane	GSH content, enzyme leakage, lipid peroxidation, organic anion accumulation	Trevisian et al., 1993
Hexachlorobutadiene	Enzyme leakage, organic ion accumulation, lipid peroxidation, gluconeogenesis, GSH content	Smith, 1988

Cephaloradine	Enzyme leakage, organic ion accumulation, lipid peroxidation, gluconeogenesis, GSH content, protein degradation	Cojocel et al., 1983b Smith, 1988
Gentamicin	Enzyme leakage, organic ion accumulation, lipid peroxidation, gluconeogenesis, GSH content, protein degradation, toxicant accumulation	Cojocel et al., 1983a Kluwe and Hook, 1978 Smith, 1988
Benzo(a)pyrene	Intracellular K^+, ATP content, gene expression, transcription factor activation	Parrish et al., 1998
Ischemia	Lipid peroxidation	McAnulty and Huang, 1997
t-Butyl hydroperoxide Hydrogen peroxide	Enzyme leakage, lipid peroxidation, organic anion accumulation, Na^+/K^+-ATPase activity	Kim and Kim, 1996
Sodium arsenite 2-amino-4,5-dichlorophenol (2A45CP)	Intracellular K^+, gene expression, transcription factor activation, heat shock protein expression, intracellular potassium, toxicant accumulation, metabolismEnzyme leakage, GSH content, glucose production, lipid peroxidation	Burton et al., 1995 Lerman and Clarkson, 1983 Parrish et al., 1999 Valentovic et al., 2002
Chloronitrobenzene	Enzyme leakage, glucose production	Hong et al., 2002

Human Kidney Slices

The introduction of kidney slice models utilizing human tissue has bridged an important gap between *in vitro* animal models and human exposures. Renal slices prepared from nontransplantable human kidney allow interspecies comparison of toxicants and correlations of animal exposure thresholds to human situations.

The first report of precision-cut human kidney slices was the identification of unique metabolic profiles of cyclosporin for human compared with rat or dog (Vickers et al., 1992). Fisher et al. characterized the use of precision-cut human kidney slices in toxicology for the comparative study of the nephrotoxic effects of mercury and cisplatin (Fisher et al., 1994). Human kidney slices show similar dose–response curves to those acquired from comparable experiments performed in renal slices obtained from animals. Interestingly, the human kidney slice model detected an increased human susceptibility to cisplatin that mimicked low-level human intoxication associated with the clinical administration of the drug.

Human kidneys are obtained from organ banks and tissue distribution sources, under strict governmental regulations. Kidneys which become available for experimental use usually fall into one of two categories: kidneys that were rejected by the recipient following transplantation, or kidneys whose condition deteriorated prior to removal from the donor or during transport and were thus unacceptable for transplantation. Tissue from these sources can vary widely in viability and functional capabilities.

Genetic heterogeneity of human populations is often responsible for introducing variability into data sets. These factors must be taken into consideration when performing experiments and interpreting data from human tissue slices. To account for these issues, larger sample sizes are often employed in human slice experiments compared with those using inbred animals. Despite this obstacle, tissue obtained from these sources continues to provide important toxicologic data for human exposures that might otherwise not be attainable.

Differences from Animal Studies

The quality and availability of human tissue for experimental purposes remains the most critical issue faced when preparing and experimenting with human renal tissue. At the same time, the continued comparison of data from animal and human *in vitro* models remains a high priority. Studies comparing exposed human renal slices with their counterparts from animal models have periodically indicated important differences in susceptibility to toxicants between humans and animals (Vickers et al., 1992; Yousif et al., 1999). Human renal slices have also been used to look at interspecies differences in metabolism (Vickers et al., 1996).

Application of Renal Slices to Toxicity Studies

Performance Analyses

Tissue slices can be analyzed using a significant number of histological, metabolic, and molecular endpoints. Renal slices have been analyzed effectively using all of the following methods:

- Histology (light and electron microscopy)
- Organic anion transport and accumulation (TEA and PAH)
- Intracellular potassium (using a flame photometer)
- Glutathione (GSH) content (fluorescence detection)
- ATP content (fluorescence detection)
- Enzyme leakage (LDH, ALP, GGT, alanine aminotransferase [ALT])
- Mitochondrial function (MTT assay)
- Toxicant accumulation (radioactive tracing, PIXE analysis)
- Lipid peroxidation (malondialdehyde formation)
- Covalent binding (precipitatable radioactivity)
- Metabolism (HPLC analysis of metabolites)
- Other molecular endpoints (immunoblotting, nucleic acid hybridization, etc.)

Advantages and Disadvantages

The major successes brought on by the use of tissue slices in mechanistic toxicology have involved cysteine conjugates (dichlorovinyl-cysteine [DCVC]), metals (mercury and cisplatin) and an aminoglycoside antibiotic (gentamicin). While much of the knowledge accumulated about gentamicin was obtained using older slice models, the onset of precision-cut tissue slices opened the door to a more detailed understanding of the nephrotoxicity of cysteine conjugates and metals. In each case, the maintenance of the cellular architecture in the renal slice facilitated identification of site-specific disposition or metabolism that corresponded with biochemical and histological necrosis of cells in the proximal tubule.

Precision-cut kidney slices have proven useful in studies of the metabolism of numerous compounds. For example, kidney slices from humans and rats have been used in metabolism studies for lidocaine, testosterone, and 7-ethoxycoumarin (De Kanter et al., 1999, 2002). Similarly, kidney slices from hamsters have been used to study the metabolism of 4-(methylnitrosamino)-1-(3-pyridyl)-1-butanone (NNK) and 4-(methylnitrosamino)-1-(3-pyridyl)-1-butanol (NNAL) (Richter et al., 2000).

The current limitation for the continued use of kidney slices for mechanistic toxicology is their heterogeneity. Cellular heterogeneity makes it difficult to pinpoint subtoxic molecular effects within a specific target cell type. Because of the cellular diversity within a single slice, subtle changes in subsets of cells are masked by the lack of an effect in the rest of the tissue. When investigating these low-level changes, it is often difficult to isolate sufficient biomass (mRNA, protein, etc.) to identify a low-level molecular change, especially when focusing on a less-frequent subset of cells.

Because of the nature of the model, a toxicant must increase specific gene or protein expression in a significant number of cells within the slice if it is to be detected. This limitation is not unique to slices, but is compounded by the heterogeneity present within the model. Furthermore, many of the most prevalent slice models (rabbit, rat, and dog) have limited databases of shared genetic information, leaving the researcher at a disadvantage when utilizing such new technologies as cDNA microarray and proteomics. These disadvantages can be avoided as researchers continue adapting and using slice technology for human tissue samples.

Future Considerations

The onset of the transgenic age has brought a new challenge to those researchers who work in the area of *in vitro* toxicology. Significant advances interfacing transgenic models of toxicology with currently available *in vitro* systems, such as tissue slicing (Catania et al., 2003), have become a reality. The ability to meld these two areas holds promise for expanding the depth and breadth of questions that can be addressed, while limiting the numbers of animals needed for *in vivo* toxicology studies. Slices from transgenic animals will surely present a new high-throughput avenue for *in vitro* and mechanistic toxicology.

Although there are currently no publications documenting the use of kidney slice models to study apoptosis, it seems reasonable to assume that these areas may present an avenue for future research. Studies examining apoptotic mechanisms by using precision-cut liver slices have been published (Chen et al., 2000).

Future advances in the use of kidney slices for toxicological study will proceed, most notably into the areas of gene expression and signal transduction. As cDNA microarray and proteomic technologies become increasingly available, slices will likely prove useful in describing genetic and proteomic disruptions that precede cell-specific necrosis within the kidney. Ongoing work in the area of tissue slices has also focused on adapting the slicing technique with the advances in live time confocal microscopy.

NEPHRON COMPONENTS

Many toxicants have been studied using *in vitro* models of isolated and purified nephron components. These systems allow investigators to pinpoint mechanisms of site-specific toxicity within the kidney. Studies using nephron components compliment *in vivo* renal toxicology studies nicely because experiments can focus on cellular effects in the absence of complicating factors, such as blood flow, endocrine effects, and toxicant disposition. All of the methods described in this section involve separation of one or more components of individual nephrons away from the remaining tissue. The methods for isolation and experimentation with nephron components summarized here have been thoroughly reviewed (Chini and Dousa, 1996; Groves and Schnellmann, 1996; Zalups and Barfuss, 1996), as a result this section focuses on the applications of these techniques as they apply to present and future issues in renal toxicology.

The main advantage presented by isolated nephron components is that the physical structure of the component remains unaltered during experimental preparation. Therefore, nephron segments can be exposed to toxicants and the effects studied just as they might happen *in vivo*.

The use of nephron components in toxicology has been largely dominated by studies involving the proximal tubule. There has been a significant number of publications utilizing isolated glomeruli. In addition to the proximal tubule and glomerulus, methods also exist to examine the remainder of the nephron's segments, including loop of Henle, distal tubule, collecting duct (Harriman et al., 2000), and juxtaglomerular apparatus (Ito, 1998).

Past Studies

The proximal tubule has been the primary segment of interest for a number of reasons. It has been shown from whole-animal studies that the majority of all renal toxicants act specifically on the proximal tubule. Because the majority of the proximal tubule is positioned in the cortical regions of the kidney and adjacent to the glomerulus, it has been technically feasible to isolate both individual tubule segments (for very specific studies) and large quantities of highly purified (≈95%) tubules (for biochemical analyses). Because very few toxicants are known to exert significant toxicological activity in either the loop of Henle or the distal tubule, and the isolation of these nephron components is more challenging, neither has been studied as thoroughly as have regions of the proximal tubule.

Aside from the proximal tubule, significant advances in renal toxicology have been made utilizing isolated glomeruli. The technique for isolating glomeruli is technically simple, for many of the same reasons as the proximal tubule.

Methods for Glomeruli Isolation, Incubation and Toxicity Studies

Isolation Techniques

Two main methods have been described for the isolation of glomerular segments of the nephron. The first technique relies upon perfusion of the kidney with an iron oxide precipitate. Iron oxide particles are trapped inside each glomerulus; kidney tissue is then diced by hand into small cubes or enzymatically digested into small pieces of tissue. Glomeruli are captured using a magnetic stir bar to attract and separate the iron oxide containing tissue from the remaining parts of the kidney (Groves and Schnellmann, 1996).

The complimentary technique takes advantage of the fact that the diameter of a glomerulus is significantly larger than that of any other nephron subunit. Because of this, glomeruli can be collected and purified by filtering with a mesh screen large enough to pass the tubular components while capturing the glomeruli (Chini and Dousa, 1996). As with the iron-oxide method, either mechanical or enzymatic methods are used to separate the tissue. Iron oxide isolation techniques are often combined with a sieve purification procedure to obtain a highly purified glomerular sample (Groves and Schnellmann, 1996). A further method has been described for isolation of glomeruli from mice, whereby magnetic 4.5 μm diameter Dynabeads are perfused through the heart (Takemoto et al., 2002).

Incubation Systems

Isolated glomeruli are usually maintained in suspension culture. Suspension cultures are prepared at concentrations ranging from 0.5 to 1.0 μg total protein/ml culture media, and maintained at 37°C under normal atmospheric conditions and with gentle agitation (100 cycles per minute) (Ishikawa and Kitamura, 1999; Schambelan et al., 1985; Sraer et al., 1979). Dulbecco's Modified Eagle/Hams F12 (DME-F12) medium containing 1% fetal calf serum (FCS) has been shown to maintain glomerular suspensions for several hours. Experiments have on occasion utilized adherent cultures of isolated glomeruli, but these are not as commonly employed as suspensions. Poly-L-lysine coated dishes have been used to improve the quality of adherent glomerular cultures (Carraro et al., 2000).

Under certain circumstances, incubation of isolated glomeruli is not required. In glomerular perfusion studies, for example, isolated tissue must only be maintained until it is tested on the perfusion apparatus.

Applications

One of the most important accomplishments made in renal toxicology using isolated glomeruli has been the thorough mechanistic description of glomerular contraction following cyclosporin exposure. Using isolated glomeruli, investigators from different groups have shown the importance of calcium, oxidative stress, prostaglandins, endothelin-1 (L'Azou et al., 1999), and the nitric oxide pathway (Potier et al., 1996b) in this contraction. Microscopic indices have been developed to detect and quantitate changes in glomerular size. In a similar study, this *in vitro* technique was coupled with proteomics for the analysis and identification of proteins that alter the permeability to albumin within isolated glomeruli (Musante et al., 2002).

In contrast with renal slices, isolated glomeruli can be used in studies that look at cellular apoptosis. Ishikawa et al. (1999) used isolated glomeruli to show that heparin inhibits the spontaneous apoptosis of glomerular mesengial cells. This group successfully used a histochemical staining technique, dUTP nick-end labeling, and agarose gel electrophoresis (DNA laddering) as endpoints.

The isolated glomerulus has been utilized for the study of many biomolecules, therapeutic agents, and toxicants. Table 4.2 lists some of the compounds whose effects have been studied in the isolated glomerulus and the technical endpoints that have been evaluated.

Methods for Isolating Nephron Segments and Toxicity Studies

A wide variety of existing methods can be used to isolate and purify different segments of the nephron. When selecting a method, the quantity of tissue needed (as for robust biochemical assays) must be balanced against the need for pure or homogeneous components (as for site-specific mechanistic studies). For example, a study involving transport of a solute in an isolated perfused tubule will require a small number of uncompromised nephrons, whereas a study aiming to identify metabolites of a toxicant specifically produced in the proximal tubule will require larger amounts of less pure components. In many cases, except when individual nephron isolation is a feasible option, density gradients of Percoll and Ficoll are used to enrich the homogeneity (Groves and Schnellmann, 1996).

Preparation of Nephron Segments (Tubules)

Two main technical approaches exist for separation and isolation of nephron segments for experimentation. Mechanical isolation involves the cutting and separating of nephron components by using a loose-fitting ground-glass or teflon homogenizer. While this technique is time-consuming and technically challenging, a skilled technician using this approach can produce the most homogeneous and uncompromised collection of nephron segments for analysis. This procedure is contrasted by harsher enzymatic separation techniques, in which digestive enzymes such as collagenase cleave cellular and extracellular proteins, separating the nephrons from one another. While digestion of the connective proteins allows for a far more effective separation of individual nephrons, the cleavage of membrane proteins leaves the cells damaged and more susceptible to a toxic exposure (Rodeheaver et al., 1990).

Table 4.2 Toxicants Studied Using Isolated Glomeruli

Toxicant	*Endpoint*	*Reference*
Uranyl nitrate	Eicosanoid synthesis	Chaudhari and Kirschenbaum, 1984
Gentamicin	Glomerular morphology, organic anion transport	Savin et al., 1985
Cyclosporin A	Glomerular morphology, enzyme leakage	Azou and Cambar, 1990 L'Azou et al., 1999 Potier et al., 1996a, 1996b Wiegmann et al., 1990
Adriamycin	Glucose metabolism, superoxide production	Barbey et al., 1989 Kastner et al., 1991 Romero et al., 1997
Metals: mercury, cadmium, copper zinc, manganese, nickel, and cobalt	Glomerular morphology, metabolism, protein and proteoglycan synthesis, glutathione levels, and fatty acid synthesis, renin release, enzyme leakage, mitochondrial activity, dye exclusion	Barrouillet et al., 1999 Kozma et al., 1996 Templeton and Chaitu, 1990 Wilks et al., 1990
Hydrogen peroxide Menadione	Antioxidant levels, superoxide production, enzyme leakage, immunohistochemical analysis protein synthesis, intracellular GSH, intracellular NADPH	Chen et al., 2000 Kawamura et al., 1991 Morgan et al., 1998 Romero et al., 1997
Lipopolysaccharride	Gene expression and immunohistochemical analysis	Sade et al., 1999 Zhou et al., 2000
Cicletanine	cGMP synthesis	Valdivielso et al., 1998
Transforming growth factor-β (TGF-β)	Albumin permeability	Sharma et al., 2000

Often, less stringent protocols for enzymatic digestion are performed in combination with mechanical separation techniques.

Incubation of Tubule Suspensions

Suspensions of renal proximal tubules have been utilized for studies lasting up to 24 hours. Cultures are usually prepared at a concentration of 1.0 mg/ml total protein in DME-F12 or other comparable cell culture media. While suspensions prepared from enzymatically and mechanically disrupted nephrons have both been reported to be functional for 24 hours, only tubules prepared without digestive enzymes appear to maintain their normal tubular morphology (Rodeheaver et al., 1990). Incubations of suspended tubules have been shown to retain the activity of metabolic enzymes, including CYP450, glutathione-*S*-transferase, β-lyase, UDP glucuronyl transferase, and sulfotransferase (Groves and Schnellmann, 1996).

Primary Culture of Tubules

Techniques used for the primary culture of tubules have been thoroughly described previously (Aleo and Kostyniak, 1996; McGuinness et al., 1993; Nowak and Schnellmann, 1996). Primary tubular culture models are reported to maintain active transport functions, cellular polarity, and brush border enzymatic activities similar to fresh isolated tubules.

The currently accepted culture conditions include the use of serum-free, hormonally defined media. Furthermore, adequate oxygen delivery and a mitochondrial fatty acid energy source (heptanoic acid, 2 m*M*) greatly improve the quality of primary tubular cultures. Because of important interaction between tubules and extracellular matrices, cultures are often grown on collagen- or similarly coated dishes. Once the primary tubule cultures have stabilized, they can be assessed by all of the same techniques used to evaluate other cell-culture models.

Viability Evaluation

The viability of nephron components is measured in much the same way as in other *in vitro* models of the kidney, but with a few important considerations. As in most *in vitro* models, the common measures of viability are enzyme leakage (e.g., LDH), dye exclusion (e.g., Trypan

Blue, neutral red), mitochondrial function (e.g., MTT, XTT), intracellular ATP, and intracellular K^+.

Special care must be employed if evaluating enzyme leakage in nephron components exposed to significant stresses, either enzymatic or manual. Such stresses can result in damage to cell membranes. When evaluating viability in the kidney using dye exclusion, it is crucial that data are interpreted with a full understanding of the kidney's ability to specifically transport the molecules of the dye being employed.

It is often important to use molecular approaches to confirm or test the purity of the isolation procedure. The three common methods for validating isolation effectiveness are cellular morphology (microscopic analysis), transport activity (using a radiolabelled substrate), and molecular markers (immunostaining). Besides validating isolation techniques, identifying a loss of these characteristics following toxicant exposure provides evidence toward potential mechanisms of toxicity.

Species Concerns

Observation has shown that distinct differences that affect the isolation of nephron components exist between particular species. Tubules from laboratory rats are generally more difficult to isolate and purify than those of other species. Rat tubules tend to be interwoven among each other in tangled bunches, making them more difficult to separate from surrounding tissue. In contrast, tubules from rabbits tend to be less intertwined with one another and are relatively simple to isolate. Proximal tubule suspensions from rabbit remain the most commonly used model for isolated tubule studies.

Use in Mechanistic Studies

Primary monolayer and suspension cultures of isolated proximal tubules have been utilized effectively in the study of many renal toxicants. Many of the toxicants that have been studied are listed in Table 4.3.

Nephron components have been used by a number of groups to investigate mechanisms of renal toxicity. Schnellmann's group has used this model extensively to investigate mechanisms of nephrotoxicity and reversibility of injury. Harriman et al. (2000) used primary cultured tubules and tubule suspensions to ascertain the importance of calpain

Table 4.3 Toxicants Studied Using Isolated Proximal Tubules

Toxicant	*Endpoint*	*Reference*
Gentamicin	Enzyme leakage, energy meatbolism	Obatomi and Plummer, 1995 Takeda et al., 1995
Atractyloside	Enzyme leakage, mitochondrial function, cellular ATP, cellular GSH, PAH transport	Obatomi and Bach, 1996a, 1996b, 2000
Linoleic acid	Oxidative stress, mitochondrial function, sodium transport	Moran et al., 1997
4-aminophenol	Mitochondrial function, cellular ATP, cellular GSH, enzyme leakage	Lock et al., 1993
Mercury, cadmium, chromium	Protein synthesis, lipid peroxidation	Wilks et al., 1990
Antimycin	Mechanism of action	Harriman et al., 2000
Dichlorovinyl-l-cysteine (DCVC)	Mitochondrial function, Na^+ transport, oxygen consumption, glucose transport	Nowak et al., 1999, 2000

activity in the molecular progression of renal injury as well as the protective mechanisms of ascorbic acid. Other studies have successfully highlighted the stages of plasma membrane damage following anoxia (Chen et al., 2001), the importance of collagen and collagen-binding integrins in repair following sublethal DCVC exposure (Nony and Schnellmann, 2001; Nony et al., 2001), mechanisms of cisplatin induced apoptosis (Cummings and Schnellmann, 2002), and the role of phospholipase A2 in oxidant-mediated cell death (Cummings et al., 2002), and presented a mechanistic description of cytoprotection by calpain inhibitors (Liu et al., 2001, 2002).

Shuprisha and colleagues (2000) used isolated proximal tubules to report on the effects of the protein kinase C (PKC) activator and known tumor promoter phorbol 12-myristate 13-acetate (PMA) in the proximal tubule. This work confirmed the importance of PKC on organic anion secretion. Isolated proximal tubules were analyzed by fluorescent microscopy to confirm that these PMA-induced effects could be stymied by staurosporine or bisindolylmaleimide I treatment.

Limitations

As with the other intact *in vitro* renal models, the short life span of the tissue is the main experimental limitation. The main technical limitation preventing more groups from utilizing isolated nephron segments as an *in vitro* model to study renal toxicants is the technical difficulty of isolating high-quality tubules. This is especially true in relation to isolation of individual tubules by manual methods, but is also a consideration with tubule suspensions.

REFERENCES

Aleo MD, and Kostyniak PJ (1996) Characterization and use of rabbit renal proximal tubular cells in primary culture for toxicology research, in *Methods in Renal Toxicology*. (Zalups RK and Lash LH, Eds) pp. 163–188. CRC Press, Boca Raton, Florida.

Azou BL, and Cambar J (1990) Protective effect of verapamil in cyclosporin-incubated isolated rat glomeruli. *Toxicol Lett* 53:247–250.

Bagate K, Develioglu L, Imbs JL, Michel B, Helwig JJ, and Barthelmebs M (1999) Vascular kinin B(1) and B(2) receptor-mediated effects in the rat isolated perfused kidney – differential regulations. *Br J Pharmacol* 128:1643–1650.

Barbey MM, Fels LM, Soose M, Poelstra K, Gwinner W, Bakker W, and Stolte, H. (1989) Adriamycin affects glomerular renal function: Evidence for the involvement of oxygen radicals. *Free Radic Res Commun* 7:195–203.

Barrouillet MP, Potier M, and Cambar J (1999) Cadmium nephrotoxicity assessed in isolated rat glomeruli and cultured mesangial cells: Evidence for contraction of glomerular cells. *Exp Nephrol* 7:251–258.

Bekersky I (1983) Use of the isolated perfused kidney as a tool in drug disposition studies. *Drug Metab Review* 14:931–960.

Benigni A, Chiabrando C, Piccinelli A, Perico N, Gavinelli M, Furci L, Patino O, Abbate M, Bertani T, and Remuzzi G (1988) Increased urinary excretion of thromboxane B2 and 2,3-dinor-TxB2 in cyclosporin A nephrotoxicity. *Kidney Int* 34:164–174.

Berndt WO (1965) The efflux of urate from rabbit renal cortex slices. *J Pharmacol Exp Ther* 150:414–419.

Burton CA, Hatlelid K, Divine K, Carter DE, Fernando Q, Brendel K, and Gandolfi AJ (1995) Glutathione effects on toxicity and uptake of mercuric chloride and sodium arsenite in rabbit renal cortical slices. *Environ Health Perspect* 103:81–84.

Cailleaux S, Lopes-Martins RA, Aimbire F, Cordeiro RS, and Tibirica E (1999) Involvement of platelet-activating factor in the modulation of vascular tone in the isolated perfused rabbit kidney. *Naunyn Schmiedebergs Arch Pharmacol* 359:505–511.

Carraro M, Mancini W, Artero M, Zennaro C, Faccini L, Candido R, Armini L, Calci M, Carretta R, and Fabris B (2000) Albumin permeability in isolated glomeruli in incipient experimental diabetes mellitus. *Diabetologia* 43:235–241.

Catania JM, Parrish AR, Kirkpatrick DS, Chitkara M, Bowden GT, Henderson CJ, Wolf CR, Clark AJ, Brendel K, Fisher RL, and Gandolfi AJ (2003) Precision-cut tissue slices from transgenic mice as an *in vitro* toxicology system. *Toxicol in vitro* 17:201–205.

Chaudhari A, and Kirschenbaum MA (1984) Altered glomerular eicosanoid biosynthesis in uranyl nitrate-induced acute renal failure. *Biochim Biophys Acta* 792:135–140.

Chen HC, Guh JY, Shin SJ, Tsai JH, and Lai YH (2000) Reactive oxygen species enhances endothelin-1 production of diabetic rat glomeruli *in vitro* and *in vivo*. *J Lab Clin Med* 135:309–315.

Chen J, Gokhale M, Schofield B, Odwin S, and Yager JD (2000) Inhibition of TGF-beta-induced apoptosis by ethinyl estradiol in cultured, precision cut rat liver slices and hepatocytes. *Carcinogenesis* 21:1205–1211.

Chen J, Liu X, Mandel LJ, and Schnellmann RG (2001) Progressive disruption of the plasma membrane during renal proximal tubule cellular injury. *Toxicol Appl Pharmacol* 171:1–11.

Chini EN, and Dousa TP (1996) Microdissection and microanalysis of specific nephron segments from mammalian kidney, in *Methods in Renal Toxicology* (Zalups RK, and Lash LH, Eds) pp. 97–108. CRC Press, Boca Raton, Florida.

Cojocel C, Dociu N, Maita K, Sleight SD, and Hook JB (1983a) Effects of aminoglycosides on glomerular permeability, tubular reabsorption, and intracellular catabolism of the cationic low-molecular-weight protein lysozyme. *Toxicol Appl Pharmacol* 68:96–109.

Cojocel C, Smith JH, Maita K, Sleight SD, and Hook JB (1983b) Renal protein degradation: A biochemical target of specific nephrotoxicants. *Fundam Appl Toxicol* 3:278–284.

Cox PG, Moons WM, Russel FG, and van Ginneken CA (1990) Renal disposition and effects of naproxen and its l-enantiomer in the isolated perfused rat kidney. *J Pharmacol Exp Ther* 255:491–496.

Cox PG, Moons WM, Russel FG, and van Ginneken CA (1991a) Indomethacin: Renal handling and effects in the isolated perfused kidney. *Pharmacology* 42:287–296.

Cox PG, Moons WM, Russel FG, and van Ginneken CA (1991b) Renal handling and effects of salicylic acid in the isolated perfused rat kidney. *Pharmacol Toxicol* 68:322–328.

Cox PG, Moons WM, Russel FG, and van Ginneken CA (1991c) Renal handling and effects of S(+)-ibuprofen and R(−)-ibuprofen in the rat isolated perfused kidney. *Br J Pharmacol* 103:1542–1546.

Cummings BS, and Schnellmann RG (2002) Cisplatin-induced renal cell apoptosis: Caspase 3-dependent and -independent pathways. *J Pharmacol Exp Ther* 302:8–17.

Cummings BS, McHowat J, and Schnellmann RG (2002) Role of an endoplasmic reticulum Ca(2+)-independent phospholipase A(2) in oxidant-induced renal cell death. *Am J Physiol* 283: F492–F498.

De Kanter R, Olinga P, De Jager MH, Merema MT, Meijer DKF, and Groothius GMM (1999) Organ slices as an *in vitro* test system for drug metabolism in human liver, lung and kidney. *Tox in vitro* 13:737–744.

De Kanter, R, De Jager, MH, Draaisma, AL, Jurva, JU, Olinga, P, Meijer, DK, and Groothuis GM (2002) Drug-metabolizing activity of human and rat liver, lung, kidney and intestine slices. *Xenobiotica* 32:349–362.

Diamond GL (1996) The isolated perfused kidney, in *Methods in Renal Toxicology* (Zalups RK, and Lash LH, Eds) pp. 59–78. CRC Press, Boca Raton, Florida.

Drapeau G, Petitclerc E, Toulouse A, and Marceau F (1992) Dissociation of the antimicrobial activity of bacitracin USP from its renovascular effects. *Antimicrob Agents Chemother* 36:955–961.

Ellis AG, and Adam WR (1991) Effects of opiates on sodium excretion in the isolated perfused rat kidney. *Clin Exp Pharmacol Physiol* 18:835–842.

Engbersen R, Moons MM, Wouterse AC, Dijkman HB, Kramers C, Smits P, and Russel FG (2000) Sulphonylurea drugs reduce hypoxic damage in the isolated perfused rat kidney. *J Pharmacol* 130:1678–1684.

Fisher RL, Sanuik JT, Gandolfi AJ, and Brendel K (1994) Toxicity of cisplatin and mercuric chloride in human kidney cortical slices. *Hum Exp Toxicol* 13:517–523.

Forster RP (1948) Use of thin kidney slices and isolated renal tubules for direct study of cellular transport kinetics. *Science* 108:65–67.

Gandolfi AJ, Brendel K, and Fernando Q (1996) Preparation and use of precision-cut renal cortical slices in renal toxicology, in *Methods in Renal Toxicology* (Zalups RK, and Lash LH, Eds) pp. 110-121. CRC Press, Boca Raton, Florida.

Groves CE, and Schnellmann RG (1996) Suspensions of rabbit renal proximal tubules, in *Methods in Renal Toxicology* (Zalups RK, and Lash LH, Eds) pp. 147–162. CRC Press, Boca Raton, Florida.

Harriman JF, Waters-Williams S, Chu DL, Powers JC, and Schnellmann RG (2000) Efficacy of Novel Calpain Inhibitors in Preventing Renal Cell Death. *J Pharmacol Exp Ther* 294:1083–1087.

Havt A, Fonteles MC, and Monteiro HS (2001) The renal effects of Bothrops jararacussu venom and the role of PLA(2) and PAF blockers. *Toxicon* 39:1841–1846.

Hong SK, Anestis DK, Ball JG, Valentovic MA, and Rankin GO (2002) *In vitro* nephrotoxicity induced by chloronitrobenzenes in renal cortical slices from Fischer 344 rats. *Toxicol Lett* 129:133–141.

Ishikawa Y, and Kitamura M (1999) Inhibition of glomerular cell apoptosis by heparin. *Kidney Int* 56:954–963.

Ito S (1998) Characteristics of isolated perfused juxtaglomerular apparatus. *Kidney Int Suppl* 67:S46–S48.

Kastner S, Wilks MF, Gwinner W, Soose M, Bach PH, and Stolte H (1991) Metabolic heterogeneity of isolated cortical and juxtamedullary glomeruli in adriamycin nephrotoxicity. *Ren Physiol Biochem* 14:48–54.

KawamuraT, Yoshioka T, Bills T, Fogo A, and Ichikawa I (1991) Glucocorticoid activates glomerular antioxidant enzymes and protects glomeruli from oxidant injuries. *Kidney Int* 40:291–301.

Keith RL, McGuinness SJ, Gandolfi AJ, Lowe TP, Chen Q, and Fernando Q (1995) Interaction of metals during their uptake and accumulation in rabbit renal cortical slices. *Environ Health Perspect* 103:77–80.

Kim YK, and Kim YH (1996) Differential effect of Ca2+ on oxidant-induced lethal cell injury and alterations of membrane functional integrity in renal cortical slices. *Toxicol Appl Pharmacol* 141:607–616.

Kirkpatrick DS, Dale KV, Catania JM, and Gandolfi AJ (2003) Low-level arsenite causes accumulation of ubiquinated proteins in rabbit renal cortical slices and HEK293 cells. *Toxicol Appl Pharmacol* 186:101–109.

Klanke B, Simon M, RocklW, Weich HA, Stolte H, and Grone HJ (1998) Effects of vascular endothelial growth factor (VEGF)/vascular permeability factor (VPF) on haemodynamics and permselectivity of the isolated perfused rat kidney. *Nephrol Dial Transplant* 13:875–885.

Klotzbach JM, and Diamond GL (1988) Complexing activity and excretion of 2,3-dimercapto-1-propane sulfonate in rat kidney. *Am J Physiol* 254:F871–F878.

Kluwe WM, and Hook JB (1978) Analysis of gentamicin uptake by rat renal cortical slices. *Toxicol Appl Pharmacol* 45:531–539.

Kozma L, Lenkey A, Varga E, and Gomba S (1996) Induction of renin release from isolated glomeruli by inorganic mercury(II). *Toxicol Lett* 85:49–54.

Krumdieck CL, dos Santos JE, and Ho KJ (1980) A new instrument for the rapid preparation of tissue slices. *Anal Biochem* 104:118–123.

L'Azou B, Medina J, Frieauff W, Cordier A, Cambar J, and Wolf A (1999) *In vitro* models to study mechanisms involved in cyclosporine A-mediated glomerular contraction. *Arch Toxicol* 73:337–345.

Leibbrandt ME, and Wolfgang GH (1995) Differential toxicity of cisplatin, carboplatin, and CI-973 correlates with cellular platinum levels in rat renal cortical slices. *Toxicol Appl Pharmacol* 132:245–252.

Lerman S, and Clarkson TW (1983) The metabolism of arsenite and arsenate by the rat. *Fundam Appl Toxicol* 3:309–314.

Liu X, Harriman JF, Rainey JJ, and Schnellmann RG (2001) Calpains mediate acute renal cell death: Role of autolysis and translocation. *Am J Physiol* 281:F728–F738.

Liu X, Harriman JF, and Schnellmann RG (2002) Cytoprotective properties of novel nonpeptide calpain inhibitors in renal cells. *J Pharmacol Exp Ther* 302:88–94.

Lo M, Medeiros IA, Mullins JJ, Ganten D, Barres C, Cerutti C, Vincent M, and Sassard J (1993) High blood pressure maintenance in transgenic mRen-2 vs. Lyon genetically hypertensive rats. *Am J Physiol* 265:R180–R186.

Lock EA, Cross TJ, and Schnellmann RG (1993) Studies on the mechanism of 4-aminophenol-induced toxicity to renal proximal tubules. *Hum Exp Toxicol* 12:383–388.

Longoni B, Migliori M, Ferretti A, Origlia N, Panichi V, Boggi U, Filippi C, Cuttano MG, Giovannini L, and Mosca F (2002) Melatonin prevents cyclosporine-induced nephrotoxicity in isolated and perfused rat kidney. *Free Radic Res* 36:357–363.

Lowe T, Chen Q, Fernando Q, Keith R, and Gandolfi AJ (1993) Elemental analysis of renal slices by proton-induced X-ray emission. *Environ Health Perspect* 101:302–308.

Maack T (1980) Physiological evaluation of the isolated perfused rat kidney. *Am J Physiol* 238:F71–F78.

Macknight AD (1968a) Water and electrolyte contents of rat renal cortical slices incubated in potassium-free media and containing ouabain. *Biochim Biophys Acta* 150:263–270.

Macknight AD (1968b) The extracellular space in rat renal cortical slices incubated at 0.5 degrees and 25 degrees. *Biochim Biophys Acta* 163:85–92.

McAnulty JF, and Huang XQ (1997) The efficacy of antioxidants administered during low temperature storage of warm ischemic kidney tissue slices. *Cryobiology* 34:406–415.

McGuinness SJ, Gandolfi AJ, and Brendel K (1993) Use of renal slices and renal tubule suspensions for *in vitro* toxicity Studies. *In vitro Toxicology* 6:1–24.

Miric G, Dallemagne C, Endre Z, Margolin S, Taylor SM, and Brown L (2001) Reversal of cardiac and renal fibrosis by pirfenidone and spironolactone in streptozotocin-diabetic rats. *Br J Pharmacol* 133:687–694.

Miura K, Goldstein RS, Pasino DA, and Hook JB (1987) Cisplatin nephrotoxicity: Role of filtration and tubular transport of cisplatin in isolated perfused kidneys. *Toxicol* 44:147–158.

Mokuda O, and Sakamoto Y (1999) Impaired parathyroid hormone action on urinary phosphorus excretion in isolated perfused kidney of streptozotocin-induced diabetic rats. *Horm Metab Res* 31:543–545.

Monteiro HS, Lima AA, and Fonteles MC (1999) Glomerular effects of cholera toxin in isolated perfused rat kidney: a potential role for platelet activating factor. *Pharmacol Toxicol* 85:105–110.

Moran JH, Weise R, Schnellmann RG, Freeman JP, Grant DF (1997) Cytotoxicity of linoleic acid diols to renal proximal tubular cells. *Toxicol Appl Pharmacol* 146:53–59.

Morgan WA, Kaler B, and Bach PH (1998) The role of reactive oxygen species in adriamycin and menadione-induced glomerular toxicity. *Toxicol Lett* 94:209–215.

Musante L, Candiano G, Bruschi M, Zennaro C, Carraro M, Artero M, Giuffrida MG, Conti A, Santucci A, and Ghiggeri GM. (2002) Characterization of plasma factors that alter the permeability to albumin within isolated glomeruli. *Proteomics* 2:197–205.

Newton JF, and Hook JB (1981) Detoxification and Drug Metabolism: Conjugation and Related Systems, in *Methods in Enzymology* (Jakoby WB, Ed.) pp. 94–105. Academic Press, London.

Nony PA, Nowak G, and Schnellmann RG (2001) Collagen IV promotes repair of renal cell physiological functions after toxicant injury. *Am J Physiol* 281:F443–F453.

Nony PA, and Schnellmann RG (2001) Interactions between collagen IV, and collagen-binding integrins in renal cell repair after sublethal injury. *Molec Pharmacol* 60:1226–1234.

Nowak G, Keasler KB, McKeller DE, and Schnellmann RG (1999) Differential effects of EGF on repair of cellular functions after dichlorovinyl-L-cysteine-induced injury. *Am J Physiol* 276:F228–F236.

Nowak G, Carter CA, and Schnellmann RG (2000) Ascorbic acid promotes recovery of cellular functions following toxicant-induced injury. *Toxicol Appl Pharmacol* 167:37–45.

Nowak G, and Schnellmann R (1996) L-ascorbic acid regulates growth and metabolism of renal cells: Improvements in cell culture. *Am J Physiol* 271:C2072–C2080.

Obatomi DK, and Bach PH (1996a) Inhibition of mitochondrial respiration and oxygen uptake in isolated rat renal tubular fragments by atractyloside. *Toxicol Lett* 89:155–161.

Obatomi, DK, and Bach PH (1996b) Selective cytotoxicity associated with *in vitro* exposure of fresh rat renal fragments and continuous cell lines to atractyloside. *Arch Toxicol* 71:93–98.

Obatomi DK, and Bach PH (2000) Atractyloside nephrotoxicity: *In vitro* studies with suspensions of rat renal fragments and precision-cut cortical slices. *In Vitro Mol Toxicol* 13:25–36.

Obatomi DK, and Plummer DT (1995) Renal damage caused by gentamicin: A study of the effect *in vitro* using isolated rat proximal tubular fragments. *Toxicol Lett* 75:75–83.

Obatomi DK, Brant S, Anthonypillai V, and Bach PH (1998a) Toxicity of atractyloside in precision-cut rat and porcine renal and hepatic tissue slices. *Toxicol Appl Pharmacol* 148:35–45.

Obatomi DK, Thanh NT, Brant S, and Bach PH (1998b) The toxic mechanism and metabolic effects of atractyloside in precision-cut pig kidney and liver slices. *Arch Toxicol* 72:524–530.

Parrish AR, Gandolfi AJ, and Brendel K (1995) Precision-cut tissue slices: Applications in pharmacology and toxicology. *Life Sci* 57:1887–1901.

Parrish AR, Fisher R, Bral CM, Burghardt RC, Gandolfi AJ, Brendel K, and Ramos KS (1998) Benzo(a)pyrene-induced alterations in growth-related gene expression and signaling in precision-cut adult rat liver and kidney slices. *Toxicol Appl Pharmacol* 152:302–308.

Parrish AR, Zheng XH, Turney KD, Younis HS, and Gandolfi AJ (1999) Enhanced transcription factor DNA binding and gene expression induced by arsenite or arsenate in renal slices. *Toxicol Sci* 50:98–105.

Parrish AR, Sallam K, Nyman DW, Orozco J, Cress AE, Dalkin BL, Nagle RB, and Gandolfi AJ. (2002) Culturing precision-cut human prostate slices as an *In vitro* model of prostate pathobiology. *Cell Biol Toxicol* 18:205–219.

Phelps JS, Gandolfi AJ, Brendel K, and Dorr R (1987) Cisplatin nephrotoxicity: *In vitro* studies with precision-cut rabbit renal cortical slices. *Toxicol Appl Pharmacol* 90:501–512.

Poola NR, Bhuiyan D, Ortiz S, Savant IA, Sidhom M, Taft DR, Kirschenbaum H, and Kalis M. (2002) A novel HPLC assay for pentamidine: comparative effects of creatinine and inulin on GFR estimation and pentamidine renal excretion in the isolated perfused rat kidney. *J Pharm Pharm Sci* 5:135–145.

Potier M, L'Azou B, and Cambar J (1996a) Isolated glomeruli and cultured mesangial cells as *in vitro* models to study immunosuppressive agents. *Cell Biol Toxicol* 12:263–270.

Potier M, Winicki J, and Cambar J (1996b) Nitric oxide (NO) donor 3-morpholinosydnonimine antagonizes cyclosporin A-induced contraction in two *in vitro* glomerular models. *Cell Biol Toxicol* 12:335–339.

Redegeld FA, Hofman GA, Koster AS, and Noordheok J (1991a) Flow-dependent extraction of 1-naphthol by the rat isolated perfused kidney. *Naunyn Schmiedebergs Arch Pharmakol* 343:330–333.

Redegeld FA, Hofman GA, van de Loo PG, Koster AS, and Noordehoek J (1991b) Nephrotoxicity of the glutathione conjugate of menadione (2-methyl-1,4-napthoquinone) in the isolated perfused rat kidney. Role of metabolism by gamma-glutamyltranspeptidase and probenecid sensitive transport. *J Pharmacol Exp Ther* 256:665–669.

Richter E, Friesenegger S, Engl J, Tricker AR (2000) Use of precision-cut tissue slices in organ culture to study metabolism of 4-(methylnitrosamino)-1-(3-pyridyl)-1-butanone (NNK) and 4-(methylnitrosamino)-1-(3-pyridyl)-1-butanol (NNAL) by hamster lung, liver and kidney. *Toxicol* 144:83–91.

Rodeheaver DP, Aleo MD, and Schnellmann RG (1990) Differences in enzymatic and mechanical isolated rabbit renal proximal tubules: Comparison in long-term incubation. *In vitro Cell Dev Biol* 26:898–904.

Romero M, Mosquera J, and Rodriguez-Iturbe B (1997) A simple method to identify NBT-positive cells in isolated glomeruli. *Nephrol Dial Transplant* 12:174–179.

Ross CR, and Farah A (1966) *p*-Aminohippurate and *N*-methylnicotinamide transport in dog renal slices–an evaluation of the counter-transport hypothesis. *J Pharmacol Exp Ther* 151:159–167.

Ross CR, Pessah NI, and Farah A (1968) Studies of uptake and runout of p-aminohippurate and *N*-methylnicotinamide in dog renal slices. *J Pharmacol Exp Ther* 160:381–386.

Ruegg CE, Gandolfi AJ, Nagle RB, and Brendel K (1987a) Differential patterns of injury to the proximal tubule of renal cortical slices following *in vitro* exposure to mercuric chloride, potassium dichromate, or hypoxic conditions. *Toxicol Appl Pharmacol* 90:261–273.

Ruegg CE, Gandolfi AJ, Nagle RB, Krumdieck CL, and Brendel K. (1987b) Preparation of positional renal slices for study of cell-specific toxicity. *J Pharmacol Methods* 17:111–123.

Ruegg CE, Wolfgang GHI, Gandolfi AJ, Brendel K, and Krumdieck CL (1989) Preparation of positional renal slices for *in vitro* nephrotoxicity studies, in *In Vitro Toxicology: Model Systems and Methods* (McQueen CA, Ed.) pp. 197–230. Tellford Press, Caldwell, NJ.

Sade K, Schwartz D, Wolman Y, Schwartz I, Chernichovski T, Blum M, Brazowski E, Keynan S, Raz I, Blantz RC, and Iaina A (1999) Time course of lipopolysaccharide-induced nitric oxide synthase mRNA expression in rat glomeruli. *J Lab Clin Med* 134:471–477.

Safirstein R, Miller P, and Guttenplan JB (1984) Uptake and metabolism of cisplatin by rat kidney. *Kidney Int* 25:753–758.

Savin V, Karniski L, Cuppage F, Hodges G, and Chonko A (1985) Effect of gentamicin on isolated glomeruli and proximal tubules of the rabbit. *Lab Invest* 52:93–102.

Schambelan M, Blake S, Sraer J, Bens M, Nivez MP, and Wahbe F (1985) Increased prostaglandin production by glomeruli isolated from rats with streptozotocin-induced diabetes mellitus. *J Clin Invest* 75:404–412.

Schurr A, and Rigor BM (1995) *Brain Slices in Basic and Clinical Research*. CRC Press, Boca Raton, Florida.

Sharma R, Khanna A, Sharma M, and Savin VJ (2000) Transforming growth factor-beta1 increases albumin permeability of isolated rat glomeruli via hydroxyl radicals. *Kidney Int* 58:131–136.

Shuprisha A, Lynch RM, Wright SH, and Dantzler WH (2000) PKC regulation of organic anion secretion in perfused S2 segments of rabbit proximal tubules. *Am J Physiol* 278:F104–F109.

Sirianni GL, and Pang KS (1999) Inhibition of esterolysis of enalapril by paraoxon increases the urinary clearance in isolated perfused rat kidney. *Drug Metab Dispos* 27:931–936.

Smith PF, Gandolfi AJ, Krumdieck CL, Putnam CW, Zukoski cf., Davis WM, and Brendel K (1985) Dynamic organ culture of precision liver slices for *in vitro* toxicology. *Life Sci* 36:1367–1375.

Smith JH (1988) The use of renal cortical slices from the Fischer 344 rat as an *in vitro* model to evaluate nephrotoxicity. *Fundam Appl Toxicol* 11:132–142.

Sraer J, Sraer JD, Chansel S, Russo-Marie F, Kouznetzova B, and Ardaillou R (1979) Prostaglandin synthesis by isolated rat renal glomeruli. *Mol Cell Endocrinol* 16:29–37.

Takeda M, Jung KY, Sekine T, Endou H, and Koide H (1995) Guanidinoacetic acid (GAA) synthesis in rat tubular suspension as a system for evaluating gentamicin (GM) nephrotoxicity. *Toxicol Lett* 81:85–89.

Takemoto M, Asker N, Gerhardt H, Lundkvist A, Johansson BR, Saito Y, and Betsholtz C (2002) A new method for large scale isolation of kidney glomeruli from mice. *Am J Pathol* 161:799–805.

Tarloff JB, and Goldstein RS (1994) *In vitro Assessment of Toxicology* (Gad SC, Ed.) pp. 149–194. Raven Press, New York.

Templeton DM, and Chaitu N (1990) Effects of divalent metals on the isolated rat glomerulus. *Toxicol* 61:119–133.

Toutain HJ, Sarsat JP, Bouant A, Hoet D, Leroy D, and Moronvalle-Halley V (1996) Precision-cut dog renal cortical slices in dynamic organ culture for the study of cisplatin nephrotoxicity. *Cell Biol Toxicol* 12:289–298.

Trevisan A, Meneghetti P, Maso S, and Troso O (1993) *In-vitro* mechanisms of 1,2-dichloropropane nephrotoxicity using the renal cortical slice. *Hum Exp Toxicol* 12:117–121.

Turney KD, Parrish AR, Orozco J, and Gandolfi AJ (1999) Selective activation in the MAPK pathway by Hg(II) in precision-cut rabbit renal cortical slices. *Toxicol Appl Pharmacol* 160:262–270.

Valdivielso JM, Morales AI, Perez-Barriocanal F, and Lopez-Novoa JM (1998) Effect of cicletanine on renal cGMP production. *Can J Physiol Pharmacol* 76:1151–1155.

Valentovic MA, Ball JG, Sun H, and Rankin GO (2002) Characterization of 2-amino-4,5-dichlorophenol (2A45CP) *in vitro* toxicity in renal cortical slices from male Fischer 344 rats. *Toxicol* 172:113–123.

Van Crugten JT, Sallustio BC, Nation RL, and Somogyi AA (1991) Renal tubular transport of morphine, morphine-6-glucuronide, and morphine-3-glucuronide in the isolated perfused rat kidney. *Drug Metab Dispos* 19:1087–1092.

Vickers AE, Fischer V, Connors S, Fisher RL, Baldeck JP, Maurer G, and Brendel K (1992) Cyclosporin A metabolism in human liver, kidney, and intestine slices. Comparison to rat and dog slices and human cell lines. *Drug Metab Dispos* 20:802–809.

Vickers AE, Fischer V, Connors MS, Biggi WA, Heitz F, Baldeck JP, and Brendel K (1996) Biotransformation of the antiemetic 5-HT3 antagonist tropisetron in liver and kidney slices of human, rat and dog with a comparison to *in vivo*. *Eur J Drug Metab Pharmacokinet* 21:43–50.

Weiss C, Passow H, and Rothstein A (1959) Autoregulation of flow in isolated rat kidney in the absence of red cells. *Am J Physiol* 196:1115–1118.

Wiegmann TB, Sharma R, Diederich DA, and Savin VJ (1990) *In vitro* effects of cyclosporine on glomerular function. *Am J Med Sci* 299:149–152.

Wilks MF, Kwizera EN, and Bach PH (1990) Assessment of heavy metal nephrotoxicity *in vitro* using isolated rat glomeruli and proximal tubular fragments. *Ren Physiol Biochem* 13:275–284.

Willinger CC, Moschen I, Kulmer S, and Pfaller W (1995a) The effect of sodium fluoride at prophylactic and toxic doses on renal structure and function in the isolated perfused rat kidney. *Toxicol* 95:55–71.

Willinger CC, Thamaree S, Schramek H, Gstraunthaler G, and Pfaller W (1995b) *In vitro* nephrotoxicity of Russell's viper venom. *Kidney Int* 47:518–528.

Wolfgang GH, Gandolfi AJ, Stevens JL, and Brendel K (1989a) *In vitro* and *in vivo* nephrotoxicity of the L- and D- isomers of *S*-(1,2-dichlorovinyl)-cysteine. *Toxicol* 58:33–42.

Wolfgang GH, Gandolfi AJ, Stevens JL, and Brendel K (1989b) *N*-acetyl *S*-(1,2-dichlorovinyl)-L-cysteine produces a similar toxicity to S-(1,2-dichlorovinyl)-L-cysteine in rabbit renal slices: Differential transport and metabolism. *Toxicol Appl Pharmacol* 101:205–219.

Wolfgang GH, Gandolfi AJ, Nagle RB, Brendel K, and Stevens JL (1990) Assessment of *S*-(1,2-dichlorovinyl)-L-cysteine induced toxic events in rabbit renal cortical slices. Biochemical and histological evaluation of uptake, covalent binding, and toxicity. *Chem Biol Interact* 75:153–170.

Yousif T, Pooyeh S, Hannemann J, Baumann J, Tauber R, and Baumann K (1999) Nephrotoxic and peroxidative potential of meropenem and imipenem/cilastatin in rat and human renal cortical slices and microsomes. *Int J Clin Pharmacol Ther* 37:475–486.

Zalups RK, and Barfuss DW (1996) *In vitro* perfusion of isolated nephron segments: A method of renal toxicology, in *Methods in Renal Toxicology* (Zalups RK, and Lash LH Eds) pp. 123–146. CRC Press, Boca Raton, Florida.

Zalups RK, and Lash LH, Eds (1996). *Methods in Renal Toxicology,* CRC Press, Boca Raton, FL.

Zhou XJ, Laszik Z, Ni Z, Wang XQ, Brackett DJ, Lerner MR, Silva FG, and Vaziri ND (2000) Down-regulation of renal endothelial nitric oxide synthase expression in experimental glomerular thrombotic microangiopathy. *Lab Invest* 80:1079–1087.

5

In Vitro Techniques in Screening and Mechanistic Studies: Cell Culture, Cell-Free Systems, and Molecular and Cell Biology

Sue M. Ford

INTRODUCTION

The evolution of cell culture in toxicology is similar to the development of *in vivo* toxicology studies in the way that it has moved from basic, descriptive toxicity studies to cutting-edge research using sophisticated analysis of cellular and molecular events. This progression in cell culture was facilitated by acceleration of several trends, including commitment of the toxicology community to using alternatives to animals, advances in molecular biology and cell signaling, and the expansion of commercial sources that provide tissue culture supplies and reagents. It was not so long ago that investigators who needed collagen-coated culture dishes would have to collect rat tails, extract the collagen, and coat and cure the dishes; now collagen-coated culture ware is commercially available. Much of the labor-intensive tissue culture housekeeping, including preparation of culture media and growth factors (Lincoln and Gabridge, 1998; Mather, 1998), sterilization of laboratory ware, testing for mycoplasma, and development of assays has shifted to the commercial sector, making cell culture a practical technique for most laboratories rather than the arcane art of tissue culturists.

Cell culture has been a valuable tool in the study of renal toxicology. There is a significant degree of differential sensitivity to toxicants among the regions in the kidney, presenting some difficulty in resolving effects on individual cell types, *in vivo* or *in vitro*. Cell culture allows investigators to select and expand populations of cells from specific sites. Cells have been grown successfully from all parts of the kidney, including proximal tubules (Chung et al., 1982; Dudas and Renfro, 2001; Kim et al., 1998; Zaki et al., 2003), glomerular mesangial cells (Mirto et al., 1999; Mühl et al., 1996; Patel et al., 2002; Vicart, 1994), interstitial cells (Johnson et al., 1999), collecting tubules (Grenier and Smith, 1978; Lee et al., 2001, 2002), and distal tubules (Cummings et al., 2000). The application of molecular biology methods to cell culture has led to innovative approaches to *in vitro* renal toxicology research.

PRIMARY CULTURES VS. CELL LINES

Isolated cells taken from a fresh kidney and placed in culture are referred to as "primary cultures." Primary cultures may be used for studies or transferred to other vessels for further cultivation ("subcultured"), at which point they are referred to as "cell lines." Most cell lines die after a finite number of cell divisions (Hayflick and Moorhead, 1965), a phenomenon known as "in vitro senescence." When variants appear that seem to be immortal, the resulting cultures are considered "continuous cell lines." (The Society for In Vitro Biology [SIVB, formerly the Tissue Culture Association] has published a guide to cell culture terminology at its website http://www.sivb.org/edu_terminology.asp; originally published in Schaeffer, 1990.)

Compared with cell lines, primary cultures are closer in nature to the intact tissue. However, a major disadvantage of primary cultures is that they must be prepared from fresh tissue each time cells are needed. In addition to being inconvenient, this introduces a source of variability. This variability is less of a problem for cells taken from genetically similar laboratory animals bred under controlled conditions. In contrast, human kidneys are obtained from random sources and from individuals who differ in age, gender, lifestyle, and other personal characteristics. Additionally, the methods of isolation, selection, and maintenance of primary cultures may not be consistent between laboratories, resulting in cultures that differ significantly. Continuous cell lines that can be maintained for generations offer greater convenience and the potential for increased reproducibility.

There are numerous cell lines for use for renal toxicology studies. Many are available from cell culture repositories, such as ATCC (American Type Culture Collection, Manassas, VA; www.atcc.org) and the European Collection of Cell Cultures (ECACC, U.K.; www.ecacc.org.uk), and can be located through the hyperlinked Cell Line Database (HyperCLDB; www.biotech.ist.unige.it/interlab/cldb.html), among others on the Worldwide Web.

One of the cell lines most commonly used in renal physiology and pharmacology is the LLC-PK$_1$ line, isolated from the renal cortex of a pig in 1958 (Hull et al., 1976). LLC-PK$_1$ has been well characterized and possesses many of the properties of the proximal tubule, including apical membrane enzymes (Gstraunthaler et al., 1985), sodium-dependent glucose transport, and microvilli (Handler, 1983), although the cells do exhibit vasopressin-responsive adenylate cyclase, characteristic of the distal nephron. The major use of LLC-PK$_1$ is to model the proximal tubule and its responses to toxicants. MDCK (Madin Darby canine kidney) is another useful cell line, which is closer in nature to distal tubules cells (Rindler et al., 1979). Both of these lines form "domes" in confluent monolayers. Domes are dynamic fluid-filled blisters that result from a combination of fluid and water transport capability, suitable degree of cell adhesion to the substratum, and the presence of tight junctions (Lever, 1979).

Other cell lines have been employed for their specific characteristics. OK (opossum kidney) cells are valuable for phosphate transport. COS-7, an African green monkey cell line originally developed as a viral host, is useful as transfection host for membrane transporters because it lacks many native transporters and it is easily transfected (Kuze et al., 1999; Zhang et al., 2002). The human embryonic line HEK-293 and its variant HEK-293T (which express the SV40 large T antigen and is resistant to G418) are useful because they are easily grown and readily transfected. They have been used for studies involving transfection with membrane transporters (Goralski et al., 2002) and calcium channels (Hajela et al., 2003), as well as for toxicity studies (Kirkpatrick et al., 2003; Zheng et al., 2003). Although the HEK-293 lines are epithelial in phenotype they do not adhere well to substrata, which may present problems for some studies. The NRK-52E cell line has been used in studies of apoptosis and cisplatin toxicity (Huang et al., 2001; Lash et al., 2002b; Tsuruya et al., 2003), and arsenite accumulation has been studied in HEK293 cells (Kirkpatrick et al., 2003).

Most of the currently available renal cell culture models (primaries and cell lines) have disadvantages that preclude their universal use in toxicology studies. The ability to develop more suitable models of primary cultures and cell lines from renal cells has been hindered by two factors: the loss of differentiated characteristics when cells are cultured and the limited lifespan of mammalian cells.

Maintenance of Differentiated Phenotype

Many ultrastructural and functional characteristics are lost or attenuated when cells are removed from the body and placed in culture. This process is called "dedifferentiation" or "anaplasia," and represents the backtracking of a committed cell to a less specialized state (Maclean and Hall, 1987; Odelberg, 2002). Cultured kidney cells differ from their *in vivo* counterparts in function as well as structure. For example, the porcine cell line LLC-PK_1 has fewer classes of glutathione-*S*-transferases compared with pig kidney (Bohets et al., 1996). Reduction in brush border enzymes (alkaline phosphatase, γ-glutamyltranspeptidase) compared with fresh tissue or primary cultures have been observed for rabbit (Taub et al., 2002) and rat (Lash et al., 2002a) cell lines. Although proximal tubule cells remain polarized, with microvilli on the apical surface, the microvilli are less abundant, as are mitochondria (Lash et al., 2002a). Differentiated characteristics are not lost uniformly. For example, cytochromes P450 and sulfotransferase activity and expression are lost within 24 hours after rat proximal tubule cells are transferred to culture, and 97% of alkaline phosphatase activity is gone within a week (Schaaf et al., 2001). In contrast, UDP-glucuronyltransferase and glutathione-*S*-transferase activities are reduced by only 25% during this interval. Haenen and coworkers (1996) observed minimal basolateral transport of glutathione (GSH) conjugates by rat proximal tubule cells cultured for four days on filters, but the apical to basolateral transport of substrates was well-maintained.

Restoration of some differentiated characteristics has been accomplished by manipulation of physical and chemical aspects of the culture environment. For example, growing proximal tubule cells on permeable filters permits transepithelial transport of endogenous substrates and xenobiotics (Genestie et al., 1995; Palmoski et al., 1992), which is not possible when the cells are grown on plastic. Increased access of LLC-PK_1 cells to oxygen via roller bottles reduces glycolytic metabolism and increases utilization of glutamine (Gstraunthaler et al., 1999). Sodium-dependent glucose transport is an important function of proximal tubules and its expression in culture is of considerable interest. Culturing cells in low-glucose media (Courjault et al., 1993; Moran et al., 1983) and assaying transport in high-sodium buffer (Schaaf et al., 2001) are two adjustments that have markedly improved the behavior of the cells with respect to glucose transport. Taub and coworkers (2002) observed that although the activity of Na^+-dependent glucose transport was reduced in immortalized rabbit proximal tubule cells, the transporters were actually present at the same level as in fresh tissue. The authors suggested that the reduced transport was due to alterations in a regulatory pathway. Another factor influencing expression of differentiated function is the stage of confluency. For example Na^+-dependent glucose transporter activity is reduced in proliferating cultured cells compared with confluent monolayers. Korn and coworkers (2001) discovered a polypeptide, RS1, that modulates upregulation of SGLT1 in LLC-PK_1 cells after the cells become confluent.

Growth factors, substrates, and other media additives also alter the expression of differentiated functions. Supplementation of the medium with cholesterol-methyl-β-cyclodextrin restored Na^+-dependent glucose transport by primary proximal tubule cultures (Runembert et al., 2002). Triiodothyronine enhanced growth as well as differentiated phenotype in rat proximal tubule cells, whereas calcitonin was effective in distal tubule cultures (Lash et al., 1995). P-glycoprotein (PGP) activity of HK-2 cells can be modulated by putative substrates, including dexamethasone, aldosterone, and 1,25 $(OH)_2$ vitamin D_3 (Romiti et al., 2002). Certain agents, such as dimethylsulfoxide, stimulate the formation of domes in renal cell cultures (Lever, 1986); these structures are the result of solute and water transport and are considered to be evidence of differentiated function in transporting epithelia.

The use of attachment factors and basement membrane components on the growth surface often improves differentiated functions. Cells grown on contractible collagen rafts, for example, exhibit transepithelial resistance and potential difference values close to those observed *in vivo* (Dudas and Renfro, 2001). Virally-immortalized proximal tubule cell lines grown on a basement membrane extract form tubule-like structures that are not seen when the cells are grown on plastic (Takeuchi et al., 2002). The formation of tubules in primary rat proximal tubule cultures may also be induced by a combination of growth factors (epidermal [EGF], fibroblast [FGF-7], and insulin-like [IGF-1]) or by hepatocyte growth factor alone (Bowes et al., 1999). Hepatocyte growth factor, along with transforming growth factor (TGF-β1), has been shown to be involved in collagen turnover in mouse proximal tubule cells (Inoue et al., 2002). Other substrata, such as alginate, have been shown to reverse dedifferentiation in epithelial cells (Eurell et al., 2003).

The above examples of modulation of cell characteristics by manipulating the culture environment illustrate that the phenomenon of dedifferentiation is likely to be a consequence of cell adaptation to the unphysiologic culture environment. Dedifferentiation may be at least partially reversible, a process referred to as "redifferentiation" (Maclean and Hall, 1987). The ability to redifferentiate cells may depend on factors such as the gene agents used for immortalization (Takeuchi et al., 2002) and the presence of other cell types in the culture. For example, serum promotes fibroblast overgrowth in rabbit proximal tubule cultures (Chung et al., 1982), and fibroblasts have been shown to prevent redifferentiation of other cell types (Dispersyn et al., 2001; Rücker-Martin et al., 2002).

Recent advances in cell and molecular biology are leading to a greater understanding of the changes in cell-cycle signaling and gene expression that may help explain the phenomena of dedifferentiation, redifferentiation, and transdifferentiation, *in vivo* and *in vitro* (Echeverri and Tanaka, 2002; Odelberg, 2002). Microarray assays have identified genes and gene products that appear at various stages in renal development or in response to hormones *in vivo* (Braam et al., 2003; Valerius et al., 2002); this knowledge may ultimately benefit cell culture technology. Developing a clearer picture of dedifferentiation, transdifferentiation, and redifferentiation at the level of gene expression will be critical for establishing cell culture models that more closely resemble the original tissue. In addition, the effects of some nephrotoxicants may involve the same processes of dedifferentiation as cultured cells (Gekle et al., 1998; Richter and Vamvakas, 1998; Vamvakas et al., 1996) so that further study of these processes will shed light on carcinogenesis and other toxicities.

Immortalization and the Development of New Cell Lines

The inconvenience and variability of primary cultures and the limitations of available cell lines necessitates the search for more suitable models. Some current lines were isolated from normal tissue and formed continuous cultures spontaneously, i.e., with no reported intervention by the investigators. Such was the case for LLC-PK_1 (Hull et al., 1976) and MDCK (Rindler et al., 1979). However, because of the rarity of this phenomenon, it is generally not a useful technique for producing a continuous renal cell line. Cell lines from normal (noncancerous) kidney cells have been difficult to produce, especially from human tissue. Cell lines have therefore been developed using embryonic cells, neoplastic tissue (Williams et al., 1976), cells treated *in vitro* with carcinogens, *in vitro* transfection with viral transforming genes, or tissue from animals transgenic with one of the viral vectors. Currently, the most reliable process of immortalizing normal cells *in vitro* involves transfection of cells with viral genes (Katakura et al., 1998), such as those from adenoviruses (Takeuchi et al., 2002), human papilloma virus (Ostrow et al., 1993), or SV40 large T antigen (SV40 Tag/SV40 LTag) (Loghman-Adham et al., 2003). Immortalization by transfection depends on the responsiveness of target cells, the efficiency of the transfection method, and the probability of a suitable chromosomal event occurring to enable cells in the transfected population to survive "crisis" (Lustig, 1999; Ray et al., 1990). The methodology for transfection continues to improve. An important advance is replacement of electroporation and calcium phosphate precipitation methods by transfection with more effective lipid reagents.

Continuous kidney cell lines have been developed from rabbit proximal tubule (Taub et al., 2002), human proximal tubule (Racusen et al., 1997; Zhao and Ford, 2002), and collecting duct cells (Lee et al., 2001). The process used to immortalize the cells may influence which differentiated characteristics are retained. Takeuchi et al. (2002) observed differences in behavior between cell lines immortalized by various genes and suggested that some may be manipulated ("reprogrammed") by extracellular factors to influence expression of differentiated characteristics. Transgenic mice, generally those with immortalizing genes derived from viruses, have been used to establish immortalized cultures from the proximal tubule (Nesbitt et al., 1995; Taher et al., 1995; Takeuchi et al., 1994; Vallet et al., 1995), mesangial cells (Vicart et al., 1994), and multiple segments of the nephron (Lee et al., 2002).

GENETICALLY MODIFIED CELLS

The choice of cell line for toxicology studies is often involves decisions based on availability, how easy the line is to work with, and which lines posses characteristics critical for the investigation. For example, Chen and coworkers (1999) compared the three renal cell lines, LLC–PK_1, MDCK, and OK, with regard to transport of nucleosides by organic cation transporters. They concluded that OK cells had a mechanism consistent with nucleoside secretion *in vivo* and would be most suitable for their studies. However, evaluation of numerous cell-culture lines for particular functions may be time-consuming and expensive, and in many cases it is more practical to insert the desired trait into a host cell via transfection with DNA or RNA vectors. This has been done frequently for cytochrome P450s and for transporters.

"Stable transfection" occurs when the transfected DNA is incorporated into the host genome, so that the daughter cells retain the transfected functions. This is a rare event and requires procedures to select successfully transfected cells from the rest of the population. The development of a stably transfected cell line takes more time than "transient transfections" but ideally provides a long-term source of cells expressing the desired characteristic. Transient transfections must be done for each batch of cells. Expression of the inserted gene lasts several days, until the DNA is degraded. With the commercial availability of improved reagents, such techniques are now routine and are suitable for automation and high-throughput (e.g., jetPEI^{TM}, Qbiogene, Carlsbad, CA). Additionally, several renal cell lines, such as HEK-293, HEK-293T, and COS-7, are easily transfected, making them good choices for transient transfections. Transient transfection can be very useful if conditions are carefully controlled to reduce variability. Several companies that sell transfection reagents (e.g., Invitrogen, Qiagen, Clontech, Stratagene, Promega) provide manuals in portable document format (pdf) on the Worldwide Web detailing important considerations for transfection, which include potential toxicity of the transfection reagent, efficacy of gene transfer, use of reporter and selection genes, quality of plasmid DNA, and effect of cell density. Promega has a "transfection assistant" interactive tool on its website (http://www.promega.com/transfectionasst/default.htm) that helps investigators select conditions and reagents for the cell line of interest. General considerations for preparing and use of transfectants in xenobiotic metabolism studies were presented in a 1998 symposium

(Townsend et al., 1999). Some useful studies for evaluating conditions for transient transfections of polarized epithelial monolayers, such as MDCK, have been detailed by Tucker and coworkers (2003).

Genetically modified cells are useful tools for evaluating mechanisms of toxicity. Using such methods, the proximal convoluted tubule cell line PKSV-PCT, derived from SV40 transgenic mice (Cartier et al., 1993), was stably transfected with tetracycline-regulated kidney androgen-regulated protein (KAP)-expressing system (Cebrián et al., 2001). Removal of tetracycline from the medium allowed KAP overexpression and reduced cyclosporin A toxicity, demonstrating a possible role for KAP in nephrotoxicity. Stable transfection of superoxide or catalase vectors into HEK-293 cells demonstrated that superoxide dismutase, but not catalase, protects against cisplatin toxicity (Davis et al., 2001). Transfection of LLC-PK_1 cells with peptide cotransporters allowed identification of a histidyl residue that can act as a binding site for the β-lactam antibiotics (Terada et al., 1997b, 1998). MDCK and HEK-293 cells are useful for transfections inasmuch as the cells are kidney-derived epithelioid cells (Davis et al., 2001). Since they lack many of the renal transporters for xenobiotics, transfection of the cells with single transporters can help elucidate characteristics of the transporters. For example, it was shown that organic cation transporters (OCT1 and OCT2) transfected into MDCK cells became localized to the basolateral membrane and had broad substrate specificities similar to *in vivo* localization and expression (Urakami et al., 1998), suggesting that OCT1/2 mediate tubular secretion of cationic drugs *in vivo*. A similar approach was taken by transfecting COS-7 cells with mouse kidney organic anion transporter (mOAT). PAH transport was sensitive to inhibition by diethyl pyrocarbonate (DEPC), suggesting involvement of a histidyl residue (Kuze et al., 1999). MDR1-MDCK represents a line of MDCK cells transfected with PGP-expressing *mdr1 gene*. The cells express PGP at a high level but form overlapping layers in culture (Braun et al., 2000). Nonetheless, they have been found to be useful in studying PGP substrate transport across monolayers (Benet, 2002; Karyekar et al., 2003).

Another interesting approach is to transfect cultured cells with antisense RNA constructs to prevent translation of mRNA, or with cDNA that encodes for the antisense RNA. This technique eliminates or diminishes the production of proteins of interest. LLC-PK_1 cells transfected with antisense oligonucleotide to the stress protein Grp78 were more sensitive to tert-butyl hydroperoxide (Liu et al., 1998), H_2O_2

(Hung et al., 2003), and disturbance of intracellular Ca^{++} (Liu et al., 1997), demonstrating the role of this protein in the cellular response to toxicants. MDCK cells transfected with antisense RNA to the molecular chaperone ORP150 were more susceptible to prolonged hypoxia (Bando et al., 2000). Initial experiments showed that c-myc transcription was increased following exposure of LLC-PK_1 cells to cadmium; however blocking translation with c-myc antisense oligodeoxynucleotide did not prevent cadmium-induced apoptosis, indicating that c-myc is not involved in this process (Ishido et al., 1998).

CELL DEATH AS A TOXIC ENDPOINT

Toxicologists aim to describe and understand the mechanisms leading to cell injury and to establish dose–response relationships between chemical exposure and adverse effects, including cell death. In the past decade, recognition of the role of apoptosis in chemical toxicity has required clarification of terminology related to cell death (Kanduc et al., 2003; Majno and Joris, 1995; Sloviter, 2002; Trump and Berezesky, 1996, 1998) and the Society of Toxicologic Pathologists has recommended standard nomenclature (Levin et al., 1999). "Oncosis" refers to cell death pathways proceeding through cell swelling, in contrast to the cell shrinkage occurring in "apoptosis" (Majno and Joris, 1995). Another variant of apoptosis, "anoikis," is a consequence of disturbances in cellular adhesion and loss of focal adhesion kinase (FAK) activity (Frisch and Francis, 1994). It is recommended that use of the term "necrosis" be limited to the changes subsequent to cell death, not to the process of cell death itself. A fundamental point is that oncosis and apoptosis are pre-lethal pathways whereas necrosis refers to the changes that occur after cell death; however, some authors use necrosis to refer to an alternative to apoptosis.

Cell culture has been crucial in defining the processes involved in apoptosis. A given xenobiotic may cause cell death by predominantly apoptosis under some conditions (e.g., low dose) and by oncosis under others. Apoptosis and oncosis in cultured cells can be distinguished by several parameters in toxicology studies. Oncotic cell death is characterized by degeneration of membranes, often resulting in cell lysis. The release of cytoplasmic constituents (e.g., lactate dehydrogenase) into culture media, the uptake of indicator dyes (e.g., trypan blue) into intact but dead cells, or the lack of uptake of others dyes (e.g., neutral red) are convenient endpoints. The plasma membrane is not grossly damaged during apoptosis *in vivo*; instead, apoptotic cell remnants (apoptotic bodies) are produced and removed by macrophages. However, apoptotic cells *in vitro* may swell and lyse, a process referred to as "secondary necrosis" (Kelly et al., 2003; Nowak et al., 2003; Padanilam, 2003; Yasuhara et al., 2003). Therefore, apoptosis *in vitro* is most clearly detected at a relatively early stage.

Xenobiotics can cause apoptosis by several pathways that have characteristic markers which can be used to evaluate the role of apoptosis in toxic cell death (Boelsterli, 2003). Activation of caspases, decreased mitochondrial membrane potential, chromatin condensation, DNA fragmentation, phosphatidylserine movement to the external plasma membrane leaflet, and hypodiploid nuclei are indications of apoptosis that can be measured in cultured cells (Loo and Rillema, 1998; Moore et al., 1998; Zhou et al., 2000). The distinctive nuclear condensation of apoptosis can be observed following staining with fluorescent dyes, such as propidium iodide, acridine orange, or bisbenzimide. DNA fragmentation is assessed with terminal deoxyribonucleotide transferase-mediated dUTP-X nick end-labeling staining (TUNEL) or *in situ* end labeling (ISEL). Phosphatidylserine movement in the plasma membrane can be assessed with annexin V, a phospholipid binding protein with a high specificity for phosphatidylserine and which can be conjugated with a fluorochrome. These methods can be used with flow cytometry to provide information on apoptosis in cell populations as well as cell-cycle distribution. The MTT (methyl-thiazolyl tertrazolium) assay measures the ability of mitochondria to reduce MTT into a colored product and therefore provides an indicator of live cell number. It is useful in following cell proliferation or, conversely, cell death, and has been conveniently adapted to multiwell plates for high-throughput screening. However, the assay does not distinguish between apoptosis and oncosis and may reflect effects on cell proliferation rather than cell death.

When investigating the actions of xenobiotics on cell death, it is often desirable to examine early and late events. Evaluation of apoptosis and necrosis in primary human proximal tubule cultures exposed to dichlorovinyl cysteine (DCVC) showed that apoptosis occurs earlier and at a lower concentration than necrosis (Lash et al., 2001); a similar relationship was noted for outer cortical collecting duct cultures exposed to cisplatin (Lee et al., 2001). Following exposure of LLC-PK_1 cells to H_2O_2, inhibition of poly(ADP-ribose)polymerase (PARP), an enzyme involved in DNA repair, was shown to protect against necrosis (measured by LDH release) but not against the DNA fragmentation characteristic of apoptosis (Filipovic et al., 1999).

Apoptosis was determined by analysis of DNA fragmentation subsequent to exposure of four renal cell lines to amphotericin B (Varlam et al., 2001). Apoptosis was inhibited by recombinant human insulin-like growth factor-1 (rhIGF-1), a known apoptosis inhibitor. These authors validated the results with *in vivo* studies in rats. Mühl and coworkers (1996) used DNA fragmentation as an index of apoptotic cell death in glomerular mesangial cells and showed that nitric oxide donors induced apoptosis but endogenous production of nitric oxide did not. Wong and coworkers (2001) also determined DNA fragmentation to show that apoptosis in primary human renal proximal tubule cultures treated with okadaic acid involves activation of caspases 9, 3, and 7. Apoptosis due to DCVC or doxorubicin treatment of LLC-PK_1 cells was assessed by annexin V staining, cell-cycle analysis, and caspase activation, and compared with LDH release as an index of necrosis (van de Water et al., 2001). In NRK52E cells, cisplatin increased production of death receptors, reduced viability, and increased caspase activity (Tsuruya et al., 2003); these effects were not seen when dimethylthiourea, a hydroxyl radical scavenger, was included in the media.

TRANSPORT

Transport of xenobiotics by the proximal tubules is a critical issue in toxicology because it is frequently a determinant of the sensitivity of the organ to toxicants and has a central role in the pharmacokinetic behavior of drugs. A basic view of renal organic ion transport has given way to an understanding of the complex nature of renal handling of drugs. This is particularly important for organic cations that may be transported by multiple mechanisms, including renal organic cation transporters (OCTs) and PGP. *In vitro* methods have proven to be valuable tools for the study of renal transport processes, particularly in unraveling polyspecific transporters.

Cultured kidney cells can be grown on permeable membrane filters, forming monolayers that allow three-compartment analysis of transport: apical, basolateral, and intracellular compartments. It is possible to study the influx or efflux of the monolayer from either side of the cell as well as the transepithelial transport of substrates in either direction. The integrity of the monolayer must be ascertained either by monitoring the transepithelial movement of marker molecules, such as inulin or mannitol, or by measuring transepithelial resistance. Transport studies are generally performed using radiolabeled substrates, and in some cases nonspecific binding to the filter may be an issue. Prefabricated filter inserts are available for 6- or 24-well tissue culture plates. Several filter types are available, including clear filters with or without attachment matrices, such as collagen; in some cases, the choice of filter may rest on which type allows growth and confluence of the cells.

LLC-PK$_1$ grown on filters have been used to study transport of numerous xenobiotics, including quinolones (Matsuo et al., 1998), nicotine and tetraethylammonium (Takami et al., 1998), and β-lactam antibiotics (Terada et al., 1997a, 1997b). In both rabbit proximal tubule suspensions and cultured rabbit proximal tubule cells, chlorotrifluoroethylcysteine is transported by organic anion transporter (OAT) in the basolateral membrane and a neutral amino acid transporter on the apical side (Groves and Morales, 1999). Schwerdt and coworkers (1997, 1998) concluded that apical uptake of ochratoxin A in MDCK cells occurred by three mechanisms: organic anion transport, proton-dipeptide transport, and diffusion. Sauvant and coworkers (1998) investigated possible causes of the decline in *p*-aminohippurate (PAH) secretion that follows ochratoxin A exposure. Incubation of OK cells with ochratoxin A for 30 min competitively inhibited PAH secretion. However long-term incubation (72 hr) of cells with the toxin decreased efflux at the apical membrane, and organic anion/dicarboxylate exchange was impaired due to effects on affinity.

Cultured cells can be transfected with genes for various transporters, including OATs, OCTs, proton-dipeptide transporters (PEPT), and PGP, which is encoded by the multidrug resistance (mdr) genes. The ability to create cells overexpressing individual transporters allows characterization of kinetic behavior and substrate specificity of each transporter (Guo et al., 2002; Hasegawa et al., 2002, 2003; Sugiyama et al., 2001). MDCK cells represent a convenient model for study of PGP because native OCT activity is low (Ito et al., 1999) because OCT also transports some PGP substrates. LLC-PK$_1$ cells transfected with human MDR1-PGP were used to demonstrate that verapamil metabolites have distinct properties (Pauli-Magnus et al., 2000). Additionally, cultured nonrenal cells, including fibroblasts (Pan et al., 1999) and *Xenopus laevis* oocytes, can be transfected with human renal transporters (Arndt et al., 2001; Guo et al., 2002) in order to study the transporters in the absence of competing reactions. Comparison with *in vivo* transporters can be made directly for validation. LLC-PK_1 cells transfected with human MDR1 or mouse mdr1a were compared with mdr1a knockout mice (Yamazaki et al., 2001). The characteristics of mouse mdr1a PGP-mediated transport were similar *in vivo* and *in vitro*, and were different to human MDR1 PGP, confirming the validity of the *in vitro* model.

MICROARRAY

The ability to analyze global gene expression in response to xenobiotic exposure opens up exciting possibilities for the elucidation of mechanisms and pathways of toxicity and for the identification of biomarkers for screening and risk assessment. Genes that respond to particular compounds can be selected for use as early-event biomarkers. These can be developed into screening tools using assays that are more suitable for rapid analysis of specific events. For example, to identify potential biomarkers for peroxisome proliferator activity, gene expression by HK-2 cells in response to known peroxisome proliferators has been examined by microarray analysis using a human genome set (Liu et al., 2003). Pyruvate dehydrogenase kinase 4 (PDK4) was strongly induced; this correlated with induction of the same enzyme in hamsters as well as with the physiological effects noted, suggesting that PDK4 may be a useful biomarker for peroxisome proliferator exposure. The ability to study temporal, species, and dose-related patterns of gene expression associated with xenobiotic exposure in cultured cells (Katsuma et al., 2002; Zheng et al., 2003) will be invaluable for understanding the progression of nephrotoxic pathology (Katsuma et al., 2002). It will also aid understanding of differences in *in vivo* and *in vitro* responses to nephrotoxicants such as cisplatin (Huang et al., 2001) and arsenite (Zheng et al., 2003).

The vast quantities of data generated by microarrays require sophisticated software and databases to dissect genetic events of interest from background noise (Dooley et al., 2003; Farr and Dunn, 1999; Glatt et al., 2001; Gracey and Cossins, 2003). An additional issue with microarray is reporting the results in conventional print publication; in some cases, provision of a web address to a downloadable data file may be appropriate (Valerius et al., 2002). However, it is possible to choose and design chips for targeted groups of genes, such as those developed for particular animal species, drug metabolism studies, or pathways related to specific cell responses (Braam et al., 2003; Katsuma et al., 2002; Zheng et al., 2003). Despite the power that microarray expression data offers, there are still issues of validation that must be considered when extrapolating from cultured cells to *in vivo*, from animals to humans, and between cultures (Zheng et al., 2003). For example, NRK-52E kidney epithelial cells were less responsive than liver cells to cisplatin, in contrast to *in vivo* results (Huang et al., 2001). Microarray technology is at an early stage of development and significant issues must be resolved before it can be used for regulatory purposes.

REFERENCES

Arndt P, Volk C, Gorboulev V, Budiman T, Popp C, Ulzheimer-Teuber I, Akhoundova A, Koppatz S, Bamberg E, and Nagel G (2001) Interaction of cations, anions, and weak base quinine with rat renal cation transporter rOCT2 compared with rOCT1. *Am J Physiol Renal Physiol* 281: F454–F468.

Bando Y, Ogawa S, Yamauchi A, Kuwabara K, Ozawa K, Hori O, Yanagi H, Tamatani M, and Tohyama M (2000) 150-kDa oxygen-regulated protein (ORP150) functions as a novel molecular chaperone in MDCK cells. *Am J Physiol (Cell Physiol)* 278:C1172–C1182.

Benet SM (2002) Can the enhanced renal clearance of antibiotics in cystic fibrosis patients be explained by P-glycoprotein transport? *Pharmaceut Res* 19:457–462.

Boelsterli UA (2003) *Mechanistic Toxicology. The Molecular Basis of How Chemicals Disrupt Biological Targets*, New York: Taylor & Francis.

Bohets HH, Nouwen EJ, De Broe ME, and Dierickx PJ(1996) Isolation and characterisation of the class alpha, mu and pi glutathione transferases in LLC-PK1 and pig kidney. *Comp Biochem Physiol B Biochem Mol Biol* 114:261–267.

Bowes RC III, Lightfoot RT, van de Water B, and Stevens JL (1999) Hepatocyte growth factor induces tubulogenesis of primary renal proximal tubular epithelial cells. *J Cell Physiol* 180:81–90.

Braam B, Allen P, Benes E, Koomans HA, Navar LG, and Hammond T (2003) Human proximal tubular cell responses to angiotensin II analyzed using DNA microarray. *Eur J Pharmacol* 464:87–94.

Braun A, Hammerle S, Suda K, Rothen-Rutishauser B, Gunthert M, Kramer SD, and Wunderli-Allenspach H (2000) Cell cultures as tools in biopharmacy. *Eur J Pharmaceut Sci* 11:S51–S60.

Cartier N, Lacave R, Vallet V, Hagege J, Hellio R, Robine S, Pringault E, Cluzeaud F, Briand P, Kahn A, and Vandewalle A (1993) Establishment of renal proximal tubule cell lines by targeted oncogenesis in transgenic mice using the L-pyruvate kinase-SV40 (T) antigen hybrid gene. *J Cell Sci* 104:695–704.

Cebrián C, Areste C, Nicolas A, Olive P, Carceller A, Piulats J, and Meseguer A (2001) Kidney androgen-regulated protein interacts with cyclophilin B and reduces cyclosporine A-mediated toxicity in proximal tubule cells. *J BiolChem* 276:29410–29419.

Chen R, Pan BF, Sakurai M, and Nelson JA (1999) A nucleoside-sensitive organic cation transporter in opossum kidney cells. *Am J Physiol (Renal Physiol)* 276:F323–F328.

Chung SD, Alavi N, Livingston D, Hiller S, and Taub M (1982) Characterization of primary rabbit kidney cultures that express proximal tubule functions in a hormonally defined medium. *J Cell Biol* 95:118–126.

Courjault F, Chevalier J, Leroy D, and Toutain H (1993) Effect of glucose and insulin deprivation on differentiation and carbohydrate metabolism of rabbit proximal tubular cells in primary culture. *Biochim Biophys Acta- Molec Cell Res* 1177:147–159.

Cummings BS, Zangar RC, Novak RF, and Lash LH (2000) Cytotoxiicty of trichloroethylene and *S*-(1,2-dichlorovonyl)-L-cysteine in primary cultures of rat renal proximal tubular and distal tubular cells. *Toxicol* 150:83–98.

Davis CA, Nick HS, and Agarwal A (2001) Manganese superoxide dismutase attenuates cisplatin-induced renal injury: Importance of superoxide. *J Am Soc Nephrol* 12:2683–2690.

Dispersyn GD, Geuens E, Ver Donck L, Ramaekers FCS, and Borgers M (2001) Adult rabbit cardiomyocytes undergo hibernation-like dedifferentiation when co-cultured with cardiac fibroblasts. *Cardiovasc Res* 51:230–240.

Dooley TP, Curto EV, Reddy SP, Davis RL, Lambert G, and Wilborn TW (2003) A method to improve selection of molecular targets by circumventing the ADME pharmacokinetic system utilizing PharmArray DNA microarrays. *Biochem Biophys Acta* 303:828–841.

Dudas PL, and Renfro JL (2001) Assessment of tissue-level kidney functions with primary cultures. *Comp Biochem Physiol- Part A: Molec Int Physiol* 128:199–206.

Echeverri K, and Tanaka EM (2002) Mechanisms of muscle dedifferentiation during regeneration. *Sem Cell Develop Biol* 13:353–360.

Eurell TE, Brown DR, Gerding PA, and Hamor RE (2003) Alginate as a new biomaterial for the growth of porcine retinal pigment epithelium. *Vet Ophthalmol* 6:237–243.

Farr S, and Dunn RT (1999) Concise review: Gene expression applied to toxicology. *Toxicol Sci* 50:1–9.

Filipovic DM, Meng X, and Reeves WB (1999) Inhibition of PARP prevents oxidant-induced necrosis but not apoptosis in LLC-PK1 cells. *Am J Physiol (Renal Physiol)* 277:F428–F436.

Frisch SM, and Francis H (1994) Disruption of epithelial cell-matrix interactions induces apoptosis. *J Cell Biol* 124:619–626.

Gekle M, Gassner B, Freudinger R, Mildenberger S, Silbernagl S, Pfaller W, and Schramek H (1998) Characterization of an ochratoxin-A-dedifferentiated and cloned renal epithelial cell line. *Toxicol Appl Pharmacol* 152:282–291.

Genestie I, Morin JP, Vannier B, and Lorenzon G (1995) Polarity and transport properties of rabbit kidney proximal tubule cells on collagen IV-coated porous membranes. *Am J Physiol* 269:F22–F30.

Glatt CM, Davis LG, Ladics GS, Ciaccio PJ, and Slusher LB (2001) An evaluation of the DNA array for use in toxicological studies. *Toxicol Meth* 11:247–275.

Goralski KB, Lou G, Prowse MT, Gorboulev V, Volk C, Koepsell H, and Sitar DS (2002) The cation transporters rOCT1 and rOCT2 interact with bicarbonate but play only a minor role for amantadine uptake into rat renal proximal tubules. *J Pharmacol Exp Ther* 303:959–968.

Gracey AY, and Cossins AR (2003) Application of microarray technology in environmental and comparative physiology. *Ann Rev Physiol* 65:231–259.

Grenier FC, and Smith WL (1978) Formation of 6-keto-PGF1[alpha] by collecting tubule cells isolated from rabbit renal papillae. *Prostaglandins* 16:759–772.

Groves CE, and Morales MN (1999) Chlorotrifluoroethylcysteine interaction with rabbit proximal tubule cell basolateral membrane organic anion transport and apical membrane amino acid transport. *J Pharmacol Exp Ther* 291:555–561.

Gstraunthaler G, Pfaller W, and Kotanko P (1985) Biochemical characterization of renal epithelial cell cultures (LLC-PK1 and MDCK). *Am J Physiol (Renal Physiol)* 248:F536-F544.

Gstraunthaler G, Seppi T, and Pfaller W (1999) Impact of culture conditions, culture media volumes, and glucose content on metabolic properties of renal epithelial cell cultures. Are renal cells in tissue culture hypoxic? *Cell Physiol Biochem: Int J Exper Cell Physiol Biochem Pharmacol* 9:150–172.

Guo A, Marinaro W, Hu P, and Sinko PJ (2002) Delineating the contribution of secretory transporters in the efflux of etoposide using Madin-Darby canine kidney (MDCK) cells overexpressing P-glycoprotein (Pgp), multidrug resistance-associated protein (MRP1), and canalicular multispecific organic anion transporter (cMOAT). *Drug Metab Dispos* 30:457–463.

Haenen HEMG, Spenkelink A, Teunissen C, Temmink JHM, Koeman JH, and van Bladeren PJ (1996) Transport and metabolism of glutathione conjugates of menadione and ethacrynic acid in confluent monolayers of rat renal proximal tubular cells. *Toxicol* 112:117–130.

Hajela RK, Peng S, and Atchison WD (2003) Comparative effects of methylmercury and Hg^{2+} on human neuronal N- and R-type high voltage activated calcium channels transiently expressed in human embryonic kidney cells (HEK293). *J Pharmacol Exp Ther* 306:1129–1136.

Handler JS (1983) Use of cultured epithelia to study transport and its regulation. *J Exp Biol* 106:55–69.

Hasegawa M, Kusuhara H, Sugiyama D, Ito K, Ueda S, Endou H, and Sugiyama Y (2002) Functional involvement of rat organic anion transporter 3 (rOat3; Slc22a8) in the renal uptake of organic anions. *J Pharmacol Exp Ther* 300:746–753.

Hasegawa M, Kusuhara H, Endou H, and Sugiyama Y (2003) Contribution of organic anion transporters to the renal uptake of anionic compounds and nucleoside derivatives in rat. *J Pharmacol Exp Ther* 305:1087–1097.

Hayflick L, and Moorhead PS (1965) The serial cultivation of human diploid cell strains. *Exp Cell Res* 25:585–621.

Huang Q, Dunn RT II, Jayadev S, DiSorbo O, Pack FD, Farr SB, Stoll RE, and Blanchard KT (2001) Assessment of cisplatin-induced nephrotoxicity by microarray technology. *Toxicol Sci* 63:196–207.

Hull RN, Cherry WR, and Weaver GW (1976) The origin and characteristics of a pig kidney cell strain, LLC-PK_1. *in vitro* 12:670–677.

Hung CC, Takaharu I, Stevens JL, and Bonventre JV (2003) Protection of renal epithelial cells against oxidative injury by endoplasmic reticulum stress preconditioning is mediated by ERK1/2 activation. *J Biol Chem* 278:29317–29326.

Inoue T, Okada H, Kobayashi T, Watanabe Y, Kikuta T, Kanno Y, Takigawa M, and Suzuki H (2002) TGF-β1 and HGF coordinately facilitate collagen turnover in subepithelial mesenchyme. *Biochem Biophys Acta* 297:255–260.

Ishido M, Tohyama C, and Suzuki T (1998) C-myc is not involved in cadmium-elicited apoptotic pathway in porcine kidney LLC-PK_1 cells. *Life Sci* 63:1195–1204.

Ito S, Woodland C, Sarkadi B, Hockmann G, Walker SE, and Koren G (1999) Modeling of P-glycoprotein-involved epithelial drug transport in MDCK cells. *Am J Physiol (Renal Physiol)* 277:F84-F96.

Johnson DW, Saunders HJ, Johnson FJ, Huq SO, Field MJ, and Pollock CA (1999) Cyclosporin exerts a direct fibrogenic effect on human tubulointerstitial cells: roles of insulin-like growth factor I, transforming growth factor β_1, and platelet-derived growth factor. *J Pharmacol Exp Ther* 289:535–542.

Kanduc D, Mittelman A, Serpico R, Sinigaglia E, Sinha AA, Natale C, Santacroce R, Di Corcia MG, Lucchese A, Dini L, Pani P, Santacroce S, Simone S, Bucci R, and Farber E (2003) Cell death: apoptosis versus necrosis. *Int J Oncol* 21:165–170.

Karyekar CS, Eddington ND, Garimella TS, Gubbins PO, and Dowling TC (2003) Evaluation of p-glycoprotein-mediated renal drug interactions in an MDR1-MDCK model. *Pharmacotherapy* 23:436–442.

Katakura Y, Alam S, and Shirahata S (1998) Immortalization by gene transfection. *Methods Cell Biol* 57:69–91.

Katsuma S, Shiojima S, Hirasawa A, Takagaki K, Kaminishi Y, Koba M, Hagidai Y, Murai M, Ohgi T, Yano J, and Tsujimoto G (2002) Global analysis of differentially expressed genes during progression of calcium oxalate nephrolithiasis. *Biochem Biophys Acta* 296:544–552.

Kelly KJ, Sandoval RM, Dunn KW, Molitoris BA, and Dagher PC (2003) A novel method to determine specificity and sensitivity of the TUNEL reaction in the quantitation of apoptosis. *Am J Physiol (Cell Physio)* 284:C1309–C1318.

Kim YK, Ko SH, Woo JS, Lee SH, and Jung JS (1998) Difference in H_2O_2 toxicity between intact renal tubules and cultured proximal tubular cells. *Biochem Pharmacol* 56:489–495.

Kirkpatrick DS, Dale KV, Catania JM, and Gandolfi AJ (2003) Low-level arsenite causes accumulation of ubiquitinated proteins in rabbit renal cortical slices and HEK293 cells. *Toxicol Appl Pharmacol* 186:101–109.

Korn T, Kuhlkamp T, Track C, Schatz I, Baumgarten K, Gorboulev V, and Koepsell H (2001) The plasma membrane-associated protein RS1 decreases transcription of the transporter SGLT1 in confluent LLC-PK1 cells. *J Biol Chem* 276:45330–45340.

Kuze K, Graves P, Leahy A, Wilson P, Stuhlmann H, and You G (1999) Heterologous expression and functional characterization of a mouse renal organic anion transporter in mammalian cells. *J Biol Chem* 274:1519–1524.

Lash LH, Hueni SE, and Putt DA (2001) Apoptosis, necrosis, and cell proliferation induced by *S*-(1,2-dichlorovinyl)-L-cysteine in primary cultures of human proximal tubular cells. *Toxicol Appl Pharmacol* 177:1–16.

Lash LH, Putt DA, Hueni SE, Cao W, Xu F, Kulidjian SJ, and Horwitz JP (2002a) Celllular energetics and glutathione status in NRK-52E cells: Toxicological implications. *Biochem Pharmacol* 64:1533–1546.

Lash LH, Tokarz JJ, and Pegouske DM (1995) Susceptibility of primary cultures of proximal tubular and distal tubular cells from rat kidney to chemically induced toxicity. *Toxicol* 103:85–103.

Lash LH, Putt DA, and Matherly LH (2002b) Protection of NRK-52E cells, a rat renal proximal tubular cell line, from chemical-induced apoptosis by overexpression of a mitochondrial glutathione transporter. *J Pharmacol Exp Ther* 303:476–486.

Lee RH, Song JM, Park MY, Kang SK, Kim YK, and Jung JS (2001) Cisplatin-induced apoptosis by translocation of endogenous Bax in mouse collecting duct cells. *Biochem Pharmacol* 62:1013–1023.

Lee WK, Jang SB, Cha SH, Lee JH, Lee KH, Kim J, Jo YH, and Endou H (2002) Different sensitivity to nephrotoxic agents and osmotic stress in proximal tubular and collecting duct cell lines derived from transgenic mice. *Toxicol in vitro* 16:55–62.

Lever JE (1979) Inducers of mammalian cell differentiation stimulate dome formation in a differentiated kidney epithelial cell line (MDCK). *Proc Natl Acad Sci USA* 76:1323–1327.

Lever JE (1986) Expression of differentiated functions in kidney epithelial cell lines. *Mineral Electrolyte Metab* 12:14–19.

Levin S, Bucci TJ, Cohen SM, Fix AS, Hardisty JF, LeGrand EK, Maronpot RR, and Trump BF (1999) The nomenclature of cell death: recommendations of an ad hoc Committee of the Society of Toxicologic Pathologists. *Toxicol Pathol* 27:484–490.

Lincoln CK, and Gabridge MG (1998) Cell culture contamination: sources, consequences, prevention, and elimination. *Methods Cell Biol* 57:49–65.

Liu H, Bowes RC III, Van De Water B, Sillence C, Nagelkerke JF, and Stevens JL (1997) Endoplasmic reticulum chaperones GRP78 and calreticulin prevent oxidative stress, Ca^{2+} disturbances, and cell death in renal epithelial cells. *J Biol Chem* 272:21751–21759.

Liu H, Miller E, Van De Water B, and Stevens JL (1998) Endoplasmic reticulum stress proteins block oxidant-induced Ca^{2+} increases and cell death. *J Biol Chem* 273:12858–12862.

Liu PCC, Huber R, Stow MD, Schlingmann KL, Collier P, Liao B, Link J, Burn TC, Hollis G, Young PR, and Mukherjee R (2003) Induction of endogenous genes by peroxisome proliferator activated receptor alpha ligands in a human kidney cell line and *in vivo*. *J Steroid Biochem Molec Biol* 85:71–79.

Loghman-Adham M, Nauli SM, Soto CE, Kariuki B, and Zhou J (2003) Immortalized epithelial cells from human autosomal dominant polycystic kidney cysts. *Am J Physiol (Renal Physiol)* 285:F397–F412.

Loo DT, and Rillema JR (1998) Measurement of cell death. *Methods Cell Biol* 57:251–264.

Lustig AJ (1999) Crisis intervention: The role of telomerase. *Proc Natl Acad Sci USA* 96:3339–3341.

Maclean N, and Hall BK (1987) *Cell Commitment and Differentiation*, Cambridge: Cambridge University Press.

Majno G, and Joris I (1995) Apoptosis, oncosis, and necrosis. An overview of cell death. *Am J Pathol* 146:3–15.

Mather JP (1998) Making informed choices: Medium, serum, and serum-free medium. How to choose the appropriate medium and culture system for the model you wish to create. *Methods Cell Biol* 57:19–30.

Matsuo Y, Yano I, Ito T, Hashimoto Y, and Inui K-I (1998) Transport of quinolone antibacterial drugs in a kidney epithelial cell line, LLC-PK1. *J Pharmacol Exp Ther* 287:672–678.

Mirto H, Barrouillet M-P, Henge-Napoli M-H, Ansoborlo E, Fournier M, and Cambar J (1999) Uranium-induced vasoreactivity in isolated glomeruli and cultured rat mesangial cells. *Toxicol in vitro* 13:707–711.

Moore A, Donahue CJ, Bauer KD, and Mather JP (1998) Simultaneous measurement of cell cycle and apoptotic cell death. *Methods Cell Biol* 57:265–78.

Moran A, Turner RJ, and Handler JS (1983) Regulation of sodium-coupled glucose transport by glucose in a cultured epithelium. *J Biol Chem* 258:15087–15090.

Mühl H, Sandau K, Brune B, Briner VA, and Pfeilschifter J (1996) Nitric oxide donors induce apoptosis in glomerular mesangial cells, epithelial cells and endothelial cells. *Eur J Pharmacol* 317:137–149.

Nesbitt T, Econs MJ, Byun JK, Martel J, Tenenhouse HS, and Drezner MK (1995) Phosphate transport in immortalized cell cultures from the renal proximal tubule of normal and Hyp mice: Evidence that the HYP gene locus product is an extrarenal factor. *J Bone Miner Res* 10:1327–1333.

Nowak G, Price PM, and Schnellmann RG (2003) Lack of a functional p21WAF1/CIP1 gene accelerates caspase-independent apoptosis induced by cisplatin in renal cells. *Am J Physiol (Renal Physiol)* 285:F440-F450.

Odelberg SJ (2002) Inducing cellular dedifferentiation: A potential method for enhancing endogenous regeneration in mammals. *Sem Cell Develop Biol* 13:335–343.

Ostrow RS, Liu Z, Schneider JF, McGlennen RC, Forslund K, and Faras AJ (1993) The products of the E5, E6, or E7 open reading frames of RhPV 1 can individually transform NIH 3T3 cells or in cotransfections with activated Ras can transform primary rodent epithelial cells. *Virology* 196:861–867.

Padanilam BJ(2003) Cell death induced by acute renal injury: A perspective on the contributions of apoptosis and necrosis. *Am J Physiol (Renal Physiol)* 284:F608–F627.

Palmoski MJ, Masters BA, Flint OP, Ford SM, and Oleson FB (1992) Characterization of rabbit primary proximal tubule kidney cell cultures grown on Millicell-HA membrane filters. *Toxicol in vitro* 6:557–567.

Pan BF, Sweet DH, Pritchard JB, Chen R, and Nelson JA (1999) A transfected cell model for the renal toxin transporter, rOCT2. *Toxicol Sci* 47:181–186.

Patel VA, Dunn MJ, and Sorokin A (2002) Regulation of MDR-1 (P-glycoprotein) by cyclooxygenase-2. *J Biol Chem* 277:38915–38920.

Pauli-Magnus C, von Richter O, Burk O, Ziegler A, Mettang T, Eichelbaum M, and Fromm MF(2000) Characterization of the major metabolites of verapamil as substrates and inhibitors of P-glycoprotein. *J Pharmacol Exp Ther* 293:376–382.

Racusen LC, Monteil C, Sgrignoli A, Lucskay M, Marouillat S, and Rhim JGS (1997) Cell lines with extended *in vitro* growth potential from human renal proximal tubule: Characterization, response to inducers, and comparison with established cell lines. *J Lab Clin Med* 129:318–329.

Ray FA, Peabody DS, Cooper JL, Cram LS, and Kraemer PM (1990) SV40 T antigen *alone* drives karyotype instability that precedes neoplastic transformation of human diploid fibroblasts. *J Cell Physiol* 42:13–31.

Richter H, and Vamvakas S (1998) *S*-(1,2-Dichlorovinyl)–cysteine-induced dedifferentiation and p53 gene mutations in LLC-PK1 cells: A comparative investigation with *S*-(2-chloroethyl)cysteine, potassium bromate, cis-platinum and styrene oxide. *Toxicol Lett* 94:145–157.

Rindler MJ, Chuman LM, Shaffer L, and Saier MH, Jr. (1979) Retention of differentiated properties in an established dog kidney epithelial cell line (MDCK). *J Cell Biol* 81:635–648.

Romiti N, Tramonti G, and Chieli E (2002) Influence of different chemicals on MDR-1 p-glycoprotein expression and activitiy in the HK-2 proximal tubular cell line. *Toxicol Appl Pharmacol* 183:83–91.

Rücker-Martin C, Pecker F, Godreau D, and Hatem SN (2002) Dedifferentiation of atrial myocytes during atrial fibrillation: Role of fibroblast proliferation *in vitro*. *Cardiovasc Res* 55:38–52.

Runembert I, Queffeulou G, Federici P, Vrtovsnik F, Colucci-Guyon E, Babinet C, Briand P, Trugnan G, Friedlander G, and Terzi F (2002) Vimentin affects localization and activity of sodium-glucose cotransporter SGLT1 in membrane rafts. *J Cell Sci* 115:713–724.

Sauvant C, Silbernagl S, and Gekle M (1998) Exposure to ochratoxin A impairs organic anion transport in proximal-tubule-derived opossum kidney cells. *J Pharmacol Exp Ther* 287:13–20.

Schaaf GJ, de Groene EM, Maas RF, Commandeur JNM, and Fink-Gremmels J (2001) Characterization of biotransformation enzyme activities in primary rat proximal tubular cells. *Chemico-Biol Interact* 134:167–190.

Schaeffer WI (1990) Terminology associated with cell, tissue and organ culture, molecular biology and molecular genetics. *In vitro Cell Dev Biol Animal* 26:97–101.

Schwerdt G, Gekle M, Freudinger R, Mildenberger S, and Silbernagl S (1997) Apical-to-basolateral transepithelial transport of ochratoxin A by two subtypes of Madin-Darby canine kidney cells. *Biochim Biophys Acta-Biomembranes* 1324:191–199.

Schwerdt G, Freudinger R, Silbernagl S, and Gekle M (1998) Apical uptake of radiolabelled ochratoxin A into Madin-Darby canine kidney cells. *Toxicol* 131:193–202.

Sloviter RS (2002) Apoptosis:a guide for the perplexed. *Trends Pharmacol Sci* 23:19–24.

Sugiyama D, Kusuhara H, Shitara Y, Abe T, Meier PJ, Sekine T, Endou H, Suzuki H, and Sugiyama Y (2001) Characterization of the efflux transport of 17β-estradiol-D-17β-glucuronide from the brain across the blood-brain barrier. *J Pharmacol Exp Ther* 298:316–322.

Taher A, Yanai N, and Obinata M (1995) Properties of incompletely immortalized cell lines generated from a line established from temperature-sensitive SV40 T-antigen gene transgenic mice. *Exp Cell Res* 219:332–338.

Takami K, Saito H, Okuda M, Takano M, and Inui K-I (1998) Distinct characteristics of transcellular transport between nicotine and tetraethylammonium in LLC-PK_1 cells. *J Pharmacol Exp Ther* 286:676–680.

Takeuchi K, Yanai N, Takahashi N, Abe T, Tsutsumi E, Obinata M, and Abe K (1994) Different cellular mechanisms of vasopressin receptor V1 and V2 subtype in vasopression-induced adenosine 3′,5′-monophosphate formation in an immortalized renal tubule cell line, TKC2. *Biochem Biophys Acta* 202:680–687.

Takeuchi K, Sakurada K, Endou H, Obinata M, and Quinlan MP (2002) Differential effects of DNA tumor virus genes on the expression profiles, differentiation, and morphogenetic reprogramming potential of epithelial cells. *Virology* 300:8–19.

Taub M, Han HJ, Rajkhowa T, Allen C, and Park JH (2002) Clonal analysis of immortalized renal proximal tubule cells: Na+/glucose cotransport system levels are maintained despite a decline in transport function. *Exp Cell Res* 281:205–212.

Terada T, Saito H, Mukai M, and Inui K-I (1997a) Characterization of stably transfected kidney epithelial cell line expressing rat H^+/peptide cotransporter PEPT1: localization of PEPT1 and transport of β-lactam antibiotics. *J Pharmacol Exp Ther* 281:1415–1421.

Terada T, Saito H, Mukai M, and Inui K-I (1997b) Recognition of β-lactam antibiotics by rat peptide transporters, PEPT1 and PEPT2, in LLC-PK_1 cells. *Am J Physiol (Renal Physiol)* 273:F706-F711.

Terada T, Saito H, and Inui K-I (1998) Interaction of β-lactam antibiotics with histidine residue of rat H^+/peptide cotransporters, PEPT1 and PEPT2. *J Biol Chem* 273:5582–5585.

Townsend AJ, Kiningham KK, St.Clair D, Tephly TR, Morrow CS, and Guengerich FP (1999) Symposium overview: Characterization of xenobiotic metabolizing enzyme function using heterologous expression systems. *Toxicol Sci* 48:143–150.

Trump BF, and Berezesky IK (1996) The role of altered $[Ca^{2+}]i$ regulation in apoptosis, oncosis, and necrosis. *Biochim Biophys Acta- Molecular Cell Research* 1313:173–178.

Trump BF, and Berezesky IK (1998) The reactions of cells to lethal injury: Oncosis and necrosis- the role of calcium, in *When Cells Die: A Comprehensive Evaluation of Apoptosis and Programmed Cell Death,* (Lockshin RA, Zakeri Z, and Tilly JL, Eds), Wiley-Liss, New York.

Tsuruya K, Tokumoto M, Ninomiya T, Hirakawa M, Masutani K, Taniguchi M, Fukuda K, Kanai H, Hirakata H, and Iida M (2003) Antioxidant ameliorates cisplatin-induced renal tubular cell death through inhibition of death receptor-mediated pathways. *Am J Physiol (Renal Physiol)* 285:F208–F218.

Tucker TA, Varga K, Bebok Z, Zsembery A, McCarty NA, Collawn JF, Schwiebert EM, and Schwiebert LM (2003) Transient transfection of polarized epithelial monolayers with CFTR, and reporter genes using efficacious lipids. *Am J Physiol (Cell Physiol)* 284:C791-C804.

Urakami Y, Okuda M, Masuda S, Saito H, and Inui KI (1998) Functional characteristics and membrane localization of rat multispecific organic cation transporters, OCT1 and OCT2, mediating tubular secretion of cationic drugs. *J Pharmacol Exp Ther* 287:800–805.

Valerius MT, Patterson LT, Witte DP, and Potter SS (2002) Microarray analysis of novel cell lines representing two stages of metanephric mesenchyme differentiation. *Mech Develop* 112:219–232.

Vallet V, Bens M, Antoine B, Levrat F, Miquerol L, Kahn A, and Vandewalle A (1995) Transcription Factors and Aldolase B Gene Expression in Microdissected Renal Proximal Tubules and Derived Cell Lines. *Exp Cell Res* 216:363–370.

Vamvakas S, Richter H, and Bittner D (1996) Induction of dedifferentiated clones of LLC-PK1 cells upon long-term exposure to dichlorovinylcysteine. *Toxicol* 106:65–74.

van de Water B, Houtepen F, Huigsloot M, and Tijdens IB (2001) Suppression of chemically induced apoptosis but not necrosis of renal proximal tubular epithelial (LLC-PK1) cells by focal adhesion kinase (FAK). Role of FAK in maintaining focal adhesion organization after acute renal cell injury. *J Biol Chem* 276:36183–36193.

Varlam DE, Siddiq MM, Parton LA, and Russmann H (2001) Apoptosis contributes to amphotericin B-induced nephrotoxicity. *Antimicrob Agents Chemother* 45:679–685.

Vicart P, Schwartz B, Vandewalle A, Bens M, Delouis C, Panthier JJ, Pournin S, Babinet C, and Paulin D (1994) Immortalization of multiple cell types from transgenic mice using a transgene containing the vimentin promoter and a conditional oncogene. *Exp Cell Res* 214:35–45.

Williams RD, Elliott AY, Stein N, and Fraley EE (1976) In vitro cultivation of human renal cell cancer. I. Establishment of cells in culture. *In vitro* 12:623–627.

Wong VY, Keller PM, Nuttall ME, Kikly K, DeWolf J, Lee D, Ali SM, Nadeau DP, and Grygielko ET (2001) Role of caspases in human renal proximal tubular epithelial cell apoptosis. *Eur J Pharmacol* 433:135–140.

Yamazaki M, Neway WE, Ohe T, Chen IW, Rowe JF, Hochman JH, Chiba M, and Lin JH (2001) In vitro substrate identification studies for P-glycoprotein-mediated transport: species difference and predictability of *in vivo* results. *J Pharmacol Exp Ther* 296:723–735.

Yasuhara S, Zhu Y, Matsui T, Tipirneni N, Yasuhara Y, Kaneki M, Rosenzweig A, and Martyn JAJ (2003) Comparison of comet assay, electron microscopy, and flow cytometry for detection of apoptosis. *J Histochem Cytochem* 51:873–885.

Zaki EL, Springate JE, and Taub M (2003) Comparative toxicity of ifosfamide metabolites and protective effect of mesna and amifostine in cultured renal tubule cells. *Toxicol in vitro* 17:397–402.

Zhang X, Evans KK, and Wright SH (2002) Molecular cloning of rabbit organic cation transporter rbOCT2 and functional comparisons with rbOCT1. *Am J Physiol (Renal Physiol)* 283:F124-F133.

Zhao B, and Ford SM (2002) Characterization of a human kidney cell line. *Int J Toxicol* 21:516.

Zheng XH, Watts GS, Vaught S, and Gandolfi AJ (2003) Low-level arsenite induced gene expression in HEK293 cells. *Toxicol* 187:39–48.

Zhou X, Zhao A, Goping G, and Hirszel P (2000) Gliotoxin-induced cytotoxicity proceeds via apoptosis and is mediated by caspases and reactive oxygen species in LLC-PK1 cells. *Toxicol Sci* 54:194–202.

II

Mechanisms of Nephrotoxicity

6

Role of Xenobiotic Metabolism

Gary O. Rankin and Monica A. Valentovic

INTRODUCTION

The kidney is a target organ for a wide range of drugs and nontherapeutic chemicals. Although some drugs (e.g., gentamicin, amphotericin B) and nontherapeutic chemicals (e.g., carbon monoxide, cyanide ion) can induce nephrotoxicity through the actions of the parent compound, a large number of toxicants require bioactivation via metabolism to produce the ultimate nephrotoxicant species. In some cases, initial biotransformation can occur in extrarenal organs so that the penultimate or ultimate toxicant species is carried to the kidney via the blood. For example, hepatic biotransformation of haloalkanes or haloalkenes can result in the formation of glutathione, cysteine and *N*-acetylcysteine conjugates of the halocarbons (Dekant et al., 1993). These conjugates are then transported via the circulation to the kidney, where further biotransformation occurs and toxicity is observed in proximal tubular cells of the S_3 segment (Dekant et al., 1993; also see Chapter 23 for an in-depth discussion of these nephrotoxicants). In other cases, the parent nephrotoxicants (e.g., chloroform) (Anders, 1989; Smith, 1986) are biotransformed completely in the kidney to their ultimate nephrotoxicant species.

The kidney has specialized features and requirements that allow it to accomplish the goals of maintaining electrolyte homeostasis while promoting the elimination of xenobiotics and their metabolites (see Chapter 1). However, these features can also lead to the accumulation of toxic levels of xenobiotics and their metabolites and can make kidney cells particularly vulnerable to toxic metabolites. For example, kidney cells possess numerous transport mechanisms that help regulate levels of endogenous compounds in blood and tissues (Bendayan, 1996; Pritchard and Miller, 1993). However, these transport features of the kidney can also allow toxicants and their metabolites to concentrate in renal tissue, with resultant nephrotoxic effects. For example, cephaloridine accumulates in proximal tubular cells via the basolateral organic anion transporter (Tune and Fernholt, 1973). Once in cells, cephaloridine can be metabolized to generate free radicals, which can cause lipid peroxidative damage and cell death as one toxic mechanism (Cojocel et al., 1985). The kidney can therefore promote the accumulation of nephrotoxicant chemicals that are substrates for renal bioactivation mechanisms.

The kidney also has a high requirement for energy production to support the many transport systems required for homeostasis of electrolytes, and protein, amino acid and glucose reabsorption from the glomerular filtrate. Any chemical or metabolite that interferes with oxygen utilization or energy production can very quickly induce nephrotoxicity. Mitochondria are thus targets of many nephrotoxicants (see Chapter 11). In some cases, parent compounds (e.g., cyanide ion) are the nephrotoxicants, but in other cases renal mitochondria metabolize toxicants to their ultimate toxic species. For example, many cysteine *S*-conjugate metabolites of halogenated alkenes or alkanes are metabolized to their ultimate nephrotoxicant species by cysteine conjugate β-lyase located at the outer membrane of mitochondria (Anders, 1989; Dekant et al., 1993; Lash et al., 1986).

From these few examples, it is clear that the kidney can be the target of extrarenally produced metabolites and can directly bioactivate chemicals or their metabolites to nephrotoxicant species. Historically, the majority of research into this mechanistic area of bioactivation of toxicants has been directed toward understanding how the liver metabolizes xenobiotics to toxicant metabolites, including nephrotoxicant metabolites. However, within the past twenty years great strides have been made in understanding the role that renal metabolism plays in the nephrotoxicity induced by drugs, nontherapeutic chemicals, and their metabolites. This chapter focuses primarily on the enzyme systems present in kidney that can contribute to the renal metabolism and bioactivation of nephrotoxicants, and considers their role in xenobiotic-induced nephrotoxicity.

BIOTRANSFORMATION ENZYMES

The kidney is an important organ in the biotransformation of drugs and nondrug chemicals (Anders, 1989; Lohr et al., 1998). Biotransformation reactions generally convert parent compounds to more polar and hydrophilic metabolites that are more rapidly excreted. However, as mentioned above, many renal biotransformation reactions bioactivate drugs and other xenobiotics to reactive chemical species that induce nephrotoxicity. The enzymes that catalyze the biotransformation of xenobiotics are generally classified as Phase I or Phase II enzymes. Phase I enzymes are responsible for catalyzing oxidation, reduction, and hydrolysis biotransformation reactions, while Phase II enzymes catalyze conjugation or synthesis reactions. In Phase II reactions, endogenous molecules (e.g., sulfate) are added to the parent compound or a metabolite via covalent bonds. Many of the oxidative and reductive enzymes, along with the Phase II glucuronosyl transferase enzymes, are located primarily in the endoplasmic reticulum within cells. Differential subcellular isolation of endoplasmic reticulum (100,000 g fraction) results in the disruption of this organelle and the formation of small membrane fragments known as microsomes. As a result, these enzymes are sometimes referred to as microsomal enzymes. Enzymes that catalyze hydrolysis reactions and most Phase II enzymes are primarily located in the cytoplasm and mitochondria within cells, and are sometimes referred to as nonmicrosomal enzymes. Numerous reviews and books or book chapters that have been published previously discuss renal biotransformation and bioactivation of xenobiotics (Anders, 1980, 1985, 1989; Commandeur and Vermeulen, 1990; Lash, 1994a, 1994b; Lock and Reed, 1998; Lohr et al., 1998; Monks and Lau, 1988; Mulder, 1990; Rush et al., 1984).

PHASE I ENZYMES

Cytochromes P450

One of the most important enzyme families for the biotransformation of xenobiotics is the cytochromes P450, also know as the P450s, CYPs, or mixed function oxidases (MFOs) (Ortiz de Montellano, 1986). This superfamily of hemoproteins is composed of a large number of members, with four gene families (CYP1 to 4) being primarily responsible for catalyzing xenobiotic biotransformation reactions. Specific enzymes (e.g., CYP2E1) are named by family (e.g., CYP2), subfamily (e.g., E) and individual (1) designations. As a family, the P450s biotransform a wide range of substrates that possess some lipophilic character. Biotransformation reactions catalyzed by P450s include aliphatic and aromatic hydroxylation, epoxide formation, dealkylation of heteroatoms (*N*-, *S*-, or *O*-), deamination, desulfuration, dehalogenation, and reduction of nitro and azo functional groups.

Another characteristic of P450 enzymes is that their activity can be induced by a wide range of substances, including drugs, environmental chemicals, tobacco smoke, and alcohols. However, the inducibility of P450 activity among subfamilies for a particular inducer or family of related compounds varies greatly. In general, CYP1A enzymes are induced by polycyclic aromatic hydrocarbons (e.g., benzo[a]-pyrene, 3-methylcholanthrene, 2,3,7,8-tetrachlorodibenzo-*p*-dioxin, β-naphthoflavone), CYP2B enzymes are induced by barbiturates (e.g., phenobarbital), CYP2E1 is induced by alcohols (e.g., ethanol) and some solvents (e.g., acetone), CYP3A enzymes are induced by steroids (e.g., pregnenolone-16α-carbonitrile), and CYP4A enzymes are induced by hypolipidemic agents and peroxisome proliferators (Lock and Reed, 1998; Lohr et al., 1998).

P450 activity is generally much lower in the kidney than in the liver, with renal concentrations of total P450 being about 10% of hepatic levels (Cummings et al., 1999; Lock and Reed, 1998). In kidney, the highest concentrations of P450 are found in the renal cortex, although smaller amounts are found in the medulla (Anders, 1980; Armbrecth et al., 1979; Mohandas et al., 1981; Zenser et al., 1978). Further studies have examined the distribution of P450s along the nephron. Endou (1983) examined different segments of isolated single nephrons from rabbits and rats and found P450 localized to proximal tubules with the S_2 and S_3 proximal tubule segments having the majority of P450 content. Baron et al. (1983) detected P450 activity in proximal and distal tubules and collecting ducts in rats using immunohistochemical techniques. Cummings et al. (1999) found CYP2B1/2, CYP2C11, CYP2E1, and CYP4A2/3 expression in enriched populations of freshly isolated renal proximal and distal tubular cells from male Fischer 344 rats. CYP3A1/2 was detected in microsomes from whole kidney homogenates but not in microsomes from proximal or distal tubular cells. Ito et al. (1998) found that expression of P450 4A (1, 2, 3 and 8) isoforms could be detected in whole kidneys from Sprague-Dawley rats, with the greatest level of 4A protein occurring in the proximal tubule. However, CYP4A1 mRNA was not detected in any nephron segment and CYP4A2 mRNA was found to be fourfold higher in kidney from male vs. female rats. The majority of P450 in kidney is therefore found in proximal tubular epithelial cells (Cummings et al., 1999; Jones et al., 1980), but the amount of activity for the individual enzymes can vary depending on species, age, and gender (Guengerich et al., 1982; Ito et al., 1998). The predominant proximal tubule location of P450 contributes to the bioactivation of many proximal tubule toxicants. There are several xenobiotics for which renal P450s play a role in bioactivation, and some of these are discussed below.

Acetaminophen

Acetaminophen (APAP) is a commonly used analgesic drug that can induce hepatotoxicity and nephrotoxicity in patients that overdose with this drug (Kaloyanides, 1991). At therapeutic levels, APAP is eliminated primarily via glucuronide and sulfate conjugate formation (Figure 6.1). However, a small amount of APAP can also be *N*-hydroxylated by CYP2E1, and possibly other enzymes, to *N*-hydroxy-APAP, which is then converted by loss of water to *N*-acetyl-*p*-benzoquinone imine (NAPQI) (Figure 6.1). NAPQI is a reactive chemical species that can be conjugated with glutathione (GSH; detoxification) or arylate renal cellular macromolecules, particularly at protein sulfhydryl groups. In overdose situations, or situations that reduce renal GSH levels, arylation of renal macromolecules sufficient to induce nephrotoxicity can occur.

In some species, deacetylation of APAP to 4-aminophenol (4AP) may also play an important role in APAP nephrotoxicity. 4AP, a known metabolite of APAP, is a more potent nephrotoxicant than APAP and induces a similar renal lesion as APAP (Newton et al., 1982). Inhibition of APAP deacetylation can also decrease APAP nephrotoxicity in rat (Newton et al., 1983). Additionally, 4AP appears to undergo a similar bioactivation pathway as APAP (Figure 6.1) (Crowe et al., 1979), however the full contribution of 4AP to APAP nephrotoxicity remains to be determined with certainty.

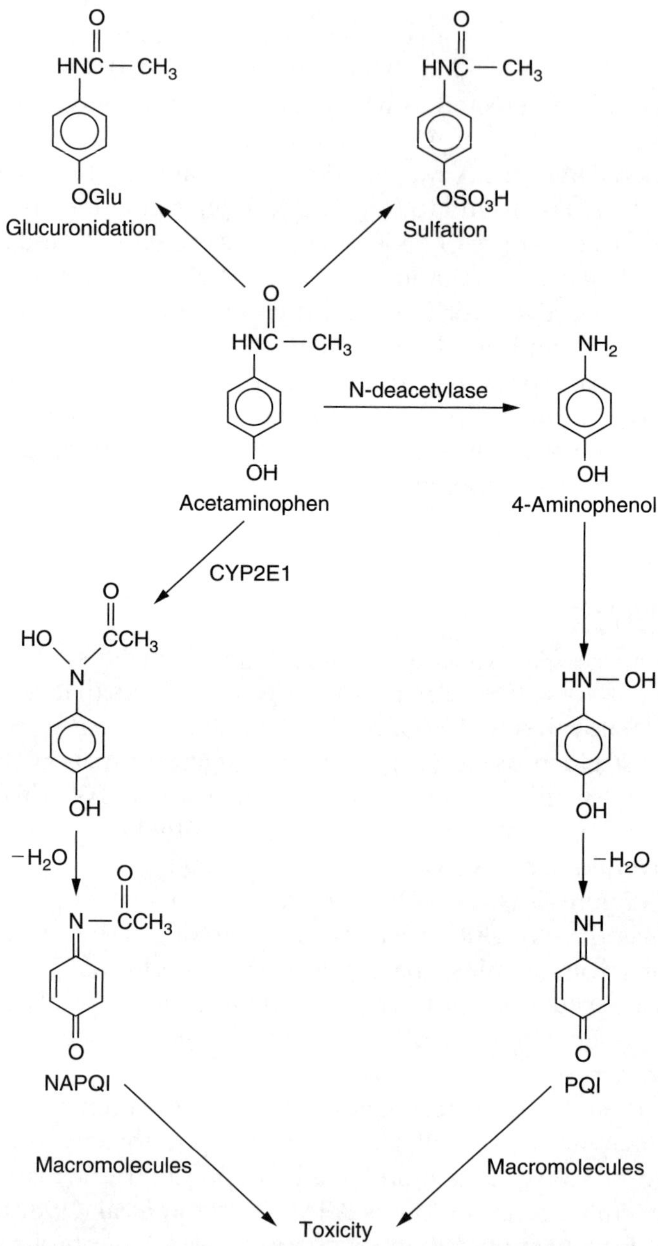

Figure 6.1 Bioactivation of acetaminophen by renal enzymes. (NAPQI = *N*-acetyl-*p*-benzoquinone imine; PQI = *p*-benzoquinone imine.)

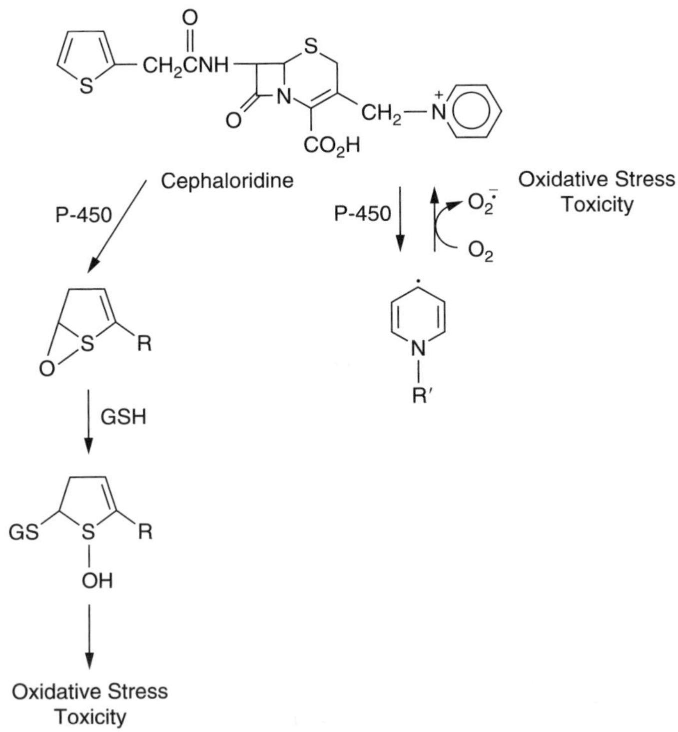

Figure 6.2 Proposed renal biotransformation pathways for the bioactivation of cephaloridine by cytochromes P450. (GSH = glutathione.)

Cephaloridine

Cephaloridine is a *β*-lactam antimicrobial drug that accumulates in renal tissues via the renal basolateral organic anion transport system (Tune and Fernholt, 1973). However, cephaloridine induces proximal tubular necrosis in mammals and so fell into disuse as an antimicrobial drug (Tune, 1986). Several mechanisms have been proposed to explain the nephrotoxicity induced by cephaloridine, including oxidative stress and mitochondrial toxicity (Tune, 1986; Tune et al., 1988). While the majority of evidence supports arylation of certain mitochondrial transporters by the *β*-lactam ring of cephalosprins as the primary mechanism of toxicity, P450-mediated biotransformation of the thiophene or pyridinium rings in cephaloridine may also contribute to the overall toxicity of this cephalosporin. P450-mediated biotransformation at either site could lead to oxidative stress (Figure 6.2), and lipid peroxidation

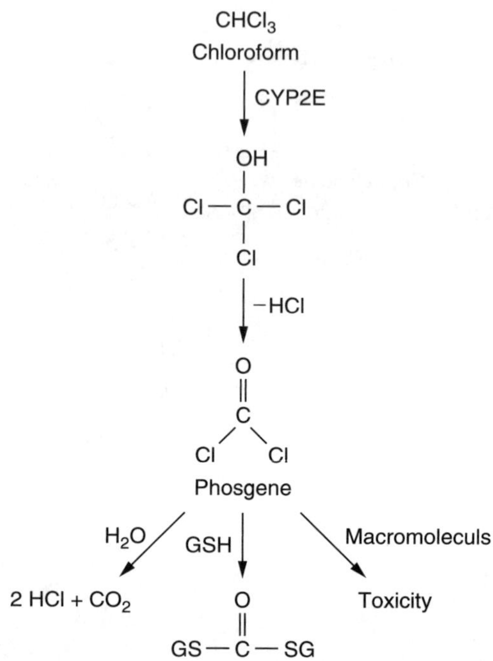

Figure 6.3 The renal bioactivation of chloroform to the reactive metabolite phosgene.

has long been considered a possible mechanism of cephaloridine nephrotoxicity (Cojocel et al., 1985; Kuo et al., 1983). Demonstration of the nephrotoxic potential of 4-(2-thienyl)butyric acid (Lash and Tokarz, 1995) and 1-(carboxymethyl)pyridine (Lash, L.H., personal communication), non-β-lactam analogs of cephaloridine, also provides evidence that P450-mediated biotransformation of cephaloridine may contribute to the overall nephrotoxic mechanism.

Chloroform

Chloroform has long been recognized as a nephrotoxicant, inducing proximal tubular necrosis. Clear evidence supports the renal bioactivation of chloroform via P450-mediated biotransformation (Anders, 1989; Smith, 1986). In the kidney, chloroform is hydroxylated by CYP2E1 to trichloromethanol, which rapidly eliminates HCl to form phosgene (Figure 6.3), the ultimate nephrotoxicant species (Lock and Reed, 1998; Pohl et al., 1977; Smith, 1986). Although phosgene can be hydrolyzed to form carbon dioxide and HCl, or be detoxified by reaction with GSH, the highly reactive nature of phosgene results in reaction with many cellular macromolecular nucleophiles to disrupt cell function and lead to cell death.

Others

A recent study demonstrated that biotransformation of the mycotoxin ochratoxin A mediated by P450 (CYPs 1A2, 2C11 and 3A) may control, in part, the bioactivation of ochratoxin A into nephrotoxic and nephrocarcinogenic metabolites (Pfohl-Leszkowicz et al., 1998). Additionally, renal CYP2E1-mediated biotransformation of chloromethane to formaldehyde and 1,1-dichloroethene to chloroacetyl chloride may contribute to the renal tumors and nephrotoxicity induced by these halocarbons (Dekant et al., 1995; Speerschneider and Dekant, 1995). Lastly, renal P450s in the CYP1 and CYP2 gene families appear to catalyze the formation of nephrotoxic monoepoxide metabolites of linoleic acid (Moran et al., 2000).

Flavin-Containing Monooxygenases

The flavin-containing monooxygenases (FMOs) are microsomal enzymes that catalyze the oxidation of heteroatoms such as nitrogen, sulfur, and selenium in xenobiotics (Zeigler, 1993). There are five isoforms of FMO(1 to 5) whose expression varies in mammalian tissues depending on gender, species, tissue, and age (Hines et al., 1994; Koukouritaki et al., 2002; Ripp et al., 1999; Yeung et al., 2000). FMO1, FMO3, and FMO5 have been quantitated in human kidney samples, with expression levels of FMO1>FMO5>FMO3 (Krause et al., 2003).

Only recently has the toxicological significance of FMO-catalyzed renal bioactivation of xenobiotics come to be recognized. Cysteine *S*-conjugates of many haloalkanes and haloalkenes are potent nephrotoxicants (Anders, 1989; Lash, 1994b), and may also be substrates for FMOs (Krause et al., 2002; 2003; Ripp et al., 1997). The resulting sulfoxide metabolites can be nephrotoxicants more potent than the parent cysteine *S*-conjugate (Lash et al., 1994). A study by Lash et al. (2001) found that methimazole, an inhibitor of FMO3, is more effective than aminooxyacetic acid, an inhibitor of cysteine conjugate β-lyase, in inhibiting *S*-(1,2-dichlorovinyl)-L-cysteine (DCVC)-induced cell death in primary cultures of human proximal tubular cells.

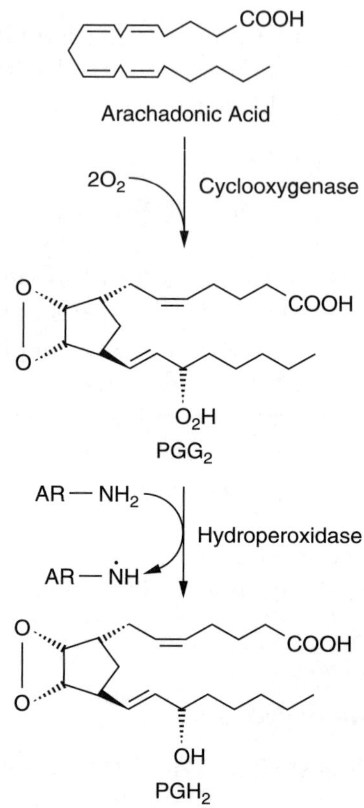

Figure 6.4 Renal cooxidation of xenobiotics by prostaglandin H synthase.

This suggests an important role for FMO in the bioactivation of DCVC. While more work is needed in this area, renal FMO-catalyzed reactions clearly may contribute to the bioactivation of some nephrotoxicants.

Prostaglandin Synthase

Prostaglandins, important modulators of renal physiology, are synthesized in the kidney from arachidonic acid (AA) released from membrane phospholipids. AA is converted to the intermediate prostaglandin G_2 (PGG_2), and then to prostaglandin H_2 (PGH_2) by prostaglandin H synthase (PHS) (Figure 6.4) (Needleman et al., 1986). PHS is a heme-containing protein with cyclooxygenase activity, which converts AA to PGG_2 and hydroperoxidase activity, which then catalyzes the formation of PGH_2. In kidney, PHS is found primarily in medullary interstitial and collecting duct cells, with smaller amounts of activity found in glomeruli, loop of Henle, and medullary thick ascending limb.

The cyclooxygenase activity of PHS is fairly specific to using polyunsaturated fatty acids as substrate, but the hydroperoxidase activity will accept many different types of reducing cosubstrates to donate electrons to the heme group of the enzyme (Eling et al., 1983; Spry et al., 1986). The result is that electron donor groups (e.g., aromatic amines, phenolic hydroxyls) can become free radicals (Figure 6.4). Numerous compounds undergo this cooxidation process, with subsequent dehydrogenation, demethylation, epoxidation, sulfoxidation, *N*-oxidation, *S*-oxidation, or dioxygenation (Anders, 1989; Eling et al., 1983). Reactive metabolites of xenobiotics that are formed by cooxidation can covalently bind to tissue macromolecules (e.g., proteins, enzymes, DNA) to alter function or induce cell death.

The list of nephrotoxicants and urinary tract carcinogens that can be substrates for cooxidation is long and includes: APAP, 4AP, benzidine, several nitrofurans, phenetidine, and diethylstilbesterol (Anders, 1989: Harvison et al., 1988; Larsson et al., 1985; Lash, 1994b; Newton et al., 1985; Ross et al., 1985; Zenser et al., 1979; 1983). Of particular interest is the nephrotoxicity induced by APAP. Acute overdose with APAP leads to proximal tubular necrosis, and the role of P450 in forming NAPQI has been established as described above. However, in chronic APAP nephrotoxicity, APAP accumulates in the inner medulla and nephrotoxicity is characterized by interstitial nephritis and papillary necrosis. In this portion of the kidney, little or no P450 is present, but PHS activity is high. Therefore, in chronic APAP nephrotoxicity, PHS is the primary bioactivator of APAP, rather than P450.

Reductases

The kidney also contains several enzymes capable of catalyzing reduction reactions, including P450 enzymes that catalyze reductive dehalogenation of haloalkanes, aldehyde, and ketone reductases, and enzymes that catalyze glutathione-dependent reduction of α-haloketones (Anders, 1980). There are few examples of renal reduction reactions resulting in bioactivation of a nephrotoxicant. However, several 5-nitrofurans are nephrotoxic and induce renal and bladder cancer.

Reduction of the 5-nitro group in these compounds to reactive chemical species (e.g., free radicals, nitroso groups) by renal enzymes with nitroreductase activity (e.g., xanthine oxidase, NADPH-cytochrome c reductase, aldehyde oxidase) and redox cycling are believed to be at least part of the mechanism of nephrotoxicity induced by the 5-nitrofurans (Anders, 1989; Commandeur and Vermeulen, 1990; Zenser et al., 1981).

Hydrolases

Carboxylesterase (A- and B-esterase) activity is found in the microsomal fraction of kidney homogenates (Pond et al., 1995; Yan et al., 1994). Carboxylesterase catalyzes the hydrolysis of carboxylic acid esters to carboxylic acids and alcohols or thiols. Such reactions appear to be detoxifying rather than bioactivating in liver, although few studies have been conducted with renal enzymes (Lohr et al., 1998). However, renal amidase activity may contribute to the deacetylation of APAP to form 4AP, which could then be bioactivated by P450 or PHS to PQI (Figure 6.1) to contribute to acute or chronic APAP nephrotoxicity (Anders, 1989; Newton et al., 1985). Epoxide hydrolase (EH) activity has been found in microsomal and cytosolic fractions of renal cortex and medulla from human kidneys and in proximal tubules from rat kidney (Lock and Reed, 1998). EH catalyzes the hydrolysis of epoxides to trans-diols, but there is little evidence that EH contributes to the bioactivation of nephrotoxicants.

PHASE II ENZYMES

UDP-Glucuronosyltransferases

Glucuronidation is one of the most common Phase II biotransformation reactions in mammals and results from the reaction of uridine 5′-diphosphoglucuronic acid (UDPGA) with nucleophilic substrates (e.g., phenols, alcohols, aromatic amines, carboxylic acids). The formation of glucuronide conjugates is catalyzed by over 15 UDP-glucuronosyltransferases (UGTs) in this superfamily of proteins (Fisher et al., 2001; Tukey and Strassburg, 2000). The major organ that catalyzes glucuronidation is the liver, but the kidney does possess significant UGT activity. The level of the individual renal UGT isoform activity varies greatly compared with hepatic activity. Some substrates (e.g., morphine, diflunisal, mycophenolic acid) are glucuronidated by human kidney microsomes at levels comparable with or higher than liver, but overall, renal glucuronidation rates are usually much lower than hepatic glucuronidation rates (Fisher et al., 2001; Mulder et al., 1990).

Glucuronidation is primarily a detoxification pathway. However, the formation of acyl glucuronides from nonsteroidal anti-inflammatory drugs (NSAIDs) and other xenobiotics converts the carboxylic acid group in a parent compound into reactive metabolites (Spahn-Langguth and Benet, 1992; Spahn-Langguth et al., 1996). Several studies have suggested that acyl glucuronide metabolites may contribute to the observed toxicity of carboxylic acid drugs; drugs that form acyl glucuronides covalently bind to plasma and organ (including kidney) proteins (Bolze et al., 2002; King and Dickinson, 1993: Smith and Liu, 1993, 1995; Spahn-Langguth and Benet, 1992; Spahn-Langguth et al., 1996). Therefore, acyl glucuronide metabolites may contribute to the acute or chronic nephrotoxicity associated with drugs such as clofibrate, suprofen, diclofenac, and other NSAIDs. However, it remains to be determined whether the origin of the acyl glucuronides that interact with renal proteins are produced renally or extrarenally.

Sulfotransferases

Sulfate conjugates are formed from the reaction of acceptor groups (e.g., alcohols, phenols) with 3′-phosphoadenosine 5′-phosphosulfate (PAPS) in the presence of a sulfotransferase (SULT). Over 40 cytosolic SULTs have been identified in mammals, with ten SULT genes identified in humans (Nagata and Yamazoe, 2000). Two primary SULT families exist: SULT1, which is a male-predominant phenol sulfotransferase family, and SULT2, which is sometimes referred to as the hydroxysteroid SULT family and is predominant in females.

The kidney possesses the ability to form sulfate conjugates of phenols and, to a lesser degree, alcohols (Aldred et al., 2000; Anders, 1989; Dunn and Klaassen, 1998; Runge-Morris, 1994). Although hepatic SULT plays a role in the bioactivation of various hepatocarcinogens (Kato and Yamazoe, 1994; Mulder and Jakoby, 1990), there is little evidence that renal sulfation of xenobiotics contributes to the formation of nephrotoxicant sulfate conjugates. For example, while sulfation plays a key role in the bioactivation of the nephrotoxicant *N*-(3,5-dichlorophenyl)-2-hydroxysuccinimide (NDHS), the sulfation of NDHS does not appears to occur in the kidney. This conclusion was based on the findings that *in vivo* administered sodium sulfate potentiates

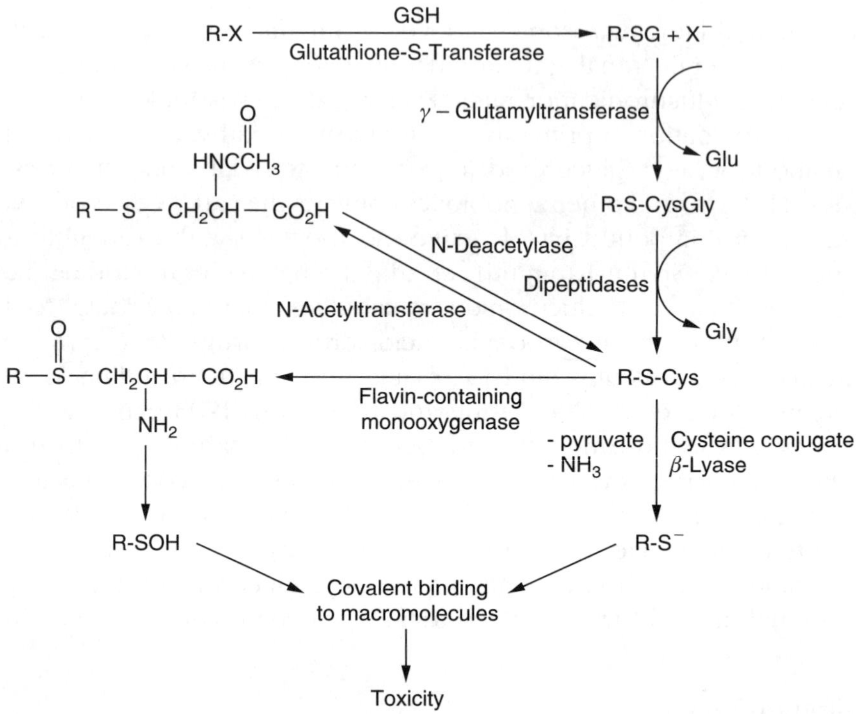

Figure 6.5 Formation of glutathione *S*-conjugates and biotransformation pathways to *N*-acetylcysteine conjugates, *S*-oxides, and reactive thiols. (R–SG = glutathione *S*-conjugate; R–S–CysGly = cysteinylglycine *S*-conjugate; R–S–Cys = cysteine *S*-conjugate; R–S = reactive thiol; R-SOH = sulfenic acid metabolite.)

NDHS nephrotoxicity, and *in vitro* the sulfate conjugate of the alcohol group of NDHS, but not unconjugated NDHS, is a potent nephrotoxicant (Aleo et al., 1991; Hong et al., 1999; Rankin et al., 2001). Therefore, while sulfation appears to play a role in the bioactivation of nephrotoxicants, the sulfation occurs extrarenally, at least for NDHS.

Glutathione-*S*-Transferases and Related Biotransforming Enzymes

The conjugation of electrophilic sites in xenobiotics or their metabolites with the tripeptide L-γ-glutamyl-L-cysteinylglycine (GSH) is catalyzed by one or more of the many (20+) glutathione *S*-transferases (GSTs) found in mammals to form a thiol ether bond between glutathione and the substrate (Figure 6.5) (Anders, 1980; Lohr et al., 1998). The liver has much more GST activity than kidney, but kidney is capable of significant GST activity.

Human kidney contains members of the α, μ, and π classes of GST, with the α-class GST comprising the majority of GST activity (Singh et al., 1987). The highest GST activity in kidney is located in the renal cortex, with proximal tubular cells generally having the most GST activity (Anders, 1989; Lohr et al., 1998). However, class distribution differences occur; α-GSTs are found predominantly in proximal tubules, while π class GSTs are found predominately in distal tubules. The inducibility of GST activity in liver by a number of xenobiotics has been well documented, but kidney GST induction has been less studied. Nonetheless, different patterns of GST induction have been observed, as phenobarbital and 3-methylcholanthrene induce hepatic but not renal GST activity (Derbel et al., 1993).

Substrates for glutathione conjugation vary widely, and include epoxides, haloalkanes, haloalkenes, haloaryls, and α,β-unsaturated carbonyl containing compounds (e.g., acrolein, ethacrynic acid). Glutathione *S*-conjugates can be further degraded to *N*-acetylcysteine (mercapturate) metabolites (Figure 6.5). Initially, glutamic acid is removed by γ-glutamyltransferase to form the cysteinylglycine conjugate. Dipeptidases (e.g., cysteinylglycine dipeptidase) remove the glycine moiety to produce the cysteine *S*-conjugate. Cysteine *S*-conjugates are then *N*-acetylated to form the mercapturate metabolite. Although the kidney is capable of forming and biotransforming glutathione *S*-conjugates to reactive species, there are only a few examples of where initial renal glutathione conjugation results in formation of the nephrotoxic metabolites (e.g., 1,2-dibromoethane and hexafluoropropene; Commandeur and Vermeulen, 1990; Koob and Dekant, 1990). Hepatic glutathione conjugation thus appears to be responsible for the initial glutathione conjugation step for many compounds, while further hepatic, intestinal, and renal biotransformation of the glutathione *S*-conjugates contribute to the relative concentrations of glutathione-derived metabolites to which the kidney is exposed. As a result, the kidney may be exposed *in vivo* to glutathione, cysteine, and *N*-acetylcysteine conjugates following renal or extrarenal biotransformation of xenobiotics and their glutathione *S*-conjugates.

Although many different types of substrates can form glutathione *S*-conjugates, nephrotoxic conjugates are primarily formed from haloalkanes and haloalkenes. The nature of the reactive metabolite and mechanism of nephrotoxicity, however, can vary depending on the nature of the xenobiotic (Anders, 1989; Lash, 1994b; Rankin, 1998). Four basic pathways have been identified for the formation of reactive metabolites from glutathione *S*-conjugates of haloalkanes and haloalkenes: intramolecular cyclization, renal accumulation of reactive quinones, cysteine conjugate β-lyase activation, and *S*-oxidation.

Intramolecular Cyclization

The 1,2-dihaloethanes (XCH_2CH_2X; X = Cl, Br) can induce nephrotoxicity characterized as proximal tubular necrosis, primarily in the juxtaglomerular regions of the kidney. Initial bioactivation of the 1,2-dihaloethanes involves conjugation with glutathione (Figure 6.6). While both 1,2-dichloroethane and 1,2-dibromoethane may be conjugated with glutathione in liver, only 1,2-dibromoethane forms a glutathione *S*-conjugate in kidney (Commandeur and Vermeulen, 1990). The glutathione *S*-conjugate of 1,2-dibromoethane is unstable and quickly cyclizes to form the episulfonium ion, a reactive metabolite that

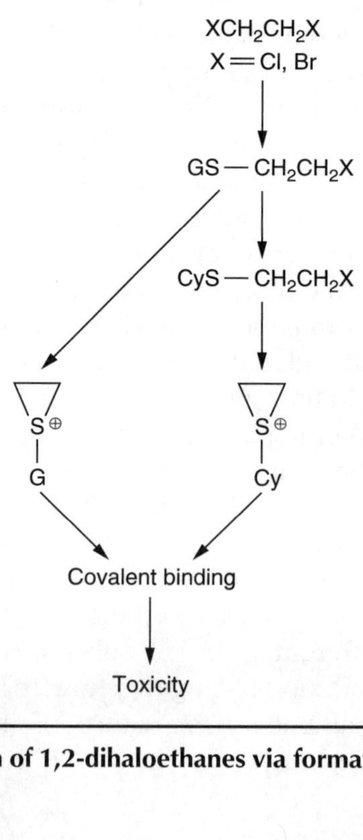

Figure 6.6 Bioactivation of 1,2-dihaloethanes via formation of episulfonium ions.

covalently binds to renal cellular macromolecules, including DNA, to induce nephrotoxicity (Figure 6.6).

It is likely that because of the instability of the glutathione *S*-conjugate, only glutathione *S*-conjugates of 1,2-dibromoethane formed in kidney contribute to the nephrotoxicity induced by this haloalkane. The glutathione *S*-conjugate of 1,2-dichloroethane formed in liver is more stable than the corresponding conjugate of 1,2-dibromoethane, and can be transported to kidney as the glutathione *S*-conjugate or related metabolite. Both the glutathione and cysteine *S*-conjugates of 1,2-dichloroethane appear to contribute to 1,2-dichloroethane nephrotoxicity through the formation of reactive episulfonium ions (Figure 6.6) (Commandeur and Vermeulen, 1990; Foureman and Reed, 1987). Although the cysteine conjugate reacts more readily with DNA, the relative contribution of the two conjugates to the carcinogenic mechanism for 1,2-dichloroethane has yet to be determined.

Cysteine Conjugate β-Lyase

The major mechanism for the bioactivation of nephrotoxic or nephrocacinogenic glutathione *S*-conjugates of haloalkenes involves the biotransformation of the cysteine *S*-conjugate by cysteine conjugate *β*-lyase to form pyruvate, ammonia, and a reactive thiol species (Lash 1994b; Rankin, 1998). Acute nephrotoxicity induced by these conjugates is characterized by proximal tubular necrosis, with the initial renal lesion appearing in the S_3 segment in most rodent models and in the S_1 and S_2 segments in dogs. Additionally, several factors contribute to susceptibility to the nephrotoxicity induced by these conjugates, including age, gender, and strain within a species.

The kidney can concentrate glutathione-derived conjugates via several mechanisms (Rankin, 1998). Glutathione *S*-conjugates can be processed at the luminal membrane of proximal tubular cells by *γ*-glutamyltransferase and dipeptidases to release cysteine *S*-conjugates, which are usually accumulated in the cytoplasm. Glutathione *S*-conjugates can also be actively transported into proximal tubular cells at the basolateral membrane by a sodium-dependent uptake mechanism and cleaved by *γ*-glutamyl cyclotransferase and peptidases to release cysteine *S*-conjugates. Cysteine *S*-conjugates can accumulate in proximal tubular cells via luminal amino-acid transporters, while *N*-acetylcysteine conjugates (mercapturates) can accumulate in proximal tubular cells via transport by basolateral organic anion transporters.

Mercapturates can then be deacetylated by amidases to release the cysteine *S*-conjugates. Therefore, even though the kidney may be exposed to varying amounts of the glutathione-derived conjugates, ultimately these conjugates can accumulate in kidney and be available as the cysteine *S*-conjugate of the halocarbon.

A primary target for these nephrotoxic cysteine conjugates is the mitochondrion, which contains a portion of the cellular cysteine conjugate β-lyase enzymes localized to the outer membrane. The mechanism of the bioactivation of cysteine conjugates of haloalkenes by β-lyase has been well described (Anders et al., 1988; Dekant et al., 1988; 1994; Lash et al., 1988) (Figure 6.5 and Figure 6.7). The ultimate nephrotoxic species following β-lyase biotransformation depends on the nature of the parent haloalkene. When the parent compound is a geminal difluoroalkene (e.g., tetrafluoroethlyene), the glutathione and cysteine *S*-conjugates are alkyl conjugates, while if the parent compound is a geminal dichloroalkene (e.g., trichloroethylene), the conjugates are alkenyl conjugates. Alkyl cysteine *S*-conjugates are bioactivated by β-lyase to a thionoacyl fluoride, which can rapidly acylate renal macromolecules to induce nephrotoxicity, while alkenyl cysteine *S*-conjugates form highly reactive thioketene metabolites following biotransformation by β-lyase (Figure 6.7). The chemical nature of the reactive metabolite also contributes to the observed nephrotoxic response as cysteine conjugates of chloroalkenes are both nephrotoxic and mutagenic, while the cysteine conjugates of fluoroalkanes are only acutely nephrotoxic (Green and Odum, 1985; Vamvakas et al., 1989).

Facilitated Accumulation

Bromobenzene induces nephrotoxicity characterized by proximal tubular necrosis, with the greatest necrosis observed in the S_3 segment. The mechanism for bioactivation of bromobenzene to nephrotoxic metabolites begins with hepatic P450-mediated oxidation of bromobenzene to 2-bromophenol and then to 2-bromohydroquinone (Figure 6.8). Renal quinol oxidase can activate 2-bromohydroquinone to 2-bromoquinone, an arylating chemical species. While 2-bromoquinone is capable of redox cycling and causing oxidative stress, arylation appears to be more important for cellular toxicity (Lash, 1994b).

Studies by Monks et al. (1985) demonstrated a role for glutathione *S*-conjugates of 2-bromohydroquinone in bromobenzene nephrotoxicity.

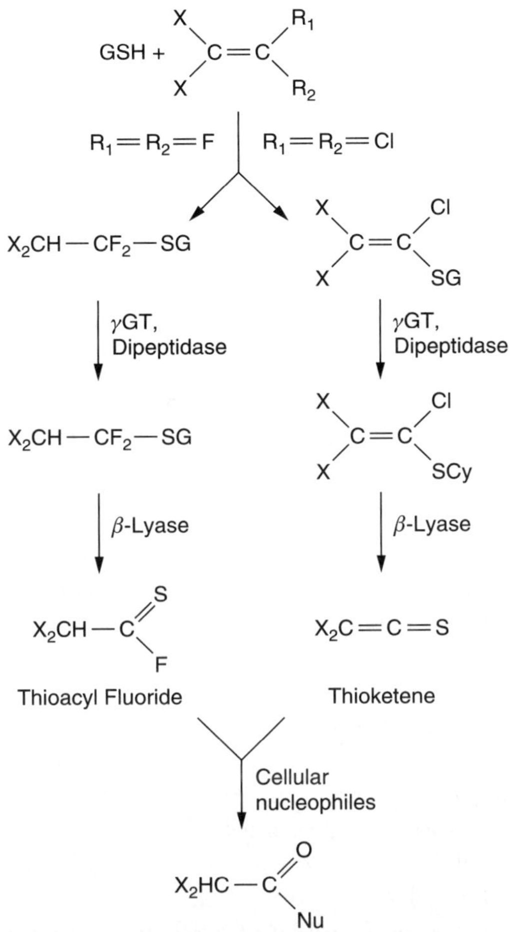

Figure 6.7 Formation of glutathione *S*-conjugates from haloalkenes and bioactivation by renal mitochondrial cysteine conjugate β-lyase. (γGT = γ-glutamyltransferase.)

Monoglutathionyl and diglutathionyl conjugates of 2-bromohydroquinone are formed *in vivo* (Figure 6.8) and diglutathionyl conjugates of 2-bromohydroquinone induce nephrotoxicity identical to bromobenzene and 2-bromohydroquinone nephrotoxicity, but at a dose that is 10 to 15 times lower than with 2-bromohydroquinone (Lau and Monks, 1993). Further *in vivo* studies have demonstrated that inhibition of renal γ-glutamyltransferase by acivicin, but not inhibition of cysteine conjugate β-lyase by aminooxyacetic acid, attenuated the nephrotoxicity

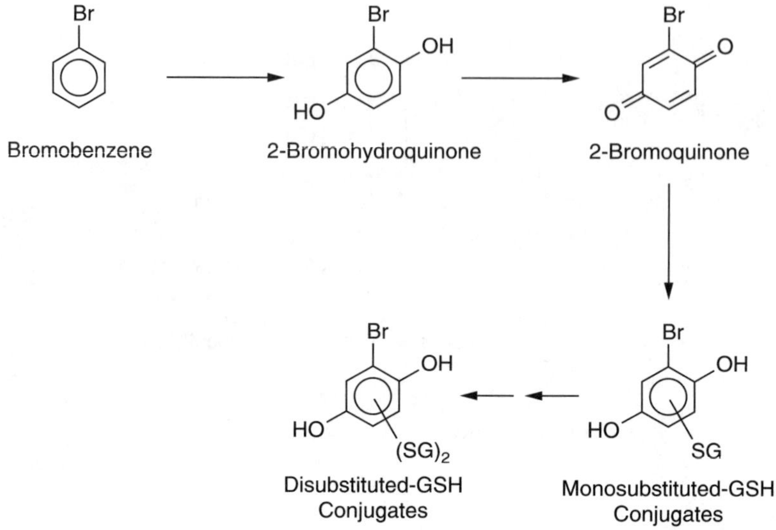

Figure 6.8 Bioactivation of bromobenzene to nephrotoxic metabolites via oxidation and glutathione conjugation.

induced by the diglutathionyl conjugate of 2-bromohydroquinone and provided partial protection against 2-bromohydroquinone-induced nephrotoxicity (Monks et al., 1985; Lau and Monks, 1993). Glutathione conjugation of 2-bromohydroquinone at extrarenal sites can thus contribute to the transport and selective accumulation in proximal tubular cells of nephrotoxic 2-bromohydroquinone metabolites. The ultimate mechanism of toxicity is related to the ability of the accumulated cysteine conjugates of 2-bromohydroquinone to induce oxidative stress.

A similar mechanism of transport and selective renal accumulation has been proposed for at least part of the mechanism for the acute nephrotoxicity induced by APAP and 4AP (Fowler et al., 1991; Klos et al., 1992; Trumper et al., 1996).

S-Oxidation

A number of cysteine *S*-conjugates are oxidized to the corresponding sulfoxide metabolites by flavin-containing monooxygenases in renal and hepatic microsomes, as discussed above. The sulfoxide metabolites may nonenzymatically release a reactive sulfenic acid metabolite (Figure 6.5) or serve as a Michael acceptor for reaction with thiol groups of renal macromolecules (Sausen and Elfarra, 1991). There may thus be more than one mechanism for induction of nephrotoxicity from the cysteine *S*-conjugate *S*-oxide metabolites. However, further work is needed to determine the relative contributions of the two pathways to cysteine *S*-conjugate nephrotoxicity.

Other Phase II enzymes

The kidney also possesses the ability to catalyze *N*-, *O*-, and *S*-methylation, glycine and glutamine conjugation with carboxylic acid or alcohol groups, and acetylation of aromatic amines (Lohr et al., 1998). These reactions are primarily detoxification reactions and generally not involved in the direct bioactivation of nephrotoxicants.

REFERENCES

Aldred S, Foster JJR, Lock EA, and Waring RH (2000) Investigation of the localization of dehydroepiandrosterone sulphotransferase in adult rat kidney. *Nephron* 86:176–182.

Aleo MD, Rankin GO, Cross TJ, and Schnellmann RG (1991) Toxicity of *N*-(3,5-dichlorophenyl)succinimide and metabolites to rat renal proximal tubules and mitochondria. *Chem-Biol Interact* 78:109–121.

Anders MW (1980) Metabolism of drugs by the kidney. *Kidney Int* 18:636–647.

Anders MW (Ed.) (1985) *Bioactivation of Foreign Compounds*. New York: Academic Press.

Anders MW (1989) Biotransformation and bioactivation of xenobiotics by the kidney, in *Intermediary Xenobiotic Metabolism in Animals: Methodology, Mechanisms, and Significance*. Hutson DH, Caldwell J, and Paulson GD (Eds), London: Taylor & Francis, pp. 81–97.

Anders MW, Lash LH, Dekant W, Elfarra AA, and Dohn DR (1988) Biosynthesis and biotransformation of glutathione S-conjugates to toxic metabolites. *CRC Crit Rev Toxicol* 18:311–341.

Armbrecth HJ, Birnbaum LS, Zenser TV, Mattamal MB, and Davis BB (1979) Renal cytochrome P450s – Electrophoretic and electroparamagnetic resonance studies. *Arch Biochem Biophys* 197:277–284.

Baron J, Kawabata TT, Redick JA, Knapp SA, Wick DG, Wallace RB, Jakoby WB, and Guengerich FP (1983) Localization of carcinogen-metabolizing enzymes in human and animal tissues, in *Extrahepatic Drug Metabolism and Chemical Carcinogenesis*. Rydstrom J, Montelius J, and Bengtsson M (Eds), Amsterdam: Elsevier, pp. 73–88.

Bendayan R (1996) Renal drug transport: A review. *Pharmacotherapy* 16:971–985.

Bolze S, Bromet N, Gay-Feutry C, Massiere F, Boulieu R, and Hulot T (2002) Development of an *in vitro* screening model for the biosynthesis of acyl glucuronide metabolites and the assessment of their reactivity toward human serum albumin. *Drug Metab Dispos* 30:404–413.

Cojocel C, Hannemann J, and Baumann K (1985) Cephaloridine-induced lipid peroxidation initiated by a reactive oxygen species as a possible mechanism of cephaloridine nephrotoxicity. *Biochim Biophys Acta* 834:402–410.

Commandeur JNM, and Vermeulen NPE (1990) Molecular and biochemical mechanisms of chemically induced nephrotoxicity: A review. *Chem Res Toxicol* 3:171–194.

Crowe CA, Young AC, Calder IC, Ham KN, and Tange JD (1979) The nephrotoxicity of p-aminophenol. I. The effect on microsomal enzymes, glutathione, and covalent binding in kidney and liver. *Chem-Biol Interact* 27:235–243.

Cummings BS, Zangar RC, Novak RF, and Lash LH (1999) Cellular distribution of cytochromes P-450 in the rat kidney. *Drug Metab Dispos* 27:542–548.

Dekant W, Anders MW, and Monks TJ (1993) Bioactivation of halogenated xenobiotics by *S*-conjugate formation, in Renal Disposition and Nephrotoxicity of Xenobiotics. Anders MW, Dekant W, Hensler D, Oberleither H, and Silbernagl S, eds. (Eds),

Dekant W, Frischmann C, and Speerschneider P (1995) Sex, organ and species specific bioactivation of chloromethane by cytochrome P4502E1. *Xenobiotica* 25:1259–1265.

Dekant W, Lash LH, and Anders MW (1988) Fate of glutathione conjugates and bioactivation of cysteine *S*-conjugates by cysteine conjugate β-lyase, in *Glutathione Conjugation: Mechanisms and Biological Significance.* Sies H, and Ketterer B (Eds), London: Academic Press, pp. 415–447.

Dekant W, Vamvakas S, and Anders MW (1994) Formation and fate of nephrotoxic and cytotoxic glutathione *S*-conjugates: cysteine conjugate β-lyase pathway, in *Conjugation-Dependent Carcinogenicity and Toxicity of Foreign Compounds.* Advances in Pharmacology, Vol. 27. Anders MW, and Dekant W (Eds), New York: Academic Press, pp. 115–162.

Derbel M, Igarashi T, and Satoh T (1993) Differential induction of glutathione-*S*-transferase subunits by phenobarbital, 3-methylcholanthrene and ethoxyquin in rat liver and kidney. *Biochim Biophys Acta* 1158:175–180.

Dunn RT II, and Klaassen CD (1998) Tissue-specific expression of rat sulfotransferase messenger RNAs. *Drug Metab Dispos* 26:598–604.

Eling T, Boyd J, Reed G, Mason R, and Sivarajah K (1983) Xenobiotic metabolism by prostaglandin endoperoxide synthase. *Drug Metab Rev* 14:1023–1053.

Endou H (1983) Distribution and some characteristics of cytochrome P-450 in the kidney. *J Toxicol Sci* 8:165–176.

Fisher MB, Paine MF, Strelevitz TJ, and Wrighton SA (2001) The role of hepatic and extrahepatic UDP-glucuronosyltransferases in human drug metabolism. *Drug Metab Rev* 33:273–297.

Foureman GL, and Reed DJ (1987) Formation of *S*-[2-(N7-guanyl)ethyl] adducts by the postulated *S*-(2-chloroethyl)cysteinyl and *S*-(2-chloroethyl)glutathionyl conjugates of 1,2-dichloroethane. *Biochemistry* 26:2028–2033.

Fowler LM, Moore RB, Foster JR, and Lock EA (1991) Nephrotoxicity of 4-aminophenol glutathione conjugate. *Human Exp Toxicol* 10:451–459.

Green T, and Odum J (1985) Structure/activity studies of the nephrotoxic and mutagenic action of cysteine conjugates of chloro- and fluoroalkenes. *Chem-Biol Interact* 54:15–31.

Guengerich FP, Wang P, and Davidson NK (1982) Estimation of isozymes of microsomal cytochrome P-450 in rats, rabbits, and human using immunochemical staining coupled with sodium dodecyl sulfate-polyacrylamide gel electrophoresis. *Biochemistry* 21:1698–1706.

Harvison PJ, Egan RW, Gale PH, Christian GD, Hill BS, and Nelson SD (1988) Acetaminophen and analogs as cosubstrates and inhibitors of prostaglandin H synthase. *Chem-Biol Interact* 64:251–266.

Hines RN, Cashman JR, Philpot RM, Williams DE, and Zeigler DM (1994) The mammalian flavin-containing monooxygenases: Molecular characterization and regulation of expression. *Toxicol Appl Pharmacol* 125:1–6.

Hong SK, Anestis DK, Ball JG, Valentovic MA, Brown PI, and Rankin GO (1999) Sodium sulfate potentiates *N*-(3,5-dichlorophenyl)-2-hydroxysuccinimide (NDHS) and *N*-(3,5-dichlorophenyl)-2-hydroxysuccinamic acid (2-NDHSA) nephrotoxicity in the Fischer 344 rat. *Toxicology* 138:165–174.

Ito O, Alonso-Galicia M, Hopp KA, and Roman RJ (1998) Localization of cytochrome P-450 4A isoforms along the rat nephron. *Am J Physiol* 274:F395–F404.

Jones DP, Orrenius S, and Jakobson SW (1980) Cytochrome P-450-linked monooxygenase systems in the kidney, in *Extrahepatic Metabolism of Drugs and Other Foreign Compounds*. Gram TE (Ed.), New York: Spectrum Publications, pp. 123–158.

Kaloyanides GJ (1991) Metabolic interations between drugs and renal tubulointerstitial cells: role in nephrotoxicity. *Kidney Int* 39:531–540.

Kato R, and Yamazoe Y (1994) Metabolic activation of *N*-hydroxylated metabolites of carcinogenic and mutagenic arylamines and arylamides by esterification. *Drug Metab Rev* 26:413–430.

King AR, and Dickinson RG (1993) Studies on the reactivity of acyl glucuronides – IV. Covalent binding of diflunisal to tissues of the rat. *Biochem Pharmacol* 45:1043–1047.

Klos C, Koob M, Kramer C, and Dekant W (1992) p-Aminophenol nephrotoxicity: biosynthesis of toxic glutathione conjugates. *Toxicol Appl Pharmacol* 115:98–116.

Koukouritaki SB, Simpson P, Yeung CK, Rettie AE, and Hines RN (2002) Human hepatic flavin-containing monooxygenase 1 (FMO1) and 3 (FMO3) developmental expression. *Ped Res* 51:236–243.

Koob M, and Dekant W (1990) Metabolism of hexafluoropropene: Evidence for bioactivation by glutathione conjugate formation in the kidney. *Drug Metab Dispos* 18:911–916.

Krause RJ, Glocke SC, and Elfarra AA (2002) Sulfoxides as urinary metabolites of *S*-allyl-L-cysteine in rats: Evidence for the involvement of flavin-containing monooxygenases. *Drug Metab Dispos* 30:1137–1142.

Krause RJ, Lash LH, and Elfarra AA (2003) Human kidney flavin-containing monooxygenases and their potential roles in cysteine *S*-conjugate metabolism and nephrotoxicity. *J Pharmacol Exp Ther* 304:185–191.

Kuo C-H, Maita K, Sleight SD, and Hook JB (1983) Lipid peroxidation: A possible mechanism of cephaloridine-induced nephrotoxicity. *Toxicol Appl Pharmacol* 67:78–88.

Larsson R, Ross D, Berlin T, Olsson LI, and Moldéus P (1985) Prostaglandin synthase catalyzed metabolic activation of p-phenetidine and acetaminophen by microsomes isolated from rabbit and human kidney. *J Pharmacol Exp Ther* 235:475–480.

Lash LH (1994a) Role of renal metabolism in risk to toxic chemicals. *Environ Health Perspect* 102 (Suppl 11):75–79.

Lash LH (1994b) Role of metabolism in chemically induced nephrotoxicity, in *Mechanisms of Injury in Renal Disease and Toxicity*, Chapter 9, Goldstein RS (Ed.), Boca Raton: CRC Press, pp. 207–234.

Lash LH, and Tokarz JJ (1995) Oxidative stress and cytotoxicity of 4-(2-thienyl)butyric acid in isolated rat renal proximal tubular and distal tubular cells. *Toxicology* 103:167–175.

Lash LH, Anders MW, and Jones DP (1988) Glutathione homeostasis and glutathione *S*-conjugate toxicity in the kidney. *Rev Biochem Toxicol* 9:29–67.

Lash LH, Elfarra AA, and Anders MW (1986) Renal cysteine conjugate β-lyase: Bioactivation of nephrotoxic cysteine *S*-conjugates in mitochondrial outer membrane. *J Biol Chem* 261:5930–5935.

Lash LH, Hueni SE, and Putt DA (2001) Apoptosis, necrosis, and cell proliferation induced by *S*-(1,2-dichlorovinyl)- L-cysteine in primary cultures of human proximal tubular cells. *Toxicol Appl Pharmacol* 177:1–16.

Lash LH, Sausen PJ, Duescher RJ, Cooley AJ, and Elfarra AA (1994) Roles of cysteine conjugate β-lyase and *S*-oxidase in nephrotoxicity: studies with *S*-(1,2-dichlorovinyl)-L-cysteine and *S*-(1,2-dichlorovinyl)-L-cysteine sulfoxide. *J Pharmacol Exp Ther* 269:374–383.

Lau SS, and Monks TJ (1993) Nephrotoxicity of bromobenzene: The role of Quinone-thioethers, in *Toxicolgy of the Kidney*, 2nd ed., Hook JB, and Goldstein RS (Eds), New York: Raven Press, pp. 415–436.

Lock EA, and Reed CJ (1998) Xenobiotic metabolizing enzymes of the kidney. *Toxicol Pathol* 26:18–25.

Lohr JW, Willsky GR, and Acara MA (1998) Renal drug metabolism. *Pharmacol Rev* 50:107–141.

Mohandas J, Duggin GG, Horvath JS, and Tiller DJ (1981) Regional differences in peroxidative activation of paracetamol (acetaminophen) mediated by cytochrome P450 and prostaglandin endoperoxide synthetase in rabbit kidney. *Res Commun Chem Path Pharmacol* 34:69–80.

Monks TJ, and Lau SS (1988) Reactive intermediates and their toxicological significance. *Toxicology* 52:1–53.

Monks TJ, Lau SS, Highet RJ, and Gillette JR (1985) Glutathione conjugates of 2-bromohydroquinone are nephrotoxic. *Drug Metab Dispos* 13:553–559.

Moran JH, Mitchell LA, Bradbury JA, Qu W, Zeldin DC, Schnellmann RG, and Grant DF (2000) Analysis of the cytotoxic properties of lineolic acid metabolites produced by renal and hepatic P450s. *Toxicol Appl Pharmacol* 168:268–279.

Mulder GJ (Ed.) (1990) *Conjugation Reactions in Drug Metabolism: An Integrated Approach*. New York: Taylor & Francis.

Mulder GJ, and Jakoby WB (1990) Sulfation, in *Conjugation Reactions in Drug Metabolism: An Integrated Approach*, Mulder GJ (Ed.), New York: Taylor & Francis, pp. 107–161.

Mulder GJ, Coughtrie MWH, and Burchell B (1990) Glucuronidation, in *Conjugation Reactions in Drug Metabolism: An Integrated Approach*, Mulder GJ (Ed.), New York: Taylor & Francis, pp. 51–105.

Nagata K, and Yamazoe Y (2000) Pharmacogenetics of sulfotransferase. *Annu Rev Pharmacol Toxicol* 40:159–176.

Needleman P, Turk J, Kakschik B, Morrison A, and Lefkowith JB (1986) Arachidonic acid metabolism. *Annu Rev Biochem* 55:69–102.

Newton JF, Bailie MB, and Hook JB (1983) Acetaminophen nephrotoxicity in the rat. Renal metabolic activation *in vitro. Toxicol Appl Pharmacol* 70:433–444.

Newton JF, Braselton WE Jr, Kuo C-H, Kluwe WM, Gemboys MW, Mudge GH, and Hook JB (1982) Metabolism of acetaminophen by the isolated perfused kidney. *J Pharmacol Exp Ther* 221:76–79.

Newton JF, Kuo C-H, DeShone GM, Hoefle D, Bernstein J, and Hook JB (1985) The role of *p*-aminophenol in acetaminophen-induced nephrotoxicity: Effect of bis(*p*-nitrophenyl)phosphate on acetaminophen and *p*-aminophenol nephrotoxicity and metabolism in Fischer 344 rats. *Toxicol Appl Pharmacol* 81:416–430.

Ortiz de Montellano PR (Ed.) (1986) *Cytochrome P-450.* New York: Plenum Press.

Pfohl-Leszkowicz A, Pinelli E, Bartsch H, Mohr U, and Castegnaro M (1998) Sex- and strain-specific expression of cytochrome P450s in ochratoxin A-induced genotoxicity and carcinogenicity in rats. *Mol Carcinog* 23:76–85.

Pohl LR, Bhooshan B, Whittaker NF, and Krishna G (1977) Phosgene: a metabolite of chloroform. *Biochem Biophys Res Commun* 79:684–691.

Pond AL, Chambers HW, and Chambers JE (1995) Organophosphate detoxification potential of various rat tissues via A-esterase and aliesterase activities. *Toxicol Lett* 78:361–369.

Pritchard JB, and Miller DS (1993) Mechanisms mediating renal secretion of organic anions and cations. *Physiol Rev* 73:765–796.

Rankin GO (1998) Kidney, in *Encyclopedia of Toxicology*, vol. 2, Wexler P (Ed.), New York: Academic Press, pp. 198–225.

Rankin GO, Hong SK, Anestis DK, Lash LH, and Miles SL (2001) In vitro nephrotoxicity induced by *N*-(3,5-dichlorophenyl)succinimide (NDPS) metabolites in isolated renal cortical cells from male and female Fischer 344 rats. Evidence for a nephrotoxic sulfate conjugate metabolite. *Toxicology* 163:73–82.

Ripp SL, Itagaki K, Philpot RM, and Elfarra AA (1999) Species and sex differences in expression of flavin-containg monooxygenase form 3 in liver and kidney microsomes. *Drug Metab Dispos* 27:46–52.

Ripp SL, Overby LH, Philpot RM, and Elfarra AA (1997) Oxidation of cysteine *S*-conjugates by rabbit liver microsomes and cDNA-expressed flavin-containing monooxygenases: studies with *S*-(1,2-dichlorovinyl)-L-cysteine, *S*-(1,2,2-trichlorovinyl)-L-cysteine, *S*-allyl-L-cysteine and *S*-benzyl-L-cysteine. *Mol Pharmacol* 51:507–515.

Ross D, Larsson R, Norbeck K, Ryhage R, and Moldéus P (1985) Characterization and mechanism of formation of reactive products formed during peroxidase-catalyzed oxidation of *p*-phenetidine: Trapping of reactive species by reduced glutathione and butylated hydroxyanisole. *Mol Pharmacol* 27:277–286.

Runge-Morris MA (1994) Sulfotransferase gene expression in rat hepatic and extrahepatic tissues. *Chem-Biol Interact* 92:67–76.

Rush GF, Smith JH, Newton JF, and Hook JB (1984) Chemically induced nephrotoxicity: Role of metabolic activation. *CRC Crit Rev Toxicol* 13:99–160.

Sausen PJ, and Elfarra AA (1991) Reactivity of cysteine *S*-conjugate sulfoxides: formation of *S*-[1-chloro-2-(*S*-glutathionyl)vinyl]-L-cysteine sulfoxide by the reaction of *S*-(1,2-dichlorovinyl)-L-cysteine sulfoxide with glutathione. *Chem Res Toxicol* 4:655–660.

Singh SV, Leal T, Ansari GAS, and Awashthi YC (1987) Purification and characterization of glutathione-*S*-transferase of human kidney. *Biochem J* 246:179–186.

Smith JH (1986) Role of renal metabolism in chloroform nephrotoxicity. *Comments Toxicol* 1:125–144.

Smith PC, and Liu JH (1993) Covalent binding of suprofen acyl glucuronide to albumin *in vitro*. *Xenobiotica* 23:337–348.

Smith PC, and Liu JH (1995) Covalent binding of suprofen to renal tissue correlates with excretion of its acyl glucuronide. *Xenobiotica* 25:531–540.

Spahn-Langguth H, and Benet LZ (1992) Acyl glucuronides revisited: Is the glucuronide process a toxification as well as a detoxification mechanism? *Drug Metab Rev* 24:5–48.

Spahn-Langguth H, Dahms M, and Hermening A (1996) Acyl glucuronides: Covalent binding and its potential relevance. *Adv Exp Med Biol* 387:313–328.

Speerschneider P, and Dekant W (1995) Renal tumorigenicity of 1,1-dichloroethene in mice: The role of male-specific expression of cytochrome P450 2E1 in the renal bioactivation of 1,1-dichloroethene. *Toxicol Appl Pharmacol* 130:48–56.

Spry LA, Zenser TV, and Davis BB (1986) Bioactivation of xenobiotics by prostaglandin H synthase in the kidney: Implications for therapy. *Comments Toxicol* 1:109–123.

Trumper L, Monasterolo LA, and Elias M (1996) Nephrotoxicity of acetaminophen in male Wistar rats: Role of hepatically derived metabolites. *J Pharmacol Exp Ther* 279:548–554.

Tukey RH, and Strassburg CP (2000) Human UDP-glucuronosyltransferases: metabolism, expression, and disease. *Annu Rev Pharmacol Toxicol* 40:581–616.

Tune BM (1986) The nephrotoxicity of cephalosporin antibiotics – structure-nephrotoxicity relationships. *Comments Toxicol* 1:145–170.

Tune BM, and Fernholt M (1973) Relationship between cephaloridine and p-aminohippurate transport in the kidney. *Am J Physiol* 225:1114–1117.

Tune BM, Sibley RK, and Hsu C-Y (1988) The mitochondrial respiratory toxicity of cephalosporin antibiotics. An inhibitory effect on substrate uptake. *J Pharmacol Exp Ther* 245:1054–1059.

Vamvakas S, Dekant W, and Henschler D (1989) Assessment of unscheduled DNA synthesis in a cultured line of renal epithelial cells exposed to cysteine *S*-conjugates of haloalkenes and haloalkanes. *Mutat Res* 222:329–335.

Yan B, Yang D, Brady M, and Parkinson A (1994) Rat kidney carboxylesterase. Cloning, sequencing, cellular localization, and relationship to rat liver hydrolase. *J Biol Chem* 269:29688–29696.

Yeung CK, Lang DH, Thummel KE, and Rettie AE (2000) Immunolocalization of FMO1 in human liver, kidney, and intestine. *Drug Metab Dispos* 28:1107–1111.

Zeigler DM (1993) Recent studies of the structure and function of multisubstrate flavin-containing monooxygenases. *Annu Rev Pharmacol Toxicol* 33:179–199.

Zenser TV, Cohen SM, Mattammal MB, Wise RW, Rapp NS, and Davis BB (1983) Prostaglandin hydroperoxidase-catalyzed activation of certain *N*-substituted aryl renal and bladder carcinogens. *Environ Health Perspect* 49: 33–41.

Zenser TV, Mattammal MB, and Davis BB (1978) Differential distribution of the mixed-function oxidase activities in rabbit kidney. *J Pharmacol Exp Ther* 207:719–725.

Zenser TV, Mattammal MB, and Davis BB (1979) Cooxidation of benzidine by renal medullary prostaglandin cyclooxygenase. *J Pharmacol Exp Ther* 211:460–464.

Zenser TV, Mattammal MB, Palmier MO, and Davis BB (1981) Microsomal nitroreductase activity in rabbit kidney and bladder: implications in 5-nitrofuran-induced toxicity. *J Pharmacol Exp Ther* 219:735–740.

7

Mechanisms of Renal Cell Death

Jay F. Harriman and Rick G. Schnellmann

INTRODUCTION

Recent advances in the determination of cell death mechanisms have not only clarified how molecular alterations lead to the morphological changes readily observed with oncosis and apoptosis, but have also revealed how these two pathways share common mediators. In many cases, the end stages of cell death are easily measured and observed, but the challenge remains to completely determine and understand the steps between the injurious insult and the end stages of cell death. Because renal dysfunction is a direct downstream result of the pathological events occurring at the cellular and subcellular levels, the mechanisms of cell death must be elucidated to prevent acute renal failure (ARF) and to develop treatments. This chapter describes the distinguishing characteristics and signaling pathways of oncosis and apoptosis as they relate to acute renal cell death.

Oncosis is a rediscovered term for necrotic cell death and is characterized by organelle and cell swelling prior to the gross breakdown of the plasma membrane (Majno and Joris, 1995). The term oncosis (from *onkos*, a Greek word meaning "swelling") was coined by von Recklinghausen (1910) over 90 years ago. Necrosis describes the morphological changes that occur after a cell is dead. These changes include the karyolysis, pyknosis, karyorrhexis, condensation of the cytoplasm and eosinophilia, loss of structure, and fragmentation. Oncosis encompasses the prelethal events leading to the lethal events (including the "point of no return"). The Cell Death Nomenclature Committee of the Society of Toxicologic Pathologists has recommended that the term necrosis be used when dead cells are observed in a histological section, whereas oncosis be used to distinguish death by cell swelling from death by apoptosis (Levin et al., 1999). Oncosis is commonly assessed *in vitro* by measuring the gross breakdown of the plasma membrane. For example, the release of a cytosolic enzyme, such as lactate dehydrogenase, from the cell and uptake of a normally impermeant fluorescent nuclear stain, such as propidium iodide, are two common markers of oncosis. Unlike apoptosis, oncosis may mediate further cell and tissue injury through the release of intracellular contents and the initiation of an inflammatory response. Most toxicants that cause ARF produce oncosis to some extent.

Apoptosis occurs physiologically to balance cell proliferation and regulate the mass of a tissue. Both toxicant and ischemic insults result in apoptosis. Compared with oncosis, apoptosis is advantageous for surrounding healthy tissue because damaged cells are removed through a process that minimizes tissue destruction. ATP-dependent signaling cascades responsive to intracellular and extracellular stimuli regulate the balance of apoptotic inducers and repressors; thus, apoptosis is an active process requiring ATP. These signaling cascades culminate in a decrease in cell volume and the activation of caspases, with the downstream caspases mediating the development of additional morphological markers (e.g., chromatin margination and cell fragmentation into apoptotic bodies) that characterize apoptosis. Apoptosis may be identified experimentally by these morphological markers and by measuring increases in caspase activities.

A unique type of apoptotic cell death experienced by the kidney and other organs is termed "anoikis" (from the ancient Greek word for homelessness) (Frisch and Francis, 1994), which denotes apoptosis that occurs in cells that detach from the extracellular matrix. In the kidney, anoikis occurs as a result of the sloughing of cells from the basement membrane due to normal turnover, renal blockage, or ischemia or toxicant injury. It has been postulated that cells may detach from the basement membrane due to disruption of integrin function and associated protein kinase signaling pathways. For example, integrin activation of focal adhesion kinase can suppress

Table 7.1 Biochemical and Morphological Hallmarks of Apoptosis and Oncosis

Parameter	*Apoptosis*	*Oncosis*
Regulation	Signaled with cellular control	Signaled without cellular control
Cell shape	Shrinkage	Swelling
Membrane	Intact	Increased permeability
Organelles	Largely intact	Swollen
Nucleus	Fragmentation	Karyolysis
Chromatin	Condensed	Compact but heterogeneous
DNA	Oligonucleosomal cleavage	Random cleavage
Energy	ATP dependent	ATP independent

Source: Modified from Dartsch, 1999.

anoikis, while a stress-activated protein kinase promotes anoikis (Frisch and Ruoslahti, 1997).

Cell death is characterized by changes in cell and organelle morphology, chromatin and DNA degradation, protease activation, ATP depletion, and mitochondrial alterations (Table 7.1). The morphological characteristics of oncotic cell death are distinctly different from apoptosis (Figure 7.1). While apoptosis is characterized by cellular shrinkage without overt damage to mitochondria, the endoplasmic reticulum (ER), or the plasma membrane, oncosis is associated with cell swelling, mitochondrial and nuclear swelling, and loss of plasma membrane integrity. Cell swelling occurs rapidly as a result of direct damage to the cell membrane or after inhibition of energy production. The loss of ATP inhibits the Na^+/K^+-ATPase, resulting in increased Na^+ and H_2O influx and K^+ efflux. Oncotic bleb formation occurs as a result of cell swelling and modifications to the cytoskeleton. The blebs in oncotic cells are usually clear and have a band of actin at the base. Budding occurs with apoptosis, with the buds (apoptotic bodies) containing cytoplasm and organelles as opposed to the watery blebs seen in oncosis. In addition, the remnants of the oncotic cell membrane most often remain attached to adjacent cell structures and the basement membrane. The release of oncotic cell remnants into surrounding tissue results in an inflammatory response that exacerbates localized tissue damage.

The kidney is uniquely susceptible to toxicant-induced cell death, either directly by the interaction of the toxicant with the renal cell or

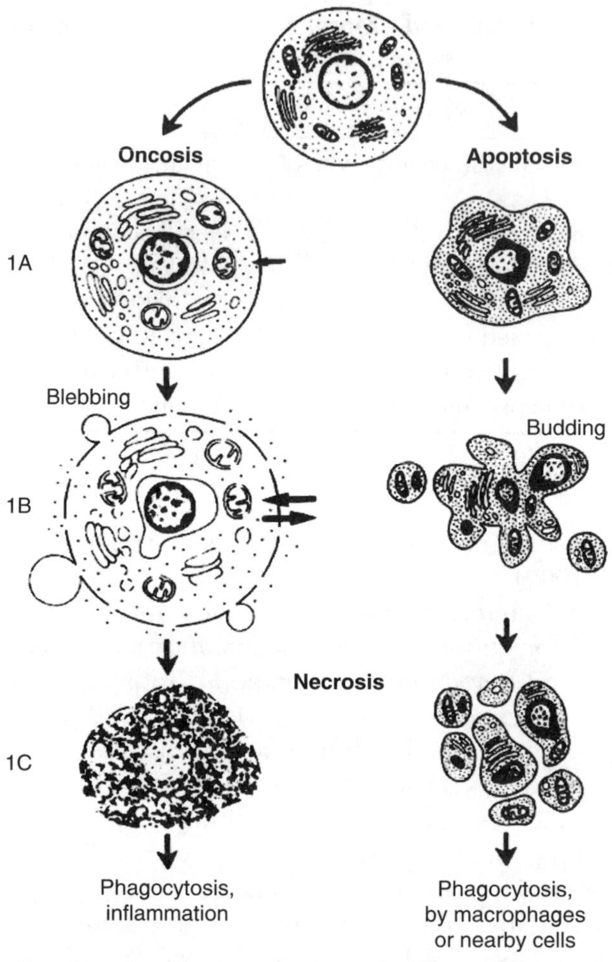

Figure 7.1 Morphological features of oncosis and apoptosis. A normal cell is shown at the top. Oncosis: (1A) swelling, (1B) vacuolization, blebbing, and increased permeability, and (1C) coagulation, shrinking, and karyolysis. Apoptosis: (2A) shrinkage and pyknosis, (2B) budding and karyorrhexis, and (2C) breakup into apoptotic bodies. (From Majno and Joris, 1995, with permission.)

indirectly due to occlusion of the renal artery and the resultant ischemia. In order to maintain its many transport functions, the kidney has an enormous ATP requirement. Consequently, the high ATP demand results in its vulnerability to a variety of toxicants and to anoxic or hypoxic conditions. Because transport functions vary with each segment of the nephron, some segments are more vulnerable to injury. For example, renal proximal tubules (RPT) have a high rate of transport and are more dependent on ATP than distal tubules, making RPT more susceptible to toxicant and ischemic injury.

Understanding the regulation and cellular changes associated with cell death processes in the kidney will enable investigators to both identify and explore toxicant-induced renal pathologies. Furthermore, this understanding will aid in the identification of pharmaceutical agents for the treatment and prevention of renal injury. This chapter discusses the characteristics of oncotic and apoptotic cell death and gives insight into the proposed mechanisms that mediate the cell death processes.

ONCOTIC CELL DEATH

Characteristics of Oncotic Cell Death

Toxicant-induced oncosis affects masses of contiguous cells; it is characterized by organelle and cell swelling, cell rupture and release of intracellular contents, and is followed by an inflammatory response. Morphologically, cell death can be visualized using hematoxylin and eosin-stained sections of paraffin-embedded tissue. Light microscopic analysis of oncotic sections of tissue reveals cellular swelling, membrane disruption, organelle disruption, and local tissue damage due to an immune response. Electron microscopy defines the finer features of cell death, including the integrity of the plasma membrane and intracellular organelles and the state of the nuclear chromatin (Figure 7.2). Watery blebs devoid of organelles also can be seen with microscopic analysis of oncotic cells.

In addition to microscopy, oncosis has been determined by measuring the release of cellular enzymes such as lactate dehydrogenase (LDH), alanine aminotransferase (ALT), or sorbitol dehydrogenase into the blood or media. LDH is a large cytosolic enzyme (140 kDa) released when the cell membrane breaks down. The percentage of LDH activity released into the media is an indication of the percentage of cells that have died. Dye exclusion can also be successfully employed to determine cell death. Propidium iodide (PI) is a small (mol wt = 668) membrane-impermeable dye that fluoresces following its intercalation into the DNA of cells with plasma membrane degradation. PI can be used in conjunction with other dyes that

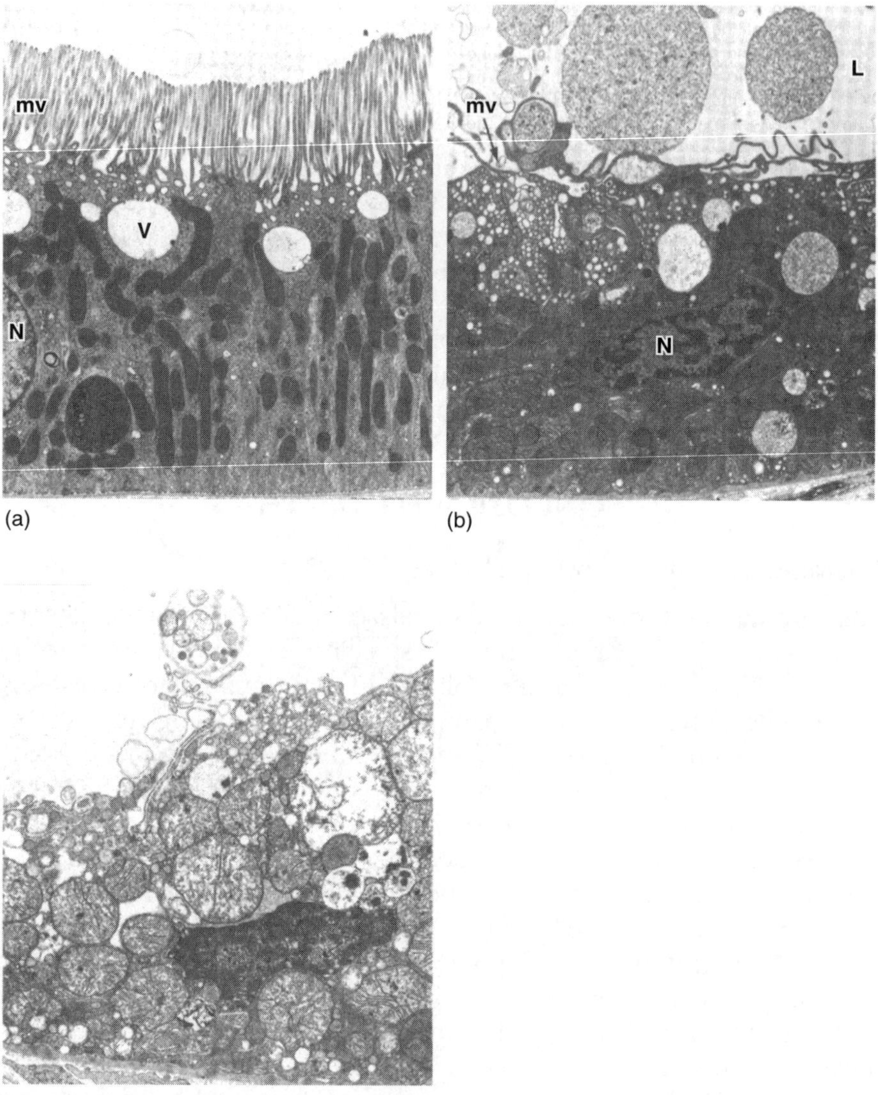

(a) (b) (c)

Figure 7.2 Electron micrograph of an RPT cell during reversible and irreversible injury. (a) Normal epithelial cell. Note the numerous microvilli (mv) (N = nucleus, v = vacuole); (b) RPT cell with reversible ischemic injury. Note the loss of microvilli, bleb formation and extrusion into the lumen (L), and swollen mitochondria; (c) RPT cell with irreversible injury. Note the swollen mitochondria with amorphous densities, membrane disruption, and dense pyknotic nucleus. (From Robbins, 1994, with permission.)

quantitate changes in intracellular pH and cations. Cell viability may also be measured by adding the membrane-impermeable, nuclear-staining dye, trypan blue. Finally, the release of endogenous or exogenous compounds (radiolabeled adenosine, chromium, or the fluorescent cytosolic dye calcein) may be measured as a determination of cell death (Groves and Schnellmann, 1996; Trump and Berezesky, 1994; Welder and Acosta, 1994). In these cases, cells are preincubated with the agent before treatment with a toxicant, and its release from the cells monitored during toxicant exposure. One potential problem with the uptake of dyes is that some dyes enter cells with an intact membrane, thus overestimating cellular membrane disruption or death. Additionally, a disadvantage of measuring the release of small compounds into the media is that they may be released via a cellular transporter. These issues are especially important considering the high rate of endogenous and exogenous chemical transport in renal cells.

Chen and colleagues (1997, 2001) and Dong et al. (1998) used fluorescein-labeled dextrans of graded molecular size to characterize porous plasma membrane defects during ATP depletion in RPT and Madin-Darby canine kidney (MDCK) cells. They observed the transition of membrane defects from small pores to larger pores during the loss of viability. The cell membrane became permeable to PI (mol wt = 668) early during anoxia, but was impermeable to phallacidin (mol wt = 1,125), 3 kDa dextran and 70 kDa dextran. Subsequently, the cell membrane became permeable to phallacidin and 3 kDa dextran, but remained impermeable to 70 kDa dextran. In the final stage, the cell membrane became permeable to 70 kDa dextran and LDH. Permeability to 70 kDa dextran and LDH was irreversible, whereas cells with increased permeability to propidium iodide were able to recover. These studies suggest that the increase in membrane permeability that occurs during oncosis is not an "all or none" phenomenon but is, rather, a graded increase in membrane permeability. Additionally, these studies demonstrate that release or uptake of small molecules in viability studies may represent reversible membrane changes, while uptake or release of larger molecules should be used for assessment of irreversible cell death.

Mechanisms of Oncosis

Many proposed mechanisms involving numerous intracellular targets are implicated in oncotic cell death, including mitochondrial dysfunction, loss of ion homeostasis (including Ca^{2+}), oxidative injury, protease activation, cytoskeletal derangements, and lysosomal disruption. The cell death process may involve a number of initial pathways feeding into several key downstream mediators. Along these pathways is a "point of no return," where the cell dies without regard to intervention.

Cellular Energetics

A large expenditure of energy is required for renal tubular reabsorption of solutes and water. Both oxidative phosphorylation and glycolysis supply this energy, although 95% of renal ATP is generated by oxidative phosphorylation (Mandel, 1985). Many toxicants act by interfering with cellular energetics. Cells with low glycolytic activity, including the S_1 and S_2 segments of the RPT, are more susceptible to toxicants that interfere with mitochondrial function. Toxicants disrupt mitochondrial function by preventing oxygen delivery to the tissue, by increasing ATP and oxygen utilization, or by interfering directly with mitochondrial energy production. The first cellular effect of ischemia, hypoxia, or mitochondrial inhibition is the prevention of ATP production. Toxicants that interfere with renal perfusion decrease oxygen delivery, resulting in decreased ATP levels in the renal cortex (Trifillis et al., 1981a, 1981b). Many nephrotoxicants interfere directly with mitochondrial respiration (Schnellmann and Griner, 1994), including mercurials (Weinberg et al., 1982), cisplatin (Gordon and Gattone, 1986), cyclosporine (Walker et al., 1986), and cysteine conjugates (Groves et al., 1993).

The induction of the mitochondria permeability transition (MPT) by the opening of the MPT pore is one method by which mitochondria lose their ability to produce ATP. Saturable inhibition by cyclosporin A suggested a channel or pore, and by using electrophysiological methods a high conductance transition pore was identified in the mitochondrial membrane (Szabo and Zoratti, 1991). The MPT channel can be defined as a voltage-dependent, cyclosporin A-sensitive, high-conductance inner membrane channel with a pore diameter of about 3 nm (Bernardi et al., 1999; Massari and Azzone, 1972). The collapse of the proton gradient and opening of the MPT pore prevents the synthesis of ATP and leads to ATP depletion. The MPT renders the inner membrane permeable to solutes less than 1,500 kDa, and is associated with mitochondrial swelling, loss of membrane potential, and release of intramitochondrial solutes, including glutathione and pyridine nucleotides (Bernardi et al., 1994). The induction of the MPT requires matrix Ca^{2+} and can be induced by a variety of agents, including oxidants, heavy metals, fatty acids, and atractyloside (Zoratti and Szabo, 1995). Additionally, various agents inhibit the MPT, including Mg^{2+}, cyclosporin A, sulfhydryl reducing agents, thiol alkylating agents, carnitine, ADP, NADH, NADPH, polyamines, and low pH (Bernardi et al., 1992, 1994; Crompton et al., 1988; Novgorodov et al., 1994; Petronilli et al., 1994).

The primary consequence of MPT channel opening is mitochondrial membrane depolarization. Common MPT assays are based on the fluorescence change of probes that accumulate in mitochondria as a result of depolarization. For example, experiments using confocal microscopy demonstrate cyclosporin-sensitive redistribution of the fluorophore marker calcein from the cytosol into the mitochondrial matrix space in cultured hepatocytes (Trost and Lemasters, 1997).

Induction of the MPT is implicated in both apoptotic and oncotic cell death (Lemasters, 1999). When the onset of the MPT is rapid, ATP is depleted and oncosis occurs. If the induction of the MPT is slower or if other ATP sources are available, less ATP is depleted and apoptotic cell death proceeds. Common death signals can result in either oncosis or apoptosis in the same tissue, and the induction of the MPT fits this combined pathway (Lemasters, 1999). The MPT also has been described in the kidney, in which the transition was induced by salicylate and was prevented by cyclosporin A in rat kidney mitochondria (Al-Nasser, 1999).

For further analysis of the ATP-dependent switch from apoptosis to oncosis, Lieberthal et al. (1998) exposed primary cultures of mouse RPT cells to the mitochondrial inhibitor antimycin A or 2-deoxyglucose, while varying the concentrations of dextrose to achieve graded ATP depletion. In this study, cells exposed to severe ATP depletion (75 to 85%) died by oncosis, while cells exposed to milder ATP depletion died by apoptosis. In another study, Leist el al. (1997) performed ATP clamping experiments by incubating oligomycin-treated Jurkat cells in varying concentrations of dextrose. In this study, two classical inducers of apoptosis (staurosporine and CD95 stimulation) resulted in oncotic cell death when ATP levels were reduced 50 to 70%. These studies suggest that when a threshold of ATP depletion is reached, the mechanism of cell death switches from apoptosis to oncosis (Figure 7.3).

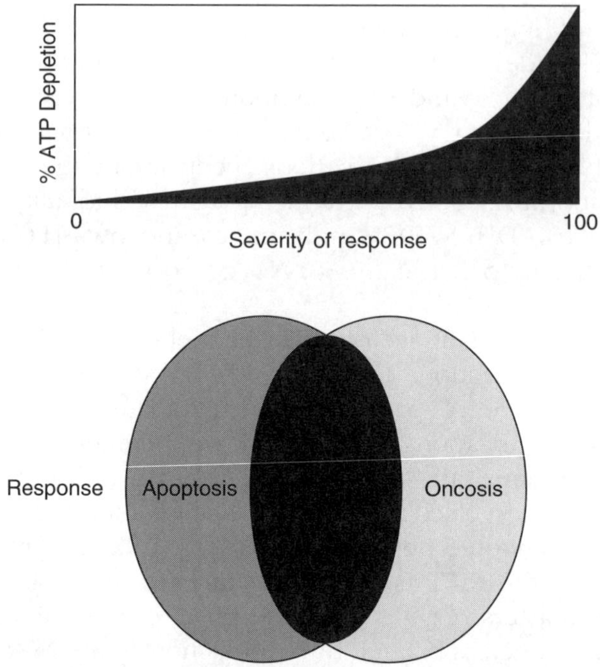

Figure 7.3 Response of renal cells following ATP depletion. The response to ATP depletion depends on its severity. Complete depletion results in oncosis, while partial depletion of ATP leads to apoptosis. Some common pathways may be shared between oncosis and apoptosis. (Adapted from Ueda and Shah, 2000, with permission.)

Clearly, ATP is critical for the normal function of a cell. When ATP production falls, concurrent with a loss of mitochondrial function, either oncotic (rapid) or apoptotic (delayed) cell death ensues. The resulting lack of ATP caused by mitochondrial dysfunction results in both the inability of protein kinases to keep phosphoproteins phosphorylated and to keep ATP-dependent ion transporters active. A decrease in protein phosphorylation has been observed during anoxia in RPT (Kobryn and Mandel, 1994). Additionally, protein phosphatase inhibitors are cytoprotective in rabbit RPT subjected to anoxia or exposed to antimycin A (Griffin and Schnellmann, 1997), suggesting that protein dephosphorylation may play an important role in cell death, secondary to mitochondrial dysfunction and ATP depletion. (See also Chapter 11 for additional discussion of the role of cellular energetics in nephrotoxicity.)

Cell Volume and Ion Homeostasis

Volume maintenance is vital to the integrity of the cell and the loss thereof results in the cell swelling that is associated with oncosis. Although some toxicants interact with the plasma membrane directly, to increase ion permeability and disrupt ion homeostasis, toxicants generally disrupt cell volume and ion homeostasis by inhibiting energy production. The loss of ATP results in the inhibition of ion transporters that maintain the differential ion gradients. The Na^+/K^+-ATPase maintains a high concentration of K^+ and a low concentration of Na^+ inside the cell. As ATP levels decrease, Na^+/K^+-ATPase activity decreases, resulting in K^+ efflux, Na^+ influx and a decrease in the normally negative membrane potential. Downstream from these events, Ca^{2+} and Cl^- entry occurs, as well as additional Na^+ and H_2O influx ,and the marked cell swelling associated with oncosis.

For example, in freshly isolated RPT suspensions exposed to the mitochondrial inhibitor antimycin A, cellular respiration ceases immediately and ATP depletion occurs within 2 min. The Na^+ and K^+ gradients are then lost. Following a lag period, extracellular Ca^{2+} and Cl^- influx occurs by an unknown mechanism, but is inhibited by blockers of Ca^{2+} channels (nifedipine) and Cl^- channels [5-nitro-2-(3-phenylpropylamino)-benzoate (NPPB), indanyloxyacetic acid (IAA-94), niflumic acid and diphenylamine-2-carboxylate (DPC)]. These Ca^{2+} and Cl^- channel blockers prevented cell death and decreased plasma membrane permeability associated with antimycin A toxicity (Miller and Schnellmann, 1993; Waters and Schnellmann, 1996; 1998; Waters et al., 1997a). In the renal epithelial cell line LLC-PK_1, oxidant-induced loss of membrane integrity was prevented by substitution of Cl^- with isethionate, Cl^- channel blockers (e.g., NPPB, niflumic acid, and DPC), and inhibitors of volume-sensitive Cl^- channels (e.g., ketoconazole and tamoxifen) (Meng and Reeves, 2000). This implies a role for volume-sensitive Cl^- channels in the progressive loss of cell membrane integrity during oxidant injury.

Oxidative Injury

Reactive oxygen species (ROS) contribute to cell injury during inflammation, during ischemia-reperfusion, and following exposure to toxicants. Patients with chronic renal failure are subjected to increased oxidative stress, as measured by plasma malondialdehyde concentration, which increased with the severity of renal dysfunction (Mimic-Oka et al., 1999). Mitochondria, through the incomplete reduction of O_2 to water, are sources of ROS in cells. A one-electron reduction of O_2 results in the production of the superoxide anion free radical, and a two-electron reduction of O_2 results in the production of H_2O_2. The other major ROS, the hydroxyl radical, is formed from H_2O_2 and the superoxide anion free radical by the iron-catalyzed Fenton reaction. The hydroxyl radical reacts immediately with adjacent molecules, while H_2O_2 and the superoxide anion are less reactive and may produce injury away from the site of formation.

Oxidative stress leads to cellular damage when the production of ROS is not compensated by the normal protective mechanisms of the cell. Toxicants may initiate oxidative stress by increasing the production of ROS. Gentamicin, glycerol, and $HgCl_2$ have been shown to increase the production of H_2O_2 (Guidet and Shah, 1989; Walker and Shah, 1987). Oxidative stress can occur also from redox cycling. Quinones undergo a one-electron reduction to a semiquinone radical and then a second one-electron reduction to a hydroquinone, followed by oxidation back to the quinone. This cycling produces superoxide anion from O_2.

Iron also plays a role in oxidant injury by catalyzing the formation of hydroxyl radical (Halliwell and Gutteridge, 1990). Iron-mediated free radical formation is responsible for the toxicity of *t*-butyl hydroperoxide in rabbit RPT because the iron chelator deferoxamine, the antioxidants promethazine and butylated hydroxytoluene, and the disulfide-reducing agent dithiothreitol prevent the lipid peroxidation, the mitochondrial dysfunction, and cell death (Schnellmann, 1988). Deferoxamine and other metal chelators are also cytoprotective during H_2O_2-mediated toxicity in LLC-PK_1 cells (Walker and Shah, 1991).

Increases in cytosolic free Ca^{2+} can activate nitric oxide synthase (NOS). NO may play a role in hypoxia/reoxygenation injury in RPT as a result of its free radical nature or due to the generation of peroxynitrite. NO generation by RPT can be stimulated by hypoxia and an inhibitor of NOS, *N*-nitro-L-arginine methylester, is cytoprotective during hypoxic cell injury in rat RPT (Yu et al., 1994). Additionally, treatment of African green monkey kidney cells with H_2O_2 resulted in increased immunodetectable iNOS, elevated NO release, nitrite production, and decreased viability. Treatment with antisense *S*-oligodeoxynucleotides to iNOS prevented its expression, nitrite production, and lethal cell damage (Goligorsky and Noiri, 1999). These data indicate that NO generated by iNOS and its metabolite, peroxynitrite, is cytotoxic to renal tubules.

ROS are very reactive and damage neighboring macromolecules. They may induce lipid peroxidation, oxidize sulfhydryl or amino groups on proteins, depolymerize polysaccharides, and induce DNA strand breaks. The interaction of free radicals with polyunsaturated fatty acid chains of membrane phospholipids results in lipid free radicals. The reaction of oxygen with this free radical results in the production of lipid hydroperoxides and toxic breakdown products. Secondary free radicals initiate a chain reaction of lipid peroxidation that can be terminated by an antioxidant or free radical dimerization. Lipid peroxidation results in alterations in membrane fluidity, permeability, cross-linking reactions, and enzyme activity. The oxidation of sulfhydryl or amino groups on proteins results in the loss of protein functions that leads to cell injury or death.

Calcium

Under physiological conditions, intracellular cytosolic free Ca^{2+} ($Ca^{2+}{}_f$) concentrations are tightly regulated. The maintenance of the $Ca^{2+}{}_f$ concentration is achieved by a variety of mechanisms. Ca^{2+} can be bound to the plasma membrane, sequestered within intracellular organelles, or can be either free or bound in the cytoplasm. The plasma membrane and ER Ca^{2+}-ATPases are perhaps most vital in the maintenance of the intracellular $Ca^{2+}{}_f$ concentration to approximately 100 n*M*, a concentration 1/10,000th the extracellular level (Weinberg, 1991). The plasma membrane Ca^{2+}-ATPase pumps excess Ca^{2+} out of the cell, while the ER Ca^{2+}-ATPase pumps Ca^{2+} into the ER. These Ca^{2+}-ATPases counteract both the Ca^{2+} leak into the cell from the extracellular space and the leak from the ER into the cytoplasm. Mitochondrial uptake of Ca^{2+} does not occur under controlled physiological conditions, but may occur when cytosolic levels reach 400 to 500 n*M*, as occurs during injury (Weinberg, 1991).

It is widely postulated that during cell injury, $Ca^{2+}{}_f$ levels rise, leading to a variety of destructive events, including cellular degradation by activated proteases, and damage to the cytoskeleton and membrane (Harman and Maxwell, 1995). The majority of studies that measure $Ca^{2+}{}_f$ increases with the high-affinity intracellular indicator fura-2 in ATP-depleted proximal tubules report increases approaching 500 nM to 1 μM (Jacobs et al., 1991; Kribben et al., 1994; Rose et al., 1994).

The initial increase in Ca^{2+}_f during cell injury may come from intracellular Ca^{2+} release. The ER has a high-affinity, low-capacity Ca^{2+}-ATPase, and contains the largest intracellular store of Ca^{2+} (Berridge, 1993). ER Ca^{2+} is released by the cell signaling mediator inositol-1,4,5-trisphosphate (IP_3) and may be depleted by the ER Ca^{2+}-ATPase inhibitors thapsigargin and cyclopiazonic acid (Sagara and Inesi, 1991; Thastrup et al., 1990). Treatments of freshly isolated rabbit RPT suspensions with thapsigargin or cyclopiazonic acid released and depleted ER Ca^{2+} (Harriman et al., 2002; Waters et al., 1997b). The release and depletion of ER Ca^{2+} prior to antimycin A exposure prevented both antimycin A-induced increases in Ca^{2+}_f and oncosis (Harriman et al. 2002; Waters et al., 1997b). These data strongly suggest the importance of ER Ca^{2+} release during cellular injury and death.

The influx of extracellular Ca^{2+} may also mediate increased Ca^{2+}_f levels during cell injury. In antimycin A-treated proximal tubules, incubation in low-Ca^{2+} medium prevented Ca^{2+}_f increases, inhibited polyphosphoinositide hydrolysis, and preserved the structure of mitochondria and microvilli (Garza-Quintero et al., 1993). Chelation of intracellular or extracellular Ca^{2+}, as well as the Ca^{2+} channel blockers verapamil or nifedipine, protected against cell death associated with anoxia and a variety of toxicants (Almeida et al., 1992; Rose et al., 1993, 1994; Wetzels et al., 1993; Waters et al., 1997a). This evidence indicates that increases in Ca^{2+}_f and extracellular Ca^{2+} influx are critical mediators of cell death. Furthermore, the extracellular Ca^{2+} influx may occur through an L-type Ca^{2+} channel. McCarty and O'Neil (1991) and Zhang and O'Neil (1996a, 1996b) have provided evidence for dihydropyridine-sensitive Ca^{2+} channels in rabbit RPT. However, the exact mechanisms responsible for Ca^{2+} influx during cell injury remain unresolved.

A pathway has been proposed by which ER Ca^{2+} stores and Ca^{2+}_f regulate extracellular entry. In nonexcitable cells, including renal cells, receptor-mediated release of IP_3 results in the release of Ca^{2+} from the ER, with a subsequent increase in Ca^{2+}_f (Berridge, 1993, 1995; Putney and Bird, 1993). This increase in Ca^{2+}_f is followed by a larger and sustained increase in Ca^{2+}_f due to extracellular Ca^{2+} entry. The extracellular Ca^{2+} entry is coupled to the depletion of the ER Ca^{2+} stores by a mechanism termed "capacitative Ca^{2+} entry" or "store-operated Ca^{2+} entry." In mast cells (Hoth and Penner, 1993) and MDCK cells (Delles et al. 1995), the whole-cell patch clamp technique has identified capacitative Ca^{2+} entry as Ca^{2+} release-activated Ca^{2+} (CRAC) channels. The CRAC channels are highly selective for Ca^{2+} and can be blocked by Cd^{2+} and La^{3+}, and the amplitude of CRAC channel current is dependent on the extracellular Ca^{2+} concentration. The sensitivity of CRAC channels to Ca^{2+} channel blockers has not been established. Store operated Ca^{2+} entry has not been characterized in differentiated adult RPT cells under physiological or pathological conditions. Additionally, store operated Ca^{2+} entry was not apparent following the ATP depletion-induced release of ER Ca^{2+} stores during antimycin A exposure in rabbit RPT suspensions (Harriman et al., 2002).

The consequences of the rise in Ca_f^{2+} levels are still debated. Numerous studies have concluded that an increase in $Ca^{2+}{}_f$ occurs immediately prior to, or concurrent with, a loss of cell viability, and is not required for oncosis (Jacobs et al., 1991; Lemasters et al., 1987; Weinberg, 1991). However, other investigators have demonstrated that Ca_f^{2+} plays an important role in oncosis. Kribben et al. (1994) demonstrated that Ca_f^{2+} increases significantly prior to hypoxia-induced cell death in rat RPT. Additionally, investigators have demonstrated that decreasing the extracellular Ca^{2+} concentration reduces anoxia- or hypoxia-induced oncosis in RPT (Rose et al., 1994; Takano et al., 1985; Waters et al., 1997a; Wetzels et al., 1993). Cytoprotection and prevention of $^{45}Ca^{2+}$ uptake have also been observed with Ca^{2+} channel blockers, intracellular Ca^{2+} chelation, and inhibitors of the Ca^{2+}-activated protease, calpain. Indeed, one of the consequences of the increase in $Ca^{2+}{}_f$ levels has been hypothesized to be activation of calpain, which then mediates further Ca^{2+} influx and oncosis.

To summarize, evidence suggests that increases in $Ca^{2+}{}_f$ levels play an important role in the oncosis cascade (Harriman et al., 2002). Prevention of $Ca^{2+}{}_f$ increases by either chelation of intracellular or extracellular Ca^{2+} or by addition of Ca^{2+} channel blockers prevents cell injury. Evidence also indicates that increases in $Ca^{2+}{}_f$ occur prior to irreversible membrane damage.

Degradative Proteins

Phospholipases

Phospholipases, particularly the phospholipase A_2 (PLA_2) family, may mediate the damage resulting from drugs, chemicals, and ischemia or reperfusion. PLA_2 hydrolyzes the acyl bond at the sn-2 position of phospholipids, resulting in the release of arachidonic acid and a lysophospholipid (Balsinde et al., 1999). The family members of this group have different substrate preferences, Ca^{2+} dependencies, and biochemical characteristics (Cummings et al., 2000; Dennis, 1994). An increase in PLA_2 activity as a result of toxicant exposure could result in the direct loss of membrane phospholipids and impairment of both plasma and mitochondrial membrane permeabilities and functions (Malis and Bonventre, 1986, 1988). The breakdown products may also contribute to injury. The lysophospholipids and free fatty acids may alter membrane permeability by acting as detergents or uncouple mitochondrial respiration. The released arachidonic acid may result in the production of vasoactive metabolites (e.g., thromboxane) and in turn may affect organ function by causing vasoconstriction and ischemia. Additionally, arachidonic acid and its metabolites are proinflammatory and chemotactic.

In renal cells, the Ca^{2+}-dependent, cytosolic PLA_2 ($cPLA_2$) is the isoform that is suggested to participate in oxidant-induced oncosis (Cummings et al., 2000). In LLC-PK_1 cells, the overexpression of $cPLA_2$, but not a secretory PLA_2 ($sPLA_2$), increased the susceptibility to H_2O_2 toxicity (Sapirstein et al., 1996). The authors suggested that oxidant-induced damage results in the translocation of $cPLA_2$ to a specific subcellular location, where it produces injury. In addition, a study employing multiple inhibitors of PLA_2 in MDCK cells implicated $cPLA_2$ in oxalate-induced renal cell oncosis (Kohjimoto et al., 1999). This evidence suggests a role for $cPLA_2$, but not ($sPLA_2$, in oxidant-induced renal injury, although the mechanisms remain unclear.

In contrast, Ca^{2+}-independent PLA_2 ($iPLA_2$) has been shown to be cytoprotective during oxidant injury in RPT. Pretreatment of rabbit RPT and cultured RPT cells with bromoenol lactone, a specific inhibitor of microsomal $iPLA_2$, potentiated oxidant-induced LDH release and PI staining (Cummings et al., 2002; McHowat et al., 1995), suggesting oxidant-induced oncosis is potentiated by inhibition of microsomal $iPLA_2$. Therefore, in oxidant injury, microsomal $iPLA_2$ may function to hydrolyze oxidized phospholipids and confer cytoprotection during oncotic cell death.

Endonucleases

DNA breakdown occurs late in oncotic cell injury, resulting in a characteristic smear observed during electrophoretic analysis of the

DNA. However, DNA "ladder formation," typical of endonuclease-mediated internucleosomal DNA cleavage during apoptosis, has also been observed during oncosis by some researchers (Schumer et al., 1992; Ueda et al., 1995; Ueda and Shah, 1992). Dong et al. (1997) noted DNA "laddering" coupled directly to plasma membrane damage in MDCK cells undergoing oncosis. In a study by Ueda et al. (1995) using rat RPT, activity of a 15 kDa endonuclease was demonstrated during hypoxia/reoxygenation injury. In contrast, Schnellmann et al. (1993) reported the lack of endonuclease activity prior to or after the onset of oncotic cell death in rabbit RPT exposed to antimycin A. These results indicate that endonuclease activation may not be a generalized mechanism following renal cell injury, and may be a consequence rather than a cause of oncosis.

Proteinases

Activation of proteinases in the cytosol or membranes could disrupt normal membrane and cytoskeletal function, leading to a loss of ion homeostasis, cell structure, and eventual cell death. Because calcium-activated neutral cysteine proteinases (calpains) are activated by Ca^{2+}, are ubiqitous in nature, and have numerous key cellular substrates, they are implicated as mediators of oncosis. Under physiological conditions, calpains are thought to exist as inactive proenzymes, termed "pro-calpains" (Croall and Demartino, 1991; Edelstein et al., 1997). Two major calpain isoforms are differentiated by their Ca^{2+} requirement for activation *in vitro*; μ-calpain is activated by micromolar concentrations of Ca^{2+} and m-calpain is activated by millimolar concentrations of Ca^{2+}. In cells, the Ca^{2+} requirement for calpain activation may be lowered by membrane-associated phospholipids (Arthur and Crawford, 1996; Suzuki et al., 1992). Calpain activation is restricted, however, by the physiological calpain inhibitor calpastatin (Ma et al., 1994; Mohan and Nixon, 1995).

Both μ- and m-calpains have two subunits, an 80 kDa catalytic subunit and a 30 kDa regulatory subunit. The 80 kDa catalytic subunit contains four domains. Domain I, on the amino terminus, is cleaved upon binding of the enzyme to Ca^{2+}, converting calpain from an 80 kDa to a 76 kDa enzyme. Domain II contains the catalytic site and is homologous to other cysteine proteases. The function of domain III is unknown. Domain IV, on the carboxyl terminus, contains four EF-hand Ca^{2+}-binding sites. Domain IV differs between μ- and m-calpain, accounting for the differences in Ca^{2+} sensitivity (Suzuki et al., 1992). Hydrolysis of the small subunit (30 kDa to 18 kDa) also occurs during calpain activation after binding of Ca^{2+} to its carboxy-terminal region. However, calpain autolysis may not be required for substrate hydrolysis.

Calpains have been implicated in toxicant-induced cell death in brain, liver, heart, and the kidney (Bronk and Gores, 1993; Croall and Demartino, 1991; Edelstein et al., 1995, 1996a, 1996b; Nicotera et al., 1986; Saido et al., 1994; Wang and Yuen, 1994). These results rely primarily on cytoprotection by inhibitors of calpain. Various calpain inhibitors are cytoprotective in RPT exposed to antimycin A (Harriman et al., 2000; Liu et al., 2002; Waters et al., 1997a). Additionally, calpain inhibitors are cytoprotective in RPT exposed to a variety of toxicants, including an alkylating quinone (bromohydroquinone), an oxidant (t-butyl hydroperoxide), and a toxicant that forms a reactive electrophile (tetrafluoroethyl-L-cysteine) (Schnellmann et al., 1994; Schnellmann and Williams, 1998). Edelstein et al. (1996a, 1996b) reported an increase in calpain activity in rat RPT subjected to hypoxia and cytoprotection by calpain inhibitors. These results suggest that calpains play a key role in cell death produced by hypoxia and a variety of toxicants. Calpain inhibitors prevented extracellular Ca^{2+} and Cl^- influx during mitochondrial inhibitor-induced oncosis (Waters et al., 1997a). These results suggest that calpain activation precedes and mediates both Ca^{2+} and Cl^- influx to trigger oncosis.

The involvement of calpains in physiological and pathological processes is a consequence of substrate modulation. Calpains are modulatory, rather than digestive, proteases and have numerous substrates, including cytoskeletal proteins, membrane proteins, and enzymes (Carafoli and Molinari, 1998). Calpains directly modify the cytoskeletal proteins fodrin, spectrin, talin, filamin, microtubule-associated proteins and the membrane proteins cadherin, Ca^{2+}-ATPase, and integrins (Saido et al., 1994). Evidence indicates that calpain mediates the hydrolysis of cytoskeleton-associated paxillin, vinculin, and talin during renal cell death (Liu and Schnellmann, 2003).

Cytoskeleton

The cytoskeleton is responsible for maintaining cell shape and polarity, intracellular transport, and for connecting organelles. During oncosis, the cell experiences a loss of apical brush border, blebbing of the plasma membrane, and loss of cell polarity. These changes may result from alterations in cytoskeletal components and interactions of the cytoskeleton with the plasma membrane. Two cytoskeletal components involved in the maintenance of cell shape are spectrin and ankyrin. The degradation of spectrin and ankyrin has been observed in rat RPT cells during ischemia (Doctor et al., 1993) and may be a consequence of the activation of calpain (Lofvenberg and Backman, 1999). In addition to studies demonstrating alterations in spectrin and ankyrin, increases in DNase-reactive actin, redistribution of pelletable actin, and loss of microvilli have been observed in rabbit RPT subjected to hypoxia (Nurko et al., 1996). Liu and Schnellmann (2003) provided evidence that calpain mediates the graded increase in rabbit RPT plasma membrane permeability and hydrolysis of cytoskeleton-associated paxillin, vinculin, and talin during renal cell death.

The tubular epithelial cell is polarized with respect to transporters and enzymes, including the Na^+/K^+-ATPase (basolateral membrane) and alkaline phosphatase (brush border). During ischemia and ATP depletion, cell polarity is lost. In rat RPT, the Na^+/K^+-ATPase dissociates from the cytoskeletal components actin, fodrin, and uvomorulin, resulting in its apical redistribution after ischemic injury (Molitoris et al., 1992).

The loss of polarity may also be a result of the degradation of cadherin–catenin complexes during epithelial ischemia. Bush et al. (2000) noted degradation of E-cadherin in both whole ischemic rat kidneys and in MDCK cells treated with antimycin A. Inhibitors of the proteasome, lysosomal proteases, or calpain did not block this degradation, suggesting a protease active at low ATP and Ca^{2+} concentrations. The authors of the study proposed that degradation of E-cadherin and prevention of cadherin–catenin complexes disrupt the adherens junction and lead to the loss of polarity, resulting in the ischemic epithelial cell phenotype.

Lysosomes

Lysosomes play an important role in the toxicity of aminoglycoside antibiotics and in α_{2u}-globulin nephropathy, but play a minor or insignificant role with other toxicants. α_{2u}-Globulin nephropathy is sex- and species-specific, occurring in some strains of male rats but not in humans, mice, or females of any species. Nephropathy occurs when compounds such as unleaded gasoline, d-limonene, 1,4-dichlorobenzene, tetrachloroethylene, decalin, and lindane bind to α_{2u}-globulin and prevent its renal proximal tubular lysosomal degradation. The size and number of lysosomes increase, ultimately resulting in single-cell necrosis.

Aminoglycosides induce lysosomal dysfunction and cause ARF. Aminoglycosides are filtered at the glomerulus, bound by phospholipids in the RPT brush border, reabsorbed by RPT cells, and accumulate in the lysosomes. The size and number of lysosomes increase and electron-dense myeloid bodies appear. Myeloid bodies contain undegraded phospholipids and are thought to occur via aminoglycoside-induced inhibition of lysosomal phospholipases. The mechanism by which lysosomal accumulation of phospholipids results in cell death is unclear. One possible explanation suggests that lysosomes become progressively distended until they rupture. The released lysosomal enzymes and aminoglycosides interact with membranes and organelles, causing cell death (Kaloyanides, 1992).

Stress Proteins

The cell has many protective mechanisms to prevent cell injury and allow recovery from injury. Antioxidants, including ascorbic acid and α-tocopherol, prevent damage associated with ROS, while glutathione can detoxify ROS and reactive electrophiles. Heat shock proteins (HSPs) and other stress proteins also prevent cell damage. HSPs are induced by elevated temperature and are classified according to their molecular weight. They enable a cell to withstand harsh conditions by protecting other proteins from degradation or structural alterations. Not only are HSPs produced as a result of an increase in temperature, but HSP production is also triggered by ROS, damaged proteins, anoxia/ischemia, and toxicants (Lovis et al., 1994; Kashgarian, 1995). In the kidney, HSP27, 32, 60, 70, and 72 all may play a role during ischemia and toxicant exposure. HSP32 lessens the nephrotoxicity due to rhabdomyolysis, possibly due to an increase in heme degradation. HSP72 is induced after stress and is localized to the apical membrane of RPT after ischemia (Van Why et al., 1992). The role of HSP72 during oncosis has not been demonstrated, but it co-immunoprecipitates with Bcl-2 and may play a role in ischemia-induced apoptosis (Wang et al., 1999). HSP27 has been localized with actin in the cytoskeleton (Zhu et al., 1994), and has also been found in the apical border of RPT cells during ischemia. HSP27 functions to stabilize and accelerate recovery of microfilaments (Lavoie et al., 1995). Nephrotoxic doses of *S*-(1,1,2,2-tetrafluoroethyl)-L-cysteine (TFEC) result in the formation of adducts to HSP60 and HSP70, although the role of these adducts in toxicity is unclear (Bruschi et al., 1993).

Grp78 and Grp94, members of the glucose regulated protein family, are chaperones in the ER that mediate protein folding and maturation (Helenius et al., 1997; Kaufman, 1999). They also bind Ca^{2+} and contribute to the maintainance of ER Ca^{2+} levels (Liu et al., 1997, 1998). Grp78 is an ER stress protein that is able to prevent oxidative stress, Ca^{2+} disturbances, and cell death in LLC-PK_1 cells exposed to iodoacetamide (Liu et al., 1997). Treatment of LLC-PK_1 cells with trans-4,5-dihydroxy-1,2-dithiane (DTTox), the intramolecular disulfide form of dithiothreitol, activated the transcription of Grp78 and rendered the cells tolerant to TFEC toxicity (Halleck et al., 1997). This preconditioning of renal epithelial cells with mild ER stress increases the expression of ER stress proteins and prevents TFEC-induced oncotic cell death. The increased expression of stress proteins has been postulated to control intracellular Ca^{2+} levels and prevent oxidative stress.

APOPTOTIC CELL DEATH

Characteristics of Apoptotic Cell Death

Apoptosis is responsible for numerous physiological and pathological events. For example, the programmed destruction of cells during embryogenesis, the hormone-dependent loss of cells that occurs during the menstrual cycle, and the cell deletion that occurs to balance cell proliferation in tissues with high cell turnover is the result of apoptosis. Additionally, apoptosis may be responsible for some of the cell death that occurs in ARF produced by both ischemia and toxicants (Ueda and Shah, 2000; Zhou et al., 1999). The biochemical characteristics and dose of the toxicant and the extent of ATP depletion all determine whether a cell dies by apoptosis or oncosis (Kern and Kehrer, 2002; Lash et al., 2001; Leist et al., 1997; Lieberthal et al., 1998). This section focuses on the mechanisms of apoptosis that may occur in renal cells exposed to injurious stimuli.

In oncosis, the loss of ATP results in the failure of the Na^+/K^+-ATPase, leading to decreased intracellular K^+ and increased intracellular Na^+ and water. As injury proceeds, Cl^- influx occurs with further cell swelling and eventual cell rupture. In apoptosis, intracellular ions such as K^+ and Cl^- are lost through ion transport systems, leading to water loss and shrunken and dense cells (Benson et al., 1996; Bortner et al., 1997). Volume-regulatory K^+ and Cl^- channel blockers prevent both the decrease in cell volume and subsequent apoptotic cell death. However, the broad-spectrum caspase inhibitor Z-Val-Ala-Asp(OMe)-CH_2F (zVAD-fmk) prevents caspase 3 activation, DNA laddering, and apoptosis, but is unable to prevent the cell shrinking (Maeno et al., 2000). Therefore, the decrease in cell volume is mediated by volume-regulatory K^+ and Cl^- channels, and precedes the numerous biochemical and morphological alterations, including cytochrome c release, caspase 3 activation, and DNA "laddering" (Figure 7.4).

Detection of Apoptotic Cell Death

Apoptotic morphology includes cell shrinkage, chromatin condensation, and the formation of cellular buds and apoptotic bodies. The apoptotic bodies are phagocytized by adjacent cells or, in the case of the kidney, may be lost to the tubular filtrate. Because of their quick disposal, the histologic detection of apoptosis can be difficult. When present, the cells appear to have an intensely eosinophilic cytoplasm when stained with hematoxylin and eosin, and dense nuclear chromatin fragments can be observed.

Apoptosis can be detected by a variety of techniques, including light or electron microscopy, flow cytometry, caspase assays, immunoblot analysis, and agarose-gel electrophoresis. To be certain of apoptotic cell death, more than one technique must be employed. Due to the distinct morphology of the apoptotic cell, observation under light or electron microscopy is the determining factor for the identification of apoptosis. The advantages of morphological detection are numerous, including the direct recognition of apoptotic (versus oncotic) processes and their regulation. For example, an inhibitor may inhibit the DNA fragmentation but not the other features of apoptosis. Additionally, engulfed and sloughed cells can be detected and identified using microscopy. Another benefit is the determination of the tissue distribution of apoptotic cells following an injury.

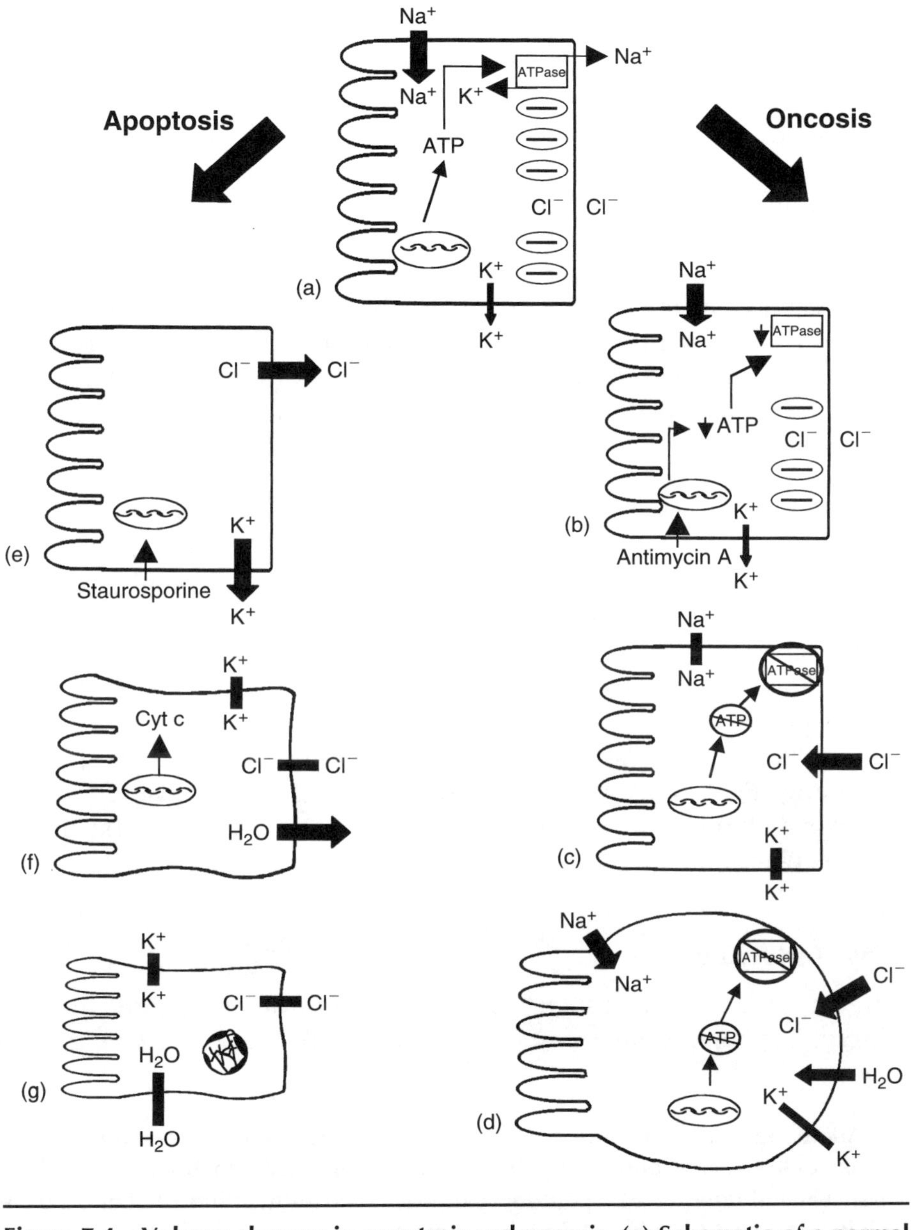

Figure 7.4 Volume changes in apoptosis and oncosis. (a) Schematic of a normal renal cell. (b–d) Oncotic pathway: (b)The addition of antimycin A prevents cell respiration, leading to ATP depletion. ATP depletion results in the loss of Na^+-K^+-ATPase activity, influx of Na^+ and efflux of K^+. (c) Increases in intracellular Ca^{2+} and influx of extracellular Cl^- follow through an unidentified pathway.

The available microscopic techniques for cell evaluation include light, electron, and fluorescence microscopy. Quantification of apoptotic cells can be determined by light microscopy, while electron microscopy provides a detailed view of the ultrastructural changes during apoptosis (but is time consuming and is not quantitative). Chromatin evaluation is performed using fluorescent nucleic acid stains, while DNA strand breaks are examined by labeling the ends of DNA with a detectable marker (e.g., TUNEL assay; see below). Fluorescence microscopy can be employed to assess mitochondrial function, including the mitochondrial membrane potential, which may be lost during oncosis or apoptosis. In *in vitro* studies, when cell detachment from plates or cell processing is required for morphological evaluation, care must be taken because cells may be damaged and morphology altered during these processes.

The binding of the 35 kDa Ca^{2+}-dependent phospholipid binding protein annexin V to phosphatidylserine (PS) may also be employed to examine apoptosis *in vitro* using microscopy or flow cytometry (Vermes et al., 1995). During early apoptosis, the charged head group of PS, normally located on the inner face of the plasma membrane, redistributes randomly to the outer face of the plasma membrane. Annexin V binds to the exposed PS head group located on the outer surface. Annexin V may be conjugated with a variety of chromophores, including FITC and biotin, to identify apoptotic cells. However, annexin V binding in late apoptotic and oncotic cells may be due to intracellular annexin V labeling.

DNA Laddering

The nuclear changes in apoptosis include pyknotic nuclei, chromatin condensation, and DNA cleavage into large fragments

(d) Additional Na^+ and water influx and cell swelling occur as a result of the Cl^- influx. Cell swelling and cytoskeletal derangements lead to breakdown of the plasma membrane. (e–g) Apoptotic pathway: (e) The addition of staurosporine results in K^+ and Cl^- efflux through volume-sensitive channels. (f) Efflux of water occurs as a result of K^+ and Cl^- efflux, resulting in cell shrinkage. Cytochrome c is released from the mitochondria and caspase activation occurs. (g) Additional morphological characteristics of apoptosis appear, including chromatin condensation. (Adapted from Schnellmann, 1997, with permission.)

(50 to 300 kilobases) followed by smaller oligonucleosomal fragments (Arends et al., 1990). During internucleosomal DNA cleavage, DNA is cleaved at linker regions between histones to form fragments of 180 to 200 base pairs and generating free 3′-hydroxyl ends. These fragments appear on electrophoretic gels as a DNA "ladder" pattern. Fragmentation can be identified and assessed using conventional agarose gel electrophoresis (CAGE) for DNA ladders, field inversion gel electrophoresis (FIGE) for large kilobase pair fragments, or *in situ* end-labeling (ISEL) to detect both DNA ladders and large fragments (Gavrieli et al., 1992). Detection of apoptosis by DNA fragmentation alone is problematic because not all apoptotic cells undergo DNA fragmentation and some oncotic cells do undergo DNA fragmentation.

TUNEL Assay

The fragmentation of nuclear DNA leading to the generation of free 3′-hydroxyl ends is the basis of the terminal deoxynucleotidyl transferase (TdT)-dUTP nick end labeling (TUNEL) assays. These procedures rely on TdT to incorporate dUTP into the 3′-hydroxyl ends of double- and single-stranded DNA. Conjugation of dUTP with biotin or fluorescent avidin enables the detection and quantitation of DNA fragmentation by either flow cytometry or fluorescence microscopy. Unfortunately, this method is not specific for apoptosis and can label oncotic cells with DNA damage (Grasl-Kraupp et al., 1995).

Flow Cytometry

In flow cytometry, apoptosis is assessed in intact cells instead of cell lysates. Flow cytometry allows rapid analysis of a large number of cells, and can identify and quantify subpopulations in heterogeneous cell populations. Flow cytometry assays used to detect apoptosis include:

- sub-G_1 DNA content method (Hotz et al., 1994)
- *in situ* nick translation assay, to detect DNA fragmentation (Gorczyca et al., 1993)
- annexin V assay, to determine PS externalization (Van Engeland et al., 1996)
- normalized mitochondrial membrane potential (NMMP/NADH) and cardiolipin (Dumas et al., 1995) assays, to assess mitochondrial function

Condensed chromatin during apoptosis can also be detected with a two-color flow cytometry assay using a combination of Hoechst 33342 and PI. The Hoechst dye stains condensed chromatin blue, while PI stains the DNA of oncotic cells red. As stated above, annexin V is a 35 kDa Ca^{2+}-dependent phospholipid binding protein that binds to externalized PS during apoptosis. Used in conjunction with PI during flow cytometry, annexin V can distinguish apoptotic (annexin V positive, PI negative) from oncotic (annexin V negative, PI positive) cells.

Caspase Activity

Another method for the detection and quantification of apoptosis is the measurement of cysteine aspartate-specific proteases, or caspases. Caspases form a large family of interleukin-1β-converting enzyme (ICE)-like proteases, presently containing 14 members (Ahmad et al., 1998). The caspases are present in the cell as inactive proenzymes and are activated following cleavage at specific aspartate cleavage sites. They have absolute specificity for aspartic acid in the P1 position and contain a conserved QACXG (where X is R, Q, or G) pentapeptide active-site motif. The active enzyme is a heterotetramer, containing two large and two small subunits. Caspase activation leads to the cleavage of structural proteins, RNA splicing proteins, DNA repair proteins, and other caspases. Cells undergoing apoptosis activate numerous caspases in a cascade that leads to the morphological and biochemical characteristics of an apoptotic cell (Thornberry, 1997).

Caspases can be detected by immunoblotting and measurement of activity. Antibodies are available for most of the established caspase family members and can be utilized for immunoblot analysis. In order to distinguish active caspases from precursor caspases, some antibodies recognize the active form. Specific fluorogenic substrates also are available for some of the caspases. Cleavage of these substrates by caspases results in the release of a fluorescent marker, which can be monitored by a fluorescent plate reader, fluorescence microscopy, or by flow cytometry. For example, Ac-Asp-Glu-Val-Asp-7-amido-4-methyl coumarin (Ac-DEVD-AMC) is used as a substrate for caspase 3 (Nicholson, 1999; Thornberry et al., 1997) and its cleavage results in the formation of the fluorescent AMC.

Although substrates are now available for caspase 1, 3, 6, 8, and 9, a lack of specificity or overlapping specificity is a concern.

Specific inhibitors of caspases, based on the DEVD and YVAD motifs, are commercially available and have been demonstrated to prevent apoptosis in some models. However, there is a report that inhibition of caspase activity significantly decreased LDH release (oncosis) in hypoxic proximal tubules (Edelstein et al., 1999), suggesting a role for caspases in oncosis as well as apoptosis. However, recent evidence in RPT cells indicates that 50% of the apoptosis induced by the dissimilar toxicants cisplatin, staurosporine, vincristine, and A23187 proceed via caspase-independent pathways (Cummings and Schnellmann, 2002).

Laminin and poly(ADP-ribose) polymerase (PARP) are intracellular substrates of caspases whose cleavage products may be used to detect caspase activity (Cohen, 1997). An anti-FITC-PARP antibody can recognize a caspase-specific cleavage site in PARP, and is suitable for flow cytometry, immunohistochemistry, and immunoblot applications (Brush, 2000). However, studies in fibroblasts (Ha and Snyder, 1999) and bovine aortic endothelial cells (Walisser and Thies, 1999) suggest that PARP cleavage may occur during oncosis.

An additional substrate of caspases is cytokeratin, an intermediate filament in epithelial cells. A study by Leers et al. (1999) demonstrated that cleavage of cytokeratin occurs early in apoptosis, before cells become TUNEL positive or display annexin V binding. In this study, an antibody to a cytokeratin hinge region compatible with caspase accessibility and exposed during early apoptosis was described, which may provide an excellent tool for the detection of apoptosis in epithelial cells.

Mechanisms of Apoptotic Cell Death

Injury to cells may result in apoptosis by modification of three major targets: DNA, the plasma membrane, and the mitochondria (Figure 7.5). Injury to DNA results in either the repair of the DNA or apoptosis, and the outcome is dependent on the nuclear phosphoprotein p53 (Bellamy, 1996). Injury to the plasma membrane activates sphingomyelinase and generates ceramide from membrane lipids. Ceramide is a second messenger that initiates apoptosis through the mitogen-activated protein kinase (MAPK) or stress-activated protein kinase (SAPK) signaling pathways (Westwick et al., 1995). Mitochondrial injury leads to induction of the MPT and the formation of the MPT pore. The MPT pore and the depolarization of the mitochondria is linked

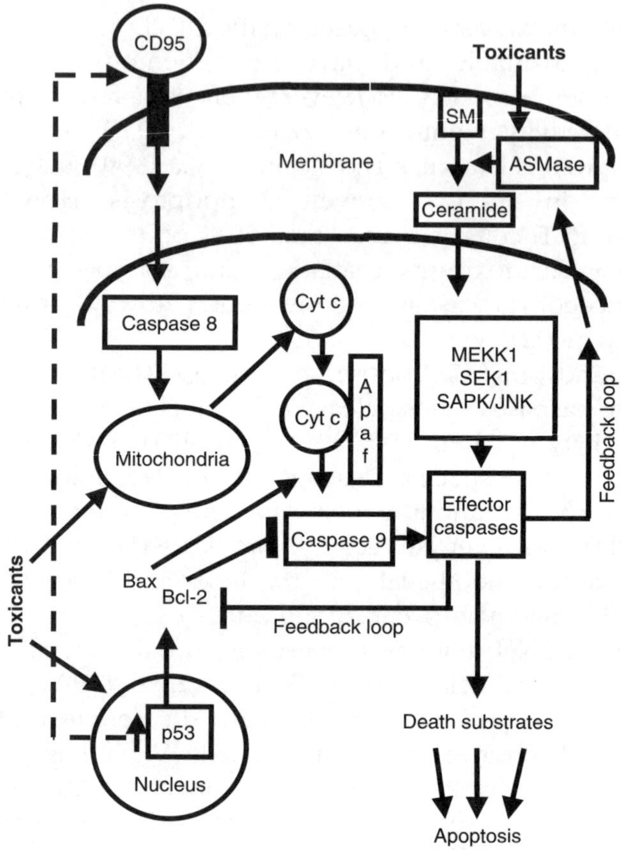

Figure 7.5 Chemical-mediated apoptotic signal transduction cascade. Toxicants may act on the nucleus, mitochondria, or membrane to initiate apoptotic signal transduction mechanisms. In the nucleus, toxicant-induced DNA damage elevates p53 levels. p53 regulates the expression of Bcl-2 family members, and may increase CD95 receptor/ligand interaction (TNF family). This interaction activates caspase 8, which cleaves and activates Bid, resulting in the release of cytochrome c from the mitochondria. Toxicants may act directly on the mitochondria to release cytochrome c. Cytochrome c binds Apaf and activates caspase 9, leading to activation of effector caspases and resulting in the cleavage of death substrates and the morphology of apoptosis. Toxicants may act on the plasma membrane and activate acid sphingomyelinase (ASMase), resulting in the cleavage of sphingomyelin and increased ceramide levels. Ceramide activates effector caspases through the stress-activated protein kinase/c-Jun amino-terminal kinase (SAPK/JNK) signaling pathway. In a feedback loop, caspase 3 (effector caspase) can result in the formation of ceramide, and can cleave Bcl-2. Bcl-2 inhibits the activation of caspase 9 when intact, but is proapoptotic when cleaved.

to the release of pro-apoptotic factors from the mitochondria, including apoptosis initiating factor (AIF) and cytochrome c, known to activate procaspase 9 and subsequently caspase 3 (Li et al., 1997). As seen from these examples, signaling cascades resulting from three different targets of toxicants all lead to apoptosis.

The signal transduction cascade of apoptosis depends on the apoptotic signal, the cell type, and factors as yet unknown. The cascade includes the signal itself, various death receptors, MAP kinases, caspases, transcription factors, and survival factors as a counterbalance. Current evidence indicates that a variety of different signals converge on the caspase proteases to produce the same set of terminal proteolytic events and apoptosis.

Death Receptors

Physiological stimuli may result in apoptosis via nuclear and cytokine receptors and intermediates related to the tumor necrosis factor (TNF) family. For example, activated lymphocytes have on their surface a CD95/apo-1/fas receptor of the TNF receptor family that binds a TNF-family ligand (Trauth et al., 1989). Binding of ligand to the receptor initiates receptor trimerization, which then facilitates the recruitment of proteins to form a death initiating signaling complex (DISC). Fas contains an amino acid sequence called a "death domain," which binds to a DISC protein called FADD (fas-associated death domain). FADD contains a death effector domain that binds to a third protein, FLICE (fas-activated protein-like ICE, also known as caspase 8). Caspase 8 can activate additional downstream caspases (Muzio et al., 1996) as well as cleave Bid (Li et al., 1998; Luo et al., 1998). The Bid cleavage fragment is targeted to the mitochondria and promotes the release of cytochrome c. Cytochrome c binds with Apaf-1 and can activate procaspase 9, subsequently activating caspase 3 (Li et al., 1997). The TNF receptor family has many family members that bind to death domains, such as the TRADD (TNF receptor-associated death domain), which can then couple to FADD and caspase 8. Additionally, many adapter proteins may positively or negatively modify the death signal. An example of negative regulation is the blockage of the TNF signal by TNF-receptor associated factor (TRAF1). TRAF1 is cleaved by Fas ligand and the cleaved product of TRAF1 is pro-apoptotic (Irmler et al., 2000).

The TNF pathway mediates apoptosis in numerous renal models during various injuries. Fumonisin B_1-induced apoptosis is mediated by the TNF pathway in African green monkey fibroblasts and neonatal kidney cells (Ciacci-Zanella and Jones, 1999). Fas ligand mRNA and protein are present in mouse and rat kidney, and are increased during glomerular injury (Lorz et al., 2000). Fas and Fas ligand also are expressed in renal tubular epithelial cells isolated from mice injected with lipopolysaccharide, and were shown to mediate the resulting apoptosis (Koide et al., 1999). Although Fas is expressed on human renal tubule epithelial cells, these cells appeared to be resistant to Fas-mediated apoptosis (Boonstra et al., 1997).

Bax and Bcl-2

The Bcl-2 protein family plays an important role in the control of apoptosis (Brown, 1996). This family of proteins appears to behave as a checkpoint between signals from the cell surface and the activation of caspases. There are currently nine members of this family. Bax is a member that is 45% homologous to Bcl-2. However, overexpression of Bax commits a cell to apoptosis while overexpression of Bcl-2 prevents apoptosis. Bcl-2 may bind to Bax to exert its apoptotic blocking effect. Bcl-2 appears to prevent the signal responsible for the activation of the apoptotic machinery from reaching the caspases. Recent evidence demonstrates the direct association of Bcl-XL, an antiapoptotic member of the Bcl family, with caspase 9 and Apaf-1 (Hu et al., 1998). The complex of caspase 9 and Apaf-1 can initiate apoptosis by caspase-3 activation (Li et al., 1997), but caspase-3 activation is prevented by the binding of Bcl-XL to this complex (Hu et al., 1998).

In the kidney, some studies report little or no Bcl-2, while Bax is predominant (Aschoff et al., 1999; Krajewski et al., 1994). However, increased expression of both Bcl-2 and Bax has been observed in rat RPT following ischemic injury (Basile et al., 1997; Gobe et al., 2000), indicating that Bcl-2 and Bax may be important regulators of apoptosis during renal tubule regeneration after ischemia. In the study by Gobe et al. (2000), distal tubules markedly increased Bcl-2, which may partly explain the resistance of distal tubules to apoptosis resulting from ischemic injury. In contrast to *in vivo* ischemic injury, addition of folate to mouse renal tubular epithelial cells increased mRNA expression of Bcl-XL and Bax, but decreased expression of Bcl-2 (Ortiz et al., 2000).

Similar results were observed in rats treated with cyclosporin A (Shihab et al., 1999). In this study, mRNA levels were increased for p53, Bax, Fas ligand, and caspase 1 while Bcl-2 mRNA levels decreased following cyclosporin A exposure. Overexpression of Bcl-2 prevented gentamicin-induced apoptosis in both MDCK and LLC-PK_1 cells (El Mouedden et al., 2000). Additionally, Bcl-2 overexpression prevented caspase 3 activation and apoptosis induced by cisplatin, whereas Bcl-2 overexpression did not prevent caspase 3 activation and apoptosis induced by *S*-(1,2-dichlorovinyl)-L-cysteine in LLC-PK_1 cells (Zhan et al., 1999). Therefore, caspase 3 activation may occur by both Bcl-2-dependent and Bcl-2-independent mechanisms in chemical-induced apoptosis of renal epithelial cells.

Evidence in receptor-mediated apoptosis indicates a positive feedback loop. Activation of caspase 3 is known to cleave Bcl-2 and Bcl-XL, inactivating the enzymes and terminating their survival function. The fragments of Bcl-2 and Bcl-XL are pro-apoptotic (Cheng et al., 1997; Clem et al., 1998) and have been found to localize to mitochondria and cause the release of cyochrome c (Kirsch et al., 1999). Cytochrome c binds to Apaf-1 and causes activation of caspase 9. Caspase 9 may initiate the processing of many additional caspases, including caspases 2, 3, 6, 7, 8, and 10 (Slee et al., 1999). The activation of caspase 3 completes the loop and ensures the death of the cell by apoptosis.

Caspases

The caspases, first described in the nematode *Caenorhabditis elegans* (*C. elegans*), are evolutionarily conserved and presently comprise 14 family members (Cohen, 1997; Ahmad et al., 1998; Thornberry and Lazebnik, 1998). Due to a conserved region (S1) between family members, aspartate is required in the P1 position of substrates. They also contain a nonconserved region (S4), allowing for more specialized function of the family members.

Activation of caspases is the common event in apoptosis produced by cellular death signals and diverse injuries. Interaction with Bax, Bcl-2, Apaf-1, and cytochrome c mediates the activation of the caspase cascade in many models. However, the cascade varies with receptor-mediated and chemical-induced apoptosis. While activation of Bid and cytochrome c release is caspase-dependent in receptor-mediated apoptosis, evidence indicates caspase-independent cytochrome c release and Bid activation in chemical-induced apoptosis

(Sun et al., 1999). Recent evidence also indicates p53-mediated caspase 3 activation in the absence of caspase 8 and caspase 9 in rabbit RPT apoptosis induced by cisplatin (Cummings and Schnellmann, 2002). The cascade may vary with the physiological or pathological stimuli, but it inevitably leads to activation of downstream caspases. Recent evidence now indicates the presence of caspase-independent apoptotic pathways in cisplatin, staurosporine, vincristine, and A23187-induced RPT cell apoptosis (Cummings and Schnellmann, 2002). However, these pathways have not yet been defined.

As the executioners of apoptosis, caspases have many significant substrates. Potential endogenous substrates for the caspases have been identified and primarily fall into two groups: those involved in cell homeostasis and repair mechanisms, and those involved in the maintenance of cell structure (Thornberry, 1997). Those involved in cell homeostasis and repair include PARP, a small ribonucleoprotein (U1, 70kDa), DNA-dependent protein kinases, and protein kinase C δ (PKCδ). Substrates involved with maintenance of cell structure include the cytoskeletal proteins fodrin, actin, and Gas2, and proteins maintaining nuclear structure, such as NuMA and lamins (Thornberry, 1997). *In vitro* work suggests that many of the homeostatic substrates appear to be cleaved by caspase 3 (Casciola-Rosen et al., 1996). Caspase 3 cleaves the inner nuclear membrane protein LAP2 and the nucleoporin Nup 153 (Buendia et al., 1999), while caspase 6 cleaves the structural proteins lamin A (Orth et al., 1996; Takahashi et al., 1996), lamin B, and lamin B receptor (Duband-Goulet et al., 1998). This cleavage of nuclear structural proteins most likely accounts for the chromatin condensation during apoptosis. Caspase 3 also cleaves the endogenous inhibitor of endonuclease, resulting in endonuclease activition and the characteristic DNA fragmentation observed during apoptosis (Widlak et al., 2000). Although examining the roles of caspases 1 and 3 during renal apoptosis has begun, few studies have explored the roles of the other caspase family members.

Ceramide

Ceramide is generated by the hydrolysis of the plasma membrane phospholipid sphingomyelin, and has been observed following environmental stressors and initiators of apoptosis, including TNFα (Obeid et al., 1993), Fas (Gulbins et al., 1995), ionizing radiation (Haimovitz-Friedman et al., 1994), and heat shock and oxidative stress

(Verheij et al., 1996). Ceramide acts as a second messenger to promote apoptosis through the stress activated protein kinase (SAPK)/c-Jun kinase (JNK) cascade (Haimovitz-Friedman et al., 1997). Toxicants induce ceramide-mediated apoptosis by both the Fas pathway (Friesen et al., 1996) and the SAPK/JNK pathway (Ham et al., 1995; Verheij et al., 1996) in renal cells (Gekle et al., 2000).

Direct targets for ceramide include ceramide-activated protein kinase (CAPK), ceramide-activated protein phosphatase, a guanine-nucleotide exchange factor, and PKC δ (Ballou et al., 1996; Spiegel et al., 1996). Ceramide generation results in the sequential activation of MEKK1, SEK1, SAPK, and c-Jun (Yan et al., 1994). It is presumed that the SAPK/JNK pathway phosphorylates targets upstream of caspase 3 to signal apoptosis (Haimovitz-Friedman et al., 1997). A study by Takeda et al. (1999) indicated that purified caspase 3 increases sphingomyelinase activity in a cell-free system, suggesting that caspase 3 may release ceramide. Caspase-dependent ceramide generation has also been proposed in several apoptosis models (Pronk et al., 1996; Hartfield et al., 1997; Genestier et al., 1998), suggesting activation of caspases both upstream and downstream of ceramide production.

Calpains

Studies implicate calpains in apoptotic cell death. Squier et al. (1994) found increased calpain activity in thymocytes undergoing apoptosis after treatment with dexamethasone and noted that the apoptosis was blocked by calpain inhibitors. Both increased calpain activity and protection by calpain inhibitors were also observed in apoptosis induced by a reovirus in murine fibroblasts (Debiasi et al., 1999), by cycloheximide in neutrophils (Squier et al., 1999), and by ischemia-reperfusion in the rat liver (Kohli et al., 1999). Bax was cleaved by calpain during drug-induced apoptosis in HL-60 cells (Wood et al., 1998). Although Bax was originally thought to be inactivated by calpain cleavage, a more recent study indicated that the cleavage product of Bax, when expressed in human embryonic kidney cells, induced apoptosis more potently than the full-length protein (Wood and Newcomb, 2000), demonstrating that an increase in calpain activity may promote apoptosis.

In contrast, other studies suggest that calpain may be a negative regulator of apoptosis. Evidence in HeLa cells indicates that upstream caspases are cleaved and inactivated by calpain, thus preventing downstream caspase processing and apoptosis (Chua et al., 2000). A similar observation was made in a neuronal model, with calpain preventing the ability of cytochrome c to activate the caspase cascade (Lankiewicz et al., 2000). This is in contrast to the previously mentioned studies showing prevention of morphological and biochemical changes, and cell death by inhibitors of calpain. In conclusion, the role of calpain in apoptotic cell death remains unclear and may be specific to particular models or to particular tissues or cells. The role of calpain in renal apoptosis has not been addressed.

TOXICANT-INDUCED RENAL CELL DEATH

The Kidney as a Susceptible Organ

The kidney is susceptible to a variety of toxicants for numerous reasons. The kidneys receive approximately 25% of the resting cardiac output, ensuring high levels of toxicant delivery over a period of time. The concentrating ability of the kidney results in high concentrations of toxicants in the tubular cells, lumen, and interstitium. The increased concentration of relatively insoluble compounds may result in intraluminal precipitation, tubular obstruction, and secondary ARF. Other factors also play a role in the susceptibility of the kidney to injury, including biotransformation enzymes that result in the formation of reactive intermediates and toxic metabolites. Additionally, renal toxicity may be potentiated by vasoconstrictive agents or by the inhibition of vasodilatory substances, as occurs with nonsteroidal anti-inflammatory drugs (NSAIDs).

Various endogenous and exogenous compounds are excreted from the kidney through organic anion and organic cation transport systems. These transport systems play a role early in the nephrotoxicity of many compounds. The first of these transport systems to be cloned was the organic cation transporter OCT1 (Grundemann et al., 1994). OCT1 has been demonstrated to excrete a variety of organic cations, such as endogenous compounds (choline, corticosterone, dopamine, epinephrine, norepinephrine, progesterone, and 5-hydroxytryptamine), drugs (cimetidine and clonidine), and the toxins nicotine and cyanine (Burckhardt and Wolff, 2000).

The organic anion transporter OAT1, first cloned in 1997 (Sekine et al., 1997; Sweet et al., 1997), also has wide substrate specificity, including endogenous substrates (cyclic nucleotides, prostaglandin), drugs (β-lactam antibiotics with anionic moieties, NSAIDs, antineoplastic drugs), and the natural mycotoxin ochratoxin A (Burckhardt and Wolff, 2000; Sekine et al., 1997; Sweet et al., 1997). An example of nephrotoxicity of a drug transported into the epithelial cell by OAT1 is the β-lactam antibiotic cephaloridine (Jariyawat et al,. 1999). Cephaloridine treatment causes acute proximal tubular oncosis in humans and laboratory animals (Goldstein et al., 1988). The β-lactam antibiotics, such as cephaloridine, accumulate in the kidney as a result of uptake by OAT1 before producing cytotoxic effects (e.g., lipid peroxidation or mitochondrial dysfunction) (Tune, 1997).

Many nephrotoxicants display segment-specific toxicity. The segments of the kidney have differences in transport and accumulation of chemicals (Endou, 1998), biotransformation enzymes (Lohr et al., 1998), and cellular energetics (Bastin et al., 1987) that account for the site-selective injury. For example, the proximal tubule is the target segment for nephrotoxic antibiotics (Mingeot-Leclercq and Tulkens, 1999), antineoplastics (Safirstein et al., 1984), halogenated hydrocarbons (Elfarra, 1993), mycotoxins (Gekle and Silbernagl, 1994), and heavy metals (Kone et al. 1990), while the glomerulus is the primary target for immune complexes (Chen et al., 1994).

Evidence for Oncosis

Nephrotoxic injury *in vivo* commonly results in acute renal failure (ARF). ARF is characterized by a decline in the glomerular filtration rate (GFR) and can be divided into three phases: the initiation phase, the maintenance phase, and the recovery phase. The initiation phase involves exposure of the kidney to the toxicant and impaired renal function. The maintenance phase results in a sustained loss of renal function. The recovery phase restores function if the damage is not too severe.

The pathogenesis of ARF is defined by oncosis or apoptosis due to direct tubular cell injury, hypoperfusion or hypofiltration due to vasoconstriction or glomerular injury, direct obstruction of the tubule, or tubulointerstitial nephritis due to immunologic or inflammatory responses. Agents known to induce ARF include mercuric chloride, potassium dichromate, uranyl nitrate, arsenite, diethylene glycol, aminoglycoside antibiotics, and cyclosporin A (Olsen and Solez, 1994).

The kidney is the primary target for the uptake, accumulation, and toxicity of Hg^{2+}. $HgCl_2$ poisoning was at one time the best-known cause of toxic acute tubular oncosis and resulted from accidental ingestion or intentional ingestion (i.e., suicide) (Olsen and Solez, 1994). Renal uptake of Hg^{2+} is rapid; as much as 50% of a nontoxic dose of Hg^{2+} is found in the kidneys of rats within a few hours of exposure (Zalups, 1993). *P*-aminohippuric acid (PAH), an organic anion transport system substrate, decreases the renal uptake of Hg^{2+} (Zalups, 1998), suggesting that a Hg^{2+} conjugate utilizes the basolateral organic anion transport system for uptake. Increased uptake of Hg^{2+} was observed when coadministered with *N*-acetylcysteine or cysteine, indicating that Hg^{2+}-*S*-conjugates of small endogenous thiols may be the primary transportable forms of Hg^{2+} (Zalups, 1998). In addition to the basolateral anion transport system, luminal γ-glutamyltransferase (γ-GT) appears to play a role in uptake of Hg^{2+}. Much of the evidence implicating γ-GT in uptake of Hg^{2+} comes from experiments in which the γ-GT inhibitor acivicin decreased the renal uptake of Hg^{2+} (Zalups, 2000).

The mechanisms of toxicity of Hg^{2+} may involve oxidative stress, mitochondrial damage, effects on Ca^{2+} homeostasis, and damage to plasma membrane proteins (Zalups, 2000). Lipid peroxidation resulting from oxidative stress was detected in renal cortical homogenates following administration of $HgCl_2$ (Fukino et al., 1984). Lund et al. (1993) demonstrated that isolated mitochondria treated with $HgCl_2$ produced elevate hydrogen peroxide formation, indicating that $HgCl_2$ has direct effects on mitochondria. Increases in ${Ca^{2+}}_f$ have also been demonstrated in rabbit RPT treated with $HgCl_2$ (Smith et al., 1987, 1991). In these studies, chelation of extracellular Ca^{2+} delayed bleb formation and oncotic cell death. Additionally, plasma membrane proteins possessing sulfhydryl groups, including the Na^+/K^+-ATPase, are bound and inactivated by Hg^{2+} (Imesch et al., 1992).

Aminoglycoside antibiotics are highly polar cations containing two or more amino sugars and a central hexose nucleus joined in a glycosidic linkage. Members of this group include streptomycin, kanamycin, neomycin, gentamicin, netilmicin, and tobramycin, and are widely used for the treatment of Gram-negative infections. Aminoglycosides are filtered in the glomerulus and absorbed by binding to anionic phospholipids in the S_1 and S_2 segments of the proximal tubule (Vandewalle et al., 1981). Following endocytosis, they are sequestered in the lysosomes, which increase in size and number and form myeloid bodies containing undegraded phospholipids. It is thought that the swollen lysosomes rupture and release degradative enzymes into the cytoplasm that damage the brush border, ER, and mitochondria, and ultimately cause oncosis (De Broe et al., 1984; Mingeot-Leclercq and Tulkens, 1999; Kaloyanides, 1992). Clinically, aminoglycoside treatment results in renal dysfunction in 5 to 26% of patients (Laurent et al., 1990; Fillastre and Godin, 1991; Kacew and Bergeron, 1990) and includes the clinical observations of polyuria, urinary brush border enzymes, glucosuria, aminoaciduria, and proteinuria (Gilbert, 1995; Patel et al., 1975).

Cyclosporin A is an immunosuppressive agent used to prevent rejection of transplanted organs and in the treatment of autoimmune-type diseases. Cyclosporin A treatment *in vivo* results in renal dysfunction that appears vascular, rather than tubular, in nature, with characteristics of decreased renal blood flow and GFR and increased blood urea nitrogen (BUN) and serum creatinine. Vasoconstriction induced by cyclosporine A may be due to increased production of thromboxane (Remuzzi and Bertani, 1989) and endothelin (Lanese and Conger, 1993), or the decreased production of nitric oxide from L-arginine (Bloom et al., 1995).

High doses of cyclosporine A result in tubular morphological changes, including vacuolization, giant mitochondria, lysosomal inclusion bodies, single-cell necroses, and microcalcifications in the proximal tubule (Mihatsch et al., 1994). This renal tubular injury may be caused by hypoxia-reoxygenation (due to vasoconstriction) because cyclosporin A increases hypoxia and free-radical production in rat kidneys (Zhong et al., 1998). Administration of an antioxidant suppressed lipid peroxidation and cell death of renal cells in both *in vivo* and *in vitro* models (Wang and Salahudeen, 1994).

Evidence for Apoptosis

Apoptosis in renal cells occurs as a result of ischemic injury and after exposure to some toxicants, including cyclosporin A, cadmium, *S*-(1,2-dichlorovinyl)-L-cysteine (DCVC), and cisplatin. Although normally associated with renal tubular oncosis *in vivo*, cyclosporin A-induced apoptosis of LLC-PK_1 cells has been observed (Takeda et al., 1998) and may be mediated by the Fas signaling cascade (Healy et al., 1998). While Ca^{2+} exposure is also associated with oncosis *in vivo*, Cd^{2+} produced apoptosis in rat kidneys, as assessed by TUNEL staining and morphology (Ishido et al., 1999). *In vitro*, $CdCl_2$ induced apoptosis in the porcine kidney cell line LLC-PK_1 (Ishido et al., 1995). DCVC, a major metabolite of trichloroethylene, causes apoptosis in cultured human RPT cells at concentrations potentially relevant to environmental exposures (Lash et al., 2001). Apoptosis in renal cells also occurres during treatment with the chemotherapeutic agent cisplatin, which initiates apoptosis in primary cultures of rabbit RPT (Cummings and Schnellmann, 2002) and mouse RPT (Lieberthal et al., 1996), a mouse S_3 proximal tubule cell line (Takeda et al., 1997), and LLC-PK_1 cells (Okuda et al., 2000). In an *in vivo* study, treatment of rats with cisplatin for five days resulted in an increased number of TUNEL-positive cells in the outer stripe of the outer medulla (Zhou et al., 1999). Cisplatin-induced apoptosis may be regulated by the induction of p53 and p21. Induction of p53 and p21 were observed after cisplatin treatment (Megyesi et al., 1996), and a p21 knockout mouse displayed a more rapid onset of cisplatin-induced acute renal failure (Megyesi et al., 1998), implicating a protective role for p21. In conclusion, toxicant-induced apoptotic cell death occurs in both *in vitro* and *in vivo* renal models.

Due to the high susceptibility of the kidney to ischemic and toxic insults, a variety of agents are nephrotoxic and induce oncosis, apoptosis, or both, by numerous mechanisms. Many of these mechanisms remain poorly characterized in the kidney (e.g., signal transduction mechanisms involved during toxicant-induced renal cell apoptosis). The key for treatment and prevention of nephrotoxicity, and thus the challenge of research, lies in the understanding of the steps between the injurious insult and the end stages of cell death.

REFERENCES

Ahmad M, Srinivasula SM, Hegde R, Mukattash R, Fernandes-Alnemri T, and Alnemri ES (1998) Identification and characterization of murine caspase-14, a new member of the caspase family. *Cancer Res* 58:5201–5205.

Almeida ARP, Bunachak D, Burnier M, Wetzels JFM, Burke TJ, and Schrier RW (1992) Time-dependent protective effects of calcium channel blockers on anoxia- and hypoxia-induced proximal tubule injury. *J Pharmacol Exper Ther* 260:526–532.

Al-Nasser IA (1999) Salicylate-induced kidney mitochondrial permeability transition is prevented by cyclosporin A. *Toxicol Lett* 105:1–8.

Arends MJ, Morris RG, and Wyllie AH (1990) Apoptosis. The role of the endonuclease. *Am J Pathol* 136:593–608.

Arthur JSC, and Crawford C (1996) Investigation of the interaction of m-calpain with phospholipids: calpain-phospholipid interactions. *Biochem Biophys Acta* 1293:201–206.

Aschoff AP, Ott U, Funfstuck R, Stein G, and Jirikowski GF (1999) Colocalization of Bax and Bcl-2 in small intestine and kidney biopsies with different degrees of DNA fragmentation. *Cell Tissue Res* 296:351–357.

Ballou LR, Laulederkind SJ, Rosloniec EF, and Raghow R (1996) Ceramide signalling and the immune response. *Biochim Biophys Acta* 1301:273–287.

Balsinde J, Balboa MA, Insel PA, and Dennis EA (1999) Regulation and inhibition of phospolipase A_2. *Annu Rev Pharmacol Toxicol* 39:175–189.

Basile DP, Liapis H, and Hammerman MR (1997) Expression of bcl-2 and bax in regenerating rat renal tubules following ischemic injury. *Am J Physiol* 272:F640–F647.

Bastin J, Cambon N, Thompson M, Lowry OH, and Burch HB (1987) Change in energy reserves in different segments of the nephron during brief ischemia. *Kidney Int* 31:1239–1247.

Bellamy COC (1996) p53 and apoptosis. *Brit Med Bull* 53:522–538.

Benson RS, Heer S, Dive C, and Watson AJ (1996) Characterization of cell volume loss in CEM-C7A cells during dexamethasone-induced apoptosis. *Am J Physiol* 270:C1190–C1203.

Bernardi P, Broekemeier KM, and Pfeiffer DR (1994) Recent progress on regulation of the mitochondrial permeability transition pore; a cyclosporin sensitive pore in the inner-mitochondrial membrane. *J Bioenerg Biomembr* 26:509–517.

Bernardi P, Scorrano L, Colonna R, Petronilli V, and Lisa FD (1999) Mitochondria and cell death, Mechanistic aspects and methodological issues. *Eur J Biochem* 264:687–701.

Bernardi P, Vassanelli S, Veronese P, Colonna R, Szabo I, and Zoratti M (1992) Modulation of the mitochondrial permeability transition pore: Effect of protons and divalent cations. *J Biol Chem* 267:2934–2939.

Berridge MJ (1993) Inositol trisphosphate and calcium signaling. *Nature* 361:315–325.

Berridge MJ (1995) Capacitative calcium entry. *Biochem J* 312:1–11.

Bloom IT, Bentley FR, Spain DA, and Garrison RN (1995) An experimental study of altered nitric oxide metabolism as a mechanism of cyclosporin-induced renal vasoconstriction. *Brit J Surg* 82:195–198.

Boonstra JG, Van der Woude FJ, Wever PC, Laterveer JC, Daha MR, and Kooten CV (1997) Expression and function of Fas (CD95) on human renal tubular epithelial cells. *J Am Soc Nephrol* 8:1517–1524.

Bortner CD, Hughes FM Jr, and Cidlowski JA (1997) A primary role for K^+ and Na^+ efflux in the activation of apoptosis. *J Biol Chem* 272:32436–32442.

Bronk SF, and Gores GJ (1993) pH-dependent nonlysosomal proteolysis contributes to lethal anoxic injury of rat hepatocytes. *Am J Physiol* 264:G744–G751.

Brown R (1996) The bcl-2 family of proteins. *Brit Med Bull* 53:466–477.

Bruschi SA, West KA, Crabb JW, Cupta RS, and Stevens JL (1993) Mitochondrial HSP60 (P1 Protein) and a HSP70-like protein (mortalin) are major targets for modification during *S*-(1,1,2,2-tetrafluoroethyl)-L-cysteine-induced nephrotoxicity. *J Biol Chem* 268:23157–23161.

Brush MD (2000) Recourse to death: A bevy of new products harnesses the power of flow cytometry for detecting apoptosis. *The Scientist* 14:25–28.

Buendia B, Santa-Maria A, and Courvalin JC (1999) Caspase-dependent proteolysis of integral and peripheral proteins of nuclear membranes and nuclear pore complex proteins during apoptosis. *J Cell Sci* 112:1743–1753.

Burckhardt G, and Wolff NA (2000) Structure of renal organic anion and cation transporters. *Am J Physiol* 278:F853–F866.

Bush KT, Tsukamoto T, and Nigam SK (2000) Selective degradation of E-cadherin and dissolution of E-cadherin-catenin complexes in epithelial ischemia. *Am J Physiol* 278:F847–F852.

Carafoli E, and Molinari M (1998) Calpain: A protease in search of a function? *Biochem Biophys Res Comm* 247:193–203.

Casciola-Rosen L, Nicholson DW, Chong T, Rowan KR, Thornberry NA, Miller DK, and Rosen A (1996) Apopain/CPP32 cleaves proteins that are essential for cellular repair: A fundamental principle of apoptotic death. *J Exp Med* 183:1957–1964.

Chen A, Wei CH, Lee WH, and Lin CY (1994) Experimental IgA nephropathy: Factors influencing IgA-immune complex deposition in the glomerulus. *Springer Sem Immunopathol* 16:97–103.

Chen J, and Mandel LJ (1997) Reversibility/irreversibility of anoxia-induced plasma membrane permeabilization. *J Am Soc Nephrol* 8:584.

Chen J, Liu X, Mandel LJ, and Schnellmann RG (2001) Progressive disruption of the plasma membrane during renal proximal tubule cellular injury. *Toxicol Appl Pharmacol* 171:1–11.

Cheng EH, Kirsch DG, Clem RJ, Ravi R, Kastan MB, Bedi A, Ueno K, and Hardwick JM (1997) Conversion of Bcl-2 to a Bax-like death effector by caspases. *Science* 278:1966–1968.

Chua BT, Guo K, and Li P (2000) Direct cleavage by the calcium-activated protease calpain can lead to inactivation of caspases. *J Biol Chem* 275:5131–5135.

Ciacci-Zanella JR, and Jones C (1999) Fumonisin B1, a mycotoxin contaminant of cereal grains, and inducer of apoptosis via the tumour necrosis factor pathway and caspase activation. *Food Chem Toxicol* 37:703–712.

Clem RJ, Cheng EH, Karp CL, Kirsch DG, Ueno K, Takahashi A, Kastan MB, Griffin DE, Earnshaw WC, Veliuona MA, and Hardwick JM (1998) Modulation of cell death by Bcl-XL through caspase interaction. *Proc Nat Acad Sci USA* 95:554–559.

Cohen GM (1997) Caspases: The executioners of apoptosis. *Biochem J* 326:1–16.

Croall DE, and Demartino GN (1991) Calcium-activated neutral protease (calpain) system: structure, function and regulation. *Physiol Reviews* 71:813–847.

Crompton M, Ellinger H, and Costi A (1988) Inhibition by cyclosporin A of a Ca^{2+}-dependent pore in heart mitochondria activated by inorganic phosphate and oxidative stress. *Biochem J* 255:357–360.

Cummings BS, McHowat J, and Schnellmann RG (2002) Role of an endoplasmic reticulum Ca^{2+}-independent phospholipase A_2 in oxidant-induced renal cell death. *Am J Physiol* 283:492–498.

Cummings BS, Mchowat J, and Schnellmann RG (2000) Phospholipase A_2s in cell injury and death. *J Pharmacol Exp Ther* 294:793–799.

Cummings BS, and Schnellmann RG (2002) Cisplatin-induced renal cell apoptosis: Caspase 3-dependent and -independent pathways. *J Pharmacol Exp Ther* 302:8–17.

Dartsch DC (1999) Mechanisms of toxicity, programmed cell death (apoptosis), in *Toxicology* Marquardt H, Schafer SG, McClellan R, Welsch F (Eds), San Diego: Academic Press, pp. 245–255.

De Broe ME, Paulus GJ, Verpooten GA, Roels F, Buyssens N, Wedeen R, and Tulkens PM (1984) Early effects of gentamicin, tobramycin and amikacin on the human kidney. *Kidney Int* 25:643–652.

Debiasi RL, Squier MK, Pike B, Wynes M, Dermody TS, Cohen JJ, and Tyler KL (1999) Reovirus-induced apoptosis is preceded by increased cellular calpain activity and is blocked by calpain inhibitors. *J Virol* 73:695–701.

Delles C, Haller T, and Dietl P (1995) A highly calcium-selective cation current activated by intracellular calcium release in MDCK cells. *J Physiol* 486:557–569.

Dennis EA (1994) Diversity of group types, regulation, and function of phospholipase A_2. *J Biol Chem* 269:13057–13060.

Doctor RB, Bennett V, and Mandel LJ (1993) Degradation of spectrin and ankyrin in the ischemic rat kidney. *Am J Physiol* 264:C1003-C1013.

Dong Z, Patel Y, Saikumar P, Weinberg JM, and Venkatachalam MA (1998) Development of porous defects in plasma membranes of adenosine triphosphate-depleted Madin-Darby Canine Kidney cells and its inhibition by glycine. *Lab Invest* 78:657–668.

Dong Z, Saikumar P, Weinberg JM, and Venkatachalam MA (1997) Internucleosomal DNA cleavage triggered by plasma membrane damage during necrotic cell death. *Am J Pathol* 151:1205–1213.

Duband-Goulet I, Courvalin JC, and Buendia B (1998) LBR, a chromatin and lamin binding protein from the inner nuclear membrane, is proteolyzed at late stages of apoptosis. *J Cell Sci* 111:1441–1451.

Dumas M, Maftah A, Bonte F, Ratinaud MH, Meybeck A, and Julien R (1995) Flow cytometric analysis of human epidermal cell ageing using two fluorescent mitochondrial probes. *C R Acad Sci III* 318:191–197.

Edelstein CL, Ling H, and Schrier RW (1997) The nature of renal cell injury. *Kidney Int* 51:1341–1351.

Edelstein CL, Shi Y, and Schrier RW (1999) Role of caspases in hypoxia-induced necrosis of rat renal proximal tubules. *J Am Soc Nephrol* 10:1940–1949.

Edelstein CL, Wieder ED, Yaqoob MM, Gengaro PE, Burke TJ, Nemenoff RA, and Schrier RW (1995) The role of cysteine proteases in hypoxia-induced rat renal proximal tubular injury. *Proc Natl Acad Sci USA* 92:7662–7666.

Edelstein CL, Yaqoob MM, Alkhunaizi AM, Gengaro PE, Nemenoff RA, Wang KK, and Schrier RW (1996a) Modulation of hypoxia-induced calpain activity in rat renal proximal tubules. *Kidney Int* 50:1150–1157.

Edelstein CL, Yaqoob MM, and Schrier RW (1996b) The role of the calcium-dependent enzymes nitric oxide synthase and calpain in hypoxia-induced proximal tubule injury. *Renal Failure* 18:501–511.

El Mouedden M, Laurent G, Mingeot-Leclercq MP, and Tulkens PM (2000) Gentamicin-induced apoptosis in renal cell lines and embryonic rat fibroblasts. *Toxicol Sci* 56:229–239.

Elfarra AA (1993) Aliphatic halogenated hydrocarbons, in *Toxicology of the Kidney*, 2nd ed., Hook JB, Goldstein RS (Eds), New York: Raven Press, pp. 387–413.

Endou H (1998) Recent advances in molecular mechanisms of nephrotoxicity. *Toxicol Lett* 102–103:29–33.

Fillastre JP, and Godin M (1991) An overview of drug-induced nephropathies, in *Mechanisms of Injury in Renal Disease and Toxicity*, Goldstein RS (Ed.), Boca Raton, Fl: CRC, pp. 123–147.

Friesen C, Herr I, Krammer PH, and Debatin KM (1996) Involvement of the CD95 (APO-1/FAS) receptor/ligand system in drug-induced apoptosis in leukemia cells. *Nature Med* 2:574–577.

Frisch SM, and Francis H (1994) Disruption of epithelial cell-matrix interactions induces apoptosis. *J Cell Biol* 124:619–626.

Frisch SM, and Ruoslahti E (1997) Integrins and anoikis. *Curr Opin Cell Biol* 9:701–706.

Fukino H, Hirai M, Hsueh YM, and Yamane Y (1984) Effect of zinc pretreatment on mercuric chloride-induced lipid peroxidation in the rat kidney. *Toxicol Appl Pharmacol* 73:395–401.

Garza-Quintero R, Weinberg JM, Ortega-Lopez J, Davis JA, and Venkatachalam MA (1993) Conservation of structure in ATP-depleted proximal tubules: Role of calcium, polyphosphoinositides, and glycine. *Am J Physiol* 265:F605–F623.

Gavrieli Y, Sherman Y, and Ben-Sasson S (1992) Identification of programmed cell death in situ via specific labeling of DNA fragmentation. *J Cell Biol* 119:493–501.

Gekle M, Schwerdt G, Freudinger R, Mildenberger S, Wilflingseder D, Pollack V, Dander M, and Schramek H (2000) Ochratoxin A induces JNK activation and apoptosis in MDCK-C7 cells at nanomolar concentrations. *J Pharmacol Exp Ther* 293:837–844.

Gekle M, and Silbernagl S (1994) The role of the proximal tubule in ochratoxin A nephrotoxicity *in vivo*: Toxodynamic and toxokinetic aspects. *Renal Physiol Biochem* 17:40–49.

Genestier L, Prigent AF, Paillot R, Quemeneur L, Durand I, Banchereau J, Revillard JP, and Bonnefoy-Bérard N (1998) Caspase-dependent ceramide production in Fas- and HLA class I-mediated peripheral T cell apoptosis. *J Biol Chem* 273:5060–5066.

Gilbert DN (1995) Aminoglycosides, in *Principles and practice of infectious diseases*, 4th ed., Mandell GL, Bennett JE, Dolin R (Eds), New York, NY: Churchill Livingstone, pp. 279–306.

Gobe G, Zhang X-J, Willgoss DA, Schoch E, Hogg NA, and Endre ZH (2000) Relationship between expression of Bcl-2 genes and growth factors in ischemic acute renal failure in the rat. *J Am Soc Nephrol* 11:454–467.

Goldstein RS, Smith PF, Tarloff JB, Contardi L, Rush GF, and Hook JB (1988) Biochemical mechanisms of cephaloridine nephrotoxicity. *Life Sci* 42:1809–1816.

Goligorsky MS, and Noiri E (1999) Duality of nitric oxide in acute renal injury. *Sem Nephrol* 19:263–271.

Gorczyca W, Melamed MR, and Darzynkiewicz Z (1993) Apoptosis of S-phase HL-60 cells induced by DNA topoisomerase inhibitors: detection of DNA strand breaks by flow cytometry using the in situ nick translation assay. *Toxicol Lett* 67:249–258.

Gordon JA, and Gattone VH (1986) Mitochondrial alterations in cisplatin-induced acute renal failure. *Am J Physiol* 250:F991-F998.

Grasl-Kraupp B, Ruttkay-Nedecky B, Koudelka H, Bukowska K, Bursch W, and Schulte-Hermann R (1995) In situ detection of fragmented DNA (TUNEL assay) fails to discriminate among apoptosis, necrosis, and autolytic cell death: A cautionary note. *Hepatology* 21:1465–1468.

Griffin JM, and Schnellmann RG (1997) Protein dephosphorylation in renal proximal tubular (RPT) injury: Effects of glycine, strychnine and calyculin A. *Toxicologist* 30:1558.

Groves CE, Hayden PJ, Lock EA, and Schnellmann RG (1993) Differential cellular effects in the toxicity of haloalkene and haloalkane cysteine conjugates to rabbit renal proximal tubules. *J Biochem Toxicol* 8:49–56.

Groves CE, and Schnellmann RG (1996) Suspensions of rabbit renal proximal tubules, in *Methods in Renal Toxicology*, Zalups RK and Lash LH (Eds), Boca Raton: CRC Press, pp. 147–162.

Grundemann D, Gorboulev V, Gambaryan S, Veyhl M, and Loepsell H (1994) Drug excretion mediated by a new prototype of polyspecific transporter. *Nature* 372:549–552.

Guidet B, and Shah SV (1989) Enhanced *in vivo* H_2O_2 generation by rat kidney in glycerol-induced renal failure. *Am J Physiol* 257:F440–F445.

Gulbins E, Bissonnette R, Mahbouti A, Martin S, Nishioka W, Brunner T, Baier G, Baier-Bitterlich G, Byrd C, and Lang F (1995) Fas-induced apoptosis is mediated by a ceramide-initiated RAS signaling pathway. *Immunity* 2:341–351.

Ha HC, and Snyder SH (1999) Poly(ADP-ribose) polymerase is a mediator of necrotic cell death by ATP depletion. *Proc Natl Acad Sci USA* 96:13978–13982.

Haimovitz-Friedman A, Kan C-C, Ehleiter D, Persaud RS, McLoughlin M, Fuks Z, and Kolesnick RN (1994) Ionizing radiation acts on cellular membranes to generate ceramide and induce apoptosis. *J Exp Med* 180:525–535.

Haimovitz-Friedman A, Kolesnick RN, and Fuks Z (1997) Ceramide signaling in apoptosis. *Brit Med Bull* 53:539–553.

Halleck MM, Liu H, North J, and Stevens JL (1997) Reduction of trans-4,5-dihydroxy-1,2-dithiane by cellular oxidoreductases activates gadd153/chop and grp78 transcription and induces cellular tolerance in kidney epithelial cells. *J Biol Chem* 272:21760–21766.

Halliwell B, and Gutteridge JM (1990) Role of free radicals and catalytic metal ions in human disease: An overview. *Meth Enzymol* 186:1–85.

Ham J, Babij C, Whitfield J, Pfarr CM, Lallemand D, Yaniv M, and Rubin LL (1995) A c-Jun dominant negative mutant protects sympathetic neurons against programmed cell death. *Neuron* 14:927–939.

Harman AW, and Maxwell MJ (1995) An evaluation of the role of calcium in cell injury. *Annu Rev Pharmacol Toxicol* 35:129–144.

Harriman JF, Liu XL, Aleo MD, Machaca K, and Schnellmann RG (2002) Endoplasmic reticulum Ca^{2+} signaling and calpains mediate renal cell death. *Cell Death Differ* 9:734–741.

Harriman JF, Waters-Williams S, Chu D-L, Powers JC, and Schnellmann RG (2000) Efficacy of novel calpain inhibitors in preventing renal cell death. *J Pharmacol Exp Ther* 294:1083–1087.

Hartfield PJ, Mayne GC, and Murray AW (1997) Ceramide induces apoptosis in PC12 cells. *FEBS Lett* 401:148–152.

Healy E, Dempsey M, Lally C, and Ryan MP (1998) Apoptosis and necrosis: mechanisms of cell death induced by cyclosporine A in a renal proximal tubular cell line. *Kidney Int* 54:1955–1966.

Helenius A, Trombetta ES, Hebert DN, and Simons JF (1997) Calnexin, calreticulin, and the folding of glycoproteins. *Trends Cell Biol* 7:193–200.

Hoth M, and Penner R (1993) Calcium release-activated calcium current in rat mast cells. *J Physiol* 465:359–386.

Hotz MA, Gong J, Traganos F, and Darzynkiewicz Z (1994) Flow cytometric detection of apoptosis: comparison of the assays of in situ DNA degradation and chromatin changes. *Cytometry* 15:237–244.

Hu Y, Benedict MA, Wu D, Inohara N, and Núñez G (1998) Bcl-XL interacts with Apaf-1 and inhibits Apaf-1-dependent caspase-9 activation. *Proc Natl Acad Sci USA* 95:4386–4391.

Imesch E, Moosmayer M, and Anner BM (1992) Mercury weakens membrane anchoring of Na-K-ATPase. *Am J Physiol* 262:F837–F842.

Irmler M, Steiner V, Ruegg C, Wajant H, and Tschopp J (2000) Caspase-induced inactivation of the anti-apoptotic TRAF1 during Fas ligand-mediated apoptosis. *FEBS Lett* 468:129–133.

Ishido M, Homma, ST, Leung PS, and Tohyama C (1995) Cadmium-induced DNA fragmentation is inhibitable by zinc in porcine kidney LLC-PK_1 cells. *Life Sci* 56:PL351–PL356.

Ishido M, Tohyama C, and Suzuki T (1999) Cadmium-bound metallothionein induces apoptosis in rat kidneys, but not in cultured kidney LLC-PK_1 cells. *Life Sci* 64:797–804.

Jacobs WR, Sgambati M, Gomez G, Vilaro P, Higdon M, Bell PD, and Mandel LJ (1991) Role of cytosolic Ca in renal tubule damage induced by anoxia. *Am J Physiol* 260:C545–C554.

Jariyawat S, Sekine T, Takeda M, Apiwattanakul N, Kanai Y, Sophasan S, and Endou H (1999) The interaction and transport of beta-lactam antibiotics with the cloned rat renal organic anion transporter 1. *J Pharmacol Exp Ther* 290:672–677.

Kacew S, and Bergeron S (1990) Pathogenic factors in aminoglycoside-induced nephrotoxicity. *Toxicol Lett* 51:241–259.

Kaloyanides GJ (1992) Drug-phospholipid interactions: Role in aminoglycoside nephrotoxicity. *Renal Failure* 14:351–357.

Kashgarian M (1995) Stress proteins induced by injury to epithelial cells. *Contemporary Issues in Nephrology* 30:75–95.

Kaufman, RJ (1999) Stress signaling from the lumen of the ER: Coordination of gene transcriptional and translational controls. *Genes Dev* 13:1211–1233.

Kern JC, and Kehrer JP (2002) Acrolein-induced cell death: a caspase-influenced decision between apoptosis and oncosis/necrosis. *Chem-Biol Interact* 139:79–95.

Kirsch DG, Doseff A, Nelson, Chau B, Lim D-S, de Souza-Pinto NC, Hansford R, Kastan MB, Lazebnik YA, and Hardwick JM (1999) Caspase-3-dependent cleavage of Bcl-2 promotes release of cytochrome c. *J Biol Chem* 274:21155–21161.

Kobryn CE, and Mandel LJ (1994) Decreased protein phosphorylation induced by anoxia in proximal renal tubules. *Am J Physiol* 267:C1073–C1079.

Kohjimoto Y, Kennington L, Scheid CR, and Honeyman RA (1999) Role of phospholipase A_2 in the cytotoxic effects of oxalate in cultured renal epithelial cells. *Kidney Int* 56:1432–1441.

Kohli V, Madden JF, Bentley RC, and Clavien PA (1999) Calpain mediates ischemic injury of the liver through modulation of apoptosis and necrosis. *Gastroenterol* 116:168–178.

Koide N, Narita K, Kato Y, Sugiyama T, Chakravortty D, Morikawa A, Yoshida T, and Yokochi T (1999) Expression of Fas and Fas ligand on mouse renal tubular epithelial cells in the generalized shwartzman reaction and its relationship to apoptosis. *Infect Immun* 67:4112–4118.

Kone BC, Brenner RM, and Gullans SR (1990) Sulfhydryl-reactive heavy metals increase cell membrane K^+ and Ca^{2+} transport in renal proximal tubule. *J Membr Biol* 113:1–12.

Krajewski S, Krajewska M, Shabaik A, Miyashita T, and Reed JC (1994) Immunohistochemical determination of *in vivo* distribution of Bax a dominant inhibitor of Bcl-2. *Am J Pathol* 145:1323–1336.

Kribben A, Wieder ED, Wetzels JF, Yu L, Gengaro PE, Burke TJ, and Schrier RW (1994) Evidence for role of cytosolic free calcium in hypoxia-induced proximal tubule injury. *J Clin Invest* 93:1922–1929.

Lanese DM, and Conger JD (1993) Effects of endothelin receptor antagonist on cyclosporine-induced vasoconstriction in isolated rat renal arterioles. *J Clin Invest* 91:2144–2149.

Lankiewicz S, Luetjens CM, Bui NT, Krohn AJ, Poppe M, Cole GM, Saido TC, and Prehn JHM (2000) Activation of calpain I converts excitotoxic neuron death into a caspase-independent cell death. *J Biol Chem* 275:17064–17071.

Lash LH, Hueni SE, and Putt DA (2001) Apoptosis, necrosis, and cell proliferation induced by *S*-(1,2-Dichlorovinyl)-L-cysteine in primary cultures of human proximal tubular cells. *Toxicol Appl Pharmacol* 177:1–16.

Laurent G, Kishore BK, and Tulkens PM (1990) Aminoglycoside-induced renal phospholipidosis and nephrotoxicity. *Biochem Pharmacol* 40:2383–2392.

Lavoie JN, Lambert H, Hickey E, Weber LA, and Landry J (1995) Modulation of cellular thermoresistance and actin filament stability accompanies phosphorylation-induced changes in the oligomeric structure of heat shock protein 27. *Mol Cell Biol* 15:505–516.

Leers MPG, Kolgen W, Bjorklund V, Bergman T, Tribbick G, Persson B, Bjorklund P, Ramaekers FCS, Bjorklund B, Nap M, Jornvall H, and Schutte B (1999) Immunocytochemical detection and mapping of a cytokeratin 18 neo-epitope exposed during early apoptosis. *J Pathol* 187:567–572.

Leist M, Single B, Castoldi AF, Kuhnle S, and Nicotera P (1997) Intracellular adenosine triphosphate (ATP) concentration: A switch in the decision between apoptosis and necrosis. *J Exp Med* 185:1481–1486.

Lemasters JJ, DiGuiseppi J, Nieminen AL, and Herman B (1987) Blebbing free Ca^{2+} and mitochondrial membrane potential preceding cell death in hepatocytes. *Nature* 325:78–81.

Lemasters JJ (1999) Necrapoptosis and the mitochondrial permeability transition: shared pathways to necrosis and apoptosis. *Am J Physiol* 276:G1–G6.

Levin S, Bucci TJ, Cohen SM, Fix AS, Hardisty JF, Legrand EK, Maronpot RR, and Trump BF (1999) The nomenclature of cell death: Recommendations of an ad hoc Committee of the Society of Toxicologic Pathologists. *Toxicol Pathol* 27:484–490.

Li H, Zhu H, Xu CJ, and Yan J (1998) Cleavage of BID by caspase 8 mediates the mitochondrial damage in the Fas pathway of apoptosis. *Cell* 94:491–501.

Li P, Nijhawan D, Budihardjo I, Srinivasula SM, Ahmad M, Alnemri ES, and Wang X (1997) Cytochrome c and dATP-dependent formation of Apaf-1/caspase-9 complex initiates an apoptotic protease cascade. *Cell* 91:479–489.

Lieberthal W, Menza SA, and Levine JS (1998) Graded ATP depletion can cause necrosis or apoptosis of cultured mouse proximal tubular cells. *Am J Physiol* 274:F315–F327.

Lieberthal W, Triaca V, and Levine J (1996) Mechanism of death induced by cisplatin in proximal tubular epithelial cells: apoptosis vs necrosis. *Am J Physiol* 270:F700–F708.

Liu H, Bowes RC III, van de Water B, Sillence C, Nagelkerke JF, and Stevens JL (1997) Endoplasmic reticulum chaperones Grp78 and calreticulin prevent oxidative stress, Ca^{2+} disturbances, and cell death in renal epithelial cells. *J Biol Chem* 272:21751–21759.

Liu H, Miller E, van de Water B, and Stevens JL (1998) Endoplasmic reticulum stress proteins block oxidant-induced Ca^{2+} increases and cell death. *J Biol Chem* 274:12858–12862.

Liu X, Harriman JF, and Schnellmann RG (2002) Cytoprotective properties of novel nonpeptide calpain inhibitors in renal cells. *J Pharmacol Exp Ther* 302:88–94.

Liu X, and Schnellmann RG (2003) Calpain mediates progressive plasma membrane permeability and proteolysis of cytoskeleton-associated paxillin, talin, and vinculin during renal cell death. *J Pharmacol Exp Ther* 304:63–70.

Lofvenberg L, and Backman L (1999) Calpain-induced proteolysis of beta-spectrins. *FEBS Lett* 443:89–92.

Lohr JW, Willsky GR, and Acara MA (1998) Renal Drug Metabolism. *Pharmacol Rev* 50:107–141.

Lorz C, Ortiz A, Justo P, Gonzalez-Cuadrado S, Duque N, Gomez-Guerrero C, and Egido J (2000) Proapoptotic Fas ligand is expressed by normal kidney tubular epithelium and injured glomeruli. *J Am Soc Nephrol* 11:1266–1277.

Lovis C, Mach F, Donati YR, Bonventre JV, and Polla BS (1994) Heat shock proteins and the kidney. *Renal Failure* 16:179–192.

Lund BO, Miller DM, and Woods JS (1993) Studies on Hg(II)-induced H_2O_2 formation and oxidative stress *in vivo* and *in vitro* in rat kidney mitochondria. *Biochem Pharmacol* 45:2017–2024.

Luo X, Budihardjo I, Zou H, Slaughter C, and Wang X (1998) Bid, a Bcl2 interacting protein, mediates cytochrome c release from mitochondria in response to activation of cell surface death receptors. *Cell* 94:481–490.

Ma H, Yang HQ, Takano E, Hatanaka M, and Maki M (1994) Amino-terminal conserved region in proteinase inhibitor domain of calpastatin potentiates its calpain inhibitory activity by interacting with calmodulin-like domain of the proteinase. *J Biol Chem* 269:24430–24436.

Maeno E, Ishizaki Y, Kanaseki T, Hazama A, and Okada Y (2000) Normotonic cell shrinkage because of disordered volume regulation is an early prerequisite to apoptosis. *Proc Natl Acad Sci USA* 97:9487–9492.

Majno G, and Joris I (1995) Apoptosis, oncosis, and necrosis. An overview of cell death. *Am J Pathol* 146:3–15.

Malis CD, and Bonventre JV (1986) Mechanism of calcium potentiation of oxygen free radical injury to renal mitochondria. A model for post-ischemic and toxic mitochondrial damage. *J Biol Chem* 261:14201–14208.

Malis CD, and Bonventre JV (1988) Susceptibility of mitochondrial membranes to calcium and reactive oxygen species: implications for ischemic and toxic tissue damage. *Prog Clin Biol Res* 282:235–259.

Mandel LJ (1985) Metabolic substrates, cellular energy production, and the regulation of proximal tubular transport. *Annu Rev Physiol* 47:85–101.

Massari S, and Azzone GF (1972) The equivalent pore radius of intact and damaged mitochondria and the mechanism of active shrinkage. *Biochim Biophys Acta* 283:23–29.

McCarty NA, and O'Neil RG (1991) Calcium-dependent control of volume regulation in renal proximal tubule cells. II. Roles of dihydropyridine-sensitive and -insensitive Ca^{2+} entry pathways. *J Membr Biol* 123:161–170.

McHowat J, Miller GW, Creer MH, and Schnellmann RG (1995) The role of calcium-independent plasmalogen-selective phospholipase A_2 (CIPS-PLA2) in oxidant injury. The *Toxicologist* 15:1485.

Megyesi J, Safirstein RL, and Price PM (1998) Induction of $p21^{WAF1/CIP1/SDI1}$ in kidney tubule cells affects the course of cisplatin-induced acute renal failure. *J Clin Invest* 101:777–782.

Megyesi J, Udvarhelyi N, Safirstein RL, and Price PM (1996) The p53-independent activation of transcription of $p21^{WAF1/CIP1/SDI1}$ after acute renal failure. *Am J Physiol* 271:F1211–F1216.

Meng X, and Reeves WB (2000) Effects of chloride channel inhibitors on H_2O_2-induced renal epithelial cell injury. *Am J Phyisol* 278:F83–F90.

Mihatsch MJ, Gudat F, Ryffel B, and Thiel G (1994) Cyclosporine nephropathy, in *Renal Pathology: With Clinical and Functional Correlations*, 2nd ed., Tisher CC, and Brenner MB (Eds), Philadelphia: J.B. Lippincott Co, pp. 1641–1681.

Miller GW, and Schnellmann RG (1993) Cytoprotection by inhibition of chloride channels: the mechanism of action of glycine and strychnine. *Life Sci* 53:1211–1215.

Mimic-Oka J, Simic T, Djukanovix L, Reljc Z, and Davicevic Z (1999) Alteration in plasma antioxidant capacity in various degrees of chronic renal failure. *Clin Nephrol* 51:233–241.

Mingeot-Leclercq M, and Tulkens PM (1999) Aminoglycosides: Nephrotoxicity. *Antimicrob Agents Chemother* 43:1003–1012.

Mohan PS, and Nixon RA (1995) Purification and properties of high molecular weight calpastatin from bovine brain. *J Neurochem* 64:859–866.

Molitoris BA, Dahl R, and Geerdes A (1992) Cytoskeletal disruption and apical redistribution of proximal tubule Na^+-K^+-ATPase during ischemia. *Am J Physiol* 263:F488–F495.

Muzio M, Chinnaiyan AM, Kischkel FC, O'Rourke K, Shevchenko A, Ni J, Scaffidi C, Bretz JD, Zhang M, Gentz R, Mann M, Krammer PH, Peter ME, and Dixit VM (1996) FLICE, a novel FADD-homologous ICE/CED-3-like protease, is recruited to the CD95 (Fas/APO-1) death-inducing signaling complex. *Cell* 85:817–827.

Nicholson DW (1999) Caspase structure, proteolytic substrates, and function during apoptotic cell death. *Cell Death Differ* 6:1028–1042.

Nicotera P, Hartzell P, Baldi C, Svensson SA, Bellomo G, and Orrenius S (1986) Cystamine induces toxicity in hepatocytes through the elevation of cytosolic Ca^{2+} and the stimulation of a nonlysosomal proteolytic system. *J Biol Chem* 261:14628–14635.

Novgorodov SA, Gudz TI, Brierley GP, and Pfeiffer DR (1994) Magnesium ion modulates the sensitivity of the mitochondrial permeability transition pore to cyclosporin A and ADP. *Arch Biochem Biophys* 311:219–228.

Nurko S, Sogabe K, Davis JA, Roeser NF, Defrain M, Chien A, Hinshaw D, Athey B, Meixner W, Venkatachalam MA, and Weinberg JM (1996) Contribution of actin cytoskeletal alterations to ATP depletion and calcium-induced proximal tubule cell injury. *Am J Physiol* 270:F39–F52.

Obeid LM, Linardic CM, Karolak LA, and Hannun YA (1993) Programmed cell death induced by ceramide. *Science* 259(5102):1769–1771.

Okuda M, Masaki K, Fukatsu S, Hashimoto Y, and Inui K (2000) Role of apoptosis in cisplatin-induced toxicity in the renal epithelial cell line LLC-PK_1. *Biochem Pharmacol* 59:195–201.

Olsen S, and Solez K (1994) Acute tubular necrosis and toxic renal injury, in *Renal Pathology: With Clinical and Functional Correlations*, 2nd ed., Tisher CC and Brenner BM (Eds), Philadelphia, PA: J.B.Lippincott Company, pp. 769–809.

Orth K, Chinnaiyan AM, Garg M, Froelich CJ, and Dixit VM (1996) The CED-3/ICE-like protease Mch2 is activated during apoptosis and cleaves the death substrate lamin A. *J Biol Chem* 271:16443–16446.

Ortiz A, Lorz C, Catalan MP, Danoff TM, Yamasaki Y, Egido J, and Neilson EG (2000) Expression of apoptosis regulatory proteins in tubular epithelium stressed in culture or following acute renal failure. *Kidney Int* 57:969–981.

Patel V, Luft FC, Yum MN, Patel B, Zeman W, and Kleit SA (1975) Enzymuria in gentamicin-induced kidney damage. *Antimicrob Agents Chemother* 7:364–369.

Petronilli V, Costantini P, Scorrano L, Colonna R, Passamonti S, and Bernardi P (1994) The voltage sensor of the mitochondrial permeability transition pore is tuned by the oxidation-reduction state of vicinal thiols. *J Biol Chem* 269:16638–16642.

Pronk GJ, Ramer K, Amiri P, and Williams LT (1996) Requirement of an ICE-like protease for induction of apoptosis and ceramide generation by REAPER. *Science* 271:808–810.

Putney JW, and Bird GSJ (1993) The signal for capacitative calcium entry. *Cell* 75:199–201.

Remuzzi G, and Bertani T (1989) Renal vascular and thrombotic effects of cyclosporine. *Am J Kidney Dis* 13:261–272.

Robbins SL (1994) *Pathologic Basis of Disease.* Coltran RS, Kumar V, and Robbins SL (Eds), Philadelphia, PA: Saunders, p. 8.

Rose UM, Bindels RJM, Jansen JWCM, and Van Os CH (1994) Effects of Ca^{2+} channel blockers, low Ca^{2+} medium and glycine on cell Ca^{2+} and injury in anoxic rabbit proximal tubules. *Kidney Int* 46:223–229.

Rose UM, Bindels RJM, Vis A, Jansen JWC, and Van Os CH (1993) The effect of L-type Ca^{2+} channel blocker on anoxia-induced increases in intracellular Ca^{2+} concentration in rabbit proximal tubule cells in primary culture. *Eur J Physiol* 423:378–386.

Safirstein R, Miller P, and Guttenplan JB (1984) Uptake and metabolism of cisplatin by rat kidney. *Kidney Int* 25:753–758.

Sagara Y, and Inesi G (1991) Inhibition of the sarcoplasmic reticulum Ca^{2+} transport ATPase by thapsigargin at subnanomolar concentrations. *J Biol Chem* 266:13503–13506.

Saido TC, Sorimachi H, and Suzuki K (1994) Calpain: New perspectives in molecular diversity and physiological-pathological involvement. *FASEB J* 8:814–822.

Sapirstein A, Spech RA, Witzgall R, and Bonventre JV (1996) Cytosolic phospholipase A_2 (PLA_2) but not secretory PLA_2, potentiates hydrogen peroxide cytotoxicity in kidney epithelial cells. *J Biol Chem* 271:21505–21513.

Schnellmann RG (1997) Pathophysiology of Nephrotoxic Cell Injury, in *Diseases of the Kidney*, Schrier RW, and Gottschalk CW (Eds), Boston: Little Brown, pp. 1049–1067.

Schnellmann RG, and Griner RD (1994) Mitochondrial mechanisms of tubular injury. Mechanisms of Injury, in *Renal Disease and Toxicity*, Goldstein RS (Ed.), Boca Raton: CRC Press.

Schnellmann RG, Swagler AR, and Compton MM (1993) Absence of endonuclease activation during acute cell death in renal proximal tubules. *Am J Physiol* 265:C485–C490.

Schnellmann RG, and Williams SW (1998) Proteases in renal cell death: Calpains mediate cell death produced by diverse toxicants. *Renal Failure* 20:679–686.

Schnellmann RG, Yang X, and Cross TJ (1994) Calpains play a critical role in renal proximal tubule (RPT) cell death. *Canad J Physiol Pharmacol* 72:602.

Schnellmann RG (1988) Mechanisms of t-butyl hydroperoxide-induced toxicity to rabbit renal proximal tubules. *Am J Physiol* 255:C28–C33.

Schumer M, Colombel MC, Sawczuk IS, Gobe G, Connor J, O'Toole KM, Olsson CA, Wise GJ, and Buttyan R (1992) Morphologic, biochemical, and molecular evidence of apoptosis during the reperfusion phase after brief periods of renal ischemia. *Am J Pathol* 140:831–838.

Sekine T, Wananabe N, Hosoyamada M, Kanai Y, and Endou H (1997) Expression cloning and characterization of a novel multispecific organic anion transporter. *J Biol Chem* 272:18526–18529.

Shihab FS, Andoh TF, Tanner AM, Yi H, and Bennett WM (1999) Expression of apoptosis regulatory genes in chronic cyclosporine nephrotoxicity favors apoptosis. *Kidney Int* 56:2147–2159.

Slee EA, Harte MT, Kluck RM, Wof BB, Casiano CA, Newmeyer DD, Wang H-G, Reed JC, Nicholson DW, Alnemri ES, Green DR, and Martin SJ (1999) Ordering the cytochrome c-initiated caspase cascade: hierarchical activation of caspases-2, -3, -6, -7, -8, and -10 in a caspase-9-dependent manner. *J Cell Biol* 144:281–292.

Smith MW, Ambudkar IS, Phelps PC, Regec AL, and Trump BF (1987) $HgCl_2$-induced changes in cytosolic Ca^{2+} of cultured rabbit renal tubular cells. *Biochim Biophys Acta* 931:130–142.

Smith MW, Phelps PC, and Trump BF (1991) Cytosolic Ca^{2+} deregulation and blebbing after $HgCl_2$ injury to cultured rabbit proximal tubule cells as determined by digital imaging microscopy. *Proc Natl Acad Sci USA* 88:4926–4930.

Spiegel S, Foster D, and Kolesnick R (1996) Signal transduction through lipid second messengers. *Curr Opin Cell Biol* 8:159–167.

Squier MK, Miller AC, Malkinson AM, and Cohen JJ (1994) Calpain activation in apoptosis. J *Cell Physiol* 159:229–237.

Squier MK, Sehnert AJ, Sellins KS, Malkinson AM, Takano E, and Cohen JJ (1999) Calpain and calpastatin regulate neutrophil apoptosis. *J Cell Physiol* 178:311–319.

Sun X-M, MacFarlane M, Zhuang J, Wolf BB, Green DR, and Cohen GM (1999) Distinct caspase cascades are initiated in receptor-mediated and chemical-induced apoptosis. *J Biol Chem* 274:5053–5060.

Suzuki K, Saido TC, and Hirai S (1992) Modulation of cellular signals by calpain. *Annals NY Acad Sci* 674:218–227.

Sweet DH, Wolff NA, and Pritchard JB (1997) Expression cloning and characterization of ROAT1. The basolateral organic anion transporter in rat kidney. *J Biol Chem* 272:30088–30095.

Szabo I, and Zoratti M (1991) The giant channel of the inner mitochondrial membrane is inhibited by cyclosporin A. *J Biol Chem* 266:3376–3379.

Takahashi A, Musy PY, Martins LM, Poirier GG, Moyer RW, and Earnshaw WC (1996) CrmA/SPI-2 inhibition of an endogenous ICE-related protease responsible for lamin A cleavage and apoptotic nuclear fragmentation. *J Biol Chem* 271:32487–32490.

Takano T, Soltoff SP, Murdaugh S, and Mandel LJ (1985) Intracellular respiratory dysfunction and cell injury in short term anoxia of rabbit renal proximal tubules. *J Clin Invest* 76:2377–2384.

Takeda M, Kobayashi M, Shirato I, and Endou H (1998) Cyclosporine induces apoptosis of mouse terminal proximal straight tubule cells. *Nephron* 80:121–122.

Takeda M, Kobayashi M, Shirato I, Osaki K, and Endou H (1997) Cisplatin-induced apoptosis of immortalized mouse proximal tubule cells is mediated by interleukin-1β converting enzyme family proteases but inhibited by overexpression of Bcl-2. *Arch Toxicol* 71:612–621.

Takeda Y, Tashima M, Takahashi A, Uchiyama T, and Okazaki T (1999) Ceramide generation in nitric oxide-induced apoptosis. Activation of magnesium-dependent neutral sphingomyelinase via caspase-3. *J Biol Chem* 274:10654–10660.

Thastrup O, Cullen PJ, Drobak B, Hanley MR, and Dawson AP (1990) Thapsigargin, a tumor promoter, discharges intracellular Ca^{2+} stores by specific inhibition of the endoplasmic reticulum Ca^{2+}-ATPase. *Proc Natl Acad Sci USA* 87:2466–2470.

Thornberry NA, and Lazebnik Y (1998) Caspases: Enemies within. *Science* 281:1312–1316.

Thornberry NA, Rano TA, Peterson EP, Rasper DM, Timkey T, Garcia-Calvo M, Houtzager VM, Nordstrom PA, Roy S, Vaillancourt JP, Chapman KT, and Nicholson DW (1997) A combinatorial approach defines specificities of members of the caspase family and granzyme B. Functional relationships established for key mediators of apoptosis. *J Biol Chem* 272:17907–17911.

Thornberry NA (1997) The caspase family of cysteine proteases. *Brit Med Bull* 53:478–490.

Trauth BC, Klas C, Peters AM, Matzku S, Möller P, Falk W, Debatin KM, and Krammer PH (1989) Monoclonal antibody-mediated tumor regression by induction of apoptosis. *Science* 245:301–305.

Trifillis AL, Kahng MW, and Trump BF (1981a) Metabolic studies of glycerol-induced acute renal failure in the rat. *Exp Mol Pathol* 35:1–13.

Trifillis AL, Kahng MW, and Trump BF (1981b) Metabolic studies of $HgCl_2$-induced acute renal failure in the rat. *Exp Mol Pathol* 35:14–24.

Trost LC, and Lemasters JL (1997) Role of the mitochondrial permeability transition in salicylate toxicity to cultured rat hepatocytes: Implications for the pathogenesis of Reye's syndrome. *Toxicol Appl Pharmacol* 147:431–441.

Trump BF, and Berezesky IK (1994) Cellular and molecular pathobiology of reversible and irreverible injury, in *Methods in Toxicology*, Tyson CA, and Frazier JM (Eds), San Diego: Academic Press, Inc., pp. 1–22.

Tune BM (1997) Nephrotoxicity of beta-lactam antibiotics: Mechanisms and strategies for prevention. *Pediatr Nephrol* 11:768–772.

Ueda N, and Shah SV (1992) Endonuclease-induced DNA damage and cell death in oxidant injury to renal tubular epithelial cells. *J Clin Invest* 90:2593–2597.

Ueda N, and Shah SV (2000) Tubular cell damage in acute renal failure-apoptosis, necrosis, or both. *Nephrol Dial Transplant* 15:318–323.

Ueda N, Walker PD, Hsu SM, and Shah SV (1995) Activation of a 15-kd endonuclease in hypoxia/reoxygenation injury without morphological features of apoptosis. *Proc Natl Acad Sci USA* 92:7202–7206.

Van Engeland M, Ramaekers FC, Schutte B, and Reutelingsperger CP (1996) A novel assay to measure loss of plasma membrane asymmetry during apoptosis of adherent cells in culture. *Cytometry* 24:131–139.

Van Why SK, Hildebrandt F, Ardito T, Mann AS, Siegel NJ, and Kashgarian M (1992) Induction and intracellular localization of HSP-72 after renal ischemia. *Am J Physiol* 263:F769–F775.

Vandewalle A, Farman AN, Morin JP, Fillastre JP, Hatt PY, and Bonvalet JP (1981) Gentamicin incorporation along the nephron: autoradiographic study on isolated tubules. *Kidney Int* 19:529–539.

Verheij M, Bose R, Lin XH, Yao B, Jarvis WD, Grant S, Birrer MJ, Szabo E, Zon LI, Kyriakis JM, Haimovitz-Friedman A, Fuks Z, and Kolesnick RN (1996) Requirement for ceramide-initiated SAPK/JNK signalling in stress-induced apoptosis. *Nature* 380:75–79.

Vermes I, Haanen C, Steffens-Nakken H, and Reutelingsperger C (1995) A novel assay for apoptosis. Flow cytometric detection of phosphatidylserine expression on early apoptotic cells using fluorescein labelled Annexin V. *J Immunol Meth* 184:39–51.

Von Recklinghausen F (1910) *Untersuchungen uber rachitis und osteomalacie.* Verlag Gustav Fischer, Jena.

Walisser JA, and Thies RL (1999) Poly(ADP-ribose) polymerase inhibition in oxidant-stressed endothelial cells prevents oncosis and permits caspase activation and apoptosis. *Exp Cell Res* 251:401–413.

Walker PD, Das C, and Shah SV (1986) Cyclosporin A induced lipid peroxidation in renal cortical mitochondria. *Kidney Int* 29:311.

Walker PD, and Shah SV (1987) Gentamicin enhanced production of hydrogen peroxide by renal cortical mitochondria. *Am J Physiol* 253:C495–C499.

Walker PD, and Shah SV (1991) Hydrogen peroxide cytotoxicity in LLC-PK_1 cells: A role for iron. *Kidney Int* 40:891–898.

Wang C, and Salahudeen AK (1994) Cyclosporin nephrotoxicity: attenuation by an antioxidant-inhibitor of lipid peroxidation *in vitro* and *in vivo. Transplantation* 58:940–946.

Wang KKW, and Yuen P-W (1994) Calpain inhibition: An overview of its therapeutic potential. *Trends Pharm Sci* 15:412–419.

Wang Y, Knowlton AA, Christensen TG, Shih T, and Borkan SC (1999) Prior heat stress inhibits apoptosis in adenosine triphospate-depleted renal tubular cells. *Kidney Int* 55:2224–2235.

Waters SL, and Schnellmann RG (1996) Extracellular acidosis and chloride channel inhibitors act in the late phase of cellular injury to prevent death. *J Pharmacol Exp Ther* 278:1012–1017.

Waters SL, Sarang SS, Wang KKW, and Schnellmann RG (1997a) Calpains mediate calcium and chloride influx during the late phase of cell injury. *J Pharmacol Exp Ther* 283:1177–1184.

Waters SL, Wong JK, and Schnellmann RG (1997b) Depletion of endoplasmic reticulum calcium stores protects against hypoxia- and mitochondrial inhibitor-induced cellular injury and death. *Biochem Biophys Res Commun* 240:57–60.

Waters SL, and Schnellmann RG (1998) Examination of the mechanisms of action of diverse cytoprotectants in renal cell death. *Toxicol Pathol* 26:58–63.

Weinberg JM, Harding PG, and Humes HD (1982) Mitochondrial bioenergetics during the initiation of mercuric chloride-induced renal injury. I. Direct effects of *in vitro* mercuric chloride on renal cortical mitochondrial function. *J Biol Chem* 257:60–67.

Weinberg JM (1991) The cell biology of ischemic renal injury. *Kidney Int* 39:476–500.

Welder AA, and Acosta D (1994) Enzyme leakage as an indicator of cytotoxicity in cultured cells, in *Methods in Toxicolog*, Tyson CA, and Frazier JM (Eds), San Diego: Academic Press, Inc., pp. 46–49.

Westwick JK, Bielawska AE, Dbaibo G, Hannun YA, and Brenner DA (1995) Ceramide activates the stress-activated protein kinases. *J Biol Chem* 270:22689–22692.

Wetzels JFM, Yu L, Wang X, Kribben A, Burke TJ, and Schrier RW (1993) Calcium modulation and cell injury in isolated rat proximal tubules. *J Pharmacol Exp Ther* 267:176–180.

Widlak P, Li P, Wang X, and Garrard WT (2000) Cleavage preferences of the apoptotic endonuclease DFF40 (caspase-activated Dnase or nuclease) on naked DNA and chromatin substrates. *J Biol Chem* 275:8226–8232.

Wood DE, and Newcomb EW (2000) Cleavage of Bax enhances its cell death function. *Exp Cell Res* 256:375–382.

Wood DE, Thomas A, Devi LA, Berman Y, Beavis RC, Reed JC, and Newcomb EW (1998) Bax cleavage is mediated by calpain during drug-induced apoptosis. *Oncogene* 17:1069–1078.

Yan M, Dai T, Deak JC, Kyriakis JM, Zon LI, Woodgett JR, and Templeton DJ (1994) Activation of stress-activated protein kinase by MEKK1 phosphorylation of its activator SEK1. *Nature* 372:798–800.

Yu L, Gengaro PE, Niederberger M, Burke TJ, and Schrier RW (1994) Nitric oxide: A mediator in rat tubular hypoxia/reoxygenation injury. *Proc Natl Acad Sci USA* 91:1691–1695.

Zalups RK (1993) Early aspects of the intrarenal distribution of mercury after the intravenous administration of mercuric chloride. *Toxicology* 79:215–228.

Zalups RK (1998) Basolateral uptake of mercuric conjugates of *N*-acetylcysteine and cysteine in the kidney involves the organic anion transport system. *J Toxicol Env Health* 55:13–29.

Zalups RK (2000) Molecular interactions with mercury in the kidney. *Pharmacol Rev* 52:113–143.

Zhan Y, Van de Water B, Wang Y, and Stevens JL (1999) The roles of caspase-3 and bcl-2 in chemically-induced apoptosis but not necrosis of renal epithelial cells. *Oncogene* 18:6505–6512.

Zhu Y, O'Neill S, Saklatvala J, Tassi L, and Mendelsohn ME (1994) Phosphorylated HSP27 associates with the activation-dependent cytoskeleton in human platelets. *Blood* 84:3715–3723.

Zhang MI, and O'Neil RG (1996a) A regulated calcium channel in apical membranes of renal proximal tubule cells. *Am J Physiol* 271:C1757–C1764.

Zhang MI, and O'Neil RG (1996b) An L-type calcium channel in renal epithelial cells. *J Membr Biol* 259–266.

Zhong Z, Arteel G, Connor HD, Yin M, Frankenberg MV, Stachlewitz RF, Raleigh JA, Mason RP, and Thurman RG (1998) Cyclosporin A increases hypoxia and free radical production in rat kidneys: Prevention by dietary glycine. *Am J Physiol* 275:F595-F604.

Zhou H, Miyaji T, Kato A, Fujigaki Y, Sano K, and Hishida A (1999) Attenuation of cisplatin-induced acute renal failure is associated with less apoptotic cell death. *J Lab Clin Med* 134:649–658.

Zoratti M, and Szabo I (1995) The mitochondrial permeability transition. *Biochim Biophys Acta* 1241:139–176.

8

Signal Transduction in Renal Cell Repair and Regeneration

Bob van de Water, Raoef Imamdi, and Marjo de Graauw

INTRODUCTION

Acute renal failure (ARF) in humans is caused by either ischemia and reperfusion (IR) or nephrotoxic chemicals, and the proximal tubular epithelium is generally the primary target. Extensive research on molecular mechanisms of acute renal tubular pathologies has led to improved insight into the molecular and cellular mechanisms of renal cell injury. Despite these efforts, ARF still remains an important clinical problem. Therefore, further basic research to better understand the molecular basis of renal cell injury and repair remains crucial for identifying potential novel therapeutic strategies, either to combat the initiation phase of ARF or to accelerate the renal regeneration process. Such strategies may interfere with any of the phases that can be recognized in the ARF and regeneration process: (sublethal) injury of proximal tubular epithelial cells, initiation of cellular repair, survival or death programs, and regeneration of damaged tubules by increased proliferation followed by differentiation.

During each phase, a specific pattern of the expression or activity of a variety of (stress response) proteins is found, which is essential for either cell repair, survival, or death programs, as well as the regeneration and differentiation pathways. Identification of these proteins and understanding their individual roles is required to identify potential novel targets for future therapeutic strategies. In this chapter, the different molecular events that take place after cell injury are discussed, in particular in the light of cell repair processes to prevent the onset of apoptosis. Additionally, the changes in the expression and activity of growth factors and their downstream pathways will be discussed with respect to cell renewal and kidney restructuring.

The clinical features of ARF are mostly related to IR or hypoxic injury. Likewise, most experimental work on molecular mechanisms of proximal tubular cell injury followed by regeneration has used renal IR as a model. Therefore, results describing molecular mechanisms of nephrotoxicity will not be limited to nephrotoxicant-induced ARF but will also include IR experiments.

SIGNAL TRANSDUCTION AND STRESS RESPONSES IN DAMAGED CELLS

Biochemical Perturbations and Cellular Stress Responses

Damage of renal epithelial cells *in vitro* as well as *in vivo* by either IR or nephrotoxicants causes increased expression of a variety of (early stress response) proteins. These include proteins that (Safirstein, 1994; Safirstein et al., 1990; Stevens et al., 2000):

- regulate gene transcription (Egr, Gro, Fos, and Jun)
- are involved in repair of damaged proteins (molecular chaperones such as heat shock proteins and glucose regulated proteins)
- are antioxidant enzymes (metallothioneins, superoxide dismutase, and glutathione *S*-transferases)
- regulate structural reorganization of cells as well as their direct environment (actin, extracellular matrix components, and metalloproteinases)
- are components of the machinery that controls mitogenesis (growth factors and their receptors, cyclins, histones)

Some proteins are often upregulated after nephrotoxicant exposure. For example, heat shock proteins are induced by a variety of different conditions, including exposure to $CdCl_2$, $HgCl_2$, cisplatin, oxidative stress, and nephrotoxic cysteine *S*-conjugates (Chen et al., 1992a; Goering et al., 2000; Hernandez-Pando et al., 1995; Liu et al., 1996a, 1996b; Satoh et al., 1994); similarly, glucose regulated proteins (GRPs) are induced by nephrotoxic cysteine *S*-conjugetes, iodoacetamide, oxidative stress, as well as tunicamycin (Halleck et al., 1997a, 1997b; Liu et al., 1997; Zinszer et al., 1998).

Induction of these diverse stress response related genes are most likely mediated by upstream biochemical perturbations, of which the correct maintenance is crucial for protein function and cellular homeostasis. The most prominent and best studied biochemical parameters in renal cells are intracellular free calcium perturbations, (non)protein thiol modification, and oxidative stress (Chen and Stevens, 1991; Chen et al., 1990; Nicotera et al., 1990; Reed, 1990; van de Water et al., 1994). These early changes in the cellular homeostasis feed signaling pathways that activate repair processes (see below), but at the same time may initiate the expression of proapoptotic proteins to enable removal of injured cells in the event that cellular repair is insufficient to prolong normal cellular homeostasis. An organelle specific perturbation (for example, depletion of calcium stores from the endoplasmic reticulum (ER) vs. increases in free calcium in the matrix of the mitochondria) is likely to activate different stress response routes or repair pathways. The exact protein expression profile will also be chemical-dependent and, likewise, largely reflect the various biochemical perturbations at a specific cellular location that are induced by a particular compound. Several studies have sorted out the relationships between biochemical perturbations in renal cells and altered gene expression of proteins (Figure 8.1).

(Non)Protein Thiol Modifications

(Non)protein thiol modifications are caused by various nephrotoxicants, including nephrotoxic cysteine conjugates, cephaloridine, and heavy metals, but also by IR (Chen et al., 1990; Nath et al., 1996; Maines et al., 1999; Tune et al., 1989). Free protein thiol group depletion may involve an overall decrease in protein thiols or protein specific thiol modification (Reed, 1990). Within the cell, these thiol changes may be site-dependent and occur specifically at places where toxicants are bioactivated or accumulate. The cause of protein thiol modification by toxicants can either be direct, through covalent modification of –SH groups of reactive intermediates, or indirect, due to oxidative stress or depletion or oxidation of glutathione (GSH). In this respect,

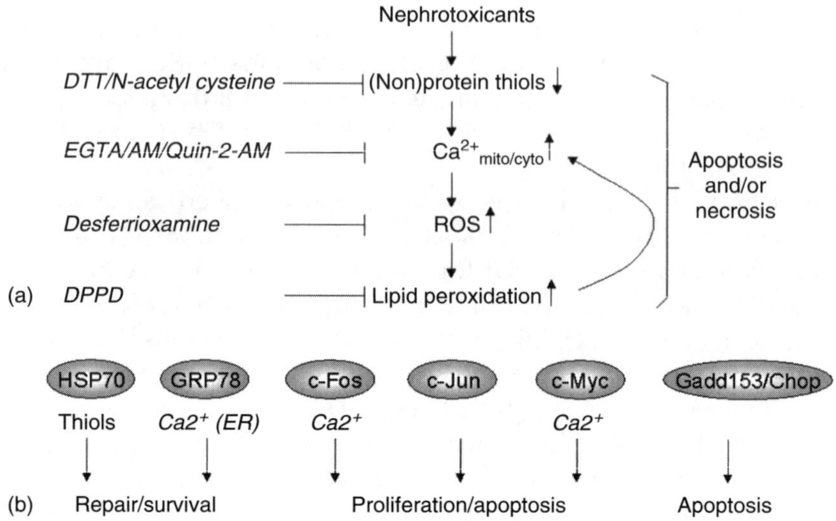

Figure 8.1 Ordering of biochemical events in relation to nephrotoxicity (A) and stress gene induction (B). DTT, dithiothreitol; DPPD, diphenyl-phenylene-diamine; ER, endoplasmic reticulum; mito, mitochondria; cyto, cytosol; ROS, reactive oxygen species.

maintainance of sufficient reduced GSH levels is essential. This is mediated by the GSH reductase enzymes, which are themselves also affected by some nephrotoxicants (van de Water et al., 1996).

To study the role of (non)protein thiol modification in the cellular stress response to nephrotoxicants, the thiol-reducing compound dithiothreitol (DTT) has often been used (Chen et al., 1991; van de Water et al., 1996). Iodoacetamide (IDAM) is an alkylating agent that depletes cellular GSH, which is followed by increase of intracellular free calcium, oxidative stress, lipid peroxidation, and finally necrotic cell death (Chen et al., 1991). IDAM-induced cell death is blocked by DTT as well as by the lipid radical scavenger diphenyl-*p*-phenylene diamine (DPPD) and the intracellular calcium chelator EGTA-AM (Chen et al., 1991). IDAM affects protein thiols, which are associated with induction of heat shock protein 70 (*hsp70*) (Liu et al., 1996b). Although DTT, EGTA-AM, and DPPD are all capable of inhibiting necrotic cell death, only DTT blocks the upregulation of *hsp70* (Liu et al., 1996b). This indicates an important role for (non)protein thiol modification in this process. Another nephrotoxicant, the cysteine *S*-conjugate

S-(1,2-dichlorovinyl)-L-cysteine (DCVC) that is often used as a model compound to study mechanisms of nephrotoxicity, both *in vitro* and *in vivo*, also causes activation of various stress genes, including *hsp70, c-myc, c-fos, c-jun, grp78* and *gadd153* (Stevens et al., 2000). The biochemical perturbations caused by DCVC are similar to those of IDAM, and also involve calcium perturbations, oxidative stress, lipid peroxidation followed by necrotic cell death, which is blocked by DTT, EGTA-AM, and DPPD. Although DTT inhibits DCVC-induced *hsp70* induction (Chen et al., 1992b), it only partially affects the expression of *c-myc, c-fos* and *c-jun* (Yu et al., 1994).

Modification of protein thiols affects protein folding. Consequently, constitutive heat shock protein 70 (HSC70) binds to these misfolded proteins. Subsequently, the transcription factor heat shock factor 1 (HSF1) that normally attaches to HSC70 is released. HSF1 then translocates to the nucleus to activate gene transcription of various heat shock proteins. IDAM activates HSF1; this is blocked by DTT but not by DPPD or EGTA-AM. This is in-line with the DTT-mediated protection against IDAM-induced *hsp70* induction (Liu et al., 1996b).

Oxidative Stress

Oxidative stress occurs after exposure of cells to a variety of nephrotoxicants, as well as after IR. It may arise from direct redox cycling, as observed for cephaloridine and various nephrotoxic hydroquinones, or indirectly, through modulation of enzyme systems that are involved in the formation or detoxification of reactive oxygen species (ROS). One of the major cellular sources of ROS is the mitochondrion; electrons may escape from the respiratory chain and react with molecular oxygen to form superoxide or hydroxyl radicals. Many nephrotoxicants, including cisplatin, nephrotoxic cysteine conjugates, and $HgCl_2$, inhibit the mitochondrial respiratory chain (van de Water et al., 1994; Kruidering et al., 1997; Lund et al., 1991). This is associated with increased formation of superoxide, which is converted to hydrogen peroxide by superoxide dismutase, and next to highly reactive hydroxyl radicals that react with proteins, lipids, and DNA. Oxidative stress mediated activation of stress response routes is most prominent for stress activated protein kinases (including JNK and p38; Kyriakis and Avruch, 2001) and the nuclear factor kappaB (NF-κB) pathways (Schreck et al., 1992). Here the NF-κB route is discussed; stress activated kinases are discussed below.

NF-κB activation occurs under several nephrotoxic conditions, including IR *in vivo* and exposure of cultured renal proximal tubular cells to nephrotoxic cysteine *S*-conjugates (Meldrum et al., 2002; Otieno and Anders, 1997). Activation of NF-κB is dependent on the formation of reactive oxygen intermediates, since *N*-acetyl cysteine and DPPD can prevent it (Otieno and Anders, 1997). Interestingly, despite the fact that $HgCl_2$ as well as arsenite activate other stress response routes, these compounds do not activate NF-κB, and even block the lipopolysaccharide-induced activation of NF-κB. This is due to inhibition of I-κB kinase activity (Dieguez-Acuna et al., 2001). Tumor necrosis factor mediated activation of NF-κB is blocked by antioxidants, including *N*-acetyl cysteine (Schulze-Osthoff et al., 1993). The ROS involved in this process seem to be derived from the mitochondrial electron transport chain since rotenone and antimycin A prevent activation of NF-κB. Whether this is also true for nephrotoxicant-induced injury is unclear.

Oxidative stress does not seem to be a major player in the induction of *grp78, gadd153, c-jun, c-fos, c-myc* and *hsp70.* For example, nephrotoxic cysteine *S*-conjugate-induced upregulation of these stress response proteins is not prevented by the antioxidant DPPD (Stevens et al., 2000). Thiol modifications are directly involved in *hsp70* induction, but oxidative stress is not. Therefore, thiol modification and oxidative stress play different roles in the regulation of stress response pathways, in spite of the fact that DTT as well as DPPD protect against renal cell necrosis caused by oxidative stress. These combined data indicate dissociation between, on the one hand, critical biochemical events that cause oxidative stress dependent cell death and, on the other hand, the activation of cellular stress response routes activated by means other than oxidative stress.

Intracellular Calcium Perturbations

The cytosolic free calcium concentration in cells is 100 to 200 n*M*, compared with approximately 1.3 m*M* in the circulation. The activity of many enzymes is tightly regulated by free calcium and, therefore, the cytosolic calcium levels require tight control. This is mediated by uptake of calcium into various intracellular calcium pools, e.g., the ER, the mitochondria, and the nucleus, and by pumping of calcium out of the cell across the plasma membrane. Sustained perturbation of intracellular calcium concentrations is associated with cytotoxicity

(Nicotera et al., 1990). This has been observed in many cell types, including renal cells. Thus, various nephrotoxicants, including $HgCl_2$, $CdCl_2$, and nephrotoxic cysteine *S*-conjugates, cause an increase in intracellular free calcium that is directly linked to oxidative stress dependent cell death (Maki et al., 1992; Smith et al., 1992; van de Water et al., 1993, 1994). Thus, chelation of intracellular free calcium using BAPTA/AM or EGTA/AM blocks the induction of lipid peroxidation and cell death (van de Water et al., 1993, 1994). Chelation of intracellular free calcium inhibites, to some extent, DCVC-induced upregulation of *c-myc* (Yu et al., 1994), suggesting a possible involvement of calcium dependent protein kinases in this process. Similarly, in primary cultured rat proximal tubular epithelial cells the oxidative stress induced upregulation of *c-fos* is inhibited by a chelator of intracellular free calcium, Quin-2/AM (Maki et al., 1992). However, no effect of chelation of calcium on the upregulation of *c-fos, hsp70,* or *gadd153* by DCVC was observed in the LLC-PK_1 renal cell line (Chen et al., 1992a, 1992b; Yu et al., 1994).

The cellular effect of toxicants is complex and deregulated calcium levels may act in concert with other perturbations to upregulate various stress response proteins. Therefore, it is of interest to determine the role of perturbation of selective intracellular calcium pools, such as that in the ER, on the upregulation of stress response proteins. For example, an inhibitor of the ER Ca^{2+}-ATPase pump, thapsigargin, or general calcium ionophores, such as A23187 or ionomycin, cause induction of *grp78* and other ER stress response genes, including *grp94*, in renal cells (Liu et al., 1997). These perturbations, however, do not affect *hsp70* expression. These data indicate at least some role for calcium perturbations in the upregulation of several stress proteins. Given the fact that many enzymes are directly or indirectly regulated by calcium, it is most likely that the calcium dependent stress response gene expressions are regulated, at least in part, by calcium dependent kinase activity.

Signal Transduction Pathways Activated In Response To Cell Injury

The activation of protein kinases is also fundamental in regulation of the expression of various genes. Protein kinases comprise (non)receptor tyrosine protein kinases, such as the epidermal growth factor (EGF) receptor family or Src-family kinases, as well as serine or threonine protein kinases, including, for example, the family of protein kinase C as well as the mitogen activated protein kinase (MAPK) family. Many protein kinases exist; each kinase has its own particular cellular function, which depends on cellular localization and substrate availability. Substrates include transcription factors that become (in)activated after phosphorylation and thereby affect gene expression, resulting in a defined biological response. Protein kinases also modulate phosphorylation of many other proteins that do not affect gene expression directly but may affect other cellular events or behavior.

Protein kinase activity is counterbalanced by the activity of protein phosphatases. The balance between various protein kinases and phosphatases that are (in)activated after cellular stress ultimately determines the biological outcome: cell survival or cell death, (de)differentiation, mitogenesis, etc. This is an exciting area of basic research, which so far has only been applied to a limited extent to the pathophysiology of ARF. Some of the novel insights of protein kinase activated pathways in acute renal cell injury will be discussed below.

Stress-Activated Protein Kinase Pathways That Facilitate Apoptosis

The family of MAPKs (Kyriakis and Avruch, 2001), including c-Jun amino-terminal kinase (JNK), p38, and extracellular signal-regulated kinases (ERKs), are all involved in stress signaling in the kidney after acute renal cell injury. After IR, JNK, p38, and ERK are activated (Di Mari et al., 1997; Kundusova et al., 2002; Pombo et al., 1994, 1997; Yin et al., 1997). The activation of JNK seems to involve an oxidative stress related pathway, since it is inhibited by *N*-acetyl cysteine (Di Mari et al., 1997). This oxidative stress during *in vivo* IR seems to involve monoamine oxidase (MAO) because an inhibitor of MAO, pargyline, prevents formation of reactive oxygen species. This effect is associated with decreased levels of JNK-P but increased levels of ERK-P (Kundusova et al., 2002). Hypoxia/reoxygenation in primary cultured human kidney cells also causes JNK activation (Garay et al., 2000). Activation of JNK in ARF seems to be involved in cell death signaling. Thus, downregulation of JNK1 in human kidney cells with antisense oligonucleotides prevents JNK activation and inhibits hypoxia/reoxygenation-induced apoptosis (Garay et al., 2000). The prevention of JNK phosphorylation *in vivo* with pargyline is also associated with reduced levels of apoptotic cells (Kundusova et al., 2002).

Mild injury that does not lead to cell death usually results in a preconditioning of cells and protection against a subsequent, more severe, stress brought about by the same stimulus; this is also the case for IR. Preconditioning by a brief ischemic period prevents the activation of JNK and p38, but not ERK, after an additional period of IR. This is associated with prevention of the activation of kinases that act upstream of JNK: MKK7, MKK4, and MKK3/6. IR-induced activation of MEK1/2 that acts upstream of ERK is not prevented by ischemia preconditioning (Park et al., 2001). In another model, preconditioning by 24 hours of unilateral ureteral obstruction (UUO) 6 to 8 days prior to IR, also prevented IR-induced activation of JNK and p38. Such a protection is related to an increased expression of Hsp27 in UUO kidneys. Importantly, increased expression of Hsp27 in LLC-PK_1 cells inhibits the activation of JNK and p38 and the onset of cell death (Park et al., 2002a).

Preconditioning and protection against injury-induced stress kinase activation is most likely mediated by proteins that are upregulated after cellular stress that directly control the stress kinase pathways. These proteins may include thioredoxin, thioredoxin reductase, or glutathione *S*-transferase μ, which control the activation of a kinase that is upstream of JNK, such as apoptosis signal related kinase (ASK1) (Saitoh et al., 1998). Glutathione *S*-transferase π may also be upregulated, thereby inhibiting the JNK pathway (Adler et al., 1999).

As described above, stress-induced JNK activation seems to be critical in controlling apoptosis of renal cells. This effect of JNK may be mediated by the activation of the transcription factor c-Jun, thereby affecting gene expression of target genes that may affect the apoptotic process (Behrens et al., 1999; Verheij et al., 1996). Alternatively, JNK may affect the phosphorylation of various apoptosis-related proteins. Indeed, JNK can phosphorylate the antiapoptotic protein Bcl-2 and thereby sensitize cells to apoptosis (Deng et al., 2001; Yamamoto et al., 1999). More work is needed to unravel the exact mechanism by which JNK affects renal cell apoptosis and whether modulation of JNK itself or its downstream substrates may be potential therapeutic targets to protect against ARF.

Activation of Survival Signaling Pathways in Response to Injury

Cellular stress in renal cells also activates signaling kinases that are associated with survival signaling pathways, such as ERK and protein kinase B (PKB). For example, cisplatin treatment causes the activation of PKB (Kaushal et al., 2001) and the nephrotoxicant 2,3,5-tris-(glutathion-*S*-yl)hydroquinone (TGHQ) activates ERK in LLC-PK_1 cells (Ramachandiran et al., 2002). Interestingly, the TGHQ-mediated activation of ERK seems to be mediated by the EGF-receptor since an inhibitor of this receptor, AG1478, prevents the activation of ERK (Ramachandiran et al., 2002). IR *in vivo* also activates ERK (Park et al., 2001).

The mechanisms involved in ERK and PKB activation remain unclear. This activation may, however, be related to a direct activation of the EGF-receptor by oxidative stress. Alternatively, the release or activation of (pro)growth factors, which are present in the plasma membrane or attached to the extracellular matrix and released after cell injury, may activate the corresponding receptors. Activation of either ERK or PKB seems directly related to survival signaling in renal cells. Thus, PD098065, an inhibitor of MEK1, inhibits ERK activation and accelerates the cell death induced by TGHQ. Inhibition of the EGF-receptor with AG1478 also facilitates cell death (Ramachandiran et al., 2002).

• In proximal tubular cells, prooxidant-induced cell death is prevented by insulin-like growth factor I (IGF-I), which activates ERK through the Ras–Raf–MEK pathways; this is blocked by the MEK1 inhibitor PD098059. In thick ascending limb cells, ERK is activated after oxidative stress; this is a requirement for survival because the MEK1 inhibitor PD098059 prevents ERK activation and facilitates cell death (Di Mari et al., 1999). Protein kinase A also seems to be related to prosurvival pathways. A cell-permeable cAMP derivative protects against cephaloridine induced oxidative stress and cell injury in renal cortical slices; this is prevented by inhibition of protein kinase A activity (Kohda and Gemba, 2001).

Proteins that are directly involved in the control of apoptosis are important phosphorylation targets for prosurvival kinases such as PKB. For example, activation of PKB by cisplatin in LLC-PK_1 cells is associated with phosphorylation of the proapoptotic protein Bad (Kaushal et al., 2001). Phosphoinositide-3 kinase (PI-3 kinase) is a key player in the activation of PKB. Activation of PKB results in translocation of phosphorylated Bad to the cytoplasm and association with the protein 14–3-3, thereby preventing a proapoptotic action of Bad (Zha et al., 1996).

Activation of PKB seems important for survival signaling in renal cells as inhibitors of PI-3 kinase, LY29002, and wortmannin prevent the activation of PKB and phosphorylation of Bad and enhance cisplatin-induced apoptosis of LLC-PK_1 cells (Kaushal et al., 2001). Although more proteins, including forkhead transcription factors, are known to be phosphorylation by PKB and involved in the control of apoptosis in other cells (Cardone et al., 1998; Dijkers et al., 2002; Kops et al., 1999), the relevance of these other pathways needs further investigation in renal epithelial cells both *in vitro* and especially *in vivo*.

Inhibition of Pro-Survival Pathways After Renal Cell Injury

Focal contacts are important sites of signal transduction. Many phosphoproteins, in particular proteins phosphorylated on tyrosine residues, are localized at focal contacts. These include both structural proteins, such as paxillin and p130Cas, and protein tyrosine kinases, such as focal adhesion kinase (FAK), Pyk2, and Src, as well as other kinases, including JNK, and ERK (Figure 8.2). The signaling mediated by these focal adhesion associated kinases is essential in various cellular processes, including cell migration, proliferation, and survival. Importantly, these kinases are also located at cell-matrix adhesions in freshly isolated proximal tubules as well as in primary cultured proximal tubular cells.

Interestingly, hypoxia in freshly isolated rabbit proximal tubules results in the loss of tyrosine phosphorylation of several focal adhesion

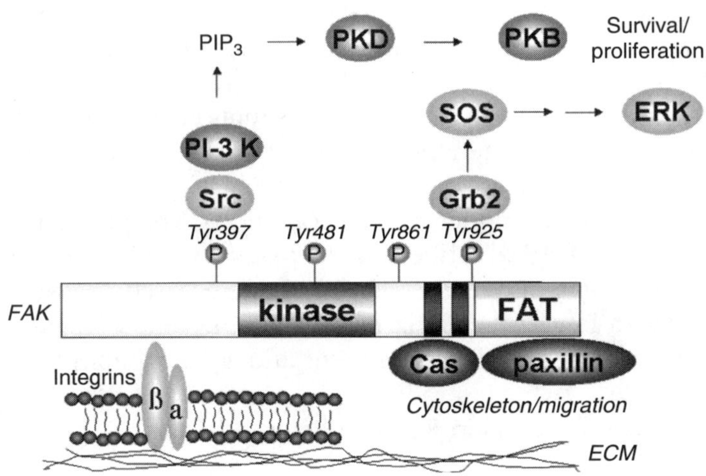

Figure 8.2 Role of focal adhesion kinase (FAK) in cell-matrix mediated survival signaling and migration.

associated proteins, including FAK and paxillin (Weinberg et al., 2001). Tyrosine phosphorylation of FAK and paxillin in primary cultured rat renal epithelial cells is also lost after treatment with the nephrotoxicant DCVC (van de Water et al., 1999a, 2001).

FAK is an important nonreceptor tyrosine kinase associated with focal adhesions, and activated in particular by adhesion of cells to the extracellular matrix. Various factors that are important in renal development and function, such as hepatocyte growth factor (HGF) and angiotensin II, also activate FAK. Activation of FAK provides survival signaling, including activation of both ERK and PKB (Sonoda et al., 1999). Loss of focal adhesions caused by forced detachment of cells is associated with apoptosis. This process, also called anoikis, has been best studied in the renal cell line MDCK. Overexpression of constitutively active FAK protects against anoikis in MDCK cells; this is related to the activation of PKB (Frisch et al., 1996; Khwaja and Downward, 1997).

FAK is also important in the onset of apoptosis in primary cultured rat renal proximal tubular cells. Thus, the nephrotoxicant DCVC causes a rapid loss of FAK phosphorylation, which is associated with loss of focal adhesions. These events occur prior to the activation of caspases, and are not prevented by caspase inhibitors (van de Water et al., 1999a). Additionally, overexpression of a dominant negative acting FAK construct prevents the localization of FAK at focal adhesions and facilitates the activation of apoptosis by DCVC (van de Water et al., 2001). These studies indicate that focal adhesion mediated signaling is critical for maintenance of survival.

Given the fact that, in the *in vivo* situation, the cell adhesion to extracellular matrix is often compromised after a renal insult (Molitoris and Marrs, 1999), FAK-mediated signaling may also be crucial for the control of cell survival under *in vivo* conditions. The above indicates an important role of focal adhesion mediated signaling complexes in proximal tubular epithelial cells. Further research is warranted to elucidate the role of other kinases that are present at focal adhesions in the process of renal cell detachment and onset of apoptosis.

Stress Response Proteins in Repair of Injured Cells

Heat Shock Proteins and Control of Apoptosis

Heat shock proteins play a critical role in the adaptive response of cells to stress conditions and are upregulated in the kidney after renal injury

(Schober et al., 1997). HSP70 is the major inducible heat shock protein and is involved in many cellular activities, including control of protein synthesis, folding, and translocation into organelles. The upregulation of HSP70 in the kidney is important for the control of renal cell survival and organ function after renal injury. In LLC-PK_1 cells, the inducible upregulation of HSP70 protects against apoptosis caused by ATP depletion (Wang et al., 2002). Upregulation of HSP70 by sodium arsenite improved the recovery from IR (Yang et al., 2001). Moreover, whole-body hyperthermia caused upregulation of HSP70 and subsequent improvement of renal function after cold ischemia and renal transplantation (Radealli et al., 2002).

The *in vivo* renal protection is also associated with protection against apoptosis (Yang et al., 2001). How does HSP70 protect cells against apoptosis? The catalytic activity of HSP70 that mediates refolding of damaged proteins is required for the protection of cells against the onset of apoptosis caused by heat stress (Mosser et al., 2000). This is associated with an inhibition of cytochrome c release from the mitochondria, indicating that HSP70 protection acts upstream of the mitochondria-dependent activation of apoptosis. The protein binding capacity of HSP70 itself is also involved in the regulation of apoptosis. It would be favorable if HSP70 interferes with proapoptotic proteins and, thereby, keeps these proteins in the inactive state. This would then allow for a full adaptive response after cell injury, and only when repair turns out to be impossible would injured cells commit suicide.

In addition to the protein recovery functions, HSP70 is also involved in the control of signaling pathways that control the onset of apoptosis, as well as protein complexes that are central in the activation of the apoptotic pathways (Figure 8.3). Thus, in NIH 3T3 cells, oxidative stress causes activation of ASK1, which is prevented by either prior mild heat shock or by overexpression of HSP70 (Park et al., 2002b). This seems to be mediated by a direct interaction between HSP70 and ASK1, thereby preventing the homo-oligomerization and activation of ASK1 as well as the downstream activation of stress kinase p38. In this way HSP70 prevents the onset of apoptosis caused by overexpression of ASK1 (Park et al., 2002b). Additionally, HSP70 binds directly to apoptosis protease activating factor 1 (APAF-1) (Beere et al., 2000; Saleh et al., 2000). APAF-1 is essential in the formation of the so-called apoptosome, a protein complex consisting of cytochrome c released from the mitochondria, APAF-1, ATP/dATP and caspase-9 that is

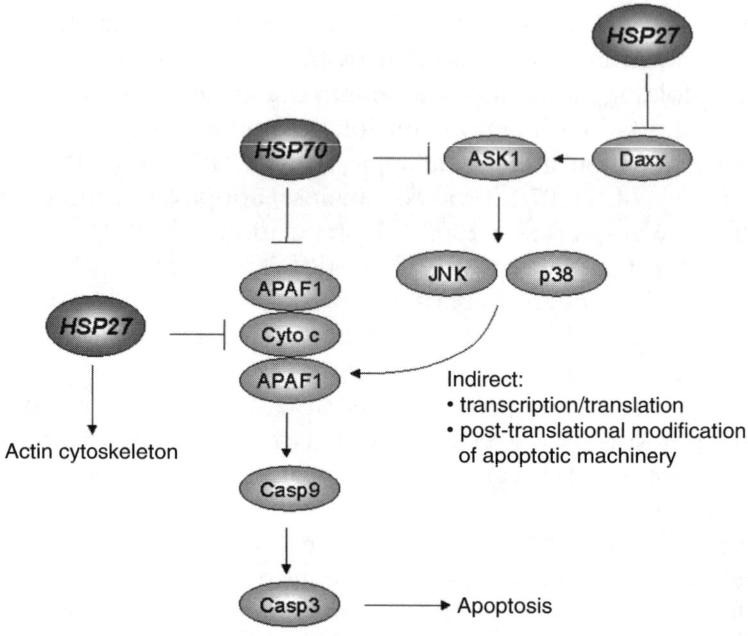

Figure 8.3 Interaction of heat shock proteins with regulators of apoptosis.

required for the activation of caspase-9 and subsequent activation of caspase-3. Binding of HSP70 to the caspase activation recruitment domain (CARD) of APAF-1 prevents the formation of the apoptosome, and thus prevents activation of caspase-9 and caspase-3.

The small heat shock protein HSP27 also has a crucial role in the regulation of renal cell survival after injury. HSP27 transiently translocates to the particulate cytoskeletal fraction at early time-points after renal IR (Schober et al., 1997). In the recovery phase after injury, increased levels of HSP27 are detected, especially in proximal tubular cells (Park et al., 2001). Increased HSP27 is also detected in the kidney after transient ureter obstruction (Park et al., 2002a). Interestingly, adenovirus-mediated overexpression of HSP27 inhibits cell death of LLC-PK_1 cells caused by either oxidative stress or ATP depletion (Park et al., 2002a).

Several mechanisms have been suggested by which HSP27 protects against cell death. For example, one branch of the Fas receptor activation pathway results in the activation of ASK1 and JNK through a protein called Daxx. Recent evidence indicates that the phosphorylated form of HSP27 binds Daxx and thereby prevents Daxx-induced apoptosis (Charette et al., 2000).

Death receptor independent apoptosis is also controlled by HSP27. This seems to be mediated through maintenance of the F-actin network organization, thereby preventing mitochondrial release of cytochrome c (Paul et al., 2002). F-actin disruption resulting from cell injury causes release of the BH3-only proapoptotic Bcl-2 family member Bmf, which activates the mitochondrial pathway of apoptosis (Puthalakath et al., 2001); such a release is likely to be prevented when HSP27 preserves F-actin organization. In the event that a release of cytochrome c might occur, HSP27 can also bind free cytochrome c and thereby still prevent formation of the apoptosome and caspase activation (Bruey et al., 2000).

Although HSP90 upregulation has been observed in several *in vitro* models of nephrotoxicity, little is known about the expression and role of HSP90 in the control of renal failure *in vivo*. Interestingly, studies with a novel inhibitor of HSP90, geldanamycin, indicate that HSP90 function is essential for cell survival; geldanamycin induces apoptosis in a variety of cell types. A decreased chaperone activity of HSP90 depletes growth factor receptors from cells that mediate survival signals, for example HER-family tyrosine kinases and Raf, and causes induction of apoptosis in tumor cells (Zheng et al., 2000). This indicates that HSP90 also has a critical role in the maintenance of proper cellular function, and that increased expression of HSP90 in renal cells may allow increased cellular repair or control of apoptosis through increased activation of growth factor mediated survival signaling pathways.

Glucose Regulated Proteins: Cellular Repair and Tolerance

Another group of molecular chaperones that is upregulated after toxicant stress are the glucose-regulated proteins (GRPs). GRPs belong to a group of ER resident chaperones that include Grp78/BiP, Grp94, Grp170, protein disulfide isomerase, and calreticulin (Kaufman, 1999). These proteins are generally upregulated as a result of an accumulation of unfolded proteins in the lumen of the ER, also called the unfolded protein response (UPR, Figure 8.4). This response is mediated by the activity of several recently discovered membrane-associated serine/threonine kinases that reside in the ER, such as PERK and Ire1 (Kaufman, 1999).

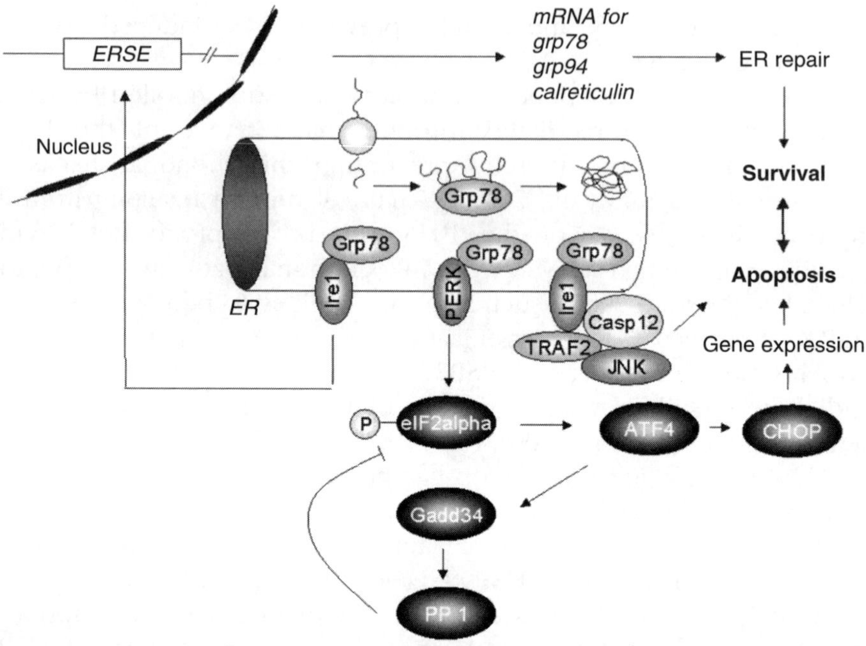

Figure 8.4 Balance of the activation of antiapoptotic and proapoptotic signaling routes from the endoplasmic reticulum (ER).

Various conditions that result in accumulation of unfolded proteins in the ER, e.g., ER calcium perturbations, reduction of protein thiols and protein glycosylation inhibition, cause activation of these ER stress kinases. While active PERK phosphorylates eIF2α, thereby inhibiting the initiation of translation, Ire1 activation results in increased expression of *grp78* (Wang et al., 1998). Activation of PERK as well as Ire1 is controlled by the protein level of Grp78. Under normal conditions, PERK and Ire1 both bind Grp78. After ER stress, Grp78 is uncoupled from these kinases to interact with misfolded proteins in the lumen of the ER. Only after proper activation of Ire1 and subsequent Grp78 upregulation, PERK and Ire1 activities are again downregulated due to reassociation of the kinases with Grp78. Thus, coordinated activation of PERK and Ire1 allows for reduction of protein synthesis workload, thereby facilitating upregulation of ER chaperones and restoration of ER function and cellular repair.

Prior upregulation of ER stress proteins provides tolerance against cell death caused by various ER stress conditions (Kaufman, 1999).

Prior upregulation of ER stress proteins in renal cells *in vitro* also provides tolerance against subsequent cell death caused either by ATP depletion (Bush et al., 1999) or by prooxidants, such as iodoacetamide and *t*-butyl-hydroperoxide (Liu et al., 1997,1998a). Grp78 is the major chaperone upregulated after ER stress and primarily responsible for the tolerant phenotype of ER stressed cells. Overexpression of Grp78 itself protects against activation of PERK and Ire1 (Bertolotti et al., 2000), which is associated with protection against cell death (Kaufman, 1999).

Overexpression of an antisense Grp78 construct prevents the upregulation of Grp78 after mild ER stress conditions; this is associated with reduced cytoprotection against prooxidant-induced cell death (Liu et al., 1997, 1998a). Prior upregulation of ER stress proteins protects renal cells against both necrosis and apoptosis (van de Water et al., 1999b), although there seems to be a fundamental difference in the mechanisms involved in the two types of cell death. Prooxidants cause necrosis, which is related to release of ER calcium into the cytoplasm followed by uptake of free calcium by the mitochondria with the subsequent induction of oxidative stress, resulting in lipid peroxidation and plasma membrane rupture (Liu et al., 1997, 1998a).

Upregulation of Grp78 prevents this process by maintaining a proper intracellular calcium homeostasis, thereby preventing oxidative stress and lipid peroxidation. This is also observed after overexpression of calreticulin, an ER resident calcium binding protein (Liu et al., 1997, 1998a). However, prooxidant-induced apoptosis in the same cells does not require an uncontrolled rise of intracellular calcium and is not prevented by calreticulin overexpression. Nevertheless, pre-ER stress provides tolerance against the induction of apoptosis (van de Water et al., 1999b). This protection seems to be related to an inhibition of the UPR, because translation inhibition after prooxidant treatment is reduced under these conditions.

The UPR in ARF *in vivo* has not been studied in much detail; this should now be possible with phospho-state-specific antibodies directed against active PERK or Ire1. Attempts have been made to elucidate a role of ER stress proteins in the control of nephrotoxicity *in vivo*. Treatment of rats with trans-4,5-dihydroxy-1,2-dithiane (i.e., oxidized dithiothreitol) caused an upregulation of Grp78 in the kidney, which was associated with a cytoprotection against acute renal failure caused by tetrafluoroethyl-L-cysteine (Stevens, J.L., personal communication). Whether such a protection is related to prevention of the oxidative stress or the UPR requires further investigation. Moreover, these studies need to be extended to other models of ARF.

CHOP/GADD153 and ER Stress-Induced Apoptotic Pathways

CHOP/GADD153 (CEBP homology protein/growth arrest and DNA damage protein 153) expression is induced by several nephrotoxicants and other forms of renal cell stress (Chen et al., 1992a; Halleck et al., 1997b; Jeong et al., 1996; Kultz et al., 1998; Zhang et al., 1999; Zinszer et al., 1998). Several pathways are involved in the induction of *chop*, including the UPR in the ER, amino acid starvation, glucose deprivation, and hypoxia.

PERK-mediated phosphorylation of eIF2α is responsible for the induction of *chop* caused by the UPR (Harding et al., 2000). Although eIF2α phosphorylation shuts down general protein synthesis, it now allows the synthesis of the transcription factor ATF4 that activates *chop* expression (Harding et al., 2000; Ma et al., 2002). Interestingly, at the same time the growth arrest and DNA damage protein GADD34 is upregulated in a PERK dependent manner, GADD34 binds the protein phosphatase 1 and thereby allows dephosphorylation of eIF2α with the subsequent attenuation of translational inhibition (Novoa et al., 2001), ultimately resulting in full cell recovery.

The effect of nephrotoxicants and IR on ATF4 and GADD34 upregulation has not been determined. However, CHOP-deficient mice have been used to evaluate a role for CHOP in regulation of renal cell injury *in vivo*. Thus, treatment of mice with an inhibitor of *N*-linked glycosylation, tunicamycin, causes ARF, which develops over several days and is associated with the occurrence of apoptosis in the renal proximal tubular epithelial cells (Zinszner et al., 1998). In CHOP-deficient mice, ARF induced by tunicamycin is diminished and apoptosis does not occur. Tunicamycin-induced activation of caspases and apoptosis are also attenuated in primary cultured, CHOP-deficient fibroblasts (Zinszner et al., 1998). These data indicate that *in vivo*, CHOP-mediated regulation of transcription in the kidney is an important contributor to the control of apoptosis of proximal tubular cells after ER stress.

What is the link between the ER and the induction of apoptosis? In part this may be an increase of cytosolic calcium that has leaked out of the ER and can accumulate in the mitochondria, thereby affecting the onset of a permeability transition resulting in activation of caspase-3

(Figure 8.4). Alternatively, increased cytosolic calcium may activate calpains, which may induce the onset of apoptosis.

A more likely explanation is the activation of caspase-12, which is located at the ER and specifically activated by ER stresses such as those caused by tunicamycin and thapsigargin. Indeed, tunicamycin-induced renal cell death is attenuated in caspase-12 deficient mice (Nakagawa et al., 2000). Activation of caspase-12 is possibly mediated by the recruitment of the tumor necrosis factor associated factor 2 (TRAF-2) to Ire1, which is activated after ER stress (Yoneda et al., 2001). TRAF-2 mediated binding and clustering of caspase-12 may result in the activation of caspase-12 and subsequent activation of downstream caspases, such as caspase-3 (Morishima et al., 2002).

Alternatively, Ire1-mediated activation of JNK through TRAF-2 may also result in induction of the apoptotic machinery at the level of the mitochondria (Urano et al., 2000). An involvement of Ire1 in the activation of caspase-12 would fit with a model of renal cell repair whereby increased levels of Grp78 in the ER would keep Ire1 in its inactive form, thereby preventing the onset of caspase-12 activation and induction apoptosis.

These data, and those described above for Grp78, suggest a tight coupling between ER stress response signaling pathways. On the one hand, these upregulated Grps enable repair of ER function and downregulation of PERK kinase activation. On the other hand, the upregulated CHOP regulates proapoptotic signaling cascades. The balance in expression and activity of the prorepair and antiapoptotic glucose regulated proteins on the one hand, and the proapoptotic effects of CHOP-dependent pathways on the other hand, most likely determines the final outcome of cellular stress: survival or death.

NF-κB: Inflammation and Apoptosis

NF-κB transcription factors regulate the activation of various genes that play a role in inflammation (cytokines and chemokines), proliferation (e.g., c-myc and cyclin D1) and cell survival or death (inhibitor of apoptosis proteins, Bcl-2 family members, TRAIL) (Ghosh and Karin, 2002). In mammals, the NF-κB group of transcription factors has five members, which form homodimers or heterodimers. NF-κB dimers are bound to members of the family of IkappaB proteins that retain NF-κB in the cytosol. After phosphorylation of I-κB by I-κB kinases

(IKKs), I-κB is degraded by the proteasome, allowing the translocation of NF-κB to the nucleus to activate gene transcription (Ghosh and Karin, 2002).

NF-κB may have an effect on both cell survival and organ function. On the one hand, activated NF-κB induces antiapoptotic Bcl-2 family members Bcl-X_L and A1, as well as the caspase inhibitors, i.e., X-1L, c-IAP-1, and c-IAP-2 (Chu et al., 1997; Chen et al., 2000; Furusu et al., 2001; Pagliari et al., 2000; Wu et al., 1998). Likewise, activation of NF-κB in cells *in vitro* is generally associated with protection against apoptosis, despite the fact that NF-κB can also upregulate the expression of the proapoptotic TRAIL receptors DR4 and DR5 (Ravi et al., 2001). On the other hand, NF-κB pathways induce the expression of cytokines, chemokines, and cell adhesion receptors, which promote the infiltration and activation of monocytes or macrophages.

Although such an infiltration is essential to the removal of injured cells, sustained presence of infiltrating mononuclear cells in the kidney is associated with nephrotoxicity (Ramesh and Reeves, 2002). Given the importance of NF-κB activation in cell injury responses, future *in vivo* studies are needed to further delineate the role of NF-κB pathways in the pathophysiology of ARF.

c-Myc

Myc is a transcription factor that acts in collaboration with the proteins Max and Mad. It is essential in the control of cell proliferation and exhibits increased expression due to gene amplification in a variety of tumor cells. Myc is also associated with the onset of apoptosis, which particularly occurs after withdrawal of growth factors (Grandori et al., 2000). Nephrotoxicant-induced upregulation of c-myc has only been demonstrated in renal cells *in vitro* and renal IR *in vivo* (Rafaat et al., 1997; Zhan et al., 1997). The nephrotoxicant DCVC and oxidative stress cause the induction of *c-myc* in LLC-PK_1 cells (Maki et al., 1992; Vamvakas et al., 1993; Zhan et al., 1997). Although calcium perturbations seem to be involved in the mechanism of myc induction, other players remain to be elucidated. Some signaling pathways that are activated after cellular stress are known to upregulate c-myc. These include β-catenin mediated signaling that is associated with loss of cell–cell adhesion (He et al., 1998), as well as NF-κB activation (Bourgarel-Rey et al., 2001; La Rosa et al., 1994). Such a link needs further elucidation in the renal cells.

DCVC-induced induction of c-myc seems relevant for the onset of apoptosis: enforced overexpression of c-myc enhances DCVC-induced apoptosis. This is not observed for transcriptionally inactive mutants of c-myc, indicating involvement of transcriptional processes and potentially upregulation of other proteins in the enhanced apoptosis induction (Zhan et al., 1997). DCVC-induced upregulation of c-myc is associated with upregulation of at least one c-myc target gene, ornithine decarboxylase. An inhibitor of this enzyme decreases the onset of cell death by DCVC (Zhan et al., 1997).

Use of serial analysis of gene expression (SAGE) in other cell types has uncovered other target genes of c-myc that are mainly involved in cell cycle regulation and genome integrity preservation: Cyclin E binding protein 1, Cyclin B1, BRCA1, and MSH2 (Menssen and Hermeking, 2002). Although it is unlikely that these genes are directly related to induction of apoptosis, blockade of c-Myc induced mitogenesis due to other cellular stresses caused by toxicant exposure may turn on the suicide program. This is well known for myc-induced apoptosis caused by enforced c-myc overexpression in the absence of growth factor survival signaling in a variety of other cell types (Evan et al., 1992). The latter apoptotic program is dependent on p53, FasL-Fas, and bcl-2 dependent pathways (Fanidi et al., 1992; Hueber et al., 1997; Juin et al., 2002; Klefstrom et al., 1997). The exact pathways involved in myc-induced apoptosis of renal cells still need further investigation. Moreover, the role for myc in ARF still requires further elucidation.

RENAL REGENERATION

Time Course of Renal Regeneration After Acute Renal Failure

ARF is associated with denudation and excretion of proximal tubular cells (often still viable), which generally involves the S_3 segment at the outer stripe of the outer medulla but may extend all the way to the outer cortex, depending on the type and dose of a particular nephrotoxicant. The time course of the injury may also differ: cysteine *S*-conjugates cause severe damage within 24 hours, comparable with IR (Monks et al., 1991; Rivera et al., 1994; Safirstein, 1999; Wallin et al., 1992). However, cisplatin- and gentamicin-induced proximal tubular cell turnover generally develops over a time course of several days (Nonclercq et al., 1992; Safirstein, 1999).

After severe loss of the proximal tubular cells, an outburst of PCNA/BrdU-positive proliferating cells is seen between days one and three after initial injury. After day seven this proliferation is complete, and the injured renal tissue is fully recovered and repopulated with differentiated cells after approximately three to four weeks. The regeneration phase involves four different processes: mitogenesis, motogenesis, morphogenesis, and differentiation. Each process requires a different set of proteins. These different protein expression patterns are probably differentially regulated by the action of receptor-mediated activation of either paracrine-activated or autocrine-activated (stress) signal transduction pathways.

Various growth factors and cytokines play an essential role in the regulation of the renal regeneration process, including adenine nucleotides, hepatocyte growth factor, epidermal growth factor, insulin growth factor, fibroblast growth factor, and transforming growth factor (Hammerman, 1998). These factors are released from injured cells, from cells already present in the interstitium, or from inflammatory cells that have infiltrated the tissue. For example, mechanical injury of LLC-PK_1 cells causes release of growth factors such as acidic fibroblast growth factor (FGF), basic FGF, and PDGF-BB (Anderson and Ray, 1998). Infiltrated mononuclear cells (i.e., macrophages) are important in the regeneration process by excretion of various cytokines and growth factors, such as FGF, transforming growth factor (TGF) α, EGF-like, IL-2, etc.

Many of the above mentioned factors affect the proliferation rate and differentiation status in regenerating areas. However, factors that regulate hemodynamics within the kidney are also released, such as endothelin and angiotensin, which may affect the renal regeneration process. Indeed, endothelin seems to play a role in chronic models of renal injury, most likely through its vasoconstrictive effect (Benigni and Remuzzi, 1995). The combination of the expression and activity of all these factors within the kidney will ultimately determine the long-term outcome of the regeneration process: complete recovery or progression towards end-stage renal disease. Some of the most studied growth factors involved in the renal regeneration process are discussed below.

Role of Individual Growth Factors in the Renal Regeneration Process

Epidermal Growth Factor

EGF is produced in the kidney in large amounts and is present in the distal convoluted tubule and the thick ascending limb as a preform (preproEGF) inserted in the membrane. EGF localizes to distal tubules, while the EGF-receptor localizes to proximal tubules. In general, acute renal injury results in downregulation of the levels of EGF. For example, tobramycin-induced injury of the convoluted proximal tubule and cisplatin-induced damage of straight proximal tubule cause a reduction in immunoreactive EGF and a decrease of tissue bound preproEGF (Leonard et al., 1994; Safirstein et al., 1989).

EGF expression is also decreased after exposure to gentamicin and $HgCl_2$, which correlates with injury to proximal straight tubules (Nonclercq et al., 1993; Verstrepen et al., 1995). Reduction in the levels of preproEGF, EGF, and EGF receptors also occurs after IR, with lowest expression 48 hours after the initial ischemic period; normal levels are only reached after complete renal regeneration (Safirstein et al., 1989; Toubeau et al., 1994). A decrease of renal EGF correlates in all cases with decreased EGF levels in the urine (Safirstein et al., 1989; Taira et al., 1994). A fall of EGF levels seems to be compensated by an induction of heparin-binding EGF (HB-EGF) mRNA, as seen in both IR and $HgCl_2$-induced injury.

Expression of HB-EGF is confined to the straight proximal tubules in the outer stripe of the outer medulla. This may be related to activation of mitogenesis since *in vitro* rhHB-EGF stimulates cellular DNA synthesis (Hise et al., 2000; Homma et al., 1995). Given the loss of EGF after acute renal injury, attempts were made to modulate renal failure by EGF treatment. Indeed, treatment with EGF accelerates the renal regeneration process after gentamicin treatment in rats (Morin et al., 1992). EGF also potentiates renal cell repair of $HgCl_2$-damaged tissue when administered 2 to 4 hours after the $HgCl_2$. This effect seems to be mediated by increased proliferation of cells of proximal tubular origin. EGF treatment resulted in lower peak levels of blood urea nitrogen and also an earlier return of blood urea nitrogen levels to control baseline levels (Coimbra et al., 1990).

These effects of EGF involve both prosurvival signaling events and increased mitogenesis. For example, EGF suppresses renal tubular apoptosis and increases proliferation in *in vivo* models of obstructive nephropathy. This effect is associated with decreased induction of stress related proteins, such as clusterin and TGFβ1, as well as markers of dedifferentiated cells, such as vimentin (Chevalier et al., 1998; Kennedy et al., 1997). This was also observed *in vitro*, where EGF stimulated the recovery of confluency of rabbit proximal tubular cells that were treated with low concentrations of either tert-butylhydroperoxide or

DCVC. Importantly, no complete recovery was observed with either IGF-I or insulin, indicating that EGF is superior to these growth factors (Kays and Schnellmann, 1995).

Hepatocyte Growth Factor

HGF plays a role in various cellular processes, including mitogenesis, motogenesis, and morphogenesis (Balkovetz and Lipschutz, 1999). In the kidney, HGF is crucial for renal development as well as for the regeneration process. This is supported by the fact that HGF transgenic mice have renal proximal hypertrophy associated with polycystic kidney disease and glomerulosclerosis. This stresses a critical role for HGF in continuous cell survival and renewal of renal proximal tubular epithelium (Takayama et al., 1997).

HGF is present in tissues in an inactivated form, but is proteolytically released by certain stimuli, including renal tissue injury (Miyazawa et al., 1994). Levels of the HGF-receptor, c-met, increase in the kidney after ARF caused by glycerol; this is associated with increased levels of free HGF in liver, spleen, and lung (Goto et al., 1997). After $HgCl_2$-induced ARF, increased expression of HGF is seen in the renal interstitial cells (Igawa et al., 1993).

HGF is essential in renal protection and facilitates renal regeneration; the effects seem more potent than for EGF. For example, gene transfer of HGF protects against unilateral ureter obstructrion; this is associated with increased proliferation and decreased apoptosis of tubular cells (Gao et al., 2002). Cisplatin- or $HgCl_2$-induced ARF in mice is also markedly reduced by treatment with HGF (Kawaida et al., 1994), and recombinant human HGF prevents renal toxicity of cyclosporine in mice associated with increased proliferation (Amaike et al., 1996).

HGF activates the c-met/HGF receptor, a tyrosine kinase that activates downstream signaling pathways involved in cell survival, mitogenesis, motogenesis, and morphogenesis. Indeed, cisplatin-induced apoptosis of renal mIMCD-3 cells is inhibited by HGF (Liu et al., 1998b). This is also seen in cisplatin-induced proximal tubule injury *in vivo*; this cytoprotection occurs in association with increased cell proliferation, thereby accelerating the renal regeneration process (Kawaida et al., 1994).

Another effect of HGF is modulation of the extracellular matrix composition. Thus, HGF inhibits increased synthesis and deposition of extracellular matrix components, thereby preventing renal fibrosis and end-stage renal disease. Such an effect of HGF has been observed in the development of renal fibrosis in a model of chronic renal failure in a spontaneous mouse model of end-stage renal disease. This was accompanied by increased proximal tubular proliferation and decreased TGFβ1 and PDGF expression levels (Mizuno et al., 1998).

Insulin-Like Growth Factor-I

IGF-I production in the kidney is primarily located in the collecting duct. IGF-I may be causative in hypertrophic cell increases of glomerular cells and proximal tubular cells during renal regeneration after ARF (Hammerman, 1991). Like EGF, the levels of IGF-I drop after IR, although autophosphorylation of the IGF-I receptor remains normal (Friedlaender et al., 1998).

Activity of IGF-I not only depends on the expression of IGF-I itself, but also on the expression of IGF binding proteins and IGF-I receptors (Tsao et al., 1995). For example, in radiocontrast agent-induced ARF, the levels of IGF-I, as well as those of IGF binding proteins, change in association with injury to medullary thick ascending limb (Symon et al., 1998). Treatment with recombinant IGF-I hastens recovery and accelerates repair or regeneration of damaged epithelia after renal IR when given either prior to or 24 hours after IR (Miller et al., 1994).

IGF-I also ameliorates acute nephrotoxicity of cyclosporine (Maestri et al., 1997). In a rat model of IR as well as an *in vitro* model of anoxia and reoxygenation, recombinant human IGF-I is cytoprotective. This is associated with increased ATP repletion, increased mitogenesis, and decreased apoptosis (Hirschberg and Ding, 1998).

Fibroblast Growth Factors

The family of FGFs consists of at least 14 members and plays an important role in cell proliferation and differentiation. Of these 14 members, only a few are expressed in the kidney. FGF-1 is present in distal tubules, cortical and medullary collecting ducts, blood vessels, and glomeruli, but is absent in normal proximal tubular cells. However, increased FGF-1 expression was observed in rat proximal tubular epithelial cells *in vitro*, which may be similar to the *in vivo* regeneration process (Zhang et al., 1993).

One day after renal injury caused by tetrafluoroethylcysteine (TFEC) *in vivo*, an increase in the percentage of proliferating cells was associated with increased expression of FGF-1. Although increased levels of FGF-1 are present in infiltrating mononuclear cells at early time points, vimentin-positive, regenerating proximal tubular epithelial cells also stain positive for FGF-1 at later time points (Ichimura et al., 1995).

FGF-7 seems to play a more prominent role in renal growth and development than FGF-1. Thus, FGF-7-null mice have kidneys that possess 30% fewer nephrons than control kidneys (Qiao et al., 1999). FGF-7 is primarily present in the interstitium and its expression increases after treatment with the nephrotoxicant TFEC. No FGF-7 was expressed in primary cultured renal proximal tubular epithelial cells (Ichimura et al., 1996).

These combined observations suggest a role for both FGF-1 and FGF-7 in the mitogenic response and morphogenic changes that occur during the renal regeneration process. However, research is still required to further validate the requirement for both FGF-1 and FGF-7 in renal regeneration after both nephrotoxicant exposure and renal IR. Evaluation of whether recombinant FGFs may be safely used to promote the renal regeneration process is also needed.

Transforming Growth Factor

TGFβ1 is the most important isoform of the TGF family; it has a prominent role in the response of the kidney to severe injury. TGFβ1 activity in the kidney is regulated by elevated glucose, angiotensin II, oxidant stress, and hemodynamic forces. Activation levels of free TGFβ1 are increased early after IR (Basile et al., 1996). An increase in TGFβ1 is also observed in nephropathy caused by nephrotoxicants. For example, in human allografts, an association is observed between the levels of TGFβ1 and the extent of cyclosporine toxicity (Pankewycz et al., 1996). An increase of TGFβ1 is also observed after treatment of cultured mouse proximal tubular cells after exposure to cyclosporine, suggesting a direct cellular effect of cyclosporine in the mechanism leading to increased TGFβ1 expression and activity (Wolf et al., 1996).

Not all toxicants cause an increase in TGFβ1. For example, $CdCl_2$ and $HgCl_2$ decrease the levels of TGFβ1 in immortalized proximal tubular cells *in vitro* (Jiang et al., 2002). Sustained elevation of TGFβ1 is related to a perturbed regeneration process, eventually resulting in fibrosis. Such a role for TGFβ1 in renal regeneration has become evident from the treatment of animal models with anti-TGFβ antibodies and antisense oligonucleotides.

In a nephrotoxic rat model of cyclosporine-induced fibrosis, anti-TGFβ1 antibody treatment prevented fibrosis (Islam et al., 2001). Anti-TGFβ2 antibody also decreases fibrosis in a streptozotocin-induced diabetic rats (Hill et al., 2001). Moreover, TGFβ1 antisense oligodeoxynucleotides prevented fibrosis in a unilateral ureter obstruction model (Isaka et al., 2000).

Besides extracellular matrix (ECM) disposition, TGFβ1 also regulates the proliferation and differentiation of injured tubule cells (Docherty et al., 2002). For example, TGFβ1 has a negative effect on cell proliferation and it inhibits regeneration of rabbit renal proximal tubular cells *in vitro* after exposure to tert-butylhydroperoxide (Kays et al., 1996). Such an effect of TGFβ is also observed in a three-dimensional culture of primary cultured proximal tubular cells from the rat; this is associated with increased apoptosis (Bowes et al., 1999).

HGF vs. TFGβ: Control of ECM, Tubulogenesis and Cell Survival

HGF and TGFβ are probably the most important growth factors that control the regeneration process in the kidney. Their effects seem opposite: HGF promotes whereas TGFβ1 antagonizes proper regeneration. The downstream effects and signaling pathways involved are described below (Figure 8.5).

HGF and TGFβ1 and Control of ECM Composition

HGF and TGFβ1 are central to the regulation of ECM composition and turnover. Inappropriate regulation of expression of TGFβ1 seems to be involved in increased production of ECM components, resulting in scarring of tissue and fibrosis. HGF seems to counteract this effect and prevent the scarring of tissue.

The effect of TGFβ1 is related to the increased expression of genes coding for the tissue inhibitor of metalloproteinases (TIMP-1), alphaIV collagen, fibronectin EIIIA, PAI-1, and decreased expression of the metalloproteinase MMP-9 (Basile et al., 1998). Such changes in expression are attenuated by treatment with TGF-β antibodies. In contrast, HGF decreases the expression of TIMP-1 and TIMP-2 and increases that of MMP-9 (Liu et al., 2000). Moreover, treatment with anti-HGF antibodies in a mouse model of renal fibrosis results in

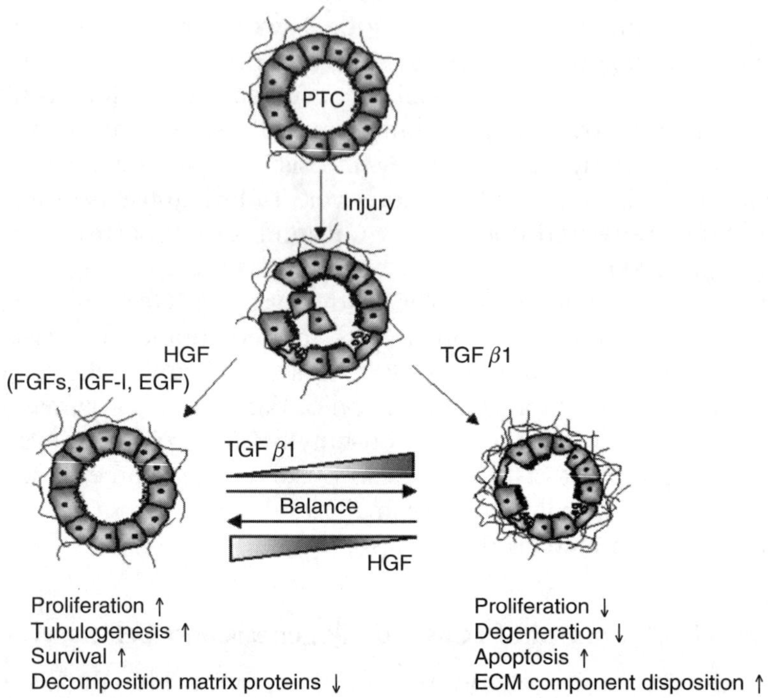

Figure 8.5 Differential roles of hepatocyte growth factor and transforming growth factor in the renal regeneration process.

markedly increased matrix deposits in the kidney (Mizuno et al., 2000). Thus, there is a requirement for a correct balance in the expression and activity of HGF and TGFβ1 during the renal regeneration process. When this balance tips towards TGFβ1, increased production of matrix proteins may lead to fibrosis, with end-stage renal disease as the worst outcome.

Morphometric and Morphogenic Effects of HGF

An important effect of HGF is its potential to induce cell scattering and tubulogenesis. These effects have been studied best in MDCK cells. HGF-induced scattering of MDCK cells is associated with loss of E-cadherin mediated adhesions at adherens junctions. This process requires the activity of several signaling pathways, including a shift in the activity of RhoGTPase Rac1 to RhoA (Hordijk et al., 1997).

In three-dimensional matrix cultures, HGF stimulates the branching of MDCK foci into structures that resemble tubules, a process called tubulogenesis (Montesano et al., 1991). This process mimics the HGF-induced outgrowth of metanephroi (van Adelsberg et al., 2001). HGF-transfected cells form branching tubules, which is associated with increased expression of the c-Met receptor (Liu et al., 1998c). Signal transduction pathways downstream of the c-Met receptor, including Grb2, PLCγ and PI-3 kinase activation, are required for branching tubulogenesis of MDCK cells. Likewise, overexpression of R-Ras or PI-3 kinase causes formation of branching tubules of MDCK cells. Moreover, inhibitors of ERK as well as PI-3 kinase block tubulogenesis, indicating a role for these signaling pathways in morphometric changes during the renal regeneration process (Khwaja et al., 1998).

HGF also activates ERK and p38 pathways, while inhibitors of these kinases prevent the development of metanephroi induced by HGF (Hida et al., 2002). TGFβ1 counteracts effects of HGF, such as tubular branching of MCDK cells (Santos and Nigam, 1993). Additionally, TGFβ1 also inhibits the tubulogenesis of primary cultured rat proximal tubular cells (Bowes et al., 1999).

Control of Survival and Proliferation by HGF and TGFβ1

Tubulogenesis is associated with increased cell proliferation. This is also observed *in vivo* after ARF when HGF is used. The increased proliferation may be related to increased expression of c-Myc and c-Fos, which promote mitogenesis (Clifford et al., 1998). HGF also protects against apoptosis *in vitro*, which is related to the activation of PI-3 kinase and its downstream signaling pathways, including protein kinase B (Liu et al., 1998b). HGF-mediated activation of the c-Met receptor activates ERK in the kidney, which is associated with protection against glycerol-induced ARF (Nagano et al., 2002). Additionally, HGF stimulates the expression of the antiapoptotic protein Bcl-2 (Gao et al., 2002). TGFβ1 counteracts the survival signaling of HGF but also induces apoptosis by itself in various cell types, including proximal tubular cells. Treatment with anti-TGFβ1 antibodies prevents apoptosis in intact kidneys after unilateral ureter obstruction (Miyajima et al., 2000).

TGFβ1-mediated signaling involves regulation of the family of Smad proteins that in their turn affect gene expression. Smad proteins, therefore, probably mediate the proapoptotic action of TGFβ1 in renal cells. Indeed, overexpression of Smad7 induces apoptosis and caspase activation in rat mesangial cells and Smad7 antisense oligonucleotides prevent TGFβ induced apoptosis. Smad2 and Smad3 do not seem to be involved in this process (Okado et al., 2002).

In conclusion, HGF and TGFβ control the renal regeneration process at the level of ECM composition and turnover, cell scattering and tubulogenesis, and mitogenesis and cell survival. Further understanding of the downstream signaling pathways for each of these effects in the kidney will be essential for a targeted therapy to accelerate the renal regeneration process and prevent renal fibrosis in patients that have encountered severe acute renal injury.

CONCLUSIONS AND FUTURE PERSPECTIVES

From the above discussions it follows that nephrotoxicants induce changes in protein expression that can in part be ascribed to general biochemical perturbations, such as calcium homeostasis, oxidative stress, and thiol modification. Of course, other more subtle and less well characterized biochemical perturbations also take place, which may be essential in the cellular stress response. These require further identification and characterization.

Only a few stress response genes have been covered in detail in this chapter. The use of cDNA and oligonucleotide microarrays and chips will allow further detailed analysis of the different stress response genes that are upregulated by various nephrotoxic medicines and chemicals, as well as by IR. Moreover, such cDNA arrays will be valuable tools for studying the relationship between the biochemical perturbations and particular stress response pathways in greater detail, as well the effect of growth factors in the control of renal toxicity.

A next step will be to comprehend the exact roles of these genes in the diversity of biological events and routes that are initiated after cellular stress, including survival or death, damage repair, mitogenesis, and cellular shape changes and tissue remodelling. The use of both transgenic and knock-out mouse models, as well as small interference RNA technologies, are required to further unravel the role of these individual genes in the induction of nephrotoxicity and the renal regeneration process. Eventually, this will lead to the identification of novel therapeutic targets that may help to prevent ARF or speed up the regeneration process.

ACKNOWLEDGMENTS

We would like to thank Gerard Mulder and Fred Nagelkerke for critical reading of the manuscript and Floor van de Water for help with manuscript preparation. Financial support was provided by the Royal Academy of Arts and Sciences (Fellowship to BvdW) and grants 902-21-229 and 902-21-217 from the Netherlands Organization for Scientific Research (to BvdW).

REFERENCES

Adler V, Yin Z, Fuchs SY, Benezra M, Rosario L, Tew KD, Pincus MR, Sardana M, Henderson CJ, Wolf CR, Davis RJ, and Ronai Z (1999) Regulation of JNK signaling by GSTp. *EMBO J* 18:1321–1334.

Amaike H, Matsumoto K, Oka T, and Nakamura T (1996) Preventive effect of hepatocyte growth factor on acute side effects of cyclosporin A in mice. *Cytokine* 8:387–394.

Anderson RJ, and Ray CJ (1998) Potential autocrine and paracrine mechanisms of recovery from mechanical injury of renal tubular epithelial cells. *Am J Physiol* 274:F463–F472.

Balkovetz DF, and Lipschutz JH (1999) Hepatocyte growth factor and the kidney: It is not just for the liver. *Int Rev Cytol* 18:6225–6260.

Basile DP, Rovak JM, Martin DR, and Hammerman MR (1996) Increased transforming growth factor-beta 1 expression in regenerating rat renal tubules following ischemic injury. *Am J Physiol* 270:F500–F509.

Basile DP, Martin DR, and Hammerman MR (1998) Extracellular matrix-related genes in kidney after ischemic injury: potential role for TGF-beta in repair. *Am J Physiol* 275:F894–F903.

Beere HM, Wolf BB, Cain K, Mosser DD, Mahboubi A, Kuwana T, Tailor P, Morimoto RI, Cohen GM, and Green DR (2000) Heat-shock protein 70 inhibits apoptosis by preventing recruitment of procaspase-9 to the Apaf-1 apoptosome. *Nature Cell Biol* 2:469–475.

Behrens A, Sibilia M, and Wagner EF (1999) Amino-terminal phosphorylation of c-Jun regulates stress-induced apoptosis and cellular proliferation. *Nature Genet* 21:326–329.

Benigni A, and Remuzzi G (1995) Endothelin in the progressive renal disease of glomerulopathies. *Miner Electrolyte Metab* 21:283–291.

Bertolotti A, Zhang Y, Hendershot LM, Harding HP, and Ron D (2000) Dynamic interaction of BiP and ER stress transducers in the unfolded-protein response. *Nature Cell Biol* 2:326–332.

Bourgarel-Rey V, Vallee S, Rimet O, Champion S, Braguer D, Desobry A, Briand C, and Barra Y (2001) Involvement of nuclear factor kappaB in c-Myc induction by tubulin polymerization inhibitors. *Mol Pharmacol* 59:1165–1170.

Bowes RC, Lightfoot RT, van de Water B, and Stevens JL (1999) Hepatocyte growth factor induces tubulogenesis of primary renal proximal tubular epithelial cells. *J Cell Physiol* 180:81–90.

Bruey JM, Ducasse C, Bonniaud P, Ravagnan L, Susin SA, Diaz-Latoud C, Gurbuxani S, Arrigo AP, Kroemer G, Solary E, and Garrido C (2000) Hsp27 negatively regulates cell death by interacting with cytochrome c. *Nature Cell Biol* 2:645–652.

Bush KT, George SK, Zhang PL, and Nigam SK (1999) Pretreatment with inducers of ER molecular chaperones protects epithelial cells subjected to ATP depletion. *Am J Physiol* 277:F211–F218.

Cardone MH, Roy N, Stennicke HR, Salvesen GS, Franke TF, Stanbridge E, Frisch S, and Reed JC (1998) Regulation of cell death protease caspase-9 by phosphorylation. *Science* 282:1318–1321.

Charette SJ, Lavoie JN, Lambert H, and Landry J (2000) Inhibition of Daxx-mediated apoptosis by heat shock protein 27. Mol Cell Biol 20:7602–7612.

Chen C, Edelstein LC, and Gelinas C (2000) The Rel/NF-kappaB family directly activates expression of the apoptosis inhibitor Bcl-x(L). *Mol Cell Biol* 20:2687–2695.

Chen Q, and Stevens JL (1991) Inhibition of iodoacetamide and t-butylhydroperoxide toxicity in LLC-PK_1 cells by antioxidants: A role for lipid peroxidation in alkylation induced cytotoxicity. *Arch Biochem Biophys* 284:422–430.

Chen Q, Jones TW, Brown PC, and Stevens JL (1990) The mechanism of cysteine conjugate cytotoxicity in renal epithelial cells. Covalent binding leads to thiol depletion and lipid peroxidation. *J Biol Chem* 265:21603–21611.

Chen Q, Yu K, Holbrook NJ, and Stevens JL (1992a) Activation of the growth arrest and DNA damage-inducible gene gadd 153 by nephrotoxic cysteine conjugates and dithiothreitol. *J Biol Chem* 267:8207–8212.

Chen Q, Yu K, and Stevens JL (1992b) Regulation of the cellular stress response by reactive electrophiles. The role of covalent binding and cellular thiols in transcriptional activation of the 70-kilodalton heat shock protein gene by nephrotoxic cysteine conjugates. *J Biol Chem* 267:24322–24327.

Chevalier RL, Goyal S, Wolstenholme JT, and Thornhill BA (1998) Obstructive nephropathy in the neonatal rat is attenuated by epidermal growth factor. *Kidney Int* 54:38–47.

Chu ZL, McKinsey TA, Liu L, Gentry JJ, Malim MH, and Ballard DW (1997) Suppression of tumor necrosis factor-induced cell death by inhibitor of apoptosis c-IAP2 is under NF-kappaB control. *Proc Natl Acad Sci USA* 94:10057–10062.

Clifford SC, Czapla K, Richards FM, O'Donoghue DJ, and Maher ER (1998) Hepatocyte growth factor-stimulated renal tubular mitogenesis: Effects on expression of c-myc, c-fos, c-met, VEGF and the VHL tumour-suppressor and related genes. *Br J Cancer* 77:1420–1428.

Coimbra TM, Cieslinski DA, and Humes HD (1990) Epidermal growth factor accelerates renal repair in mercuric chloride nephrotoxicity. *Am J Physiol* 259:F438–F443.

Deng X, Xiao L, Lang W, Gao F, Ruvolo P, and May WS Jr (2001) Novel role for JNK as a stress-activated Bcl2 kinase. *J Biol Chem* 276:23681–23688.

Dieguez-Acuna FJ, Ellis ME, Kushleika J, and Woods JS (2001) Mercuric ion attenuates nuclear factor-kappaB activation and DNA binding in normal rat kidney epithelial cells: implications for mercury-induced nephrotoxicity. *Toxicol Appl Pharmacol* 173:176–187.

Dijkers PF, Birkenkamp KU, Lam EWF, Thomas NSB, Lammers JWJ, Koenderman L, and Coffer PJ (2002) FKHR-L1 can act as a critical effector of cell death induced by cytokine withdrawal: protein kinase B-enhanced cell survival through maintenance of mitochondrial integrity. *J Cell Biol* 156:531–542.

Di Mari J, Megyesi J, Udvarhelyi N, Price P, Davis R, and Safirstein RN (1997) Acetyl cysteine ameliorates ischemic renal failure. *Am J Physiol* 272:F292–F298.

Di Mari JF, Davis R, and Safirstein RL (1999) MAPK activation determines renal epithelial cell survival during oxidative injury. *Am J Physiol* 277:F195–F203.

Docherty NG, Perez-Barriocanal F, Balboa NE, and Lopez-Novoa JM (2002) Transforming growth factor-beta1 (TGF-beta1): A potential recovery signal in the post-ischemic kidney. *Renal Failure* 24:391–406.

Evan GI, Wyllie AH, Gilbert CS, Littlewood TD, Land H, Brooks M, Waters CM, Penn LZ, and Hancock DC (1992) Induction of apoptosis in fibroblasts by c-myc protein. *Cell* 69:119–128.

Fanidi A, Harrington EA, and Evan GI (1992) Cooperative interaction between c-myc and bcl-2 proto-oncogenes. *Nature* 359:554–556.

Friedlaender MM, Fervenza FC, Tsao T, Hsu F, and Rabkin R (1998) The insulin-like growth factor-I axis in acute renal failure. *Renal Failure* 20:343–348.

Frisch SM, Vuori K, Ruoslahti E, and Chan-Hui PY (1996) Control of adhesion-dependent cell survival by focal adhesion kinase. *J Cell Biol* 134:793–799.

Furusu A, Nakayama K, Xu Q, Konta T, Sugiyama H, and Kitamura M (2001) Expression, regulation, and function of inhibitor of apoptosis family genes in rat mesangial cells. *Kidney Int* 60:579–586.

Gao X, Mae H, Ayabe N, Takai T, Oshima K, Hattori M, Ueki T, Fujimoto J, and Tanizawa T (2002) Hepatocyte growth factor gene therapy retards the progression of chronic obstructive nephropathy. *Kidney Int* 62:1238–1248.

Garay M, Gaarde W, Monia BP, Nero P, and Cioffi CL (2000) Inhibition of hypoxia/reoxygenation-induced apoptosis by an antisense oligonucleotide targeted to JNK1 in human kidney cells. *Biochem Pharmacol* 59:1033–1043.

Ghosh S, and Karin M (2002) Missing pieces in the NF-kappaB puzzle. *Cell* 109 (Suppl):S81–S96.

Goering PL, Fisher BR, Noren BT, Papaconstantinou A, Rojko JL, and Marler RJ (2000) Mercury induces regional and cell-specific stress protein expression in rat kidney. *Toxicol Sci* 53:447–457.

Goto T, Sugimura K, Harimoto K, Kasai S, Kim T, and Kishimoto T (1997) Hepatocyte growth factor in glycerol-induced acute renal failure. *Nephron* 77:440–444.

Grandori C, Cowley SM, James LP, and Eisenman RN (2000) The Myc/Max/Mad network and the transcriptional control of cell behavior. *Annu Rev Cell Dev Biol* 16:653–699.

Halleck MM, Holbrook NJ, Skinner J, Liu H, and Stevens JL (1997a) The molecular response to reductive stress in LLC-PK_1 renal epithelial cells: coordinate transcriptional regulation of gadd153 and grp78 genes by thiols. *Cell Stress Chaperones* 2:31–40.

Halleck MM, Liu H, North J, and Stevens JL (1997b) Reduction of trans-4,5-dihydroxy-1, 2-dithiane by cellular oxidoreductases activates gadd153/chop and grp78 transcription and induces cellular tolerance in kidney epithelial cells. *J Biol Chem* 272:21760–21766.

Hammerman MR (1991) The renal growth hormone/insulin-like growth factor I axis. *Am J Kidney Dis* 17:644–666.

Hammerman MR (1998) Growth factors and apoptosis in acute renal injury. *Curr Opin Nephrol Hypertens* 7:419–424.

Harding HP, Novoa I, Zhang Y, Zeng H, Wek R, Schapira M, and Ron D (2000) Regulated translation initiation controls stress-induced gene expression in mammalian cells. *Mol Cell* 6:1099–1108.

He TC, Sparks AB, Rago C, Hermeking H, Zawel L, da Costa LT, Morin PJ, Vogelstein B, and Kinzler KW (1998) Identification of c-MYC as a target of the APC pathway. *Science* 281:1509–1512.

Hernadez-Pando R, Pedraza-Chaverri J, Orozco-Estevez H, Silva-Serna P, Moreno I, Rondan-Zarate A, Elinos M, Correa-Rotter R, and Larriva-Sahd J (1995) Histological and subcellular distribution of 65 and 70 kD heat shock proteins in experimental nephrotoxic injury. *Exp Toxicol Pathol* 47:501–508.

Hida M, Omori S, and Awazu M (2002) ERK and p38 MAP kinase are required for rat renal development. *Kidney Int* 61:1252–1262.

Hill C, Flyvbjerg A, Rasch R, Bak M, and Logan A (2001) Transforming growth factor-beta2 antibody attenuates fibrosis in the experimental diabetic rat kidney. *J Endocrinol* 170:647–651.

Hirschberg R, and Ding H (1998) Mechanisms of insulin-like growth factor-I-induced accelerated recovery in experimental ischemic acute renal failure. *Miner Electrolyte Metab* 24:211–219.

Hise MK, Liu L, Salmanullah M, Drachenberg CI, Papadimitriou JC, and Rohan RM (2000) mRNA expression of transforming growth factor-alpha and the EGF receptor following nephrotoxic renal injury. *Renal Failure* 22:423–434.

Homma T, Sakai M, Cheng HF, Yasuda T, Coffey RJ, and Harris RC (1995) Induction of heparin-binding epidermal growth factor-like growth factor mRNA in rat kidney after acute injury. *J Clin Invest* 96:1018–1025.

Hordijk PL, ten Klooster JP, van der Kammen RA, Michiels F, Oomen LC, and Collard JG (1997) Inhibition of invasion of epithelial cells by Tiam1-Rac signaling. *Science* 278:1464–1466.

Hueber AO, Zornig M, Lyon D, Suda T, Nagata S, and Evan GI (1997) Requirement for the CD95 receptor-ligand pathway in c-Myc-induced apoptosis. *Science* 278:1305–1309.

Ichimura T, Maier JA, Maciag T, Zhang G, and Stevens JL (1995) FGF-1 in normal and regenerating kidney: Expression in mononuclear, interstitial, and regenerating epithelial cells. *Am J Physiol* 269:F653–F662.

Ichimura T, Finch PW, Zhang G, Kan M, and Stevens JL (1996) Induction of FGF-7 after kidney damage: a possible paracrine mechanism for tubule repair. *Am J Physiol* 271:F967–F976.

Igawa T, Matsumoto K, Kanda S, Saito Y, and Nakamura T (1993) Hepatocyte growth factor may function as a renotropic factor for regeneration in rats with acute renal injury. *Am J Physiol* 265:F61–F69.

Isaka Y, Tsujie M, Ando Y, Nakamura H, Kaneda Y, Imai E, and Hori M (2000) Transforming growth factor-beta 1 antisense oligodeoxynucleotides block interstitial fibrosis in unilateral ureteral obstruction. *Kidney Int* 58:1885–1892.

Islam M, Burke JF Jr, McGowan TA, Zhu Y, Dunn SR, McCue P, Kanalas J, and Sharma K (2001) Effect of anti-transforming growth factor-beta antibodies in cyclosporine-induced renal dysfunction. *Kidney Int* 59:498–506.

Jeong JK, Stevens JL, Lau SS, and Monks TJ (1996) Quinone thioether-mediated DNA damage, growth arrest, and gadd153 expression in renal proximal tubular epithelial cells. *Mol Pharmacol* 50:592–598.

Jiang J, McCool BA, and Parrish AR (2002) Cadmium- and mercury-induced intercellular adhesion molecule-1 expression in immortalized proximal tubule cells: Evidence for a role of decreased transforming growth factor-beta1. *Toxicol Appl Pharmacol* 179:13–20.

Juin P, Hunt A, Littlewood T, Griffiths B, Swigart LB, Korsmeyer S, and Evan G (2002) c-Myc functionally cooperates with Bax to induce apoptosis. *Mol Cell Biol* 22:6158–6169.

Kaufman RJ (1999) Stress signaling from the lumen of the endoplasmic reticulum: Coordination of gene transcriptional and translational controls. *Genes Dev* 13:1211–1233.

Kaushal GP, Kaushal V, Hong X, and Shah SV (2001) Role and regulation of activation of caspases in cisplatin-induced injury to renal tubular epithelial cells. *Kidney Int* 60:1726–1736.

Kawaida K, Matsumoto K, Shimazu H, and Nakamura T (1994) Hepatocyte growth factor prevents acute renal failure and accelerates renal regeneration in mice. *Proc Natl Acad Sci USA* 91:4357–4361.

Kays SE, and Schnellmann RG (1995) Regeneration of renal proximal tubule cells in primary culture following toxicant injury: Response to growth factors. *Toxicol Appl Pharmacol* 132:273–280.

Kays SE, Nowak G, and Schnellmann RG (1996) Transforming growth factor-beta 1 inhibits regeneration of renal proximal tubular cells after oxidant exposure. *J Biochem Toxicol* 11:79–84.

Kennedy WA, Buttyan R, Garcia-Montes E, D'Agati V, Olsson CA, and Sawczuk IS (1997) Epidermal growth factor suppresses renal tubular apoptosis following ureteral obstruction. *Urology* 49:973–980.

Khwaja A, and Downward J (1997) Lack of correlation between activation of Jun-NH2-terminal kinase and induction of apoptosis after detachment of epithelial cells. *J Cell Biol* 139:1017–1023.

Khwaja A, Lehmann K, Marte BM, and Downward J (1998) Phosphoinositide 3-kinase induces scattering and tubulogenesis in epithelial cells through a novel pathway. *J Biol Chem* 273:18793–18801.

Klefstrom J, Arighi E, Littlewood T, Jaattela M, Saksela E, Evan GI, and Alitalo K (1997) Induction of TNF-sensitive cellular phenotype by c-Myc involves p53 and impaired NF-kappaB activation. *EMBO J* 16:7382–7392.

Kohda Y, and Gemba M (2001) Modulation by cyclic AMP and phorbol myristate acetate of cephaloridine-induced injury in rat renal cortical slices. *Jpn J Pharmacol* 85:54–59.

Kops GJ, de Ruiter ND, De Vries Smits AM, Powell DR, Bos JL, and Burgering BM (1999) Direct control of the Forkhead transcription factor AFX by protein kinase B. *Nature* 398:630–634.

Kruidering M, van de Water B, de Heer E, Mulder GJ, and Nagelkerke JF (1997) Cisplatin-induced nephrotoxicity in porcine proximal tubular cells: Mitochondrial dysfunction by inhibition of complexes I to IV of the respiratory chain. *J Pharmacol Exp Ther* 280:638–649.

Kultz D, Madhany S, and Burg MB (1998) Hyperosmolality causes growth arrest of murine kidney cells. Induction of GADD45 and GADD153 by osmosensing via stress-activated protein kinase 2. *J Biol Chem* 273:13645–13651.

Kunduzova OR, Bianchi P, Pizzinat N, Escourrou G, Seguelas MH, Parini A, and Cambon C (2002) Regulation of JNK/ERK activation, cell apoptosis, and tissue regeneration by monoamine oxidases after renal ischemia-reperfusion. *FASEB J* 16:1129–1131.

Kyriakis JM, and Avruch J (2001) Mammalian mitogen-activated protein kinase signal transduction pathways activated by stress and inflammation. *Physiol Rev* 81:807–869.

La Rosa FA, Pierce JW and Sonenshein GE (1994) Differential regulation of the c-myc oncogene promoter by the NF-kappa B rel family of transcription factors. *Mol Cell Biol* 14:1039–1044.

Leonard I, Zanen J, Nonclercq D, Toubeau G, Heuson-Stiennon JA, Beckers JF, Falmagne P, Schaudies RP, and Laurent G (1994) Modification of immunoreactive EGF and EGF receptor after acute tubular necrosis induced by tobramycin or cisplatin. *Renal Failure* 16:583–608.

Liu J, Squibb KS, Akkerman M, Nordberg GF, Lipsky M, and Fowler BA (1996a) Cytotoxicity, zinc protection, and stress protein induction in rat proximal tubule cells exposed to cadmium chloride in primary cell culture. *Renal Failure* 18:867–882.

Liu H, Lightfoot R, and Stevens JL (1996b) Activation of heat shock factor by alkylating agents is triggered by glutathione depletion and oxidation of protein thiols. *J Biol Chem* 271:4805–4812.

Liu H, Bowes RC, van de Water B, Sillence C, Nagelkerke JF, and Stevens JL (1997) Endoplasmic reticulum chaperones GRP78 and calreticulin prevent oxidative stress, Ca^{2+} disturbances, and cell death in renal epithelial cells. *J Biol Chem* 272:21751–21759.

Liu H, Miller E, van de Water B, and Stevens JL (1998a) Endoplasmic reticulum stress proteins block oxidant-induced $Ca2^{+}$ increases and cell death. *J Biol Chem* 273:12858–12862.

Liu Y, Sun AM, and Dworkin LD (1998b) Hepatocyte growth factor protects renal epithelial cells from apoptotic cell death. *Biochem Biophys Res Commun* 246:821–826.

Liu Y, Centracchio JN, Lin L, Sun AM, and Dworkin LD (1998c) Constitutive expression of HGF modulates renal epithelial cell phenotype and induces c-met and fibronectin expression. *Exp Cell Res* 242:174–185.

Liu Y, Rajur K, Tolbert E, and Dworkin LD (2000) Endogenous hepatocyte growth factor ameliorates chronic renal injury by activating matrix degradation pathways. *Kidney Int* 58:2028–2043.

Lund BO, Miller DM, and Woods JS (1991) Mercury-induced H_2O_2 production and lipid peroxidation *in vitro* in rat kidney mitochondria. *Biochem Pharmacol* 42 (Suppl):S181–S187.

Ma Y, Brewer JW, Diehl JA, and Hendershot LM (2002) Two distinct stress signaling pathways converge upon the CHOP promoter during the mammalian unfolded protein response. *J Mol Biol* 318:1351–1365.

Maestri M, Dafoe DC, Adams GA, Gaspari A, Luzzana F, Innocente F, Rademacher J, Dionigi P, Barbieri A, Zonta F, Zonta A, and Rabkin R (1997) Insulin-like growth factor-I ameliorates delayed kidney graft function and the acute nephrotoxic effects of cyclosporine. *Transplantation* 64:185–190.

Maines MD, Raju VS, and Panahian N (1999) Spin trap (*N*-t-butyl-alpha-phenylnitrone)-mediated suprainduction of heme oxygenase-1 in kidney ischemia/reperfusion model: Role of the oxygenase in protection against oxidative injury. *Pharmacol Exp Ther* 291:911–919.

Maki A, Berezesky IK, Fargnoli J, Holbrook NJ, and Trump BF (1992) Role of [$Ca2^{+}$]i in induction of c-fos, c-jun, and c-myc mRNA in rat PTE after oxidative stress. *FASEB J* 6:919–924.

Meldrum KK, Hile K, Meldrum DR, Crone JA, Gearhart JP, and Burnett AL (2002) Simulated ischemia induces renal tubular cell apoptosis through a nuclear factor-kappaB dependent mechanism. *J Urol* 168:248–252.

Menssen A, and Hermeking H (2002) Characterization of the c-MYC-regulated transcriptome by SAGE: Identification and analysis of c-MYC target genes. *Proc Natl Acad Sci USA* 99:6274–6279.

Miller SB, Martin DR, Kissane J, and Hammerman MR (1994) Rat models for clinical use of insulin-like growth factor I in acute renal failure. *Am J Physiol* 266:F949–F956.

Miyajima A, Chen J, Lawrence C, Ledbetter S, Soslow RA, Stern J, Jha S, Pigato J, Lemer ML, Poppas DP, Vaughan ED, and Felsen D (2000) Antibody to transforming growth factor-beta ameliorates tubular apoptosis in unilateral ureteral obstruction. *Kidney Int* 58:2301–2313.

Miyazawa K, Shimomura T, Naka D, and Kitamura N (1994) Proteolytic activation of hepatocyte growth factor in response to tissue injury. *J Biol Chem* 269:8966–8970.

Mizuno S, Kurosawa T, Matsumoto K, Mizuno-Horikawa Y, Okamoto M, and Nakamura T (1998) Hepatocyte growth factor prevents renal fibrosis and dysfunction in a mouse model of chronic renal disease. *J Clin Invest* 101:1827–1834.

Mizuno S, Matsumoto K, Kurosawa T, Mizuno-Horikawa Y, and Nakamura T (2000) Reciprocal balance of hepatocyte growth factor and transforming growth factor-beta 1 in renal fibrosis in mice. *Kidney Int* 57:937–948.

Molitoris BA, and Marrs J (1999) The role of cell adhesion molecules in ischemic acute renal failure. *Am J Med* 106:583–592.

Monks TJ, Jones TW, Hill BA, and Lau SS (1991) Nephrotoxicity of 2-bromo-(cystein-S-yl) hydroquinone and 2-bromo-(*N*-acetyl-L-cystein-S-yl) hydroquinone thioethers. Toxicol *Appl Pharmacol* 111:279–298.

Montesano R, Matsumoto K, Nakamura T, and Orci L (1991) Identification of a fibroblast-derived epithelial morphogen as hepatocyte growth factor. *Cell* 67:901–908.

Morin NJ, Laurent G, Nonclercq D, Toubeau G, Heuson-Stiennon JA, Bergeron MG, and Beauchamp D (1992) Epidermal growth factor accelerates renal tissue repair in a model of gentamicin nephrotoxicity in rats. *Am J Physiol* 263:F806-F811.

Morishima N, Nakanishi K, Takenouchi H, Shibata T, and Yasuhiko Y (2002) An endoplasmic reticulum stress-specific caspase cascade in apoptosis. Cytochrome c-independent activation of caspase-9 by caspase-12. *J Biol Chem* 277:34287–34294.

Mosser DD, Caron AW, Bourget L, Meriin AB, Sherman MY, Morimoto RI, and Massie B (2000) The chaperone function of hsp70 is required for protection against stress-induced apoptosis. *Mol Cell Biol* 20:7146–7159.

Nagano T, Mori-Kudo I, Tsuchida A, Kawamura T, Taiji M, and Noguchi H (2002) Ameliorative effect of hepatocyte growth factor on glycerol-induced acute renal failure with acute tubular necrosis. *Nephron* 91:730–738.

Nakagawa T, Zhu H, Morishima N, Li E, Xu J, Yankner BA, and Yuan J (2000) Caspase-12 mediates endoplasmic-reticulum-specific apoptosis and cytotoxicity by amyloid-beta. *Nature* 403:98–103.

Nath KA, Croatt AJ, Likely S, Behrens TW, and Warden D (1996) Renal oxidant injury and oxidant response induced by mercury. *Kidney Int* 50:1032–1043.

Nicotera P, Bellomo G, and Orrenius S (1990) The role of Ca^{2+} in cell killing. *Chem Res Toxicol* 3:484–494.

Nonclercq D, Wrona S, Toubeau G, Zanen J, Heuson-Stiennon JA, Schaudies RP, and Laurent G (1992) Tubular injury and regeneration in the rat kidney following acute exposure to gentamicin: a time-course study. *Renal Failure* 14:507–521.

Nonclercq D, Toubeau G, Laurent G, Schaudies RP, Zanen J, and Heuson-Stiennon JA (1993) Immunocytological localization of epidermal growth factor in the rat kidney after drug-induced tubular injury. *Eur J Morphol* 31:65–71.

Novoa I, Zeng H, Harding HP, and Ron D (2001) Feedback inhibition of the unfolded protein response by GADD34-mediated dephosphorylation of eIF2alpha. *J Cell Biol* 153:1011–1022.

Okado T, Terada Y, Tanaka H, Inoshita S, Nakao A, and Sasaki S (2002) Smad7 mediates transforming growth factor-beta-induced apoptosis in mesangial cells. *Kidney Int* 62:1178–1186.

Otieno MA, and Anders MW (1997) Cysteine *S*-conjugates activate transcription factor NF-kappa B in cultured renal epithelial cells. *Am J Physiol* 273:F136-F143.

Pagliari LJ, Perlman H, Liu H, and Pope RM (2000) Macrophages require constitutive NF-kappaB activation to maintain A1 expression and mitochondrial homeostasis. *Mol Cell Biol* 20:8855–8865.

Pankewycz OG, Miao L, Isaacs R, Guan J, Pruett T, Haussmann G, and Sturgill BC (1996) Increased renal tubular expression of transforming growth factor beta in human allografts correlates with cyclosporine toxicity. *Kidney Int* 50:1634–1640.

Park KM, Chen A, and Bonventre JV (2001) Prevention of kidney ischemia/reperfusion-induced functional injury and JNK, p38, and MAPK kinase activation by remote ischemic pretreatment. *J Biol Chem* 276:11870–11876.

Park KM, Kramers C, Vayssier-Taussat M, Chen A, and Bonventre JV (2002a) Prevention of kidney ischemia/reperfusion-induced functional injury, MAPK and MAPK kinase activation, and inflammation by remote transient ureteral obstruction. *J Biol Chem* 277:2040–2049.

Park HS, Cho SG, Kim CK, Hwang HS, Noh KT, Kim MS, Huh SH, Kim MJ, Ryoo K, Kim EK, Kang WJ, Lee JS, Seo JS, Ko YG, Kim S, and Choi EJ (2002b) Heat shock protein hsp72 is a negative regulator of apoptosis signal-regulating kinase 1. *Mol Cell Biol* 22:7721–7730.

Paul C, Manero F, Gonin S, Kretz-Remy C, Virot S, and Arrigo A-P (2002) Hsp27 as a negative regulator of cytochrome C release. *Mol Cell Biol* 22:816–834.

Pombo CM, Bonventre JV, Avruch J, Woodgett JR, Kyriakis JM, and Force, T. (1994) The stress-activated protein kinases are major c-Jun amino-terminal kinases activated by ischemia and reperfusion. *J Biol Chem* 269:26546–26551.

Pombo CM, Tsujita T, Kyriakis JM, Bonventre JV, and Force T (1997) Activation of the Ste20-like oxidant stress response kinase-1 during the initial stages of chemical anoxia-induced necrotic cell death. Requirement for dual inputs of oxidant stress and increased cytosolic [Ca^{2+}]. *J Biol Chem* 272:2937–2939.

Puthalakath H, Villunger A, O'Reilly LA, Beaumont JG, Coultas L, Cheney RE, Huang DC, and Strasser A (2001) Bmf: A proapoptotic BH3-only protein regulated by interaction with the myosin V actin motor complex, activated by anoikis. *Science* 293:1829–1832.

Qiao J, Uzzo R, Obara-Ishihara T, Degenstein L, Fuchs E, and Herzlinger D (1999) FGF-7 modulates ureteric bud growth and nephron number in the developing kidney. *Development* 126:547–554.

Raafat AM, Murray MT, McGuire T, DeFrain M, Franko AP, Zafar RS, Palmer K, Diebel L, and Dulchavsky SA (1997) Calcium blockade reduces renal apoptosis during ischemia reperfusion. *Shock* 8:186–192.

Radaelli CA, Tien YH, Kubulus D, Mazzucchelli L, Schilling MK, and Wagner AC (2002) Hyperthermia preconditioning induces renal heat shock protein expression, improves cold ischemia tolerance, kidney graft function and survival in rats. *Nephron* 90:489–497.

Ramachandiran S, Huang Q, Dong J, Lau SS, and Monks TJ (2002) Mitogen-Activated Protein Kinases Contribute to Reactive Oxygen Species-Induced Cell Death in Renal Proximal Tubule Epithelial Cells. *Chem Res Toxicol* 15:1635–1642.

Ramesh G, and Reeves WB (2002) TNF-alpha mediates chemokine and cytokine expression and renal injury in cisplatin nephrotoxicity. *J Clin Invest* 110:835–842.

Ravi R, Bedi GC, Engstrom LW, Zeng Q, Mookerjee B, Gelinas C, Fuchs EJ, and Bedi A (2001) Regulation of death receptor expression and TRAIL/Apo2L-induced apoptosis by NF-kappaB. *Nature Cell Biol* 3:409–416.

Reed DJ (1990) Review of the current status of calcium and thiols in cellular injury. *Chem Res Toxicol* 3:495–502.

Rivera MI, Jones TW, Lau SS, and Monks TJ (1994) Early morphological and biochemical changes during 2-Br-(diglutathion-*S*-yl)hydroquinone-induced nephrotoxicity. *Toxicol Appl Pharmacol* 128:239–250.

Safirstein R (1994) Gene expression in nephrotoxic and ischemic acute renal failure. *J Am Soc Nephrol* 4:1387–1395.

Safirstein RL (1999) Lessons learned from ischemic and cisplatin-induced nephrotoxicity in animals. *Renal Failure* 21:359–364.

Safirstein R, Zelent AZ, and Price PM (1989) Reduced renal prepro-epidermal growth factor mRNA and decreased EGF excretion in ARF. *Kidney Int* 36:810–815.

Safirstein R, Price PM, Saggi SJ, and Harris RC (1990) Changes in gene expression after temporary renal ischemia. *Kidney Int* 37:1515–1521.

Saitoh M, Nishitoh H, Fujii M, Takeda K, Tobiume K, Sawada Y, Kawabata M, Miyazono K, and Ichijo H (1998) Mammalian thioredoxin is a direct inhibitor of apoptosis signal-regulating kinase (ASK) 1. *EMBO J* 17:2596–2606.

Saleh A, Srinivasula SM, Balkir L, Robbins PD, and Alnemri ES (2000) Negative regulation of the Apaf-1 apoptosome by Hsp70. *Nature Cell Biol* 2:476–483.

Santos OF, and Nigam SK (1993) HGF-induced tubulogenesis and branching of epithelial cells is modulated by extracellular matrix and TGF-beta. *Dev Biol* 160:293–302.

Satoh K, Wakui H, Komatsuda A, Nakamoto Y, Miura AB, Itoh H, and Tashima Y (1994) Induction and altered localization of 90-kDa heat-shock protein in rat kidneys with cisplatin-induced acute renal failure. *Renal Failure* 16:313–323.

Schober A, Muller E, Thurau K, and Beck FX (1997) The response of heat shock proteins 25 and 72 to ischaemia in different kidney zones. *Pflugers Arch* 434:292–299.

Schreck R, Albermann K, and Baeuerle PA (1992) Nuclear factor kappa B: An oxidative stress-responsive transcription factor of eukaryotic cells (a review). *Free Radic Res Commun* 17:221–237.

Schulze-Osthoff K, Beyaert R, Vandevoorde V, Haegeman G, and Fiers W (1993) Depletion of the mitochondrial electron transport abrogates the cytotoxic and gene-inductive effects of TNF. *EMBO J* 12:3095–3104.

Smith MW, Phelps PC, and Trump BF (1992) Injury-induced changes in cytosolic Ca^{2+} in individual rabbit proximal tubule cells. *Am J Physiol* 262:F647–F655.

Sonoda Y,Watanabe S, Matsumoto Y, Aizu-Yokota E, and Kasahara T (1999) FAK is the upstream signal protein of the phosphatidylinositol 3-kinase-Akt survival pathway in hydrogen peroxide-induced apoptosis of a human glioblastoma cell line. *J Biol Chem* 274:10566–10570.

Stevens JL, Liu H, Halleck M, Bowes RC, Chen QM, and van de Water B (2000) Linking gene expression to mechanisms of toxicity. *Toxicol Lett* 112–113:479–486.

Symon Z, Fuchs S, Agmon Y, Weiss O, Nephesh I, Moshe R, Brezis M, Flyvbjerg A, and Raz I (1998) The endogenous insulin-like growth factor system in radiocontrast nephropathy. *Am J Physiol* 274:F490–F497.

Taira T, Yoshimura A, Inui K, Oshiden K, Ideura T, Koshikawa S, and Solez K (1994) Immunochemical study of epidermal growth factor in rats with mercuric chloride-induced acute renal failure. *Nephron* 67:88–93.

Takayama H, LaRochelle WJ, Sabnis SG, Otsuka T, and Merlino G (1997) Renal tubular hyperplasia, polycystic disease, and glomerulosclerosis in transgenic mice overexpressing hepatocyte growth factor/scatter factor. *Lab Invest* 77:131–138.

Toubeau G, Nonclercq D, Zanen J, Laurent G, Schaudies PR, and Heuson-Stiennon JA (1994) Renal tissue expression of EGF and EGF receptor after ischaemic tubular injury: an immunohistochemical study. *Exp Nephrol* 2:229–239.

Tsao T, Wang J, Fervenza FC, Vu TH, Jin IH, Hoffman AR, and Rabkin R (1995) Renal growth hormone–insulin-like growth factor-I system in acute renal failure. *Kidney Int* 47:1658–1668.

Tune BM, Fravert D, and Hsu CY (1989) Oxidative and mitochondrial toxic effects of cephalosporin antibiotics in the kidney. A comparative study of cephaloridine and cephaloglycin. *Biochem Pharmacol* 38:795–802.

Urano F, Wang X, Bertolotti A, Zhang Y, Chung P, Harding HP, and Ron D (2000) Coupling of stress in the ER to activation of JNK protein kinases by transmembrane protein kinase IRE1. *Science* 287:664–666.

Vamvakas S, Bittner D, and Koster U (1993) Enhanced expression of the protooncogenes c-myc and c-fos in normal and malignant renal growth. *Toxicol Lett* 67:161–172.

van Adelsberg J, Sehgal S, Kukes A, Brady C, Barasch J, Yang J, and Huan Y (2001) Activation of hepatocyte growth factor (HGF) by endogenous HGF activator is required for metanephric kidney morphogenesis *in vitro*. *J Biol Chem* 276:15099–15106.

van de Water B, Zoetewey JP, de Bont HJ, Mulder GJ, and Nagelkerke JF (1993) The relationship between intracellular Ca^{2+} and the mitochondrial membrane potential in isolated proximal tubular cells from rat kidney exposed to the nephrotoxin 1,2-dichlorovinyl-cysteine. *Biochem Pharmacol* 45: 2259–2267.

van de Water B, Zoeteweij JP, de Bont HJ, Mulder GJ, and Nagelkerke JF (1994) Role of mitochondrial Ca^{2+} in the oxidative stress-induced dissipation of the mitochondrial membrane potential. Studies in isolated proximal tubular cells using the nephrotoxin 1,2-dichlorovinyl-L-cysteine. *J Biol Chem* 269:14546–14552.

van de Water B, Zoeteweij JP, and Nagelkerke JF (1996) Alkylation-induced oxidative cell injury of renal proximal tubular cells: Involvement of glutathione redox-cycle inhibition. *Arch Biochem Biophys* 327:71–80.

van de Water B, Nagelkerke JF, and Stevens JL (1999a) Dephosphorylation of focal adhesion kinase (FAK) and loss of focal contacts precede caspase-mediated cleavage of FAK during apoptosis in renal epithelial cells. *J Biol Chem* 274:13328–13337.

van de Water B, Wang Y, Asmellash S, Liu H, Zhan Y, Miller E, and Stevens JL (1999b) Distinct endoplasmic reticulum signaling pathways regulate apoptotic and necrotic cell death following iodoacetamide treatment. *Chem Res Toxicol* 12:943–951.

van de Water B, Houtepen F, Huigsloot M, and Tijdens IB (2001) Suppression of chemically induced apoptosis but not necrosis of renal proximal tubular epithelial (LLC-PK_1) cells by focal adhesion kinase (FAK). Role of FAK in maintaining focal adhesion organization after acute renal cell injury. *J Biol Chem* 276:36183–36193.

Verheij M, Bose R, Lin XH, Yao B, Jarvis WD, Grant S, Birrer MJ, Szabo E, Zon LI, Kyriakis JM, Haimovitz-Friedman A, Fuks Z, and Kolesnick RN (1996) Requirement for ceramide-initiated SAPK/JNK signalling in stress-induced apoptosis. *Nature* 380:75–79.

Verstrepen WA, Nouwen EJ, Zhu MQ, Ghielli M, and De Broe ME (1995) Time course of growth factor expression in mercuric chloride acute renal failure. *Nephrol Dial Transplant* 10:1361–1371.

Wallin A, Zhang G, Jones TW, Jaken S, and Stevens JL (1992) Mechanism of the nephrogenic repair response. Studies on proliferation and vimentin expression after ^{35}S-1,2-dichlorovinyl-L-cysteine nephrotoxicity *in vivo* and in cultured proximal tubule epithelial cells. *Lab Invest* 66:474–484.

Wang XZ, Harding HP, Zhang Y, Jolicoeur EM, Kuroda M, and Ron D (1998) Cloning of mammalian Ire1 reveals diversity in the ER stress responses. *EMBO J* 17:5708–5717.

Wang YH, Knowlton AA, Li FH, and Borkan SC (2002) Hsp72 expression enhances survival in adenosine triphosphate-depleted renal epithelial cells. *Cell Stress Chaperones* 7:137–145.

Weinberg JM, Venkatachalam MA, Roeser NF, Senter RA, and Nissim I (2001) Energetic determinants of tyrosine phosphorylation of focal adhesion proteins during hypoxia/reoxygenation of kidney proximal tubules. *Am J Pathol* 158:2153–2164.

Wolf G, Zahner G, Ziyadeh FN, and Stahl RA (1996) Cyclosporin A induces transcription of transforming growth factor beta in a cultured murine proximal tubular cell line. *Exp Nephrol* 4:304–308.

Wu MX, Ao Z, Prasad KV, Wu R, and Schlossman SF (1998) i.e.,X-1L, an apoptosis inhibitor involved in NF-kappaB-mediated cell survival. *Science* 281 998–1001.

Yamamoto K, Ichijo H, and Korsmeyer SJ (1999) Bcl-2 is phosphorylated and inactivated by an ASK1/Jun N-terminal protein kinase pathway normally activated at G(2)/M. *Mol Cell Biol* 19:8469–8478.

Yang CW, Kim BS, Kim J, Ahn HJ, Park JH, Jin DC, Kim YS, and Bang BK (2001) Preconditioning with sodium arsenite inhibits apoptotic cell death in rat kidney with ischemia/reperfusion or cyclosporine-induced injuries. The possible role of heat-shock protein 70 as a mediator of ischemic tolerance. *Exp Nephrol* 9:284–294.

Yin T, Sandhu G, Wolfgang CD, Burrier A, Webb RL, Rigel DF, Hai T, and Whelan J (1997) Tissue-specific pattern of stress kinase activation in ischemic/reperfused heart and kidney. *J Biol Chem* 272:19943–19950.

Yoneda T, Imaizumi K, Oono K, Yui D, Gomi F, Katayama T, and Tohyama M (2001) Activation of caspase-12, an endoplastic reticulum (ER) resident caspase, through tumor necrosis factor receptor-associated factor 2-dependent mechanism in response to the ER stress. *J Biol Chem* 276:13935–13940.

Yu K, Chen Q, Liu H, Zhan Y, and Stevens JL (1994) Signaling the molecular stress response to nephrotoxic and mutagenic cysteine conjugates: differential roles for protein synthesis and calcium in the induction of c-fos and c-myc mRNA in LLC-PK_1 cells. *J Cell Physiol* 161:303–311.

Zha J, Harada H, Yang E, Jockel J, and Korsmeyer SJ (1996) Serine phosphorylation of death agonist BAD in response to survival factor results in binding to 14–3-3 not BCL-X(L). *Cell* 87:619–628.

Zhan Y, Cleveland JL, and Stevens JL (1997) A role for c-myc in chemically induced renal-cell death. *Mol Cell Biol* 17:6755–6764.

Zhang G, Ichimura T, Maier JA, Maciag T, and Stevens JL (1993) A role for fibroblast growth factor type-1 in nephrogenic repair. Autocrine expression in rat kidney proximal tubule epithelial cells *in vitro* and in the regenerating epithelium following nephrotoxic damage by *S*-(1,1,2,2-tetrafluoroethyl)-L-cysteine *in vivo*. *J Biol Chem* 268:11542–11547.

Zhang Z, Yang XY, and Cohen DM (1999) Urea-associated oxidative stress and Gadd153/CHOP induction. *Am J Physiol* 276:F786–F793.

Zheng FF, Kuduk SD, Chiosis G, Munster PN, Sepp-Lorenzino L, Danishefsky SJ, and Rosen N (2000) Identification of a geldanamycin dimer that induces the selective degradation of HER-family tyrosine kinases. *Cancer Res* 60:2090–2094.

Zinszner H, Kuroda M, Wang X, Batchvarova N, Lightfoot RT, Remotti H, Stevens JL, and Ron D (1998) CHOP is implicated in programmed cell death in response to impaired function of the endoplasmic reticulum. *Genes Dev* 12: 982–995.

9

Chemical-Induced Nephrocarcinogenicity in the Eker Rat: A Model of Chemical-Induced Renal Carcinogenesis

Terrence J. Monks and Serrine S. Lau

THE MODEL

The Eker Rat and the Tuberous Sclerosis 2 (*Tsc2*) Tumor Suppressor Gene

Several potential genetic targets for hereditary and sporadic renal cell carcinoma (RCC) have been identified. Three genes, the von Hippel-Lindau and Tuberous Sclerosis-2 (*Tsc2*) tumor suppressor genes (Eker and Mossige, 1961; Gnarra et al., 1994) and the c-met protooncogene (Walker, 1998), act as determinants of susceptibility for RCC in humans and rodents, and are targets for somatic events that lead to the development of spontaneous and carcinogen-induced renal tumors. Although familial predisposition to renal cancer has led to the identification of genes involved in spontaneous RCC in humans, few genetically susceptible animal models are available to study the induction of this disease by chemical carcinogens (Paulson et al., 1985).

Spontaneous renal cell carcinoma is rare in rats, occurring in most strains with a frequency of $<0.05\%$ (Yeung et al., 1995). However, the Eker rat (Eker and Mossige, 1961) carries a single autosomal mutation that predisposes them to the development of spontaneous renal cell tumors at a very high incidence. A germline insertion of an endogenous retrovirus in the *Tsc2* gene, on rat chromosome 10q syntenic with human chromosome 16p, is responsible for the predisposing "Eker" mutation (Hino et al., 1995; Yeung et al., 1994). Alterations in the *Tsc2* gene may be responsible for the development of renal tumors (Walker, 1998; Walker et al., 1992; Yeung et al., 1994). Loss of heterozygosity (LOH) at the *Tsc2* locus has been demonstrated in renal tumors in humans (Carbonara et al., 1996; Green et al., 1994) and in spontaneous or chemically induced RCCs in rats (Kubo et al., 1994; Urakami et al., 1997; Yeung et al., 1995).

The finding that *Tsc2* knockout mice develop RCCs provides further compelling evidence that this gene functions as a tumor suppressor of renal carcinogenesis (Kobayashi et al., 1999; Onda et al., 1999). Preneoplastic lesions in the renal tubules begin to appear in rats carrying the Eker mutation at around two to 3 months of age, and by the age of 1 year the incidence of renal cell tumors in gene carriers approaches 100%. The majority of renal cell tumors observed in the Eker rat originate from the renal proximal tubules and are histologically similar to renal tumors in humans (Everitt et al., 1992; Walker, 1998). In humans, tuberous sclerosis is an autosomal dominant genetic disease that leads to the development of renal lesions, including RCC (Green et al., 1994).

Following exposure to known renal carcinogens, the multiplicity of renal tumors in susceptible Eker rats (i.e., those carrying the germline mutation) is greatly increased, with those carrying the mutation ($Tsc2^{Ek/+}$) being greater than seventy times more susceptible to the induction of RCC than their homozygous wild-type litter mates ($Tsc2^{+/+}$) (Walker et al., 1992). Loss of the second allele of this gene as a somatic event leads to the development of RCC in these animals, fulfilling Knudson's "two-hit" hypothesis for the loss of tumor suppressor gene function in tumorigenesis (Yeung et al., 1995). Thus, the Eker rat model system offers a unique opportunity to investigate mechanisms of chemical induced renal carcinogenesis in kidney epithelial cells that are predisposed to tumor development.

Smoking, Hydroquinone and Renal Cancer

Smoking non-filter cigarettes and long term cigarette smoking (≥ 30 years) both correlate with a high incidence of renal cell carcinoma risk in men (Muscat et al., 1995; Randerath and Randerath, 1993; Tavani and La Vecchia, 1997). Although the basis for the increased incidence of renal tumors in cigarette smokers is not known, cigarette smoke contains high concentrations of oxidants and free radicals, the principal radical in the tar phase being the 1,4-benzoquinone/hydroquinone redox couple (Church and Pryor, 1985). Hydroquinone (HQ) was nominated for study by the National Cancer Institute based on its high levels of production, the potential for human exposure, and the lack of adequate carcinogenicity data (Kari, 1989). Renal tubular cell degeneration was subsequently reported in male and female F344 rats receiving 1.82 mmol HQ/kg (13 week gavage studies) and long-term studies demonstrated that HQ caused marked increases in renal tubular cell adenomas (Kari, 1989). Shibata and colleagues (Shibata et al., 1991) also reported the induction of renal cell tumors in rats and mice following exposure to HQ in the diet (0.8%) for 2 years. HQ also modifies second-stage carcinogenesis, an effect dependent on the target organ, and on the initiation protocol used (Yamaguchi et al., 1989).

Mutagenicity of Hydroquinone and Generation of Reactive Oxygen Species

Although HQ is generally nonmutagenic in short-term bacterial mutagenicity assays (Florin et al., 1980; Sakai et al., 1985), and no mutagenic activity has been found in mouse cells *in vivo* (Gocke et al., 1983), it causes base-pair changes in the TA1535 *Salmonella* tester strain (Gocke et al., 1981) and is mutagenic in oxidant-sensitive (TA104 and TA2637) *Salmonella* tester strains (Hakura et al., 1996). This is consistent with the mutagenicity of 1,4-benzoquinone in several Ames bacterial tester strains (Hakura et al., 1995). HQ also induces sister chomatid exchange, is clastogenic (Gocke et al., 1981; Kari et al., 1992; Tsutsui et al., 1997), catalyzes the *in vitro* formation of 8-oxogaunine (8-oxo-dG) (Lau et al., 1996; Leanderson and Tagesson, 1990) and causes DNA single strand breaks in isolated hepatocytes (Walles, 1992).

The cellular response to oxygen free radicals is largely influenced by the type and concentration of oxygen radical generated. A single exposure to xanthine/xanthine oxidase predominantly increases smooth muscle cell proliferation, whereas frequent exposures to high levels of xanthine/xanthine oxidase result in cell death (Li et al., 1997b). Cotreatment studies with superoxide dismutase and catalase suggest that xanthine/xanthine oxidase-dependent superoxide anion ($O_2^{\cdot -}$) is mitogenic, while hydrogen peroxide (H_2O_2) is cytotoxic (Li et al., 1997b). In the presence of transition metal ions, and iron in particular, H_2O_2 is converted to the extremely reactive hydroxyl radical ($^{\bullet}OH$) which is thought to be the primary toxic species responsible for DNA damage and cell death (Dahm-Daphi et al., 2000; Mertens et al., 1995; Polla, 1999).

The mutagenicity of $^{\bullet}OH$ has been documented in a variety of experimental systems. For example, when the pZ189 plasmid is exposed to $^{\bullet}OH$, under cell-free conditions, and replicated in host cells, high numbers of deletions and base substitutions are observed (Akman et al., 1991; Moraes et al., 1989). When $^{\bullet}OH$ is generated via ionizing radiation, G:C to A:T transitions are observed; human cells appear particularly susceptible to this form of genetic damage (Waters et al., 1991). Additionally, exposure of pZ189 to ultraviolet (UV) B light (313 nm) followed by replication in CV-1 monkey kidney cells, leads mainly to G:C to A:T transitions (Keyse et al., 1988). When pZ189, containing the *supF* gene as a target, is exposed to γ-irradiation and replicated in human lymphoblastoid GM606 cells, base substitutions also occur, predominantly at G : C base pairs, with G : C to A : T transitions the most common mutations observed (Sikpi et al., 1991). G:C to T:A transversions and G:C to C:G transversions are also found. In contrast to these findings, although base substitutions also occur predominantly at G:C base pairs, when pSP189 is exposed to γ-irradiation and replicated in human embryonic kidney AD293 cells, the most common mutations are G:C to T:A transversions (Jeong et al., 1998).

METABOLISM DEPENDENT NEPHROTOXICITY OF HYDROQUINONE

Activity of γ-Glutamyl Transpeptidase (γ-GT)

HQ is acutely toxic to the kidney of Fischer 344 rats in a manner requiring the activity of γ-GT (Peters et al., 1997), suggesting that HQ-mediated nephrotoxicity is dependent on the formation of a metabolite that requires processing by this enzyme (Figure 9.1). Consistent with this view, glutathione (GSH) conjugates of HQ,

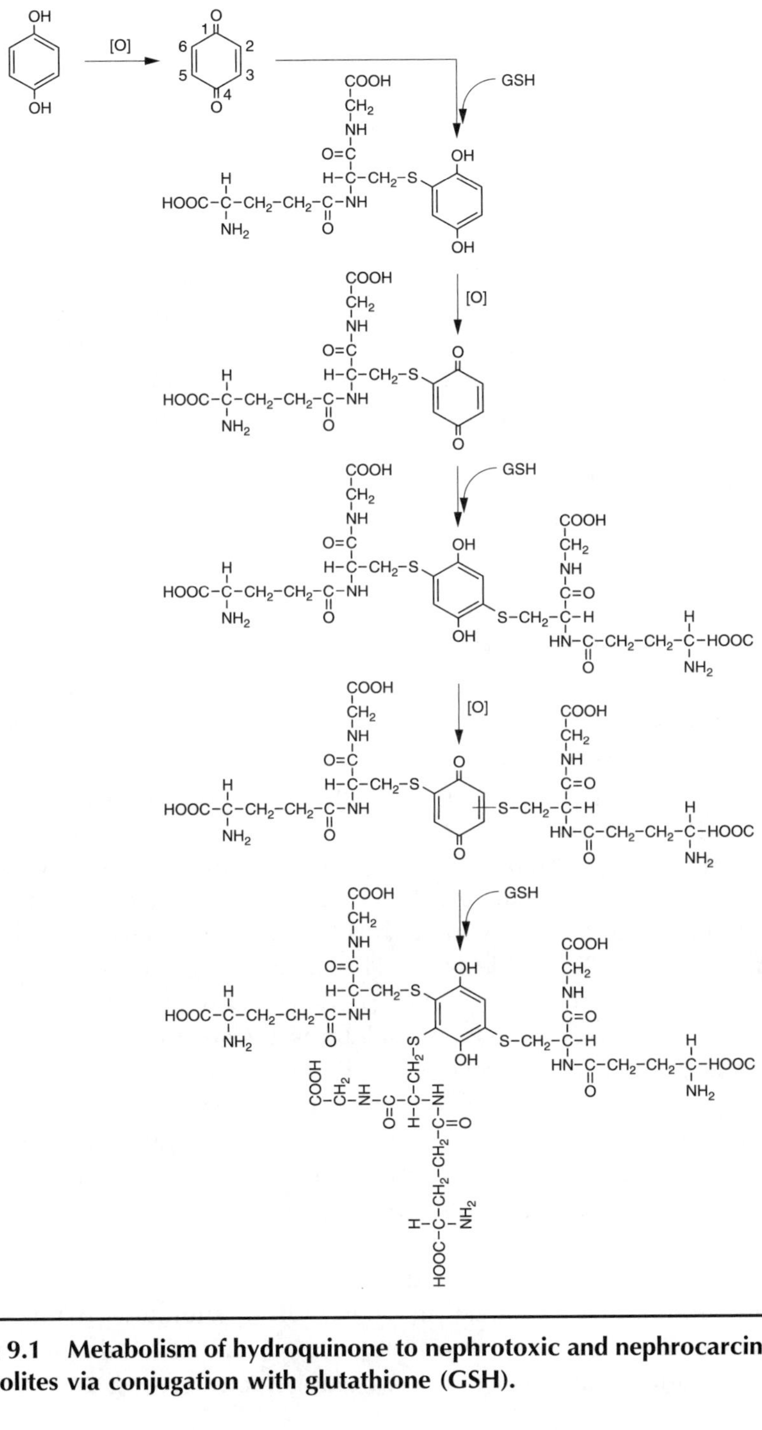

Figure 9.1 Metabolism of hydroquinone to nephrotoxic and nephrocarcinogenic metabolites via conjugation with glutathione (GSH).

in particular 2,3,5-tris(glutathion-*S*-yl)HQ (TGHQ), are potent nephrotoxicants (Lau et al., 1988; Peters et al., 1997), and HQ is metabolized *in vivo* to GSH conjugates in amounts sufficient to support their role in the acute nephrotoxic effects of HQ (Hill et al., 1993). TGHQ causes a similar pattern of renal injury in male Fischer 344 and Eker rats. However, TGHQ is approximately 600 times more potent than HQ, indicating that only a small fraction of HQ requires metabolism to TGHQ in order to produce toxicity. TGHQ (7.5 μmol/kg) causes selective toxicity to cells localized in the outer stripe of the outer medulla (OSOM) of the kidney, which progresses with time along the medullary rays. The site-selectivity of this toxicity is probably a consequence of the susceptibility of this area to oxidative stress, and to the high concentrations of γ-GT in the brush border membrane of proximal tubular cells (Monks and Lau, 1994). Indeed, the site-selective toxicity is reflected in the rapid excretion of γ-GT into urine in the absence of gross kidney dysfunction.

Maintenance of Redox Activity

The cytotoxicity of TGHQ in renal proximal tubular epithelial cell cultures (LLC-PK_1 cells) is markedly reduced by catalase and deferoxamine, scavengers of H_2O_2 and Fe^{3+}/Fe^{2+}, respectively, suggesting that the cytotoxic response is mediated by oxygen free radicals, in particular the $^{\bullet}$OH (Towndrow et al., 2000). Moreover, when the shuttle vector pSP189, containing the *supF* gene, was treated with TGHQ and replicated in both human AD293 cells and *Eschericha coli* MBL50 cells the mutation frequency increased 4.6-fold and 2.6-fold, respectively (Jeong et al., 1999) (Table 9.1). The major type of mutations were base substitutions, which occurred predominantly at G : C sites in both cell types. A high frequency of deletions (30%), including those less than and greater than 10 bp, were observed in AD293-replicated plasmids. The most common types of mutations in AD293 cells were G : C to A : T transitions (33.8%), and G : C to T : A (29.4%) and G : C to C : G transversions (19.1%). In MBL50 cells, the major mutations were G : C to T : A (33.8%) and G : C to C : G transversions (31.3%), and G : C to A : T transitions (27.5%).

The mutation spectra are thus similar to those reported for $^{\bullet}$OH-induced mutations, suggesting that $^{\bullet}$OH generated from TGHQ conjugates not only play a role in cytotoxicity, but also provide a basis for their mutagenicity and carcinogenicity (Jeong et al., 1999).

Table 9.1 Frequency of Base Substitution Mutations Induced by TGHQ in pSP189 Replicated in Human AD293 and Bacterial MBL50 Cells

	Number of mutations (% of total)			
Sequence alteration	*Human AD293*		*Bacterial MBL50*	
Transversions				
GC⇒TA	20	(29.4)	19	(37.3)
GC⇒CG	13	(19.1)	16	(31.3)
AT⇒TA	4	(5.9)	0	(-0-)
AT⇒CG	6	(8.8)	1	(2.0)
Transitions				
GC⇒AT	23	(33.8)	14	(27.5)
AT⇒GC	2	(2.9)	1	(2.0)
Total mutations	68	(100)	51	(100)

Source: Jeong et al. (1999) *Cancer Res* 59:3641–3645 (with permission).

The mutagenic and cytotoxic properties of quinone-thioethers, therefore, provide a basis for their ability to behave as complete carcinogens. Mutations induced by TGHQ may be propagated by its ability to cause a sustained regenerative hyperplasia within renal proximal tubular epithelial cells (see below), the site at which tumors eventually develop.

Signaling Pathways Associated with Cell Transformation

TGHQ-mediated cytotoxicity is associated with a prominent oxidative stress component (Jeong et al., 1999; Towndrow et al., 2000) and oxygen free radicals have been implicated in the carcinogenic process (Klaunig et al., 1998). Acute oxidative injury is thought to produce cell death and a compensatory increase in cell proliferation, resulting in the clonal expansion of preinitiated cells. Alternatively, an acute exposure to a sublethal concentration of reactive oxygen species (ROS) could result in damage to DNA, which, if misrepaired, would lead to the generation of newly initiated cells. Chronic oxidative injury may contribute to carcinogenic processes through the modulation of growth-related signal transduction pathways, such as those associated with the activator protein-1 (AP-1) and nuclear factor kappa B (NF-κB) transcription factors. AP-1 and NF-κB are consistently activated by ROS and widely implicated in transformation responses (Barchowsky et al., 1995; Ding et al., 1999; Dong et al., 1994; Guyton et al., 1996; Hsu et al., 2000; Li et al., 1997a; Meyer et al., 1993; Schmidt et al., 1996) indicative of a role for ROS-dependent regulation of these transcription factors in carcinogenesis. AP-1 is a heterodimeric complex composed of *c-jun* (c-Jun, JunB, JunD) and *c-fos* (c-Fos, FosB, Fra-1) protooncogene family members, as either a Jun : Jun homodimer or Jun : Fos heterodimer that specifically binds to the 12-O-tetradecanoyl phorbol-13-acetate responsive element (TRE) (Curran and Franza, 1988). NF-κB DNA-binding activity is associated with at least five different NF-κB family members: NF-κB1 (p105/p50), NF-κB2 (p100/p52), RelA (p65), RelB, and c-Rel (Wahl et al., 1998). The most common NF-κB dimers consist of RelA (p65) and NF-κB1 (p50) or NF-κB2 (p52) subunits (Siebenlist et al., 1994). Hydrogen peroxide, but not superoxide anion, is thought to regulate NF-κB DNA-binding activity (Schmidt et al., 1996). In contrast, the AP-1-dependent induction of *gadd153* following H_2O_2 treatment is inhibited by *o*-phenanthroline and mannitol, raising the possibility that iron-generated $^{\bullet}OH$ regulates AP-1-related signal transduction (Guyton et al., 1996).

In addition to ROS, a number of growth- and stress-related signal transduction pathways, including the protein kinase C and mitogen activated protein kinase pathways, regulate AP-1 and NF-κB activity (Barchowsky et al., 1995; Forrest and Curran, 1992; Weber et al., 1997). Protein kinase C represents a family of at least 11 isoforms that transduce signals from a wide variety of stimuli, and are the receptors for the tumor promoting phorbol esters (Nishizuka, 1992). The mitogen activated protein kinase cascade is firmly established in the transformation response to chemical and polypeptide tumor promoters (Dong et al., 1994; Huang et al., 1998; Li et al., 1997a; Watts et al., 1998), and are typically defined on the basis of extracellular signal regulated kinase (ERK1/ERK2) activity (Watts et al., 1998; Huang et al., 1999). TGHQ modulates TRE- and NF-κB-binding activity in renal epithelial cells via the generation of an oxidative stress (Weber et al., 2001). However, TGHQ-mediated TRE- and NF-κB-binding activity exhibit different ROS-dependent regulation, TRE-binding activity being transcriptionally-dependent, regulated predominantly in an ERK-dependent fashion, and apparently secondary to $^{\bullet}OH$-dependent cytotoxicity (Weber et al., 2001). In contrast, TGHQ-induced NF-κB-binding activity appears to be regulated by H_2O_2, and negatively regulated by ERK activity (Weber et al., 2001).

TGHQ Transformation of Primary Renal Epithelial Cells

HQ and TGHQ induce cell transformation in primary renal epithelial cells derived from the Eker rat (Yoon et al., 2001). Treatment of primary Eker rat renal epithelial cells with HQ (25 μM and 50 μM) or TGHQ (100 μM and 300 μM) induces a twofold to fourfold and sixfold to twentyfold increase in cell transformation, respectively. Cell lines established from TGHQ-induced transformed colonies exhibit a broad range of numerical cytogenetic alterations, LOH at the *Tsc2* gene locus, and loss of expression of tuberin, the protein encoded by the *Tsc2* gene. Only heterozygous ($Tsc2^{EK/+}$) kidney epithelial cells are susceptible to transformation by HQ and TGHQ, as wild-type cells ($Tsc2^{+/+}$) show no increase in transformation frequency following chemical exposure over background levels (Yoon et al., 2001). These data indicate that TGHQ and HQ are capable of directly transforming rat renal epithelial cells and that the *Tsc2* tumor suppressor gene is an important target of TGHQ-mediated renal epithelial cell transformation.

TOXICITY-INDUCED CELL PROLIFERATION

Regenerative Hyperplasia

As noted above, HQ belongs to a class of carcinogens whose mechanism of action is somewhat unclear, as the precise properties of a nongenotoxic compound that acts as a carcinogen to induce tumor formation are unknown. In the case of HQ, it is metabolized to a potent yet quantitatively minor metabolite (Hill et al., 1993), TGHQ, that is both nephrotoxic and genotoxic (Jeong et al., 1999; Lau et al., 1988; Peters et al., 1997). Certain chemicals may induce carcinogenesis by mechanisms involving cytotoxicity followed by sustained regenerative hyperplasia and, ultimately, tumor formation, and HQ has been proposed to belong to such a class of nongenotoxic carcinogens (Florin et al., 1980; Gocke et al., 1983; Sakai et al., 1985). Both HQ and TGHQ are acutely nephrotoxic, producing a proliferative response subsequent to tissue damage (Peters et al., 1997). Although it is unlikely that proliferation *per se* initiates tumorigenesis, additional rounds of DNA synthesis associated with compensatory cell proliferation in the kidney could increase the opportunity for sustaining genetic alterations relevant for tumor initiation. In this regard, the ability of HQ and its metabolites to act as clastogens, inducing both numerical and structural chromosome alterations, is well established (Eastmond et al., 1994; Gocke et al., 1983; Rupa et al., 1997; Stillman et al., 1997; Tsutsui et al., 1997).

It is important that HQ and TGHQ induced cell proliferation is confined to the proximal tubular cells in the area of necrosis, and this presumably occurs in an attempt to compensate for proximal tubular cell loss (Peters et al., 1997). Moreover, the localization of cell proliferation to the area of toxicity, and the absence of cell proliferation distal to the site of toxicity, suggest that cell proliferation is a response to cytolethality, and not to a mitotic stimulus *per se*. Unlike the liver, the kidney does not appear to possess a stem-cell population (Laurent et al., 1988), and necrotic tubules are thought to be repopulated by neighboring cells, which are stimulated to reenter the cell cycle from G_0 and proceed through DNA-replication (S) into mitosis (M).

After proximal tubular epithelial cell necrosis, and exfoliation of dead cells, the basement membrane is repopulated with new cells. These regenerative cells initially lack a differentiated phenotype, as indicated by their basophilicity, noncolumnar appearance and absence of a brush border. Such regenerating cells also express vimentin (Ward et al., 1992), a characteristic of the embryonic kidney. The dedifferentiation of regenerative cells during nephrogenic repair has been referred to as the "epithelial-mesenchymal transition," a reversal of the transition that occurs during organogenesis, and that only occurs at sites of cell proliferation, correlating with necrosis of tubular epithelium (Stevens and Jones, 1990). A fully differentiated epithelium eventually replaces the regenerative cells after cessation of proliferation. Thus, the kidney can survive a toxic insult if it can repopulate with sufficient fidelity to maintain function. However, during the process of tissue regeneration, a variety of factors may adversely influence the remodeling process. The nature of the proliferative response is important since mitogenic agents that directly stimulate cell proliferation in the absence of necrosis (Schulte-Hermann et al., 1988; Schroter et al., 1987) do not transform hepatocytes into preneoplastic foci, even though proliferation is similar to that induced by cytotoxicants (Columbano et al., 1987). In the kidney, a relationship between cell type selective toxicity, cell proliferation, and ultimate tumor formation has been demonstrated for a number of carcinogens.

Nephrocarcinogenic Effects of TGHQ

To determine whether TGHQ is a carcinogen in the kidney, TGHQ (2.5 μmol/kg, i.p.) was administered to Eker rats (2 months of age) for 4 or 10 months. As early as 4 months after initiation of treatment, TGHQ-treated rats developed numerous toxic tubular dysplasias of a form rarely present in vehicle-treated rats (Lau et al., 2001). These preneoplastic lesions probably represent early transformation within tubules undergoing regeneration after injury by TGHQ, and adenomas subsequently arose within these lesions. After 10 months treatment (2.5 μmol/kg for 4 months followed by 3.5 μmol/kg for 6 months) there were sixfold, sevenfold, and tenfold more basophilic dysplasias, adenomas, and renal cell carcinomas, respectively, in TGHQ-treated animals than in controls. Most of these lesions were in the region of TGHQ-induced acute renal injury, the OSOM, illustrating a clear link between TGHQ-induced nephrotoxicity and nephrocarcinogenicity. Thus, acute toxicity after both HQ and TGHQ exposure occurs in the OSOM of the kidney and progresses with time along the medullary rays (Peters et al., 1997). This region is the site of the vast majority of preneoplastic and neoplastic lesions observed in TGHQ-exposed kidneys (Lau et al., 2001). LOH of *Tsc2* within intoxicated tubule cells, isolated by laser capture microdissection, is consistent with the high frequency LOH observed in TGHQ-associated tumors (12/12), providing a direct molecular link between this preneoplastic lesion and frank tumors (Figure 9.2).

Thus, although HQ is generally considered a nongenotoxic carcinogen, HQ nephrocarcinogenesis is probably mediated by the formation of the quantitatively minor yet potent nephrotoxic metabolite, TGHQ, which induces sustained regenerative hyperplasia, loss of tumor suppressor gene function and the subsequent formation of renal adenomas and carcinomas. The relationship between HQ and TGHQ emphasizes that assumptions regarding mechanisms of action of nongenotoxic carcinogens need to be considered carefully in the absence of data on the profiles of metabolites generated by these compounds in specific target organs for tumor induction. It is, therefore, important that mechanistic studies in specific target tissues be conducted for carcinogens whose mechanism of action is unknown.

In the absence of data on the carcinogenicity of specific metabolites, especially quantitatively minor metabolites derived from so-called

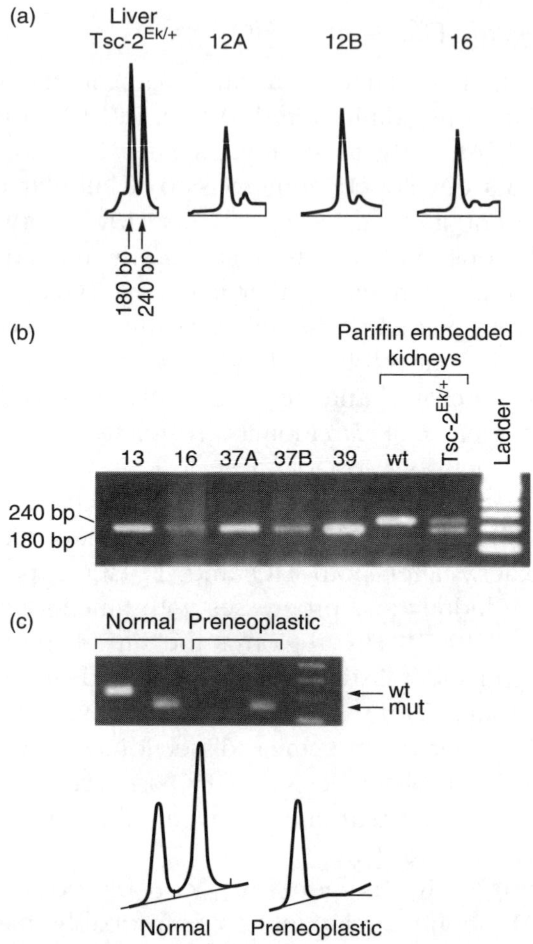

Figure 9.2 TGHQ-induced loss of heterozygosity (LOH) of the *Tsc2* gene. (A) LOH of the *Tsc2* wild-type allele determined by HPLC-WAVE DNA fragment analysis; analysis of three tumors obtained from two TGHQ-treated *Tsc2*$^{EK/+}$ rats (two tumors, 12A and 12B, from one animal, the third, 16, from a second animal). (B) agarose gel (1%) analysis of the wild-type and mutant allelles of the *Tsc2* gene; gel stained with ethidium bromide, DNA from paraffin-embedded *Tsc2*$^{+/+}$ and *Tsc2*$^{EK/+}$ vehicle-treated kidneys used as controls to illustrate *Tsc2* heterozygosity, tumors from TGHQ-treated rats (samples 13, 16, 37A, 37B, 39) exhibiting LOH. (C) laser capture microdissection followed by PCR-based LOH analysis of toxic tubule cells. (From Lau et al. (2001) *Chem Res Toxicol* 14:25–33, with permission.)

Table 9.2 TGHQ-Induced Lesions in the Eker Rat

	Lesions/animal	
Lesions	*Control (n = 26)*	*TGHQ treated (n = 26)*
Dysplasias		
Basophilic	3.0 ± 0.4	18.4 ± 1.6† (6.1)
Eosinophilic	0.7 ± 0.1	0.9 ± 0.2
Total	3.8 ± 0.4	19.3 ± 1.9† (5.0)
Adenomas		
Basophilic	0.6 ± 0.2	4.1 ± 0.7† (6.8)
Eosinophilic	0.3 ± 0.1	0.6 ± 0.2
Total	0.9 ± 0.2	4.7 ± 0.7† (5.2)
Carcinomas		
Basophilic	0.1 ± 0.1	1.0 ± 0.2† (10.0)
Eosinophilic	0.3 ± 0.1	0.2 ± 0.1
Total	0.4 ± 0.1	1.2 ± 0.2* (3.0)
Total lesions	5.0 ± 0.6	25.2 ± 2.3† (5.0)

The lesions were quantitated from a single kidney section and scored by established methods. The data are means ± SE. Values significantly different from control, as analyzed by HSD comparison of means, are indicated with †($p < 0.0001$) and *($p = 0.0035$). The values in parentheses are fold increases over control values.
Source: Lau et al. (2001) *Chem Res Toxicol* 14:25–33 (with permission).

nongenotoxic carcinogens, the designation "nongenotoxic" would appear premature. The *in vivo* formation of a quantitatively minor yet potent genotoxic metabolite from a nongenotoxic precursor such as HQ has important implications for risk assessment. The value of incorporating sensitive, genetically defined strains of rodents into the two year, two species carcinogen bioassay has been considered (Tennant, 1998; Tennant et al., 1998). Assumptions regarding mechanisms of action could ultimately affect the selection of genetically engineered animals for risk assessment and the interpretation of bioassay results obtained with these animals. That there are species, sex, tissue and interindividual differences in the ability to metabolize a nongenotoxic compound to a genotoxic carcinogen underscores the importance of understanding the mechanism of action of potential carcinogens.

The similar ratios (1:4) of adenomas to dysplasias in control and TGHQ-treated Eker rats (Table 9.2) suggests that TGHQ acts at the initiation stage of the carcinogenic process and is possibly a complete carcinogen. It is clear that carcinogens can act at multiple stages of carcinogenesis: initiation, promotion, and progression. Besides directly initiating this process, via the induction of gene mutations for example, carcinogens can also act to promote existing preneoplastic lesions via both genotoxic and nongenotoxic mechanisms.

In the case of TGHQ, the similar adenoma:dysplasia ratio suggests that this compound is not simply acting via promotion of the growth of preexisting preneoplastic lesions (dysplasias), as this ratio remains the same (i.e., progression occurs at a constant rate) in both treated and control rats. This may be a result of synergism between the clastogenic and DNA-damaging effects of TGHQ and cell proliferation induced in response to nephrotoxicity. The ability of TGHQ to significantly increase mutations in the *supF* gene replicated in human AD293 or bacterial cells (Jeong et al., 1999) is consistent with this view. Although the mutation spectrum induced by TGHQ *in vitro* in short-term assays is consistent with the participation of •OH in quinone-thioether mediated DNA damage and cytotoxicity, several mutations occur with a frequency indicative of the involvement of additional mutagenic species, most likely quinone-thioether DNA adducts (Jeong et al., 1999). TGHQ-mediated tumorigenesis may, therefore, involve both genotoxic and nongenotoxic mechanisms. Indeed, reactive electrophilic metabolites of TGHQ become covalently adducted to proteins in the same region of the kidney that ultimately gives rise to tumors (Kleiner et al., 1998). Thus, the proteins necessary for both the bioactivation and transport of TGHQ are at the site of the initial tissue injury and cell proliferation.

Tsc2 Tumor Suppressor Function

Loss of the normal allele of the *Tsc2* tumor suppressor gene in TGHQ-associated tumors suggests that loss of tuberin function is involved in tumor development, but the precise cellular and molecular mechanism by which TGHQ induces renal tumors in the Eker rat is not known. The *Tsc2* gene encodes a protein of 1784 amino acids, tuberin (European Chromosome 16 Tuberous Sclerosis Consortium, 1993), which is widely expressed in most adult tissues (Xiao et al., 1997). A 58 amino acid region near the carboxyl terminus of tuberin exhibits homology with a portion of the catalytic domain of the GTPase activating protein (GAP) for Rap1 (Wienecke et al., 1997) and, as predicted, tuberin exhibits specific GAP activity toward Rap1 (Wienecke et al., 1997). Rap1 is a member of the *ras* superfamily of small GTP-binding proteins, and appears to function as a transducer of mitogenic signals from the cell membrane to the nucleus (Ohtsuka et al., 1996; York et al., 1998; Yoshida et al., 1992).

Four Rap1GAPs have been identified, including tuberin (Cullen et al., 1995; Kurachi et al., 1997; Rubinfeld et al., 1991), but it is not known whether tuberin represents a major source of Rap1GAP activity in the kidney. However, with the exception of tuberin, there are no studies reported on expression of Rap1GAPs in the kidney. Rap1GAPs are likely expressed in a tissue- and cell-specific manner. Consistent with this view, tuberin colocalizes with Rap1 in Golgi apparatus of cultured human cell lines (Wienecke et al., 1996), and has similar patterns of expression with Rap1 in tissues such as the kidney, skin, and adrenal gland (Wienecke et al., 1997). Tuberin may, therefore, be the predominant Rap1GAP in the kidney.

The GAP homology domain is highly conserved between humans and rat (European Chromosome 16 Tuberous Sclerosis Consortium, 1993; Kobayashi et al., 1995) and introduction of the wild-type *Tsc2* gene or the carboxy-terminal tuberin construct, including the GAP homology domain, into tuberin-negative cell lines suppresses cell proliferation and tumorigenicity. Thus, the *Tsc2* gene functions as a tumor suppressor, and the tumor suppressor function resides at the carboxyl terminus (Jin et al., 1996; Orimoto et al., 1996).

Rap1GAP negatively regulates Rap1 activity by stimulating the hydrolysis of active, GTP-bound Rap1 to the inactive, GDP-bound form (Rubinfeld et al., 1991). The Eker germline insertion in the *Tsc2* gene results in premature termination of tuberin synthesis upstream of the catalytic domain, leading to the loss of the Rap1GAP-like domain. Loss of tuberin expression in cells would, therefore, maintain Rap1 in the active GTP-bound state, contributing to the development of RCC. In addition to Rap1, tuberin also exhibits GAP activity toward Rab5, implicating a role for regulating the docking and fusion process of the endocytic pathway, a process critical for normal functioning of renal epithelial cells. The GTPase activity of Rab5, which is regulated by tuberin, modulates endosome fusion (Barbieri et al., 1998; Christoforidis et al., 1999; Gournier et al., 1998; Simonsen et al., 1998), and loss of *Tsc2* gene function is associated with disrupted endocytosis (Xiao et al., 1997), but the role of this disruption in tumor pathogenesis requires further investigation. Tuberin is also involved in cell cycle control (Barbieri et al., 1998; Gournier et al., 1998; Field et al., 1998), but the precise physiological or cellular roles of tuberin and its role in renal carcinogenesis still remain elusive.

Tumor Formation

Although TGHQ (2.5 μmol/kg, i.p.) markedly increases cell proliferation within the OSOM of the kidney in both *Tsc2*$^{EK/+}$ and *Tsc2*$^{+/+}$ rats (Figure 9.3), only TGHQ-treated *Tsc2*$^{EK/+}$ rats develop renal tumors (Yoon et al., 2002). Therefore, in this model of chemical carcinogenesis, cell proliferation may be necessary but it is certainly not sufficient for tumor development (Yoon et al., 2002). Genetic alterations must, therefore, also be an important factor in TGHQ-induced nephrocarcinogenesis. However, it is unclear whether TGHQ causes genetic alterations directly; in *Tsc2*$^{EK/+}$ rats, a highly sustained mitotic environment might indirectly lead to inactivation of the remaining wild-type allele of the *Tsc2* gene via mechanisms such as chromosome nondisjunction. In this respect, TGHQ-induced sustained regenerative cell proliferation within the OSOM, without directly interacting with the genetic machinery, may therefore promote LOH at the *Tsc2* locus in *Tsc2*$^{EK/+}$ rats. Alternatively, as noted above, TGHQ may be a complete carcinogen (Jeong et al., 1999; Yoon et al., 2001).

Tuberin and the Cell Cycle

Basal tuberin expression was not observed in the OSOM of "naïve" animals. In contrast, i.p. saline injections to otherwise "untreated" animals slightly increased tuberin expression, suggesting that physiological stress may be able to induce a modest upregulation of tuberin (Yoon et al., 2002). In contrast, tuberin expression was initially induced within the OSOM following TGHQ treatment, but was lost within TGHQ-induced renal tumors (Yoon et al., 2002). Loss of *Tsc2* occurs as an early, perhaps initiating event, in preneoplastic lesions (toxic tubular dysplasias) (Lau et al., 2001), leading to the development of renal tumors in TGHQ-treated *Tsc2*$^{EK/+}$ rats. The early upregulation of tuberin expression observed at 4 months of age in response to TGHQ suggests that tuberin is induced in response to acute cellular stress or injury or acts as a brake to dampen subsequent cell proliferation, or both. Findings from primary renal epithelial cells support this view. Establishment of primary cell culture is stressful to

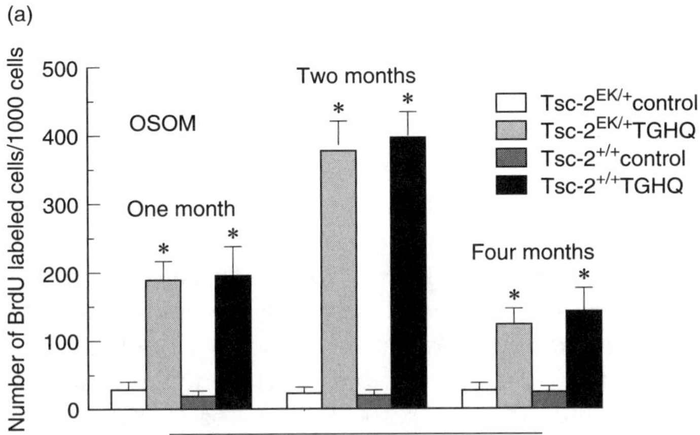

Number of tumors/animal @4 months		
	Tsc-2$^{EK/+}$	Tsc-2$^{+/+}$
Control	1/11	0/11
TGHQ	3/11	0/11

(b)

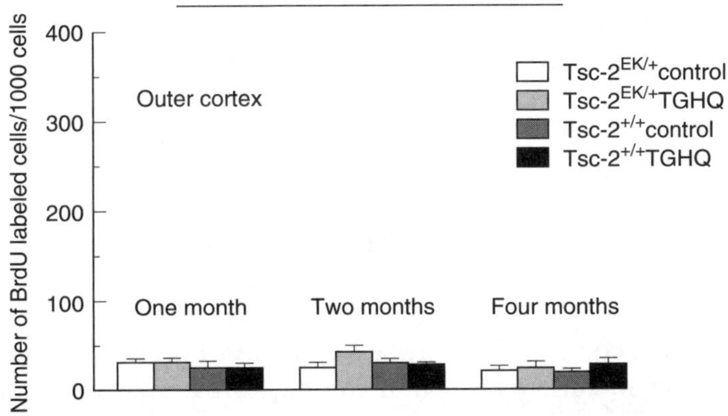

Figure 9.3 Quantitation of TGHQ-induced cell proliferation within the outer stripe of the outer medulla of Eker rat kidney. Positively immunostained nuclei of tubular epithelial cells were counted. Units represent the number of proliferating cells per 1000 cells. Values represent mean ±SE ($n=4$ to 5). A significant difference was found between control and TGHQ treatment groups at $^*p<0.01$. (From Yoon et al. (2002) *Am J Physiol* 283:F262–F270, with permission.)

cells and although tuberin expression is not detected in normal kidney, primary renal epithelial cells isolated from the OSOM exhibit high tuberin expression in culture (Yoon et al., 2001).

The precise function of the upregulation of tuberin expression in response to TGHQ is unclear. Tuberin is involved in cell cycle control (Ito and Rubin, 1999; Soucek et al., 1997, 1998). For example, loss of tuberin expression induces quiescent G_0-arrested cells to reenter the cell cycle, and also prevents them from reentering G_0 (Ito and Rubin, 1999). Reducing levels of tuberin with antisense oligonucleotides also increases cyclin D1 protein expression, and causes a transition from G_0/G_1 to S phase (Ito and Rubin, 1999). Therefore, the upregulation of tuberin expression in response to tissue injury may function to limit the mitogenic repair response, thereby counteracting increases in cell proliferation. The finding that tuberin induction in OSOM tissue obtained from rats treated with TGHQ for 4 months appears to be inversely correlated with the cell proliferation supports this view (Lau et al., 2001).

Interestingly, there is no significant difference in ERK activity between control ($Tsc2^{EK/+}$ and $Tsc2^{+/+}$) rats or rats treated with TGHQ for either 1 or 2 months, despite clear increases in cell proliferation at these times. Thus, cell proliferation in response to TGHQ-induced tissue injury is not driven by the upregulation of ERK. However, following 4 months of TGHQ treatment, $Tsc2^{EK/+}$ rats exhibit higher ERK activity than either TGHQ-treated or control $Tsc2^{+/+}$ rats and tuberin expression in these animals is lower than in their $Tsc2^{+/+}$ counterparts. Renal tumors derived from TGHQ-treated $Tsc2^{EK/+}$ rats also exhibit significantly higher ERK and cyclin D1 activity compared with normal kidney (Yoon et al., 2002). Therefore, while changes in ERK activity do not correlate with changes in cell proliferation *per se*, highest ERK activity is associated with decrease in tuberin function under comparable exposure conditions. Consistent with these findings, constitutive activation of ERKs causes cell transformation (Huang et al., 1998; Okazaki and Sagata, 1995), leading to a reduced dependence of cellular growth on mitogens (Brunet et al., 1994). Additionally, inhibition of the ERK pathway reverts tumor cells to a nontransformed phenotype *in vitro*, arrests tumor growth *in vivo* (Sebolt-Leopold et al., 1999), and inhibits the growth of Ras-transformed cells *in vitro* (Nishio et al., 1999). Reexpression of *Tsc2* in tuberin-negative cells decreased ERK activity, consistent with the growth suppressive effects of this tumor suppressor gene, suggesting that LOH at the *Tsc2* locus in TGHQ-induced renal tumors may contribute to high ERK activity in developing renal tumors (Yoon 2004).

In summary, stimulation of cell proliferation during the early stages of toxicant exposure in the Eker rat model of chemical-induced nephrocarcinogenesis is insufficient for tumor formation following renal injury. Indeed, tuberin is initially upregulated following TGHQ treatment in response to tissue injury and the compensatory cell proliferation, suggesting that *Tsc2* is an acute-phase response gene that helps limit the proliferative response after injury. Subsequent inactivation of the *Tsc2* gene following longer periods of exposure to TGHQ is accompanied by the eventual loss of tuberin expression, which is associated with cell cycle deregulation, perhaps via increases in ERK activity and cyclin D1 expression.

REPAIR OF OXIDATIVE DNA DAMAGE

Constitutive 8-Oxoguanine-DNA Glycosylase Expression

The *Tsc2* gene appears to influence constitutive 8-oxoguanine-DNA glycosylase (OGG1) expression and the ability of OGG1 to respond to an oxidative stress, consistent with the proposal that *Tsc2* is an acute-phase response gene. Constitutive expression of OGG1 in $Tsc2^{EK/+}$ rats is threefold lower than in wild-type $Tsc2^{+/+}$ rats (Habib et al., 2003). The basis for the suppression of renal OGG1 in $Tsc2^{EK/+}$ rats is not known. However, the decrease in OGG1 expression in $Tsc2^{EK/+}$ rats has important functional consequences, compromising the ability of these animals to respond to oxidative stress.

The nephrotoxicity of TGHQ is dependent on the generation of ROS (Towndrow et al., 2000). Therefore, factors that influence the ability of cells and tissues to respond to ROS would be expected to have a major impact on the response to TGHQ. Consistent with this view, $Tsc2^{EK/+}$ rats, with lower constitutive renal OGG1 expression (Figure 9.4), experience substantially higher levels of 8-oxo-deoxyguanosine (8-oxo-dG) than do wild type $Tsc2^{+/+}$ rats (Habib et al., 2003) (Figure 9.5). Moreover, the poor ability of $Tsc2^{EK/+}$ rats to cope with the TGHQ-induced oxidative stress is exacerbated by their inability to rapidly upregulate this enzyme. Indeed, for at least 12 hours after challenge with TGHQ, OGG1 expression is suppressed by as much as 80% in $Tsc2^{EK/+}$ rats (Habib et al., 2003) (Figure 9.5). The combination of the higher constitutive expression of OGG1 in $Tsc2^{+/+}$ rats, and its rapid induction in response to TGHQ treatment, coupled to the initial decrease in OGG1 expression in $Tsc2^{EK/+}$ rats, results in $Tsc2^{EK/+}$ OGG1 protein levels just 5% of those seen in $Tsc2^{+/+}$ rats 8 hours after treatment. Coincidentally, 8-oxo-dG levels in $Tsc2^{+/+}$ rats 8 hours after

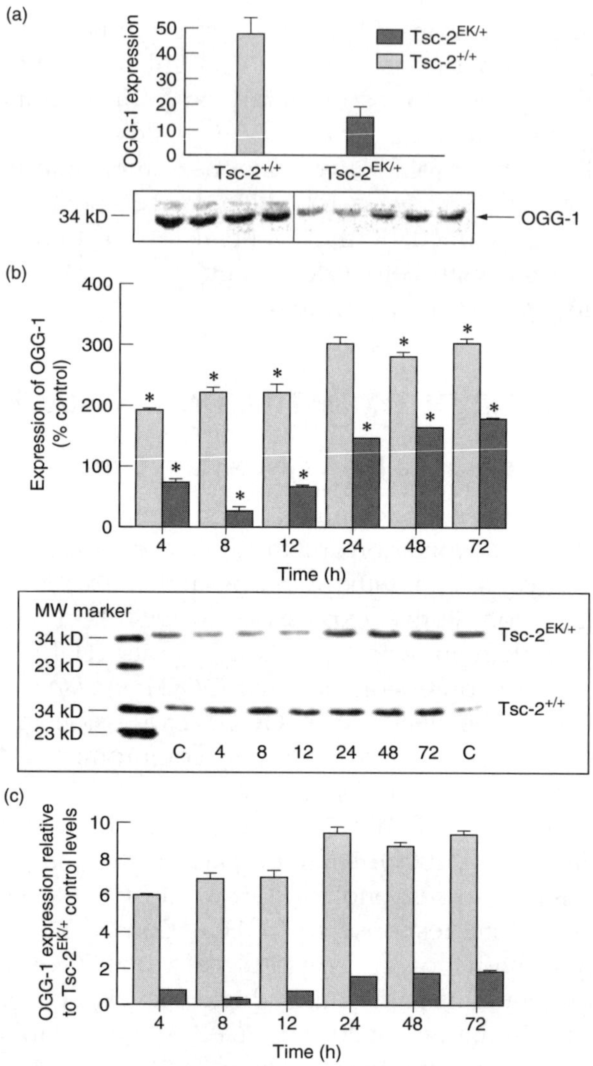

Figure 9.4 Constitutive OGG1 expression is lower in *Tsc2*$^{Ek/+}$ rats and its induction is attenuated following administration of TGHQ (7.5 μmol/kg, i.v.): (A) constitutive OGG1 expression in *Tsc2*$^{+/+}$ and *Tsc2*$^{Ek/+}$ rats determined by Western analysis; (B) kinetics of OGG1 expression in *Tsc2*$^{+/+}$ and *Tsc2*$^{Ek/+}$ rats following TGHQ treatment; (C) comparative expression of OGG1 in *Tsc2*$^{+/+}$ and *Tsc2*$^{Ek/+}$ rats relative to a control value of 1 for constitutive expression in the *Tsc2*$^{Ek/+}$ rats. Values represent the mean ± SE ($n=4$). *Significantly different from control at $p<0.01$. (From Habib et al. (2003) *Carcinogenesis* 24:573–582, with permission.)

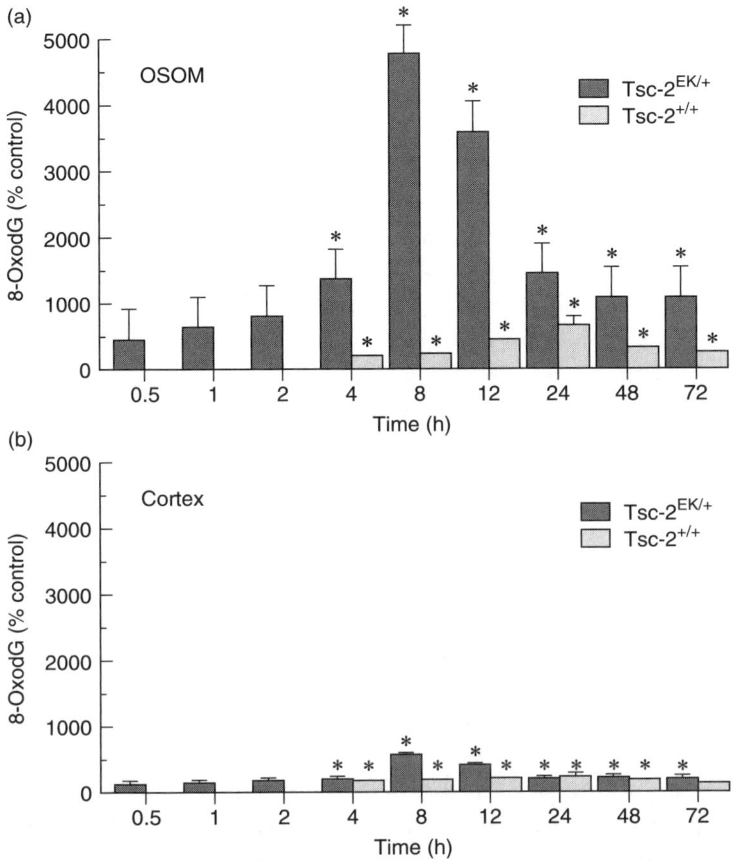

Figure 9.5 8-Oxo-dG formation is greater in the *Tsc2*$^{EK/+}$ rats than in the *Tsc2*$^{+/+}$ rats following administration of TGHQ (7.5 μmol/kg, i.v.). Values for 8-oxo-dG represent the mean ± SE, and are expressed as a percentage of the corresponding control: (A) outer stripe of outer medulla (OSOM); (B) cortex. *Significantly different from controls at $p<0.01$. Control values were 0.38 ± 0.26 and 0.93 ± 0.4 in OSOM, and 1.4 ± 0.37 and 0.82 ± 0.14 8 oxo dG/10^{-5} × dG in the cortex of *Tsc2*$^{Ek/+}$ and *Tsc2*$^{+/+}$ rats, respectively. (From Habib et al. (2003) *Carcinogenesis* 24:573–582, with permission.)

treatment (Figure 9.4) with TGHQ are just 5% of those that occur in *Tsc2*$^{EK/+}$ rats. Other prooxidants, such as nitric oxide and peroxynitrite, inhibit OGG1 activity without affecting protein expression (Jaiswal et al., 2001). Moreover, sodium dichromate decreases OGG1 mRNA and protein levels in human A549 lung carcinoma cells, an effect that appears independent of its ability to generate hydrogen peroxide

(Hodges and Chipman, 2002). Thus, addition of hydrogen peroxide to A549 cells had no effect on *OGG1* mRNA levels.

Tuberin Function and the DNA Damage Repair Response

The basis for the lack of induction of *OGG1* in *Tsc2*$^{EK/+}$ rats is not known. In humans and rats, the *Tsc2* (16p13.3 [European Chromosome 16 Tuberous Sclerosis Consortium, 1993] and 10q [Yeung et al., 1993], respectively) and *OGG1* (3p26.5 [Arai et al., 1997]) and 4 [Masuda et al., 2001], respectively) genes are located on different chromosomes. Clearly, however, the presence of a germline insertion in one allele of the *Tsc2* gene somehow influences the regulation of the *OGG1* gene. This appears to be the first evidence linking the *Tsc2* and *OGG1* genes. A single treatment of primary renal epithelial cells derived from *Tsc2*$^{EK/+}$ mutant or *Tsc2*$^{+/+}$ wild-type rats with TGHQ for 4 hours induces cell transformation only in mutant *Tsc2*$^{EK/+}$ cells (Yoon et al., 2001).

Transformation is associated with loss of tuberin expression, suggesting that a single exposure to TGHQ has the potential to induce LOH at the *Tsc2* locus. Such a genetic event may subsequently influence the regulation of *OGG1* expression. Tuberin is a structurally complex protein (Henry et al., 1998; Kobayashi et al., 1997a), and in addition to possessing a Rap1GAP homology domain, tuberin also contains potentially important domains for gene expression and protein–protein interactions: two transcriptional activation domains (Henry et al., 1998), a zinc finger like region and a potential src-homology 3 binding domain (Kobayashi et al., 1997b).

Although most studies have focused on the Rap1GAP function of tuberin, detailed functional analysis of the other regions has not been performed, and such domains may yet provide important functional roles for the tumor suppressor activity of tuberin. For example, tuberin behaves as a transcriptional coregulator, binding and selectively modulating members of the steroid receptor superfamily of genes, including the retinoid X receptor, the peroxisome proliferator receptor, the vitamin D receptor, and the glucocorticoid receptor (Henry et al., 1998). Tuberin translocation to the nucleus and modulation of steroid receptor mediated transcription may be regulated by phosphorylation (Lou et al., 2001). It is possible that the lower levels of tuberin expression in *Tsc2*$^{EK/+}$ rat kidney (Yoon et al., 2002) contribute to the lower constitutive expression of OGG1 in these animals.

In contrast to *Tsc2*$^{EK/+}$ rats, OGG1 is rapidly induced in *Tsc2*$^{+/+}$ rats, and this seems to effectively limit increases in 8-oxo-dG levels in these animals. OGG1 is also rapidly upregulated in mouse brain following ischemia-reperfusion injury (Lin et al., 2000). As noted, despite the efficient and prolonged induction of OGG1 in *Tsc2*$^{+/+}$ rats, 8-oxo-dG levels continue to rise, and reach a maximum at 24 hours after administration of TGHQ. This suggests that the repair process is saturated in *Tsc2*$^{+/+}$ rats despite the upregulation of OGG1. Consistent with this view, OGG1 protein alone may not be rate limiting for the repair of 8-oxo-dG, at least under normal growth conditions (Hollenbach et al., 1999). The repair of much larger quantities of 8-oxo-dG in the *Tsc2*$^{EK/+}$ rats, despite the somewhat attenuated induction of OGG1, may then be due to the coordinated induction of additional DNA repair proteins triggered by the extensive oxidative stress in these animals. Interestingly, mutations in the *OGG1* gene occur in human lung and kidney tumors (Chevillard et al., 1998; Audebert et al., 2000). Thus, the combination of chemical-induced loss of tuberin in TGHQ-treated *Tsc2*$^{EK/+}$ rats coupled to the subsequent deregulated expression of OGG1 may together selectively predispose these animals to TGHQ-induced renal tumors (Lau et al., 2001).

The time course of 8-oxo-dG formation in *Tsc2*$^{EK/+}$ and *Tsc2*$^{+/+}$ rats is intriguing. TGHQ is rapidly eliminated from the circulation following administration to rats, being undetectable 3 hours after dosing (Lau et al., 2001). 8-oxo-dG levels are maximal 8 hours after TGHQ administration in *Tsc2*$^{EK/+}$ rats, and at 24 hours in *Tsc2*$^{+/+}$ rats (Habib et al., 2003). The source of the ROS required to catalyze the generation of 8-oxo-dG at these later time-points is unclear. Although TGHQ is rapidly cleared from the circulation, a major fraction of the dose reaching the kidney is sequestered there, a much smaller fraction of metabolites being excreted in the urine (Hill et al., 1994). In addition to being redox active, oxidation of TGHQ and its metabolites gives rise to electrophilic intermediates capable of alkylating tissue macromolecules (Kleiner et al., 1998). Thus, following infusion of 10 μmol TGHQ directly into the kidney, a significant fraction (35.6%, greater at lower doses) becomes covalently bound to protein (Hill et al., 1994) and the nucleus seems to be a preferred target of reactive quinone-thioether metabolites (Kleiner et al., 1998).

It is thus likely that metabolites of TGHQ bind to nuclear proteins, where they continue to redox cycle and generate ROS. Following depletion of nuclear antioxidant defenses, nuclear generated ROS would then cause extensive oxidative DNA damage; the more extensive oxidative DNA damage in *Tsc2*$^{EK/+}$ rats being reflected in more extensive tissue necrosis compared with *Tsc2*$^{+/+}$ rats. Thus, although both groups of animals initially shed brush border membrane into the tubular lumen, the ability of *Tsc2*$^{+/+}$ rats to contain the oxidative DNA damage results in less cell death, and less overt tissue damage.

Constitutive renal OGG1 expression in rats carrying a germline mutation in the tuberous sclerosis tumor suppressor gene (*Tsc2*$^{EK/+}$) is lower than in wild-type *Tsc2*$^{+/+}$ rats, and the ability of OGG1 to respond to oxidative DNA damage is impaired in *Tsc2*$^{EK/+}$ rats, leading to excessive levels of 8-oxo-dG. The inability to efficiently repair this DNA damage results in more extensive cell death and tissue necrosis in *Tsc2*$^{EK/+}$ rats compared to the wild type *Tsc2*$^{+/+}$ rats. The deregulation of OGG1 in *Tsc2*$^{EK/+}$ rats may be coupled to the ability of tuberin, the *Tsc2*$^{EK/+}$ gene product, to behave as a transcriptional coactivator.

SUMMARY

The Eker rat model, with a germline mutation in the *Tsc2* tumor suppressor gene, has proven to be a useful model with which to dissect the molecular and genetic events that culminate in HQ-induced renal carcinogenesis. In particular, the model has provided important insights into the role of cell proliferation in carcinogenesis. The question of whether or not cell proliferation *per se* is sufficient to produce tumors is one that has been debated for many years. Data from TGHQ in the Eker rat has incontrovertibly demonstrated that cell proliferation, while perhaps necessary, is insufficient for chemical-induced renal carcinogenesis in the Eker rat.

In addition, the model has revealed that assumptions about the mechanism of action of nongenotoxic carcinogens must be weighed cautiously in the absence of information on the mutagenicity and carcinogenicity profiles of metabolites generated by "nongenotoxicants," especially in specific target organs for tumor induction. Mechanistic studies in specific target tissues should, therefore, be performed for carcinogens for which the mechanism of action remains unclear. Without data on the mutagenicity and carcinogenicity of specific metabolites, especially quantitatively minor metabolites derived from "nongenotoxic" carcinogens, the designation "nongenotoxic" would appear premature, with subsequent implications for risk assessment.

Finally, it has been shown that the *Tsc2* tumor suppressor gene behaves as an acute-phase response gene that also influences both constitutive and inducible renal *OGG1* gene expression.

REFERENCES

European Chromosome 16 Tuberous Sclerosis Consortium (1993) Identification and characterization of the tuberous sclerosis gene on chromosome 16. *Cell* 75:1305–1315.

Akman SA, Forrest GP, Doroshow JH, and Dizdaroglu M (1991) Mutation of potassium permanganate- and hydrogen peroxide-treated plasmid pZ189 replicating in CV-1 monkey kidney cells. *Mutat Res* 261:123–130.

Arai K, Morishita K, Shinmura K, Kohno T, Kim SR, Nohmi T, Taniwaki M, Ohwada S, and Yokota J (1997) Cloning of a human homolog of the yeast OGG1 gene that is involved in the repair of oxidative DNA damage. *Oncogene* 14:2857–2861.

Audebert M, Chevillard S, Levalois C, Gyapay G, Vieillefond A, Klijanienko J, Vielh P, El Naggar AK, Oudard S, Boiteux S, and Radicella JP (2000) Alterations of the DNA repair gene OGG1 in human clear cell carcinomas of the kidney. *Cancer Res* 60:4740–4744.

Barbieri MA, Hoffenberg S, Roberts R, Mukhopadhyay A, Pomrehn A, Dickey BF, and Stahl PD (1998) Evidence for a symmetrical requirement for Rab5-GTP in *in vitro* endosome-endosome fusion. *J Biol Chem* 273:25850–25855.

Barchowsky A, Munro SR, Morana SJ, Vincenti MP, and Treadwell M (1995) Oxidant-sensitive and phosphorylation-dependent activation of NF-kappa B and AP-1 in endothelial cells. *Am J Physiol* 269:L829-L836.

Brunet A, Pages G, and Pouyssegur J (1994) Constitutively active mutants of MAP kinase kinase (MEK1) induce growth factor-relaxation and oncogenicity when expressed in fibroblasts, *Oncogene* 9:3379–3387.

Carbonara C, Longa L, Grosso E, Mazzucco G, Borrone C, Garre ML, Brisigotti M, Filippi G, Scabar A, Giannotti A, Falzoni P, Monga G, Garini G, Gabrielli M, Riegler P, Danesino C, Ruggieri M, Magro G, and Migone N (1996) Apparent preferential loss of heterozygosity at TSC2 over TSC1 chromosomal region in tuberous sclerosis hamartomas. *Genes Chromosomes Cancer* 15:18–25.

Chevillard S, Radicella JP, Levalois C, Lebeau J, Poupon MF, Oudard S, Dutrillaux B, and Boiteux S (1998) Mutations in OGG1, a gene involved in the repair of oxidative DNA damage, are found in human lung and kidney tumours. *Oncogene* 16:3083–3086.

Christoforidis S, McBride HM, Burgoyne RD, and Zerial M (1999) The Rab5 effector EEA1 is a core component of endosome docking. *Nature* 397:621–625.

Church DF, and Pryor WA (1985) Free-radical chemistry of cigarette smoke and its toxicological implications. *Environ Health Perspect* 64:111–126.

Columbano A, Ledda-Columbano GM, Lee G, Rajalakshmi S, and Sarma DS (1987) Inability of mitogen-induced liver hyperplasia to support the induction of enzyme-altered islands induced by liver carcinogens. *Cancer Res* 47:5557–5559.

Cullen PJ, Hsuan JJ, Truong O, Letcher AJ, Jackson TR, Dawson AP, and Irvine RF (1995) Identification of a specific Ins(1,3,4,5)P4-binding protein as a member of the GAP1 family. *Nature* 376:527–530.

Curran T, and Franza BR Jr (1988) Fos and Jun: The AP-1 connection. *Cell* 55:395–397.

Dahm-Daphi J, Sass C, and Alberti W (2000) Comparison of biological effects of DNA damage induced by ionizing radiation and hydrogen peroxide in CHO cells. *Int J Radiat Biol* 76:67–75.

Ding M, Li JJ, Leonard SS, Ye JP, Shi X, Colburn NH, Castranova V, and Vallyathan V (1999) Vanadate-induced activation of activator protein-1: Role of reactive oxygen species. *Carcinogenesis* 20:663–668.

Dong Z, Birrer MJ, Watts RG, Matrisian LM, and Colburn NH (1994) Blocking of tumor promoter-induced AP-1 activity inhibits induced transformation in JB6 mouse epidermal cells. *Proc Natl Acad Sci USA* 91:609–613.

Eastmond DA, Rupa DS, and Hasegawa LS (1994) Detection of hyperdiploidy and chromosome breakage in interphase human lymphocytes following exposure to the benzene metabolite hydroquinone using multicolor fluorescence in situ hybridization with DNA probes. *Mutat Res* 322:9–20.

Eker R, and Mossige J (1961) A dominant gene for renal adenomas in the rat. *Nature* 189:858–859.

Everitt JI, Goldsworthy TL, Wolf DC, and Walker CL (1992) Hereditary renal cell carcinoma in the Eker rat: A rodent familial cancer syndrome. *J Urol* 148:1932–1936.

Field H, Farjah M, Pal A, Gull K, and Field MC (1998) Complexity of trypanosomatid endocytosis pathways revealed by Rab4 and Rab5 isoforms in Trypanosoma brucei. *J Biol Chem* 273:32102–32110.

Florin I, Rutberg L, Curvall M, and Enzell CR (1980) Screening of tobacco smoke constituents for mutagenicity using the Ames' test. *Toxicology* 15:219–232.

Forrest D, and Curran T (1992) Crossed signals: Oncogenic transcription factors. *Curr Opin Genet Dev* 2:19–27.

Gnarra JR, Tory K, Weng Y, Schmidt L, Wei MH, Li H, Latif F, Liu S, Chen F, Duh FM, and et al. (1994) Mutations of the VHL tumour suppressor gene in renal carcinoma. *Nat Genet* 7:85–90.

Gocke E, King MT, Eckhardt K, and Wild D (1981) Mutagenicity of cosmetics ingredients licensed by the European Communities. *Mutat Res* 90:91–109.

Gocke E, Wild D, Eckhardt K, and King MT (1983) Mutagenicity studies with the mouse spot test. *Mutat Res* 117:201–212.

Gournier H, Stenmark H, Rybin V, Lippe R, and Zerial M (1998) Two distinct effectors of the small GTPase Rab5 cooperate in endocytic membrane fusion. *Embo J* 17:1930–1940.

Green AJ, Smith M, and Yates JR (1994) Loss of heterozygosity on chromosome 16p13.3 in hamartomas from tuberous sclerosis patients. *Nat Genet* 6:193–196.

Guyton KZ, Xu Q, and Holbrook NJ (1996) Induction of the mammalian stress response gene GADD153 by oxidative stress: role of AP-1 element. *Biochem J* 314:547–554.

Habib SL, Phan MN, Patel SK, Li D, Monks TJ, and Lau SS (2003) Reduced constitutive 8-oxoguanine-DNA glycosylase expression and impaired induction following oxidative DNA damage in the tuberin deficient Eker rat. *Carcinogenesis* 24:573–582.

Hakura A, Mochida H, Tsutsui Y, and Yamatsu K (1995) Mutagenicity of benzoquinones for Ames Salmonella tester strains. *Mutat Res* 347:37–43.

Hakura A, Tsutsui Y, Mochida H, Sugihara Y, Mikami T, and Sagami F (1996) Mutagenicity of dihydroxybenzenes and dihydroxynaphthalenes for Ames Salmonella tester strains. *Mutat Res* 371:293–299.

Henry KW, Yuan X, Koszewski NJ, Onda H, Kwiatkowski DJ, and Noonan DJ (1998) Tuberous sclerosis gene 2 product modulates transcription mediated by steroid hormone receptor family members. *J Biol Chem* 273:20535–20539.

Hill BA, Davison KL, Dulik DM, Monks TJ, and Lau SS (1994) Metabolism of 2-(glutathion-*S*-yl)hydroquinone and 2,3,5-(triglutathion-*S*-yl)hydroquinone in the in situ perfused rat kidney: relationship to nephrotoxicity. *Toxicol Appl Pharmacol* 129:121–132.

Hill BA, Kleiner HE, Ryan EA, Dulik DM, Monks TJ, and Lau SS (1993) Identification of multi-*S*-substituted conjugates of hydroquinone by HPLC-coulometric electrode array analysis and mass spectroscopy. *Chem Res Toxicol* 6:459–469.

Hino O, Kobayashi E, Hirayama Y, Kobayashi T, Kubo Y, Tsuchiya H, Kikuchi Y, and Mitani H (1995) Molecular genetic basis of renal carcinogenesis in the Eker rat model of tuberous sclerosis (Tsc2). *Mol Carcinog* 14:23–27.

Hodges NJ, and Chipman JK (2002) Down-regulation of the DNA-repair endonuclease 8-oxo-guanine DNA glycosylase 1 (hOGG1) by sodium dichromate in cultured human A549 lung carcinoma cells. *Carcinogenesis* 23:55–60.

Hollenbach S, Dhenaut A, Eckert I, Radicella JP, and Epe B (1999) Overexpression of Ogg1 in mammalian cells: Effects on induced and spontaneous oxidative DNA damage and mutagenesis. *Carcinogenesis* 20:1863–1868.

Hsu TC, Young MR, Cmarik J, and Colburn NH (2000) Activator protein 1 (AP-1)- and nuclear factor kappaB (NF-kappaB)-dependent transcriptional events in carcinogenesis. *Free Radic Biol Med* 28:1338–1348.

Huang C, Ma WY, and Dong Z (1999) The extracellular-signal-regulated protein kinases (Erks) are required for UV-induced AP-1 activation in JB6 cells. *Oncogene* 18:2828–2835.

Huang C, Ma WY, Young MR, Colburn N, and Dong Z (1998) Shortage of mitogen-activated protein kinase is responsible for resistance to AP-1 transactivation and transformation in mouse JB6 cells. *Proc Natl Acad Sci USA* 95:156–161.

Ito N, and Rubin GM (1999) gigas, a Drosophila homolog of tuberous sclerosis gene product-2, regulates the cell cycle. *Cell* 96:529–539.

Jaiswal M, LaRusso NF, Nishioka N, Nakabeppu Y, and Gores GJ (2001) Human Ogg1, a protein involved in the repair of 8-oxoguanine, is inhibited by nitric oxide. *Cancer Res* 61:6388–6393.

Jeong JK, Juedes MJ, and Wogan GN (1998) Mutations induced in the supF gene of pSP189 by hydroxyl radical and singlet oxygen: relevance to peroxynitrite mutagenesis. *Chem Res Toxicol* 11:550–556.

Jeong JK, Wogan GN, Lau SS, and Monks TJ (1999) Quinol-glutathione conjugate-induced mutation spectra in the supF gene replicated in human AD293 cells and bacterial MBL50 cells. *Cancer Res* 59:3641–3645.

Jin F, Wienecke R, Xiao GH, Maize JC Jr, DeClue JE, and Yeung RS (1996) Suppression of tumorigenicity by the wild-type tuberous sclerosis 2 (Tsc2) gene and its C-terminal region. *Proc Natl Acad Sci USA* 93:9154–9159.

Kari FW (1989) Toxicology and carcinogenesis of hydroquinone in F344/N rats ad B6C3F1 mice (Gavage Studies). In *National Toxicology Program Technical Report 366*, PHS, NIH: U.S. Department of Health and Human Services.

Kari FW, Bucher J, Eustis SL, Haseman JK, and Huff JE (1992) Toxicity and carcinogenicity of hydroquinone in F344/N rats and B6C3F1 mice. *Food Chem Toxicol* 30:737–747.

Keyse SM, Amaudruz F, and Tyrrell RM (1988) Determination of the spectrum of mutations induced by defined-wavelength solar UVB (313-nm) radiation in mammalian cells by use of a shuttle vector. *Mol Cell Biol* 8:5425–5431.

Klaunig JE, Xu Y, Isenberg JS, Bachowski S, Kolaja KL, Jiang J, Stevenson DE, and Walborg EF Jr (1998) The role of oxidative stress in chemical carcinogenesis. *Environ Health Perspect* 106 (Suppl 1):289–295.

Kleiner HE, Jones TW, Monks TJ, and Lau SS (1998) Immunochemical analysis of quinol-thioether-derived covalent protein adducts in rodent species sensitive and resistant to quinol-thioether-mediated nephrotoxicity. *Chem Res Toxicol* 11:1291–1300.

Kobayashi T, Minowa O, Kuno J, Mitani H, Hino O, and Noda T (1999) Renal carcinogenesis, hepatic hemangiomatosis, and embryonic lethality caused by a germ-line Tsc2 mutation in mice. *Cancer Res* 59:1206–1211.

Kobayashi T, Mitani H, Takahashi R, Hirabayashi M, Ueda M, Tamura H, and Hino O (1997a) Transgenic rescue from embryonic lethality and renal carcinogenesis in the Eker rat model by introduction of a wild-type Tsc2 gene. *Proc Natl Acad Sci USA* 94:3990–3993.

Kobayashi T, Nishizawa M, Hirayama Y, Kobayashi E, and Hino O (1995) cDNA structure, alternative splicing and exon-intron organization of the predisposing tuberous sclerosis (Tsc2) gene of the Eker rat model. *Nucleic Acids Res* 23:2608–2613.

Kobayashi T, Urakami S, Cheadle JP, Aspinwall R, Harris P, Sampson JR, and Hino O (1997b) Identification of a leader exon and a core promoter for the rat tuberous sclerosis 2 (Tsc2) gene and structural comparison with the human homolog. *Mamm Genome* 8:554–558.

Kubo Y, Mitani H, and Hino O (1994) Allelic loss at the predisposing gene locus in spontaneous and chemically induced renal cell carcinomas in the Eker rat. *Cancer Res* 54:2633–2635.

Kurachi H, Wada Y, Tsukamoto N, Maeda M, Kubota H, Hattori M, Iwai K, and Minato N (1997) Human SPA-1 gene product selectively expressed in lymphoid tissues is a specific GTPase-activating protein for Rap1 and Rap2. Segregate expression profiles from a rap1GAP gene product. *J Biol Chem* 272:28081–28088.

Lau SS, Hill BA, Highet RJ, and Monks TJ (1988) Sequential oxidation and glutathione addition to 1,4-benzoquinone: Correlation of toxicity with increased glutathione substitution. *Mol Pharmacol* 34:829–836.

Lau SS, Monks TJ, Everitt JI, Kleymenova E, and Walker CL (2001) Carcinogenicity of a nephrotoxic metabolite of the "nongenotoxic" carcinogen hydroquinone. *Chem Res Toxicol* 14:25–33.

Lau SS, Peters MM, Kleiner HE, Canales PL, and Monks TJ (1996) Linking the metabolism of hydroquinone to its nephrotoxicity and nephrocarcinogenicity. *Adv Exp Med Biol* 387:267–273.

Laurent G, Toubeau G, Heuson-Stiennon JA, Tulkens P, and Maldague P (1988) Kidney tissue repair after nephrotoxic injury: biochemical and morphological characterization. *Crit Rev Toxicol* 19:147–183.

Leanderson P, and Tagesson C (1990) Cigarette smoke-induced DNA-damage: Role of hydroquinone and catechol in the formation of the oxidative DNA-adduct, 8-hydroxydeoxyguanosine. *Chem Biol Interact* 75:71–81.

Li JJ, Westergaard C, Ghosh P, and Colburn NH (1997a) Inhibitors of both nuclear factor-kappaB and activator protein-1 activation block the neoplastic transformation response. *Cancer Res* 57:3569–3576.

Li PF, Dietz R, and von Harsdorf R (1997b) Differential effect of hydrogen peroxide and superoxide anion on apoptosis and proliferation of vascular smooth muscle cells. *Circulation* 96:3602–3609.

Lin LH, Cao S, Yu L, Cui J, Hamilton WJ, and Liu PK (2000) Up-regulation of base excision repair activity for 8-hydroxy-2′-deoxyguanosine in the mouse brain after forebrain ischemia-reperfusion. *J Neurochem* 74:1098–1105.

Lou D, Griffith N, and Noonan DJ (2001) The tuberous sclerosis 2 gene product can localize to nuclei in a phosphorylation-dependent manner. *Mol Cell Biol Res Commun* 4:374–380.

Masuda K, Miyamoto T, Jung CG, Ding M, Cheng JM, Tsumagari T, Manabe T, and Agui T (2001) Linkage mapping of the rat 8-oxoguanine DNA glycosylase gene to chromosome 4. *Exp Anim* 50:353–354.

Mertens JJ, Gibson NW, Lau SS, and Monks TJ (1995) Reactive oxygen species and DNA damage in 2-bromo-(glutathion-*S*-yl) hydroquinone-mediated cytotoxicity. *Arch Biochem Biophys* 320:51–58.

Meyer M, Schreck R, and Baeuerle PA (1993) H_2O_2 and antioxidants have opposite effects on activation of NF-kappa B and AP-1 in intact cells: AP-1 as secondary antioxidant-responsive factor. *Embo J* 12:2005–2015.

Monks TJ, and Lau SS (1994) Glutathione conjugation as a mechanism for the transport of reactive metabolites. *Adv Pharmacol* 27:183–210.

Moraes EC, Keyse SM, Pidoux M, and Tyrrell RM (1989) The spectrum of mutations generated by passage of a hydrogen peroxide damaged shuttle vector plasmid through a mammalian host. *Nucleic Acids Res* 17:8301–8312.

Muscat JE, Hoffmann D, and Wynder EL (1995) The epidemiology of renal cell carcinoma. A second look. *Cancer* 75:2552–2557.

Nishio K, Fukuoka K, Fukumoto H, Sunami T, Iwamoto Y, Suzuki T, Usuda J, and Saijo N (1999) Mitogen-activated protein kinase antisense oligonucleotide inhibits the growth of human lung cancer cells. *Int J Oncol* 14:461–469.

Nishizuka Y (1992) Intracellular signaling by hydrolysis of phospholipids and activation of protein kinase C. *Science* 258:607–614.

Ohtsuka T, Shimizu K, Yamamori B, Kuroda S, and Takai Y (1996) Activation of brain B-Raf protein kinase by Rap1B small GTP-binding protein. *J Biol Chem* 271:1258–1261.

Okazaki K, and Sagata N (1995) MAP kinase activation is essential for oncogenic transformation of NIH3T3 cells by Mos. *Oncogene* 10:1149–1157.

Onda H, Lueck A, Marks PW, Warren HB, and Kwiatkowski DJ (1999) Tsc2(+/−) mice develop tumors in multiple sites that express gelsolin and are influenced by genetic background. *J Clin Invest* 104:687–695.

Orimoto K, Tsuchiya H, Kobayashi T, Matsuda T, and Hino O (1996) Suppression of the neoplastic phenotype by replacement of the Tsc2 gene in Eker rat renal carcinoma cells. *Biochem Biophys Res Commun* 219:70–75.

Paulson DE, Perez CA, and Anderson T (1985) Cancer of the kidney and ureter, in *Cancer: Principles and Practice of Oncology*, De Vita VTJ, Hellman S, and Rosenberg SA (Eds), Philadelphia: J.B. Lippincott.

Peters MM, Jones TW, Monks TJ, and Lau SS (1997) Cytotoxicity and cell-proliferation induced by the nephrocarcinogen hydroquinone and its nephrotoxic metabolite 2,3,5-(tris-glutathion-*S*-yl)hydroquinone. *Carcinogenesis* 18:2393–2401.

Polla BS (1999) Therapy by taking away: The case of iron. *Biochem Pharmacol* 57:1345–1349.

Randerath E, and Randerath K (1993) Monitoring tobacco smoke-induced DNA damage by ^{32}P-postlabelling. *IARC Sci Publ* 305–314.

Rubinfeld B, Munemitsu S, Clark R, Conroy L, Watt K, Crosier WJ, McCormick F, and Polakis P (1991) Molecular cloning of a GTPase activating protein specific for the Krev-1 protein p21rap1. *Cell* 65:1033–1042.

Rupa DS, Schuler M, and Eastmond DA (1997) Detection of hyperdiploidy and breakage affecting the 1cen-1q12 region of cultured interphase human lymphocytes treated with various genotoxic agents. *Environ Mol Mutagen* 29:161–167.

Sakai M, Yoshida D, and Mizusaki S (1985) Mutagenicity of polycyclic aromatic hydrocarbons and quinones on Salmonella typhimurium TA97. *Mutat Res* 156:61–67.

Schmidt KN, Amstad P, Cerutti P, and Baeuerle PA (1996) Identification of hydrogen peroxide as the relevant messenger in the activation pathway of transcription factor NF-kappaB. *Adv Exp Med Biol* 387:63–68.

Schroter C, Parzefall W, Schroter H, and Schulte-Hermann R (1987) Dose-response studies on the effects of alpha-, beta-, and gamma-hexachlorocyclohexane on putative preneoplastic foci, monooxygenases, and growth in rat liver. *Cancer Res* 47:80–88.

Schulte-Hermann R, Ochs H, Bursch W, and Parzefall W (1988) Quantitative structure-activity studies on effects of sixteen different steroids on growth and monooxygenases of rat liver. *Cancer Res* 48:2462–2468.

Sebolt-Leopold JS, Dudley DT, Herrera R, Van Becelaere K, Wiland A, Gowan RC, Tecle H, Barrett SD, Bridges A, Przybranowski S, Leopold WR, and Saltiel AR (1999) Blockade of the MAP kinase pathway suppresses growth of colon tumors *in vivo*. *Nat Med* 5:810–816.

Shibata MA, Asakawa E, Hagiwara A, Kurata Y, and Fukushima S (1991) DNA synthesis and scanning electron microscopic lesions in renal pelvic epithelium of rats treated with bladder cancer promoters. *Toxicol Lett* 55:263–272.

Siebenlist U, Franzoso G, and Brown K (1994) Structure, regulation and function of NF-kappa B. *Annu Rev Cell Biol* 10:405–455.

Sikpi MO, Freedman ML, Ziobron ER, Upholt WB, and Lurie AG (1991) Dependence of the mutation spectrum in a shuttle plasmid replicated in human lymphoblasts on dose of gamma radiation. *Int J Radiat Biol* 59:1115–1126.

Simonsen A, Lippe R, Christoforidis S, Gaullier JM, Brech A, Callaghan J, Toh BH, Murphy C, Zerial M, and Stenmark H (1998) EEA1 links PI(3)K function to Rab5 regulation of endosome fusion. *Nature* 394:494–498.

Soucek T, Pusch O, Wienecke R, DeClue JE, and Hengstschlager M (1997) Role of the tuberous sclerosis gene-2 product in cell cycle control. Loss of the tuberous sclerosis gene-2 induces quiescent cells to enter S phase. *J Biol Chem* 272:29301–29308.

Soucek T, Yeung RS, and Hengstschlager M (1998) Inactivation of the cyclin-dependent kinase inhibitor p27 upon loss of the tuberous sclerosis complex gene-2. *Proc Natl Acad Sci USA* 95:15653–15658.

Stevens JL, and Jones TW (1990) The role of damage and proliferation in renal carcinogenesis. *Toxicol Lett* 53:121–126.

Stillman WS, Varella-Garcia M, Gruntmeir JJ, and Irons RD (1997) The benzene metabolite, hydroquinone, induces dose-dependent hypoploidy in a human cell line. *Leukemia* 11:1540–1545.

Tavani A, and La Vecchia C (1997) Epidemiology of renal-cell carcinoma. *J Nephrol* 10:93–106.

Tennant RW (1998) Evaluation and validation issues in the development of transgenic mouse carcinogenicity bioassays. *Environ Health Perspect* 106 (Suppl 2):473–476.

Tennant RW, Tice RR, and Spalding JW (1998) The transgenic Tg.AC mouse model for identification of chemical carcinogens. *Toxicol Lett* 102–103:465–471.

Towndrow KM, Mertens JJ, Jeong JK, Weber TJ, Monks TJ, and Lau SS (2000) Stress- and growth-related gene expression are independent of chemical-induced prostaglandin E(2) synthesis in renal epithelial cells. *Chem Res Toxicol* 13:111–117.

Tsutsui T, Hayashi N, Maizumi H, Huff J, and Barrett JC (1997) Benzene-, catechol-, hydroquinone- and phenol-induced cell transformation, gene mutations, chromosome aberrations, aneuploidy, sister chromatid exchanges and unscheduled DNA synthesis in Syrian hamster embryo cells. *Mutat Res* 373:113–123.

Urakami S, Tokuzen R, Tsuda H, Igawa M, and Hino O (1997) Somatic mutation of the tuberous sclerosis (Tsc2) tumor suppressor gene in chemically induced rat renal carcinoma cell. *J Urol* 158:275–278.

Wahl C, Liptay S, Adler G, and Schmid RM (1998) Sulfasalazine: A potent and specific inhibitor of nuclear factor kappa B. *J Clin Invest* 101:1163–1174.

Walker C (1998) Molecular genetics of renal carcinogenesis. *Toxicol Pathol* 26:113–120.

Walker C, Goldsworthy TL, Wolf DC, and Everitt J (1992) Predisposition to renal cell carcinoma due to alteration of a cancer susceptibility gene. *Science* 255:1693–1695.

Walles SA (1992) Mechanisms of DNA damage induced in rat hepatocytes by quinones. *Cancer Lett* 63:47–52.

Ward JM, Stevens JL, Konishi N, Kurata Y, Uno H, Diwan BA, and Ohmori T (1992) Vimentin metaplasia in renal cortical tubules of preneoplastic, neoplastic, aging, and regenerative lesions of rats and humans. *Am J Pathol* 141:955–964.

Waters LC, Sikpi MO, Preston RJ, Mitra S, and Jaberaboansari A (1991) Mutations induced by ionizing radiation in a plasmid replicated in human cells. I. Similar, nonrandom distribution of mutations in unirradiated and X-irradiated DNA. *Radiat Res* 127:190–201.

Watts RG, Huang C, Young MR, Li JJ, Dong Z, Pennie WD, and Colburn NH (1998) Expression of dominant negative Erk2 inhibits AP-1 transactivation and neoplastic transformation. *Oncogene* 17:3493–3498.

Weber TJ, Huang Q, Monks TJ, and Lau SS (2001) Differential regulation of redox responsive transcription factors by the nephrocarcinogen 2,3,5-tris(glutathion-*S*-yl)hydroquinone. *Chem Res Toxicol* 14:814–821.

Weber TJ, Monks TJ, and Lau SS (1997) PGE_2-mediated cytoprotection in renal epithelial cells: Evidence for a pharmacologically distinct receptor. *Am J Physiol* 273:F507-F515.

Wienecke R, Maize JC Jr, Reed JA, de Gunzburg J, Yeung RS, and DeClue JE (1997) Expression of the TSC2 product tuberin and its target Rap1 in normal human tissues. *Am J Pathol* 150:43–50.

Wienecke R, Maize JC Jr, Shoarinejad F, Vass WC, Reed J, Bonifacino JS, Resau JH, de Gunzburg J, Yeung RS, and DeClue JE (1996) Co-localization of the TSC2 product tuberin with its target Rap1 in the Golgi apparatus. *Oncogene* 13:913–923.

Xiao GH, Shoarinejad F, Jin F, Golemis EA, and Yeung RS (1997) The tuberous sclerosis 2 gene product, tuberin, functions as a Rab5 GTPase activating protein (GAP) in modulating endocytosis. *J Biol Chem* 272:6097–6100.

Yamaguchi S, Hirose M, Fukushima S, Hasegawa R, and Ito N (1989) Modification by catechol and resorcinol of upper digestive tract carcinogenesis in rats treated with methyl-*N*-amylnitrosamine. *Cancer* Res 49:6015–6018.

Yeung RS, Buetow KH, Testa JR, and Knudson AG Jr (1993) Susceptibility to renal carcinoma in the Eker rat involves a tumor suppressor gene on chromosome 10. *Proc Natl Acad Sci USA* 90:8038–8042.

Yeung RS, Xiao GH, Everitt JI, Jin F, and Walker CL (1995) Allelic loss at the tuberous sclerosis 2 locus in spontaneous tumors in the Eker rat. *Mol Carcinog* 14:28–36.

Yeung RS, Xiao GH, Jin F, Lee WC, Testa JR, and Knudson AG (1994) Predisposition to renal carcinoma in the Eker rat is determined by germ-line mutation of the tuberous sclerosis 2 (TSC2) gene. *Proc Natl Acad Sci USA* 91:11413–11416.

Yoon HS, Monks TJ, Everitt JI, Walker CL, and Lau SS (2002) Cell proliferation is insufficient, but loss of tuberin is necessary, for chemically induced nephrocarcinogenicity. *Am J Physiol* 283:F262–F270.

Yoon HS, Monks TJ, Walker CL, and Lau SS (2001) Transformation of kidney epithelial cells by a quinol thioether via inactivation of the tuberous sclerosis-2 tumor suppressor gene. *Mol Carcinog* 31:37–45.

Yoon HS, Ramachandiran S, Monks TJ, and Lau SS (2004) The tuberous sclerosis-2 tumor suppressor modulates ERK and B-raf activity in transformed renal epithelial cells. *Am J Physiology* 286:F417–F424.

York RD, Yao H, Dillon T, Ellig CL, Eckert SP, McCleskey EW, and Stork PJ (1998) Rap1 mediates sustained MAP kinase activation induced by nerve growth factor. *Nature* 392:622–626.

Yoshida Y, Kawata M, Miura Y, Musha T, Sasaki T, Kikuchi A, and Takai Y (1992) Microinjection of smg/rap1/Krev-1 p21 into Swiss 3T3 cells induces DNA synthesis and morphological changes. *Mol Cell Biol* 12:3407–3414.

10

The Role of Endocytosis in Nephrotoxicity

David P. Sundin and Bruce A. Molitoris

INTRODUCTION

The kidney plays a major role in removal of low molecular weight proteins (LMWPs) and other low molecular weight solutes from the circulation. Following glomerular filtration, substances present in the glomerular filtrate are subsequently processed by the proximal tubule. Under normal physiological conditions, many are almost completely reclaimed from the tubular fluid via an intensive process of reabsorption. Much of this reabsorptive process involves uptake and catabolism of LMWPs and other substances via receptor-mediated endocytosis. Therefore, the purpose of this review is to describe the general importance and relevance of the endocytic process to the kidney specifically and the animal in general.

THE KIDNEY'S ROLE IN HOMEOSTASIS OF LOW MOLECULAR WEIGHT SUBSTANCES

The mammalian kidney plays a significant role in homeostasis of LMWPs, roughly those proteins with a molecular weight of $\leq$50 kDa. Although this class of proteins comprises a minority fraction of total circulating protein, it represents a diverse and biologically important component of serum proteins. Examples include hormones and growth factors (e.g., insulin, parathyroid hormone, epidermal growth factor), enzymes (e.g., lysozyme, ribonuclease, cytochrome c), and immunoproteins (e.g., immunoglobulin light chains, β_2-microglobulin).

The serum half-life of many LMWPs is measured in terms of minutes or hours, rather than days or weeks as with medium and large molecular weight proteins that are retained by the glomerular barrier. By way of illustration, the half-life and serum levels of many hormones and growth factors are short and stringently regulated. Their regulation is critical to the organism if it is to respond to and recover from various stimuli effectively. The process of filtration, endocytosis, and proximal tubule cell (PTC) catabolism effectively removes these compounds from the circulation and helps to maintain the appropriate half-life and plasma levels of this class of proteins.

Along with the liver, the kidney plays an important role in removal and inactivation of drugs and toxic substances from blood. Inherent properties of the kidney, such as its large blood supply and filtration activity, make it an ideal organ to help rid the body of these compounds. However, once filtered, the intrinsic characteristics of the tubule, such as its large surface area, its many transport and uptake mechanisms, and its efficient concentrating mechanisms, all make the kidney particularly susceptible to injury from these substances.

Immunoglobulin light chains (LCs) in their monomer (22 kDa) and dimer (44 kDa) forms are among the most common proteins present in the circulation. Other immunoproteins, such as β_2-microglobulin (11.8 kDa, the invariant light chain of the class I major histocompatibility antigens), are also present at significant levels. All these proteins are LMWPs and are rapidly and efficiently filtered by the kidney. In the case of β_2-microglobulin, there are well known pathological consequences if it is not cleared from the circulation. If not efficiently removed, as in the case of long-term dialysis, patients can develop β_2-microglobulin-induced carpal tunnel syndrome, bone cysts, development of joint inflammations, and vertebral collapse (Alfrey, 1989; Ullian et al., 1989).

There has been much speculation as to the significance of efficient recovery of LMWPs from the tubular fluid. It is clear these proteins are taken up, catabolized, and the constituent amino acids subsequently returned to the circulation. It is also commonly acknowledged that the body has developed this mechanism to conserve and recycle amino acids. However, hard data to unequivocally establish this fact are lacking.

Role of Endocytosis in Homeostasis of Low Molecular Weight Substances

LMWPs are removed from the circulation by glomerular filtration and subsequently reclaimed and degraded in the proximal tubule. An essential component of this homeostatic process is receptor-mediated endocytosis. The endocytic process makes it possible to reclaim these proteins from the tubular fluid. Most, if not all, uptake of LMWPs from the tubular fluid occurs in the proximal tubule. This, in fact, may be protective of the distal tubule. For instance, in circumstances of elevated protein loads reaching the distal tubule, precipitation and formation of protein casts, resulting in intratubular obstruction, can occur. Myeloma-induced light chain cast formation is an example of this phenomenon that has been well described. The low pH of the distal tubule and interaction with factors not yet uncharacterized are thought to be responsible for the precipitation and cast formation.

Relevance of Endocytosis to Nephrotoxicity

The PTC is one of the most active and efficient cells in the mammalian body involved in endocytosis. It expends abundant energy on the process of endocytosis to minimize protein loss in the urine. It is also apparent that alterations in the endocytic activity of PTCs can lead to pathological consequences. In addition, the proximal tubule can be particularly sensitive to toxicants since sensitivity requires their binding, uptake, and concentration. The endocytic activity of the proximal tubule provides a direct mechanism for concentration of toxicants and subsequent development of toxicity. All these factors imply that PTC endocytosis plays an important role in the normal functioning of the mammalian kidney as well as in the development of toxic cell injury.

This chapter provides a concise review of the general process of endocytosis, the cellular components involved, and the similarities and differences in basolateral and apical endocytosis. This is followed by a more specific discussion of PTC endocytosis. Next, studies using the aminoglycoside antibiotic gentamicin, which describe a model for development of toxicity resulting directly from endocytosis, are discussed. Finally, exploitation of endocytosis by other pathogenic substances and how this may provide a general paradigm for development of toxicity are briefly discussed.

ENDOCYTOSIS

General Overview of Endocytosis

What is Endocytosis and What does it do?

Different forms of endocytosis have been recognized for at least 100 years. Metchnikoff (1893) first described the regulated process of phagocytosis. Somewhat later, Lewis (1931) described the constitutive process of pinocytosis as an invagination of plasma membrane enclosing fluid droplets. In 1963, de Duve proposed the term "endocytosis" to describe these and similar processes. Today, endocytosis is recognized as being a variety of mechanisms that eucaryotic cells have evolved to selectively internalize plasma membrane (PM) and extracellular material. This broad definition of endocytosis has come to include mechanisms for phagocytosis (engulfment of very large extracellular particles), pinocytosis (also referred to as "cell drinking"), clathrin-dependent receptor-mediated endocytosis, and clathrin-independent endocytosis. These processes have been shown to internalize much more membrane than cells are capable of synthesizing, which led to the proposal that membrane must recycle between the PM and other intracellular compartments (Schneider et al., 1979a, 1979b; Steinman et al., 1983). In addition, it is now recognized that protein receptors mediate the selective uptake of extracellular material, especially with respect to protein ligands bound to cell surface receptors (Goldstein et al., 1985; Steinman et al., 1983; Wileman et al., 1985).

All eukaryotic cells are thought to utilize one or more of the mechanisms involved in endocytosis. Which specific forms are used primarily depends on the individual needs or functions of the specific cell type. These needs include uptake of extracellular nutrients, transmission of signals from the extracellular environment, regulation of a variety of important cell-surface proteins (e.g., receptors, channels, and pores), maintenance of cellular organelle integrity, maintenance of cell polarity, and the ability to mount an immune response (e.g., antigen presentation). As these activities illustrate, endocytosis is crucial for cellular homeostasis and, in a larger context, organismal homeostasis.

Classic Examples of Receptor-Mediated Endocytosis

Perhaps the two best-characterized examples of receptor-mediated endocytosis are those involving the uptake of low-density lipoproteins (LDLs) and the serum protein transferring (Tf). Much of the original and classical work to characterize uptake of an LDL by its specific cell-surface receptor (LDL-R) was performed by Brown and Goldstein (Goldstein et al., 1985). They unequivocally showed that cellular, as well as whole body, cholesterol homeostasis was dependent on the cell-surface binding and subsequent uptake and intracellular degradation of LDLs. This was postulated by biochemical studies and corroborated by morphological studies (Farquhar et al., 1995; Fath and Burgess, 1993). They ultimately confirmed the biological significance of this process by demonstrating that mutations in the LDL-R resulted in reduced uptake of the LDL, which lead to hypercholesterolemia and an increased incidence of heart attacks (Brown and Goldstein, 1984). These studies revolutionized the study of receptor-mediated endocytosis.

The transferrin receptor (Tf-R) has been shown to play a central role in the homeostasis of cellular and whole-body iron. The seminal characterization experiments of the Tf-R were performed in the early 1980s (Ciechaniover et al., 1983; Harding et al., 1983; Karin and Mintz, 1981; Octave et al., 1981). Prior to these experiments it was thought that iron-containing Tf bound to cell surface receptors, released the iron to the cells via an unknown mechanism, and then apo-transferrin (apo-Tf) was released from the cell without being internalized. These experiments showed that Tf was internalized, the iron released to the inside of the cell, and then Tf and its receptor recycled to the cell surface where both were able to undergo further rounds of internalization.

Overview of Steps Involved in Endocytosis

The general paradigm for receptor-mediated uptake of many ligands includes (see Figure 10.1):

1. Sequestering of receptor–ligand complexes in coated pits
2. Internalization into coated vesicles
3. Loss of vesicle coat
4. Fusion of uncoated vesicles with the early endosomal system
5. Segregation of receptor and ligand into different compartments

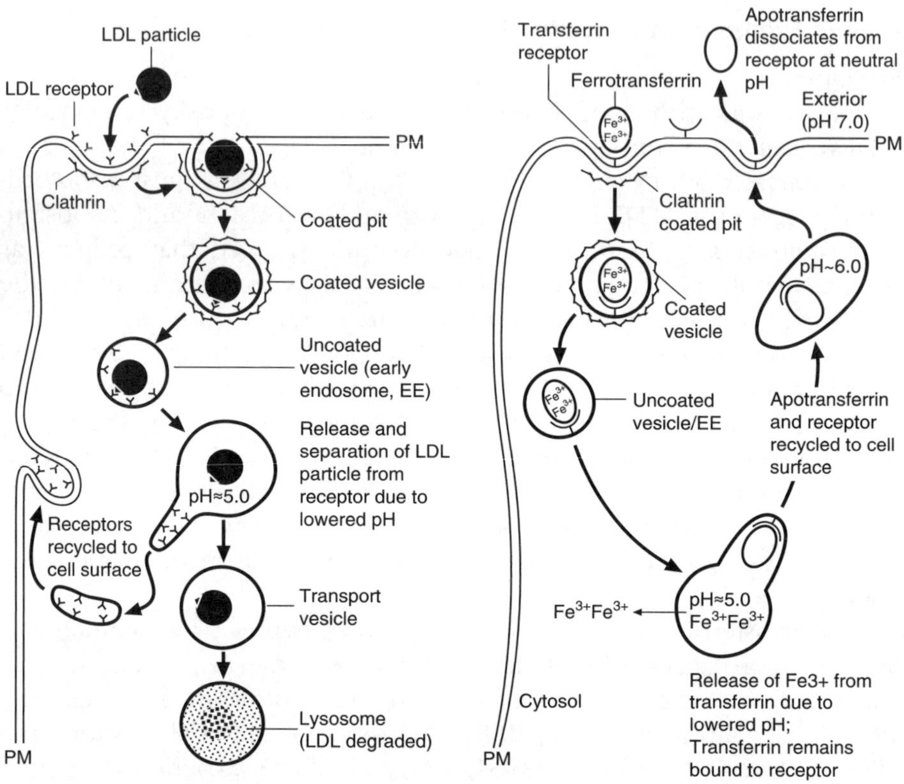

Figure 10.1 The general endocytic pathway. (a) In the case of the LDL-R, LDL is internalized via coated pits into coated vesicles. These vesicles lose their coat and fuse with the early endosome compartment. The receptor–ligand complex is then exposed to lower pH, which induces dissociation of ligand. This allows segregation of ligand and receptor making it possible for the receptor to recycle back to the cell surface, while the ligand is transported to the lysosome for degradation. (b) The Tf-R follows the same pathway except the low pH causes release of iron (Fe3+) from Tf rather than release of Tf from its receptor. Tf remains bound to its receptor and is recycled back to the cell surface where it dissociates at the pH of the extracellular milieu.

6. Recycling of receptor to the cell surface and degradation of ligand in lysosomes

A crucial step in this scheme is the dissociation and segregation of ligand and receptor. Dissociation and subsequent segregation is achieved by lowering the pH of the endosomal system before delivery to the lysosomal system at an even lower pH. In many cases, this decrease in pH causes a decrease in the affinity of ligand for receptor and induces dissociation of the receptor–ligand complex (Brown et al., 1983; Helenius et al., 1983). Acidification is achieved and maintained by an ATP-dependent proton pump (i.e., vacuolar H^{+}-ATPase) that is a member of a family of multisubunit proton pumps (Al Awqati, 1986; Gluck, 1993; Nelson, 1987).

Although this general pathway is followed by many ligands taken up via receptor-mediated endocytosis, the specific details can vary depending on the specific receptor–ligand system involved. In Figure 10.1 both the LDL-R and Tf-R pathways are graphically characterized. The LDL-R and its ligand (Figure 10.1a) follow the exact pathway just described. In the case of the Tf-R and its ligand (Figure 10.1b), the details are different but the result is similar. At the decreased pH of the endosomal system, iron is released from Tf rather than Tf being released from its receptor. Indeed, it has been shown that the Tf-R has a higher affinity for apo-Tf at the pH of the endosome (Dautry-Varsat et al., 1983; Klausner et al., 1983). Release of apo-Tf from its receptor then occurs at the higher pH of the extracellular milieu when the complex is recycled back to the cell surface.

Other examples that deviate from the general paradigm include the receptors for polymeric IgA and IgG that are transcytosed across cells (Abrahamson and Rodewald, 1981; Kuhn and Kraehenbuhl, 1979; Mostov and Blobel, 1982) and the epidermal growth factor (EGF), insulin, and macrophage Fc receptors, which do not recycle and so do not get degraded with their ligand (Carpenter and Cohen, 1976; Mellman and Plunter, 1984).

Endocytosis involves a number of functionally discrete steps. These steps have been characterized by a variety of methods that include mutational analysis (Goldstein et al., 1985), temperature blocks (Dunn et al., 1980, 1986; Hopkins and Trowbridge, 1983; Marsh et al., 1983; van Duers et al., 1987), and specific inhibitors (Heuser and Anderson, 1989; Larkin et al., 1983, 1985, 1986; Sandvig et al., 1987, 1988, 1989b). The initial step in endocytosis is binding of ligand to receptor. Binding is thought to initiate a series of events that lead to internalization of the receptor–ligand complex and its subsequent intracellular trafficking. A well-documented event initiated by binding is the phosphorylation of serine, threonine, or tyrosine residues in the cytoplasmic tail domain of the receptor (Brown et al., 1983). As an example, phosphorylation of tyrosine residues leads to putative and demonstrated interactions of the receptor with cytosolic and membrane-associated factors (Koch et al., 1991). These interactions lead to further downstream events resulting in the internalization and trafficking of receptor and ligand to their specific cellular destinations.

Most types of receptor-mediated endocytosis share a common pathway. Receptors, either with or without ligand, migrate into coated pits where they are clustered or concentrated and then internalized into coated vesicles. This process of "clustering" receptors into coated pits before internalization results in a very rapid and efficient mechanism for internalization of receptor–ligand complexes. Some receptors spontaneously "cluster" in coated pits and are internalized even in the absence of ligand. The LDL-R and Tf-R fall into this category (Anderson et al., 1982; Basu et al., 1981; Hopkins and Trowbridge, 1983; Hopkins, 1985). After recycling back to the PM, LDL-Rs are thought to remain clustered so they can be rapidly incorporated into coated pits, while Tf-Rs seem to be transiently dispersed before reclustering in coated pits (Hopkins, 1985; Robenek and Hesz, 1983). Other receptors of this type are the receptors for α_2-macroglobulin (Hopkins, 1982; Via et al., 1982), asialoglycoproteins (Berg et al., 1983; Wall et al., 1980), and insulin (Krupp and Lane, 1982). Alternatively, EGF and Fc receptors are diffusely distributed over the PM until ligand binding, whereupon the receptors become clustered in coated pits and are internalized (Dunn and Hubbard, 1984; Miettinen et al., 1992; Schlessinger, 1980). Therefore, even though clustering of receptors in coated pits for internalization is common to many types of receptor-mediated endocytosis, considerable variability exists in the details of how specific receptors are handled.

The mechanisms involved in clustering and internalization involve receptor-induced recruitment, association, and interaction of membrane and cytosolic factors. This targeting of receptors to coated pits occurs due to interactions of sequences in the cytoplasmic domain of the receptor and the aforementioned factors. These so called "internalization motifs" consist of conserved amino acids that vary depending on the specific receptor and motif. There are at least two known tyrosine internalization motifs, NPXY for the LDL-R (Davis et al., 1987; Mellman, 1996; Trowbridge et al., 1993) and YXRF for the Tf-R (Jing et al., 1990; McGraw and Maxfield, 1990; Mellman, 1996; Trowbridge et al., 1993). Other internalization motifs have been recognized and include a dileucine motif (Miettinen et al., 1989), receptor ubiquitination at lysine residues (Hicke and Riezman, 1996), and other less-well characterized motifs (Mellman, 1996; Mukherjee et al., 1996; Trowbridge et al., 1993). Thus, there is a variety of mechanisms for inducing internalization of receptor–ligand complexes, and many appear to be initiated by the initial event of binding.

Once the receptor–ligand complex has been internalized, the complex must undergo a series of recognition, sorting, and recycling steps if the receptor and ligand are to reach their appropriate cellular destinations. These events are collectively referred to as trafficking. The first event to occur is the loss of clathrin coat from internalized clathrin-coated vesicles (Mellman, 1996; Mukherjee et al., 1996; Trowbridge et al., 1993). This step allows fusion of vesicles with each other, to form endosomes, or with existing endosomes, where the acidification occurs it induces release of ligand from its receptor. It is at this point that sorting or segregation of receptor from ligand begins (Mellman, 1996; Mukherjee et al., 1996; Tojo and Endou, 1992). There are putative, mostly uncharacterized, factors that cause the receptors to concentrate or "cluster" in tubular elements of the endosomal system, while the ligands remain behind in the lumen of the large vesicular element of the endosomal system (see Figure 10.1 and Mellman, 1996; Mukherjee et al., 1996; Trowbridge et al., 1993). Following the "sorting process," receptors are recycled back to their membrane of origin and ligands continue further into the endosomal system (see Figure 10.1 and Mellman, 1996; Mukherjee et al., 1996; Trowbridge et al., 1993).

Cellular Components Involved in Endocytosis

In order to render specificity to each step of the endocytic pathway, a growing list of cytosolic and membrane associated factors have been shown to be important. Recently characterized and redefined organelles have also been shown to be required for this pathway. Therefore, this section will introduce some of the best-characterized components involved.

Clathrin and Adaptors

Depending on cell type, as much as 3.8% to as little as 0.4% of the cell surface area is occupied by coated pits (Goldberg et al., 1987; Nilsson et al., 1983). Many of the small transport vesicles involved in transport of proteins from the cell surface to intracellular organelles are derived from these structures. Biochemical characterization of enriched preparations of these "coated vesicles" has identified the protein, clathrin, as one of the major constituents. Calthrin is the protein that forms the well-known triskelion molecule, which assembles to form the basket-like lattice structure that envelops coated pits (Pearse and Crowther, 1987; Pearse and Robinson, 1990; Schmid, 1997). Another major coat constituent is the protein known as adaptor protein-2 (AP-2), which is thought to mediate the interaction of receptors and the clathrin lattice at the cell surface within the coated pit (Pearse and Robinson, 1990; Robinson, 1994; Schmid, 1997).

Clathrin is an oligomeric protein consisting of three ≈190 kDa heavy chains (HCs) and either of two ≈30 kDa LCs (Pearse and Crowther, 1987; Schmid, 1997). These three LCs and three HCs form the classic three-legged triskelion structure of clathrin. Under the appropriate conditions, clathrin can self-assemble into the characteristic lattice of hexagons and pentagons that form the basket-like structure of coated pits and vesicles. This self-assembly property and the fact that clathrin is found free and unassembled in the cytosol suggest there are probably additional factors that regulate clathrin assembly and recruitment to the appropriate membrane. The adaptor protein AP-2 most certainly is a factor (see below), but accumulating observations imply there may be additional, as yet uncharacterized, factors (Robinson, 1994; Schmid, 1997; Schmid and Damke, 1995). Thus, clathrin appears to provide a scaffolding structure whose assembly and membrane recruitment can be regulated by a number of factors. In addition, when attached to a membrane, the assembled scaffold could provide the structural support to influence the membrane to invaginate and form a budding coated vesicle.

AP-2 is a heterotetrameric protein consisting of two ≈100 kDa proteins called adaptins, a medium chain of ≈50 kDa, and a small chain of ≈20 kDa (Robinson, 1994; Schmid, 1992). These subunits assemble into structures consisting of a large "head-like" region containing the bulk of the adaptins that is flanked by two smaller "ear-like" regions (Heuser and Keen, 1988; Robinson, 1994; Schmid, 1992). Another closely related adaptor complex associates with the clathrin coat of the *trans*-Golgi network (TGN) and is called AP-1. It is thought to perform essentially the same function as the PM form, but is TGN-specific (Robinson, 1994; Schmid, 1992). AP-2 is a cytosolic protein that, like clathrin, must be recruited onto the membrane. *In vitro* studies have shown that adaptors can interact with the cytoplasmic tails of various receptors (Glickman et al., 1989; Pearse, 1988). It is thought that both adaptors (AP-1 and AP-2) provide membrane-binding sites for soluble clathrin. *In vitro* studies showed that adaptors bind to clathrin and promote its assembly into coats (Zaremba and Keen, 1983). Although implied but not yet confirmed, it is likely that specific "receptor-like" activities reside in membranes that target the right adaptor to the right membrane, i.e., PM vs. TGN (Robinson, 1994; Schmid, 1992). Thus, clathrin and AP-2 provide much of the specificity and structural framework to generate an endocytic vesicle.

Dynamin

Dynamin was originally identified as a microtubule-binding protein (Shpetner and Vallee, 1989) and has only more recently been shown to be involved in the endocytic pathway (Herskovitz et al., 1993; van der Bliek et al., 1993). It is a $\approx$100 kDa protein that binds guanosine triphosphate (GTP), has GTPase activity, and can self-assemble into helical stacks of rings that are localized to the necks of invaginated coated pits (Koenig and Ikeda, 1989; Liu and Robinson, 1995; Takel et al., 1995; Warnock and Schmid, 1996). Mutations in the *Drosophila* homologue of the dynamin gene, *shibire*, have been shown to block endocytosis (Kosaka and Ikeda, 1983). Coated pits at the PM were able to form normally but failed to pinch off and became deeply invaginated. It has been suggested that dynamin associates with the clathrin lattice, constricts the necks of coated pits, and eventually contributes to formation of coated vesicles through its GTP binding and hydrolysis activities (Schmid, 1997, and references therein). However, for complete detachment of coated vesicles *in vitro*, adenosine triphosphate (ATP) hydrolysis is required, suggesting dynamin function alone is not enough (Lamaze et al., 1993; Schmid and Smythe, 1991). Nevertheless, it is clear that dynamin plays a pivotal role in the fission and release of an endocytic vesicle.

A simplistic model is presented Figure 10.2. Although only one molecule of each factor is shown, many of each would be present. Step one involves adaptor protein recruitment to the PM. Although depicted in the diagram, the "AP-Receptor" is only assumed to be present. It has not been isolated nor characterized. Binding of ligand to receptor with a coordinate receptor modification and generation

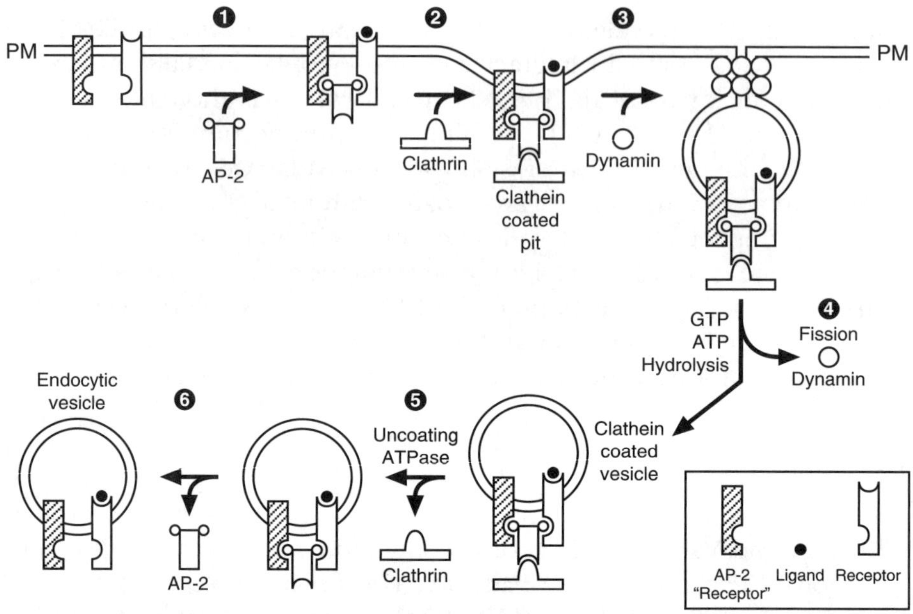

Figure 10.2 Generation of endocytic vesicles. In the first step (1), which may be induced by ligand binding, AP-2 is recruited to the membrane, where it interacts with its putative receptor and the receptor–ligand complex. The "membrane docked" AP-2 interacts with the receptor–ligand complex and acts as a nucleation site for the assembly of clathrin in the second step (2). Steps three and four involve recruitment of dynamin to the membrane and growing complex (3), dynamin-mediated constriction of the coated pit neck, and fission of a coated vesicle (4). In steps five and six, the clathrin coat is lost (5) followed by release of AP-2 (6).

of a putative signal may induce recruitment of AP-2 to the membrane. Evidence suggests that phosphorylation and dephosphorylation of the AP-2 complex may regulate its interaction with the PM and clathrin (Wilde and Brodsky, 1996). Step two is the recruitment and assembly of clathrin on the membrane and initiation of coated pit formation. The now "membrane-docked" AP-2 interacts with the receptor and acts as a nucleation site for the assembly of clathrin triskelions. Clathrin interacts directly with AP-2 and through AP-2 with the membrane. The signal responsible for this is not known. It is also not clear whether the AP-2 receptor would remain associated with the complex or not. Steps three and four involve recruitment of dynamin to the membrane, constriction of the coated pit neck, and coated vesicle fission. Although dynamin's precise role in these steps is not understood, a recently developed model suggests the GTP binding and GTPase activity of dynamin plays an important role (Hinshaw and Schmid, 1995). Dynamin would bind to the clathrin lattice in its GDP-bound form. Exchange of GDP for GTP would then trigger redistribution of dynamin such that it would self-assemble into the helical rings that play a role in constricting the necks of coated pits. GTP hydrolysis would then induce a conformational change in the dynamin complex, leading to tightening of the collar and the final events required for vesicle fission. These final events are not well characterized, however it is clear that dynamin function alone is not enough to drive the vesicle detachment reaction.

Generation of fully mature coated vesicles requires ATP hydrolysis (Lamaze et al., 1993; Schmid and Smythe, 1991) and phosphorylation/dephosphorylation of dynamin itself may play a role as well, implicating involvement of other factors (Robinson et al., 1993). Step 5 involves loss of clathrin coat from the vesicle. This reaction is mediated *in vitro* by hsc70 (heat shock protein with a molecular weight of 70 kDa), the uncoating ATPase (Greene and Eisenberg, 1990; Schlossman et al., 1984). Hsc70 is a member of the chaperonin family involved in protein folding in the endoplasmic reticulum (ER) and translocation processes of the ER, mitochondria, lysosomes, and perhaps the nucleus (Rassow et al., 1995). Although hsc70 releases clathrin from coated vesicles, it does not participate in the release of AP-2. Step 6, the last step, is release of AP-2 from the "uncoated" vesicle. Factors involved in this step remain uncharacterized.

v- and t-SNAREs

The next step in the ongoing process is recognition and docking of newly uncoated endocytic vesicles with their target organelle. This would be followed by fusion of the vesicle with the organelle. Present understanding of specific membrane fusion has been developed in large part by Rothman and colleagues and is referred to as the "SNARE hypothesis" (Rothman, 1994; Söllner et al., 1993a). It is based on interactions between soluble cytosolic factors and integral membrane factors. The integral membrane factors function essentially as receptors for the cytosolic factors. Unfortunately, the terminology for these factors can be confusing. Definition of some of the more important factors is necessary. NSF (*N*-ethylmaleimide-sensitive factor) is a soluble protein known to be required for vesicle fusion (Malhotra et al., 1988; Orci et al., 1989). SNAPs (soluble NSF attachment proteins) are also soluble proteins that are required for fusion (Clary et al., 1990; Clary and Rothman, 1990). They will bind to membranes in the absence of NSF, while the converse is not true. As their name implies, SNAPs are required for binding of NSF to membranes and must bind to membranes before NSF (Clary et al., 1990; Clary and Rothman, 1990). Receptors for SNAPs, called SNAREs, are integral membrane proteins that bind SNAPs. It is only when one or more SNAPs are bound to their receptor (i.e., SNARE) that they mediate binding of NSF to the growing complex.

The SNARE hypothesis suggests that each transport vesicle would contain at least one v-SNARE derived from its parental membrane (i.e., donor membrane), while each "target" membrane (i.e., acceptor membrane) or organelle would contain at least one matching t-SNARE. A simple model displaying these events, which is largely adapted from models developed by others (Novick and Brennwald, 1993; Rothman, 1994; Söllner et al., 1993b; Wilson et al., 1991), is depicted in Figure 10.3. Binding of matching v-SNARE and t-SNARE pairs would achieve specific targeting of the uncoated endocytic vesicle. According to this hypothesis, the specificity of each fusion event would be determined by its specific set of matching v- and t-SNARES, whereas NSF and the SNAPs would be common to all fusion events (Rothman, 1994; Söllner et al., 1993a, 1993b). The whole complex is then disassembled when NSF cleaves ATP, which is thought to occur before membrane fusion (Söllner et al., 1993a). The mechanism and details of the fusion step are only vaguely understood at this time. Alhough predictions of the SNARE hypothesis have largely been accurate, there is still much that remains to be explained.

Rab Proteins

Rab proteins belong to the Ras superfamily of low molecular weight GTPases (Chavrier et al., 1992; Takai et al., 1992; Valencia et al., 1991). Like all Ras family members, Rab proteins are able to bind GTP as well as hydrolyze it. Many of the Rab family members are ubiquitously expressed in all cell types examined. Although Rab proteins have been predominantly found on the cytoplasmic face of cellular membranes, a significant portion is also found in the cytosol in obligatory association with an accessory protein (Araki et al., 1990; Pfeiffer, 1994; Soldati et al., 1993; Ullrich et al., 1993). Observations in yeast that Rab-like proteins

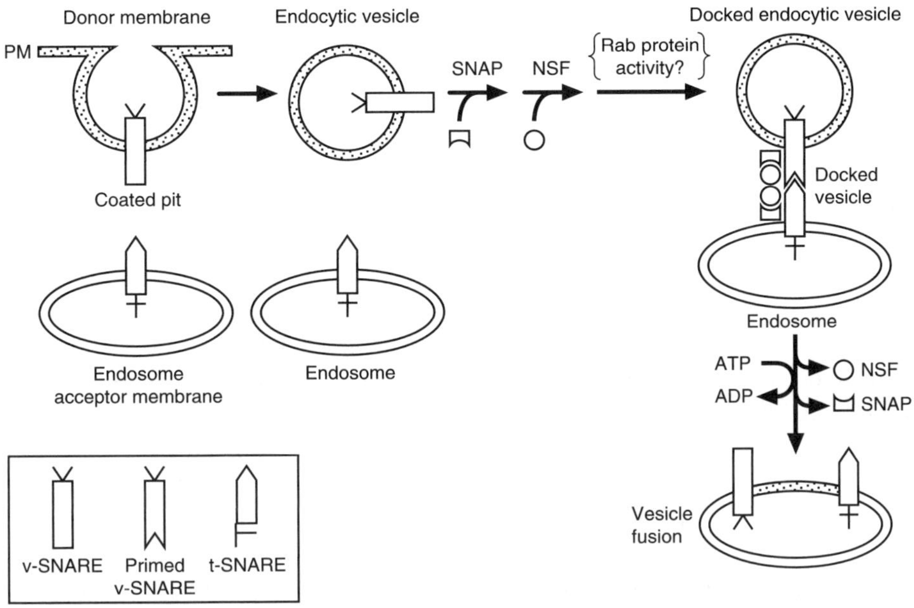

Figure 10.3 Docking and fusion of endocytic vesicles. The endocytic vesicle is derived from the "donor membrane" and contains a v-SNARE. The endosome is the "acceptor membrane" and contains a t-SNARE. Once fission of the nascent endocytic vesicle is complete, first the SNAPs and then the NSF molecules interact with the vesicle and the endosome membranes. Rab proteins may be involved in the timing of interaction of all these factors. Once the vesicle is docked, ATP hydrolysis occurs, the SNAPs and NSF fall off, and the vesicle and endosome membranes fuse and mix via a relatively uncharacterized mechanism. The stoichiometry of the individual elements is not necessarily accurate.

are required at different steps of the secretory pathway (Salminen and Novick, 1987) led to the hypothesis that each step in vesicular traffic may be regulated by at least one specific Rab (Bourne, 1988). This model implied that many of these proteins would be expressed and that each would be associated with distinct intracellular compartments. With more than 30 members of the Rab family now identified, members of the Rab family have been implicated in essentially all steps of membrane trafficking (Pfeiffer, 1994; Simons and Zerial, 1993; von Mollard et al., 1994; Zerial and Stenmark, 1993). In addition, the distinctive subcellular distribution of Rab proteins has become a defining characteristic. Although the mechanism of action of Rab proteins in membrane trafficking is still not completely understood, current models suggest specific accessory proteins play an important role in the normal cellular functioning of Rab proteins (Novick and Brennwald, 1993; Pfeiffer, 1994; Zerial and Stenmark, 1993).

Rab proteins are thought to catalyze membrane fusion in their GTP-bound form (Pfeiffer, 1994). Either during or after fusion, a Rab-specific accessory protein stimulates GTP hydrolysis. A GTPase-activating protein (GAP) is responsible for stimulating the GTP hydrolysis rate (Strom et al., 1993). This step would take place at the target membrane because fusion of transport vesicles occurs there. The accessory protein responsible for extracting Rab proteins in their GDP-bound form from target membranes and recycling them back to their membrane of origin is a cytosolic protein named guanine-nucleotide dissociation inhibitor (GDI) (Novick and Brennwald, 1993; Pfeiffer, 1994). These proteins have a higher affinity for the GDP-bound form of GDI than the GTP form and they have a preference for the prenylated form of mature Rab proteins (Soldati et al., 1994; Ullrich et al., 1994). Lastly, the accessory protein responsible for stimulating nucleotide exchange is a guanine-nucleotide exchange factor (GEF) (Goud and McCaffrey, 1991; Novick and Brennwald, 1993; Pfeiffer, 1994). The GEF may work in concert with a hypothesized GDI-displacement factor (GDF), such that targeting of Rabs to specific intracellular membranes is accompanied by release of GDI followed by nucleotide exchange (Soldati et al., 1994; Ullrich et al., 1994). This step would take place at the vesicle-forming or donor membrane. A simple model illustrating the steps involved in this process is given in Figure 10.4. This model was adapted from those developed by others (Novick and Brennwald, 1993; Pfeiffer, 1994; Simons and Zerial, 1993).

Endosomal System

To this point, a number of the early steps and important participants involved in the endocytic pathway have been described. The succeeding organelle that a receptor–ligand complex sees after internalization, the endosome, will be described.

Because of the pleiomorphic nature of the endosomal system, there is no consensus for terminology used to describe these compartments. As a result, references to these structures can be very confusing. In this section, the definitions of this complex system of organelles

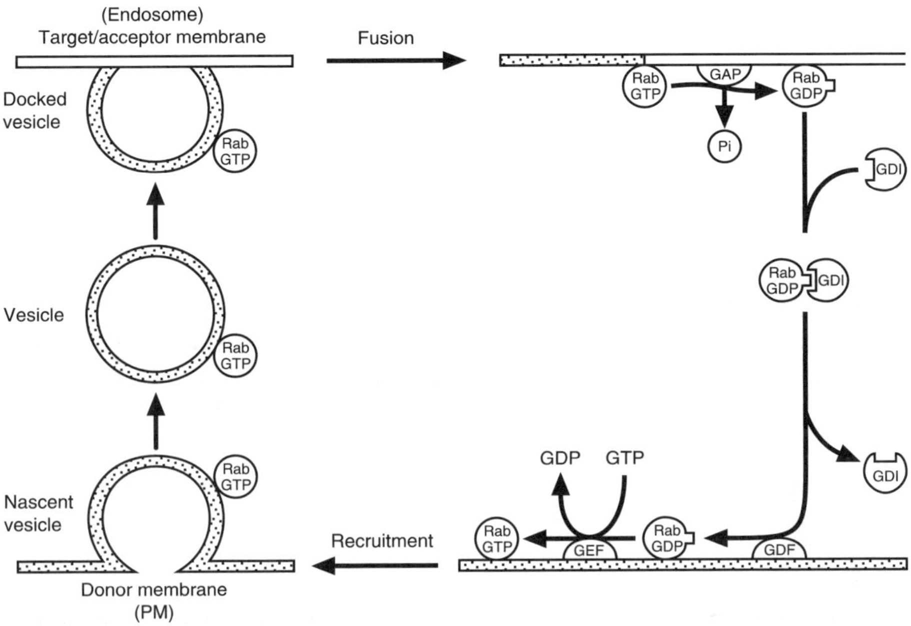

Figure 10.4 The Rab protein cycle. The docked vesicle in this figure is equivalent to the docked vesicle in Figure 10.3. Either during or following fusion of the vesicle with its target or "acceptor membrane," an activating protein called GAP stimulates GTP hydrolysis. The resultant form of the Rab protein, Rab-GDP, is then extracted from the acceptor membrane by GDI and transported through the cytosol back to its membrane of origin. At the "donor Membrane," the GDI–Rab-GDP complex is acted on by two factors: GDF (which displaces the GDI) and GEF (which stimulates nucleotide exchange). At this point, the Rab protein is ready for another round of the cycle.

along the endocytic pathway will be kept as simple as possible. First, the various structures will be defined, and then discussion will focus on what is thought to occur in the individual compartments.

At least two models have been proposed to describe the passage of material through the endosomal system and finally to the lysosome (Dunn and Maxfield, 1992; Griffiths and Gruenberg, 1991; Gruenberg et al., 1989; Helenius et al., 1983; Steinman et al., 1983; Stoorvogel et al., 1991; van Duers et al., 1989). In the "maturation model," early endosomes (EEs) are continually being formed *de novo* from homologous fusion of incoming endocytic vesicles. These EEs mature and are transformed into late endosomes (LEs), which then either fuse with or mature into lysosomes. In the "vesicle shuttle model," EEs, LEs, and lysosomes are all relatively stable organelles with defined and constant functions. Transfer of material between the various organelles is mediated by a vesicle shuttle mechanism. Incoming endocytic vesicles fuse with and pinch off from EEs. Vesicles that pinch off from the EEs fuse with the LEs, and the same scenario is repeated between the LEs and lysosomes. An alternative model for which there is evidence has also been proposed to reconcile some of the differences between these two models (Hopkins et al., 1990).

EEs have been operationally defined as the first element of the endosomal system with which uncoated endocytic vesicles interact (Helenius et al., 1983; Mellman, 1996; Mukherjee et al., 1996; Wileman et al., 1985). These structures tend to be peripherally oriented and exhibit a dynamic and complex morphology consisting of tubular-vesicular networks. Endocytic vesicles reach and fuse with this compartment as quickly as one to two minutes (Dunn et al., 1989; Griffiths et al., 1989; Mukherjee et al., 1996; Wileman et al., 1985). Material within the early endosome then undergoes acidification, which results in dissociation of the receptor–ligand complex (see Figure 10.1). Acidification and dissociation initiates the primary function of EEs, which is the physical sorting of receptor and ligand. Indeed, EEs are referred to as "sorting endosomes" (Dunn et al., 1989; Mayor et al., 1993; Mukherjee et al., 1996). The geometric shape of the EEs (i.e., tubular and vesicular) may be the primary factor involved in the process of sorting of receptor from ligand. Ligands accumulate predominantly in the large vesicular regions (where the greatest fractional volume is), and recycling receptors accumulate in the tubular extensions (where the greatest fractional surface area lies) (Dunn et al., 1989; Geuze et al., 1983; Marsh et al., 1986; Mukherjee et al., 1996). It is generally accepted that receptors recycling back to the cell surface must traffic through tubular elements of the EEs. These tubular elements eventually pinch off vesicles enriched in receptors, which are transported back to the cell surface. The vesicular element then communicates with the LE either with transport vesicles that pinch off, or by migrating en masse deeper into the cell and maturing into a LE (see models described above).

Although there is evidence that suggests EEs can be subdivided into two physically distinct compartments, the sorting and recycling endosomes (Dunn et al., 1989; Dunn and Maxfield, 1992; Marsh et al., 1986; Mayor et al., 1993; Mukherjee et al., 1996), for the purposes of this text they will be considered as one organelle. It is also important to note that the distal limit of the EE compartment has been defined by other observations as well. Penetration of recycling receptors and fluid phase tracers beyond the EE compartment has been shown to be inhibited when the temperature was lowered to 15 to 20°C, or microtubules were depolymerized (Gruenberg et al., 1989; Marsh et al., 1983). In addition, EEs and LEs have been separated by density gradient centrifugation (Branch et al., 1987; Kindberg et al., 1984; Storrie et al., 1984; Wall and Hubbard, 1985), immunoisolation (Gruenberg et al., 1989; Mueller and Hubbard, 1986), and free-flow electrophoresis (Schmid et al., 1988). Lastly, the Rab proteins Rab4 and Rab5 have been shown to be enriched predominately on EE (Bucci et al., 1992; Daro et al., 1996; Gorvel et al., 1991).

LEs are the next stop on the endocytic itinerary. They can be operationally defined as large, vesicular/vacuolar structures beyond the EEs that accumulate and concentrate their internalized contents (Mellman, 1996; Wileman et al., 1985). In addition to the distinguishing characteristics of the LE that were mentioned above, the pH of the LE is lower than the EE (Kornfield and Mellman, 1989; Maxfield and Yamashiro, 1991). There is also an abrupt change in protein composition (Beaumelle et al., 1990; Schmid et al., 1988). LEs have been shown to be enriched in the cationic-dependent mannose-6-phosphate receptor (MPR) and the lysosomal membrane glycoprotein, lgp 120 (Geuze et al., 1989; Griffiths et al., 1988), as well as Rab7 and Rab9 (Chavrier et al., 1990; Feng et al., 1995; Simons and Zerial, 1993). Movement of material from EE to LE generally involves microtubule dependent movement from more-peripheral regions of the cell to the interior or perinuclear region of the cell (Gruenberg and Howell, 1989; Mellman, 1996; Wileman et al., 1985). This movement involves a class of vesicles that are larger and longer-lived than the endocytic vesicles that fuse with the EE (Gruenberg et al., 1989; Mellman, 1996). These vesicles seem to be accompanied by an increase in the amount of internal membranes they contain. The internal membrane has been suggested to be a collection of material from more than one EE destined to be degraded (Gruenberg and Howell, 1989) and invaginations of the limiting endosomal membrane (Cohn et al., 1966; McKanna et al., 1979). Again, whether these larger "LE vesicles" mature into LEs or fuse with existing LEs is not resolved at this time. In any case, the mature or existing LE acquires the classical multivesicular appearance that is thought to be indicative of a prelysosomal compartment whose contents are destined to be degraded. This compartment would then either fuse with a lysosome or mature into one.

Microfilaments and Microtubules

In polarized epithelial cells, vesicular transport in the cytoplasm is thought to be largely microtubule dependent (Achler et al., 1989). Microtubules are required both for movement of internalized material from the EE to the perinuclear region (see above) and for movement of newly synthesized material to the apical surface (Achler et al., 1989; Eilers et al., 1989; Matlin and Simons, 1983; Pavelka et al., 1983; Pfeiffer et al., 1985). As has been described above, a 15 to 20°C temperature block and microtubule depolymerizing reagents specifically inhibit movement of internalized proteins from EE to LE. These treatments also block the movement of newly synthesized proteins from the TGN to the apical surface. Importantly, internalization and vesicular movement from the plasma membrane to EE (i.e., endocytosis) remains unaffected under these conditions. The implication is that internalization and movement from the surface to the EE must occur via a microtubule independent mechanism. It has been suggested that vesicular movement along actin filaments may play an important role in transport across the terminal web for apically targeted newly synthesized proteins as well as in apical endocytosis (Bornens, 1991; Fath and Burgess, 1993). Indeed, recent studies have shown that disruption of the actin cytoskeleton with cytochalasin D, an actin-depolymerizing agent, inhibits endocytosis in polarized epithelial cells (Jackman et al., 1994; Lamaze et al., 1997). It was also shown that endocytosis at the basolateral membrane (BLM), where the actin cytoskeleton is not as complex, was not inhibited (Jackman et al., 1994). Therefore, it appears that in polarized epithelial cells where the apical membrane actin cytoskeleton is complex and well developed and there are no appreciable microtubules, i.e., the brush border membrane(BBM), movement of vesicles may occur along actin microfilaments in conjunction with a myosin-based motor. In these cells, microtubules seem to be most involved in moving vesicles from EEs to LEs and lysosomes, to and from the BLM, and up to the terminal web. From this point on, the actin-based movement would predominate.

Lysosomes

Lysosomes are generally considered to be the end point of the endocytic pathway. They have the highest concentration of lysosomal enzymes in the endocytic pathway; a result have been called the main site of degradation in the endocytic pathway (Kornfield and Mellman, 1989). However, it has also been shown that significant degradation of endocytosed substrates may also take place in prelysosomal compartments (Berg et al., 1985; Blum et al., 1991; Tjelle et al., 1996). Along with the higher lysosomal enzyme content, lysosomes can be discriminated from LEs by virtue of their luminal pH (LE, pH ~5.5; lysosome, pH 4 to 5) (Gruenberg and Howell, 1989; Mellman et al., 1986), the lysosomes' higher density (de Duve, 1975; Helenius et al., 1983; Kornfield and Mellman, 1989), the lysosomes' lack of MPR (Helenius et al., 1983; Kornfield and Mellman, 1989), and the lysosomes' inherently more negative charge (Helenius et al., 1983; van Duers et al., 1989). In addition to degradation of materials ingested via the endocytic pathway, lysosomes can also fuse with and digest large particles taken up via phagocytosis (Mellman et al., 1986) and obsolete intracellular organelles via autophagy (Dunn, 1990).

Whether lysosomes are relatively stable structures or whether they are continuously maturing from LEs is not resolved at this time. However, in either model there is fusion and fission of structures. In the maturation model it would be LEs fusing with each other or already mature lysosomes (Helenius et al., 1983; Stoorvogel et al., 1991; van Duers et al., 1989). In the vesicle shuttle model it would be transport vesicles from stable or pre-existing LEs fusing with stable or pre-existing lysosomes (Griffiths and Gruenberg, 1991; Gruenberg et al., 1989; Helenius et al., 1983; van Duers et al., 1989). Interestingly, there is also evidence that lysosomes may communicate with LEs via a retrograde fusion step (Griffiths et al., 1988, 1990; Jahraus et al., 1994). A fusion of these models may provide the most accurate representation of the true situation.

Basolateral and Apical Endocytosis

The plasma membrane of polarized epithelial cells is partitioned into two functionally and biochemically distinct, but physically continuous, domains: the apical membrane domain and the basolateral membrane domain. These membrane domains have distinctly different lipid and protein compositions. Generation and maintenance of these distinct domains provide the means by which epithelial cells perform vectorial transport and act as a barrier to the extracellular milieu (for reviews see Matter and Mellman, 1994; Rodriguez-Boulan and Nelson, 1989; Simons and Wandlinger-Ness, 1990). The tight junction provides a physical barrier that inhibits mixing of apical and basolateral membrane constituents that would occur as a result of diffusion within the membrane. Specialized membrane trafficking and sorting mechanisms are necessary to generate and maintain the unique composition of apical and basolateral domains. Many overlap with the sorting mechanisms for endocytic uptake that has been discussed above. The biosynthetic pathway provides the initial mechanism for polarized sorting of apical and basolateral components at the level of the TGN. Numerous studies have defined targeting signals for polarized targeting via the TGN (Matter and Mellman, 1994; Mostov, 1995; Rindler et al., 1984). Although the mechanisms of endocytic uptake that have already been discussed are analogous to the ones used in a polarized epithelium, there can be differences in the specifics of endocytosis from the apical and basolateral domains. In addition, it also should be noted that there are inherent differences between cultured cells grown on solid or permeable supports and the *in vivo* situation. Conclusions drawn from cell culture studies therefore must be considered carefully before directly applying them to the *in vivo* situation.

Apical Surface Endocytic Activity

One of the most important functions of the proximal tubule is to reclaim protein filtered by the glomerulus (Maack et al., 1979, 1985). Most proteins are almost quantitatively removed from the filtrate within the proximal tubule. This is accomplished rapidly and efficiently by endocytosis at the apical surfaces of PTCs. Much of this endocytosis is via the classical clathrin-mediated uptake pathway. Considering the large amounts of protein filtered on a daily basis and the minute quantities excreted in the urine, this implies a very endocytically active apical membrane. Once internalized, the vast majority of protein is delivered to lysosomes and degraded into constituent amino acids (Christensen and Nielsen, 1991; Maack et al., 1979, 1985).

Although endocytosis is thought to occur at the basolateral surfaces of PTCs (Nielsen and Christensen, 1985; Ottosen and Maunsbach, 1973; Tisher and Kokko, 1974; Venkatachalam and Karnovsky, 1972 and reviewed in Christensen and Nielsen, 1991), for most proteins basolateral uptake in the proximal tubule is thought to be insignificant (Christensen and Nielsen, 1991; Christensen et al., 1998a; Maunsbach and Christensen, 1992). However, specific binding of a number of hormones and growth factors to the basolateral membrane of the proximal tubule has been identified (Hammerman and Gavin, 1984; Hensley et al., 1989; Milton et al., 1988; Nielsen et al., 1987; Nielsen et al., 1989; Rogers and Hammerman, 1989). For these substances, basolateral uptake may be of quantitative and physiologic importance, even though the overall protein uptake is quite low. Therefore, despite the fact that the proximal tubule is a polarized epithelium, its specific needs may differ enough from other polarized epithelia such that trafficking and sorting mechanisms are somewhat different from other examples of polarized epithelia.

Similarities, Differences, and Convergence of the Apical and Basolateral Pathways

When discussing the similarities of the apical and basolateral endocytic pathways, it is important to establish whether the discussion relates to cultured kidney cells or the *in vivo* situation. Although cultured cells provide a valuable tool for exploring specific cellular processes, care must be taken when directly comparing the two models. As an example, in opossum kidney (OK) cells and Madin-Darby canine kidney (MDCK) cells, basolateral endocytosis was shown to be two times and four to six times higher than apical endocytosis in OK and MDCK cells, respectively (Bomsel et al., 1989; Rabkin et al., 1989; von Bonsdorff et al., 1985). This differs dramatically from proximal tubules, where basolateral endocytosis contributes only a very small proportion of total endocytic uptake (Bourdeau and Carone, 1973; Bourdeau et al., 1973; Nielsen et al., 1987, 1989). In addition, only 10% of internalized ligand is presented to lysosomes in MDCK cells (Bomsel et al., 1989), while in the proximal tubule probably greater than 95% of internalized ligand is delivered to the lysosome (Christensen and Nielsen, 1991; Maack et al., 1979, 1985). The fact that MDCK cells are derived from a more distal region of the tubule may contribute to some of the differences observed between these cells and *in vivo* proximal tubule models.

Both apical and basolateral membranes exhibit clathrin-coated pit mediated endocytic uptake, in cultured cells as well as in proximal tubules. These pathways follow the basic endocytic itinerary we have already described of internalization, fusion with early endosomes where ligands are released and receptors recycled, ligands are trafficked to late endosomes and eventually deposited in lysosomes. Experiments using MDCK and LLC-PK_1 cells have shown colocalization of ligands internalized independently from the apical and basolateral surfaces (Bomsel et al., 1989; Cohen et al., 1995; Ford et al., 1994; Parton et al., 1989). This colocalization was shown to be at the level of the late-endosomal compartment. Interestingly, in the MDCK model, it has been shown that elements of the apical early endosomal compartment do not fuse with elements of the basolateral early endosomal compartment (Bomsel et al., 1990; Bucci et al., 1994). The implication from these experiments seems to be that each early-endosomal system must have its own specific and different recognition system. In practice, this could prevent the potential mixing of apical and basolateral membrane proteins. Similar colocalization of apical and basolateral applied ligands has also been reported using a polarized epithelial cell line derived from the small intestine (Hughson and Hopkins, 1990). In contrast, only very limited colocalization was seen in experiments performed using isolated perfused proximal tubules (Nielsen et al., 1985). It is possible that little or no colocalization was observed due to the decreased endocytic activity at the basolateral membrane *in vivo*. These studies are again an example of the sometimes-divergent observations between the *in vivo* and cell culture models.

Even though the basic mechanisms of basolateral and apical endocytic uptake are in general the same, there can be distinct and interesting differences. As was mentioned before, treatment with the drug cytochalasin D, which results in depolymerization of the actin cytoskeleton, inhibits apical membrane endocytosis much more dramatically than basolateral endocytosis (Gottlieb et al., 1993; Jackman et al., 1994; Lamaze et al., 1997). Furthermore, activation of protein kinase C has been shown to specifically stimulate the endocytic rate of the apical membrane (Holm et al., 1995). It has been suggested that these differences may be because significantly more nonclathrin-mediated uptake occurs at the apical membrane than at the basolateral membrane (Gottlieb et al., 1993). Again, it is important to consider the model systems used when trying to assimilate these data into a coherent context.

KIDNEY ENDOCYTOSIS

The significance of the endocytic process in general and the specific activities related to the proximal tubule have been discussed. This section expands on the specific activity and function of endocytosis as it relates to the proximal tubule.

Significance of Proximal Tubule Endocytosis

Proximal Tubule Endocytosis

It is clear that intensive reabsorption of protein filtered at the glomerulus occurs via endocytosis in the proximal tubule (for reviews see Christensen and Nielsen, 1991; Maack et al., 1979, 1985; Maunsbach and Christensen, 1992). This is the predominant mechanism for reclaim or scavenging of filtered protein. It is generally thought, though not rigorously proven, that this "protects" the organism from losing these potential sources of amino acids. The activity of the recently described multiligand receptor megalin, also called gp330, provides a good example of this.

Megalin is a very large (≈600 kDa) protein, originally described as the Heymann nephritis autoantigen (Kerjaschki and Farquhar, 1982). The proximal tubule has the highest concentration of megalin in all tissues tested, and it constitutes a significant proportion of total proximal tubule protein. Interestingly, it is present only at the apical surface in the cell types where it is present (see reviews below). It has been shown to bind many different protein ligands and is thought to mediate their uptake and delivery to lysosomes (Christensen et al., 1998a, 1998b; Farquhar et al., 1995; Moestrup, 1994). Although binding of some of the ligands is thought to be physiologically irrelevant because they would not be filtered, many are filtered and would be very relevant. In addition to its putative role in the scavenging of filtered protein, recent studies have suggested that megalin may also play an important role in the handling of filtered calcium and in the renal recovery of vitamins A and B_{12} (Christensen et al., 1999; Lundgren et al., 1994; Moestrup et al., 1996).

Lysosomal Pathway Predominates

Since Straus first demonstrated that endocytosed protein is transferred to lysosomes (Straus et al., 1964), many studies have shown that the predominant pathway of endocytosis in the proximal tubule is delivery to the lysosome for degradation (Christensen and Nielsen, 1991; Maack et al., 1979, 1985). It is clear from these types of studies that intracellular degradation is the major way in which filtered proteins are handled by PTC. In fact, it is the only renal mode that results in complete degradation of proteins into their constituent amino acids. It is possible that this pool of amino acids represents a specific and significant pool for PTC protein synthesis.

In summary, the proximal tubule plays a major role in the reclamation and preservation of nutrients from the glomerular filtrate. The very active endocytic process that is found in PTC, in large part makes this possible.

Endocytic Activity in the Proximal Tubule

Many investigators have worked to characterize the mammalian proximal tubule, functionally and morphologically. In many cases, tubule function has correlated nicely with morphology. The mammalian proximal tubule, which begins at the glomerulus and ends at the thin descending loop of Henle, is made up of three distinct segments. These three segments are more or less involved in all the processes that have been discussed. Aspects of the specific segments as they relate to this discussion will now be reviewed briefly. Extensive reviews on the function and morphology of the proximal tubule can be found elsewhere (Christensen and Nielsen, 1991; Hebert and Kriz, 1993; Maunsbach, 1966; Maunsbach and Christensen, 1992).

Endocytosis in Specific Segments of the Tubule

The mammalian proximal tubule can be subdivided into three segments. These segments are more easily distinguishable in mice and rats than in humans. The first segment, or S_1, corresponds to the earliest part of the tubule that has its origins at the urinary pole of Bowman's capsule. It has taller cells than the other segments, has tall, well-developed microvilli, many mitochondria oriented perpendicular to the basement membrane, and many lysosomes that are relatively "electron lucent" by electron microscopy. The second segment, or S_2, begins at the distal part of the convoluted section of the proximal tubule and extends to the straight section. It has cells that are somewhat shorter than the S_1 cells, shorter and less-developed microvilli, many mitochondria as in S_1, and many lysosomes that are more electron-dense than in S_1. The third segment, or S_3, corresponds to the distal or straight section of the proximal tubule. It has cells that are shorter still than the previous segments, has tall, well-developed microvilli, as in S_1, fewer and more randomly oriented mitochondria, and noticeably smaller lysosomes (Christensen and Nielsen, 1991; Hebert and Kriz, 1993; Maunsbach, 1966; Maunsbach and Christensen, 1992).

The endocytic apparatus is well developed in all three segments of the proximal tubule, though somewhat less so in S_3. All three segments have numerous clathrin-coated pits and vesicles at the apical or luminal surface, although the S_3 segment has visibly fewer apical vesicular structures. Furthermore, all three segments have many of the larger early- and late-endosomal structures that are intimately involved in the intracellular trafficking and processing of endocytosed ligands. Again, somewhat fewer of these structures are present in the S_3 segment (Christensen and Nielsen, 1991; Hebert and Kriz, 1993; Maunsbach, 1966; Maunsbach and Christensen, 1992). The general morphology of these segments correlates very well with the functional aspects of protein uptake and degradation in the proximal tubule. Protein uptake in the nephron is clearly the greatest in the proximal tubule, with uptake being greatest in S_1 and the least in S_3 for most LMWPs (Cortney et al., 1970; Cui et al., 1993; Maack et al., 1979, 1985; Tojo and Endou, 1992). Degradation of protein also seems to follow this same pattern with lysosomal hydrolase activity and protein degradation being greatest in the initial segments and least in S_3 (Clapp et al., 1988; Madsen and Park, 1987; Yokota and Kato, 1988a, 1988b).

Basolateral Endocytic Activity

Only limited basolateral endocytosis is thought to occur in the proximal tubule. Most, if not all, of this uptake is thought to involve hormones or growth factors (Hammerman, 1985; Hammerman and Gavin, 1984; Hensley et al., 1989; Milton et al., 1988; Nielsen et al., 1987, 1989; Rogers and Hammerman, 1989). Consistent with these data, the basolateral endocytic apparatus is not nearly as developed as that seen at the apical surface (Hammerman, 1985; Hammerman and Gavin, 1984; Hensley et al., 1989; Milton et al., 1988; Nielsen et al., 1987, 1989; Rogers and Hammerman, 1989). Coated pits and vesicles are intermittently found and larger components of the endosomal system are occasionally seen close to the basolateral membrane, but there are clearly many fewer of these structures present. Although microvilli are not present at the basolateral surface of PTCs, elaborate infoldings of the PM are seen. These infoldings dramatically increase the surface area and are thought to provide a larger zone for enzymes, such as Na^+/K^+-ATPase, that are active in various transport processes (Ernst and Schreiber, 1981).

Insertion of Aquaporins, Proximally and Distally

Aquaporins, formerly known as water channels, are present and active throughout most tissues of the body (Brown et al., 1995; King and Agre, 1996). These proteins play a particularly important role in reabsorption of water in the mammalian kidney, where the proximal tubule and descending thin limbs of the Loop of Henle reabsorb more than 150 liters of water per day. Aquaporins present in the collecting duct play an essential role in the concentration of urine. These channels are acutely regulated in response to vasopressin, while other less-well characterized factors are involved in their long-term regulation (Marples et al., 1999). Targeting and insertion of these channels into the appropriate membrane is thought to utilize the same mechanisms as are involved in the exocytic and endocytic pathways, which are intimately connected (Brown and Stow, 1996, and references therein). In fact, disruption of microtubules has been shown to alter the distribution and activity of water channels present in proximal tubules and collecting ducts in a manner similar to the way the endocytic pathway is affected by microtubule disruption (Dousa and Barnes, 1974; Gutmann et al., 1989; Phillips and Taylor, 1989). Therefore, the machinery and components of the endocytic pathway and its regulation appear to play a significant role in the normal activity of regions of the nephron that are and are not considered to be actively involved in endocytosis. This reiterates the role the endocytic pathway plays in the normal process of membrane homeostasis and provides the mechanism by which this pathway can be used to regulate specific cellular functions.

Consequences of Altered Endocytic Activity

The consequences referred to here result from situations in which the endocytic process of the proximal tubule is inhibited, overwhelmed, or caused to operate in the context of other pathology. Because the protein uptake system in the proximal tubule is functioning at close to its capacity under physiological conditions, situations that dramatically increase the protein load to the proximal tubule can lead to rapid and abrupt effects in the proximal as well as other regions of the tubule. Several circumstances thought to impinge on the protein uptake system of the proximal tubule will be discussed briefly.

Heavy Metal Poisoning

Heavy metals are used in many industries and have been widely studied with respect to their renal toxicity (Bernard and Lauwerys, 1991; Lauwerys et al., 1992). Many heavy metals accumulate in the kidney, chiefly the proximal tubule, which seems to be most affected. The damage to the proximal tubule is usually irreversible. Workers that have been exposed to heavy metals, such as cadmium (Cd), develop a persistent and initially specific LMWP proteinuria that can be as much as 100- to 1000-times greater than that seen in controls. The severity of the proteinuria is directly related to the duration of exposure (Kjellström et al., 1977; Tschuchiya, 1976). It has been suggested that this biological change, resulting from toxic effects on the primary region of the nephron involved in endocytosis, is predictive of kidney degeneration and should be considered an adverse effect (Lauwerys et al., 1992). In addition, Ca^{2+} wasting, manifested as ureteral calculi, is commonly induced (Kazantzis, 1979) and is another indication that disruption of normal endocytic events (see megalin involvement in renal and body Ca^{2+} above) in the proximal tubule can lead to pathological effects.

Rhabdomyolysis

Rhabdomyolysis can arise due to a number of reasons, perhaps the most common of which are traumatic muscle injury and drug overdose or abuse (Flamenbaum et al., 1983; Gabow et al., 1982). During clinical rhabdomyolysis, large quantities of myoglobin are released into the circulation. Small enough to be freely filtered, large amounts reach the proximal tubule where, if severe enough, the uptake capacity can be overwhelmed with resultant myoglobinuria. Increased levels of myoglobin reaching the distal tubule can lead to precipitation, cast formation, and tubule obstruction (Sanders et al., 1988). Furthermore, in the setting of increased tissue catabolism and resultant ischemic acute renal failure, increased proximal tubular uptake has been shown to exacerbate ischemic renal injury (Zager et al., 1987). Therefore, by overwhelming the endocytic capacity and increasing the metabolic requirements of the proximal tubule, serious consequences can arise.

Light Chain Disease

This is a well-described condition that arises from overproduction and enhanced excretion of immunoglobulin light chains, usually as a result of myeloma. As many as 50% of patients with multiple myeloma develop renal involvement (de Fronzo et al., 1975, 1978; Ganeval et al., 1992) and it can be the leading cause of death in 15 to 25% of patients with myeloma (Ganeval et al., 1992; Oken, 1984). Both a direct effect of light chains proximally (Batuman et al., 1986; Preuss et al., 1968; Sanders et al., 1988) and an indirect effect by formation of light-chain casts distally (Hill et al., 1983; Levi et al., 1968) have been proposed as pathogenetic mechanisms involved in the observed nephrotoxicity. Direct interaction with PTCs via endocytic mechanisms is thought to be necessary to induce the proximal effects, while precipitation, cast formation, and tubular obstruction are thought to cause the distal effects. Again, endocytosis appears to be either directly or indirectly involved in the pathology of the renal disease.

Aminoglycoside-Induced Nephrotoxicity

Aminoglycosides antibiotics are widely used for the treatment and prevention of Gram-negative bacterial infections. Although they have been, and continue to be, successfully used, they are associated with well-known and well-described complications of ototoxicity and nephrotoxicity (for comprehensive reviews see Cojocel, 1996, and Kaloyanides, 1993). Uptake and accumulation of aminoglycosides occur as a result of adsorptive endocytosis at the apical membrane of PTCs (Just et al., 1977; Pastoriza-Munoz et al., 1979; Sastrasinh et al., 1982; Silverblatt and Kuehn, 1979). This uptake follows a classical receptor-mediated endocytic pathway with eventual deposition of the antibiotics in lysosomes. Because uptake and concentration of aminoglycosides occur primarily in PTCs of the kidney (also cells of the inner ear) and the observed toxicity correlates with this, it is widely accepted that interaction with the PM and the endocytic process must play an important role in development of toxic effects. Subsequent to uptake, the aminoglycosides are thought to interact with, and disrupt a variety of, intracellular functions (Ali, 1995; Cojocel, 1996; Humes et al., 1982; Kaloyanides, 1993). Alterations in cellular function, which result of aminoglycoside endocytosis, lead to the many observed toxic effects. These include, similar to above: cast formation distally from BBM fragments, extruded myeloid bodies and membrane vesicles, and cellular debris from dead and dying cells.

GENTAMICIN-INDUCED NEPHROTOXICITY

This section expands on the effects and mechanisms involved in aminoglycoside- induced nephrotoxicity. Most of the discussion will revolve around studies that have used gentamicin. Both short-term as well as long-term studies are discussed, and references is made to work that implicate effects on the mechanism of endocytosis and intracellular trafficking.

Renal Handling of Gentamicin

The susceptibility of the kidney to aminoglycoside-induced toxicity derives from the fact that the kidney is the exclusive route for excretion of this class of antibiotics. Essentially all of an injected dose of gentamicin is rapidly filtered by the kidney, resulting in a plasma half-life of 30 to 90 minutes in rats (Fabre et al., 1976; Luft et al., 1975a) and 1.5 to 4 hours in humans (Just and Habermann, 1977; Schentag and Jusko, 1977; Schentag et al., 1977). Once filtered, a small, but significant fraction ($\leq$5%) is taken up and concentrated in PTCs, while the vast majority is excreted (Pastoriza-Munoz et al., 1979; Senekjian et al., 1981). During the process of excretion, the small proportion of the antibiotic taken up and concentrated in PTCs is thought to subsequently lead to the observed toxicity.

Mechanism of Uptake

Involvement of the proximal tubule in the uptake of aminoglycosides was first implied by studies that showed excretion to be significantly less than the filtered load (Chiu et al., 1976, Schentag and Jusko, 1977). Direct evidence has since been obtained using microinjection and micropuncture techniques (Frommer et al., 1983; Pastoriza-Munoz et al., 1979, 1984; Senekjian et al., 1981; Sheth et al., 1981). A preponderance of the evidence indicates that aminoglycosides are taken up at the apical surface of PTCs, with very little being taken up via the basolateral surface (Collier et al., 1979; Just et al., 1977; Pastoriza-Munoz et al., 1979; Senekjian et al., 1981; Silverblatt and Kuehn, 1979; Weeden et al., 1983). Binding of aminoglycosides to the apical membrane is mediated, in part, by electrostatic interactions with acidic phospholipids, primarily phosphatidylinositol (PI) (Feldman et al., 1982; Laurent et al., 1982; Sastrasinh et al., 1982). In addition, recent evidence suggests involvement of the large ($\approx$600 kDa) multiligand receptor megalin, also known as gp330, in uptake of aminoglycosides (Moestrup et al., 1995). Megalin is a member of the LDL-R supergene family that includes the LDL-R, the very low density lipoprotein receptor, and the α_2-macroglobulin receptor (Christensen et al., 1998a; Moestrup, 1994). The current working model involves initial binding of aminoglycosides to PI at the "tips" of microvilli followed by diffusion of the PI–aminoglycoside complex within the plane of the membrane to the coated pit region at the base of microvilli (Moestrup et al., 1995). Once the complex is in coated pits, transfer of the aminoglycoside to megalin is thought to occur via a mechanism that is as yet undescribed, which is followed by uptake via the receptor-mediated endocytic pathway.

Intracellular Trafficking

Aminoglycosides are taken up by receptor-mediated endocytosis. Accordingly, coated pits, coated and uncoated vesicles, endosomes, and lysosomes are all expected to be involved in the intracellular itinerary of aminoglycoside antibiotics. This has been documented in large part by autoradiographic techniques following *in vivo* administration of radioactive gentamicin (Just et al., 1977; Silverblatt and Kuehn, 1979). Although a qualitative rather than a quantitative technique, it has provided compelling evidence that gentamicin does indeed transit through and accumulate in these structures. It has been suggested that lysosomal accumulation may result, at least in part, from nonendocytic or alternative mechanisms (Weeden et al., 1983). Corroborating studies from other investigators have duplicated these results. However, cellular uptake and intracellular trafficking of gentamicin has more recently been investigated using immunocytochemical techniques (Molitoris et al., 1993). These studies, which were probably more sensitive than the autoradiographic ones, effectively corroborated uptake via the receptor-mediated endocytic pathway as the earlier results had shown.

Short-Term Metabolic Effects of Gentamicin

Although acute effects of aminoglycosides can be seen in time periods as short as hours, the normal clinically recognized nephrotoxic effects usually take time periods of several days. Due to the rather sudden disruption of renal function and the gross necrosis that then occurs as a result of aminoglycoside-induced nephrotoxicity, it has been difficult to distinguish between primary and secondary causative events. For these reasons, and to characterize the early events responsible for development of toxicity, investigators have used relatively short time points of exposure (one to three days). This section discusses studies that look at early effects of gentamicin as a method to characterize the mechanisms involved in the development of gentamicin-induced nephrotoxicity.

Short-Term Effects on Protein Metabolism

The bactericidal activity of aminoglycosides is derived from their ability to bind prokaryotic ribosomes, thereby blocking the ribosomal initiation complex, or by causing mistranslation of message (Davies et al., 1965, 1968). This effectively inhibits or deranges protein synthesis, leading eventually to bacterial death. Recent evidence suggests that these antibiotics may also inhibit protein synthesis by eukaryotic ribosomes. Elevated concentrations of aminoglycosides have been shown to cause mistranslation or block incorporation of amino acids by eukaryotic ribosomes in *in vitro* protein synthesis assays (Buss and Piatt, 1985; Buss et al., 1984; Wilhelm et al., 1978). It has also been shown that concentrations of aminoglycosides attained in the rat renal cortex during treatment can inhibit *in vitro* microsomal protein synthesis following *in vivo* gentamicin treatment (Bennett et al., 1988; Buss and

Piatt, 1985; Buss et al., 1984). This inhibition was seen at treatment time points that were before the times when significant alterations in cellular function were observed. In addition, further studies have shown in *in vivo* experiments that gentamicin treatment significantly inhibits rat protein synthesis following two days of treatment and continues for the third day (Sundin et al., 2001). These studies suggest that inhibition of PTC protein synthesis may play a direct and prominent role in the observed PTC toxicity, even though the vast majority of aminoglycoside accumulates in lysosomes. With respect to the accumulation of gentamicin in lysosomes, evidence also suggests degradation of protein may be inhibited by short-term gentamicin (STG) treatment (Olbricht et al., 1991).

Short-Term Effects on Phospholipid Metabolism

Alterations in rat renal phospholipid metabolism following aminoglycoside treatment have been described by a number of laboratories (Feldman et al., 1982; Josepovitz et al., 1985; Laurent et al., 1982; Ramsammy et al., 1989; Sundin et al., 1997). Short-term gentamicin treatment (hours to several days) has been shown to cause alterations in the phospholipid content of renal cortical homogenates and isolated BBM (Feldman et al., 1982; Josepovitz et al., 1985; Knauss et al., 1983; Ramsammy et al., 1988). These changes are thought to result, at least in part, from the inhibition of phospholipases, since it is known that gentamicin can inhibit phospholipases *in vitro* (Carlier et al., 1983; Hostetler and Hall, 1982; Laurent et al., 1982; Tulkens, 1986). Also, it is generally accepted that the characteristic myeloid bodies found in lysosomes after gentamicin treatment consist of undegraded phospholipid-containing membranes. A decrease in membrane fluidity has been observed in a STG model (Moriyama et al., 1989), while in a long-term gentamicin (LTG) treatment or "recovered" model no significant changes in membrane fluidity could be detected (Sundin et al., 1997). It has also been shown that STG treatment significantly increased phospholipid content of renal cortical homogenates and BBM, while incorporation of ^{32}P into total cellular phospholipid (i.e. synthesis) was reduced (Sundin et al., 2001). It was concluded that inhibition of phospholipid degradation was the major contributing factor to the STG effects on phospholipid metabolism, though there was also an apparent decrease in synthesis of total cellular phospholipid.

Short-Term Effects on Endocytosis

It is clear that gentamicin treatment can have a rapid effect on the uptake of protein in the proximal tubule. The increased urinary excretion of LMWPs is one of the earliest and most common expressions of aminoglycoside-induced nephrotoxicity (Bernard et al., 1986; Gibey et al., 1981; Schentag et al., 1978; Schentag, 1983). Although aminoglycosides have been shown to compete with LMWPs for binding and uptake in PTC (Bernard et al., 1986; Just et al., 1977; Neuhaus, 1986), the fact the induced proteinuria persists after the drug has been excreted or discontinued suggests a more complex sequence of events than a simple competitive inhibition of uptake. Indeed, it has been suggested that multiple minor alterations in combination may induce the observed effects (Kaloyanides, 1984).

Experiments have in large part concured with this general concept, when applied to the process of endocytosis. In addition to short-term effects on protein and phospholipid metabolism, there is evidence to suggest that gentamicin may be progressively trafficked to intracellular destinations other than the lysosome (Sundin et al., 2001). Additionally, preliminary experiments, designed to directly characterize effects of gentamicin on endocytosis, have suggested that STG treatment causes a decline in its own endocytosis, induces alterations in its intracellular trafficking, and induces a subcellular redistribution of Rab5 (Sundin et al., unpublished data).

All three of these effects were seen by three days of treatment. These types of "multiple endocytic and trafficking alterations" in combination could certainly provide a putative mechanism for induction of nephrotoxicity.

Implications of Gentamicin-Induced Altered Trafficking – An Hypothesis

Many hypotheses have been proposed to describe the mechanisms involved in aminoglycoside-induced nephrotoxicity (Ali, 1995; Bennett, 1989; Cojocel, 1996; Kaloyanides, 1993; Tulkens, 1986). Although there are data implicating many alternatives, the precise mechanisms remain unknown. The authors' hypothesis involves increasing disruption of normal trafficking and fusion processes, which eventually result in disruption of many cellular activities, and, in many cases, cell death. In agreement with this concept, several investigators have proposed gentamicin-induced alterations in membrane traffic

(Giurgea-Marion et al., 1986; Josepovitz et al., 1985; Klausner et al., 1983; Olbricht et al., 1991; Ramsammy et al., 1989). There is evidence that gentamicin treatment inhibits vesicle/lysosome fusion *in vivo* and homotypic endosome fusion *in vitro* (Giurgea-Marion et al., 1986; Hammond et al., 1997), both of which would reinforce the authors' hypothesis. Further, a long-term model of gentamicin treatment in which rats recover has shown that an apparent reduction in the endosomal compartment occurs, which again suggests effects on fusion capabilities (Sundin et al., 1997). This type of hypothesis provides a mechanism of gentamicin-induced toxicity with far-reaching consequences not only for endocytosis and trafficking, but also for many other cellular activities and organelles. It could also account for the many effects described by many investigators on various cellular activities and organelles.

Long-Term Effects of Gentamicin on Endocytosis

When administered at relatively high doses, gentamicin induces a well-characterized and reproducible initiation and development of induced nephrotoxicity in the rat model (Cuppage et al., 1977; Feldman et al., 1982; Gilbert et al., 1979; Houghton et al., 1976, 1986; Knauss et al., 1983; Laurent et al., 1982; Luft et al., 1975b; Sundin et al., 1997). There is convincing evidence in the rat LTG model that after 10 to 14 days of treatment, recovery of the proximal tubule epithelium commences, even with the continued administration of gentamicin (Cuppage et al., 1977; Gilbert et al., 1979; Luft et al., 1975b; Sundin et al., 2001). Serum creatinine values, cellular enzyme activities, and histology all return to close to normal by 18 to 21 days. Although it is clear that recovery from the toxic events does occur, it is probably not complete since at least some alterations do persistent (Cuppage et al., 1977; Houghton et al., 1986; Sundin et al., 2001). However, the mechanism of recovery or development of resistance to toxicity is of great interest, which has made this an active area of investigation. This section discusses the putative effects and implications of this model on and for endocytosis and trafficking.

Inhibition of Gentamicin Endocytosis

Studies have shown that during the course of gentamicin administration, accumulation of the drug increased until the point at which serum creatinine values began to rise significantly (Gilbert et al., 1979). Very soon after this point, gross cellular necrosis and death occurred, which was thought to be largely responsible for the reduced uptake of gentamicin that was subsequently seen. The reduction in uptake and accumulation of gentamicin continued until the recovery phase observed in this model had been achieved. At this point, accumulation of gentamicin resumed and eventually attained similar levels as before, without the previously observed toxicity, although it was not determined that gentamicin was internalized as efficiently as before treatment (Gilbert et al., 1979; Houghton et al., 1986). It was suggested that recently regenerated or relatively immature PTCs might not initially have the capacity to accumulate gentamicin, although they were able to later without recurring gross cellular toxicity.

This issue was investigated in a series of experiments using a similar model. Rats were injected once daily with vehicle or gentamicin (100 mg/kg) for 18 days (i.e., recovery had occurred), then sacrificed on day 19 and BBM prepared. Binding studies performed on BBMs isolated from control or LTG-treated rats showed that gentamicin binding was increased more than twofold in the LTG-treated rats (Sundin et al., 2001). However, when internalization of gentamicin was morphologically determined in the recovered rats, it was clearly reduced compared with vehicle injected control rats. In these rats, gentamicin uptake appeared to be specifically reduced compared with two other markers of endocytosis, horseradish peroxidase and β_2-microlobulin; markers of fluid-phase and receptor-mediated endocytosis, respectively. The uptake of these two markers was, if anything, increased compared with gentamicin. It was concluded that gentamicin uptake was being selectively inhibited by LTG treatment. Further, since the uptake of horseradish peroxidase and β_2-microglobulin was not decreased by LTG treatment, it was suggest that not all endocytic events or processes are affected to the same extent by gentamicin treatment. In the case of β_2-microglobulin and gentamicin, this may imply these ligands are taken up via different receptors.

Effects of LTG Treatment on Factors Involved in Gentamicin Endocytosis

Because gentamicin endocytosis was inhibited by its own administration, effects on the elements thought to be involved in its uptake needed to be determined. As has been mentioned previously, phosphatidylinositol

(PI) and megalin are thought to be involved in gentamicin internalization (Moestrup et al., 1995). When effects on these elements in the LTG model were determined, PI levels were shown to be increased twofold in total cellular homogenates and in BBM, while no significant alteration in levels of BBM-associated megalin was observed by Western blot or by immunofluorescence (Sundin et al., 2001). The increase in gentamicin binding to BBM correlated very well with the twofold increase in BBM PI levels that were observed in this model. Megalin levels might have been expected to decrease to explain the reduced uptake of gentamicin, but this was not the case. Therefore, the decrease in gentamicin internalization did not result from a reduction in either of the factors thought to be involved in its uptake.

Effect of LTG Treatment on the Endocytic Process

The studies just described indicate that LTG treatment and development of apparent resistance to gentamicin may involve a selective inhibition of gentamicin uptake. It has been postulated that normal trafficking of megalin and megalin-containing structures of megalin is reduced in LTG-treated rats (Sundin et al., 2001). This could be a mechanism whereby the proximal tubule is able to develop the apparent relative tolerance to gentamicin in the LTG model. The fact that megalin levels were unaffected suggests that its processing may be altered in some way such that internalization is less efficient. All these factors imply involvement of the endocytic and trafficking mechanisms that have been described in this review. For instance, if endocytic recognition sequences on megalin were modified or no longer recognized by the endocytic machinery, internalization could clearly be effected. Alternatively, in this same scenario, normal intracellular trafficking after internalization could also be altered such that gentamicin would be trafficked to alternative intracellular destinations. These putative alternate destinations could be either toxic or protective. Even if endocytosis and trafficking were only moderately affected, their cumulative effects could certainly contribute to the observed developments in the LTG model.

EXPLOITATION OF ENDOCYTOSIS BY TOXICANTS

Many plant and bacterial protein toxicants exert their toxic effects on cells following endocytosis, in many cases receptor-mediated endocytosis (Sandvig and van Duers, 1996; Sandvig et al., 1992a).

After internalization and translocation of an enzymatically active part of the molecule into the cytosol, most of these toxicants kill cells by inhibiting protein synthesis. In fact, it is now clear that endocytosis and at least some intracellular trafficking through specific organelles are required to induce their toxic effects (Sandvig and Olsnes, 1982; Sandvig and van Duers, 1996; Sandvig et al., 1992a). Viruses also use similar mechanisms to gain entry into cells and the cytosol (Kielian and Jungerwirth, 1990; Marsh and Helenuis, 1989). The sequences, signals, and intracellular components used by these toxicants are the same ones that have been discussed in this review. This section briefly reviews the mechanisms of some of these toxicants and then develops a general model of toxicity that results from endocytosis.

Bacterial and Plant Toxins

A number of bacterial protein toxins are important in the development of widespread and severe diseases, such as diphtheria, cholera, dysentery, and hemolytic uremic syndrome (for reviews see Fishman and Orlandi, 1994; Karmali, 1989; O'Brien et al., 1992; Sandvig and Olsnes, 1991). Better understanding of these toxins and their mechanism of action are important for understanding these diseases and their disease processes, and also for improving putative treatments. Plants also produce substances that can have significant and relevant toxicological effects (Barbieri et al., 1993; Lord et al., 1994). Because of the extreme toxicity of many of these toxins, they have been used to construct immunotoxins for antibody-targeted treatment of cancer and other diseases (Brinkman and Pastan, 1994; Ghetie and Vitetta, 1994; Vitetta et al., 1993). In addition, these toxins have proven to be valuable tools in modern cell biology for studying the processes of endocytosis, intracellular trafficking, exocytosis, and translocation of proteins across membranes (Olsnes et al., 1991; Sandvig and van Duers, 1994, 1996; Sandvig et al., 1992a; Südhoff et al., 1993; Theuer et al., 1993). Two of these toxins are discussed below.

Shiga Toxin

Shiga toxin is produced by *Shigella dysenteriae* and consists of five small B subunits in noncovalent association with a larger A subunit (Sandvig and van Duers, 1994). It is structurally related to a larger group of protein toxins that all generally consist of an enzymatically active

A subunit and multiple copies of a binding or B subunit (Keusch et al., 1986). It has been shown to bind to the neutral glycolipid globotriosylceramide (Lindberg et al., 1987) and is internalized via clathrin-coated pits (Sandvig et al., 1989a, 1991).

Shiga toxin is the first lipid-binding ligand that has been shown to be internalized via clathrin-coated pits but the mechanism behind its clustering in these structures is not understood very well. Further, it is the first molecule shown to undergo retrograde transport from the cell surface, through the Golgi, to the ER and nuclear membrane, which has demonstrated a new intracellular transport pathway (Sandvig et al., 1992b, 1994). Interestingly, as with gentamicin, the majority of internalized toxin is trafficked to lysosomes where it is degraded very slowly (Sandvig and van Duers, 1996; Sandvig et al., 1989a, 1991). Only a small fraction of the toxin is trafficked to the Golgi and ER, but these steps are essential for toxicity (Sandvig et al., 1991, 1994). It is thought that an uncharacterized Golgi processing step may be required for intoxication while release into the cytosol occurs from the ER.

Ricin

Ricin is a plant protein toxin that comes from the seeds of *Ricinus communis* and has several different 63 to 66 kDa isoforms (Barbieri et al., 1993; Lord et al., 1994). Like Shiga toxin, ricin consists of two polypeptide chains: an A chain, which carries the enzymatic activity, and a B chain attached to the A chain via disulfide linkage, which contains the binding activity. Unlike Shiga toxin, ricin binds to both glycolipids and glycoproteins that contain terminal galactose residues, which are important for efficient intoxication (Lord et al., 1994; Newton et al., 1995; Sandvig and van Duers, 1994, 1996; van Duers et al., 1989). Studies of ricin uptake have shown it is internalized by both clathrin-dependent and clathrin-independent mechanisms, which may reflect its wider binding specificity.

Continued uptake following inhibition of clathrin-coated pit internalization provided some of the first direct evidence that more than one mechanism of endocytosis exists (Moya et al., 1985; Sandvig and van Duers, 1991, 1996; Sandvig et al., 1987). Like Shiga toxin, ricin is trafficked to the Golgi and this transport and subsequent translocation are required for a strong toxic effect (Sandvig and van Duers, 1996). However, ricin has not so far been observed to traffic to the ER.

As with Shiga toxin, the majority of ricin is transported to the lysosome. The amount of ricin that reaches the Golgi has been estimated to be approximately 5% of the total amount internalized (van Duers et al., 1988).

Exploitation of Endocytosis: a General Paradigm for Development of Toxicity

This review has endeavored to describe the relative importance of endocytic activity to the kidney proximal tubule specifically, and to the kidney and entire animal in general. The endocytic process is an essential component of proximal tubule activity, and when it is disrupted, pathological consequences result. It is clear that there are many natural and synthetic elements that can affect this process; not only in the kidney, but also in general. In fact, it appears that toxins, such as ricin and Shiga toxin, or preparations of antibiotics, such as gentamicin, utilize this pathway to elicit their induced toxicity.

Endocytosis is an essential process that would have evolved early in development of life. It is reasonable to assume that organisms would have evolved mechanisms to take advantage of this ancient and essential process for their benefit or protection. This seems to be the case for both the bacterial and plant toxins discussed, as well as for the antibiotic gentamicin. To induce their toxic effects they all must be internalized, at least initially. Furthermore, they may need to undergo specific trafficking steps within the endocytic pathway to induce their toxic effects. Therefore, the requirement of internalization for activity of many toxins may imply a common mechanism. Indeed, the authors postulate exploitation of the endocytic pathway as a general paradigm whereby many toxicants induce toxicity. If true, it becomes even more imperative to better characterize the process of endocytosis and trafficking in order to develop strategies for potential therapies for toxicant induced diseases.

REFERENCES

Abrahamson DR, and Rodewald R (1981) Evidence for the sorting of endocytic vesicle contents during the receptor-mediated transport of IgG across the newborn rat intestine. *J Cell Biol* 91:270–280.

Achler C, Filmer D, Merte C, and Drenckhahn D (1989) Role of microtubules in polarized delivery of apical membrane proteins to the brush border of the intestinal epithelium. *J Cell Biol.* 109:179–189.

Al Awqati Q (1986) Proton translocating ATPases. *Ann Rev Cell Biol* 2:179–199.
Alfrey AC (1989) Beta2-microglobulin amyloidosis. *AKF Nephrology Letter* 6:27–34.
Ali BH (1995) Gentamicin nephrotoxicity in humans and animals: Some recent research. *Gen Pharmac* 26:1477–1487.
Anderson RGW, Brown MS, Beisiegel U, and Goldstein JL (1982) Surface distribution and recycling of the LDL receptor as visualized by anti-receptor antibodies. *J Cell Biol* 93:523–531.
Araki S, Kikuchi A, Hata Y, Isomura M, and Takai Y (1990) Regulation of reversible binding of Smg p25A, a Ras p21-like GTP binding protein, to synaptic plasma membranes and vesicles by its specific regulatory protein, GDP dissociation inhibitor. *J Biol Chem* 265:13007–13015.
Barbieri LM, Battelli MG, and Stirpe F (1993) Ribosome-inactivating proteins from plants. *Biochim Biophys Acta* 1154:237–282.
Basu SK, Goldstein JL, Anderson RGW, and Brown MS (1981) Monensin interrupts the recycling of low density lipoprotein receptors in human fibroblasts. *Cell* 24:493–502.
Batuman V, Sastrasinh M, and Sastrasinh S (1986) Light chain effects on alanine and glucose uptake by renal brush border membranes. *Kidney Int* 30:662–665.
Beaumelle BD, Gibson A, and Hopkins CR (1990) Isolation and preliminary characterization of the major membrane boundaries of the endocytic pathway in lymphocytes. *J Cell Biol* 111:1811–1823.
Bennett WM (1989) Mechanisms of aminoglycoside nephrotoxicity. *Clin Exp Pharmacol Physiol* 16:1–6.
Bennett WM, Mela-Riker LM, Houghton DC, Gilbert DN, and Buss WC (1988) Microsomal protein synthesis inhibition: An early manifestation of gentamicin nephrotoxicity. *Am J Physiol* 255:F265-F269.
Berg T, Blomhoff R, Naess L, Tolleshaug H, and Drevon CA (1983) Monensin inhibits receptor-mediated endocytosis of asialoglycoproteins in rat hepatocytes. *Exp Cell Res* 148:319–330.
Berg T, Kindberg GM, Ford T, and Blomhoff R (1985) Intracellular transport of asialoglycoproteins in rat hepatocytes: Evidence for two subpopulations of lysosomes. *Exp Cell Res* 161:285–296.
Bernard A, and Lauwerys RR (1991) Proteinuria: Changes and mechanisms in toxic nephropathies. *Crit Rev Toxic* 21:373–405.
Bernard A, Viau C, Ouled A, Tulkens P, and Lauwerys R (1986) Effects of gentamicin on the renal uptake of endogenous and exogenous protein in conscious rats. *Toxicol Appl Pharmacol* 84:431–438.
Blum JS, Fiani ML, and Stahl PD (1991) Proteolytic cleavage of ricin A chain in endosomal vesicles. Evidence for the action of endosomal proteases at both neutral and acidic pH. *J Biol Chem* 266:22091–22095.
Bomsel M, Parton RG, Kuznetsov SA, Schroer TA, and Gruenberg J (1990) Microtubule- and motor-dependent fusion *in vitro* between apical and basolateral endocytic vesicles from MDCK cells. *Cell* 62:719–731.
Bomsel M, Prydz K, Parton RG, and Simons K (1989) Endocytosis in filter-grown Madin-Darby canine kidney cells. *J Cell Biol* 109:3243–3258.
Bornens M (1991) Cell polarity: Intrinsic or externally imposed? *New Biol* 3:627–636.
Bourdeau JE, and Carone F (1973) Contraluminal serum albumin uptake in isolated perfused renal tubules. *Am J Physiol* 224:399–404.

Bourdeau JE, Chen ERY, and Carone F (1973) Insulin uptake in the renal proximal tubule. *Am J Physiol* 225:1399–1404.

Bourne HR (1988) Do GTPases direct membrane traffic in secretion? *Cell* 53:669–671.

Branch WJ, Mullock BM, and Luzio JP (1987) Rapid subcellular fractionation of the rat liver endocytic compartments involved in transcytosis of polymeric immunoglobulin A, and endocytosis of asialofetuin. *Biochem J* 244:311–315.

Brinkman U, and Pastan I (1994) Immunotoxins against cancer. *Biochim Biophys Acta* 1198:27–45.

Brown D, and Stow JL (1996) Protein trafficking and polarity in kidney epithelium: From cell biology to physiology. *Physiol Rev* 76:245–297.

Brown D, Katsura T, Kawashima M, Verkman AS, and Sabolic I (1995) Cellular distribution of the aquaporins: A family of water channel proteins. *Histochem Cell Biol* 104:1–9.

Brown MS, Anderson RGW, and Goldstein JL (1983) The round-trip itinerary of migrant membrane proteins. *Cell* 32:663–667.

Brown MS, and Goldstein JL (1984) How LDL receptors influence cholesterol and atherosclerosis. *Sci Am* 251:58–66.

Bucci C, Parton RG, Mather IH, Stunnenberg H, Simons K, Hoflack B, and Zerial M (1992) The small GTPase rab5 functions as a regulatory factor in the early endocytic pathway. *Cell* 70:715–728.

Bucci C, Wandinger-Ness A, Lutcke A, Chiariello M, Bruni CB, and Zerial M (1994) Rab5a is a common component of the apical and basolateral endocytic machinery in polarized epithelial cells. *Proc Natl Acad Sci USA* 91:5061–5065.

Buss WC, and Piatt MK (1985) Gentamicin administered *in vivo* reduces protein synthesis in microsomes subsequently isolated from rat kidney but not from rat brain. *J Antimicrob Chemother* 15:715–721.

Buss WC, Piatt MK, and Kauten R (1984) Inhibition of mammalian microsomal protein synthesis by aminoglycoside antibiotics. *J Antimicrob Chemother* 14:231–241.

Carlier MB, Laurent G, Claes PJ, Vanderhaeghe HJ, and Tulkens PM (1983) Inhibition of lysosomal phospholipases by aminoglycoside antibiotics: *In vitro* comparative studies. *Antinicrob Agents Chemother* 23:440–449.

Carpenter G, and Cohen S (1976) ^{125}I-labeled human epidermal growth factor. Binding, internalization, and degradation in human fibroblast. *J Cell Biol* 71:159–171.

Chavrier P, Parton RG, Hauri HP, Simons K, and Zerial M (1990) Localization of low molecular weight GTP binding proteins to exocytic and endocytic compartments. *Cell* 62:317–329.

Chavrier P, Simnos K, and Zerial M (1992) The complexity of the Rab and Rho GTP-binding protein subfamilies revealed by a PCR cloning approach. *Gene* 112:261–264.

Chiu PJ, Brown A, Miller G, and Long JF (1976) Renal extraction of gentamicin in anesthetized dogs. *Antimicrob Agents Chemother* 10:277–282.

Christensen EI, and Nielsen S (1991) Structural and functional features of protein handling in the kidney proximal tubule. *Sem Nephrol* 11:414–439.

Christensen EI, Birn H, Verroust, P, and Moestrup SK (1998a) Membrane receptors for endocytosis in the renal proximal tubule. *Int Rev Cyto* 180:237–283.

Christensen EI, Birn H, Verroust P, and Moestrup SK (1998b) Megalin-mediated endocytosis in renal proximal tubule. *Renal Failure* 20:191–199.

Christensen EI, Moskaug JØ, Vorum H, Jacobsen C, Gundersen TE, Nykjaer A, Blomhoff R, Willnow TE, and Moestrup SK (1999) Evidence for an essential role of megalin in transepithelial transport of retinol. *J Am Soc Nephrol* 10:685–695.

Ciechanover A, Schwartz AL., Dautry-Varsat A, and Lodish HF (1983) Kinetics of internalization and recycling of transferrin and the transferrin receptor in a human hepatoma cell line. Effect of lysosomotropic Agents. *J Biol Chem* 258:9681–9689.

Clapp WL, Park CH, Madsen KM, and Tisher CC (1988) Axial heterogeneity in the handling of albumin by the rabbit proximal tubule. *Lab Invest* 58:549–558.

Clary DO, and Rothman JE (1990) Purification of three related peripheral membrane proteins needed for vesicular transport. *J Biol Chem* 265:10109–10117.

Clary DO, Griff IC, and Rothman JE (1990) SNAPs, a family of NSF attachment proteins involved in intracellular membrane fusion in animals and yeast. *Cell* 61:709–721.

Cohen M, Sundin DP, Dahl R, and Molitoris BA (1995) Convergence of apical and basolateral endocytic pathways for β_2-microglobulin in LLC-PK_1 cells. *Am J Physiol* 268:F829–838.

Cohn ZA, Hirsch JG, and Fedorko ME (1966) The *in vitro* differentiation of mononuclear phagocytes. V. The formation of macrophage lysosomes. *J Exp Med* 123:757–766.

Cojocel C (1996) in *Comprehensive Toxicology* (Sipes G, McQueen C, and Gandolfi AJ, Eds) pp. 495–523, Elsevier Sci. Inc., New York.

Collier VU, Lietman PS, and Mitch WE (1979) Evidence for luminal uptake of gentamicin in the perfused rat kidney. *J Pharmacol Exp Ther* 210:247–251.

Cortney MA, Sawin LL, and Weiss DD (1970) Renal tubular protein absorption in the rat. *J Clin Invest* 49:1–4.

Cui S, Flyvbjerg A, Nielsen S, Kiess W, and Christensen EI (1993) IGF-II/Man-6-P receptors in rat kidney: Apical localization in proximal tubule cells. *Kidney Int* 43:796–807.

Cuppage FE, Setter K, Sullivan LP, Reitzes EJ, and Melnykovych AO (1977) Gentamicin nephrotoxicity: II. Physiological, biochemical, and morphological effects of prolonged administration to rats. *Virchows Arch B Cell Pathol* 24:121–138.

Daro E, Slujis PVD, Galli T, and Mellman I (1996) Rab4 and cellubrevin define different early endosome populations on the pathway of transferrin receptor recycling. *Proc Natl Acad Sci USA* 93:9559–9564.

Dautry-Varsat A, Ciechanover A, and Lodish HF (1983) pH, and the recycling of transferrin during receptor-mediated endocytosis. *Proc Natl Acad Sci USA* 80:2258–2262.

Davis BD, Weisblum B, and Davies J (1968) Antibiotic inhibitors of the bacterial ribosome. *Bacteriol Rev* 32:493–528.

Davis CG, Goldstein JL, Sudhof TC, Anderson RGW, Russell DW, and Brown MS (1987) Acid-dependent ligand dissociation and recycling of LDL receptor mediated by growth factor homology region. *Nature* 326:760–765.

Davies J, Gorini L, and Davis BD (1965) Misreading of RNA code words induced by aminoglycoside antibiotics. *Mol Pharmacol* 1:93–106.

de Duve C (1963) Lysosomes, in *Ciba Foundation Symposium* (de Reuck AVS, and Cameron MP, Eds), pp. 411–412, Churchill, London.

de Duve C (1975) Exploring cells with a centrifuge. *Science* 189:186–194.

de Fronzo RA, Cooke CR, Wright JR, and Humphrey RL (1978) Renal function in patients with multiple myeloma. *Medicine* 57:151–166.

de Fronzo RA, Humphrey RL, and Wright JR (1975) Acute renal failure in multiple myeloma. *Medicine* 54:209–223.

Dousa TP, and Barnes LD (1974) Effects of colchicine and vinblastine on the cellular action of vasopressin in the mammalian kidney. A possible role of microtubules. *J Clin Invest* 54:252–262.

Dunn WA (1990) Studies on the mechanism of autophagy: Formation of the autophagic vacuole. *J Cell Biol* 110:1923–1933.

Dunn WA, and Hubbard AL (1984) Receptor-mediated endocytosis of epidermal growth factor by hepatocytes in the perfused rat liver: Ligand and receptor dynamics. *J Cell Biol* 89:2148–2159.

Dunn KW, and Maxfield FR (1992) Delivery of ligands from sorting endosomes to late endosomes occurs by maturation of sorting endosomes. *J Cell Biol* 117:301–310.

Dunn WA, Connolly TP, and Hubbard AL (1986) Receptor-mediated endocytosis of epidermal growth factor by rat hepatocytes: Receptor pathway. *J Cell Biol* 102:24–36.

Dunn WA, Hubbard AL, and Aronson NN (1980) Low temperature selectively inhibits fusion between pinocytic vesicles and lysosomes during heterophagy of ^{125}I-asialofetuin by the perfused rat liver. *J Biol Chem* 255:5971–5978.

Dunn KW, McGraw TE, and Maxfield FR (1989) Iterative fractionation of recycling receptors from lysosomally destined ligands in an early sorting endosome. *J Cell Biol* 109:3303–3314.

Eilers U, Klumperman J, and Hauri HP (1989) Nocodazole, a microtubule-active drug, interferes with apical protein delivery in cultured intestinal epithelial cells (Caco-2). *J Cell Biol* 108:13–22.

Ernst SA, and Schreiber JH (1981) Ultrastructural localization of Na^+K^+-ATPase in rat and rabbit kidney medulla. *J Cell Biol* 91:803–813.

Fabre J, Rudhardt M, Blanchard DP, and Regamey C (1976) Persistence of sisomicin and gentamicin in renal cortex and medulla compared to other organs and serum of rats. *Kidney Int* 10:444–449.

Farquhar MG, Saito A, Kerjaschki D, and Orlando RA (1995) The Heyman nephritis antigenic complex: Megalin (gp330) and RAP. *J Am Soc Nephrol* 6:35–47.

Fath KR, and Burgess DR (1993) Golgi-derived vesicles from developing epithelial cells bind actin filaments and posses myosin-I as a cytoplasmically oriented peripheral membrane protein. *J Cell Biol* 120:117–127.

Feldman S, Wang M-Y, and Kaloyanides GJ (1982) Aminoglycosides induce a phospholipidosis in the renal cortex of the rat: An early manifestation of nephrotoxicity. *J Pharmacol Exp Ther* 220:514–520.

Feng Y, Press B, and Wandinger NA (1995) Rab7: An important regulator of late endocytic membrane traffic. *J Cell Biol* 131:1435–1452.

Fishman PH, and Orlandi PA (1994) Mechanism of the interaction of cholera toxin and *Escherichia coli* heat-labile enterotoxin with cells. *Trends Glycosci Glycotechnol* 6:387–406.

Flamenbaum W, Gehr M, and Gross M (1983) in *Acute Renal Failure* (Brenner BM, and Lazarus JM, Eds), pp. 269, Saunders, Philadelphia.

Ford DM, Dahl R, Lamp CA, and Molitoris BA (1994) Apically and basolaterally internalized aminoglycosides colocalize in LLC-PK_1 lysosomes and alter cell function. *Am J Physiol* 266:C52–C57.

Frommer JP, Senekjian HO, Babino H, and Weinman EJ (1983) Intratubular microinjection study of gentamicin transport in the rat. *Mineral Electro. Metab* 9:108–112.

Gabow PA, Kaehny WD, and Kelleher SP (1982) The spectrum of rhabdomyolysis. *Medicine* 61:141–152.

Ganeval D, Rabian C, Guerin V, Pertuiset N, Landais P, and Jungers P (1992) Treatment of multiple myeloma with renal involvement. *Adv Nephrol Necker Hospit* 21:347–370.

Geuze HJ, Slot JW, Strous GJAM, and Schwartz AL (1983) Intracellular site of asialoglycoprotein receptor-ligand uncoupling: Double immunoelectron microscopy during receptor-mediated endocytosis. *Cell* 32:277–287.

Geuze HJ, Stoorvogel W, Strous GJ, Slot J, Zijderhand-Bleekemolen J, and Mellman I (1989) Sorting of mannose-6-phosphate receptors and lysosomal membrane proteins in endocytic vesicles. *J Cell Biol* 107:2491–2501.

Ghetie MA, and Vitetta ES (1994) Recent developments in immunotoxin therapy. *Curr Opin Immunol* 6:707–714.

Gibey R, Dupond JL, Alber D, Leconte des Floris R, and Henry JC (1981) Predictive value of urinary *N*-acetyl-beta-D-glucosaminidase (NAG), alanine-aminopeptidase (AAP) and beta-2-microglobulin (beta 2M) in evaluating nephrotoxicity of gentamicin. *Clin Chim Acta* 116:25–34.

Gilbert DN, Houghton DC, Bennett WM, Plamp CE, Reger K, and Porter GA (1979) Reversibility of gentamicin nephrotoxicity in rats: Recovery during continuous drug administration. *Proc Soc Exp Biol Med* 160:99–103.

Giurgea-Marion L, Toubeau G, Laurent G, Heuson-Stiennon JA, and Tulkens PM (1986) Impairment of lysosome-pinocytotic vesicle fusion in rat kidney proximal tubules after treatment with gentamicin at low doses. *Toxicol Appl Pharmacol* 86:271–285.

Glickman JN, Conibear E, and Pearse BMF (1989) Specificity of binding of clathrin adaptors to signals on the mannose-6-phosphate/insulin-like growth factor II receptor. *EMBO J* 8:1041–1047.

Gluck SL (1993) The Vacuolar H^+-ATPases: Versatile proton pumps participating in constitutive and specialized functions of eucaryotic cells. *Int Rev Cytol* 137:105–137.

Goldstein JL, Brown MS, Anderson RGW, Russel DW, and Schneider WJ (1985) Receptor-mediated endocytosis: Concepts emerging from the LDL receptor system. *Ann Rev Cell Bio*l 1:1–39.

Goldberg RL, Smith RM, and Jarrett L (1987) Insulin and α_2-macroglobulin-methylamine undergo endocytosis by different mechanisms in rat adipocytes. I. Comparison of cell surface events. *J Cell Physiol* 133:203–212.

Gorvel J-P, Chavrier P, Zerial M, and Gruenberg J (1991) Rab5 controls early endosome fusion *in vitro*. *Cell* 64:915–925.

Gottlieb TA, Ivanov IE, Adesnik M, and Sabatini DD (1993) Actin microfilaments play a critical role in endocytosis at the apical but not the basolateral surface of polarized epithelial cells. *J Cell Biol* 120:695–710.

Goud B, and McCaffrey M (1991) Small GTP-binding proteins and their role in transport. *Curr Opin Cell Biol* 3:626–633.

Greene LE, and Eisenberg E (1990) Dissociation of clathrin from coated vesicles by the uncoating ATPase. *J Biol Chem* 265:6682–6687.

Griffiths G, and Gruenberg J (1991) The arguments for pre-existing early and late endosomes. *Trends Cell Biol* 1:5–9.

Griffiths G, Back R, and Marsh M (1989) A quantitative analysis of the endocytic pathway in baby hamster kidney cells. *J Cell Biol* 109:2703–2720.

Griffiths G, Hoflack B, Simons K, Mellman I, and Kornfeld S (1988) The mannose-6-phosphate receptor and the biogenesis of lysosomes. *Cell* 52:329–341.

Griffiths G, Matteoni R, Back R, and Hoflack B (1990) Characterization of the cation-independent mannose-6-phosphate receptor-enriched prelysosomal compartment in NRK cells. *J Cell Sci* 95:441–461.

Gruenberg J, and Howell KE (1989) Membrane traffic in endocytosis: Insights from cell-free assays. *Ann Rev Cell Biol* 5:453–481.

Gruenberg J, Griffiths G, and Howell KE (1989) Characterization of the early endosome and putative endocytic carrier vesicles *in vivo* and with an assay of vesicle fusion *in vitro*. *J Cell Biol* 108:1301–1316.

Gutmann EJ, Niles JL, McCluskey RT, and Brown D (1989) Colchicine-induced redistribution of endogenous apical membrane glycoprotein (gp 330) in kidney proximal tubule epithelium. *Am J Physiol* 257:C397-C407.

Hammerman MR (1985) Interaction of insulin with the renal proximal tubular cell. *Am J Physiol* 249:F1–F11.

Hammerman MR, and Gavin III Jr (1984) Insulin-stimulated phosphorylation and insulin binding in canine renal basolateral membranes. *Am J Physiol* 247:F408–F417.

Hammond TG, Majewski RR, Kaysen JH, Goda FO, Navar GL, Pnotillon F, and Verroust PJ (1997) Gentamicin inhibits rat renal cortical homotypic fusion: Role of megalin. *Am J Physiol* 272:F117-F23.

Harding C, Heuser J, and Stahl P (1983) Receptor-mediated endocytosis of transferrin and recycling of the transferrin receptor in rat reticulocytes. *J Cell Biol* 97:329–339.

Hebert SC, and Kriz W (1993) in *Diseases of the Kidney*, 5th edition (Schrier RW, and Gottschalk CW, Eds), pp. 3–63, Little, Brown and Co., Boston.

Helenius A, Mellman I, Wall D, and Hubbard A (1983) Endosomes. *Trends Biochem Sci* 8:245–250.

Hensley CB, Bradley ME, and Mircheff AK (1989) Parathyroid hormone-induced translocation of Na-H antiporters in rat proximal tubules. *Am J Physiol* 257:C637–C645.

Herskovitz JS, Burgess JS, Obar RA, and Vallee RB (1993) Effects of mutant rat dynamin on endocytosis. *J Cell Biol* 122:565–578.

Heuser J, and Anderson RGW (1989) Hypertonic media inhibit receptor-mediated endocytosis by blocking clathrin-coated pit formation. *J Cell Biol* 108:389–400.

Heuser JE, and Keen JH (1988) Deep-etch visualization of proteins involved in clathrin assembly. *J Cell Biol* 107:877–886.

Hicke L, and Riezman H (1996) Ubiquitination of a yeast plasma membrane receptor signals its' ligand-stimulated endocytosis. *Cell* 84:277–287.

Hill GS, Morel-Maroger L, Mery JP, Brouet JC, and Mignon F (1983) Renal lesions in multiple myeloma: Their relationship to associated protein abnormalities. *Am J Kid Dis* 2:423–438.

Hinshaw JE, and Schmid SL (1995) Dynamin self-assembles into rings suggesting a mechanism for coated vesicle budding. *Nature* 374:190–192.

Holm PK, Eker P, Sandvig K, and van Duers B (1995) Phorbol myristate acetate selectively stimulates apical endocytosis via protein kinase C in polarized MDCK cells. *Exp Cell Res* 217:157–168.

Hopkins CR (1982) Membrane recycling. *Ciba Foundation Symposium* 92:239–242.

Hopkins CR (1985) The appearance and internalization of transferrin receptors at the margins of spreading human tumor cells. *Cell* 40:199–208.

Hopkins CR, and Trowbridge IS (1983) Internalization and processing of transferrin and transferrin receptors in human carcinoma cells. *J Cell Biol* 97:508–521.

Hopkins CR, Gibson A, Shipman M, and Miller K (1990) Movement of internalized ligand-receptor complexes along a continuous endosomal reticulum. *Nature* 346:335–339.

Hostetler KY, and Hall LB (1982) Inhibition of kidney phospholipases A, and C by aminoglycoside antibiotics: Possible mechanism of aminoglycoside toxicity. *Proc Natl Acad Sci USA* 79:1663–1667.

Houghton DC, Hartnett M, Campbell-Boswell M, Porter G, and Bennett W (1976) A light and electron microscopic analysis of gentamicin nephrotoxicity in rats. *Am J Pathol* 82:589–612.

Houghton DC, Lee D, Gilbert DN, and Bennett WM (1986) Chronic gentamicin nephrotoxicity:Continued tubular injury with preserved glomerular filtration function. *Am J Pathol* 123:183–194.

Hughson EJ, and Hopkins CR (1990) Endocytic pathways in polarized CaCO-2 cells: Identification of an endosomal compartment accessible from both apical and basolateral surfaces. *J Cell Biol* 110:337–348.

Humes HD, Weinberg JM, and Knauss TC (1982) Clinical and pathophysiological aspects of aminoglycoside nephrotoxicity. *Am J Kid Dis* 2:5–29.

Jackman MR, Shurety W, Ellis JA, and Luzio JP (1994) Inhibition of apical but not basolateral endocytosis of ricin and folate in Caco-2 cells by cytochalasin D. *J Cell Sci* 107:2547–2556.

Jahraus A, Storrie B, Griffiths G, and Desjardins M (1994) Evidence for retrograde traffic between terminal lysosomes and the prelysosome/late endosome compartment. *J Cell Sci* 107:145–157.

Jing SQ, Spencer T, Miller K, Hopkins CR, and Trowbridge IS (1990) Role of the human transferrin receptor cytoplasmic domain in endocytosis: Localization of a specific signal sequence for internalization. *J Cell Biol* 110:283–294.

Josepovitz C, Forruggelia T, Levine R, Lane B, and Kaloyanides GJ (1985) Effects of netilmicin on phospholipids: Composition of subcellular fractions of rat renal cortex. *J Pharmacol Exp Ther* 235:810–819.

Just M, and Habermann E (1977) The renal handling of polybasic drugs. 2. *In vitro* studies with brush border and lysosomal preparations. *Nauyn Schmied Arch Pharmacol* 300:67–76.

Just M, Erdmann G, and Habermann T (1977) The renal handling of polybasic drugs: 1. Gentamicin and aprotinin in intact animals. *Nauyn Schmied Arch Pharmacol* 300:57–66.

Kaloyanides GJ (1984) Aminoglycosidcce-induced functional and biochemical defects in the renal cortex. *Fund Appl Toxicol* 4:930–943.

Kaloyanides GJ (1993) in *Diseases of the Kidney*, 5th Edition (Schrier RW, and Gottschalk CW, Eds), pp. 1131–1164, Little, Brown and Co., Boston.

Karin M, and Mintz B (1981) Receptor-mediated endocytosis in developmentally totipotent mouse teratocarcinoma stem cells. *J Biol Chem* 256:3245–3252.

Karmali MA (1989) Infection by verocytotoxin-producing *Escherichia coli*. *Clin Microbiol Rev* 2:15–38.

Kazantzis G (1979) Renal tubular dysfunction and abnormalities of calcium metabolism in cadmium workers. *Environ Health Perspect* 28:155–159.

Kerjaschki, D. and Farquhar, M.G. (1982) The pathogenic antigen of Heymann nephritis is a membrane glycoprotein of the renal proximal tubule brush border. *Proc Natl Acad Sci USA* 79:5557–5561, 1982.

Keusch GT, Donohue-Rolfe A, and Jacewicz M (1986) in *Microbial Lectins and Agglutinins Properties and Biological Activity*, pp. 271–295, John Wiley and Sons, New York.

Kielian M, and Jungerwirth S (1990) Mechanisms of enveloped virus entry into cells. *Mol Biol Med* 7:17–31.

Kindberg GM, Ford T, Blomhoff R, Rickwood D, and Berg T (1984) Separation of endocytic vesicles in nycodenz gradients. *Anal Biochem* 142:455–462.

King LS, and Agre P (1996) Pathophysiology of the aquaporin water channels. *Ann Rev Physiol* 58:619–648.

Kjellström T, Evrin PE, and Rahnster B (1977) Dose-response analysis of cadmium-induced tubular proteinuria:a study of urinary beta-2-microglobulin excretion among workers in a battery factory. *Environ Res* 13:303–317.

Klausner RD, Ashwell G, Renswoude JV, Harford JB, and Bridges KR (1983) Binding of apotransferrin to K562 Cells: Explanation of the transferrin cycle. *Proc Natl Acad Sci USA* 80:2263–2266.

Knauss TC, Weinberg JM, and Humes HD (1983) Alterations in renal cortical phospholipid content induced by gentamicin: Time course, specificity, and subcellular localization. *Am J Physiol* 244:F535–F546.

Koch CA, Anderson D, Moran MF, Ellis C, and Pawson T (1991) SH2 and SH3 domains: Elements that control interactions of cytoplasmic signaling proteins. *Science* 252:668–674.

Koenig JH, and Ikeda K (1989) Disappearance and reformation of synaptic vesicle membrane upon transmitter release observed under reversible blockage of membrane retrieval. *J Neurosci* 11:3844–3860.

Kornfield S, and Mellman I (1989) The biogenesis of lysosomes. *Ann Rev Cell Biol* 5:483–525.

Kosaka T, and Ikeda K (1983) Reversible blockage of membrane retrieval and endocytosis in the Garland cell of the temperature-sensitive mutant of *Drosophila melanogaster*, shibire. *J Cell Biol* 97:499–507.

Krupp MN, and Lane MD (1982) Evidence for different pathways for the degradation of insulin and insulin receptor in the chick liver cell. *J Biol Chem* 257:1372–1377.

Kuhn LC, and Kraehenbuhl JP (1979) The membrane receptor for polymeric immunoglobulin is structurally related to secretory component. Isolation and characterization of membrane secretory component from rabbit liver and mammary gland. *J Biol Chem* 254:11066–11071.

Lamaze C, Baba T, Redelmeier TE, and Schmid SL (1993) Recruitment of epidermal growth factor receptor and transferrin receptors into coated pits *in vitro*: Differing biochemical requirements. *Mol Biol Cell* 3:1181–1194.

Lamaze C, Fujimoto LM, Yin HL, and Schmid SL (1997) The actin cytoskeleton is required for receptor-mediated endocytosis in mammalian cells. *J Biol Chem* 272:20332–20335.

Larkin JM, Brown MS, Goldstein JL, and Anderson RGW (1983) Deletion of intracellular potassium arrests coated pit function and receptor-mediated endocytosis in fibroblasts. *Cell* 33:273–285.

Larkin JM, Donzell WC, and Anderson RGW (1985) Modulation of intracellular potassium and ATP: Effects on coated-pit function in fibroblasts and hepatocytes. *J Cell Physiol* 124:372–378.

Larkin JM, Donzell WC, and Anderson RGW (1986) Potassium dependent assembly of coated pits: New-coated pits form as planar clathrin lattices. *J Cell Biol* 103:2619–2627.

Laurent G, Carlier M-B, Rollman B, van Hoff F, and Tulkens P (1982) Mechanism of aminoglycoside-induced lysosomal phospholipidosis: *In vitro* and *in vivo* studies with gentamicin and amikacin. *Biochem Pharmacol* 31:3861–3870.

Lauwerys R, Bernard A, and Cardenas A (1992) Monitoring of early nephrotoxic effects of industrial chemicals. *Toxicol Lett* 64–65:33–42.

Levi DF, Williams RC Jr and Lindström FD (1968) Immunofluorescent studies of the myeloma kidney with special reference to light chain disease. *Am J Med* 44:922–933.

Lewis WH (1931) Pinocytosis. *Bull Johns Hopkins Hosp* 49:17–27.

Lindberg AA, Brown JE, Strømberg M, Wesling-Ryd M, Schultze JE, and Karlsson KA (1987) Identification of the carbohydrate receptor for Shiga toxin produced by *Shigella dysenteriae* type 1. *J Biol Chem* 262:1779–1785.

Liu J-P, and Robinson PJ (1995) Dynamin and endocytosis. *Endocrine Rev* 16:590–607.

Lord JM, Roberts LM, and Robertus JD (1994) Ricin: Structure, mode of action, and some current applications. *FASEB J* 8:201–208.

Luft FC, Patel V, Yum MN, Patel B, and Kleit SA (1975a) Experimental aminoglycoside nephrotoxicity. *J Lab Clin Med* 86:213–220.

Luft FC, Rankin LI, Sloan RS, and Yum MN (1975b) Recovery from aminoglycoside nephrotoxicity with continued drug administration. *Antimicrob Agents Chemother* 14:284–287.

Lundgren S, Hjälm G, Hellman P, Ek B, Juhlin C, Rastad J, Klareskog L, Åkerström G, and Rask L (1994) A protein involved in calcium sensing of the human parathyroid and placental cytotrophoblast cells belongs to the LDL-receptor protein superfamily. *Exp Cell Res* 212:344–350.

Maack T, Johnson V, Kau ST, Figueiredo J, and Sigulem D (1979) Renal filtration, transport, and metabolism of low-molecular-weight proteins. *Kidney Int* 16:251–271.

Maack T, Park CH, and Camargo MJF (1985), in *The kidney: Physiology and pathophysiology* (Seldon DW, and Giebisch G, Eds), pp. 1773–1803, Raven Press, New York.

Madsen KM, and Park CH (1987) Lysosome distribution and cathepsin B, and L activity along the rabbit proximal tubule. *Am J Physiol* 253:F1290–F1301.

Malhotra V, Orci L, Glick BS, Block MR, and Rothman JE (1988) Role of an *N*-ethylmaleimide-sensitive transport component in promoting the fusion of transport vesicles with cisternae of the Golgi stack. *Cell* 54:221–227.

Marples D, Frokiaer J, and Nielsen S (1999) Long-term regulation of aquaporins in the kidney. *Am J Physiol* 276:F331–F339.

Marsh M, and Helenius A (1989) Virus entry into animal cells. *Adv Virus Res* 36:107–151.

Marsh M, Bolzau E, and Helenius A (1983) Penetration of Semliki Forest virus from acidic prelysosomal vacuoles. *Cell* 32:931–940.

Marsh M, Griffiths G, Dean GE, Mellman I, and Helenius A (1986) Three-dimensional structure of endosomes in BHK-21 cells. *Proc Natl Acad Sci USA* 83:2899–2903.

Matlin, KS, and Simons K (1983) Reduced temperature prevents transfer of a membrane glycoprotein to the cell surface but does not prevent terminal glycosylation. *Cell* 34:233–243.

Matter K, and Mellman I (1994) Mechanisms of cell polarity:Sorting and transport in epithelial cells. *Curr Opin Cell Biol* 6:545–554.

Maunsbach AB (1966) Absorption of ^{125}I-labeled homologous albumin by rat kidney proximal tubule cells. A study of microperfused single proximal tubules by electron microscopic autoradiography and histochemistry. *J Ultrastruct Res* 15:197–241.

Maunsbach AB, and Christensen EI (1992) in *Handbook of Physiology: Renal Physiology*, 2nd ed., Windhager EE (Ed.), pp. 41–107, Oxford University Press, Washington, D.C.

Maxfield FR, and Yamashiro JD (1991) in *Trafficking of Membrane Proteins*, Steer JC, and Harford J (Ed.), pp. 103–165, Academic Press, New York.

Mayor S, Presley JF, and Maxfield FR (1993) Sorting of membrane components from endosomes and subsequent recycling to the cell surface occurs by a bulk flow process. *J Cell Biol* 121:1257–1269.

McGraw TE, and Maxfield FR (1990) Human transferrin receptor internalization is partially dependent upon an aromatic amino acid on the cytoplasmic domain. *Cell Regul* 1:369–377.

McKanna JA, Haigler HT, and Cohen S (1979) Hormone receptor topology and dynamics:morphological analysis using ferritin-labeled epidermal growth factor. *Proc Natl Acad Sci USA* 76:5689–5693.

Mellman I (1996) Endocytosis and molecular sorting. *Ann Rev Cell Biol* 12:575–625.

Mellman I, and Plunter H (1984) Internalization and degradation of macrophage Fc receptors bound to polyvalent immune complex. *J Cell Biol* 98:1170–1177.

Mellman I, Fuchs R, and Helenius A (1986) Acidification of the endocytic and exocytic pathways. *Ann Rev Biochem* 55:663–700.

Metchnikoff E (1893) *Lectures on the comparative pathology of inflammation.* Paul, Kegan, Trench, & Trabner, London.

Miettinen HM, Matter K, Hunziker W, Rose JK, and Mellman I (1992) Fc receptor endocytosis is controlled by a cytoplasmic domain determinant that actively prevents coated pit localization. *J Cell Biol* 116:875–888.

Miettinen HM, Rose JK, and Mellman I (1989) Fc receptor isoforms exhibit distinct abilities for coated pit localization as a result of cytoplasmic domain heterogeneity. *Cell* 58:317–327.

Milton A, Hellfritzsch M, and Christensen EI (1988) Basolateral binding and uptake of ^{125}I-insulin in proximal tubule cells after peritubular extraction in the avian kidney. *Diabetes Res* 7:189–195.

Moestrup SK (1994) The α_2-macroglobulin receptor and epithelial glycoprotein-330: Two giant receptors mediating endocytosis of multiple ligands. *Biochim Biophys Acta* 1197:197–213.

Moestrup SK, Birn H, Fisher PB, Petersen CM, Verroust PJ, Sim RB, Christensen EI, and Nexø E (1996) Megalin-mediated endocytosis of transcobalamin-vitamin-B12 complexes suggests a role of the receptor in vitamin-B12 homeostasis. *Proc Natl Acad Sci USA* 93:8612–8617.

Moestrup SK, Cui S, Vorum H, Bregengård C, Bjørn E, Norris K, Gliemann J, and Christensen EI (1995) Evidence that epithelial glycoprotein 330/megalin mediates uptake of polybasic drugs. *J Clin Invest* 96:1404–1413.

Molitoris BA, Meyer C, Dahl R, and Geerdes A (1993) Mechanism of ischemia-enhanced aminoglycoside binding and uptake by proximal tubule cells. *Am J Physiol* 264:F907–F916.

Moriyama T, Nakahama H, Fukuhara Y, Horio M, Yanase M, Kamada T, Kanashiro M, and Miyake Y (1989) Decrease in the fluidity of brush border membranes vesicles induced by gentamicin. *Biochem Pharmacol* 38:1169–1174.

Mostov KE (1995) Regulation of protein traffic in polarized epithelial cells. *Histol Histopathol* 10:423–431.

Mostov KE, and Blobel G (1982) A transmembrane precursor of secretory component. The receptor for transcellular transport of polymeric immunoglobulins. *J Biol Chem* 257:11816–11821.

Moya M, Dautry-Varsat A, Goud B, Louvard D, and Boquet P (1985) Inhibition of coated pit formation in Hep2 cells blocks the cytotoxicity of diphtheria toxin but not that of ricin toxin. *J Cell Biol* 101:548–559.

Mueller SC, and Hubbard AL (1986) Receptor-mediated endocytosis of asialoglycoproteins by rat hepatocytes: Receptor-positive and receptor-negative endosomes. *J Cell Biol* 102:932–942.

Mukherjee S, Ghosh RN, and Maxfield FR (1996) Endocytosis. *Physiol Rev* 77:759–803.

Nelson N (1987) The vacuolar proton-ATPase of eucaryotic cells. *Bioessays* 7:251–254.

Neuhaus OW (1986) Renal absorption of low molecular weight proteins in adult male rats:alpha 2u-globulin. *Proc Soc Exp Biol Med* 182:531–539.

Newton DL, Wales R, Richardson PT, Walbridge S, Sax SK, Ackerman EJ, Roberts LM, Lord JM, and Youle R (1995) Cell surface and intracellular functions for ricin galactose binding. *J Biol Chem* 11917–11922.

Nielsen JT, and Christensen EI (1985) Basolateral endocytosis of protein in isolated perfused proximal tubules. *Kidney Int* 27:39–45.

Nielsen JT, Nielsen S, and Christensen EI (1985) Transtubular transport of proteins in rabbit proximal tubules. *J Ultrastruct Res* 92:133–145.

Nielsen S, Nexø E, and Christensen EI (1989) Absorption of epidermal growth factor and insulin in rabbit renal proximal tubules. *Am J Physiol* 256:E55–E63.

Nielsen S, Nielsen JT, and Christensen EI (1987) Luminal and basolateral uptake of insulin in isolated, perfused, proximal tubules. *Am J Physiol* 253:F857–F867.

Nilsson J, Thyberg J, Heldin C-H, Westermark B, and Wasteson Å (1983) Surface binding and internalization of platelet-derived growth factor in human fibroblasts. *Proc Natl Acad Sci USA* 80:5592–5596.

Novick P, and Brennwald P (1993) Friends and family: The role of the Rab GTPases in vesicular traffic. *Cell* 75:597–601.

O'Brien AD, Tesh VL, Donohue-Rolfe A, Jackson MP, Olsnes S, Sandvig K, Lindberg AA, and Keusch GT (1992) in *Pathogenesis of Shigellosis*, Sansonetti PJ (Ed.), pp. 65–94, Springer-Verlag, Berlin.

Octave JN, Schneider Y, Crichton RR, and Trouet A (1981) Transferrin uptake by cultured rat embryo fibroblasts. The influence of lysosomotropic agents, iron chelators, and colchicine on the uptake of iron and transferrin. *Eur J Biochem* 115:611–618.

Oken MM (1984) Multiple myeloma. *Med Clin North Am* 68:757–787.

Olbricht CJ, Fink M, and Gutjahr E (1991) Alterations in lysosomal enzymes of the proximal tubule in gentamicin nephrotoxicity. *Kidney Int* 39:639–646.

Olsnes S, Kozlov JV, van Duers B, and Sandvig K (1991) Bacterial protein toxins acting on intracellular targets. *Semin. Cell Biol* 2:7–14.

Orci L, Malhotra V, Amherdt M, SerafiniT, and Rothman JE (1989) Dissection of a single round of vesicular transport: Sequential intermediates for intercisternal movement in the Golgi stack. *Cell* 58:357–368.

Ottosen PD, and Maunsbach AB (1973) Transport of peroxidase in flounder kidney tubules studied by electron microscope histochemistry. *Kidney Int* 3:315–326.

Parton RG, Prydz K, Bomsel M, Simons K, and Griffiths G (1989) Meeting of the apical and basolateral endocytic pathways of the Madin-Darby canine kidney cell in late endosomes. *J Cell Biol* 109:3259–3272.

Pastoriza-Munoz E, Bowman RL, and Kaloyanides GJ (1979) Renal tubular transport of gentamicin in the rat. *Kidney Int* 16:440–450.

Pastoriza-Munoz E, Timmerman D, and Kalyonides GJ (1984) Renal transport of netilmicin in the rat. *J Pharmacol Exp Ther* 228:65–72.

Pavelka M, Ellinger A, and Gangl A (1983) Effect of colchicine on rat small intestinal absorptive cells. I. Formation of basolateral microvillus borders. *J Ultrastr Res* 85:249–259.

Pearse BMF (1988) Receptors compete for adaptors found in plasma membrane coated pits. *EMBO J* 7:3331–3336.

Pearse BMF, and Crowther RA (1987) Structure and assembly of coated vesicles. *Ann Rev Biophys Biophys Chem* 16:49–68.

Pearse BMF, and Robinson MS (1990) Clathrin, adaptors, and sorting. *Ann Rev Cell Biol* 6:151–171.

Pfeffer SR (1994) Rab GTPases:master regulators of membrane trafficking. *Curr Opin Cell Biol* 6:522–526.

Pfeiffer S, Fuller SD, and Simons K (1985) Intracellular sorting and basolateral appearance of the G protein of vesicular stomatitis virus in MDCK cells. *J Cell Biol* 101:470–476.

Phillips ME, and Taylor A (1989) Effect of nocodazole on the water channel permeability response to vasopressin in rabbit collecting tubules *in vitro*. *J Physiol Lond* 411:529–544.

Preuss HG, Hammack WJ, and Murdaugh HV (1968) The effect of Bence Jones protein in the *in vitro* function of rabbit renal cortex. *Nephron* 5:210–216.

Rabkin R, Yagil C, and Frank B (1989) Basolateral and apical binding, internalization, and degradation of insulin by cultured kidney epithelial cells. *Am J Physiol* 257:E895-E902.

Ramsammy LS, Josepovitz C, and Kaloyanides GJ (1988) Gentamicin inhibits agonist stimulation of the phosphatidylinositol cascade in primary cultures of rabbit proximal tubular cells and in rat renal cortex. *J Pharmacol Exp Ther* 247:989–996.

Ramsammy LS, Josepovitz C, Lane B, and Kaloyanides GJ (1989) Effect of gentamicin on phospholipid metabolism in cultured rabbit proximal tubular cells. *Am J Physiol* 256:C204-C213.

Rassow J, Voos W, and Pfanner N (1995) Trends *Cell Biol* 5:207–212.

Rindler MJ, Ivanov IE, Plesken H, Rodriguez-Boula E, and Sabatini DD (1984) Viral glycoproteins destined for apical or basolateral plasma membrane domains transverse the same Golgi apparatus during their intracellular transport in Madin-Darby canine kidney cells. *J Cell Biol* 98:1304–1319.

Robenek H, and Hesz A (1983) Dynamics of low density lipoprotein recetors in the plasma membrane of cultured human skin fibroblasts as visualized by colloidal gold in conjunction with surface replicas. *Eur. J Cell Biol* 31:275–282.

Robinson MS (1994) The role of clathrin, adaptors, and dynamin in endocytosis. *Curr Opin Cell Biol* 6:538–544.

Robinson PJ, Sontag J-M, Liu J-P, Fykse EM, Slaughter C, McMahon H, and Sudhof TC (1993) Dynamin GTPase regulated by protein kinase C phosphorylation in nerve terminals. *Nature* 365:163–166.

Rodriguez-Boulan E, and Nelson WN (1989) Morphogenesis of the polarized epithelial cell phenotype. *Science* 245:718–725.

Rogers SA, and Hammerman MR (1989) Growth hormone activates phospholipase C in proximal tubular basolateral membranes from canine kidney. *Proc Natl Acad Sci USA* 86:6363–6366.

Rothman JE (1994) Mechanisms of intracellular protein transport. *Nature* 372:55–63.

Salminen A, and Novick PJ (1987) A ras-like protein is required for a post-Golgi event in yeast secretion. *Cell* 49:527–538.

Sanders PW, Herrera GA, Chen A, Booker BB, and Galla JH (1988) Differential nephrotoxicity of low molecular weight proteins including Bence Jones proteins in the perfused rat nephron *in vivo*. *J Clin Invest* 82:2086–2096.

Sandvig K, and Olsnes S (1982) Entry of the toxic proteins abrin, modeccin, ricin, and diphtheria toxin into cells. II. Effect of pH, metabolic inhibitors, and ionophores and evidence for toxin penetration from endocytic vesicles. *J Biol Chem* 257:7504–7513.

Sandvig K, and Olsnes S (1991) in *Sourcebook of Bacterial Protein Toxins* (Alouf JE, and Freer JH, Eds) pp. 57–73, Academic Press, London.

Sandvig K, and van Duers B (1991) Endocytosis without clathrin (a minireview). *Cell Biol Int Rep* 15:3–8.

Sandvig K, and van Duers B (1994) Endocytosis and intracellular sorting of ricin and Shiga toxin. *FEBS Lett* 346:99–102.

Sandvig K, and van Duers B (1996) Endocytosis, intracellular transport, and cytotoxic action of Shiga toxin and ricin. *Physiol Rev* 76:949–966.

Sandvig K, Garred Ø, Holm PK, and van Duers B (1992a) Endocytosis and intracellular transport of protein toxins. *Biochem Soc Trans* 21:707–710a.

Sandvig K, Garred Ø, Prydz K, Kozzlov JV, Hansen SH, and van Duers B (19992b) Retrograde transport of endocytosed Shiga toxin to the endoplasmic reticulum. *Nature Lond* 358:510–512.

Sandvig K, Olsnes S, Petersen OW, and van Duers B (1987) Acidification of the cytosol inhibits endocytosis from coated pits. *J Cell Biol* 105:679–689.

Sandvig K, Olsnes S, Petersen OW, and van Duers B (1988) Inhibition of endocytosis from coated pits by acidification of the cytosol. *J Cell Biochem* 36:73–81.

Sandvig K, Olsnes S, Brown JE, Petersen OW, and van Duers B (1989a) Endocytosis from coated pits of Shiga toxin: A glycolipid-binding protein from Shigella dysenteriae 1. *J Cell Biol* 108:1331–1343.

Sandvig K, Olsnes S, Petersen OW, and van Duers B (1989b) Control of coated-pit formation by cytoplasmic pH. *Meth Cell Biol* 32:365–382.

Sandvig K, Prydz K, Ryd M, and van Duers B (1991) Endocytosis and intracellular transport of the glycolipid-binding ligand Shiga toxin in polarized MDCK cells. *J Cell Biol* 113:553–562.

Sandvig K, Ryd M, Garred Ø, Schweda E, Holm PK, and van Duers B (1994) Retrograde transport from the Golgi complex to the ER of both Shiga toxin and the nontoxic Shiga B-fragment is regulated by butyric acid and cAMP. *J Cell Biol* 126:53–64.

Sastrasinh M, Knauss TC, Weinberg JM, and Humes HD (1982) Identification of the aminoglycoside binding site in rat renal brush border membranes. *J Pharmcol Exp Ther* 222:350–358.

Schentag JJ (1983) Specificity of renal tubular damage criteria for aminoglycoside nephrotoxicity in critically ill patients. *J. Clin. Pharmacol.* 23:473–483.

Schentag JJ, and Jusko WJ (1977) Renal clearance and tissue accumulation of gentamicin. *Clin Pharmacol Ther* 22:364–370.

Schentag JJ, Jusko WJ, Plaut ME, Cumbo TJ, Vance JW, and Abrutyn E (1977) Tissue persistence of gentamicin in man. *JAMA* 238:327–329.

Schentag JJ, Sutfin TA, Plaut ME, and Jusko WJ (1978) Early detection of aminoglycoside nephrotoxicity with urinary beta-2-microglobulin. *J Med* 9:201–210.

Schlessinger J (1980) The mechanism and role of hormone-induced clustering of membrane receptors. *Trends Biochem Sci* 5:210–214.

Schlossman DM, Schmid SL, Braell WA, and Rothman JE (1984) A role for clathrin light chains in the recognition of clathrin cages by 'uncoating ATPase'. *J Cell Biol* 99:723–733.

Schmid SL (1992) The mechanism of receptor-mediated endocytosis: More questions than answers. *BioEssays* 14:589–596.

Schmid SL (1997) Clathrin-coated vesicle formation and protein sorting: An integrated process. *Ann Rev Biochem* 66:511–548.

Schmid SL, and Damke H (1995) Coated vesicles: A diversity of form and function. *FASEB J* 9:1445–1453.

Schmid SL, and Smythe E (1991) Stage-specific assays for coated pit formation and coated vesicle budding *in vitro*. *J Cell Biol* 114:869–880.

Schmid SL, Fuchs R, Male P, and Mellman I (1988) Two distinct subpopulations of endosomes involved in membrane recycling and transport to the lysosomes. *Cell* 52:73–83.

Schneider YJ, Tulkens P, de Duve C, and Trouet A (1979a) Fate of plasma membrane during endocytosis. I. Uptake and processing of anti-plasma membrane and control immunoglobulins by cultured fibroblasts. *J Cell Biol* 82:449–465.

Schneider YJ, Tulkens P, de Duve C, and Trouet A (1979b) Fate of plasma membrane during endocytosis. II. Evidence for recycling (shuttle) of plasma membrane constituents. *J Cell Biol* 82:466–474.

Senekjian HO, Knight TF, and Weinman EJ (1981) Micropuncture study of the handling of gentamicin by the rat kidney. *Kidney Int* 19:416–423.

Sheth AU, Senekjian HO, Babino H, Knight TF, and Weinman EJ (1981) Renal handling of gentamicin by the Munich-Wistar rat. *Am J Physiol* 241:F645–F648.

Shpetner HS, and Vallee RB (1989) Identification of dynamin, a novel mechanochemical enzyme that mediates interactions between microtubules. *Cell* 59:421–432.

Silverblatt FJ, and Kuehn C (1979) Autoradiography of gentamicin uptake by the rat proximal tubule cell. Kidney Int 15:335–345.

Simons K, and Wandinger-Ness A (1990) Polarized sorting in epithelia. *Cell* 62:207–210.

Simons K, and Zerial M (1993) Rab proteins and the road maps for intracellular transport. *Neuron* 11:789–799.

Soldati T, Riederer MA, and Pfeffer SR (1993) Rab GDI: A solubilizing and recycling factor for Rab9 protein. *Molec Biol Cell* 4:425–434.

Soldati T, Shapiro AD, Dirac-Svejstrup AB, and Pfeffer SR (1994) Membrane targeting of the small GTPase Rab9 is accompanied by nucleotide exchange. *Nature* 369:76–78.

Söllner T, Bennett MK, Whiteheart SW, Scheller RH, and Rothman JE (1993a) A protein assembly-disassembly pathway *in vitro* that may correspond to sequential steps of vesicle docking, activation, and fusion. *Cell* 75:409–418.

Söllner T, Whiteheart SW, Brunner M, Erdjument-Bromage H, Geromanos S, Tempst P, and Rothman JE (1993b) SNAP receptors implicated in vesicle targeting and fusion. *Nature* 362:318–423.

Steinman RM, Mellman IS, Muller WA, and Cohm ZA (1983) Endocytosis and the recycling of plasma membrane. *J Cell Biol* 96:1–27.

Stoorvogel W, Strous GJ, Geuze HJ, Oorschot V, and Schwartz AL (1991) Late endosomes derive from early endosomes by maturation. *Cell* 65:417–427.

Storrie B, Pool RR Jr, Sachdeva M, Maurey KM, and Oliver C (1984) Evidence for both prelysosomal and lysosomal intermediates in endocytic pathways. *J Cell Biol* 98:108–115.

Straus W (1964) Cytochemical observations on the relationship between lysosomes and phagosomes in kidney and liver by combined staining for acid phosphatase in intravenously injected horseradish peroxidase. *J Cell Biol* 20:497–507.

Strom M, Vollmer P, Tan TJ, and Gallwitz D (1993) A yeast GTPase-activating protein that interacts specifically with a member of the Ypt/Rab family. *Nature* 361:736–739.

Südhoff TC, de Camilli, P, Niemann H, and Jahn R (1993) Membrane fusion machinery: Insights from synaptic proteins. *Cell* 75:1–4.

Sundin DP, Meyer C, Dahl R, Geerdes A, Sandoval R, and Molitoris BA (1997) Cellular mechanism of aminoglycoside tolerance in long-term gentamicin treatment. *Am J Physiol* 272:C1309-C1318.

Sundin DP, Sandoval R, and Molitoris BA (2001) Gentamicin inhibits renal protein and phospholipid metabolism in the rat: Implications involving intracellular trafficking. *J Am Soc Nephrol* 12:114–23.

Takai Y, Kaibuchi K, Kikuchi A, and Kawata M (1992) Small GTP-binding proteins. *Int Rev Cytol* 133:187–231.

Takel K, McPherson PS, Schmid SL, and de Camilli P (1995) Tubular membrane invaginations coated by dynamin rings are induced by GTP- gamma S in nerve terminals. *Nature* 374:186–190.

Theuer CP, Buchner J, Fitzgerald D, and Pasten I (1993) The N-terminal region of the 37-kDa translocated fragment of Pseudomonas exotoxin A aborts translocation by promoting its own export after microsomal membrane insertion. *Proc. Natl Acad. Sci. USA* 90:7774–7778.

Tisher CC, and Kokko JP (1974) Relationship between peritubular oncotic pressure gradients and morphology in isolated proximal tubules. *Kidney Int* 6:146–156.

Tjelle TE, Brech A, Juvet LK, Griffiths G, and Berg TJ (1996) Isolation and characterization of early endosomes, late endosomes and terminal lysosomes: their role in protein degradation. *J Cell Sci* 109:2905–2914.

Tojo A, and Endou H (1992) Intrarenal handling of proteins in rats using fractional micropuncture technique. *Am J Physiol* 263(4 Pt 2):F601-F606.

Trowbridge IS, Collawn JF, and Hopkins CR (1993) Signal-dependent membrane protein trafficking in the endocytic pathway. *Ann Rev Cell Biol* 9:129–161.

Tschuchiya K (1976) Proteinuria of cadmium workers. *J Occup Med* 18:463–470.

Tulkens PM (1986) Experimental studies on nephrotoxicity of aminoglycosides at low doses. *Am J Med* 80:105–114.

Ullian ME, Hammond WS, Alfrey AC, Schultze A, and Molitoris BA (1989) Beta-2-microglobulin-associated amyloidosis in chronic hemodialysis patient with carpal tunnel syndrome. *Medicine* 68:107–115.

Ullrich O, Horiuchi H, Bucci C, and Zerial M (1994) Membrane association of Rab5 mediated by GDP-dissociation inhibitor and accompanied by GDP/GTP exchange. *Nature* 368:157–160.

Ullrich O, Stenmark H, Alexandrov K, Huber LA, Kaibuchi K, Sasaki T, Takai Y, and Zerial M (1993) Rab GDP dissociation inhibitor as a general regulator for the membrane association of Rab proteins. *J Biol Chem* 268:18143–18150.

Valencia A, Chardin P, Wittinghofer A, and Sander C (1991) The *ras* protein family: Evolutionary tree and role of conserved amino acids. *Biochemistry* 30:4637–4648.

van der Bliek AM, Redelmeier TE, Damke H, Tisdale EJ, Meyerowitz EM, and Schmid S (1993) Mutations in human dynamin block an intermediate stage in coated vesicle function. *J Cell Biol* 122:553–563.

van Duers B, Petersen OW, Olsnes S, and Sandvig K (1987) Delivery of internalized ricin from endosomes to cisternal Golgi elements is a discontinuous, temperature-sensitive process. *Exp Cell Res* 171:137–152.

van Duers B, Petersen OW, Olsnes S, and Sandvig K (1989) The ways of endocytosis. *Int Rev Cytol* 117:131–177.

van Duers B, Sandvig K, Petersen OW, Olsnes S, Simons K, and Griffiths G (1988) Estimation of the amount of internalized ricin that reaches the *trans*-Golgi network. *J Cell Biol* 106:253–267.

Venkatachalam MA, and Karnovsky MJ (1972) Extravascular protein in the kidney. *Lab Invest* 27:435–444.

Via DP, Willingham MC, Pastan I, Gotto AM Jr and Smith LC (1982) Co-clustering and internalization of low density lipoproteins and α_2-macroglobulin in human skin fibroblasts. *Exp Cell Res* 141:15–22.

Vitetta ES, Thorpe PE, and Uhr JW (1993) Immunotoxins: Magic bullets or misguided missiles? *Trends Pharmacol Sci* 14:148–154.

von Bonsdorff C-H, Fuller SD, and Simons K (1985) Apical and basolateral endocytosis in Madin-Darby canine kidney (MDCK) cells grown on nitrocellulose filters. *EMBO J* 4:2781–2792.

von Mollard GF, Stahl B, Li C, Südhof TC, and Jahn R (1994) Rab proteins in regulated exocytosis. *Trends Biochem Sci* 19:164–168.

Wall DA, and Hubbard AL (1985) Receptor-mediated endocytosis of asialoglycoproteins by rat liver hepatocytes: Biochemical characterization of the endosomal compartments. *J Cell Biol* 101:2104–2112.

Wall DA, Wilson G, and Hubbard AL (1980) The galactose specific recognition system of mammalian liver: The route of ligand internalization in rat hepatocytes. *Cell* 21:79–93.

Warnock DE, and Schmid SL (1996) Dynamin GTPase, a force-generating molecular switch. *BioEssays* 18:885–893.

Weeden RP, Batuman V, Cheeks C, Marquet E, and Sobel H (1983) Transport of gentamicin in rat proximal tubule. *Lab Invest* 48:212–223.

Wilde A, and Brodsky FM (1996) *In vivo* phosphorylation of adaptors regulates their interaction with clathrin. *J Cell Biol* 135:635–646.

Wileman T, Harding C, and Stahl P (1985) Receptor-mediated endocytosis. *Biochem J* 232:1–14.

Wilhelm JM, Jessop JJ, and Pettit SE (1978) Aminoglycoside antibiotics and eukaryotic synthesis: Stimulation of errors in the translation of natural messengers in extracts of cultured human cells. *Biochemistry* 17:1149–1153.

Wilson DW, Whiteheart SW, Orci L, and Rothman JE (1991) Intracellular membrane fusion. *Trends Biochem Sci* 16:334–337.

Yokota S, and Kato K (1988a) Involvement of cathepsins B, and H in lysosomal degradation of horseradish peroxidase endocytosed by the proximal tubule cells of the rat kidney: I. Histochemical and immunohistochemical studies. *Anat Rec* 221:783–790.

Yokota S, and Kato K (1988b) Involvement of cathepsins B, and H in lysosomal degradation of horseradish peroxidase endocytosed by the proximal tubule cells of the rat kidney: II. Immunocytochemical studies using protein A-gold technique applied to conventional and serial sections. *Anat Rec* 221:791–801.

Zager RA, Teuben EJ, and Adler S (1987) Low molecular weight proteinuria exacerbates experimental ischemic renal injury. *Lab Invest* 56:180–188.

Zaremba S, and Keen JH (1983) Assembly polypeptides from coated vesicles mediate reassembly of unique clathrin coats. *J Cell Biol* 97:1339–1347.

Zerial M, and Stenmark H (1993) Rab GTPases in vesicular transport. *Curr Opin Cell Biol* 5:613–620.

11

Role of Cellular Energetics in Nephrotoxicity

Lawrence H. Lash

RENAL ENERGY METABOLISM AND TISSUE FUNCTION

Energy Requirements for Renal Function

To assess the importance and role of cellular energetics in processes leading to chemically or pathologically induced nephrotoxicity, it is first necessary to consider the role of energy metabolism in maintaining normal renal function. Energy-requiring processes in the kidneys can be divided into three categories: glomerular filtration, membrane transport processes, and biosynthetic reactions. For the tubular epithelial cells, only the last two processes are relevant. Membrane transport includes primary and secondary active transporters on renal brush-border and basolateral plasma membranes for inorganic ions and metabolites, and encompasses both reabsorptive and secretory processes. The energy-dependent, biosynthetic reactions include the biosynthesis of macromolecules (i.e., protein, DNA, RNA, lipid), the biosynthesis of specialized products, such as erythropoietin and vitamin D, and drug metabolism reactions. All of the above mentioned processes are either directly or indirectly dependent on an adequate supply of ATP (Soltoff, 1986), and are illustrated schematically in Figure 11.1. The large array of transport and biosynthetic processes in the renal tubular epithelium

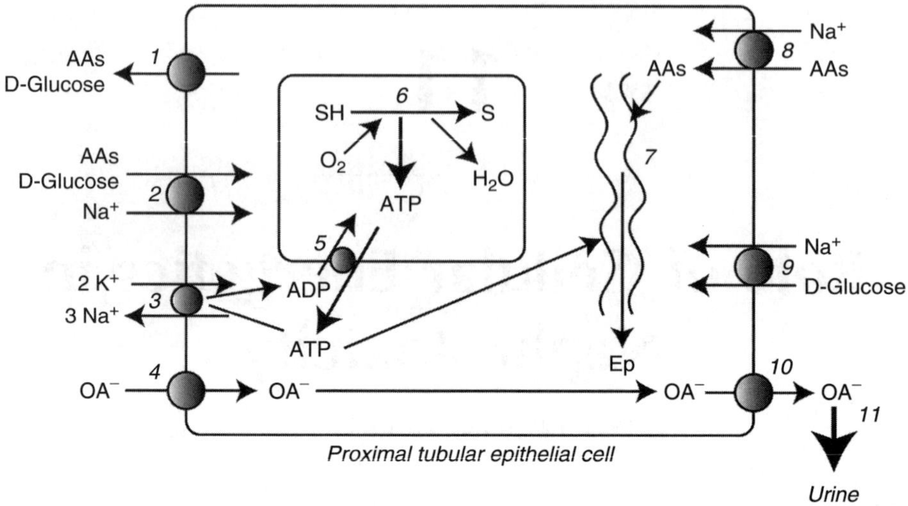

Figure 11.1 Energy-dependent processes in renal proximal tubular cells. Generalized scheme of some of the major energy-requiring and energy-yielding processes in the renal proximal tubule cell: (1) reabsorption of amino acids (AAs) and D-glucose by basolateral efflux; (2) Na^+-dependent uptake of AAs and D-glucose; (3) ($Na^+ + K^+$)-stimulated ATPase; (4) organic anion (OA^-) uptake across the basolateral membrane; (5) ATP/ADP translocase; (6) substrate-dependent ATP generation through citric acid cycle; (7) erythropoietin (Ep) synthesis in rough endoplasmic reticulum; (8) Na^+-coupled AA uptake at brush-border membrane; (9) Na^+-coupled D-glucose uptake at brush-border membrane; (10) OA^- efflux into lumen; (11) OA^- excretion into urine.

that require ATP implies that maintenance of an adequate supply of ATP is essential for proper renal function and that conditions that deplete ATP or otherwise interfere with its production or utilization will lead to a loss of renal function and toxicity.

Inasmuch as the mitochondria are the primary sites within the renal cell where ATP is produced, these organelles are an important target for nephrotoxicants. For the mitochondria, or any organelle for that matter, to be a target, two criteria should be met. First, there must be target sites at which the chemicals or metabolites of the chemicals can react to produce some change in the organelle. Second, the organelle should have an appropriate response that leads to toxicity. With respect to the first criterion, the mitochondria are replete with potential target sites whose modification can lead to mitochondrial dysfunction and toxicity. For example, a large number of critical proteins that are involved in membrane transport or the electron transport chain contain cysteinyl residues with critical sulfhydryl groups that must be maintained in the reduced form to maintain structure or function (Jones and Lash, 1993; Lash and Jones, 1996). These sulfhydryl-containing proteins include the various dehydrogenases, acylases, transport ATPases, and other transporters, such as the dicarboxylate carrier. With respect to the second criterion, several well-studied responses that lead to organelle dysfunction and toxicity, which occur after alteration of mitochondrial redox status or depletion of critical intermediates, are known. These responses include organelle swelling, ATP depletion with the consequent inhibition of energy-dependent processes (e.g., biosynthetic reactions, drug metabolism reactions, active transport), induction of the mitochondrial permeability transition, and initiation of apoptosis.

An illustration of how mitochondrial respiratory function and cellular energy requirements are coordinated in renal cells is shown by the data in Figure 11.2. Suspensions of freshly isolated renal cortical cells from rats that are incubated with substrates that deplete or enhance cellular ATP, or inhibitors or uncouplers that alter cellular utilization of ATP, lead to a corresponding alteration in mitochondrial oxygen consumption (QO_2). Thus, depletion of cellular ATP by addition of fructose or addition of adenosine as a precursor for ATP, which stimulates ATP synthesis, stimulates QO_2, whereas inhibition of the Na^+/K^+-ATPase with ouabain decreases QO_2, and uncoupling of the Na^+/K^+-ATPase with nystatin depletes cellular ATP content and stimulates QO_2. These effects on cellular ATP content and respiration demonstrate a coordinated, regulated response such that mitochondrial function is upregulated or downregulated in response to the need for ATP. This coordinated regulation of ATP-generating and ATP-requiring processes has become a guiding principle in renal biochemistry (Kiil, 1977; Mandel, 1986; Soltoff, 1986).

Many of the responses of renal mitochondria to toxicant exposure and the processes that ensue (as outlined above) will be discussed in other chapters in this book in some detail. The focus here will be to highlight the role of mitochondria in the renal cellular and organ response to chemical toxicants and pathological conditions. Examples of toxicants that specifically target renal mitochondria and pathological conditions that strongly affect renal mitochondria will also be given, to illustrate the scope and significance of the mitochondria and cellular energetics in numerous forms of nephrotoxicity.

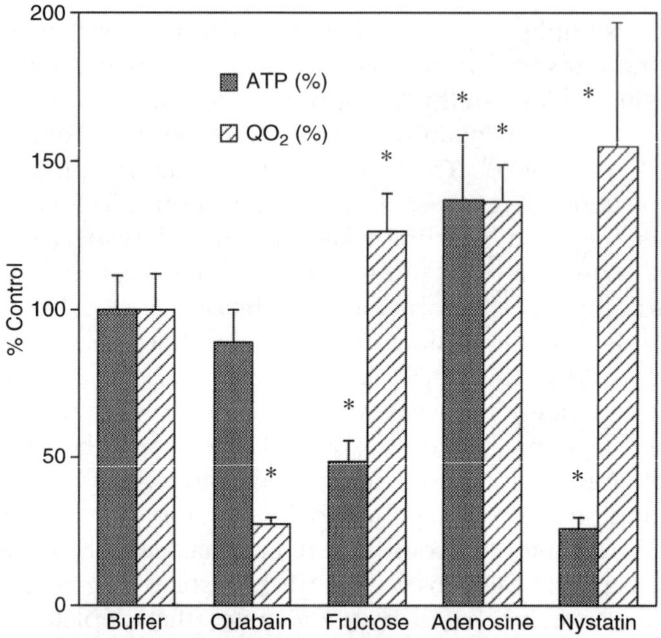

Figure 11.2 Manipulation of mitochondrial function in renal PT cells with metabolic inhibitors. Isolated proximal tubule cells (1×10^6/ml) were isolated from male Sprague-Dawley rats and were incubated for 15 min with either buffer (= control), 0.1 m*M* ouabain, 10 m*M* fructose, 10 m*M* adenosine, or 25 mg nystatin/ml for 15 min. Subsequently, aliquots of cells were taken for determination of cellular ATP contents (in perchloric acid extracts by high-performance liquid chromatography, HPLC) or cellular oxygen consumption rates (QO_2; in a Gilson 5/6H Oxygraph). Results are means ± SEM of measurements from four separate cell preparations. Source: L.H. Lash and M.W. Anders, unpublished data.

Nephron Heterogeneity and Cellular Energetics

Although most tissues, even those that are seemingly homogeneous, possess some diversity of cell populations, the kidneys are especially heterogeneous, being comprised of more than 15 distinct, identifiable cell types (Horster, 1978; Jacobson, 1981). A cursory consideration of key energy-related processes in selected nephron segments reveals some profound differences between the segments (Table 11.1). As would be readily expected, many of these differences correlate with segment-specific physiology and biochemistry, thus illustrating structure–function correspondence along the segments of the nephron.

Table 11.1 Nephron Heterogeneity of Cellular Energetics in Mammalian Kidney

Nephron segment	*Glucose metabolism*	*Mitochondrial function*	*Oxygenation*	*Energy-dependent reactions*
Proximal tubule	Gluconeogenesis high Glycolysis low	High activity	High	Erythropoietin, vitamin D synthesis; primary active transport: Na^+, K^+-ATPase; secondary active transport: organic anion, amino acids, D-glucose; drug metabolism
Distal convoluted tubule	Gluconeogenesis low Glycolysis high	High activity	High	Na^+, divalent cation transport
Medullary thick ascending limb	Gluconeogenesis low Glycolysis high	Low activity	Low	Na^+, K^+, Cl^- transport

Hence, for example, the proximal tubular epithelium possesses a high level of mitochondrial function, high rates of gluconeogenesis, but low rates of glycolysis, and these capacities are accompanied by high rates of several energy-requiring processes, in particular active transport and macromolecular synthesis (Cohen, 1986; Guder and Ross, 1984; Guder et al., 1986; Klein et al., 1981; Tessitore et al., 1986; Uchida and Endou, 1988). Besides differences in activities of the major pathways of intermediary metabolism, substrate utilization, and mitochondrial respiratory activity, each nephron segment or kidney region possesses a characteristic degree of oxygenation. Thus, there is a gradient of oxygenation from outer cortex to inner medulla, such that the inner medulla has been described as being on the brink of anoxia (Cohen, 1979; Epstein, 1997). Besides helping to explain differences in substrate utilization and the derivation of the energy supply for biosynthetic and energy-requiring processes, these segment differences in energetics and oxygenation have both physiological and toxicological consequences.

To probe the underlying susceptibility of different nephron segments to inhibition of cellular ATP generation, the influence of either treatment with 20 μM iodoacetate (IAA) + KCN or hypoxia on acute cytotoxicity, adenine nucleotide status, and thiol status were studied in suspensions of freshly isolated renal proximal tubular (PT) and distal tubular (DT) cells (Lash et al., 1996). Inhibition of glycolytic and mitochondrial ATP generation by 20 μM IAA and 1 mM KCN, respectively, led to rapid and marked depletion of cellular content of ATP and increases in cellular contents of ADP and AMP in both PT and DT cells (Figure 11.3a). While the extent of ATP depletion was slightly higher in DT cells, acute cellular necrosis, as indicated by release of lactate dehydrogenase (LDH), was markedly higher in DT cells than in PT cells (Figure 11.3b).

In another set of studies, suspensions of freshly isolated rat renal PT and DT cells were exposed to hypoxia, and effects on cellular thiol and adenine nucleotide status and acute cellular necrosis were examined (Lash et al., 1993, 1996). Rat renal PT or DT cells were incubated in an atmosphere of either 95% O_2/5% CO_2 (normoxia) or 95% N_2/5% CO_2 (hypoxia) and cellular contents of glutathione (GSH) and glutathione disulfide (GSSG), and protein sulfhydryl groups (PrSH) were measured (Figure 11.4a). Only modest effects on GSH/GSSG and PrSH contents were observed in both cell types. In contrast, while PT cells exhibited only a small increase in LDH release (7.1%), DT cells exhibited a much larger increase in LDH release (21.0%) (Figure 11.4a). Examination

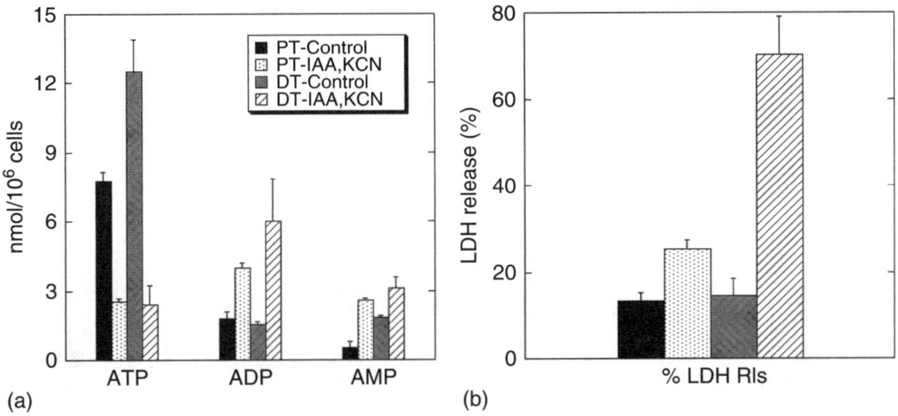

Figure 11.3 Acute cytotoxicity induced by energetic failure in isolated renal PT and DT cells: (a) cellular content of ADP, ADP, and AMP; (b) release of lactate dehydrogenase (LDH). Isolated renal PT and DT cells (2 to 3 × 10^6 cells/ml) were incubated with either buffer or 20 μM IAA + 1 mM KCN for 30 min (adenine nucleotides) or 60 min (LDH release) at 37°C under an atmosphere of 95% O_2/5% CO_2 before measurements were made. Results are the means ± SEM of measurements from three separate cell preparations.

of cellular contents of adenine nucleotides showed that DT cells exhibited only a slightly greater amount of ATP depletion than did PT cells (80.4% vs. 67.4% in DT and PT cells, respectively) and a similar increase in hypoxanthine (HXan)(3.6- and 4.1-fold in DT and PT cells, respectively), indicating a similar amount of degradation of adenine nucleotides in the two cell types (Figure 11.4b). Taken together, the results with (IAA + KCN)-treatment and hypoxia suggest that although PT and DT cells respond similarly with regard to the status of thiols and adenine nucleotides, DT cells are significantly more sensitive to acute cellular necrosis from multiple modes of energetic dysfunction.

Cellular Energetics and Redox Status

Mitochondrial energetics and redox status are tightly regulated and interrelated processes in renal cells. Besides the large number of mitochondrial proteins containing critical sulfhydryl groups mentioned above, mitochondria in renal proximal tubules also contain a large pool of GSH that represents the majority of the nonprotein thiols in the

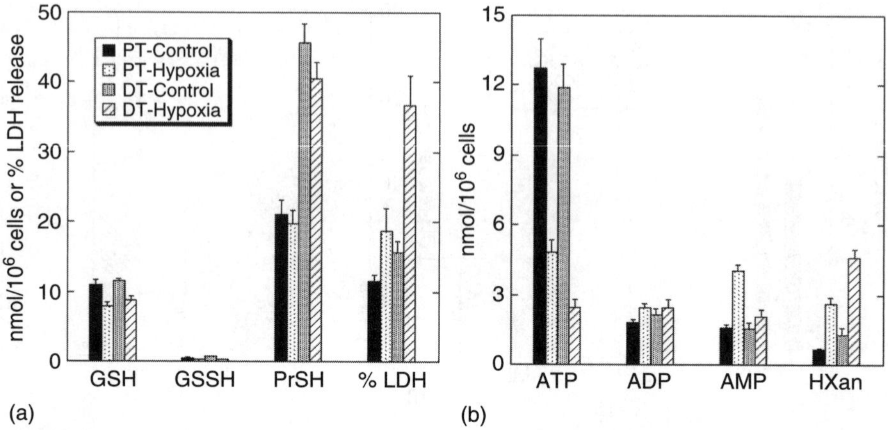

Figure 11.4 Effects of hypoxia on PT and DT cellular content of adenine nucleotides, thiols and disulfides, and cytotoxicity. Isolated renal PT and DT cells (2–3 × 106 cells/ml) were incubated with either buffer or 20 *μM* IAA + 1 m*M* KCN for 30 min (all assays except LDH release) or 60 min (LDH release) at 37°C under an atmosphere of 95% O_2/5% CO_2. After the appropriate incubation time, adenine nucleotides and GSH/GSSG were measured by HPLC, PrSH was measured by a colorimetric method, and LDH release was measured spectrophotometrically. Results are the means ± SEM of measurements from three separate cell preparations.

organelle (Lash et al., 1998; Schnellmann et al., 1988). This pool of GSH is derived, not from *in situ* synthesis from precursors as the cytoplasmic pool is, but rather from transport across the mitochondrial inner membrane from the cytoplasm (McKernan et al., 1991; Schnellmann, 1991; Smith et al., 1996).

While this GSH pool contributes to regulation of mitochondrial energetics and functional integrity by virtue of its role as a reductant and being the primary nonprotein thiol in mitochondria (Beatrice et al., 1984; Lê-Quôc and Lê-Quôc, 1985, 1989; Yagi and Hatefi, 1984), a relationship between mitochondrial energetics and GSH status is also suggested by the mechanism of GSH transport across the inner membrane. As illustrated in Figure 11.5, studies by Lash and colleagues (Chen and Lash, 1998; Chen et al., 2000; McKernan et al., 1991) demonstrated that GSH enters renal cortical mitochondria primarily, if not entirely, by counter-transport that is mediated by the dicarboxylate carrier (DCC) and the 2-oxoglutarate carrier (OGC). A consequence of

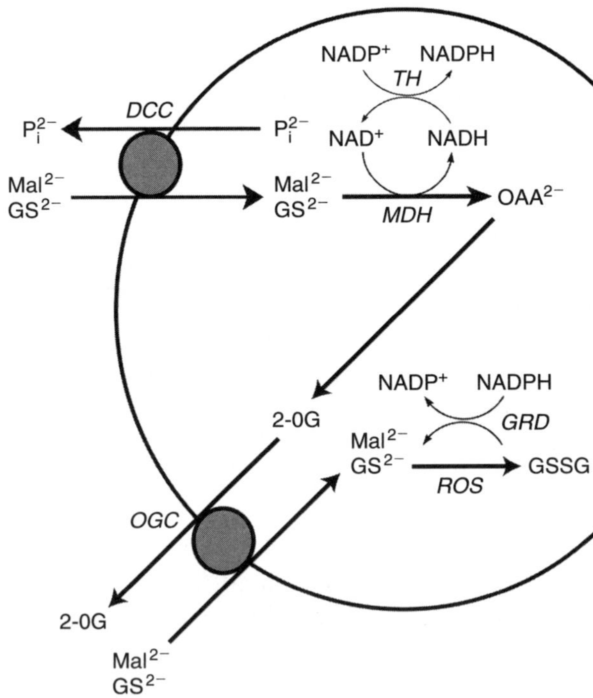

Figure 11.5 Function of the dicarboxylate carrier (DCC) and 2-oxoglutarate carrier (OGC) in the transport of GSH into mitochondria. Data are consistent with the function of the DCC and OGC in the uptake of GSH into renal mitochondria. GSH, in the form of the thiolate anion (GS^{2-}), or a dicarboxylate such as malate (mal^{2-}), is transported by the DCC in exchange for inorganic phosphate (P_i^{2-}) or by the OGC in exchange for 2-oxoglutarate (2-OG). Once inside the mitochondrial matrix, GSH can serve as a reductant for reactive oxygen species (ROS) in reactions that generate GSSG. GSH is regenerated from GSSG by GSH reductase (GRD). GSH transport interacts indirectly with the transport of other dicarboxylates, such as malate, because the malate–aspartate shuttle generates NADH, and through the action of the transhydrogenase (TH), the NADPH necessary for GRD to reduce GSSG. Other abbreviations: (MDH) malate dehydrogenase; (AST) aspartate aminotransferase; (OAA^{2-}) oxaloacetate. Reprinted with permission from Chen et al. (2000).

this coupling is that the transport and metabolism of citric acid cycle intermediates in mitochondrial matrix may be influenced by mitochondrial GSH status. Thus, coupling of GSH transport to that of dicarboxylates such as malate will influence equilibrium concentrations of substrates for various dehydrogenases in mitochondrial matrix, thereby influencing reduced and oxidized pyridine nucleotide ratios.

The toxicological significance of renal mitochondrial GSH transport processes has been demonstrated by transfecting NRK-52E cells (an immortalized cell line derived from normal rat kidney proximal tubular cells) with the cDNA for either the DCC or the OGC (Lash et al., 2002a). Wild-type NRK-52E cells exhibit only modest activity of mitochondrial GSH transport as compared with freshly isolated rat PT cells (Lash et al., 2002b). Therefore, these cells provide a low background for transfection studies. Accordingly, cells that transiently overexpressed the DCC exhibited higher rates of uptake of GSH (three- to tenfold) as compared with the nontransfected, wild-type cells. These higher rates of mitochondrial GSH uptake were accompanied by a markedly diminished susceptibility to apoptosis, induced by either tert-butyl hydroperoxide (tBH), which is an oxidant and mitochondrial toxicant (McKernan et al., 1991), or *S*-(1,2-dichlorovinyl)-L-cysteine (DCVC), which is a nephrotoxic cysteine conjugate of the environmental contaminant trichloroethylene and has the mitochondria as an early cellular target (Lash and Anders, 1986, 1987; also see below). As shown in Figure 11.6, wild-type NRK-52E cells incubated with either 10 μM tBH or 50 μM DCVC exhibited large increases in the fraction of cells undergoing apoptosis. In contrast, NRK-52E cells that overexpressed the DCC exhibited little increases in the proportion of apoptotic cells after exposure to tBH or DCVC. Hence, mitochondrial redox status plays a significant role in determining sensitivity to chemically induced apoptosis.

The protective role of enhanced mitochondrial GSH in renal PT cells is related, at least in part, to the function of the GSH redox cycle (Figure 11.7). The cycle involves coordinated function of GSH peroxidase, GSSG reductase, and substrate-level NADPH generation by various dehydrogenases. Besides providing GSH for the redox cycle, enhanced mitochondrial uptake of GSH by overexpression of the DCC or OGC may also provide GSH to maintain the reduced state of critical sulfhydryl groups of mitochondrial proteins, as described above.

The redox status of GSH and other thiols has, therefore, long been known to be critical for proper mitochondrial function (e.g., Hunter et al., 1964; Jocelyn, 1975; Jocelyn and Dickson, 1980; Jocelyn and Kamminga, 1974). Alterations in GSH concentration and redox status are an early effect of oxidative stress in mitochondria from a variety of tissues, including kidney, liver, brain, and tumor cells

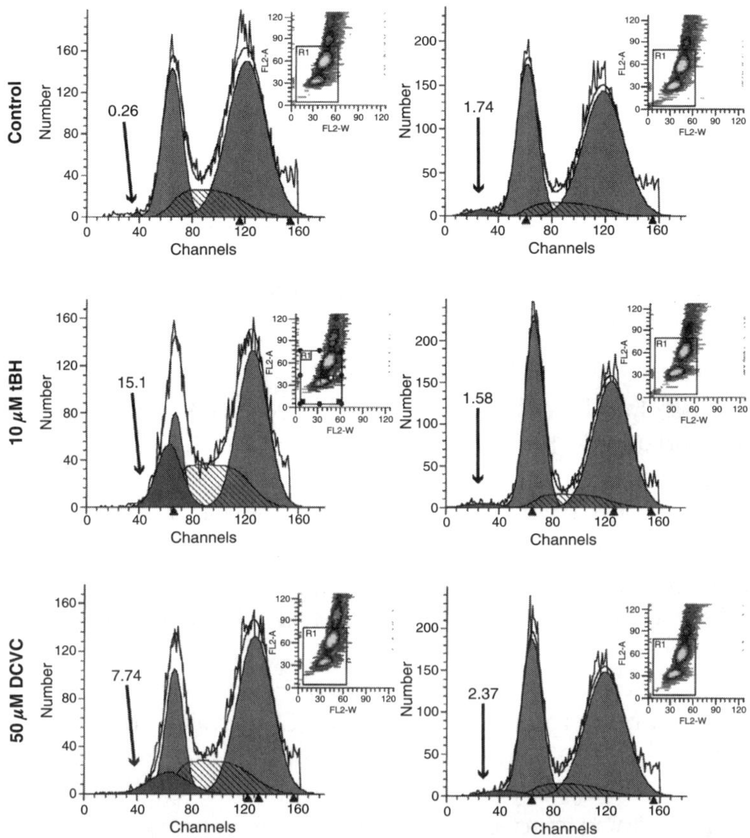

Figure 11.6 Flow cytometry analysis of propidium iodide-stained, NRK-52E cells treated with media, tBH, or DCVC. Confluent, wild-type NRK-52E cells (NRK-52E-WT) and cells transiently transfected to overexpress the DCC (NRK-52E-rDCC+) grown on 35 mm culture dishes were pretreated for 24 hours with 1 m*M* GSH and then treated for up to 4 hours with either media, 10 μ*M* tBH, or 50 μ*M* DCVC in the presence of 20 μ*M* GSH. Cells were then harvested by trypsin/EDTA treatment, washed in sterile PBS, and fixed overnight in ethanol. Cells were then stained with propidium iodide and analyzed by flow cytometry using a Becton Dickinson FACSCalibur flow cytometer. Panels represent DNA histograms, with cell number (*y*-axis) plotted against DNA content or channel number (*x*-axis). Peaks from left to right represent apoptotic cells, cells in G_0/G_1, cells in S, and cells in G_2/M. Values above the left peaks indicate the percentage of apoptotic cells. Insets are scatter profiles showing distribution of cells according to propidium iodide fluorescence intensity; *y*-axis = forward scatter; *x*-axis = side scatter. Cells outside the box are those that were excluded from the analysis due to aggregation. Adapted from Lash et al. (2001b).

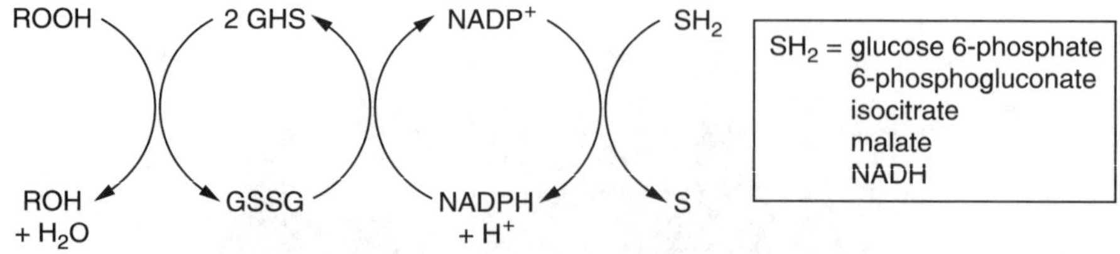

Figure 11.7 Glutathione redox cycle. (ROOH) lipid hydroperoxide; (ROH) lipid alcohol; (SH_2) and (S), subtrate and product, respectively, of various dehydrogenase reactions.

(Brodie and Reed, 1992; Maddaiah, 1990; McKernan et al., 1991; Meredith and Reed, 1983; Olafsdottir and Reed, 1988; Ravindrath and Reed, 1990; Shertzer et al., 1994; Sies and Moss, 1978), and leads to a range of mitochondrial and cellular responses. Altered GSH status modulates several processes in mitochondria, including regulation of mitochondrial Ca^{2+} ion distribution and pyridine nucleotide oxidation status (Beatrice et al., 1984; Lotscher et al., 1979; Moore et al., 1983; Olafsdottir et al., 1988; Savage et al., 1991), integrity of mitochondrial DNA (de la Asuncion et al., 1996; Esteve et al., 1999), and induction of the membrane permeability transition (Chernyak and Bernardi, 1996; Costantini et al., 1996; Halestrap et al., 1997; Masini et al., 1992; Reed and Savage, 1995; Savage and Reed, 1994; Scarlett et al., 1996). Ultimately, these effects can lead to cell death by either necrosis or apoptosis, as summarized in Figure 11.8, demonstrating how maintenance of proper mitochondrial GSH status plays a critical and early role in the renal response to oxidative stress.

Besides reactive oxygen species (ROS), recent attention has focused on the role of reactive nitrogen species (RNS), in particular nitric oxide (NO) and peroxynitrite ($ONOO^-$), in the regulation of mitochondrial and cellular function and in mediating certain forms of chemically induced and pathological injury. NO inhibits mitochondrial respiration (Bellamy et al., 2002; Beltrán et al., 2002) by direct inhibition of cytochrome oxidase (Barone et al., 2003; Borutaite et al., 2000). In renal proximal tubules, NO can regulate Na^+/K^+-ATPase activity (Liang and Knox, 2000; Zhang and Mayeux, 2001) and induces expression of heme oxygenase-1 (Liang et al., 2000). Moreover, RNS have been identified as important mediators of injury during renal ischemia (Walker et al., 2000, 2001). These effects of NO appear to derive from activation of a constitutively expressed and membrane-bound, mitochondrial NO synthase (mtNOS)

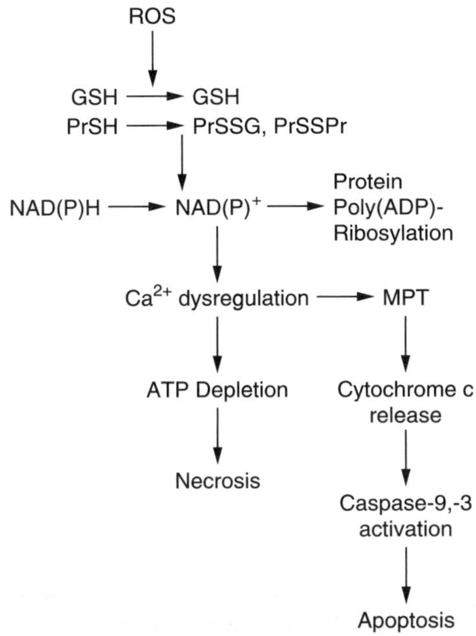

Figure 11.8 Summary scheme of selected processes involved in cell death induced by mitochondrial GSH oxidation. (MPT) membrane permeability transition; (PrSH) protein sulfhydryl; (PrSSG) glutathionylated protein; (PrSSPr) protein disulfide; (ROS) reactive oxygen species.

(Elfering et al., 2002; Ghafourifar et al., 1999), which generates $ONOO^-$. Besides direct reactions with cellular molecules, effects of NO may also be mediated by formation of *S*-nitrosoglutathione (GSNO) (Ji et al., 1999; Sandau et al., 1999; Sarkela et al., 2001; Wong and Fukuto, 1999). Figure 11.9 illustrates reactions by which NO reacts with GSH to form GSNO. Figure 11.10 illustrates reactions by which NO is subsequently generated from GSNO, showing how GSNO can be an NO donor. The normally high concentration of GSH in the mitochondrial matrix and the presence of a constitutively expressed NOS in the organelle suggests that formation of GSNO plays an important physiological role.

MITOCHONDRIA AS PRIMARY INTRACELLULAR TARGETS OF TOXICANTS AND IN PATHOLOGICAL CONDITIONS

The preceding discussion outlined the importance of mitochondrial energetics in maintaining renal cellular function, briefly discussed the

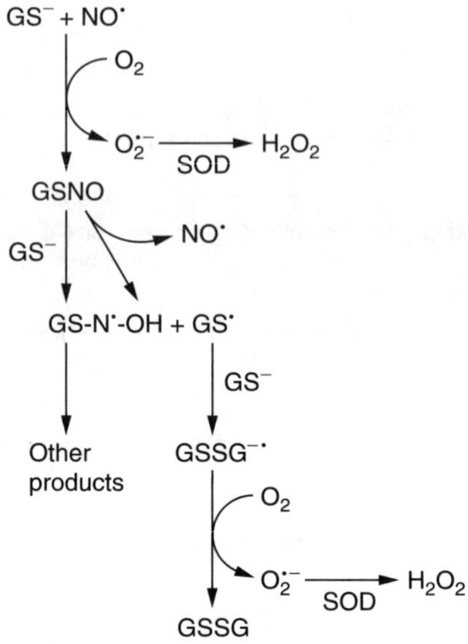

Figure 11.9 Reactions by which GSH reacts with NO to form GSNO. GSH, as the thiolate, reacts with NO in the presence of O_2 and forms GSNO. GSNO can release NO (function of GSNO as an NO donor) or it may react in the presence of the thiolate to form a species that can glutathionylate protein sulfhydryl groups. (SOD) superoxide dismutase. Adapted from Sarkela et al. (2001).

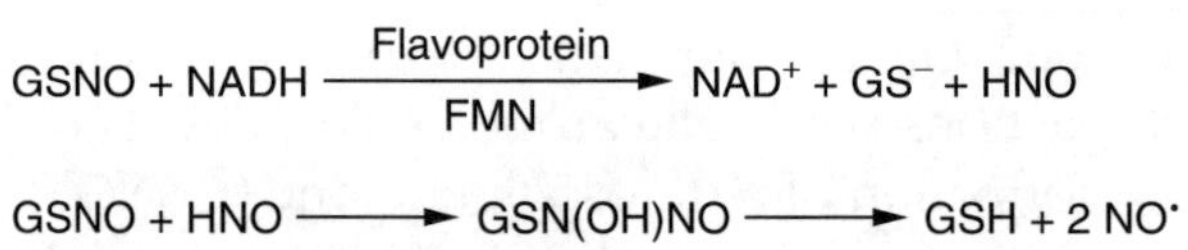

Figure 11.10 NO formation from GSNO. GSNO acts as an NO donor in two reactions, the first of which is mediated by a flavoprotein containing flavin mononucleotide. The HNO generated from the first reaction can react with another molecule of GSNO to form an intermediate that decomposes to GSH and

influence of nephron heterogeneity in determining patterns of susceptibility to nephrotoxicants and pathological conditions, and discussed the importance of regulation of mitochondrial redox status, in particular that of GSH, in maintaining mitochondrial function and the cellular responses to toxicants. The following sections briefly present selected, key examples of nonspecific toxicants, specific nephrotoxicants, and pathological conditions that exhibit mitochondrial toxicity as an early and prominent effect in the sequelae of events that lead from exposure to nephrotoxicity. As noted by Wallace in his introduction to a Society of Toxicology Symposium Review (Wallace et al., 1997), all cells except erythrocytes are vulnerable to mitochondrial poisoning. Determinants of organ specificity of toxic responses to mitochondrial toxicants include pharmacokinetics and the number and size of mitochondria in specific cell types. Nephrotoxicity due to mitochondrial toxicants, with the PT cells as primary targets along the nephron, can thus be largely attributable to both pharmacokinetics and to the high concentration of mitochondria within those cells.

Classical Mitochondrial Poisons

Several so-called "metabolic inhibitors," chemicals that inhibit mitochondrial electron transport by interacting with specific components of the respiratory chain or cause a generalized increase in inner membrane permeability, have been used extensively to investigate the role of mitochondrial energetics in renal cellular function and have been important tools for study of cellular energetics. These chemicals, such as antimycin A, 2,4-dinitrophenol, and rotenone, are potent inhibitors of mitochondrial respiration, leading to alterations in renal cellular energy metabolism and energy-dependent functions, such as active transport (Dickman and Mandel, 1989, 1990; Gullans et al., 1982). There is also coordination between mitochondrial and glycolytic ATP generation, as inhibition of glycolysis enhances sensitivity of rabbit renal proximal tubules to mitochondrial metabolic inhibitors, including antimycin A (Griner and Schnellmann, 1994). Similarly, there is linkage between the function of mitochondria and the endoplasmic reticulum in renal PT cells in regulating cellular calcium ion homeostasis; depletion of endoplasmic reticulum calcium stores protects renal PT cells from both hypoxia- and mitochondrial inhibitor-induced cell injury and death (Waters et al., 1997). This provides further evidence that mitochondrial processes that regulate energy and ion status function in concert with, and can be influenced by, processes in other subcellular organelles.

This theme of concerted action is illustrated further by the observations that glycine, which protects renal PT cells from a diverse array of toxicants, also protects from inhibitors of mitochondrial ATP production (Aleo and Schnellmann, 1992; Moran and Schnellmann, 1997;

Weinberg et al., 1990). Moreover, Miller and Schnellmann (1993a, 1993b, 1994) demonstrated that the mechanism of glycine-dependent protection involves the binding of glycine to plasma membrane receptors.

Besides producing acute cellular necrosis, mitochondrial inhibitors such as antimycin A can also cause renal PT cells to undergo apoptosis (Kaushal et al., 1997). This finding highlights the complexities in the response of cells to alterations in mitochondrial function. The mitochondrial complex I inhibitor rotenone can also induce apoptosis by a mechanism involving enhanced mitochondrial generation of reactive oxygen species (Li et al., 2003). Using HL-60 cells (a human promeylocytic leukemia cell line), apoptosis was demonstrated by cytochrome c release, caspase 3 activation, DNA fragmentation, and DNA staining with Hoechst 33342. Notably, rotenone-induced apoptosis was inhibited by treatment with several antioxidants (e.g., GSH, L-cysteine, ascorbate) and was markedly lower in HT180 cells, which overexpress superoxide dismutase.

Schnellmann and colleagues (Waters et al., 1997) emphasized that although the early events that result from inhibition of mitochondrial respiration due to either hypoxia/anoxia or metabolic inhibitors are generally agreed to include rapid depletion of intracellular ATP and K^+ and a rise in intracellular Na^+ concentrations, those that follow in the later stages of cell injury are not well understood. Although this statement was made several years ago and the knowledge of cellular responses to mitochondrial dysfunction has greatly increased over those years, many unanswered questions remain to be investigated.

Oxidants and Alkylating Agents

As indicated elsewhere in this volume, the kidneys are primary targets for a vast array of toxic chemicals. Moreover, due to the high degree of energy dependence of the PT cell, many of these chemicals target the mitochondria during the early phases of the processes leading from exposure to cell injury and death. A major class of these chemicals includes various toxicants that act as either oxidants or alkylating agents. Below are described just a few of the prominent examples. Chapters elsewhere in this volume review the nephrotoxicity of several of these chemicals in greater detail. Here, the focus is on their effects on renal mitochondria.

Cysteine Conjugates of Halogenated Alkanes and Alkenes

Several halogenated alkanes and alkenes are bioactivated, rather than detoxified, by conjugation with GSH. Because of the manner in which these compounds are metabolized and transported, they are accumulated in renal PT cells and are selectively nephrotoxic in most mammalian species. The alkanes include chemicals such as 1,1,2,2-tetrafluoroethylene and 1-chloro-1,2,2-trifluoroethylene, and the alkenes include chemicals such as trichloroethylene, perchloroethylene, and hexachlorobutadiene. All the nephrotoxic halogenated alkanes and alkenes have in common the presence of a good leaving group (i.e., a halogen atom) or an electrophilic carbon–carbon double bond (or both), thus providing a site for GSH nucleophilic addition. Once the GSH conjugate is formed, which is catalyzed by GSH *S*-transferases predominantly in the liver, it undergoes processing to the cysteine conjugate by reactions that occur in either the biliary or intestinal tract or the kidneys, and ultimately is absorbed by the renal proximal tubules. Once inside the renal PT cell, the cysteine conjugate is either *N*-acetylated or can be a substrate for either the cysteine conjugate β-lyase or flavin-containing monooxygenases (see Lash et al., 2000 for review of nephrotoxic cysteine conjugate metabolism). It is metabolism of the cysteine conjugate by the β-lyase or the flavin-containing monooxygenase that forms the reactive species that can elicit nephrotoxicity and, in some cases, nephrocarcinogenicity. More detailed discussions of the mechanisms of GSH conjugate-dependent bioactivation and nephrotoxicity are presented in Chapters 6, 7, 8, and 23, while this section focuses on the targeting of mitochondria in the early stages of renal cell injury.

Mitochondrial toxicity of cysteine conjugates arises in part because cysteine conjugate β-lyase activity is present within the organelle (Cooper et al., 2001, 2002a; Lash et al., 1986; Stevens et al., 1988). Consequently, reactive metabolite is generated in close proximity to target molecules. In support of this suggestion, several groups have isolated and chemically identified adducts of reactive metabolites from cysteine conjugates with mitochondrial DNA, RNA, protein, and phospholipids (Banki and Anders, 1989; Hayden and Stevens, 1990; Hayden et al., 1991, 1992).

Inhibition of mitochondrial oxygen consumption in isolated renal proximal tubules and PT cells and in isolated mitochondrial suspensions by various cysteine conjugates provides the clearest demonstration that these chemicals inhibit mitochondrial function. Inhibition of mitochondrial respiration has been demonstrated with 1-chloro-1,2,2-trifluoroethyl-L-cysteine (Banki et al., 1986), 1,2,3,4,5,5-pentachlorobutadienyl-L-cysteine, the cysteine conjugate of hexachlorobutadiene (Jones et al., 1986; Schnellmann et al., 1989; Wallin et al., 1987), DCVC (Lash and Anders, 1986, 1987; Lash et al., 1995, 2001a), DCVC sulfoxide (Lash et al., 1994a, 2003), and *S*-(1,1,2-trichlorovinyl)glutathione (Lash et al., 2002c). In all cases, potent inhibition of state 3 respiration was demonstrated, whereas only with 1,2,3,4,5,5-pentachlorobutadienyl-L-cysteine has an increase in state 4 respiration, indicative of uncoupling, been found (Schnellmann et al., 1989).

The ability of DCVC to inhibit mitochondrial respiration depends on the respiratory substrate that is used. As shown in Figure 11.11, DCVC selectively inhibits mitochondrial respiration in both intact renal PT cells and renal cortical mitochondria when succinate, a coupling site II-linked substrate, is used; little or no inhibition was observed when either site I-linked or site III-linked substrates were used (Lash and Anders, 1986, 1987). DCVC sulfoxide, the immediate product of the catalytic action of the flavin-containing monooxygenase on DCVC, was also the most potent inhibitor of mitochondrial respiration when succinate was the respiratory substrate (Lash et al., 1994a, 2003).

Another functional assay for mitochondrial integrity is based on the ability of the organelle to sequester calcium ions. Besides inhibition of respiration, DCVC inhibits the ability of mitochondria in renal PT cells to retain and sequester calcium ions (Vamvakas et al., 1990, 1992; van de Water et al., 1993, 1994). Time-dependent inhibition of calcium retention in the mitochondrial compartment of rat renal PT cells by DCVC is illustrated in Figure 11.12. It is likely that reactive acylating species generated by the bioactivation of DCVC targets mitochondrial calcium transporters among other proteins.

Besides targeting of calcium transporters, cysteine conjugates, including DCVC, 1,2,3,4,5,5-pentachlorobutadienyl-L-cysteine, and tetrafluoroethyl-L-cysteine, acylate or oxidize critical sulfhydryl or amino groups or otherwise inhibit several mitochondrial dehydrogenases and GSSG reductase (Cooper et al., 2002b; James et al., 2002; Lash and Anders, 1987; Lock and Schnellmann, 1990; van de Water et al., 1995). Adducts of DCVC and tetrafluoroethyl-L-cysteine with cysteinyl sulfhydryl groups and ε-amino groups of lysyl residues of these various enzymes have been isolated (Hayden and Stevens, 1990; Hayden et al., 1991). Potent targeting of mitochondrial dehydrogenases and other

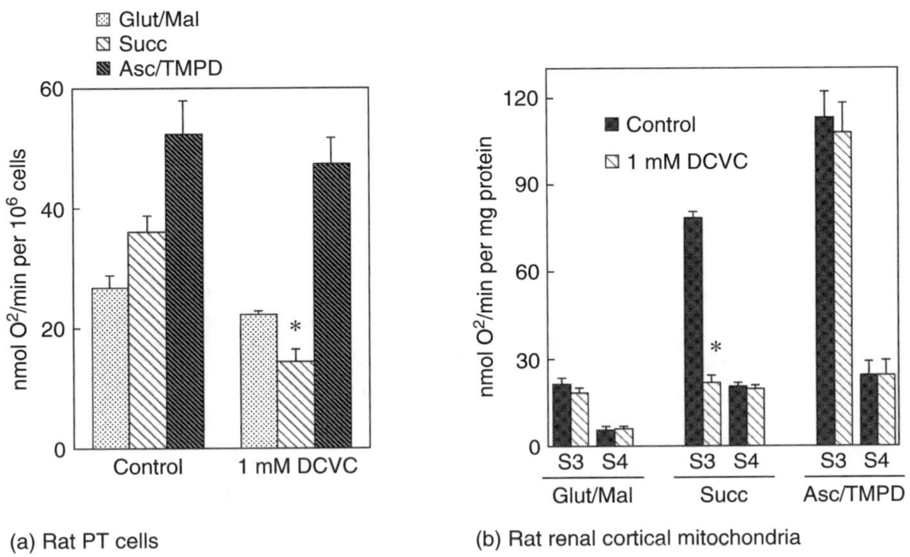

Figure 11.11 Inhibition of cellular and mitochondrial respiration in rat kidney by DCVC: (a) Isolated renal PT cells (1×10^6 cells/ml) from male Fischer 344 rats were incubated for 30 min with either Krebs-Henseleit buffer (= control) or 1 m*M* DCVC at 37°C. Cellular oxygen consumption was measured with either 4 m*M* glutamate + 2 m*M* malate (Glut/Mal), 3.3 m*M* succinate in the presence of 5 μ*M* rotenone, or 1 m*M* ascorbate + 0.2 m*M* *N,N,N′,N′*-tetramethyl-*p*-phenylenediamine (TMPD) as respiratory substrates. (b) Suspensions of isolated mitochondria (1 mg protein/ml) from renal cortex of male Fischer 344 rats were incubated for 15 min with either buffer (= control) or 1 mM DCVC at 25°C. Mitochondrial oxygen consumption was measured with respiratory substrates as in cells. Rates of state 3 (S3) and state 4 (S4) oxygen consumption were measured by addition of substrate and ADP to mitochondria and after exhaustion of ADP, respectively. Oxygen contents of incubation mixtures were measured in a Gilson 5/6H Oxygraph using a Clark-type oxygen electrode. Results are means ± SEM of measurements from three separate cell or mitochondrial preparations. *Significantly different ($p < 0.05$) from corresponding control incubations.

enzymes of the citric acid cycle are consistent with the substrate specificity pattern of cysteine conjugate induced inhibition of mitochondrial respiration.

It has been shown that DCVC induces apoptosis in a renal PT cell line (LLC-PK_1) (Chen et al., 2001) and in primary cultures of human PT cells (Lash et al., 2001b) by the mitochondrial-dependent pathway, as evidenced by induction of a membrane permeability transition

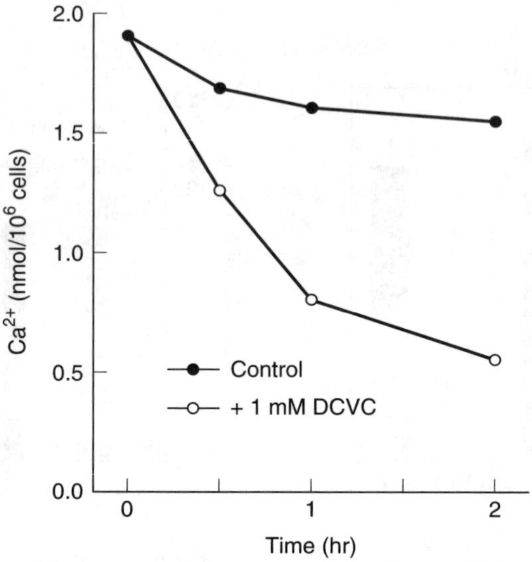

Figure 11.12 Inhibition of mitochondrial calcium sequestration by DCVC. Isolated renal PT cells (4 to 5 × 10^6/ml) from male Fischer 344 rats were separated from extracellular medium by rapid centrifugation through Ca^{2+}- and Mg^{2+}-free Hanks solution containing 20% (v/v) Percoll. Intracellular Ca^{2+} compartmentation was determined by dual-wavelength spectroscopy (654–685 nm) using the Ca^{2+}-sensitive dye arsenazo III (2,2′-(1,8-dihydroxy-3,6-disulfo-2,7-naphthalene-bisazo)-dibisbenzenearsonic acid; 30 *μM* final concentration). The protonophore carbonyl cyanide *p*-trifluoromethoxyphenylhydrazone (FCCP; 10 *μM*) and the Ca^{2+}-ionophore A23187 (15 μM) were then added sequentially, and the absorbance changes were recorded. The FCCP-releasable Ca^{2+} pool is equivalent to Ca^{2+} sequestered within mitochondria, and Ca^{2+} released by addition of A23187, after FCCP-induced Ca^{2+} release, represents mainly the pool sequestered by the endoplasmic reticulum. Results are means of measurements from three separate cell preparations for the mitochondrial pool.

(MPT) (Chen et al., 2001) and caspase-3 activation (Chen et al., 2001; Lash et al., 2001b). DCVC sulfoxide similarly induced apoptosis in primary cultures of human PT cells by the mitochondrial-dependent pathway, as evidenced by a decrease in mitochondrial membrane potential (Lash et al., 2003). 1,2,3,4,5,5-Pentachlorobutadienyl-L-cysteine also induced the MPT, although it has been concluded that this occurred by a cysteine conjugate *β*-lyase-independent mechanism (Brown et al., 1996).

Inorganic Mercury

Inorganic salts of mercury (Hg) act by binding to sulfhydryl groups and are selectively nephrotoxic. Although other subcellular sites, such as the plasma membranes, may be affected at an earlier stage after exposure of the renal PT cell to Hg, mitochondria are also prominent and relatively early targets and play a significant role in Hg-induced nephrotoxicity (Zalups and Lash, 1994, and see Chapter 22). Studies by Weinberg and colleagues (Weinberg et al., 1982a, 1982b) showed that Hg is directly toxic to isolated mitochondria in an *in vitro* model of suspensions of rat kidney mitochondria, and in an *ex vivo* model in which animals were first treated with $HgCl_2$ and then suspensions of renal cortical mitochondria were isolated. This suggests that many of the effects of Hg on mitochondria are irreversible. This suggestion is supported by a finding that treatment with dithiothreitol reversed the Hg-induced inhibition of state 3 respiration in the *in vitro* model but not in the *ex vivo* model.

Similar findings were obtained by Lund et al. (1991, 1993), who showed that Hg caused an increase in production of H_2O_2 at both the ubiquinone-cytochrome b region and the NADH dehydrogenase region, depletion of mitochondrial GSH, lipid peroxidation, and oxidative stress in both suspensions of isolated mitochondria from rat kidney cortex treated *in vitro* with $HgCl_2$ and in kidney mitochondria isolated from rats treated *in vivo* with $HgCl_2$. Mitochondrial toxicity of Hg should likely occur by formation of Hg-thiol group adducts with cysteinyl residues of the various thiol-containing proteins in the mitochondria. In support of this suggestion, Chavez et al. (1991) found that captopril [1-(3-mercapto-2-methyl-1-oxopropyl)-proline], an angiotensin converting enzyme inhibitor and thiol-containing compound, protected against Hg-induced toxicity in rat kidney mitochondria both in *in vitro* and *in vivo* experiments.

Aminoglycosides

Aminoglycoside antibiotics, such as gentamicin, are limited in their clinical use primarily due to nephrotoxicity and ototoxicity (see Cojocel, 1997, for a review). Their mechanism of nephrotoxicity has been attributed primarily to effects on the plasma membranes and lysosomes, with the polycationic aminoglycoside interacting with various anionic groups, such as phosphatidylinsolitol and other anions. Although ionic interactions play a major role in the mechanism of action of aminoglycosides, other cellular effects occur that do not seem to be directly correlated with charge. Thus, while plasma membranes (especially the brush-border membranes of PT cells) and lysosomes appear to be the major subcellular sites of interaction, significant effects on mitochondrial function are also observed.

Gentamicin inhibits mitochondrial oxidative phosphorylation in suspensions of isolated mitochondria from rat kidney cortex (Weinberg and Humes, 1980; Simmons et al., 1980). Continuous exposure of rats to gentamicin over a period of 21 days and subsequent isolation of renal cortical mitochondria, also showed impaired state 3 respiration and calcium ion accumulation (Mela-Riker et al., 1986). Shah and colleagues (Ueda et al., 1993; Walker and Shah, 1987) provided data that implicate oxidative processes in gentamicin-induced toxicity in renal cortical mitochondria; they found that gentamicin caused increased mobilization of iron from mitochondria, with the ensuing metal ion-dependent oxidative stress, and caused increased production of H_2O_2.

It is clear, therefore, that gentamicin causes mitochondrial dysfunction. However, the correlation between mitochondrial toxicity and nephrotoxicity, either in experimental animals or in the clinic, is rather poor. Hence, the significance of mitochondria as a target site in gentamicin- and aminoglycoside-induced nephrotoxicity is uncertain.

Cisplatin

Although cisplatin is used extensively as a chemotherapeutic agent for several types of cancer, its clinical use is also limited by nephrotoxicity. Studies by Gullans and colleagues (Brady et al., 1990) using suspensions of rabbit proximal tubules showed that cisplatin caused a dose-dependent inhibition of respiration. Dissection of the respiratory effect with several inhibitors and other agents suggested that mitochondrial injury was a very early effect of cisplatin and preceded other cellular effects of the drug. Zhang and Lindup (1993, 1994) studied the mitochondrial toxicity of cisplatin in renal cortical slices and found that oxidative effects play a major role in the mechanism of mitochondrial dysfunction. Effects included increased lipid peroxidation, depletion of mitochondrial and total cellular GSH, decreases in mitochondrial protein sulfhydryl content, and decreased calcium uptake.

More-recent studies on the mechanism of cisplatin-induced apoptosis have focused on its ability to cause renal PT cell death by both apoptotic and nonapoptotic mechanisms. Studies in LLC-PK_1 cells (Park et al., 2002) and rabbit proximal tubule suspensions (Cummings and Schnellmann, 2002) showed a prominent role for the mitochondrial (i.e., caspase 3/9-dependent) pathway in cisplatin-induced apoptosis. However, Cummings and Schnellmann (2002) also found that at least 50% of cisplatin-induced apoptosis was actually due to a p53-dependent mechanism that was independent of either caspase 3 or mitochondrial dysfunction. In a recent review, Gonzalez et al. (2001) concluded that cisplatin-induced cell death might also occur by necrosis. Whether this applies only to the action of cisplatin in tumor cells or whether it also applies to effects in the renal proximal tubule remains to be determined. This conclusion, taken together with that of Cummings and Schnellmann (2002), indicates that cisplatin probably produces cytotoxicity in multiple and complex ways. For nephrotoxicity, mitochondria are only one target for cisplatin.

Cyclosporine A

Cyclosporine A (CsA) is a fungal-derived peptide with immunosuppressant properties. Although it has become a critical drug in the prevention of organ transplant rejection, nephrotoxicity is a serious and potentially limiting side effect. Renal PT cells are the primary target. The nephrotoxicology of CsA and other immunosuppressive agents is discussed in great detail in Chapter 17 of this volume. Therefore, this section focuses only briefly on the role of mitochondrial dysfunction in CsA-induced nephrotoxicity.

Several studies have demonstrated prominent mitochondrial toxicity for CsA. For example, Jung and Pergande (1985) showed marked effects of CsA in isolated rat kidney mitochondria: with site I substrates (i.e., glutamate and malate), CsA both increased state 4 respiration and inhibited state 3 respiration, indicating both a direct inhibitory effect on the electron transport chain and an uncoupling effect; in contrast, when a site II coupling substrate (i.e., succinate) was used, the major effect was inhibition of state 3 respiration with only a very modest effect on state 4 respiration. Elzinga et al. (1989) showed that renal cortical mitochondria isolated from CsA-treated rats exhibited decreased pyruvate/malate-stimulated state 3 respiration and decreased calcium ion accumulation. Strzelecki et al. (1988), however, showed that the effects of CsA on mitochondria differ depending on the range of doses used and on whether renal cortical mitochondria are exposed *in vitro* (i.e., direct exposure) or whether mitochondria are isolated from rats previously treated with CsA. Thus, they found that direct treatment of renal mitochondria with CsA resulted in marked inhibition of succinate-dependent state 3 respiration. However, respiration was not significantly affected in renal mitochondria subsequently isolated from rats treated with an immunosuppressive dose of CsA (i.e., 25 mg/kg per day p.o.) but was inhibited in mitochondria isolated from rats treated with a higher dose of CsA (i.e., 75 mg/kg per day).

The findings described above bring into question the relationship between the mitochondrial and immunosuppressive effects of CsA. Moreover, CsA is also a well-characterized inhibitor of the mitochondrial MPT (Broekemeier et al., 1989), which usually protects many cell types from several forms of chemically induced injury.

How do the inhibitory effects on mitochondrial function, which ultimately produce toxicity, relate to the protective effects that are linked with the MPT? Although this issue has not been specifically addressed in direct experimentation, the various effects of CsA would seem to vary depending on dose, such that relatively lower doses are immunosuppressive and protective on a cellular level, whereas relatively higher doses produce overt nephrotoxicity via inhibition of mitochondrial function. In support of this suggestion are the findings of Fournier et al. (1987), who showed in rat liver mitochondria that CsA blocked efflux of calcium. The authors suggested that this would lead to decreased levels of calcium in the cytoplasm, which may correlate with the inhibitory effect of CsA on the mitogenic stimulation of T lymphocytes.

In contrast with the conclusion stated above regarding dose-dependence of mitochondrial toxicity, Justo et al. (2003) studied mechanisms of CsA-induced apoptosis in a murine PT cell line (MCT cells) and arrived at a different conclusion. MCT cells were treated with relatively low doses of CsA (up to 20 μg/ml). Although CsA increased the expression of Fas, cells were not sensitized to FasL-induced apoptosis. Moreover, while CsA caused endoplasmic reticulum stress, as indicated by induction of GADD153, caspase-12 was not activated. Rather, the primary pathway of apoptosis induction involved Bax translocation to the mitochondria and activation of caspase-2, -3, and -9. These authors concluded that the primary pathway for CsA induction of PT cell apoptosis involves the mitochondrial pathway. Whether or not this result is an anomaly of the cell culture model or is applicable to the *in vivo* proximal tubule remains to be determined.

Cephalosporin Antibiotics

The first-generation cephalosporins were *β*-lactam antibiotics, whose therapeutic utility was limited by nephrotoxicity (Tune, 1986). Among the cephalosporins that were potently nephrotoxic is cephaloridine (CPH). CPH is unique among this class of compounds in that it possesses two other functional groups that may be sites of metabolism besides the *β*-lactam ring (Figure 11.13). On one end of the molecule is a thiophene ring that can undergo metabolism by cytochrome P450 to yield a reactive epoxide. On the other end of the molecule is a pyridinium ring that may undergo redox cycling in a manner analogous to that of paraquat, yielding a superoxide anion and causing oxidative stress. One important factor in determining the nephrotoxicity of CPH is that it is readily taken up into the renal PT cells by organic anion transporters on the basolateral membrane. Once inside the cell, however, CPH is a poor substrate for efflux into the lumen by transport across the brush-border membrane. Consequently, CPH tends to accumulate to relatively high concentrations within the PT cell.

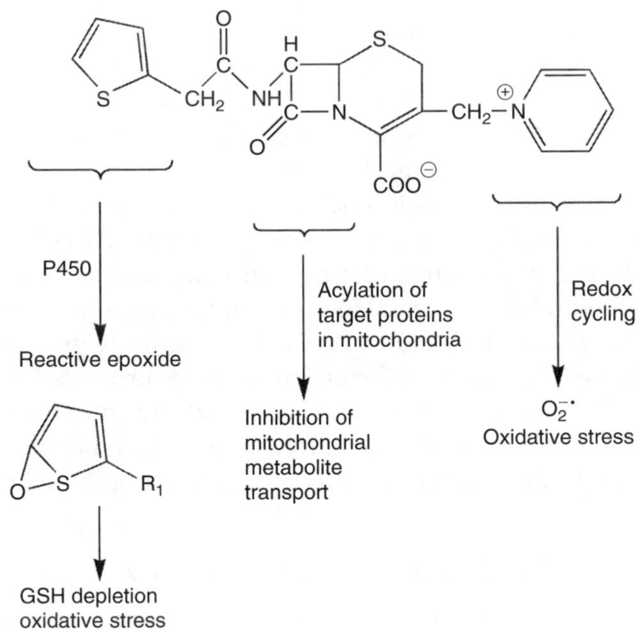

Figure 11.13 Structure and putative reactivity of functional groups of cephaloridine.

Numerous studies on cephalosporin- and CPH-induced nephrotoxicity have identified mitochondrial dysfunction as a prominent effect. For example, Tune et al. (1988, 1989) showed that CPH caused oxidative and mitochondrial injury to the rat kidney and was more potent than other cephalosporins. Similarly, Lash and colleagues (Lash et al., 1994b), studying determinants of cell type-selective toxicity in freshly isolated renal cells derived from different nephron segments, showed that CPH was a potent inhibitor of succinate-dependent respiration in PT cells, did not inhibit respiration in DT cells, and was much more potent than two other cephalosporin antibiotics (i.e., cephalothin and cephalexin).

In a series of studies, Tune and Hsu (1990, 1994, 1995a, 1995b) established that a major mechanism by which CPH produces mitochondrial dysfunction is by inhibition of substrate transport across the mitochondrial inner membrane. In particular, CPH potently inhibits transport of anionic substrates, including monocarboxylates and fatty acids. The proposed mechanism of inhibition involves acylation of key mitochondrial proteins by the β-lactam ring of the CPH molecule (see Figure 11.13).

More-recent studies suggest additional mechanisms of CPH-induced mitochondrial toxicity. Kiyomiya et al. (2000) found that cytochrome c oxidase is a potent target for CPH in LLC-PK_1 cells, and they suggest that this is the primary site of action in PT cells. These authors also did not observe any one-electron reduction of CPH by microsomal NADPH-cytochrome P450 reductase and no generation of superoxide anions by redox cycling. Whether their findings are unique to the cell culture model they were using or apply in general to renal PT cells needs to be evaluated. Kohda and Gemba (2002) found increases in protein kinase C activity and chemiluminescence intensity in the mitochondria from renal cortex of rats previously treated with CPH. These findings suggest a mode of action that is completely different from those proposed by Tune and colleagues. Additional study is needed to sort out these various modes of action and determine which one predominates and the importance of mitochondrial dysfunction in CPH-induced nephrotoxicity.

PATHOLOGICAL CONDITIONS

While there are numerous cardiovascular, metabolic, and genetic diseases that affect the kidneys in various ways, few are selective in altering renal energetics and mitochondrial function. Two pathological states that are notable for selectively or prominently affecting renal cellular energetics and mitochondrial function are oxygen deprivation and reduced renal mass (such as occurs after uninephrectomy). This section briefly highlights studies on these two pathologies as they relate to renal mitochondria.

Ischemia-Reperfusion and Hypoxia

Ischemia-reperfusion or other forms of oxygen deprivation (i.e., hypoxia or anoxia) are prominently studied causes of renal injury, both in the clinical as well as in the experimental setting. Weinberg (1991) provided an extensive review of the biochemistry of the renal cellular response to ischemia and reperfusion. Mitochondrial respiratory dysfunction and alterations in cellular levels of ATP and calcium ions (Malis and Bonventre, 1986; Snowdowne et al., 1985; Takano et al., 1985; Weinberg et al., 2000) are obvious effects of oxygen deprivation, but other effects that may result from the mitochondrial dysfunction have also been documented. For example, studies have demonstrated an early effect of oxygen deprivation on cytoskeletal proteins in proximal tubules (Chen et al., 1994; Doctor et al., 1997), which may lead to the cell surface blebbing that is often observed. Although effects of hypoxia in renal PT cells on cellular and mitochondrial GSH status appear to be secondary (Lash et al., 1996), augmentation of cellular GSH can protect PT cells from cell death due to ischemia or hypoxia (Lash et al., 1996; Scaduto et al., 1988).

An important aspect of renal injury induced by oxygen deprivation that relates to the subject of nephron heterogeneity discussed above is that not all nephron segments exhibit the same susceptibility to such injury (Epstein, 1997). Hence, renal epithelial cells of the medullary thick ascending limb, for example, exhibit much more injury from a given duration of ischemia than cells of the proximal tubules (Brezis et al., 1985; Shanley et al., 1986). These susceptibility differences can be largely attributed to the workload or degree of energy dependence of each nephron segment, such that an increase in mitochondrial activity or an energy-consuming process such as transport exacerbates injury from anoxia or ischemia (Bastin et al., 1987; Brezis et al., 1985, 1986).

Reduced Renal Mass

When functional renal mass is significantly reduced, such as occurs by surgical removal of one kidney (uninephrectomy) or progressively with aging, a series of compensatory physiological, morphological, and biochemical changes occur in the remaining renal tissue (Fine, 1986; Shirley and Walter, 1991). This compensatory response is predominantly a cellular hypertrophy and occurs primarily in the proximal tubules. Some components of the hypertrophic response include increases in cellular protein content (both generally and some specific proteins in particular), increases in brush-border and basolateral membrane surface area, increases in overall cellular size, and alterations in cellular energetics.

Figure 11.14 summarizes results of a study by Lash et al. (2001c) that investigated the biochemical effects of compensatory renal cellular hypertrophy in isolated mitochondria from rats that had undergone uninephrectomy and were allowed to undergo the compensatory growth response for 10 days prior to isolation of renal cortical mitochondria. Overall rates of respiration and electron flow increased in mitochondria from the hypertrophied kidney. Along with this increase in oxygen consumption, an increase in release of ROS and lipid peroxidation were observed. In spite of this, however, there were marked increases in rates of GSH transport by the DCC and OGC, leading to increased concentrations of GSH within the mitochondrial matrix. These findings are consistent with previous work by Harris and colleagues (Harris and Tay, 1997; Harris et al., 1988), who showed that compensatory renal cellular hypertrophy or other conditions that increase the workload of the kidney, such as proteinuria or iron overload, are associated with a hypermetabolic state. Hence, adaptive changes in mitochondrial energetics play important roles in the compensatory hypertrophy response or to large increases in workload for the kidneys. It is important to realize, however, that this hypermetabolic state also can predispose the renal PT cell to chemically induced injury due to, among other factors, a higher baseline level of ROS than is found in normal mitochondria.

MITOCHONDRIAL PERMEABILITY TRANSITION AND RENAL CELL DEATH

Although the topic of the membrane permeability transition and its role as an early response to a vast array of nephrotoxicants is considered in more detail elsewhere in this volume (e.g., see Chapters 7 and 8), no discussion of the role of mitochondrial energetics in renal toxicology would be complete without at least some consideration of

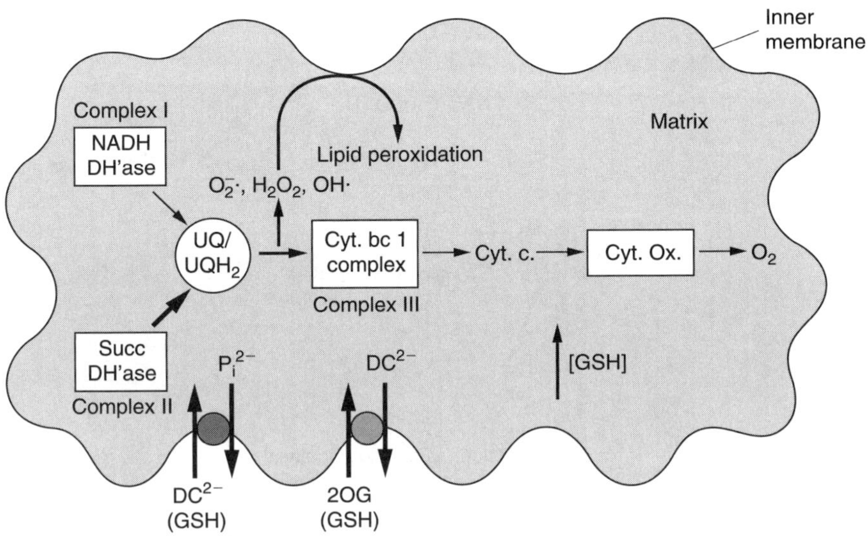

Figure 11.14 Processes in isolated renal mitochondria that are altered in uninephrectomized rats. The scheme summarizes the processes of metabolite and electron transport in renal mitochondria. The thickness of arrows is used to indicate relative flux of each step, with the thicker arrows indicating higher flux. Data from Lash et al. (2001c) show that rates of metabolite and GSH transport into renal mitochondria and rates of electron flow, particularly through complexes I to III, are significantly increased after uninephrectomy and compensatory renal growth. The increased electron flow results in increased release of ROS, presumably at the ubiquinone to cytochrome bc_1 step, resulting in increased rates of lipid peroxidation. Abbreviations: (DH'ase) dehydrogenase; (Succ) succinate; (UQ) and (UQH_2) oxidized and reduced ubiquinone; (Cyt) cytochrome; (Ox) oxidase; (DC^{2-}) dicarboxylate; (2OG) 2-oxoglutarate; (P_i^{2-}) inorganic phosphate.

this intriguing and critical process. MPT is viewed as a critical early response that occurs initially in a small proportion of mitochondria and progresses to occur in a higher proportion as the duration of exposure or dose of the toxicant or pathological condition increases (Lemasters et al., 1998).

A generalized scheme showing the progression of mitochondrial effects due to exposure of cells to tBH, based on discussion in Lemasters et al. (1998), is presented in Figure 11.15. Initial effects of tBH exposure include depletion of reduced pyridine nucleotides, by both oxidation and degradation, and oxidation of matrix GSH. This depletion of

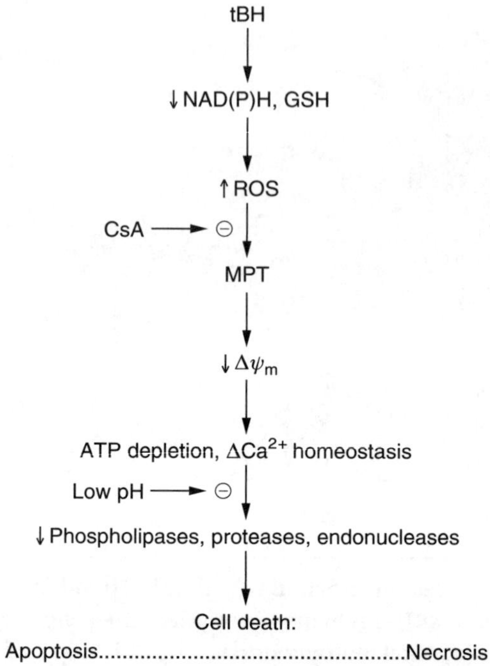

Figure 11.15 Sequence of events leading from tert-butyl hydroperoxide exposure to cell death. CsA and low pH inhibit the process.

reductants leads to an increase in production of ROS, which can then cause the MPT. Although the initial stages of the MPT are generally not associated with a loss of mitochondrial membrane potential ($\Delta\psi_m$), progression of the MPT to include a larger proportion of mitochondria decreases the $\Delta\psi_m$, which in turn inhibits respiratory function leading to ATP depletion and altered Ca^{2+} ion homeostasis. The disturbances in ATP and Ca^{2+} ion homeostasis cause activation of several degradative enzymes that can progress to cell death. The precise form of cell death (i.e., apoptosis or necrosis) is determined by degree of the disruption in mitochondrial function. CsA is a well-established inhibitor of the MTP and low pH inhibits activation of the various degradation enzymes, thereby protecting the cells and preventing cell death.

SUMMARY AND CONCLUSIONS

This chapter highlights the central importance of mitochondrial energetics in renal cellular function and provides examples of nephrotoxicants and pathological conditions whose mode of action involves primarily interference with mitochondrial function. Mitochondrial energetics are critical to renal function by virtue of the large number and wide array of energy-dependent functions that are in turn critical to renal function. An understanding of the role and importance of energetics in renal function is dependent on recognition of factors such as the biochemical, physiological, and morphological heterogeneity of the nephron cell populations and on the variations in oxygenation of the various nephron cell types. Redox homeostasis also plays a central role in the regulation of mitochondrial function by virtue of the ability of mitochondria to generate ROS and RNS and because of the prevalence of critical sulfhydryl groups on mitochondrial proteins. Alterations in mitochondrial redox status, particularly that of GSH, can lead to the MPT and apoptosis or necrosis, depending on the severity of the toxic or pathologic insult.

Although not discussed in this chapter, another facet of renal toxicity and disease involves genetic defects in either the mitochondrial DNA or the nuclear DNA, as mitochondrial proteins are encoded by both mitochondrial and nuclear genes (Buemi et al., 1997; Rötig and Munnich, 2003). Defects in these genes are associated with specific tissue pathologies and diseases, affecting those organs that have a high degree of dependence on mitochondrial ATP generation for their function. Therefore, certain skeletal muscle, neural, and renal diseases have an etiology involving mutations or defects in genes that encode mitochondrial proteins.

For many of the examples of nephrotoxicants that exhibit prominent effects on renal mitochondria, there is debate as to how critical the role of mitochondrial dysfunction is in the progression of the nephrotoxicity. In some cases, the *in vivo* relevance of oxidative stress and the ensuing damage in mitochondria has been questioned (Anson et al., 2000). Therefore, it is important to carefully define experimental models and to control for potential artifacts. Nonetheless, it is clear that maintenance of proper mitochondrial function plays a central role in the etiology of numerous toxicologic and pathologic processes in the kidneys.

REFERENCES

Aleo MD, and Schnellmann RG (1992) The neurotoxicants strychnine and bicuculline protect renal proximal tubules from mitochondrial inhibitor-induced cell death. *Life Sci* 51:1783–1787.

Anson RM, Hudson E, and Bohr VA (2000) Mitochondrial endogenous oxidative damage has been overestimated. *FASEB J* 14:355–360.

Banki K, Elfarra AA, Lash LH, and Anders MW (1986) Metabolism of *S*-(2-chloro-1,1, 2-trifluoroethyl)-L-cysteine to hydrogen sulfide and the role of hydrogen sulfide in *S*-(2-chloro-1,1,2-trifluoroethyl)-L-cysteine-induced mitochondrial toxicity. *Biochem Biophys Res Commun* 138:707–713.

Banki K, and Anders MW (1989) Inhibition of rat kidney mitochondrial DNA, RNA and protein synthesis by halogenated cysteine *S*-conjugates. *Carcinogenesis* 10:767–772.

Barone MC, Darley-Usmar VM, and Brookes PS (2003) Reversible inhibition of cytochrome c oxidase by peroxynitrite proceeds through ascorbate-dependent generation of nitric oxide. *J Biol Chem* 278:27520–27524.

Bastin J, Cambon N, Thompson M, Lowry OH, and Burch HB (1987) Change in energy reserves in different segments of the nephron during brief ischemia. *Kidney Int* 31:1239–1247.

Beatrice MC, Stiers DL, and Pfeiffer DR (1984) The role of glutathione in the retention of Ca^{2+} by liver mitochondria. *J Biol Chem* 259:1279–1287.

Bellamy TC, Griffiths C, and Garthwaite J (2002) Differential sensitivity of guanylyl cyclase and mitochondrial respiration to nitric oxide measured using clamped concentrations. *J Biol Chem* 277:31801–31807.

Beltrán B, Quintero M, Garcia-Zaragozá E, O'Connor E, Esplugues JV, and Moncada S (2002) Inhibition of mitochondrial respiration by endogenous nitric oxide: A critical step in Fas signaling. *Proc Natl Acad Sci USA* 99:8892–8897.

Borutaite V, Budriunaite A, and Brown GC (2000) Reversal of nitric oxide-, peroxynitrite- and *S*-nitrosothiol-induced inhibition of mitochondrial respiration or complex I activity by light and thiols. *Biochim Biophys Acta* 1459:405–412.

Brady HR, Kone BC, Stromski ME, Zeidel ML, Giebisch G, and Gullans SR (1990) Mitochondrial injury: An early event in cisplatin toxicity to renal proximal tubules. *Am J Physiol* 258: F1181–F1187.

Brezis M, Shanley P, Silva K, Spokes K, Lear S, Epstein FH, and Rosen S (1985) Disparate mechanisms for hypoxic cell injury in different nephron segments: Studies in the isolated perfused rat kidney. *J Clin Invest* 76:1796–1806.

Brezis M, Rosen S, Silva P, Spokes K, and Epstein FH (1986) Mitochondrial activity: A possible determinant of anoxic injury in renal medulla. *Experientia* 42:570–572.

Brodie AE, and Reed DJ (1992) Glutathione disulfide reduction in tumor mitochondria after t-butyl hydroperoxide treatment. *Chem-Biol Interact* 84:125–132.

Broekmeier KM, Dempsey ME, and Pfeiffer DR (1989) Cyclosporin A is a potent inhibitor of the inner membrane permeability transition in liver mitochondria. *J Biol Chem* 264:7826–7830.

Brown PC, Sokolove PM, McCann DJ, Stevens JL, and Jones TW (1996) Induction of a permeability transition in rat kidney mitochondria by pentachlorobutadienyl cysteine: A β-lyase-independent process. *Arch Biochem Biophys* 331:223–231.

Buemi M, Allegra A, Rotig A, Gubler MC, Aloisi C, Corica F, Pettinato G, Frisina N, and Niaudet P (1997) Renal failure from mitochondrial cytopathies. *Nephron* 76:249–253.

Chavez E, Zazueta C, Osornio A, Holguin JA, and Miranda ME (1991) Protective behavior of captopril on Hg^{++}-induced toxicity on kidney mitochondria. *In vivo* and *in vitro* experiments. *J Pharmacol Exp Ther* 256:385–390.

Chen J, Doctor RB, and Mandel LJ (1994) Cytoskeletal dissociation of ezrin during renal anoxia: Role in microvillar injury. *Am J Physio*l 267:C784–C795.

Chen Y, Cai J, Anders MW, Stevens JL, and Jones DP (2001) Role of mitochondrial dysfunction in *S*-(1,2-dichlorovinyl)-L-cysteine-induced apoptosis. *Toxicol Appl Pharmacol* 170:172–180.

Chen Z, and Lash LH (1998) Evidence for mitochondrial uptake of glutathione by dicarboxylate and 2-oxoglutarate carriers. *J Pharmacol Exp Ther* 285:608–618.

Chen Z, Putt DA, and Lash LH (2000) Enrichment and functional reconstitution of glutathione transport activity from rabbit kidney mitochondria: Further evidence for the role of the dicarboxylate and 2-oxoglutarate carriers in mitochondrial glutathione transport. *Arch Biochem Biophys* 373:193–202.

Chernyak BV, and Bernardi P (1996) The mitochondrial permeability transition pore is modulated by oxidative agents through both pyridine nucleotides and glutathione at two separate sites. *Eur J Biochem* 238:623–630.

Cohen JJ (1979) Is the function of the renal papilla coupled exclusively to an anaerobic pattern of metabolism? *Am J Physiol* 236:F423–F433.

Cohen JJ (1986) Relationship between energy requirements for Na^+ reabsorption and other renal functions. *Kidney Int* 29:32–40.

Cojocel C (1997) Aminoglycoside nephrotoxicity, Chapter 7.26, in *Comprehensive Toxicology, Renal Toxicology*, vol. 7, Goldstein RS (Ed.), Elsevier Science Ltd., Cambridge, UK, pp. 495–523.

Cooper AJL, Wang J, Gartner CA, and Bruschi SA (2001) Co-purification of mitochondrial HSP70 and mature protein disulfide isomerase with a functional rat kidney high-Mr cysteine *S*-conjugate β-lyase. *Biochem Pharmacol* 62:1345–1353.

Cooper AJL, Bruschi SA, Iriarte A, and Martinez-Carrion M (2002a) Mitochondrial aspartate aminotransferase catalyses cysteine S-conjugate β-lyase reactions. *Biochem J* 368:253–261.

Cooper AJL, Bruschi SA, and Anders MW (2002b) Toxic, halogenated cysteine *S*-conjugates and targeting of mitochondrial enzymes of energy metabolism. *Biochem Pharmacol* 64:553–564.

Costantini P, Chernyak BV, Petronilli V, and Bernardi P (1996) Modulation of the mitochondrial permeability transition pore by pyridine nucleotides and dithiol oxidation at two separate sites. *J Biol Chem* 271:6746–6751.

Cummings BS, and Schnellmann RG (2002) Cisplatin-induced renal cell apoptosis: Caspase 3-dependent and -independent pathways. *J Pharmacol Exp Ther* 302:8–17.

de la Asuncion JG, Millan A, Pla R, Bruseghini L, Esteras A, Pallardo FV, Sastre J, and Vina J (1996) Mitochondrial glutathione oxidation correlates with age-associated damage to mitochondrial DNA. *FASEB J* 10:333–338.

Dickman KG, and Mandel LJ (1989) Glycolytic and oxidative metabolism in primary renal proximal tubule cultures. *Am J Physiol* 257:C333–C340.

Dickman KG, and Mandel LJ (1990) Differential effects of respiratory inhibitors on glycolysis in proximal tubules. *Am J Physiol* 258:F1608–F1615.

Doctor RB, Zhelev DV, and Mandel LJ (1997) Loss of plasma membrane structural support in ATP-depleted renal epithelia. *Am J Physiol* 272:C439–C449.

Elfering SL, Sarkela TM, and Giulivi C (2002) Biochemistry of mitochondrial nitric-oxide synthase. *J Biol Chem* 277:38079–38086.

Elzinga LW, Mela-Riker LM, Widener LL, and Bennett WM (1989) Renal cortical mitochondrial integrity in experimental cyclosporine nephrotoxicity. *Transplantation* 48:102–106.

Epstein FH (1997) Oxygen and renal metabolism. *Kidney Int* 51:381–385.

Esteve JM, Mompo J, de la Asuncion JG, Sastre J, Asensi M, Boix J, Vina JR, and Pallardo FV (1999) Oxidative damage to mitochondrial DNA and glutathione oxidation in apoptosis: Studies *in vivo* and *in vitro*. *FASEB J* 13:1055–1064.

Fine L (1986) The biology of renal hypertrophy. *Kidney Int* 29:619–634.

Fourner N, Ducet G, and Crevat A (1987) Action of cyclosporine on mitochondrial calcium fluxes. J *Bioenerg Biomembr* 19:297–303.

Ghafourifar P, Schenk U, Klein SD, and Richter C (1999) Mitochondrial nitric-oxide synthase stimulation causes cytochrome c release from isolated mitochondria: Evidence for intramitochondrial peroxynitrite formation. *J Biol Chem* 274:31185–31188.

Gonzalez VM, Fuertes MA, Alonso C, and Perez JM (2001) Is cisplatin-induced cell death always produced by apoptosis? *Mol Pharmacol* 59:657–663.

Griner RD, and Schnellmann RG (1994) Decreasing glycolysis increases sensitivity to mitochondrial inhibition in primary cultures of renal proximal tubule cells. *In Vitro Cell Dev Biol* 30A:30–34.

Guder WG, and Ross BD (1984) Enzyme distribution along the nephron. *Kidney Int* 26:101–111.

Guder WG, Wagner S, and Wirthensohn G (1986) Metabolic fuels along the nephron: Pathways and intracellular mechanisms of interaction. *Kidney Int* 29:41–45.

Gullans SR, Brazy PC, Soltoff SP, Dennis VW, and Mandel LJ (1982) Metabolic inhibitors: Effects on metabolism and transport in the proximal tubule. *Am J Physiol* 243:F133–F140.

Halestrap AP, Woodfield K-Y, and Connern CP (1997) Oxidative stress, thiol reagents, and membrane potential modulate the mitochondrial permeability transition by affecting nucleotide binding to the adenine nucleotide translocase. *J Biol Chem* 272:3346–3354.

Harris DCH, and Tay Y-C (1997) Mitochondrial function in rat renal cortex in response to proteinuria and iron. *Clin Exp Pharmacol Physiol* 24:916–922.

Harris DCH, Chan L, and Schrier RW (1988) Remnant kidney hypermetabolism and progression of chronic renal failure. *Am J Physiol* 254:F267–F276.

Hayden PJ, and Stevens JL (1990) Cysteine conjugate toxicity, metabolism, and binding to macromolecules in isolated rat kidney mitochondria. *Mol Pharmacol* 37:468–476.

Hayden PJ, Ichimura T, McCann DJ, Pohl LR, and Stevens JL (1991) Detection of cysteine conjugate metabolite adduct formation with specific mitochondrial proteins using antibodies raised against halothane metabolite adducts. *J Biol Chem* 266:18415–18418.

Hayden PJ, Welsh CJ, Yang Y, Schaefer WH, Ward AJI, and Stevens JL (1992) Formation of mitochondrial phospholipid adducts by nephrotoxic cysteine conjugate metabolites. *Chem Res Toxicol* 5:231–237.

Horster M (1978) Principles of nephron differentiation. *Am J Physiol* 235:F387–F393.

Hunter FE Jr, Scott A, Hoffsten PE, Gebicki JM, Weinstein J, and Schneider A (1964) Studies on the mechanism of swelling, lysis, and disintegration of isolated liver mitochondria exposed to mixtures of oxidized and reduced glutathione. *J Biol Chem* 239:614–621.

Jacobson HR (1981) Functional segmentation of the mammalian nephron. *Am J Physiol* 241:F203–F218.

James EA, Gygi SP, Adams ML, Pierce RH, Fausto N, Aebersold RH, Nelson SD, and Bruschi SA (2002) Mitochondrial aconitase modification, functional inhibition, and evidence for a supramolecular complex of the TCA cycle by the renal toxicant *S*-(1,1,2,2-tetrafluoroethyl)-L-cysteine. *Biochemistry* 41:6789–6797.

Ji Y, Akerboom TP, Sies H, and Thomas JA (1999) *S*-Nitrosylation and *S*-glutathionylation of protein sulfhydryls by *S*-nitrosoglutathione. *Arch Biochem Biophys* 362:67–78.

Jocelyn PC (1975) Some properties of mitochondrial glutathione. *Biochim Biophys Acta* 369:427–436.

Jocelyn PC, and Dickson J (1980) Glutathione and the mitochondrial reduction of hydroperoxides. *Biochim Biophys Acta* 590:1–12.

Jocelyn PC, and Kamminga A (1974) The non-protein thiol of rat liver mitochondria. *Biochim Biophys Acta* 343:356–362.

Jones DP, and Lash LH (1993) Introduction: Criteria for assessing normal and abnormal mitochondrial function, in *Mitochondrial Dysfunction*, Lash LH, and Jones DP (Eds), Academic Press, San Diego, pp. 1–7.

Jones TW, Wallin A, Thor T, Gerdes RG, Ormstad K, and Orrenius S (1986) The mechanism of pentachlorobutadienyl-glutathione nephrotoxicity studied with isolated rat renal epithelial cells. *Arch Biochem Biophys* 251:504–513.

Jung K, and Pergande M (1985) Influence of cyclosporin A on the respiration of isolated rat kidney motochondria. *FEBS Lett* 183:167–169.

Justo P, Lorz C, Sanz A, Egido J, and Ortiz A (2003) Intracellular mechanisms of cyclosporin A-induced tubular cell apoptosis. *J Am Soc Nephrol* 14:3072–3080.

Kaushal GP, Ueda N, and Shah SV (1997) Role of caspases (ICE/CED 3 proteases) in DNA damage and cell death in response to a mitochondrial inhibitor, antimycin A. *Kidney Int* 52:438–445.

Kiil F (1977) Renal energy metabolism and regulation of sodium reabsorption. *Kidney Int* 11:153–160.

Klein KL, Wang M-S, Torikai S, Davidson WD, and Kurokawa K (1981) Substrate oxidation by isolated single nephron segments of the rat. *Kidney Int* 20:29–35.

Kiyomiya K-i, Matsushita N, Matsuo S, and Kurebe M (2000) Cephaloridine-induced inhibition of cytochrome c oxidase activity in the mitochondria of cultured renal epithelial cells (LLC-PK_1) as a possible mechanism of its nephrotoxicity. *Toxicol Appl Pharmacol* 167:151–156.

Kohda Y, and Gemba M (2002) Enhancement of protein kinase C activity and chemiluminescence intensity in mitochondria isolated from the kidney cortex of rats treated with cephaloridine. *Biochem Pharmacol* 64:543–549.

Lash LH, and Anders MW (1986) Cytotoxicity of *S*-(1,2-dichlorovinyl)glutathione and *S*-(1,2-dichlorovinyl)-L-cysteine in isolated rat kidney cells. *J Biol Chem* 261:13076–13081.

Lash LH, and Anders MW (1987) Mechanism of *S*-(1,2-dichlorovinyl)-L-cysteine- and *S*-(1,2-dichlorovinyl)-L-homocysteine-induced renal mitochondrial toxicity. *Mol Pharmacol* 32:549–556.

Lash LH, and Jones DP (1996) Mitochondrial toxicity in renal injury, in *Methods in Renal Toxicology*, Zalups RK, and Lash LH (Eds), CRC Press, Boca Raton, pp. 299–329.

Lash LH, Elfarra AA, and Anders MW (1986) Renal cysteine conjugate *β*-lyase: Bioactivation of nephrotoxic cysteine *S*-conjugates in mitochondrial outer membrane. *J Biol Chem* 261:5930–5935.

Lash LH, Fisher JW, Lipscomb JC, and Parker JC (2000) Metabolism of Trichloroethylene. *Environ Health Perspec* 108 (Suppl. 2):177–200.

Lash LH, Qian W, Putt DA, Hueni SE, Elfarra AA, Krause RJ, and Parker JC (2001a) Renal and hepatic toxicity of trichloroethylene and its glutathione-derived metabolites in rats and mice: Sex-, species-, and tissue-dependent differences. *J Pharmacol Exp Ther* 297:155–164.

Lash LH, Hueni SE, and Putt DA (2001b) Apoptosis, necrosis and cell proliferation induced by *S*-(1,2-dichlorovinyl)-L-cysteine in primary cultures of human proximal tubular cells. *Toxicol Appl Pharmacol* 177:1–16.

Lash LH, Putt DA, Horky III SJ, and Zalups RK (2001c) Functional and toxicological characteristics of isolated renal mitochondria: Impact of compensatory renal growth. *Biochem Pharmacol* 62:383–395.

Lash LH, Putt DA, and Matherly LH (2002a) Protection of NRK-52E cells, a rat renal proximal tubular cell line, from chemical induced apoptosis by overexpression of a mitochondrial glutathione transporter. *J Pharmacol Exp Ther* 303:476–486.

Lash LH, Putt DA, Hueni SE, Cao W, Xu F, Kulidjian SJ, and Horwitz JP (2002b) Cellular energetics and glutathione status in NRK-52E cells: Toxicological implications. *Biochem Pharmacol* 64:1533–1546.

Lash LH, Putt DA, Hueni SE, Krause RJ, and Elfarra AA (2003) Roles of necrosis, apoptosis, and mitochondrial dysfunction in *S*-(1,2-dichlorovinyl)-L-cysteine sulfoxide-induced cytotoxicity in primary cultures of human renal proximal tubular cells. *J Pharmacol Exp Ther* 305:1163–1172.

Lash LH, Qian W, Putt DA, Desai K, Elfarra AA, Sicuri AR, and Parker JC (2002c) Renal toxicity of perchloroethylene and *S*-(1,2,2-trichlorovinyl)glutathione in rats and mice: Sex- and species-dependent differences. *Toxicol Appl Pharmacol* 179:163–171.

Lash LH, Sausen PJ, Duescher RJ, Cooley AJ, and Elfarra AA (1994a) Roles of cysteine conjugate *β*-lyase and *S*-oxidase in nephrotoxicity: Studies with *S*-(1,2-dichlorovinyl)-L-cysteine and *S*-(1,2-dichlorovinyl)-L-cysteine sulfoxide. *J Pharmacol Exp Ther* 269:374–383.

Lash LH, Tokarz JJ, Woods EB, and Pedrosi BM (1993) Hypoxia and oxygen dependence of cytotoxicity in renal proximal tubular and distal tubular cells. *Biochem Pharmacol* 45:191–200.

Lash LH, Tokarz JJ, and Woods EB (1994b) Renal cell type specificity of cephalosporin-induced cytotoxicity in suspensions of isolated proximal tubular and distal tubular cells. *Toxicology* 94:97–118.

Lash LH, Tokarz JJ, Chen Z, Pedrosi BM, and Woods EB (1996) ATP depletion by iodoacetate and cyanide in renal distal tubular cells. *J Pharmacol Exp Ther* 276:194–205.

Lash LH, Visarius TM, Sall JM, Qian W, and Tokarz JJ (1998) Cellular and subcellular heterogeneity of glutathione metabolism and transport in rat kidney cells. *Toxicology* 130:1–15.

Lash LH, Xu Y, Elfarra AA, Duescher RJ, and Parker JC (1995) Glutathione-dependent metabolism of trichloroethylene in isolated liver and kidney cells of rats and its role in mitochondrial and cellular toxicity. *Drug Metab Dispos* 23:846–853.

Lemasters JJ, Nieminen A-L, Qian T, Trost LC, Elmore SP, Nishimura Y, Crowe RA, Cascio WE, Bradham CA, Brenner DA, and Herman B (1998) The mitochondrial permeability transition in cell death: a common mechanism in necrosis, apoptosis and autophagy. *Biochim Biophys Acta* 1366:177–196.

Lê-Quôc K, and Lê-Quôc D (1985) Critical role of sulfhydryl groups in the mitochondrial inner membrane structure. *J Biol Chem* 260:7422–7428.

Lê-Quôc D, and Lê-Quôc K (1989) Relationships between the NAD(P) redox state, fatty acid oxidation, and inner membrane permeability in rat liver mitochondria. Arch *Biochem Biophys* 273:466–478.

Li N, Ragheb K, Lawler G, Sturgis J, Rajwa B, Melendez JA, and Robinson JP (2003) Mitochondrial complex I inhibitor rotenone induces apoptosis through enhancing reactive oxygen species production. *J Biol Chem* 278:8516–8525.

Liang M, and Knox FG (2000) Production and functional roles of nitric oxide in the proximal tubule. *Am J Physiol* 278:R1117–R1124.

Liang M, Croatt AJ, and Nath KA (2000) Mechanisms underlying induction of heme oxygenase-1 by nitric oxide in renal tubular epithelial cells. *Am J Physiol* 279:F728–F735.

Lock EA, and Schnellmann RG (1990) The effect of haloalkene cysteine conjugates on rat renal glutathione reductase and lipoyl dehydrogenase activities. *Toxicol Appl Pharmacol* 104:180–190.

Lotscher HR, Winterhalter KH, Carafoli E, and Richter C (1979) Hydroperoxides can modulate the redox state of pyridine nucleotides and the calcium balance in rat liver mitochondria. *Proc Natl Acad Sci USA* 76:4340–4344.

Lund B-O, Miller DM, and Woods JS (1991) Mercury-induced H_2O_2 production and lipid peroxidation *in vitro* in rat kidney mitochondria. *Biochem Pharmacol* 42 Suppl:S181–S187.

Lund B-O, Miller DM, and Woods JS (1993) Studies on Hg(II)-induced H_2O_2 formation and oxidative stress *in vivo* and *in vitro* in rat kidney mitochondria. *Biochem Pharmacol* 45:2017–2024.

Mandel LJ (1986) Primary active sodium transport, oxygen consumption, and ATP: Coupling and regulation. *Kidney Int* 29:3–9.

Maddaiah VT (1990) Glutathione correlates with lipid peroxidation in liver mitochondria of triiodothyronine-injected hypophysectomized rats. *FASEB J* 4:1513–1518.

Malis CD, and Bonventre JV (1986) Mechanism of calcium potentiation of oxygen free radical injury to renal mitochondria: A model for post-ischemic and toxic mitochondrial damage. *J Biol Chem* 261:14201–14208.

Masini A, Ceccarelli D, Trenti T, Gallesi D, and Muscatello U (1992) Mitochondrial inner membrane permeability changes induced by octadecadienoic acid hydroperoxide: Role of mitochondrial GSH pool. *Biochim Biophys Acta* 1101:84–89.

McKernan TB, Woods EB, and Lash LH (1991) Uptake of glutathione by renal cortical mitochondria. *Arch Biochem Biophys* 288:653–663.

Mela-Riker LM, Widener LL, Houghton DC, and Bennett WM (1986) Renal mitochondrial integrity during continuous gentamicin treatment. *Biochem Pharmacol* 35:979–984.

Meredith MJ, and Reed DJ (1983) Depletion *in vitro* of mitochondrial glutathione in rat hepatocytes and enhancement of lipid peroxidation by adriamycin and 1,3-bis(2-chloroethyl)-1-nitrosourea (BCNU). *Biochem Pharmacol* 32:1383–1388.

Miller GW, and Schnellmann RG (1993a) A novel low-affinity strychnine binding site on renal proximal tubules: Role in toxic cell death. *Life Sci* 53:1203–1209.

Miller GW, and Schnellmann RG (1993b) Cytoprotection by inhibition of chloride channels: The mechanism of action of glycine and strychnine. *Life Sci* 53:1211–1215.

Miller GW, and Schnellmann RG (1994) A putative cytoprotective receptor in the kidney: Relation to the neuronal strychnine-sensitive glycine receptor. *Life Sci* 55:27–34.

Moran JH, and Schnellmann RG (1997) Diverse cytoprotectants prevent cell lysis and promote recovery of respiration and ion transport. *Biochem Biophys Res Commun* 234:275–277.

Moore GA, Jewell SA, Bellomo G, and Orrenius S (1983) On the relationship between Ca^{2+} efflux and membrane damage during t-butyl hydroperoxide metabolism by liver mitochondria. *FEBS Lett* 153:289–292.

Olafsdottir K, and Reed DJ (1988) Retention of oxidized glutathione by isolated rat liver mitochondria during hydroperoxide treatment. *Biochim Biophys Acta* 964:377–382.

Olafsdottir K, Pascoe GA, and Reed DJ (1988) Mitochondrial glutathione status during Ca^{2+} ionophore-induced injury to isolated hepatocytes. *Arch Biochem Biophys* 263:226–235.

Park MS, De Leon, M, and Devarajan P (2002) Cisplatin induces apoptosis in LLC-PK_1 cells via activation of mitochondrial pathways. *J Am Soc Nephrol* 13:858–865.

Ravindranath V, and Reed DJ (1990) Glutathione depletion and formation of glutathione-protein mixed disulfide following exposure of brain mitochondria to oxidative stress. *Biochem Biophys Res Commun* 169:1075–1079.

Reed DJ, and Savage MK (1995) Influence of metabolic inhibitors on mitochondrial permeability transition and glutathione status. *Biochim Biophys Acta* 1271:43–50.

Rötig A, and Munnich A (2003) Genetic features of mitochondrial respiration chain disorders. *J Am Soc Nephrol* 14:2995–3007.

Sandau KB, Callsen D, and Brüne B (1999) Protection against nitric oxide-induced apoptosis in rat mesangial cells demands mitogen-activated protein kinases and reduced glutathione. *Mol Pharmacol* 56:744–751.

Sarkela TM, Berthiaume J, Elfering S, Gybina AA, and Giulivi C (2001) The modulation of oxygen radical production by nitric oxide in mitochondria. *J Biol Chem* 276:6945–6949.

Savage MK, and Reed DJ (1994) Release of mitochondrial glutathione and calcium by a cyclosporin A-sensitive mechanism occurs without large amplitude swelling. *Arch Biochem Biophys* 315:142–152.

Savage MK, Jones DP, and Reed DP (1991) Calcium- and phosphate-dependent release and loading of glutathione by liver mitochondria. *Arch Biochem Biophys* 290:51–56.

Scaduto RC Jr, Fattone VH II, Grotyohann LW, Wertz J, and Martin LF (1988) Effect of an altered glutathione content on renal ischemic injury. *Am J Physiol* 255:F911–F921.

Scarlett JL, Packer MA, Porteus CM, and Murphy MP (1996) Alterations to glutathione and nicotinamide nucleotides during the mitochondrial permeability transition induced by peroxynitrite. *Biochem Pharmacol* 52:1047–1055.

Schnellmann RG (1991) Renal mitochondrial glutathione transport. *Life Sci* 49:393–398.

Schnellmann RG, Cross TJ, and Lock EA (1989) Pentachlorobutadienyl-L-cysteine uncouples oxidative phosphorylation by dissipating the proton gradient. *Toxicol Appl Pharmacol* 100:498–505.

Schnellmann RG, Gilchrist SM, and Mandel LJ (1988) Intracellular distribution and depletion of glutathione in rabbit renal proximal tubules. *Kidney Int* 34:229–233.

Shanley PF, Rosen MD, Brezis M, Silva P, Epstein FH, and Rosen S (1986) Topography of focal proximal tubular necrosis after ischemia with reflow in the rat kidney. *Am J Pathol* 122:462–468.

Shertzer HG, Bannenberg GL, Zhu H, Liu R-M, and Moldéus P (1994) The role of thiols in mitochondrial susceptibility to iron and tert-butyl hydroperoxide-mediated toxicity in cultured mouse hepatocytes. *Chem Res Toxicol* 7:358–366.

Shirley DG, and Walter SJ (1991) Acute and chronic changes in renal function following unilateral nephrectomy. *Kidney Int* 40:62–68.

Sies H, and Moss KM (1978) A role of mitochondrial glutathione peroxidase in modulating mitochondrial oxidations in liver. *Eur J Biochem* 84:377–383.

Simmons CF Jr, Boguski RT, and Humes HD (1980) Inhibitory effect of gentamicin on renal mitochondrial oxidative phosphorylation. *J Pharmacol Exp Ther* 214:709–715.

Smith CV, Jones DP, Guenthner TM, Lash LH, and Lauterburg BH (1996) Compartmentation of glutathione: Implications for the study of toxicity and disease. *Toxicol Appl Pharmacol* 140:1–12.

Snowdowne KW, Freudenrich CC, and Borle AB (1985) The effects of anoxia on cytosolic free calcium, calcium fluxes, and cellular ATP levels in cultured kidney cells. *J Biol Chem* 260:11619–11626.

Soltoff SP (1986) ATP and the regulation of renal cell function, *Annu Rev Physiol* 48:9–31.

Stevens JL, Ayoubi N, and Robbins JD (1988) The role of mitochondrial matrix enzymes in the metabolism and toxicity of cysteine conjugates. *J Biol Chem* 263:3395–3401.

Strzelecki T, Kumar S, Khauli R, and Menon M (1988) Impairment by cyclosporine of membrane-mediated functions in kidney mitochondria. *Kidney Int* 34:234–240.

Takano T, Soltoff SP, Murdaugh S, and Mandel LJ (1985) Intracellular respiratory dysfunction and cell injury in short-term anoxia of rabbit renal proximal tubules. *J Clin Invest* 76:2377–2384.

Tessitore N, Sakhrani LM, and Massry SG (1986) Quantitative requirement for ATP for active transport in isolated renal cells. *Am J Physiol* 251:C120–C127.

Tune BM (1986) The nephrotoxicity of cephalosporin abtibiotics – Structure-activity relationships. *Comments Toxicol* 1:145–170.

Tune BM, and Hsu C-Y (1990) The renal mitochondrial toxicity of cephalosporins: Specificity of the effect on anionic substrate uptake. *J Pharmacol Exp Ther* 252:65–69.

Tune BM, and Hsu C-Y (1994) Toxicity of cephaloridine to carnitine transport and fatty acid metabolism in rabbit renal cortical mitochondria: Structure-activity relationships. *J Pharmacol Exp Ther* 270:873–880.

Tune BM, and Hsu C-Y (1995a) Effects of nephrotoxic β-lactam antibiotics on the mitochondrial metabolism of monocarboxylic substrates. *J Pharmacol Exp Ther* 274:194–199.

Tune BM, and Hsu C-Y (1995b) Toxicity of cephalosporins to fatty acid metabolism in rabbit renal cortical mitochondria. *Biochem Pharmacol* 49:727–734.

Tune BM, Sibley RK, and Hsu C-Y (1988) The mitochondrial respiratory toxicity of cephalosporin antibiotics. An inhibitory effect on substrate uptake. *J Pharmacol Exp Ther* 245:1054–1059.

Tune BM, Fravert D, and Hsu C-Y (1989) Oxidative and mitochondrial toxic effects of cephalosporin antibiotics in the kidney: A comparative study of cephaloridine and cephaloglycin. *Biochem Pharmacol* 38:795–802.

Uchida S, and Endou H (1988) Substrate specificity to maintain cellular ATP along the mouse nephron. *Am J Physiol* 255:F977–F983.

Ueda N, Guidet B, and Shah SV (1993) Gentamicin-induced mobilization of iron from renal cortical mitochondria. *Am J Physiol* 265:F435–F439.

Vamvakas S, Sharma VK, Sheu S-S, and Anders MW (1990) Perturbations of intracellular calcium distribution in kidney cells by nephrotoxic haloalkenyl cysteine *S*-conjugates. *Mol Pharmacol* 38:455–461.

Vamvakas S, Bittner D, Dekant W, and Anders MW (1992) Events that precede and that follow *S*-(1,2-dichlorovinyl)-L-cysteine-induced release of mitochondrial Ca^{2+} and their association with cytotoxicity to renal cells. *Biochem Pharmacol* 44:1131–1138.

van de Water B, Zietbey JP, de Bront HJGM, Mulder GJ, and Nagelkerke JF (1993) The relationship between intracellular Ca^{2+} and the mitochondrial membrane potential in isolated proximal tubular cells from rat kidney exposed to the nephrotoxin 1,2-dichlorovinyl-cysteine. *Biochem Pharmacol* 45:2259–2267.

van de Water B, Zoeteweij JP, de Bont HJGM, Mulder GJ, and Nagelkerke JF (1994) Role of mitochondrial Ca^{2+} in the oxidative stress-induced dissipation of the mitochondrial membrane potential: Studies in isolated proximal tubular cells using the nephrotoxin 1,2-dichlorovinyl-L-cysteine. *J Biol Chem* 269:14546–14552.

van de Water B, Zoeteweij JP, de Bont HJGM, and Nagelkerke JF (1995) Inhibition of succinate:ubiquinone reductase and decrease of ubiquinol in nephrotoxic cysteine *S*-conjugate-induced oxidative cell injury. *Mol Pharmacol* 48:928–937.

Walker PD, and Shah SV (1987) Gentamicin enhanced production of hydrogen peroxide by renal cortical mitochondria. *Am J Physiol* 253:C495–C499.

Walker LM, Walker PD, Imam SZ, Ali SF, and Mayeux PR (2000) Evidence for peroxynitrite formation in renal ischemia-reperfusion injury: Studies with the inducible nitric oxide synthase inhibitor L-N^6-(1-iminoethyl)lysine. *J Pharmacol Exp Ther* 295:417–422.

Walker LM, York JL, Iman SZ, Muldrew KL, and Mayeux PR (2001) Oxidative stress and reactive nitrogen species generation during renal ischemia. *Toxicol Sci* 63:143–148.

Wallace KB, Eells JT, Madeira VMC, Cortopassi G, and Jones DP (1997) Mitochondria-mediated cell injury. *Fund Appl Toxicol* 38:23–37.

Wallin A, Jones TW, Vercesi AE, Cotgreave I, Ormstad K, and Orrenius S (1987) Toxicity of *S*-pentachlorobutadienyl-L-cysteine studied with isolated rat renal cortical mitochondria. *Arch Biochem Biophys* 258:365–372.

Waters SL, Wong JK, and Schnellmann RG (1997) Depletion of endoplasmic reticulum calcium stores protects against hypoxia- and mitochondrial inhibitor-induced cellular injury and death. *Biochem Biophys Res Commun* 240:57–60.

Weinberg JM (1991) The cell biology of ischemic renal injury. *Kidney Int* 39:476–500.

Weinberg JM, and Humes HD (1980) Mechanisms of gentamicin-induced dysfunction of renal cortical mitochondira. I. Effects on mitochondrial respiration. *Arch Biochem Biophys* 205:221–231.

Weinberg JM, Harding PG, and Humes HD (1982a) Mitochondrial bioenergetics during the initiation of mercuric chloride-induced renal injury. I. Direct effects of *in vitro* mercuric chloride on renal cortical mitochondrial function. *J Biol Chem* 257:60–67.

Weinberg JM, Harding PG, and Humes HD (1982b) Mitochondrial bioenergetics during the initiation of mercuric chloride-induced renal injury. II. Functional alterations of renal cortical mitochondria isolated after mercuric chloride treatment. *J Biol Chem* 257:68–74.

Weinberg JM, Davis JA, Abarzua M, Kiani T, and Kunkel R (1990) Protection by glycine of proximal tubules from injury due to inhibitors of mitochondrial ATP production. *Am J Physiol* 258:C1127–C1140.

Weinberg JM, Venkatachalam MA, Roeser NF, and Nissim I (2000) Mitochondrial dysfunction during hypoxia/reoxygenation and its correction by anaerobic metabolism of citric acid cycle intermediates. *Proc Natl Acad Sci USA* 97:2826–2831.

Wong PS-Y, and Fukuto JM (1999) Reaction of organic nitrate esters and *S*-nitrosothiols with reduced flavins: A possible mechanism of bioactivation. *Drug Metab Dispos* 27:502–509.

Yagi T, and Hatefi Y (1984) Thiols in oxidative phosphorylation: Inhibition and energy-potentiated uncoupling by monothiol and dithiol modifiers. *Biochemistry* 23:2449–2455.

Zalups RK, and Lash LH (1994) Recent advances in understanding the renal transport and toxicity of mercury. *J Toxicol Environ Health* 42:1–44.

Zhang C, and Mayeux PR (2001) NO/cGMP signaling modulates regulation of Na^+-K^+-ATPase activity by angiotensin II in rat proximal tubules. *Am J Physiol* 280:F474–F479.

Zhang J-G, and Lindup W (1993) Role of mitochondria in cisplatin-induced oxidative damage exhibited by rat renal cortical slices. *Biochem Pharmacol* 45:2215–2222.

Zhang J-G, and Lindup WE (1994) Cisplatin nephrotoxicity: Decreases in mitochondrial protein sulfhydryl concentration and calcium uptake by mitochondria from rat renal cortical slices. *Biochem Pharmacol* 47:1127–1135.

12

Adhesion Molecules in Renal Physiology and Pathology

Alan R. Parrish

INTRODUCTION

The importance of cell adhesion molecules in the context of tissue architecture has long been recognized and is a critical area of cell biology. Interest in this area has exploded in the past decade with the identification of specific molecules mediating cell–cell adhesion and, importantly, the organization and structure of distinct cell adhesion complexes. This chapter focuses on cell adhesion molecules in the kidney, including discussion of both cell–extracellular matrix and homotypic and heterotypic adhesion complexes. While it is apparent that, in addition to a critical structural role, cell adhesion molecules may also be important signaling modulators (reviewed in Juliano, 2002), a focus will be given to the cell–extracellular matrix or cell–cell adhesion role of these complexes.

A great deal of progress has been made in our understanding of the cell adhesion molecules mediating cell–extracellular or cell–cell interactions in the kidney. Most of our knowledge is derived from epithelial cells in the proximal and distal tubule; this chapter therefore focuses in large part on tubular epithelial cell adhesion, although some information on glomerular complexes is included.

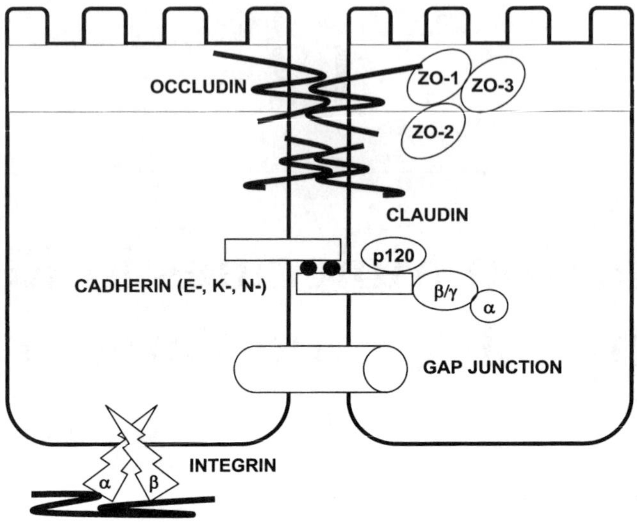

Figure 12.1 Representation of the cell adhesion complexes present in proximal tubular epithelial cells. Most proximal to the apical membrane is the tight junction, composed of transmembrane (occludin and claudins) and cytosolic proteins (ZO-1, ZO-2, ZO-3). ZO-1 binds directly to occludin, and is turn bound by ZO-2 and ZO-3. The adherens junction (cadherin/catenin complex) is located basal to the tight junction. Cadherins (E-, K-, N-) are single transmembrane proteins whose cytoplasmic domain is bound by either *β*- or *γ*-catenin. *α*-Catenin does not directly bind cadherins, but links the complex to the actin cytoskeleton (directly or indirectly) via *β*- or *γ*-catenin. $p120^{ctn}$ is thought to mediate the lateral clustering of cadherins. Ksp-cadherin also mediates cell–cell adhesion, but does not interact with catenins. Desmosomes are also cadherin-based adhesion complexes that also mediate homotypic cell adhesion (not shown). Gap junctions are also critical for cell–cell interactions, and require functional adhesion complexes to mediate intercellular communication between adjacent cells. Integrin receptors are heterodimers of *α*- and *β*-subunits that are linked to the actin cytoskeleton and bind extracellular matrix proteins within the basement membrane.

Tubular epithelial cells are attached to the basement membrane via integrin receptors and are linked to neighboring cells via several homotypic cell–cell adhesion complexes (Figure 12.1). These epithelial cells, despite several unique segment-specific functions, share a common need for establishing and maintaining cell polarity (reviewed in Wagner and Molitoris, 1999). Unique apical and basolateral domains separate the glomerular filtrate from the blood and allow for the movement of ions, water, and macromolecules to and from each compartment. These domains can be separated by structural, biochemical, and physiological parameters. It is thought that attachments to the basement membrane, as well as cell–cell adhesion complexes, are critical for cell polarity. Tight junctions, cadherin and catenin complexes, and gap junctions mediate tubular epithelial cell interactions and have distinct localization in cells.

While not discussed in this chapter, both cell–extracellular and cell–cell adhesion complexes are linked to the cytoskeleton, which also has an important role in cell polarity. Apical surface membrane amplification is actin-dependent; cell–cell/cell–matrix adhesion requires actin and intermediate filaments; and intracellular transport requires both microtubules and actin. It has become clear that both integrin and cell adhesion molecules are components of highly involved cell complexes that also contain cytoskeletal and signaling related proteins. The role that disruption of cell adhesion complexes may play in renal injury is also receiving increased attention (reviewed in Molitoris and Marrs, 1999; Wagner and Molitoris, 1999) and is discussed. Important in this aspect is the recent attention given the importance of adhesion molecules mediating inflammatory target cell interactions (heterotypic cell adhesion) in the kidney. This chapter discusses what is currently understood about the localization of cell adhesion molecules in the kidney. Additionally, the disruption of cell adhesion during renal injury is addressed in detail.

RENAL CELL ADHESION MOLECULES

Cell–Extracellular Matrix

Integrin superfamily members are cell-surface glycoproteins that predominately act as receptors for extracellular matrix proteins such as collagen, fibronectin, and laminin. Integrin receptors are heterodimers, each consisting of an α and β subunit. Each of these subunits has a large extracellular domain, a single transmembrane domain, and a short cytoplasmic domain (except $\beta4$) that is ultimately linked to the actin cytoskeleton. At least 18 α and eight β subunits have been identified in vertebrates (reviewed in Juliano, 2002). Interestingly, the ligand affinity of integrin receptors is regulated by extracellular (divalent cations) or intracellular (small GTPases such as R-Ras and Rap1) signals.

The cytoplasmic domain of integrin receptors is part of a large complex of signaling- and cytoskeletal-related proteins. Clustering of integrins and these proteins at sites of cell–extracellular matrix adhesion forms a complex, specialized structure commonly referred to as a focal adhesion or focal contact that attaches the cell to the extracellular matrix. Focal adhesion kinase (FAK) is a nonreceptor tyrosine kinase that is tyrosine phosphorylated and activated by integrin-mediated adhesion, although the protein does not directly associate with integrins (Hanks et al., 1992; Schaller et al., 1992). Conversely, dephosphorylation occurs rapidly when cells are detached from the extracellular matrix. FAK is capable of binding to a number of other proteins, including c-src, phosphatidylinositol (PI)-3-kinase, paxillin, talin, and p130cas. Recently, FAK has been shown to play a critical role in the regulation of cell motility and apoptosis (reviewed in Juliano, 2002).

The spatial pattern of integrin subunit expression in the normal human kidney has been determined (Patey et al., 1994). The $\beta 1$ integrin subunit is expressed ubiquitously, while $\beta 3$ is expressed only in the epithelium of glomeruli and tubules. $\beta 2$ and $\beta 4$ are not detected in the normal human kidney. The $\alpha 1$ subunit is expressed in parietal epithelial and mesangial cells in the glomeruli, and in both the proximal and distal tubular epithelium. $\alpha 2$ is expressed in mesangial cells and distal tubules, while $\alpha 3$ is expressed in all glomerular cells and the distal tubules. $\alpha 5$ expression is mainly confined to endothelial cells in the kidney, while $\alpha 6$ is expressed in all cells. The αv subunit is expressed in both proximal and distal tubules and in the glomerular epithelium. Of the many possible subunit combinations, only a few have been shown to play a significant role in renal development and function, mainly $\alpha 1\beta 1$, $\alpha 2\beta 1$, $\alpha 3\beta 1$, $\alpha 6\beta 1$, $\alpha 8\beta 1$, and $\alpha v\beta 3$ (reviewed in Hamerski and Santoro, 1999).

The distribution of extracellular matrix proteins in the normal human kidney has also been investigated. The $\alpha 1$ and $\alpha 2$ chains of collagen IV were found throughout the kidney, while the $\alpha 3$ chain of collagen IV was not present in the mesangial space or the basement membrane of proximal tubules. Laminin was also ubiquitously expressed, while fibronectin and collagen I were limited to the mesangial space of the glomeruli.

Significant progress has been made in identifying the spatial pattern of integrin receptors and their ligands (extracellular matrix proteins) along the nephron. While the relevance of the distinct localization of both integrins and extracellular matrix protein in regards to functional differences between nephron segments is not clear, this information provides a basis to investigate the signaling process regulating, and regulated by, distinct integrin receptor–ligand interactions in the kidney.

Homotypic Cell–Cell

Tight Junctions

Tight junctions (zonula occludens) were first described as a dense condensation at the apical end of polarized cells where the plasma membranes of adjacent cells appeared to fuse. The tight junction has two critical functions; the "fence" function prevents intradomain movement of membrane proteins and phospholipids, while the "gate" function blocks the paracellular movement of solutes. More recently, much effort has focused on identifying the components of tight junctions. In addition to the structural components of the tight junction, a wide variety of proteins involved in cell signaling, cytoskeleton modulation, and vesicular trafficking are concentrated at the tight junction (reviewed in Lapierre, 2000).

Occludin, claudins, and junctional adhesion molecule (JAM) are all transmembrane components of tight junctions. Occludin has four transmembrane domains, two extracellular domains, and a long cytoplasmic (carboxy-terminal) domain. Occludin is a phosphoprotein, and it is the hyperphosphorylated form that is the main component within a functional tight junction (Wong, 1997). Claudins are a multigene family of transmembrane proteins that have a similar structure as occludin (four transmembrane domains and two extracellular domains). At least eight claudins have been identified, although the potential for additional claudins exists (Tsukita and Furuse, 1999). All carboxy-terminal tails of claudins contain a PDZ binding motif that may promote interaction with cytoplasmic protein constituents of tight junctions. JAM is a member of the IgG superfamily and is structurally distinct from occludin and claudins. Based on homology with IgG family members, it is thought that JAM may play a role in the transmigration of inflammatory cells during an immune reaction (Martin-Padura et al., 1998). ZO-1, ZO-2, and ZO-3 are cytoplasmic proteins that are members of the membrane-associated guanylate kinase (MAGUK) gene superfamily; common structural features are three PDZ domains, an SH3 domain, and a guanylate kinase homology region (GUK) (Fanning et al., 1996). ZO-1 can bind occludin, while both ZO-2 and ZO-3 bind to ZO-1 but not each other. It is thought that ZO-1 interacts with actin to link the tight junction complex to the cytoskeleton.

In the kidney, the proximal tubule has a "leaky" tight junction and the transepithelial electrical resistance (TER) and complexity of the tight junction increases from the proximal tubule to the collecting duct. Therefore, efforts have focused on determining the spatial expression pattern of tight junction proteins along the nephron.

In the rabbit kidney, immunfluorescence was used to detect occludin, ZO-1, and ZO-2 expression (Gonzalez-Mariscal et al., 2000). Occludin staining was quite weak in the proximal tubules, with much more intense staining in the distal tubules. ZO-1 was found in all tubules, but was expressed at higher levels in distal segments than in proximal. ZO-2 was diffusely expressed in the proximal tubules with weak cell border staining, but was highly expressed at cell borders from loop of Henle to collecting ducts. In mouse kidney, claudins-1 and -2 are expressed in the Bowman's capsule, while claudins-2, -10, and -11 are found in proximal tubules (Enck et al., 2001; Kiuchi-Saishin et al., 2002). In the thin descending limb of Henle, claudin-2 is expressed, and claudins-3, -4, and -8 are localized to the thin ascending limb of Henle. Claudins-3, -10, -11, and -16 are expressed in the thick ascending limb of Henle, while claudins-3 and -8 are present in the distal tubule and claudins-3, -4, and -8 in the collecting duct.

It is postulated that claudin-2 might form a "leakier" tight junction than claudin-3, based on localization along the nephron. The finding that enforced expression of claudin-2 is associated with the conversion from "tight" to "leaky" junctions in Madin-Darby canine kidney (MDCK) cells supports this hypothesis (Furuse et al., 2001). However, the higher expression of occludin, ZO-1, and ZO-2 in distal tubules than in proximal tubules probably also contributes to differences in the "strength" of tight junctions along the nephron.

While the localization of tight junction components has been well defined in the kidney, and the pattern of protein expression appears to dictate the relative "tightness" of the junction, much research remains to be done. For example, little is known about whether the signaling proteins that are concentrated at tight junctions are affected by the composition of the tight junction. In addition, it can be hypothesized that the strength of the tight junction in the distal tubules as compared with the proximal tubules may mediate, in part, the relative susceptibility of these different regions to injury.

Cadherin/Catenin Complexes

The cadherin gene superfamily encodes for transmembrane proteins that regulate Ca^{2+}-dependent cell–cell adhesion (Wu and Maniatis, 1999). Type-I (classical) cadherins comprise a large family of proteins that share significant structural conservation and include E-cadherin (epithelial) (L-cell adhesion molecule [CAM], uvomorulin), N-cadherin (neuronal) (A-CAM), and P-cadherin (placental) (reviewed in Kemler, 1992). The extracellular domain contains four highly conserved Ca^{2+}-binding "cadherin repeats" (EC1–4) and one membrane proximal extracellular domain (EC5). Type-II (atypical) cadherins have a similar structure, but lack a conserved His–Ala–Val (HAV) sequence in the EC1 domain. The desmocollins and desmogleins are localized at desmosomal junctions, while the remaining cadherin molecules have low homology to type-I cadherins (less than 44% at the amino acid level) and have from 6 to 34 cadherin repeats. Antibodies to type-I cadherins immunoprecipitate several cytoplasmic proteins; α-, β-, and γ-catenin (plakoglobin), with molecular weights of 102, 94, and 86 kDa, respectively. α-Catenin is linked to the cytoplasmic domain of cadherins via β- or γ-catenin; α-catenin does not directly bind to cadherin. This finding is supported by studies demonstrating two mutually exclusive cadherin/catenin complexes in cells; β-/α- or γ-/α-catenin (Nathke et al., 1994). Due to homology with vinculin, it is suggested that α-catenin links the cadherin/catenin complex to the cytoskeleton (Nagafuchi et al., 1991; Tsukita et al., 1992). A more recently identified protein, $p120^{ctn}$, binds to the cadherin cytoplasmic domain and shares sequence homology with β- and γ-catenin but does not bind α-catenin (Reynolds et al., 1994).

The renal expression of cadherin molecules is complex, with reports of at least six cadherins present in the kidney (E-, K-, Ksp-, N-, P-, and T-cadherin). In addition to the developmental regulation of cadherin expression, evidence is accumulating that cadherins and catenins are differentially expressed along the nephron.

In the developing human kidney, N-cadherin is expressed in the proximal tubule and thin limb and E-cadherin is expressed in the thin limb, distal tubule, and collecting tubule (Nouwen et al., 1993). In the developing mouse kidney, the proximal tubule progenitors express K-cadherin (a type-II cadherin), while the distal tubule expresses E-cadherin, and the glomeruli are positive for P-cadherin (Cho et al., 1998). K-cadherin has low homology with N- (38%), E- (35%), and

P-cadherin (32%), and is expressed at high levels in fetal human kidney and at low levels in adult kidney (Xiang et al., 1994). In adult mouse kidney, E-cadherin is detected everywhere but the initial segment where the proximal tubule joins Bowman's capsule (Piepenhagen et al., 1995). This pattern of expression contrasts with that of adult human kidney, where E-cadherin is not detected in the proximal tubule (Nouwen et al., 1993). Both α- and β-catenin are expressed in all nephron segments (adult mouse), while γ-catenin is only detected in distal part of nephron (Piepenhagen and Nelson, 1995). It has been shown that $p120^{ctn}$ is localized in the proximal, but not distal, tubules (Parrish, unpublished observation).

Aronson and colleagues have recently cloned Ksp-cadherin, a 130 kDa protein that lacks the prosequence and HAV adhesion recognition sequence and possesses a truncated cytoplasmic domain that does not interact with the catenins (Thomson et al., 1995). Although Ksp-cadherin is expressed in the developing mouse kidney (Wertz and Herrmann, 1999) and in the developing and adult rabbit kidney (Thomson and Aronson, 1999), the specific role of this cadherin remains unclear.

Given that E- and N-cadherin are highly homologous, the functional relevance of the localization of N-cadherin in the proximal tubules and E-cadherin in the distal tubules is unclear. As specific interactions between components of tight junctions and cadherin/catenin complexes, specifically ZO-1/ZO-2 and α-catenin, have been identified (Itoh et al., 1997, 1999), it is possible that, along with distinct tight junction component expression, the cadherin and catenin expression pattern may also dictate the relative permeability of proximal and distal tubules.

Gap Junctions

Gap junctions are specialized structures comprised of connexins (Cx). Connexins are a multigene family with over 12 members, ranging in size from 25 to 50 kDa. Each connexin has four transmembrane domains and two extracellular loops, each containing three critical cysteine residues involved in disulfide bond formation. Six connexins assemble into a hemichannel (connexon), which interacts with a connexon on a neighboring cell to form a channel (reviewed in Yeager et al., 1998). The channel (gap junction) allows for intercellular communication via the exchange of nutrients, metabolites, ions, and small molecules (less than 1 kDa).

In the kidneys of 4-day-old mice, Cx45 was detected in glomeruli and distal tubules, while Cx32 and 26 were coexpressed in proximal tubules (Butterweck et al., 1994). Cx43 was not detected at this stage of development. However, Cx43 mRNA is expressed in adult rat kidney.

Proximal convoluted tubules and proximal straight tubules express significantly less Cx43 than inner (Guo et al., 1998) medullary collecting ducts, glomeruli, and cortical collecting ducts. Although it is expressed in the adult rodent kidney, Cx43 is not critical for renal development given that Cx43 knockout mice develop normal kidneys (Silverstein et al., 2001). At this time, it is unknown if there are segment-specific differences in gap junction expression and function.

Heterotypic Cell–Cell (Inflammation)

Ig CAM superfamily members (ICAMs, PECAM, VCAM) and selectins, as well as integrins, are leukocyte adhesion molecules. Inflammation is a key component of many nephropathies, including toxicant-induced acute renal failure. Interest in these adhesion molecules was initiated when the antigen presentation function of renal tubular epithelial cells was described by Kelley and Singer (1993). Tubular epithelial cells express both ICAM-1 and VCAM-1, which mediate adherence to inflammatory cells. ICAMs are members of the Ig CAM superfamily and are ligands for the $\beta2$ integrin present on leukocytes (reviewed in Hubbard and Rothlein, 2000). ICAMs have two or more Ig-like repeats in the extracellular domain, a single transmembrane domain, and a short cytoplasmic tail. Selectins play a major role in the adhesion of leukocytes to endothelial cells and platelets during inflammation. L-selectin (leukocyte), E-selectin (endothelial), and P-selectin (platelet) have conserved structures and mediate heterotypic cell adhesion via a calcium-dependent mechanism (reviewed in Juliano, 2002).

A complex pattern of expression of adhesion molecules is seen in the kidney, with multiple components of cell–matrix and cell–cell (both homotypic and heterotypic) expressed at distinct sites along the nephron (Figure 12.2). While this is a major advance in the understanding of cell adhesion in the kidney, many important questions remain to be addressed. The relationship between expression of distinct adhesion

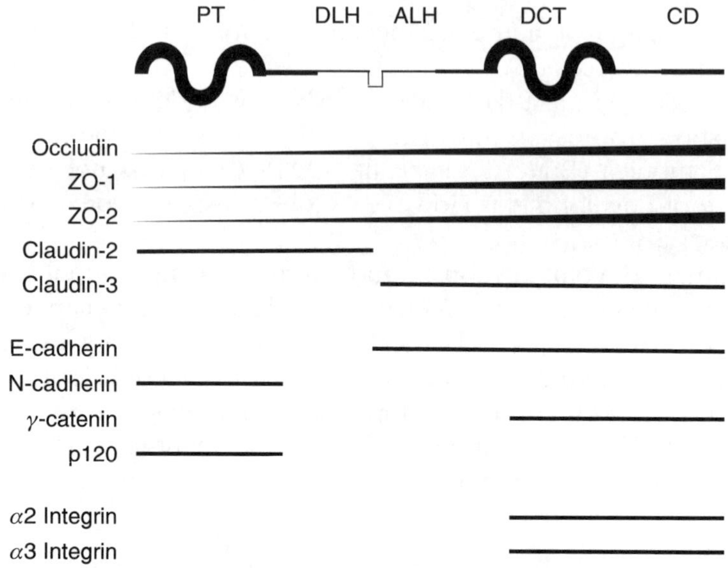

Figure 12.2 Schematic representation of differential expression of cell adhesion proteins along the nephron. The tight junction components occludin, ZO-1, and ZO-2 are expressed in the proximal tubules (PT), but the levels are much higher in the distal convoluted tubules (DCT) and collecting ducts (CD). In addition, claudin-2 appears to be expressed in the proximal tubules, while claudin-3 expression is limited to the distal tubules and collecting ducts. A number of other of claudins are also expressed in the kidney in a complex pattern. N-cadherin and $p120^{ctn}$ are expressed in the proximal tubules, while E-cadherin and γ-catenin are primarily expressed in the distal tubules. Ksp-cadherin and α-/β-catenin are expressed all along the nephron. The $\alpha 2$ and $\alpha 3$ integrin subunits are expressed only in the distal tubules. The $\beta 1$, $\beta 3$, $\alpha 1$, $\alpha 6$, and αv integrin subunits are expressed at the same levels in proximal and distal tubules. (DLH) descending loop of Henle, (ALH) ascending loop of Henle (thin and thick). (Adapted from Piepenhagen et al. (1995).)

molecules and critical transporters along the nephron has yet to be defined. Furthermore, it is increasingly clear that adhesion complexes such as integrins, tight junctions, and cadherin/catenin complexes also regulate critical signaling pathways. The possibility that distinct cell adhesion complexes differentially regulate these signaling processes in a segment-specific manner also exists. A great deal, therefore, remains to be defined about the role of cell adhesion in normal renal development and function.

DISRUPTION OF RENAL CELL ADHESION MOLECULES

Cell-Matrix

It has been over a decade since the first report suggesting that acute injury in the kidney is associated with disruption of integrin receptors. In a pair of seminal manuscripts, Goligorsky and colleagues laid the groundwork for a number of studies investigating the role of integrin receptors in renal injury. It was first demonstrated that nonlethal oxidative stress disrupts focal adhesions and is associated with a redistribution of the $\alpha 3$ subunit from the basal to apical domain of cultured renal epithelial cells (Gailit et al., 1993). It was hypothesized that renal tubular obstruction was mediated, in part, by detachment of viable tubular cells from the basement membrane and adhesion of detached cells to tubular cells via redistribution of integrin receptors from the basal to the apical surface (Goligorsky et al., 1993). The subsequent finding that cyclic RGD peptides attenuate ischemic-reperfusion injury in rats added strength to this hypothesis (Noiri et al., 1994). The major mechanism of protection was related to inhibition of cell–cell adhesion, while inhibition of cell–extracellular matrix adhesion was of less significance. Since these initial findings, a number of studies have investigated the role of integrins in renal pathologies.

The impact of injury on integrins in renal cells continues to be an area of interest. Primary cultures of mouse proximal tubular cells subjected to ATP depletion via cyanide in the absence of dextrose (1 hr) have significant changes, including disrupted cell–extracellular matrix adhesion (Kroshian et al., 1994). In addition, mannitol permeability was significantly increased, suggesting that cell–cell adhesion was also disrupted. In a proximal tubule cell line (JTC-12), 10 min exposure to hydrogen peroxide (10 to 50 m*M*) inhibited cell–matrix adhesion (Nigam et al., 1998). During recovery from toxicant exposure, $\alpha 6$ expression increased and correlated with increased cell–extracellular matrix adhesion.

More-recent work has investigated the molecular pathways associated with injury-induced disruption of renal integrins. ATP depletion of mouse inner medullary collecting duct-3 cells resulted in decreased cell–extracellular matrix adhesion that was reversible by the addition of exogenous hepatocyte growth factor (HGF) (Liu et al., 2001). Interestingly, HGF induced extracellular signal-related kinase (ERK) phosphorylation and blockade of ERK activation with U0126 completely prevented the HGF-induced restoration of cell–extracellular matrix adhesion. Primary cultures of rat proximal tubular cells were challenged by oxidative stress, and integrins were antagonized during the recovery phase by a GRGD peptide or monoclonal $\beta 1$ antibody (Wijesekera et al., 1997). Inhibition of integrins during recovery was associated with increased apoptosis and it was demonstrated that activation of protein kinase C prevented the GRGD peptide-induced increase in apoptosis. Nephrotoxicant-induced apoptosis has been demonstrated to involve alterations in FAK activity that precede activation of caspase-3 (van de Water et al., 1999, 2001). Therefore, intracellular signaling pathways regulated by kinases may play a role in the integrin-related cell death and the recovery of integrin function following renal injury.

The delocalization of integrin receptor subunits from the basal to the apical domain of cells also remains a focus of research efforts. ATP depletion of mouse proximal tubular cells causes loss of focal contacts and redistribution of the $\beta 1$ integrin to the apical domain (Lieberthal et al., 1997). Interestingly, the integrin subunit is functional, as assessed by binding of RGD-coated beads. In contrast, another study demonstrated that during reperfusion following ischemia, $\beta 1$ integrin subunits appear on the lateral borders of epithelial cells, but not on the apical surface (Zuk et al., 1998). The $\beta 1$ subunit is not detected in exfoliated cells following injury, however, fibronectin is detected in the tubular spaces of the S_3 segment of the proximal tubule and at high levels in the distal tubule. The contribution of fibronectin to acute renal failure is unknown, but it is postulated that the appearance of functional integrins on the apical surface of tubular epithelial cells coupled with both the exfoliation of viable cells and the appearance of extracellular matrix proteins in the tubular lumen could contribute to tubular blockage during acute renal failure.

The role of integrin ligands (extracellular matrix proteins) in renal injury has also been investigated. In renal proximal tubular cells, *S*-(1,2-dichlorovinyl)-L-cysteine was associated with a decrease in the basal localization of integrin subunits, and an appearance of integrins on the apical membrane (Nony and Schnellmann, 2001). Interestingly, exogenous collagen IV, but not function-stimulating antibodies to integrin subunits, promoted the basal relocalization of integrin subunits. This suggests that outside-in signaling via ligand binding may also play a role in the recovery of integrin function following injury.

Another integrin ligand that has been studied during renal injury is osteopontin (OPN), a secreted glycoprotein (reviewed in Xie et al., 2001). OPN is expressed primarily in the loop of Henle and distal tubules, however it is upregulated in all tubular segments and glomeruli following renal injury. The role of OPN in renal injury is unclear; although it has a role in macrophage-mediated injury, OPN may also be protective via induction of tolerance to acute ischemia, reduction of cellular peroxide levels, prevention of apoptosis, and stimulation of regeneration following injury. It has been shown that OPN expression is not sufficient to serve as the principal mediator of monocyte or macrophage influx in cyclosporine toxicity in the human kidney (Hudkins et al., 2001), although a role for OPN in renal inflammation cannot be excluded.

Taken together, studies over the past decade have provided evidence for a critical role of integrin receptors and ligands in both renal injury and recovery. Most of the work has, however, focused on determining expression and localization of these proteins during injury and recovery and much remains to be done on the molecular mechanisms that may regulate the role of integrins and their ligands in renal pathologies.

Homotypic Cell–Cell

Tight Junctions and Cadherin/Catenin Complexes

A "backleak" of glomerular filtrate into the interstitium and venous system has been demonstrated in animal models of renal injury (Donohoe et al., 1978) and in human patients with renal failure (Moran and Myers, 1985; Myers et al., 1979). In addition, Racusen et al. (1991) demonstrated that acute renal failure (ARF) patients shed viable renal tubular cells in their urine.

More-recent work has investigated the impact of ARF on cell–cell adhesion proteins. In allografts classified as sustained ARF ($n = 19$), 57% of the filtrate (inulin) was lost to transtubular backleak (Kwon et al., 1998). No backleak was detected in recovering ARF patients ($n = 20$). Histochemical analysis revealed significant abnormalities of proximal tubular epithelial cells, but not distal tubules. Staining for ZO-1 and α-, β-, and γ-catenin revealed diminished intensity and redistribution from the apico-lateral domain. These data provide compelling evidence for a disruption of cell–cell adhesion in renal failure and provide a rationale for investigating the mechanisms underlying disruption of adhesion in nephrotoxicity.

A number of studies have investigated both tight junctions and the cadherin/catenin complex following renal injury in cell lines. ATP depletion of mouse proximal tubular cells causes a significant increase in the tyrosine phosphorylation of β- and γ-catenin and diminished staining of E-cadherin at the basolateral membrane (Schwartz et al., 1999). This can be prevented by genistein, a tyrosine kinase inhibitor, and replicated with vanadate, a protein tyrosine phosphatase inhibitor. Genistein also attenuated the reduction of TER associated with ATP depletion. Cadmium has been shown to disrupt E-cadherin dependent cell adhesion in a number of cell lines (reviewed in Prozialeck, 2000), putatively via displacement of Ca^{2+} from the extracellular binding domain. This hypothesis is supported by the finding that cadmium can bind E-cadherin, with a dissociation constant approximating 20 μM (Prozialeck et al., 1996) and that cadmium can bind the Ca^{2+} binding site on a synthetic polypeptide corresponding to the Ca^{2+} binding region of E-cadherin (Prozialeck and Lamar, 1999). Zimmerhackl et al. (1998) have shown that cadmium is more toxic to LLC-PK_1 cells (proximal tubule) than to MDCK cells (distal tubule and collecting duct), putatively via disruption of E-cadherin linkage to the actin cytoskeleton.

Diatrizoate and ioxaglate are ionic radiocontrast agents that elicit E-cadherin, ZO-1, and occludin redistribution in parallel with a loss of TER in MDCK cells (Schick et al., 2002). Although these compounds reduced extracellular Ca^{2+}, the same pattern of injury was not replicated by EGTA, and raising the extracellular Ca^{2+} level to control levels in toxicant-challenged cultures did not abrogate toxicity, suggesting that the mechanism for disruption of cell–cell adhesion was due to a mechanism distinct from Ca^{2+} displacement from the cadherin extracellular domain. Crude venom from the snake *Bothrops moojeni* caused time- and dose-dependent loss of TER in MDCK monolayers (Collares-Buzato et al., 2002). Although distributions of E-cadherin, occludin, and ZO-1 were not affected, cytoskeleton alterations were seen at cell–matrix contacts.

In vivo evidence for disruption of cell–cell adhesion during renal injury in animal models is lacking, although some evidence suggests that it may be associated with renal injury. Bush et al. (2000) showed that ischemia caused a loss of cell-surface E-cadherin and degradation of E-cadherin to an approximately 80 kDa fragment, suggesting that disruption of the cadherin/catenin complex in the kidney may represent a critical event in ischemia-induced renal injury. In addition, bismuth, a nephrotoxic metal, was shown to specifically alter the cellular localization of N-cadherin, but not E-cadherin, in the mouse proximal tubular epithelium (Leussink et al., 2001). Interestingly, no changes in tight junction constituents were seen.

While evidence suggests that disruption of cell–cell adhesion is associated with renal failure in patients, the understanding of the cause and effect of this relationship is limited by the lack of *in vivo* studies. It is clear that a number of toxic insults (metals, ischemia, toxins, etc.) disrupt cell–cell contacts *in vitro*, however the *in vivo* relevance of these findings in uncertain. Many of the *in vitro* studies have focused on E-cadherin despite that fact that expression of this cadherin is mostly limited to distal tubules. The large gaps in the understanding of cell adhesion complexes and the role that disruption of homotypic cell adhesion may play in renal injury await further investigation.

Gap Junctions

Several studies have demonstrated disruption of gap junctional intercellular communication (GJIC) by toxicants in renal cell lines. Methylmercury inhibits GJIC in primary cultures of rat proximal tubular cells, as assessed by dye coupling (Yoshida et al., 1998). In primary cultures of rat proximal tubular cells exposed to 100 μM cadmium chloride for 60 min, dye coupling revealed a decrease in GJIC, which was associated with an increase in intracellular Ca^{2+} (Fukumoto et al., 2001). A rapid, yet reversible, inhibition of GJIC is seen in a dolphin kidney epithelial cell line in response to a number of environmental contaminants (perfluorooctane sulfonic acid, perfluorooctane sulfonamide, and perfluorohexane sulfonic acid, but not perfluorobutane sulfonic acid) (Hu et al., 2002).

It is proposed that decreased GJIC also represents an early event in mercuric chloride toxicity in MDCK cells (Aleo et al., 2002). In this study, 0.1 μM mercuric chloride (4 hr) was shown to reduce GJIC by approximately 50%, as measured by dye-coupled cells. While these *in vitro* studies suggest that GJIC may be disrupted in renal injury, *in vivo* studies are required to further define the role that GJIC may play during renal injury.

Other Factors

Kidney injury molecule-1 (KIM-1) is a recently identified member of the immunoglobin gene superfamily. The protein contains a novel six cysteine immunoglobin-like domain and a mucin domain and is upregulated in cells that are dedifferentiated and undergoing replication (Ichimura et al., 1998). It has been shown that KIM-1 protein is cleaved by a metalloprotease and shed. It is proposed that this allows cells to scatter and reconstitute the epithelial barrier (Bailly et al., 2002). The identification of KIM-1 is exciting given that it is rapidly upregulated following injury. Furthermore, fragments of the protein may be detected in the urine early during renal failure, providing a reliable and early indicator of renal damage in the clinical setting.

Neural cell adhesion molecule (NCAM) is expressed during renal development, however expression is low in the adult kidney. In rats, during recovery from ischemia (5 days following reperfusion) NCAM is abundant at basal and lateral borders in S_3 cells and expression persists for up to 7 weeks (Abbate et al., 1999). This recapitulation of developmental expression of NCAM may provide some insight into the molecular pathways that regulate recovery from injury in the kidney.

While the literature suggests an important role for disruption of homotypic cell–cell adhesion during renal injury, much work remains to be done to determine the cause and effect relationship between disruption of cell adhesion and renal injury *in vivo*. Importantly, much of the knowledge concerning disruption of cell adhesion during injury is taken from *in vitro* models, and the significance of these findings is unclear. The relationship between tight junctions, cadherin/catenin complexes, and gap junctions during injury also remains uncertain, while a distinct possibility exists that renal injury related to disruption of cell–cell adhesion involves a concerted action on each of the complexes. The mechanisms underlying disruption of these complexes remain unclear, as does the impact of disrupted cell–cell adhesion on critical cellular signaling pathways.

Heterotypic Cell–Cell (Inflammation)

In the mid 1990s, an important study suggested that ICAM-1 was the key mediator of injury in the kidney (Kelly et al., 1994). In this study, it was demonstrated that an anti-ICAM-1 antibody protects the rat kidney against ischemia-induced damage. This protective effect can be duplicated by antisense for ICAM-1 (Haller et al., 1996), and an anti-ICAM-1 antibody is also protective against cisplatin-induced nephrotoxicity (Kelly et al., 1999). Interestingly, an antibody against ICAM-1 cannot attenuate mercuric chloride induced ARF (Ghielli et al., 2000), suggesting that inflammation may play a more significant role in ischemia- and cisplatin-induced ARF than in mercuric chloride induced ARF. A role for ICAM-1 in renal pathophysiology is definitively proven by the fact that ICAM-1 knockout mice are protected against ischemic renal injury (Kelly et al., 1996).

Leukocytes can bind target cells via the interaction of the $\beta 2$ integrin with ICAM-1. As such, pretreatment with antibodies against $\beta 2$ integrins protects the rat kidney against ischemia-reperfusion injury (Rabb et al., 1994), as assessed by serum creatinine and renal histopathology. This was supported by the finding that multiple antibodies against the $\beta 2$ integrin subunit (three antibodies that recognize heavy or light chains) attenuated ischemia-reperfusion injury (Tajra et al., 1999). Therefore, blockage of adhesion molecules that mediate heterotypic cell–cell adhesion during inflammation on either the infiltrating ($\beta 2$ integrin) or target (ICAM-1) cells attenuates renal injury in several models.

More-recent studies have focused on the molecular pathways that may regulate the overexpression of ICAM-1 associated with renal injury. Interleukin-10 has been demonstrated to attenuate increased ICAM-1 expression following either ischemia or cisplatin, and subsequently to have protective effects against the renal injury induced by these insults (Deng et al., 2001). In addition, statins (hydroxy-3-methylglutaryl coenzyme A reductase inhibitors) significantly reduced monocyte and macrophage infiltration of rat kidneys during ischemia-induced acute renal failure, putatively via reducing the ICAM-1 expression associated with ischemia (Gueler et al., 2002). It has also been shown that cadmium and mercury increase ICAM-1 expression in immortalized proximal tubule epithelial cells (Jiang et al., 2002). Importantly, both metals cause a decrease in secreted TGFβ-1 and addition of exogenous TGFβ-1 to the culture media attenuated metal-induced increases of ICAM-1, suggesting that TGFβ-1 signaling may regulate inducible ICAM-1 levels in the kidney.

While work has principally focused on ICAM-1, a role for P-selectin in the infiltration of macrophages during renal injury has also been demonstrated (Naruse et al., 2002). Unilateral ureteral obstruction in rats is associated with significant macrophage infiltration in the interstitium. Selective expression of P-selectin was seen early during unilateral ureteral obstruction in the vasa recta, while no change in E- or L-selectin was seen. This suggests that the vasa recta, via expression of P-selectin, may contribute to macrophage infiltration during renal injury.

Taken together, these data suggest a critical role for ICAM-1 in renal injury in multiple models. Importantly, ICAM-1 represents a compelling candidate for therapeutic intervention in acute renal failure.

CONCLUSIONS

It is clear that a great deal of progress has been made in the understanding of the specific proteins and complexes that mediate cell adhesion in the kidney. Importantly, the identification of the segment-specific expression of adhesion proteins has provided a rationale for some of the functional differences (e.g., strength of tight junction) between the proximal and distal tubules. Given the emerging data showing that cell adhesion components may also have critical signaling roles, it will be interesting to investigate whether the tightly regulated pattern of expression in the kidney influences other differences between epithelial cells along the nephron.

Table 12.1 Toxicant-Induced Disruption of Renal Cell Adhesion

Toxicant	*Effect*
Bismuth	*In vivo* disruption of N-cadherin localization
Cadmium	*In vitro* disruption of E-cadherin; *in vitro* decrease in GJIC; *in vitro* induction of ICAM-1
Cisplatin	*In vivo* induction of ICAM-1
DCVC	*In vitro* delocalization of integrins and dephosphorylation of FAK
Ischemia/oxidative stress	*In vitro* disruption of integrin localization; *in vivo* disruption of E-cadherin, *in vivo* induction of ICAM-1
Mercury	*In vitro* disruption of GJIC, *in vitro* induction of ICAM-1
Radiocontrast	*In vitro* disruption of cell–cell junctions
Venom	*In vitro* disruption of cell–cell adhesion/actin contacts

A potential role for disruption of cell adhesion in renal pathologies is also emerging, and a number of toxicants have been suggested to disrupt various cell adhesion molecules (Table 12.1). However, it is presently unclear whether the *in vitro* findings will correlate with *in vivo* renal dysfunction following injury. In addition, it remains to be determined whether disruption of cell adhesion is a cause or an effect of renal injury. The *in vivo* investigation of these questions is made difficult by the complex and overlapping patterns of adhesion molecule expression. However, data suggest that cell adhesion molecules may be important biomarkers of renal injury (KIM-1) and may also a focal point for therapeutic intervention (ICAM-1). Many critical issues concerning the role of adhesion molecules in renal physiology and pathophysiology remain to be addressed.

REFERENCES

Abbate M, Brown D, and Bonventre JV (1999) Expression of NCAM recapitulates tubulogenic development in kidneys recovering from acute ischemia. *Am J Physiol* 277:F454–463.

Aleo MF, Morandini F, Bettoni F, Tanganelli S, Vezzola A, Giuliania R, Steimberg N, Apostoli P, and Mazzoleni G (2002) Antioxidant potential and gap junction-mediated intercellular communication as early biological markers of mercuric chloride toxicity in the MDCK cell line. *Toxicol in vitro* 16:457–465.

Bailly V, Zhang Z, Meier W, Cate R, Sanicola M, and Bonventre JV (2002) Shedding of kidney injury molecule-1, a putative adhesion protein involved in renal regeneration. *J Biol Chem* 277:39739–39748.

Bush KT, Tsukamoto T, and Nigam SK (2000) Selective degradation of E-cadherin and dissolution of E-cadherin-catenin complexes in epithelial ischemia. *Am J Physiol* 278:F847–F852.

Butterweck A, Gergs U, Elfgang C, Willecke K, and Traub O (1994) Immunochemical characterization of the gap junction protein connexin45 in mouse kidney and transfected human HeLa cells. *J Membr Biol* 141:247–256.

Cho EA, Patterson LT, Brookhiser WT, Mah S, Kintner C, and Dressler GR (1998) Differential expression and function of cadherin-6 during renal epithelium development. *Development* 125:803–812.

Collares-Buzato CB, de Paula le Sueur L, and da Cruz-Hofling MA (2002) Impairment of the cell-to-matrix adhesion and cytotoxicity induced by *Bothrops moojeni* snake venom in cultured renal tubular epithelia. *Toxicol Appl Pharmacol* 181: 124–132.

Deng J, Kohda Y, Chiao H, Wang Y, Hu X, Hewitt SM, Miyaji T, McLeroy P, Nibhanupudy B, Li S, and Star RA (2001) Interleukin-10 inhibits ischemic and cisplatin-induced acute renal failure. *Kidney Int* 60:2118–2128.

Donohoe JF, Venkatachalam MA, Bernard DB, and Levinsky NG (1978) Tubular leakage and obstruction in acute ischemic renal failure. *Kidney Int* 13: 208–222.

Enck AH, Berger UV, and Yu AS (2001) Claudin-2 is selectively expressed in proximal nephron in mouse kidney. *Am J Physiol* 281:F966–974.

Fanning AS, Lapierre LA, Brecher AR, van Itallie CM, and Anderson JM (1996) Protein interactions in the tight junction: The role of MAGUK proteins in regulating tight junction organization and function, in *Current Topics in Membranes*, Nelson WJ (Ed.), Academic Press, San Diego, pp. 211–235.

Fukumoto M, Kujiraoka T, Hara M, Shibasaki T, Hosoya T, and Yoshida M (2001) Effect of cadmium on gap junctional intercellular communication in primary cultures of rat renal proximal tubular cells. *Life Sci* 69:247–254.

Furuse M, Furuse K, Sasaki H, and Tsukita S (2001) Conversion of zonulae occludentes from tight to leaky strand type by introducing claudin-2 into Madin-Darby Canine Kidney I cells. *J Cell Biol* 153:263–272.

Gailit J, Colflesh D, Rabiner I, Simone J, and Goligorsky MS (1993) Redistribution and dysfunction of integrins in cultured renal epithelial cells exposed to oxidative stress. *Am J Physiol* 264:F149–157.

Ghielli M, Verstrepen WA, de Greef KE, Helbert MH, Ysebaert DK, Nouwen EJ, and De Broe ME (2000) Antibodies to both ICAM-1 and LFA-1 do not protect the kidney against toxic (HgCl2) injury. *Kidney Int* 58: 1121–1134.

Goligorsky MS, Lieberthal W, Racusen LC, and Simon EE (1993) Integrin receptors in renal tubular epithelium: New insights into pathophysiology of acute renal failure. *Am J Physiol* 264:F1–F8.

Gonzalez-Mariscal L, Namorado MC, Martin D, Luna J, Alarcon L, Islas S, Valencia L, Muriel P, Ponce L, and Reyes JL (2000) Tight junction proteins ZO-1, ZO-2, and occludin along isolated renal tubules. *Kidney Int* 57:2386–2402.

Gueler F, Rong S, Park JF, Fiebeler A, Menne J, Elger M, Mueller DN, Hampich F, Dechend R, Kunter U, Luft FC, and Haller H (2002) Postischemic acute renal failure is reduced by short-term statin treatment in a rat model. *J Am Soc Nephrol* 13: 2288–2298.

Guo R, Liu L, and Barajas L (1998) RT-PCR study of the distribution of connexin 43 mRNA in the glomerulus and renal tubular segments. *Am J Physiol* 275: R439–F447.

Haller H, Dragun D, Miethke A, Park JK, Weis A, Lippoldt A, Gross V, and Luft FC (1996) Antisense oligonucleotides for ICAM-1 attenuate reperfusion injury and renal failure in the rat. *Kidney Int* 50:473–480.

Hamerski DA, and Santoro SA (1999) Integrins and the kidney: Biology and pathobiology. *Curr Opin Nephrol Hypertens* 8:9–14.

Hanks SK, Calalb MB, Harper MC, and Patel SK (1992) Focal adhesion protein tyrosine kinase phosphorylated in response to cell spreading on fibronectin. *Proc Natl Acad Sci USA* 89:8487–8491.

Hu W, Jones PD, Upham BL, Trosko JE, Lau C, and Giesy JP (2002) Inhibition of gap junctional intercellular communication by perfluorinated compounds in rat liver and dolphin kidney epithelial cell lines *in vitro* and Sprague-Dawley rats *in vivo*. *Toxicol Sci* 68:429–436.

Hubbard AK, and Rothlein R (2000) Intercellular adhesion molecule-1 (ICAM-1) expression and cell signaling cascades. *Free Radic Biol Med* 28:1379–1386.

Hudkins KL, Le Qc, Segerer S, Johnson RJ, Davis CL, Giachelli CM, and Alpers CE (2001) Osteopontin expression in human cyclosporine toxicity. *Kidney Int* 60:635–640.

Ichimura T, Bonventre JV, Bailly V, Wei H, Hession CA, Cate RL, and Sanicola M (1998) Kidney injury molecule-1 (KIM-1), a putative epithelial cell adhesion molecule containing a novel immunoglobin domain, is up-regulated in renal cells after injury. *J Biol Chem* 273:4135–4142.

Itoh M, Morita K, and Tsukita S (1999) Characterization of ZO-2 as a MAGUK family member associated with tight as well as adherens junctions with a binding affinity to occludin and α-catenin. *J Biol Chem* 274:5981–5986.

Itoh M, Nagafuchi A, Moroi S, and Tsukita S (1997) Involvement of ZO-1 in cadherin-based cell adhesion through its direct binding to α-catenin and actin filaments. *J Cell Biol* 138:181–192.

Jiang J, McCool BA, and Parrish AR (2002) Cadmium- and mercury-induced intercellular adhesion molecule-1 expression in immortalized proximal tubule cells: Evidence for a role of decreased transforming growth factor-beta1. *Toxicol Appl Pharmacol* 179:13–20.

Juliano RL (2002) Signal transduction by cell adhesion receptors and the cytoskeleton: Functions of integrins, cadherins, selectins, and immunoglobulin-superfamily members. *Annu Rev Pharmacol Toxicol* 42:283–323.

Kelley VR, and Singer GG (1993) The antigen presentation function of renal tubular epithelial cells. *Exp Nephrol* 1:102–111.

Kelly KJ, Meehan SM, Colvin RB, Williams WW, and Bonventre JV (1999) Protection from toxicant-mediated renal injury in the rat with anti-CD54 antibody. *Kidney Int* 56:922–931.

Kelly KJ, Williams WW Jr, Colvin RB, and Bonventre JV (1994) Antibody to intercellular adhesion molecule-1 protects the kidney against ischemic injury. *Proc Natl Acad Sci USA* 91:812–816.

Kelly KJ, Williams WW Jr, Colvin RB, Meehan SM, Springer TA, Gutierrez-Ramos JC, and Bonventre JV (1996) Intercellular adhesion molecule-1-deficient mice are protected against ischemic renal injury. *J Clin Invest* 97:1056–1063.

Kemler R (1992) Classical cadherins. *Semin Cell Biol* 3:149–155.

Kiuchi-Saishin, Gotoh S, Furuse M, Takasuga A, Tano Y, and Tsukita S (2002) Differential expression patterns of claudins, tight junction membrane proteins, in mouse nephron segments. *J Am Soc Nephrol* 13:875–886.

Kroshian VM, Sheridan AM, and Lieberthal W (1994) Functional and cytoskeletal changes induced by sublethal injury in proximal tubular epithelial cells. *Am J Physiol* 266: F21–F30.

Kwon O, Nelson WJ, Sibley R, Huie P, Scandling JD, Dafoe D, Alfrey E, and Myers BD (1998) Backleak, tight junctions, and cell-cell adhesion in postischemic injury to the renal allograft. *J Clin Invest* 101:2054–2064.

Lapierre LA (2000) The molecular structure of the tight junction. *Adv Drug Deliv Rev* 41: 255–264.

Leussink BT, Litvinov SV, de Heer E, Slikkerveer A, van der Voet GB, Bruijn JA, and de Wolff FA (2001) Loss of homotypic epithelial cell adhesion by selective N-cadherin displacment in bismuth nephrotoxicity. *Toxicol Appl Pharmacol* 175:54–59.

Lieberthal W, McKenney JB, Kiefer CR, Synder LM, Kroshian VM, and Sjaastad MD (1997) Beta1 integrin-mediated adhesion between renal tubular cells after anoxic injury. *J Am Soc Nephrol* 8:175–183.

Liu ZX, Nickel CH, and Cantley LG (2001) HGF promotes adhesion of ATP-depleted renal tubular in a MAPK-dependent manner. *Am J Physiol* 281:F62–F70.

Martin-Padura I, Lostaglio S, Schneemann M, Williams L, Romano M, Fruscell P, Panzeri C, Stoppacciaro A, Ruco L, Villa A, Simmons D, and Dejana E (1998) Junctional adhesion molecule, a novel member of the immunoglobulin superfamily that distributes at intercellular junctions and modulates monocyte transmigration. *J Cell Biol* 142:117–127.

Molitoris BA, and Marrs J (1999) The role of cell adhesion molecules in ischemic acute renal failure. *Am J Med* 106:583–592.

Moran SM, and Myers BD (1985) Pathophysiology of protracted acute renal failure in man. *J Clin Invest* 76:1440–1448.

Myers BD, Chui F, Hilberman M, and Michaels AS (1979) Transtubular leakage of glomerular filtrate in human acute renal failure. *Am J Physiol* 237:F319–F325.

Nagafuchi A, Takeichi M, and Tsukita S (1991) The 102 kd cadherin-associated protein: Similarity to vinculin and posttranscriptional regulation of expression. *Cell* 65:849–857.

Naruse T, Yuzawa Y, Akahori T, Mizuno M, Maruyama S, Kannagi R, Hotta N, and Matsuo S (2002) P-selectin-dependent macrophage migration into the tubulointerstitium in unilateral ureteral obstruction. *Kidney Int* 62:94–105.

Nathke IS, Hinck L, Swedlow JR, Papkoff J, and Nelson WJ (1994) Defining interactions and distributions of cadherin and catenin complexes in polarized epithelial cells. *J Cell Biol* 125:1341–1352.

Nigam S, Weston CE, Liu CH, and Simon EE (1998) The actin cytoskeleton and integrin expression in the recovery of cell adhesion after oxidant stress to a proximal tubule cell line (JTC-12). *J Am Soc Nephrol* 9:1787–1797.

Noiri E, Gailit J, Sheth D, Magazine H, Gurrath M, Muller G, Kessler H, and Goligorsky MS (1994) Cyclic RGD peptides ameliorate ischemic acute renal failure in rats. *Kidney Int* 46:1050–1058.

Nony PA, and Schnellmann RG (2001) Interactions between collagen IV, and collagen-binding integrins in renal cell repair after sublethal injury. *Mol Pharmacol* 60:1226–1234.

Nouwen EJ, Dauwe S, van der Biest I, and de Broe ME (1993) Stage- and segment-specific expression of cell-adhesion molecules N-CAM, A-CAM, and L-CAM in the kidney. *Kidney Int* 44:147–158.

Patey N, Balbwachs-Mecarelli L, Droz D, Lesavre P, and Noel LH (1994) Distribution of integrin subunits in normal human kidney. *Cell Adhes Commun* 2: 159–167.

Piepenhagen PA, and Nelson WJ (1995) Differential expression of cell-cell and cell-substratum adhesion proteins along the kidney nephron. *Am J Physiol* 269: C1433–C1449.

Piepenhagen PA, Peters LL, Lux SE, and Nelson WJ (1995) Differential expression of Na^+-K^+-ATPase, ankyrin, fodrin, and E-cadherin along the kidney nephron. *Am J Physiol* 269:C1417–C1432.

Prozialeck WC (2000) Evidence that E-cadherin may be a target for cadmium toxicity in epithelial cells. *Toxicol Appl Pharmacol* 164:231–249.

Prozialeck WC, and Lamar PC (1997) Cadmium (Cd2+) disrupts E-cadherin-dependent cell-cell junctions in MDCK cells. *In Vitor Cell Dev Biol Anim* 33:516–520.

Prozialeck WC, Lamar PC, and Ikura M (1996) Binding of cadmium (Cd2+) to E-CADI, a calcium-binding polypeptide analog of E-cadherin. *Life Sci* 58:PL325–PL330.

Rabb H, Mendiola CC, Dietz J, Saba SR, Issekutz TB, Abanilla F, Bonventre JV, and Ramirez G (1994). Role of CD11a and CD11b in ischemic acute renal failure in rats. *Am J Physiol* 267:F1052–1058.

Racusen LC, Fivush BA, Li YL, Slatnik I, and Solez K (1991) Dissociation of tubular cell detachment and tubular cell death in clinical and experimental "acute tubular necrosis." *Lab Invest* 64:546–556.

Reynolds AB, Daniel J, McCrea PD, Wheelock MJ, Wu J, and Zhang Z (1994) Identification of a new catenin: The tyrosine kinase substrate $p120^{cas}$ associates with E-cadherin complexes. *Mol Cell Biol* 14:8333–8342.

Schaller MD, Borgman CA, Cobb BS, Vines RR, Reynolds AB, and Parsons JT (1992) pp125FAK, a structurally distinctive protein tyrosine-kinase associated with focal adhesions. *Proc Natl Acad Sci USA* 89:192–196.

Schick CS, Bangert R, Kubler W, and Haller C (2002) Ionic radiocontrast media disrupt intercellular contacts via an extracellular calcium-independent mechanism. *Exp Nephrol* 10:209–215.

Schwartz JH, Shih T, Menza SA, and Lieberthal W (1999) ATP depletion increases tyrosine phosphorylation of beta-catenin and plakoglobin in renal tubular cells. *J Am Soc Nephrol* 10:2297–2305.

Silverstein DM, Urban M, Gao Y, Mattoo TK, Spray DC, and Rozental R (2001) Renal morphology in connexin43 knockout mice. *Pediatr Nephrol* 16:467–471.

Tajra LC, Martin X, Margonari J, Blanc-Brunat N, Ishibashi M, Vivier G, Panaye G, Stephens JP, Kawashima H, Miyasaka M, Treille-Ritouet D, Dubernard JM, and Revillard JP (1999) In vivo effects of monoclonal antibodies against rat beta(2) integrins on kidney ischemia-reperfusion injury. *J Surg Res* 87:32–38.

Thomson RB, and Aronson PS (1999) Immunolocalization of Ksp-cadherin in the adult and developing rabbit kidney. *Am J Physiol* 277:F146–F156.

Thomson RB, Igarashi P, Biemesderfer D, Kim R, Abu-Alfa A, Soleimani M, and Aronson PS (1995) Isolation and cDNA cloning of Ksp-cadherin, a novel kidney-specific member of the cadherin multigene family. *J Biol Chem* 270:17594–17601.

Tsukita SH, and Furuse M (1999) Occludin and claudins in tight junction strands: Leading or supporting players? *Trends Cell Biol* 9:268–273.

Tsukita S, Tsukita S, Nagafuchi A, and Yonemura S (1992) Molecular linkage between cadherins and actin filaments in cell-cell adherens junctions. *Curr Opin Cell Biol* 4:834–839.

Van de Water B, Nagelkerke JF, and Stevens JL (1999) Dephosphorylation of focal adhesion kinase (FAK) and loss of focal contacts precede caspase-mediated cleavage of FAK during apoptosis in renal epithelial cells. *J Biol Chem* 274:13328–13337.

Van de Water B, Houtepen F, Huigsloot M, and Tijdens IB (2001) Suppression sof chemically induced apoptosis but not necrosis of renal proximal tubular epithelium (LLC-PK1) cells by focal adhesion kinase (FAK). Role of FAK in maintaining focal adhesion organization after acute renal cell injury. *J Biol Chem* 276:36183–16193.

Wagner MC, and Molitoris BA (1999) Renal epithelial polarity in health and disease. *Pediatr Nephrol* 13:163–170.

Wertz K, and Herrmann BG (1999) Kidney-specific cadherin (cdh16) is expressed in embryonic kidney, lung, and sex ducts. *Mech Dev* 84:185–188.

Wijesekera DS, Zarama MJ, and Paller MS (1997) Effects of integrins on proliferation and apoptosis of renal epithelial cells after acute injury. *Kidney Int* 52:1511–1520.

Wong V (1997) Phosphorylation of occludin correlates with occludin localization and function at the tight junction. *Am J Physiol* 273:C1859–C1867.

Wu Q, and Maniatis T (1999) A striking organization of a large family of human neural cadherin-like cell adhesion genes. *Cell* 97:779–790.

Xiang Y-Y, Tanaka M, Suzuki M, Igarashi H, Kiyokawa E, Naito Y, Ohtawara Y, Shen Q, Sugimura H, and Kino I (1994) Isolation of complementary DNA encoding K-cadherin, a novel cadherin preferentially expressed in fetal kidney and kidney carcinoma. *Cancer Res* 54:3034–3041.

Xie Y, Sakatsume M, Nishi S, Narita I, Arakawa M, and Gejyo F (2001) Expression, roles, receptors, and regulation of osteopontin in the kidney. *Kidney Int* 60: 1645–1657.

Yeager M, Unger VM, and Falk MM (1998) Synthesis, assembly and structure of gap junction intercellular channels. *Curr Opin Struct Biol* 8:517–524.

Yoshida M, Kujiraoka T, Hara M, Nakazawa H, and Sumi Y (1998) Methylmercury inhibits gap junctional intercellular communication in primary cultures of rat proximal tubular cells. *Arch Toxicol* 72:192–196.

Zimmerhackl LB, Momm F, Wiegle G, and Brandis M (1998) Cadmium is more toxic to LLC-PK1 cells than to MDCK cells acting on the cadherin-catenin complex. *Am J Physiol* 275:F143–F153.

Zuk A, Bonventre JV, Brown D, and Matlin KS (1998) Polarity, integrin and extracellular matrix dynamics in the postischemic rat kidney. *Am J Physiol* 275: C711–731.

13

Oxidant Mechanisms in Toxic Acute Renal Failure

Alexei G. Basnakian, Gur P. Kaushal, Norishi Ueda, and Sudhir V. Shah

INTRODUCTION

The notion that reactive oxygen metabolites (ROMs) may be important in inflammation was initiated by a publication in 1969 in which McCord and Fridovich described an enzyme, superoxide dismutase, which scavenges superoxide anion (McCord and Fridovich, 1969). They reasoned that since phagocytizing neutrophils (the effector cells of the acute inflammatory response) release large amounts of superoxide extracellularly and superoxide dismutase possesses anti-inflammatory activity, the superoxide anion and other oxygen metabolites may be important chemical mediators of the inflammatory process (McCord, 1974). This hypothesis has received considerable support from a large number of studies over the past decade, which indicate that partially reduced oxygen metabolites are important mediators of ischemic, toxic, and immune-mediated tissue injury (McCord and Fridovich, 1969; Halliwell and Gutteridge, 1990).

In this review, current evidence for a role of ROMs in toxic acute renal failure induced by cisplatin, gentamicin, and by rhabdomyolysis, is summarized. Some of the cellular mechanism involved in renal tubular injury, including the role of the apoptotic pathway, are then discussed.

Oxygen normally accepts four electrons and is converted directly to water. However, partial reduction of oxygen can, and does, occur in biological systems. Therefore, the sequential reduction of oxygen along the univalent pathway leads to the generation of superoxide anion, hydrogen peroxide, hydroxyl radical, and water (Fridovich, 1978; Halliwell and Gutteridge, 1990). Superoxide and hydrogen peroxide appear to be the primary species generated. These species may then play roles in the generation of additional and more reactive oxidants, including the highly reactive hydroxyl radical (or a related highly oxidizing species) in which iron salts play a catalytic role in a reaction, commonly referred to as the metal-catalyzed Haber-Weiss reaction (Halliwell and Gutteridge, 1990).

Additional ROMs can be formed as a result of the metabolism of hydrogen peroxide by neutrophil-derived myeloperoxidase (MPO), the enzyme responsible for the green color of pus, to produce highly reactive toxic products, including hypochlorous acid. These oxygen metabolites, including the free radical species, superoxide and hydroxyl radical, and other metabolites, such as hydrogen peroxide and hypohalous acids, are often collectively referred to as ROMs.

ROLE OF REACTIVE OXYGEN METABOLITES IN CISPLATIN-INDUCED NEPHROTOXICITY

Cisplatin is a widely used antineoplastic agent, which has nephrotoxicity as a major side effect. The underlying mechanism of this nephrotoxicity is not well understood. There is much evidence supporting a role for ROMs in cisplatin-induced nephrotoxicity (see Table 13.1). Cisplatin has been shown to induce *in vitro* generation of hydrogen peroxide in proximal tubular cells (Kruidering et al., 1997; Tsutsumishita et al., 1998). Multiple studies indicated that scavengers of ROMs are protective *in vitro* and *in vivo* (Baliga et al., 1998a; Davis et al., 2001; Gemba et al., 1988; Kim et al., 1997; Matsushima et al., 1998; Sugihara and Gemba, 1986; Tsutsumishita et al., 1998). The catalytic iron content and the effect of iron chelators have been examined in an *in vitro* model of cisplatin-induced cytotoxicity in LLC-PK$_1$ cells (renal tubular epithelial cell) and in an *in vivo* model of cisplatin-induced acute renal failure in rats (Baliga et al., 1999). Exposure of LLC-PK$_1$ cells to cisplatin resulted in a significant increase in bleomycin-detectable iron (iron capable of catalyzing free radical reactions) released into the medium. Concurrent incubation of LLC-PK$_1$ cells with iron chelators, including deferoxamine and

Table 13.1 Evidence Suggesting a Role of ROM in Cisplatin-induced Acute Renal Failure

- Cisplatin enhances *in vitro* generation of hydrogen peroxide in proximal tubule cells (Kruidering et al., 1997; Tsutsumishita et al., 1998).
- Cisplatin increases catalytic iron in LLC-PK_1 cells *in vitro* (Baliga et al., 1998a) and iron chelators are protective *in vitro* (Baliga et al., 1998a; Kim et al., 1997) and *in vivo* (Baliga et al., 1998a; Matsushima et al., 1998).
- Scavengers of ROMs, including catalase (Kim et al., 1997; Tsutsumishita et al., 1998) and superoxide dismutase (Davis et al., 2001; Matsushima et al., 1998), and hydroxyl radical scavengers (Baliga et al., 1998a; Gemba et al., 1988; Matsushima et al., 1998; Sugihara and Gemba, 1986) are protective *in vitro* and *in vivo.*
- Cisplatin increases renal cytochrome P450 content (Jollie and Maines, 1985) and bleomycin iron (Baliga et al., 1998b) and cytochrome P450 inhibitors prevent increase in bleomycin iron (Baliga et al., 1998b) and are protective in *in vitro* and *in vivo* (Baliga et al., 1998b; Hannemann and Baumann, 1988; Liu et al., 2002).
- Cisplatin induces rapid peroxidation *in vitro* and *in vivo* (Hannemann and Baumann, 1988).
- Cisplatin induces reduces levels of GSH and thiol *in vitro* (Kruidering et al., 1997) and *in vivo* (Mistry et al., 1991; Sugiyama et al., 1989), and glutathione ester (Anderson et al., 1990; Babu et al., 1995) is protective and inhibition of glutathione (Babu et al., 1995) and deficiency of selenium (Satoh et al., 1987) accelerate nephrotoxicity *in vivo.*
- Cisplatin reduces the levels of antioxidants in the cancer patient (Weijl et al., 1998) and antioxidants (Bull et al., 1988; Gemba et al., 1988; Sugihara and Gemba, 1986) are protective *in vivo* and *in vitro* (Gemba and Fukuishi, 1991).
- Cisplatin induces heme oxygenase (Agarwal et al., 1995) and heme protein (Jollie and Maines, 1985), however inhibition of heme oxygenase accelerates cisplatin nephrotoxicity (Agarwal et al., 1995).

1,10-phenanthroline, significantly attenuated cisplatin-induced cytotoxicity as measured by lactate dehydrogenase release. Bleomycin-detectable iron content was also markedly increased in the kidney of rats treated with cisplatin. Similarly, the administration of deferoxamine in rats provided marked functional (as measured by blood urea nitrogen and creatinine) and histological protection against cisplatin-induced acute renal failure.

A separate study examined the role of the hydroxyl radical in cisplatin-induced nephrotoxicity. Incubation of LLC-PK_1 cells with cisplatin caused an increase in hydroxyl radical formation. Hydroxyl radical scavengers dimethyl sulfoxide, mannitol, and benzoic acid significantly reduced cisplatin-induced cytotoxicity, and treatment with dimethyl sulfoxide or dimethylthiourea provided significant protection against cisplatin-induced acute renal failure. Taken together, the data from these studies strongly support a critical role for iron in mediating tissue injury via hydroxyl radical (or a similar oxidant) in this model of nephrotoxicity.

ROLE OF REACTIVE OXYGEN METABOLITES IN GENTAMICIN NEPHROTOXICITY

Gentamicin is widely used in the treatment of Gram-negative infections. A major complication of the use of aminoglycoside antibiotics, including gentamicin, is nephrotoxicity, which accounts for 10 to 15% of all cases of acute renal failure (Humes and Weinberg, 1986). The precise mechanisms of gentamicin nephrotoxicity remain unknown, however studies have provided strong evidence to suggest a role of ROMs in gentamicin-induced renal failure (see Table 13.2). Both *in vitro* and *in vivo* studies indicate enhanced generation of hydrogen peroxide and release of iron in response to gentamicin. Most, if not all, of the hydrogen peroxide generated by mitochondria is derived from the dismutation of superoxide. Therefore, the enhanced generation of hydrogen peroxide by gentamicin suggests that superoxide anion production is also increased. Superoxide and hydrogen peroxide may interact (with trace metals such as iron as the redox agent) to generate highly reactive and unstable oxidizing species, including the hydroxyl radical. It has been demonstrated that hydroxyl radical scavengers and iron chelators provide a marked protective effect on renal function in gentamicin-induced acute renal failure in rats (Nakajima et al., 1994).

Additionally, several interventional agents markedly reduced histological evidence of damage. Other studies have provided support for these observations. Administration of superoxide dismutase or the oxidant scavenger dimethylthiourea provided a marked protection against gentamicin-induced impairment of renal function and lipid peroxidation, and dimethylthiourea attenuated the tubular damage (Nakajima et al., 1994). In contrast, it was reported that amelioration of gentamicin-induced lipid peroxidation by treatment with the antioxidant diphenylphenyenediamine failed to prevent nephrotoxicity (Ramsammy

Table 13.2 Evidence Suggesting a Role of ROM in Gentamicin-induced Acute Renal Failure

- Gentamicin enhances the generation of superoxide anion, hydrogen peroxide, and hydroxyl radical by renal cortical mitochondria (Cuzzocrea et al., 2002; Du and Yang, 1994; Guidet and Shah, 1989b; Walker and Shah, 1987; Yang et al., 1995a).
- There is enhanced *in vivo* generation of hydrogen peroxide in renal cortex in the gentamicin-treated rat (Guidet and Shah, 1989b).
- Gentamicin enhances the release of iron from renal cortical mitochondria (Ueda et al., 1993).
- Gentamicin–iron complex causes lipid peroxidation *in vitro* and is a potent catalyst for free radical formation (Priuska and Schacht, 1995).
- Hydroxyl radical scavengers are protective in gentamicin-induced acute renal failure in rats (Ali and Mousa, 2001; Nakajima et al., 1994; Walker and Shah, 1988; Zurofsky and Haber, 1995).
- Iron chelators are protective in gentamicin-induced hydroxyl radical formation by renal cortical mitochondria (Yang et al., 1995a) and acute renal failure in the rat (Walker and Shah, 1988).
- Administration of superoxide dismutase provides a marked protection against gentamicin-induced impairment of renal function (Ali, 1995; Nakajima et al., 1994).
- Iron supplementation enhances gentamicin nephrotoxicity *in vivo* (Ben Ismail et al., 1994; Kays et al., 1991).
- Pretreatment with zinc prevents gentamicin nephrotoxicity by inducing metallothionein that can scavenge ROMs (Du and Yang, 1994).
- Coadministration of antioxidants, vitamins E and C, and probucol, as well as selenium, is protective against gentamicin-induced nephrotoxicity (Abdel-Naim et al., 1999; Ademuyiwa et al., 1990; Kavutcu et al., 1996; Zurovsky and Haber, 1995).

et al., 1986). However, it was also demonstrated that coadministration of antioxidants, vitamin E and selenium, is protective against gentamicin-induced nephrotoxicity (Ademuyiwa et al., 1990). It is not clear why the contradictory results were obtained, however one explanation is that it may be due to the difference in the mechanisms of the protective effect of the various antioxidants.

Additional support for a role of iron-catalyzed free radical generation has been provided by demonstrating that gentamicin-induced generation of hydroxyl radicals is reduced by iron chelators *in vitro* (Yang et al., 1995a) and iron supplementation enhances gentamicin nephrotoxicity *in vivo* (Ben Ismail et al., 1994; Kays et al., 1991). Taken together, it appears that ROMs are among the mediators responsible for gentamicin nephrotoxicity.

ROLE OF REACTIVE OXYGEN METABOLITES IN MYOGLOBINURIC ACUTE RENAL FAILURE

During the Battle of Britain, Bywaters and Beall (1941) described the first causative association between acute renal failure and skeletal muscle injury, with the release of muscle cell contents, including myoglobin, into plasma (rhabdomyolysis). Since then the spectrum of etiologies for rhabdomyolysis, myoglobinuria, and renal failure has been markedly expanded, with the recognition of both traumatic and, more recently, nontraumatic causes (Gabow et al., 1982; Grossman et al., 1974; Koffler et al., 1976). The most widely used model of myoglobinuric acute renal failure is produced by subcutaneous or intramuscular injection of hypertonic glycerol (Hostetter et al., 1983). Current evidence suggesting a role of ROMs in myoglobinuric acute renal failure is shown in Table 13.3. Guidet and Shah (1989a) demonstrated enhanced generation of hydrogen peroxide in glycerol-induced acute renal failure. Zager (1996a, 1996b) provided evidence for mitochondria as a critical site of heme-induced free radical formation. When heme-laden proximal tubular segments were exposed to mitochondrial respiratory chain inhibitors, there was a marked alteration in lipid peroxidation: blockade at site 2 or site 3 prevented heme-induced lipid peroxidation whereas blockade at site 1 increased oxidative damage.

The recognition that hydrogen peroxide is produced in excessive amounts in this model motivated the examination of the potential efficacy of pyruvate, an α-ketoacid (Salahudeen et al., 1991). A property shared by a wide range of α-ketoacids is the ability of these metabolites to scavenge hydrogen peroxide through a nonenzymatic, oxidative decarboxylation reaction (Bunton, 1949). The administration of pyruvate following the intramuscular injection of glycerol improved renal function, as measured by serum creatinine determinations accompanied by a marked reduction in structural injury (Salahudeen et al., 1991). A property of pyruvate that perhaps contributes to its protective effect is its facile distribution across plasma and mitochondrial membranes (Halestrap et al., 1980; Murer and Burckhardt, 1983). This attribute

Table 13.3 Evidence Suggesting a Role of ROM in Myoglobinuric Acute Renal Failure

- There is enhanced *in vivo* generation of hydrogen peroxide in renal cortex in rats with glycerol-induced acute renal failure (Guidet and Shah, 1989a).
- Pyruvate, a scavenger of hydrogen peroxide, is protective in glycerol-induced acute renal failure (Salahudeen et al., 1991).
- Hydroxyl radical scavengers and iron chelators are protective in glycerol-induced acute renal failure in the rat (Shah and Walker, 1988).
- Iron chelator is protective in two experimental models of pigment-induced acute renal failure, intramuscular glycerol injection, and intravenous hemoglobin infusion without and with concurrent ischemia in the rat (Paller, 1988).
- Iron chelator and mannitol can each protect against myohemoglobinuric acute renal failure (Zager, 1992).
- The protective effect of glutathione (GSH) and the detrimental effect of either depletion of GSH or interference with recycling of glutathione disulfide (GSSG) into GSH indicate an important role of glutathione in glycerol-induced acute renal failure (Abul-Ezz et al., 1991).
- There is increase in both the heme-oxygenase mRNA and the enzyme activity in glycerol-induced acute renal failure (Nath et al., 1992). It appears to serve a protective role because inhibiting the enzyme worsens renal failure.
- A lipid peroxidation inhibitor, 21-aminosteroid, prevent heme protein-induced lipid peroxidation and cytotoxicity in *in vitro* and in *in vivo* models of glycerol-induced acute renal failure (Nath et al., 1995; Salahudeen et al., 1996).
- Antioxidative bioflavonoid proanthocyanidin-BP1 is protective against glycerol-induced acute renal failure in the rat (Avramovic et al., 1999).

delivers pyruvate widely within the intracellular compartment and to subcellular sites at which potentially damaging peroxides are produced.

Shah and Walker (1988) also examined the effect of hydroxyl radical scavengers and iron chelators in glycerol-induced acute renal failure in rats. Dimethylthiourea, a hydroxyl radical scavenger, provided marked protection against glycerol-induced acute renal failure. In contrast to the effect of dimethylthiourea, urea (which is not a hydroxyl radical scavenger and served as a control) failed to provide any protection. A second hydroxyl radical scavenger, sodium benzoate, and an iron chelator, deferoxamine, had similar protective effects on renal function. The interventional agents were also associated with a marked reduction in histological evidence of renal damage. Paller (1988) has also demonstrated that deferoxamine treatment was protective in three models of myoglobinuric renal injury, namely hemoglobin-induced nephrotoxicity, glycerol-induced acute renal failure, and a combined renal ischemia hemoglobin insult. Similarly, Zager (1992) demonstrated the protective effect of an iron chelator in myohemoglobinuric injury. Taken together, the histological and functional protective effects of the hydroxyl radical scavengers and an iron chelator implicate a role for the hydroxyl radical in glycerol-induced acute renal failure.

MECHANISMS OF RENAL TUBULAR EPITHELIAL CELL INJURY

The modern study of cell death began with the landmark publication by Kerr, Wyllie and Currie in 1972 (Kerr et al., 1972), in which they coined the term apoptosis and made a distinction between necrosis and apoptosis based on morphological criteria. In necrosis there is swelling of cell organelles, a loss of plasma membrane integrity, and rupture of the cell, invoking an inflammatory response (Kerr et al., 1972; Lieberthal and Levine, 1996; Thompson, 1998). In contrast, in apoptosis, cells shrink, lose microvilli and cell junctions, and explode into a series of membrane-bound condensed apoptotic bodies and affected cells are phagocytized by adjacent viable cells with little leakage of cellular contents, thus invoking no inflammation. Apoptosis has been shown to be important, for example, in embryogenesis and normal tissue turnover, in which it is only appropriate because apoptosis evokes no inflammation (Kerr et al., 1972; Lieberthal and Levine, 1996; Thompson, 1998). In contrast, accidental cell death that might occur after severe insults, such as ischemic or toxic injury, is generally assumed to result in a catastrophic breakdown of regulated cellular homeostasis, i.e., necrosis.

One of the major advances in the understanding of cell death has been the recognition that the pathways traditionally associated with apoptosis may be critical in determining the form of cell injury associated with necrosis. It is now recognized that the same insult may result in apoptosis or necrosis, with the mild injury generally resulting in apoptosis and severe injury in necrosis (Lieberthal et al., 1998; Schumer et al., 1992; Thompson, 1998). Thus, the pathway that is followed by the cell is dependent on both the nature and severity of insults. It appears likely that the cascades that lead to the apoptotic and necrotic modes of cell death are activated almost simultaneously, and that there are common pathways that are shared and regulated in the two modes of cell death (Lieberthal and Levine, 1996).

DNA Fragmentation and Endonucleases in Renal Tubular Epithelial Cell Injury

Several *in vivo* studies have demonstrated chromatin condensation, the morphological hallmark of apoptosis, in models of acute renal failure, including ischemia/reperfusion injury. However, much of the evidence for the role of apoptotic mechanisms in renal tubular epithelial (RTE) cell injury relates to the demonstration of endonuclease activation, resulting in oligonucleosome-length DNA fragmentation (approximately 200 bp), which has been regarded as one of the biochemical hallmarks of apoptosis.

A key question is whether or not endonuclease activation is related to cell death. Additionally, it is important to know whether the endonuclease activation and subsequent DNA fragmentation is at a point of no return so that any attempt to halt this process would not prevent cell death. The 200 bp ladder, which is commonly used because of its simplicity, actually measures very late events. Clearly, the chromatin does not need to be cut to the 200 bp fragments to induce cell death. Approximately 40 double-strand DNA breaks per cell have been shown to be lethal (Ueda and Shah, 1992). Beyond this level, the repair of DNA breaks is no longer effective. The sensitivity of the method used to measure DNA strand breaks largely determines the time point at which the DNA fragmentation can be detected. Methods that are aimed at quantifying rare DNA breaks, e.g., pulse-field electrophoresis and the random oligonucleotide-primed synthesis (ROPS) assay, provide more-accurate timing of DNA fragmentation.

The authors' interest in the role of endonucleases in RTE cell injury began with attempts to understand the biochemical and cellular mechanisms of oxidant injury. It was generally accepted that hydrogen peroxide gets to the DNA and, in the presence of iron in the DNA, results in the site-specific generation of hydroxyl radicals, which cause the DNA damage. It was possible to show that, in fact, in hydrogen peroxide-induced injury to RTE cells, endonuclease activation occurs as an early event leading to DNA fragmentation, and that endonuclease inhibitors prevented hydrogen peroxide-induced DNA strand breaks, DNA fragmentation, and cell death (Baliga et al., 1997). Based on the demonstration of the role of endonucleases in oxidant injury, it was surmised that endonuclease activation might be important in those forms of acute renal tubular injury where ROMs have been implicated (Ueda et al., 1995).

Multiple studies by several groups have shown in different systems that there is a direct link between endonuclease-generated DNA breaks and subsequent cell death (Basnakian et al., 2002a; Krieser and Eastman, 1998; Nagata, 2000; Polzar et al., 1993; Zhang et al., 1999). There is also direct evidence that overexpression of DNase I (Zhang et al., 1999), DNase II (Nagata, 2000), and caspase-activated DNase (Krieser and Eastman, 1998) cause DNA fragmentation and irreversible cell death. Acting alone, each of these DNases is capable of causing cell death. However, it should be recognized that a significant portion of DNA breaks occurs after cell death as a part of the "clean-up" of the debris from dead cells.

Endonucleases in the Kidney

Only imited information is available regarding the endonucleases responsible for the DNA fragmentation in the kidney. Studies have shown the presence of two major endonucleases in kidneys and kidney cells, the 15 kDa endonuclease mentioned above and the 30 to 34 kDa DNase I-like endonuclease (Baliga et al., 1997; Enari et al., 1998; Hagar et al., 1996a; Takeshita et al., 2000) (Figure 13.1). The 30 kDa enzyme is mainly a cytoplasmic enzyme, whereas the 15 kDa endonuclease is located in the nuclei. This enzyme was similar to a DNase I by its biochemical

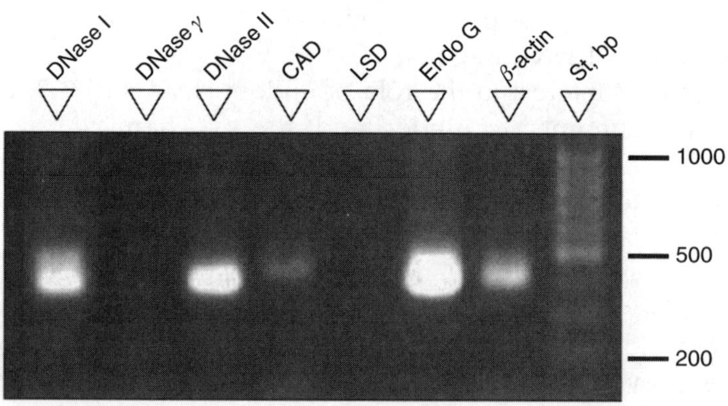

Figure 13.1 Expression of different DNases in mouse kidney. The RT-PCR reaction was performed using primers designed based on GenBank sequences, and the identity of the products were confirmed by DNA sequencing. β-Actin was used as internal control.

characteristics. DNase I is found in all studied species and tissues (Lacks, 1981) and is expressed principally in tissues of the digestive system (Wang et al., 1998). In nondigestive tissues (including kidney), the role of DNase I is not known. Among various organs and tissues, the kidney has one of the highest levels of DNase I activity as measured using DNA-substrate gel electrophoresis (Lacks, 1981; Wang et al., 1998).

One study to examine the role of DNase I in renal injury (Takeshita et al., 2000) showed that DNase I in rat kidney can be regulated by alternative pre-mRNA splicing in 5′ untranslated region (Basnakian et al., 2002b). Some DNase I isoforms can be generated by posttranslational modification, namely mannose-type glycosylation of the protein (Lacks, 1981). Caspase-activated deoxyribonuclease (CAD) activity of the 40 kDa enzyme has been identified by Nagata's group in the cytoplasmic fraction of mouse lymphoma cells (Enari et al., 1998). The specific protein inhibitor of CAD, named ICAD, is a substrate for caspase 3. CAD is the best documented example of an apoptotic endonuclease. This enzyme is present in mouse and human kidney, whereas some other tissues were found to be CAD-negative (Mukae et al., 1998).

Activation of DNases by Reactive Oxygen Metabolites

The activation of endogenous endonucleases in response to oxidative stress has been known since Skalka's work in the 1960s (Skalka and Matyasova, 1967). The exact sequence of events leading to the activation of endonuclease by oxidants is unclear. Data has indicated that different reactive oxygen species contribute to the activation of endonuclease and the enzymatic DNA damage induced by chemical hypoxic injury to renal tubular epithelial cells (Hagar et al., 1996b). Significant protection against DNA strand breaks induced by chemical hypoxia was provided by superoxide dismutase (a scavenger of the superoxide radical), by pyruvate (a scavenger of hydrogen peroxide), by hydroxyl radical scavengers (such as dimethylthiourea, salicylate, and sodium benzoate), and by the metal chelators deferoxamine and 1,10-phenanthroline. The association of endonuclease activation with different ROMs can be suggestive of the absence of a direct link between the type of ROM species and the type of endonuclease. Human recombinant DNase I can be directly activated by metal-ion catalyzed oxidation *in vitro* (Basnakian et al., 2002c) (Figure 13.2). However, it has yet to be investigated whether this pathway takes place *in vivo*, and whether other DNases can be activated in a similar fashion.

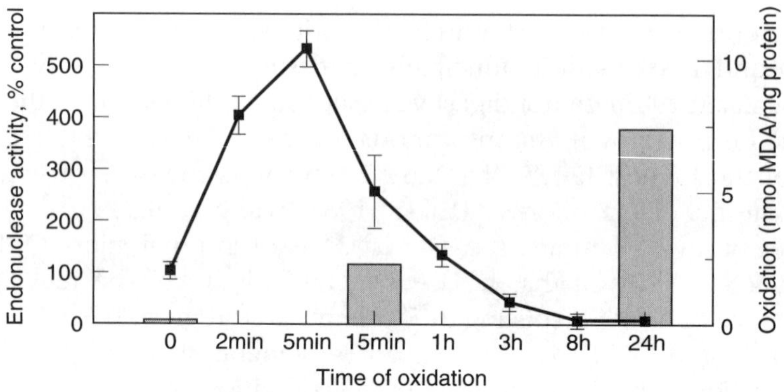

Figure 13.2 In vitro activation of human recombinant DNase I by mild metal ion-catalyzed oxidation. DNase I (Pulmozyme, Genentech) was exposed to 5 m*M* cupric sulfate at 37°C for various periods of time. The reaction was stopped by the addition of EDTA, and the excess cupric sulfate was removed by dialysis. Endonuclease activity was measured using the plasmid incision assay as previously described (Matsushima et al., 1998). Oxidation was assessed using the thiobarbituric acid reactive substances (TBARS) assay.

Another pathway that has to be explored further is the contribution of DNA repair endonucleases to the generation of DNA strand breaks during the repair of oxidatively modified DNA (Lu et al., 2001). Although these DNA strand breaks are very specific to modified residues, it is likely that endonuclease-generated scissions may contribute to the pool of DNA breaks generated by ROM (Figure 13.3).

Mitochondria: Armed and Lethal

Although the paper by Kerr, Wyllie and Currie (Kerr et al., 1972) stated that several organelles, such as mitochondria, did not undergo major modifications during apoptosis, recent evidence suggests that mitochondria act as central coordinators of the downstream execution phase of cell death (Hengartner, 2000; Ravagnan et al., 2002; Zimmermann and Green, 2001). Several proapoptotic signal transduction and damage pathways converge on mitochondria to induce mitochondrial membrane permeabilization (MMP) and this phenomenon is under the control of Bcl-2-related proteins. The inner membrane is characterized by a transmembrane potential ($\Delta\psi_m$) generated through the activity of proton pumps of the respiratory chain. $\Delta\psi_m$ dissipates after the cells are induced to die. The outer mitochondrial member becomes completely

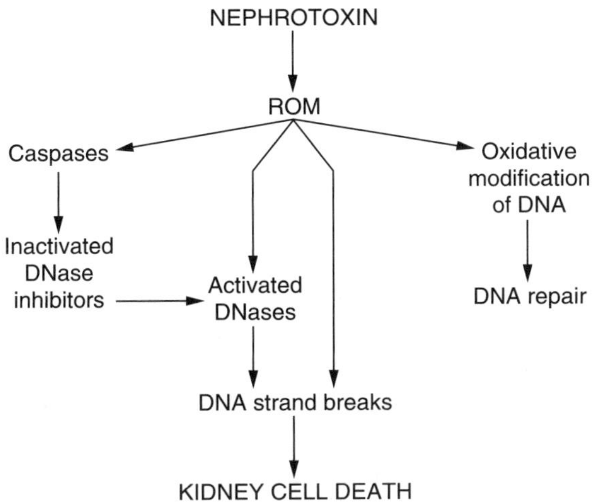

Figure 13.3 Hypothetical scheme for the ROM-mediated mechanisms of cell death induced by nephrotoxicants. ROM, induced by nephrotoxicants, activate caspases and DNases, introduce oxidative DNA damage (mainly 8-hydroxyguanosine), and direct DNA strand breaks in DNA. Caspases inactivate protein inhibitors of DNases, which further promotes enzymatic DNA fragmentation. Endonucleases participating in excision DNA repair of oxidatively damaged DNA may contribute to the total pool of DNA breaks. Failure of DNA breaks repair system leads to cell death.

permeabilized to proteins, resulting in the leakage of potentially toxic mitochondrial intermembrane proteins that orchestrate the degradation phase of cell death.

The notion that a specific class of proteases, the "caspases" (cysteine aspartate specific proteases) are involved in apoptosis emerged from genetic studies of the nematode *Caenorrhabditis elegans*, which revealed that 131 out of 1,093 cells die during the development of the worm and that a specific set of genes are required for this process to occur. Caspases form a family of structurally related cysteine proteases that play a central role in the execution of apoptosis (Earnshaw et al., 1999; Thornberry and Lazebnik, 1998; Wolf and Green, 1999). On receiving a proapoptotic stimulus, the caspases are proteolytically processed to their active forms from their normally synthesized inactive proenzymes. At least fourteen caspases encoded by distinct genes have been cloned and sequenced to date in mammals (Wolf and Green, 1999). Caspase-2, -8, -9, and -10 have large prodomains and initiate the activation of downstream caspases. Caspase-3, -6, and -7, with smaller domains, are identified as effecter or executioner caspases (Earnshaw et al., 1999; Thornberry and Lazebnik, 1998; Wolf and Green, 1999).

The executioner caspases are the major active caspases detected in apoptotic cells and are widely regarded as mediating the execution of apoptosis by cleaving and inactivating intracellular proteins that are essential for cell survival and proliferation (Cryns and Yuan, 1998; Earnshaw et al., 1999; Thornberry and Lazebnik, 1998; Wolf and Green, 1999). The specificity of downstream executioner caspases to cleave cellular proteins is unique because of their different primary sequences and different recognition sites on the target proteins. For example, following activation, caspase-3 primarily recognizes DEVD or DMQD tetrapeptide sequences, whereas caspase-6 recognizes the VEID tetrapeptide sequence for cleavage after the aspartate residue on the target proteins (Earnshaw et al., 1999; Thornberry and Lazebnik, 1998; Wolf and Green, 1999).

Two cell death pathways that result in the activation of the downstream executioner caspase-3 have been relatively well characterized. One is receptor-mediated (Strasser et al., 2000) and the other is mitochondrial-dependent. The receptor-mediated pathway is initiated by activation of cell death receptors (Fas and tumor necrosis factor α) leading to activation of procaspase-8, which in turn cleaves and activates procaspase-3. The mitochondrial-dependent pathway is triggered by cytochrome c release from the mitochondria. In 1996, Xiadong Wang and colleagues (Liu et al., 1996) reported the surprising observation that holo-cytochrome c (but not apo-cytochrome c) is required for the activation of caspase-3 in a cell-free system. Cytochrome c, once present in the cytoplasm, drives the assembly of a high molecular weight caspase activating complex, termed the "apoptosome." Cytochrome c binds to Apaf-1, within its carboxy-terminal region rich in WD motifs. Initially believed to be required only transiently for caspase-9 activation, the Apaf-1/caspase-9 complex is now thought to actually represent the true active form of caspase-9. Thus, Apaf-1 must be viewed not simply as a caspase-9 activator, but rather as an essential regulatory subunit of a caspase-9 holoenzyme. This holoenzyme, often referred to as the apoptosome, is a very large complex that might well contain several additional proteins (Hengartner, 2000).

Some studies have demonstrated ROM-induced upregulation of Fas/FasL system (Bauer et al., 1998; Huang et al., 2000; Kwon et al., 2001; Kasahara et al., 1997; Vogt et al., 1998), while others have shown that H_2O_2-induced apoptosis is Fas-independent (Dumont et al., 1999; Hildeman et al., 1999). The receptor-dependent pathway is initiated by activation of cell death receptors, such as Fas and tumor necrosis factor. After activation, the death receptor recruits procaspase-8 and the adaptor protein FADD to form a death-inducing signaling complex (DISC), which facilitates the activation of procaspase-8. Active caspase-8 in turn cleaves and activates downstream executioner caspases, caspases-3, -6, and -7 (Boldin et al., 1996; Kischkel et al., 1995). Receptor-induced cell death and caspase-8 activation is inhibited by the cowpox virus protein Crm A (Miura et al., 1995; Takahashi et al., 1997; Tewari and Dixit, 1995) but not by Bcl-2 (Newton and Strasser, 1998). Several studies have provided evidence that H_2O_2 induces apoptosis through the activation of caspase-3 (Matsura et al., 1999; Nie et al., 1998), which may be mediated through both the mitochondrial- and receptor-mediated pathways.

After induction of apoptosis, mitochondria release several potentially lethal proteins that either participate in caspase activation (cytochrome c, Smac/DIABLO, HtrA2) or can induce cell death in a caspase-independent fashion (AIF, Endo G, HtrA2). It is difficult to weigh the relative contribution of each of these factors to apoptosis.

Caspases and Cell Death in Toxic Acute Renal Failure

The effect of cisplatin, one of the chemotherapeutic agents that cause nephrotoxicity, on proximal tubular epithelial cell injury has been extensively studied. The primary targets of cisplatin in the kidney are the proximal tubular epithelial cells, where it accumulates and promotes the damage of these cells (Safirstein et al., 1987). The cellular and molecular mechanisms responsible for drug-induced nephrotoxicity to renal tubular epithelial cells are not well understood.

Cisplatin induces cell death in renal tubular epithelial cells (Okuda et al., 2000; Takeda et al., 1996). Caspase-3 is activated in renal proximal tubular cells by cisplatin treatment, suggesting that cisplatin-induced cell death is mediated by caspases. Data have demonstrated that cisplatin induces selective and differential activation of caspases, including executioner caspase-3 and initiator caspase-8 and -9, but not proinflammatory caspase-1 (Kaushal et al., 2001). The selective activation of these caspases was markedly inhibited by their respective peptide inhibitors, suggesting that these caspases may play an important role in cisplatin-induced injury to renal tubular epithelial cells. DEVD-CHO or LEHD-CHO, inhibitors of caspase-3 and caspase-9, respectively, provided marked protection against cisplatin-induced cell death and partial protection against DNA damage in LLC-PK_1 cells, as revealed by an alkaline unwinding assay and by agarose gel electrophoresis (Takeda et al., 1997).

The specific role of caspase-3 and its more direct involvement in cisplatin-induced injury has been identified by studies utilizing the baculovirus protein p35, a potent inhibitor of caspase-3 (Takeda et al., 1997). Overexpression of p35 blocks the induction of apoptosis in insect and mammalian cells. A stably transfected LLC-PK_1 cell line developed to overexpress p35 was therefore capable of providing protection against cisplatin-induced injury, demonstrating that cisplatin injury involves the participation of caspases (Kaushal et al., 2001). Overexpression of crmA, a cowpox viral gene known to inhibit caspase-8, also provided protection against cisplatin-induced apoptosis in mouse proximal tubular cells (Takeda et al., 1997). Cisplatin-induced activation of caspase-8 and caspase-9 in renal proximal tubules indicates that both receptor and mitochondrial pathways participate in the activation process.

REGULATORY MECHANISMS

In living cells, mitochondrial changes are predominantly prevented by antiapoptotic members of the Bcl-2 family of proteins. Bcl-2 was first discovered as a protooncogene in follicular B-cell lymphoma. Subsequently, it was identified as a mammalian homolog to the apoptosis repressor ced-9 in *Caenorhabditis elegans*. Since then, at least 19 Bcl-2 family members have been identified in mammalian cells, which possess at least one of four conserved motifs; known as Bcl-2 homology domains (BH1 to BH4). The Bcl-2 family members can be subdivided into three categories, according to their function and structure: antiapoptotic members, such as Bcl-2, Bcl-XL, Mcl-1, and A1 (Bfl-1), proapoptotic molecules, such as Bax, Bak, and Bok (Mtd), and the BH3-only proteins, Bid, Bad, and Bim (these are called BH3-only proteins because of four Bcl-2 homology regions, they share only the third).

Recent studies from several laboratories have recognized PI-3 kinase/Akt phosphorylation as one of the signaling pathways that block apoptosis and promote cell survival in response to diverse apoptotic stimuli in different cell types (Datta et al., 1997, 1999; Franke et al., 1997; Yang et al., 1995b). Akt (also known as protein kinase B) was originally identified as the cellular homolog of the transforming oncogene of the AKT8 retrovirus (Bellacosa et al., 1993; Staal, 1987). Akt is a serine/threonine kinase, one of the downstream targets of PI-3 kinase (Datta et al., 1999; Franke et al., 1997), which phosphorylates Akt and activates it.

Several pathways downstream of PI-3/Akt phosphorylation have been proposed for cell survival (Datta et al., 1999; Franke et al., 1997). One of the well-studied molecules that mediate cell survival by Akt phosphorylation is the proapoptotic Bcl-2 family member Bad. Bad has the ability to directly interact and bind to antiapoptotic Bcl-2 and Bcl-XL and blocks their survival function (Gross et al., 1999; Vaux and Korsmeyer, 1999; Yang et al., 1995b; Zha et al., 1996). Phosphorylated Akt can directly phosphorylate Bad both *in vitro* and *in vivo* (Datta et al., 1997; del Peso et al., 1997; Yang et al., 1995b), and may render Bad incapable of binding to Bcl-XL and restore the antiapoptotic function of Bcl-2 (Datta et al., 1999; Gross et al., 1999). Sequestering phosphorylated Bad by 14-3-3 proteins (Yang et al., 1995b; Zha et al., 1996) may also participate in the Akt survival pathway by making it unavailable to bind to Bcl-2 or prevent it from damaging the mitochondria.

Many studies have provided evidence that Bcl-2 family members regulate activation of caspases through control of cytochrome c release from the mitochondria (Gross et al., 1999; Yang et al., 1997) and by directly binding to Apaf 1, thus preventing the activation of procaspase-9 and subsequently caspase-3 (Hu et al., 1998). Evidence has been presented showing that PI-3 kinase-mediated Akt phosphorylation is associated with Bad phosphorylation and suppression of caspase-9 and caspase-3 activation in cisplatin-induced injury to renal tubular epithelial cells (Kaushal et al., 2001). Wortmannin and LY294002, inhibitors of PI-3 kinase, have been shown to block cisplatin-induced phosphorylation of both Akt and Bad and enhance activation of caspase-9 and caspase-3 (Kaushal et al., 2001). However, these inhibitors had no effect on the activation of proinflammatory caspase-1 and receptor-dependent initiator caspase-8. Similar results on activation of caspase-3 and caspase-9 were obtained on inhibition of PI-3 kinase in hypoxia-induced injury to renal tubular epithelial cells. A further study has shown that Akt can also phosphorylate human caspase-9, resulting in reduction of caspase-9 activity (Cardone et al., 1998). Based on these studies, the inhibition of Akt phosphorylation as well as Bad phosphorylation by wortmannin and LY294002 in hypoxia or cisplatin-induced injury may contribute to enhanced activation of mitochondrial dependent caspase-3 and caspase-9, but not to receptor-mediated activation of caspase-8 or proinflammatory caspase-1.

CONCLUSION

There is compelling evidence for the implication of ROMs in toxic acute renal failure induced by cisplatin, gentamicin, cyclosporin, or glycerol. Although toxic acute renal failure has been associated with the necrotic form of cell death, many questions regarding the mechanisms of ROM-mediated nephrotoxicity remain to be elucidated. In particular, the studies described above indicate that the apoptotic mode of cell death is also very important in toxic tubular epithelial cell injury. The signal transduction pathways that link ROMs with the activation of downstream caspases and endonucleases have yet to be defined. In the final analysis, it is obviously important to determine whether or not these mechanisms are applicable to acute renal failure in humans. The collective body of evidence suggests an important role for ROMs in toxic acute renal failure and may provide therapeutic opportunities for preventing or treating acute renal failure in humans.

REFERENCES

Abdel-Naim AB, Abdel-Wahab NH, and Attia FF (1999) Protective effects of vitamin e and probucol against gentamicin-induced nephrotoxicity in rats. *Pharmacol Res* 40:183–187.

Abul-Ezz SR, Walker PD, and Shah SV (1991) Role of glutathione in an animal model of myoglobinuric acute renal failure. *Proc Natl Acad Sci USA* 88: 9833–9837.

Ademuyiwa O, Ngaha EO, and Ubah FO (1990) Vitamin E and selenium in gentamicin nephrotoxicity. *Human Exp Toxicol* 9:281–288.

Agarwal A, Balla J, Alam J, Croatt, AJ, and Nath KA (1995) Induction of heme oxygenase in toxic renal injury: A protective role in cisplatin nephrotoxicity in the rat. *Kidney Int* 48:1298–1307.

Ali BH (1995) Gentamicin nephrotoxicity in humans and animals: some recent research. *Gen Pharmacol* 26:1477–1487.

Ali BH, and Mousa HM (2001) Effect of dimethyl sulfoxide on gentamicin-induced nephrotoxicity in rats. *Hum Exp Toxicol* 20:199–203.

Anderson ME, Naganuma A, and Meister A (1990) Protection against cisplatin toxicity by administration of glutathione ester. *FASEB J* 4:3251–3255.

Avramovic V, Vlahovic P, Mihailovic D, and Stefanovic V (1999) Protective effect of a bioflavonoid proanthocyanidin-BP1 in glycerol-induced acute renal failure in the rat: Renal stereological study. *Ren Fail* 21:627–634.

Babu E, Gopalakrishnan VK, Sriganth IN, Gopalakrishnan R, and Sakthisekaran D (1995) Cisplatin induced nephrotoxicity and the modulating effect of glutathione ester. *Mol Cell Biochem* 144:7–11.

Baliga R, Ueda N, Walker PD, and Shah SV (1997) Oxidant mechanisms in toxic acute renal failure. *Am J Kidney* Dis 29:465–477.

Baliga R, Zhang Z, Baliga M, Ueda N, and Shah SV (1998a) *In vitro* and *in vivo* evidence suggesting a role for iron in cisplatin-induced nephrotoxicity. *Kidney Int* 53:394–401.

Baliga R, Zhang Z, Baliga M, Ueda N, and Shah SV (1998b) Role of cytochrome P-450 as a source of catalytic iron in cisplatin-induced nephrotoxicity. *Kidney Int* 54:1562–1569.

Baliga R, Ueda N, Walker PD, and Shah SV (1999) Oxidant mechanisms in toxic acute renal failure. *Drug Metab Rev* 31:971–979.

Basnakian AG, Ueda N, Kaushal GP, Mikhailova MV, and Shah SV (2002a) DNase I-like endonuclease in rat kidney cortex and activation during ischemia/reperfusion injury. *J Am Soc Nephrol* 13:1000–1007.

Basnakian AG, Singh AB, and Shah SV (2002b) Identification of novel alternatively spliced segments in rat DNase I gene expressed in the kidney. *Gene* 289:87–96.

Basnakian AG, Yin X, and Shah SV (2002c) Activation of DNase I by metal-ion catalyzed oxidation *in vitro*. *J Am Soc Nephrol* 13:325A.

Bauer MKA, Vogt M, Los M, Siegel J, Wesselborg S, and Schulze-Osthoff K (1998) Role of reactive oxygen intermediates in activation-induced CD95 (Apo-1/Fas) ligand expression. *J Biol Chem* 273:8048–8055.

Bellacosa A, Franke TF, Gonzalez-Portal ME, Datta K, Taguchi T, Gardner J, Cheng JQ, Testa JR, and Tsichlis PN (1993) Structure, expression and chromosomal mapping of c-akt: Relationship to v-akt and its implications. *Oncogene* 8:745–754.

Ben Ismail TH, Ali BH, and Bashir AA (1994) Influence of iron, deferoxamine and ascorbic acid on gentamicin-induced nephrotoxicity in rats. *Gen Pharmacol* 25:1249–1252.

Boldin M, Goncharov T, and Goltsev Y (1996) Involvement of MACH, a novel MORT1/FADD-interacting protease, in Fas/APO-1- and TNF receptor-induced cell death. *Cell* 85:803–815.

Bull JM, Strebel FR, Sunderland BA, Bulger RE, Edwards M, Siddik ZH, and Newman RA (1988) o-(beta-Hydroxyethyl)-rutoside-mediated protection of renal injury associated with cis-diamminedichloroplatinum(II)/hyperthermia treatment. *Cancer Res* 48:2239–2244.

Bunton CA (1949) Oxidation of α-diketones and α-keto-acids by hydrogen peroxide. *Nature* 163:444.

Bywaters EGL, and Beall D (1941) Crush injuries with impairment of renal function. *Br Med J* 1:427–432.

Cardone MH, Roy N, Stennicke HR, Salvesen GS, Franke TF, Stanbridge E, Frisch S, and Reed JC (1998) Regulation of cell death protease caspase-9 by phosphorylation. *Science* 282:1318–1321.

Cryns V, and Yuan J (1998) Proteases to die for. *Genes and Develop* 12:1551–1570.

Cuzzocrea S, Mazzon E, Dugo L, Serraino I, Di Paolo R, Britti D, De Sarro A, Pierpaoli S, Caputi A, Masini E, and Salvemini D (2002) A role for superoxide in gentamicin-mediated nephropathy in rats. *Eur J Pharmacol* 450:67–76.

Datta SR, Dudek H, Tao X, Masters S, Fu H, Gotoh Y, and Greenberg ME (1997) Akt phosphorylation of BAD couples survival signals to the cell-intrinsic death machinery. *Cell* 91:231–241.

Datta SR, Brunet A, and Greenberg ME (1999) Cellular survival: A play in three akts. *Genes and Develop* 13:2905–2927.

Davis CA, Nick HS, and Agarwal A (2001) Manganese superoxide dismutase attenuates cisplatin-induced renal injury: Importance of superoxide. *J Am Soc Nephrol* 12:2683–2690.

del Peso L, Gonzalez-Garcia M, Page C, Herrera R, and Nunez G (1997) Interleukin-3-induced phosphorylation of BAD through the protein kinase Akt. *Science* 278:687–689.

Du XH, and Yang CL (1994) Mechanism of gentamicin nephrotoxicity in rats and the protective effect of zinc-induced metallothionein synthesis. *Nephrol Dial Transplant* 9:135–140.

Dumont A, Hehner SP, Hofmann TG, Ueffing M, Droge W, and Schmitz ML (1999) Hydrogen peroxide-induced apoptosis is CD95-independent, requires the release of mitochondria-derived reactive oxygen species and the activation of NF-kappaB. *Oncogene* 18:747–757.

Earnshaw WC, Martins LM, and Kaufmann SH (1999) Mammalian caspases: Structure, activation, substrates, and functions during apoptosis. *Annu Rev Biochem* 68:383–424.

Enari M, Sakahara H, Yokohama H, Okawa K, Iwamatsu A, and Nagata S (1998) A caspase-activated DNase that degrades DNA during apoptosis, and its inhibitor ICAD. *Nature* 391:43–50.

Franke TF, Kaplan DR, and Cantley LC (1997) PI3K downstream of AKT blocks apoptosis. *Cell* 88:435–437.

Fridovich I (1978) The biology of oxygen radicals. The superoxide radical is an agent of oxygen toxicity; superoxide dismutases provide an important defense. *Science* 201:875–880.

Gabow PA, Kaehny WD, and Kelleher SP (1982) The spectrum of rhabdomyolysis. *Medicine* 61:141–152.

Gemba M, and Fukuishi N (1991) Amelioration by ascorbic acid of cisplatin-induced injury in cultured renal epithelial cells. *Contrib Nephrol* 95:138–142.

Gemba M, Fukuishi N, and Nakano S (1988) Effect of N,N′-diphenyl-p-phenylenediamine pretreatment on urinary enzyme excretion in cisplatin nephrotoxicity in rats. *Jpn J Pharmacol* 46:90–92.

Gross A, McDonnell JM, and Korsmeyer S (1999) Bcl-2 family members and the mitochondria in apoptosis. *Genes Dev* 13:1899–1911.

Grossman RA, Hamilton RW, Morse BM, Penn AS, and Goldberg M (1974) Nontraumatic rhabdomyolysis and acute renal failure. *N Engl J Med* 291:807–811.

Guidet B, and Shah SV (1989a) Enhanced *in vivo* H_2O_2 generation by rat kidney in glycerol-induced renal failure. *Am J Physiol* 257:F440–F445.

Guidet BR, and Shah SV (1989b) *In vivo* generation of hydrogen peroxide by rat kidney cortex and glomeruli. *Am J Physiol* 256:F158–F164.

Hagar H, Ueda N, and Shah SV (1996a) Endonuclease induced DNA damage and cell death in chemical hypoxic injury to LLC-PK_1 cells. *Kidney Int* 49:355–361.

Hagar H, Ueda N, and Shah SV (1996b) Role of reactive oxygen metabolites in DNA damage and cell death in chemical hypoxic injury to LLC-PK_1 cells. *Am J Physiol* 271:F209–F215.

Halestrap AP, Scott RD, and Thomas AP (1980) Mitochondrial pyruvate transport and its hormonal regulation. *Int J Biochem* 11:97–105.

Halliwell B, and Gutteridge JMC (1990) Role of free radicals and catalytic metal ions in human disease: An overview. *Meth Enzymol* 186:1–85.

Hannemann J, and Baumann K (1988) Cisplatin-induced lipid peroxidation and decrease of gluconeogenesis in rat kidney cortex: Different effects of antioxidants and radical scavengers. *Toxicology* 51:119–132.

Hengartner MO (2000) The biochemistry of apoptosis. *Nature* 407:770–776.

Hildeman DA, Mitchell T, Teague TK, Henson P, Day BJ, Kappler J, and Marrack PC (1999) Reactive oxygen species regulate activation-induced T cell apoptosis. *Immunity* 10:735–744.

Hostetter TH, Wilkes BM, and Brenner BM (1983) Renal circulatory and nephron function in experimental acute renal failure, in *Acute Renal Failure*, Brenner BM, and Lazarus JM (Eds), W.B. Saunders Company, Philadelphia. pp. 99–115.

Hu Y, Benedict MA, Wu D, Inohara N, and Nunez G (1998) Bcl-XL interacts with Apaf-1 and inhibits Apaf-1-dependent caspase-9 activation. *Proc Natl Acad Sci USA* 95:4386–4391.

Huang C, Li J, Zheng R, and Cui K (2000) Hydrogen peroxide-induced apoptosis in human hepatoma cells is mediated by CD95(APO-1/Fas) receptor/ligand system and may involve activation of wild-type p53. *Mol Biol Rep* 27:1–11.

Humes HD, and Weinberg JM (1986) Toxic Nephropathies, in *The Kidney*, Brenner BM, and Rector FC (Eds), W.B. Saunders Company, Philadelphia. pp. 1491–1532.

Jollie DR, and Maines MD (1985) Effect of cis-platinum on kidney cytochrome P-450 and heme metabolism: Evidence for the regulatory role of the pituitary hormones. *Arch Biochem Biophys* 240:51–59.

Kasahara Y, Iwai K, Yachie A, Ohta K, Konno A, Seki H, Miyawaki T, and Taniguchi N (1997) Involvement of reactive oxygen intermediates in spontaneous and CD95 (Fas/APO-1)-mediated apoptosis of neutrophils. *Blood* 89:1748–1753.

Kaushal GP, Kaushal V, Hong X, and Shah SV (2001) Role and regulation of activation of caspases in cisplatin-induced injury to renal tubular epithelial cells. *Kidney Int* 60:1726–1736.

Kavutcu M, Canbolat O, Ozturk S, Olcay E, Ulutepe S, Ekinci C, Gokhun IH, and Durak I (1996) Reduced enzymatic antioxidant defense mechanism in kidney tissues from gentamicin-treated guinea pigs: Effects of vitamins E and C. *Nephron* 72:269–274.

Kays SE, Crowell WA, and Johnson MA (1991) Iron supplementation increases gentamicin nephrotoxicity in rats. *J Nutr* 121:1869–1875.

Kerr JFR, Wyllie AH, and Currie AR (1972) Apoptosis: A basic biological phenomenon with wide-ranging implications in tissue kinetics. *Br J Cancer* 26:239–257.

Kim YK, Jung JS, Lee SH, and Kim YW (1997) Effects of antioxidants and Ca^{2+} in cisplatin-induced cell injury in rabbit renal cortical slices. *Toxicol Appl Pharmacol* 146:261–269.

Kischkel F, Hellbardt S, and Behrmann I (1995) Cytotoxicity-dependent APO-1 (Fas/CD95)-associated proteins form a death-inducing signaling complex (DISC) with the receptor. *EMBO J* 14:5579–5588.

Koffler A, Friedler RM, and Massry SG (1976) Acute renal failure due to nontraumatic rhabdomyolysis. *Ann Intern Med* 85:23–28.

Krieser RJ, and Eastman A (1998) The cloning and expression of human deoxyribonuclease II. A possible role in apoptosis. *J Biol Chem* 273:30909–30914.

Kruidering M, van de Water B, de Heer E, Mulder GJ, and Nagelkerke JF (1997) Cisplatin-induced nephrotoxicity in porcine proximal tubular cells: Mitochondrial dysfunction by inhibition of complexes I to IV of the respiratory chain. *J Pharmacol Exp Ther* 280:638–649.

Kwon D, Choi C, Lee J, Kim KO, Kim JD, Kim SJ, and Choi IH (2001) Hydrogen peroxide triggers the expression of Fas/FasL in astrocytoma cell lines and augments apoptosis. *J Neuroimmunol* 113:1–9.

Lacks SA (1981) Deoxyribonuclease I in mammalian tissues. Specificity of inhibition by actin. *J Biol Chem* 256:2644–2648.

Lieberthal W, and Levine JS (1996) Mechanisms of apoptosis and its potential role in renal tubular epithelial cell injury. *Am J Physiol* 271:F477–F488.

Lieberthal W, Menza SA, and Levine JS (1998) Graded ATP depletion can cause necrosis or apoptosis of cultured mouse proximal tubular cells. *Am J Physiol* 274:F315–F327.

Liu X, Kim CN, Yang J, Jemmerson R, and Wang X (1996) Induction of apoptotic program in cell-free extracts: Requirement for dATP and cytochrome c. *Cell* 86:147–157.

Liu H, Baliga M, and Baliga R (2002) Effect of cytochrome P450 2E1 inhibitors on cisplatin-induced cytotoxicity to renal proximal tubular epithelial cells. *Anticancer Res* 22:863–868.

Lu AL, Li X, Gu Y, Wright PM, and Chang DY (2001) Repair of oxidative DNA damage: Mechanisms and functions. *Cell Biochem Biophys* 35:141–170.

Matsura T, Kai M, Fujii Y, Ito H, and Yamada K (1999) Hydrogen peroxide induced apoptosis in HL-60 requires caspase-3 activation. *Free Rad Res* 30:73–83.

Matsushima H, Yonemura K, Ohishi K, and Hishida A (1998) The role of oxygen free radicals in cisplatin-induced acute renal failure in rats. *J Lab Clin Med* 131:518–526.

McCord JM (1974) Free radicals and inflammation: Protection of synovial fluid by superoxide dismutase. *Science* 185:529–531.

McCord JM, and Fridovich I (1969) Superoxide dismutase. An enzymic function for erythrocuprein (hemocuprein). *J Biol Chem* 244:6049–6055.

Mistry P, Merazga Y, Spargo DJ, Riley PA, and McBrien DC (1991) The effects of cisplatin on the concentration of protein thiols and glutathione in the rat kidney. *Cancer Chemother Pharmacol* 28:277–282.

Miura M, Friedlander R, and Yuan J (1995) Tumor necrosis factor-induced apoptosis is mediated by a CrmA-sensitive cell death pathway. *Proc Natl Acad Sci USA* 92:8318–8322.

Mukae N, Enari M, Sakahira H, Fukuda Y, Inazawa J, Toh H, and Nagata S (1998) Molecular cloning and characterization of human caspase-activated DNase. *Proc Natl Acad Sci USA* 95:9123–9128.

Murer H, and Burckhardt G (1983) Membrane transport of anions across epithelia of mammalian small intestine and kidney proximal tubule. *Rev Physiol Biochem Pharmacol* 96:1–51.

Nagata S (2000) Apoptotic DNA fragmentation. *Exp Cell Res* 256:12–18.

Nakajima T, Hishida A, and Kato A (1994) Mechanisms for protective effects of free radical scavengers on gentamicin-mediated nephropathy in rats. *Am J Physiol* 266:F425–F431.

Nath KA, Balla G, Vercellotti GM, Balla J, Jacob HS, Levitt MD, and Rosenberg ME (1992) Induction of heme oxygenase is a rapid, protective response in rhabdomyolysis in the rat. *J Clin Invest* 90:267–270.

Nath KA, Balla J, Croatt AJ, and Vercellotti GM (1995) Heme protein-mediated renal injury: A protective role for 21-aminosteroids *in vitro* and *in vivo*. *Kidney Int* 47:592–602.

Newton K, and Strasser A (1998) The Bel-2 family and cell death regulation. *Curr Op Genet Dev* 8:68–75.

Nie Z, Mei Y, Ford M, Rybak L, Marcuzzi A, Ren H, Stiles GL, and Ramkumar V (1998) Oxidative stress increases A1 adenosine receptor expression by activating nuclear factor kappa B. *Mol Pharmaco*l 53:663–669.

Okuda M, Masaki K, and Fukatsu S (2000) Role of apoptosis in cisplatin-induced toxicity in the renal epithelial cell line LLC-PK_1: Implication of the functions of apical membranes. *Biochem Pharmacol* 59:95–201.

Paller MS (1988) Hemoglobin- and myoglobin-induced acute renal failure in rats: Role of iron in nephrotoxicity. Am *J Physiol* 255:F539–F544.

Polzar B, Peitsch MC, Loos R, Tschopp J, and Mannherz HG (1993) Overexpression of deoxyribonuclease I (DNase I) transfected into COS-cells: its distribution during apoptotic cell death. *Eur J Cell Biol* 62:397–405.

Priuska EM, and Schacht J (1995) Formation of free radicals by gentamicin and iron and evidence for an iron/gentamicin complex. *Biochem Pharmacol* 50:1749–1752.

Ramsammy LS, Josepovitz C, Ling KY, Lane BP, and Kaloyanides GJ (1986) Effects of diphenyl-phenylenediamine on gentamicin-induced lipid peroxidation and toxicity in rat renal cortex. *J Pharmacol Exp Ther* 238:83–88.

Ravagnan L, Roumier T, and Kroemer G (2002) Mitochondria, the killer organelles and their weapons. *J Cell Physiol* 192:131–137.

Safirstein R, Winston J, Moel D, Dikman S, and Guttenplan J (1987) Cisplatin nephrotoxicity: Insights into mechanism. *Intl J Androl* 10:325–346.

Salahudeen AK, Clark EC, and Nath KA (1991) Hydrogen peroxide-induced renal injury. A protective role for pyruvate *in vitro* and *in vivo*. *J Clin Invest* 88:1886–1893.

Salahudeen AK, Wang C, Bigler SA, Dai Z, and Tachikawa H (1996) Synergistic renal protection by combining alkaline-diuresis with lipid peroxidation inhibitors in rhabdomyolysis: Possible interaction between oxidant and non-oxidant mechanisms. *Nephrol Dial Transplant* 11:635–642.

Satoh M, Naganuma A, and Imura N (1987) Deficiency of selenium intake enhances manifestation of renal toxicity of cis-diamminedichloroplatinum in mice. *Toxicol Lett* 38:155–160.

Schumer M, Colombel MC, Sawczuk IS, Gobe G, Connor J, O'Toole KM, Olsson CA, Wise GJ, and Buttyan R (1992) Morphologic, biochemical, and molecular evidence of apoptosis during the reperfusion phase after brief periods of renal ischemia. *Am J Physiol* 140:831–838.

Shah SV, and Walker PD (1988) Evidence suggesting a role for hydroxyl radical in glycerol-induced acute renal failure. *Am J Physiol* 255:F438–F443.

Skalka M, and Matyasova J (1967) The effect of radiation on deoxyribonucleoproteins in animal tissue. *Folia Biol (Praha)* 13:457–464.

Staal SP (1987) Molecular cloning of the akt oncogene and its human homologs Akt1 and Akt2: Amplification of Akt1 in a primary human gastric adenocarcinoma. *Proc Natl Acad Sci USA* 84:5034–5037.

Strasser A, O'Connor L, and Dixit VM (2000) Apoptosis signaling. *Annu Rev Biochem* 69:217–245.

Sugihara K, and Gemba M (1986) Modification of cisplatin toxicity by antioxidants. *Jpn J Pharmacol* 40:353–355.

Sugiyama S, Hayakawa M, Kato T, Hanaki Y, Shimizu K, and Ozawa T (1989) Adverse effects of anti-tumor drug, cisplatin, on rat kidney mitochondria: Disturbances in glutathione peroxidase activity. *Biochem Biophys Res Commun* 159:1121–1127.

Takahashi A, Hirata H, Yonehara S, Imai Y, Lee K-K, Moyer RW, Turner PC, Mesner PW, Okazaki T, Sawai H, Kishi S, Yamamoto K, Okuma M, and Sasada M (1997) Affinity labeling displays the stepwise activation of ICE-related proteases by Fas, staurosporine, and CrmA-sensitive caspase-8. *Oncogene* 14:2741–2752.

Takeda M, Fukuoka K, and Endou H (1996) Cisplatin-induced apoptosis in mouse proximal tubular cell line. *Contr Nephrol* 118:24–28.

Takeda M, Kobayashi M, Shirato I, Osaki T, and Endou H (1997) Cisplatin-induced apoptosis of immortalized mouse proximal tubule cells is mediated by interleukin-1b converting enzyme (ICE) family of proteases but inhibited by overexpression of bcl-2. *Arch Toxicol* 71:612–621.

Takeshita H, Mogi K, Yasuda T, Nakajima T, Nakajima Y, Mori S, Hoshino T, and Kishi K (2000) Mammalian deoxyribonucleases I are classified into three types: Pancreas, parotid, and pancreas-parotid (mixed), based on differences in their tissue concentrations. *Biochem Biophys Res Commun* 269:481–484.

Tewari M, and Dixit VM (1995) Fas and tumor necrosis factor-induced apoptosis is inhibited by the poxvirus crmA gene product. *J Biol Chem* 270:3255–3260.

Thompson EB (1998) Special topic: Apoptosis. *Annu Rev Physiol* 60:525–532.

Thornberry NA, and Lazebnik Y (1998) Caspases: Enemies within. *Science* 281:1312–1316.

Tsutsumishita Y, Onda T, Okada K, Takeda M, Endou H, Futaki S, and Niwa M (1998) Involvement of H_2O_2 production in cisplatin-induced nephrotoxicity. *Biochem Biophys Res Commun* 242:310–312.

Ueda N, and Shah SV (1992) Endonuclease-induced DNA damage and cell death in oxidant injury to renal tubular epithelial cells. *J Clin Invest* 90:2593–2597.

Ueda N, Guidet B, and Shah SV (1993) Gentamicin-induced mobilization of iron from renal cortical mitochondria. *Am J Physiol* 265:F435–F439.

Ueda N, Walker PD, Hsu SM, and Shah SV (1995) Activation of a 15-kDa endonuclease in hypoxia/reoxygenation injury without morphologic features of apoptosis. *Proc Natl Acad Sci USA* 92:7202–7206.

Vaux DL, and Korsmeyer SJ (1999) Cell death in development. *Cell* 96:245–254.

Vogt M, Bauer MK, Ferrari D, and Schulze-Osthoff K (1998) Oxidative stress and hypoxia/reoxygenation trigger CD95 (APO-1/Fas) ligand expression in microglial cells. *FEBS Lett* 429:67–72.

Walker PD, and Shah SV (1987) Gentamicin enhanced production of hydrogen peroxide by renal cortical mitochondria. *Am J Physiol* 253:C495–C499.

Walker PD, and Shah SV (1988) Evidence suggesting a role for hydroxyl radical in gentamicin-induced acute renal failure in rats. *J Clin Invest* 81:334–341.

Wang CC, Lu SC, Chen HL, and Liao TH (1998) Porcine spleen deoxyribonuclease II. Covalent structure, cDNA sequence, molecular cloning, and gene expression. *J Biol Chem* 273:17192–17198.

Weijl NI, Hopman GD, Wipkink-Bakker A, Lentjes e.g., Berger HM, Cleton FJ, and Osanto S (1998) Cisplatin combination chemotherapy induces a fall in plasma antioxidants of cancer patients. *Ann Oncol* 9:1331–1337.

Wolf BB, and Green DR (1999) Suicidal tendencies: Apoptotic cell death by caspase family proteinases. *J Biol Chem* 274:20049–20052.

Yang CL, Du XH, and Han YX (1995a) Renal cortical mitochondria are the source of oxygen free radicals enhanced by gentamicin. *Ren Fail* 17:21–26.

Yang E, Zha J, Jockel J, Boise LH, Thompson CB, and Korsmeyer SJ (1995b) Bad, a heterodimeric partner for Bcl-XL and Bcl-2, displaces Bax and promotes cell death. *Cell* 80:285–291.

Yang J, Liu X, Bhalla K, Kim CN, Ibrado AM, Cai J, Peng T-I, Jones DP, and Wang X (1997) Prevention of apoptosis by Bcl-2: Release of cytochome c from mitochondria blocked. *Science* 275:1129–1132.

Zager RA (1992) Combined mannitol and deferoxamine therapy for myohemoglobinuric renal injury and oxidant tubular stress. Mechanistic and therapeutic implications. *J Clin Invest* 90:711–719.

Zager RA (1996a) Rhabdomyolysis and myohemoglobinuric acute renal failure. *Kidney Int* 49:314–326.

Zager RA (1996b) Mitochondrial free radical production induces lipid peroxidation during myohemoglobinuria. *Kidney Int* 49:741–751.

Zha J, Harada H, Yang E, Jockel J, and Korsmeyer SJ (1996) Serine phosphorylation of death agonist BAD in response to survival factor results in binding to 14-3-3 not BCL-X. *Cell* 87:619–628.

Zhang S, Demirs JT, Bove KE, and Xu M, (1999) DNA fragmentation factor 45-deficient cells are more resistant to apoptosis and exhibit different dying morphology than wild-type control cells. *J Biol Chem* 274:37450–37454.

Zimmermann KC, and Green DR (2001) How cells die: Apoptosis pathways. *J Allergy Clin Immunol* 108:S99–S103.

Zurovsky Y, and Haber C (1995) Antioxidants attenuate endotoxin-gentamicin induced acute renal failure in rats. *Scand J Urol Nephrol* 29:147–154.

III

Clinical Nephrotoxicity and Specific Classes of Nephrotoxicants

14

Vasoactive and Inflammatory Substances

Norberto Perico, Marina Noris, and Giuseppe Remuzzi

INTRODUCTION

The functioning of the kidney, at both glomerular and tubular levels, is regulated by a complex network of circulating and locally produced hormones. These hormones comprise a chemically heterogeneous group, which includes proteins, lipids, nucleosides, and amino acid-derived molecules. Besides influencing physiological determinants of renal function, these substances, when produced in excess, may participate through their vasoactive and proinflammatory properties to the pathogenic processes, leading to acute or chronic renal dysfunction. They are also involved in the mechanisms by which drugs and chemicals can cause glomerular and tubular structural and functional changes, both in laboratory animals and in humans.

In this chapter, the physiologic actions of the substances that modulate renal function are reviewed. Where appropriate, recent advances of their roles in the pathophysiology of progressive kidney diseases are emphasized. Moreover, the potential relevance of mechanisms and mediators of renal injury are discussed, including the current state of knowledge on the mechanisms of toxicity by which drugs such as cyclosporine, gentamicin, amphotericin B, nonsteroidal anti-inflammatory drugs, and radiocontrast media impair renal function.

THE RENIN–ANGIOTENSIN SYSTEM

Regulation of Renin Production and Secretion

The renin–angiotensin system is a phylogenetically old system. It involves multiple organ systems to control the systemic blood pressure and fluid and electrolyte homeostasis. Renin, a glycoprotein synthesized in the juxtaglomerular apparatus of the kidney, converts its substrate, the $\alpha 2$-globulin angiotensinogen synthesized and released by liver cells, to the decapeptide angiotensin I (Ang I). This conversion is believed to take place in plasma. Angiotensin converting enzyme (ACE), a nonspecific dipeptidyl carboxypeptidase present primarily on endothelial cells, acts on Ang I by cleaving the octapeptide angiotensin II (Ang II) from its carboxy-terminal dipeptide. The major effector of the system is actually Ang II, whose actions include vasoconstriction, stimulation of aldosterone secretion, increased sodium reabsorption by the macula densa, stimulation of thirst by interacting with sympathetic nerve transmission in the central nervous system, and direct inhibition of renin release in the kidney.

In addition to the conversion of Ang I to Ang II, ACE inactivates two vasodilator peptides, bradykinin and kallidin. Inhibition of ACE thus lowers blood pressure through two mechanisms: prevention of the formation of Ang II and potentiation of the hypotensive properties of bradykinin. The availability of specific inhibitors and antagonists of Ang II receptors has helped in defining the concept that not all the actions of endogenous Ang II are carried out by the circulating form of the peptide. ACE inhibitors control blood pressure for periods of time that greatly exceed their plasma half-lives (Naftilan, 1994).

There is also a discrepancy between the antihypertensive effect of ACE inhibitors and plasma ACE activity. These observations prompted investigators to look for possible local formation of Ang II. The finding that bilateral nephrectomy virtually eliminated plasma renin activity without reducing arterial wall renin concentrations in normotensive and hypertensive rats suggested that the arterial renin concentration was independent of renal renin (Naftilan, 1994). Moreover, isolated and perfused vascular tissues generate Ang II, which could be taken to indicate that all components of the cascade are present in vascular tissues (Naftilan, 1994).

Molecular biology techniques allowed cloning and DNA sequencing of all the components of the renin–angiotensin system, including renin, angiotensinogen, ACE, and the Ang II receptors (Dzau et al., 1988).

Renin is expressed in various tissues and organs in the mouse and rat, including vascular tissue, brain, testes, heart, submandibular gland and kidney. Angiotensinogen mRNA is present in the heart and vascular tissue, and angiotensinogen expression has been detected in the vascular smooth muscle cells of the aorta. *In vivo* autoradiography methods and mRNA have located ACE in the heart, with higher signals in the atria than in ventricles. These studies confirmed earlier immunocytochemical observations that ACE activity can be detected in the endothelial cell layer, and further indicate the importance of an intact endothelium for the local generation of Ang II in the vascular wall.

It recently became clear that the renin–angiotensin system is much more complex than previous research suggested. The new story of renin–angiotensin system began in 2000, with the discovery of an enzyme similar to ACE, named ACE2 (Donoghue et al., 2000; Tipnis et al., 2000). ACE2 is expressed predominately in vascular endothelial cells of the heart and kidney. Although both are carboxypeptidases, ACE cleaves two amino acids at a time, whereas ACE2 shortens peptides by only one amino acid. The result is that ACE and ACE2 have different biochemical activities. Therefore, Ang I is thought to be converted to angiotensin 1–9 (with nine amino acids) by ACE2, but to the Ang II (with eight amino acids) by ACE. Angiotensin 1–9 has no known effects and cannot be converted to Ang II by ACE2, but can be converted to angiotensin 1–7 (a blood-vessel dilator) by ACE (Boehm and Nabel, 2002). Therefore, it has been suggested that ACE2 prevents the formation of the vasoconstrictor Ang II. The role of ACE2 in blood pressure control is still unclear. However, studies undertaken in *ace2* gene knockout mice showed that loss of ACE2 does not alter blood-pressure homeostasis but does severely impair cardiac function (Boehm and Nabel, 2002; Crackower et al., 2002).

In the kidney, renin is produced and stored in granular juxtaglomerular cells, which are modified aortic smooth muscle cells found in the media of afferent arterioles (Griendling et al., 1993). Genomic analysis of the renin gene identified a single locus in humans and rats, but mice have two renin genes, designed *Ren-1* and *Ren-2* (Griendling et al., 1993), the latter corresponding to the renin produced in mouse kidney. Renin is synthesized in an inactive precursor form, preprorenin. Cleavage of the signal peptide from the carboxyl terminal of preprorenin results in prorenin, which is also biologically inactive. Subsequent glycosylation and proteolytic cleavage leads to formation of renin, a 37 to 40 kDa molecule. Both prorenin and renin are secreted from juxtaglomerular cells. Because prorenin is the major circulating form, it is postulated that significant conversion of prorenin to renin follows secretion.

Stimulation of renin release by juxtaglomerular cells is mediated by increased intracellular cyclic adenosine monophosphate (cAMP), while a rise in cytosolic free calcium is inhibitory (Kurtz, 1986). Physiologic regulators of renin secretion include the urinary NaCl concentration, sensed by macula densa cells in the distal tubule. Decreased NaCl delivery to macula densa cells stimulates renin secretion, whereas increased urinary NaCl exerts an opposite effect (Lorenz et al., 1993). Changes in luminal Cl concentration alter the rate of Na^+-K^+-$2Cl^-$ transport in macula densa cells (Schlatter, 1989). The precise mechanism by which variation in the activity of this transport translates into a signal that regulates renin release by adjacent juxtaglomerular granular cells is not entirely clear. Postulated mediators include adenosine, which inhibits renin secretion through activation of adenosine 1 (A_1) receptors on juxtaglomerular cells, and alterations in interstitial osmolality, which may affect renin secretion directly (Lorenz et al., 1993). Experimental evidence also suggests that nitric oxide (NO) produced by macula densa and endothelial cells regulates renin secretion (Lorenz et al., 1993; Sigmon et al, 1992).

The activity of the renal sympathetic nervous system is well recognized to control renin secretion. Stimulation of post-junctional β-adrenergic receptors increases renin release, whereas the role of α-adrenergic receptors is controversial (Koppa and DiBona, 1993).

Changes in intrarenal perfusion pressure are associated with alterations in renin release (Hackenthal et al., 1990). Elevation of perfusion pressure inhibits renin release and induces the so-called "pressure natriuresis" phenomenon. It has been postulated that increased renal perfusion pressure elevates intracellular calcium in juxtaglomerular cells to inhibit renin secretion (Hackenthal et al., 1990; Kurtz, 1986). Increased perfusion pressure also stimulates NO production by endothelial cells, which in turn suppresses renin secretion (Lorenz et al., 1993). Conversely, decreased renal perfusion enhances the production of prostacyclin (PGI_2), which increases renin release (Henrich, 1981).

Several endocrine and paracrine hormones regulate renin secretion by the kidney. Beside atrial natruiretic peptide (ANP), inhibitory hormones include vasopressin (AVP), endothelin (ET), adenosine and Ang II, the latter probably being the most physiologically relevant by inhibiting both renin gene expression and peptide secretion in a negative feedback loop (Burns et al., 1993; Hackenthal et al., 1990; Kurtz et al., 1986; Lorenz et al., 1993).

Arachidonate metabolites produced in the kidney also play an important role in renin secretion (Lorenz et al., 1993). Intrarenal infusion of arachidonic acid increases, and indomethacin decreases, plasma renin activity in rabbits (Larsson et al., 1974). Moreover, several studies have confirmed that prostaglandins stimulate and lipoxygenase products inhibit renin release (Heinrich et al., 1990; Lorenz et al., 1993).

Vasoactive Properties of Angiotensin II

Circulating and locally produced Ang II exerts its effects by binding to specific receptors (Siragy, 2002). Recently, two subtypes of Ang II receptors have been described in various tissues, based on their affinity for recently developed Ang II antagonists, namely the nonpeptide biphenylimidazoles, typified by losartan, and the tetrahydroimidapyridines (PD), typified by PD123177 and PD121981, or CGP42112A, a modified peptide analogue of Ang II (Edwards and Aiyar, 1993; Burnier and Brunner, 1994). Inhibition by losartan characterizes the angiotensin I receptor (AT_1 receptor), whereas the PD compounds and CGP42112A identify the AT_2 receptor. Using inhibitors of these receptor subtypes, it has been shown that the AT_1 receptor is the predominant receptor subtype in the vasculature, liver, and kidney of adult rats, but expression of mRNA for AT_1 receptor has also been found in adrenal gland and lung. cDNA clones for AT_1 receptors have been isolated from mice, rabbits, and humans, and show a high degree of homology with the original rat clone. These nucleotide sequences encode for a 359 amino acid protein that has the seven hydrophobic transmembrane domains typical of G protein-coupled receptors.

The rat and mouse have a second form of AT_1 receptor, the AT_{1B} receptor, with 96% amino acid homology with the original rat AT_1 receptor (now designed as AT_{1A} receptor). These receptor subtypes are very similar in ligand specificity and signal transduction mechanisms. Vascular smooth muscle and lung express primarily the AT_{1A} mRNA, the adrenal and pituitary glands express mainly AT_{1B}, and kidney expresses both. These two receptor isoforms may differ more in the regulation of their expression than in their functional properties.

The AT_2 receptor is a seven transmembrane receptor with a molecular mass of approximately 41,000 da, but it exhibits only approximately 34% homology with the protein sequence of the AT_1 receptor (Carey et al., 2000). In fetal tissues, the AT_2 receptor is widely expressed, predominantly in areas of mesenchymal differentiation (Siragy and Carey, 2001). In the adult rat, the AT_2 receptor is expressed in the vascular smooth muscle cells of mesenteric vessels (Matrougui et al., 1999). In general, sheep and humans have a much higher level of expression of the AT_2 receptor than rodents. The AT_2 receptor protein is present in the heart, coronary arteries, atrial myocytes, and the ventricular myocardium in rat (Wang et al., 1998b). In humans, AT_2 receptor expression can be equal to, or even exceed, that of AT_1 receptors, but the functional significance of the AT_2 receptor in humans is still unknown (Carey et al., 2000).

Binding of Ang II to the AT_1 receptor activates a number of different signal transduction pathways, which include stimulation of phospholipase C and, as a consequence, activation of phosphoinositide turnover and calcium mobilization, and inhibition of adenylate cyclase activity (Burnier and Brunner, 1994; Edwards and Aiyar, 1993). Unlike other receptor systems, such as α-adrenergic receptors or AVP receptors, the AT_1 receptor is coupled to multiple signal transduction pathways through different G proteins (Gq or Gi).

Much less is known about the signaling mechanisms coupled to the AT_2 receptor. However, this receptor does not appear to operate through the classical signal transduction pathways and in most tissues is not coupled to G proteins. Based on studies of activation of the AT_2 receptor in membrane preparations from rat adrenal glomerulosa cells, it has been suggested that Ang II, through the AT_2 receptor, activates a phosphotyrosine phosphatase that may regulate ATP binding to the kinase domain of particulate guanylate cyclase (GC). This view has been challenged by the observation that in cells permanently transfected with a plasmid harboring AT_2 complementary DNA, activation of this receptor is linked to the inhibition of protein phosphotyrosine phosphatase activity. In some cells, AT_1 receptors have been shown to inhibit GC (Burnier and Brunner, 1994). More recently, it has been shown that AT_2 signaling activates protein phosphatase 2A (PP2A) and mitogen-activated protein (MAP) kinase phosphatase-1 (MKP-1) and has variable effects on extracellular signal-related protein kinase (ERK)1/2 (Luft, 2002). Moreover, AT_2 uses arachidonate as second messenger, and stimulates the production of bradykinin that, in turn, stimulates the NO system (Luft, 2002). Finally AT_2 signaling activates NF-kB by mechanisms that are only now being explored (Ruiz-Ortega et al., 2000, 2001a, 2001b).

Ang II is though to be central in the regulation of systemic blood pressure. This is achieved either by binding of circulating Ang II to its receptors on vascular smooth muscle cells, or by locally generated Ang II, which exerts tonic effects on the vessel wall. Indeed, through direct action on smooth muscle cells, Ang II significantly increases arteriolar resistance in renal, mesenteric, coronary, and cerebral vascular beds (Forsyth et al., 1971). Ang II also exerts indirect pressor effects via the central and peripheral nervous systems. Its effects on the central nervous system include increased sympathetic discharge and decreased vagal tone (Reid, 1984). Peripherally, Ang II augments the vasoconstrictive response to renal nerve stimulation in dogs (Wong et al., 1991), and its inhibition attenuates the pressor response to norepinephrine in humans (Smith et al., 1991).

Although the vasoconstrictor activity of Ang II is well known, it is now becoming clear that this hormone may play an equally important role in numerous physiological processes, including vascular remodeling and smooth muscle cell growth (Naftilan, 1994). Structural changes in the vasculature are in fact consistent features of hypertension, and may involve local tissue renin–angiotensin systems.

Until recently the majority of the biological actions of Ang II were believed to be mediated by the AT_1 receptor. AT_1 actions induce vasoconstriction, aldosterone release, sodium reabsorption, thirst and salt appetite, growth, and inflammation (Luft, 2002). However, knowledge about the functional roles of AT2 receptors has been expanded. The roles now include embryonic growth regulation, inhibition of cell proliferation, neuronal development and cell repair, apoptosis induction, and production of bradykinin, NO and prostanoids (Luft, 2002; Siragy and Carey, 2001). It is also apparent that the AT_2 receptor provides important counter-regulatory vasodilation, off-setting the pressor action of Ang II at the AT_1 receptor (Gross et al., 2000; Israel et al., 2000; Tamura et al., 2000; Tsutsumi et al., 1999).

Studies have reported that AT_2 receptors may inhibit ACE. If so, this may account for part of the increase in bradykinin mediated by the AT_2 receptor. However, AT_2 receptor blockade with PD1234319 decreases renal Ang II levels, suggesting that this may not be the case (Siragy and Carey, 2001).

Role of Angiotensin II in the Regulation of Renal Function

Ang II is a potent vasoconstrictor of the renal vascular bed, with potential targets that regulate the glomerular microcirculation, i.e., the afferent and efferent arterioles as well as mesangial cells (which dictate glomerular size and filtration area through their contractile properties). *In vivo* studies have shown that Ang II contracts both afferent and efferent arterioles, modulating intraglomerular capillary pressure and, ultimately, glomerular filtration.

How Ang II contributes to the regulation of renal perfusion and glomerular filtration has been clarified by pharmacological manipulations that interfere with the renin–angiotensin axis. In animals and humans, particularly in conditions of activation of the renin–angiotensin system, inhibition of ACE normally increases renal blood flow (RBF) but does not change glomerular filtration rate (GFR) (Brunner, 1992). It thus appears that Ang II is a predominant vasoconstrictor of efferent arteriole (Brunner, 1992). Besides their effect on the renin–angiotensin system, ACE inhibitors block the hydrolysis of other peptides, including bradykinin. This contributes to enhanced renal vasodilatation. The actual contribution of Ang II to renal hemodynamics is limited by the fact that ACE catalyzes the formation of Ang II and the hydrolysis of bradykinin.

Ang II receptor antagonists have been used to define the role of endogenous Ang II in modulating glomerular microcirculation (Edwards and Aiyar, 1993; Burnier and Brunner, 1994). In the isolated perfused rat kidney, Ang II receptor blockade with losartan completely prevents the increase in renal vascular resistance induced by Ang II. Using the hydronephrotic rabbit kidney model, which enables direct access to renal microvessels, Ang II receptor antagonist fully blocks the effect of the peptide on both the afferent and efferent arterioles. In normotensive rats, short-term Ang II receptor blockade does not change GFR, but in most cases increases renal plasma flow (RPF) (Burnier and Brunner, 1994). This was also observed in dogs given the AT_1 receptor antagonist losartan; there was a selective increase in RPF or a combined increase of RPF and GFR, depending on whether animals were normotensive or hypertensive, respectively. There are also studies in humans that have explored the effect of Ang II receptor blockade on renal function. In normotensive subjects on a low- or high-sodium diet, losartan did not change RPF or GFR (Burnier and Brunner, 1994). However, Ang II receptor blockade enhanced RPF but not GFR in hypertensive patients with moderate chronic renal failure. In all these studies AT_1 receptor blocking agents were used, thus providing evidence of the renal vascular and glomerular effects of Ang II through its AT_1 receptor. The effects of AT_2 receptor antagonists have also been examined, but in most cases no changes in renal hemodynamics were found (Edwards and Aiyar, 1993), which excludes a major role of these receptors in the control of glomerular function. Moreover, Ang II induces mesangial cell contraction, leading to decreased glomerular ultrafiltration coefficient (K_f) *in vivo* (Ausiello et al., 1980). This effect, however, is attenuated by the concomitant production of prostaglandins by mesangial cells (Foidart and Mahieu, 1986).

In addition to its vascular effects, Ang II participates in the regulation of urinary sodium excretion (Burns et al., 1993; Edwards and Aiyar,1993). Increasing evidence suggests that Ang II controls renal sodium excretion not only by affecting renal hemodynamics and aldosterone biosynthesis but also by directly regulating epithelial sodium transport. Ang II acts predominantly in the S_1 segment of the proximal convoluted tubule to reduce intracellular cAMP and hence increases Na^+/H^+ antiport activity. Both the basolateral and luminal membranes have Ang receptors, which have a biphasic effect on sodium transport. Therefore, both *in vitro* and *in vivo* experiments have shown that at low concentrations (10^{-12} to 10^{-10} M) Ang II stimulates sodium and water reabsorption, whereas at higher concentrations (10^{-8} to 10^{-6} M) it inhibits sodium transport in the microperfused subcortical rabbit proximal convoluted tubule. A reduction in Ang II concentration also reduces sodium transport both in the loop of Henle (due to vasodilatation of the medullary circulation) and in the cortical collecting tubule (secondary to a reduction in plasma aldosterone levels).

The receptor subtype involved in the complex and unique mechanism of action of Ang II on the proximal tubule has not been clearly established, although AT_1 receptors appear to play a role. However, in sodium-depleted anesthetized dogs the AT_2 receptor antagonist PD123319 increases urine volume and free water clearance in a dose-dependent manner, without affecting renal hemodynamics. Because PD123319 does not affect circulating AVP, a direct tubular effect mediated by the AT_2 receptor has been suggested (Edwards and Aiyar, 1993).

In addition to the direct effect mediated by Ang II binding to proximal tubule receptors, Ang II may also indirectly regulate proximal tubule transport. Therefore, proximal tubular cells are intensively innervated by sympathetic nerves. Denervation of the proximal tubule resulted in marked attenuation of the stimulatory effect of Ang II on sodium reabsorption.

There is now convincing evidence that Ang II is a hypertrophogenic hormone that participates in the regulation of renal cell growth (Mezzano et al., 2001; Wolf and Neilson, 1993). Tubular hypertrophy is an important factor in the adaptive response of the kidney to various physiological and pathological stimuli. The important observation that activation of the intrarenal renin–angiotensin system is altered in situations associated with renal growth, and the fact that ACE inhibitors block compensatory renal hypertrophy in many models, provided a basis for establishing the influence of Ang II on the growth of various renal cell lines. Ang II has no effect on the growth of murine fibroblasts isolated from the renal interstitium, but induces proliferation of a murine mesangial cell line and induces cellular enlargement in proximal tubular cells after several days. Other investigators found evidence of Ang II-mediated hypertrophy in primary cultures of rabbit proximal tubules. These stimulatory effects were transduced through AT_1 receptors because losartan abolished the Ang II-induced hypertrophy.

Ang II-induced hypertrophy in proximal tubular cells is characterized by elevated secretion of collagen type IV, but not type I. Since type IV collagen is an integral part of the basement membrane, an increase in its biosynthesis looks like an appropriate consequence of cellular enlargement to accommodate the increased cellular mass. The *in vitro* hypertrophic activity of Ang II has been confirmed by the *in vivo* observation in rats that infusion of Ang II into the renal artery leads to a significant increase in the renal expression of c-fos and Egr-1, immediate early genes, which is blocked by saralasin, and is independent of blood pressure. Although these findings suggest a direct cellular effect of Ang II on the expression of these early genes, the effect of Ang II may be through the induction of ischemia, which is also associated with an increase in immediate early genes.

Role of the Renin–Angiotensin System in Abnormal Protein Traffic through the Glomerular Barrier and the Progression of Renal Disease

Chronic renal disease evolves to end-stage renal failure (ESRF) through a series of events that cause progressive parenchymal damage, relatively independent of the initial insult (Remuzzi et al., 2002b). Among the several theories on the pathophysiology of progressive nephropathies (Remuzzi and Bertani, 1990), the most convincing one suggests that the initial reduction in nephron number progressively damages the remaining nephrons, which suffer the consequences of adaptive increases in glomerular pressure and flow (Hostetter et al., 1981). Glomerular capillary hypertension is normally accompanied by enhanced transglomerular protein traffic (Anderson et al., 1985) and both are prevented in laboratory animals by antihypertensive drugs (Zatz et al., 1986). In the past, most nephrologists considered the amount of protein found in urine (taken as an indicator of the underlying abnormality in glomerular permeability) simply as a marker of the severity of renal lesions. Today the results of many studies (Bertani et al., 1986; Eddy et al., 1991; Remuzzi and Bertani, 1990) indicate that proteins filtered through the glomerular capillary may have intrinsic renal toxicity, which together with other independent risk factors, such as hypertension, can play a contributory role in the progression of renal damage (Remuzzi and Bertani, 1998; Remuzzi et al., 2002b).

According to the most widely used models of glomerular size selectivity, the glomerular capillary can be considered as being perforated by hypothetical cylindrical pores with different radii (Remuzzi et al., 1987). Different pore-size distribution probabilities have been proposed. Those that best replicate the curves of tracer molecules are used to predict the sieving coefficient of the membrane in human glomerular diseases. Enhanced transglomerular passage of dextrans in diabetes (Myers et al., 1991) and in a number of different proteinuric conditions (Myers and Guasch, 1994) defined the functional nature of glomerular permeability defects in nephrotic patients and served to establish that immune and nonimmune mediated glomerular diseases with massive proteinuria all share abnormalities in glomerular barrier size-selective function. Alterations in local hemodynamics play a decisive role (Myers and Guasch, 1994; Remuzzi, 1995) in the charge- and size-selective function of the barrier, independently of the different structural components of the capillary that participate. In a model of renal vein obstruction (Yoshioka et al., 1986), acute increases in glomerular capillary pressure are accompanied by a sudden impairment of barrier size-selectivity. In another study, aimed at mathematically dissociating hemodynamics from the intrinsic permeability property of the glomerular membrane (Yoshioka et al., 1987), structural modifications of the barrier were required to explain changes in dextran data obtained upon an increase of glomerular pressure in passive Heymann nephritis (PHN). It therefore appears that local generation of Ang II, possibly formed in excessive amounts in response to hemodynamic injury to the endothelium (Lee et al., 1995), shifts glomerular permselective pores toward larger dimensions. There is experimental and human evidence that ACE inhibitors tend to enhance glomerular size selectivity, but how they affect the functional properties of the glomerular barrier remains speculative.

In a study in rats with spontaneous age-dependent proteinuria (Iordache et al., 1994), no difference in the morphometrical parameters examined, including podocytes and frequency of epithelial slit-diaphragms, was found in untreated and ACE inhibitor-treated animals, despite the fact that the latter had considerably less proteinuria. These findings have been interpreted as indicating that improvement in the selective properties of the glomerular capillary wall induced by the ACE inhibitor reflected differences in macromolecular organization of the protein matrix in the glomerular basement membrane or in the slit-diaphragms of the podocytes. This is consistent with other findings that ACE inhibitors reduce mRNA expression for extracellular matrix components in experimental diabetes (Nakamura et al., 1995c) as well as in immunological models of massive proteinuria (Ruiz-Ortega et al., 1995). Moreover, the renoprotective effects of ACE inhibitors in rat models of proteinuric renal disease can be associated with the preservation of glomerular distribution of slit diaphragm components, such as the tight function protein zonula occudens-1 (ZO-1) (Macconi et al., 2000) or nephrin (Benigni et al., 2001b), which are essential for maintaining the filtration barrier.

A 90-day course of enalapril, given to humans with insulin-dependent diabetes with nephropathy reduced systemic blood pressure, lowered the fractional clearances of albumin and immunoglobulin G (IgG), and uniformly lowered the dextran sieving profile (Morelli et al., 1990). The transmembrane dextran flux achieved during enalapril therapy could only be explained by a reduction in the mean radius of the functional pores perforating the glomerular capillary. In ten patients with insulin-dependent diabetes and more severe renal insufficiency, a low dose of enalapril also lowered fractional clearance of albumin and IgG and uniformly reduced the dextran sieving profile over a wide range of molecular radii (Remuzzi et al., 1993).

To evaluate intrinsic changes in the glomerular membrane permeability properties independently of the potential influence of hemodynamics, dextran fractional clearance data were analyzed using a theoretical model that evaluated the permeability of the membrane as a function of size of the hypothetical pores that perforate it. Analysis of the sieving curves with this computational tool indicated that enalapril has a negligible effect on the log-normal component of the assumed pore-size distribution, but that it reduces to a remarkable extent mean pore dimensions of a nonselective shunt pathway that allows large macromolecules to penetrate the glomerular capillary wall. Therefore, enalapril, by reducing the relative importance of the shunt pathway, modulates intrinsic membrane permeability of the glomerular capillary wall, an effect that might explain its antiproteinuric properties (Remuzzi et al., 1993).

Evidence that calculated changes in membrane pore radii are unrelated to assumed changes in hydraulic pressure gradient across the glomerular membrane suggests that the reduced filtration of circulating macromolecules associated with ACE inhibition therapy does not necessarily result from a hypothetical reduction in glomerular capillary pressure, but must derive from changes in intrinsic membrane permselective properties (Remuzzi et al., 1993). Similar results have been achieved with Ang II receptor antagonists that improve glomerular size-selectivity in patients with nondiabetic chronic nephropathy (Remuzzi et al., 1999).

Several studies in recent years have addressed the possibility that albumin or other proteins that accumulate in the lumen of proximal tubular cells because of glomerular permeability dysfunction, cause renal injury (Remuzzi et al., 1997). Indeed, experimental evidence is available that protein overabsorption has deleterious effects in proximal tubules and the renal interstitium (Remuzzi and Bertani, 1998). Proximal tubular cells can thus change their phenotype in response to protein overload (Benigni et al., 1995). In cultured proximal tubular cells, increasing concentrations of delipidated albumin, IgG, or transferrin cause concentration-dependent increases in the rate of synthesis of vasoactive and inflammatory substances, including endothelin-1 (ET-1), monocyte chemoattractant protein-1 (MCP-1) and RANTES (regulated on activation, normal, T-cell expressed, and secreted) (Wang et al., 1997; Zoja et al., 1995, 1998a).

Tubular epithelial cells *in vivo* are organized as a continuous polarized layer with highly specialized luminal and basolateral compartments. In cultured cells that maintain their polarized organization, ET-1, MCP-1 and RANTES are mainly secreted into the basolateral compartment in response to protein overloading (Wang et al., 1997; Zoja et al., 1995, 1998a). If a similar pattern of secretion occurs *in vivo*, these substances could be released into the interstitium to promote the migration of macrophages and T lymphocytes (Wenzel and Abboud, 1995). The interstitial accumulation of chemokines could induce proliferation of fibroblasts, increased synthesis of extracellular matrix, and inflammation.

If the interstitial inflammatory reaction and consequent fibrosis in chronic proteinuric nephropathies are caused by protein overloading, limiting protein filtration and reabsorption should prevent activation of the tubular cells and renal injury. This is what happens in animals given drugs that improve the size-selective function of the glomerular membrane. ACE inhibitors were found to have a greater renoprotective effect than other antihypertensive drugs in rats with a remnant kidney (Anderson et al., 1986). Most subsequent studies in rats with diabetes mellitus (Zatz et al., 1986), puromycin- and doxorubicin-induced nephropathies (Anderson et al., 1988), age-induced changes (Remuzzi et al., 1990), and Heymann's nephritis (Zoja et al., 1996) confirmed the antiproteinuric and renoprotective properties of this class of compounds. In two studies, however, ACE inhibition failed to reduce proteinuria or protect the kidneys from injury (Fogo et al., 1988; Marinides et al., 1987). The fact that ACE inhibitors may not limit renal damage when no substantial antiproteinuric effect is achieved suggests that the renoprotection conferred by this class of drugs is a result of their effect on membrane permeability rather than a result of the morphogenic action of Ang II.

Pharmacological manipulation to reduce urinary protein excretion in humans also limits the progressive loss of renal function. Antihypertensive drugs have been used in humans to slow the progression of renal disease in diabetic and nondiabetic glomerulopathies. For a given level of blood pressure control, ACE inhibitors give more renal protection than other antihypertensives used in human nephropathies. This appears to be linked to the enhanced ability to lower urinary proteins (Bjorck et al., 1992). During four years of follow up, patients with insulin-dependent diabetes mellitus given captopril showed a lower incidence of doubling of serum creatinine concentrations than those on conventional therapy (Lewis et al., 1993). Systemic blood pressure reduction was comparable in the two groups, whereas urinary protein actually decreased with captopril but was higher than baseline in the conventional pretreatment group (Breyer, 1995). The ACE Inhibition in Progressive Renal Insufficiency (AIPRI) study (Maschio et al., 1996) also found a lower risk of doubling baseline serum creatinine in patients given ACE inhibition treatment with benazepril than in those given conventional therapy. However, a difference in systolic/diastolic blood pressure between the two treatments left open the question of whether the renoprotective effect of the active drug is related to its antiproteinuric property or to better blood-pressure control.

The Ramipril Efficacy in Nephropathy (REIN) study (GISEN Group, 1997) supports the concept that the renal protection conferred by ACE inhibitors exceeds their antihypertensive effect. The REIN study was a randomized, double-blind, placebo-controlled trial, designed to test whether glomerular protein traffic and its modification by an ACE inhibitor (ramipril) influenced renal disease progression in 352 patients with chronic nondiabetic nephropathies. A prestratification strategy recognized two levels of proteinuria (>1 but <3 g/24 hr, and >3 g/24 hr) in patients randomly assigned to ramipril or conventional antihypertensive therapy. Treatments were aimed to achieve the same level of blood pressure control (diastolic blood pressure <90 mmHg). The REIN core study found that in patients with proteinuria of >3 g/24 hr who were fast progressors, ramipril safely lowered the rate of GFR decline and reduced by half the combined risk of doubling serum creatinine or end-stage renal disease (ESRD). These effects were accompanied by a substantial lowering of the urinary protein excretion rate, which exceeded that expected from the degree of blood pressure reduction, indicating that the renoprotection was linked to reduction of protein traffic. At the end of the REIN core study, patients with proteinuria of >3 g/24 hr, who either continued on ramipril or were shifted to ramipril, entered the REIN follow-up study. The REIN follow-up study indicated that in patients originally randomized to ramipril or conventional therapy, ramipril slowed the rate of GFR decline and limited progression to ESRD even better than in the core study. The novel finding of the follow-up study was that GFR became almost stable in patients originally randomized to ramipril and who continued with the active drug for more than 36 months, indicating that if the treatment period is long enough the ACE inhibitor can reverse the tendency of GFR to decline.

Long-term ACE inhibition therapy may even help achieve remission of renal disease in humans. Results from the whole REIN study (core and follow-up) showed that in chronic nephropathies, the tendency for GFR to decline with time can not only be halted but even reversed, even in patients with very severe disease (Ruggenenti et al., 1999). This is in agreement with the findings of Hovind and coworkers (2001) who showed remission of nephrotic-range albuminuria in type 1 diabetic patients with continued long-term ACE inhibition therapy and intensified blood-pressure control.

Other large studies have also indicated renal benefits from Ang II receptor blocker treatment beyond the reduction of blood pressure in patients with diabetic nephropathy (Brenner et al., 2001; Lewis et al., 2001; Parving et al., 2001). The Reduction of Endpoints in NIDDM with the Ang II-Antagonist Losartan (RENAAL) study was conducted in patients with type 2 diabetes (noninsulin dependent diabetes mellitus, NIDDM) and nephropathy who received losartan (in addition to usual antihypertensive treatment). The study found that losartan reduced the risk of ESRD (28% risk reduction) and doubling of serum creatinine (25% risk reduction) compared with placebo (Brenner et al., 2001). The level of proteinuria declined by 35% with losartan.

The Irbesartan Diabetic Nephropathy trial (IDNT) studied the effects of the Ang II receptor blocker irbesartan compared with amlodipine or placebo in hypertensive patients with type 2 diabetes and nephropathy. The trial reported that the risk of doubling of the serum creatinine concentration was 33% less in the irbesartan group than in the placebo group, and was 37% less in the irbesartan group than in the amlodipine group. The relative risk of ESRD was 23% less in the irbesartan than in other groups (Lewis et al., 2001). Similar benefits were seen in the Irbesartan Microalbuminuria (IRMA II) trial, in which irbesartan was compared with placebo in patients with type 2 diabetes and microalbuminuria (Parving et al., 2001).

Perhaps the more interesting question is whether or not the combination of ACE inhibitors and Ang II receptor antagonists is more effective than either drug alone. In diabetic nephropathy, there are small studies with conflicting results (Rossing et al., 2002; Agarwal, 2001).

Together, experimental and clinical studies indicate that interstitial inflammation and progression of disease can be limited by drugs that strengthen the glomerular permeability barrier to proteins, thus limiting proteinuria and filtered protein-dependent signaling for mononuclear cell infiltration and extracellular matrix deposition.

THE RENAL KALLIKREIN–KININ SYSTEM

Renal Actions of Kinins

The kallikrein–kinin system participates in circulatory homeostasis by generating the peptide bradykinin, a potent vasodilator (Carretero et al., 1993). Kinins generated within the kidney do not gain access to the systemic circulation, so the systemic and renal kallikrein–kinin systems act in two separate compartments. The kallikrein–kinin system consists of four components: kallikreins, kininogens, kinins, and kininases (Coyne and Morrison, 1991).

Kallikreins are serine protease enzymes that release the biologically active kinins from kininogen, the precursor glycoprotein. The enzymatic activity of these proteases depends on a serine residue at their catalytic site. There are two distinct enzymes in this family: the circulating (plasma) form and the tissue (glandular) form. Plasma kallikrein participates in the clotting cascade. Tissue kallikreins are present in the kidney, small intestine, pancreas, and salivary glands, where they are responsible for local production of kinins. Renal kallikrein activity is predominantly found in the cortex, where it has been detected in glomeruli and proximal and distal tubules. The majority of kallikrein found in the urine is of the tissue form and is secreted by the kidney.

Renal kallikrein acts on low and high molecular weight kininogens to generate the decapeptide lys-bradykinin (kallidin), which is converted into bradykinin by an aminopeptidase that cleaves the lysyl group of the peptide. Bradykinin is rapidly inactivated by kininase I and II (also known as angiotensin converting enzyme) by hydrolysis of the carboxy-terminal arginine or the dipeptide phenylalanine-arginine, respectively, in the vascular endothelium and renal tubular cells. In addition to kininases, the proximal tubule brush border contains endopeptidases II (enkephalases), which also inactivate kallidin and bradykinin (Ura et al., 1987). Several serine protease inhibitors, including aprotinin and α1-antitrypsin, block the enzymatic activity of renal and other tissue kallikrein (Scicli and Carretero, 1986).

Studies in normal human subjects (Gill et al., 1965), anesthetized dogs (Webster and Gilmore, 1964), or isolated perfused kidneys (McGiff et al., 1975) have established that either intravenously or intra-arterially administered bradykinin or kallidin cause renal arteriolar vasodilatation. However, the fact that pharmacologic doses of kinins produce vasodilatation does not prove that the renal-kallikrein system performs this function *in situ*. Nevertheless, data have been obtained that are consistent with this possibility. Inferential data support a direct correlation between renal kallikrein-kinin activity and RBF. Kininase II inhibition increases RBF with concomitant increases in renal venous and urinary kinins (McCaa et al., 1978).

Kallikrein excretion is also well correlated with RBF in both normal and hypertensive human subjects (Levy et al., 1977). However, evidence is available that kinin infusion does not reduce renal vascular resistance in rats on a chronic low-sodium diet, but would do so if these animals had been pretreated with aprotinin (Johnston et al., 1981). This suggests that the activity of the renal kallikrein system is already maximal in the low-sodium condition. In contrast, in rats on chronic high-sodium diet, an Ang II receptor antagonist failed to reduce renal vascular resistance, but the ACE inhibitor captopril did. The effect of captopril could be prevented by aprotinin, suggesting that with the high-sodium diet, captopril is acting by preventing kinin catabolism. Furthermore, in these high-sodium rats, kinins are capable of reducing renal vascular resistance. Collectively, these data infer that renal kinins are exerting some level of control over renal vascular resistance, at least in rats on a low-sodium diet.

Whether the tissue kallikrein–kinin system participates in the regulation of electrolyte and water excretion as a natriuretic (or diuretic) or an antinatriuretic (or antidiuretic) influence is uncertain. The observations by Webster and Gilmore (1964) and Gill et al. (1965) established that injected kinins increase sodium and water excretion. Bradykinin antiserum from rabbit, given intravenously to rats, decreased sodium and water excretion, while control serum had no such effects (Marin-Grez, 1974). Some studies of bradykinin injected into the renal artery suggeste reduced proximal reabsorption of sodium and water during stop-flow in the dog (Capelo and Alzamora, 1977). Others have shown that intra-arterial kinin does not alter sodium reabsorption in micropunctured, superficial proximal tubules, suggesting that any kinin-induced natriuresis could be the result of changes in proximal reabsorption in deeper cortical nephrons not available for micropuncture (Stein et al., 1972). Moreover, acute infusion of bradykinin induces significant natriuresis and diuresis in the absence of GFR alterations (Granger and Hall, 1985).

In experiments in which bradykinin was administered for several days, the acute rise in salt and water excretion was not sustained, while renal vasodilatation persisted (Granger and Hall, 1985). Although these studies suggest a natriuretic role for bradykinin, they do not elucidate the physiological role of the renal kallikrein–kinin system. Several groups have attempted to block bradykinin action *in vivo* in order to determine these physiologic functions. Inhibition of endogenous bradykinin, by means of specific antibodies or aprotinin, has been shown to blunt the natriuretic and diuretic effect of saline infusion (Scicli and Carretero, 1986). It has also been demonstrated that infusion of a bradykinin antagonist into the renal arteries of dogs on low-sodium diets causes antidiuresis and significant decreases in the fractional excretion of sodium (Siragy, 1993). There were no changes in GFR, plasma aldosterone concentration, plasma renin activity, or systemic arterial pressure during intrarenal administration of the antagonist. These results suggest that endogenous kinins can act as natriuretic substances.

The mechanism by which kinins induce natriuresis remains ill defined. Several studies suggest that kinins directly regulate epithelial transport of ions and water (Scicli and Carretero, 1986). The renal target of these actions is most likely the collecting duct, which, at least in rabbits, has a high density of specific bradykinin binding sites (Scicli and Carretero, 1986). In addition, renal kallikrein and kininogen have been localized to the distal tubule, connecting segment, and collecting duct, confirming the presence of a complete kallikrein–kinin axis in the distal part of the nephron (Scicli and Carretero, 1986). However, the complex interaction of kinins with other hormonal systems makes it difficult to discriminate whether their renal effects are indeed direct. Therefore, kinins activate phospholipase A_2, resulting in increased prostaglandin synthesis (Coyne and Morrison, 1991). In a rat model of unilateral ureteral obstruction, it was shown that indomethacin abolished bradykinin-induced natriuresis and diuresis in the contralateral kidney (Kopp and Smith, 1993). Kallidin has been reported to inhibit the hydroosmotic action of AVP on isolated rabbit cortical collecting duct (Schuster et al., 1984). On the other hand, bradykinin stimulates AVP release when infused into the renal artery but not when administered systemically, suggesting that this effect is mediated by afferent renal nerves (Yamamoto et al., 1992). Interactions *in vivo* between the kallikrein–kinin and renin–angiotensin systems are unclear so far. Because kininase II and ACE are the same entity, the hemodynamic and renal actions of ACE inhibitors could be in part mediated by accumulation of kinins.

ENDOTHELIN

Biochemistry, Synthesis and Receptor Biology

In search of endogenous factors with vasoactive properties, Yanagisawa and coworkers in 1988 found that porcine cells generated a 21-amino-acid peptide, endothelin (ET), one of the most potent endogenous vasoconstrictors yet identified (Yanagisawa et al., 1988). Endothelin-1 (ET-1) was the first peptide of the ET family to be discovered (Remuzzi and Benigni, 1993; Remuzzi et al., 2002b). Within a year, two further ET isoforms, ET-2 and ET-3, were described, the former differing from ET-1 by two amino acids, the latter by six (Simonson and Dunn, 1992).

ETs are synthesized by proteolytic cleavage of pre-proETs, large isopeptide precursors (200 amino acids) encoded by the three separate genes on distinct chromosomal loci, by processing endopeptidases that recognize paired basic amino acids (Arg–Arg or Lys–Arg) which form biologically inactive intermediates known as Big-ET-1, -2, and -3. These are then converted to the responsive mature peptides by cleavage of the Trp_{21}–Val_{22} bond via a highly specific ET converting enzyme (ECE) (Remuzzi and Benigni, 1993; Yanagisawa et al., 1988). Alternative splicing produces three ECE-1 isoform of this membrane bound metalloproteinase ET-converting enzyme from a single gene that comprises 19 exons and is located on human chromosome 1p36 (Valdenaire et al., 1995).

ECE-1a and ECE-1c are localized at the cell surface, whereas ECE-1b resides intracellularly (Schweizer et al., 1997). The extracellular form of ECE-1 cleaves externally supplied BIG-ET-1 from outside the cell, but endogenously produced BIG-ET-1 is converted to active peptide by the intracellular enzyme (Xu et al., 1994). A fourth isoform of ECE-1 (ECE-1d) has been cloned, which is generated from the same *ECE1* gene as the three other isoforms but through an alternative promoter located upstream of the third exon (Valdenaire et al., 1999).

The biological activity of ET-1 is regulated by a complex process of synthesis and breakdown. The peptide is mainly metabolized by neutral endopeptidase EC3.4.24.11, a membrane-bound enzyme widely involved in the degradation of peptide hormones, and to a large degree represented in the brush border vesicles of the proximal tubules (Abassi et al., 1993). ET-1 breakdown catalyzed by the metallopeptidase EC3.4.24.30 (called coccolysin) produced by an *Enterococcus faecalis* has been reported (Makinen and Makinen, 1994).

Besides the vascular endothelium, circulating and resident cells of different organs, including lung, gut, and kidney, synthesize ET-1 (MacCumber et al., 1989; Ehrenreich et al., 1990). ET-1 and ET-3 are both formed in the kidney, the former mainly in glomeruli and inner medullary collecting ducts and the latter predominantly at the tubular level (Marsen et al., 1994). ET-2 mRNA has been found in the renal medulla (Marsen et al., 1994). Most of the ET formed is secreted towards the basolateral side of the cell (Wagner et al., 1992), indicating ET-1 as a local hormone with paracrine activity.

ET expression and release in cultured cells is enhanced by a variety of proinflammatory and vasoactive agents, including thrombin, transforming growth factor β (TGF-β) and Ang II. Mechanical forces modulate

ET-1 synthesis and release differently, depending on the duration and level of flow (Kuchan and Frangos, 1993). Transfection experiments with ET-1 promoter constructs revealed the presence of a shear stress responsive element on the ET-1 gene, distinct from the activator protein-1 (AP-1) consensus sequence and GATA-2 motif sites essential for basal ET-1 transcription (Malek et al., 1993).

ETs act on two pharmacologically distinct subtypes of G protein-coupled receptors, termed ETA and ETB due to their different affinity for ET isopeptides (Malek et al., 1993). ETA binds ET-1 more strongly, while ETB has similar affinity for all isopeptides. ETA and ETB receptors possess two separable ligand subdomains (Sakamoto et al., 1993) within the region spanning the transmembrane helices IV–VI (and the adjacent loop regions for the ETA receptor subtype) and at the transmembrane helices I, II, III, and VII (plus the intervening loop for ETA receptor subtype). ETA receptors reside in vascular smooth muscle cells, mediating vasoconstriction and smooth muscle cell proliferation, while the ETB receptors present on endothelial cells mediate vasodilatory effect via NO. A further subtype of ETB receptors, with affinity in the picomolar range, has been identified and characterized in rat brain and cardiac atrium (Sokolovsky et al., 1992).

It is not known whether, of the two ETB receptors, the super high affinity receptor type is related to the vasodilatory property of ET, or the high affinity receptor type is related to the vasoconstrictor action of the peptide. Several reports have confirmed that ETB receptors on smooth muscle cells mediate vasoconstriction if challenged with low concentrations of ET. Clearance of ET-1 from the circulation depends on pulmonary ETB receptors (Fukuroda et al., 1994), which also regulate ET-1 gene expression by enhancing pre-proET-1 transcription and mRNA stability (Buchan et al., 1994; Saijonmaa et al., 1992).

In situ hybridization studies contributed to identifying the precise location of ET receptors within the kidney (Hori et al., 1992). The ETA receptor type is mainly expressed in renal arteries and in glomerular afferent and efferent arterioles (Hori et al., 1992). ETB receptors are present in abundance on the surface of glomerular endothelial and mesangial cells, as well as epithelial podocytes and in vasa recta bundles (Hori et al., 1992). Weak hybridization signals corresponding to ETB mRNA were also present over epithelial cells in thin segments of loop of Henle, probably in interstitial and capillary endothelial cells, but not in epithelial cells of collecting ducts (Kohan et al., 1992). Reverse transcription and polymerase chain reaction studies demonstrated the presence of both receptor subtypes in inner medullary collecting ducts. Cultured mesangial cells express two classes of ET binding sites, but the molecular subtypes have not been analyzed (Badr et al., 1989b). Relative abundances of ETA and ETB receptors in the kidney varies from species to species. In human kidney, ETB receptors are more abundant (Nambi et al., 1992a), while in the rat kidney, abundances of ETA and ETB receptors are comparable (Nambi et al., 1992b). Multiple intracellular signal transduction pathways mediate ET effects. ET stimulates inositol triphosphate (IP3) formation through phospatidyinositole (PI) hydrolysis, by phospholipase C (PLC) and intracellular calcium mobilization (Wilkes et al., 1991). In smooth muscle and bovine mesangial cells, ET also activates phospholipase A2 (PLA2) (Barnett et al., 1994). Rat renal medullary interstitial cells possess ETA receptors that are linked to phosphatidylinositol-specific PLC (PI-PLC) and PLA2, and to phosphatidylcholine-specific phospholipase D (PC-PLD) (Friedlaender et al., 1993).

Biological Effects of Endothelin in the Kidney

Renal vessels are particularly sensitive to the vasoconstrictive effect of ET-1 (Pernow et al., 1988). Infusion of ET-1 into the renal artery increases renal vascular resistance with a consequent decrease in RBF (Perico et al., 1990b). As in the systemic circulation, renal vasoconstriction follows a transient phase of vasodilatation due to induction of endothelium-derived vasodilators through ETB receptors on endothelial cells (Simonson and Dunn, 1993). By binding to the ETA receptor, ET-1 contracts afferent and efferent arterioles and increases the tone of the arcuate and interlobular arteries, giving rise to a long-lasting vasoconstriction (Simonson and Dunn, 1993).

The effect of pharmacological doses of ET-1 on renal hemodynamics has been clearly defined by systemic infusion studies, but the effects of physiological concentrations of ET-1 have not been determined. This has only been clarified by experiments with specific receptor antagonists. Injection of ET-1 in anesthetized rats raises blood pressure and renal vascular resistance and reduces RBF and GFR (Simonson and Dunn, 1993). Systemic but not renal pressor effects of ET-1 in rats are abolished by BQ123, a selective ETA receptor antagonist (Pollock and Opgenorth, 1993). Both systemic and renal vasoconstriction are completely prevented by PD145065, a nonselective ETA and ETB receptor antagonist, suggesting a relevant role for ETB receptor in the renal response to ET-1 infusion in the rat. Data from healthy volunteers infused with as much as 2.5 ng/kg/min ET-1, whose systemic blood pressure was modestly affected in the face of a substantial decrease in renal function and excretion of water and sodium, concur to indicate that renal vessels in humans are uniquely sensitive to ET (Sorensen et al., 1994).

It is not known which receptor subtypes mediate renal vasoconstriction in response to ET-1 infusion in humans. There are data suggesting that ET-1 is involved in mesangial cell contraction and mitogenesis (Badr et al., 1989b). This latter phenomenon is associated with upregulation of several genes, including c-fos and jun proto-oncogenes (Badr et al., 1989b), which convert short-term transmembrane signals into long-term responses. ET-1 also increases expression of mRNA for collagen types I, III, IV, and laminin in cultured mesangial cells (Ishimura et al., 1991). ET-1 directly inhibits water reabsorption by an intrarenal mechanism (Badr et al., 1989b) and through the inhibition of AVP-mediated cAMP accumulation (Tomita et al., 1990).

ET-1 has complex direct effects on sodium homeostasis and water balance, which add to those mediated by other hormones, including Ang II, aldosterone, and ANP (Simonson and Dunn, 1993). In laboratory animals, low-dose systemic ET-1 has a natriuretic effect despite a fall in GFR and RBF (Perico et al., 1991b). Natriuresis may be due to stimulation of ANP or to higher blood pressure during ET infusion. In the isolated perfused rat kidney, sodium excretion also increases after ET-1 administration (Perico et al., 1991b). Unlike in the rat, systemic infusion of high doses of ET-1 in dogs reduces sodium excretion because of a decrease in filtered load or renin–angiotensin stimulation (Miller et al., 1989). In humans, systemic administration of ET-1 at concentrations comparable with those measured in physiological conditions, such as during upright positioning, causes renal sodium retention. However, higher concentrations of ET-1, like those found in pathological conditions such as heart failure, hepatorenal syndrome, and renal failure, cause strong sodium retention and, in addition, renal vasoconstriction (Sorensen et al., 1994). Because sodium retention continues after discontinuation of low-dosage ET-1 in the absence of changes in renal hemodynamics, it has been suggested that ET-1 may directly stimulate tubular sodium reabsorption.

Role of Endothelin-1 in Renal Disease Progression

A large body of evidence has consistently documented that ET-1 is deeply involved in the process of progressive renal injury, particularly through its chemotactic property and its stimulation of mesangial and interstitial cell proliferation and deposition of extracellular matrix deposition (Benigni and Remuzzi, 1997; Benigni et al., 2001a). ET-1 has no direct effect on glomerular permeability properties, but rather it perpetuates the renal damage once alteration of filtration of macromolecules has already taken place. Proteins abnormally filtered through the glomerular capillary have an intrinsic renal toxicity (Remuzzi and Bertani, 1998). There is evidence that proximal tubular cells (which in basal conditions produce negligible amounts of ET-1), when exposed to high molecular weight proteins such as albumin, IgG, and transferrin, are stimulated to synthesize and release the ET-1 toward the basolateral compartment (Zoja et al., 1995). This evidence has raised interest on the possible contribution of tubular derived ET-1 to tubulointerstitial damage. ET-1 is mainly secreted toward the basolateral membrane of the cells, thus a prolonged increase in tubular derived ET-1, such as it may occur in nephrosis, could lead to ischemia of peritubular capillaries, interstitial fibroblast proliferation, matrix deposition, and infiltration of active macrophages (Remuzzi and Bertani, 1998).

Additional findings supporting the role of tubular-derived ET-1 in mediating tubulointerstitial injury arise from a study in which lipoproteins enhance ET-1 synthesis in human proximal tubular cells in culture (Ong et al., 1994). Should this occur *in vivo* one might expect to find increased renal synthesis of ET-1 in proximal tubules in proteinuric progressive renal disease eventually correlating with renal damage. It has been shown that in rats with PHN (an immunological model of renal disease characterized by glomerulosclerosis and tubulointerstitial damage) urinary ET-1 time-dependently increased and correlated with renal damage (Zoja et al., 1996). Studies on location of ET-1 mRNA and peptide by *in situ* hybridization and immunohistochemistry showed that, unlike controls, PHN rats exhibited intense staining in proximal tubular cells at the basolateral side of cytoplasm, and in the vicinity of interstitial fibroblasts and macrophages (Zoja et al., 1998b). Combined treatment with an ACE inhibitor and an Ang II receptor antagonist, which efficiently stops progression of renal disease, reduced urinary protein excretion and lowered renal ET-1 mRNA and synthesis to control levels (Zoja et al., 1998b).

Evidence that ET effectively plays a role in the progressive renal injury of chronic renal diseases is available from studies in which selective pharmacological manipulation of ET receptors has been performed. In rats undergoing extensive renal mass reduction (RMR), a model of chronic renal disease characterized by hypertension, severe proteinuria, and structural damage in the kidney, there is time-dependent upregulation of renal pre-proET-1 gene expression and synthesis of the corresponding peptide which can be quantified in urine (Benigni et al., 1991a). In this model, a significant correlation between proteinuria, renal injury, and urinary ET-1 has been demonstrated. Administration of a selective ETA receptor antagonist, FR139317, to RMR rats has been shown to normalize blood pressure, reduce proteinuria and improve renal function and histology (Benigni et al., 1993). Treated animals show very low average frequency of segmental glomerulosclerosis, and renal expression of c-fos proto-oncogene, an early marker for cell proliferation that was upregulated in untreated remnant kidney animals, was suppressed.

In the same model, bosentan, an orally active ETA and ETB receptor antagonist, reduced blood pressure, partially prevented proteinuria and, most importantly, protected animals from renal failure and death (Benigni et al., 1996). Furthermore, lupus mice chronically treated with FR139317 show reduced blood pressure, lowered proteinuria, and improved renal function and histology (Nakamura et al., 1995b). Northern blot analysis has revealed that treatment with the ETA receptor antagonist is associated with decrease in renal gene expression of c-myc, c-fos, and c-jun, early markers of cell proliferation, as well as of extracellular matrix components, metalloproteinases, and the tissue inhibitor of metalloproteinase-1 (Nakamura et al., 1995b). Similarly, FR139317 given to streptozotocin-diabetic rats significantly attenuates extracellular matrix component gene expression and mRNA levels for growth factors, including tumor necrosis factor α (TNF-α), platelet-derived growth factor B (PDGF-B), TGF-β, and basic fibroblast growth factor (FGF) (Nakamura et al., 1995a). Further data suggest that a nonselective ET receptor antagonist, PD142,893 (a peptidic mixed antagonist of ET receptors), when chronically administered to diabetic animals with overt proteinuria, was equally effective as an ACE inhibitor at lowering blood pressure and controlling protein excretion (Benigni et al., 1998).

Studies in transgenic animals further support the hypothesis that enhanced renal ET formation may favor the development of renal lesions. Mice overexpressing human ET-1 promoter (Theuring et al., 1995) form excessive ET-1 in their kidneys and have renal lesions. Moreover, rats transgenic for the human ET-2 gene that did not develop hypertension were characterized by severe renal lesions, including interstitial fibrosis (Hocher et al., 1996). Data from experimental progressive renal injury have also shown a superior renoprotective effect of combining an agent that limits glomerular protein traffic, such as an ACE inhibitor, with a single therapy alone, opening novel therapeutic perspectives for the treatment of renal disease.

The role of ET-1 in progressive renal diseases in humans is the subject of much research. Urinary excretion of ET-1 is increased in patients with chronic progressive nephropathies (Ohta et al., 1991). It has been shown that urinary ET-1 levels were significantly higher in patients with unilateral nephrectomy compared with subjects who underwent adrenalectomy (Takeda et al., 1994). However, no evidence of a cause–effect relationship between the development of sclerotic lesions and abnormal ET generation in humans is available.

The possibility of administering nonpeptidic ET receptor antagonists or ECE inhibitors to humans in the near future will hopefully clarify the actual role of ET in the progression of renal disease in humans. The exciting developments in cell biology and pharmacology of ETs should soon have a decisive impact on those diseases of the kidney for which evidence of excessive local formation of the hormone is already overwhelming.

NITRIC OXIDE

Biochemistry

Nitric oxide (NO) is a lipophilic, highly reactive, free radical gas with diverse biomessenger functions (Wallace and Leiper, 2002). Evidence reveals that NO is involved in a remarkable array of key physiological processes, including regulation of vascular tone, platelet aggregation, host-defense, inflammation, neurotransmission, learning and memory, penile erection, gastric emptying, hormone release, cell differentiation, cell migration, and apoptosis (Lane and Gross, 1999). The synthesis of NO by vascular endothelium is responsible for the vasodilator tone, essential for the regulation of blood pressure (Rees et al., 1989). In 1980, Furchgott and Zawadski showed that the relaxation of arteries by pharmacological agents such as acetylcholine was dependent on the presence of an intact endothelium, a phenomenon that was attributed to an endothelium derived relaxing factor (EDRF). In 1987, Palmer and coworkers demonstrated that NO accounted for EDRF activity and that the precursor for its formation was the semi-essential aminoacid L-arginine (Palmer et al., 1988).

In the central nervous system, NO is a neurotransmitter that underpins several function, including the formation of memory (Garthwaite et al., 1988). In the periphery, there is a widespread network of nerves, previously recognized as nonadrenergic and noncholinergic, that operate through a NO-dependent mechanism to mediate some forms of neurogenic vasodilatation and regulate various gastrointestinal, respiratory, and genitourinary tract functions (Sanders et al., 1992). NO also contributes to the control of platelet aggregation (Radomski et al., 1990) and the regulation of cardiac contractility (Moncada and Higgs, 1995). These actions are all regulated by the activation of soluble guanylate cyclase and the consequent increase in the concentration of cyclic guanosine monophosphate (cGMP) in target cells (Wang and Marsden, 1995).

Another area in which NO has been shown to play a central role is as a mediator of macrophage cytotoxicity during host defense and immunologic reactions. Macrophages can induce oxidative injury to tumor cells and mycobacteria that is associated with the release of NO (Stuehr and Nathan, 1989). Further work demonstrated that the analogs of L-arginine (e.g., *N*-nitro-L-arginine-methyl ester, L-NAME), which inhibit NO synthesis, could block this activity, and that NO could mimic the pattern of macrophage cytotoxic effects (Stuehr and Nathan, 1989).

NO is derived from the amino acid L-arginine in an unusual reaction catalyzed by a family of enzymes, termed NO synthases, that convert arginine and oxygen into citrulline and NO (Moncada and Higgs, 1995; Wang and Marsden, 1995). Mammalian NO synthases (NOSs) have been characterized from various cell types and found to comprise three distinct isoforms. The isoforms are 50% to 60% homologous, and are distinguished by their histological expression, susceptibility to arginine-based inhibitors, intracellular localization, NO output, and mode of regulation (Sessa, 1994). These NOS isoforms are products of distinct genes (Marsden et al., 1993) and are functionally categorized by whether their expression is constitutive (cNOSs) or inducible (iNOSs). Notably, two isoforms are constitutively expressed in specific cell types, in which activity is regulated by changing levels of intracellular calcium.

The two cNOSs are neuronal NOS (nNOS) and endothelial NOS (eNOS), named for the tissues from which they were initially isolated. These cNOSs are transiently activated by agonist-induced elevations in intracellular calcium, resulting in the evanescent binding of calcium/calmodulin. This pulsatile activation is ideally suited to a regulatory role by puffs of NO in cells and tissues. Inducible NOS (iNOS) can be induced by immunostimulants in most cell types; activity is continuous, high output, and calcium-independent. iNOS was originally isolated from murine macrophages as an enzyme that contains tightly bound calmodulin (Cho et al., 1992), and accordingly it is fully active at low resting levels of intracellular calcium. iNOS produces a large continuous flux of NO, which may be limited only by substrate availability. Transcriptional upregulation of the iNOS gene occurs in response to various inflammatory cytokines (e.g., interleukin-1 [IL-1], TNF-α, or immunostimulants, such as bacterial lipopolysaccharide [LPS]), by a mechanism that is synergistic with interferon-γ (Nathan and Xie, 1994). Although the high output of NO that derives from iNOS plays an important role in host defense against intracellular parasitism (Green et al., 1990), it can be toxic to the host, contributing to inflammatory dysfunctions and potentially lethal hypotension (Petros et al., 1991). Evidence suggests that iNOS may be expressed in some tissues under normal conditions; these include the fetal and adult lung (Kobzik et al., 1993) and the kidney. This possibility challenges the common view that iNOS is exclusively inducible.

All three NOS isoforms catalyze a five-electron oxidation of one of the equivalent guanido nitrogens of L-arginine to yield one molecule each of NO and L-citrulline, at the cost of 1.5 molecules of nicotinamide adenine dinucleotide phosphate (NADPH) and two molecules of dioxygen (Stuehr and Griffith, 1992). The reaction involves two successive monooxygenation reactions, with N^{w}-hydroxy-L-arginine produced as an isolatable intermediate (Stuehr and Griffith, 1992). All NOS isoforms contain four prosthetic groups: flavin adenine dinucleotide (FAD), flavin adenine mononucleotide (FMN), iron protoporphyrin IX (heme), and tetrahydrobiopterin (BH_4). The flavins are involved in electron storage and delivery, accepting two electrons from NADPH and then delivering single electrons to the heme group within the active site (Bredt et al., 1991). The heme has an established redox role in the activation of molecular oxygen (Stuehr and Ikeda-Saito, 1992). Although essential to NO production, the role for BH_4 is presently unclear.

Two reports have suggested the possibility of additional enzymes capable of synthesizing NO, which have very distinct properties from the known NOS isoforms. In one study, an enzymatic activity capable of producing citrulline and nitrate was identified in rat kidney. The activity is not inhibited by conventional arginine analogs, is not dependent on calcium or calmodulin, does not bind adenosine diphosphate (ADP)-sepharose under standard condition, and does not immunoreact with an anti-NOS antibody (Singh et al., 1997). (The molecular identity of this activity remains to be established.) In the second study, a protein purified from rat cerebellum was shown to convert bradykinin to NO (Chen and Rosazza, 1996). With bradykinin as a substrate, both substituted L-arginine derivatives and specific bradykinin 2 (B2) receptor antagonists inhibited the activity. However, with L-arginine as a substrate, the protein synthesized NO in a calmodulin-dependent reaction that is inhibited by L-arginine derivatives. The expression of this protein in other organs has not been examined, and its primary structure has not been elucidated. Studies are needed to determine whether these activities belong to the NOS, and whether the bradykinin-to-NO pathway is physiologically relevant.

Earlier attempts to localize NOS activity in the kidney were based on indirect evidence for NO biosynthesis obtained by the measurement of nitrites and nitrates, the stable end products of NO, and cGMP levels. For the measurements, tissue slices of dissected renal zones, as well as isolated glomeruli and various lines of cultured cells, were used (Brezis et al., 1991). It was suggested from these *in vitro* studies that renal NO production is not only derived from endothelial cells, but also from smooth muscle cells, mesangial cells, and tubular epithelial cells. For general histochemical detection of NOS, the nitroblue tetrazolium reaction has been used (Schmidt et al., 1992). The intensity of this reaction varies with the NADPH diaphorase activity of NOS, thereby indicating NOS activity irrespective of the particular isoform. Immunohistochemical methods, using isoform-specific antibodies and *in situ* hybridization technique, have been applied to assign the diaphorase labeling to a particular NOS isoform (Schmidt et al., 1992).

All the three known NOS isoforms have been detected in the kidney. The constitutive nNOS has been localized strictly to the cells of the macula densa of the rat and mouse kidney (Mundel et al., 1992), thus a strong NADPH-diaphorase reaction was observed in the macula densa of the juxtaglomerular apparatus which colocalized with immunostaining for nNOS (Mundel et al., 1992). An immunoelectron microscopy study localized nNOS to the cytoplasm and to cytoplasmic vesicles (Tojo et al., 1994). However, using reverse transcriptase and polymerase chain reaction (RT-PCR) in individual, microdissected rat nephron segments, Terada and coworkers (1992) localized nNOS mRNA in the inner medullary collecting duct and, to a lesser extent, in the glomerulus, inner medullary thin limb, and cortical and outer medullary collecting duct, as well as in parts of the renal vasculature. RT-PCR has demonstrated mRNA for eNOS in the glomerulus and afferent and efferent arterioles (Ujiie et al., 1994). The distribution of eNOS has also been studied immunohistochemically in the rat. eNOS was found in the glomerular endothelium, with the efferent arteriolar endothelium showing stronger staining than the afferent arteriolar endothelium (Aiello et al., 1997). These two forms of NOS are crucially involved in renal pathophysiology and local release of NO serves to control RBF and modulate tubuloglomerular feedback (Ito, 1995).

The mesangial cell is able to express inducible NOS after cytokine stimulation, and this has been shown for rat, bovine, and human mesangial cells (Pfeilschifter et al., 1993). In rat mesangial cells, iNOS is induced by IL-1β, TNF-α, and endotoxin (Pfeilschifter et al., 1993). Forskolin, an activator of adenylate cyclase, synergizes with cytokines to increase iNOS in mesangial cells (Saura et al., 1995), indicating that cAMP stimulates an independent pathway of induction. In human mesangial cells, iNOS is induced by IL-1β and interferon-γ in combination (Pfeilschifter et al., 1993); the effect is augmented by TNF-α. Using the PCR technique on mRNA extracted from control and cytokine-stimulated proximal tubular cells, McLay and coworkers (1994) found iNOS product, indicating that proximal tubular cells express iNOS in basal condition and respond to immune challenge through the induction of iNOS. Other studies in the rat have shown that glomerular mesangium and afferent arteriole, as well as various segments of the nephron, express iNOS under basal conditions and that renal iNOS expression is augmented after immune activation with cytokines or endotoxin (Morrissey et al., 1994). Consistently, immunoperoxidase analysis, with specific antisera to iNOS, shows iNOS signal localized to afferent arteriole, collecting ducts, outer medullary tubules, and medullary thick ascending limbs of rat kidney (Aiello et al., 1997; Tojo et al., 1994). The steady-state amount of iNOS, mRNA, and protein in normal rat kidney is highest in tubules of the outer medulla, and *in vitro* studies on rat homogenates have shown that the specific activity of NOS in the medulla was three times that of the cortex (McKee et al., 1994). Altogether, these data suggest that NO exerts its effects not only on glomerular function, but also on tubular function.

Renal Actions of Nitric Oxide

The renal vascular bed produces large amounts of NO whose basal production maintains RBF and GFR (Lahera et al., 1991). Evaluation of the role of NO in the regulation of RBF and GFR has relied heavily on the use of various NOS inhibitors, with different degrees of specificity with respect to each NOS isoforms. Examples of these nonspecific NOS inhibitors include N^G-monomethyl-L-arginine (L-NMMA), *N*-nitro-L-arginine methylester (L-NAME), and N(omega)-nitro-L-arginine (L-NNA). By allowing evaluation of the functional effects of local blockade of NO synthesis, NOS blockers have been very helpful in providing semiquantitative evaluation of NO production and have enabled investigators to bypass limitations to accurate measurement of NO production in specific target cells. Studies have been performed in different strains of animals, including normal human volunteers.

The kidney seems more sensitive to acute inhibition of NO synthesis than other organs. Intravenous infusion of the NO synthesis inhibitor L-NAME, at a dose that does not modify systemic blood pressure, reduces RBF and GFR in normal rats (Lahera et al., 1991). Baylis and coworkers (1992) have shown that chronic administration of an inhibitor of NO synthesis in rats results in sustained systemic and glomerular hypertension, accompanied by glomerular injury. These data suggest that reduced renal NO synthetic capacity may contribute to the initiation or maintenance of intraglomerular hypertension. Micropuncture studies in rat kidney (Zatz and De Nucci, 1991) revealed that NO blockade leads to complex changes in glomerular hemodynamics, with increases in both preglomerular and efferent arteriolar resistance such that glomerular blood pressure increases significantly. Using a juxtamedullary rat nephron preparation, Imig and coworkers (1992) found that NO primarily alters afferent vascular tone, thereby modifying the ability of the preglomerular vasculature to autoregulate glomerular capillary pressure. Similarly, Deng and Baylis (1993) stated that local NO controls afferent arteriolar resistance, whereas efferent resistance is not under the tonic control of NO. However, studies in the avascular hydronephrotic kidney have indicated a preferential effect of NO blockade in the efferent arteriole (Hoffend et al., 1993). Tolins and Raij (1991) suggested that part of the intrarenal hemodynamic effects of NO synthesis inhibition could be due to unmasking the effect of intrarenal Ang II. Indeed, concomitant administration of L-NAME with an Ang II receptor blocker attenuated the hemodynamic effect of NO synthesis inhibition.

Ito et al. (1991) demonstrated in preparations of isolated glomeruli, that NO modulates the effects of Ang II on the afferent arteriole. De Nicola et al. (1992) have shown in micropuncture studies that inhibition of NO synthesis results in glomerular arteriolar vasoconstriction, decreased glomerular ultrafiltration coefficient, and reduced single-nephron GFR. In the same experiments, they showed that these changes could be prevented to a great extent by simultaneous administration of an Ang II receptor antagonist. Therefore, it appears that within the kidney, NO is the natural antagonist of vasoconstrictive agents such as Ang II and ET-1. It is likely that NO and Ang II continuously interact in the normal kidney modulating renal hemodynamics and, to a certain extent, sodium excretion.

The interaction between NO and ET-1 may be more important in pathological situations that result in renal ischemia. Indeed, endothelium-dependent relaxation in response to NO agonists is impaired in kidneys with post-ischemic acute renal failure (Conger et al., 1988). At the same time, intrarenal synthesis of ET-1 is markedly and persistently increased after ischemia (Firth and Ratcliffe, 1992).

There is substantial evidence that NO is involved in tubuloglomerular feedback and renin release. The macula densa, which plays a crucial role in both, has by far the highest NOS levels in normal kidney (Mundel et al. 1992; Tojo et al., 1994). *In vivo* micropuncture studies have shown that L-NAME infused into the proximal tubule or peritubular capillary augments the tubuloglomerular feedback-mediated decrease in single-nephron GFR (Wilcox et al., 1992). Using isolated microperfused afferent arterioles and attached macula densa, Ito and Ren (1993) demonstrated that selective inhibition of NO at the macula densa causes constriction of the afferent arteriole when the NaCl concentration at the macula is high, but has no effect when NaCl is low. It has also been shown that high NaCl intake increases NOS activity in the kidney, which contributes to the rightward shift of the tubuloglomerular feedback response curve and renal vasodilatation (Deng et al., 1993).

In studies using kidney slices, stimulation of NO synthesis and release with NO agonists reduced renin release (Vidal et al., 1988). Henrich et al. (1988) have shown that the inhibitory effect of NO on renin release is mediated by cGMP. Similarly, evidence is available that other agents that increase cGMP accumulation, such as sodium nitroprusside and 8-bromo-cGMP, also inhibit renin release (Henrich et al., 1988). Chronic inhibition of NO synthesis has been shown to impair sodium excretion and result in hypertension (Ribeiro et al., 1992). Such effects may be due, at least in part, to inappropriate inhibition of the renin–angiotensin system, insufficient rightward shift of tubuloglomerular feedback, or both.

It has been demonstrated that in coincubation experiments NO release from glomerular endothelial cells increases cGMP within mesangial cells (Marsden et al., 1990). Furthermore, NO produced by endothelial cells inhibits Ang II-induced mesangial cell contraction (Shultz et al., 1990). Therefore, NO may be an important signaling molecule in the cross-communication between glomerular endothelial and mesangial cells. Besides its relaxant effects, NO also inhibits mesangial cell proliferation induced by serum (Garg and Hassid, 1989) and by growth factors such as PDGF (Pfeilschifter et al., 1993). This growth-inhibitory effect of NO may help to preserve the structure of the glomerulus under conditions of increased production of growth factors as it is typically found in certain forms of glomerulonephritis.

The significance of the high expression of iNOS in normal kidney has not been completely elucidated. It has been proposed that NO production by iNOS in medullary thick ascending limbs serves to maintain and regulate medullary blood flow and oxygenation (Mattson et al., 1994). Proximal tubular cells contain large amounts of cGMP and iNOS, which can be induced with various cytokines (Guzman et al., 1995). However, it is difficult to assume that the characteristic prolonged activity of iNOS can modulate the minute-to-minute changes in proximal reabsorption through its effect on Na^+/K^+-ATPase.

Experiments performed in anesthetized rats confirm initial observations in awake chronically catheterized rats that, as well as increasing systemic blood pressure and decreasing RPF, nonspecific NOS blockers increase urinary sodium concentration and water excretion (Baylis et al., 1991). This suggests an important role for NO in proximal tubule function. In support of these observations, studies by Wang (1997) using *in situ* microperfusion techniques showed that L-NAME decreases water and bicarbonate reabsorption in the proximal tubule, suggesting an important role for NO dependent, cGMP-linked sodium and bicarbonate transport in this tubular segment.

Wang et al. (2000) reported that proximal tubules from nNOS knockout mice exhibited lower fluid and bicarbonate absorption rates than proximal tubules from wild-type mice. Significant but less pronounced decreases in fluid and bicarbonate absorption were also observed in iNOS knockout mice (Wang et al., 2002). Together, these data indicate that NO produced within proximal tubules stimulates fluid and bicarbonate reabsorption.

In contrast to the above findings, studies by several groups of investigators have found that NO inhibits the Na^+/K^+-ATPase activity in the proximal tubule (Roczniak and Burns, 1996). *In vivo* studies also support these findings, insofar as NOS blockade is associated with reduced sodium and water excretion. Consistent with the inhibitory effect of endogenous NO on proximal tubule transport, Vallon et al., (2001) observed that *in vivo* microperfused proximal tubules of nNOS knockout mice exhibit higher fluid and chloride absorption rates compared with proximal tubules of normal mice.

The explanation for such differences between experimental findings is not clear. Some of the difference may relate to the species used or dose and type of nonspecific NOS blockers. In addition, in nNOS and iNOS knockout mice the specific NOS isoform is genetically deleted in all tissues, not just the proximal tubule. Therefore, the difference between wild-type and knockout mice may be due to the effect of deleting nNOS or iNOS from other organs. Overall, the published data favor an inhibitory effect of NO on proximal tubule transport, most likely caused by a decrease in apical Na^+–H^+ exchanger activity and Na^+/K^+-ATPase.

All the mRNAs encoding NOS isoforms are present in the outer and inner medulla. By immunohistochemistry, positive nNOS, eNOS and iNOS immunostaining has been found in almost all segments of rat renal tubule, but no immunostaining has been observed in the thin limbs of Henle. Data are also available that show NO modulates the function of cortical and inner medullary collecting ducts (Roczniak et al., 1998). In these segments of the tubule, NO-inhibited tubular reabsorption by inhibiting the Na^+/K^+-ATPase activity. Constitutive-type nNOS may play an important role in the regulation of sodium excretion in the inner medullary collecting duct because rats exposed to high salt intake for three days had a significant increase in nNOS protein, as shown by Western blotting (Roczniak et al., 1998).

Consistent with the above observations, administration of the NOS inhibitor L-NMMA to normal human volunteers is associated with a significant reduction in sodium and water excretion (Dijkhorst-Oei and Koomans, 1998). This effect is susceptible to changes in sodium intake because a high-salt diet enhances this response whereas a low-salt diet suppresses the effects of L-NMMA on fractional excretion of sodium (Bech et al., 1998).

NO and prostaglandins may closely interact in the regulation of sodium excretion. Indeed, the blockade of sodium excretion resulting from inhibition of NO synthesis is greatly magnified if prostanoid synthesis is concomitantly stopped by cyclooxygenase inhibitors (Romero et al., 1992).

Nitric Oxide and Renal Disease

In addition to its role on renal hemodynamics, NO has potentially harmful effects, possibly by direct toxicity or indirectly by interacting with leukocyte-derived oxygen radicals. Cattel and coworkers (1990) first demonstrated increased activity of the L-arginine–NO pathway in experimental glomerulonephritis. In culture, glomeruli from rats with nephrotoxic nephritis synthesize increased amounts of nitrate, the stable end product of NO, and this can be prevented by the NOS inhibitor L-NMMA. Nitrate generation is seen in glomeruli isolated at four hours after the induction of glomerulonephritis, is maximal by 24 hours, and persists for up to three weeks. Normal glomeruli do not synthesize nitrate under these conditions. The inducible NOS message is just detectable in normal glomeruli and is increased several hundredfold six hours after induction of glomerulonephritis, with high levels persisting for seven days. It has also been shown that there is nitrite production from glomeruli as well as increased inducible NOS gene expression in other experimental models of glomerulonephritis, including *in situ* immune complex glomerulonephritis, membranous glomerulonephritis, and mesangial proliferative glomerulonephritis induced by an antibody to the mesangial cell antigen, Thy-1 (Blantz and Munger, 2002; Cattel et al., 1991, 1993). In these models, the levels of nitrite generation correlates with glomerular macrophage infiltration. In active Heymann nephritis, nitrite synthesis is reduced by whole-body irradiation, which suppresses macrophage infiltration (Cattel et al., 1991), implying that macrophages are of vital importance in glomerular NO synthesis. Inhibition by dexamethasone suggests that NO synthesis occurred through the induction of inducible NOS.

What role does enhanced NO synthesis play in glomerulonephritis? Analogy with injury at other sites and knowledge of the action of NO in the kidney suggest a range of possible effects. Beneficial effects could include maintenance of RBF and GFR, inhibition of thrombosis, superoxide scavenging with reduction of adhesion molecule expression, downregulation of the expression of proinflammatory cytokines, and inhibition of mesangial cell proliferation. Detrimental effects might include toxicity to intrinsic glomerular cells, either directly or through the formation of peroxynitrite with subsequent peroxidative damage. NO may also activate cyclooxygenase to stimulate glomerular prostaglandin production.

There is a major obstacle to establishing the role of NO in glomerulonephritis: it is difficult to achieve inhibition of inducible NOS without also inhibiting endothelial NOS, thus causing systemic hypertension which itself tends to exacerbate the course of glomerulonephritis. A second problem in some experimental models is that of separating the local effects of NO synthesis in the glomerulus from the effect of systemic NOS inhibition on the generation of the immune response. Weinberg et al. (1994) studied the role of NO in MLR-lpr/mpr mice, which spontaneously develop an autoimmune disease characterized by autoantibody production, glomerulonephritis, vasculitis, and arthritis, all of which mimic human systemic lupus erythematosus. During the course of the disease there is an increase in urinary excretion of the NO products, nitrites and nitrates, and an increased expression of mRNA for inducible NOS in kidney and spleen. Chronic treatment with L-NMMA in the drinking water over this period preventes the development of proteinuria and markedly improves the histological changes of glomerulonephritis and arthritis (Weinberg et al., 1994). The levels of circulating anti-double-stranded DNA antibody did not differ between the treatment and control groups, suggesting that this is not due to an effect on the immune mechanism.

In a more recent report (Reilly et al., 2002), chronic administration of 1-N^6-1-imimoethyl-lysine (L-NIL), a selective iNOS inhibitor, to MLR-lpr mice blocked the development of pathologic renal disease more efficiently than L-NMMA, providing evidence for a harmful role of iNOS in this disease. Narita et al (1995) reported that administering a single dose of L-NMMA one hour before anti-thymocyte serum (ATS) injection in rats, preventes mesangial cell lysis and subsequent accumulation of extracellular matrix. The ATS-induced proteinuria and the increase in urinary nitrite excretion are also prevented (Narita et al., 1995).

However, a beneficial effect for NO has been reported in other experimental models of glomerulonephritis. In Heymann nephritis, L-NAME administration aggravates proteinuria and increases interstitial infiltration of mononuclear cells (Tikkanen et al., 1994). The effects can, however, be reversed by returning blood pressure to normal with captopril, suggesting that they may be due to the hypertensive effect of L-NAME. In a nonimmune model of chronic glomerular disease in the rat induced by adriamycin administration, treatment with L-NIL exacerbates the progression of the disease, as evidenced by reduced creatinine clearance, increased tubular atrophy, interstitial volume and monocyte infiltration (Rangan et al., 2001). From these observations it appears that in certain models of experimental inflammatory renal disease, induction of iNOS occurs within the kidney. However, the role of NO still remains to be elucidated fully; there is evidence of both beneficial and detrimental effects depending on the model of glomerulonephritis, the timing of NOS inhibition and the agents used to inhibit NO synthesis (Cook and Cattel, 1995).

A role for NO in the pathogenesis of human glomerulonephritis has been suggested. Among glomerular cells, the exact type that expresses NOS and the NOS isoforms and contributes to the local production of NO in the diseased kidney of humans has not been identified. Furusu et al. (1998) examined the expression of iNOS, eNOs, and nNOS in kidney biopsies obtained from patients with lupus nephritis, membranous nephropathy, immunoglobulin A (IgA) nephropathy, and minimal change nephrotic syndrome. Normal portions of surgically resected kidney were used as controls. There was expression of eNOS in glomerular endothelial cells and in the endothelium of cortical vessels, both in control and diseased kidneys. The iNOS was present in mesangial cells, podocytes, and infiltrating macrophages in the diseased glomeruli, whereas immunostaining of iNOS was less detectable in control kidneys. The expression pattern of eNOS in each glomerulus was the reverse of that of iNOS in IgA nephropathy and lupus nephritis. The degree of staining for eNOS correlated negatively with the degree of glomerular injury, whereas the expression of iNOS correlated positively with the degree of glomerular injury. No detectable nNOS was found in normal or diseased glomeruli. In renal biopsies of patients with IgA nephropathy and mesangial proliferative glomerulonephritis, iNOS was consistently detected. This was not the case in normal kidney tissue removed because of malignancy (Kashem et al., 1996). Immunohistochemical studies of these samples revealed that cells expressing iNOS were predominantly infiltrating monocytes and macrophages in the tubulointerstitial compartment. At the same sites, TNF-α mRNA expression was also found. Although iNOS expression in the diseased kidney is predominantly caused by macrophage infiltration (Kashem et al., 1996), studies *in vitro* found that mesangial cells and tubulointerstitial cells respond to immune stimuli by producing substantial amounts of iNOS and NO (Mohaupt et al., 1994; Kunz et al., 1994).

EICOSANOIDS

Eicosanoid is the generic term that refers to a group of locally acting hormones or autacoids derived from dietary polyunsaturated fatty acids.

In humans, arachidonic acid, an essential fatty acid esterified into cellular membrane phospholipids, is the most abundant and important precursor. After deesterification by phospholipases, free arachidonic acid may rapidly reesterify into membrane lipids, avidly bind intracellular proteins, or undergo enzymatic oxygenation to yield the various biologically active molecules referred to as eicosanoids. The biosynthetic pathway leading to the formation of the unstable intermediate endoperoxides from arachidonic acid seems to be identical in all tissues investigated (Lands, 1979). Subsequent conversion of the endoperoxides follows a tissue-specific pattern resulting in the formation of different prostaglandins by different tissues. The endoperoxides are synthesized by a glycoprotein enzyme with two enzymatic sites, cyclooxygenase and hydroperoxidase (Lands, 1979). The endoperoxides can then be transformed by prostacyclin synthase to PGI2, by thromboxane synthase to TxA2, by prostaglandin (PG) isomerases to PGE2 and PGD2, and by endoperoxide reductase to yield PGF2α. There is, in addition, some nonenzymatic decomposition of PGH2 into PGE2 and PGD2 (Lands, 1979).

Cyclooxygenase Products

Biosynthesis and Metabolism

In both animals and humans, two separate cyclooxygenase (COX) enzymes have been identified, which are encoded in two separate genes: COX-1 (Yokohama and Tanabe, 1989) and COX-2 (Jones et al., 1993). The COX pathway is the major pathway for arachidonic acid metabolism in the kidney (Morrison, 1986). The human COX-1 enzyme is a 68.5 kDa protein that is constitutively present in arterial and arteriolar endothelial cells (Satoh and Satoh, 1984), mesangial cells (Schlondorff et al., 1985), glomerular epithelial cells (Smith and Bell, 1978), renal interstitial cells (Brown et al., 1980), and along most segments of the tubule, although in markedly varying concentrations (Alavi et al., 1987). The COX-2 enzyme is a 603 amino acid protein whose expression has been demonstrated in the macula densa and renal papillae under normal conditions (Harris et al., 1994) and in mesangial cells subjected to continuous stretch-relaxation in culture (Akai et al., 1992). The COX-2 isoform is constitutive in some tissues, but unlike COX-1 this isoenzyme is markedly induced by bacterial endotoxins, cytokines, and growth factors, and catalyzes the synthesis of proinflammatory PGs (Khan et al., 2002). The antagonism of inflammation and pain by nonsteroidal anti-inflammatory drugs (NSAIDs) is believed to result from the inhibition of COX-2, while the antagonism of gastric mucosal defense and platelet aggregation are the results of COX-1 inhibition.

The biological role of COX-2 extends beyond inflammation and pain. Early female reproductive function is dependent on COX-2 and certain forms of precancerous lesions and cancers, including renal and bladder cancer, overexpress COX-2 (Khan et al., 2002).

The PGs and TxA_2 undergo rapid destruction and inactivation within the kidney by cytosolic degradative enzymes (Samuelsson, 1987). Elimination of PGE_2, $PGF_{2\alpha}$, and PGI_2 proceeds through enzymatic oxidation and nonenzymatic hydrolysis, while that of TxA_2 is exclusively nonenzymatic. The initial degradative step is catalyzed by 15-hydroxyprostaglandins dehydrogenase, with formation of biologically inactive 15-keto-PGs. These metabolites are further degraded by a PG reductase. PGI_2 and TxA_2 undergo rapid nonenzymatic degradation to 6-keto-$PGF_{1\alpha}$ and TxB_2, respectively. As the kidney metabolizes, and also excretes, circulating PGs, it is difficult to evaluate the net rate of renal production of a particular PG. Intact PGs and stable hydrolysis products are excreted in the urine and largely reflect the rate of renal PG production (Lote and Haylor, 1989). Urinary excretion of 6-keto-$PGF_{1\alpha}$, the hydrolytic product of PGI_2, may reflect both augmented systemic and renal PGI_2 synthesis (Patrono and Dunn, 1987).

Vasoactive and Inflammatory Actions of COX Products

PGs have diverse actions, in part related to their site of synthesis and the cells on which they act. Their principal physiologic role is mediation or modulation of hormone action (Patrono and Dunn, 198; Schlondorff et al., 19857). Moreover, under pathophysiologic conditions, such as inflammatory injury, local release of prostanoids may mediate some of the functional derangements that characterize these conditions (Lianos et al., 1983).

Prostanoids act through specific and distinct receptors (Coleman and Humphrey, 1993) that have been cloned and sequenced. All these receptors are members of the G-protein coupled family of receptors. Multiple subtypes of each of these prostanoid receptors may exist, as in the case with the PGE_2 receptor, for whom three receptor subtypes (EP-1, EP-2, EP-3) have been identified, thus explaining the apparently contrasting effects mediated by PGE_2 on smooth muscle cells (Hebert et al., 1993). The TxA_2 receptor (TP receptor) appears to signal via phosphatidylinositol hydrolysis, leading to increased intracellular calcium (Hirata et al., 1991) and a vasoconstrictor response.

There is pharmacological evidence for the existence of TP receptors in the glomerulus (Menè et al., 1989). The PGI_2 receptor (IP receptor) signals via stimulation of cAMP generation (Veis et al., 1990). PGI_2 has been demonstrated to play an important vasodilatory role in the glomerular microvasculature, where the effects of PGI_2 and PGE_2 to stimulate cAMP generation were distinct and additive (Chaudhari et al., 1990).

Renal Actions of COX Products

Unlike the gastrointestinal tract and platelets, in which physiological function is dependent on COX-1, the kidneys of laboratory species (rat and dog) and primates (monkeys and humans) constitutively express both COX-1 and COX-2 (Khan et al., 2002). COX-1 immunoreactivity is abundant and appears uniformly across species in the collecting ducts, renal vasculature, glomeruli, and papillary interstitial cells. In contrast, basal COX-2 immunoreactivity in kidney is less intense and exhibits some interspecies differences in localization. The intrarenal distribution of COX-2 in rats and dogs includes the macula densa, thick ascending limbs, and papillary interstitial cells. In normal humans, COX-2 is co-localized with COX-1 in glomeruli and small blood vessels, but is not detectable in the macula densa (Khan et al., 2002).

The physiological role of prostaglandins in modifying various renal functions has become increasingly apparent over the past two decades. At present there is convincing evidence that prostaglandins influence RBF and GFR, the release of renin, and the urinary concentrating mechanism. A contribution of prostaglandins to phosphate and hydrogen-ion handling by the kidney is likely, but not fully established. Finally, a role for prostaglandins in the control of renal erythropoietin production has been postulated.

In general, the importance of renal prostaglandins for the maintenance of blood flow and glomerular filtration can be demonstrated only under conditions where the vasoconstrictor system is activated (Henrich et al., 1978). Similar to the vasculature in other organs, PGE_2 and PGI_2 are vasodilators in the kidney (Jackson et al., 1982). Occasional vasoconstriction observed after PGE_2 or PGI_2 administration may be secondary to an increase in renin release and, consequently, Ang II formation (Scharschmidt et al., 1983). When the formation of Ang II is blocked, both PGE_2 and PGI_2 act as renal vasodilators. Conversely, thromboxane is a potent vasoconstrictor that also contracts isolated glomeruli (Scharschmidt et al., 1983).

The physiological role of the vasodilator prostaglandins is most likely to counteract the effect of vasoconstrictors. Renal vasoconstrictors increase synthesis of vasodilatory prostaglandins, which thus act as a negative feedback. Prostaglandins may affect RBF to varying degrees in different regions of the kidney (Kirschenbaum et al., 1974). Inhibition of endogenous prostaglandin synthesis will preferentially decrease medullary blood flow, possibly indicating that the high renal medullary PG synthesis maintains blood flow to this poorly oxygenated and hypertonic region of the kidney.

Prostaglandins also play a contributory role in renal autoregulation (Schnermann et al., 1984). As modulators of renal vascular tone and mesangial contraction, prostaglandins also influence GFR (Scharschmidt et al., 1983; Schnermann et al., 1984). They may do so by changing afferent and efferent arteriolar resistance (Schnermann et al., 1984), by influencing the renin–angiotensin system (Schnermann et al., 1984), and by directly affecting glomerular filtration characteristics through changes in mesangial contraction (Scharschmidt et al., 1983). Micropuncture experiments have demonstrated that prostaglandins may also contribute to tubuloglomerular feedback regulation (Schnermann et al., 1984).

Prostaglandins, particularly PGE_2 and PGI_2, have been shown to enhance plasma renin activity *in vivo* and directly increase renin release *in vitro*, and thus represent one of the multiple factors controlling renin (Henrich, 1981). *In vitro* studies have shown that locally produced PGs act on renin secretion (Henrich, 1981). Stimulation of renin release also occurs in isolated glomeruli superfused with arachidonic acid, PGE_2 or PGI_2 (Beierwaltes et al., 1982), and this is blocked by PG synthesis inhibition.

Ever since the original observations that PGE_1 inhibits the effect of AVP in the toad urinary bladder (Orloff et al., 1965) and in the mammalian collecting duct (Grantham and Orloff, 1968), the interaction of AVP with PGs has attracted much interest. Multiple studies have confirmed that PGE_1 and PGE_2 antagonize the antidiuretic action of AVP both *in vitro* and *in vivo* (Gross et al., 1981). Therefore, the major effect of PGs in the concentrating mechanism is their interference with the cellular mechanism of action of AVP. It has also been proposed that thromboxane may directly enhance the action of AVP (Burch and Halushka, 1980), although these findings remain controversial (Ludens and Taylor, 1982).

Infusion of arachidonic acid or of PGE_2 or PGI_2 directly into the renal artery results in natriuresis (Hebert et al., 1991). While hemodynamic changes probably contribute, the natriuresis is largely a direct tubular phenomenon originating in the distal nephron (Hebert et al., 1991). PGE_2 has mild or no effects on Na transport in the proximal tubule and most segments of the ascending limb of Henle, with the exception of the medullary thick ascending limb in some species.

Published reports on the effects of PGs on renal phosphate transport are also conflicting. It has been shown that PGE_1 alone is mildly antiphosphaturic but that it markedly inhibites the phosphaturic response to parathyroid hormone (PTH) in intact dog (Beck et al., 1972). In contrast, others have reported that PGE_1 increases the fractional excretion of phosphate in intact dogs, whereas PGE_2 is inactive (Strandhoy et al., 1974). Thus, some evidence supports a role for PGs in inhibition of tubular phosphate transport. It should be recalled, however, that the proximal tubule, the major site of phosphate transport, has very little capacity to produce PGs.

A role for PG in the production and release of erythropoietin by the kidney has been postulated based on some indirect experimental evidence (Radte et al., 1980).

Role of Prostaglandins and Thromboxane in Renal Disease

Many studies have investigated the signaling molecules that attract inflammatory cells into the kidney upon immune complex deposition (Noris et al., 1995) and the mediators actually involved in the subsequent cascade of events leading to tissue injury. There is evidence that abnormalities in the metabolism of arachidonic acid in the kidney take part in the dynamics of tissue injury and modulate the severity of the inflammatory reactions. TxA_2 is one of the arachidonic-acid metabolites widely implicated as a mediator of damage in renal disease (Remuzzi et al., 1992). In an experimental model of immune-mediated mesangial cell injury in the rat, glomerular synthesis of TxA_2 was increased, and pretreatment with a TxA_2 synthase inhibitor prevented the drop in GFR (Stahl et al., 1990).

In a murine model of lupus nephritis, TxA_2 was produced within the kidney in much larger amounts than in control mice, and correlated with the amount of protein in the urine and with the severity of renal pathology (Kelley et al., 1986). In the same model, long-term pharmacologic blocking of TxA_2 synthesis limited proteinuria and renal lesions and prolonged survival (Salvati et al., 1995). Similarly, in patients studied during the active phase of lupus nephritis, renal TxA_2 synthesis was enhanced, as reflected by increased urinary excretion of its stable breakdown product TxB_2. The levels of urinary TxB_2 correlated with the severity of renal lesions and deterioration of renal function (Patrono et al., 1985). In addition, both RBF and GFR were increased by pharmacologic blockade of TxA_2 receptor in patients with lupus nephritis (Pierucci et al., 1989). However, in lupus nephritis, renal function depends on vasodilatory prostaglandins and, in fact, inhibiting the prostaglandin-forming enzyme cyclooxygenase with NSAIDs was detrimental, as indicated by a lowering of RBF and GFR in patients with lupus nephritis who were given aspirin or ibuprofen (Patrono et al., 1985). It has been found that TxB_2 synthesis and COX-2 gene expression were higher in peripheral blood mononuclear cells from patients with active lupus nephritis compared with patients with the inactive form of the disease and with healthy subjects (Tomasoni et al., 1998). In addition, immunohistochemistry studies have found an intense expression of COX-2 protein in the glomeruli of patients with lupus nephritis, which was localized in infiltrating macrophages (Tomasoni et al., 1998). Unlike COX-2, levels of COX-1 mRNA were comparable in lupus patients and control subjects and were not influenced by the disease activity.

This finding of a selective upregulation of COX-2 isoenzyme in circulating monocytes and kidney macrophages indicates a novel pathway of inflammatory injury in human lupus nephritis. If this interpretation is correct, COX-2 may become a target for therapeutic intervention in this disease. That this may be the case is supported by finding that administration of a selective COX-2 inhibitor, combined with the immunosuppressant mycophenolate mofetil, to NZB/W lupus mice delays the onset of renal disease and prolongs survival (Zoja et al., 2001).

Another aspect of eicosanoid involvement in glomerulonephritis is the role of cyclooxygenase products in the proliferative response to glomerular immune injury and the resultant crescent formation. Cybulsky et al. (1992) studied the role of prostaglandins and thromboxane in the proliferative response of glomerular epithelial cells. They showed that cell proliferation induced by epidermal growth factor (EGF) decreases in the presence of cyclooxygenase or thromboxane synthase inhibitors. These inhibitors decrease the EGF-dependent tyrosine phosphorylation of growth factor receptor, suggesting that eicosanoids enhance

EGF-induced receptor activation and mitogenic response in glomerular epithelial cells.

The role of COX products in mediating diabetic nephropathy remains controversial. Vasodilator prostaglandins may contribute to the hyperfiltration that occurs in early stages of the disease, whereas TxA_2 may play a role in the subsequent development of albuminuria and basement membrane changes (DeRubertis and Craven, 1993). Glomerular synthesis and urinary excretion of TxB_2 increase in streptozotocin-treated diabetic rats. Enhanced urinary excretion of TxB_2 in human type 1 diabetes mellitus has also been described.

Although selective inhibitors of thromboxane synthesis have been reported to ameliorate proteinuria in diabetic nephropathy, the cellular source of urinary TxB_2 has not been clear. A study documented that platelet depletion failed to significantly reduce production of TxB_2 in streptozotocin-treated diabetic rats. However, a selective thromboxane synthase inhibitor given in a dose that suppressed both platelet and glomerular TxB_2 production prevented the development of albuminuria in these rats (DeRubertis and Craven, 1992). These results suggest that platelets are not the major source of TxB_2 in this model and emphasize the role of glomerular thromboxane synthesis.

In humans with diabetic nephropathy, a thromboxane synthase inhibitor produced dose- and time-dependent reductions in both serum TxB_2 and urinary 11-dehydro-TxB_2 (Alessandrini et al., 1990). A possible role of COX-2 in mediating diabetic nephropathy has been suggested. Cheng and coworkers (2002) documented that COX-2 expression in the renal cortex was significantly increased in rats made diabetic by streptozotocin administration, as compared with control rats. Immunohistochemical localization indicated increased COX-2 expression in cells of the macula densa and surrounding cortical thick ascending limb of Henle. Treatment with a selective COX-2 inhibitor decreased expression of plasminogen activator inhibitor-1 (PAI-1), vascular endothelial growth factor and fibronectin, and also decreased biochemical, functional, and structural markers of renal injury (Cheng et al., 2002).

Lipoxygenase Products

Biosynthesis and Metabolism

Arachidonic acid can also be metabolized via lipoxygenase pathways. Enzymatic lipoxygenation of arachidonic acid leads to the generation of leukotrienes (LTs), lipoxin (LXs), and hydroxyeicosatetraenoic acids (HETEs) (Samuelsson et al., 1987). Formation of these compounds is initiated by 5-, 12-, or 15-lipoxygenase, whereby a hydroperoxy group is introduced into arachidonic acid at carbon 5, carbon 12 or carbon 15, respectively, to yield the corresponding 5-, 12-, or 15-hydroperoxytetraenoic acid (HPETE). HPETEs are unstable compounds that are transformed into the corresponding 5-, 12-, and 15-HETE, which in turn undergo enzymatic modification leading to the generation of the various LTs and LXs.

Leukotriene A_4 (LTA_4) is an early pivotal intermediate in the 5-lipoxygenase pathway whose metabolism lead to the production of the LT series of metabolites (Samuelsson et al., 1987). Formation of LTB_4 requires LTA_4 hydrolase activity, whereas generation of the peptidyl-leukotrienes (LTC_4, LTD_4, and LTE_4) requires the enzymatic action of glutathione-*S*-transferase (Samuelsson et al., 1987).

The 15-lipoxygenase enzyme catalyzes the production of 15-HETE and initiates another major pathway of arachidonic-acid metabolism, which leads to trihydroxy derivatives, the LXs (Samuelsson et al., 1987). The main LXs derived from 15-HETE are designed LXA_4, LXB_4, and 7-*cis*-11-*trans*-LXA_4 (Serhan, 1994).

In the kidney, lipoxygenase products are largely generated by infiltrating leukocytes or resident cells of macrophage-monocyte origin, but renal cells are capable of generating LTs and LXs either directly or through metabolism of intermediates by resident leukocytes (Lefkowith et al., 1988).

Renal Actions of Lipoxygenase Products

The LTs are potent proinflammatory molecules (Samuelsson et al., 1987). LTB_4 has no significant effects on renal hemodynamics in normal animals, but amplifies glomerular inflammation and proteinuria in animals with glomerulonephritic injury (Yared et al., 1991). LTC_4 and LTD_4 exert potent effects on glomerular hemodynamics. In rats, systemic administration of LTC_4 leads to reduction of RBF and GFR (Badr, 1984). Similarly, infusion of either LTC_4 or LTD_4 in the isolated perfused kidney results in dramatic increases in renal vascular resistance and reductions in GFR (Rosenthal and Pace-Asciak, 1983). LTD_4 mediates these effects by causing a significant increase in efferent arteriolar resistance, leading to a fall in GPF, and a rise in glomerular capillary hydraulic pressure (Badr et al., 1987). In addition, LTD_4 markedly reduces the glomerular capillary ultrafiltration coefficient, and therefore its overall effect is to decrease single-nephron GFR. Moreover, LTC_4 and LTD_4 contract mesangial cells (Barnett et al., 1986). Specific mesangial cell LTD_4 receptors have been identified in both rats and humans (Simonson et al., 1988).

Different lipoxins display distinct effects on renal hemodynamics (Serhan, 1994). In rats, LXA_4 causes a selective decrease in afferent arteriolar resistance, thereby increasing RBF, intracapillary hydraulic pressure, and GFR (Katoh, 1992). The LXA_4-induced increase in GFR, however, is partially offset by its mild effect in decreasing the ultrafiltration coefficient (Katoh, 1992). The vasodilator actions of LXA_4 are mediated by prostaglandins (Hebert et al., 1993). LXB_4 and 7-*cis*-11-*trans*-LXA_4 display vasoconstrictive effects on renal hemodynamics in rats that are independent of COX activity (Katoh, 1992).

Lipoxygenase Products in Renal Disease

LTs are increasingly recognized as being major mediators of glomerular hemodynamic and structural deterioration during early phases of experimentally induced glomerulonephritis (Badr, 1992b). Increased glomerular generation of LTB_4 and peptidyl-LTs has been demonstrated in several models of glomerular injury (Badr, 1992a, 1992b). LTB_4 probably worsens glomerular injury by augmenting leukocyte recruitment and activation, and the peptidyl-LTs, by depressing the ultrafitration coefficient and GFR (Badr et al., 1987; Rosenthal and Pace-Asciak, 1983; Yared et al., 1991). Indeed, *in vivo* administration of LTB_4 to rats with mild nephrotoxic serum-induced injury has been associated with an increase in polymorphonucleocyte infiltration and a marked exacerbation of the fall in GFR, the latter correlating with the number of infiltrating polymorphonucleocytes per glomerulus (Yared et al., 1991). Selective blockade of the 5-lipoxygenase pathway in the course of glomerular injury is associated with significant amelioration of the deterioration of renal hemodynamic and structural parameters (Badr et al., 1988).

LTs are probably involved in the pathophysiology of human glomerulonephritis. In this regard, 5-lipoxygenase has been detected in kidney biopsy specimens from some patients with IgA nephropathy and mesangial proliferative glomerulonephritis, and associated with clinically worse renal status (Rifai et al., 1993). In MRL-lpr/lpr mice, which develop overt manifestations of autoimmune disease with nephritis similar to human systemic lupus erythematosus, impaired renal hemodynamic function has been associated with enhanced ionophore-stimulated production of LTB_4 and LCB_4 from preparations of renal cortex (Spurney et al., 1991). Significant inverse correlation was observed between GFR and *in vitro* production of both LTB_4 and LTC_4 in kidneys from MLR lpr/lpr mice but not from control mice. Administration of the specific peptidoleukotriene receptor antagonist SKF-104353 to MRL lpr/lpr mice significantly improved both GFR and RBF, whereas this agent had no effect on renal hemodynamics in control mice. Furthermore, urinary LTE_4 levels have been found to be elevated in patients with active systemic lupus erythematosus (Hackshaw et al., 1992).

A pathophysiologic role for LTs has also been described in experimental acute allograft rejection, cyclosporine toxicity, and acute ureteral obstruction. In the latter case, in glomeruli isolated from rats after unilateral release of bilateral ureteral obstruction, there is a significant increase in LTB_4 production as compared with control rats given sham operations. Animals treated with a 5-lipoxygenase inhibitor prior to bilateral ureteral obstruction show attenuation in the fall of GFR observed in the untreated animals (Reyes et al., 1992).

Nonenzymatic Oxidative Products of Arachidonic Acid

A series of novel prostaglandin F_2-like compounds produced by a non-cyclooxygenase mechanism involving free radical catalyzed lipid peroxidation has also been described (Morrow et al., 1990). These compounds, termed isoprostanes, are easily detected in normal human plasma and urine, and their production is enhanced markedly in animal models of free radical induced injury (Morrow et al., 1990). One of these compounds, 8-epi-prostaglandin $F_{2\alpha}$, is the most potent renal vasoconstrictor eicosanoid reported to date (Morrow et al., 1990). In normal rats, intrarenal arterial infusion of 8-epi-prostaglandin $F_{2\alpha}$ induces a dose-dependent reduction in GFR and RBF, with renal function ceasing at the highest dose (Takahashi et al., 1992). Micropuncture measurements revealed a predominant increase in afferent compared with efferent arteriolar resistance, resulting in a significant fall in transcapillary hydraulic pressure difference and reducing single-nephron GFR and plasma flow (Takahashi et al., 1992). Changes induced in renal hemodynamics by 8-epi-prostaglandin $F_{2\alpha}$ are completely absent in the presence of a TxA_2 receptor antagonist (Takahashi et al., 1992). These findings indicate that 8-epi-prostaglandin $F_{2\alpha}$ is a potent preglomerular vasoconstrictor acting principally through activation of TxA_2 receptors.

ROLE OF VASOACTIVE AND INFLAMMATORY MEDIATORS IN DRUG-INDUCED RENAL NEPHROTOXICITY

Cyclosporine

Immunosuppression with cyclosporine (CsA) has resulted in improved allograft survival in renal, liver, heart, and pancreas transplantation (Remuzzi and Perico, 1995). CsA has also been advocated as a promising agent for treatment of some autoimmune diseases, including uveitis, psoriasis, systemic lupus erythematosus, rheumatoid arthritis, and various forms of glomerulonephritis (Remuzzi and Perico, 1995). Whereas the efficacy of CsA as an immunosuppressant is undisputed, its therapeutic potential is often limited by its concomitant nephrotoxicity, which may manifest as an acute decline in GFR, rapidly reversible, or a chronic form of renal vascular damage. These are probably the consequences of the toxic effect of CsA on renal vessels, which has been consistently documented in laboratory animals and humans (Kahan, 1989; Perico et al., 1991b). Indeed CsA promotes physiologic responses leading to an increase in renal vascular resistance, as shown in the isolated perfused rat kidney in which CsA, but not its vehicle, cause a dose-dependent fall in renal perfusate flow (Perico et al., 1990a). These findings, suggesting CsA has a direct effect on renal circulation, have been confirmed by *in vivo* studies, which show a decline in RBF associated with a reduction in GFR after intravenous or oral CsA administration to normal rats (Kon et al., 1990; Perico et al., 1990a).

The mechanisms responsible for this effect are still matter of speculation. Activation of the renin–angiotensin system and increased sympathetic nervous traffic to the kidney has been implicated, as well as an alteration in arachidonate metabolism and ET synthesis (Dieperink et al., 1998). Siegl et al. (1983) introduced the theory that CsA stimulates the sympathetic nervous system, giving rise to increased activity of renin, angiotensin, and aldosterone. Murray et al. (1985) confirmed the importance of the renal sympathetic nervous system by observing that unilateral renal denervation of rats relieves CsA-induced vasoconstriction. Moss and coworkers (1985) demonstrated that denervated kidneys are protected from the decrease in GFR observed in innervated kidneys after CsA infusion. They found that CsA infusion increases efferent and afferent renal and genitofemoral nerve activity in parallel, suggesting generalized sympathetic activation. Increased renal nerve activity can cause vasoconstriction, but may also increase proximal tubular reabsorption with or without decreased RBF or GFR (Abildgaard et al., 1986). It has been argued that transplanted kidneys are denervated and should therefore be protected from the nephrotoxic effect of CsA. In reality, however, allograft recipients have increased renal vascular resistance (Curtis et al., 1986), and transplanted kidneys are vulnerable to CsA nephrotoxicity. Perhaps this discrepancy can be explained by increased sensitivity of the denervated organ to nonnervous stimulation, as previously proposed (Moss et al., 1985).

The renin–angiotensin system is not of primary importance in acute CsA nephrotoxicity. Plasma renin concentration in the rat has a bimodal correlation with CsA dosage; increased renin levels were seen with normal CsA doses while plasma renin decreased with high CsA doses (Diepernick et al., 1983). CsA increases renin release from renal cortical slices *in vitro* (Moss et al., 1985), suggesting a direct effect on the juxtaglomerular apparatus. However, improvements in renal function are not uniformly observed after inhibition of converting enzyme (Diepernick et al., 1986). These experimental observations indicate that acute CsA-induced nephrotoxicity can occur independently of activation of the sympathetic nervous and renin–angiotensin systems. Data from a chronic model produced by sodium depletion have, however, implicated the renin–angiotensin system in the tubulointerstitial fibrosis and arteriolopathy produced by CsA, independent of renal hemodynamics (Elzinga et al., 1993).

Arachidonate metabolites, through the enzyme cyclooxygenase, have been reported to participate in the regulation of glomerular function, and might contribute to the deterioration of renal function induced by CsA. Therefore, in laboratory animals, CsA consistently augments the generation of TxA_2, a potent renal vasoconstrictor (Perico et al., 1986), while its effect on vasodilatory prostaglandins is controversial (Lau et al., 1989; Zoja et al., 1986). The renal hemodynamic significance of the increased TxA_2 generation during CsA administration has been supported by the fact that a TxA_2 synthase inhibitor (Perico et al., 1986) or a specific TxA_2 receptor antagonist (Perico et al., 1991a) partially prevent the CsA-induced acute decline in GFR and RPF in normal rats. The beneficial effects of the antagonist on CsA nephrotoxicity have been confirmed in a rat model of kidney isograft (Perico et al., 1992). Studies in humans have also shown that CsA alters the balance between the various vasoactive products of arachidonate metabolism. In patients with recent renal transplant who were receiving CsA as a part of their immunosuppressive regimen and who had evidence of nephrotoxicity, the urinary excretion of TxA_2 metabolites was greater than in healthy subjects (Smith et al., 1992). Short-term administration of a specific TxA_2 synthase inhibitor caused selective and nearly complete inhibition of TxB_2 excretion in all patients, associated with 9% and 33% improvement of GFR and RPF, respectively, indicating that renal vasoconstriction caused by CsA in these patients was partially mediated by enhanced production of TxA_2. Moreover, two studies have shown that a dietary regimen with omega-3 polyunsaturated fatty acids, known to inhibit renal eicosanoid synthesis (Elzinga et al., 1987), favorably influenced renal function, both in the early period and one year after transplantation (van der Heide et al., 1992).

The fact that after pharmacological manipulation of TxA_2 synthesis or biological activity renal function in both experimental animals and humans improved but did not return to normal suggests that factors other than TxA_2 are involved in the acute renal vasoconstriction caused by CsA. Attention has been focused on the potential contribution of ET to CsA-associated vasoconstriction (Perico et al., 1990a). It has been shown that CsA stimulates ET production *in vitro* and *in vivo*, increasing ET binding in renal tissue as well as the urinary excretion of the peptide (Awazu et al., 1991; Bunchman and Brookshire, 1991; Nambi et al., 1990; Perico et al., 1990a). Exposure of bovine aortic endothelial cells to CsA was associated with a concentration-dependent increase in ET release into the incubation medium (Zoja et al., 1986). Similarly, Bunchman and Brookshire (1991) showed a direct effect of CsA on ET release from cultured human endothelial cells. These findings support persistent activation of ET gene or receptor regulation within the kidney that contributes to CsA-associated injury.

Further evidence for the role of endogenous ET derives from *in vivo* studies showing that an anti-ET antibody or an ET receptor antagonist markedly prevents the decline in GFR induced by giving CsA in normal rats (Perico et al., 1990a). Experimental use of ET_A or ET_A/ET_B receptor antagonists attenuated CsA-induced decrease in GFR and RBF (Fogo et al., 1992), vasoconstriction of large preglomerular arteries and reduction in glomerular blood flow (Cavarape et al., 1998), afferent arteriole vasoconstriction (Lanese and Conger, 1993), myosin light chain phosphorylation in glomerular mesangial cells (Takeda et al., 1992), and calcium rise in smooth muscle cells (Meyer-Lehnert et al., 1997).

When ET_A and ET_A/ET_B receptor antagonists were compared, additional ET_B blockade did not attenuate CsA effects further (Cavarape et al., 1998).

However, other studies with ET receptor blockade led to conflicting results. Indeed ETA receptor antagonism attenuated the CsA-induced fall in GFR and RBF only when infused into the renal artery, but not systemically, before CsA administration (Fogo et al., 1992). Others (Davis et al., 1994) showed that a selective ET_A antagonist or a combination of an ET_A/ET_B receptor blocker does not prevent CsA-induced renal vasoconstriction in rats. Moreover a study in 10 healthy volunteers found that the ET_A/ET_B receptor antagonist bosentan limited CsA-induced renal hypoperfusion without effects on the systemic blood pressure (Binet et al., 2000).

Given the hemodynamic changes caused by CsA, it is not surprising that a growing body of data has emerged linking acute CsA nephrotoxicity with disturbances in NO pathway. Studies have consistently shown that CsA impairs endothelium-dependent vasodilation mediated by NO of *in vitro* human subcutaneous vessels and *in vivo* forearm vessels of heart transplant recipients (Bracht et al., 1999; Richards et al., 1989). This has also been shown in numerous *in vitro*, *ex vivo* and *in vivo* experimental studies evaluating mesenteric arteries, aortic rings, renal arteries, afferent and efferent arterioles, femoral arteries and thoracic aorta of rats (Diederich et al., 1992; Gallego et al., 1993, 1994; Lee et al., 1999; Roullet et al., 1994; Takenaka et al., 1992). However, studies assessing CsA effect on tissue, plasma, and urinary NO levels and on tissue expression of NOS isoforms have provided contradictory results. *In vivo* experiments in rats have shown that CsA does not change the urinary excretion of NO metabolites (Bobadilla et al., 1994; Vaziri et al., 1998). Studies in healthy volunteers found that CsA increased NOS activity (Stroes et al., 1997), whereas studies in renal transplant recipients showed that CsA induced impaired basal and stimulated NO production (Morris et al., 2000) or produced no significant reductions in urinary excretion of NO metabolites after the first dose (Grossman et al., 2001).

Enhancement of NO production by administration of L-arginine improves, while blockade of NO generation by L-NAME worsens, the changes in the endothelium-dependent vasodilation and in renal and glomerular hemodynamics induced by CsA in laboratory animals (Andoh et al., 1997; Bloom et al., 1995; De Nicola et al., 1993; Gallego et al., 1993; Gardner et al., 1996; Lee et al., 1999). Clinical trials assessing the effects of L-arginine supplementation in CsA-treated renal or heart transplant patients were generally negative, showing no improvement in renal function or hemodynamics (Gaston et al., 1995; Koller-Strametz et al., 1999). Other investigators (Andres et al., 1997), however, have reported increases in GFR, RBF and natriuresis after administration of L-arginine to stable renal transplant recipients.

Although acute renal hypoperfusion induced by CsA is no longer a problem, chronic CsA nephrotoxicity remains a major challenge for transplant physicians. Lesions of chronic CsA nephropathy consist of vascular and tubulointerstitial changes, including degenerative hyaline alterations extending from preglomerular areas proximally up the afferent arteriole, with progressive narrowing and eventual occlusion of the lumen (Mihatsch et al., 1995). Tubulointerstitial changes include narrow stripes of atrophy and fibrosis, mostly corresponding to areas of cortex with afferent arteriole lesions (Mihatsch et al., 1988), which can be diffuse.

The precise cause of chronic CsA renal toxicity is not known. It may be that sustained glomerular afferent arteriolar vasoconstriction eventually promotes structural alteration and eventual occlusion of preglomerular vessels (Myers and Newton, 1991; Remuzzi and Perico, 1995). This causes an ischemic damage of downstream glomeruli and interstitium that culminates in glomerular obsolescence and interstitial fibrosis. Although prolonged vasoconstriction could contribute to chronic CsA nephropathy by producing chronic ischemia, this relation has been difficult to demonstrate (Andoth and Bennett, 1998). How tubulointerstitial injury induced by CsA develops remains ill defined.

In an animal model of chronic CsA nephrotoxicity produced by one week of salt depletion, which mimics the human renal pathology of chronic CsA nephropathy, interstitial fibrosis is mediated, at least partly, through Ang II induction of TGF-β1 expression (Shihab et al., 1997). Indeed, in these rats, ACE inhibition, as well as Ang II receptor blockade, decreased the expression of transforming growth factor-β1 and markedly reduced tubulointerstitial fibrosis. However, the structural changes in this model are probably related to salt depletion that activates Ang II-dependent growth factors for interstitial fibroblasts, lymphokines, and cytokines, thus questioning the role of Ang II in chronic CsA interstitial injury in humans, who are mostly in a normal-sodium diet.

Intense staining for ET-1, RANTES, and MCP-1 mRNAs selectively localized at tubular epithelial cells was found in biopsies taken from patients with proven CsA nephropathy, but not in the chronic graft-rejection patients, whose tubules had only minimal staining for RANTES mRNA (Benigni et al., 1999). It is tempting to speculate that CsA exerts its renal toxicity on tubular epithelial cells by inducing excessive formation of vasoactive and chemotactic mediators that are then secreted toward the basolateral compartment of tubular cells, thus contributing to the inflammatory reaction and subsequent fibrosis of the interstitium. In proximal tubular cells in culture, ET-1 synthesis is enhanced when the cells are exposed to CsA (Nakahama, 1990). Moreover, *in vivo* in rats, CsA administration is accompanied by an increase of urinary ET excretion (Benigni et al., 1991b). Conversely, no data are available on the ability of CsA to stimulate RANTES and MCP-1 generation by tubular epithelial cells. Because it has been shown that CsA accumulates in tubular cells (Von Willebrand and Hayry, 1983), the high intracellular concentration of the drug may activate the synthetic machinery for ET-1 and possibly for other chemokines. This possibility is supported by the demonstration that CsA is an inhibitor of the multidrug resistance transporter, p-glycoprotein, which is constitutively expressed in the epithelial cells of renal tubules (Garcia del Moral et al., 1995). Inhibition of this transporter could allow the accumulation of CsA into tubular cells that theoretically would result in enhanced generation of mediators of tubulointerstitial injury. Further research should clarify this potential mechanism of CsA-induced chronic nephrotoxicity and may disclose possible strategies for new medical treatments.

Some studies have raised the possibility that vascular endothelial growth factor (VEGF) plays a role in chronic CsA nephrotoxicity. VEGF is a potent endothelial cell mitogen that mediates endothelial cell proliferation and survival, induces angiogenesis, participates in vascular remodeling and repair, and causes vasodilation and increased vascular permeability by enhancing NO production (Shihab et al., 2001). Evidence is available of VEGF upregulation in salt-depleted CsA-treated rats, but not in normal-salt-diet animals (Shihab et al., 2001). Interestingly, VEGF administration to rats with established chronic CsA nephropathy improves interstitial fibrosis, decreases osteopontin expression, macrophage infiltration and collagen III deposition, as well as reducing blood pressure (Kang et al., 2001).

Other mediators have been experimentally implicated in the genesis of chronic CsA nephrotoxicity, including reactive oxygen species that, besides their effects on renal function, mediate tissue injury favoring fibrosis. The use of the antioxidant vitamin E preventes the increase in TGF-β and osteopontin mRNA and the development of renal fibrosis in CsA-treated rats (Jenkins et al., 2001).

Aminoglycosides

Nephrotoxic injury is a common complication of aminoglycoside antibiotic therapy. Studies that have used well-defined measures of nephrotoxicity indicate an incidence rate of 7 to 36% (Kumin, 1980). This variability reflects differences with respect to nephrotoxicity potentials of aminoglycoside antibiotics in clinical use as well as differences among patients receiving these drugs. A survey of clinical studies revealed that the average incidence of nephrotoxicity caused by specific aminoglycoside antibiotics was: gentamicin 14%, tobramycin 12.9%, amikacin 9.4%, and netilmicin 8.9% (Kahlmeter and Dahlagers, 1984). In critically ill patients, the incidence of aminoglycoside nephrotoxicity may rise twofold (Plaut et al., 1979; Schentag et al., 1982).

The earliest and most common expression of aminoglycoside nephrotoxicity is increased urinary excretion of low molecular weight proteins (Schentag, 1983) and lysosomal and brush border membrane enzymes (Beck et al., 1977). These changes may be detected within 24 hours of initiating drug therapy, and the frequency and magnitude of these changes increase as a function of dose and duration of therapy. Indeed, the incidence of renal toxicity, as well as the clinical and bacteriologic efficacy, of the once-daily dosing of aminoglycosides compared with conventional therapy have been well investigated over the past decade, in both immunocompetent and immunosuppressed patients. Meta-analyses by different investigators (Ali and Goetz, 1997; Bailey et al., 1997; Fisman and Kaye, 2000) have shown at least a similar clinical efficacy with no additional risk of renal toxicity. Two studies have reported a significantly lower risk of nephrotoxicity with once-daily dosing (Barza et al., 1996; Ferriols-Lisart and Alos-Aminana, 1996).

Unfortunately, the urinary excretion of low molecular weight proteins and tubular enzymes do not predict which patients will progress to acute renal failure. Nonoliguric renal failure is a common expression of aminoglycoside nephrotoxicity (Anderson et al., 1977) and may reflect a direct inhibitory effect on solute transport along the thick ascending limb of loop of Henle (Kidwell, 1994) or possibly tubulointerstitial cell injury (Rosen et al., 1994), which results in impaired ability to maintain a hypertonic medullary interstitium. Inhibition of adenylate cyclase may also contribute to the polyuria (Humes and Weinberg, 1983). Neither mechanism, however, adequately explains the maintenance of normal to high urine output even in the face of severe depression of whole-kidney GFR.

The slow evolution of acute renal failure, which has been attributed to a variable susceptibility of renal proximal tubular cells to aminoglycoside toxicity (Kaloyanides and Pastoiza-Munoz, 1980), may allow for the development of maximal compensatory adaptation by residual intact nephrons. In addition, micropuncture experiments (Safirstein et al., 1983) implicate a marked depression of solute and water transport along the proximal tubule such that the large increase in the fraction of filtrate escaping reabsorption along the proximal tubule may overwhelm the reabsorptive capacity of the distal nephron and contribute to the pattern of nonoliguric renal failure. When oliguria occurs, it usually signifies the influence of one or more complicating factors, i.e., ischemia or another nephrotoxicant.

Depression of GFR is a relatively late manifestation of aminoglycoside nephrotoxicity. In humans, depression of GFR typically does not occur before five to seven days of therapy have been completed (Lietman and Smith, 1983), unless there has been a major complicating factor such as renal ischemia. Studies in animal models of aminoglycoside nephrotoxicity have implicated activation of the renin–angiotensin system (Schor, 1981), tubular obstruction (Neugarten et al., 1983), and release of platelet activating factor from mesangial cells (Rodriguez-Barbero et al., 1995) as pathogenic factors causing depression of GFR. Indeed activation of the renin–angiotensin system and the ensuing local vasoconstriction appear to be primarily responsible for the decrease in GFR (Hishida et al., 1994). This explains very well the aggravating effect of NSAIDs on aminoglycoside nephrotoxicity, since these drugs inhibit the production of the vasodilatory prostaglandin, PGE_2 (Assael et al., 1985). An increase in proximal intratubular free-flow pressure of single nephrons, most likely related to necrotic obstruction, has also been observed (Aynedjian et al., 1988), suggesting that the decline of glomerular filtration has a multifactoral origin and involves a combination of tubular and nontubular mechanisms.

A growing body of evidence supports the view that the pathogenesis of aminoglycoside toxicity is causally related to the capacity of these cationic drugs to bind to and perturb the function and structure of biologic membranes (Beauchamp and Labrecque, 2001; Mingeot-Leclercq and Tulkens, 1999). Aminoglycosides have been shown to bind to anionic (Lullman and Vollmer, 1982; Ramsammy and Kaloyanides, 1987) but not to neutral phospholipids (Kirschblaum, 1984) of the brush border membrane and are rapidly transferred into endocytic vacuoles by megalin, a transmembrane protein (Moestrup et al., 1995).

In fact, shortly after their administration, aminoglycosides can be localized within endocytic vacuoles and lysosomes of the proximal tubular cells (Beauchamp and Labrecque, 2001). Among the anionic phospholipids, aminoglycosides bind most avidly to PI-4,5-bisphosphate (PIP2) (Schacht, 1979; Au et al., 1987).

Several approaches have been used to gain insight into the molecular interaction between aminoglycosides and anionic phospholipids (Reid and Gajjar, 1987). All models indicate an electrostatic interaction between a protonated amine group and the anionic phosphate head group. It is well established that these drugs interact with and perturb the function of plasma membranes (Williams et al., 1981), lysosomes (Hostetler and Hall, 1982), mitochondria (Simmons et al., 1980), and microsomes (Buss and Piatt, 1985). It remains unclear, however, whether toxicity results from disruption of a single critical membrane function or multiple membrane functions. It is possible that the injury cascade is triggered by the rupture of lysosomes engorged with aminoglycoside antibiotic and with myeloid bodies. The resultant release of potent acid hydrolases and high concentrations of drug into the cytoplasm might cause disruption of a number of critical intracellular processes, including mitochondrial respiration (Simmons et al., 1980), microsomal protein synthesis (Buss and Piatt, 1985), intracellular signaling via the PI cascade (Ramsammy et al., 1988), as well as the generation of hydroxyl radicals (Walker and Shah; 1988,Yang et al., 1995), all of which have been observed in experimental models of aminoglycoside toxicity. Similar effects of aminoglycoside nephrotoxicity have been also observed in humans (De Broe et al., 1984).

Further insight into the pathogenesis of aminoglycoside nephrotoxicity has been derived from studies of interventions that modify the severity of this disorder in laboratory animals. It has thus been shown that polyasparagine and polyaspartic acid inhibit binding of gentamicin to rat renal brush-border membrane *in vitro*, and when injected *in vivo* confers protection against the development of aminoglycoside nephrotoxicity without inhibition of the renal cortical accumulation of drug (Williams et al., 1986). Daptomycin, a lipopeptide antibiotic containing three Asp residues, has also been shown to protect against aminoglycoside toxicity (Beauchamp et al., 1990; Couture et al., 1994).

The clinical application of polyaspartic acid needs further development, while daptomycin is not currently available. Subsequently, other compounds capable of forming electrostatic complexes with aminoglycosides have been reported to protect against nephrotoxicity (Kacew, 1989). It has been demonstrated that hydroxyl radical scavengers and iron chelators provide a marked protective effect on renal function in gentamicin-induced acute renal failure in rats (Walker et al., 1999). Antioxidants, such as vitamin E and C, selenium, probucol, and particularly deferoxamine, are effective in protecting the kidney against aminoglycoside toxicity. More recently, carvedilol, an antihypertensive agent with antioxidant activity (Kumar et al., 2000), and taurine, the major intracellular free β-amino acid (known to be an endogenous antioxidant and a membrane-stabilizing agent) (Erdem et al., 2000), were also shown to protect rats against gentamicin-induced nephrotoxicity. However, the clinical application of these interesting compounds has not been reported. The once-daily dosing of aminoglycosides is the only approach initially studied in rats and used successfully clinically to reduce aminoglycoside nephrotoxicity.

Amphotericin B

Amphotericin B is a member of the polyene macrolide class of antibiotics. The most restrictive adverse effect associated with amphotericin B therapy is its potential to induce nephrotoxicity, manifested as disturbance in both glomerular and tubular function (Deray, 2002; Fanos and Cataldi, 2000). The incidence of amphotericin B nephrotoxicity is very high and acute renal failure is common. Reported rates of acute renal failure for patients on amphotericin B are in the range 49 to 65% (Luke and Boyle, 1998; Walsh et al., 1999; Wingard et al., 1999). The clinical manifestations usually include azotemia, renal tubular acidosis, decreased concentrating ability of the kidney, and electrolyte disturbances such as urinary potassium wasting and hypokalemia, and magnesium wasting and hypomagnesemia (Sabra and Branch, 1990).

Amphotericin B is known to cause acute renal vasoconstriction and to damage preferentially the distal tubular epithelium (Deray, 2002). These alterations are responsible for the decrease in GFR and tubular dysfunction. It has been known for some years that amphotericin B, when given to animals, decreases RBF. This can happen as quickly as 45 minutes after infusion of the drug. The same effect has been reported in humans (Bell et al., 1962). The exact mechanisms mediating its nephrotoxicity have not been clearly defined. The initial event is thought to involve binding of amphotericin B to membrane sterols in the renal vasculature and renal epithelial cells to alter membrane permeability. This interaction may trigger other cellular events that result in activation of second messenger systems, release of mediators, or activation of renal homeostatic mechanisms. It is, therefore, possible that the membrane effect per se is not the sole factor that determines the extent of change in renal function. Furthermore, factors that interact with these secondary responses and mechanisms may modify the net effect of amphotericin B on renal function.

The mechanisms of the renal vasoconstriction responses to amphotericin B have not been identified. Amphotericin B has a vasoconstrictive effect on afferent renal arterioles, which reduces RBF and GFR (Sawaya et al., 1995). Theoretically, the drug can act either directly on the vascular smooth muscle or through release of secondary mediators. Neither renal denervation nor Ang II receptor blockade prevents the renal vasoconstriction or the reduction in GFR (Tolins and Raij, 1988). ET does not appear to be involved in the acute responses to amphotericin B (Heyman et al., 1992). A role for TxA_2 has been suggested, based upon partial inhibition of the amphotericin B-induced vasoconstriction and reduction in GFR by pretreatment with ibuprofen or a thromboxane receptor antagonist (Hardie et al., 1993).

It has also been suggested that activation of tubuloglomerular feedback may play a role in the acute renal effects of this compound. According to this hypothesis, the tubular toxicity of amphotericin B results in impaired reabsorption of sodium and chloride by the proximal tubule that increases distal tubular delivery of these ions, thus activating tubuloglomerular feedback (Branch, 1988). Indirect evidence in support of a role for tubuloglomerular feedback was derived from studies that demonstrated inhibition of the acute renal effects of amphotericin B by physiological and pharmacological interventions that also blocked tubuloglomerular feedback, namely salt loading and administration of furosemide, theophylline, or calcium channel blockers (Gerkens et al., 1983; Heidemann et al., 1991; Tolins and Raij, 1988). Later studies, however, argued against a role for tubuloglomerular feedback in acute amphotericin B nephrotoxicity, and suggested that the drug has a direct vasoconstrictor effect.

Other studies also suggest a protective effect of pentoxyphylline, a vascular decongestant and antagonist of TNF-α and IL-1α, against amphotericin B-induced acute and chronic nephrotoxicity, indicating a role for these factors in the renal effects of the drug (Luke et al., 1991; Wasan et al., 1990).

Nonsteroidal Anti-Inflammatory Drugs

NSAIDs have been established as valuable therapeutic agents in modern clinical practice. These drugs have become increasingly popular because of their proven effectiveness in a broad range of common clinical disorders, such as arthritis, overuse injuries such as tendonitis, orthopedic injuries, and dysmenorrhea. The increase in use has been accompanied by a growing recognition that these drugs are capable of causing an acute reduction in renal function. In fact, NSAID-related acute renal failure is reported to rank behind aminoglycoside antibiotics as a cause of nephrotoxic acute renal failure. In descending order of clinical frequency, the primary NSAID related abnormalities of renal function are: fluid and electrolyte disturbances, acute deterioration of renal function, nephrotic syndrome with interstitial nephritis, and papillary necrosis.

Sodium chloride and water retention are the most commonly encountered side effects of the use of NSAIDs (Schlondorff, 1993). This should not be considered, however, a toxicity of the drugs since it represents a modification of a physiologic control mechanism without the production of a true functional disorder within the kidney. In fact, all individuals who ingest NSAID will manifest a fractional increase in extracellular fluid, but this is of a magnitude not clinically apparent except in individuals who are otherwise prone to sodium chloride and water retention. In adults, the formation of detectable edema, related to NSAID use in the absence of obvious renal functional impairment, is typically seen in less than 5% of such individuals. The concurrent use of NSAID and diuretics, in particular loop diuretics, can lead to a drug–drug interaction that blunts the usual response of a patient to the diuretic. NSAID-induced fluid and electrolyte retention is typically benign, rapidly responds to discontinuation of the drug, and is easily managed in those who require continuing NSAID therapy. The effect on sodium excretion may be related to an aldosterone-like effect, to redistribution of renal medullary blood flow, or to a direct antinatriuretic effect of prostaglandin synthesis inhibition (Schlondorff, 1993). Other electrolyte abnormalities are also induced by NSAID and the most important of these is potassium retention with resultant hyperkalemia (Whelton and Hamilton, 1991). The inhibition of prostaglandins synthesis with NSAIDs results in a decrement in renin release (Schlondorff, 1993). This decrease in renin has led to hypoaldosteronism and hyperkalemia in susceptible patients, particularly those with preexisting renal insufficiency. Most reports of cases with NSAID-induced hyperkalemia implicate indomethacin (Goldszer et al., 1981); however, hyperkalemia remains a potential adverse effect in patients exposed to any of the NSAIDs.

From the clinical point of view, perhaps the most worrisome renal related functional abnormality is the induction of hemodynamically mediated acute renal failure that occurs in individuals with preexisting reductions in blood perfusion to their kidneys (Schlondorff, 1993). It appears that this form of acute deterioration in renal function occurs in less than 1% of individuals who ingest NSAIDs on a chronic basis (Whelton and Watson, 1998). Acute renal deterioration in this setting can be attributed to interruption of the delicate balance between hormonally mediated pressor mechanisms and prostaglandin-associated vasodilatory effects. In the "at risk" patients, which include those with congestive heart failure, cirrhosis, nephrotic syndrome, chronic renal diseases, and dehydration, volume contraction triggers pressor responses via adrenergic and renin–angiotensin pathways. Typically, vasodilatory renal prostaglandins counterbalance the vasoconstrictive effects of norepinephrine and Ang II. The addition of NSAIDs increases the risk of hemodynamically mediated ischemic damage to the kidney by removing the protective effects of vasodilatory prostaglandins and allowing unopposed vasoconstriction.

The greatest number of reported cases of renal insufficiency has occurred with indomethacin use (Schlondorff, 1993), which probably is partly related to the length of time that the drug has been available for use compared with newer preparations. Acute renal failure has, however, been associated with other drugs in the carbo-hetero-cyclic acetic acid class (for example, zomepirac) (Fellner and Arraha, 1981), as well as with drugs in other chemical derivative classes of NSAID, such as propionic acid derivatives (ibuprofen and fenoprofen) (Fong and Cohen, 1982), and the fenamic acid derivative (meclofenamic acid) (Husserl et al., 1979). Moreover, salicylic acid (aspirin) and the enolic acid derivative phenylbutazone are capable of inducing renal ischemic lesions, particularly in patients with preexisting renal disease (Kimberly et al., 1978). Some reports have suggested that sulindac may not induce the renal insufficiency caused by other NSAIDs (Brunning and Barth, 1982), but other communications have questioned the "renal-sparing" effect of this drug (Roberts et al., 1984).

Essentially all NSAIDs can cause another type of renal dysfunction associated with various levels of functional impairment and characterized by the nephrotic syndrome together with interstitial nephritis

(Bender et al., 1984). The features of this NSAID-induced renal syndrome are somewhat variable. The patient may experience edema, oliguria, and clinical signs indicative of significant proteinuria (Levin, 1988). Systemic signs of allergic interstitial nephritis, such as fever, drug rash, peripheral eosinophilia, and eosinophiluria, are typically absent. The urine sediment contains microscopic hematuria and cellular elements reported as pyuria (Levin, 1988). Proteinuria is usually in the nephrotic range. The functional extent of renal deterioration can range from minimal to severe.

Characteristically, the histology of this form of NSAID-induced nephrotic syndrome consists of minimal change glomerulonephritis with tubulointerstitial nephritis. This is an unusual combination of findings and, when noted in the clinical setting of protracted NSAID use, is virtually pathognomic of NSAID-related nephrotic syndrome.

While the pathogenesis of the lesions is not known, it has been postulated that prostaglandin inhibition may lead to a cycle of immunologic events that culminate in enhanced lymphokine production, thereby allowing a delayed hypersensitivity reaction to proceed unchecked (Torres, 1982). Alternatively, a direct glomerular or tubular toxicity may be operating in some cases. An intriguing possible cause of the syndrome is the notion that by inhibiting cyclooxygenase, arachidonic-acid products are stimulated via the lipooxygenase pathways to form polyenoic acids, including leukotrienes. Thus, the production of a number of synthetic products with known vascular effects may actually be increased by some NSAIDs, at least locally. Finally, the possibility that a delayed hypersensitivity reaction may be present in some cases also has been postulated.

The onset of NSAID-induced nephrotic syndrome is usually delayed, having a mean time of onset of 5.4 months after initiation of NSAID therapy and ranging from 2 weeks to 18 months (Abraham and Keane, 1984). NSAID-induced nephritic syndrome is usually reversible between one month and one year after discontinuation of NSAID therapy. During the recovery period, some patients may require dialysis.

It appears that NSAIDs and analgesic drugs can cause papillary necrosis as an acute or chronic event. The chronic progression of events that lead to NSAID-related papillary necrosis have been well known since the days of the first descriptions of chronic phenacetin abuse nephropathy and the subsequent extensive investigations that defined the consequences of chronic (5 to 20 years) exposure of the kidney to high doses of analgesic combinations, such as salicylate and acetaminophen (the metabolite of phenacetin), often with the addition of caffeine (Kincaid-Smith, 1986). Fortunately, the incidence of this form of chronic analgesic abuse nephropathy has diminished because of a better understanding of the drugs involved, patient education, and, in some countries, efficient regulatory measures.

The mechanism of NSAID-induced acute papillary necrosis is often not clear and the causative role of the NSAID in question may be difficult to delineate because of the presence of confounding factors such as underlying disease, urinary tract infection, and concomitant medications. Selected NSAIDs may exert a direct toxic effect on renal papillae and may become highly concentrated in the medullary and papillary regions of the kidney. Aspirin depletes cellular glutathione, which would otherwise detoxify the acetaminophen metabolite, *N*-acetyl-benzo-quinonemine. Without glutathione, this highly reactive metabolite could lead to cell death (Kincaid-Smith, 1986). Prostaglandin inhibition may also play a role (Clive and Stoff, 1984). Medullary ischemia, a possible precipitating factor in development of papillary necrosis, results from NSAID-induced reduction of blood into the renal medulla in experimental models (Kirschenbaum et al., 1974; Stein and Fadem, 1978). Although clinical toxicity is exceedingly rare it has been reported for ibuprofen (Shah et al., 1981), mefenamic acid (Roberston et al., 1980), and according to prescribing information, several other NSAIDs.

Because NSAIDs inhibit the general activity of cyclooxygenase, this action, in large part, accounts for both their therapeutic effectiveness and their toxic side effects. Given the propensity of NSAIDs to cause renal side effects, there is keen interest in the renal safety of COX-2 specific drugs. The idea that COX-isoform functions are mutually exclusive, with COX-1 involved in maintenance of normal renal physiology and COX-2 rapidly induced in response to inflammation, inspired hopes of exciting new therapeutic possibilities. However studies with two selective COX-2 inhibitors, celecoxib and rofeoxib, in healthy elderly subjects have shown moderate and transient sodium retention that reiterated the effect seen with naproxen and indomethacin (Khan et al., 2002). However, neither COX-2 specific drug affected glomerular filtration, in contrast with the decreases seen with both nonselective NSAIDs. Thus, it is concluded that COX-2-specific inhibitors produce some disturbance of sodium excretion that is characteristic of conventional NSAIDs, but unlike NSAIDs, they spare glomerular filtration. However, two cases of acute renal failure associated with selective COX-2 inhibitor therapy in elderly subjects have been reported (Morales and Mucksavage, 2002; Papaioannides et al., 2001), which dampen the enthusiasm for these drugs. Similarly, a first case of allergic interstitial nephritis associated with nephritic syndrome occurred in a patient treated with the selective COX-2 inhibitor, celecoxib (Alper et al., 2002).

Renal papillary necrosis was not observed in preclinical toxicology studies with celecoxib in sentinel species (Khan et al., 2002). These data, together with findings that COX-1 is the principal COX isoform in human papilla, have suggested that there may be little risk for renal papillary necrosis with COX-2 selective drugs in humans. However, a report of a case of celecoxib-related renal papillary necrosis (Akhund et al., 2003) should alert physicians on the possibility that this complication could occur with COX-2 specific inhibitors as well.

RADIOCONTRAST MEDIA

The administration of radiographic contrast agents remains an important cause of hospital-acquired acute renal failure, which contributes to morbidity and mortality during hospitalization, prolongs hospital stay, and increases the incidence of chronic end-stage renal disease and costs of health care (Gruberg et al., 2000; McCullough et al., 1997). The incidence of nephropathy associated with contrast media depends on the patient cohort studied. Indeed, chronic renal insufficiency, diabetes mellitus, contrast media volume, and recurrent administration are considered important risk factors (Rudnick et al., 1996; Erley, 1999; Mueller et al., 2002). Kidney damage induced by radiocontrast agents includes a hemodynamic response and a self-standing tubulotoxicity (Erley, 1999).

Vascular ischemia has particular appeal as a major contributor to contrast-associated nephropathy since acute changes in renal hemodynamics have been consistently reported using a variety of experimental protocols (Lund et al., 1984; Talner and Davidson, 1968). The renal circulatory response is characterized by initial transient vasodilation followed by prolonged vasoconstriction (Larson et al., 1983; Lund et al., 1984; Talner and Davidson, 1968), although this effect is a feature that is restricted to high osmolar contrast agents (El Sayed et al., 1991).

Many mediators have been suggested to explain the vasoconstriction, including calcium, adenosine, reactive oxygen species, renin–angiotensin system, and renal PGs and ET. In addition, direct tubular toxicity from the iodinated contrast media has also been documented (Humes et al., 1987; Ziegler et al., 1975), which undoubtedly contributes to contrast-associated nephropathy. Activation of tubuloglomerular feedback by the excessive osmotic load delivered to the macula densa during the initial circulation of contrast media should trigger renin release and hence a reduction in GFR. The reflex renin release from the juxtaglomerular apparatus requires the integrity of the tubuloglomerular feedback arc (Navar et al., 1986).

The role of the renin–angiotensin system in contrast associated nephropathy has been evaluated in the salt-restricted dog model (Larson et al., 1983). By stimulating the renin–angiotensin system with salt depletion, an exaggerated vascular response to intra-arterially injected contrast media occurs during both the vasodilatory and vasoconstriction phase. Saralasin pretreatment diminished, but did not eliminate, the vasoconstrictor response, while sodium repletion blunted both phases of the renal hemodynamic effects of contrast media (Larson et al., 1983). Other investigators, however, have been unable to confirm any contribution of the renin–angiotensin system to the contrast induced hemodynamic changes (Katzberg et al., 1977).

Superoxide radicals are a favorite mechanism for explaining renal cell injury (Baud and Ardaillou, 1986). Bakris (1993) reported measurable increases in the production of reactive oxygen species when human mesangial cells were exposed to diatrizoate sodium. Furthermore, in a salt-depleted animal model the same investigators (Bakris et al., 1990) found that the oxygen free radical scavenger superoxide dismutase prevents the contrast-induced fall in GFR without affecting the changes in RBF. Renal biopsies performed three hours after contrast injection confirmed an influx of polymorphonuclear leukocytes and macrophages, cells that release oxygen free radicals.

Confounding this interpretation are the findings of Yoshioka et al. (1992) that volume-depleted rats have a reduced cell content of superoxide dismutase compared with their euvolemic mates. Since volume depletion is a well-known risk factor for contrast-associated nephropathy, depletion of superoxide dismutase may play a role in increasing the risk of cell damage in clinical situations.

Contrast media could also induce renal hemodynamic changes secondary to preferential action of endogenous vasoconstrictor substances on the arteriolar network of the glomerulus (Schor and Brenner, 1980). Norepinephrine-induced vasoconstriction involves both afferent and efferent arterioles, while the vasoconstrictor effect of Ang II is limited to the efferent arteriole (Remuzzi et al., 1990). Renal prostaglandins counteract this vasoconstrictive action and restore the hemodynamic balance. When the intrinsic vasodilating effects of PGs are eliminated through cyclooxygenase inhibition, glomerular filtration pressure is maintained while post-glomerular blood flow is reduced.

Based on experimental models, the most adverse impact of contrast media on renal function occurs when both sets of endogenous vasomotor mediators are stimulated. Heyman et al. (1988) indicated that the thick ascending limb of Henle is the segment most vulnerable to oxygen deprivation. Whether contrast media induce reversible renal dysfunction or histologic tubular necrosis is probably the result of the combined effects of the intensity of renal vasoconstriction and the magnitude of the oxygen deprivation. This may help to explain the clinical observation that the presentation of contrast-associated nephropathy can vary from mild, often unrecognized, dysfunction to acute oliguric renal failure requiring dialysis support for survival. Experimental evidence supporting a role for ET as a mediator of contrast-induced renal vasoconstriction also continues to accumulate (Oldroyd et al., 1994) with advent of ET receptor antagonists (Oldroyd et al., 1995). However, a multicenter double-blind randomized trial in high-risk patients undergoing coronary angiography has shown that intravenous administration of the ET_A/ET_B receptor antagonist bosentan does not prevent the nephrotoxicity that follows radiocontrast administration, questioning the validity of ET-receptor antagonism prophylaxis in these patients (Wang et al., 1998a).

Beside renal hemodynamic changes, direct tubular toxicity of contrast media, as well as intraluminal obstruction secondary to co-precipitation of contrast media with various urinary proteins, may also play a role in the induction of contrast nephropathy. The intrinsic mechanisms involved, however, remain ill-defined.

REFERENCES

Abassi ZA, Klein H, Golomb E, and Keiser HR (1993) Urinary endothelin: A possible biological marker of renal damage. *Am J Hypertens* 6:1046–1054.

Abildgaard U, Holstein-Rathlou N, and Leyssac P (1986) Effect of renal nerve activity on tubular sodium and water reabsorption in dog kidneys as determined by the lithium clearance method. *Acta Physiol Scand* 126:251–257.

Abraham PA, and Keane WF (1984) Glomerular and interstitial disease induced by nonsteroidal anti-inflammatory drugs. *Am J Nephrol* 4:1–6.

Agarwal R (2001) Add-on angiotensin receptor blockade with maximized ACE inhibition. *Kidney Int* 59:2282–2289.

Aiello S, Noris M, Todeschini M, Zappella S, Foglieni C, Benigni A, Corna D, Zoja C, Cavallotti D, and Remuzzi G (1997) Renal and systemic nitric oxide synthesis in rats with renal mass reduction. *Kidney Int* 52:171–181.

Akai Y et al., (1992) Mechanical stretch relaxation stimulates release of arachidonic acid and induces transient expression of the mitogen sensitive PGH synthase (PGHS-2) gene in cultured rat mesangial cells. *J Am Soc Nephrol* 3:450 (Abstract).

Akhund L, Quinet RJ, and Ishaq S (2003) Celecoxib-related renal papillary necrosis. *Arch Intern Med* 163:114–115.

Alavi N, Lianos EA, and Bentzel CJ (1987) Prostaglandin and thromboxane synthesis by highly enriched rabbit proximal tubular cells in culture. *J Lab Clin Med* 110:338–345.

Alessandrini P, Salvati P, Pugliese F, Ciabattoni G, and Patrono C (1990) Inhibition by FCE22178 of platelet and glomerular thromboxane synthase in animal and human kidney disease. *Adv Prostaglandin Thromboxane Leukot Res* 21:707.

Ali MZ, and Goetz MB (1997) A meta-analysis of the relative efficacy and toxicity of single daily dosing versus multiple daily dosing of aminoglycosides. *Clin Infect Dis* 24:796–809.

Alper AB, Meleg-Smith S, and Krane NK (2002) Nephrotic syndrome and interstitial nephritis associated with celecoxib. *Am J Kidney Dis* 40:1086–1090.

Anderson RJ, Linas SL, Berns AS, Henrich WL, Miller TR, Gabow PA, and Schrier RW (1977) Nonoliguric acute renal failure. *N Engl J Med* 296:1134–1138.

Anderson S, Diamond JR, Karnovsky MJ, and Brenner BM (1988) Mechanisms underlying transition from acute glomerular injury to late glomerular sclerosis in a rat model of nephrotic syndrome. *J Clin Invest* 82:1757–1768.

Anderson S, Meyer TW, Rennke HG, and Brenner BM (1985) Control of glomerular hypertension limits glomerular injury in rats with reduced renal mass. *J Clin Invest* 76:612–619.

Anderson S, Rennke HG, and Brenner BM (1986) Therapeutic advantage of converting enzyme inhibitors in arresting progressive renal disease associated with systemic hypertension in the rat. *J Clin Invest* 77:1993–2000.

Andoh TF, Gardner MP, and Bennett WM (1997) Protective effects of dietary L-arginine supplementation on chronic cyclosporine nephrotoxicity. *Transplantation* 64:1236–1240.

Andoth TF, and Bennett WM (1998) Chronic cyclosporine nephrotoxicity. *Curr Opin Nephrol Hypertens* 7:265–270.

Andres A, Morales JM, Praga M, Campo C, Lahera V, Garcia-Robles R, Rodicio JL, and Ruilope LM (1997) L-arginine reverses the antinatriuretic effect of cyclosporin in renal transplant patients. *Nephrol Dial Transplant* 12:1437–1440.

Assael BM, Chiabrando C, Gagliardi L, Noseda A, Bamonte F, and Salmona M (1985) Prostaglandins and aminoglycoside nephrotoxicity. *Toxicol Appl Pharmacol* 78:386–394.

Au S, Weiner ND, and Schacht J (1987) Aminoglycosides preferentially increase permeability of phosphoinositide-containing membranes: A study with carboxyfluorescein in liposomes. *Biochim Biophys Acta* 902:80–86.

Ausiello DA, Kreisberg JI, Roy C, and Karnovsky MJ (1980) Contraction of cultured rat glomerular cells of apparent mesangial origin after stimulation with angiotensin II, and arginine vasopressin. *J Clin Invest* 65:754–760.

Awazu M, Sugiura M, Inagami T, Ichikawa I, and Kon V (1991) Cyclosporine promotes glomerular endothelial binding *in vivo*. *J Am Soc Nephrol* 1:1253–1258.

Aynedjian HS, Nguyen D, Lee HY, Sablay LB, and Bank N (1988) Effects of dietary electrolyte supplementation on gentamicin nephrotoxicity. *Am J Med Sci* 295:444–452.

Badr KF (1992a) 15-lipoxygenase products as leukotriene antagonists: Therapeutic potential in glomerulonephritis. *Kidney Int Suppl* 38:S101–S108.

Badr KF (1992b) Five-lipoxygenase products in glomerular immune injury. *J Am Soc Nephrol* 3:907–915.

Badr KF, Baylis C, Pfeffer JM, Pfeffer MA, Soberman RJ, Lewis RA, Austen KF, Corey EJ, and Brenner BM (1984) Renal and systemic hemodynamic responses to intravenous infusion of leukotriene C4 in the rat. *Circ Res* 54:492–499.

Badr KF, Brenner BM, and Ichikawa I (1987) Effects of leukotriene D4 on glomerular dynamics in the rat. *Am J Physiol* 253:F239–F243.

Badr KF, Murray JJ, Breyer MD, Takahashi K, Inagami T, and Harris RC (1989b) Mesangial cell, glomerular, and renal vascular response to endothelin in the kidney. *J Clin Invest* 83:336–342.

Badr KF, Schreiner GF, Wasserman M, and Ichikawa I (1988) Preservation of the glomerular capillary ultrafiltration coefficient during rat nephrotoxic serum nephritis by a specific leukotriene D4 receptor antagonist. *J Clin Invest* 81:1702–1709.

Bailey TC, Little JR, Littenberg B, Reichley RM, and Dunagan WC (1997) A meta-analysis of extended-interval dosing versus multiple daily dosing of aminoglycosides. *Clin Infect Dis* 24:786–795.

Bakris GL (1993) Pathogenesis and therapeutic aspects of radiocontrast-induced renal dysfunction, in *Toxicology of the kidney*, Hook JB, and Goldstein RS (Eds), Raven Press Ltd, New York., p. 361.

Bakris GL, Lass NA, Gaber CA, Jones JD, and Burnett JC Jr (1990) Radiocontrast medium-induced declines in renal function: a role for oxygen free radicals. *Am J Physiol* 258:F115–F120.

Barnett R, Goldwasser P, Scharschmidt LA, and Schlondorff D (1986) Effects of leukotrienes on isolated rat glomeruli and cultured mesangial cells. *Am J Physiol* 250:F838–F844.

Barnett TL, Ruffini L, Hart D, and Nord EP (1994) Mechanism of endothelin activation of phospholipase A2 in rat renal medullary interstitial cells. *Am J Physiol* 266:F46–F56.

Barza M, Ioannidis JP, Cappelleri JC, and Lau J (1996) Single or multiple daily doses of aminoglycosides: a meta-analysis. *Brit Med J* 312:338–345.

Baud L, and Ardaillou R (1986) Reactive oxygen species: Production and role in the kidney. *Am J Physiol* 251:F765–F776.

Baylis C, Harton P, and Engels K (1991) Endothelial-derived relaxing factor control renal hemodynamics in the normal rat kidney. *J Am Soc Nephrol* 1:875–881.

Baylis C, Mitruka B, and Deng A (1992) Chronic blockade of nitric oxide synthesis in the rat produces systemic hypertension and glomerular damage. *J Clin Invest* 90:278–281.

Beauchamp D, and Labrecque G (2001) Aminoglycoside nephrotoxicity: do time and frequency of administration matter? *Curr Opin Crit Care* 7:401–408.

Beauchamp D, Pellerin M, Gourde P, Pettigrew M, and Bergeron MG (1990) Effects of daptomycin and vancomycin on tobramycin nephrotoxicity in rats. *Antimicrob Agents Chemother* 34:139–147.

Bech JN, Nielsen CB, Ivarsen P, Jensen KT, and Pedersen EB (1998) Dietary sodium affects systemic and renal hemodynamic response to NO inhibition in healthy humans. *Am J Physiol* 274:F914–F923.

Beck NP, DeRubertis FR, Michelis MF, Fusco RO, Fleld JB, and Davis BB (1972) Effect of prostaglandin E1 on certain renal actions of parathyroid hormone. *J Clin Invest* 51:2352 (Abstract).

Beck PR, Thomson RB, and Chaudhuri AKR (1977) Aminoglycoside antibiotics and renal function: Changes in urinary gamma-glutamyltransferase excretion. *J Clin Pathol* 30:432–437.

Beierwaltes WH, Schryver S, Sanders E, Strand J, and Romero JC (1982) Renin release selectively stimulated by prostaglandin I2 in isolated rat glomeruli. *Am J Physiol* 243:F276–F283.

Bell NH, Andriole VT, Sabesin SM, and Utz JP (1962) On the nephrotoxicity of amphotericin B in man. *Am J Med* 33:64–69.

Bender WL, Whelton A, Beschorner WE, Darwish MO, Hall-Craggs M, and Solez K (1984) Interstitial nephritis, proteinuria, and renal failure caused by nonsteroidal anti-inflammatory drugs: Immunologic characterization of the inflammatory infiltrate. *Am J Med* 76:1006–1012.

Benigni A, and Remuzzi G (1997) The renoprotective potential of endothelin receptor antagonists. *Exp Opin Ther Patents* 7:139.

Benigni A, Bruzzi I, Mister M, Azzollini N, Gaspari F, Perico N, Gotti E, Bertani T, and Remuzzi G (1999) Nature and mediators of renal lesions in kidney transplant patients given cyclosporine for more than one year. *Kidney Int* 55:674–685.

Benigni A, Colosio V, Brena C, Bruzzi I, Bertani T, and Remuzzi G (1998) Unselective inhibitor of endothelin receptors reduces renal dysfunction in experimental diabetes. Diabetes 47:450–456.

Benigni A, Perico N, and Remuzzi G (2001a) Research on renal endothelin in proteinuric nephropathies dictates novel strategies to prevent progression. *Curr Opin Nephrol Hypertens* 10:1–6.

Benigni A, Perico N, Gaspari F, and Remuzzi G (1991a) Increased renal endothelin production in rats with reduced renal mass. *Am J Physiol* 260:F331–F339.

Benigni A, Perico N, Ladny JR, Imberti O, Bellizzi L, and Remuzzi G (1991b) Increased urinary excretion of endothelin-1 and its precursor big endothelin-1 in rats chronically treated with cyclosporine. *Transplantation* 52:175–177.

Benigni A, Tomasoni S, Gagliardini E, Zoja C, Grunkemeyer JA, Kalluri R, and Remuzzi G (2001b) Blocking angiotensin II synthesis/activity preserves glomerular nephrin in rats with severe nephrosis. *J Am Soc Nephrol* 12:941–948.

Benigni A, Zoja C, and Remuzzi G (1995) The renal toxicity of sustained glomerular protein traffic. *Lab Invest* 73:461–468.

Benigni A, Zoja C, Corna D, and Remuzzi G (1993) A specific endothelin subtype A receptor antagonist protects against injury in renal disease progression. *Kidney Int* 44:440–444.

Benigni A, Zoja C, Corna D, and Remuzzi G (1996) Blocking both type A, and B endothelin receptors in the kidney attenuates renal injury and prolongs survival in rats with remnant kidney. *Am J Kidney Dis* 27:416–423.

Bertani T, Cutillo F, Zoja C, Broggini M, and Remuzzi G (1986) Tubulointerstitial lesions mediate renal damage in adriamycin glomerulopathy. *Kidney Int* 30:488–496.

Binet I, Wallnofer A, Weber C, Jones R, and Thiel G (2000) Renal hemodynamics and pharmacokinetics of bosentan with and without cyclosporine A. *Kidney Int* 57:224–231.

Bjorck S, Mulec H, Johnsen SA, Norden G, and Aurel M (1992) Renal protective effect of enalapril in diabetic nephropathy. *Br Med J* 304:339–343.

Blantz R, and Munger K (2002) Role of nitric oxide in inflammatory conditions. *Nephron* 90:373–378.

Bloom IT, Bentley FR, Spain DA, and Garrison RN (1995) An experimental study of altered nitric oxide metabolism as a mechanism of cyclosporin-induced renal vasoconstriction. *Br J Surg* 82:195–198.

Bobadilla NA, Tapia E, Franco M, Lopez P, Mendoza S, Garcia-Torres R, Alvarado JA, and Herrera-Acosta J (1994) Role of nitric oxide in renal hemodynamic abnormalities of cyclosporin nephrotoxicity. *Kidney Int* 46:773–779.

Boehm M, and Nabel e.g., (2002) Angiotensin-converting enzyme 2 – A new cardiac regulator. *N Engl J Med* 347:1795–1797.

Bracht C, Yan XW, LaRocca HP, Sutsch G, and Kiowski W (1999) Cyclosporine A, and control of vascular tone in the human forearm: influence of post-transplant hypertension. *J Hypertens* 17:357–363.

Branch RA (1988) Prevention of amphotericin B-induced renal impairment: A review on the use of sodim supplementation. Arch Intern Med 148:2389–2394.

Bredt DS, Hwang PM, and Glatt CE (1991) Cloned and expressed nitric oxide synthase structurally resembles cytochrome P-450 reductase. *Nature* 351:714–718.

Brenner BM, Cooper ME, de Zeeuw D, Keane WF, Mitch WE, Parving HH, Remuzzi G, Snapinn SM, Zhang Z, Shahinfar S; RENAAL Study Investigators (2001) Effects of losartan on renal and cardiovascular outcomes in patients with type 2 diabetes and nephropathy. *N Engl J Med* 345:861–869.

Breyer JA (1995) Medical management of nephropathy in type I diabetes mellitus: Current recommendations. *J Am Soc Nephrol* 6:1523–1529.

Brezis M, Heyman SN, Dinour D, Epstein FH, and Rosen S (1991) Role of nitric oxide in renal medullary oxygenation. Studies in isolated and intact rat kidneys. *J Clin Invest* 88:390–395.

Brown CA, Zusman RM, and Haber E (1980) Identification of an angiotensin receptor in rabbit renomedullary interstitial cells in culture. Correlation with prostaglandin biosynthesis. *Circ Res* 46:802–807.

Brunner HR (1992) ACE inhibitors in renal diseases. *Kidney Int* 42:463–479.

Buchan KW, Alldus C, Christodoulou C, Clark KL, Dykes CW, Sumner MJ, Wallace DM, White DG, and Watts IS (1994) Characterization of three nonpeptide endothelin receptor ligands using human cloned ETA, and ETB receptors. *Br J Pharmacol* 112:1251–1257.

Bunchman TE, and Brookshire CA (1991) Cyclosporine-induced synthesis of endothelin by cultured human endothelial cells. *J Clin Invest* 88:310–314.

Bunning RD, and Barth WF (1982) Sulindac: A potentially renal-sparing nonsteroidal anti-inflammatory drug. *JAMA* 248:2864–2867.

Burch RM, and Halushka PV (1980) Thromboxane and stable prostaglandin endoperoxide analogs stimulate water permeability in the toad urinary bladder. *J Clin Invest* 66:1251. (Abstract)

Burnier M, and Brunner HR (1994) Angiotensin II receptor antagonists and the kidney. *Curr Opin Nephrol Hypertens* 3:537–545.

Burns KD, Homma T, and Harris RC (1993) The intrarenal renin-angiotensin system. *Semin Nephrol* 13:13–30.

Buss UC, and Piatt MK (1985) Gentamicin administered *in vivo* reduces protein synthesis in microsomes subsequently isolated from rat kidneys but not from rat brains. *J Antimicrob Chemother* 15:715–721.

Capelo LR, and Alzamora F (1977) A stop-flow analysis of the effects of intrarenal infusion of bradykinin. *Arch Int Pharmacodyn Ther* 230:156–165.

Carey RM, Wang Z-Q, and Siragy HM (2000) Role of the angiotensin type 2 receptor in the regulation of blood pressure and renal function. *Hypertension* 35:155–163.

Carretero DA, Cabrini LA, and Scicli AG (1993) The molecular biology of the kallikrein-kinin system: I. General description, nomenclature and the mouse gene family. *J Hypertens* 7:693.

Cattel V, Cook T, and Moncada S (1990) Glomeruli synthesize nitrite in experimental nephrotoxic nephritis. *Kidney Int* 38:1056–1060.

Cattel V, Largen P, de Heer E, and Cook T (1991) Glomeruli synthesize nitrite in active Heymann nephritis: the source is infiltrating macrophages. *Kidney Int* 40:847–851.

Cattel V, Lianos E, Largen P, and Cook T (1993) Glomerular NO synthase activity in mesangial cell immune injury. *Exp Nephrol* 1:36–40.

Cavarape A, Endlich K, Feletto F, Parekh N, Bartoli E, and Steinhausen M (1998) Contribution of endothelin receptors in renal microvessels in acute cyclosporine-mediated vasoconstriction in rats. *Kidney Int* 53:963–939.

Chaudhari A, Gupta S, and Kirschenbaum MA (1990) Biochemical evidence for PG12 and PGE2 receptors in the rabbit renal preglomerular microvasculature. *Biochim Biophys Acta* 1053:156–161.

Chen Y, and Rosazza JP (1996) Oligopeptides as substrates and inhibitors for a new constitutive nitric oxide synthase from rat cerebellum. *Biochem Biophys Res Commun* 224:303–308.

Cheng HF, Wang CJ, Moeckel GW, Zhang MZ, McKanna JA, and Harris RC (2002) Cyclooxygenase-2 inhibitor blocks expression of mediators of renal injury in a model of diabetes and hypertension. *Kidney Int* 62:929–939.

Cho HJ, Xie QW, and Calaycay J (1992) Calmodulin is a subunit of nitric oxide synthase from macrophages. *J Exp Med* 176:599–604.

Clive DM, and Stoff JS (1984) Renal syndromes associated with nosteroidal anti-inflammatory drugs. *N Engl J Med* 310:563–572.

Coleman RA, and Humphrey PPA (1993) Prostanoid receptors: their function and classification, in *Therapeutic Applications of Prostaglandin*, Vane JR, and O'Grady J (Eds), Edward Arnold, London.

Conger JD, Robinette JB, and Schrier RW (1988) Smooth muscle calcium and endothelium-derived relaxing factor in the abnormal vascular responses of acute renal failure. *J Clin Invest* 82:532–537.

Cook HT, and Cattel V (1995) Nitric oxide synthetic pathway and regulation: its potential role in immune and inflammatory reactions in the glomerulus. *J Nephrol* 8:247.

Couture M, Simard M, Gourde P, Lessard C, Gurnani K, Lin L, Carrier D, Bergeron MG, and Beauchamp D (1994) Daptomycin may attenuate experimental tobramycin nephrotoxicity by electrostatic complexation to tobramycin. *Antimicrob Agents Chemother* 38:742–749.

Coyne DW, and Morrison AR (1991) Kinins: Biotransformation and cellular mechanisms of action. In: Contemporary Issues, in *Nephrology: Hormones,*

Autacoids, and the Kidney, Goldfarb S, and Ziyadeh FN (Eds), Churchill Livingstone, New York.

Crackower MA, Sarao R, Oudit GY, Yagil C, Kozieradzki I, Scanga SE, Oliveira-dos-Santos AJ, da Costa J, Zhang L, Pei Y, Scholey J, Ferrario CM, Manoukian AS, Chappell MC, Backx PH, Yagil Y, and Penninger JM (2002) Angiotensin-converting enzyme 2 is an essential regulator of heart function. *Nature* 417:822–828.

Curtis JJ, Luke RG, and Dubovsky E (1986) Cyclosporin in therapeutic doses increases renal allograft vascular resistance. *Lancet* 2:477–479.

Cybulsky AV, Goodyer PR, Cyr MD, and McTavish AJ (1992) Eicosanoids enhance epidermal growth factor receptor activation and proliferation in glomerular epithelial cells. *Am J Physiol* 262:F639–F646.

Davis LS, Haleen SJ, Doherty AM, Cody WL, and Keiser JA (1994) Effects of selective endothelin antagonists on the hemodynamic response to cyclosporin A. *J Am Soc Nephrol* 4:1448–1454.

De Broe ME, Paulus GJ, Verpooten GA, Roels F, Buyssens N, Wedeen R, Van Hoof F, and Tulkens PM (1984) Early effects of gentamicin, tobramycin, and amikacin on the human kidney. Kidney Int 25:643–652.

De Nicola L, Blantz RC, and Gabbai FB (1992) Nitric oxide and angiotensin II glomerular and tubular interaction in the rat. *J Clin Invest* 89:1248–1256.

De Nicola L, Thomsom SC, Wead LM, Brown MR, and Gabbai FB (1993) Arginine feeding modifies cyclosporine nephrotoxicity in rats. *J Clin Invest* 92:1859–1865.

Deng A, and Baylis C (1993) Locally produced EDRF controls preglomerular resistance and ultrafiltration coefficient. *Am J Physiol* 264:F212–F215.

Deng X, Welch WJ, and Wilcox CS (1993) Nitric oxide confers selective renal vasodilation during dietary salt loading. *J Am Soc Nephrol* 4:547 (Abstract).

Deray G (2002) Amphotericin B nephrotoxicity. *J Antimicrob Chemother* 49 (Suppl) S1:37–41.

DeRubertis FR, and Craven PA (1992) Contribution of platelet thromboxane production to enhanced urinary excretion and glomerular production of thromboxane and to the pathogenesis of albuminuria in the streptozotocin-diabetic rat. *Metabolism* 41:90–96.

DeRubertis FR, and Craven PA (1993) Eicosanoids in the pathogenesis of the functional and structural alterations of the kidney in diabetes. *Am J Kidney Dis* 22:727–735.

Diederich D, Yang Z, and Luscher TF (1992) Chronic cyclosporine therapy impairs endothelium-dependent relaxation in the renal artery of the rat. *J Am Soc Nephrol* 2:1291–1297.

Dieperink H, Leyssac P, Starklint H, Jorgensen K, and Kemp E (1986) Antagonist capacities of nifedipine, captopril, phenoxybenzamine, prostacyclin and indomethacin on cyclosporin A induced impairment of rat renal function. *Eur J Clin Invest* 16:540–548.

Dieperink H, Perico N, Nielsen FT, and Remuzzi G (1998) Cyclosporine/tacrolimus (FK-506), in *Clinical Nephrotoxins. Renal injury from drugs and chemicals*, De Broe ME, Porter GA, Bennett WM, and Verpooten GA (Eds), Kluwer Academic Publishers, Boston, pp. 275.

Dieperink H, Starklint H, and Leyssac PP (1983) Nephrotoxicity of cyclosporine – an animal model: Study of the nephrotoxic effect of cyclosporine on overall and renal tubular function in conscious rats. *Transplant Proc* 15:2736.

Dijkhorst-Oei LT, and Koomans HA (1998) Effects of a nitric oxide synthesis inhibitors on renal sodium handling and diluting capacity in humans. *Nephrol Dial Transplant* 13:587–593.

Donoghue M, Hsieh F, Baronas E, Godbout K, Gosselin M, Stagliano N, Donovan M, Woolf B, Robison K, Jeyaseelan R, Breitbart RE, and Acton S (2000) A novel angiotensin-converting enzyme-related carboxypeptidase (ACE2) converts angiotensin I to angiotensin 1–9. *Circ Res* 87:E1–9.

Dzau VJ, Burt DW, and Pratt RE (1988) Molecular biology of the renin-angiotensin system. *Am J Physiol* 255:F563–F573.

Eddy AA, McCulloch L, Liu E, and Adams J (1991) A relationship between proteinuria and acute tubulointerstitial disease in rats with experimental nephrotic syndrome. *Am J Pathol* 138:1111–1123.

Edwards RM, and Aiyar N (1993) Angiotensin II receptor subtypes in the kidney. *J Am Soc Nephrol* 3:1643–1653.

Ehrenreich H, Anderson RW, and Fox CH (1990) Endothelins, peptides with potent vasoactive properties, are produced by human macrophages. *J Exp Med* 172:1741–1748.

El Sayed AA, Haylor J, El Nahas M, Salzano S, and Morcos SK (1991) Haemodynamic effects of water-soluble contrast media on the isolated perfused rat kidney. *Brit J Radiol* 64:435–439.

Elzinga L, Kelley VE, Houghton DC, Barrett LV, Bennett WM, and Strom TB (1987) Modification of experimental nephrotoxicity with fish oil as the vehicle of cyclosporine. *Transplantation* 43:271–274.

Elzinga L, Rosen S, and Bennett WM (1993) Dissociation of glomerular filtration rate from tubulointerstitial fibrosis in experimental chronic cyclosporine nephropathy: Role of sodium intake. *J Am Soc Nephrol* 4:214–221.

Erdem A, Gundogan NU, Usubutun A, Kilinc K, Erdem SR, Kara A, and Bozkurt A (2000) The protective effect of taurine against gentamicin-induced acute tubular necrosis in rats. Nephrol Dial Transplant15:1175–1182.

Erley CM (1999) Nephrotoxicity: focusing on radiocontrast nephropathy. *Nephrol Dial Transplant* 14 (Suppl. 4):13–15.

Fanos V, and Cataldi L (2000) Amphotericin B-induced nephrotoxicity: A review. *J Chemother* 12:463–470.

Fellner SK, and Arraha HB (1981) Wolfe C. Acute renal failure and zomepirac. *Arch Intern Med* 141:1846.

Ferriols-Lisart R, and Alos-Aminana M (1996) Effectiveness and safety of once-daily aminoglycosides: A meta-analysis. *Am J Health Syst Pharm* 53:1141–1150.

Firth JD, and Ratcliffe PJ (1992) Organ distribution of the three rat endothelin messenger RNAs and the effects of ischemia on renal gene expression. *J Clin Invest* 90:1023–1031.

Fisman DN, and Kaye KM (2000) Once-daily dosing of aminoglycosides antibiotics. *Infect Dis Clin North Am* 14:475–487.

Fogo A, Hellings SE, Inagami T, and Kon V (1992) Endothelin receptor antagonism is protective *in vivo* acute cyclosporine toxicity. *Kidney Int* 42:770–774.

Fogo A, Yoshida Y, Glick AD, Homma T, and Ichikawa I (1988) Serial micropuncture analysis of glomerular function in two rat models of glomerular sclerosis. *J Clin Invest* 82:322–330.

Foidart JB, and Mahieu P (1986) Glomerular mesangial cell contractility *in vitro* is controlled by an angiotensin-prostaglandin balance. *Mol Cell Endocrinol* 47:163–173.

Fong HJ, and Cohen AH (1982) Ibuprofen-induced acute renal failure with acute tubular necrosis. *Am J Nephrol* 2:28–31.

Forsyth RP, Hoffbrand BI, and Melmon KL (1971) Hemodynamic effects of angiotensin in normal and environmentally stressed monkeys. *Circulation* 44:119–129.

Friedlaender MM, Jain D, and Ahmed Z (1993) Endothelin activation of phospholipase D: Dual modulation by protein kinase C, and Ca^{2+}. *Am J Physiol* 264:F845–F853.

Fukuroda T, Fujikawa T, Ozaki S, Ishikawa K, Yano M, and Nishikibe M (1994) Clearance of circulating endothelin-1 by ETB receptors in rats. *Biochem Biophys Res Commun* 199:1461–1465.

Furchgott RF, and Zawadski JV (1980) The obligatory role of endothelial cells in the relaxation of smooth muscle by acetylcholine. *Nature* 288:373–376.

Furusu A, Miyazaki M, Abe K, Tsukasaki S, Shioshita K, Sasaki O, Miyazaki K, Ozono Y, Koji T, Harada T, Sakai H, and Kohno S (1998) Expression of endothelial and inducible nitric oxide synthase in human glomerulonephritis. *Kidney Int* 53:1760–1768.

Gallego MJ, Garcia Villalon AL, Lopez Farre AJ, Garcia JL, Garron MP, Casado S, Hernando L, and Caramelo CA (1994) Mechanisms of the endothelial toxicity of cyclosporin A. Role of nitric oxide, cGMP, and Ca^{2+}. *Circ Res* 74:477–484.

Gallego MJ, Lopez Farre A, Riesco A, Monton M, Grandes SM, Barat A, Hernando L, Casado S, and Caramelo CA (1993) Blockade of endothelium-dependent responses in conscious rats by cyclosporin A: Effect of L-arginine. *Am J Physiol* 264:H708–H714.

Garcia del Moral R, O'Valle F, Andujar M, Aguilar M, Lucena MA, Lopez-Hidalgo J, Ramirez C, Medina-Cano MT, Aguilar D, and Gomez-Morales M (1995) Relationship between p-glucoprotein expression and cyclosporin A in kidney: An immunohistological and cellular culture study. *Am J Pathol* 146:398–408.

Gardner MP, Houghton DC, Andoh TF, Lindsley J, and Bennett WM (1996) Clinically relevant doses and blood levels produce experimental cyclosporine nephrotoxicity when combined with nitric oxide inhibition. *Transplantation* 61:1506–1512.

Garg U, and Hassid A (1989) Inhibition of rat mesangial cell mitogenesis by nitric oxide-generating vasodilators. *Am J Physiol* 257:F60–F66.

Garthwaite J, Charles SL, and Chess-Williams R (1988) Endothelium-derived relaxing factor release on activation of NMDA receptors suggests a role as intracellular messenger in the brain. *Nature* 336:385–388.

Gaston RS, Schlessinger SD, Sanders PW, Barker CV, Curtis JJ, and Warnock DG (1995) Cyclosporine inhibits the renal response to L-arginine in human kidney transplant recipients. *J Am Soc Nephrol* 5:1426–1433.

Gerkens JF, Heidemann HT, Jackson EK, and Branch RA (1983) Aminophylline inhibits renal vasoconstriction induced by intrarenal hypertonic saline. *J Pharmacol Exp Ther* 224:609–613.

Gill JR Jr, Melmon KL, Gillespie L Jr and Bartter FC (1965) Bradykinin and renal function in normal man: Effects of adrenergic blockade. *Am J Physiol* 209:844–848.

GISEN Group (1997) Randomised placebo-controlled trial of effect of ramipril on decline in glomerular filtration rate and risk of terminal renal failure in proteinuric, non-diabetic nephropathy. *Lancet* 349:1857–1863.

Goldszer RC, Coodley EL, Rosner MJ, Simons WM, and Schwartz AB (1981) Hyperkalemia associated with indomethacin. *Arch Intern Med* 141:802–804.

Gossmann J, Radounikli A, Bernemann A, Schellinski O, Raab HP, Bickeboller R, and Scheuermann EH (2001) Pathophysiology of cyclosporine-induced nephrotoxicity in humans: a role for nitric oxide? *Kidney Blood Press Res* 24:111–115.

Granger JP, and Hall JE (1985) Acute and chronic actions of bradykinin on renal function and arterial pressure. *Am J Physiol* 248:F87–F92.

Grantham JJ, and Orloff J (1968) Effect of prostaglandin E1 on the permeability response of the isolated collecting tubule to vasopressin, adenosine 3′5′-monophosphate, and theophylline. *J Clin Invest* 47:1154–1161.

Green SJ, Mellouk S, and Hoffman SL (1990) Cellular mechanisms of nonspecific immunity to intracellular infection: Cytokine-induced synthesis of toxic nitrogen oxides from L-arginine by macrophages and hepatocytes. *Immunol Lett* 25:15–19.

Griendling KK, Murphy TJ, and Alexander RW (1993) Molecular biology of the renin-angiotensin system. *Circulation* 87:1816–1828.

Gross PA, Schrier RW, and Anderson RJ (1981) Prostaglandins and water metabolism: A review with emphasis on *in vivo* studies. *Kidney Int* 19:839 (Abstract).

Gross V, Milia AF, Plehm R, Inagami T, and Luft FC (2000) Long-term blood pressure telemetry in AT2 receptor-disrupted mice. *J Hypertens* 18:955–961.

Gruberg L, Mintz GS, Mehran R, Gangas G, Lansky AJ, Kent KM, Pichard AD, Satler LF, and Leon MB (2000) The prognostic implications of further renal function deterioration within 48 h of interventional coronary procedures in patients with pre-existent chronic renal insufficiency. *J Am Coll Cardiol* 36:1542–1548.

Guzman NJ, Fang MZ, and Tang SS (1995) Autocrine inhibition of $Na^+/K^{(+)}$ATPase by nitric oxide in mouse proximal tubule epithelial cells. *J Clin Invest* 95:2083–2088.

Hackenthal E, Paul M, Ganten D, and Taugner R (1990) Morphology, physiology, and molecular biology of renin secretion. *Physiol Rev* 70:1067–1116.

Hackshaw KV, Voelkel NF, Thomas RB, and Westcott JY (1992) Urine leukotriene E4 levels are elevated in patients with active systemic lupus erythematosus. *J Rheumatol* 19:252–258.

Hardie W, Ebert J, Takahashi K, and Badr KF (1993) Thromboxane A2 receptor antagonism reverses amphotericin B-induced renal vasoconstriction in the rat. *Prostaglandins* 45:47–56.

Harris RC, McKanna JA, Akai Y, Jacobson HR, Dubois RN, and Breyer MD (1994) Cyclooxygenase-2 is associated with the macula densa of rat kidney and increases with salt restriction. *J Clin Invest* 94:2504–2510.

Hebert RL, Jacobson HR, and Breyer MD (1991) Prostaglandin E2 inhibits sodium transport in the rabbit CCD by raising intracellular calcium. *J Clin Invest* 87:1992–1998.

Hebert RL, Jacobson HR, Fredin D, and Breyer MD (1993) Evidence that separate PGE2 receptors modulate water and sodium transport in rabbit cortical collecting duct. *Am J Physiol* 265:F643–F650.

Heidemann HT, Bolten M, and Inselmann G (1991) Effect of chronic theophylline administration on amphotericin B nephrotoxicity. *Nephron* 59:294–298.

Heinrich WL, Falck JR, and Campbell WB (1990) Inhibition of release by 14,15-epoxyeicosatrienoic acid in renal cortical slices. *Am J Physiol* 258:E269–E274.

Henrich WL (1981) Role of prostaglandins in renin secretion. *Kidney Int* 19:822–830.

Henrich WL, Berl T, McDonald KM, Anderson RJ, and Schrier RW (1978) Angiotensin II, renal nerves and prostaglandins in renal hemodynamics during hemorrhage. *Am J Physiol* 235:F46–F51.

Henrich WL, McAllister EA, Smith PB, and Campbell WB (1988) Guanosine 3′,5′-cyclic monophosphate as a mediator of inhibition of renin release. *Am J Physiol* 255:F474–F478.

Heyman SN, Brezis M, Reubinoff CA, Greenfeld Z, Lechene C, Epstein FH, and Rosen S (1988) Acute renal failure with selective medullary injury in the rat. *J Clin Invest* 82:401–412.

Heyman SN, Clark BA, Kaiser N, Epstein FH, Spokes K, Rosen S, and Brezis M (1992) *In-vivo* and *in-vitro* studies on the effect of amphotericin B on endothelin release. *J Antimicrob Chemother* 29:69–77.

Hirata M, Hayashi Y, Ushikubi F, Yokota Y, Kageyama R, Nakanishi S, and Narumiya S (1991) Cloning and expression of cDNA for a human thromboxane A2 receptor. *Nature* 349:617–620.

Hishida A, Nakajiama M, Yamada A, Kato A, and Honda N (1994) Roles of hemodynamic and tubular factor in gentamicin-mediated nephropathy. *Ren Fail* 16:109–116.

Hocher B, Liefeldt L, and Thone-Reineke C (1996) Characterization of the renal phenotype of transgenic rats expressing the human endothelin-2 gene. *Hypertension* 28:196–201.

Hoffend J, Cavarape A, Endlich K, and Steinhausen M (1993) Influence of endothelium-derived relaxing factor on renal microvessels and pressure-dependent vasodilation. *Am J Physiol* 265:F285–F292.

Hori S, Komatsu Y, Shigemoto R, Mizuno N, and Nakanishi S (1992) Distinct tissue distribution and cellular localization of two messenger ribonucleic acids encoding different subtypes of rat endothelin receptors. *Endocrinology* 130:1885–1895.

Hostetler KY, and Hall LB (1982) Inhibition of kidney lysosomal phospholipases A, and C by aminoglycoside toxicity. *Proc Natl Acad Sci USA* 79:1663–1667.

Hostetter TH, Olson JL, Rennke HG, Venkatachalam MA, and Brenner BM (1981) Hyperfiltration in remnant nephrons: A potentially adverse response to renal ablation. *Am J Physiol* 241:F85–F93.

Hovind P, Rossing P, Tarnow L, Toft H, Parving J, and Parving HH (2001) Remission of nephrotic-range albuminuria in type 1 diabetic patients. *Diabetes Care* 24:1972–1977.

Humes HD, and Weinberg JM (1983) The effect of gentamicin on antidiuretic hormone-stimulated osmotic water flow in the toad urinary bladder. *J Lab Clin Med* 101:472–478.

Humes HD, Hunt DA, and White MD (1987) Direct toxic effect of the radiocontrast agent diatrizoate on renal proximal tubule cells. *Am J Physiol* 252:F246–F255.

Husserl FE, Lange RK, and Kantrow CM (1979) Renal papillary necrosis and pyelonephritis accompanying fenoprofen therapy. *JAMA* 242:1896–1898.

Imig JD, and Roman RJ (1992) Nitric oxide modulates vascular tone in preglomerular arterioles. *Hypertension* 19:770–774.

Iordache BE, Imberti O, Foglieni C, Remuzzi G, and Bertani T (1994) Effects of angiotensin-converting enzyme inhibition on glomerular capillary wall ultrastructure in MWF/Ztm rats. *J Am Soc Nephrol* 5:1378–1384.

Ishimura E, Shouji S, Nishizawa Y, Morii H, and Kashgarian M (1991) Regulation of mRNA expression for extracellular matrix (ECM) by cultured rat mesangial cells. *J Am Soc Nephrol* 2:546. (Abstract)

Israel A, Cierco M, and Sosa B (2000) Angiotensin AT2 receptors mediate vasodepressor response to foot shock in rats. Role of kinins, nitric oxide and prostaglandins. *Eur J Pharmacol* 394:103–108.

Ito S (1995) Nitric oxide in the kidney. *Curr Opin Nephrol Hypertens* 4:23–30.

Ito S, and Ren Y (1993) Evidence for the role of nitric oxide in macula densa control of glomerular hemodynamics. *J Clin Invest* 92:1093–1098.

Ito S, Johnson CS, and Carretero OA (1991) Modulation of angiotensin II-induced vasoconstriction by endothelium-derived relaxing factor in the isolated microperfused rabbit afferent arteriole. *J Clin Invest* 87:1656–1663.

Jackson EK, Heidemann HT, Branch RA, and Gerkens JF (1982) Low dose intrarenal infusions of PGE2, PGI2 and 6-keto-PGE1 vasodilate the *in vivo* rat kidney. *Circ Res* 51:67–72.

Jenkins JK, Huang H, Ndebele K, and Salahudeen AK (2001) Vitamin E inhibits renal mRNA expression of COX II, HO I, TGF-beta, and osteopontin in the rat model of cyclosporine nephrotoxicity. *Transplantation* 71:331–334.

Johnston PA, Bernard DB, Perrin NS, Arbeit L, Lieberthal W, and Levinsky NG (1981) Control of rat vascular resistance during alterations in sodium balance. *Circ Res* 48:728–733.

Jones DA, Carlton DP, McIntyre TM, Zimmerman GA, and Prescott SM (1993) Molecular cloning of human prostaglandin endoperoxide synthase II, and demonstration of expression in response to cytokines. *J Biol Chem* 268:9049–9054.

Kacew S (1989) Inhibition of gentamicin-induced nephrotoxicity by pyridoxal-5′-phosphate in the rat. *J Pharmacol Exp Ther* 248:360–366.

Kahan BD (1989) Cyclosporine. *N Engl J Med* 321:1725–1738.

Kahlmeter G, and Dahlagers J (1984) Aminoglycoside toxicity - a review of clinical studies published between 1975 and 1982. *J Antimicrob Chemother* 13 (suppl A):9–22.

Kaloyanides GJ, and Pastoriza-Munoz E (1980) Aminoglycoside nephrotoxicity. *Kidney Int* 18:571–582.

Kang DH, Kim YG, Andoh TF, Gordon KL, Suga S, Mazzali M, Jefferson JA, Hughes J, Bennett W, Schreiner GF, and Johnson RJ (2001) Post-cyclosporine-mediated hypertension and nephropathy: Amelioration by vascular endothelial growth factor. *Am J Physiol* 280:F727–F736.

Kashem A, Endoh M, Yano N, Yamauchi F, Nomoto Y, and Sakai H (1996) Expression of inducible NOS in human glomerulonephritis: The possible source is infiltrating monocytes/macrophages. *Kidney Int* 50:392–399.

Katoh T, Takahashi K, DeBoer DK, Serhan CN, and Badr KF (1992) Renal hemodynamic actions of lipoxins in rats: A comparative physiological study. *Am J Physiol* 263:F436–F442.

Katzberg RW, Morris TW, Bengener FA, Kamm DE, and Fischer HW (1977) Renal renin and hemodynamic responses to selective renal artery catheterization and angiography. *Invest Radiol* 12:381–388.

Kelley VE, Sneve S, and Musinski S (1986) Increased renal thromboxane production in murine lupus nephritis. *J Clin Invest* 77:252–259.

Khan KN, Paulson SK, Verburg KM, Lefkowith JB, and Maziasz TJ (2002) Pharmacology of cyclooxygenase-2 inhibition in the kidney. *Kidney Int* 61:1210–1219.

Kidwell DT, McKeown JW, Grider JS, McCombs GB, Ott CE, and Jackson BA (1994) Acute effects of gentamicin on thick ascending limb function in the rat. *Eur J Pharmacol* 270:97–103.

Kimberly RP, Bowden RE, and Keiser HR (1978) Reduction of renal function by newer nonsteroidal anti-inflammatory drugs. *Am J Med* 64:804–807.

Kincaid-Smith P (1986) Effects of non-narcotic analgesics on the kidney. *Drugs* 32 (suppl 4):109–128.

Kirschblaum BB (1984) Interactions between renal brush border membranes and polyamines. *J Pharmacol Exp Ther* 229:409–416.

Kirschenbaum MA, White N, Stein JH, and Ferris TF (1974) Redistribution of renal cortical blood flow during inhibition of prostaglandin synthesis. *Am J Physiol* 227:801–805.

Kobzik L, Bredt DS, and Lowenstein CJ (1993) Nitric oxide synthase in human and rat lung: Immunocytochemical and histochemical localization. *Am J Respir Cell Mol Biol* 9:371–377.

Kohan DE, Hughes AK, and Perkins SL (1992) Characterization of endothelin receptors in the inner medullary collecting duct of the rat. *J Biol Chem* 267:12336–12340.

Koller-Strametz J, Wolzt M, Fuchs C, Putz D, Wisser W, Mensik C, Eichler HG, Laufer G, and Schmetterer L (1999) Renal hemodynamic effects of L-arginine and sodium nitroprusside in heart transplant recipients. *Kidney Int* 55:1871–1877.

Kon V, Sugiura M, Inagami T, Harvie BR, Ichikawa I, and Hoover RL (1990) Role of endothelin in cyclosporine-induced glomerular dysfunction. *Kidney Int* 37:1487–1491.

Kopp UC, and Smith LA (1993) Role of prostaglandins in renal sensory receptor activation by substance P, and bradykinin. *Am J Physiol* 265:R544–R551.

Koppa UC, and DiBona GF (1993) Neural regulation of renin secretion. *Semin Nephrol* 13:543–551.

Kuchan MJ, and Frangos JA (1993) Shear stress regulates endothelin-1 release via protein kinase C, and cGMP in cultured endothelial cells. *Am J Physiol* 264:H150–H156.

Kumar KV, Shifow AA, Naidu MUand Ratnakar KS (2000) Carvedilol: A beta blocker with antioxidant property protects against gentamicin-induced nephrotoxicity in rats. *Life Sci* 66:2603–2611.

Kumin GD (1980) Clinical nephrotoxicity of tobramycin and gentamicin: A prospective study. JAMA 244:1808–1810.

Kunz D, Muhl H, Walker G, and Pfeilschifter J (1994) Two distinct signaling pathways triggers the expression of inducible nitric oxide synthase in rat mesangial cells. *Proc Natl Acad Sci USA* 91:5387–5391.

Kurtz A (1986) Intracellular control of renin release – an overview. *Klin Wochenschr* 64:838–846.

Kurtz A, Della Bruna R, Pfeilschifter J, Taugner R,and Bauer C (1986) Atrial natriuretic peptide inhibits renin release from isolated renal juxtaglomerular cells by a cGMP-mediated process. *Proc Natl Acad Sci USA* 83:4769–4773.

Lahera V, Salom MG, Miranda-Guardiola F, Moncada S, and Romero JC (1991) Effects of NG-nitro-L-arginine methylester on renal function and blood pressure. *Am J Physiol* 261:F1033–1037.

Lands WEM (1979) The biosynthesis and metabolism of prostaglandins. *Annu Rev Physiol* 41:633–652.

Lane P, and Gross SS (1999) Cell signaling by nitric oxide. *Semin Nephrol* 19:215–229.

Lanese DM, and Conger JD (1993) Effects of endothelin receptor antagonist on cyclosporine-induced vasoconstriction in isolated rat renal arterioles. *J Clin Invest* 91:2144–2149.

Larson TS, Hudson K, Mertz JI, Romero JC, and Knox FG (1983) Renal vasoconstrictive response to contrast media. The role of sodium balance and the renin-angiotensin system. *J Lab Clin Med* 101:385–391.

Larsson C, Weber P, and Anggard E (1974) Arachidonic acid increases and indomethacin decreases plasma renin activity in the rabbit. *Eur J Pharmacol* 28:391–394.

Lau DCW, Wong KL, and Hwang WS (1989) Cyclosporine toxicity on cultured rat microvascular endothelial cells. *Kidney Int* 35:604–613.

Lee J, Kim SW, Kook H, Hang DG, Kim NH, and Choi KC (1999) Effects of L-arginine on cyclosporin-induced alterations of vascular NO/cGMP generation. *Nephrol Dial Transplant* 14:2634–2638.

Lee LK, Meyer TW, Pollock AS, and Lovett DH (1995) Endothelial cell injury initiates glomerular sclerosis in the rat remnant kidney. *J Clin Invest* 96:953–964.

Lefkowith JB, Morrison AR, and Schreiner GF (1988) Murine glomerular leukotriene B4 synthesis. Manipulation by (n-6)fatty acid deprivation and cellular origin. *J Clin Invest* 82:1655–1660.

Levin ML (1988) Patterns of tubulo-interstitial damage associated with nonsteroidal anti-inflammatory drugs. *Semin Nephrol* 8:55–61.

Levy SB, Lilley JJ, Frigon RP, and Stone RA (1977) Urinary kallikrein and plasma renin activity as determinants of renal blood flow. *J Clin Invest* 60:129–138.

Lewis EJ, Hunsicker LG, Bain RP, and Rohde RD (1993) The effect of angiotensin-converting-enzyme inhibition on diabetic nephropathy. *N Engl J Med* 329:1456–1462.

Lewis EJ, Hunsicker LG, Clarke WR, Berl T, Pohl MA, Lewis JB, Ritz E, Atkins RC, Rohde R, Raz I; Collaborative Study Group (2001) Renoprotective effect of the angiotensin-receptor antagonist irbesartan in patients with nephropathy due to type 2 diabetes. *N Engl J Med* 345:851–860.

Lianos EA, Andres GA, and Dunn MJ (1983) Glomerular prostaglandin and thromboxane synthesis in rat nephrotoxic serum nephritis. Effects on renal hemodynamics. *J Clin Invest* 72:1439–1448.

Lietman PS, and Smith CR (1983) Aminoglycoside nephrotoxicity in man. *Rev Infect Dis* 5 (suppl 2):284.

Lorenz JN, Grenberg SG, and Griggs JP (1993) The macula densa mechanism for control of renin secretion. *Semin Nephrol* 13:531–542.

Lote CJ, and Haylor J (1989) Eicosanoids, in renal function, in *Prostaglandins, Leukotrienes and Essential Fatty Acids*, Churchill Livingstone, New York.

Ludens JH, and Taylor CJ (1982) Inhibition of ADH-stimulated water flow by stable prostaglandin endoperoxide analogues. *Am J Physiol* 242:F119–F125.

Luft FC (2002) Angiotensin II, the AT2 receptor, and nuclear factor-kB activation. *Kidney Int* 61:2272–2273.

Luke DR, Wasan KM, Verani RR, Brunner LJ, Vadiel K, and Lopez-Berestein G (1991) Attenuation of amphotericin B nephrotoxicity in the candidiasis rat model. *Nephron* 59:139–144.

Luke RG, and Boyle JA (1998) Renal effects of amphotericin B lipid complex. *Am J Kidney Dis* 31:780–785.

Lullmann H, and Vollmer B (1982) An interaction of aminoglycoside antibiotics with Ca binding to lipid monolayers and to biomembranes. *Biochem Pharmacol* 31:3769–3773.

Lund G, Einzig S, Rysavy J, Borgwardt B, Salomonowitz E, Cragg A, and Amplatz K (1984) Role of ischemia in contrast-induced renal damage: An experimental study. *Circulation* 69:783–789.

MacCumber MW, Ross CA, Glaser BM, and Snyder SH (1989) Endothelin: visualization of mRNAs by *in situ* hybridization provides evidence for local action. *Proc Natl Acad Sci USA* 86:7285–7289.

Macconi D, Ghilardi M, Bonassi ME, Mohamed EI, Abbate M, Colombi F, Remuzzi G, and Remuzzi A (2000) Effect of angiotensin-converting enzyme inhibition on glomerular basement membrane permeability and distribution of zonula occludens-1 in MWF rats. *J Am Soc Nephrol* 11:477–489.

Makinen PL, and Makinen KK (1994) The *Enterococcus faecalis* extracellular metalloendopeptidase (EC 3.4.24.30; coccolysin) inactivates human endothelin at bonds involving hydrophobic amino acid residues. *Biochem Biophys Res Commun* 200:981–985.

Malek A, Greene AL, and Izumo S (1993) Regulation of endothelin 1 gene by fluid shear stress is transcriptionally mediated and independent of protein kinase C, and cAMP. *Proc Natl Acad Sci USA* 90:5999–6003.

Marin-Grez M (1974) The influence of antibodies against bradykinin on isotonic saline diuresis in the rat. *Pflugers Arch* 350:231–239.

Marinides GN, Groggel GC, and Cohen AH (1987) Failure of angiotensin converting enzyme inhibition to affect the course of chronic puromycin aminonucleoside nephropathy. *Am J Pathol* 129:394–401.

Marsden PA, Brock TA, and Ballermann BJ (1990) Glomerular endothelial cells respond to calcium-mobilizing agonists with release of EDRF. *Am J Physiol* 258:F1295–F1303.

Marsden PA, Heng HH, and Scherer SW (1993) Structure and chromosomal localization of the human constitutive endothelial nitric oxide synthase gene. *J Biol Chem* 268:17478–17488.

Marsen TA, Schramek H, and Dunn MJ (1994) Renal actions of endothelin: linking cellular signaling pathways to kidney disease. *Kidney Int* 45:336–344.

Maschio G, Alberti D, Janin G, Locatelli F, Mann JFE, and Motolese M (1996) Effect of the angiotensin-converting-enzyme inhibitor benazepril on the progression of chronic renal insufficiency: The Angiotensin-Converting-Enzyme Inhibition in Progressive Renal Insufficiency Study Group. *N Engl J Med* 334:939–945.

Matrougui K, Loufrani L, Heymes C, Levy BI, and Henrion D (1999) Activation of AT2 receptors by endogenous angiotensin II is involved in flow-induced dilation in rat resistance arteries. *Hypertension* 34:659–665.

Mattson DL, Shanhong L, Nakanishi K, Papanek PE, and Cowley AW Jr (1994) Effect of chronic renal medullary nitric oxide inhibition on blood pressure. *Am J Physiol* 266:H1918–H1926.

McCaa RE, Hall JE, and McCaa CS (1978) The effects of angiotensin I-converting enzyme inhibitors on arterial blood pressure and urinary sodium excretion. Role of the renin-angiotensin and kallikrein-kinin systems. *Circulation* 43 (suppl 1):32.

McCullough PA, Wolyn R, Rocher LL, Levin RN, and O'Neill WW (1997) Acute renal failure after coronary intervention: incidence, risk factors, and relationship to mortality. *Am J Med* 103:368–375.

McGiff JC, Itskovitz HD, and Terragno NA (1975) The actions of bradykinin and eledoisin in the canine isolated kidney: Relationship to prostaglandins. *Clin Sci Mol Med* 49:125–131.

McKee M, Scavone C, and Nathanson JA (1994) Nitric oxide, cGMP, and hormone regulation of active sodium transport. *Proc Natl Acad Sci USA* 91:12056–12060.

McLay JS, Chatterjee P, Nicolson AG, Jardine AG, McKay NG, Ralston SH, Grabowski P, Haites NE, MacLeod AM, and Hawksworth GM (1994) Nitric oxide production by human proximal tubular cells: a novel immunomodulatory mechanism? *Kidney Int* 46:1043–1049.

Menè P, Simonson MS, and Dunn MJ (1989) Physiology of the mesangial cell. *Physiol Rev* 69:1347–1424.

Meyer-Lehnert H, Bokemeyer D, Friedrichs U, Backer A, and Kramer HJ (1997) Cellular mechanisms of cyclosporine A-associated side-effects: Role of endothelin. *Kidney Int Suppl* 61:S27–S31.

Mezzano SA, Ruiz-Ortega M, and Egido J (2001) Angiotensin II, and renal fibrosis. *Hypertension* 38:635–638.

Mihatsch MJ, Ryffel B, and Gudat F (1995) The differential diagnosis between rejection and cyclosporine toxicity. *Kidney Int* 48 (suppl 52):S63–S69.

Mihatsch MJ, Thiel G, and Ryffel B (1988) Histopathology of cyclosporine nephrotoxicity. *Transplant Proc* 20 (suppl 3):759–771.

Miller WL, Redfield MM, and Burnett JC (1989) Integrated cardiac, renal, and endocrine actions of endothelin. *J Clin Invest* 83:317–320.

Mingeot-Leclercq M-P, and Tulkens PM (1999) Aminoglycosides: Nephrotoxicity. *Antimicrob Agents Chemother* 43:1003–1012.

Moestrup SK, Cui S, Vorum H, Bregengard C, Bjorn SE, Norris K, Gliemann J, and Christensen EI (1995) Evidence that epithelial glycoprotein 330/megalin mediates uptake of polybasic drugs. *J Clin Invest* 96:1404–1413.

Mohaupt MG, Elzie JL, Ahn KY, Clapp WL, Wilcox CS, and Kone BC (1994) Differential expression and induction of mRNAs encoding two inducible nitric oxide synthases in rat kidney. *Kidney Int* 46:653–665.

Moncada S, and Higgs EA (1995) Molecular mechanisms and therapeutic strategies related to nitric oxide. *FASEB J* 9:1319–1330.

Morales E, and Mucksavage JJ (2002) Cycoloxygenase-2 inhibitor-associated acute renal failure: Case report with rofecoxib and review of the literature. *Pharmacotherapy* 22:1317–1321.

Morelli E, Loon N, Peters W, and Myers BD (1990) Effects of converting-enzyme inhibition on barrier function in diabetic glomerulopathy. *Diabetes* 39:76–82.

Morris ST, McMurray JJ, Rodger RS, Farmer R, and Jardine AG (2000) Endothelial dysfunction in renal transplant recipients maintained on cyclosporine. *Kidney Int* 57:1100–1106.

Morrison AR (1986) Biochemistry and pharmacology of renal arachidonic acid metabolism. *Am J Med* 80(1A):3–11.

Morrissey JJ, McCracken R, Kaneto H, Vehaskari M, Montani D, and Klahr S (1994) Location of an inducible nitric oxide synthase mRNA in the normal kidney. *Kidney Int* 45:998–1005.

Morrow JD, Hill KE, Burk RF, Nammour RM, Badr KF, and Roberts JL (1990) A series of prostaglandin F2-like compounds are produced *in vivo* in humans by a non-cyclooxygenase free radical-catalyzed mechanism. *Proc Natl Acad Sci USA* 87:9383–9387.

Moss N, Powell S, and Falk R (1985) Intravenous cyclosporine activates afferent and efferent renal nerves and causes sodium retention in innervated kidneys in rats. *Proc Natl Acad Sci USA* 82:8222–2226.

Mueller C, Buerkle G, Buettner HJ, Petersen J, Perruchoud AP, Eriksson U, Marsch S, and Roskamm H (2002) Prevention of contrast media-associated nephropathy. Randomized comparison of 2 hydration regimens in 1620 patients undergoing coronary angioplasty. *Arch Intern Med* 162:329–336.

Mundel P, Bachmann S, and Bader M (1992) Expression of nitric oxide synthase in kidney macula densa cells. *Kidney Int* 42:1017–1019.

Murray BM, Paller MS, and Ferris TF (1985) Effect of cyclosporine administration on renal hemodynamics in conscious rat. *Kidney Int* 28:767–774.

Myers B, and Newton L(1991) Cyclosporine-induced chronic renal nephropathy: An obliterative microvascular renal injury. *J Am Soc Nephrol* 2:S45–S52.

Myers BD, and Guasch A (1994) Mechanisms of massive proteinuria. *J Nephrol* 7:254.

Myers BD, Nelson RG, Williams GW, Bennett PH, Hardy SA, Berg RL, Loon N, Knowler WC, and Mitch WE (1991) Glomerular function in Pima Indians with noninsulin-dependent diabetes mellitus of recent onset. *J Clin Invest* 88:524–530.

Naftilan AJ (1994) Role of tissue renin-angiotensin system in vascular remodeling and smooth muscle cell growth. *Curr Opin Nephrol Hypertens* 3:218–227.

Nakahama H (1990) Stimulatory effect of cyclosporine A on endothelin secretion by a cultured renal epithelial cell line, LLC-PK1 cells. *Eur J Pharmacol* 180:191–192.

Nakamura T, Ebihara I, Fukui M, Tomino Y, and Koide H (1995a) Effect of a specific endothelin receptor A antagonist on mRNA levels for extracellular matrix components and growth factors in diabetic glomeruli. *Diabetes* 44:895–899.

Nakamura T, Ebihara I, Tomino Y, and Koide H (1995b) Effect of a specific endothelin A receptor antagonist on murine lupus nephritis. *Kidney Int* 47:481–489.

Nakamura T, Takahashi T, Fukui M, Ebihara I, Osada S, Tomino Y, and Koide H (1995c) Enalapril attenuates increased gene expression of extracellular matrix components in diabetic rats. *J Am Soc Nephrol* 5:1492–1497.

Nambi P, Pullen M, and Wu HL (1992a) Identification of endothelin receptor subtypes in human renal cortex and medulla using subtype-selective ligands. *Endocrinology* 131:1081–1086.

Nambi P, Pullen M, Contino LC, and Brooks DP (1990) Up-regulation of renal endothelin receptors in rats with cyclosporine A-induced nephrotoxicity. *Eur J Pharmacol* 187:113–116.

Nambi P, Wu HL, and Pullen M (1992b) Identification of endothelin receptor subtypes in rat kidney cortex using subtype-selective ligands. *Mol Pharmacol* 42:336–339.

Narita I, Border WA, Ketteler M, and Noble NA (1995) Nitric oxide mediates immunologic injury to kidney mesangium in experimental glomerulonephritis. *Lab Invest* 72:17–24.

Nathan C, and Xie QW (1994) Nitric oxide synthases: Roles, tolls, and controls. *Cell* 78:915–918.

Navar LG, Champion WJ, and Thomas CE (1986) Effects of calcium channel blockade on renal vascular resistance response to changes in perfusion pressure and angiotensin converting enzyme inhibition in dogs. *Circ Res* 58:874–881.

Neugarten J, Aynedjian HS, and Bank N (1983) Role of tubular obstruction in acute renal failure due to gentamicin. *Kidney Int* 24:330–335.

Noris M, Bernasconi S, Casiraghi F, Sozzani S, Gotti E, Remuzzi G, and Mantovani A (1995) Monocyte chemoattractant protein-1 is excreted in excessive amounts in the urine of patients with lupus nephritis. *Lab Invest* 73:804–809.

Ohta K, Hirata Y, and Shichiri M (1991) Urinary excretion of endothelin-1 in normal subjects and patients with renal disease. *Kidney Int* 39:307–311.

Oldroyd S, Slee SJ, Haylor J, Morcos SK, and Wilson C (1994) Role of endothelin in the renal responses to radiocontrast media in the rat. *Clin Sci* 87:427–434.

Oldroyd SD, Haylor JL, and Morcos SK (1995) Bosentan, an orally active endothelin antagonist: Effect on the renal response to contrast media. *Radiology* 196:661–665.

Ong ACM, Jowett TP, Moorhead JF, and Owen JS (1994) Human high density lipoproteins stimulate endothelin-1 release by cultured human renal proximal tubular cells. *Kidney Int* 46:1315–1321.

Orloff J, Handler JS, and Bergstrom S (1965) Effect of prostaglandin PGE1 on the permeability response of toad bladder to vasopressin, theophylline and adenosine 3′,5′-monophosphate. *Nature* 205:397–398.

Palmer RMJ, Ferrige AG, and Moncada S (1987) Nitric oxide release accounts for the biological activity of endothelium-derived relaxing factor. *Nature* 327:524–526.

Palmer RMJ, Rees DD, Ashton DS, and Moncada S (1988) L-arginine is the physiological precursor for the formation of nitric oxide in endothelium-dependent relaxation. *Biochem Biophys Res Commun* 153:1251–1256.

Papaioannides D, Bouropoulos C, Sinapides D, Korantzopoulos P, and Akritidis N (2001) Acute renal dysfunction associated with selective COX-2 inhibitor therapy. *Int Urol Nephrol* 33:609–611.

Parving HH, Lehnert H, Bröchner-Mortensen J, Gomis R, Andersen S, and Arner P (2001) The effect of irbersartan on the development of diabetic nephropathy in patients with type 2 diabetes. *N Engl J Med* 345:870–878.

Patrono C, and Dunn MJ (1987) The clinical significance of inhibition of renal prostaglandin synthesis. *Kidney Int* 32:1–12.

Patrono C, Ciabattoni G, Remuzzi G, Gotti E, Bombardieri S, Di Munno O, Tartarelli G, Cinotti GA, Simonetti BM, and Pierucci A (1985) Functional significance of renal prostacyclin and thromboxane A2 production in patients with systemic lupus erythematosus. *J Clin Invest* 76:1011–1018.

Perico N, Benigni A, Zoja C, Delaini F, and Remuzzi G (1986) Functional significance of exaggerated renal thromboxane A2 synthesis induced by cyclosporin A. *Am J Physiol* 20:F581.

Perico N, Dadan J, and Remuzzi G (1990a) Endothelin mediates the renal vasoconstriction induced by cyclosporine in the rat. *J Am Soc Nephrol* 1:76–83.

Perico N, Dadan J, Gabanelli M, and Remuzzi G (1990b) Cyclooxygenase products and atrial natriuretic peptide modulate renal response to endothelin. *J Pharmacol Exp Ther* 252:1213–1220.

Perico N, Pasini M, Gaspari F, Abbate M, and Remuzzi G (1991a) Co-participation of thromboxane A2 and leukotriene C4 and D4 in mediating cyclosporine-induced acute renal failure. *Transplantation* 52:873–878.

Perico N, Plata Cornejo R, and Remuzzi G (1991b) Endothelin induces diuresis and natriuresis in the rat by acting on proximal tubular cells through a mechanism mediated by lipoxygenase. *J Am Soc Nephrol* 2:57–69.

Perico N, Rossini M, Imberti O, Malanchini B, Cornejo RP, Gaspari F, Bertani T, and Remuzzi G (1992) Thromboxane receptor blockade attenuates chronic cyclosporine nephrotoxicity and improves survival in rats with renal isograft. *J Am Soc Nephrol* 2:1398–1404.

Pernow J, Boutier JF, Franco-Cereceda A, Lacroix JS, Matran R, Lundberg and JM (1988) Potent selective vasoconstrictor effects of endothelin in the pig kidney *in vivo*. *Acta Physiol Scand* 134:573–574.

Petros A, Bennett D, and Vallance P (1991) Effect of nitric oxide synthase inhibitors on hypotension in patients with septic shock. *Lancet* 338:1557–1558.

Pfeilschifter J, Kunz D, and Muhl H (1993) Nitric oxide: An inflammatory mediator of glomerular mesangial cells. *Nephron* 64:518–525.

Pierucci A, Bianca MD, and Simonetti M (1989) Improvement of renal function with selective thromboxane antagonist in lupus nephritis. *N Engl J Med* 320:273.

Plaut ME, Schentag JJ, and Jusko WJ (1979) Aminoglycoside nephrotoxicity: Comparative assessment in critically ill patients. *J Med* 10:257–266.

Pollock DM, and Opgenorth TJ (1993) Evidence for endothelin-induced renal vasoconstriction independent of ETA receptor activation. *Am J Physiol* 264:R222–R226.

Radomski MW, Palmer RMJ, and Moncada S (1990) An L-arginine/nitric oxide pathway present in human platelets regulates aggregation. *Proc Natl Acad Sci USA* 87:5193–5197.

Radte HW, Jubiz W, Smith JB, and Fisher JW (1980) Albuterol-induced erythropoietin production and prostaglandins release in the isolated perfused dog kidney. *J Pharmacol Exp Ther* 214:467–471.

Ramsammy LS, and Kaloyanides G (1987) Effect of gentamicin on the transition temperature and permeability to glycerol of phosphatidylinositol-containing liposomes. *Biochem Pharmacol* 36:1179–1181.

Ramsammy LS, Josepovitz C, and Kaloyanides GJ (1988) Gentamicin inhibits agonist stimulation of the phosphatidylinositol cascade in primary cultures of rabbit proximal tubular cells and in rat renal cortex. *J Pharmacol Exp Ther* 247:989–996.

Rangan GK, Wang Y, and Harris DC (2001) Pharmacological modulators of nitric oxide exacerbate tubulointerstitial inflammation in proteinuric rats. *J Am Soc Nephrol* 12:1696–1705.

Rees DD, Palmer RM, and Moncada S (1989) Role of endothelium-derived nitric oxide in the regulation of blood pressure. *Proc Natl Acad Sci USA* 86:3375–3378.

Reid DG, and Gajjar K (1987) A proton and carbon 13 nuclear magnetic resonance study of neomycin B, and its interactions with phosphatidylinositol 4,5-bisphosphate. *J Biol Chem* 262:7967–7972.

Reid IA (1984) Actions of angiotensin II on the brain: mechanisms and physiologic role. *Am J Physiol* 246:F533–F543.

Reilly CM, Farrelly LW, Viti D, Redmond ST, Hutchison F, Ruiz P, Manning P, Connor J, and Gilkeson GS (2002) Modulation of renal disease in MLR/lpr mice by pharmacological inhibition of inducible nitric oxide synthase. *Kidney Int* 61:839–846.

Remuzzi A (1995) Mathematical description of transport of water and macromolecules through the glomerular capillary wall. *Curr Opin Nephrol Hypertens* 4:343–348.

Remuzzi A, Battaglia C, Rossi L, Zoja C, and Remuzzi G (1987) Glomerular size selectivity in nephrotic rats exposed to diets with different protein content. *Am J Physiol* 253:F318–F327.

Remuzzi A, Perico N, Sangalli F, Vendramin G, Moriggi M, Ruggenenti P, and Remuzzi G (1999) ACE inhibition and Ang II receptor blockade improve glomerular size-selectivity in IgA nephropathy. *Am J Physiol* 276:F457–F466.

Remuzzi A, Puntorieri S, Battaglia C, Bertani T, and Remuzzi G (1990) Angiotensin converting enzyme inhibition ameliorates glomerular filtration of macromolecules and water and lessens glomerular injury in the rat. *J Clin Invest* 85:541–549.

Remuzzi A, Ruggenenti P, Mosconi L, Pata V, Viberti G, and Remuzzi G (1993) Effect of low-dose enalapril on glomerular size-selectivity in human diabetic nephropathy. *J Nephrol* 6:36.

Remuzzi G, and Bertani T (1998) Pathophysiology of progressive nephropathies. *N Engl J Med* 339:1448–1456.

Remuzzi G, and Benigni A (1993) Endothelins in the control of cardiovascular and renal function. *Lancet* 342:589–593.

Remuzzi G, and Bertani T (1990) Is glomerulosclerosis a consequence of altered glomerular permeability to macromolecules? *Kidney Int* 38:384–394.

Remuzzi G, and Perico N (1995) Cyclosporine-induced renal dysfunction in experimental animals and humans. *Kidney Int* 48 (suppl 52):S70–S74.

Remuzzi G, Fitzgerald PA, and Patrono C (1992) Thromboxane synthesis and action within the kidney. *Kidney Int* 41:1483–1493.

Remuzzi G, Ruggenenti P AND Benigni A (1997) Understanding the nature of renal disease progression. *Kidney Int* 51:2–15.

Remuzzi G, Ruggenenti P, and Perico N (2002b) Chronic renal diseases: Renoprotective benefits of renin-angiotensin system inhibition. *Ann Intern Med* 136:604–615.

Reyes AA, Lefkowith J, Pippin J, and Klahr S (1992) Role of the 5-lipoxygenase pathway in obstructive nephropathy. *Kidney Int* 41:100–106.

Ribeiro MO, Antunes E, De Nucci G, Lovisolo SM, and Zatz R (1992) Chronic inhibition of nitric oxide synthesis: A new model of arterial hypertension. *Hypertension* 20:298–303.

Richards NT, Poston L, and Hilton PJ (1989) Cyclosporine A inhibits relaxation but does not induce vasoconstriction in human subcutaneous resistance vessels. *J Hypertens* 7:1–3.

Rifai A, Sakai H, and Mitsunori Y (1993) Expression of 5-lipoxygenase and 5-lipoxygenase activation protein in glomerulonephritis. *Kidney Int* 39:S95–S99.

Roberts DG, Gerber JG, and Nies AS (1984) Comparative effects of sulindac and indomethacin in humans. *Clin Res* 32:72A (Abstract).

Robertson CE, Ford MJ, Van Someren V, Dlugolecka M, and Prescott LF (1980) Mefenamic acid nephropathy. *Lancet* 2:232–233.

Roczniak A, and Burns KD (1996) Nitric oxide stimulates guanylate cyclase and regulates sodium transport in rabbit proximal tubule. *Am J Physiol* 270:F106–F115.

Roczniak A, Zimpelmann J, and Burns KD (1998) Effect of dietary salt on neuronal nitric oxide synthase in the inner medullary collecting duct. *Am J Physiol* 275:F46–F54.

Rodriguez-Barbero A, Rodriguez-Lopez AM, Gonzalez-Sarmiento R, and Lopez-Novoa JM (1995) Gentamicin activates rat mesangial cells – a role for platelet activating factor. *Kidney Int* 47:1346–1353.

Romero JC, Lahera V, Salom MG, and Biondi ML (1992) Role of the endothelium-dependent relaxing factor nitric oxide on renal function. *J Am Soc Nephrol* 2:1371–1387.

Rosen S, Brezis M, and Stillman I (1994) The pathology of nephrotoxic injury: A reappraisal. *Miner Electrolyte Metab* 20:174–180.

Rosenthal A, and Pace-Asciak CR (1983) Potent vasoconstriction of the isolated perfused kidney by leukotrienes C4 and D4. *Can J Physiol Pharmacol* 61:325–328.

Rossing K, Christensen PK, Jensen BR, and Parving HH (2002) Dual blockade of the renin-angiotensin system in diabetic nephropathy: a randomized double-blind crossover study. *Diabetes Care* 25:95–100.

Roullet JB, Xue H, McCarron DA, Holcomb S, and Bennett WM (1994) Vascular mechanisms of cyclosporin-induced hypertension in the rat. *J Clin Invest* 93:2244–2250.

Rudnick MR, Berns JS, Cohen RM, and Goldfarb S (1996) Contrast media-associated nephrotoxicity. *Curr Opin Nephrol Hypertens* 5:127–133.

Ruggenenti P, Perna A, Benini R, Bertani T, Zoccali C, Maggiore Q, Salvadori M, and Remuzzi G (1999) In chronic nephropathies prolonged ACE inhibition can induce remission: Dynamics of time-dependent changes in GFR. *J Am Soc Nephrol* 10:997–1006.

Ruiz-Ortega M, Gonzalez S, Seron D, Condom E, Bustos C, Largo R, Gonzalez E, Ortiz A, and Egido J (1995) ACE inhibition reduces proteinuria, glomerular lesions and extracellular matrix production in a normotensive rat model of immune complex nephritis. *Kidney Int* 48:1778–1791.

Ruiz-Ortega M, Lorenzo O, Ruperez M, Blanco J, and Egido J (2001a) Systemic infusion of angiotensin II into normal rats activates nuclear factor-kappa B, and AP-1 in the kidney: Role of AT(1) and AT(2) receptors. *Am J Pathol* 158:1743–1756.

Ruiz-Ortega M, Lorenzo O, Ruperez M, Konig S, Wittig B, and Egido J (2000) Angiotensin II activates nuclear transcription factor kB through AT1 and AT2 in vascular smooth muscle cells: Molecular mechanism. *Circ Res* 86:1266–1272.

Ruiz-Ortega M, Lorenzo O, Ruperez M, Suzuki Y, and Egido J (2001b) Angiotensin II activates nuclear transcription factor-kB in aorta of normal rats and in vascular smooth muscle cells of AT1 knockout mice. *Nephrol Dial Transplant* 16 (suppl 1):27–33.

Sabra R, and Branch RA (1990) Amphotericin B nephrotoxicity. *Drug Saf* 5:94–108.

Safirstein R, Miller P, and Kahn T (1983) Cortical and papillary absorptive defects in gentamicin nephrotoxicity. *Kidney Int* 24:526–533.

Saijonmaa O, Nyman T, and Fyhrquist F (1992) Endothelin-1 stimulates its own synthesis on human endothelial cells. *Biochem Biophys Res Commun* 188:286–291.

Sakamoto A, Yanagisawa M, and Sawamura T (1993) Distinct subdomains of human endothelin receptors determine their selectivity to endothelin A-selective antagonist and endothelin B-selective agonists. *J Biol Chem* 268:8547–8553.

Salvati P, Lamberti E, Ferrario R, Ferrario RG, Scampini G, Pugliese F, Barsotti P, and Patrono C (1995) Long-term thromboxane-synthase inhibition prolongs survival in murine lupus nephritis. *Kidney Int* 47:1168–1175.

Samuelsson B (1987) An elucidation of the arachidonic acid cascade. Discovery of prostaglandins, thromboxane, and leukotrienes. *Drugs* 33:2–9.

Samuelsson B, Dahlen SE, Lindgren JA, Rouzer CA, and Serhan CN (1987) Leukotrienes and lipoxins: Structures, biosynthesis, and biological effects. *Science* 237:1171–1176.

Sanders KM, Ward SM, Thornbury KD, Dalziel HH, Westfall DP, and Carl A (1992) Nitric oxide as a non-adrenergic, non-cholinergic neuro-transmitter in the gastrointestinal tract. *Jpn J Pharmacol* 58 (suppl 2):220P–225P.

Satoh H, and Satoh S (1984) Prostaglandin E2 and I2 production in isolated dog renal arteries in the absence of presence of vascular endothelial cells. *Biochem Biophys Res Commun* 118:873–876.

Saura M, Lopez S, Puyol MR, Puyol DR, and Lamas S (1995) Regulation of inducible nitric oxide synthase expression in rat mesangial cells and isolated glomeruli. *Kidney Int* 47:500–509.

Sawaya BP, Briggs JP, and Schermann J (1995) Amphotericin B membrane properties. *J Am Soc Nephrol* 6:154–164.

Schacht J (1979) Isolation of an aminoglycoside receptor from guinea pig inner ear tissúes and kidneys. *Arch Otorhinolaryngol* 224:129–134.

Scharschmidt LA, Lianos E, and Dunn MJ (1983) Arachidonate metabolites and the control of glomerular function. *Fed Proc* 42:3058–3063.

Schentag JJ (1983) Specificity of tubular damage criteria for aminoglycoside nephrotoxicity in critically ill patients. *J Clin Pharmacol* 23:473–483.

Schentag JJ, Cerra FB, and Plaut ME (1982) Clinical and pharmacokinetic characteristics of aminoglycoside nephrotoxicity in 201 critically ill patients. *Antimicrob Agents Chemother* 21:721–726.

Schlatter E (1989) Macula densa cells sense luminal NaCl concentration via furosemide sensitive $Na_2^+Cl^-K^+$ cotransport. *Pflugers Arch* 414:286–290.

Schlondorff D (1993) Renal complications of nonsteroidal anti-inflammatory drugs. *Kidney Int* 44:643–653.

Schlondorff D, Perez J, and Satriano JA (1985) Differential stimulation of PGE2 synthesis in mesangial cells by angiotensin and A23187. *Am J Physiol* 248:C119–C126.

Schmidt HHHW, Gagne GD, Nakane M, Pollock JM, Miller MF, and Murad F (1992) Mapping of neuronal nitric oxide synthase in the rat suggests frequent co-localization with NADPH-diaphorase but not with soluble guanylyl cyclase, and novel paraneural function for nitrinergic signal transduction. *J Histochem Cytochem* 40:1439–1456.

Schnermannn J, Briggs JP, and Weber PC (1984) Tubuloglomerular feedback, prostaglandins and angiotensin in the autoregulation of glomerular filtration rate. *Kidney Int* 25:3–64.

Schor N (1981) Pathophysiology of altered glomerular function in aminoglycoside-treated rats. *Kidney Int* 19:288–296.

Schor N, and Brenner BM (1980) Humoral regulation of glomerular filtration, in *Humoral regulation of sodium excretion* Lichardus B, Schier RW, and Pance J (Eds), Elsevier/North Holland, Amsterdam, p 28.

Schuster VO, Kokko JP, and Jacobson HR (1984) Interactions of lysylbradykinin and antidiuretic hormone in the rabbit collecting tubule. *J Clin Invest* 73:1659–1667.

Schweizer A, Valdenaire O, Nelbock P, Deuschle U, Dumas Milne Edwards JB, Stumpf JG, and Loffler BM (1997) Human endothelin-converting enzyme (ECE-1): Three isoforms with distinct subcellular localization. *Biochem J* 328:871–877.

Scicli AG, and Carretero OA (1986) Renal kallikrein-kinin system. *Kidney Int* 29:120–130.

Serhan CN (1994) Lipoxin biosynthesis and its impact in inflammatory and vascular events. *Biochim Biophys Acta* 1212:1–25.

Sessa WC (1994) The nitric oxide synthase family of proteins. *J Vasc Res* 31:131–143.

Shah GM, Muhalwas KK, and Winer RL (1981) Renal papillary necrosis due to ibuprofen. *Arthritis Rheum* 24:1208–1210.

Shihab FS, Bennett WM, Tanner AM, and Andoth TF (1997) Angiotensin II blockade decreases TGF-beta-1 and matrix proteins in cyclosporine nephropathy. *Kidney Int* 52:660–673.

Shihab FS, Bennett WM, Yi H, and Andoh TF (2001) Expression of vascular endothelial growth factor and its receptors Flt-1 and KDR/Flk-1 in chronic cyclosporine nephrotoxicity. *Transplantation* 72:164–168.

Shultz PJ, Schorer AE, and Raij L (1990) Effects of endothelium-derived relaxing factor and nitric oxide on rat mesangial cells. *Am J Physiol* 258:F162–F167.

Siegel H, Ryffel B, and Petric R (1983)Cyclosporin A, RAAS, and renal adverse reactions. *Transplant Proc* 15:2719.

Sigmon DH, Carretero OA, and Beierwaltes WH (1992) Endothelium-derived relaxing factor regulates renin release *in vivo*. *Am J Physiol* 263:F256–F261.

Simmons cf., Bogusky RT, and Humes HD (1980) Inhibitory effects of gentamicin on renal mitochondrial oxidative phosphorylation. *J Pharmacol Exp Ther* 214:709–715.

Simonson MS, and Dunn MJ (1992) The molecular mechanisms of cardiovascular and renal regulation by endothelin peptides. *J Lab Clin Med* 119:622–639.

Simonson MS, and Dunn MJ (1993) Renal actions of endothelin peptides. *Curr Opin Nephrol Hypertens* 2:51–60.

Simonson MS, Menè P, Dubyak GR, and Dunn MJ (1988) Identification and transmembrane signaling of leukotriene D4 receptors in human mesangial cells. *Am J Physiol* 255:C771–C780.

Singh R, Pervin S, and Rogers NE (1997) Evidence for the presence of an unusual nitric oxide- and citrulline-producing enzyme in rat kidney. *Biochem Biophys Res Commun* 232:672–677.

Siragy HM (1993) Evidence that intrarenal bradykinin plays a role in regulation of renal function. *Am J Physiol* 265:E648–E654.

Siragy HM (2002) Angiotensin receptor blockers: How important is selectivity? *Am J Hypertens* 15:1006–1014.

Siragy HM, and Carey RM (2001) Angiotensin type 2 receptors: Potential importance in the regulation of blood pressure. *Curr Opin Nephrol Hypertens* 10:99–103.

Smith DHG, Neutel JM, and Weber MA (1991) Effects of angiotensin II on pressor responses to norepinephrine in humans. *Life Sci* 48:2413–2421.

Smith SR, Creech EA, Schaffer AV, Martin LL, Rakhit A, Douglas FL, Klotman PE, and Coffman TM (1992) Effects of thromboxane synthase inhibition with CGS 13080 in human cyclosporine nephrotoxicity. *Kidney Int* 41:199–205.

Smith WL, and Bell TG (1978) Immunohistochemical localization of the prostaglandin-forming cyclooxygenase in renal cortex. *Am J Physiol* 235:F451–F457.

Sokolovsky M, Ambar I, and Galron R (1992) A novel subtype of endothelin receptors. *J Biol Chem* 267:20551–20554.

Sorensen SS, Madsen JK, and Pedersen EB (1994) Systemic and renal effect of intravenous infusion of endothelin-1 in healthy human volunteers. *Am J Physiol* 266:F411–F418.

Spurney RF, Riuz P, Pisetsky DS, and Coffman TM (1991) Enhanced renal leukotriene production in murine lupus: role of lipoxygenase metabolites. *Kidney Int* 39:95–102.

Stahl RAK, Thaiss F, Kahf S, Schoeppe W, and Helmchen UM (1990) Immune-mediated mesangial cell injury: Biosynthesis and function of prostanoids. *Kidney Int* 38:273–281.

Stein JH, and Fadem SZ (1978) The renal circulation. *JAMA* 239:1308–1312.

Stein JH, Congbalay RC, Karsh DL, Osgood RW, and Ferris TF (1972) The effect of bradykinin on proximal tubular sodium reabsorption in the dog: Evidence for functional nephron heterogeneity. *J Clin Invest* 51:1709–1721.

Strandhoy JW, Ott CE, Schneider e.g., Willis LR, Beck NP, Davis BB, and Knox FG (1974) Effects of prostaglandins E1 and E2 on renal sodium reabsorption and Starling forces. *Am J Physiol* 226:1015–1021.

Stroes ES, Luscher TF, de Groot FG, Koomans HA, and Rabelink TJ (1997) Cyclosporin A increases nitric oxide activity *in vivo*. *Hypertension* 29:570–575.

Stuehr DJ, and Griffith OW (1992) Mammalian nitric oxide synthases. *Adv Enzymol Relat Area Mol Biol* 65:287.

Stuehr DJ, and Ikeda-Saito M (1992) Spectral characterization of brain and macrophage nitric oxide synthase. Cytochrome P-450-like hemeproteins that contain a flavin semiquinone radical. *J Biol Chem* 267:20547–20550.

Stuehr DJ, and Nathan CF (1989) Nitric oxide: a macrophage product responsible for cytostasis and respiratory inhibition in tumor target cells. *J Exp Med* 169:1543–1255.

Takahashi K, Nammour TM, Fukunaga M, Ebert J, Morrow JD, Roberts LJ 2nd, Hoover RL, and Badr KF (1992) Glomerular actions of a free radical-generated novel prostaglandin, 8-epi-PGF2a, in the rat: Evidence for interaction with thromboxane A2 receptors. *J Clin Invest* 90:136–141.

Takeda M, Breyer MD, Noland TD, Homma T, Hoover RL, Inagami T, and Kon V (1992) Endothelin-1 receptor antagonist: Effects on endothelin- and cyclosporine-treated mesangial cells. *Kidney Int* 41:1713–1719.

Takeda M, Komeyama T, and Tsutsui T (1994) Changes in urinary excretion of endothelin-1-like immunoreactivity before and after unilateral nephrectomy in humans. *Nephron* 67:180–184.

Takenaka T, Hashimoto Y, and Epstein M (1992) Diminished acetylcholine-induced vasodilation in renal microvessels of cyclosporine-treated rats. *J Am Soc Nephrol* 3:42–50.

Talner LB, and Davidson AJ (1968) Effect of contrast media on renal extraction of para-amino-hippurate. *Invest Radiol* 3:301–309.

Tamura M, Takagi T, Howard EF, Landon EJ, Steimle A, Tanner M, Myers PR (2000) Induction of angiotensin II subtype 2 receptor-mediated blood pressure regulation in synthetic diet-fed rats. *J Hypertens* 18:1239–1246.

Terada Y, Tomita K, Nonoguchi H, and Mammo F (1992) Polymerase chain reaction localization of constitutive nitric oxide synthase and soluble guanylyl cyclase messenger RNAs in microdissected rat nephron segments. *J Clin Invest* 90:659–665.

Theuring F, Schmager F, and Thone-Reinicke C (1995) Transgenic mice in endothelin research. *Fourth Int Conf on Endothelin*;London UK:P146 (Abstract).

Tikkanen I, Uhlenius N, and Tikkanen T (1994) Renoprotective role of nitric oxide (NO) in Heymann nephritis (HN). *J Am Soc Nephrol* 5:594 (Abstract).

Tipnis SR, Hooper NM, Hyde R, Karran E, Christie G, and Turner AJ (2000) A human homolog of angiotensin-converting enzyme: cloning and functional expression as a captopril-insensitive carboxypeptidase. *J Biol Chem* 275:33238–33243.

Tojo A, Gross SS, and Zhang L (1994) Immunocytochemical localization of distinct isoforms of nitric oxide synthase in the juxtaglomerular apparatus of the normal kidney. *J Am Soc Nephrol* 4:1438–1447.

Tolins JP, and Raij L (1988) Adverse effect of amphotericin B on renal hemodynamics in the rat: Neurohumoral mechanisms and influence of calcium channel blockade. *J Pharmacol Exp Ther* 245:594–599.

Tolins JP, and Raij L (1991) Effects of aminoacid infusion on renal hemodynamics: Role of endothelium-derived relaxing factor. *Hypertension* 17:1045–1051.

Tomasoni S, Noris M, Zappella S, Gotti E, Casiraghi F, Bonazzola S, Benigni A, and Remuzzi G. (1998) Upregulation of renal and systemic cyclooxygenase-2 in patients with active lupus nephritis. *J Am Soc Nephrol* 9:1202–1212.

Tomita K, Nonoguchi H, and Marumo F (1990) Effects of endothelin on peptide-dependent cyclic adenosine monophosphate accumulation along the nephron segments of the rat. *J Clin Invest* 85:2014–2018.

Torres VE (1982) Present and future of the non-steroidal anti-inflammatory in nephrology. *Mayo Clin Proc* 57:389–393.

Tsutsumi Y, Matsubara H, Masaki H, Kurihara H, Murasawa S, Takai S, Miyazaki M, Nozawa Y, Ozono R, Nakagawa K, Miwa T, Kawada N, Mori Y, Shibasaki Y, Tanaka Y, Fujiyama S, Koyama Y, Fujiyama A, Takahashi H, and Iwasaka T (1999) Angiotensin II type 2 receptor overexpression activates the vascular kinin system, and causes vasodilatation. *J Clin Invest* 104:925–935.

Ujiie K, Yuen J, Hogarth L, Danziger R, and Star RA (1994) Localization and regulation of endothelial NO synthase mRNA expression in rat kidney. *Am J Physiol* 267:F296–F302.

Ura N, Carretero OA, and Erdos e.g., (1987) Role of renal endopeptidase 24.11 in kinin metabolism *in vitro* and *in vivo*. *Kidney Int* 32:507–513.

Valdenaire O, Lapailleur-Enouf D, Egidy G, Thouard A, Barret A, Vranckx R, Tougard C, and Michel JB (1999) A fourth isoform of endothelin-converting enzyme (ECE-1) is generated from an additional promoter. Molecular cloning and characterization. *Eur J Biochem* 264:341–349.

Valdenaire O, Rohrbacher E, and Mattei MG (1995) Organization of the gene encoding the human endothelin-converting enzyme (ECE-1). *J Biol Chem* 270:29794–29798.

Vallon V, Traynor T, Barajas L, Huang YG, Briggs JP, and Schnermannn J (2001) Feedback control of glomerular vascular tone in neuronal nitric oxide synthase knockout mice. *J Am Soc Nephrol* 12:1599–1606.

van der Heide J, Bilo H, Donker A, Wilmink J, Sluiter W, and Tegzess A (1992) The effect of dietary supplementation with fish oil on renal function and the course of early postoperative rejection episodes in cyclosporine-treated renal transplant recipients. *Transplantation* 54:257–263.

Vaziri ND, Ni Z, Zhang YP, Ruzics EP, Maleki P, and Ding Y (1998) Depressed renal and vascular nitric oxide synthase expression in cyclosporine-induced hypertension. *Kidney Int* 54:482–491.

Veis JH, Dillingham MA, and Berl T (1990) Effects of prostacyclin on the cAMP system in cultured rat inner medullary collecting duct cells. *Am J Physiol* 258:F1218–1223.

Vidal MJ, Romero JC, and Vanhoutte PM (1988) Endothelium-derived relaxing factor inhibits renin release. *Eur J Pharmacol* 149:401–402.

Von Willebrand E, and Hayry P (1983) Cyclosporine A deposits in renal allografts. *Lancet* 2:189–192.

Wagner OF, Christ G, and Wojta J (1992) Polar secretion of endothelin-1 by cultured endothelial cells. *J Biol Chem* 267:16066–16068.

Walker PD, and Shah SV (1988) Evidence suggesting a role for hydroxyl radical in gentamicin-induced acute renal failure in rats. *J Clin Invest* 81:334–341.

Walker PD, Barri Y, and Shah SV (1999) Oxidant mechanisms in gentamicin nephrotoxicity. *Ren Fail* 21:433–442.

Wallace P, and Leiper J (2002) Blocking NO synthesis: how, where and why? *Nature Rev Drug Discov* 1:939.

Walsh TJ, Finberg RW, Arndt C, Hiemenz J, Schwartz C, Bodensteiner D, Pappas P, Seibel N, Greenberg RN, Dummer S, Schuster M, and Holcenberg JS (1999) Liposomal amphotericin B for empirical therapy in patients with persistent fever and neutropenia. *N Engl J Med* 340:764–771.

Wang A, Bashore T, and Holcslaw T (1998a) Randomized prospective double blind multicentre trial of endothelin receptor antagonist in the prevention of contrast nephrotoxicity. *J Am Soc Nephrol* 9:137A (Abstract).

Wang T (1997) Nitric oxide regulates HCO_3^- and Na^+ transport by a cGMP-mediated mechanism in the kidney proximal tubule. *Am J Physiol* 272:F242-F248.

Wang T, Giebisch G, and Aronson PS (2002) Use of transgenic animals to study renal acid–base transport. *J Nephrol* 15 (Suppl. 5):S151–S160.

Wang T, Inglis FM, and Kalb RG (2000) Defective fluid and $HCO3^-$ adsorption in proximal tubule of neuronal nitric oxide synthase-knockout mice. *Am J Physiol* 279:F518–F524.

Wang Y, and Marsden PA (1995) Nitric oxide synthases: biochemical and molecular regulation. *Curr Opin Nephrol Hypertens* 4:12–22.

Wang Y, Chen J, Chen L, Tay Y-C, Rangan GK, and Harris DCH (1997) Induction of monocyte chemoattractant protein-1 in proximal tubule cells by urinary protein. *J Am Soc Nephrol* 8:1537–1545.

Wang Z-Q, Moore AF, Ozono R, Siragy HM, and Carey RM (1998b) Immunolocalization of subtype 2 angiotensin II (AT2) receptor protein in the rat heart. *Hypertension* 32:78.–83.

Wasan KM, Vadiel K, Lopez-Berenstein G, Verani RR, and Luke DR (1990) Pentoxifylline in amphotericin B toxicity rat model. *Antimicrob Agents Chemother* 34:241–244.

Webster ME, and Gilmore JP (1964) Influence of kallidin-10 on renal function. *Am J Physiol* 206:714.

Weinberg JB, Granger DL, Pisetsky DS, Seldin MF, Misukonis MA, Mason SN, Pippen AM, Ruiz P, Wood ER, and Gilkeson GS (1994) The role of nitric oxide in the pathogenesis of spontaneous murine autoimmune disease: increased nitric oxide production and nitric oxide synthase expression in MLR-lpr/lpr mice, and reduction of spontaneous glomerulonephritis and arthritis by orally administered NG-monomethyl-L-arginine. *J Exp Med* 179:651–660.

Wenzel UO, and Abboud HE (1995) Chemokines and renal disease. *Am J Kidney Dis* 26:982–994.

Whelton A, and Hamilton CW (1991) Nonsteroidal anti-inflammatory drugs: Effects on kidney function. *J Clin Pharmacol* 31:588–598.

Whelton A, and Watson AJ (1998) Nonsteroidal anti-inflammatory drugs: Effects on kidney function in *Clinical Nephrotoxins*, De Broe ME, Porter GA, Bennett WM, Verpooten GA (Eds), Kluwer Academic Publishers, Dordrecht, p. 203.

Wilcox CS, Welch WJ, and Murad F (1992) Nitric oxide synthase in macula densa regulates glomerular capillary pressure. *Proc Natl Acad Sci USA* 89:11993–11997.

Wilkes BM, Ruston AS, and Mento P (1991) Characterization of the endothelin-1 receptor and signal transduction mechanisms in rat renal medullary interstitial cells. *Am J Physiol* 260:F579–F589.

Williams PD, Hattendorf GH, and Bennett DB (1986) Inhibition of renal membrane binding and nephrotoxicity of aminoglycosides. *J Pharmacol Exp Ther* 237:919–925.

Williams PD, Holohan PD, and Ross CR (1981) Gentamicin nephrotoxicity. I Acute biochemical correlates in the rat. *Toxicol Appl Pharmacol* 61:234–242.

Wingard JR, Kubilis P, Lee L, Yee G, White M, Walshe L, Bowden R, Anaissie E, Hiemenz J, and Lister J (1999) Clinical significance of nephrotoxicity in patients treated with amphotericin B for suspected or proven aspergillosis. *Clin Infect Dis* 29:1402–1407.

Wolf G, and Neilson EG, (1993) Angiotensin II as a hypetrophogenic cytokine for proximal tubular cells. *Kidney Int* 43 (Suppl 39):S100–S107.

Wong PC, Hart SD, and Timmermans P (1991) Effect of angiotensin II antagonism on canine renal sympathetic nerve function. *Hypertension* 17:1127–1134.

Xu D, Emoto N, and Giaid A (1994) ECE-1: A membrane-bound metalloprotease that catalyses the proteolytic activation of big endothelin-1. *Cell* 78:473–485.

Yamamoto A, Keil LC, and Reid IA (1992) Effect of intrarenal bradykinin infusion on vasopressin release in rabbits. *Hypertension* 19:799–803.

Yanagisawa M, Kurihara H, and Kimura S (1988) A novel potent vasoconstrictor peptide produced by vascular endothelial cells. *Nature* 332:411–415.

Yang CL, Du XH, and Han YX (1995) Renal cortical mitochondria are the source of oxygen free radicals enhanced by gentamicin. *Ren Fail* 17:21–26.

Yared A, Albrightson-Winslow C, Griswold D, Takahashi K, Fogo A, and Badr KF (1991) Functional significance of leukotriene B4 in normal and glomerulonephritic kidneys. *J Am Soc Nephrol* 2:45–56.

Yokohama C, and Tanabe T (1989) Cloning of human gene encoding prostaglandin endoperoxide synthase and primary structure of the enzyme. *Biochem Biophys Res Commun* 165:888–894.

Yoshioka T, Fogo A, and Beckman JK (1992) Reduced activity of antioxidant enzymes underlies contrast media-induced renal injury in volume depletion. *Kidney Int* 41:1008–1015.

Yoshioka T, Mitarai T, Kon V, Deen WM, Rennke HG, and Ichikawa I (1986) Role for angiotensin II in an overt functional proteinuria. *Kidney Int* 30:538–545.

Yoshioka T, Rennke HG, Salant DJ, Deen WM, and Ichikawa I (1987) Role of abnormally high transmural pressure in the permselectivity defect of glomerular capillary wall: A study in early passive Heymann nephritis. *Circ Res* 61:531–538.

Zatz R, and De Nucci G (1991) Effects of acute nitric oxide-inhibition on rat glomerular microcirculation. *Am J Physiol* 261:F360–F363.

Zatz R, Dunn BR, and Meyer TW (1986) Prevention of diabetic glomerulopathy by pharmacological amelioration of glomerular capillary hypertension. *J Clin Invest* 77:1925–1930.

Ziegler TW, Ludens JH, Fanesti DD, and Talner LB (1975) Inhibition of active sodium transport by radiographic contrast media. *Kidney Int* 7:68–76.

Zoja C, Benigni A, Noris M, Corna D, Casiraghi F, Pagnoncelli M, Rottoli D, Abbate M, and Remuzzi G (2001) Mycophenolate mofetil combined with a cyclooxygenase-2 inhibitor ameliorates murine lupus nephritis. *Kidney Int* 60:653–663.

Zoja C, Corna D, Bruzzi I, and Remuzzi G (1996) Passive Heymann nephritis: Evidence that angiotensin-converting enzyme inhibition reduces proteinuria and retards renal structural injury. *Exp Nephrol* 4:213–221.

Zoja C, Donadelli R, Colleoni S, and Remuzzi G (1998a) Protein overload stimulates RANTES production by proximal tubular cells depending on NF-kB activation. *Kidney Int* 53:1608–1615.

Zoja C, Furci L, Ghilardi F, Zilio P, Benigni A, and Remuzzi G (1986) Cyclosporine-induced endothelial cell injury. *Lab Invest* 55:455–462.

Zoja C, Liu XH, Abbate M, Corna D, Schiffrin EL, Remuzzi G, and Benigni A (1998b) Angiotensin II blockade limits tubular protein overreabsorption and the consequent up-regulation of endothelin 1 gene in experimental membranous nephropathy. *Exp Nephrol* 6:121–131.

Zoja C, Morigi M, Figliuzzi M, and Remuzzi G (1995) Proximal tubular cell synthesis and secretion of endothelin-1 on challenge with albumin and other proteins. *Am J Kidney Dis* 26:934–941.

15

Analgesic Nephropathy

Marc E. de Broe

INTRODUCTION

Classic analgesic nephropathy is a slowly progressive disease resulting from the daily use over many years of mixtures containing at least two antipyretic analgesics, and usually including caffeine or codeine (or both), which may lead to psychological dependence. This type of nephropathy has never been described after the intake of a single analgesic substance. The nephropathy is characterized by renal papillary necrosis and chronic interstitial nephritis, with an insidious progression to renal failure, sometimes in association with transitional-cell carcinoma of the uroepithelium (De Broe et al., 1996; Duggin, 1996; Gloor, 1961; Henrich et al., 1996; Kincaid-Smith and Nanra, 1993; Spuehler and Zollinger, 1953).

In the early stages of the disease, the clinical symptoms are limited to polyuria, sometimes associated with sterile pyuria, and renal colic, which is occasionally associated with acute renal failure due to bilateral obstruction of the urinary tract. Macroscopic hematuria and microscopic hematuria are seen within the setting of sloughing and elimination of necrotic papillae or as a result of transitional-cell carcinoma. Further progression of the disease is accompanied by the nonspecific symptoms of advanced renal failure.

Analgesic nephropathy is part of a broad spectrum of clinical findings, as discussed in this chapter. In the early 1990s, the incidence of analgesic nephropathy among patients receiving dialysis was 0.8% in the U.S., 3% in Europe, and 9% in Australia. In addition to the "classical analgesic nephropathy," there are several epidemiological reports in the literature that suggest that excessive exposure to analgesics and nonsteroidal anti-inflammatory drugs (NSAIDs) may contribute to the progression of a chronic renal disease of mixed etiology towards endstage renal failure (Fored et al., 2001; Morlans, 1990; Pommer (1993); Sandler et al., 1989; Perneger et al., 1994).

This chapter focuses on the clinical aspects, diagnosis, differential diagnosis, and prevention of classical analgesic nephropathy.

CLINICAL MANIFESTATION OF CLASSICAL ANALGESIC NEPHROPATHY

The clinical features of the analgesic syndrome (Duggan, 1974) are many and highly varied and correspond to those of a multisystem disease. Women are affected five to seven times more often than men. The diagnosis is made more frequently with increasing age, and is very rare in patients younger than 30 years old.

Renal Manifestations

Analgesic nephropathy is frequently asymptomatic for years, until the late stages of renal insufficiency. Disordered tubular function, such as impaired urinary concentration and acidification, is the earliest renal manifestation. Renal salt losses and clinically manifest tubular acidosis occur in 10% of the patients with advanced renal insufficiency (Nanra, 1992; Nanra et al., 1978). Sterile leucocyturia and slight proteinuria may be present, although their diagnostic predictive value is limited (Elseviers et al., 1992). A protein excretion of more than 3 g/24 hr indicates a concomitant glomerular lesion. Hematuria must be considered as indicating a fresh renal papillary necrosis with sequestration or a concomitant complication, such as a urinary tract infection or (in later stages) an uroepithelial carcinoma. Bacterial urinary tract infections occur frequently, particularly in the later stages of the disease (Gsell, 1974; Murray and Goldberg, 1978; Nanra et al., 1978).

Arterial hypertension is present in about 50% of patients, sometimes in association with atheromatosis or renal artery stenosis, and this can determine the clinical picture and rate of progression of the renal insufficiency. Acute papillary necrosis can elicit a hypertensive crisis in which a paradoxical water and sodium depletion may be present (Nanra et al., 1978). Papillary necrosis itself mostly has a clinically silent course, only causing symptoms when papillary tissue is shed into the efferent urinary tract, giving rise to hemorrhage, ureteral colic, or ureteral obstruction. Calcified papillary necrosis may manifest as urinary calculi. With a decreasing glomerular filtration rate, all the metabolic consequences of renal insufficiency occur. Kincaid-Smith (1988) estimated that only about 10% of patients with manifest analgesic nephropathy reach the terminal stage of renal insufficiency. Progression is frequently slow and the individual symptoms often show no clinical difference from other kidney diseases. In patients with analgesic nephropathy and terminal renal insufficiency, complex urinary tract infections occur more frequently.

Urinary Tract Malignancy

Transitional cell carcinomas of the renal pelvis, ureter, and bladder (which may be multiple and bilateral) all occur with increased frequency in this setting (Blohme and Johansson, 1981; McCredie et al., 1986; Piper et al., 1985). The incidence of renal cell carcinoma may also be enhanced, but this remains controversial (Chow et al., 1994).

It is estimated that a urinary tract malignancy will develop in as many as 8 to 10% of patients with analgesic nephropathy (Blohme and Johansson, 1981; McCredie, 1986; Piper et al., 1985), but in well under 1% of users of phenacetin-containing analgesic without kidney disease (Dubach et al., 1991). In women under the age of 50, for example, analgesic abuse is the most common cause of bladder cancer, an otherwise unusual disorder in young women (Piper et al., 1985). The potential magnitude of this problem has also been illustrated by histologic examination of nephrectomy specimens obtained prior to renal transplantation; the incidence of urothelial atypia in this setting approaches 50% (Blohme and Johansson, 1981).

The tumors generally become apparent after 15 to 25 years of analgesic abuse (Blohme and Johansson, 1981), usually (but not always) in patients with clinically evident analgesic nephropathy (McCredie et al., 1986). Most patients are still taking the drug at the time of diagnosis, but clinically evident disease can first become apparent several years after cessation of analgesic intake and even after renal transplantation has been performed (Blohme and Johansson, 1981). In Australia, for example, the incidence of analgesic nephropathy declined progressively in the first 10 years after phenacetin-containing compounds were removed from over-the-counter analgesic combinations and 5 years after over-the-counter sales of analgesic mixtures were banned (McCredie et al., 1989). By comparison, the incidence of urinary tract malignancy continued to rise (at a greater rate than other malignancies), a possible reflection of late phenacetin-induced injury (McCredie et al., 1989).

It is presumed that the induction of malignancy results from the intrarenal accumulation of *N*-hydroxylated phenacetin metabolites that have potent alkylating action (McCredie et al., 1986). Because of urinary concentration, the highest concentration of these metabolites will be in the renal medulla, ureters, and bladder, possibly explaining the predisposition to carcinogenesis at these sites.

The pathogenetic importance of phenacetin metabolites is suggested indirectly from the observation that there appears to be no association between tumor formation and the prolonged ingestion of analgesics that can cause papillary necrosis but do not form these metabolites, such as acetaminophen and the NSAIDs (Castelao et al., 2000; Jensen et al., 1989; McCredie and Stewart, 1988; Nanra, 1993).

The major presenting symptom of urinary tract malignancy in analgesic nephropathy is microscopic or gross hematuria. Therefore, continued monitoring is essential. New hematuria should be evaluated by urinary cytology, and, if indicated, cystoscopy with retrograde pyelography (Blohme and Johansson, 1981). It may also be prudent to obtain yearly urine cytology for the first several years if analgesics are discontinued, or indefinitely if drug intake persists. The incidence of urothelial carcinoma after renal transplantation in patients with analgesic nephropathy is comparable with the general incidence of up to 10% of urothelial carcinomas in end-stage renal failure patients with analgesic nephropathy. Removal of the native kidneys prior to renal transplantation has also been suggested, but the efficacy of this regimen has not been proven (Blohme and Johansson, 1981).

Extrarenal Manifestations

Most of these extrarenal findings were obtained in clinical and autopsy studies in which consumers of analgesics containing phenacetin and paracetamol were investigated. Very few findings have been verified in controlled studies.

Cardiovascular Complications

Accelerated atherosclerosis, arterial hypertension, renal artery stenoses, and increased cardiovascular mortality are frequently associated with chronic analgesic consumption (Dubach et al., 1983; Kaladelfos and Edwards, 1976; Kincaid-Smith, 1979; Mihatsch et al., 1982; Nanra, 1980; Schwarz et al., 1984). Most of these studies are retrospective, and individual findings have almost never been controlled with regard to other cardiovascular risk factors.

In the only prospective cohort study, increased cardiovascular mortality was demonstrated in female consumers of phenacetin-containing analgesics (Dubach et al., 1983), but no additional risk factors were investigated. While subjects were stratified by urine concentration of *N*-acetyl-*p*-aminophenol, which may derive from either phenacetin or paracetamol, the etiological role of the anilides (and other analgesic components) remains unknown. In an autopsy study, an increased prevalence of myocardial infarction in phenacetin misusers compared with a control population could not be confirmed. Differences in the frequency of hypertension became less obvious when the degree of restricted kidney function was considered (Mihatsch et al., 1982). In a controlled study, an increased incidence of fatal cerebral hemorrhage was found in dialysis patients with analgesic nephropathy (Chachati et al., 1987).

Gastrointestinal Complications

Peptic ulcers and erosive gastritis are regarded as the most common concomitant diseases in chronic analgesic consumption and are frequently related to the acetylsalicylic acid component of combination analgesics (Gault and Wilson, 1978; Gsell, 1974; Mihatsch et al., 1982; Nanra, 1980; Prescott, 1982). In the Basle autopsy study, gastric ulcers, but not duodenal ulcers, were more frequent in analgesic misusers (Mihatsch et al., 1982). In a prospective clinical investigation, chronic pancreatitis was found more frequently in patients with analgesic nephropathy than in the control group with other renal diseases (Hangartner et al., 1987).

Hematological Complications

Formerly, hematological changes, such as agranulocytosis, were an early manifestation of the analgesic syndrome (Prescott, 1982; Sarre, 1958). Currently, with the altered spectrum of analgesic preparations they are likely to be less common. An anemia inappropriate to the degree of renal insufficiency and mild hemolysis, partly in combination with splenomegaly and methemoglobinemia, were ascribed to phenacetin. However, in the comparative autopsy study mentioned above, there was no significant enlargement of the spleen in phenacetin misusers (Mihatsch et al., 1982). Chronic hemorrhagic anemia due to the acetylsalicylic acid component of mixed analgesics has been observed, especially in English-speaking countries (Carro-Ciampi, 1978).

Skeletal Complications

Skeletal problems are much more frequently symptomatic in analgesic misusers than in other patients with renal insufficiency (Fassett et al., 1982; Fellner and Tuttle, 1969; Jaeger et al., 1982). At comparable stages and duration of renal failure, patients with analgesic nephropathy have lower blood calcium and consequently higher parathyroid hormone and alkaline phosphatase levels when compared with patients with interstitial nephritis of other origin, suggesting more severe vitamin D deficiency and osteodystrophy.

Psychosomatic Aspects

A large number of psychopathological alterations have been described in chronic analgesic consumers (Carro-Ciampi, 1978; Gsell, 1974; Murray, 1978; Prescott, 1982). In one of the few controlled psychometric investigations, in women from Swiss watch factories, heavy analgesic consumers were described by the Freiburg Personality Inventory as significantly more nervous and sensitive to irritation, depressive, more easily frustrated, and emotionally labile (Ladewig et al., 1979). In the period preceding an analgesic dependence, disturbances of subjective well-being are frequently associated with headaches, sleep disorders, gastrointestinal complaints, cardiac symptoms, and lumbar, as well as shoulder and neck, pain. Factors giving rise to the intake of analgesics are emotional tension and overstrain. Additionally, specific occupational demands resulting from shiftwork and piecework in various industrial production plants, and access to analgesics at the workplace, played a role in the establishment of their misuse (Gsell, 1974; Sarre et al., 1958). Family habits are likely to be important for the inception of chronic analgesic consumption (Carro-Ciampi, 1978;

Gsell, 1974; Murray, 1978). The psychotropic effect of analgesic substances and the centrally active additives in combination analgesics may lead on the one hand to perpetuation of the chronic intake and on the other possibly to chronic pain states (Diener, 1988; Prescott, 1982; Woerz, 1983). It has been argued that these effects are restricted to phenacetin-containing preparations (Feinstein et al., 2000).

DIAGNOSIS AND DIFFERENTIAL DIAGNOSIS

Until recently, the diagnosis of analgesic nephropathy was based on nonspecific clinical and biochemical signs, certain findings on renal imaging with an unknown diagnostic yield, and a history of analgesic abuse. Despite the nonspecific nature of the renal presentation, frequently there are other findings that point towards the presence of analgesic nephropathy (Murray and Goldberg, 1978; Nanra et al., 1978). Most patients are women between the ages of 30 and 70. Careful questioning often reveals a history of chronic headaches or lower-back pain that leads to the analgesic use. Also common are other somatic complaints (such as malaise and weakness), and ulcer-like symptoms or a history of peptic ulcer disease due to chronic aspirin ingestion, with or without associated exposure to NSAIDs. Most patients have no symptoms referable to the urinary tract, although flank pain or hematuria from a sloughed or obstructing papilla may occur. Urinary tract infection is also somewhat more common in women with this disorder.

A history of analgesic abuse is, however, often difficult to obtain, since most patients deny having such a history. This lack of reliable criteria and the high prevalence of analgesic nephropathy in the dialysis population in the 1980s in Belgium (17.9% in 1984) led to the undertaking of three prospective, multicenter, controlled studies to define and validate diagnostic criteria for this disease (Elseviers et al., 1992, 1995a).

In the first study, all 273 patients at 13 Belgian dialysis centers who entered the dialysis program were divided into case patients and controls (Elseviers et al., 1992). Case patients consisted of 85 patients who were each found to have a history of analgesic abuse based on an interview, a medical-chart analysis, and a review of additional information provided by the patient's nephrologist. The interview focused on the use of medications for the relief of symptoms such as headache and joint pain. To help identify the medications that were being used, the patients were shown a book containing color photographs of the 12 analgesics most frequently sold in Belgium. Analgesic abuse was defined as daily use of analgesics for at least 5 years, with a minimum total dose of 3000 units (1 unit = 1 tablet or 1 dose of powder). The controls consisted of all 188 other patients with no history of analgesic abuse. Findings of a decrease in the length of the kidneys, irregular ("bumpy") contours, and papillary calcifications on sonography and conventional tomography had a sensitivity of 72% and a specificity of 97%, given a positive predictive value of 92%. In contrast to what is written in several textbooks of medicine and nephrology, the sensitivity and specificity of nonspecific signs, such as hypertension, anemia, sterile pyuria, bacteriuria, and slight proteinuria, were insufficient to be helpful in the diagnosis of analgesic nephropathy in patients with end-stage renal disease (Figure 15.1) (Elseviers et al., 1992). The results this study were corroborated by the second study, conducted in 12 European countries and Brazil (Elseviers et al., 1995b).

The third study compared the usefulness of sonography, conventional tomography, and computed tomography (CT) without contrast medium for the diagnosis of analgesic nephropathy. In 40 patients with end-stage

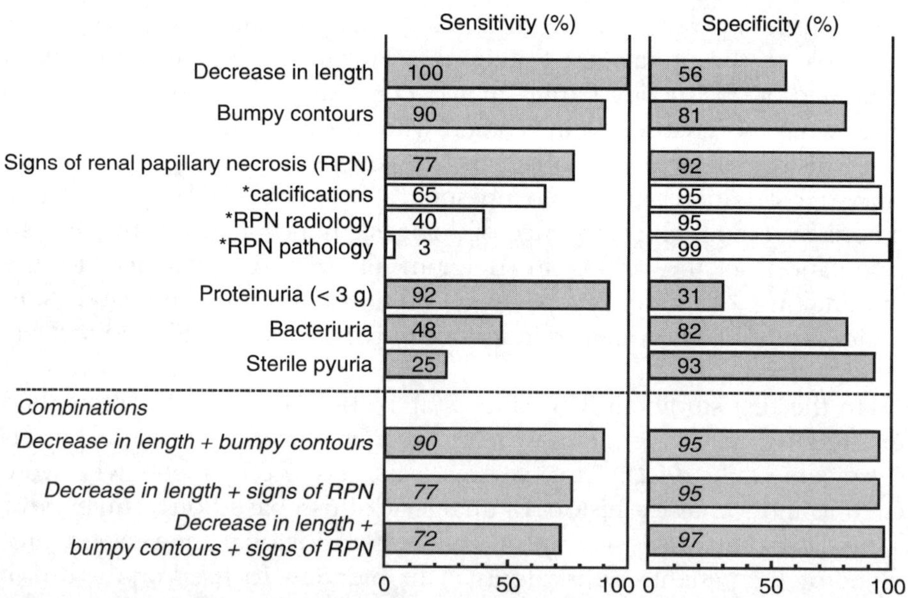

Figure 15.1 Sensitivity and specificity of the criteria of analgesic nephropathy (Elseviers et al., 1992).

Table 15.1 Sensitivity and Specificity of CT Imaging Criteria Used to Diagnose Analgesic Nephropathy

Finding on renal imaging	*Sensitivity (%)*	*Specificity (%)*
Decrease in length		
Patients with ESRD	95	10
Patients with RF	77	86
Bumpy contours		
Patients with ESRD	50	90
Patients with RF	62	93
Papillary calcifications		
Patients with ESRD	87	97
Patients with RF	92	100
Decrease in length and either bumpy contours or papillary calcifications		
Patients with ESRD	90	90
Patients with RF	77	100

Notes: 40 patients with end-stage renal disease (ESRD) and 53 patients with incipient, mild, or moderate renal failure (RF). CT was performed without contrast medium. In patients with incipient, mild, or moderate renal failure, serum creatinine concentrations ranged from 1.5 to 4 mg/dl (133 to 354 μmol/l).
Source: Data were adapted from Elseviers et al. (1995a).

renal disease, renal size should be evaluated with similar accuracy with all three techniques, but CT scanning was the best method for detecting papillary calcifications (sensitivity, 87%, specificity, 97%) (Table 15.1). The results were similar for the group of 53 patients with incipient, mild or moderate renal failure (serum creatinine concentration of 1.5 to 4 mg/dl (133 to 354 μmol/l). The sensitivity of CT scanning for papillary calcifications was 92%, with a specificity of 100% (Table 15.1). The finding of papillary calcifications in the early stages of analgesic nephropathy corroborates experimental observations in rats (Prescott, 1982).

On the basis of the results of these three studies, CT scanning without contrast medium is recommended to diagnose or rule out analgesic nephropathy as the possible cause of renal disease in patients with end-stage renal disease as well as in patients with mild and moderate renal failure, even in the absence of reliable information on the use of analgesics.

The finding of a bilateral decrease in length of the kidneys combined with bilateral bumpy contours or papillary calcifications (papillary necrosis) has a high diagnostic performance for that particular disease.

Renal biopsy is of little use in verification of the diagnosis; the sample contains mainly cortical tissue, and the pathognomonic medullary alterations, renal papillary necrosis, and capillary sclerosis can only be detected in exceptional cases that contain medullary tissue.

Papillary necrosis is not diagnostic of analgesic nephropathy, because similar changes can occur in a variety of other diseases. These include diabetes mellitus (particularly during an episode of acute pyelonephritis), urinary tract obstruction, sickle cell anemia or trait, and renal tuberculosis. However, the history and appropriate laboratory tests can usually differentiate these disorders from analgesic-induced disease (Figure 15.2).

COURSE OF RENAL DISEASE

The course of the renal disease depends both on the severity of the renal damage at the time of presentation and on whether drug therapy is discontinued (Buckalew and Schey, 1986; Gault and Wilson, 1978; Murray and Goldberg, 1978). The decline in renal function can be expected to progress if analgesics are continued. Even aspirin, which is generally not nephrotoxic when given alone (Sandler et al., 1989), can promote further renal damage in analgesic nephropathy (Buckalew and Schey, 1986; Gault and Wilson, 1978).

Conversely, renal function stabilizes or mildly improves in most patients if analgesic consumption is discontinued (Buckalew and Schey, 1986; Gault and Wilson, 1978). If, however, the renal disease is already advanced, progression may occur in the absence of drug intake, presumably due to secondary hemodynamic and metabolic changes associated with nephron loss (Garber et al., 1999).

The late course of analgesic nephropathy may also be complicated by two additional problems, malignancy and atherosclerotic disease.

PREVENTION OF RENAL DISEASE

Analgesic nephropathy is one of the few renal diseases for which primary prevention is suitable.

Information campaigns that were focused on the population at risk did not solve the problem of analgesic nephropathy. In Belgium, it was clearly demonstrated that in most abusers, sustained analgesic

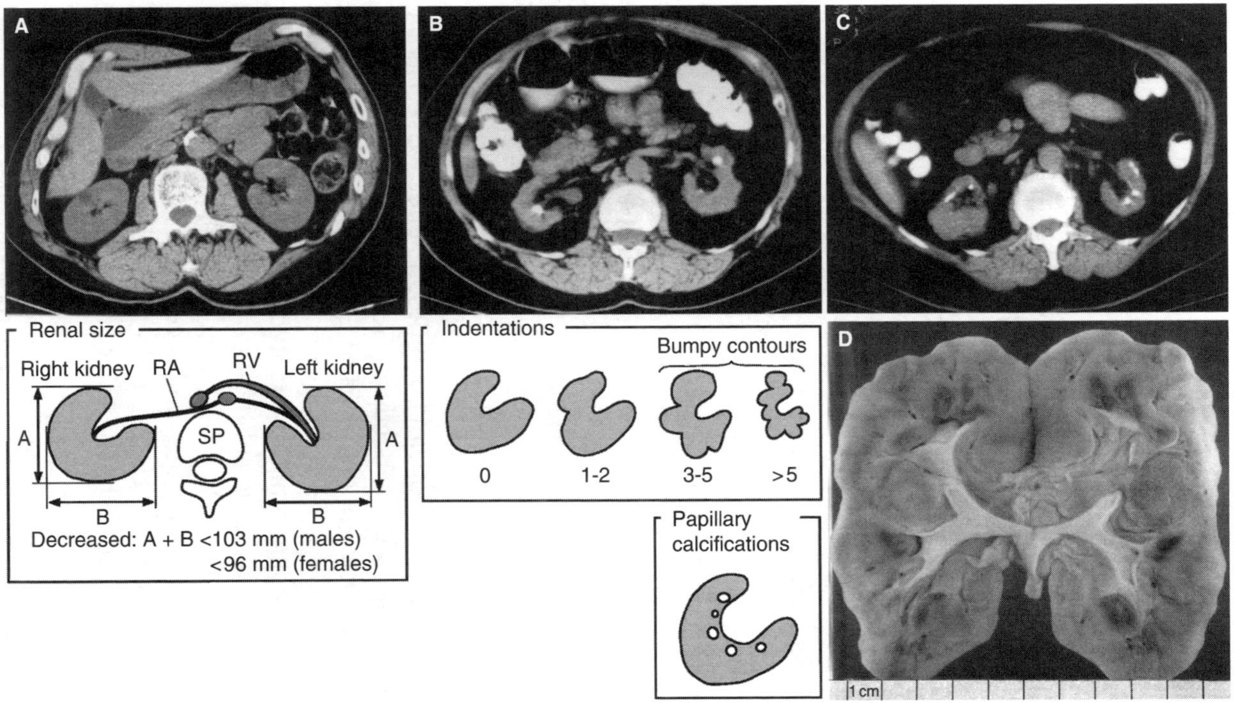

Figure 15.2 Renal imaging criteria of analgesic nephropathy as observed in CT scan without contrast medium, including a decreased renal size, bumpy contours, and papillary calcifications. (A) Normal kidney. (B) Incipient renal failure (serum creatinine = 2.3 mg/dl). (C) End-stage renal failure. (D) Postmortem kidney of an analgesic nephropathy patient with moderate renal failure. Irregular contours of the kidney, fibrotic-necrotic foci of papillae. (RA) renal artery, (RV) renal vein, (SP) spine. (Elseviers et al., 2003.)

consumption was no longer related to a physical complaint. Instead, analgesics were mainly taken for their mood-altering capacities. Most analgesic abusers admitted to having been informed of the health risks related to long-standing analgesic abuse, and even if renal impairment occurred, only a few of the cases stopped their analgesic abuse (Elseviers and De Broe, 1994).

The withdrawal of phenacetin from analgesic mixtures also did not solve the problem. When phenacetin was withdrawn from most analgesic mixtures in Australia (in the 1970s), no decline in the occurrence of analgesic nephropathy could be observed (Kincaid-Smith, 1979, 1986). A declining incidence rate was only observed after restriction of the over-the-counter sales of all analgesic mixtures (between 1979 and 1980) (ANZDATA, 2000; Nanra and Kincaid-Smith, 1993; Stewart et al., 1994). Some countries in Europe, particularly Sweden, have succeeded in controlling the disease after legislative measures were taken. As early as the 1960s, Sweden drafted legislation that only a few years later had become very effective.

The legislation was simple and clear: all analgesics containing even the slightest dose of phenacetin became prescription limited. This resulted in a prescription status for almost all combined analgesics, hence a dramatic drop in their sale. In spite of the substantial total increase of consumption of single analgesics between 1980 and 1990, analgesic nephropathy in Sweden now belongs to the history of medicine and nephrology (less than 1% of the Swedish dialysis population) (Noels et al., 1995).

In contrast, in many other European countries, no effective legislative measures have been imposed. In Belgium, Germany, and Switzerland, the pharmaceutical industry spontaneously removed phenacetin from their products. Phenacetin was replaced by another analgesic substance, such as pyrazole, maintaining a high volume of analgesic mixtures that containing two or more analgesic substances. In Belgium, the Ministry of Health decided in 1988 that when obtaining analgesic mixtures in the pharmacy, users had to sign a request and receive an information sheet warning for possible renal consequences of extensive analgesic consumption. This resulted in a obvious fall of the sale and consumption of analgesic mixtures. Although these measures were only effective during one year, their indirect effect was more important. After 1988, several pharmaceutical companies modified their analgesic products, resulting in a reduction of the mixtures from two analgesic components to one analgesic plus caffeine or codeine (or both).

Analgesic nephropathy gained recognition in recent years in several Central and Eastern European countries. Abuse of analgesic mixtures is also reported in several developing countries, where there is little knowledge about the extent of the problem of analgesic nephropathy. Moreover, in many countries there are no legislative restrictions on the introduction of analgesic mixtures containing two analgesic substances combined with caffeine or codeine into the market.

In view of prevention, it would be advisable to obtain legislative measures worldwide in order to limit the over-the-counter availability of all analgesics containing two analgesic components plus caffeine or codeine. This has been formally proposed in Europe (Elseviers and De Broe, 1996) as well as in the U.S. (Henrich et al., 1996) by a large group of investigators active in the field.

REFERENCES

ANZDATA (2000) *Primary renal disease.* ANZDATA Registry Report.

Blohme I, and Johansson S (1981) Renal pelvic neoplasms and atypical urothelium in patients with end-stage analgesic nephropathy. *Kidney Int* 20:671–675.

Buckalew VM Jr, and Schey HM (1986) Renal disease from habitual antipyretic analgesic consumption: An assessment of the epidemiologic evidence. *Medicine* 65:291–303.

Carro-Ciampi G (1978) Phenacetin abuse: A review. *Toxicology* 10:311–339.

Castelao JE, Yuan JM, Gago-Dominguez M, Yu MC, and Ross RK (2000) Non-steroidal anti-inflammatory drugs and bladder cancer prevention. *Br J Cancer* 82:1364–1369.

Chachati A, Dechenne C, and Godon JP (1987) Increased-incidence of cerebral hemorrhage mortality in patients with analgesic nephropathy on hemodialysis. *Nephron* 45:167–168.

Chow WH, McLaughlin JK, Linet MS, Niwa S, and Mandel JS (1994) Use of analgesics and risk of renal cell cancer. *Int J Cancer* 59:467–470.

De Broe ME, and Elseviers MM (1998) Analgesic nephropathy. *New Eng J Med* 338:446–452.

De Broe ME, Elseviers MM, Bengtsson U, Mihatsch MJ, Molzahn M, Pommer W, Ritz E, and Schwarz A (1996) Analgesic nephropathy. *Nephrol Dialysis Transplant* 11:2407–2408.

Diener HC (1988) Klinik des Analgetikakopfschmerzes. *Deutsche Medizinische Wochenschrift* 113:472–476.

Dubach UC, Rosner B, and Pfister E (1983) Epidemiologic study of abuse of analgesics containing phenacetin: Renal morbidity and mortality (1968–1979). *New Eng J Med* 308:357–362.

Dubach UC, Rosner B, and Sturmer T (1991) An epidemiologic study of abuse of analgesic drugs: Effects of phenacetin and salicylate on mortality and cardiovascular morbidity. *New Eng J Med* 324:155–160.

Duggan JM (1974) The analgesic syndrome. *Aust NZ J Med* 4:365–372.

Duggin GG (1996) Combination analgesic-induced kidney disease: The Australian experience. *Am J Kidney Dis* 28 (Suppl 1):S39–S47.

Elseviers MM, and De Broe ME (1994) Analgesic nephropathy in Belgium is related to the sales of particular analgesic mixtures. *Nephrol Dialysis Transplant* 9:41–46.

Elseviers MM, and De Broe ME (1996) Combination analgesic involvement in the pathogenesis of analgesic nephropathy. *Am J Kidney Dis* 28:958–962 (Letter).

Elseviers MM, Bosteels V, Cambier P, De Paepe M, Godon JP, Lins R, Lornoy W, Matthys E, Moeremans C, Roose R, Theelen B, Van Caesbroeck D, Verbanck J, and De Broe ME (1992) Diagnostic criteria of analgesic nephropathy in patients with end-stage renal failure – results of the Belgian study. *Nephrol Dialysis Transplant* 7:479–486.

Elseviers MM, De Schepper A, Corthouts R, Bosmans JL, Cosyn L, Lins RL, Lornoy W, Matthys E, Roose R, Van Caesbroeck D, Waller I, Horackova M, Schwarz A, and De Broe ME (1995a) High diagnostic performance of CT scan for analgesic nephropathy in patients with incipient to severe renal failure. *Kidney Int* 48:1316–1323.

Elseviers MM, Waller I, Nenov D, Levora J, Matousovic K, Tanquerel T, Pommer W, Schwarz A, Keller E, Thieler H, Köhler H, Lemoniatou H, Cresseri D, Bonnucchi D, and De Broe ME (1995b) Evaluation of diagnostic criteria for analgesic nephropathy in patients with end-stage renal failure: Results of the ANNE study. *Nephrol Dialysis Transplant* 10:808–814.

Elseviers MM, Schwarz A, and De Broe ME (2003) Analgesics and 5-aminosalicylic acid, in *Clinical nephrotoxins – renal injury from drugs and chemicals*, 2nd ed., De Broe ME, Porter GA, Bennett WM, Verpooten GA (Eds), Dordrecht: Kluwer Academic Publishers, pp. 263–278.

Fassett RG, Lien JWK, Mathew TH, and McClure J (1982) Bone disease in analgesic nephropathy. *Clin Nephrol* 18:273–279.

Feinstein AR, Heinemann LAJ, Dalessio D, Fox JM, Goldstein J, Haag G, Ladewig D, and OÇBrien CP (2000) Do caffeine-containing analgesics promote dependence? A review and evaluation. *Clin Pharmacol Ther* 68:457–467.

Fellner SK, and Tuttle EP (1969) The clinical syndrome of analgesic abuse. *Arch Int Med* 124:379–382.

Fored CM, Ejerblad E, Lindblad P, Fryzek JP, Dickman PW, Signobello LB, Lipworth L, Elinder C-G, Blot WJ, McLaughlin JK, Zack MM, and Nyrén O (2001) Acetaminophen, aspirin, and chronic renal failure. *New Eng J Med* 345:1801–1808.

Garber SL, Mirochnik Y, Arruda JA, and Dunea G (1999) Evolution of experimentally induced papillary necrosis to focal segmental glomerulosclerosis and nephrotic proteinuria. *Am J Kidney Dis* 33:1033–1039.

Gault MH, and Wilson DR (1978) Analgesic nephropathy in Canada: Clinical syndrome, management, and outcome. *Kidney Int* 13:58–63.

Gloor F (1961) Die doppelseitige chronische nichtobstruktive interstitielle Nephritis. *Ergebnisse der allgemeinen Pathologie und Pathologischen Anatomie* 41:63–207.

Gsell O (1974) Nephropathie durch Analgetika. *Ergebnisse der inneren Medizin und Kinderheilkunde* 35:68–175.

Hangartner PJ, Buehler H, Muench R, Zaruba K, Stamm B, and Ammann R (1987) Chronische Pankreatitis als wahrscheinliche Folge eines Analgetikaabusus. *Schweizerische Medizinische Wochenschrift* 117:638–642.

Henrich WL, Agodoa LE, Barrett B, Bennett WM, Blantz RC, Buckalew VM Jr, D'Agati VD, De Broe ME, Duggin GG, and Eknoyan G (1996) Analgesics and the kidney: Summary and recommendations to the Scientific Advisory Board of the National Kidney Foundation from an ad hoc committee of the National Kidney Foundation. *Am J Kidney Dis* 27:162–165.

Jaeger P, Burckardt P, and Wauters JP (1982) Ca, P metabolism is particularly disturbed in analgesic abuse nephropathy. *Kidney Int* 21:903–904.

Jensen OM, Knudsen JB, Tomasson H and Sørensen BL (1989) The Copenhagen case-control study of renal pelvis and ureter cancer, role of analgesics. *Int J Cancer* 44:965–968.

Kaladelfos G, and Edwards KDG (1976) Increased prevalence of coronary heart disease in analgesic nephropathy: Relation to hypertension, hypertriglyceridemia and combined hyperlipidemia. *Nephron* 16:388–400.

Kincaid-Smith P (1979) Analgesic nephropathy in Australia. *Contrib Nephrol* 16:57–64.

Kincaid-Smith P (1986) Renal toxicity of non-narcotic analgesics – at-risk patients and prescribing applications. *Med Toxicology* 1 (Suppl 1):14–22.

Kincaid-Smith P (1988) Analgesic-induced renal disease, in *Diseases of the kidney*, 5th ed., Vol. 2, Schrier RW, and Gottschalk CW (Eds), Boston: Little, Brown, pp. 1202–1216.

Kincaid-Smith P, and Nanra RS (1993) Lithium-induced and analgesic-induced renal diseases, in *Diseases of the kidney*, 5th ed., Vol. 2, Schrier RW, and Gottschalk CW (Eds), Boston: Little, Brown, pp. 1099–1129.

Ladewig D, Dubach UC, Ettlin G, and Hobi V (1979) Zur Psychologie des Analgetikakonsums bei berufstaetigen Frauen. Ergebnisse einer epidemiologischen Perspektivstudie. *Nervenarzt* 50:219–224.

McCredie M, and Stewart JH (1988) Does paracetamol alone cause urothelial cancer or renal papillary necrosis? *Nephron* 49:296–300.

McCredie M, Stewart JH, Carter JJ, Turner J, and Mahony JF (1986) Phenacetin and papillary necrosis: Independent risk factors for renal pelvic cancer. *Kidney Int* 30:81–84.

McCredie M, Stewart JH, Mathew TH, Disney AP, and Ford JM (1989) The effect of withdrawal of phenacetin-containing compounds on the incidence of kidney and urothelial cancer and renal failure. *Clin Nephrol* 31:35–39.

Mihatsch MJ, Kernen R, and Zollinger HU (1982) Phenacetinabusus VI: eine Autopsiestatistik unter besonderer Beruecksichtigung extrarenaler Befunde. *Schweizerische Medizinische Wochenschrift* 112:1383–1388.

Morlans M (1990) End-stage renal disease and non-narcotic analgesics: A case–control study. *Br J Clin Pharmacol* 30:717–723.

Murray RM (1978) Genesis of analgesic nephropathy in the United Kingdom. *Kidney Int* 13:50–57.

Murray TG, and Goldberg M (1978) Analgesic-associated nephropathy in the U.S.A.: Epidemiologic clinical and pathogenetic features. *Kidney Int* 13:64–71.

Nanra RS (1980) Clinical and pathological aspects of analgesic nephropathy. *Br J Clin Pharmacol* 10 (Suppl 2):S359–S368.

Nanra RS (1992) Pattern of renal dysfunction in analgesic nephropathy: comparison with glomerulonephritis. *Nephrol Dialysis Transplant* 7:384–390.

Nanra RS (1993) Analgesic nephropathy in the 1990s: An Australian perspective. *Kidney Int* 42 (Suppl):S86–S92.

Nanra RS, and Kincaid-Smith P (1993) Experimental evidence for nephrotoxicity of analgesics, in *Analgesic and NSAID-Induced Kidney Disease*, Stewart JH (Ed.), Oxford: Oxford University Press, pp. 17–31.

Nanra RS, Stuart-Taylor J, de Leon AH, and White K (1978) Analgesic nephropathy: Etiology, clinical syndrome, and clinicopathologic correlations in Australia. *Kidney Int* 13:79–92.

Noels LM, Elseviers MM, and De Broe ME (1995) Impact of legislative measures on the sales of analgesics and the subsequent prevalence of analgesic nephropathy: A comparative study in France, Sweden and Belgium. *Nephrol Dialysis Transplant* 10:167–174.

Perneger TV, Whelton PK, and Klag MJ (1994) Risk of kidney failure associated with the use of acetaminophen, aspirin, and nonsteroidal antiinflammatory drugs. *New Eng J Med* 331:1675–1679.

Piper JM, Tonascia J, and Matanoski GM (1985) Heavy phenacetin use and bladder cancer in women aged 20 to 49 years. *New Eng J Med* 313:292–295.

Pommer W (1993) Clinical presentation of analgesic-induced nephropathy, in *Analgesic and NSAID-Induced Kidney Disease*, Stewart JH (Ed.), Oxford University Press, pp. 108–118.

Prescott LF (1982) Analgesic nephropathy: A reassessment of the role of phenacetin and other analgesics. *Drugs* 23:75–149.

Sandler DP, Smith JC, Weinberg CR, Buckalew VM Jr, Dennis VW, Blythe WB, and Burgess WP (1989) Analgesic use and chronic renal disease. *New Eng J Med* 320:1238–1243.

Sarre H, Moench A, and Kluthe R (Eds) (1958) *Phenacetinabusus und Nierenschaedigung*, Stuttgart: Thieme.

Schwarz A, Pommer W, Keller F, Kuehn-Freitag G, Offermann G, and Molzahn M (1984) Morbidity of patients with analgesic-associated nephropathy and end-stage renal failure. *Proc Eur Dialysis Transplant Assoc–Eur Renal Assoc* 21:311–316.

Spuehler O, and Zollinger HU (1953) Die chronisch-interstitielle Nephritis. *Zeitschrift für Klinische Medicine* 151:1–50.

Stewart JH, McCredie M, Disney APS, and Mathew TH (1994) Trends in incidence of end-stage renal failure in Australia, 1972–1991. *Nephrol Dialysis Transplant* 9:1377–1382.

Woerz R (1983) Effects and risks of psychotropic and analgesic combinations. *Am J Med* 75 (Suppl 5A):139–140.

16

Antibiotic-Induced Nephrotoxicity

Hélène Servais, Marie-Paule Mingeot-Leclercq, and Paul M. Tulkens

INTRODUCTION

Antibiotics have long been, and remain, a major cause of drug-related renal toxicity, causing both acute renal failure and tubulointerstitial disease, depending on the drug. Most cases of acute renal failure are related to acute tubular necrosis, for which aminoglycosides have long been one of the leading causes. Tubulointerstitial disease is more commonly seen with β-lactams (hypersensitivity nephropathy), but can also be caused indirectly by aminoglycosides (tubulointerstitial injury).

Two complicating factors make the nephrotoxicity of antibiotics more common than expected. First, many of these drugs are given in combination, either between themselves or with other drugs. Consequently, the toxicity of one agent may be aggravated by the other one, as exemplified by vancomycin–aminoglycosides combinations, or the coadministration of aminoglycosides and nonsteroidal anti-inflammatory agents. Second, many antimicrobials are removed from the body essentially, or at least predominantly, through the renal route. The serum levels of these drugs will therefore increase as renal function becomes impaired, either as a consequence of the drug toxicity itself or because of concomitant renal damage caused by another drug or by another cause of nephrotoxic reaction. This applies both to drugs eliminated by glomerular filtration and those that are removed from the body by tubular secretion if renal function is severely compromised (creatinine clearance of 20 ml/min or less).

These situations are common in severely ill patients for whom effective antimicrobial chemotherapy must be utilized. Thus, toxic levels of aminoglycosides and vancomycin are often observed in these patients unless close monitoring of both renal function and the drugs is performed. The same may occur with certain penicillins, fluoroquinolones, and tetracyclines, for which drug monitoring is not routinely performed.

These considerations explain many cases of antibiotic-related toxicities, and necessitate careful adjustment of doses, based on age, sex, body weight, and renal function. Clinicians must, however, be warned against inappropriately moving towards low doses of antimicrobials for fear of toxicity (a trend that was frequent in the late 1980s and early 1990s). It is indeed now well established that inefficient antimicrobial therapy leads to clinical failure as well as to risk of resistance (Craig, 2001), thereby exposing the patient to additional toxicities related to the persistence of the infection and the need to prolong the therapy. Efforts must be directed at selecting the most appropriate antibiotic and administering it at the maximal acceptable dose as defined by the peak serum concentration, the area under the 24 hour serum concentration, and the time during which the serum concentration will remain above the critical threshold (Amsden et al., 2000).

This chapter reviews the data available on the three classes of antimicrobial agents that have been most commonly associated with renal toxicities: aminoglycosides, β-lactams, and vancomycin. Other antibiotics will be only briefly touched upon because data are scanty and the mechanisms often less clear.

AMINOGLYCOSIDES

Aminoglycoside antibiotics are polar, cationic molecules that are all rapidly excreted by the kidney without significant metabolism. The first group comprises streptomycin and its closely related derivative dihydrostreptomycin, which were widely used from the late 1940s through the mid 1960s to treat Gram-negative infections and tuberculosis. Emergence of resistance and the desire to enlarge the spectrum towards "difficult to treat bacteria" such as *Pseudomonas aeruginosa* led to the development and introduction of kanamycins (represented mostly by kanamycin A and tobramycin, and, in some countries, by dibekacin), sisomicin, and gentamicin (which is a mixture of three major components referred to as gentamicin C_1, C_{1a}, and C_2 in approximately equimolar amounts). These were extensively used up to the mid 1980s. Further development of resistance to these naturally-occurring molecules eventually triggered the design of semisynthetic derivatives (amikacin [from kanamycin A], netilmicin [from sisomicin], isepamicin [gentamicin B], and arbekacin [from dibekacin]), which were made to resist bacterial inactivating enzymes responsible for this resistance (Mingeot-Leclercq et al., 1999).

Despite all these developments, ototoxicity and nephrotoxicity remained of concern to clinicians and resulted in a marked limitation in the use of these otherwise potent and life-saving antibiotics. Because it proved very difficult, if not impossible, to separate antibacterial efficacy and toxicity (Price, 1986), the pharmaceutical industry largely, if not entirely, abandoned this field of research in the late 1980s. In parallel, there has been an intense effort in laboratory and clinical research (more than 1,500 publications are referenced in public databases since 1969). Therefore, there is now a deep knowledge of the epidemiology, pathology, molecular mechanisms, and clinical significance of these toxicities, but no safe compounds. This explains why efforts at reducing aminoglycoside toxicity have met with such interest since the mid 1990s, and even today these remain the only means of protecting patients.

Epidemiology

The reported incidence of aminoglycoside nephrotoxicity varies from 0 to 50%, with most reports in the 5 to 25% range (Bertino, Jr. et al., 1993; Lane et al., 1977; Lerner et al., 1983). This variability results from differences in definition of nephrotoxicity, nature and frequency of the criteria used to asses renal function, and, perhaps most importantly, the clinical setting in which the drugs are used. The incidence in aged patients suffering from multisystem diseases and exposed to other potential nephrotoxicants ranges as high as 35 to 50%, whereas this figure may be close to zero in young healthy volunteers (Appel, 1990; Lane et al., 1977). In prospective randomized studies with definitions of nephrotoxicity that reflect a substantive decrement of glomerular filtration rate in seriously ill patients, the reported incidence of nephrotoxicity varies between 5 and 10% of patient courses (Lane et al., 1977;

Smith et al., 1977). In surveys of the etiology of acute renal failure in hospitalized patients, about half of the drug-induced cases of renal toxicity were attributable to aminoglycosides.

Clinical Features

Nephrotoxicity induced by aminoglycosides manifests clinically as non-oliguric renal failure, with a slow rise in serum creatinine concentration and a hypoosmolar urinary output developing after several days of treatment. The first signs of tubular dysfunctions or alterations are: release of brush border and lysosomal enzymes, decreased reabsorption of filtered proteins, wasting of potassium, magnesium, calcium, and glucose, and phospholipiduria. Progression to dialysis-dependent oliguric-anuric renal failure is unusual unless other risk factors are present (Appel, 1990). The renal failure is generally reversible. In a few patients there has been documented recovery of renal function despite continued administration of the aminoglycoside (Trollfors, 1983). Conversely, cases of fatal anuria have been reported. Occasionally, a Fanconi's syndrome (Casteels-Van Daele et al., 1980) or a Bartter's-like syndrome (Landau and Kher, 1997) has been observed.

Risk Factors for Aminoglycoside Nephrotoxicity

Based on clinical observations, there appear to be a variety of factors that predispose to the development of renal dysfunction with aminoglycoside therapy (Appel, 1990). It is important to distinguish between patient- and drug-related factors since this will guide the caregiver in determining the most appropriate course of action.

Patient-Related Factors

Age

The incidence of amikacin nephrotoxicity rises with advancing age, from 7% in patients under age of 30 years to 25% in patients over 75. The most likely mechanisms are twofold. First, the number of active nephrons decreases with age, leaving the patient with less reserve upon injury of a large proportion of nephrons. Second, because of the falling renal function, dosages may often be excessive if only based on insensitive renal function tests such as the measurement of serum urea nitrogen or serum creatinine concentrations, which rise significantly only when a large proportion of active nephrons is damaged (Moore et al., 1984).

Pre-existing Renal Diseases

In patients with pre-existing renal disease, as estimated by serum creatinine concentration greater than 2 mg/dl, the study of Moore et al. (1984) found no increase in toxicity risk if the dose was carefully adjusted. However, pre-existing renal failure clearly exposes the patient to inadvertent overdosing. In addition, kidneys from patients with pre-existing renal disease may have decreased ability to recover from ischemic or toxic insults (Beauchamp et al., 1992b; Manian et al., 1990)).

Gender

Female gender was identified as a risk factor in one study but not confirmed in others (Kahlmeter and Dahlager, 1984; Moore et al., 1984). Conversely, male gender was also reported in a retrospective analysis as a risk factor (Bertino et al., 1993). The matter is, therefore, unsettled. Animal studies are of little value in this context since most are performed with rats, and it is known that male laboratory rats tend to be spontaneously proteinuric, which in itself may be a risk factor.

Volume Depletion/Hypotension

Depletion of intravascular volume is an important risk factor for aminoglycoside-induced nephrotoxicity, whether induced by sodium depletion, hypoalbuminemia, or diuretics, even when systemic acid–base and electrolyte–volume status are maintained (Gamba et al., 1990). Hypokalemia and hypomagnesemia may be both predisposing risk factors or consequences of aminoglycoside-induced damage (Nanji and Denegri, 1984; Zaloga et al., 1984).

Liver Diseases

Liver dysfunction was identified as a risk factor in a retrospective analysis of two large clinical trials and was then validated in two additional prospective trials (Lietman, 1988). This is particularly true for patients with biliary obstruction, cholangitis, or both, rather than for other causes of liver disease, such as alcoholic cirrhosis (Desai and Tsang, 1988). There is no simple explanation to this observation.

Sepsis

Because of the unique role of aminoglycosides in treating patients with difficult Gram-negative sepsis, the hemodynamic and metabolic perturbations of the sepsis syndrome were often associated with an increase in drug-induced nephrotoxicity. Acute or chronic endotoxemia amplifies the nephrotoxic potential and renal uptake of gentamicin in rats (Auclair et al., 1990; Ngeleka et al., 1990; Tardif et al., 1990). There is a lack of definitive studies in humans, although cases of renal failure that could be associated with the administration of batches of endotoxin-contaminated gentamicin have been observed. The mechanism can be complex, but is likely to result from increase in the release of oxygen intermediates in renal tubular cells mediated by endotoxins (or other bacterial toxins or virulence factors). The latter may then be additive to the membrane damage produced in the same cells by the aminoglycosides themselves (Joly et al., 1991).

Fever

Fever *per se* increases renal metabolic demands. When associated with shock, ischemia, and the development of foci of tissue necrosis, this enhances aminoglycoside nephrotoxicity by accelerating the course and severity of the toxic insult (Spiegel et al., 1990; Zager, 1988).

Drug-Related Factors

Duration of Aminoglycoside Therapy

A large array of clinical data supports the notion that duration of therapy is a critical factor for developing the clinical manifestation of aminoglycoside-induced nephrotoxicity. The mechanism is probably that the kidney does not use all of its nephrons at each time (leaving a subpopulation unaltered which can then be recruited once the main population becomes less functional), and proximal tubules are capable of regeneration. If regeneration is insufficient, renal function will be affected once a sufficiently large proportion of all available nephrons have been recruited and intoxicated, which typically may take about four to seven days.

Drug Choice

Much dispute has been heard in this area. In the rat model, tobramycin was clearly found to be less toxic than gentamicin, both in animals and in humans (Gilbert et al., 1978). The respective positions of amikacin and netilmicin (to limit the discussion to the most-commonly used aminoglycosides) have been subject to many more controversies. Commercial interests have largely fueled these, but genuine differences in scientific and practical approaches have also been important. Thus, high-dose studies, which are necessary in rats to cause overt renal dysfunction, showed that netilmicin was considerably safer than gentamicin. Likewise, amikacin was also declared much less toxic. The problem may relate to the dosages and criteria used. Indeed, when low, clinically relevant, doses are used, netilmicin appears as toxic as gentamicin, whereas amikacin appears a mild or nontoxic drug. (Lerner et al., 1983; Smith et al., 1977).

The question that has not been adequately answered is whether these differences are sufficient to translate into differences in clinical toxicity, in view of the abundance of risk factors discussed above that may make any comparison quite hazardous. As an example, one clinical study found netilmicin less toxic than tobramycin, but the level of toxicity of tobramycin itself in that study was much lower than was usually reported, suggesting that risks factors had been minimized in the study population. With respect to amikacin, many studies used this antibiotic as a second-line drug after failure with another aminoglycoside. Amikacin, indeed, is active against many strains resistant to gentamicin or tobramycin. In case of resistance to the former drugs, the change to amikacin will often have taken place after three or four days of exposure to the first drug. Amikacin-induced toxicity may have been largely due to the first treatment and by the greater length of these successive treatments. In studies where amikacin was used as first-line agent, toxicities are usually very mild.

Frequent Dosing Intervals

Based on numerous clinical trials, the results of which have been bundled into an impressive series of meta-analysis, a multiple daily dosing of aminoglycosides clearly appears as a safe and efficacious treatment method (Gilbert, 1997). This mode of treatment does not prevent drug toxicity but may reduce the risk.

Concomitant Administration of Drugs

The toxicity of aminoglycosides can be enhanced by the coadministration of other drugs and, conversely, other nephrotoxic drugs can amplify the nephrotoxic potential of aminoglycosides. This has been clearly demonstrated for combinations of aminoglycosides and the glycopeptide antibiotic vancomycin (Kibbler et al., 1989; Wood et al., 1986) or, to a lesser extent, teicoplanin (Wilson, 1998), the antifungal amphotericin B (Gilbert, 2000; Kibbler et al., 1989), the anesthetic methoxyflurane (Barr et al., 1973), the immunosuppressant cyclosporine (Whiting et al., 1982), and the anticancer drug cisplatin (Jongejan et al., 1989; Salem et al., 1982).

Handling of Aminoglycosides by the Kidney

The first step towards understand the pathophysiology of aminoglycoside nephrotoxicity was made in the 1970s by the demonstration of accumulation of aminoglycosides in the renal cortex (Fabre et al., 1976; Luft and Kleit, 1974). This finding was first documented in animals, but later repeatedly confirmed in the human kidney (De Broe et al., 1984, 1991; Edwards et al., 1976; Verpooten et al., 1989). Autoradiographic, micropuncture and immunocytochemical studies have shown that aminoglycosides are primarily taken up and concentrated by S_1 and S_2 proximal tubule cells (Molitoris et al., 1993; Pastoriza-Munoz et al., 1984; Silverblatt and Kuehn, 1979; Wedeen et al., 1983). The amount of aminoglycoside taken up by the renal cortex is only a small fraction of the total administered dose (about 2 to 5%) (Fabre et al., 1976), but it must be emphasized that aminoglycosides are not metabolized by mammalian cells. Any quantity of the drug that is retained by the kidney therefore remains chemically unmodified.

Mechanism of Cortical Uptake

Much attention was focused on the identification of pathway responsible for aminoglycoside uptake, and the mechanism proposed remained controversial for a long time. The drug can be taken up into the cell from both the luminal and basolateral membranes, although binding and uptake by brush border membrane predominates (Bennett, 1989). Nowadays, the consensus is that megalin, an endocytic receptor expressed on the apical surface of the proximal tubular epithelium, represents the major route of entry of aminoglycosides through the brush border of proximal tubular cells. This has been elegantly demonstrated with mice having genetic or functional megalin deficiency. These mice do not accumulate aminoglycosides in their proximal tubular cells and are protected against aminoglycoside-induced nephrotoxicity (Schmitz et al., 2002).

Intracellular Handling

Following internalization, aminoglycoside antibiotics traffic via the endocytic system and accumulate primarily in lysosomes. Inside the lysosomes, the aminoglycosides accumulate in very large amounts (reaching concentrations that exceed 10 to 100 times the serum concentration). Information obtained both from cell cultures (LLC-PK_1 cells) and *in vivo* (rat kidney) indicates that a small but quantifiable amount of the internalized gentamicin (5 to 10%) traffics directly and rapidly from the surface membrane to the Golgi apparatus in both (Molitoris, 1997; Sandoval et al., 1998, 2000; Sundin et al., 2001). This finding for gentamicin is consistent with movement along the endocytic pathway. It is also consistent with the previously known movement of Shiga toxin and ricin from the surface membrane to the Golgi apparatus (Johannes and Goud, 1998; Lord and Roberts, 1998; Sandvig and van Deurs, 1996). Because aminoglycosides are polar molecules, most of the drug taken up into proximal tubular cells (whether in lysosomes or in the Gogi apparatus) will remain for a considerable length of time. The half-life of the drug in kidney tissue may amount to several days (Fabre et al., 1976) as opposed to the short serum half-life (2 to 3 hr), mainly reflecting cell turnover with a small contribution of true exocytosis.

Mechanisms of Toxicity

A major difficulty and a point of many controversies has been, and still is, ascertainment of which changes of the numerous that have been described are truly responsible for toxicity. Several hypotheses have been suggested. An intriguing aspect of aminoglycoside nephrotoxicity is that very large amount of drugs (usually 10 times the therapeutic dose) must be administered to animals in order to cause clear-cut acute tubular necrosis and concomitant alteration of renal function (Gilbert, 2000; Parker, 1982). This is in sharp contrast to the clinical situation, in which a sizeable fraction of patients experience a loss in renal function upon treatment with clinically acceptable doses (Smith et al., 1980).

The question has therefore been raised as which subclinical features seen in animal studies are responsible for toxicity, and how these pertain to the further development of clinical toxicities (Tulkens, 1986).

However, the demonstration of a sequence of events, from a subclinical alteration to overt toxicity, is not necessarily a proof of a cause and effect relationship. It is indeed possible that only the more drastic changes, such as a decrease in glomerular filtration (Baylis et al., 1977) or extended necrosis (Parker, 1982) (both seen in animals only at high doses), really cause toxicity. Taking account of this caveat, the following sections review the various hypotheses that have received strong experimental support over the past 20 years.

Lysosomal Alterations

When in lysosomes, aminoglycosides induce a marked phospholipidosis that has been demonstrated in cell culture models (Aubert-Tulkens et al., 1979), experimental animals (Feldman et al., 1982; Giuliano et al., 1984; Josepovitz et al., 1985; Knauss et al., 1983), and humans (De Broe et al., 1984). This phospholipidosis develops rapidly and involves all major phospholipids, with, however, a predominant increase in phosphatidylinositol on a relative basis (Feldman et al., 1982; Knauss et al., 1983). Accumulation of phospholipids within lysosomes is responsible for the formation of the so-called "myeloid bodies" that were described and linked to toxicity as early as in the mid 1970s (Kosek et al., 1974; Watanabe, 1978).

It has been proposed that phospholipidosis induced by aminoglycosides result primarily from impaired phospholipid degradation due to inhibition of lysosomal phospholipases A and C and sphingomyelinase (Aubert-Tulkens et al., 1979; Laurent et al., 1990), and that it is related to the binding of the cationic aminoglycosides to phospholipids at the acid pH prevailing in lysosomes. The link between the lysosomal phospholipidosis and further cell damage has, however, remained largely indirect so far (Laurent et al., 1990). Moreover, some aminoglycosides, such as netilmicin, induce a conspicuous phospholipidosis (Toubeau et al., 1986) without marked necrosis (Luft et al., 1976). This lack of a link between subclinical alterations and overt renal toxicity in animals may, however, be related to differences in dose–effect relationships between netilmicin and other aminoglycosides (Hottendorf et al., 1981).

Mitochondrial Alterations

Besides metabolic alterations in lysosomes, aminoglycosides also induce changes in mitochondria, namely a competitive interaction with magnesium resulting in reduced mitochondrial respiration (Bendirdjian et al., 1982; Simmons et al., 1980; Weinberg and Humes, 1980), which has been considered as a cause of toxicity.

This hypothesis highlighted the lack of evidence showing that aminoglycosides could reach mitochondria before cell necrosis and postmortem redistribution of the drug stored in lysosomes. Yet, release of oxygen radical species triggered by aminoglycosides at the level of mitochondria has been proposed as a potentially important mechanism (Ueda et al., 1993; Walker and Shah, 1988).

Traffic of part of the cellular gentamicin to mitochondria *in vivo* has been observed (Sundin et al., 2001). This observation may have a particular significance since polyamines are known to be able to activate mitochondria and cause the release of cytochrome c, an important step leading to apoptosis (Mather and Rottenberg, 2001). Proteomic analysis following gentamicin administration has also indicated an energy production impairment and a mitochondrial dysfunction occurring in parallel with the onset of nephrotoxicity (Charlwood et al., 2002).

Inhibition of Protein Synthesis

Aminoglycosides act as antibiotics by inhibiting prokaryotic protein synthesis through binding to ribosomes, blocking peptide synthesis initiation, and causing mistranslation (Carter et al., 2000; Davies et al., 1965; Tai and Davis, 1979). Studies indicate that gentamicin administration *in vivo* reduces renal cortical endoplasmic reticulum protein synthesis *ex vivo* very rapidly (Bennett et al., 1988; Buss and Piatt, 1985). More recent data point to a major *in vivo* inhibitory effect of gentamicin on protein synthesis after only two days of antibiotic administration (Sundin et al., 2001). The mechanism by which this occurs is unknown, but could involve inhibition of nuclear transcription or alteration of endoplasmic reticulum- or Golgi-mediated posttranslational modifications.

An intriguing hypothesis raised by the study of the influence of gentamicin on the expression of specific proteins is that failure to translate high levels of mRNA into proportionally high levels of protein could attenuate the expression of stress response gene products, and thus diminish the possibility of recovery in gentamicin intoxication (Dominguez et al., 1996). Gentamicin has been found to significantly reduce Na^+/glucose cotransporter (SGLT1)-dependent glucose transport and to downregulated mRNA and protein levels of the SGLT1 in pig proximal tubular LLC-PK_1 cells (Takamoto et al., 2003).

Inhibition of Na^+/K^+-ATPase

Aminoglycosides inhibit Na^+/K^+-ATPase activity in basolateral membranes (Cronin et al., 1982; Williams et al., 1984), and the binding of the drugs to these membranes has been correlated with toxicity (Williams et al., 1987). Inhibition is, however, seen only when aminoglycosides are presented to the cytoplasmic face of the membrane (Williams et al., 1984). It is not known whether this occurs *in vivo* at therapeutic doses.

Apoptosis

The observation that aminoglycosides induce apoptosis *in vivo* at therapeutically relevant doses (El Mouedden et al., 2000a) shed light on the mechanisms of the early stages of nephrotoxicity. Apoptosis, also called programmed cell death, was first described in 1972 (Kerr et al., 1972) from studies of tissue development kinetics and differentiation, but has now been demonstrated to be a key determinant in cell response to many environmental signals leading to cell death. It is characterized by specific features, such as cell shrinkage, increased cytoplasmic density, condensation of chromatin, and fragmentation of DNA. It can be triggered in the kidney by a very large array of toxic agents (Davis and Ryan, 1998).

In the context of aminoglycosides, rats show a clearly detectable apoptotic reaction in proximal tubules after only four days of treatment, which becomes conspicuous after 10 days (see Figure 16.1A). This reaction is dose-dependent and occurs in the absence of necrosis (El Mouedden et al., 2000a). Gentamicin-induced apoptosis can be also demonstrated using cultured renal cells (LLC-PK_1 - see Figure 16.1B, MDCK) and nonrenal cells (embryonic fibroblasts) (El Mouedden et al., 2000b). Current work suggest that lysosome destabilization may be a key triggering event in the onset of apoptosis in LLC-PK_1 cells upon incubation with gentamicin, setting the link between lysosomal accumulation of the drug and the ensuing toxic events. Other pathways, such as those involving mitochondria, may also be involved. It is also possible that both pathways are interrelated; the lysosomal destabilization eventually causes mitochondrial depolarization and opening of the

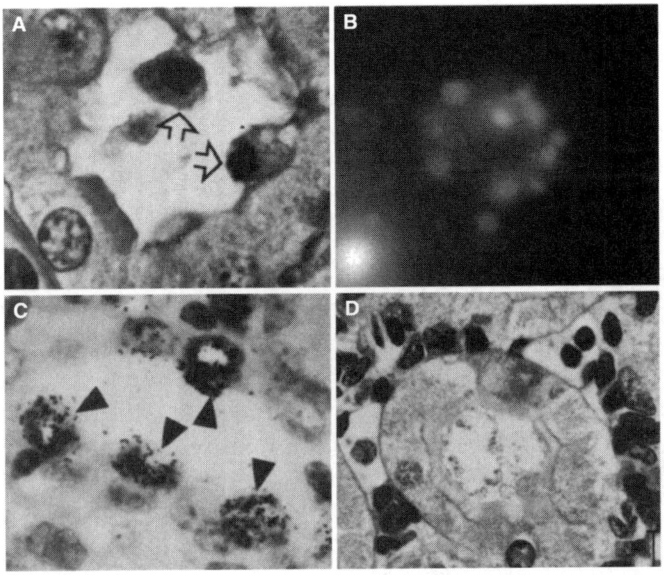

Figure 16.1 Morphological changes in rat renal cortex (A, C, D) upon treatment with gentamicin at low doses (10 mg/kg, 10 days) and in cultured LCC-PK$_1$ renal cells (B) upon incubation with gentamicin under conditions causing a drug accumulation similar to that observed in rat renal cortex of the animals treated as indicated in A, B, and C (approximately 10 μg/g tissue) (El Mouedden et al., 2000b). (A) typical image of apoptosis (shrinkage necrosis; open arrows) in a seemingly normal proximal tubule; (B) typical image of nuclear fragmentation in a single cell; (C) autoradiographic demonstration of thymidine incorporation in nuclei of proximal tubular cells (arrowheads); (D) peritubular infiltration by endothelial and fibroblasts-like cells. (A, D) hematoxylin-eosin/periodic acid Schiff staining; (B) 4′,6-diamidino-2-phenylindole staining; (C) animals injected with ^{3}H-thymidine one hour before sacrifice and tissue sections processed for autoradiographic detection of radioactively-labeled structures followed by light hematoxylin/eosin counterstaining. (A, C, D: reproduced from Laurent et al. (1983) with permission; B, unpublished data from H. Servais.)

mitochondrial permeability transition pore, a known triggering event of not only apoptosis but also necrosis, depending upon the severity of this change (Halestrap et al., 2000; Lemasters et al., 1998).

Regeneration

The kidney has a large capacity to regenerate and thereby to compensate for tubular insults. This explains why necrosis and other related processes may occur without being detected by functional deficits. The importance of this fact is best demonstrated in rats in which fairly high doses (10 times the human dose) can be given for periods as long as 40 days. After a first episode of acute renal failure, related to the synchronous necrosis of a large proportion of the proximal tubules, renal function returns to normal, as if the animals had become refractory to the toxic effects of the drug (Elliott et al., 1982a, 1982b). This "resistance" to gentamicin is a state of persistent tubular cell injury obscured functionally by preservation of the glomerular filtration rate and histologically by asynchrony of cell necrosis and regeneration (Houghton et al., 1986).

Tubular regeneration has been extensively studied by means of ^{3}H-thymidine incorporation and other methods, and has been shown to occur very early on during treatment, even at low, clinically significant, doses (see Figure 16.1C) (Laurent et al., 1983, 1988; Toubeau et al., 1986). It may therefore be speculated that all patients exposed to aminoglycosides experience focal losses of tubular tissue (through apoptosis or necrosis), and that the quality and extent of regeneration determines whether or not active renal function is maintained during treatment.

After about 10 days of exposure to gentamicin in subtoxic doses, however, the kidney will show signs of mild peritubular inflammation and fibroblast proliferation (see Figure 16.1D). This will eventually lead to chronic tubulointerstitial nephritis with progressive renal failure. Cessation of treatment is associated with microcystic and inflammatory changes, suggesting that the renal response to tubular injury can be dissociated from the amount of toxin in the renal cortex (Elliott et al., 1982a; Houghton et al., 1988).

Both regeneration and assessment of fibrosis and other signs of tubulointerstitial inflammation have therefore been used as surrogate markers in many studies evaluating and comparing aminoglycosides at low doses (Hottendorf and Gordon, 1980; Tulkens, 1986). The whole process of regeneration and fibrosis induced by aminoglycosides has also been examined within the context of the release of growth factors (Leonard et al., 1994; Morin et al., 1992). While studies aiming at stimulating regeneration have been disappointing, it has now been recognized that macrophages, myofibroblasts, transforming growth factor β, endothelin, and angiotensin II may contribute to the development of renal fibrosis in gentamicin-treated rats (Geleilete et al., 2002).

Means of Protection

Reducing or protecting against aminoglycoside nephrotoxicity has attracted much effort and attention over the last decade, based either on purely clinical approaches, or on various experimental methods. The former will be discussed in detail because they are the only ones so far that have been shown to decrease aminoglycoside toxicity in patients.

Clinical Approaches

The Once-Daily Schedule

Up until the late 1980s, aminoglycosides were commonly recommended for administration in divided doses over a 24 hour period, and typically administered on 12 or 8 hour schedules (i.e., the total daily dose divided in two or three administrations at 12 or 8 hour intervals, respectively). Kinetic and toxicodynamic studies revealed, however, that this mode of administration resulted in enhanced drug uptake and toxicity as compared with a once-daily schedule (Bennett et al., 1979; Giuliano et al., 1986). In parallel, pharmacodynamic studies examining the antibacterial activity of aminoglycosides *in vitro* (Blaser et al., 1985), in experimental infections (Gerber et al., 1989; Leggett et al., 1990), and in patients (Moore et al., 1987) indicated that a large serum peak concentration to minimum inhibitory concentration ratio was a key determinant in their therapeutic efficacy.

Aminglycosides are indeed typically concentration-dependent antibiotics and have a large post-antibiotic effect (Amsden et al., 2000). This led a series of investigators to test the once-daily schedule, first in limited clinical situations from the mid and late 1980s (Powell et al., 1983; ter Braak et al., 1990; Tulkens, 1991), then to a more wide usage in the mid 1990s (Nicolau et al., 1995; Prins et al., 1993), and eventually leading to more than 300 publications to date and several meta-analyses (Blaser and Konig, 1995; Ferriols-Lisart and Alos-Alminana, 1996; Munckhof et al., 1996). This schedule is now recommended in most cases for reasons of both potentially improved efficacy and potentially decreased toxicity (Gilbert, 1997, 2000).

The bottom line of all these efforts is that nephrotoxicity is usually delayed, but not suppressed, with the once-daily schedule. This is particularly well illustrated by the results presented in Figure 16.2, where it can be seen that both multiple-daily and once-a-day dosing

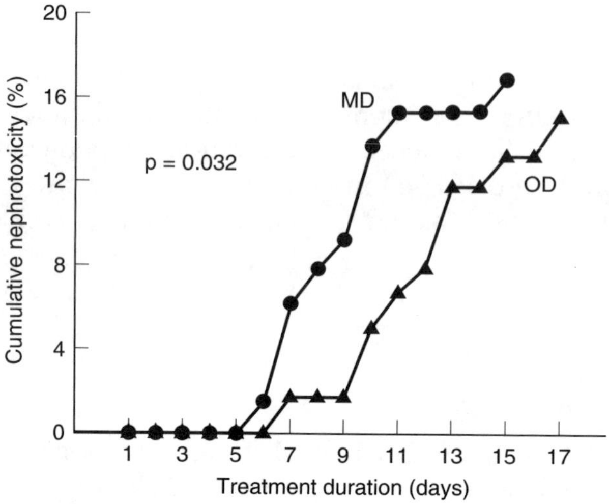

Figure 16.2 Appearance of nephrotoxic reaction in patients given netilmicin once-daily (OD) or on 12 hour or 8 hour multiple doses schedules (MD)(ter Braak et al., 1990). Nephrotoxicity was defined as an increase in serum creatinine concentration of more than 50% over baseline. Patients (141) were predominantly elderly subjects with severe bacterial infections and received simultaneously 2 g ceftriaxone/day. Netilmicin treatment did not differ significantly in mean daily dose per kg body weight (average 6.6 mg/day) nor duration of therapy between the two treatment arms except for schedule. Compared with patients receiving conventional doses, patients treated with a once-a-day dose had higher serum peak netilmicin levels and lower trough levels. (Reproduced with minor modifications with permission.)

of netilmicin will eventually lead to a similar proportion of patients experiencing nephrotoxic reactions. Patients receiving the drug once daily, however, will be safer for two to three days more than those treated with the conventional schedule. This is understandable since the once-daily schedule will not abolish but will only reduce drug uptake. Therefore, treating patients for an extended period of time will eventually cause the cortical drug level to reach in both cases the critical threshold that results in tubular insult and ensuing functional damage. This is the basis for the current recommendations to limit aminoglycoside treatments to a maximum of seven days unless there is clear and defined medical reason to the contrary. It is indeed important to note that pharmacodynamic considerations predict that the once daily schedule will be at least as effective, and even perhaps more effective than the divided dose schedule (Blaser et al., 1985, 1987; Amsden et al., 2000), so that a short course will be effective.

When all studies are considered at a fixed time point, the once-daily schedule appears almost never more, and often less, nephrotoxic than the divided-dose schedules (with a global risk factor calculated for eight studies and a total of 802 patients of 0.9, with a confidence interval of 0.63 to 1.31; Verpooten et al., 2003), while being as or more efficacious (Blaser and Konig, 1995) and offering obvious practical advantages, including less necessity of monitoring serum levels (see below). A survey made in the late 1990s among 500 acute care hospitals in the U.S. revealed that 74.7% of them use the once-a-day schedule or a close variant of it (Chuck et al., 2000).

Individualized Pharmacokinetic Monitoring

Individualized pharmacokinetic monitoring represents another approach that has been used by clinicians and clinical pharmacists in attempts to minimize aminoglycoside toxicity (Dahlgren et al., 1975; Pancorbo et al., 1982) while ensuring sufficient serum levels to obtain maximal therapeutic effects (Zaske et al., 1982). This has led to intense efforts at designing the most appropriate models and approaches to define "optimal" peak and trough levels, which were to be considered as gold standards and used for comparing drug toxicities (Smith et al., 1980). This, however, sometimes led to contradictory results (Bertino et al., 1993; Burton et al., 1991; Pancorbo et al., 1982), but did allow identification of patients at risk (Bertino et al., 1993; Mullins et al., 1987), reassessment of accepted "normal" therapeutic ranges (McCormack and Jewesson, 1992; Watling and Dasta, 1993), and led to minimized costs (Bertino et al., 1994).

When examined in large patient populations, individualized pharmacokinetic monitoring eventually demonstrated a significant *negative* risk factor of up to 0.42 vs. controls, together with substantial savings (Streetman et al., 2001). The outcomes, however, remained blurred by the misconception that a high peak was associated with toxicity, while a minimum trough level was necessary for activity. As has been shown, the converse is true, but can only be demonstrated if comparing different schedules of administration. Comparing schedules is essential because changing the dose but not the schedule modifies simultaneously, and to a similar direction, the C_{max}, the AUC and the C_{min}. These are indeed covariables with respect to the dose, making impossible to ascribe toxicity or efficacy to one of them independently of the others by simply manipulating the dose.

The introduction of once-daily dosing largely modified the way monitoring was carried out, by concentrating more on the peak levels (for efficacy), while trough levels (often vanishingly low at 24 hours) tended not to be recorded (Cronberg, 1994). Currently, optimal true peak levels (i.e., extrapolated at time = 0) are set at around 20 mg/l for gentamicin, tobramycin, and netilmicin, and 60 mg/l for amikacin, with trough (24 hour) levels lower than 1 and 3 mg/l, respectively, for patients with normal renal function (Gilbert, 2000). These levels will be obtained in most patients, decreasing the necessity of systematic monitoring. However, surveillance may be useful in patients at risk. A nomogram using the eight hour serum value, as an indicator of both the potential peak level (and thereby allowing adjustment of the dose) and the elimination constant (to detect impending renal failure), has been developed (Nicolau et al., 1995). It seems fairly popular in the U.S. as about one-third of the hospitals using the once-daily schedule have adopted it (Chuck et al., 2000).

This nomogram, as well as others, may not be optimal (Wallace et al., 2002). Application of the principles of individualized pharmacokinetic monitoring to the once-daily schedule (by adjusting the dose interval) may still ensure a further reduction of toxicity while allowing the use of larger doses to potentially increase efficacy (Bartal et al., 2003). Additional reduction of toxicity of the once-daily schedule by selecting the most appropriate time of the day for administration has been reported (Rougier et al., 2003) with early afternoon appearing as optimal. The practicability of this approach needs, however, to be critically assessed.

Experimental Approaches

Reduction of Cortical Uptake

Reduction of cortical uptake has been attempted by complexing aminoglycosides with polyanionic substances, such as dextran sulfate (Kikuchi et al., 1991) or inositol hexasulfate (Kojima et al., 1990b), or anionic β-lactams, such as piperacillin (Hayashi et al., 1988) or moxalactam [latamoxef] (Kojima et al., 1990a), fosfomycin (Fujita et al., 1983), or even simply with pyridoxal-5′-phosphate (one coenzyme form of vitamin B6; Smetana et al., 1992). Other attempts have been directed at: decreasing the electrostatic attraction of aminoglycosides and their brush border binding sites by infusion of bicarbonate (Chiu et al., 1979) to raise urine pH and decrease the polycationic character of aminoglycosides (the pKa of the aminofunctions of the common aminoglycosides span from 5.5 though to approximately 8), or competing with aminoglycosides by means of calcium (Humes et al., 1984) or lysine (Malis et al., 1984). Lack of marked efficacy or intrinsic toxicity has, however, prevented all these approaches from being developed in the clinics.

Stimulation of exocytosis by fleroxacin (a fluoroquinolone antibiotic) has also been attempted (Beauchamp et al., 1997). Pasufloxacin, an experimental fluoroquinolone, also has shown a protective effect against arbekacin-induced nephrotoxicity, and that this was attributable to a suppression of uptake of arbekacin in cortical renal tubules (Kizawa et al., 2003). The concomitant use of an aminoglycoside and a fluoroquinolone to prevent toxicity runs, however, against the rules of the safe and restricted use of antibiotics.

Prevention of Phospholipidosis

Polyaspartic acid has been shown to protect against both morphological and functional signs of aminoglycoside-induced nephrotoxicity in the rat (Beauchamp et al., 1990a; Gilbert et al., 1989). Further studies showed that this polyanionic peptide enters cells by endocytosis and reaches lysosomes where it forms ion-pair complexes with aminoglycosides, thereby preventing them from interacting with phospholipids (Kishore et al., 1990). Interestingly, this results in an increased cortical accumulation of gentamicin but does not alter the pharmacokinetic parameters relevant to the therapeutic effect of aminoglycosides (Whittem et al., 1996). Although potentially promising, the patent holder has barred by proprietary considerations and by lack of initiatives for the clinical development of polyaspartic acid. Daptomycin (a lipopeptide antibiotic active against Gram-positive organisms) also protects against lysosomal phospholipidosis (Beauchamp et al., 1990b) and nephrotoxicity (Wood et al., 1989) in experimental animals. Similarly to polyaspartic acid, daptomycin binds to phospholipid bilayers. However, the contribution of daptomycin to the membrane charge density and its effects on the lipid packing both combine to counteract the inhibition of phospholipase activity due to aminoglycosides (Gurnani et al., 1995), and result in an activation of phospholipases when aminoglycosides are present (Carrier et al., 1998). Daptomycin has been recently approved in the U.S. for clinical usage and evaluations of its potential to reduce aminoglycoside toxicity in patients at therapeutic doses is therefore awaited with interest.

Reduction of Cell Death and Increase of Cell Repair

This has been attempted with a number of compounds. Compounds have included, on the one hand, desferroxamine (Ben Ismail et al., 1994), methimazole (Elfarra et al., 1994), vitamin E plus selenium (Ademuyiwa et al., 1990), and lipoic acid (Sandhya et al., 1995), and on the other hand, ulinastatin (Nakakuki et al., 1996), fibroblast growth factor 2 (Leonard et al., 1994), and the heparin-binding epidermal growth factor (Sakai et al., 1997). Unfortunately, no clinical data are available.

Protection against Vascular and Glomerular Insults

Such protection can be obtained using calcium channel blockers (Lortholary et al., 1993), but the intrinsic pharmacological properties of these agents have prevented any clinical development. Trapidil, an antiplatelet and vasodilator agent, may also protect by antagonism of platelet-derived growth factor, vasodilation, inhibition of thrombocyte aggregation, and nitric oxide release (Buyukafsar et al., 2001).

Concluding Remarks on Aminoglycoside Nephrotoxicity

Although having attracted so much attention from laboratory and clinical researchers, the basic mechanisms of aminoglycoside nephrotoxicity, and especially the biochemical events leading to cell damage and glomerular dysfunction, still remain poorly understood. Yet, this has not prevented the generation of a vast amount of useful information concerning risk factors and the design of various means of protection. It now remains to be seen whether drug design can make use of this knowledge to obtain truly less-toxic compounds.

β-LACTAMS

β-Lactam antibiotics (penicillins, cephalosporins, carbapenems, and monobactams) are among the most important antimicrobials. As shown in Figure 16.3, all these drugs share the same pharmacophore, a five-membered ring (six-membered in the case of cephalosporins)

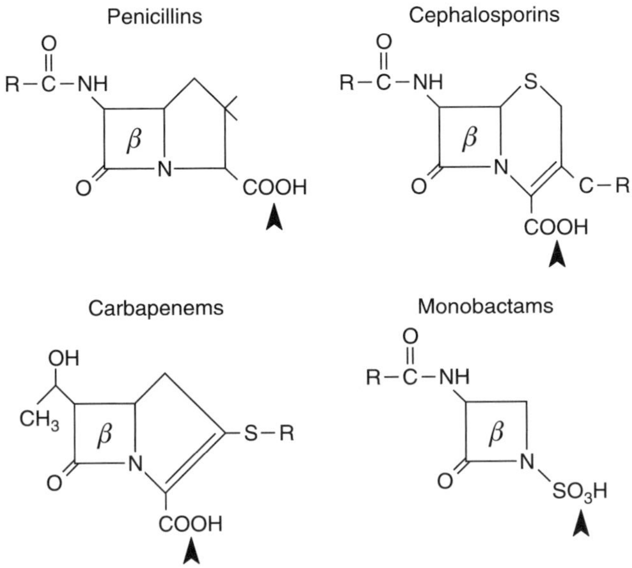

Figure 16.3 Structural formulae of the four main classes of β-lactam antibiotics used in clinics. All these antibiotics share the same pharmacophore consisting in a β-lactam ring (denoted β in the figure) fused with a five- or six-membered ring carrying a carboxylic acid function (shown by the vertical arrowheads. Monobactams, in which the pharmacophore consists in a β-lactam ring only, carry an hydrogenosulphate group (equivalent proton donor) function at the same distance from the β-lactam rings as the carboxylic acid function in the other β-lactams. This pharmacophore mimics and has the same configuration as the dipeptide D–Ala–D–Ala (from left to right: NX_2-CO-NX-CX_2-COOH, in which X are the corresponding substituents; for carbapenems, the first N atom is absent), which explains their antibacterial activity (Ghuysen et al., 1965). Mimicking a peptide has direct consequences on β-lactam handling by the kidney.

carrying a free carboxyl function fused with a β-lactam ring (monobactams possess only the β-lactam ring and the carboxyl function is replaced by an hydrogenophosphate group acting as an equivalent proton donor). This common pharmacophore mimics the structure of a D–Ala–D–Ala dipeptide (Tipper and Strominger, 1965), which is the basis for the mode of action of this class of antibiotics (Ghuysen et al., 1981). As shall be indicated, this common structure also explains some aspects in the renal handling, and therefore the potential toxicity, of β-lactams. Differences among the very numerous derivatives relate primarily to the side chains, which govern critical properties related to spectrum, resistance to β-lactamases, and, to some extent, pharmacokinetic properties (Petri, 2001).

Severe nephrotoxicity related to tubular necrosis was quickly recognized as a potential hazard of some β-lactams after the introduction of cephaloridine in the mid 1960s (Hinman and Wolinsky, 1967; Seneca, 1967). The fact that such toxicity had not been observed, at least to the same extent, with penicillins and other cephalosporins already in clinical usage at that time indicated that it was related to specific structural determinants distinct from the pharmacophore. This led to systematic screening of all new compounds under development for alterations of renal function. The result has been that, in contrast to aminoglycosides, antimicrobial activity and nephrotoxicity were quickly dissociated, at least in a blinded fashion. Most of the new β-lactams developed and commercialized over the past 20 years have, therefore, no or only very limited gross nephrotoxicity. Conversely, derivatives that were overtly nephrotoxic were quickly halted in their development based on simple screening studies. In parallel, the allergenic potentials of β-lactams were also taken into consideration, and compounds prone to induce allergic reactions (which are the main cause of the interstitial nephritis toxicity caused by β-lactams) were eliminated from the development programs. As a result, the molecules now in use are intrinsically nontoxic or only mildly toxic.

While this is obviously to the advantage of patients, it has largely cloaked the structure–relationship data (because of proprietary restrictions) and has limited the number of published mechanistic studies. Furthermore, little or no effort was spent in assessing means of protection as they were not considered essential in view of the availability of nontoxic derivatives. The study of β-lactams has nevertheless been rewarding in terms of a better understanding of renal physiology and of renal drug transport, on the one hand, and of basic mechanisms of cell toxicity on the other hand.

Pathophysiology

β-Lactam antibiotics are among the most important groups of drugs excreted by the kidney (Kamiya et al., 1983; Nightingale et al., 1975; Tune, 1997). This predictably results in increased exposure of susceptible targets within the kidney to potential toxic drugs. With penicillins, renal toxicity has varied, ranging from allergic angitis to interstitial nephritis, the latter of which is most commonly observed with methicillin, and to tubular necrosis. Administration of massive doses of any penicillin, but most often carbenicillin and ticarcillin, may result in hypokalemia owing to the large amounts of nonreabsorbable anion presented to the distal renal tubules, which alters proton excretion and secondarily results in potassium loss.

Interstitial Nephritis

Interstitial nephritis is the most frequent pattern of drug-induced immunologically mediated renal injury (Kleinknecht et al., 1978). It occurs as an apparent hypersensitivity response to β-lactams and is usually ranked among the late immunological reactions to these drugs (Levine, 1966). Indeed, β-lactams behave like haptens, which may bind to serum or cellular proteins to be subsequently processed and presented by major histocompatibility complex molecules as hapten-modified peptides. The most common form of haptenization for penicillins is the penicilloyl configuration, which arises from the opening of the strained β-lactam ring, yielding an additional carboxylic function that allows the molecule to covalently bind to the lateral and terminal aminofunctions of proteins. Serum molecules thus facilitate haptenization. This reaction occurs with the prototype benzylpenicillin and virtually all semisynthetic penicillins, but other derivatives (called minor determinants) can be formed in small quantities and stimulate variable immune responses.

Since all β-lactams share the same basic structure, they are all disposed to give rise to haptenization. Variations in the side chains, and the corresponding differences in the chemical nature of the haptens, however, explain why clinical consequences are variable from one class of β-lactams to another (Pham and Baldo, 1996; Zhao et al., 2002). Cross-reactivity between penicillins and cephalosporins is accordingly far from complete (Romano et al., 2000) and is, even, rare (Novalbos et al., 2001). The potential relation of β-lactam allergy to other medication reactions involving structurally unrelated drugs needs to be further explored.

The basic characteristic of the immunoallergic acute interstitial nephritis is the presence of edema and focal or diffuse cellular infiltrates in the renal interstitium, particularly in the corticomedullary region, the subcapsular cortex, and around the glomerulus. In β-lactam-induced acute interstitial nephritis, the presence of eosinophils is frequently observed. The majority of the infiltrate cells are mononuclear, fundamentally T-lymphocytes, although there are also monocyte-macrophages that may appear as epithelial cells.

The clinical syndrome is one of fever, macular rash, eosinophilia, proteinuria, eosinophiluria, and hematuria (Appel and Neu, 1977). Functionally, the reaction to β-lactam-induced interstitial nephritis is one of nonoliguric renal failure with a decrease in creatinine clearance and a rise in serum urea nitrogen and serum creatinine concentrations, which can progress to anuria and renal failure (Baldwin et al., 1968; Olsen and Asklund, 1976; Woodroffe et al., 1975). The incidence of acute interstitial nephritis induced by β-lactams is three times more frequent in males than in females. It can appear at any age but is most usual in young adults. The incidence and severity of nephrotoxicity associated with β-lactams are potentiated by aminoglycoside antibiotics (Dejace and Klastersky, 1987; Yver et al., 1976), renal ischemia (Browning et al., 1983), and endotoxemia (Tune et al., 1988; Tune and Hsu, 1985). Differential diagnosis from other causes of acute renal failure may be difficult, but coincident evidence of an acute allergic reaction to the β-lactam in use may help, as may the detection of eosinophils in the urine. Definitive diagnosis usually requires renal biopsy, and is important because a change in antibiotic therapy will usually result in rapid improvement in renal function (Linton et al., 1980).

Acute Tubular Necrosis

For the reasons discussed above, acute tubular necrosis is currently rare with β-lactams. It is encountered only with high-dose therapies (Atkinson et al., 1966), with the following rank order (as defined by studies in animals or humans): cephaloglycin, cephaloridine >> cefaclor, cefotiam, cefazoline, cephalothin >>> cephalexin, cefoperazone, cefotaxime and cetazidime (Boyd et al., 1973; Cojocel et al., 1988; Hottendorf et al., 1987). No tubulotoxic reaction has been reported for with monobactams, but it must be emphasized that only one molecule in this class (aztreonam) is in wide clinical use and that most of the toxicology data on compounds that have not been developed remains unpublished.

Carbapenems present a particular and specific situation. Imipenem, which was the first carbapenem introduced in clinical usage, is quickly hydrolyzed by a dehydropeptidase found in the brush border of proximal tubular cells. This not only causes a loss of drug (and a lack of efficacy in case of urinary tract infection) but also the release of degradation products that are nephrotoxic in certain animal species (Moellering et al., 1989). Imipenem is, therefore, always administered in combination with cilastatin. The latter inhibits the dehydropeptidase and allows for both a sustained concentration of active antibiotic in urine and a sparing of the toxic effects on proximal tubular cells. Meropenem, and the other carbapenems, which are not susceptible to this degradation by renal dehydropeptidase, are intrinsically devoid of such toxicity (Topham et al., 1989).

Two factors account for the development of renal tubular damage caused by β-lactams, namely the accumulation of the drug in the proximal tubular cells via active transport mechanisms, and the ability of the drug to trigger cytotoxic events.

Active Transport

As recently reviewed by Inui et al. (2000), β-lactam antibiotics are not only filtered but also actively transported and secreted by the kidney at the level of the proximal tubular cells. An overview of the transporters involved is presented in Figure 16.4. The role of organic anionic transporters (OATs) in the renal handling of cephalosporins and carbapenems and the ensuing tubular toxicity was primarily based upon experiments using inhibitors. First, probenecid (Craig, 1997; Tune, 1972; Tune et al., 1974), *p*-aminohippurate (Tune, 1975), piperacillin (Hayashi et al., 1988), *N*-acyl amino acids (Hirouchi et al., 1993), and other organic anions (Craig, 1997) were found to reduce the uptake and the nephrotoxicity of cephaloridine in experimental animals. Second, betamipron (*N*-benzoyl-3-propionic acid), an inhibitor of OATs, was shown to inhibit the uptake of panipenem and imipenem into proximal cells and to prevent their nephrotoxicities (Hirouchi et al., 1994). This demonstrated that uptake by proximal tubule cells from the blood via a renal OAT at the basolateral membrane was apparently a necessary step for toxicity. At present, it is known that OAT1 and OAT3 are the major anion exchangers in rats (Cha et al., 2001; Jariyawat et al., 1999; Kusuhara et al., 1999; Sekine et al., 1997) and in humans (Sekine et al., 1997, 2000; Sweet et al., 2000; Tojo et al., 1999). OAT2, also localized at the basolateral membrane, may interact with cephalosporins, but with a lower affinity (Enomoto et al., 2002; Sekine et al., 1998).

Transport of β-lactams across the luminal membrane to the urine is primarily mediated by OAT4 (Babu et al., 2002a; Cha et al., 2000). This transport is slower for cephaloridine, which results in the accumulation

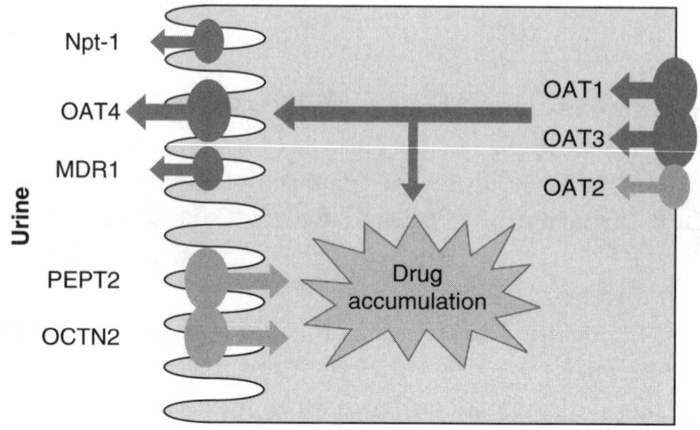

Figure 16.4 The main transport systems of β-lactam antibiotics in the kidney and the role in tubular toxicity. The most important transporters are members of the organic anion transporter (OAT) family (part of the major facilitator superfamily) that act as ion uniports or ion/H^+ symports. Influx into cells occurs primarily through the OAT1 and OAT3 transporters located at basolateral membrane, which are inhibited by probenecid, *p*-aminohippurate, and other organic anions. OAT2 has only a low affinity for β-lactams. Once in cells, β-lactams are secreted into the urine through the activity of OAT4. Imbalance between influx and efflux may result in intracellular drug accumulation and triggering of cell toxicity. The peptide transporter 2 (PEPT2) and the organic cation transporter 5 (OCTN2) are responsible for luminal reabsorption of aminocephalosporins and cephalosporins carrying quaternary nitrogen, respectively. They may contribute to the intracellular retention of these cephalosporins. The inorganic phosphate/Na^+ transporter (Npt-1) is involved in the clearance of faropenem. The role of other transporters, such as the P-glycoprotein (MDR1), is suspected but not proven. (Based on data reviewed in Inui et al., 2000, and Van Bambeke et al., 2003.)

of high intracellular drug concentrations and selective damage of these cells in the proximal tubule (Takeda et al., 2002; Tune, 1997).

Other important transporters include the H^+-coupled peptide symporters PEPT2 and the organic cation transporters OCTN2. PEPT2 will transport β-lactams in a inward fashion, from tubular fluid into the tubular epithelial cells, recognizing primarily aminocephalosporins, such as cephalexin, but not aminopenicillins (Daniel and Adibi, 1993; Ganapathy et al., 1995). Although of the same subfamily as the intestinal PEPT1 transporter, the renal PEPT2 is quite distinct with respect to recognition of β-lactam antibiotics; anionic compounds such as cefixime or ceftibuten have a much greater affinity for PEPT1 than for PEPT2, whereas the opposite is observed for zwitterionic β-lactams (Ganapathy et al., 1995, 1997). OCTN2, an organic cation/carnitine transporter, responsible for Na^+-coupled transport of carnitine in the kidney and other tissues, has been shown to transport cephaloridine, cefepime, and in general cephalosporins carrying a quaternary nitrogen in their side chain (as does carnitine), from the luminal medium into cells. Thus, OCTN2 will contribute to further increase the intracellular concentration of cephaloridine, and will cause inhibition of carnitine reabsorption in the kidney.

Interestingly, many of the β-lactam antibiotics that are not recognized by OCTN2 are good substrates for PEPT2 and *vice versa.* The mouse inorganic phosphate transporter Npt-1, which operates in the hepatic sinusoidal membrane to transport benzylpenicillin and mevalonic acid, is probably involved in the renal secretion of faropenem (an oral carbapenem) and is inhibited by anionic β-lactams, such as benzylpenicillin, ampicillin, cephalexin, and cefazolin, and other anionic drugs, such as indomethacin and furosemide (Uchino et al., 2000). The involvement of other transporters, such as the P-glycoprotein (MDR1, PGP), is suspected but not proven (Susanto and Benet, 2002).

Cytotoxicity

While site-specific transport and accumulation are a first and necessary step, these are not sufficient to cause proximal tubular nephrotoxicity. Several parameters seem involved, such as:

- the presence of the pyridinium ring (cephaloridine) and of the substituted 3-methylthiotetrazole (cefotiam) (Cojocel et al., 1988)
- the rate and extent of desacetylation of cephalosporins (Hottendorf et al., 1987; Williams et al., 1988)
- an activation of protein kinase C with an enhancement of superoxide anion generation (Kohda and Gemba, 2001)
- the activation of adenine receptors (Minami et al., 1994)
- the interference with carnitine-dependent fatty acid oxidation in mitochondria or reabsorption of carnitine in the kidney (Tune and Hsu, 1994, 1995).

Unfortunately, many of these studies have been made using only a limited number of β-lactams, so that it is not possible to determine in-depth structure–activity relationships and generalizations are difficult to draw. It is also possible that the mechanisms are genuinely multiple and depend on the chemical nature of each of the drugs studied, i.e., their side chains. Further studies combining detailed examination of the distribution, expression levels and activity of β-lactam transporters along the nephron, together with a molecular knowledge of the cytotoxic effects of each antibiotic accumulated by cells, will be needed for useful prediction of the nephrotoxicity of new compounds (Jariyawat et al., 1999).

Concluding Remarks on β-lactams

β-Lactams have shown at least two modes of renal toxicities, namely interstitial nephritis and tubular necrosis. While the former toxicity is intrinsically linked to their allergenic properties and cannot really be predicted, it remains rare and associated with only a few compounds in clinical practice. Differential diagnostic with other causes of interstitial nephritis is, however, desirable to avoid continuing exposure, or reexposure to the same drug. Tubular necrosis has been a problem with cephaloridine and a few other β-lactams when administered at high doses. The safety margin introduced by the systematic screening of new molecules should be sufficient for this toxicity to remain exceptional in clinical practice. It remains, however, a problem to be seriously considered in the development of all new β-lactams since it appears to result from a combination of common factors to this whole pharmacological class together with specific and idiosyncratic properties of each drug.

VANCOMYCIN

Vancomycin is a glycopeptide antibiotic of large molecular weight. This causes the drug not to be absorbed from the gastrointestinal tract, and to be unable to penetrate Gram-negative bacteria. Originally introduced in the clinics in the late 1950s as an agent active in case of staphylococcal and streptococcal infections, vancomycin has been notorious for early toxicity related to impurities and to histamine release (causing the so-called "red-man syndrome"). These adverse effects have been markedly reduced through better purification procedures and by giving the drug as a slow infusion over at least one hour. With these precautions, vancomycin is considered as safe, causing only relatively mild and self-limiting general toxicity (Elting et al., 1998;

Sorrell and Collignon, 1985), with the main side effects remaining phlebitis at the site of injection as well as nephrotoxicity (see below) and ototoxicity. The latter remains the most problematic because it can be irreversible (Feketi, 2000).

For many years, clinical usage of vancomycin has remained limited because of the availability of other antibiotics, including of β-lactamase-resistant penicillins and cephalosporins, β-lactamase inhibitors such as clavulanic acid (in Europe) or sulbactam (in the U.S. and some other countries), capable of counteracting the rise of β-lactamase producing staphylococci. Yet, the pandemic of nosocomial methicillin-resistant *Staphylococcus aureus* infections which started in the mid 1970s (and the fact that these strains were resistant not only to all β-lactams, including those resistant to β-lactamase, but often also to aminoglycosides, macrolides, lincosamines, and fluoroquinolones; Livermore, 2000), heralded the comeback of vancomycin for systemic use in the mid-1980s. The only real alternative has been teicoplanin, which can be administered less frequently and is claimed to be less nephrotoxic. However, uncertainties about the necessary dosages of teicoplanin in severe infections have resulted in a lack of approval in the U.S. and many other countries outside Europe. The introduction of quinupristin/dalfopristin and linezolid in the late 1990s has not really changed this situation as both drugs have potentially worrying side effects and are facing emergence of resistance.

Vancomycin elimination is predominantly by glomerular filtration and clearance amounts to 90% of inulin clearance and 80% of creatinine clearance (due to protein binding) with only a minor portion of the drug being eliminated by nonrenal mechanisms (Krogstad et al., 1980). Clearance of vancomycin and creatinine are highly correlated among patients, enabling construction of nomograms for vancomycin dosage adjustment based on creatinine clearance (Moellering, Jr. et al., 1981) and body weight (Brown and Mauro, 1988).

There have been many disputes about the intrinsic nephrotoxic character of vancomycin and the role of impurities in this context. The incidence of nephrotoxicity associated with vancomycin appears indeed low (5 to 15%) when the drug is used alone (Goetz and Sayers, 1993; Rybak et al., 1990). In a recent study in China based on 84 patients suffering from Gram-positive infections, the prevalence of nephrotoxicity ranged from 11% to 14%, depending on which criteria were used. There was no significant difference in terms of nephrotoxicity prevalence whichever criteria of nephrotoxicity were applied. Nephrotoxicity could be reversed, either during or after treatment, in 22 to 44% of patients. The development of nephrotoxicity was associated with lower respiratory tract infections and poor bacteriological response.

There is ample clinical evidence that vancomycin can enhance the nephrotoxic potential of aminoglycosides. Risk factors include the length of treatment with vancomycin longer than 21 days and vancomycin trough serum concentrations greater than 10 mg/l (Rybak et al., 1990). In a study comparing continuous and intermittent infusion of vancomycin, a significant rise in serum creatinine concentration was only observed in patients receiving other antibiotics, including aminoglycosides (Wysocki et al., 2001). Conversely, a prospective study evaluating the effect of aminoglycoside dosing regimens on rates of observed nephrotoxicity found by multivariate logistic regression analysis that concomitant use of vancomycin was a significant predictor of nephrotoxicity (Rybak et al., 1999). The same conclusion was reached from a study with oncologic patients (Elting et al., 1998). In this context, there are data that suggest that individualized pharmacokinetic monitoring results in less aminoglycoside-associated nephrotoxicity and fewer associated costs in patients exposed to vancomycin (Streetman et al., 2001). Pediatric patients are, however, at lower risk (Nahata, 1987), perhaps in relation to their better regenerating capabilities (Laurent et al., 1988).

Experimental studies support the concept that vancomycin and aminoglycosides potentiate their respective nephrotoxicities. Rats treated concomitantly with tobramycin and vancomycin show, indeed, more extensive tubular necrosis than those treated with either of these drugs alone, and also demonstrated more intense regeneration (Wood et al., 1986). This heralds early insult to cortical tissue (Laurent et al., 1988). A similar conclusion pointing to a more severe toxicity of the gentamicin-vancomycin combination compared to each of these drugs alone was reached in a study using enzymuria (alanyl aminopeptidase, γ-glutamyltransferase, *N*-acetyl-β-hexosaminidase) as criterion of tubular damage in vancomycin–gentamicin combinations (Fauconneau et al., 1992, 1997). Intriguingly enough, renal cortical levels of vancomycin were lowered in rats receiving both vancomycin and gentamicin as compared with rats receiving vancomycin alone. One possible explanation is the fact that increased tubular necrosis will result in cell shedding and ensuing decrease of cortical levels of the drug.

Unfortunately, very little is known about the uptake and handling of vancomycin in kidney. One study concluded that mediated transport for vancomycin occurred across the basolateral membrane, but not across the brush border membrane. It implied that the nephrotoxicity of vancomycin is due to entry from this pole with absence of mediated egress at the brush border membrane (Sokol et al., 1989). The transporters have, however, not been identified. Moreover, the subcellular localization of vancomycin in kidney is largely unknown. One study examined the subcellular localization of tobramycin and vancomycin in the renal cortices by immunogold labeling. Tobramycin was detected over the lysosomes of proximal tubular cells as expected, together with vancomycin (Beauchamp et al., 1992a). The significance of this colocalization in molecular and cytopathological terms has, however, not been further investigated.

Another potentiation of toxicity that may have clinical significance is that of endotoxin. While endotoxin did not cause additional nephrotoxicity in rats treated with vancomycin alone, it caused marked toxicity (as assessed by the increase in blood urea nitrogen concentration, decrease in creatinine clearance, and rise in renal cortical DNA synthesis) in animals treated with vancomycin and gentamicin. This occurred to a more severe extent than in animals receiving gentamicin alone. It was concluded that endotoxin amplified the nephrotoxic potential of gentamicin alone as well as that of gentamicin plus vancomycin (Ngeleka et al., 1990).

The usefulness of vancomycin serum concentrations, the determination of a therapeutic range of values, and their correlation to antibacterial efficacy and drug toxicity in the clinical setting are controversial. Old reports of dose-related toxicities were, indeed, confronted with the problems of impurities. Actually, only the antibacterial efficacy of vancomycin and its correlation with reported therapeutic ranges may justify obtaining a vancomycin trough concentration in certain groups of patients (Freeman et al., 1993). Evaluation by decision analysis over a range of assumptions, varying probabilities, and costs reveals that pharmacokinetic monitoring and vancomycin dosage adjustment to prevent nephrotoxicity are not cost-effective for all patients. However, such dosage adjustments demonstrate cost-effectiveness for patients receiving concomitant nephrotoxicants, intensive care patients, and probably oncology patients (Darko et al., 2003). Serum level determination may also be helpful in patients with increased volume of distribution, in patients with decreased renal function, in children or neonates, and in elderly patients (Cunha, 1995).

As mentioned earlier, teicoplanin, a glycopeptide antibiotic similar to vancomycin except for its much longer half-life (due to an hydrophobic side chain), was claimed to be devoid of nephrotoxicity. This proved correct when tested in volunteers at 6 mg/kg/day (Pierre et al., 1988), and was confirmed by clinical trials using this dose or lower ones (Wilson, 1998; Wood, 1996, 2000). However, high doses and prolonged treatment have been associated with toxicity, which appeared to consist of interstitial nephritis (Frye et al., 1992). More importantly, clinical failures were seen in the early 1990s in patients treated with the 6 mg/kg doses for severe infections, such as endocarditis (Gilbert et al., 1991), and concerns were also raised about the possible selection of resistance in *Staphylococcus aureus* and in coagulase-negative staphylococci. Unfortunately, none of the previously published animal studies had reported systematically on dose–effect relationships for teicoplanin toxicity in comparison with vancomycin (which is typically dosed in humans at 30 mg/kg/day). The introduction of teicoplanin in Japan, however, triggered such additional studies (Yoshiyama et al., 2000). These showed that the nephrotoxicity of teicoplanin was only one-fourth that of vancomycin in rats. Extrapolating these data to humans would suggest that teicoplanin, given at the high doses that may be necessary, could be as nephrotoxic as vancomycin.

Because of the renewed interest in glycopeptides, and of the mounting resistance against vancomycin, new glycopeptides have been obtained and are presently under development. Among them, oritavancin is active against vancomycin-resistant enterocococci and therefore attracts much interest. The only safety data published so far, however, concern the phase I studies, in which oritavancin was well-tolerated (Barrett, 2001). Another derivative is dalbavancin, which is characterized by a very long half-life, making once-weekly administration possible. Preclinical studies in rats and dogs show that dalbavancin is well tolerated upon intravenous bolus administrations at doses several times higher than those expected to be used in humans. Phase I or phase II clinical trials have not reported major side effects in the range of concentrations where dalbavancin is effective (Seltzer et al., 2003). No dosage adjustments are necessary in case of mild renal insufficiency (Dowell et al., 2003).

Three new antibiotics of other pharmacochemical classes (quinupristin/dalfopristin, linozolid, and daptomcyin) have recently become available as treatment options for infections caused by drug-resistant Gram-positive cocci. Although apparently devoid of nephrotoxicity, these drugs have other adverse effects and are costly. Nevertheless, quinupristin/dalfopristin and linozolid may be useful in selected patients who cannot tolerate vancomycin, or in the case of infection with resistant organisms. Studies with daptomycin, which protects against aminoglycoside nephrotoxicity in animals, are awaited with interest.

Few studies have examined protective means against nephrotoxicity induced by glycopeptide antibiotics. Yet, fosfomycin (an antibiotic with limited usage) was found to decrease not only gentamicin but also vancomycin and teicoplanin toxicities (Yoshiyama et al., 2001). The usefulness and practicability of such antibiotic associations solely for a toxicological reason is, however, subject to criticism in view of the risk of emergence of resistance.

Concluding Remarks for Vancomycin

The nephrotoxicity of vancomycin and other glycopeptide antibiotics seems relatively minor, but is not clearly understood in molecular terms. The reasons are mainly a lack of basic studies with vancomycin, which is bearing its age, and the uncertainties concerning dosages and comparative, dose-dependence studies for teicoplanin. New glycopeptides are still in development and little public data is available. Despite those difficulties, it is fair to say that glycopeptide nephrotoxicity should not be of major concern to clinicians, except in the case of co-administration of aminoglycosides.

OTHER ANTIBIOTICS

Data on nephrotoxicity of other antibiotics are scanty and most available data consist of case reports with few, if any, systematic experimental studies. The field is also made difficult to review as data concerning established antibiotics are often old (or, for mechanistic studies, poor), while those related to new compounds are often not in the public domain. The following antibiotics may, however, be briefly discussed.

Colistin is a cationic polypeptide antibiotic from the polymyxin family. These antibiotics interact with membrane phospholipids, causing destruction of the bacterial membranes by detergent-like mechanisms and, thereby, by increasing cellular permeability. Colistin, introduced in 1962, was abandoned in the early 1970s because of initial reports of severe toxicities (Ito et al., 1969). However, the difficulties created by the emergence of multidrug resistant *Pseudomonas aeruginosa* in cystic fibrosis patients have forced clinicians to resume its usage. Few data, however, are available, even though some researchers consider that the frequency of nephrotoxicity may be substantially less than previously believed (Beringer, 2001). Colistin is currently usually administered by aerosol, which tends to minimize its potential nephrotoxic effects.

The nephrotoxicity of quinolones has been linked to the development of crystalluria in experimental animals. However, crystalluria is unlikely to occur in humans at the clinical doses and renal damage has not been noted for ciprofloxacin. Premarketing clinical trials in humans ($n = 5{,}308$) of temafloxacin (which eventually was withdraw for reason of uremic hemolytic anemia; Blum et al., 1994) reported no crystalluria nor clinically significant nephrotoxicity (Krasula and Pernet, 1991). Only a few reports of acute renal failure due to oral fluoroquinolone therapy have appeared, with signs of acute interstitial nephritis (Famularo and De Simone, 2002; Ramalakshmi et al., 2003). All patients had nonoliguric renal failure, which was completely reversed after discontinuation of the fluoroquinolone. Eventually, a Medline search was conducted over the period 1985 to 1999 on ciprofloxacin, norfloxacin, levofloxacin, ofloxacin, trovafloxacin, enoxacin, sparfloxacin, grepafloxacin, gatifloxacin, clinafloxacin, and moxifloxacin to ascertain the incidence and features of fluoroquinolone nephrotoxicity. The search failed to reveal more than case reports and temporally related events, with hard-to-estimate incidences and the possibility of multifactorial aspects (Lomaestro, 2000). Interestingly, fleroxacin and pasufloxacin (an experimental fluoroquinolone) have been shown to reduce the nephrotoxic potential of gentamicin.

Nephrotoxicity was recognized as a risk factor of the combination of sulfamethoxazole and trimethoprim (cotrimoxazole) since its introduction on the European market in the late 1970s (Richmond et al., 1979). It is potentially related to tubuloobstructive effect caused by precipitation of sulfamethoxazole, as first observed after treatment with sulfonamides in the 1940s. It is particularly concerning in transplant patients (Ringden et al., 1984) as cotrimoxazole toxicity is synergistic with that of cyclosporine (Pak et al., 1988). Surprisingly, however, animal studies showed a protective effect of cotrimozole on gentamicin-induced nephrotoxicity in rats (Izzettin et al., 1994). Further basic studies are probably needed in this context.

Tetracyclines may aggravate uremia in patients with underlying renal diseases, in relation to their capacity to impair eucaryotic as well as procaryotic synthesis. In addition, cases of nonoliguric renal failures were reported as early as in the mid 1960s (Lew and French, 1966).

Ascorbic acid, isoascorbic acid, and mannitol were found to be protective (Polec et al., 1971), which suggests that osmotic diuresis was contributing to eliminate a toxic agent, however the mechanism of this toxicity has remained largely unexplored.

More-recent studies showed that the OAT1 and OAT4 transporters mediated the efflux of tetracycline, whereas human (h) OAT2 and hOAT3 did not (Babu et al., 2002b). These results have been interpreted as suggesting that OAT1 mediates the basolateral uptake and efflux of tetracycline, whereas hOAT4 is responsible for the reabsorption as well as the efflux of tetracycline at the apical side of the proximal tubule. However, the link to tetracycline-induced nephrotoxicity in the human kidney remains to be established. A Fanconi's syndrome, associated with nausea, vomiting and polyuria, has also been repeatedly observed in patients receiving degraded tetracyclines since its original description in the mid 1960s (Brodehl et al., 1968) and may result from a direct insult to the proximal renal tubules.

Rifampin has been linked with renal insufficiency associated with heavy glomerular proteinuria. Acute interstitial nephritis together with effacement of glomerular epithelial cell foot processes, electron-dense deposits in mesangial matrix and subendothelial and paramesangial sites seemed to be the cause of the disorder, the clinical signs of which were reversible upon discontinuation of therapy (Neugarten et al., 1983). Other cases of acute tubular nephritis have been described (Gallieni et al., 1999). Toxicity was shown to involve tubular cells, as demonstrated by potassium wasting, acidifying defect, high fractional excretion of uric acid, and glucosuria (Cheng and Kahn, 1984). Progressive increases in enzymuria with no changes in glomerular filtration rate were also observed, and were additive to that caused by streptomycin (Kumar et al., 1992).

CONCLUSIONS

Nephrotoxicity remains of concern for many antibiotics, but progress in the understanding of the basic underlying mechanism coupled with the optimization of the clinical use of existing agents and the effective selection of less toxic derivatives has considerably reduced the incidence of this adverse effect. It is very rewarding to see how cooperation between chemistry, biology, and clinical sciences has been so effective in this context. Efforts should, however, be maintained because nephrotoxic reactions significantly contribute to endangering patients.

They also represent an important medical and economic burden in treatment of infectious diseases.

ACKNOWLEDGMENTS

Hélène Servais is Boursier of the Belgian Fonds pour l'Encouragement de la Recherche dans l'Industrie et l'Agriculture (F.RI.A.). The experimental work of the authors is supported by the Belgian Fonds de la Recherche Scientifique Médicale (F.R.SM.), the Fonds National de la Recherche Scientifique (F.N.S.R.), and the Fonds Spécial de Recherches (F.S.R.) of the Université catholique de Louvain.

REFERENCES

Ademuyiwa O, Ngaha EO, and Ubah FO (1990) Vitamin E and selenium in gentamicin nephrotoxicity. *Hum Exp Toxicol* 9:281–288.

Amsden GW, Ballow CH, and Bertino JS (2000) Pharmacokinetics and Pharmacodynamics of Anti-infective Agents, in *Principles and Practices of Infectious Diseases*, Mandell GL, Bennett JE, and Dolin R (Eds), Churchill Livingstone, New York.

Appel GB (1990) Aminoglycoside nephrotoxicity. *Am J Med* 88:16S–20S.

Appel GB, and Neu HC (1977) The nephrotoxicity of antimicrobial agents (first of three parts). *N Engl J Med* 296:663–670.

Atkinson RM, Currie JP, Davis B, Pratt DA, Sharpe HM, and Tomich EG, (1966) Acute toxicity of cephaloridine, an antibiotic derived from cephalosporin C. *Toxicol Appl Pharmacol* 8:398–406.

Aubert-Tulkens G, Van Hoof F, and Tulkens P (1979) Gentamicin-induced lysosomal phospholipidosis in cultured rat fibroblasts. Quantitative ultrastructural and biochemical study. *Lab Invest* 40:481–491.

Auclair P, Tardif D, Beauchamp D, Gourde P, and Bergeron MG (1990) Prolonged endotoxemia enhances the renal injuries induced by gentamicin in rats. *Antimicrob Agents Chemother* 34:889–895.

Babu E, Takeda M, Narikawa S, Kobayashi Y, Enomoto A, Tojo A, Cha SH, Sekine T, Sakthisekaran D, and Endou H (2002a) Role of human organic anion transporter 4 in the transport of ochratoxin A. *Biochim Biophys Acta* 1590:64–75.

Babu E, Takeda M, Narikawa S, Kobayashi Y, Yamamoto T, Cha SH, Sekine T, Sakthisekaran D, and Endou H (2002b) Human organic anion transporters mediate the transport of tetracycline. *Jpn J Pharmacol* 88:69–76.

Baldwin DS, Levine BB, McCluskey RT, and Gallo GR (1968) Renal failure and interstitial nephritis due to penicillin and methicillin. *N Engl J Med* 279:1245–1252.

Barr GA, Mazze RI, Cousins MJ, and Kosek JC (1973) An animal model for combined methoxyflurane and gentamicin nephrotoxicity. *Br J Anaesth* 45:306–312.

Barrett JF (2001) Oritavancin. Eli Lilly & Co. *Curr Opin Investig Drugs* 2:1039–1044.

Bartal C, Danon A, Schlaeffer F, Reisenberg K, Alkan M, Smoliakov R, Sidi A, and Almog Y (2003) Pharmacokinetic dosing of aminoglycosides: A controlled trial. *Am J Med* 114:194–198.

Baylis C, Rennke HR, and Brenner BM (1977) Mechanisms of the defect in glomerular ultrafiltration associated with gentamicin administration. *Kidney Int* 12:344–353.

Beauchamp D, Gourde P, Simard M, and Bergeron MG (1992a) Subcellular localization of tobramycin and vancomycin given alone and in combination in proximal tubular cells, determined by immunogold labeling. *Antimicrob Agents Chemother* 36:2204–2210.

Beauchamp D, Gourde P, Theriault G, and Bergeron MG (1992b) Age-dependent gentamicin experimental nephrotoxicity. *J Pharmacol Exp Ther* 260:444–449.

Beauchamp D, Laurent G, Grenier L, Gourde P, Zanen J, Heuson-Stiennon JA, and Bergeron MG (1997) Attenuation of gentamicin-induced nephrotoxicity in rats by fleroxacin. *Antimicrob Agents Chemother* 41:1237–1245.

Beauchamp D, Laurent G, Maldague P, Abid S, Kishore BK, and Tulkens PM (1990a) Protection against gentamicin-induced early renal alterations (phospholipidosis and increased DNA synthesis) by coadministration of poly-L-aspartic acid. *J Pharmacol Exp Ther* 255:858–866.

Beauchamp D, Pellerin M, Gourde P, Pettigrew M, and Bergeron MG (1990b) Effects of daptomycin and vancomycin on tobramycin nephrotoxicity in rats. *Antimicrob Agents Chemother* 34:139–147.

Ben Ismail TH, Ali BH, and Bashir AA (1994) Influence of iron, deferoxamine and ascorbic acid on gentamicin-induced nephrotoxicity in rats. *Gen Pharmacol* 25:1249–1252.

Bendirdjian J, Gillaster J, and Foucher B (1982) Mitochondrial modifications with aminoglycosides, in *The Aminoglycosides, Microbiology, Clinical Use and Toxicology*, Marcel Dekker, New-York.

Bennett WM (1989) Mechanisms of aminoglycoside nephrotoxicity. C*lin Exp Pharmacol Physiol* 16:1–6.

Bennett WM, Mela-Riker LM, Houghton DC, Gilbert DN, and Buss WC (1988) Microsomal protein synthesis inhibition: An early manifestation of gentamicin nephrotoxicity. *Am J Physiol* 255:F265–F269.

Bennett WM, Plamp CE, Gilbert DN, Parker RA, and Porter GA (1979) The influence of dosage regimen on experimental gentamicin nephrotoxicity: Dissociation of peak serum levels from renal failure. *J Infect Dis* 140:576–580.

Beringer P (2001) The clinical use of colistin in patients with cystic fibrosis. *Curr Opin Pulm Med* 7:434–440.

Bertino JS, Jr., Booker LA, Franck PA, Jenkins PL, Franck KR, and Nafziger AN (1993) Incidence of and significant risk factors for aminoglycoside-associated nephrotoxicity in patients dosed by using individualized pharmacokinetic monitoring. *J Infect Dis* 167:173–179.

Bertino JS, Jr., Rodvold KA, and Destache CJ (1994) Cost considerations in therapeutic drug monitoring of aminoglycosides. *Clin Pharmacokinet* 26:71–81.

Blaser J, and Konig C (1995) Once-daily dosing of aminoglycosides. *Eur J Clin Microbiol Infect Dis* 14:1029–1038.

Blaser J, Stone BB, and Zinner SH (1985) Efficacy of intermittent versus continuous administration of netilmicin in a two-compartment *in vitro* model. *Antimicrob Agents Chemother* 27:343–349.

Blaser J, Stone BB, Groner, MC, and Zinner SH (1987) Comparative study with enoxacin and netilmicin in a pharmacodynamic model to determine importance of ratio of antibiotic peak concentration to MIC for bactericidal activity and emergence of resistance. *Antimicrob Agents Chemother* 31:1054–1060.

Blum MD, Graham DJ, and McCloskey CA (1994) Temafloxacin syndrome: Review of 95 cases. *Clin Infect Dis* 18:946–950.

Boyd JF, Butcher BT, and Stewart GT (1973) The nephrotoxic effect of cephaloridine and its polymers. *Int J Clin Pharmacol* 7:307–315.

Brodehl J, Gellissen K, Hagge W, and Schumacher H (1968) [Reversible renal Fanconi syndrome due to a toxic catabolism product of tetracycline]. *Helv Paediatr Acta* 23:373–383.

Brown DL, and Mauro LS (1988) Vancomycin dosing chart for use in patients with renal impairment. *Am J Kidney Dis* 11:15–19.

Browning MC, Hsu CY, Wang PL, and Tune BM (1983) Interaction of ischemic and antibiotic-induced injury in the rabbit kidney. *J Infect Dis* 147:341–351.

Burton ME, Ash CL, Hill DP, Jr., Handy T, Shepherd MD, and Vasko MR (1991) A controlled trial of the cost benefit of computerized bayesian aminoglycoside administration. *Clin Pharmacol Ther* 49:685–694.

Buss WC, and Piatt MK (1985) Gentamicin administered *in vivo* reduces protein synthesis in microsomes subsequently isolated from rat kidneys but not from rat brains. *J Antimicrob Chemother* 15:715–721.

Buyukafsar K, Yazar A, Dusmez D, Ozturk H, Polat G, and Levent A (2001) Effect of trapidil, an antiplatelet and vasodilator agent on gentamicin-induced nephrotoxicity in rats. *Pharmacol Res* 44:321–328.

Carrier D, Bou KM, and Kealey A (1998) Modulation of phospholipase A2 activity by aminoglycosides and daptomycin: A Fourier transform infrared spectroscopic study. *Biochemistry* 37:7589–7597.

Carter AP, Clemons WM, Brodersen DE, Morgan-Warren RJ, Wimberly BT, and Ramakrishnan V (2000) Functional insights from the structure of the 30S ribosomal subunit and its interactions with antibiotics. *Nature* 407:340–348.

Casteels-Van Daele M, Corbeel L, Van de CW, and Standaert L (1980) Gentamicin-induced Fanconi syndrome. *J Pediatr* 97:507–508.

Cha SH, Sekine T, Fukushima JI, Kanai Y, Kobayashi Y, Goya T, and Endou H (2001) Identification and characterization of human organic anion transporter 3 expressing predominantly in the kidney. *Mol Pharmacol* 59:1277–1286.

Cha SH, Sekine T, Kusuhara H, Yu E, Kim JY, Kim DK, Sugiyama Y, Kanai Y, and Endou H (2000) Molecular cloning and characterization of multispecific organic anion transporter 4 expressed in the placenta. *J Biol Chem* 275:4507–4512.

Charlwood J, Skehel JM, King N, Camilleri P, Lord P, Bugelski P, and Atif U (2002) Proteomic Analysis of Rat Kidney Cortex Following Treatment with Gentamicin. *Journal of Proteome Research* 1:73–82.

Cheng JT, and Kahn T (1984) Potassium wasting and other renal tubular defects with rifampin nephrotoxicity. *Am J Nephrol* 4:379–382.

Chiu PJ, Miller GH, Long JF, and Waitz JA (1979) Renal uptake and nephrotoxicity of gentamicin during urinary alkalinization in rats. *Clin Exp Pharmacol Physiol* 6:317–326.

Chuck SK, Raber SR, Rodvold KA, and Areff D (2000) National survey of extended-interval aminoglycoside dosing. *Clin Infect Dis* 30:433–439.

Cojocel C, Gottsche U, Tolle KL, and Baumann K (1988) Nephrotoxic potential of first-, second-, and third-generation cephalosporins. *Arch Toxicol* 62:458–464.

Craig WA (1997) The pharmacology of meropenem, a new carbapenem antibiotic. *Clin Infect Dis* 24 Suppl 2:S266–S275.

Craig WA (2001) Does the dose matter? *Clin Infect Dis* 33 Suppl 3:S233–S237.

Cronberg S (1994) Simplified monitoring of aminoglycosides. *J Antimicrob Chemother* 4:819–827.

Cronin RE, Nix KL, Ferguson ER, Southern PM, and Henrich WL (1982) Renal cortex ion composition and Na-K-ATPase activity in gentamicin nephrotoxicity. *Am J Physiol* 242:F477–F483.

Cunha BA (1995) Vancomycin. *Med Clin North Am* 79:817–831.

Dahlgren JG, Anderson ET, and Hewitt WL (1975) Gentamicin blood levels: A guide to nephrotoxicity. *Antimicrob Agents Chemother* 8:58–62.

Daniel H, and Adibi SA (1993) Transport of beta-lactam antibiotics in kidney brush border membrane. Determinants of their affinity for the oligopeptide/H+ symporter. *J Clin Invest* 92:2215–2223.

Darko W, Medicis JJ, Smith A, Guharoy R, and Lehmann DE (2003) Mississippi mud no more: Cost-effectiveness of pharmacokinetic dosage adjustment of vancomycin to prevent nephrotoxicity. *Pharmacotherapy* 23:643–650.

Davies J, Gorini L, and Davis BD (1965) Misreading of RNA codewords induced by aminoglycoside antibiotics. *Mol Pharmacol* 1:93–106.

Davis MA, and Ryan DH (1998) Apoptosis in the kidney. *Toxicol Pathol* 26:810–825.

De Broe ME, Paulus GJ, Verpooten GA, Roels F, Buyssens N, Wedeen R, Van Hoof F, and Tulkens PM (1984) Early effects of gentamicin, tobramycin, and amikacin on the human kidney. *Kidney Int* 25:643–652.

De Broe ME, Verbist L, and Verpooten GA (1991) Influence of dosage schedule on renal cortical accumulation of amikacin and tobramycin in man. *J Antimicrob Chemother* 27 Suppl C:41–47.

Dejace P, and Klastersky J (1987) [A comparative review of combination therapy: 2 beta-lactams versus beta-lactam plus aminoglycoside]. *Infection* 15 Suppl 4:S158–S167.

Desai TK, and Tsang TK (1988) Aminoglycoside nephrotoxicity in obstructive jaundice. *Am J Med* 85:47–50.

Dominguez JH, Hale CC, and Qulali M (1996) Studies of renal injury. I. Gentamicin toxicity and expression of basolateral transporters. *Am J Physiol* 270:F245–F253.

Dowell JA., Seltzer E, Stogniew M, Dorr MB, Fayokavitz S, Krauze D, and Henkel T (2003) Dalbavancin dosage adjustements not required for patients with mild renal impairment. *13th European Congress of Clinical Microbiology and Infectious Diseases*, Glasgow, UK, P1224.

Edwards CQ, Smith CR, Baughman KL, Rogers JF, and Lietman PS (1976) Concentrations of gentamicin and amikacin in human kidneys. *Antimicrob Agents Chemother* 9:925–927.

El Mouedden M, Laurent G, Mingeot-Leclercq MP, Taper HS, Cumps J, and Tulkens PM (2000a) Apoptosis in renal proximal tubules of rats treated with low doses of aminoglycosides. *Antimicrob Agents Chemother* 44:665–75.

El Mouedden M, Laurent G, Mingeot-Leclercq MP, and Tulkens PM (2000b) Gentamicin-induced apoptosis in renal cell lines and embryonic rat fibroblasts. *Toxicol Sci* 56:229–239.

Elfarra AA, Duescher RJ, Sausen PJ, O'Hara TM, and Cooley AJ (1994) Methimazole protection of rats against gentamicin-induced nephrotoxicity. *Can J Physiol Pharmacol* 72:1238–1244.

Elliott WC, Houghton DC, Gilbert DN, Baines-Hunter J, and Bennett WM (1982a) Gentamicin nephrotoxicity. I. Degree and permanence of acquired insensitivity. *J Lab Clin Med* 100:501–512.

Elliott WC, Houghton DC, Gilbert DN, Baines-Hunter J, and Bennett WM (1982b) Gentamicin nephrotoxicity. II. Definition of conditions necessary to induce acquired insensitivity. *J Lab Clin Med* 100:513–525.

Elting LS, Rubenstein EB, Kurtin D, Rolston KV, Fangtang J, Martin CG, Raad II, Whimbey EE, Manzullo E, and Bodey GP (1998) Mississippi mud in the 1990s: Risks and outcomes of vancomycin- associated toxicity in general oncology practice. *Cancer* 83:2597–2607.

Enomoto A, Takeda M, Shimoda M, Narikawa S, Kobayashi Y, Kobayashi Y, Yamamoto T, Sekine T, Cha SH, Niwa T, and Endou H (2002) Interaction of human organic anion transporters 2 and 4 with organic anion transport inhibitors. *J Pharmacol Exp Ther* 301:797–802.

Fabre J, Rudhardt M, Blanchard P, and Regamey C (1976) Persistence of sisomicin and gentamicin in renal cortex and medulla compared with other organs and serum of rats. *Kidney Int* 10:444–449.

Famularo G, and De Simone C (2002) Nephrotoxicity and purpura associated with levofloxacin. *Ann Pharmacother* 36:1380–1382.

Fauconneau B, De Lemos E, Pariat C, Bouquet S, Courtois P, and Piriou A (1992) Chrononephrotoxicity in rat of a vancomycin and gentamicin combination. *Pharmacol Toxicol* 71:31–36.

Fauconneau B, Favreliere S, Pariat C, Genevrier A, Courtois P, Piriou A, and Bouquet S (1997) Nephrotoxicity of gentamicin and vancomycin given alone and in combination as determined by enzymuria and cortical antibiotic levels in rats. *Ren Fail* 19:15–22.

Feketi R (2000) Vancomycin, teicoplanin, and the streptogramins: quinupristin and dalfopristin, in *Principles and practice of infectious diseases,* Mandell GE, Bennett JE, and Dolin R (Eds), Churchill Livingstone, Philadelphia, pp. 382–392.

Feldman S, Wang MY, and Kaloyanides GJ (1982) Aminoglycosides induce a phospholipidosis in the renal cortex of the rat: An early manifestation of nephrotoxicity. *J Pharmacol Exp Ther* 220:514–520.

Ferriols-Lisart R, and Alos-Alminana M (1996) Effectiveness and safety of once-daily aminoglycosides: A meta-analysis. *Am J Health Syst Pharm* 53:1141–1150.

Freeman CD, Quintiliani R, and Nightingale CH (1993) Vancomycin therapeutic drug monitoring: Is it necessary? *Ann Pharmacother* 27:594–598.

Frye RF, Job ML, Dretler RH, and Rosenbaum BJ (1992) Teicoplanin nephrotoxicity: First case report. *Pharmacotherapy* 12:240–242.

Fujita K, Fujita HM, and Aso Y (1983) [Protective effect of fosfomycin against renal accumulation of aminoglycoside antibiotics]. *Jpn J Antibiot* 36:3392–3394.

Gallieni M, Braidotti P, Cozzolino M, Romagnoli S, and Carpani P (1999) Acute tubulo-interstitial nephritis requiring dialysis associated with intermittent rifampicin use: Case report. *Int J Artif Organs* 22:477–481.

Gamba G, Contreras AM, Cortes J, Nares F, Santiago Y, Espinosa A, Bobadilla J, Jimenez SG, Lopez G, Valadez A, and. (1990) Hypoalbuminemia as a risk factor for amikacin nephrotoxicity. *Rev Invest Clin* 42:204–209.

Ganapathy ME, Brandsch M, Prasad PD, Ganapathy V, and Leibach FH (1995) Differential recognition of beta-lactam antibiotics by intestinal and renal peptide transporters, PEPT 1 and PEPT 2. *J Biol Chem* 270:25672–25677.

Ganapathy ME, Prasad PD, Mackenzie B, Ganapathy V, and Leibach FH (1997) Interaction of anionic cephalosporins with the intestinal and renal peptide transporters PEPT 1 and PEPT 2. *Biochim Biophy.Acta* 1324:296–308.

Geleilete TJ, Melo GC, Costa RS, Volpini RA, Soares TJ, and Coimbra TM (2002) Role of myofibroblasts, macrophages, transforming growth factor-beta endothelin, angiotensin-II, and fibronectin in the progression of tubulointerstitial nephritis induced by gentamicin. *J Nephrol* 15:633–642.

Gerber AU, Kozak S, Segessenmann C, Fluckiger U, Bangerter T, and Greter U (1989) Once-daily versus thrice-daily administration of netilmicin in combination therapy of Pseudomonas aeruginosa infection in a man-adapted neutropenic animal model. *Eur J Clin Microbiol Infect Dis* 8:233–237.

Ghuysen JM, Frère JM, and Leyh-Bouille M (1981) The D-alanyl-D-Ala peptidases. Mechanism of action of penicillins and delta-3-cephalosporins, in *Beta-Lactam Antibiotics: Mode of Action, New Development, and Future Prospects*, Salton M, and Shockman GD (Eds), Academic Press, Inc., New York, pp. 127–152.

Gilbert D.N. (2000) Aminoglycosides, in *Principles and Practices of Infectious Diseases*, Mandell GL, Bennett JE, and Dolin R (Eds), Churchill Livingstone, New-York.

Gilbert DN (1997) Meta-analyses are no longer required for determining the efficacy of single daily dosing of aminoglycosides. *Clin Infect Dis* 24:816–819.

Gilbert DN, Plamp C, Starr P, Bennet WM, Houghton DC, and Porter G (1978) Comparative nephrotoxicity of gentamicin and tobramycin in rats. *Antimicrob Agents Chemother* 13:34–40.

Gilbert DN, Wood CA, and Kimbrough RC (1991) Failure of treatment with teicoplanin at 6 milligrams/kilogram/day in patients with Staphylococcus aureus intravascular infection. The Infectious Diseases Consortium of Oregon. *Antimicrob Agents Chemother* 35:79–87.

Gilbert DN, Wood CA, Kohlhepp SJ, Kohnen PW, Houghton DC, Finkbeiner HC, Lindsley J, and Bennett WM (1989) Polyaspartic acid prevents experimental aminoglycoside nephrotoxicity. *J Infect Dis* 159:945–953.

Giuliano RA, Paulus GJ, Verpooten GA, Pattyn VM, Pollet DE, Nouwen EJ, Laurent G, Carlier MB, Maldague P, and Tulkens PM (1984) Recovery of cortical phospholipidosis and necrosis after acute gentamicin loading in rats. *Kidney Int* 26:838–847.

Giuliano RA, Verpooten GA, Verbist L, Wedeen RP, and De Broe ME (1986) In vivo uptake kinetics of aminoglycosides in the kidney cortex of rats. *J Pharmacol Exp Ther* 236:470–475.

Goetz MB, and Sayers J (1993) Nephrotoxicity of vancomycin and aminoglycoside therapy separately and in combination. *J Antimicrob Chemother* 32:325–334.

Gurnani K, Khouri H, Couture M, Bergeron MG, Beauchamp D, and Carrier D (1995) Molecular basis of the inhibition of gentamicin nephrotoxicity by daptomycin; an infrared spectroscopic investigation. *Biochim Biophys Acta* 1237:86–94.

Halestrap AP, Doran E, Gillespie JP, and O'Toole A (2000) Mitochondria and cell death. *Biochem Soc Trans* 28:170–177.

Hayashi T, Watanabe Y, Kumano K, Kitayama R, Yasuda T, Saikawa I, Katahira J, Kumada T, and Shimizu K (1988) Protective effect of piperacillin against nephrotoxicity of cephaloridine and gentamicin in animals. *Antimicrob Agents Chemother* 32:912–918.

Hinman AR, and Wolinsky E (1967) Nephrotoxicity associated with the use of cephaloridine. *JAMA* 200:724–726.

Hirouchi Y, Naganuma H, Kawahara Y, Okada R, Kamiya A, Inui K, and Hori R (1993) Protective effect of *N*-acyl amino acids (NAAs) on cephaloridine (CER) nephrotoxicity in rabbits. *Jpn J Pharmacol* 63:487–493.

Hirouchi Y, Naganuma H, Kawahara Y, Okada R, Kamiya A, Inui K, and Hori R (1994) Preventive effect of betamipron on nephrotoxicity and uptake of carbapenems in rabbit renal cortex. *Jpn J Pharmacol* 66:1–6.

Hottendorf GH, Barnett D, Gordon LL, Christensen EF, and Madissoo H (1981) Nonparallel nephrotoxicity dose-response curves of aminoglycosides. *Antimicrob Agents Chemother* 19:1024–1028.

Hottendorf GH, and Gordon LL (1980) Comparative low-dose nephrotoxicities of gentamicin, tobramycin, and amikacin. *Antimicrob Agents Chemother* 18:176–181.

Hottendorf GH, Laska DA, Williams PD, and Ford SM (1987) Role of desacetylation in the detoxification of cephalothin in renal cells in culture. *J Toxicol Environ Health* 22:101–111.

Houghton DC, English J, and Bennett WM (1988) Chronic tubulointerstitial nephritis and renal insufficiency associated with long-term "subtherapeutic" gentamicin. *J Lab Clin Med* 112:694–703.

Houghton DC, Lee D, Gilbert DN, and Bennett WM (1986) Chronic gentamicin nephrotoxicity. Continued tubular injury with preserved glomerular filtration function. *Am J Pathol* 123:183–194.

Humes HD, Sastrasinh M, and Weinberg JM (1984) Calcium is a competitive inhibitor of gentamicin-renal membrane binding interactions and dietary calcium supplementation protects against gentamicin nephrotoxicity. *J Clin Invest* 73:134–147.

Inui KI, Masuda S, and Saito H (2000) Cellular and molecular aspects of drug transport in the kidney. *Kidney Int* 58:944–958.

Ito J, Johnson WW, and Roy S, III (1969) Colistin nephrotoxicity: Report of a case with light and electron microscopic studies. *Acta Pathol Jpn* 19:55–67.

Izzettin FV, Ayca B, Uras F, Uysal V, Cevikbas U, Yardimci T, and Stohs SJ (1994) Nephrotoxicity of gentamicin and co-trimoxazole combination in rats. *Gen Pharmacol* 25:1185–1189.

Jariyawat S, Sekine T, Takeda M, Apiwattanakul N, Kanai Y, Sophasan S, and Endou H (1999) The interaction and transport of beta-lactam antibiotics with the cloned rat renal organic anion transporter 1. *J Pharmacol Exp Ther* 290:672–677.

Johannes L, and Goud B (1998) Surfing on a retrograde wave: How does Shiga toxin reach the endoplasmic reticulum? *Trends Cell Biol* 8:158–162.

Joly V, Bergeron Y, Bergeron MG, and Carbon C (1991) Endotoxin-tobramycin additive toxicity on renal proximal tubular cells in culture. *Antimicrob Agents Chemother* 35:351–357.

Jongejan HT, Provoost AP, and Molenaar JC (1989) Potentiated nephrotoxicity of cisplatin when combined with amikacin comparing young and adult rats. *Pediatr Nephrol* 3:290–295.

Josepovitz C, Farruggella T, Levine R, Lane B, and Kaloyanides GJ (1985) Effect of netilmicin on the phospholipid composition of subcellular fractions of rat renal cortex. *J Pharmacol Exp Ther* 235:810–819.

Kahlmeter G, and Dahlager JI (1984) Aminoglycoside toxicity – a review of clinical studies published between 1975 and 1982. *J Antimicrob Chemother* 13 Suppl A:9–22.

Kamiya A, Okumura K, and Hori R (1983) Quantitative investigation on renal handling of drugs in rabbits, dogs, and humans. *J Pharm Sci* 72:440–443.

Kerr JF, Wyllie AH, and Currie AR (1972) Apoptosis: A basic biological phenomenon with wide-ranging implications in tissue kinetics. *Br J Cancer* 26:239–257.

Kibbler CC, Prentice HG, Sage RJ, Hoffbrand AV, Brenner MK, Mannan P, Warner P, Bhamra A, and Noone P (1989) A comparison of double beta-lactam combinations with netilmicin/ureidopenicillin regimens in the empirical therapy of febrile neutropenic patients. *J Antimicrob Chemother* 23:759–771.

Kikuchi S, Aramaki Y, Nonaka H, and Tsuchiya S (1991) Effects of dextran sulphate on renal dysfunctions induced by gentamicin as determined by the kidney perfusion technique in rats. *J Pharm Pharmacol* 43:292–293.

Kishore BK, Kallay Z, Lambricht P, Laurent G, and Tulkens PM (1990) Mechanism of protection afforded by polyaspartic acid against gentamicin-induced phospholipidosis. I. Polyaspartic acid binds gentamicin and displaces it from negatively charged phospholipid layers *in vitro*. *J Pharmacol Exp Ther* 255:867–874.

Kizawa K, Miyazaki M, Nagasawa M, Ogake N, Nagai A, Sanzen T, and Kawamura Y (2003) [Attenuation of arbekacin-induced nephrotoxicity in rats by pazufloxacin mesilate]. *Jpn J Antibiot* 56:44–54.

Kleinknecht D, Kanfer A, Morel-Maroger L, and Mery JP (1978) Immunologically mediated drug-induced acute renal failure. *Contrib Nephrol* 10:42–52.

Knauss TC, Weinberg JM, and Humes HD (1983) Alterations in renal cortical phospholipid content induced by gentamicin: time course, specificity, and subcellular localization. *Am J Physiol* 244:F535–F546.

Kohda Y, and Gemba M (2001) Modulation by cyclic AMP, and phorbol myristate acetate of cephaloridine- induced injury in rat renal cortical slices. *Jpn J Pharmacol* 85:54–59.

Kojima R, Ito M, and Suzuki Y (1990a) Studies on the nephrotoxicity of aminoglycoside antibiotics and protection from these effects (6): A mechanism for the suppressive action of latamoxef on intrarenal tobramycin level. *Jpn J Pharmacol* 53:111–120.

Kojima R, Ito M, and Suzuki Y (1990b) Studies on the nephrotoxicity of aminoglycoside antibiotics and protection from these effects (9): Protective effect of inositol hexasulfate against tobramycin-induced nephrotoxicity. *Jpn J Pharmacol* 53:347–358.

Kosek JC, Mazze RI, and Cousins MJ (1974) Nephrotoxicity of gentamicin. *Lab Invest* 30:48–57.

Krasula RW, and Pernet AG (1991) Comparison of organ-specific toxicity of temafloxacin in animals and humans. *Am J Med* 91:38S–41S.

Krogstad DJ, Moellering RC, Jr., and Greenblatt DJ (1980) Single-dose kinetics of intravenous vancomycin. *J Clin Pharmacol* 20:197–201.

Kumar BD, Prasad CE, and Krishnaswamy K (1992) Detection of rifampicin-induced nephrotoxicity by *N*-acetyl-3-D-glucosaminidase activity. *J Trop Med Hyg* 95:424–427.

Kusuhara H, Sekine T, Utsunomiya-Tate N, Tsuda M, Kojima R, Cha SH, Sugiyama Y, Kanai Y, and Endou H (1999) Molecular cloning and characterization of a new multispecific organic anion transporter from rat brain. *J Biol Chem* 274: 13675–13680.

Landau D, and Kher KK (1997) Gentamicin-induced Bartter-like syndrome. *Pediatr Nephrol* 11:737–740.

Lane AZ, Wright GE, and Blair DC (1977) Ototoxicity and nephrotoxicity of amikacin: An overview of phase II, and phase III experience in the United States. *Am J Med* 62:911–918.

Laurent G, Kishore BK, and Tulkens PM (1990) Aminoglycoside-induced renal phospholipidosis and nephrotoxicity. *Biochem Pharmacol* 40:2383–2392.

Laurent G, Maldague P, Carlier MB, and Tulkens PM (1983) Increased renal DNA synthesis *in vivo* after administration of low doses of gentamicin to rats. *Antimicrob Agents Chemother* 24:586–593.

Laurent G, Toubeau G, Heuson-Stiennon JA, Tulkens P, and Maldague P (1988) Kidney tissue repair after nephrotoxic injury: Biochemical and morphological characterization. *CRC Crit Rev Toxicol* 19:147–183.

Leggett JE, Ebert S, Fantin B, and Craig WA (1990) Comparative dose-effect relations at several dosing intervals for beta-lactam, aminoglycoside and quinolone antibiotics against gram-negative bacilli in murine thigh-infection and pneumonitis models. *Scand J Infect Dis Suppl* 74:179–184.

Lemasters JJ, Nieminen AL, Qian T, Trost LC, Elmore SP, Nishimura Y, Crowe RA, Cascio WE, Bradham CA, Brenner DA, and Herman B (1998) The mitochondrial permeability transition in cell death: A common mechanism in necrosis, apoptosis and autophagy. *Biochim Biophys Acta* 1366:177–196.

Leonard I, Zanen J, Nonclercq D, Toubeau G, Heuson-Stiennon JA, Beckers JF, Falmagne P, Schaudies RP, and Laurent G (1994) Modification of immunoreactive EGF, and EGF receptor after acute tubular necrosis induced by tobramycin or cisplatin. *Ren Fail* 16:583–608.

Lerner AM, Reyes MP, Cone LA, Blair DC, Jansen W, Wright GE, and Lorber RR (1983) Randomised, controlled trial of the comparative efficacy, auditory toxicity, and nephrotoxicity of tobramycin and netilmicin. *Lancet* 1:1123–1126.

Levine BB (1966) Immunologic mechanisms of penicillin allergy. A haptenic model system for the study of allergic diseases of man. *N Engl J Med* 275:1115–1125.

Lew HT, and French SW (1966) Tetracycline nephrotoxicity and nonoliguric acute renal failure. *Arch Intern Med* 118:123–128.

Lietman PS (1988) Liver disease, aminoglycoside antibiotics and renal dysfunction. *Hepatology* 8:966–968.

Linton AL, Clark WF, Driedger AA, Turnbull DI, and Lindsay RM (1980) Acute interstitial nephritis due to drugs: Review of the literature with a report of nine cases. *Ann Intern Med* 93:735–741.

Livermore DM (2000) Antibiotic resistance in staphylococci. *Int J Antimicrob Agents* 16 Suppl 1:S3–S10.

Lomaestro BM (2000) Fluoroquinolone-induced renal failure. *Drug Saf* 22:479–485.

Lord JM, and Roberts LM (1998) Toxin entry: Retrograde transport through the secretory pathway. *J Cell Biol* 140:733–736.

Lortholary O, Blanchet F, Nochy D, Heudes D, Seta N, Amirault P, and Carbon C (1993) Effects of diltiazem on netilmicin-induced nephrotoxicity in rabbits. *Antimicrob Agents Chemother* 37:1790–1798.

Luft FC, and Kleit SA (1974) Renal parenchymal accumulation of aminoglycoside antibiotics in rats. *J Infect Dis* 130:656–659.

Luft FC, Yum MN, and Kleit SA (1976) Comparative nephrotoxicities of netilmicin and gentamicin in rats. *Antimicrob Agents Chemother* 10:845–849.

Malis CD, Racusen LC, Solez K, and Whelton A (1984) Nephrotoxicity of lysine and of a single dose of aminoglycoside in rats given lysine. *J Lab Clin Med* 103:660–676.

Manian FA, Stone WJ, and Alford RH (1990) Adverse antibiotic effects associated with renal insufficiency. *Rev Infect Dis* 12:236–249.

Mather M, and Rottenberg H (2001) Polycations induce the release of soluble intermembrane mitochondrial proteins. *Biochim Biophys Acta* 1503:357–368.

McCormack JP, and Jewesson PJ (1992) A critical reevaluation of the "therapeutic range" of aminoglycosides. *Clin Infect Dis* 14:320–339.

Minami T, Nakagawa H, Nabeshima M, Kadota E, Namikawa K, Kawaki H, and Okazaki Y (1994) Nephrotoxicity induced by adenine and its analogs: Relationship between structure and renal injury. *Biol Pharm Bull* 17:1032–1037.

Mingeot-Leclercq MP, Glupczynski Y, and Tulkens PM (1999) Aminoglycosides: Activity and resistance. *Antimicrob Agents Chemother* 43:727–737.

Moellering RC, Jr., Eliopoulos GM, and Sentochnik DE (1989) The carbapenems: New broad spectrum beta-lactam antibiotics. *J Antimicrob Chemother* 24 Suppl A:1–7.

Moellering RC, Jr., Krogstad DJ, and Greenblatt DJ (1981) Vancomycin therapy in patients with impaired renal function: A nomogram for dosage. *Ann Intern Med* 94:343–346.

Molitoris BA (1997) Cell biology of aminoglycoside nephrotoxicity: Newer aspects. *Curr Opin Nephrol Hypertens* 6:384–388.

Molitoris BA, Meyer C, Dahl R, and Geerdes A (1993) Mechanism of ischemia-enhanced aminoglycoside binding and uptake by proximal tubule cells. *Am J Physiol* 264:F907–F916.

Moore RD, Lietman PS, and Smith CR (1987) Clinical response to aminoglycoside therapy: Importance of the ratio of peak concentration to minimal inhibitory concentration. *J Infect Dis* 155:93–99.

Moore RD, Smith CR, Lipsky JJ, Mellits ED, and Lietman PS (1984) Risk factors for nephrotoxicity in patients treated with aminoglycosides. *Ann Intern Med* 100:352–357.

Morin NJ, Laurent G, Nonclercq D, Toubeau G, Heuson-Stiennon JA, Bergeron MG, and Beauchamp D (1992) Epidermal growth factor accelerates renal tissue repair in a model of gentamicin nephrotoxicity in rats. *Am J Physiol* 263:F806–F811.

Mullins RE, Lampasona V, and Conn RB (1987) Monitoring aminoglycoside therapy. *Clin Lab Med* 7:513–529.

Munckhof WJ, Grayson ML, and Turnidge JD (1996) A meta-analysis of studies on the safety and efficacy of aminoglycosides given either once daily or as divided doses. *J Antimicrob Chemother* 37:645–663.

Nahata MC (1987) Lack of nephrotoxicity in pediatric patients receiving concurrent vancomycin and aminoglycoside therapy. *Chemotherapy* 33:302–304.

Nakakuki M, Yamasaki F, Shinkawa T, Kudo M, Watanabe M, and Mizota M (1996) Protective effect of human ulinastatin against gentamicin-induced acute renal failure in rats. *Can J Physiol Pharmacol* 74:104–111.

Nanji AA, and Denegri JF (1984) Hypomagnesemia associated with gentamicin therapy. *Drug Intell Clin Pharm* 18:596–598.

Neugarten J, Gallo GR, and Baldwin DS (1983) Rifampin-induced nephrotic syndrome and acute interstitial nephritis. *Am J Nephrol* 3:38–42.

Ngeleka M, Beauchamp D, Tardif D, Auclair P, Gourde P, and Bergeron MG (1990) Endotoxin increases the nephrotoxic potential of gentamicin and vancomycin plus gentamicin. *J Infect Dis* 161:721–727.

Nicolau DP, Freeman CD, Belliveau PP, Nightingale CH, Ross JW, and Quintiliani R (1995) Experience with a once-daily aminoglycoside program administered to 2,184 adult patients. *Antimicrob Agents Chemother* 39:650–655.

Nightingale CH, Greene DS, and Quintiliani R (1975) Pharmacokinetics and clinical use of cephalosporin antibiotics. *J Pharm Sci* 64:1899–1926.

Novalbos A, Sastre J, Cuesta J, De Las HM, Lluch-Bernal M, Bombin C, and Quirce S (2001) Lack of allergic cross-reactivity to cephalosporins among patients allergic to penicillins. *Clin Exp Allergy* 31:438–443.

Olsen S, and Asklund M (1976) Interstitial nephritis with acute renal failure following cardiac surgery and treatment with methicillin. *Acta Med Scand* 199:305–310.

Pak K, Tomoyoshi T, Nomura Y, and Okabe H (1988) [Studies on nephrotoxicity of cyclosporin. II. Nephrotoxicity in rats receiving cyclosporin and sulfamethoxazole-trimethoprim]. *Hinyokika Kiyo* 34:1723–1731.

Pancorbo S, Compty C, and Heissler J (1982) Comparison of gentamicin and tobramycin nephrotoxicity in patients receiving individualized-pharmacokinetic dosing regimens. *Biopharm Drug Dispos* 3:83–88.

Parker RA (1982) Animal models in the study of aminoglycoside nephrotoxicity, in *The Aminoglycosides: Microbiology, Clinical use and Toxicology.* pp. 235–267, Marcel Dekker, Inc.

Pastoriza-Munoz E, Timmerman D, and Kaloyanides GJ (1984) Renal transport of netilmicin in the rat. *J Pharmacol Exp Ther* 228:65–72.

Petri WA Jr (2001) Penicillins, cephalosporins and other β-lactam antibiotics, in *Goodman & Gilman's The Pharmacological Basis of Therapeutics,* Hardman, JG, and Limbird LE (Eds), McGraw-Hill Medical Publishing Division, New York, pp. 1189–1218.

Pham NH, and Baldo BA (1996) beta-Lactam drug allergens: fine structural recognition patterns of cephalosporin-reactive IgE antibodies. *J Mol Recognit* 9:287–296.

Pierre C, Blanchet F, Seta N, Chaigne P, Labarre C, Sterkers O, Amiel C, and Carbon C (1988) Tolerance of once-daily dosing of netilmicin and teicoplanin, alone or in combination, in healthy volunteers. *Clin Pharmacol Ther* 44:458–466.

Polec RB, Yeh SD, and Shils ME (1971) Protective effect of ascorbic acid, isoascorbic acid and mannitol against tetracycline-induced nephrotoxicity. *J Pharmacol Exp Ther* 178:152–158.

Powell SH, Thompson WL, Luthe MA, Stern RC, Grossniklaus DA, Bloxham DD, Groden DL, Jacobs MR, DiScenna AO, Cash HA, and Klinger JD (1983) Once-daily vs. continuous aminoglycoside dosing: Efficacy and toxicity in animal and clinical studies of gentamicin, netilmicin, and tobramycin. *J Infect Dis* 147:918–932.

Price KE (1986) Aminoglycoside research 1975–1985: Prospects for development of improved agents. *Antimicrob Agents Chemother* 29:543–548.

Prins JM, Buller HR, Kuijper EJ, Tange RA, and Speelman P (1993) Once versus thrice daily gentamicin in patients with serious infections. *Lancet* 341:335–339.

Ramalakshmi S, Bastacky S, and Johnson JP (2003) Levofloxacin-induced granulomatous interstitial nephritis. *Am J Kidney Dis* 41:E7.

Richmond JM, Whitworth JA, Fairley KF, and Kincaid-Smith P (1979) Co-trimoxazole nephrotoxicity. *Lancet* 1:493.

Ringden O, Myrenfors P, Klintmalm G, Tyden G, and Ost L (1984) Nephrotoxicity by co-trimoxazole and cyclosporin in transplanted patients. *Lancet* 1:1016–1017.

Romano A, Mayorga C, Torres MJ, Artesani MC, Suau R, Sanchez F, Perez E, Venuti A, and Blanca M (2000) Immediate allergic reactions to cephalosporins: Cross-reactivity and selective responses. *J Allergy Clin Immunol* 106:1177–1183.

Rougier F, Claude D, Maurin M, Sedoglavic A, Ducher M, Corvaisier S, Jelliffe R, and Maire P (2003) Aminoglycoside nephrotoxicity: Modeling, simulation, and control. *Antimicrob Agents Chemother* 47:1010–1016.

Rybak MJ, Abate BJ, Kang SL, Ruffing MJ, Lerner SA, and Drusano GL (1999) Prospective evaluation of the effect of an aminoglycoside dosing regimen on rates of observed nephrotoxicity and ototoxicity. *Antimicrob Agents Chemother* 43:1549–1555.

Rybak MJ, Albrecht LM, Boike SC, and Chandrasekar PH (1990) Nephrotoxicity of vancomycin, alone and with an aminoglycoside. *J Antimicrob Chemother* 25:679–687.

Sakai M, Zhang M, Homma T, Garrick B, Abraham JA, McKanna JA, and Harris RC (1997) Production of heparin binding epidermal growth factor-like growth factor in the early phase of regeneration after acute renal injury. Isolation and localization of bioactive molecules. *J Clin Invest* 99:2128–2138.

Salem PA, Jabboury KW, and Khalil MF (1982) Severe nephrotoxicity: A probable complication of cis-dichlorodiammineplatinum (II) and cephalothin-gentamicin therapy. *Oncology* 39:31–32.

Sandhya P, Mohandass S, and Varalakshmi P (1995) Role of DL alpha-lipoic acid in gentamicin induced nephrotoxicity. *Mol Cell Biochem* 145:11–17.

Sandoval R, Leiser J, and Molitoris BA (1998) Aminoglycoside antibiotics traffic to the Golgi complex in LLC-PK1 cells. *J Am Soc Nephrol* 9:167–174.

Sandoval RM, Dunn KW, and Molitoris BA (2000) Gentamicin traffics rapidly and directly to the Golgi complex in LLC- PK(1) cells. *Am J Physiol Renal Physiol* 279: F884–F890.

Sandvig K, and van Deurs B (1996) Endocytosis, intracellular transport, and cytotoxic action of Shiga toxin and ricin. *Physiol Rev* 76:949–966.

Schmitz C, Hilpert J, Jacobsen C, Boensch C, Christensen EI, Luft FC, and Willnow TE (2002) Megalin deficiency offers protection from renal aminoglycoside accumulation. *J Biol Chem* 277:618–622.

Sekine T, Cha SH, and Endou H (2000) The multispecific organic anion transporter (OAT) family. *Pflugers Arch* 440:337–350.

Sekine T, Cha SH, Tsuda M, Apiwattanakul N, Nakajima N, Kanai Y, and Endou H (1998) Identification of multispecific organic anion transporter 2 expressed predominantly in the liver. *FEBS Lett* 429:179–182.

Sekine T, Watanabe N, Hosoyamada M, Kanai Y, and Endou H (1997) Expression cloning and characterization of a novel multispecific organic anion transporter. *J Biol Chem* 272:18526–18529.

Seltzer E, Goldstein B, Dorr MB, Dowell J, Perry M, and Henkel T (2003) Dalbavancin: Phase 2 demonstration of efficacy of a novel, weekly dosing regimen in skin and soft tissue infections. *13th European Congress of Clinical Microbiology and Infectious Diseases*, Glasgow, UK, O143.

Seneca H (1967) Nephrotoxicity from cephaloridine. *JAMA* 201:640–641.

Silverblatt FJ, and Kuehn C (1979) Autoradiography of gentamicin uptake by the rat proximal tubule cell. *Kidney Int* 15:335–345.

Simmons CF, Jr., Bogusky RT, and Humes HD (1980) Inhibitory effects of gentamicin on renal mitochondrial oxidative phosphorylation. *J Pharmacol Exp Ther* 214:709–715.

Smetana S, Khalef S, Kopolovic G, Bar-Khayim Y, Birk Y, and Kacew S (1992) Effect of interaction between gentamicin and pyridoxal-5-phosphate on functional and metabolic parameters in kidneys of female Sprague-Dawley rats. *Ren Fail* 14:147–153.

Smith CR, Baughman KL, Edwards CQ, Rogers JF, and Lietman PS (1977) Controlled comparison of amikacin and gentamicin. *N Engl J Med* 296:349–353.

Smith CR, Lipsky JJ, Laskin OL, Hellmann DB, Mellits ED, Longstreth J, and Lietman PS (1980) Double-blind comparison of the nephrotoxicity and auditory toxicity of gentamicin and tobramycin. *N Engl J Med* 302:1106–1109.

Sokol PP, Huiatt KR, Holohan PD, and Ross CR (1989) Gentamicin and verapamil compete for a common transport mechanism in renal brush border membrane vesicles. *J Pharmacol Exp Ther* 251:937–942.

Sorrell TC, and Collignon PJ (1985) A prospective study of adverse reactions associated with vancomycin therapy. *J Antimicrob Chemother* 16:235–241.

Spiegel DM, Shanley PF, and Molitoris BA (1990) Mild ischemia predisposes the S3 segment to gentamicin toxicity. *Kidney Int* 38:459–464.

Streetman DS, Nafziger AN, Destache CJ, and Bertino AS, Jr. (2001) Individualized pharmacokinetic monitoring results in less aminoglycoside-associated nephrotoxicity and fewer associated costs. *Pharmacotherapy* 21:443–451.

Sundin DP, Sandoval R, and Molitoris BA (2001) Gentamicin inhibits renal protein and phospholipid metabolism in rats: Implications involving intracellular trafficking. *J Am Soc Nephrol* 12:114–123.

Susanto M, and Benet LZ (2002) Can the enhanced renal clearance of antibiotics in cystic fibrosis patients be explained by P-glycoprotein transport? *Pharm Res* 19:457–462.

Sweet DH, Miller DS, and Pritchard JB (2000) Basolateral localization of organic cation transporter 2 in intact renal proximal tubules. *Am J Physiol Renal Physiol* 279:F826–F834.

Tai PC, and Davis BD (1979) Triphasic concentration effects of gentamicin on activity and misreading in protein synthesis. *Biochemistry* 18:193–198.

Takamoto K, Kawada M, Usui T, Ishizuka M, and Ikeda D (2003) Aminoglycoside antibiotics reduce glucose reabsorption in kidney through down-regulation of SGLT1. *Biochem Biophys Res Commun* 308:866–871.

Takeda M, Babu E, Narikawa S, and Endou H (2002) Interaction of human organic anion transporters with various cephalosporin antibiotics. *Eur J Pharmacol* 438:137–142.

Tardif D, Beauchamp D, and Bergeron MG (1990) Influence of endotoxin on the intracortical accumulation kinetics of gentamicin in rats. *Antimicrob Agents Chemother* 34:576–580.

ter Braak EW, de Vries PJ, Bouter KP, van der Vegt SG, Dorrestein GC, Nortier JW, van Dijk A, Verkooyen RP, and Verbrugh HA (1990) Once-daily dosing regimen for aminoglycoside plus beta-lactam combination therapy of serious bacterial infections: Comparative trial with netilmicin plus ceftriaxone. *Am J Med* 89:58–66.

Tipper DJ, and Strominger JL (1965) Mechanism of action of penicillins: A proposal based on their structural similarity to acyl-D-alanyl-D-alanine. *Proc Natl Acad Sci USA* 54:1133–1141.

Tojo A, Sekine T, Nakajima N, Hosoyamada M, Kanai Y, Kimura K, and Endou H (1999) Immunohistochemical localization of multispecific renal organic anion transporter 1 in rat kidney. *J Am Soc Nephrol* 10:464–471.

Topham JC, Murgatroyd LB, Jones DV, Goonetilleke UR, and Wright J (1989) Safety evaluation of meropenem in animals: Studies on the kidney. *J Antimicrob Chemother* 24 Suppl A:287–306.

Toubeau G, Laurent G, Carlier MB, Abid S, Maldague P, Heuson-Stiennon JA, and Tulkens PM (1986) Tissue repair in rat kidney cortex after short treatment with aminoglycosides at low doses. A comparative biochemical and morphometric study. *Lab Invest* 54:385–393.

Trollfors B (1983) Gentamicin-associated changes in renal function reversible during continued treatment. *J Antimicrob Chemother* 12:285–287.

Tulkens PM (1986) Experimental studies on nephrotoxicity of aminoglycosides at low doses. Mechanisms and perspectives. *Am J Med* 80:105–114.

Tulkens PM (1991) Pharmacokinetic and toxicological evaluation of a once-daily regimen versus conventional schedules of netilmicin and amikacin. *J Antimicrob Chemother* 27 Suppl C:49–61.

Tune BM (1972) Effect of organic acid transport inhibitors on renal cortical uptake and proximal tubular toxicity of cephaloridine. *J Pharmacol Exp Ther* 181:250–256.

Tune BM (1975) Relationship between the transport and toxicity of cephalosporins in the kidney. *J Infect Dis* 132:189–194.

Tune BM (1997) Nephrotoxicity of beta-lactam antibiotics: Mechanisms and strategies for prevention. *Pediatr Nephrol* 11:768–772.

Tune BM, Fernholt M, and Schwartz A (1974) Mechanism of cephaloridine transport in the kidney. *J Pharmacol Exp Ther* 191:311–317.

Tune BM, and Hsu CY (1985) Augmentation of antibiotic nephrotoxicity by endotoxemia in the rabbit. *J Pharmacol Exp Ther* 234:425–430.

Tune BM, and Hsu CY (1994) Toxicity of cephaloridine to carnitine transport and fatty acid metabolism in rabbit renal cortical mitochondria: Structure-activity relationships. *J Pharmacol Exp Ther* 270:873–880.

Tune BM, and Hsu CY (1995) Effects of nephrotoxic beta-lactam antibiotics on the mitochondrial metabolism of monocarboxylic substrates. *J Pharmacol Exp Ther* 274:194–199.

Tune BM, Sibley RK, and Hsu CY (1988) The mitochondrial respiratory toxicity of cephalosporin antibiotics. An inhibitory effect on substrate uptake. *J Pharmacol Exp Ther* 245:1054–1059.

Uchino H, Tamai I, Yabuuchi H, China K, Miyamoto K, Takeda E, and Tsuji A (2000) Faropenem transport across the renal epithelial luminal membrane via inorganic phosphate transporter Npt1. *Antimicrob Agents Chemother* 44:574–577.

Ueda N, Guidet B, and Shah SV (1993) Gentamicin-induced mobilization of iron from renal cortical mitochondria. *Am J Physiol* 265:F435–F439.

Van Bambeke F, Michot JM, and Tulkens PM (2003) Antibiotic efflux pumps in eukaryotic cells: Occurrence and impact on antibiotic cellular pharmacokinetics, pharmacodynamics and toxicodynamics. *J Antimicrob Chemother* 51: 1067–1077.

Verpooten GA, Giuliano RA, Verbist L, Eestermans G, and De Broe ME (1989) Once-daily dosing decreases renal accumulation of gentamicin and netilmicin. *Clin Pharmacol Ther* 45:22–27.

Verpooten GA, Tulkens PM, and Molitoris BA (2003) Aminoglycosides and vancomycin, in *Clinical Nephrotoxins*, 2nd ed., De Broe ME, Porter GA, Bennet WM, and Verpooten GA (Eds), New York - Kluwer Academic Publishers, pp. 301–321.

Walker PD, and Shah SV (1988) Evidence suggesting a role for hydroxyl radical in gentamicin-induced acute renal failure in rats. *J Clin Invest* 81:334–341.

Wallace AW, Jones M, and Bertino JS, Jr. (2002) Evaluation of four once-daily aminoglycoside dosing nomograms. *Pharmacotherapy* 22:1077–1083.

Watanabe M (1978) Drug-induced lysosomal changes and nephrotoxicity in rats. *Acta Pathol Jpn* 28:867–889.

Watling SM, and Dasta JF (1993) Aminoglycoside dosing considerations in intensive care unit patients. *Ann Pharmacother* 27:351–357.

Wedeen RP, Batuman V, Cheeks C, Marquet E, and Sobel H (1983) Transport of gentamicin in rat proximal tubule. *Lab Invest* 48:212–223.

Weinberg JM, and Humes HD (1980) Mechanisms of gentamicin-induced dysfunction of renal cortical mitochondria. I. Effects on mitochondrial respiration. *Arch Biochem Biophys* 205:222–231.

Whiting PH, Simpson JG, Davidson RJ, and Thomson AW (1982) The toxic effects of combined administration of cyclosporin A, and gentamicin. *Br J Exp Pathol* 63:554–561.

Whittem T, Parton K, and Turner K (1996) Effect of polyaspartic acid on pharmacokinetics of gentamicin after single intravenous dose in the dog. *Antimicrob Agents Chemother* 40:1237–1241.

Williams PD, Bennett DB, Gleason CR, and Hottendorf GH (1987) Correlation between renal membrane binding and nephrotoxicity of aminoglycosides. *Antimicrob Agents Chemother* 31:570–574.

Williams PD, Laska DA, Tay LK, and Hottendorf GH (1988) Comparative toxicities of cephalosporin antibiotics in a rabbit kidney cell line (LLC-RK1). *Antimicrob Agents Chemother* 32:314–318.

Williams PD, Trimble ME, Crespo L, Holohan PD, Freedman JC, and Ross CR (1984) Inhibition of renal Na+, K+-adenosine triphosphatase by gentamicin. *J Pharmacol Exp Ther* 231:248–253.

Wilson AP (1998) Comparative safety of teicoplanin and vancomycin. *Int J Antimicrob Agents* 10:143–152.

Wood CA, Finkbeiner HC, Kohlhepp SJ, Kohnen PW, and Gilbert DN (1989) Influence of daptomycin on staphylococcal abscesses and experimental tobramycin nephrotoxicity. *Antimicrob Agents Chemother* 33:1280–1285.

Wood CA, Kohlhepp SJ, Kohnen PW, Houghton DC, and Gilbert DN (1986) Vancomycin enhancement of experimental tobramycin nephrotoxicity. *Antimicrob Agents Chemother* 30:20–24.

Wood MJ (1996) The comparative efficacy and safety of teicoplanin and vancomycin. *J Antimicrob Chemother* 37:209–222.

Wood MJ (2000) Comparative safety of teicoplanin and vancomycin. *J Chemother* 12 Suppl 5:21–25.

Woodroffe AJ, Weldon M, Meadows R, and Lawrence JR (1975) Acute interstitial nephritis following ampicillin hypersensitivity. *Med J Aust* 1:65–68.

Wysocki M, Delatour F, Faurisson F, Rauss A, Pean Y, Misset B, Thomas F, Timsit JF, Similowski T, Mentec H, Mier L, and Dreyfuss D (2001) Continuous versus intermittent infusion of vancomycin in severe Staphylococcal infections: Prospective multicenter randomized study. *Antimicrob Agents Chemother* 45:2460–2467.

Yoshiyama Y, Yazaki T, Kanke M, and Beauchamp D (2000) Nephrotoxicity of teicoplanin in rats. *Jpn J Antibiot* 53:660–666.

Yoshiyama Y, Yazaki T, Wong PC, Beauchamp D, and Kanke M (2001) The effect of fosfomycin on glycopeptide antibiotic-induced nephrotoxicity in rats. *J Infect Chemother* 7:243–246.

Yver L, Becq-Giraudon B, Pourrat O, and Sudre Y (1976) [The nephrotoxicity of the methicillin-gentamycin combination. Apropos of 5 cases]. *Sem Hop* 52:1903–1907.

Zager RA (1988) A focus of tissue necrosis increases renal susceptibility to gentamicin administration. *Kidney Int* 33:84–90.

Zaloga GP, Chernow B, Pock A, Wood B, Zaritsky A, and Zucker A (1984) Hypomagnesemia is a common complication of aminoglycoside therapy. *Surg Gynecol Obstet.* 158:561–565.

Zaske DE, Bootman JL, Solem LB, and Strate RG (1982) Increased burn patient survival with individualized dosages of gentamicin. *Surgery* 91:142–149.

Zhao Z, Baldo BA, and Rimmer J (2002) beta-Lactam allergenic determinants: Fine structural recognition of a cross-reacting determinant on benzylpenicillin and cephalothin. *Clin Exp Allergy* 32:1644–1650.

17

Nephrotoxicity of Cyclosporine and Other Immunosuppressive and Immunotherapeutic Agents

Douglas Charney, Kim Solez, and Lorraine C. Racusen

INTRODUCTION

Solid organ and bone marrow transplantation have been revolutionized by the availability of immunosuppressive agents developed over the past few decades. The first and most widely used is cyclosporin A (CsA), a fungal metabolite with immunosuppressive properties first described by Borel in 1972. The drug acts on the immune system to selectively inhibit T-cell activation and interleukin-2-mediated events. Tacrolimus (FK506), a more recently discovered macrolide fungal product, also inhibits T-cell activation. Both of these agents, which act via inhibition of calcineurin, can produce acute and chronic renal toxicity, and therapeutic and toxic levels may be tightly linked. Additionally, antibody-based therapies are being more widely used in immunosuppressive protocols to destroy effector T-cells or block receptors and prevent or treat cell-mediated or antibody-mediated acute rejection. These agents also may have effects on renal function.

Newer immunosuppressive protocols have reduced toxic side effects, in part by adding newer agents, including sirolimus (rapamycin) and mycophenolate mofetil (MMF; Cellcept), and reducing doses of the calcineurin inhibitors. These newer agents have primarily nonrenal toxicities, although they have some potential interactions with CsA and tacrolimus. Calcineurin inhibitors have proven so useful and effective in a variety of solid organ allografts, however, that they are still widely used, and toxicity still remains a significant problem. In this chapter, the effects of CsA and tacrolimus on renal structure and function in humans and in experimental models are reviewed. Additionally, clinical and experimental toxicity of other, newer immunosuppressive agents, including antibody-based therapies, are considered.

CLINICAL NEPHROTOXICITY OF IMMUNOSUPPRESSIVE AGENTS

Cyclosporine

Pharmacokinetics and Mechanisms of Action

CsA is a macrolide antibiotic, structurally a cyclic polypeptide (undecapeptide), composed of 11 amino acids. It is produced as a fermentation product of the fungus *Beauveria nivea* (Mihatsch et al., 1998; Neoral insert). CsA is a highly lipophilic molecule that distributes widely into blood, plasma, and tissue compartments, with preferential accumulation in fat-rich organs such as liver and adipose tissue. Its lipophilic nature allows it to rapidly penetrate the plasma membrane of its target cell, the T-cell (Aklaghi et al., 2002; Campistol and Sacks, 2000). In blood, it distributes predominantly into plasma (33 to 47%) and red blood cells (41 to 58%) in a manner dependent on temperature, hematocrit, and concentration of plasma proteins. In plasma, it preferentially binds to lipoproteins such as LDL, HDL, and VLDL and, to a lesser extent, albumin. As it is a lipophilic substance, it is excreted in human milk; mothers receiving CsA should avoid breast-feeding. CsA is eliminated primarily through the biliary system, with only 6% of each dose excreted in urine, 0.1% excreted in urine as unchanged parent compound. It is extensively metabolized into at least 30 metabolites by the hepatic cytochrome P450-3A4 system, although bioactivity of its metabolites is minimal (Dunn et al., 2001).

CsA interacts with a wide variety of other drugs, either by induction of or competition for the hepatic cytochrome P450-3A4 system.

Additionally, its binding to P-glycoprotein (an ATP-dependent drug efflux transporter that pumps drugs out of the cell) is another mechanism of drug interaction. Drugs that increase CsA plasma or whole-blood concentrations include diltiazem, nicardipine, verapamil, glipizide, erythromycin, and ketoconazole, among others. Those that decrease cyclosporine concentrations include rifampicin, nafcillin, intravenous bactrim, phenytoin, and phenobartial, among others (Dunn et al., 2001). The herbal pharmaceutical St. John's Wort decreases cyclosporine levels via both cytochrome P450-3A and P-glycoprotein interaction mechanisms. This interaction is of particular importance since many patients do not view herbal supplements as "drugs" and thus do not report them to their physicians (Ernst, 2002). Additionally, grapefruit juice increases blood cyclosporine levels and should be avoided.

CsA exerts its immunosuppressive effect by inhibiting activation of calcineurin, a calcium/calmodulin-dependent phosphatase. When CsA enters the cell, it binds to cyclophilin. This complex then interacts with calcineurin to inhibit calcineurin enzyme activity. The most relevant substrate of calcineurin is nuclear factor of activated T-cells (NFAT) which, under normal circumstances, regulates transcription of interleukin-2 (IL-2), tumor necrosis factor α (TNFα), GMCSF, and other genes after it translocates to the cell nucleus (Campistol and Sacks, 2000; Dunn et al., 2001; Mihatsch et al., 1998). Other genes whose transcription is altered by calcineurin (either via NFAT or other substrates) include those coding for the interleukin-2 receptor (IL-2R), nitric oxide (NO) synthase, transforming growth factor β (TGFβ), endothelin-1, collagen types I and IV, and the bcl-2 protein (which prevents cellular apoptosis) (Campistol et al., 2000). By inhibiting the gene regulatory function of calcineurin, particularly the IL-2 gene, cyclosporine prevents T-helper (and, to a lesser extent, T-suppressor) cells from progressing from the G_0 to the G_1 phase of the cell cycle (Campistol and Sacks, 2000). Suppression of the TNFα gene also attenuates MHC class II molecule expression (Dunn et al., 2001). CsA, however, exerts no effects on cellular phagocytic functions, such as enzyme secretion or granulocyte or macrophage migration.

The original CsA formulation (Sandimmune, Novartis Pharmaceuticals, Basel, Switzerland) consisted of an oil-based suspension. The main shortcoming of this formulation was the wide pharmacokinetic variability within and between patients. This seems to result from the extreme variation in ability of patients to absorb this particular CsA formulation from the gut, hence the often-used reference to patients as either good or poor "absorbers." More recently, however, a microemulsion CsA formulation (Neoral, Novartis Pharmaceuticals, Basel, Switzerland) was devised to improve on the limited and unpredictable intestinal absorption of the oil-based product (Coukell and Plosker, 1998; Dunn et al., 2001; Frei, 1999; Helderman, 1998; Holt and Johnston, 2000; Taler et al., 1999). Neoral consists of CsA suspended in a uniform, fine ($<0.15\,\mu$m) microemulsion by a combination of lipophilic solvent, hydrophilic solvent, and surfactant. Absorption is more rapid and complete as compared with Sandimmune, in both normal subjects and transplant recipients (Keown and Primmett, 1998). Adverse drug effects are not affected by the CsA formulation and thus Neoral shows no significant difference in safety or tolerability profile as compared with Sandimmune (Coukell and Plosker 1998; Frei, 1999; Helderman, 1998; Keown and Primmett, 1998; Nashan et al., 2002; Shah et al., 1999). Although Neoral does not seem to affect absorption much in good absorbers, it substantially improves absorption of CsA in poor absorbers and in patients in need of high doses per kilogram of body weight, which include children (Frei, 1999; Hoyer, 1998), African Americans, and diabetics. Additionally, significant absorption improvement has been shown in recipients of combined kidney and pancreas transplants (Frei, 1999). Although Neoral has resulted in undisputed ease of patient management as compared with Sandimmune, the subject of its effect on acute allograft rejection is controversial. Nevertheless, most investigators conclude that it does indeed result in fewer acute allograft rejection episodes compared with Sandimmune, mainly due to more consistent blood concentration, and therapeutic drug monitoring values that are more predictive of actual drug exposure (Holt et al., 2000; Keown and Primmett, 1998; Shah et al., 1999).

Numerous investigators have reported overwhelming evidence that the two-hour post-dose blood level is the best and most accurate single time-point value for therapeutic drug monitoring of Neoral (Belitsky et al., 2000; Canadian Neoral Renal Transplantation Study Group, 2001; Dunn et al., 2001; Levy et al., 2002; Mahalati et al., 2001; Nashan et al., 2002). This is because the AUC_{0-4} (area under the concentration–time curve, from zero to four hours post-dose) represents the period of greatest blood concentration variability between patients. The greatest immunosuppressive effect of Neoral also occurs during this time (Levy et al., 2002), and patients who achieve their target AUC_{0-4}

CsA levels show significantly fewer acute rejection episodes (Canadian Neoral Renal Transplantation Study Group 2001; Levy et al., 2002; Mahalati et al., 2001). As measurement of AUC_{0-4} is unwieldy and unmanageable in routine patient management, a single time-point value that most accurately represents AUC_{0-4} is required. The two-hour post-dose CsA blood concentration level (C_2) has consistently proven to most accurately represent AUC_{0-4}, while trough blood levels (C_0) reflect AUC_{0-4} poorly (Canadian Neoral Renal Transplantation Study Group 2001; Dunn et al., 2001; Holt and Johnston, 2000; Levy et al., 2002; Nashan et al., 2002). It remains to be seen whether or not C_2 CsA blood level monitoring replaces C_0 (trough) CsA level monitoring in routine clinical practice.

Clinical Toxicity

Nonrenal Toxicity

Nonrenal toxic effects of CsA include hypertension, hepatotoxicity, neurotoxicity (encephalopathy), tremor, gingival hyperplasia, hypertrichosis, lipid abnormalilties, and low frequencies of such effects as nausea, fatigue, and leukopenia (DeMattos et al., 2000; Frei et al., 1999; Neoral prescribing information, 2002). CsA increases both the incidence and severity of hypertension in transplant patients; the highest rates occur in heart transplant and lung transplant recipients. Such hypertensive patients lose, at least initially, their normal circadian nocturnal decrease in blood pressure, often suffering consequent nocturnal headaches and urinary frequency. Even late posttransplantation, when CsA and corticosteroid doses are usually substantially reduced, hypertension usually persists (Taler et al., 1999).

Renal Toxicity

Four discrete syndromes of clinically significant renal dysfunction caused by CsA in the native or grafted kidney have been identified: acute reversible renal functional impairment (usually occurring early in the course of treatment), delayed renal allograft function, acute vasculopathy (thrombotic microangiopathy), and chronic nephropathy with interstitial fibrosis. Various combinations of these syndromes may occur, and they can also be combined with transplant rejection or other types of renal disease not directly related to CsA therapy.

Acute Reversible Renal Functional Impairment

In CsA-mediated acute reversible renal functional impairment, no morphological renal abnormalities are found. Effects include (in order of decreasing frequency): elevation in serum creatinine levels, hyperuricemia, significant hyperkalemia (sometimes with hyperchloremic metabolic acidosis), and hypomagnesemia (Mihatsch et al., 1998; Neoral prescribing information, 2002). Sodium retention and inability to concentrate urine may also ensue (Rodicio, 2000). The most severe form of functional toxicity manifests as acute renal failure with poly-, oligo-, or anuria (Mihatsch et al., 1998). Functional nephrotoxicity is usually dose-related and reversible after dose reduction or complete withdrawal (Ader and Rostaing, 1998; Campistol and Sacks, 2000). Most of these effects are secondary to alterations in renal hemodynamics and glomerular filtration rate that begin soon after cyclosporine therapy is initiated (Campistol and Sacks, 2000). The most likely mediators of cyclosporine-induced vasoconstriction include endothelin, thromboxane A2, NO synthase inhibition, and sympathetic nervous system activation (Andoh and Bennett, 1998).

This syndrome is probably most similar to acute nephrotoxicity models in animals. It was first reported in patients receiving CsA parenterally in high doses (up to 20 mg/kg/day), following bone marrow transplantation (Powles et al., 1978); patients were often also receiving other known nephrotoxic drugs. Acute renal failure (ARF) was sometimes dramatic in onset and protracted. Patients treated with lower doses of cyclosporine have a less dramatic onset of renal failure and a lower incidence of oliguria. Urine sodium is characteristically less than 10 mEq/l and the oliguria is normally self-limiting, with normal urine output returning within days (Greenberg et al., 1987; Keown et al., 1986). Characteristic of this syndrome of acute reversible renal functional impairment brought about by CsA is the very rapid return of renal function when CsA levels are reduced.

In heart allograft recipients treated with CsA, acute oliguric renal failure frequently occurs within the first four days after transplant (e.g., Greenberg et al., 1987). Although peak serum creatinine levels tend to coincide with the highest trough serum CsA levels during the first week after transplant, the absolute magnitude of the renal functional impairment correlates more closely with the degree of impaired renal function observed prior to cardiac transplantation than with serum CsA levels. This suggests that patients with inadequate renal reserve are more susceptible to acute reversible renal functional impairment brought about by CsA (Keown et al., 1986). Similar episodes of ARF have been reported following liver transplantation, particularly when intravenous CsA was used (Powell-Jackson et al., 1983).

Myers et al. (1988a, 1988b) proposed that the acute reversible renal functional impairment is caused by afferent arteriolar vasoconstriction that with time leads to hyaline accumulation. Eventually, this arteriolar lesion leads to a "striped" interstitial fibrosis and more chronic renal functional impairment (see below). There is some evidence that the renin–angiotensin system may be involved in the arteriolar vasoconstriction, although this is a less constant finding clinically than in experimental models (Lassila, 2002). In the early phases, the vasoconstriction can be reversed by dopamine and nifedipine (e.g., Conte et al., 1989; Propper et al., 1989).

The return of renal function tends to coincide with the normalization of serum CsA levels, with the slope of declining levels and declining serum creatinine concentrations being approximately equal. The relationship between acute CsA nephrotoxicity and circulating levels of the drug is controversial. However, nephrotoxicity is generally observed in association with rising or elevated serum levels exceeding 200 ng/ml and is common in patients with levels exceeding 400 ng/ml. Toxicity due to CsA alone is rarely observed with serum CsA levels of less than 200 ng/ml (Kahan et al., 1983, Keown and Primmett, 1989).

Effects on the Renal Tubule

Proximal tubular epithelial cells are uniquely sensitive to the toxic effects of calcineurin inhibitors such as CsA. Morphologic effects on renal tubular epithelial cells include isometric cytoplasmic vacuolization, necrosis with or without subsequent calcification, inclusion bodies (which correspond ultrastructurally to giant mitochondria), and giant lysosomes (which are more commonly seen with CsA than tacrolimus therapy) (Davies et al., 2000; Mihatsch et al., 1998; Neoral prescribing information, 2002). Immunohistochemical and *in situ* hybridization have revealed strong osteopontin protein and mRNA expression by tubular epithelium in biopsies with CsA toxicity (Hudkins et al., 2001), which could in turn lead to recruitment of monocyte and macrophages.

"Isometric" vacuolization of proximal tubular cells, i.e., many small equally-sized vacuoles within the tubular cytoplasm (Figure 17.1),

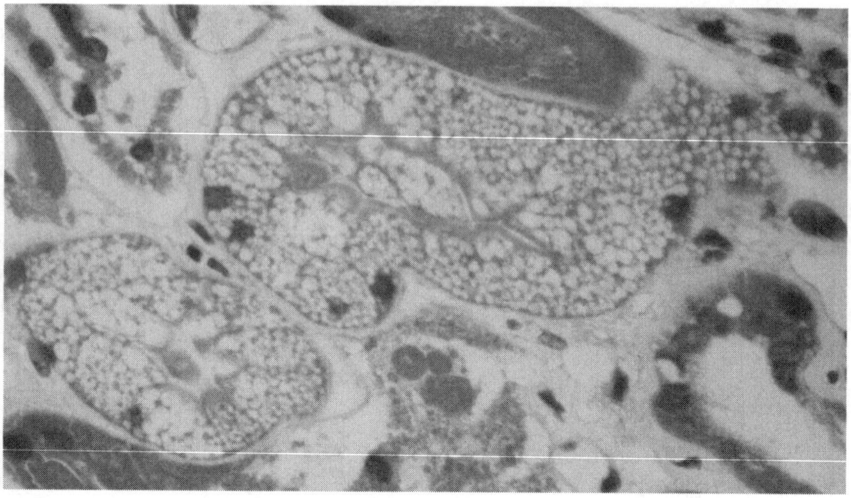

Figure 17.1 Proximal tubular profiles showing numerous, similarly sized, minute cytoplasmic vacuoles (i.e., isometric vacuolization). Trichrome stain. (Original magnification ×400.)

is a feature seen with high CsA levels and is the most common morphological finding. This feature appears to be even more prominent in fine-needle aspirates than in biopsies; this change has been reported to be present in 11 out of 12 cases of CsA toxicity studies by fine-needle aspiration (Nast et al., 1989). However, although tubular vacuolization is characteristic of CsA therapy, it is not specific for CsA nephrotoxicity. Clinical correlation with determination of drug levels is often useful.

Delayed Graft Function

There is some evidence that CsA can prolong the duration of ischemic posttransplant ARF in renal allografts (Belitsky, 1995; Shiel et al., 1984). With the availability of several agents for induction therapy, many centers avoid calcineurin inhibitor therapy during the period of renal dysfunction.

Acute Vasculopathy (Thrombotic Microangiopathy)

The earliest form of arteriolopathy seen in renal biopsies in patients taking CsA consists of arteriolar smooth muscle and endothelial cell

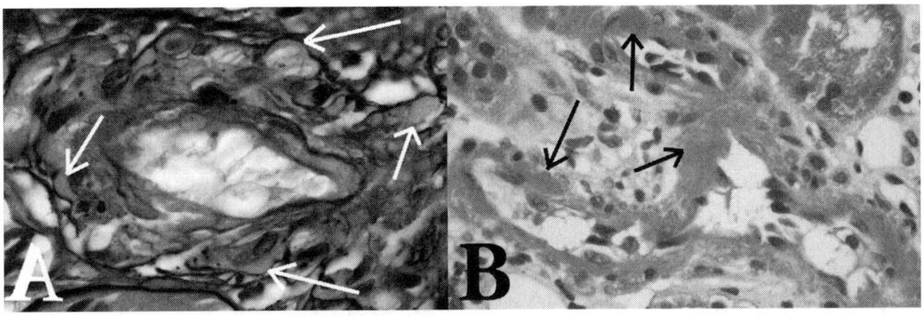

Figure 17.2 Arterioles showing peripherally located nodules of hyaline (arrows), which appear to have replaced medial myocytes. (A) Periodic acid methenamine silver (original magnification ×1000); (B) Trichrome stain (original magnification ×400).

vacuolization. The vacuolated smooth muscle fibers may later undergo cell necrosis and are replaced by insudated plasma proteins ("hyaline"), which impart a "necklace" or "ring of pearls" appearance to the outer adventitial aspect of arterioles (Davies et al., 2000; Mihatsch et al., 1998) (Figure 17.2). Arteriolar hyalinosis is frequently observed in CsA nephrotoxicity, although is not seen in all cases. In a biopsy study, this was the only lesion clearly differentiating cases of CsA nephrotoxicity from control cases with normal function (Solez et al., 1993). Of course, arteriolar hyalinosis is not specific for CsA toxicity; these changes may be due to hypertension or diabetes in the recipient, or be pre-existing in the donor kidney. Baseline graft biopsy at the time of transplant provides a baseline evaluation of extent of arteriolar change in the donor, enabling detection of significant increases in this morphological finding.

The most serious form of acute vasculopathy resulting from CsA administration, first described by Shulman et al. (1981) in bone marrow allograft recipients, is a hemolytic uremic syndrome (HUS) or thrombotic thrombocytopenic purpura (TTP)-like form of thrombotic microangiopathy (TMA), which can result in graft loss if not promptly addressed. While the TMA lesion predominantly affects arterioles, it can extend to include portions of the glomerular tuft and arteries with up to two layers of smooth muscle cells (Davies et al., 2000). This form of vasculopathy is accompanied by avid platelet consumption within the graft, as demonstrated by ^{111}In labeled platelet studies, as well as a microangiopathic hemolytic anemia and decline in renal function.

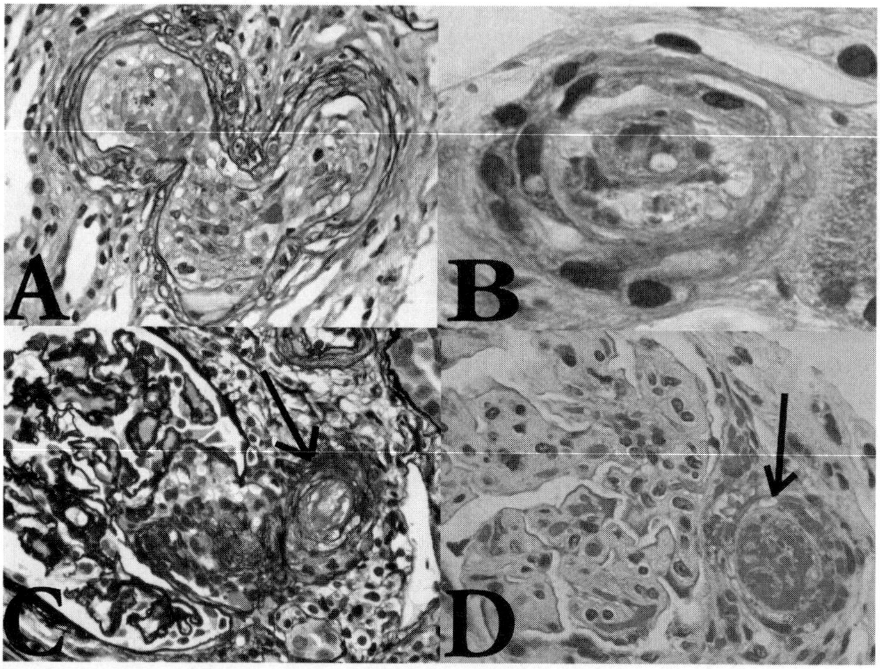

Figure 17.3 Calcineurin inhibitor induced thrombotic microangiopathy (TMA). (A) Widened arteriole with intimal expansion secondary to edema and extravasation of red blood cells, some of which are fragmented. Focal medial myocyte injury and attenuation are also present at 3′ and 10 o'clock positions. Periodic acid methenamine silver (original magnification ×400); (B) Arteriole showing intimal edema with extravasated red blood cells and subsequent luminal occlusion. Trichrome stain (original magnification ×1000); (C, D) Severe afferent arteriolar TMA (arrows) with medial and intimal red blood cells, fibrin, and edema. Consequent ischemic retraction of glomerular capillary tufts (left side of each figure) has ensued. (C) Periodic acid methenamine silver (original magnification ×400), (D) trichrome stain (original magnification ×400).

Morphologically, the lesion consists of thrombosis of the renal microvasculature, with fibrin-platelet thrombi and fragmented red blood cells occluding glomerular capillaries and afferent arterioles (Figure 17.3).

The pathogenesis of CsA-induced TMA is not completely understood, but it is thought that damaged endothelial cells release Von Willebrand factor multimers and platelet activating factor (PAF), leading to platelet aggregation. Increased platelet aggregation has been demonstrated in patients and normal volunteers treated with CsA (Grace et al., 1987). Additionally, endothelial injury results in decreased levels of prostacyclin (PGI_2), a vasodilator and platelet aggregation inhibitor, and increased levels of thromboxane, a vasoconstrictor and platelet aggregator (Grupp et al., 1998). Decreased levels of vasodilatory NO and increased levels of the vasoconstrictor endothelin also result from endothelial cell injury (Grupp et al., 1998).

Thrombotic microangiopathy occurs in approximately 1 to 5% of kidney allograft recipients and, although treatment with CsA may be resumed following clinical recovery (Grupp et al., 1998; Wiener et al., 1997; Zent et al., 1997), some advocate switching to tacrolimus (Abdalla et al., 1994; McCauley et al., 1989). Others advocate plasmapheresis or methylprednisolone pulses (Grupp et al., 1998). Risk factors for CsA-induced TMA include diarrhea-negative HUS (D-negative HUS) as the primary renal disease, a high class 2 HLA antigen (HLA-DR) mismatched graft and a shock-damaged renal graft.

There has been some confusion between the changes of cyclosporine-induced arteriolopathy and the vascular changes of acute rejection. The intimal arteritis of acute rejection involves arcuate and interlobular arteries, in which there is a cellular intimal thickening with lymphocytes incorporated into the intimal mass. CsA arteriolopathy (Figure 17.1), conversely, affects arterioles and glomerular capillaries and has no intrinsic inflammatory cell component. Arterioles show various combinations of hyaline and fibrinoid change, often with identifiable fibrinogen deposition by immunofluorescence. Glomerular capillaries show thrombosis with endothelial cell desquamation or nuclear pyknosis. Small vessel and capillary thrombosis can also occur in antibody-mediated rejection(AbAR), but coincident with other features of AbAR (Racusen et al., 2003), including inflammatory cells marginating in capillaries, and capillary immunostaining for antibodies or complement, and especially the complement split products C3d and C4d (Mauiyyedi and Colvin, 2002). Certainly, vascular lesions can occur, which represent a combination of rejection and CsA toxicity. Laboratory evaluation of drug levels and of possible antidonor antibody are useful in diagnosing these processes.

Chronic Nephropathy with Interstitial Fibrosis

Many renal allograft recipients on chronic CsA therapy show some evidence of long-term nephrotoxicity with serum creatinine levels of

2.0 mg/dl or higher. Interstitial fibrosis and tubular atrophy are well-known, albeit nonspecific, markers and consequences of chronic CsA nephrotoxicity. The pattern in which these occur has been termed "striped," as areas of interstitial fibrosis and tubular atrophy alternate with viable (or even hypertrophied) renal tubules (Andoh and Bennett, 1998; Ader and Rostaing, 1998; Davies et al., 2002). These striped tubulointerstitial changes begin in the outer medulla and progress outward toward to the medullary rays of the cortex. Such changes occur in renal and nonrenal transplant recipients, autoimmune disease patients (who generally receive lower CsA doses), and experimental animals (Andoh and Bennett, 1998). Hyaline arteriolosclerosis secondary to CsA can be viewed as an inherent element of this scarring process as it contributes a significant component of ischemia to downstream tubules and glomeruli (Ader and Rostaing, 1998; Porter et al., 1999).

In renal allografts, it is often difficult to differentiate the changes of chronic CsA toxicity from those of chronic rejection. Chronic nephrotoxicity is easier to detect in recipients of other organ grafts who are transplanted at a time when they have healthy native kidneys. In cardiac allografts, chronic nephrotoxicity with gradual decline in glomerular filtration rate (GFR) occurs with some frequency (Myers, 1986; Parry et al., 2000). Other features of chronic nephropathy in CsA-treated heart transplant recipients include severe hypertension, mild proteinuria, and tubular dysfunction. A similar chronic CsA toxicity has also been detected in liver transplant recipients (McCauley, 1990) and in pulmonary transplantation (Tsimaratos et al., 2000). The markedly impaired GFR in stable, long-surviving renal transplant patients receiving CsA is probably a manifestation of the same type of chronic nephrotoxicity. The renal failure brought about by chronic CsA nephrotoxicity may not reverse when CsA is discontinued (Rao et al., 1985).

In studies of cardiac allograft recipients, Greenberg et al. (1987) have emphasized that although both acute reversible and chronic CsA nephrotoxicity occur with high prevalence, the acute form does not appear to be a specific risk factor for the chronic form. In their study, two thirds of CsA-treated patients had renal dysfunction at one year in both the group that had experienced posttransplant ARF and the group that had maintained normal renal function in the perioperative period. In renal allograft recipients, there is a relationship between cumulative CsA dose and chronic nephrotoxicity, whereas acute nephrotoxicity is related to trough CsA blood level). One would therefore not necessarily expect the same patients to be affected by both types of toxicity. In many patients, the sustained mild-to-moderate decrease in GFR brought about by CsA therapy does not appear to progress over time (Bantle et al., 1990).

Myers et al. (1988a) and Bertani et al. (1991) have drawn attention to the significant effects of CsA on glomerular structure and function in cardiac allograft recipients. Apparently, as a result of the obliterative hyaline arteriolopathy caused by CsA, there is a progressive downstream collapse and sclerosis of glomeruli in biopsies from cardiac allograft recipients. Analysis of dextran transport with an isoporous membrane model suggested that CsA significantly reduced glomerular capillary pressure (Myers et al., 1988a). Both groups noted that a bimodal distribution of glomerular size developed, with an increased number of both abnormally small and abnormally large glomeruli; the latter probably represent compensatory hypertrophy. As end-stage renal failure sets in, there is a shift in the glomerular population, with a greater and greater number becoming abnormally small. Hyperplasia of the juxtaglomerular apparatus was also observed (Bertani et al., 1991). This is less consistently observed in renal allografts treated with CsA (Myers et al., 1988b), suggesting that the finding in cardiac allograft recipients cannot necessarily be generalized to other situations.

A sizeable number of risk factors have been purported as predictors of chronic CsA nephrotoxicity. This list includes number of CsA-induced episodes of renal functional deterioration, high CsA trough levels, number of unexplained episodes of renal functional deterioration, number of other nephrotoxic drugs administered, number of acute rejection episodes, number of acute rejection episode treatments (DeMattos et al., 2000), and high variability in levels of CsA exposure (Levy et al., 2002).

Ponticelli (1998), however, argues that chronic change attributed to CsA is overemphasized and that chronic "rejection" due to underdosage is actually the primary nemesis of prolonged renal allograft survival. He and his colleagues studied 121 cadaveric renal allograft recipients whose grafts were still functioning after 10 years of uninterrupted CsA administration. All patients underwent renal biopsy; biopsies of 15 patients showed chronic rejection, three showed *de novo* membranous nephropathy, and five revealed recurrence of primary glomerulonephritis. Patients experienced only a minimal decline in mean creatinine clearance between year one (59 ± 14.9 ml/min) and year 10 after transplant (57 ± 25.3 ml/min). He therefore concludes that prolonged CsA use favors, rather than impedes, long-term renal allograft survival.

Along similar lines, Remuzzi et al. (1998) applied a logistic regression model to retrospectively collected data from 135 consecutive renal transplant recipients on triple immunosuppressive therapy (Sandimmune, azathioprine, and steroids). Whole blood CsA trough concentrations were measured by radioimmunoassay and highest and lowest trough values were entered for each of two posttransplant periods, postoperative days (POD) 0 to 9 and POD 10 to 30. The model showed that whole blood trough concentrations of 330 to 430 ng/ml from days 0 to 9 and 260 to 390 ng/ml from days 10 to 30 predicted an acute rejection incidence of 22% and 12%, respectively. The authors conclude that this model does not support the tendency to administer low CsA doses in order to avoid early posttransplant renal complications. Rather, higher concentrations of CsA during the initial 30 days posttransplant decreases the incidence of acute rejection and may afford better preservation of renal function.

Clinical Effects in Autoimmune Disease Patients

As patients with autoimmune diseases generally receive lower CsA doses than transplant patients, it is relevant to briefly explore experience with the renal effects of CsA in this patient population. Vercauteren et al. (1998) conducted a meta-analysis and morphological review of CsA-induced nephrotoxicity in autoimmune diseases. Their results showed an unequivocal ability of CsA to induce the chronic renal lesions of interstitial fibrosis, tubular atrophy, and arteriolar hyalinosis. The authors stress studies comparing pre- and posttreatment renal morphology to prove that CsA caused the chronic renal damage rather than the autoimmune diseases themselves. This occurred despite CsA doses of 5 mg/kg/day or less. Zachariae (1999) also conducted a literature analysis of CsA in autoimmune diseases. These authors found that *de novo* morphologic chronic renal damage encompassing interstitial fibrosis, tubular atrophy, arteriolar hyalinosis, and glomerulosclerosis can be induced by as little as 12 months of low dose (≤5 mg/kg/day) CsA therapy. Careful assessment of the risk : benefit ratio is therefre strongly recommended when considering CsA for patients suffering from autoimmune diseases (Zachariae, 1999).

Tacrolimus

Pharmacokinetics and Mechanisms of Action

Tacrolimus, an immunosuppressant developed in Japan and formerly termed FK505 (James, 1996), is a 23-membered macrolactam isolated by fermenting the broth of *Streptomyces tsukubaensis*. It consists of 23 aliphatic heterocyclic rings and possesses a molecular weight of 822.05 (Christians et al., 2002; Finn 1999). Tacrolimus is 100 times more potent than CsA (Andoh et al., 1997a) and its mechanism of action is exerted by its requisite binding to the immunophilin FK binding protein 12 (FKBP12). The FKBP12–FK506 complex is then able to inhibit calcineurin by insinuating itself into a hydrophobic groove between the calcineurin A catalytic and regulatory subunits, as demonstrated by x-ray crystallography (Hemenway and Heitman, 1999). The phosphatase activity of calcineurin, a Ca^{2+}-dependent serine/threonine protein phosphatase conserved from yeast to humans, is then disrupted and phosphatase-controlled translocation of several transcription factors from the cell cytoplasm to the nucleus is then blocked. One of the most important of these is Nuclear Factor of Activated T-cells (NFAT) (Christians et al., 2002; Dumont, 2000; Finn, 1999; Suthanthiran et al., 1996). Normally, NFAT, once dephosphorylated by calcineurin, can translocate to the cell nucleus where it associates with an additional subunit and binds a key functional element within the IL-2 promoter, thus activating the IL-2 gene (Ruhlman and Nordheim, 1997). However, IL-2 is not the only cytokine whose transcription is impeded. Tacrolimus, once bound to FKBP12, also inhibits transcriptional activation of early T-cell activation genes involved in producing GM-CSF, IL-3, IL-4, IL-5, IFNγ, and TNFα (Christians et al., 2002; James, 1996; Tocci et al., 1989). Moreover, even transcription of the IL-2 receptor gene is disrupted (Suthanthiran et al., 1996).

At drug levels of 5 ng/ml, tacrolimus distributes mainly within erythrocytes (95 to 98%), which contain abundant FKBP12 (Christians et al., 2002; Nagase et al., 1994; Winkler et al., 1994). Plasma proteins to which tacrolimus binds include lipoproteins, globulins, α-1-acid glycoprotein, and albumin (Nagase et al., 1994). Tacrolimus for use in general clinical transplant practice is usually administered orally in capsules, dispersed in hydroxypropylmethylcellulose. Absorption occurs within the small intestine, mostly the duodenum and jejunum (Christians et al., 2002; van Hooff et al., 1999), but is erratic, displaying wide inter- and intraindividual variability (Kahan et al., 2002). In fact, variability between patients can range anywhere from 6% to 43% (van Hooff et al., 1999). Peak blood concentrations after oral administration occur between 1.6 and 2.3 hours after ingestion. Food decreases both the rate and extent of absorption so that bioavailability is greatest when tacrolimus is ingested in the fasting state. If oral tacrolimus cannot be taken on an empty stomach, it should at least be consumed in a consistent manner relative to meals (Bekersky et al., 2001).

Tacrolimus is metabolized by cytochrome P450 3A4. Therefore, relevant interactions occur with drugs that either induce or compete for this enzyme (Christians et al., 2002; Finn, 1999; van Hooff and Christiaans, 1999; van Hooff et al., 1999). Drugs that inhibit tacrolimus metabolism and increase its circulating concentration include CsA, erythromycin, clarithromycin, clotrimazole, fluconazole, ketoconazole, danazol, and corticosteroids (Mignat, 1997; Prograf prescribing information, 2002). Drugs that induce cytochrome P450 3A4, decreasing circulating tacrolimus concentrations, include phenytoin, phenobarbital, and rifampin (Christians et al., 2002; Prograf prescribing information, 2002). St. John's Wort (*Hypericum perforatum*) also induces CYP3A4 and has recently been added to the list of drug interactions with tacrolimus (Prograf prescribing information, 2002).

After intravenous or oral administration, the majority of the administered dose can be recovered in the feces, indicating that bile is the principal elimination route. Less than 3% of an administered dose is recovered in the urine (Moller et al., 1999). The reactions involved in metabolism of tacrolimus include hydroxylation, demethylation, and oxidation, such that four first-generation and four second-generation metabolites are generated. The only metabolite with any degree of activity is 31-O-desmethyl tacrolimus and it is produced in minimal amounts. It thus seems that tacrolimus metabolites contribute only negligibly to its activity (Christians et al., 2002; Iwasaki et al., 1993).

Due to its poor correlation between dose and blood concentration, as well as wide inter- and intraindividual variations in pharmacokinetics, therapeutic drug monitoring of tacrolimus trough concentrations is required. For renal transplant patients, a whole blood trough concentration of 7 to 20 ng/ml at months 1 to 3, and 5 to 15 ng/ml at months 4 to 12 is recommended by the manufacturer (Prograf prescribing information). van Hooff et al. (1999) recommend administering the first dose prior to transplant surgery so that the whole blood trough level will be equal to or greater than 10 ng/ml by day 7, minimizing the risk of early acute rejection. Pediatric patients clear tacrolimus twice as rapidly as adults and, therefore, require higher relative doses to achieve similar blood concentrations (Finn, 1999).

Clinical Effects

Reported clinical benefits of tacrolimus over CsA include lower levels of patient total cholesterol, low density lipoprotein (LDL), and apolipoprotein B (Claesson et al., 1998; McCune et al., 1998; Neylan 1998), lower incidence of hirsutism (Mihatsch et al., 1998; Spencer et al., 1997), and cessation or even reversal of gingival hyperplasia (Hernandez et al., 2000). Radermacher et al. (1998) report lower renal vascular resistance indices up to two months after transplant, although such values equalized with those of CsA after two months. Rescue therapy rates for refractory rejection when tacrolimus is substituted for CsA range from 60 to 98% (Boch et al., 2001; Laskow et al., 1998).

Nonrenal Toxicity

Reported nonrenal (or at least not-directly renal) toxic effects or complications of tacrolimus comprise the broad categories of neurotoxicity, posttransplantation diabetes mellitus or hyperglycemia and hypertension (Christian et al., 2002; Jindal et al., 1997; Mayer et al., 1997; Pirsch et al., 1997; Elmer et al., 1996; Laskow et al., 1998). Other adverse effects include hypomagnesemia, increased blood uric acid levels, diarrhea, infection, and posttransplant lymphoproliferative disorder (Mihatsch et al., 1998; Shapiro 1998). Needless to say, most or all of these nonrenal effects could secondarily involve the kidneys.

Neurotoxic effects of tacrolimus include tremor, paresthesia, weakness (Pirsch et al., 1997; Elmer et al., 1996), headache (Neylan 1998), aphasia, psychosis, coma (Mihatsch et al., 1998) and demyelinating peripheral neuropathy (Peltier and Russell, 2002). The neuropathy described by Peltier and Russell (2002) appears within two weeks of initiation of tacrolimus therapy and appears to be a chronic inflammatory demyelinating polyneuropathy with sensory and motor involvement. They hypothesize the cause to be immune-mediated, and possibly due to alteration of T-cell subsets. With regards to neurotoxic effects of tacrolimus, simple dose dependency does not seem to be the case (Mihatsch et al., 1998).

Posttransplant diabetes is a problematic complication of tacrolimus therapy and is generally regarded to be more severe as compared with that caused by CsA (Knoll and Bell, 1999; Mayer et al., 1997; Mihatsch et al., 1998; Neylan, 1998; Pirsch et al., 1997; Spencer et al., 1997). In fact, Greenspan et al. (2002) found that in pediatric patients, the odds ratio for developing posttransplant diabetes on tacrolimus vs. CsA was 9.1. This effect is compounded in African American transplant patients (Marchetti and Navalesi, 2000; Neylan, 1998). Calcineurin inhibitors cause posttransplant diabetes via several possible mechanisms, including decreasing insulin secretion, increasing peripheral insulin resistance (Greenspan et al., 2002) or by a direct toxic effect on the beta cell (Jindal et al., 1997). The diabetogenicity of tacrolimus may be organ specific, as the incidence of posttransplant diabetes is higher in kidney transplant patients on tacrolimus than in liver transplant patients on tacrolimus. In the latter, the incidence of posttransplant diabetes is similar to that of liver transplant patients on CsA (Jindal et al., 1997). However, both Jindal and Berloco point out that the incidence of posttransplant diabetes may have been spuriously high in early reports of kidney transplant patients on tacrolimus, as experience with it was limited and overly high doses were inadvertently administered (Jindal et al., 1997, Berloco et al., 2001). Hyperglycemia in tacrolimus-treated patients appears to be dose dependent, as a reduction in tacrolimus dose reduces this effect (Mihatsch et al., 1998).

Hypertensive effects of tacrolimus may be secondary to alteration in sympathetic outflow (Bechstein, 2000). This complication appears to be less severe in patients receiving tacrolimus than those receiving CsA (Spencer et al., 1997).

Radermacher et al. (1998) report higher incidence of pneumonia with tacrolimus, as compared with CsA. In a review of published reports of *Cryptococcus neoformans* infection in transplant recipients, Husain et al. (2001) have found that tacrolimus patients are more likely to experience cryptococcal infection of the skin, soft tissue and joints, while nontacrolimus patients are more likely to suffer central nervous system involvement.

Renal Effects

Like CsA, tacrolimus can cause nephrotoxic damage even in patients whose whole blood trough concentrations have been maintained at acceptable levels (DeMattos et al., 2000). The mechanism by which tacrolimus exerts such effects has been attributed to drug-induced alteration of prostaglandin metabolism, favoring the production of vasoconstrictor prostanoids, thus resulting in increased renal vascular resistance, reduced renal blood flow, and decreased glomerular filtration rate (Mignat, 1997). A direct glomeruloconstrictive effect has been demonstrated (DeMattos et al., 2000). Morphologic manifestations of these effects include tubular cytoplasmic vacuolization and single-cell calcification, myocyte vacuolization, necrotizing arteriolitis, thrombotic microangiopathy, striped (or diffuse) interstitial fibrosis, arteriolar hyalinosis, and even focal segmental glomerular sclerosis (Brown and Neild, 1990; Japanese FK506 Study Group, 1991; Mihatsch et al., 1998; Randhawa et al., 1993). Acute tacrolimus nephrotoxicity is reversible and caused by reduction in renal blood flow related to afferent arteriolar vasoconstriction. Chronic nephrotoxicity, in the form of striped interstitial fibrosis, is irreversible (DeMattos et al., 2000).

Randhawa et al. (1993) compared the histopathologic changes in renal allograft biopsies from patients on tacrolimus vs. those on CsA. Such changes included tubular vacuolization, myocyte vacuolization, small focal calcifications of tubular epithelial cells, tubular basement membranes or interstitium, striped interstitial fibrosis, vessel hyalinosis, and thrombotic microangiopathy. Generally speaking, toxic drug effects occurred sooner posttransplant in CsA-treated patients but affected a higher percentage of tacrolimus patients. Tubular vacuoles were small, focally confluent and involved both proximal and distal tubules, but did not differ between patients on tacrolimus vs. patients on CsA. In contrast, Morozumi et al. (1996) described a slight difference in tubular vacuolization between tacrolimus and CsA treated kidneys. They described tacrolimus-related vacuoles as foamy, nonisometric and present in the straight and convoluted portions of the proximal tubule. Conversely, CsA-related vacuoles were described as isometric and limited to the straight portion of the proximal tubule.

Higher incidences of urinary tract infections and pyelonephritis have been reported with tacrolimus compared to CsA, perhaps reflecting more potent immunosuppressive effects dose for dose (Radermacher et al., 1998). Polyoma virus infection of the kidney is also more common in the era of tacrolimus-containing immunosuppressive protocols (Ramos et al., 2002). Long known to infect the urothelium, polyoma virus infection of the kidney has been reported by many centers over the past six to seven years. The virus produces pale purple intranuclear inclusions in infected cells. Infected cells can be detected in situ or exfoliated into the tubular lumen. There may be associated inflammation in early phases, followed by zonal fibrosis and tubular atrophy as the infection progresses (Drachenburg, 2001). Treatment is problematic, requiring reduction of immunosuppressant dose, which may result in rejection.

Rapamycin

Pharmacokinetics and Mechanism of Action

Sirolimus, or Rapamycin, is a macrocyclic triene antibiotic lactone antibiotic produced by fermentation of the fungus *Streptomyces hygroscopicus*, a soil actinomycete recovered from the Vai Atari region of the Easter Islands (Hardinger et al., 2002; Johnson, 2002; Kahan, 2001). It functions in preventing organ transplantation rejection by first binding to a cytosolic immunophilin which is, interestingly, the identical immunophilin to which tactolimus binds, namely FKBP12 (Johnson et al., 2002), a peptide-prolyl isomerase functioning as a folding catalyst (Kahan and Camardo, 2001). Binding to FKBP12 occurs via its C-7 methoxy group. This C-7 methoxy group is then used to crosslink FKB12 to the multifunctional serine-threonine kinases, mammalian target of Rapamycin 1 and 2 (mTOR 1 and 2) (Johnson, 2002; Kahan and Camardo, 2001). The blockade of mTORs by sirolimus has important ramifications during cell-cycle progression. First, the lymphocyte response to costimulatory signal 2 during the G_0 to G_1 transition is impaired. Second, the lymphocyte response to cytokine signal 3 between G_1 activation and G_1 progression is attenuated. This mechanism of action renders sirolimus ideal for administration with calcineurin inhibitors because calcineurin inhibitors block signal 1 during the G_0 to G_1 transition. Sirolimus is also well suited for clinical use with anti IL-2R antibodies since anti IL-2R antibodies prevent IL-2 from triggering signal 3 between G_1 activation and progression (Hong and Kahan, 2000; Kahan and Camardo, 2001).

Sirolimus is available in both oral solution and tablet form. Its oral absorption is poor and it is distributed widely into body tissues. It is excreted predominantly in the feces, with only minor amounts excreted into the urine (Kahan and Camardo, 2001). Absorption is only mildly delayed by simultaneous ingestion of a high-fat meal, although oral bioavailability is actually minimally increased by it. Therefore, patients may take sirolimus with or without food, but should do so consistently (Zimmerman et al., 1999). Inter- and intra-patient variability in drug clearance is wide and correlation between dose, whole-blood concentration and patient demographics is poor (Mahalati and Kahan, 2001). For these reasons as well as toxicity issues (discussed below), therapeutic drug monitoring (TDM) of sirolimus is indicated. The gold standard by which all TDM methods are evaluated is high performance liquid chromatography (HPLC) coupled to mass spectrometry. However, for widespread ability of hospital and reference laboratories to perform TDM of sirolimus, less unwieldy methods are available. These include radioreceptor assay, microparticle enzyme immunoassay, and liquid chromatography with ultlraviolet detection (Mahalati and Kahan, 2001; Maleki et al., 2000). Metabolites of sirolimus, measured by HPLC/electrospray-mass spectrometry, include hydroxyl, dihydroxy, demethyl, and didemethyl sirolimus (Streit et al., 1996).

The time to peak blood concentration after oral administration of sirolimus is 1.4 ± 1.2 hours and its terminal half-life ($t_{1/2}$) is 62 ± 16 hours (Zimmerman and Kahan, 1997). Sirolimus partitions extensively into formed blood elements (blood : plasma ratio = 36 : 1) (MacDonald et al., 2000) and there is excellent correlation between the steady state trough whole-blood concentration and the area under the concentration time curve (AUC) (Kahan et al., 2000; Mahalati and Kahan, 2001; Zimmerman and Kahan, 1997). Therefore, trough whole-blood concentrations should be used as the index for TDM. If CsA is being administered concurrently, trough concentrations of sirolimus should be in the range 5 to 15 μg/l, assuming that CsA is administered at a trough concentration of 75 to 150 μg/l (Mahalati and Kahan, 2001). While TDM may not be entirely necessary for patients at no increased risk of rejection who maintain a regimen consisting of full dose CsA, sirolimus 2 mg/day and steroids, it is strongly recommended for certain groups of renal transplant patients. These include those with liver impairment, patients in whom the CsA dose is markedly reduced, patients at high risk for rejection, patients receiving concurrent doses of CYP3A4 inhibitors or inducers, and young children (MacDonald et al., 2000). Indeed, in pediatric transplant patients, $t_{1/2}$ of sirolimus is shorter and clearance is greater, necessitating higher relative doses when adjusted for body surface area (Ettenger and Grimm, 2001).

Sirolimus is metabolized by CYP3A4, which presents difficulty in terms of drug interactions, particularly with calcineurin inhibitors. Exposure to sirolimus is, thus, enhanced by concurrent administration of diltiazem, CsA, and ketoconazole, while exposure is decreased by concurrent administration of rifamycin. Exposure is unchanged by acyclovir, digoxin, or nifedipine (Kahan, 2001). For this reason, the optimal dosing interval between sirolimus and CsA is four hours. Interestingly, McAlister and colleagues (McAlister et al., 2002), using an immunoassay to measure sirolimus and tacrolimus blood levels, found no interaction between simultaneously administered sirolimus and tacrolimus, despite the fact that tacrolimus is also metabolized by CYP3A4.

Grapefruit juice is well known to increase the bioavailability of CYP3A4 substrates, such as diltiazem (Christensen et al., 2002). While no studies examining the interaction between sirolimus and grapefruit juice have been undertaken, patients might be advised against taking the two concurrently pending emergence of definitive conclusions.

Clinical Effects

Nonrenal Toxicity

Nonrenal adverse effects of sirolimus include hyperlipidemia (Flechner et al., 2002; Kahan et al., 1999a; Meier-Kriesche and Kaplan, 2000; Podbielski and Schoenberg, 2001), hypertriglyceridemia (Brattstrom et al., 1998; Flechner et al., 2002; Groth et al., 1999; Hoogeveen et al., 2001), thrombocytopenia (Groth et al., 1999; Hong and Kahan, 2000; Johnson et al., 2001; Kahan et al., 1999a; Meier-Kriesche and Kaplan, 2000; Podbielski and Schoenberg, 2001), leukopenia (Groth et al., 1999; Hong and Kahan, 2000; Johnson et al., 2001; Kahan et al., 1999a; Podbielski and Schoenberg, 2001), elevated liver function test values, hypokalemia (Groth et al., 1999; Johnson et al., 2001), and leukocytoclastic vasculitis (Hardinger et al., 2002). Additionally, CsA produces a pharmacodynamic effect that augments sirolimus-induced myelosuppression and hyperlipidemia (Podder et al., 2001). Interestingly, reports of pulmonary infiltrates with interstitial pneumonitis have also recently emerged (Morelon et al., 2001; Singer et al., 2000). While most such adverse effects can be attenuated by dosage adjustment, some, such as leukocytoclastic vasculitis (reported in one patient) (Hardinger et al., 2002) and interstitial pneumonitis (reported in at least 34 patients) (Morelon et al., 2001; Singer et al., 2000), required withdrawal of sirolimus from the immunosuppressive regimen.

Although unrelated to renal transplantation, risk of hepatic artery thrombosis in liver transplants is such that Wyeth Ayerst has added a black box warning to the sirolimus package insert warning that the safety and efficacy of sirolimus has not been established in liver transplant patients and, thus, is not recommended. This risk is a particular threat when sirolimus is administered with tacrolimus (letter from Wyeth Ayerst to AST and ASTS members, April 10, 2002).

Renal Toxicity

Clinical reports of sirolimus-induced renal toxicity are few. Theoretically, knowing that TMA/HUS is a well-known complication of calcineurin inhibitor based immunosuppression, competition for CYP3A4 by sirolimus would be expected to increase the incidence of CsA-induced TMA/HUS by enhancing systemic exposure to CsA. Langer et al. (2002) report HUS in 10 out of 672 renal transplant patients on a CsA/sirolimus (SRL)/steroid regimen. In none of these 10 patients was HUS the native renal disease. CsA was discontinued in 9 of the 10 patients and sirolimus was discontinued in all because it is known to cause thrombocytopenia. In only two of the patients could the original CsA/SRL/steroid regimen be resumed. In one additional patient, SRL/steroids were reinstituted without a calcineurin inhibitor. In the remaining patients, SRL was replaced with mycophenolate mofetil (MMF) and some received tacrolimus instead of CsA. Five of the patients switched to MMF required plasmapheresis for their HUS (Langer et al., 2002). Florman et al. (2002) attempted an SRL-based calcineurin inhibitor (CNI)-free regimen in two pediatric patients whose native kidneys were lost to atypical HUS, hoping that avoidance of CNIs would prevent HUS recurrence. However, both patients nonetheless lost their allografts to recurrent HUS.

Langer and Kahan (2002) report a higher incidence of perinephric fluid collections or lymphoceles in renal transplant patients on SRL/CsA/prednisone than in those on azathioprine/CsA/prednisone. Additionally, such patients required more-aggressive treatment of their perinephric fluid collections (such as percutaneous drainage or surgery) than did the non-SRL patients with fluid collections.

Schwarz et al. (2001) reported that patients on a CsA/SRL/prednisone regimen (with CsA later withdrawn in a subset of this study group) showed a significantly lower renal phosphate reabsorption than patients in a CsA/mycophenolate mofetil/prednisone group or in healthy controls. However, lower renal phosphate reabsorption normalized by week 28, at which time no difference was detected among the groups.

Mycophenolate Mofetil

Pharmacokinetics and Mechanism of Action

Mycophenolate mofetil (MMF), originally named RS-61443, is the mopholinoethyl ester of mycophenolic acid (MPA), which is actually the active compound. MPA is an antibiotic fermentation product and a weak organic acid isolated from *Penicillium stoloniferum*. MPA has been esterified to produce MMF due to its markedly improved bioavailability compared with its parent compound (Allison and Eugui, 2000; Eugui et al., 1991; Lee et al., 1990; Mele and Halloran, 2000). The bioavailability of mycophenolic acid from orally ingested MMF is 94% (Lee et al., 1990; Mele and Halloran, 2000). Once ingested, MMF is rapidly hydrolyzed into active MPA, a potent inhibitor of *de novo* purine synthesis. MPA functions principally to inhibit inosine monophosphate dehydrogenase (IMPDH), the rate-imiting enzyme in *de novo* guanosine nucleotide synthesis. As IMPDH is inhibited, inosine monophosphate levels fall, inhibiting guanosine nucleotide synthesis. Deoxyguanosine triphosphate (dGTP), required for DNA synthesis, is eventually depleted secondary to this inhibition, effectively blocking DNA synthesis via feedback inhibition by excess adenosine nucleotides (Allison and Eugui, 2000; Becker, 1999; Eugui et al., 1991; Mele and Halloran, 2000). The reason for the preferential inhibition of purine synthesis by MPA in lymphoid cells is that lymphocytes rely on *de novo* purine synthesis, while other cell types are able to utilize the salvage pathway of purine synthesis. In addition, MPA shows fivefold higher binding affinity for the type II (inducible) isoform of IMPDH than for the type I (constitutive) isoform. Since the type II isoform is upregulated in activated lymphocytes, they are more susceptible to the inhibitory effects of MPA (Allison and Eugui, 2000; Becker, 1999; Eugui et al., 1991; Mele and Halloran, 2000; Natsumeda and Carr, 1993). MPA also exerts its immunosuppressive effects by the following additional mechanisms: induction of apoptosis in activated T-cells (Cohn et al., 1999), inhibition of transfer of fucose and mannose residues to glycoproteins (the most relevant of which include many adhesion molecules) (Allison and Eugui, 2000; Mele and Halloran, 2000), and MPA depletion of tetrahydrobiopterin, a cofactor of inducible NO synthase (iNOS) via depleting guanosine nucleotides. The resultant suppression of NO production prevents tissue damage by peroxynitrite (Allison and Eugui, 2000).

By inhibiting *de novo* purine synthesis, MPA effectively attenuates lymphocyte proliferative and productive activity. Nagy et al. (1993)

proved that at therapeutic concentrations, MPA inhibits synthesis of the following cytokines after 48 hours following activation by *Staphylococcus aureus* endotoxin A: interleukin (IL)-1α, IL-1β, IL-2, IL-3, IL-4, IL-5, IL-6, IL-10, interferon-γ (IFN-γ), tumor necrosis factor-γ (TNF-γ), tumor necrosis factor-β (TNF-β), and granulocyte-macrophage colony stimulating factor (GM-CSF).

Following oral ingestion of MMF, it is rapidly and completely absorbed, after which it undergoes deesterification, yielding active MPA. This process does not appear to be affected by renal or hepatic impairment to any significant degree. Greater than 90% of a single dose of MMF is excreted in the urine, predominantly as mycophenolic acid glucuronide (MPAG). There is significant enterohepatic circulation and recirculation of mycophenolic acid glucuronide, in which intestinal bacterial deconjugation of MPAG plays a major role. Dosage adjustment of MMF is not necessary in renal impairment or dialysis (Bullingham et al., 1998). Therapeutic drug monitoring of MPA is recommended, particularly during the early transplant period, in order to optimize dosage in preventing acute rejection. HPLC with ultraviolet detection yields extensive pharmacokinetic information on MPA, but commercially available immunoassay kits provide ease of use and ready availability for routine laboratory use (Armstrong et al., 2001). Concomitant immunosuppression with CsA appears to lower MPA trough levels (Gregoor et al., 1999).

Clinical Effects

MMF has proven extremely useful in prevention of acute rejection in heart, kidney and liver allograft recipients. By enabling reduction of calcineurin inhibitor doses, MMF can attenuate complications secondary to their nephrotoxicity (Mele and Halloran, 2000; Yang et al., 2002a); Yang et al. (2002a) showed that complete withdrawal of calcineurin inhibitor and institution of MMF monotherapy may be most efficacious. Additionally, evidence exists for amelioration of renal impairment due to chronic allograft nephropathy, both clinically and experimentally. Gonzalez-Molina et al. (2002) reported a significant decrease in renal functional impairment, or even variable degrees of improvement, when MMF was added to a CsA-based regimen in patients with chronic allograft nephropathy (CAN). Ojo et al. (2000), utilizing data from the U.S. renal transplant scientific registry, found that MMF decreased the relative risk for development of CAN by 27% and this effect was independent of its ability to prevent acute rejection episodes.

Nonrenal Toxicity

A brief summary of nonrenal toxic side effects of MMF is included here as they may secondarily affect the kidney, particularly when prerenal azotemia enters into the differential diagnosis of acute renal failure in the allograft. The most clinically relevant information garnered about MMF toxicity can be traced to three large randomized, double-blind clinical studies, the European MMF Study Group (1995), the U.S. Renal Transplant MMF Study Group (Sollinger, 1995) and the Tricontinental MMF Renal Transplant Study Group (1996; Halloran et al., 1997); although similar manifestations of its adverse effects can be found in other studies. Most such effects appear to be dose-dependent and reversible, particularly with dosage adjustment. The most common nonrenal side effects include:

- gastrointestinal effects (nausea, vomiting, and diarrhea; less often cholestasis, hemorrhagic gastritis, pancreatitis, and large bowel perforation) (Becker 1999; Behrend 2001; Behrend et al., 1997; Carl et al., 1997; European MMF Study Group, 1995; Gallagher and Andrews, 2001; Mathew, 1998; Mele and Halloran, 2000; Moreso et al., 1998; Sollinger, 1995; Tricontinental MMF Study Group, 1996)
- myelosuppression (leukopenia, thrombocytopenia, and anemia) (Becker, 1999; Behrend, 2001; Behrend et al., 1997; Carl et al., 1997; European MMF Study Group, 1995; Gallagher and Andrews, 2001; Mele and Halloran, 2000; Mathew, 1998; Moreso et al., 1998; Sollinger, 1995; Tricontinental MMF Study Group, 1996)
- increased incidence of cytomegalovirus (CMV) infection, some of which may be tissue invasive (Carl, 1997; European MMF Study Group, 1995; Mathew, 1998; Mele and Halloran, 2000; Moreso et al., 1998; Sollinger, 1995; Tricontinental MMF Study Group, 1996).

However, CMV infection is less of a problem with doses lower than 3 g/day (Moreso et al., 1998). While transient elevations in liver function test levels have been reported, overt hepatotoxicity has not (Sollinger et al., 1992). Results of the three pivotal multicenter trials mentioned above also included reports of nonmelanoma skin cancer and posttransplant lymphoproliferative disorder (PTLD) in MMF patients. However, their incidence was no greater than that in groups taking immunosuppressants other than MMF, and most PTLDs were polyclonal (Mele and Halloran, 2002).

Renal Toxicity

Direct renal toxicity as a result of MMF has not been reported (Mele and Halloran, 2002; Vanrenterghem, 1997).

Polyclonal Antibodies

Introduction and Mechanism of Action

Polyclonal antibodies in the treatment and prophylaxis of renal allograft rejection have been extant since the early 20th century and have been used in humans since the 1970s. They are composed of many different immunoglobulins directed against a host of molecules expressed on the surface of lymphocytes (Brennan, 2001; Bock et al., 1995; Peddi et al., 2002; Prin-Mathieu et al., 1997). Until relatively recently, only horse-derived polyclonal antibodies were available for clinical use in the U.S. These included Minnesota antilymphocyte globulin (MALG) and antithymocyte gamma globulin (ATGAM). In contrast, a third horse-derived antibody (lymphoglobuline) and two rabbit-derived polyclonal antibodies (thymoglobuline and F-ATG) were available in Europe. In 1992 the U.S. Food and Drug Administration prohibited the use of MALG, leaving ATGAM as the only commercially available polyclonal antibody preparation. This continued until 1999, when Thymoglobulin was approved for use in the U.S. (Brennan, 2001).

Polyclonal antithymocyte globulin (ATG) preparations are obtained by immunizing an animal (usually rabbit or horse) with human thymocytes. The antibodies thus generated are intended to exert a cytotoxic effect upon human T-lymphocytes, although they contain idiotypic specificity against B-cell and other surface markers as well. The T-cell surface markers against which antithymocyte globulin preparations contain antibodies include CD2, CD3, T-cell receptor, CD4, CD5, CD7, CD8, CD11a, CD18, CD44, and CD25 (anti-IL-2Rα chain). B-cell surface markers against which they are directed include CD19, CD 20, CD21, and CD40. They also contain antibodies against accessory, adhesion, co-stimulatory and NK-cell markers including CD6, CD44, CD45, CD56, CD58, LFA-1, CD80 (B7-1), CD86 (B7-2), CD95 (Fas), Fas-ligand, CTLA-4 (CD152), CD40-ligand (CD154), and MHC class I and II antigens and $\beta 2$ microglobulin (Brennan, 2001; Bock et al., 1995; Peddi et al., 2002; Prin-Mathieu et al., 1997).

Polyclonal antibody preparations exert their intended action via lymphocyte depletion by complement or FC-dependent opsonization and lysis (or phagocytosis) (Brennan, 2001; Prin-Mathieu et al., 1997) although other related mechanisms are thought to be responsible for induction of host tolerance to the allograft. T-suppressor cell activity against host antiallograft immune effects has been postulated as one such mechanism. Muller et al. (1997) showed that, as compared with patients receiving either no antibody-based induction therapy or OKT3 induction therapy, patients receiving antithymocyte globulin induction therapy experienced long-term (>5 years) changes in T-cell subsets, including a significantly decreased CD4:CD8 ratio and a marked relative increase in the CD8+/CD57+ T-cell subgroup. Additionally, activation of certain cell surface markers that may induce apoptosis of donor-reactive T-cells may, in turn, induce tolerance. Such markers include Fas (CD95), TNFR1, or the T-cell receptor. Rowinski et al. (2002) showed that ATG induces Fas, Fas-ligand, and CD69 expression on the surfaces of both CD4+ and CD8+ cells, with Fas-ligand being expressed in much higher concentrations on CD8+ cells. Posttranscriptional inhibition of CD25 (α chain of the IL-2 receptor) is another means by which ATG blocks proliferation of host T lymphocytes (Prin-Mathieu et al., 1997). Additionally, polyclonal antibodies may be particularly suited for presensitized patients or for treating steroid-resistant rejection because they are capable of eliminating preactivated noncycling memory T-cells (Brennan, 2001).

Gaber et al. (1998) observed that compared with ATGAM (horse preparation), thymoglobulin (rabbit preparation) is associated with significantly greater depletion of CD2, CD3, CD4, and CD8 cells and results in one-year patient and graft survival rates at least matching those of ATGAM. Other comparisons of thymoglobulin with ATGAM have shown fewer rejection episodes, higher six-month graft survival rates and lower rates of cytomegalovirus infection and posttransplant lymphoproliferative disorder. These effects may be due to more profound and lasting T-lymphocyte depletion (Brennan, 2001).

Polyclonal antibody preparations are supplied as sterile lyophilized powders that are reconstituted with sterile diluent and administered intravenously through a high-flow vein. Quality control testing of antithymocyte preparations is performed on peripheral blood lymphocytes of healthy human subjects. However, Shenton et al. (1994) point out that not only is there a significant difference in potency amongst different commercially available preparations, but that cell surface marker blockade by and cytotoxicity of ATGs differ significantly between peripheral blood lymphocytes of renal patients and those of healthy humans (generally speaking, blockade and cytotoxicity is more potent in renal patients). Monitoring of peripheral blood T-cell counts in patients undergoing ATG therapy is therefore essential.

Clinical Effects

Nonrenal Effects

The most common adverse effects of antithymocyte preparations include fever, chills, rash, arthralgias, leukopenia, thrombocytopenia, pain, headache, edema, hypertension, and nausea (Bock et al., 1995; Rossi et al., 1993). Such effects are generally manageable and reversible. Minor anaphylactic reactions can be prevented by premedication with steroids, antihistamines, and acetominophen (Brennan, 2001; Rossi et al., 1993), and by infusing the ATG preparation over a minimum of six hours into a high-flow vein. The infusion rate may also be slowed to reduce some of these reactions. Acott et al. (2001) found a significantly higher reported incidence of unexplained fever, pulmonary edema or fluid retention, thrombocytopenia, and reactivation of Epstein-Barr and HHV-6 viruses in patients administered ATG induction vs. basiliximab (see below) induction.

Benoist et al. (1998) reported two cases of a false positive increase in serum C-reactive protein in two pediatric renal allograft patients who received antithymocyte globulin, which they attributed to heterophilic antibodies. They were able to remove these antibodies by subtotal IgG adsorption to protein A or protein G or by immunoadsorption using rabbit antilymphocyte globulin or total IgG in nonimmune rabbit serum. They stress that since C-reactive protein is elevated during acute allograft rejection, this false positive increase due to antithymocyte globulin is a potential pitfall in clinical diagnosis of acute rejection.

Schaffner et al. (1991) described transient pure red cell aplasia with reticulocytopenia in patients receiving rabbit antilymphoblast globulin (RATG). This became evident six to nine days following thrombocytopenia and lymphopenia, suggesting that antibodies were directed against red blood cell precursors. Such transient pure red cell aplasia was noted by these authors initially in three patients, which prompted a larger and more systematic study. They found that 90% of their patients treated with RATG developed reticulocytopenia and that 65% suffered complete disappearance of reticulocytes, requiring transfusion. This was confirmed by voluntary bone marrow examination in four patients.

A potentially serious adverse effect of the polyclonal antithymocyte and antilymphocyte preparations is the generation of antirabbit or antihorse antibodies in the patient. During the Thymoglobulin phase III trial, antirabbit antibodies developed in 68% of the thymoglobulin-treated patients and antihorse antibodies developed in 78% of the ATGAM-treated patients (Gaber et al., 1998). While theoretically, such host-derived antibodies would be expected to minimize efficacy of the ATG preparations, the effect of such antibodies on repeat use of ATG preparations has not been studied in controlled trials and is not entirely clear. Prin-Mathieu et al. (1997) studied pre- and postinduction serum levels of antirabbit and antihorse IgG, IgA, and IgM in kidney transplant recipients treated with horse antilymphocyte globulin or rabbit antithymocyte globulin. Amongst primary transplant recipients, they noted a significant increase in the number of patients with antirabbit IgM and antihorse IgG, IgA, and IgM after induction therapy. Sixteen of 89 patients developed serum sickness. Those patients who developed serum sickness showed a significant increase in the IgM isotype, but those who developed antihorse or antirabbit antibodies without serum sickness showed a significant increase in the IgA isotype. Even patients who were never exposed to antithymocyte or antilymphocyte antibodies may demonstrate antibodies to them, due to sensitization to environmental or food antigens. The authors therefore recommend testing patient sera for antirabbit or antihorse antibodies prior to administration of antilymphocyte preparations. Enzyme-linked immunosorbent assay (ELISA) is a relatively simple and sensitive method for detecting such antibodies (Prin-Mathieu et al., 1997; Tatum et al., 1984).

Renal Effects

The major concern for polyclonal antithymocyte or antilymphocyte preparations lies in development of acute glomerulonephritis as a manifestation of serum sickness occurring in patients generating antihorse or antirabbit antibodies to such preparations (Cunningham et al., 1987; Prin-Mathieu et al., 1997).

Monoclonal Anti-CD3 Antibodies

Introduction and Mechanism of Action

OKT3 is a murine monoclonal antibody of the IgG2a subclass directed against the CD3 antigen of human T-cells. Its ability to reverse allograft rejection stems from its ability to deplete T-cells and modulate or remove T-cell receptor complexes from T-cell surfaces (First et al., 1993; Norman, 1993). In doing so, it necessarily activates T-cells, resulting in massive cytokine release (see discussion below). OKT3 can reverse those acute rejection episodes that are unresponsive to steroids and even some that are unresponsive to polyclonal antibodies. Approximately 95% of first acute rejections are responsive to OKT3 (Norman, 1993).

Other anti-CD3 monoclonal antibodies include WT32, a monoclonal antibody of the IgG2a subclass, WT31, a monoclonal antibody of the IgG1 subclass, and BMA031, a monoclonal antibody of the IgG2b subclass. WT31 is devoid of a mitogenic effect in 30% of patients tested. BMA031 does not produce as massive a release of cytokines after the first administered dose ("first dose response") as does OKT3. WT32 demonstrates effects similar to those of OKT3 (First et al., 1993). The remainder of the monoclonal antibody discussion will focus on OKT3 as the prototype of this class of antibody-based therapy.

OKT3 is administered intravenously as a bolus and should be injected within one minute. While receiving OKT3, plasma levels should be monitored to maintain a plasma concentration of $>0.80\,\mu g/ml$ or to maintain a CD3+ T-cell concentration of $<25\,cells/mm^3$. Vasquez and Pollak (1997) reported a clinically significant interaction between OKT3 and CsA in adult renal transplant recipients. They noted a significant increase in CsA trough levels on day five (of a seven day postoperative course) in patients receiving OKT3 for transplant induction therapy. The authors postulate a cytokine-induced inhibition of CYP3A4 (responsible for CsA metabolism) as the mechanism for this interaction, as numerous cytokines are released following OKT3-induced T-cell activation.

Clinical Effects

Nonrenal Effects

Interaction of anti-CD3 monoclonal antibodies, and OKT3 in particular, with their T-cell targets results in a predictable and reproducible response characterized by febrile illness, often accompanied by diarrhea, pulmonary edema, and even a meningitis-like syndrome. Other accompanying symptoms may include headache, tremor, nausea, vomiting, abdominal pain, malaise, muscle and joint pain, generalized malaise, dyspnea, wheezing, hypotension, hypertension, chest pain, and even profound shock or cardiac or respiratory arrest (First et al., 1993; Jeyarajah and Thistlethwaite, 1993; Loertscher, 2002; Norman, 1993). This was originally thought to be due to release of cytokines from T-cells undergoing lysis, however the theory has been negated as T-cells do not contain preformed cytokines within their cytoplasm. Rather, cytokines are only manufactured by T-cells on activation (First et al., 1993; Jeyarajah and Thistlethwaite, 1993). This response is thus due to massive cytokine release by T-cells, activated by binding of OKT3 to its T-cell surface CD3 molecule target. It is known as "cytokine release syndrome," and since it is most severe after administration of the first one or two OKT3 doses it is also known as the "first dose response" (Jeyarajah and Thistlethwaite, 1993). Cytokines released during this syndrome include TNFα, IFN-γ, IL-1, IL-2, IL-3, IL-4, IL-6, IL-10, and GM-CSF (Abramowics et al., 1989; First et al., 1993; Gaston et al., 1991; Jeyarajah and Thistlethwaite, 1993; Loertscher, 2002). TNFα appears to be the cytokine most responsible for symptoms of pulmonary edema, as well as for the most severe cardiovascular effects, including angina, tachycardia, and changes in blood pressure. Intravascular thrombosis has also been reported secondary to OKT3 administration and is likewise attributed to effects of TNFα, specifically its activation of the extrinsic pathway of the clotting cascade (Jeyarajah and Thistlethwaite, 1993).

Parlevliet et al. (1995) describe this response as dose-dependent because low dose administration (0.5 mg BID) results in significantly decreased complement activation and neutrophil degranulation as compared with the standard dose (5 mg daily). Measures to alleviate symptoms of cytokine release syndrome include pretreatment with diphenhydramine, low-dose steroids, and acetaminophen, as well as rigid fluid management. More-aggressive measures to counteract sequelae of cytokine release include administration of high-dose steroids, indomethacin, pentoxifylline, and even anti-TNFα antibodies (Rossi et al., 1993).

Another serious complication of OKT3 administration is generation of human antimouse antibodies (HAMA) in 80% of patients. These antimouse antibodies rapidly neutralize OKT3, dramatically reducing its efficacy, although simultaneous administration of low-dose CsA may reduce the HAMA response to as low as 15% of patients (Loertscher, 2002; Rossi et al., 1993). The HAMA response can target either the isotypic (Fc) or idiotypic (Fv) region of the OKT3 molecule (Loertscher, 2002). Because of this HAMA response, OKT3 can be administered as pulse therapy only (Norman, 1993) and may not be effective for repeat courses of therapy.

As with other forms of immunosuppression, OKT3 therapy is accompanied by an increased risk of infectious (particularly Epstein-Barr virus and cytomegalovirus) and neoplastic (particularly posttransplant lymphoproliferative disorder, other lymphomas and skin caner) (Orthoclone package insert) complications. Hibberd et al. (1992) report a significantly increased risk of CMV disease in previously seropositive patients who were given OKT3 either for acute rejection or prophylactically, as compared with those not receiving antibody-based therapy (prednisone and CsA, with or without azathioprine).

Renal Effects

Treatment of acute renal allograft rejection with OKT3 is frequently accompanied by initial continued deterioration in renal function accompanied by an even further rise in serum creatinine levels (Batiuk et al., 1993; First et al., 1993; Jeyarajah and Thistlethwaite, 1993). This reaction is known as "cytokine nephropathy" and is believed to be secondary to decreased renal prostaglandin synthesis or by direct cytokine-induced renal damage (First et al., 1993), which may include tubular toxicity, hemodynamic compromise or both (Jeyarajah and Thistlethwaite, 1993). Renal function generally improves spontaneously. Morphologic manifestations of cytokine nephropathy are by and large unimpressive and include either no changes or mild interstitial edema (Batiuk et al., 1993).

Another potentially grave complication of OKT3 therapy is thrombotic microangiopathy (TMA), which is a particular risk in patients whose primary renal disease was a form of TMA. This has also been attributed to effects of TNFα (Doutrelepont et al., 1992; Pisoni et al., 2001). As with polyclonal antibodies, OKT3 may induce serum sickness caused by the HAMA response. Glomerulonephritis secondary to glomerular deposition of the resulting immune complexes may ensue.

Antiinterleukin-2R α Monocloncal Antibodies

Introduction and Mechanism of Action

Basiliximab and daclizumab are chimeric and humanized monoclonal antibodies, respectively, directed against the α subunit of the interleukin-2 receptor (IL-2Rα, p55, CD25, Tac subunit), which is expressed only on activated T-lymphocytes. Basiliximab was constructed by fusing the entire variable region (Fv) of a mouse anti-CD25 antibody (RFT5) to the constant region of a human IgG1 antibody. The resulting chimeric monoclonal antibody thus has a murine content of only 25% (and a human content of 75%), but a conserved binding affinity to the IL-2R (1×10^{-9} mol/l). Continuous production of this antibody was enabled by genetically engineering a mouse myeloma cell line to express a plasmid capable of producing this chimeric gene product (Cibrik et al., 2001; Loertscher et al., 2002; Pascual et al., 2001); Vilalta et al., 2002. Daclizumab is a humanized monoclonal antibody, genetically synthesized from oligonucleotides, consisting of 90% human and 10% murine antibody sequences. The human sequences are derived from the constant region of human IgG1 and the variable region of the Eu myeloma antibody. The murine sequences are derived from the complementarity-determining region of a mouse anti-Tac antibody (Loertscher et al., 2002; Pascual et al., 2001). However, the humanization of this antibody has reduced its binding affinity to the IL-2R to 3×10^{-9} mol/l (Loertscher et al., 2002).

The rationale behind the inception of these chimeric, humanized monoclonal antibodies was to target the system used by T-cells to proliferate after antigenic exposure with an antibody composed of as many human determinants and as few murine determinants as possible. IL-2 binding to its receptor is a key trigger for clonal expansion and viability of activated T-cells and blocking this receptor will thus inhibit amplification of the immune response.

The IL-2 receptor is composed of three subunits, α (CD25), β (CD122), and γ (CD132). The α subunit is essential for converting the IL-2 receptor to its functional state. Additionally, unlike the CD3/T-cell receptor complex, the IL-2Ra subunit possesses no cytoplasmic tail, cannot internalize, and cannot transduce signals. Therefore, binding to the IL-2Rα results in no cytokine release, explaining the lack of side effects of basiliximab and daclizumab (Loertscher et al., 2002; Nashan et al., 1999; Pascual et al., 2001; Vilalta et al., 2002; Vincenti, 2001; Wiseman and Faulds, 1999). Because these monoclonal antibodies contain only small murine components, they maintain a long half-life within the circulation (7.2 ± 3.2 days for basiliximab, approximately 20 days for daclizumab) since neutralization by a HAMA response is not a concern. The immunogenicity of these antibodies in human patients is minimal and does not appear to interfere with the action of the drug. Basiliximab is administered intravenously on day 0 and day 4 posttransplant, while daclizumab is administered in divided doses over three to five days (usually every other day for five days) (Carswell et al., 2001; Cibrik et al., 2001; Loertscher et al., 2002; Pascual et al., 2001; Pisani et al., 2001; Vincenti, 2001). Numerous clinical trials, including pivotal phase III trials and pediatric clinical trials, have established the safety and efficacy of these monoclonal antibodies; demonstrating minimal adverse effects and no increase in infectious or malignant complications (Acott et al., 2001; Boletis et al., 2001; Garcia-Meseguer et al., 2002; Henry et al., 2001; Kahan et al., 1999b; Nashan et al., 1997; Soulillou et al., 1990). Their main uses in transplantation include induction therapy, sequential use in delayed graft function, elimination of corticosteroid induction, and sparing of calcineurin inhibitors during induction. Hong and Kahan (1999) achieved excellent results when using anti-IL-2R monoclonal antibodies in combination with sirolimus during induction, delaying initiation of calcineurin inhibitor administration until mean serum creatinine fell below 3.0 mg/dl.

Strehlau et al. (2000) reported drug interaction between CsA and basiliximab, with substantial increase in CsA trough levels in pediatric patients receiving basiliximab. They postulate that circulating IL-2, increased by blockade of its receptors on activated T-cells, is present at increased levels and may then alter metabolism by CYP3A4. Prompted by this report, Kovarik et al. (2001) studied 39 pediatric patients receiving basiliximab, CsA, and prednisone, who were enrolled in an international multicenter trial. They found no evidence that basiliximab had any influence on CsA levels when administered at pharmacologically active concentrations and concluded that their results were in agreement with those of the adult renal transplant trials of basiliximab. However, Sifontis et al. (2002) reviewed the medical records of 12 renal transplant patients receiving basiliximab induction in conjunction with tacrolimus-based immunosuppression and compared them with the records of patients receiving ATG induction. They reported a 63% increase in tacrolimus blood levels in basiliximab-treated patients on day 3 compared with ATG-treated controls, half of which showed acute tubular injury on renal biopsy. The authors postulate that circulating IL-2, unable to bind to its receptor on activated T-cells, is thus capable of binding to IL-2 receptors on hepatocytes and enterocytes, resulting in CYP3A4 downregulation.

Clinical Effects

Nonrenal Effects

Most adverse effects of anti-IL-2Rα monoclonal antibodies, such as constipation, nausea, vomiting, diarrhea, dyspepsia, pain, headache, and insomnia, have been demonstrated to show incidences nearly identical to those of placebo. However, potentially severe hypersensitivity and anaphylactic reactions can occur with either monoclonal antibody, manifested by symptoms such as hypotension, tachycardia, wheezing, bronchospasm, respiratory failure, urticaria, rash, pruritis, and sneezing. If this occurs, therapy should be immediately discontinued and reexposure may be precluded.

Another potential adverse effect of anti-IL-2Rα monoclonal antibodies is a phenomenon termed "escape rejection." Because they only block the IL-2 receptor, allograft rejection may still occur, mediated by binding of IL-7 and/or IL-15 to activated T-cells (Vincenti, 2001).

Renal Effects

Reported adverse renal effects of these monoclonal antibodies occur at frequencies similar to placebo and cannot easily be separated from effects of concurrent immunosuppressive agents, immunosuppression in general, or other factors (i.e., renal tubular injury, urinary tract infection, hydronephrosis).

Intravenous Immunoglobulins (IVIg)

Introduction and Mechanism of Action

Therapeutic uses of intravenous immune globulin (IVIg) are myriad and include: antiviral prophylaxis, treatment of a wide variety of autoimmune diseases, and prevention of production or rebound of antiallograft antibodies (Saydain et al., 2002). Immunomodulatory action of IVIg is complex, and may vary somewhat between formulations (see Sewell and Jolles, 2002, for a review). Seven different formulations of IVIg are now licensed in the U.S. Formulations of intravenous immunoglobulin are developed from pooled sera with a variety of different vehicles. There are differences in their production and composition that affect their efficacy, tolerability, and side-effects profile (see Lemm, 2002, for a review).

Clinical Effects

Nonrenal Effects

Pulmonary toxicity with respiratory symptoms has been reported (e.g., Rault et al., 1991). Immune hemolysis, disseminated intravascular coagulation, and serum sickness have been reported in a child being treated for Kawasaki disease, which are probably due to antibodies to blood-group antigens in the preparation (Comenzo et al., 1992). Meningitis and hepatitis have been reported as well, resolving over several days following therapy (Shorr and Kester, 1996). In experimental animals, long-lasting hypotension has been reported, with evidence that IgG dimers in the preparations may activate macrophages and neutrophils via Fc receptors, causing release of inflammatory mediators such as PAF (Bleeker et al., 2000).

Renal Effects

Acute renal failure has been reported with use of these agents (Decocq et al., 1996; Cantu et al., 1995; Rault et al., 1991). Patients may present with mild renal insufficiency or with oligoanuria, with marked rise in serum creatinine. Recovery follows discontinuation of the therapy. The best-documented renal lesion is a marked vacuolization and hydropic swelling of tubular epithelial cells, with preservation of PAS-positive brush border. Resemblance of this lesion to osmotic injury has focused attention on the sucrose vehicle in which (at least some of) these preparations are suspended and administered. Less-striking forms of tubular vacuolization may occur, but must be differentiated from tubular injury due to other causes.

Interleukin-2, Tumor Necrosis Factor, and Interferon

The relative effectiveness of IL-2 in treating metastatic renal-cell carcinoma might suggest the possibility of direct renal tubular toxicity. However, it appears that most of the nephrotoxic effects of this compound are actually indirect. The renal functional impairment frequently encountered after IL-2 therapy has usually been attributed to a systemic capillary leak with hypotension and prerenal azotemia (Belldegrun et al., 1984; Hamblin, 1990; Rosenstein et al., 1986). Rhabdomyolysis secondary to convulsions has also been cited as a cause (Sarna et al., 1989). IL-2 really has no direct effect on endothehium, but stimulates release of other cytokines, including IL-1, TNF, lymphocytotoxin, and interferon-γ, that cause endothelial-cell activation and vascular leakiness to macromolecules (Butler et al., 1989; Cotran et al., 1988). Samlowski et al. (1990) reported heavy shedding of renal tubular cells into the urine of patients two to four days after initiation of IL-2 therapy. Onset of these cytologic features of tubular injury preceded significant hypotension or elevation in serum creatinine, suggesting a direct tubular toxic effect.

The systemic administration of recombinant human tumor necrosis factor is characterized by side effects very similar to those of IL-2, largely attributed to a capillary leak syndrome and prerenal azotemia (Moritz et al., 1989; Steinmetz et al., 1988). In reported series of studies, 13 to 21% of patients developed some element of renal insufficiency. The presence of enzymuria in over half the patients provided some evidence of at least subclinical tubular toxicity (Steinmetz et al., 1988).

Mercatello et al. (1991) reported on the characteristics of the renal functional impairment encountered in patients treated with a combination of lL-2 and interferon-α. In 16 patients treated for renal cell carcinoma with this combined regimen, renal plasma flow (RPF) remained constant despite a 25% decrease in GFR, 20% decrease in systemic blood pressure, and 50% decrease in urine output. Despite the preservation of renal blood how, the renal failure was associated with a decrease in fractional excretion of sodium (from 1.37% to 0.56%) and a 30% increase in heart rate, features suggestive of "prerenal azotemia."

Microalbuminuria was also observed. The authors postulate that the renal functional impairment is due either to a decrease in the efferent to afferent resistance ratio, leading to a decrease in glomerular capillary pressure, or to a decrease in ultrafiltration coefficient (K_f) or both. They point out that nonsteroidal anti-inflammatory agents, frequently used to minimize side effects of the immunotherapy, could considerably worsen the renal functional impairment observed in these protocols.

Interferon

The acute renal failure occasionally encountered in patients treated with interferon-α or -γ appears to be different from that seen with other types of immunotherapy. Proteinuria is frequently a prominent part of the presentation (Ault et al., 1988; Cooper et al., 1987), and the renal failure is often severe (Fiedler et al., 1991), occasionally leading to death (e.g., Mahaley et al., 1988). An acute interstitial nephritis was observed in one patient treated with interferon-α for mycosis fungoides (Ault et al., 1988). In another report, a patient treated with interferon-γ for acute lymphoblastic leukemia developed proteinuria and renal failure with a combination of focal glomerular sclerosis and acute tubular necrosis on biopsy, with cast formation and extensive regenerative changes in tubular epithelium (Averbuch et al., 1984). Both proteinuria and renal failure resolved after interferon therapy was terminated.

FROM THE CLINIC TO THE LABORATORY

Clearly, important observations have been made on the renal effects of immunotherapeutic agents in patients receiving these regimens, and the pathogenesis of renal pathology and pathophysiology have been addressed to some extent in clinical studies. However, experimental model systems are necessary for complete elucidation of mechanisms of nephrotoxicity, enabling careful control of experimental variables that is not possible clinically. Invasive techniques enable direct study of hemodynamics and glomerular and tubular function *in vivo* in experimental animals or in isolated perfused kidney. Cell-culture systems can be used to directly assess effects of the immunotherapeutic agent on specific cell populations in a carefully controlled milieu. Numerous *in vivo* and *in vitro* model systems have been used to study mechanisms of nephrotoxicity of immunotherapeutic agents.

Experimental Toxicity of Immunotherapeutic Agents

Cyclosporine

In experimental studies, CsA has effects on structure and function of the renal tubule, renal vasculature and hemodynamics, and renal interstitium. Which of these effects is most relevant clinically has been debated, and doubtless varies with the type of clinical CsA-induced renal dysfunction being considered and with an array of clinical variables, including acute and cumulative dose, patient age, the presence of preexisting or concomitant ischemic injury, and other factors.

Renal Tubule

Early studies using rat models suggested that CsA is a primary tubular toxin, and that tubular effects of the drug explained its effects on renal function. Blair et al. (1982) described isometric vacuolization of tubular cells, accumulation of eosinophilic bodies often representing giant mitochondria, and microcalcification in proximal tubules after up to 21 days of high dose CsA; findings confirmed by other investigators (Ryffel et al., 1983; Siegl et al., 1983; Thomson et al., 1984). However, tubular cell vacuolization was also described following treatment with the parenteral CsA vehicle (Chou et al., 1986; Kone et al., 1986), which could be dissociated from a reduction in GFR (Cunningham et al., 1985; Siegl et al., 1983; Thomson et al., 1981), especially in acute toxic injury.

Effects on the proximal tubule tend to be most prominent in the S_3 segment (Duncan et al., 1986), becoming more widespread at very high doses. However, functional defects in the proximal tubule are relatively slight, and some experimental studies have shown increased reabsorption of sodium and reduction of fractional excretion, probably as a result of compensatory mechanisms (Schwass et al., 1986). In micropuncture studies of rats treated with a low dose of CsA for five to seven days, electrolyte transport also remained normal in the loop of Henle as well as in the whole kidney (Muller-Suur and Davis, 1986). A hyperkalemic renal tubular acidosis is seen in the distal tubule. Hypomagnesemia occurs due to renal magnesium wasting.

With chronic administration of CsA in experimental animals, tubular changes are seen in both cortex and medulla. Rosen et al. (1990), in studies of rats treated for three to ten weeks with CsA (12.5 mg/kg/day), described striking morphological alterations predominantly in the medullary rays of the cortex. Changes included vacuolization in proximal tubular and thick ascending limb cells. These changes progressed with time, with decrease in medullary ray cellular mass. In the medulla, there was atrophy of medullary thick ascending limb (mTAL) cells and cast formation, most prominent in the inner stripe of the outer medulla. At ten weeks, large numbers of calcified casts were seen in occasional animals, especially in the deep outer stripe. Using electron microscopy, atrophic cells showed ultrastructural simplification with associated obliteration of tubular lumina. Histiocytes were observed infiltrating the walls of mTALs. Salt depletion exacerbated the injury, although uninephrectomized animals appeared to be protected from this combined effect. The investigators emphasized that these changes were localized and best seen on perfusion-fixed material on l-micron plastic sections of the kidney; similar injury may have been under-appreciated in other studies of chronic CsA-induced renal injury (Rosen et al., 1990).

Micropuncture studies in rats have demonstrated significantly higher electrolyte concentrations in thick ascending limb fluid in rats treated with acute infusion of CsA (5 mg/kg) or short-term CsA (15 mg/kg/day for five to seven days), suggesting a possible mTAL transport defect (Gnutzmann et al., 1986). Increased calcium accumulation in renal cells in CsA-treated rodents is most marked in the renal medulla (Borowitz, 1988). The morphological and functional alterations in the medulla and mTAL may be primary, contributing to reduction of GFR perhaps in part via tubuloglomerular feedback, or may be secondary to cortical alterations and/or hemodynamic changes with hypoperfusion.

In the intact animal or isolated perfused organ, it is difficult to separate direct tubular toxic effects from tubular effects due to hemodynamic changes. *In vitro* studies of tubular cells or segments provide an opportunity to study direct effects of the drug on tubular epithelium. Wilson and Hreniuk (1988) screened effects of CsA on cells from proximal convoluted tubule, proximal straight tubule, cortical thick ascending limb, and cortical collecting tubules. Time-dependent toxicity, measured by lactate dehydrogenase release and nigrosin uptake, is seen at doses of 100 ng/ml CsA and above; proximal tubular cells seemed most sensitive to the drug. Toxic effects appear calcium-dependent and mediated by nonlysosomal cysteine proteases.

Oxidants may play a role in tubular cell injury in CsA toxicity *in vitro*. Polte et al. (2002) showed that a cytoprotective effect of atrial natriuretic peptide in an LLC-PK_1 model appears to occur via heme oxygenase induction. CsA-induced renal lipid peroxidation can be demonstrated both *in vivo* and *in vitro*, and may be an early, important event in the pathogenesis of CsA nephrotoxicity. Antioxidants such as vitamin E and melatonin (Longoni et al., 2002) protect against CsA nephrotoxicity in experimental models.

Scoble et al. (1989) documented cellular uptake of CsA and found increasing cell vacuolization at low doses in LLC-PK_1 cells; at higher doses, they also noted inhibition of cell growth and of activity of the sodium-glucose cotransporter. CsA binds to a polypeptide component of the renal cotransporter, competing at the Na^+-dependent, high-affinity phlorizin binding site (Ziegler et al., 1990). Additional data suggested a CsA-induced alteration of transporter turnover. The authors hypothesized that CsA binds to a newly synthesized component of the transporter, inhibiting its assembly so that components, which accumulate in subapical vessels and lysosomes, build up in the cell. Moreover, in proximal straight tubules, the authors also observed vacuolization of endoplasmic reticulum (ER), and in these cells, monoclonal antibody to the cotransporter reacted with membranes of this dilated ER as well as Golgi vesicles. The authors suggest that these results might explain the concentration of CsA in renal tissue, as it binds to this polypeptide.

CsA also inhibits rat renal cortical mitochondrial respiration in a dose-dependent manner (Scoble et al., 1989). These results are comparable with those reported by Jung and Pergande (1985) in isolated rat kidney mitochondria, as well as in human kidney mitochondria (Jung et al., 1987). These studies suggest that CsA can cause impaired mitochondrial function, though only at very high concentrations. Strzelecki et al. (1988) reported that oxidation of succinate by rat kidney cortical mitochondria is markedly inhibited by toxic levels of CsA (25 to 50 nmol/mg protein). Mitochondria isolated from rats treated with CsA *in vivo* showed impaired adenosine diphosphate (ADP)-stimulated respiration at a toxic (75 mg/kg/day p.o.) but not an immunosuppressive (25 mg/kg/day p.o.) dose. Of note, Simon et al. (1997) showed that prednisolone and azathioprine worsen the decrease in oxidative phosphorylation of kidney mitochondria induced by CsA, emphasizing the potential importance of interactions among immunosuppressive agents in clinical regimens.

Elzinga et al. (1989) found only minor effects of CsA on respiration (and only pyruvate-malate supported respiration) of mitochondria isolated from these animals, and a slight depression of Ca^{2+} accumulation, despite severe depression in GFR and typical renal morphological changes, arguing against an important primary toxic effect of CsA on renal tubule cells. CsA is known to be a potent inhibitor of the inner membrane permeability transition in liver mitochondria induced by a variety of agents, with complete inhibition at 150 pmol CsA/mg protein. In liver, CsA did not inhibit Ca^{2+} uptake or mitochondrial phospholipase A, and results were consistent with CsA interaction with a membrane pore or pore regulator (Broekemeier et al., 1989). In model membrane bilayers, CsA reduces both temperature and maximum heat capacity of the dipalmitoyl phosphatidyl-choline, gel-to-liquid crystalline phase transition, and markedly reduces the amplitude of the pretransition to less than 20% compared with values in the pure lipid (O'Leary et al., 1986; Wiedmann et al., 1990). However, in an erythrocyte preparation, CsA did not cause erythrocyte hemolysis or inhibit hypotonic hemolysis, suggesting that the compound does not necessarily cause functional changes in cell membranes (O'Leary et al., 1986). The changes in mitochondrial function noted above at very high doses of the drug are probably due to alterations in mitochondrial membrane structure (Jackson et al., 1988).

Potential reduction in oxidative phosphorylation can lead to a shift to nonoxidative pathways. Indeed, *in vitro* exposure of LLC-PK_1 cells to CsA increases glucose consumption and lactate production, signaling a shift to glycolysis (Dominguez, 2002). Increased glucose consumption and glycolysis were accompanied by higher rates of GLUT1 gene transcription, and higher GLUT1 mRNA and protein. Interruption of glucose influx and glycolysis, in turn, magnified CsA cytoxicity, as measured by the percentage of lactate dehydrogenase release.

Some studies have suggested that CsA may interfere with normal protein synthesis and processing in renal cells. Buss et al. (1989) reported inhibition of renal microsomal protein chain elongation following *in vivo* CsA treatment. Cytoplasm of renal cells from CsA-treated rats inhibited microsomal protein synthesis in control microsomes. However, since these effects reach their nadir after renal dysfunction is established and recover well before renal functional recovery, alterations in renal microsomal protein synthesis are unlikely to be the direct cause of acute nephrotoxicity, although they may slow cellular recovery from injury and contribute to chronic tubular changes (Bennett et al., 1991). The discovery that the ubiquitous intracellular CsA-binding protein cyclophilin is identical to peptidyl-prolyl-*cis-trans*-isomerase has led to speculation that protein folding during synthesis, which is facilitated by this isomerase, may be perturbed by CsA (Fischer et al., 1989; Takahashi et al., 1989). These effects on protein conformation and function could be responsible for its immunosuppressive, and potentially its nephrotoxic, properties. Kidney androgen-regulated protein (KAP), the product of the most abundant gene expressed in mouse kidney proximal tubular cells, has recently been shown to specifically interact with cyclophilin B. Cebrian et al. (2001) showed that KAP levels are decreased in CsA-treated mice. Tetracycline-controlled overexpression of KAP in stably transfected proximal tubular cells significantly decreases toxic effects of CsA, and the authors suggest that this protein may represent a stress-response gene of relevance in this model of toxicity. Walker et al. (1989, 1990) showed that CsA significantly inhibits protein kinase C activity in renal cells *in vivo* and *in vitro*. Since this enzyme is involved in phosphorylation of cellular proteins and enzymes, these effects on the function of this enzyme, perhaps combined with a demonstrated inhibition of mixed-function oxidase complex in renal cells produced by the drug (Walker et al., 1990), could interfere with regulation of intracellular functions and cellular response to injury (Walker et al., 1989).

CsA has been shown to disrupt gene transcription (Morris et al., 1992). Effects of CsA on DNA and protein synthesis in renal epithelial cells may interfere with recovery from injury and development of compensatory hypertrophy and hyperplasia; concerns particularly in the transplant setting. Animal models of reduced renal mass are particularly useful for studying the effects of CsA on compensatory renal adaptation because renal changes in these models are not complicated by ischemia or rejection, which may supervene in the transplant setting.

Following *in vivo* unilateral nephrectomy, a number of investigators have shown suppression of compensatory changes in CsA-treated animals (e.g., Battle et al., 1990; Schurek et al., 1986). Whiting et al. (1985) demonstrated worsening of renal function and increased tubular morphological changes in uninephrectomized animals, despite tissue CsA levels similar to controls, suggesting that the kidney undergoing adaptive hemodynamic and growth changes may be more susceptible to CsA nephrotoxicity. Inhibition of compensatory renal growth by CsA was recently investigated *in vivo* with careful pair-feeding to control for effects of reduced oral intake. In these studies, Battle et al. (1990) found that kidneys of uninephrectomized animals treated for the three weeks following surgery with CsA increased in weight to the level of those in pair-fed controls, but renal cortical RNA and protein content were markedly reduced. The increase in weight in these studies was felt to be due to tissue edema. In long-term studies, Brunner et al. (1986) reported that a low dose of CsA begun several weeks after ablation was actually beneficial in preventing glomerular sclerosis and vasculopathy, perhaps due to a decreased blood pressure.

Lally et al. (1999) investigated the effect of CsA on the proliferation of LLC-PK_1 cells. By measuring ^{3}H-thymidine incorporation, they found that CsA significantly inhibited DNA synthesis in a time- and dose-dependent manner. DNA damage occurred as an early event, as inhibition of DNA synthesis could be observed prior to loss of cell viability (as measured by 3-(4,5-dimethylthiazol-2-yl)-2,5-diphenyltetrazolium incorporation or by neutral red uptake). Flow cytometric analysis revealed that low CsA doses (4.2 to 21 μM) arrested the cell cycle at the G_0/G_1 phase, while higher doses (21 to 42 μM) arrested the cell cycle at the G_2/M phase. Coincident with cell cycle arrest was elevation in p53 protein levels. The authors postulate that p53 transactivates p21 and thereby halts cell-cycle progression, preventing propagation of damaged DNA (Lally et al., 1999).

CsA appears to interfere with renal tubular Na^+/K^+-ATPase-dependent activity as well, which is logical considering that calcineurin regulates both baseline and receptor-activated Na^+/K^+-ATPase activity (Ader et al., 1998). As calcineurin is physically located within the proximal tubular, cortical collecting duct, and medullary thick ascending limb epithelial cells (Tumlin, 1997), calcineurin inhibitors affect widely separated nephron segments. Within the cortical collecting duct, CsA decreases Na^+/K^+-ATPase activity by 35% (Ader et al., 1998). Tubular calcification is commonly described in CsA-mediated tubular toxicity, particularly within the outer medulla (Andoh and Bennett, 1998). Elucidating the mechanism behind this phenomenon in rats, Aicher et al. (1997) showed that CsA treatment strongly reduces renal calbindin-D 28,000 da protein levels, increasing urinary calcium excretion and promoting intratubular calcifications.

Studies have shown that exogenous vascular endothelial growth factor (VEGF) protects against CsA renal toxicity. While some of this effect could be related to vascular effects (see below), recent studies evaluating the role of endogenous VEGF demonstrate that blockade of VEGF intensifies tubular injury and apoptosis *in vivo*. This effect is associated with strikingly increased tubular VEGF and Bcl-xL proteins. *In vitro*, Western blotting reveals autocrine production of VEGF by mouse tubular cells, with significant increase in CsA toxicity with VEGF blocking antibody (Alvarez-Arroyo et al., 2002a).

Overt tubular cell injury, including frank necrosis, with CsA treatment is more commonly seen in the ischemically injured kidney. The combination of mild ischemia and CsA treatment has been used by a number of investigators as a strategy for developing experimental models of CsA-induced renal injury (e.g., Bia and Tyler, 1987; Jablonski et al., 1986; Kone et al., 1986; Kanzai et al., 1986). These models have particular relevance to renal transplantation, where some level of ischemic injury is probably present in all transplanted organs, especially from cadaveric donors. The tendency of CsA to exacerbate ischemic injury contributes to the prolonged primary nonfunction as well as the more severe and prolonged "acute tubular necrosis" seen in transplanted kidneys (see below). Even a low dose of parenteral CsA (5 mg/kg) can exacerbate renal ischemic injury when given after, although not before, the ischemic insult (Bia and Tyler, 1987).

It is not clear to what extent this interaction of CsA and ischemia is due to direct effects on renal tubular cells with enhanced mitochondrial dysfunction, substrate depletion, and interference with cell repair mechanisms, or to hemodynamic effects. Walker et al. (1990) showed that CsA reduces renal content of glutathione and produces a significant increase in lipid peroxidation in renal cortical membrane fractions *in vitro*. In this setting, hypoxia and no-flow or reflow could lead to increased oxidant injury. As noted above, cellular repair in response to injury could be impaired by CsA inhibition of intracellular protein and enzyme synthesis and function. Alternatively, or perhaps additionally, it has been suggested that the vasoconstriction induced by CsA could contribute to renal ischemia, increasing the chance of developing acute tubular cell injury.

Renal Vasculature and Hemodynamics

Effects of CsA on the renal vasculature are well established. Hemodynamic effects are probably responsible for, or contribute to, all of the various clinical syndromes recognized as CsA nephrotoxicity, including acute toxicity, chronic nephropathy, and hypertension (reviewed by Campistol and Sacks, 2000, and Rodicio, 2000). Experimental models have proven very useful in defining vascular effects of CsA and in defining the mechanisms producing these effects. Decreased blood flow can be demonstrated in many organs with CsA administration, including heart, lung, liver, spleen, and kidney (McKenzie et al., 1986). Many investigators have documented acute and chronic reductions in renal blood flow and increase in renal vascular resistance in experimental animals *in vivo* in response to CsA. Marked renal vasoconstriction is also seen in the isolated perfused kidney when CsA is added to the perfusate (Besarab et al., 1987). Gillum et al. (1989) showed that partial ablation of renal mass and high-protein feeding partially protected against the fall in GFR and RPF, and the morphological changes induced by CsA given over 28 days, presumably by inducing relative renal vasodilation.

Micropuncture studies of glomerular hemodynamics following acute infusion of CsA have revealed increases in both afferent and efferent arteriolar resistance, and proportionate reductions in glomerular plasma flow (GPF) and whole-kidney and single-nephron GFR. Because there was a greater increment in efferent arteriolar resistance, filtration fraction and mean glomerular capillary hydraulic pressure increased in these studies. Despite a 33% increase in hydrostatic pressure gradient (ΔP), however, single-nephron GFR (SNGFR) fell, in part because glomerular K_f was decreased by approximately 70% (Besarab et al., 1987). Glomerular hemodynamics have also been studied after "short-term chronic" treatment with CsA (30 mg/kg s.c. for eight days). Thomson et al. (1989) found reduction in SNGFR due to decreased in single-nephron plasma flow (SNPF) and ΔP, but with no decrease in glomerular membrane hydraulic permeablity. Afferent arteriolar resistance increased disproportionately to efferent resistance in CsA-treated animals, contributing to the lower glomerular capillary hydrostatic pressures. Discrepancies between this study and that of Barros et al. (1987) described above are almost surely due to different dose and duration of treatment.

Schor et al. (1981) reported findings similar to those of Thomson et al. (1989), confirming that glomerular hemodynamic patterns after acute and chronic CsA are different. Using vascular casting, English et al. (1987) provided an elegant morphological demonstration of progressive afferent arteriolar vasoconstriction in perfused kidneys removed from rats treated with CsA. Winston et al. (1989) studied glomerular hemodynamics during compensatory renal growth in a uninephrectomy model in rats. They found that reductions in whole kidney and SNGFR in CsA-treated animals in this model were due solely to reduction in K_f with no evidence of renal vasoconstriction.

Using isolated glomeruli and acute exposure to high concentrations of CsA, Wiegmann et al. (1990) documented a dose-dependent decrease in glomerular volume, consistent with mesangial contraction, and highly significant decreases in K_f and glomerular hydraulic conductivity. Potier et al. (1998) compared the contractile effects of CsA and cyclosporine G in isolated Sprague-Dawley rat glomeruli and cultured mesangial cells using image analysis to enable accurate and precise morphometry. The authors found that at noncytotoxic doses, CsA and cyclosporine G resulted in a similar degree of mesangial cell contraction, while CsA caused significantly more contraction in isolated rat glomeruli. They propose that such effects might exemplify why CsA causes a greater compromise in GFR *in vivo*, leading to more severe nephrotoxicity.

Zimmerhackl et al. (1990) were able to assess effects of CsA on the entire renal vasculature by using a hydronephrotic rat kidney model. The kidneys of these rats lack tubular structures, enabling visualization of vascular structures and removing tubular injury as an experimental variable. Using television microscopy, they observed that acute infusion of CsA caused severe constriction of arcuate arteries in CsA-treated rats. Application of CsA directly to the kidney surface did not produce vasoconstriction, suggesting that this was not a direct effect of CsA but was mediated via other mechanisms.

What are the mechanisms of these altered hemodynamics? Depletion of effective circulating volume may contribute to the hemodynamic effects of CsA, at least in experimental models. Gerkens et al. (1984) documented an effect of salt intake on CsA induced renal functional impairment in rats, suggesting a possible prerenal component, and other investigators have suggested that decreased circulatory volume may play a role in reduction of renal function in CsA-treated animals (e.g., Kaskel et al., 1987; Ryffel et al., 1986). Devarajan et al. (1989) found that acute volume expansion (10% body weight as saline) at the end of a low dose CsA treatment period in young rats completely normalized renal blood flow (RBF) and GFR, and largely restored renal autoregulatory ability. Chronic volume expansion (10% body weight /day saline i.p.) beginning three days before and extending through CsA treatment, partially prevented the fall in GFR and RBF and the increase in renal vascular resistance produced by CsA, as well as improving autoregulation. However, in hemodynamic studies in rats treated with CsA, Thomson et al. (1989) found that plasma volume expansion caused SNGFR, ΔP, and SNPF to rise in both CsA and pair-fed vehicle-treated controls, but without eliminating differences in these parameters between animal groups. Therefore, despite a potential role for prerenal mechanisms in CsA-induced alterations in renal blood flow and function, there are clearly additional direct renal effects of CsA on determinants of renal hemodynamics and GFR.

Sympathetic nervous system activation is another important mechanism of CsA-mediated renal hypertension and alteration in renal hemodynamics (Cavarape et al., 1998; Churchill et al., 1993; Ryuzaki et al., 1997; Porter et al., 1999; Rodicio, 2000). Adrenergic nerves were initially implicated in CsA toxic effect by studies of renal denervation and use of adrenergic antagoinists (Moss et al., 1985; Murray et al., 1985; Xue et al., 1987). This may occur due to induction of norepinephrine release from nerve terminals or by potentiation of norepinephrine release via calcium influx (Cavarape et al., 1998). Nevertheless, catecholamine levels in transplant patients are not elevated (Rodicio 2000). One would expect that transplanted kidneys, which are denervated, would escape such sympathetic stimulation. However, Churchill et al. (1993) showed that Lewis rats with one native and one transplanted kidney each had no difference in function between the native and the transplanted kidney (in terms of GFR and RPF) after two to four weeks of CsA therapy. Therefore, if sympathetic stimulation is involved in hypertensive renal effects of CsA, denervated transplanted kidneys are not spared (Churchill et al., 1993). Indeed, renal allografts are significantly reinnervated within a few months of transplant (Gazdar and Dammin, 1970).

The renin–angiotensin system has also been implicated in the changes in renal hemodynamics and function in CsA-treated animals (reviewed in Lassila, 2002; Lee, 1997). Indeed, it is likely that the hemodynamic effects of renal nerve stimulation are mediated, at least in part, by angiotensin II (Blantz et al., 1976; Kopp et al., 1981; Tucker et al., 1987). In experimental animals, plasma renin rises in response to acute CsA infusion (Caterson et al., 1986; Murray et al., 1985), and may also rise with chronic administration (Murray et al., 1985; Perico et al., 1986a, 1986b); the latter effect, however, may be due to volume depletion in the CsA-treated animals. In humans, peripheral renin is reported to be unchanged or depressed (e.g., Bantle et al., 1985; Myers et al., 1984), although if total renin rather than active renin is measured, total renin is increased due to increased circulating prorenin, the activation of which appears to be depressed (Myers et al., 1988a, 1988b).

Renin release is increased on *in vitro* incubation with CsA in renal cortical slices (Duggin et al., 1986) and in primary cultures of isolated juxtaglomerular cells (Kurtz et al., 1988). Renin synthesis and release in JGA cells both appear to be elevated, with a later increase in prorenin levels, suggesting inhibition of conversion or release. A number of investigators have noted hyperplasia of the juxtaglomerular apparatus (JGA) in experimental animals during CsA therapy (e.g., Gillum et al., 1988; Nitta et al., 1987). However, this is an inconsistent finding in humans. Angiotensin-converting enzyme inhibitors abolish the renal vasoconstriction or hypertenison and reduction of kidney function produced by CsA (e.g., Barros et al., 1987; Mervaala and Lassila, 2002). Conversely, Murray et al. (1985) and Kaskel et al. (1987) found no effect of captopril pretreatment of renal dysfunction induced by short-term treatment with CsA. Interaction of CsA and renin–angiotensin system has been recently reviewed (Lassila, 2002).

The loss of renal autoregulatory ability in CsA-treated animals appears to be due to a direct effect on preglomerular myogenic mechanisms rather than to impairment of tubuloglomerular feedback (Kaskel et al., 1989). Although systemic arterial pressure is unchanged or reduced in experimental animals following CsA administration, particularly after high doses, hypertension is frequently seen clinically in patients receiving the drug. The only animal model in which CsA produces hypertension is in the spontaneously hypertensive rat; in these animals, CsA accelerates the development of hypertension, with evidence of direct arteriolar injury (Ryffel et al., 1983). This model may be relevant to the arteriolopathy seen in some patients treated with CsA, but is probably not completely relevant to the hypertension developing in patients without this lesion. Some experimental studies by Golub et al. (1989), however, showed an altered vascular response of systemic vessels in CsA treated spontaneously hypertensive rats. Following CsA, systemic blood pressure and *in vitro* tail artery contractile response to nerve stimulation and, in the high-dose group, to norepinephrine, increased. Xue et al. (1987) reported a slow contraction of rings of isolated rat aorta on exposure to CsA (5×10^{-6}M) an effect significantly inhibited by verapamil as well as phenoxybenzamine. In studies by Muller-Schweinitzer et al. (1988, 1989), CsA appeared to activate renin or a renin-like substance in peripheral and renal veins and arteries, increasing local angiotensin formation, which may contribute to CsA-induced hypertension.

Antidiuretic hormone has also been implicated in the acute glomerular functional impairment produced by CsA. Barros et al. (1987), in studies using acute infusion of CsA, found that GFR and

GPF decreased significantly less in Brattleboro rats (lacking endogenous antidiuretic hormone [ADH]) than in Munich-Wistar rats, although there was still a significant effect of CsA. ADH is known to produce renal vasoconstriction and to reduce K_f, presumably by mesangial cell contraction (e.g., Schor et al., 1981), both effects seen in CsA-treated rats. Kremer et al. (1989) used rat glomerular mesangial cells in primary culture to assess CsA effects on an ADH-induced rise in intracellular Ca^{2+} and stimulation of prostaglandin production. They found that 1 to 10 pg CsA/ml for up to 90 minutes enhanced the rise in intracellular Ca^{2+} and inhibited both basal and ADH-stimulated production of the vasodilatory prostaglandin PGE_2. In cultured rat mesangial cells *in vitro*, preincubation with CsA significantly increased the AVP-induced rise in intracellular Ca^{2+} and the percentage of contracting mesangial cells (as assessed by digital imaging analysis). Pretreatment with CsA appeared to increase Ca^{2+} uptake into ADH-sensitive stores, such as the ER; verapamil, a calcium-channel blocker, had no effect on this CsA-induced uptake (Meyer-Lehnert et al., 1988).

Platelet activating factor (PAF) causes vasoconstriction of both pre- and postglomerular arterioles and stimulates mesangial cell contraction (Rodicio, 2000). Cultured rat mesangial cells and isolated glomeruli have also been used to examine the role of PAF in producing mesangial cell contraction. Rodriguez-Puyol et al. (1989) found that PAF antagonists completely inhibited CsA-induced decrease in planar surface area and inhibited decrease in whole-glomerular cross-sectional area. In contrast, verapamil was only partially effective in inhibiting effects of CsA. Additionally, CsA increased glomerular PAF production more than twofold, an effect not inhibited by verapamil. Dos Santos et al. (1989) studied effects of a PAF antagonist *in vivo* using micropuncture techniques, and found that pretreatment of rats with the antagonist prevented renal vasoconstriction and the fall in SNGFR and glomerular K_f following acute infusion of CsA; total GFR was unaffected, suggesting significant effects of the PAF antagonist only in superficial nephrons. Another PAF antagonist has been reported to lessen the interstitial fibrosis developing in rats receiving chronic CsA (Pirotzsky et al., 1988).

The role of the vasoregulatory prostaglandins in CsA nephrotoxicity has been the subject of intense investigation, with results suggesting an important role for an imbalance of vasodilatory and vasoconstrictor prostaglandins in the pathogenesis of CsA-induced renal dysfunction. As noted above, Kremer et al. (1989) found that an ADH-stimulated increase in PGE_2 production was inhibited by CsA in mesangial cell cultures. Stahl and Kudelka (1986) reported reduction of (basal) PGE_2 formation in isolated glomeruli and papillae of rat kidneys with chronic treatment with oral CsA. Stahl et al. (1989) also found inhibition of angiotensin II- or calcium ionophore-induced PGE_2 formation in cultured rat mesangial cells, at CsA concentrations as low as 800 ng/ml; this effect appeared to be due to inhibition of release of arachidonic acid substrate from cell membranes rather than to inhibition of cyclooxygenase. Decreased synthesis of PGE_2 and $PGF_1\alpha$ was also reported in isolated glomeruli from CsA-treated rats (Bunke et al., 1988). Exogenous PGE_2 (or a variant) reduced or prevented CsA-induced nephrotoxicity in rat models (Makowka et al., 1986; Ryffel et al., 1986).

Other investigators, however, found increases in vasodilatory prostaglandins, presumably as a compensatory mechanism, with CsA treatment. Murray et al. (1985) reported that CsA, given acutely or on a more chronic regimen *in vivo*, increased urinary excretion of 6-keto-$PGF_1\alpha$, and that cyclooxygenase inhibitors exacerbated the CsA-induced decrease in RBF and increase in renal vascular resistance (RVR). A similar increase in 6-keto-$PGF_1\alpha$ was reported with acute CsA administration to conscious rabbits (Caterson et al., 1986). In a chronic postischemic, denervated rat model of CsA-induced nephrotoxicity, Coffman et al. (1987) found increased urinary excretion of 6-keto-$PGF_1\alpha$ and increased tissue production of PGE_2 as well as 6-keto-$PGF_1\alpha$ in kidneys, either perfused *in vitro* or removed from CsA-intoxicated animals.

Discrepancies in the results of analysis of vasodilatory prostaglandins in these various models are due in part to differences in dose and duration of CsA treatment. Since there are a number of potential sites of prostaglandin synthesis, including endothelial and smooth muscle cells, glomerular cells, tissue macrophages and lymphocytes, monocytes, and platelets, results with isolated glomeruli or in cultured cells may be different than with renal slices or the intact animal. Moreover, urinary excretion of prostaglandins may not necessarily reflect localized tissue levels.

In rat models, CsA causes a significant increase in urinary thromboxane B_2 (TXB_2), which is derived from thromboxane A2 (TXA_2), with ensuing decline in renal function (Campistol and Sacks, 2000; Porter et al., 1999). The primary mechanism is afferent arteriolar vasoconstruction (Hayashi et al., 1997). Coffman et al. (1987), besides finding increased production of vasodilatory prostaglandin with CsA treatment, also reported increased production and excretion of thromboxane (TX) in their model. Elzinga et al. (1987) looked at renal cortical TXB_2 content in studies on the modification of CsA toxicity by the use of fish oil rather than olive oil as the vehicle for the drug. They found significant preservation of GFR and reduction in morphological changes in tubular cells of animals given CsA in fish oil. These results were associated with significantly reduced levels of cortical TXB_2, presumably due to the omega-3 fatty acids in fish oil that are known to inhibit cyclooxygenase. These investigators also reported selective enhancement of TX in kidneys and in peritoneal macrophages in CsA-treated rats; modest reductions in PGE_2 and prostacyclin were also seen. These effects were suppressed by a diet rich in fish oil, and the decreased GFR and tubular vacuolization seen in these animals were also decreased on this regimen (Rogers et al., 1988). In clinical studies, van der Heide et al. (1990) found that three months of dietary fish oil supplementation improved GFR and reduced renal vascular resistance in patients with stable renal function on a stable dose of CsA for at least three months.

Perico et al. (1986a) specifically assessed the functional significance of the increased synthesis of TXA_2 induced by CsA. Using a selective TX-synthetase inhibitor, they found a marked reduction in urinary TXB_2 and a significant increase in GFR in CsA-treated rats (40 mg/kg q.o.d. for six weeks). No effects on RPF were seen, leading to the conclusion that TXA_2 may be reducing GFR by altering K_f, presumably via mesangial cell contraction. Perico et al. (1991) also reported that a TXA_2 receptor antagonist partially prevented the decline in GFR and RPF induced by acute CsA administration (50 mg/kg i.v.). In the same model, an antagonist to the receptor for LTC_4/D_4, sulfidopeptide leukotrienes, shown to decrease GFR and increase RVR *in vivo* and *in vitro*, also partially prevented CsA-induced hemodynamic effects. A combination of the TXA_2 and LTC_4/D_4 receptor antagonists completely abolished acute CsA-induced effects in rats.

Effects of vasoconstrictor compounds such as TX on renal hemodynamics may be mediated via substances released from endothelial cells. In particular, the most potent vasoconstrictor substance yet identified, endothelin, is a peptide released from endothelial cells. Endothelin has powerful systemic and renal vasoconstrictor effects (Fukuda et al., 1988; Hirata et al., 1988) and has been implicated in postischemic renal vasoconstriction. Badr et al. (1989) characterized renal effects of endothelin; these include increases in afferent and efferent arteriolar resistances, decrease in SNGFR, and a decrease in K_f, changes similar to those produced by CsA. Renal transplant patients have an increase in plasma levels of endothelin-1 following CsA administration, probably from injured endothelial cells (Campistol and Sacks, 2000). Kon et al. (1990) examined the role of endothelin in CsA-induced glomerular dysfunction. Using continuous infusion of rabbit antiporcine endothelin into a first-order branch of the main renal artery of Munich-Wistar rats, they then performed micropuncture on glomeruli infused vs. not infused with antiendothelin in the same kidney. CsA reduced SNGFR and glomerular plasma flow rate associated with increased afferent arteriolar resistance and decreased glomerular capillary pressure; these changes were markedly attenuated by the antiendothelin antibody, results confirmed by others (reviewed in Rodicio, 2000, and Kohan et al., 1997). In additional animals, Kon et al. (1990) demonstrated marked elevations of endothelin with CsA treatment.

Perico et al. (1990) also found that endothelin appears to mediate CsA-induced renal vasoconstriction in the rat. Awazu et al. (1991), using isolated glomerular membranes, showed that CsA promotes glomerular endothelin binding *in vivo*. CsA also diminishes acetylcholine-mediated vasodilatation in microvessels of rat kidneys, perhaps by decreasing NO synthesis (Takenaka et al., 1992). Endothelin A receptors mediate vasoconstriction, while endothelin B receptors are responsible for clearance of endothelin. CsA-induced vasocontrictive effects on intrarenal artreries and arterioles can be ameliorated by endothelin receptor A but not endothelin B receptor antagonists (Cavarape et al., 1998). CsA increases the number of endothelin A receptors and decreases the number of endothelin B receptors (Rodicio 2000).

Altered release of vasoactive substances from endothelial cells is probably due to endothelial injury. There is evidence that CsA may induce endothelial and vascular injury in experimental models, analogous to the CsA-induced clinical arteriolopathy reported in occasional renal transplant patients and the HUS-like syndrome in rare patients (see above). Neild et al. (1983) described a Schwartzman-like reaction in CsA-treated rabbits with serum sickness; these investigators later demonstrated depression of prostacyclin synthesis *in vitro* by the vasculature of CsA-treated rabbits, apparently due to decreased levels of PGI_2-stimulating factor. CsA also exacerbates development of arteriolopathy in the spontaneously hypertensive rat (Ryffel et al., 1983). Faraco et al. (1991) infused lipopolysaccharide (LPS) from *Escherischia coli* into rabbits pretreated for 10 days with parenteral CsA or its vehicle. Kidneys were harvested after 5 hr infusion, and 84.8% of glomeruli in CsA pretreated animals contained fibrin deposits compared with 5.4% of glomeruli in vehicle pretreated rats, and none in vehicle or CsA pretreated rats not given LPS. By light and electron microscopy, lesions in glomeruli in CsA pretreated animals resembled the HUS-like lesions described clinically. Because both LPS and CsA stimulate procoagulant activity in endothelial cells, the authors suggest that these agents, and potentially other inflammatory mediators induced by LPS may interact in vessels and glomeruli in the clinical setting as well as in this experimental model.

Using cultured endothelial cells from bovine aorta, Zoja et al. (1986) demonstrated that exposure to CsA produced time- and dose-dependent cell injury with early detachment followed by cell lysis; they also demonstrated time- and dose-dependent increase in prostacyclin and TXA_2. Lau et al. (1989) also described CsA-induced changes in endothelial cells of the microvasculature of the rat. Cell replication in culture was depressed significantly at 250 and 1,000 ng CsA/ml, but with no evidence of increased cell death. There were morphological changes in the cells, including nucleolar changes and formation of cytoplasmic vesicles, and prostacyclin release was increased as well.

CsA may have additional vascular effects in vessels with significant endothelial injury, with implications for the development of chronic vascular insufficiency. Ferns et al. (1990) showed that in rabbits with deendothelialized carotid arteries, treatment with CsA was associated with an increase in total intimal thickening in these vessels due to numerous foam cells, incorporation of ^{3}H-thymidine by neointimal monocytes/macrophages and intimal smooth muscle vacuolization. The authors suggested that CsA may contribute to graft-related atherosclerosis. The renal allograft, especially from a cadaveric donor, is very likely to sustain endothelial injury due to a variety of factors including altered hemodynamics and/or vascular rejection. CsA could certainly contribute to vascular pathology in such a setting.

Endothelin and a number of other vasoconstrictor hormones act at least in part via elevated levels of intracellular Ca^{2+}, and there is some experimental evidence that Ca^{2+} may mediate renal effects of CsA. Dieperink et al. (1986) found that concomitant treatment of CsA-treated rats with nifedipine improved inulin clearance, RBF, and renal arterial pressure, and improved proximal and distal tubular reabsorption;

acute infusion of the calcium channel blocker after the 13 day protocol did not reverse CsA-induced changes. Barros et al. (1987), in studies of glomerular hemodynamics with acute CsA infusion, found that verapamil significantly decreased the effects of CsA on GFR and RPF. In studies in the hydronephrotic rat kidney, the vasoconstrictor effect of CsA on the arcuate arteries was partially reversed by the calcium channel blocker nitrendipine (Zimmerhackl et al., 1990). As noted above, experimental studies have shown increased Ca^{2+} mobilization in vasopressin-treated glomerular mesangial cells exposed to CsA (Kremer et al., 1989), an effect not blocked by verapamil. In contrast, Scoble et al. (1989) reported that verapamil *potentiated* the increased CsA toxicity produced by increasing Ca^{2+} concentrations in the medium bathing LLC-PK_1 cells. CsA uptake (at 2 μg/ml in the medium) could be blocked by verapamil, but only at very high toxic doses of the calcium channel blocker. Therefore, it is likely that the functionally significant alterations in cellular calcium induced by CsA and modulation of toxicity by calcium channel blockers takes place at the level of renal vessels and glomeruli rather than at the renal tubule, although Wilson and Hreniuk (1988) found that toxic effects of CsA on cultured tubular cells were Ca^{2+}-dependent.

The relative importance of cyclophilin- vs. calcineurin-mediated mechanisms in the effect of CsA on endothelial cells has been investigated. In endothelial cell cultures, CsA was cytotoxic or proapoptotic at high concentrations, and cytoprotective or antiapoptotic at low concentrations. CsA analogs that bind to cyclophilin closely reproduced the effects of CsA. The actions of these compounds were shifted from a protective to a cellular-injury pattern in the presence of a specific anti-VEGF monoclonal antibody, consistent with a role for VEGF in CsA-induced cytoprotection. An antibody to VEGF receptor 2 also abolished the cytoprotective effect of low-dose CsA (Alvarez-Arroyo et al., 2002b).

In summary, the acute or subacute effects of CsA on renal hemodynamics may be mediated by a number of factors, including an imbalance in vasoconstrictor and vasodilator prostaglandins. The role of the renin–angiotensin system in CsA toxicity is complex. Vasoconstrictor effects may well be mediated by endothelin or PAF release (or both) and at the cellular level by altered intracellular calcium and CsA interaction with cyclophilin. Potential protective mechanisms include actions of VEGF at the endothelium. Prerenal factors, renal nerves, and interactions among these various factors may also play a role.

The relevance of these mechanisms to chronic CsA-induced nephrotoxicity as seen in humans and in some experimental models remains to be established. The interstitial fibrosis seen in chronic nephrotoxicity, which may at least in part be due to vascular insufficiency, is discussed below.

Chronic CsA Nephrotoxicity and the Renal Interstitium

A number of investigators have developed models of long-term CsA administration, usually given over several months, with progressive deterioration in renal function and the histologic pattern of striped interstitial fibrosis described in patients with evidence of chronic toxicity (Bertani et al., 1987; Gillum et al., 1988; Rosen et al., 1990). The sodium-depletion model, in particular, has been very useful in defining mechanisms of chronic CsA nephrotoxicity. Salt-depletion in rats (Porter et al., 1999) or mice (Khanna et al., 1997) provide reproducible animal models in which chronic effects of CsA resemble those seen in humans.

At high doses of CsA, focal areas of tubular atrophy and interstitial fibrosis may be seen with less than two weeks of treatment. Jackson et al. (1987) noted these changes, as well as an increase in interstitial cells and reduction of RBF and GFR, within four to ten days on CsA doses of 100 mg/kg/day s.c. in rats with low-sodium diets. There was increased incorporation of ^{3}H-thymidine into DNA in inner and outer cortex and inner and outer medulla. Using histoautoradiography and morphometric analysis, they demonstrated that interstitial cell proliferation accounted for most of the increased labeling. Rosen et al. (1990), using a rat model and lower doses of CsA (12.5 mg/kg for three to ten weeks), emphasized thick ascending limb cell atrophy with fibroblast proliferation and collagen formation in the inner stripe of the outer medulla and the medullary rays, with S_2 to S_3 segment degenerative changes as well. Morphological changes predicted renal failure, though a causal effect was not established. The authors emphasized that injury occurred in areas of limited oxygen availability and was aggravated by salt depletion, suggesting a role for chronic hypoperfusion. The actual extracellular matrix proteins deposited in rat models of chronic CsA toxicity include biglycan, decorin, type I collagen (Ader and Rostaing, 1998) and fibronectin (Porter et al., 1999).

The interstitial fibrosis seen with chronic treatment may be related to a direct stimulatory effect of CsA. Wolf et al. (1990) showed that

CsA stimulates transcription and procollagen secretion in cultured tubulointerstitial fibroblasts and proximal tubular cells. Similarly, Nast et al. (1991), in studies of rats treated with 25 mg CsA/kg/day s.c. for one or four weeks, found elevated cortical procollagen-a_1 (I) mRNA levels at one week, before morphological evidence of the focal cortical tubular atrophy and interstitial fibrosis, which had developed by four weeks in these animals. Procollagen-a_1 (Ill) and a_1 (IV) and β-actin mRNA levels were not increased. Of possible relevance to the fibrosis seen in the kidney, CsA increases connective tissue in nonrenal tissue, notably gingival tissue (Pisanty et al., 1990; Thompson et al., 1991).

Yang et al. (2001) demonstrated stimulation of the intrarenal renin–angiotensin system in the salt-depletion model. Angiotensin II secretion, stimulated by CsA, could certainly be expected to induce renal parenchymal ischemia due to afferent arteriolar constriction. However, its induction of TGFβ1 expression appears to be much more influential in causing the interstitial fibrosis and tubular atrophy characteristic of chronic CsA nephrotoxicity (Campistol and Sacks, 2000; Ruiz-Ortega and Egido, 1997; Shihab et al., 1996). Investigators confirmed this by ameliorating the experimental fibrogenic effects of CsA with angiotensin converting enzyme (ACE) inhibitors (Shihab et al., 1996, 1997b; Johnson, 2002). Sodium-depleted mice treated with CsA produce increased levels of renal TGFβ1 mRNA (Khanna et al., 1997; Porter et al., 1999). In this model, after CsA therapy for one week, typical morphologic changes of chronic CsA toxicity, such as striped tubulointerstitial fibrosis and arteriolar hyalinosis, accompany the increased levels of TGFβ1 mRNA (Khanna et al., 1997). This apparently occurs via activation of the angiotensin II type 1 (AT1) receptor (Porter et al., 1999; Ruiz-Ortega and Egido, 1997).

In vivo studies by Shihab et al. (2002b) demonstrated marked amelioration of CsA-induced fibrosis with administration of the antifibrotic compound pirfenidone. This agent markedly reduced the increase in TGFβs, PAI-1, and biglycan mRNA in this model, and markedly reduced TGFβ1 protein expression.

In addition to TGFβ1, other stimulators of extracellular matrix protein production upregulated in experimental models of chronic CsA toxicity include insulin-like growth factor-I (Johnson, 2002) and plasminogen activator inhibitor type I (PAI-1) (Ader and Rostaing, 1998; Campistol and Sacks, 2000; Porter et al., 1999; Shihab et al., 1996, 1997a, 1997b). Shihab et al. (1997a, 1997b) demonstrated decreased expression of TGFβ1 and PAI-1, measured both by Northern blot and ELISA, with angiotensin II blockade produced by losartan or enalapril administration. Johnson et al. (1999) used primary cultures of human proximal tubular cells and renal cortical fibroblasts to function as an *in vitro* model of the renal tubulointerstitium. With this model, they were able to demonstrate that clinically relevant CsA concentrations promote fibrogenesis by a combination of matrix metalloproteinase activity suppression and increased fibroblast collagen synthesis. CsA-stimulated the cultured fibroblasts to secrete insulin-like growth factor-I via an autocrine mechanism, and the cultured proximal tubular cells to secrete TGFβ1 by a paracrine mechanism, resulting in the observed interstitial fibrosis and tubular atrophy. In a rat low-salt model, correction of hypomagnesemia, which is induced clinically and experimentally by CsA, almost completely abolished CsA-induced fibrosis, with abrogation of increases in TGFβ mRNA, PAI-1, matrix proteins, and matrix metalloproteinase-1 (Miura et al., 2002).

VEGF expression is upregulated in the salt-depletion model of chronic CsA toxicity in rats (Shihab et al., 2001). Shihab et al. (2002a) demonstrated that angiotensin II receptor blockade with enalapril or losartan ameliorated toxicity, with significant reduction in VEGF mRNA and protein expression. VEGF receptor (KDR/Flk-1) mRNA expression was lower with angiotensin II blockade. The authors speculated that the actions of VEGF in this model might relate to its effects on macrophage infiltration or matrix deposition.

Osteopontin may also play a role in the early phase of fibrogenic events. Osteopontin is an adhesion molecule and macrophage chemoattractant. Increased osteopontin expression has been demonstrated in the salt-depletion model of chronic CsA toxicity, associated with increased macrophage infiltration (Pichle et al., 1995). Young et al. (1995) showed that tubular and interstitial cell proliferation and macrophage influx precede interstitial fibrosis, and correlate with decreased creatinine clearance and decreased medullary concentrating ability. Macrophages attracted to sites of tubular injury join tubular cells in secreting TGFβ1 (Campistol and Sacks, 2000). Angiotensin II may be responsible for osteopontin overproduction in addition to inflammatory cells, which normally secrete it at sites of injury (Lee, 1997). Osteopontin-null mice have less arteriolopathy, cortical macrophage infiltration, and interstitial collagen deposition, with less cortical interstitial cell proliferation; these results are consistent with osteopontin as a partial mediator of these early CsA-induced effects (Mazzali et al., 2002).

Activation of angiotensin II and decreased NO production, in addition to hemodynamic effects, increase the expression of proapoptotic genes

(Thomas et al., 1998). Accelerated apoptosis, in turn, characterizes the interstitial fibrosis in a rat model of chronic CsA toxicity (Thomas et al., 1998), with increased apoptosis of tubular and interstitial cells documented, with apoptotic cells documented in macrophages. Apoptosis was significantly decreased in animals treated with losartan, while apoptosis was significantly increased when animals were treated with *N*-nitro-L-arginine-methyl ester (L-NAME) to block NO. There was significant reduction in interstitial fibrosis when L-arginine was added to increase NO. There was a significant correlation between apoptosis and interstitial fibrosis. Yang et al. (2001) demonstrated in low-salt rats that angiotensin II blockade protects epidermal growth factor expression, which fell in CsA-treated rats without losartan; endothelial growth factor expression was well correlated with interstitial fibrosis score and number of apoptotic (TUNEL-positive) cells.

The expression of apoptosis-related genes has recently been investigated in a low-salt mouse model (Yang et al., 2002b) in which CsA administration increases apoptosis; the number of apoptotic cells correlated well with interstitial fibrosis. There were significant increases in Fas-ligand mRNA, Fas protein expression, ICE mRNA and CPP32 mRNA, and in the enzymatic activity of ICE and CPP32 protease. The ratio between bax and bcl-2 protein and levels of p53 protein increased as well in the CsA group. Immunohistochemistry revealed strong expression of Fas, Fas-ligand, ICE, and CPP32 in renal tubular cells in areas of structural injury. In salt-depleted rats, colchicine also decreases CsA-induced apoptotic cell death associated with improved creatinine clearance and decreased tubulointerstitial fibrosis (Li et al., 2002).

Tacrolimus

Experimental animal models of tacrolimus-induced renal structural and functional impairment have unequivocally proven its nephrotoxicity, both acute and chronic, by several different mechanisms. Some studies suggest that tacrolimus (FK506) is less nephrotoxic than CsA. For example, in studies in a Lewis rat renal ischemia model, Nalesnik et al. (1990) reported less nephrotoxicity with FK506 than with CsA.

Kumano et al. (1990) assessed functional and morphological changes induced by FK506 in the kidneys of heminephrectomized rats. FK506 was given at 1, 2.5, or 5 mg/kg by gavage for 21 days. Creatinine clearances were reduced in a dose-dependent manner (10 to 25%

decrease). There was no change in FeNa, but FeK decreased. Inulin and *p*-aminohippuric acid (PAH) clearances at 21 days revealed decreased GFR and RPF, with an increase in RVR. Histology revealed vacuolization in proximal tubules and arteriolar smooth muscle, as well as eosinophilic "inclusions" in the media of vessels and in tubules. Tubular atrophy and focal interstitial fibrosis were also seen. Daily doses of the calcium-channel blocker diltiazem improved renal functional parameters, whereas captopril and prazosine did not. Diltiazem also lessened vascular lesions, assessed semiquantitatively. In glomeruli isolated from normal rats, these authors also found an increase in the ratio of vasoconstrictor to vasodilator prostaglandins released on exposure to FK506. These studies suggest a role for alterations in renal hemodynamics, perhaps via altered intracellular calcium and glomerular prostaglandin synthesis, in the renal response to FK506 (Kumano et al., 1990, 1991).

The acute effects of FK506 in a salt-depleted Sprague-Dawley rat model have been extensively studied (Andoh and Bennett, 1998; Andoh et al., 1997b). They used pair-fed rats given FK506 vs. vehicle for 21 days in either a normal or low-salt diet, maintaining whole-blood FK506 trough levels similar to those recommended in humans. In the salt-depleted rats, FK506 decreased GFR, urine osmolarity, and plasma magnesium, while increasing plasma creatinine levels, fractional excretion of magnesium, urine volume, and plasma renin activity, as compared with the rats on a normal-salt diet. Structural changes in the affected rats included collapse and vacuolization of proximal tubular cells, discrete or confluent zones of tubulointerstitial edema, and a mononuclear cell infiltrate. Other chronic effects seen by 21 days included interstitial scarring associated with elevated plasma renin activity.

In the same experimental model performed by some of the same investigators, the FK506 plus low-salt diet group of rats showed discrete zones of tubular atrophy and interstitial fibrosis involving medullary rays and the inner stripe, corresponding to areas of low oxygen tension. Additionally, in this group of rats, JGA granularity was strikingly increased as compared with FK506-fed rats on a normal-salt diet. Curiously, this granularity did not correspond to circulating plasma renin activity, suggesting local renin–angiotensin induced-effects (Stillman et al., 1995). Nielsen et al. (1995), also using a rat model, clearly demonstrated decreased proximal end-delivery, as measured by lithium clearance, in the FK506-treated group. This was accompanied by a GFR that was 23% of controls. Morphologically, the FK506-treated rat kidneys showed increased tubular basophilia and atrophy. The authors attribute these effects to constriction of preglomerular vessels.

Morris et al. (1991) also used a male Sprague-Dawley rat model to examine certain nephrotoxic effects of tacrolimus and CsA, both of which were administered intramuscularly to rats maintained on a liquid diet. Compared with pair-fed controls, FK506-treated rats experienced a 50% reduction in GFR within ten days. The calculated fractional citrate excretion rate in FK506-treated rats was 7.5%, compared with 2.2% in pair-fed controls. As citrate is reabsorbed in the proximal tubule, increased citrate excretion in the face of decreased GFR indicates proximal tubular dysfunction. The investigators noted markedly decreased phosphoenolpyruvate carboxykinase (PEPCK) levels in isolated tubules from CsA-treated rats and verified decreased PEPCK mRNA levels in both CsA- and FK506-treated rats, while no such decrease was seen in levels of other proximal tubular proteins. The authors propose that this highly selective inhibition of the PEPCK gene may represent a critical step in development of calcineurin inhibitor-mediated nephrotoxicity.

Lieberman et al. (1991) used electrical resistance pulse sizing (ERPS) to measure degree of glomeruloconstriction in isolated fixed rat glomeruli, proceeding with the logical assumption that constricted glomeruli are smaller than nonconstricted glomeruli. They found a proportional increase in glomeruloconstriction with increasing concentrations of tacrolimus and proposed that contracted glomeruli necessarily have less available surface area for ultrafiltration, reducing their K_f and, consequently, single-nephron and total GFR.

McCauley et al. (1991), using the established porcine proximal tubular epithelial cell line LLC-PK_1 exposed to increasing concentrations of FK506 or CsA, found that both FK506 and CsA inhibited renal epithelial cell proliferation in a dose-dependent manner, with cell proliferation indices measured by fluorescent DNA staining. However, significant cell killing, as measured by a lactate dehydrogenase release assay and trypan blue exclusion, was not detected in either the FK506 or CsA groups, compared with vehicle-treated cells. The authors propose this inhibition of epithelial cell proliferation as one possible explanation for both acute and chronic toxic effects of calcineurin inhibitors. Acute renal failure will be prolonged secondary to this inhibition and, consequently, preferential stimulation of interstitial fibroblast proliferation can ensue.

Rapamycin

Much attention is currently focused on sirolimus (rapamycin), not only for avoidance of calcineurin inhibitor-related nephrotoxicity but also for the beneficial effects that it is purported to confer. One of the beneficial effects is the prevention, and possibly even reversal, of chronic allograft nephropathy, particularly through its effects on the cellular cytokine milieu. Indeed, sirolimus blocks cytokine-induced vascular and smooth muscle cell proliferation and inhibits the antiapoptotic influence of bcl-2. This latter feature might actually confer a tolerogenic proapoptotic effect upon effector T-cells (Kahan, 2001). Ikonen et al. (2000) found that sirolimus halted progression of graft vascular disease in primates, and actually caused partial regression of it in a subset of their subjects. Sirolimus decreases the synthesis of primary monocyte-derived cytokines, such as IL-6 and monocyte chemotactic protein-1 (MCP-1), which should, at least theoretically, promote allograft tolerance (Oliveira et al., 2002).

Additionally, in an ischemia-reperfusion rat model in which sirolimus was compared with CsA, Jain et al. (2001) found that sirolimus-treated rats showed a significantly lower incidence of proteinuria, as well as lower serum creatinine levels. TGFβ mRNA and tissue inhibitors of matrix metalloproteinases were expressed in significantly lower amounts in homogenized renal cortical tissue of sirolimus-treated rats, both of which function to inhibit extracellular matrix degradation, contributing to chronic allograft nephropathy. Wang et al. (2001) exposed cultured sirolimus vs. FK506-treated mesangial cells to platelet-derived growth factor-BB (PDGF-BB), a potent mediator of mesangial cell proliferation. They found that concentrations of sirolimus as low as 10^{-1} nmol/l inhibit PDGF-induced mesangial cell DNA synthesis with no evidence of cytotoxicity (as measured by cellular lactate dehydrogenase release). In contrast, potentially nephrotoxic concentrations of 10^4 nmol/l or more of FK506 are required for the same inhibitory effect on cultured mesangial cells.

Experimental Renal Toxicity

In contrast to the findings of Jain et al. (2001) (see above), several investigators have reported increased expression of TGFβ, both in the kidney and in lymphocytes of rodent models, which theoretically increases the risk of interstitial fibrosis, tubular atrophy, and vascular proliferation seen in chronic allograft nephropathy.

Swinford et al. (2002) found increased TGFβ mRNA expression in immortalized rat proximal tubular cells exposed to sirolimus and an additive effect was imparted by concurrent CsA administration. Dodge et al. (2000) reported that sirolimus-treated splenic leukocytes of BALB/c mice produced significantly more TGFβ (as measured by ELISA) than those of untreated controls, potentially increasing extracellular matrix production and stimulating fibroblast growth. In a Sprague-Dawley rat model, Andoh et al. (1996b) studied the structural and functional renal effects of sirolimus, CsA, or sirolimus plus CsA, in addition to effects on blood glucose for up to 28 days. They found that sirolimus alone did not increase GFR or cause any detectable renal fibrosis, nor did it cause a significant elevation in plasma glucose. However, sirolimus potentiated the hyperglycemia, GFR decrease, and renal fibrosis caused by CsA. As chronic hyperglycemia is known to accelerate fibrosis of various tissue sites, the authors postulated that the synergistic fibrogenic effect of concomitant CsA and sirolimus administration is due to their combined effects on glycemia.

Functional renal impairment by sirolimus, although demonstrated experimentally at high doses, appears negligible compared with that caused by calcineurin inhibitors. Sabbatini et al. (2000), in micropuncture experiments using intact rats, observed and reported that, compared with vehicle-treated controls, single-nephron GFR but not total GFR, was significantly reduced in rats treated with 5 mg sirolimus/kg i.v. This is in contrast to rats treated with 20 mg CsA/kg i.v., where both single-nephron and total GFR were significantly reduced at immunosuppressive doses. Additionally, glomerular plasma flow was decreased in both sirolimus and CsA treated rats due to increased afferent and efferent arteriolar constriction. The authors, thus, concluded that very high doses of sirolimus are required to produce only marginal effects on renal and glomerular hemodynamics. Similarly, Golbaekdal et al. (1994), using a Lancaster/Yorkshire female pig model with renal structure and function very similar to those of human, found that in therapeutic or subtherapeutic doses, intravenous infusion of sirolimus caused no deleterious effects on GFR, renal plasma flow, water and sodium excretion, or tubular function up to two hours after infusion. However, in contrast to the findings of Sabbatini et al. (2000) in rats, Golbaekdal et al. (1994) noted an increase in GFR and renal plasma flow when supratherapeutic doses of sirolimus were infused.

Lieberthal et al. (2001) reported that sirolimus severely impairs renal recovery after ischemia-reperfusion injury in rats (produced by renal artery occlusion) but had no effect on GFR of rats after sham surgery. This hindrance of renal recovery appears to be mediated by renal tubular cell apoptosis and inhibition of tubular epithelial cell regenerative response. These effects are possibly the result of sirolimus-induced inactivation of the 70 kDa S6 protein kinase, which the authors observed in both cultured mouse proximal tubular cells and kidney tissue of rats subjected to renal artery occlusion.

Renal functional and structural effects of sirolimus administration were studied by Andoh et al. (1996a) in rats (on a low-salt diet) and were compared with effects of CsA, tacrolimus, and vehicle. Although the calcineurin inhibitors CsA and tacrolimus decreased urinary excretion of NO, renal blood flow, and GFR, sirolimus did not. However, all three drugs caused significant hypomagnesemia due to high fractional excretion of magnesium. Surprisingly, the investigators reported similar morphological tubular findings with all three drugs, specifically tubular collapse, vacuolization, and nephrocalcinosis, which they logically attribute to a calcineurin-independent mechanism.

In summary, although sirolimus may enhance the effects of calcineurin inhibitors and vice versa due to competition for hepatic CYP3A4 metabolism, definitive evidence of nephrotoxicity in therapeutic doses is lacking, except for those potentiated by calcineurin inhibitor induced hyperglycemia.

Mycophenolate Mofetil

Wang et al. (1999) demonstrated that MPA inhibits PDGF-BB-induced mesangial cell and vascular smooth muscle cell proliferation in a dose-dependent manner. They have further elucidated that PDGF exerts its physiologic effects via induction of osteopontin overexpression. Such effects include mesangial and smooth muscle cell proliferation, chemoattraction, inflammatory cell activation, and cell adhesion. They then went on to prove that MPA blockes PDGF-induced osteopontin overexpression, providing evidence of a more directed mechanism of action.

Yang et al. (2002a) compared effects of CsA withdrawal and MMF treatment on the progression of chronic CsA toxicity in a salt-depleted rat model. Addition of MMF to a CsA regimen does not improve decreased renal function, nor the increases in tubulintersititial fibrosis, arteriolopathy, macrophage influx, renin-positive glomeuli, TUNEL-positive cells, and mRNA for osteopontin or TGF. These parameters are improved with CsA withdrawal, with further improvement with addition of MMF. The authors conclude that while combined CsA and MMF do not prevent chronic toxicity, MMF treatment after CsA withdrawal does improve toxicity, supporting the clinical use of MMF in chronic CsA toxicity.

Interleukin-2

Acute renal failure is an important complication of IL-2 therapy. Clinically, ARF has generally been ascribed to hypotension and prerenal azotemia developing in the setting of a generalized capillary leak syndrome (see above). Anderson and Hayes (1989) reported toxic effects of human recombinant IL-2 in rats. Mortality in treated animals was due to severe anemia or hepatic damage. Mild elevations in blood urea nitrogen (BUN) were detected; no other tests of renal function were performed. Renal histopathologic lesions occurred at high doses (60 and 150×10^6 units IL-2/kg/day i.v.) and consisted of cytoplasmic inclusions in proximal tubules, felt to be hemoglobin. Doses of 3.75×10^6 units IL-2/kg/injection (i.p. or i.v.) bid produced cytoplasmic inclusions and a lymphocytic infiltrate; the same total daily dose given as a single bolus produced only mild lymphoid infiltrate. Because of the paucity of histopathologic findings in the kidney, the rise in BUN was attributed to prerenal azotemia.

Welbourn et al. (1990), in studies of multiorgan edema produced by IL-2, documented significant edema of the kidney in rats when IL-2 (10×10^6 units) was given as a one hour infusion. Plasma thromboxane increased significantly as well, and polymorphonuclear leukocyte (PMNL) sequestration was demonstrated in the lung. These effects were not seen with rapid bolus injection. Renal function and renal or urinary prostanoids were not measured. These investigators subsequently showed that phalloidin, which stabilizes actin filaments, and antamanide, an analogue of phalloidin, reduce edema formation, perhaps by stabilizing the microvascular permeability barrier. Antamanide reduced kidney edema, although phalloidin, which reduced lung edema, did not. Both agents also lowered plasma TXB_2, but did not prevent neutrophil sequestration in the lung. No renal functional studies were performed (Welbourn et al., 1991). In studies of a model of IL-2 toxicity in rats, Rubinger et al. (1990) reported increases in urinary TX associated with decreased GFR and increased fractional excretion of Na, an effect exacerbated by high-dose indomethacin.

IL-2 alone does not increase endothelial permeability *in vitro* (Damle and Doyle, 1989; Klausner et al., 1989). Vascular effects seen with IL-2 therapy, which may lead directly or indirectly to renal dysfunction, appear at this time to be most likely mediated by neutrophils and thromboxane, and perhaps other cytokines (Butler et al., 1989; Cotran et al., 1988).

Other Agents

In experimental studies, Morel-Maroger et al. (1978) described the development of a progressive glomerulonephritis in newborn Swiss mice receiving daily injections of murine interferon. This finding may relate to the proteinuria observed in humans (see above). Maessen et al. (1989) looked at direct cytotoxic effects of interferon and recombinant IL-1 on cultured human kidney epithelial cells. They found that both agents caused cell injury as measured by lactate dehydrogenase release and inhibition of protein synthesis, suggesting a direct toxic effect in the tubules.

REFERENCES

Abdalla AH, Al Sulaiman MH, and Al Khader AA (1994) FK506 as an alternative in cyclosporine-induced hemolytic uremic syndrome in a kidney transplant recipient. Transplant Int 7:382–384.

Abramowicz D, Schandene L, Goldman M, Crusiaux A, Vereerstraeten P, De Pauw L, Wybran J, Kinnaert P, Dupont E, and Toussaint C (1989) Release of tumor necrosis factor, interleukin-2, and gamma-interferon in serum after injection of OKT3 monoclonal antibody in kidney transplant recipients. Transplantation 47:606–608.

Acott PD, Lawen J, Lee S, and Crocker JFS (2001) Basiliximab versus ATG/ALG induction in pediatric renal transplants: Comparison of herpes virus profile and rejection rates. Transplant Proc 33:3180–3183.

Ader JL, and Rostaing L (1998) Cyclosporin nephrotoxicity: Pathophysiology and comparison with KF506. Curr Opin Nephrol Hypertens 7:539–545.

Aicher L, Meier G, Norcross AJ, Jakubowski J, Varela MC, Cordier A, and Steiner S (1997) Decrease in kidney calbindin-D 28 kDa as a possible mechanism mediating cyclosporine A- and FK506-induced calciuria and tubular mineralization. Biochem Pharmacol 53:723–731.

Akhlaghi F, and Trull AK (2002) Distribution of cyclosporin in organ transplant recipients. Clin Pharmacokinet 41:615–637.

Allison AC, and Eugui EM (2000) Mycophenolate mofetil and its mechanism of action. Immunopharmacology 47:85–118.

Alvarez-Arroyo MV, Suzuki Y, Yague S, Lorz A, Jiminez S, Soto C, Barat A, Belda E, Gonzalez-Pacheco FR, Deudero JJ, Castilla MA, Egido J, Ortiz A, and Caramelo C

(2002a) role of endogenous vascular endothelila growth factor in tubular cell protection against acute cyclosporine toxicity. Transplantation 74:1618–1624.

Alvarez-Arroyo MV, Yague S, Wenger RM, Pereira DS, Jiminez S, Gonzalez-Pacheco FR, Castilla MA, Deudero JJ, and Caramelo C (2002b) Cyclophilin-mediated pathways in the effect of cyclosporin A on endothelial cells: Role of vascular endothelial growth factor. Circ Res 91:202–209.

Anderson TD, and Hayes TJ (1989) Toxicity of human recombinant interleukin-2 in rats. Lab Invest 60:331–346.

Andoh TF, and Bennett WM (1998) Chronic cyclosporine nephrotoxicity. Curr Opin Nephrol Hypertens 7:265–270.

Andoh TF, Burdmann EA, Fransechini N, Houghton DC, and Bennett WM (1996a) Comparison of acute Rapamycin nephrotoxicity with cyclosporine and FK506. Kidney Int 50:1110–1117.

Andoh TF, Lindsley J, Franceschini N, and Bennett WM (1996b) Synergistic effects of cyclosporine and rapamycin in a chronic nephrotoxicity model. Transplantation 62:311–316.

Andoh TF, Burdmann EA, and Bennett WM (1997a) Nephrotoxicity of immunsuppressive drugs: Experimental and clinical observations. Semin Nephrol 17:34–45.

Andoh TF, Gardner MP, and Bennett WM (1997b) Protective effects of dietary L-arginine supplementation on chronic cyclosporine nephropathy. Transplantation 64:1236–1240.

Armstrong VW, Shipkova M, Schutz E, Weber L, Tonshoff B, and Oellerich M (2001) German Study Group on MMF therapy in pediatric renal transplant recipients. Monitoring of mycophenolic acid in pediatric renal transplant recipients. Transplant Proc 33:1040–1043.

Ault BH, Stapleton F-B, Gaber L, Martin A, Roy S Ill, and Murphy SB (1988) Acute renal failure during therapy with recombinant human gamma interferon. N Engl J Med 3l9:1397–1400.

Averbuch SD, Austin HA III, Sherwin SA, Antonovych T, Bunn PA Jr, and Longo DL (1984) Acute interstitial nephritis with the nephrotic syndrome following recombinant leukocyte alpha-interferon therapy for mycosis fungoides. N Engl J Med 310:32–35.

Awazu M, Sugiura M, Inagarni T, Ichikawa I, and Kon V (1991) Cyclosporine promotes glomerular endothelin binding *in vivo*. J Am Soc Nephrol 1:1253–1258.

Badr KF, Murray JJ, Breyer MD, Takahashi K, Inagami T, and Harris RC (1989) Mesangial cell, glomerular and renal vascular responses to endothelin in the rat kidney. Elucidation of signal transduction pathways. J Clin Invest 83:336–342.

Bantle IF, Paller MS, Boudreau Ri, Olivari MT, and Ferris TF (1990) Long-term effects of cyclosporine on renal function in organ transplant recipients. J Lab Clin Med 115:233–240.

Bantle JP, Nath KA, Sutherlant DE, Najarian IS, and Ferris TF (1985) Effects of cyclosporine on the renin-angiotensin-aldosterone system and potassium excretion in renal transplant recipients. Arch Intern Med 145:505–508.

Barros RIG, Boim MA, Ajzen H, Raimas OL, and Schor N (1987) Glomerular hemodynamics and hormonal participation in cyclosporine nephrotoxicity Kidney Int 32:19–25.

Batiuk TD, Bennett WM, and Norman DJ (1993) Cytokine nephropathy during antilymphocyte therapy. Transplant Proc 25 (Suppl 1):27–30.

Battle DC, Gutterman C, Keilanj T, Peces R, and LaPointe M (1990) Effect of cyclosporin A on renal function and kidney growth in the uninephrectomlzed rat. Kidney Int 37:21–28.

Bechstein WO (2000) Neurotoxicity of calcineurin inhibitors: Impact and clinical management. Transpl Int 13:313–326.

Becker BN (1999) Mycophenolate mofetil. Transplant Proc 31:2777–2778.

Behrend M (2001) Adverse gastrointestinal effects of mycophenolate mofetil: Aetiology, incidence and management. Drug Safety 24:645–663.

Behrend M, Lueck R, and Pichlmayr R (1997) Long-term experience with mycophenolate mofetil in the prevention of renal allograft rejection. Transplant Proc 29:2927–2929.

Bekersky I, Dressler D, and Mekki Q (2001) Effect of time of meal consumption on bioavailability of a single oral 5 mg tacrolimus dose. J Clin Pharmacol 41:289–297.

Belitsky P, for the Canadian Transplant Study Group. (1985) Initial nonfunction of cadaver renal allografts preserved by simple cold storage. Transplant Proc 7:1485–1488.

Belitsky P, Dunn S, Johnston A, and Levy G (2000) Impact of absorption profiling on efficacy and safety of cyclosporin therapy in transplant recipients. Clin Pharmacokinet 39:117–125.

Belldegrun A, Webb DE, Austin HA III, et al. (1984) Effect of interleukin-2 on renal function in patients receiving immunotherapy for advanced cancer. Ann Intern Med 106:1006.

Bennett WM (1991) Clinical aspects of cyclosporine nephrotoxicity, in *Nephrology: Proceedings of the XIth international Congress of Nephrology*, Hitano M (Ed.), Springer-Verlag, Tokyo, pp. 564–575.

Bennett WM, Houghton DC, and Buss WC (1991) Cyclosporine-induced renal dysfunction: Correlations between cellular events and whole kidney function. J Am Soc Nephrol 1:1212–1219.

Benoist JF, Orbach D, and Biou D (1998) False increase in C-reactive protein attributable to heterophilic antibodies in two renal transplant patients treated with rabbit antilymphocyte globulin. Clin Chem 44:1980–1988.

Berloco P, Rossi M, Pretagostini R, Sociu-Foca Cortesini N, and Cortesini R (2001) Tacrolimus as cornerstone immunosuppressant in kidney transplantation. Transplant Proc 33:994–996.

Bertani T, Ferrazzi P, Schieppati A, et al. (1991) Nature and extent of glomerular injury induced by cyclosporine in heart transplant patients. Kidney Int 40:243–250.

Bertani T, Perico N, Abbate M, Battaglia C, and Remuzzi G (1987) Renal injury induced by long-term administration of cyclosporin A to rats. Am J Pathol 127:569–579.

Besarab A, Iarrell BE, Hirsch S, Carabasi RA, Cressman MD, and Green P (1987) Use of the isolated perfused kidney model to assess the acute pharmacologic effects of cyclosporine and its vehicle cremophor. Transplantation 44:195–201.

Bia MJ, and Tyler KA (1987) Effect of cyclosporine on renal ischemic injury. Transplantation 43:800–804.

Blair IT, Thompson AW, Whiting PH, Davidson RJL, and Simpson JG (1982) Toxicity of the immune suppressant cyclosporin A in the rat. J Pathol 138:163–177.

Blantz RC, Konnen KS, and Tucker BJ (1976) Angiotension II effects upon the glomerular microcirculation and ultrafiltration coefficient of the rat. J Clin Invest 57:419–434.

Bleeker WK, Teeling JL, Verhoeven AJ, Rigter GM, Agterberg J, Tool AT, Koenderman AH, Kuijpers TW, and Hack CE (2000) Vasoactive side effects of intravenous immunoglobulin preparations in a rat model and their treatment with recombinant platelet-activating factor acetylhydrolase. Blood 95:1856–1861.

Boch HA (2001) Steroid-resistant kidney transplant rejection: Diagnosis and treatment. J Am Soc Nephrol 12 (S17):S48–S52.

Bock P, Hobler N, Caudrelier P, and Alberici G (1995) Treatment of acute rejection in renal graft recipients by Thymoglobuline: A retrospective multicenter analysis. Transplant Proc 27:1058–1059.

Boletis JN, Theodoropoulou H, Hiras T, Stamatiadis D, Darema M, Psimenou E, Stathakis C, and Kostakis A (2001) Monoclonal antibody basiliximab with low cyclosporine dose as initial immunosuppression. Transplant Proc 33:3184–3186.

Borowitz JL (1988) Cyclosporine increases calcium in the kidney medulla. Life Sci 42:1215–1222.

Brattstrom C, Wilczek H, Tyden G, Bottinger Y, Sawe J, and Groth CG (1998) Hyperlipidemia in renal transplant recipients treated with sirolimus (rapamycin). Transplantation 65:1272–1274.

Brennan DC (2001) Polyclonal antibodies in immunosuppression. Transplant Proc 33:1002–1004.

Broekemeier KM, Dempsey ME, and Pfeiffer DR (1989) Cyclosporin A is a potent inhibitor of the inner membrane permeability transition in liver mitochondria. J Biol Chem 264:7826–7830.

Brown Z, and Neild GH (1990) FK506 and haemolytic uremic syndrome. Lancet 412.

Brunner FP, Hermle M, Mihatsch MI, and Thiel G (1986) Effect of cyclosporine in rats with reduced renal mass. Clin Nephrol 25:S148–S154.

Bullingham RE, Nicholls AJ, and Kamm BR (1998) Clinical pharmacokinetics of mycophenolate mofetil. Clin Pharmacokinet 34:429–455.

Bunke M, Wilder L, and McLeish K (1988) Effect of cyclosporine on glomerular prostaglandin production. Transplant Proc 20:646–649.

Buss WC, Steparek I, and Bennett WM (1989) A new proposal for the mechanism of cyclosporine nephrotox icily: Inhibition of renal microsomal protein chain elongation following *in vivo* cyclosporin A. Biochem Pharmacol 38:4085–4093.

Butler LD, Mohler KM, Layman NK, Cain RL, Riedl PE, Puckett LD, and Bendele AM (1989) Interleukin-2 induced systemic toxicity: Induction of mediators and immunopharmacologic intervention. Immunopharmacol Immunotoxicol 11:445.

Campistol JM, and Sacks SH (2000) Mechanisms of nephrotoxicity. Transplantation 69 Supplement:SS5–SS10.

Canadian Neoral Renal Transplantation Study Group (2001) Absorption profiling of cyclosporine microemulsion (Neoral) during the first two weeks after renal transplantation. Transplantation 72:1024–1032.

Cantu TG, Hoehn-Saric EW, Burgess KM, Racusen L, and Scheel PJ (1995) Acute renal failure associated with immunoglobulin therapy. Am J Kidney Dis 25:228–234.

Carl S, Wiesel M, and the European Mycophenolate Mofetil Co-operative Study Group (1997) Mycophenolate Mofetil (Cellcept) in renal transplantation: The European Experience. Transplant Proc 29:2932–2935.

Carswell CI, Plosker GL, and Wagstaff AJ (2001) Daclizumab: a review of its use in the management of organ transplantation. BioDrugs 15:745–773.

Caterson RI, Duggin GG, Critchley L, Baxter C, Horvath JS, Hall BM, and Tiller DJ (1986) Renal tubular transport of cyclosporine A (CsA) and associated changes in renal function. Clin Nephrol 25:S30–S33.

Cavarape A, Endlich K, Feletto F, Parekh N, Bartoli E, and Steinhausen M (1998) Contribution of endothelin receptors in renal microvessels ini acute cyclosporine-medated vasoconstriction in rats. Kidney Int 53:963–969.

Cebrian C, Areste C, Nicolas A, Olive P, Carceller A, Piulats J, and Meseguer A (2001) Kidney anddrogen-related protein interacts with cyclophilin b and reduces cyclosporine A-mediated toxicity in proximal tubule cells. J Biol Chem 276:29410–29419.

Christensen H, Asberg A, Homboe AB, and Berg KG (2002) Coadministration of grapefruit juice increases systemic exposure of diltiazem in healthy volunteers. Eur J Clin Pharmacol 58:515–520.

Christians U, Jacobsen W, Benet LZ, and Lampen A (2002) Mechanisms of clinically relevant drug interactions associated with tacrolimus. Clin Pharmacokinet 41:813–851.

Churchill MC, Churchill PC, and Bidani AK (1993) The effects of cyclosporine in Lewis rats with native and transplanted kidneys. Transplantation 55:1256–1260.

Cibrik DM, Kaplan B, and Meier-Kiesche HU (2001) Role of anti-interleukin-2 receptor antibodies in kidney transplantation. BioDrugs 15:655–666.

Claesson K, Mayer AD, Squifflet JP, Grabensee B, Eigler FW, Behrend M, Vanrenterghem Y, van Hooff J, Morales JM, Johnson RW, Buchholz B, Land W, Forsythe JL, Neumayer HH, Ericzon BG, and Muhlbacher F (1998) Lipoprotein patterns in renal transplant patients: A comparison between FK506 and cyclosporine A patients. Transplant Proc 30:1292–1294.

Coffman TM, Carr DR, Farger WE, and Klotman P (1987) Evidence that renal prostaglandin production and thromboxane production is stimulated in chronic cyclosporine nephrotoxicity Transplantation 43:282–285.

Cohn RG, Mirkovich A, Dulap B, Burton P, Chiu SH, Eugui E, and Caulfield JP (1999) Mycophenoic acid increases apoptosis, lysosomes and lipid droplets in human lymphoid and monocytic cell lines. Transplantation 68:411–418.

Comenzo RL, Malachowski ME, Meissner HC, Fulton DR, and Berkman EM (1992) Immune hemolysis, disseminated intravascular coagulation, and serum sickness after large doses of immune globulin given intravenously for Kawasaki disease. J Pediatr 120:926–928.

Conte G, Dal Canton A, Sabbatini M, Napodano P, De Nicola L, Gigliotti G, Fuiano G, Testa A, Esposito C, Russo D, et al. (1989) Acute cyclosporine renal dysfunctio0 reversed by dopamine infusion in healthy subjects. Kidney Int 36:1086–1092.

Cooper MR, Fefer A, Thompson I, Case DC Jr, Kempf R, Sacher R, Neefe J, Bickers J, Scarffe JH, Spiegel RJ, et al. (1987) Interferon alpha 2h/melphalan/prednisone in previously untreated patients with multiple myeloma. A phase I-II trial. Invest New Drugs 5:41–46.

Cotran RS, Pober IS, Gimbrone MA Jr, Springer TA, Wiebke EA, Gaspari AA, Rosenberg SA, and Lotze MT (1988) Endothelial activation during interleukin-2 immunotherapy: A possible mechanism for the vascular leak syndrome. J Immunol 140:1883–1888.

Coukell AJ, and Plosker GL (1998) Cyclosporin microemulsion (Neoral). A pharmacoeconomic review of its use compared with standard cyclosporine in renal and hepatic transplantation. Pharmacoeconomoics 14:691–708.

Cunningham C, Gavin MP, Whiting PH, Burke MD, Maclntyre F, Thomson AW, and Simpson JG (1985) Serum cyclosporine levels, hepatic drug metabolism and renal tubulotoxicity. Biochem Pharmacol 33:2857–2861.

Cunningham E, Chi Y, Brentjens J, and Venuto R (1987) Acute serum sickness with glomerulonephritis induced by antithymocyte globulin. Transplantation 43:309–312.

Damle N, and Doyle L (1989) IL-2-activated human killer lymphocytes but not their secreted products mediate increase in albumin flux across cultured endothelial monolayers. J Immunol 142:2660–2669.

Davies DR, Bittmann I, and Pardo J (2000) Histopathology of calcineurin inhibitor-induced nephrotoxicity. Transplantation 69 (Suppl 12):SS11–SS13.

Decocq G, de Cagny B, Andrejak M, and Desablens B (1996) Acute kidney failure secondary to intravenous immunoglobulin administration. Four cases and review of the literature. Therapie 51:516–526.

DeMattos AM, Olyaei AJ, and Bennett WM (2000) Nephrotoxicity of immunosuppressive drugs: long-term consequences and challenges for the future. Am J Kidney Dis 35:333–346.

Devarajan P, Kaskel Fi, Arbeit LA, and Moore LC (1989) Cyclosporine nephrotoxicity: Blood volume, sodium conservation, and renal hemodynamics. Am J Physiol 256:F71–F78.

Dieperink H, Leyssac PO, Starklint H, Jorgensen KA, and Kemp E (1986) Antagonist capacities of nifedipine, captopril, phenoxybenzamine, prostacyclin and indomethacin on cyclosporin A induced impairment of rat renal function. Eur J Clin Invest 16:540–548.

Dodge IL, Demirci G, Strom TB, and Li XC (2000) Rapamycin induces transforming growth factor-beta production by lymphocytes. Transplantation 70:1104–1106.

Dominguez JH, Soleimani M, and Batiuk T (2002) studies of renal injury IV: The GLUT1 gene protects renal cells from cyclosporine A toxicity. Kidney Int 62:127–136.

dos Santos OF, Boim MA, Bregman R, Draibe SA, Barros EJ, Pirotzky E, Schor N, and Braquet P (1989) Effect of Platelet-activating factor antagonist on cyclosporine nephrotoxicity. Transplantation 47:592–595.

Doutrelepont JM, Abramowicz D, Florquin S, de Pauw L, Goldman M, Kinnaert P, and Vereerstraeten P (1992) Early recurrence of hemolytic uremic syndrome in a renal transplant recipient during prophylactic OKT3 therapy. Transplantation 53:1378–1379.

Drachenberg RC, Drachenberg CB, Papadimitrious JC, Ramos E, Fink JC, Wali R, Weir MR, Cangro CB, Klassen DK, Khaled A, Cunningham R, and Bartlett ST (2001) Morphological spectrum of polyoma virus disease in renal allografts: Diagnostic accuracy of urine cytology. Am J Transplant 1:373–381.

Duggin GG, Baxter C, Hail BM, Horvath JS, and Tiller DJ (1986) Influence of cyclosporine A on intrarenal control of GFR. Clin Nephrol 25:S43–S45.

Dumont FJ (2000) FK506, an immunosuppressant targeting calcineurin function. Curr Med Chem 7:731–748.

Duncan JJ, Thomson AW, Aldridge RD, Simpson IG, and Whiting PH (1986) Cyclosporine-induced renal structural damage: Influence of dosage, strain, age and sex with reference to the rat and guinea pig. Clin Nephrol 25:S14–S17.

Dunn CJ, Wagstaff AJ, Perry CM, Plosker GL, and Goa KL (2001) Cyclosporin: an updated review of the pharmacokinetic properties, clinical efficacy and tolerability of a microemulsion-based formulation (Neoral) in organ transplantation. Drugs 61:1957–2016.

Elmer DS, Nyman T, Hathaway DK, Alloway R, and Gaber AO (1996) Use of FK506 immunosuppressive therapy in pancreas transplantation. J Trans Coord 6:122–127.

Elzinga L, Kelley VE, Houghton DC, and Bennett WM (1987) Modification of experimental nephrotoxicity with fish oil as the vehicle for cyclosporine. Transplantation 43:271–274.

Elzinga LW, Mela-Riker LM, Widener LL, and Bennett WM (1989) Renal cortical mitochondrial integrity in experimental cyclosporine nephrotoxicity. Transplantation 48:102–106.

English J, Evan A, Houghton DC, and Bennett WM (1987) Cyclosporine-induced acute renal dysfunction in the rat. Transplantation 44:135–141.

Ernst E (2002) St. John's Wort supplements endanger the success of organ transplantation. Arch Surg 137:316–319.

Ettenger RB, and Grimm EM (2001) Safety and Efficacy of TOR inhibitors in pediatric renal transplant recipients. Am J Kidney Dis 38 (4S2):S22–S28.

Eugui EM, Almquist SJ, Muller CD, and Allison AC (1991) Lymphocyte selective cytostatic and immunosuppressive effects of mycophenolic acid *in vitro*: Role of deoxyguanosine nucleotide depletion. Scand J Immunol 33:161–173.

European Mycophenolate Mofetil Study Group (1995) Placebo-controlled study of mycophenolate mofetil combined with cyclosporine and corticosteroids for prevention of acute rejection. Lancet 345:1321–1325.

Faraco PR, Hewitson TD, and Kincaid-Smith P (1991) An animal model for the study of the microangiopathic form of cyclosporine nephrotoxicity. Transplantation 51:1129–1131.

Ferns G, Reidy M, and Ross R (1990) Vascular effects of cyclosporine A *in vivo* and *in vitro*. Am J Pathol 137:403–413.

Fiedler W, Zellor W, Peimann CJ, Weh HJ, and Hossfeld DK (1991) A phase II combination trial with recombinant human tumor necrosis factor and gamma interferon in patients with colorectal cancer. Klin Wochenschr 59:261–268.

Finn WF (1999) FK506 nephrotoxicity. Ren Fail 21:319–329.

First MR, Schroeder TJ, and Hariharan S (1993) OKT3-induced cytokine-release syndrome: Renal effects (cytokine nephropathy). Transplant Proc 25 (2 Suppl 1):25–26.

Fischer G, Wittmann-Liebold B, Lang K, Kiefhaber T, and Schmid FX (1989) Cyclophilin and peptidylprolyl cis-trans isomerase are probably identical proteins. Nature 337:476–478.

Flechner SM, Goldfarb D, Modlin C, Feng J, Krishnamurthi V, Mastroianni B, Savas K, Cook D, and Novick AC (2002) Kidney transplantation without calcineurin inhibitor drugs: A prospective, randomized trial of sirolimus versus cyclosporine. Transplantation 74:1070–1076.

Florman S, Bechimol C, Lieberman K, Burrows L, and Bromberg JS (2002) Fulminant recurrence of atypical hemolytic uremic syndrome during a calcineurin inhibitor-free immunosuppression regimen. Pediatr Transplant 6:352–355.

Frei U (1999) Overview of the clinical experience with Neoral in transplantation. Transplant Proc 31:1669–1674.

Fukuda Y, Hirata Y, Yoshimi H, Kojima T, Kobayashj Y, Yanigasawa M, and Masairi T (1988) Endothelin is a potent secretagogue for atrial natriuretic peptide in cultured rat atrial myocytes. Biochem Biophys Res Commun 155:167–172.

Gaber AO, First MR, Tesi RJ, Gaston RS, Mendez R, Mulloy LL, Light JA, Gaber LW, Squiers E, Taylor RJ, Neylan JF, Steiner RW, Knechtle S, Norman DJ, Shihab F, Basadonna G, Brennan DC, Hodge EE, Kahan BD, Kahan L, Steinberg S, Woodle ES, Chan L, Ham JM, Schroeder TJ, et al. (1998) Results of the double-blind, randomized multicenter, phase III clinical trial of thymoglobulin versus atgam in the treatment of acute graft rejection episodes after renal transplantation. Transplantation 66:29–37.

Gallagher H, and Andrews PA (2001) Cytomegalovirus infection and abdominal pain with mycophenolate mofetil: Is there a link? Drug Safety 24:405–412.

Garcia-Meseguer C, Roldan M, Melgosa M, Alonso A, Pena A, Espinosa L, and Navarro M (2002) Efficacy and safety of basiliximab in pediatric renal transplantation. Transplant Proc 34:102–103.

Gaston RS, Deierhoi MH, Patterson T, Prasthofer E, Julian BA, Barber WH, Laskow DA, Diethelm AG, and Curtis JJ (1991) OKT3 first-dose reaction: Association with T cell subsets and cytokine release. Kidney Int 39:141–148.

Gazdar AF, and Dammin GJ (1970) Neural degeneration and regeneration in human renal transplants. N Engl J Med 283:222–224.

Gerkens IF, Bhagwandken SH, Dosen PJ, and Smith AJ (1984) The effect of salt intake on cyclosporineinduced impairment of renal function in rats. Transplantation 38:412–417.

Gillum DM, Truong L, Tasby J, Migliore P, and Suki WN (1988) Chronic cyclosporine nephrotoxicity. A rodent model. Transplantation 46:285–292.

Gillum DM, Truong L, and Suki WN (1989) Effects of umnephiectomy and high protein feeding in cyclosporine nephropathy. Kidney Int 36:194–200.

Gnutzmann KH, Hering K, and Gutsche H-U (1986) Effect of cyclosporine on the diluting capacity of the rat kidney. Clin Nephrol 25:S31–S36.

Golbaekdal K, Nielsen CB, Djurhuus JC, and Pedersen EB (1994) Effects of rapamycin on renal hemodynamics, water and sodium excretion, and plasma levels of angiotensin II, aldosterone, atrial natriuretic peptide, and vasopressin in pigs. Transplantation 58:1153–1157.

Golub MS, Lustig S, Berger ME, and Lee DBN (1989) Altered vascular responses in cyclosporine-treated rats. Transplantation 48:116–118.

Gonzalez-Molina M, Seron D, Garcia del Moral R, Carrera M, Sola E, Gomez Ullate P, Capdevila L, and Gentil MA, for the Spanish Mycophenolate Mofetil and Chronic Graft Nephropathy Study Group (2002) Treatment of chronic allograft nephropathy with mycophenolate mofetil after kidney transplantation: A Spanish multicenter study. Transplant Proc 34:335–337.

Grace AA, Barradas MA, Mikhailidis DP, Jeremy JY, Moorhead IF, Sweny P, and Dandora P (1987) Cyclosporine A enhances platelet aggregation. Kidney Int 32:889–895.

Greenberg A, Egel JW, Thompson ME, Hardesty RL, Griffith BP, Bahnson HT, Bernstein RL, Hastillo A, Hess ML, and Puschett JB (1987) Early and late forms of cyclosporine nephrotoxicity: studies in cardiac transplant recipients. Am J Kidney Dis 9:12–22.

Greenspan LC, Gitelman SE, Leung MA, Glidden DV, and Mathias RS (2002) Increased incidence in post-transplant diabetes mellitus in children: A case-control analysis. Pediatr Neprhol 17:1–5.

Gregoor PJ, de Sevaux RG, Hene RJ, Hesse CJ, Hillbrands LB, Vos P, van Gelder T, Hoitsman AJ, and Weimar W (1999) Effect of cyclosporine on mycophenolic acid trough levels in kidney transplant recipients. Transplantation 68:1603–1606.

Groth CG, Backman L, Morales JM, Calne R, Kreis H, Lang P, Touraine JL, Claesson K, Campistol JM, Durand D, Wramner L, Brattstrom C, and Charpentier B (1999) Sirolimus (rapamycin)-based therapy in human renal transplantation: Similar efficacy and different toxicity compared to cyclosporine. Sirolimus European Renal Transplant Study Group. Transplantation 67:1036–1042.

Grupp C, Schmidt F, Braun F, Lorf T, Burckhardt R, and Muller GA (1998) Haemolytic uraemic syndrome (HUS) during treatment with cyclosporin A after renal transplantation – is tacrolimus the answer? Nephrol Dial Transplant 13:1629–1631.

Hamblin TI (1990) Interleukin-2: Side effects are acceptable. Br Med J 30:275–276.

Halloran P, Mathew T, Tomlanovich S, Groth C, Hooftman L, and Barker C (1997) Mycophenolate mofetil in renal allograft recipients: A pooled efficacy analysis of three randomized, double-blind, clinical studies in prevention of rejection. Transplantation 63:39–47.

Hardinger KL, Cornelius LA, Trulock EP III, and Brennan DC (2002) Sirolimus-induced leukocytoclastic vasculitis. Transplantation 74:739–740.

Hayashi K, Loutzenhiser R, and Epstein M (1997) Direct evidence that thromboxane mimetic U44069 preferentially constricts the afferent arteriole. J Am Soc Nephrol 8:25–31.

Helderman JH (1998) Lessons from the Neoral Global Database. Transplant Proc 30:1721–1722.

Hemenway CS, and Heitman J (1999) Calcineurin. Structure, function and inhibition. Cell Biochem Biophys 30:115–151.

Henry ML, Pelletier RP, Elkhammas EA, Bumgardner GL, and Ferguson RM (2001) A single center experience with basiliximab induction therapy in renal transplantation. Transplant Proc 33:3178–3179.

Hernandez G, Arriba L, Lucas M, and de Andres A (2000) Reduction of severe gingival overgrowth in a kidney transplant patient by replacing cyclosporine A with tacrolimus. J Periodontol 71:1630–1636.

Hibberd PL, Tolkoff-Rubin NE, Cosimi AB, Schooley RT, Isaacson D, Doran M, Delvecchio A, Delmonico FL, Auchincloss H Jr, and Rubin RH (1992) Symptomatic cytomegalovirus disease in the cytomegalovirus seropositive renal transplant recipient treated with OKT3. Transplantation 53:68–72.

Hirata Y, Yoshimi H, Takata S, Watanabe X, Kumagi S, Nakajima K, and Sakakibara S (1988) Cellular mechanism of action by a novel vasoconstrictor endothelin in cultured vascular smooth muscle cells. Biochem Biophys Res Commun 154:868–875.

Holt DW, and Johnston A (2000) The impact of cyclosporin formulation on clinical outcomes. Transplant Proc 32:1552–1555.

Hong JC, and Kahan BD (1999) Use of anti-CD25 monoclonal antibody in combination with rapamycin to eliminate cyclosporine treatment during the induction phase of immunosuppression. Transplantation 68:701–704.

Hong JC, and Kahan BD (2000) Sirolimus-induced thrombocytopenia and leukopenia in renal transplant recipients: Risk factors, incidence, progression and management. Transplantation 69:2085–2090.

Hoogeveen RC, Ballantyne CM, Pownall HJ, Opekun AR, Hachey DL, Jaffe JS, Opperman S, Kahan BD, and Morrisett JD (2001) Effect of Sirolimus on the metabolism of apoB100- containing lipoproteins in renal transplant patients. Transplantation 72:1244–1250.

Hoyer PF (1998) Cyclosporin A (Neoral) in pediatric organ transplantation. Neoral Pediatric Study Group. Pediatr Transplant 2:35–39.

Hudkins KL, Le QC, Segerer S, Johnson RJ, Davis CL, Giachelli CM, and Alpers CE (2001) Osteopontin expression in human cyclosporine toxicity. Kidney Int 60:635–640.

Husain S, Wagener MM, and Singh N (2001) Cryptococcus neoformans infection in organ transplant recipients: Variables influencing characteristics and outcome. Emerg Infect Dis 7:375–381.

Ikonen TS, Gummert JF, Hayase M, Honda Y, Hausen B, Christians U, Berry GJ, Yock P, and Morris RE (2000) Sirolimus (rapamycin) halts and reverses progression of allograft vascular disease in non-human primates. Transplantation 70:969–975.

Iwasaki K, Shiraga T, Nagase K, Tozuka Z, Noda K, Sakuma S, Fujitsu T, Shimatani K, Sato A, and Fujioka M (1993) Isolation, identification, and biological activities of oxidative metabolites of FK506, a potent immunosuppressive macrolide lactone. Drug Metab Dispos 21:971–977.

Jablonski P, Harrison C, Howden BH, Rae D, Tavalis G, Marshall VC, and Tange ID (1986) Cyclosporjne and the ischemic rat kidney. Transplantation 4l:147–151.

Jackson NM, Hsu C-H, Visscher GE, Venkatachalam MA, and Humes HD (1987) Alterations in renal structure and function in a rat model of cyclosporine nephrotoxicity. J Pharmacol Exp Ther 242:749–756.

Jackson NM, O'Connor RP, and Humes HD (1988) Interactions of cyclosporine with renal proximal tubule cells and cellular membranes. Transplantation 46:109–114.

Jain S, Bicknell GR, Whiting PH, and Nicholson ML (2001) Rapamycin reduces expression of fibrosis-associated genes in an experimental model of renal ischaemia reperfusion injury. Transplant Proc 33:556–558.

James DG (1996) A new immunosuppressant: Tacrolimus. Postgrad Med J 72:586.

Japanese FK506 Study Group (1991) Clinicopathologic evaluation of kidney transplants in patients given a fixed dose of FK506. Transplant Proc 23:3111–3115.

Jeyarajah DR, and Thistlethwaite JR (1993) General aspects of cytokine-release syndrome: Timing and incidence of symptoms. Transplant Proc 25:16–20.

Jindal RM, Sidner RA, and Milgrom ML (1997) Post-transplant diabetes mellitus. The role of immunosuppression. Drug Safety 16:242–257.

Johnson DW, Saunders HJ, Johnson FJ, Huq SO, Field MJ, and Pollack CA (1999) Fibrogenic effects of cyclosporine A on the tubulointerstitium: Role of cytokines and growth factors. Exp Nephrol 7:470–478.

Johnson RW, Kreis H, Oberbauer R, Brattstrom C, Claesson K, and Eris J (2001) Sirolimus allows early cyclosporine withdrawal in renal transplantation resulting in improved renal function and lower blood pressure. Transplantation 72:777–786.

Johnson RW (2002) Sirolimus (Rapamune) in renal transplantation. Curr Opin Nephrol Hypertens 11:603–607.

Jung K, and Pergande M (1985) Influence of cyclosporin A on the respiration of isolated rat kidney mitochondria. FEBS Lett 183:167–169.

Jung K, Reinholdt C, and Scholz D (1987) Inhibitory effects of cyclosporine on the respiratory efficiency of isolated human kidney mitochondria. Transplantation 43:162–163.

Kahan BD (2001) Sirolimus: A comprehensive review. Expert Opin Pharmocother 2:1903–1917.

Kahan BD, and Camardo JS (2001) Rapamycin: Clinical results and future opportunities. Transplantation 72:1181–1193.

Kahan B, Reid M, and Newberger J (1983) Pharmacokinetics of cyclosporin A in human transplantation. Transplant Proc 15:446–453.

Kahan BD, Julian BA, Pescovitz MD, Vanrenterghem Y, and Neylan J for the Rapamune Study Group (1999a) Sirolimus reduces the incidence of acute rejection episodes despite lower cyclosporine doses in Caucasian recipients of mismatched primary renal allografts: A phase II trial. Transplantation 68:1526–1532.

Kahan BD, Rajagopalan PR, and Hall M (1999b) Reduction of the occurrence of acute cellular rejection among renal allograft recipients treated with basiliximab, a chimeric anti-interleukin-2-receptor monoclonal antibody. United States Simulect Renal Study Group. Transplantation 67:276–284.

Kahan BD, Napoli KL, Kelly PA, Podbielski J, Hussein I, Urbauer DL, Katz SH, and Van Buren CT (2000) Therapeutic drug monitoring of sirolimus: correlations with efficacy and toxicity. Clin Transplant 14:97–109.

Kahan BD, Keown P, Levy GA, and Johnston A (2002) Therapeutic drug monitoring of immunosuppressant drugs in clinical practice. Clin Ther 24:330–350.

Kanzai G, Stowe N, Steinmuller D, Ho-Hsieh H, and Novick A (1986) Effect of cyclosporine upon the function of ischemically damaged kidneys in the rat. Transplantation 4l:782.

Kaskel Fl, Devarajan P, Arbeit LA, Partin JS, and Moore LC (1987) Cyclosporine nephrotoxicity: sodium excretion, autoregulation, and angiotensin II. Am J Physiol 252:F733–F742.

Kaskel Fl, Devarajan P, Birzgalis A, and Moore LC (1989) Inhibition of myogenic autoregulation in cyclosporine nephrotoxicity in the rat. Renal Physiol Biochem 12:250–259.

Keown PA, Stiller CR, and Wallace AC (1986) Nephrotoxicity of cyclosporin A, in *Kidney Transplant Rejection: Diagnosis and Treatment,* Williams GM, Burdick JF, Solez K (Eds.), Marcel Dekker, New York, pp. 423–457.

Keown PA, and Primmett DR (1998) Cyclosporine: The principal immunosuppressant for renal transplantation. Transplant Proc 30:1712–1715.

Khanna A, Kapur S, Sharma V, Li B, and Suthanthiran M (1997) In vivo hyperexpression of transforming growth factor-beta-1 in mice: Stimulation by cyclosporine. Transplantation 63:1037–1039.

Klausner JM, Morel N, Paterson IS, Kobzik L, Valeri CR, Eberlein TJ, Shepro D, and Hechtman HB (1989) The rapid induction by interleukin-2 of pulmonary microvascular permeability. Ann Surg 209:119–128.

Knoll GA, and Bell RC (1999) Tacrolimus versus cyclosporine for immunosuppression in renal transplantation: Meta-analysis of randomized trials. Br Med J 318:1104–1107.

Kohan DE (1997) Endothelins in the normal and diseased kidney. Am J Kidney Dis 29:2–26.

Kon V, Sugiura M, Inagami T, Harvie BR, Ichikawa I, and Hoover RL (1990) Role of endothelin in cyclosporine-induced glomerular dysfunction. Kidney Int 37:1487–1491.

Kone BC, Racusen LC, Whelton A, and Solez K (1986) Acute renal failure produced by combining cyclosporine and brief renal ischemia in the Munich Wistar rat. Clin Nephrol 25:S171–S174.

Kopp U, Aurell M, Sjolander M, and Ablad B (1981) The role of prostaglandins in the alpha- and beta adrenoreceptor mediated renin release response to graded renal nerve stimulation. Pflugers Arch 391:1–8.

Kovarik JM, Korn A, and Chodoff L (2001) Within-patient controlled assessment of the influence of basiliximab on cyclosporine in pediatric de novo renal transplant recipients. Transplant Proc 33:3172–3173.

Kremer S, Margolis B, Harper P, and Skorecki K (1989) Cyclosporine induced alterations in vasopressin signalling in the glomerular mesangial cell. Clin Invest Med 12:201–206.

Kumano K, Wang G, and Sakai T (1990) Functional and morphological changes in rat kidney induced by FK506 (FK). J Am Soc Nephrol 1:613.

Kumano K, Wang G, Endo T, and Kuwao S (1991) FK506-induced nephrotoxicity in rats. Transplant Proc 23:512–515.

Kurtz A, Bruna RD, and Kuhn K (1988) Cyclosporine A enhances renin secretion and production in isolated juxtaglomerular cells. Kidney Int 33:947–953.

Lally C, Healy E, and Ryan M (1999) Cyclosporine A-induced cell cycle arrest and cell death in renal epithelial cells. Kidney Int 56:1254–1257.

Langer RM, and Kahan BD (2002) Incidence, therapy and consequences of lymphocele after sirolimus-cyclosporine-prednisone immunosuppression in renal transplant recipients. Transplantation 74:804–808.

Langer RM, Van Buren CT, Katz SM, and Kahan BD (2002) De novo hemolytic uremic syndrome after kidney transplantation in patients treated with cyclosporine-sirolimus combination. Transplantation 73:756–760.

Laskow DA, Neylan JF III, Shapiro RS, Pirsch JD, Vergne-Marini PJ, and Tomlanovich SJ (1998) The role of tacrolimus in adult kidney transplantation: A review. Clin Transplant 12:489–503.

Lassila M (2002) Interaction of cyclosporine A and the renin-angiotensin system; new perspectives. Curr Drug Metab 3:61–71.

Lau DCW, Wong K-L, and Hwang WS (1989) Cyclosporine toxicity on cultured rat microvascular endothelial cells. Kidney Int 35:604–613.

Lee DBN (1997) Cyclosporine and the renin-angiotensin axis. Kidney Int 52: 248–260.

Lee WA, Gu L, Miksztal AR, Chu N, Leung K, and Nelson PH (1990) Bioavailability improvement of mycophenolic acid through amino ester derivitization. Pharmacol Res 7:161–166.

Lemm G (2002) Composition and properties of IVIg preparations that affect tolerability and therapeutic efficacy. Neurology 59 (12 Suppl 6):S28–S32.

Levy G, Thervet E, Lake J, and Uchida K (2002) Patient management by Neoral C_2 monitoring: An international consensus statement. Transplantation 73 (9 Suppl):S12–S18.

Li C, Yang CW, Ahn HJ, Kim WY, Park CW, Park JH, Lee MJ, Yang JH, Kim YS, and Bang BK (2002) Colchicine decreases apoptotic cell death in chronic cyclosporin nephrotoxicity. J Lab Clin Med 139:364–371.

Lieberman KV, Lin WG, and Reisman L (1991) FK 506 is a direct glomeruloconstrictor, as determined by electrical pulse sizing (ERPS). Transplant Proc 23:3119–3120.

Lieberthal W, Fuhro R, Andry CC, Rennke H, Abernathy VE, Koh JS, Valeri R, and Levine JS (2001) Rapamycin impairs recovery from acute renal failure: role of cell-cycle arrest and apoptosis of tubular cells. Am J Physiol 281:F693–F706.

Loertscher R (2002) The utility of monoclonal antibody therapy in renal transplantation. Transplant Proc 34:797–800.

Longoni B, Migliori M, Ferretti A, Origlia N, Panichi V, Boggi U, Filipp C, Cuttano MG, Giovannini L, and Mosca F (2002) Melatonin prevents cyclosporine-induced nephrotoxicity in isolated and perfused rat kidney. Free Rad Res 36:357–363.

McAlister VC, Mahalati K, Peltekian KM, Fraser A, and MacDonald AS (2002) A clinical pharmocokinetic study of tacrolimus and sirolimus combination immunosuppression comparing simultaneous to separated administration. Ther Drug Monit 24:346–350.

McCauley J, Bronsther O, Fung J, Todo S, and Starzl TE (1989) Treatment of cyclosporine-induced haemolytic uraemic syndrome with FK506. Lancet 2:1516.

McCauley I, Van Thiel DH, Starzl TE, and Puschett JB (1990) Acute and chronic renal failure in liver transplantation. Nephron 55:121–128.

McCauley J, Farkus Z, Prasad SJ, Plummer HA, and Murray SA (1991) Cyclosporine and FK 506 induced inhibition of renal epithelial cell proliferation. Transplant Proc 6:2829–2830.

McCune TR, Thacker LR II, Peters TG, Mulloy L, Rohr MS, Adams PA, Yium J, Light A, Pruett T, Gaber AO, Selman SH, Jonsson J, Hayes JM, Wright FH Jr, Armata T, Blanton J, and Burdick JF (1998) Effects of tacrolimus on hyperlipidemia after successful renal transplantation: A Southeastern Organ Procurement Foundation multicenter clinical study. Transplantation 65:87–92.

MacDonald A, Scarola J, Burke JT, and Zimmerman JJ (2000) Clinical pharmacokinetics and therapeutic drug monitoring of sirolimus. Clin Ther 22SB:B101–B121.

McKenzie N, Devineni R, Vezina W, Keown P, and Stiller C (1986) The effect of cyclosporine on organ blood flow. Transplant Proc 17:1973–1975.

Maessen JG, Buurman WA, and Knotstra G (1989) Direct cytotoxic effect of cytokines in kidney parenchyma: a possible mechanism of allograft destruction. Transplant Proc 21:300–310.

Mahalati K, Belitsky P, West K, Kiberd B, Fraser A, Sketris I, Macdonald AS, McAlister V, and Lawen J (2001) Approaching the therapeutic window for cyclosporine in kidney transplantation: A prospective study. J Am Soc Nephrol 12:828–833.

Mahalati K, and Kahan BD (2001) Clinical Pharmacokinetics of sirolimus. Clin Pharmacokinet 40:573–585.

Mahaley MS, Bertsch L, Cash S, and Gillespie GY (1988) Systemic gamma-interferon therapy for recurrent glioma. J Neurosurg 69:826–829.

Makowka L, Lopatjn W, Gilas L, Falk I, Phillips Ml, and Falic R (1986) Prevention of cyclosporjne (CyA) nephrotoxicity by synthetic prostaglandins. Clin Nephrol 25:589–594.

Maleki S, Graves S, Becker S, Horwatt R, Hicks D, Stroshane RM, and Kincaid H (2000) Therapeutic monitoring of sirolimus in human whole-blood samples by high-performance liquid chromatography. Clin Ther 22SB:B25–B37.

Marchetti P, and Navalesi R (2000) The metabolic effects of cyclosporine and tacrolimus. J Endocrinol Invest 23:482–490.

Mathew TH, for the Tricontinental Mycophenolate Mofetil Renal Transplantation Study Group (1998) A blinded, long-term, randomized multicenter study of mycophenolate mofetil in cadaveric renal transplantation: Results at three years. Transplantation 65:1450–1454.

Mauiyyedi S, and Colvin RB (2002) Humoral rejection in kidney transplnatation: New concepts in diagnosis and treatment. Curr Opin Nephrol Hypertens 11:609–618.

Mayer AD, Dmitrewski J, Squifflet JP, Besse T, Grabensee B, Klein B, Eigler FW, Heemann U, Pichlmayr R, Behrend M, Vanrenterghem Y, Donck J, van Hooff J, Christiaans M, Morales JM, Andres A, Johnson RW, Short C, Buchholz B, Rehmert N, Land W, Schleibner S, Forsythe JL, Talbot D, Pohanka E, et al. (1997) Multicenter randomized trial comparing tacrolimus (FK506) and cyclosporine in the prevention of renal allograft rejection: A report of the European Tacrolimus Multicenter Renal Study Group. Transplantation 64:436–443.

Mazzali M, Hughes J, Dantas M, Liaw L, Steitz S, Alpers CE, Pichler RH, Lan HY, Giachelli CM, Shankland SJ, Couser WG, and Johnson RJ (2002) Effects of cyclosporine in ostropontin null mice. Kidney Int 62:78–85.

Meier-Kriesche HU, and Kaplan B (2000) Toxicity and efficacy of sirolimus: Relationship to whole-blood concentrations. Clin Ther 22SB:B93–B100.

Mele TS, and Halloran PF (2000) The use of mycophenolate mofetil in transplant recipients. Immunopharmacology 47:215–245.

Mercatello A, Aoumeur H-A, Negrier S, Allaouchiche B, Coronel B, Tognet E, Bret M, Favrot M, Pozet N, Moskovtchenko JF, et al. (1991) Acute renal failure with preserved renal plasma flow induced by cancer immunotherapy. Kidney Int 40:309–314.

Mervaala E, Lassila M, Vaskonen T, Krogerus L, Lahteenmaki T, Vapaatalo H, and Karppanen H (1999) Effects of ACE inhibition on cyclosporine A-induced hypertension and nephrotoxicity in spontaneously hypertensive rats on a high sodium diet. Blood Press 8:49–56.

Meyer-Lehnert H, and Schrier RW (1988) Cyclosporine A enhances vasopressin induced Ca^{2+} mobilization and contraction in mesangial cells. Kidney Int 34:89–97.

Mignat C (1997) Clinically significant drug interactions with new immunosuppressive agents. Drug Safety 16:267–278.

Mihatsch MJ, Kyo M, Morozumi K, Yamaguchi Y, Nickeleit V, and Ryffel B (1998) The side-effects of ciclosporine-A and tacrolimus. Clin Nephrol 49:356–363.

Miura K, Nakatani T, Asai T, Yamanaka S, Tamada S, Tashiro K, Kim S, Okamura M, and Iwao H (2002) Role of hypomagnesemia in chronic cyclosporine nephropathy. Transplantation 73:340-347.

Moller A, Iwasaki K, Kawamura A, Teramura Y, Shiraga T, Hata T, Schafer A, and Undre NA (1999) The disposition of ^{14}C-labeled tacrolimus after intravenous and oral administration in healthy human subjects. Drug Metab Dispos 27:633–636.

Morel-Maroger L, Slopes IC, Vinter J, Woodrow D, and Gresser I (1978) An ultrastructural study of the development of nephritis in mice treated with interferon in the neonatal period. Lab Invest 39:513–522.

Morelon E, Stern M, Israel-Biet D, Correreas JM, Daniel C, Mamzer-Bruneel MF, Peraldi MN, and Kreis H (2001) Characteristics of sirolimus-associated interstitial pneumonitis in renal transplant patients. Transplantation 72:773–774.

Moreso F, Seron D, Morales JM, Cruzado JM, Gil-Vernet S, Perez JL, Fulladosa X, Andres A, and Grinyo JM (1998) Incidence of leukopenia and cytomegalovirus disease in kidney transplants treated with Mycophenolate mofetil combined with low cyclosporine and steroid doses. Clin Transplant 12:198–205.

Moritz T, Niederle N, Baumann I, et al. (1989) Phase I study of recombinant human tumor necrosis factor a in advanced malignant disease. Cancer Immunol Immunorher 19:144–150.

Morozumi K, Sugito K, Oda A, Takeuchi O, Fukuda M, Usami T, Oikawa T, Fujinami T, Koyama K, et al. (1996) A comparative study of morphological characteristics of renal injuries of tacrolimus (FK506) and cyclosporine (CsA) in renal allografts: Are the morphologic characteristics of FK506 and CyA nephrotoxicity similar? Transplant Proc 28:1076–1078.

Morris SM, Kepla-Lenhart D, Curthoys NP, McGill RL, Marcus RJ, and Adler S (1991) Disruption of renal function and gene expression by FK506 and Cyclosporine. Transplant Proc 23:3116–3118.

Morris SM, Kepka-Lenhart D, McGill RL, et al. (1992) Specific disruption of renal function and gene transcription by cyclosporin A. J Biol Chem 267:13768–13771.

Moss NG, Powell SL, and Falk RJ (1985) Intravenous cyclosporine activates afferent and efferent renal nerves and causes sodium retention in innervated kidneys in rats. Proc Natl Acad Sci USA 82:8222–8226.

Muller TF, Grebe SO, Neumann MC, Heymanns J, Radsak K, Sprenger H, and Lange H (1997) Persistent long-term changes in lymphocyte subsets induced by polyclonal antibodies. Transplantation 64:1432–1437.

Muller-Schweinitzer E (1988) Changes in the venous compliance in bradykinin and angiotensin II and its significance for the vascular effects of cyclosporine A. Naunyn Schmiedebergs Arch Pharmacol 338:699–703.

Muller-Schweinitaer E (1989) Interaction of cyclosporine A with the renin-angiotensin system in canine veins. Naunyn Schmiedebergs Arch Pharmacol 340:252–257.

Muller-Suur R, and Davis SD (1986) Effect of cyclosporine A on renal electrolyte transport: whole kidney and Henle loop study. Effect of cyclosporine on the diluting capacity of the rat kidney. Clin Nephrol 25:S57–S6l.

Murray BM, Paller MS, and Ferns TF (1985) Effect of cyclosporine administration on renal hemodynamics in conscious rats. Kidney Int 28:767–774.

Myers BD (1986) Cyclosporine nephrotoxicity (review). Kidney Int 30:964–974.

Myers BD, Ross I, Newton L, Leutscher I, and Perlroth M (1984) Cyclosporine-associated chronic nephropathy. N Engl J Med 311:699–705.

Myers BD, Newton L, Boshkos C, Macoviak JA, Frist WH, Derby GC, Perlroth MG, and Sibley RK (1988a) Chronic injury of human renal microvessels with low-dose cyclosporine therapy. Transplantation 46:694–703.

Myers BD, Sibley R, Newton L, Tomlanovich SJ, Boshkos C, Stinson E, Luetscher JA, Whitney DJ, Krasny D, Coplon NS, et al. (1988b) The long-term course of cyclosporine-associated chronic nephropathy. Kidney Int 33:590–600.

Nagase K, Iwasaki K, Nozaki K, and Noda K (1994) Distribution and protein binding of FK506, a potent immunosuppressive macrolide lactone, in human blood and its uptake by erythrocytes. J Pharm Pharmacol 46:113–117.

Nagy SE, Andersson JP, and Anderson UG (1993) Effect of Mycophenolate Mofetil (RS-61443) on cytokine production: Inhibition of superantigen-induced cytokines. Immunophamacology 26:11–20.

Nalesnik M, Lai HS, Murase N, Todo S, and Starzl TB (1990) The effect of FK506 and cyclosporine A on the Lewis rat renal ischemia model. Transplant Proc 22:87–89.

Nashan B, Moore R, Amlot P, Schmidt AG, Abeywickrama K, and Soulillou JP (1997) Randomized trial of basiliximab versus placebo for control of acute cellular rejection in renal allograft recipients. CHIB201 International Study Group. Lancet 350:1193–1198.

Nashan B, Light S, Hardie IR, Lin A, and Johnson JR (1999) Reduction of acute renal allograft rejection by daclizumab. Daclizumab Double Therapy Study Group. Transplantation 67:110-115.

Nashan B, Cole E, Levy G, and Thervet E (2002) Clinical validation studies of neoral C2 monitoring: a review. Transplantation 73 (9 Suppl):S3–S11.

Nast C, Adler SG, Artishevsky A, Kresser CT, Ahmed K, and Anderson PS (1991) Cyclosporine induces elevated procollagen alpha 1 (I) mRNA levels in the rat renal cortex. Kidney Int 39:631–638.

Nast CC, Blifeld C, Danovitch GM, Fine RN, and Ettenger RB (1989) Evaluation of cyclosporine nephrotoxicity by renal transplant fine needle aspiration. Mod Pathol 2:577–582.

Natsumeda Y, and Carr SF (1993) Human type I and II IMP dehydrogenase as drug targets. Ann NY Acad Sci 696:88–93.

Neild GH, Rocchi G, Imberti L, Fumagalhi F, Brown Z, Remuzzi G, and Williams DG (1983) Effect of cyclosporine on prostacyclin synthesis by vascular tissue in rabbits. Transplant Proc 15:2398–2400.

Neoral prescribing information, Novartis Pharmaceuticals Corporation, East Janover, New Jersey 07936, August 2002.

Neylan JF (1998) Racial differences in renal transplantation after immunosuppression with tacrolimus versus cyclosporine. Transplantation 65:515–523.

Nielsen FT, Leyssac PP, Kemp E, Starklint H, and Dieperink H (1995) Nephrotoxicity of FK-506 in the rat. Studies on glomerular and tubular function, and on the relationship between efficacy and toxicity. Nephrol Dial Transplant 10:334–340.

Nitta K, Friedman AL, Nicastri AD, Paik S, and Friedman EA (1987) Granular juxtaglomerular cell hyperplasia caused by cyclosporine. Transplantation 44:417–421.

Norman DJ (1993) Rationale for OKT3 monoclonal antibody treatment in transplant patients. Transplant Proc 25 (2 Suppl 1):1–3.

Ojo AO, Meier-Kriesche HU, Hanson JA, Leichmann AB, Cibrik D, Magee JC, Wolf RA, Agodoa LY, and Kaplan B (2000) Mycophenolate mofetil reduces late renal allograft loss independent of acute rejection. Transplantation 69:2405–2409.

O'Leary TI, Ross PD, Lieber MR, and Levin IW (1986) Effects of cyclosporine A on biomembranes: Vibrational spectroscopic, calorinienic and hemolysis studies. Biophys J 49:795–801.

Oliveira JG, Xavier P, Sampaio SM, Heneriques C, Tavares I, Mendes AA, and Pestana M (2002) Compared to mycophenolate mofetil, rapamycin induces significant changes on growth factors and growth factor receptors in the early days post-kidney transplantation. Transplantation 73:915–920.

Parlevliet KJ, Bemelman FJ, Yong SL, Hack CE, Surachno J, Wilmink JM, ten Berge IJ, and Schellekens PT (1995) Toxicity of OKT3 increases with dosage: A controlled study in renal transplant recipients. Transplant Int 8:141–146.

Parry G, Meiser B, and Rabago G (2000) the clinical impact of cyclosporine nephrotoxicity in heart transplantation. Transplantation 69 (12 Suppl):SS23–SS26.

Pascual J, Marcen R, and Ortuno J (2001) Anti-interleukin-2 receptor antibodies: basiliximab and daclizumab. Nephrol Dial Transplant 16:1756–1760.

Peddi VR, Bryant M, Roy-Chaudhury P, Woodle ES, and First MR (2002) Safety, efficacy and cost analysis of thymoglobulin induction therapy with intermittent dosing based on CD^{3+} lymphocyte counts in kidney and kidney-pancreas transplant recipients. Transplantation 73:1514–1518.

Peltier A, and Russell JW (2002) Recent Advances in drug-induced neuropathies. Curr Opin Neurol 15:633–638.

Perico N, Benigni A, Zoja C, Delaini F, and Remuzzi G (1986a) Functional significance of exaggerated renal thromboxane A2 synthesis induced by cyclosporin A. Am J Physiol 251:F581–F587.

Perico N, Dadan I, and Remuzzi G (1990) Endothelin mediates the renal vasoconstriction induced by cyclosporine in the rat. J Am Soc Nephrol 1:76–83.

Perico N, Pasuni M, Gaspan F, Abbate M, and Remuzzi G (1991) Co-participation of thromboxane A2 and leukotriene C4 and D4 in mediating acute renal failure. Transplantation 52:873–878.

Perico N, Zoja C, Benigm A, Ghilardi F, Gualandris L, and Remuzzi G (1986b) Effect of short-term cyclosporine administration in rats on renin-angiotensi.n and thromboxane A2: Possible relevance to the reduction in glomerular filtration rate. J Pharmacol Exp Ther 239:229–235.

Pichler RH, Franceschini N, Young BA, Hugo C, Andoh TF, Burdmann EA, Shankland SJ, Alpers CE, Bennett WM, Couser WG, et al. (1995) Pathogenesis of cyclosporine nephropathy; Roles of angiotensin II and osteopontin. J Am Soc Nephrol 6:1186–1196.

Pirotzsky B, Colljez P, Guijmard C, Schaeverbeke I, and Broquet P (1988) Cyclosporin-induced nephrotoxicity: preventive effect of a PAF-acether antagonist BN52063. Transplant Proc 20:665–669.

Pirsch JD, Miller J, Deierhoi MH, Vincenti F, and Filo RS (1997) A comparison of tacrolimus (FK506) and cyclosporine for immunosuppression after cadaveric renal transplantation. FK506 Kidney Transplant Study Group. Transplantation 63:977–983.

Pisani F, Buonomo O, Laria G, Tisone G, Mazzaella V, Pollicita S, Camplone C, Piazza A, Valeri M, Famulari A, and Casciani CU (2001) Preliminary results of a prospective randomized study of basiliximab in kidney transplantation. Transplant Proc 33:2032–2033.

Pisanty S, Rahamim E, Ben-Ezra D, and Shoshan S (1990) Prolonged systemic administration of cyclosporin A affects gingival epithelium. J Periodontol 61:138–141.

Pisoni R, Ruggenenti P, and Remuzzi G (2001) Drug-induced thrombotic microangiopathy: Incidence, prevention and management. Drug Safety 24:491–501.

Podbielski J, and Schoenberg L (2001) Use of sirolimus in kidney transplantation. Prog Transplant 11:29–32.

Podder H, Stepkowski SM, Napoli KL, Clark J, Verani RR, Chou TC, and Kahan BD (2001) Pharmacokinetic interactions augment toxicities of sirolimus/cyclosporine combinations. J Am Soc Nephrol 12:1069–1071.

Polte T, Hemmerle A, Berndt G, Grosser N, Abate A, and Schroder H (2002) Atrial natriuretic peptide reduces cyclosporin toxicity in renal cells: Role of cGMP and heme oxygenase-1. Free Rad Biol Med 32:56–63.

Ponticelli C (1998) Optimization of cyclosporine therapy in renal transplantation. Transplant Proc 30:1718–1720.

Porter GA, Andoh TF, and Bennett WM (1999) An animal model of chronic cyclosporine nephrotoxicity. Ren Fail 21:365–368.

Potier M, Wolf A, and Cambar J (1998) Comparative study of cyclosporin A, cyclosporin G and the novel cyclosporin derivative IMM 125 in isolated glomeruli and cultured rat mesangial cells: A morphometric analysis. Nephrol Dial Transplant 13:1406–1411.

Powell-Jackson PR, Young B, Caine RY, and Williams R (1983) Nephrotoxicity of parenterally administered cyclosporine after orthotopic liver transplantation. Transplantation 36:505–508.

Powles RL, Barrett AJ, Clink H, Kay HEM, Sloane J, and McElwain TJ (1978) Cyclosporin A for the treatment of graft-versus-host disease in man. Lancet 2:1327–1331.

Prin-Mathieu C, Renoult E, De March AK, Bene MC, Kessler M, and Faure GC (1997) Serum anti-rabbit and anti-horse IgG, IgA and IgM in kidney transplant recipients. Nephrol Dial Transplant 112:2133–2139.

Prograf Prescribing Information, revised May 2002, Fujisawa Healthcare, Inc., Deefield, IL and Killorgin, Co. Kerry, Ireland.

Propper DJ, Whiting PH, Power DA, Edward N, and Catto GRD (1989) The effect of nifedipine on graft function in renal allograft recipients treated with cyclosporin A. Clin Nephrol 32:62–67.

Racusen LC, Colvin RB, Solez K, Mihatsch MJ, Halloran PF, Campbell PM, Cecka MJ, Cosyns JP, Demetris AJ, Fishbein MC, Fogo A, Furness P, Gibson IW, Glotz D, Hayry P, Hunsickern L, Kashgarian M, Kerman R, Magil AJ, Montgomery R, Morozumi K, Nickeleit V, Randhawa P, Regele H, Seron D, Seshan S, Sund S, and Trpkov K (2003). Antibody-mediated rejection criteria – an addition to the Banff 97 classification of renal allograft rejection. *Am J Transplant* 3:708–714.

Radermacher J, Meiners M, Bramlage C, Kleim V, Behrend M, Schlitt JH, Pichlmayr R, Koch KM, and Brunkhorst R (1998) Pronounced renal vasoconstriction and systemic hypertension in renal transplant patients treated with cyclosporine A versus FK506. Transplant Int 11:3–10.

Rahman MA, and Ing TS (1989) Cyclosporine and magnesium metabolism. Lab Clin Med 114:211–214.

Ramos E, Drachenberg CB, Papadimitrious JC, Hamze O, Fink JC, Klassen DK, Drachenberg RC, Wiland A, Wali R, Cangro CB, Schweitzer E, Bartlett ST, and Weir MR (2002) Clinical course of polyoma virus nephropathy in 67 renal transplant patients. J Am Soc Nephrol 13:2145–2151.

Randhawa PS, Shapiro R, Jordan ML, Starzl TE, and Demetris AJ (1993) The histopathological changes associated with allograft rejection and drug toxicity in renal transplant recipients maintained on FK506. Am J Surg Pathol 17:60–68.

Rao Ky, Crasson JT, and Kjellstrand CM (1985) Chronic irreversible nephrotoxicity from cyclosporin A. Nephron 41:75–77.

Rault R, Piraino B, Johnston JR, and Oral A (1991) Pulmonary and renal toxicity of intravenous immunoglobulin. Clin Nephrol 36:83–86.

Remuzzi G, Perico N, and Gaspari F (1998) Optimization of cyclosporine therapy in kidney transplantation. Transplant Proc 30:1673–1676.

Rodicio JL (2000) Calcium antagonists and renal protection from cyclosporine nephrotoxicity: Long-term trial in renal transplantation patients. J Cardiovasc Pharmacol 35 (3 Suppl 1):S7–S11.

Rodriguez-Puyol D, Lamas S, Olivera A, Lopez-Farre A, Ortega G, Hernando L, Lopez-Novoa JM (1989) Actions of cyclosporin A on cultured rat mesangial cells. Kidney Int 35:632–637.

Rogers TS, Elzinga L, Bennett WM, and Kelley VE (1988) Selective enhancement of thromboxane in macrophages and kidneys in CsA nephrotoxicity. Transplantation 45:153–156.

Rosen S, Greenfeld Z, and Brezis M (1990) Chronic cyclosporine-induced nephropathy in the rat. Transplantation 49:445–452.

Rosenstein M, Ettinghausen SE, and Rosenberg SA (1986) Extravasation of intravascular fluid mediated by the systemic administration of recombinant interleukin-2. J Immunol 137:1735–1742.

Rossi SJ, Schroeder TJ, Hariharan S, and First MR (1993) Prevention and management of the adverse effects associated with immunosuppressive therapy. Drug Safety 9:104–131.

Rowinski W, Korczak-Kowalska G, Samsel R, Zderska M, Chmura A, Wlodarczyk Z, Pliszczynski J, Wyzgal J, Cieciura T, Lagiewska B, Walaszewski J, Paczek L, Lao M, and Gorski A (2002) Can the immunosuppressive effect of perioperative single high-dose antithymocyte globulin administration in kidney allograft recipients be due to apoptosis of activated lymphocytes? Transplant Proc 34:1622–1624.

Rubinger D, Cohen E, Shiloni E, Rathaus M, Bernheim I, and Popovtzer MM (1990) Increased thromboxane (TXB2) mediates the adverse renal effects of interleukin-2 (IL-2) in rats. J Am Soc Nephrol 1:448.

Ruhlman A, and Nordheim A (1997) Effects of the immunosuppressive drugs CsA and FK506 on intracellular signaling and gene regulation. Immunobiol 198:192–206.

Ruiz-Ortega M, and Egido J (1997) Angiotensin II modulates cell growth-related events and synthesis of matrix proteins in renal interstitial fibroblasts. Kidney Int 52:1497–1510.

Ryffel B, Donatsch P, Hiestand P, and Mihatsch MJ (1986) Prostaglandin E2 reduces nephrotoxicity and immunosuppression of cyclosporine in rats. Clin Nephrol 25:595–599.

Ryffel B, Donatsch P, Madorin M, Matter BE, Ruttimann G, Schon H, Stoll R, and Wilson J (1983) Toxicological evaluation of cyclosporin A. Arch Toxicol 53:107–141.

Ryffel B, and Mihatsch MJ (1986) Cyclosporine nephrotoxicity. Toxicol Pathol 14:73–92.

Ryuzaki M, Stahl LK, Lyson T, Victor RG, and Bishop VS (1997) Sympathoexcitatory response to cyclosporine A and baroreflex resetting. Hypertension 29:576–582.

Sabbatini M, Sansone G, Uccello F, De Nicola L, Nappi F, and Andreucci VE (2000) Acute effects of rapamycin on glomerular dynamics: A micropuncture study in the rat. Transplantation 69:1946–1990.

Samlowski W, Schumann JL, Birchfield GR, Ward JH, and Schumann GB (1990) Cytologic evidence of nephrotoxicity in patients receiving high-dose interleukin-2. In Abstracts of the XIth international Congress of Nephrotoxicity. Tokyo, pp. 436A.

Sarna GP, Figlin RA, Pertckeck M, Altrock B, and Kradjian SA (1989) Systemic administration of recombinant methionyl human interleukin 2 (Ala 125) to cancer patients. Clinical results. J Biol Response Mod 8:l6–24.

Saydain G, George L, and Raoof S (2002) New therapies: Plasmapheresis, intravenous immunoglobulin, and monoclonal antibodies. Crit Care Clin 18:957–975.

Schaffner A, Thomann B, Zala GF, Ruegg R, Keusch G, Fehr J, and Gmur J (1991) Transient pure red cell aplasia caused by antilymphoblast globulin after cadaveric renal transplantation. Transplantation 51:1018–1023.

Schor N, Ichilcawa I, and Brenner BM (1981) Mechanisms of action of various hormones and ultrafiltration in the rat. Kidney Int 20:442–45l.

Schurek HJ, Neumann KH, Jessinghaus WP, Aeikens B, and Wonigeit K (1986) Influence of cyclopsorone A on adaptive hypertrophy after unilateral nephrectomy in the rat. Clin Nephrol 25:S144–S147.

Schwarz C, Bohmig GA, Steininger R, Mayer G, and Oberbauer R (2001) Impaired phosphate handling of renal allografts is aggravated under Rapamycin-based immunosuppression. Nephrol Dial Transplant 16:378–382.

Schwass DE, Sasaici AW, Houghton DC, Benner KE, and Bennett WM (1986) Effect of phenobarbital and cimetidine on experimental cyclosporine nephrotoxicity: Preliminary observations. Clin Nephrol 25:S117–S120.

Scoble JE, Senior JCM, Chan P, Varghese Z, Sweny P, and Moorhead JF (1989) In vitro cyclosporine toxicity. The effect of verapamil. Transplantation 47:647–650.

Sewell WA, and Jolles S (2002) Immunomodulatory action of intravenous immunoglobulin. Immunology 107:387–393.

Shah MB, Martin JE, Schroeder TJ, and First MR (1999) The evaluation of the safety and tolerability of two formulations of cyclosporine: Neoral and sandimmune. A meta-analysis. Transplantation 67:1411–1417.

Shapiro R (1998) Tacrolimus in pediatric renal transplantation: A review. Pediatr Transplant 2:270–276.

Shenton BK, White MD, Bell AE, Clark K, Rigg KM, Forsythe JL, Proud G, and Taylor RM (1994) The paradox of ATG monitoring in renal transplantation. Transplant Proc 26:3177–3180.

Shiel AGR, Hall BM, Tiller DI, et al. (1984) Australian trial of cyclosporine (CsA) in cadaveric donor renal transplantation, in *Cyclosporine: Biological Activity*

and Clinical Applications, Kahan BD (Ed.), Grime & Stratton, Orlando, FL, pp. 307.
Shihab FS, Andoh TF, Tanner AM, and Bennett WM (1996) Angiotensin II (AII) blockade ameliorates fibrosis and decreases TGF-[beta]-1 expression in chronic cyclosporine (CsA) nephropathy. J Am Soc Nephrol 7:1846.
Shihab FS, Bennett WM, Tanner AM, and Andoh TF (1997a). Angiotensin II blockade decreases TGF-[beta]-1 and matrix proteins in cyclosporine nephropathy. Kidney Int 52:660–673.
Shihab FS, Bennett WM, Tanner AM, and Andoh TF (1997b) Mechanism of fibrosis in experimental tacrolimus nephrotoxicity. Transplantation 64:1829–1837.
Shihab FS, Bennett WM, Yi H, and Ando TF (2001) Expression of vascular endothelial growth factor and its receptors Flt-1 and KDR/Flk-1 in chronic cyclosporine nephrotoxicity. Transplantation 72:164–168.
Shihab FS, Bennett WM, Isaac J, Yi H, and Andoh TF (2002a) Angiotensin II regulation of vascular endothelial growth factor and receptors Flt-1 and kDR/Flk-1 in cyclosporine nephrotoxocity. Kidney Int 62:422–433.
Shihab FS, Bennett WM, Yi H, and Andoh TF (2002b) Perfenidone treatment decreases transforming growth factor beta1 and matrix proteins and meliorates fibrosis in chronic cyclosporine nephrotoxicity. Am J Transplant 2:111–119.
Shorr AF, and Kester KE (1996) Meningitis and hepatitis complication intrvenous immunoglobulin therapy. Ann Pharmacother 30:1115–1116.
Shulman H, Striker G, Deeg RI, Kennedy M, Storb R, and Thomas ED (1981) Nephrotoxicity of cyclosporin A after allogeneic marrow transplantation. Glomerular thromboses and tubular injury. N Engl J Med 305:l392–1395.
Siegl H, Ryffel B, Petric R, Shoemaker P, Muller A, Donatsch P, and Mihatsch M (1983) Cyclosporine, the renin-angiotensin-aldosterone system, and renal adverse reactions. Transplant Proc 15:2719–2725.
Sifontis NM, Benedetti E, and Vasquez EM (2002) Clinically significant drug interaction between basiliximab and tacrolimus in renal transplant recipients. Transplant Proc 34:1730–1732.
Simon N, Zini R, Morin C, et al (1997) Prednisolone and azathioprine worsen the cyclosporine A-induced oxiative phosphorylation decrease of kidney mitochondria. Life Sci 61:659–666.
Singer SJ, Tiernan R, and Sullivan EJ (2000) Interstitial pneumonitis associated with sirolimus therapy in renal transplant recipients. N Engl J Med 14:1815–1816.
Solez K, Racusen L, Marcussen N, Slatnik 1, Keown P, Burdjck IF, and Olsen S (1993) Morphology of ischemic acute renal failure, normal function and cyclosporine nephrotoxicity in cyclosporine treated renal allograft recipients. Kidney Int 43:1058–1067.
Sollinger HW (1995) U.S. Renal Transplant Mycophenolate Mofetil Study Group. Mycophenolate mofetil for prevention of acute rejection in primary cadaveric renal allograft recipients. Transplantation 60:225–232.
Sollinger HW, Deierhoi MH, Belzer FO, Diethelm AG, and Kauffman RS (1992) RS-61443 – A phases I clinical trial and pilot rescue study. Transplantation 53:428–432.
Soulillou JP, Cantarovich D, Le Mauff B, Giral M, Robillard N, Hourmant M, Hirn M, and Jacques Y (1990) Randomized controlled trial of a monoclonal antibody against the

interleukin-2 receptor (33B3.1) as compared with rabbit antithymocyte globulin for prophylaxis against rejection of renal allografts. N Engl J Med 26:1224–1226.

Spencer CM, Goa KL, and Gillis JC (1997) Tacrolimus. An update of its pharmacology and clinical efficacy in the management of organ transplantation. Drugs 54:925–975.

Stahl RAK, and Kudelka S (1986) Chronic cyclosporine A treatment reduces prostaglandin B2 formation in isolated glomeruli and papilla of rat kidneys. Clin Nephrol 25:578–582.

Stahl RA, Adler S, Baker PJ, Johnson RJ, Chen YP, Pritzl P, Couser WG (1989) Cyclosporin A inhibits prostaglandin E2 formation by rat mesangial cells in culture. Kidney Int 35:1161–1167.

Steinmetz T, Schaadt M, Gahl R, Schenk V, Diehl V, and Pfreundschuh M (1988) Phase I study of 24-hour continuous infusion of recombinant human tumor necrosis factor. J Biol Response Mod 7:417–423.

Stillman TE, Andoh TF, Burdmann EA, Bennett WM, and Rosen S (1995) FK506 nephrotoxicity: Morphologic and physiologic characterization of a rat model. Lab Invest 73:794–803.

Strehlau J, Pape L, Offner G, Nashan B, and Ehrich JH (2000) Interleukin-2 receptor antibody-induced alterations of ciclosporine dose requirements in paediatric transplant recipients. Lancet 356:1327–1328.

Streit F, Christians U, Scheibel HM, Napoli KL, Ernst L, Linck A, Kahan BD, and Sewing KF (1996) Sensitive and specific quantitation of sirolimus (rapamycin) and its metabolites in blood of kidney graft recipients by HPLC/electrospray-mass spectrometry. Clin Chem 42:1417–1425.

Strzelecki T, Kuman 5, Khauli R, and Menon M (1988) Impairment by cyclosporine of membrane-mediated functions in kidney mitochondria. Kidney Int 34:234–240.

Suthanthiran M, Morris RE, and Strom TB (1996) Immunosuppressants: Cellular and molecular mechanisms of action. Am J Kidney Dis 28:159–172.

Swinford RD, Pascual M, Diamant D, Tang SS, and Ingelfinger JR (2002) Rapamycin increases transforming growth factor-beta mRNA expression in immortalized rat proximal renal tubular cells. Transplantation 73:319–320.

Takahashi N, Hayano T, and Suzuki M (1989) Peptidyl-prolyl cis-trans isomerase is the cyclosporin A-binding protein cyclophilin. Nature 337:473–475.

Takenaka T, Hashimoto Y, and Epstein M (1992) Diminished acetylcholine-induced vasodilatation in renal micro-vessels of cyclosporin-treated rats. J Am Soc Nephrol 3:42–50.

Taler SJ, Textor SC, Canzanello VJ, and Schwartz L (1999) Cyclosporine-induced hypertension: incidence, pathogenesis and management. Drug Safety 20:437–449.

Tatum AH, Bollinger RR, and Sanfilippo F (1984) Rapid serologic diagnosis of serum sickness from antithymocyte globulin therapy using enzyme immunoassay. Transplantation 38:582–586.

Thomas SE, Andoh TF, Pichler RH, Shankland SJ, Couser WG, BennettWM, and Johnson RJ (1998) Accelerated apoptosis characterizes cyclosporine-associated interstitial fibrosis. Kidney Int 53:897–908.

Thompson HL, Burbelo PD, Gabriel G, Yamada Y, and Metcalfe DD (1991) Murine mast cells synthesize basement membrane components: A potential role in early fibrosis. J Clin Invest 87:619–623.

Thomson AW, Whiting PH, and Simpson JG (1984) Cyclosporine: Immunology, toxicity and pharmacology in experimental animals. Agents Actions 15:306–327.

Thomson AW, Whiting PH, Blair IT, Davidson RJL, and Simpson IG (1981) Pathological changes developing in the rat during a 3-week course of high dosage cyclosporin A and their reversal following drug withdrawal. Transplantation 32:271–277.

Thomson SC, Tucker BJ, Gabbaj F, and Blantz RC (1989) Functional effects on glomerular hemodynamics of short term chronic cyclosponne in male rats J Clin Invest 83:960–969.

Tocci MJ, Matkovich DA, Collier KA, Kwok P, Dumont F, Lin S, Siekierka JJ, Chin J, and Hutchinson NI (1989) The immunosuppressant FK506 selectively inhibits expression of early T-cell activation genes. J Immunol 143:718.

Tricontinental Mycophenolate Mofetil Renal Transplant Study Group (1996) A blinded randomized clinical trial of Mycophenolate mofetil for the prevention of acute rejection in cadaveric renal transplantation. Transplantation 61:1029–1033.

Tsimaratos M, Viard L, Kreitmann B, Remediani C, Picon G, Camboulives J, Sarles J, and Metras D (2000) Kidney function in cyclosporine-treated paediatric pulmonary transplant recipients. Transplantation 69:2055–2059.

Tucker BJ, Mundy CA, and Blantz RC (1987) Adrenergic and angiotensin II influences on renal vascular tone in chronic sodium depletion. Am J Physiol 251:F811–F817.

Tumlin JA (1997) Expression and function of calcineurin in the mammalian nephron: Physiological roles, receptor signaling and ion transport. Am J Kidney Dis 30:884–895.

van der Heide JJH, Bib HIG, Tegzess AM, and Donker AIM (1990) The effects of dietary supplementation with fish oil on renal function in cyclosporine-treated renal transplant recipients. Transplantation 49:523–527.

van Hooff JP, and Christiaans MHL (1999) Use of tacrolimus in renal transplantation. Transplant Proc 31:3298–3299.

van Hooff JP, Boots JMM, van Duijnhoven EM, and Christiaans MHL (1999) Dosing and management guidelines for tacrolimus in renal transplant patients. Transplant Proc 31S7A:54S–57S.

Vanrenterghem Y (1997) The use of mycophenolate mofetil (Cellcept) in renal transplantation. Nephron 76:392–399.

Vasquez EM, and Pollak R (1997) OKT3 therapy increases cyclosporine blood levels. Clin Transplant 11:38–41.

Vercauteren SB, Bosmans JL, Elseviers MM, Verpooten GA, and de Broe ME (1998) A meta-analysis and morphological review of cyclosporine-induced nephrotoxicity in auto-immune diseases. Kidney Int 54:536–545.

Vilalta R, Vila A, Nieto J, Espanol T, Caragol I, and Callis L (2002) Experience with basiliximab in pediatric renal transplantation. Transplant Proc 34:100–101.

Vincenti F (2001) The role of newer monoclonal antibodies in renal transplantation. Transplant Proc 33:1000–1001.

Walker RJ, Lazzaro VA, Duggin GG, Horvath IS, and Tiller DJ (1990) Cyclosporin~induced alterations in renal metabolism: The role of lipid peroxidation and enzyme inhibition in cyclosporine nephrotoxicity. Transplantation 50:487–492.

Walker RI, Lazzaro VA, Duggin GG, Horvath IS, and Tiller DJ (1989) Cyclosposin A inhibits protein kinase C activity: A contributing mechanism in the development of nephrotoxicity. Biochem Biophys Res Commun 160:409–415.

Wang W, Mo S, and Chan L (1999) Mycophenolic acid inhibits PDGF-induced osteopontin expression in rat mesangial cells. Transplant Proc 31:1176–1177.

Wang W, Chan YH, Lee W, and Chan L (2001) Effect of rapamycin and FK506 on mesangial cell proliferation. Transplant Proc 33:1036–1037.

Welbourn R, Goldman G, Kobzik L, Valeri CR, Shepro D, and Hechtman HB (1990) Involvement of thromboxane and neutrophils in multiple-system organ edema with interleukin-2. Ann Surg 212:728–733.

Welbourn R, Goldman G, Kobzik L, Valeri CR, Hechtman HB, and Shepro D (1991) Attenuation of IL-2-induced multisystem organ edema with phalloidin and antamanide. J AppI Physiol 70:1364–1368.

Whiting PH, Duncan JI, Gavin MP, Heys SD, Simpson JG, Asfar SK, and Thomson AW (1985) Renal function in rats treated with cyclosporin following unilateral nephrectomy. Br J Exp Pathol 66:535–542.

Wiedmann TS, Trouard T, Shekar SC, Polikandritou M, and Raliinan Y-E (1990) Interaction of cyclosporin A with dipalmitoylphospfiatidylcholine. Biochim Biophys Acta 1023:12–18.

Wiegmann TB, Shanna R, Diederich DA, and Savin V (1990) In vitro effects of cyclosporine on glomerular function. Am J Med Sci 299:149–152.

Wiener Y, Nakhleh RE, Lee MW, Escobar FS, Venkat KK, Kupin WL, and Mozes MF (1997) Prognostic factors and early resumption of cyclosporine A in renal allograft recipients with thrombotic microangiopathy and hemolytic uremic syndrome. Clin Transplant 11:157–162.

Wilson PD, and Hreniuk D (1988) Nephrotoxicity of cyclosporine in renal tubule cultures and attenuation by calcium restriction. Transplant Proc 20:709–711.

Winkler M, Ringe B, Baumann J, Loss M, Wonigeit K, and Pichlmayr R (1994) Plasma vs. whole blood for therapeutic drug monitoring of patients receiving FK506 for immunosuppression. Clin Chem 40:2247–2253.

Winston I, Feingold R, and Safirstein R (1989) Glomerular hemodynamics in cyclosporine nephrotoxicity following nephrectomy. Kidney Int 35:1175–1182.

Wiseman LR, and Faulds D (1999) Daclizumab: A review or its use in the prevention of acute rejection in renal transplant recipients. Drugs 58:1029–1042.

Wolf G, Killen PD, and Neilson EG, (1990) Cyclosporin A stimulates transcription and procollagen secretion in tubulointerstitial fibroblasts and proximal tubular cells. J Am Soc Nephrol 1:918–922.

Xue H, Bukoski D, McCarron DA, and Bennett WM (1987) Induction of contraction in isolated rat aorta by cyclosporine. Transplantation 43:715–718.

Yang CW, Ahn HJ, Kim WY, Shin MJ, Kim SK, Park HJ, Kim YO, Kim YS, Kim J, and Bang BK (2001) Influence of the renin-angiotensin system on epidermal growth factor expression in normal and cyclosporine-treated rat kidneys. Kidney Int 60:847–857.

Yang CW, Ahn HJ, Kim WY, Li C, Kim HW, Choi BS, Cha JH, Kim YS, Kim J, and Bang BK (2002a) Cyclosporine withdrawl and mycophenolate mofetil treatment effects on the progression of chronic cyclosporine nephrotoxicity. Kidney Int 62:20–30.

Yang CW, Faulkner GR, Wahba IM, Christianson TA, Bagby GC, Jin DC, Abboud HE, Andoh TF, and Bennett WM (2002b) Expression of apoptosis-related genes in chronic cyclosporine nephrotoxicity in mice. Am J Transplant 2:391–399.

Young BA, Burdmann EA, Hjohnson RJ, alpers CE, Giachelli CM, Eng E, Andoh T, Bennett WM, and Couser WG (1995) Cellular proliferation and macrohage influx precede interstitial fibrosis in cyclosporine nephrotoxicity. Kidney Int 48:439–448.

Zachariae H (1999) Renal toxicity of long-term ciclosporin. Scan J Rheumatol 28:65–68.

Zent R, Katz A, Quaggin S, Cattran D, Wade J, and Cardella C (1997) Thrombotic microangiopathy in renal transplant recipients treated with cyclosporine A. Clin Nephrol 47:181–186.

Ziegler K, Frimmer M, Fritzsch G, and Koepsell H (1990) Cyclosporin binding to a protein component of the renal Na^+-D-glucose cotransporter. J Biol Chem 265:3270–3277.

Zimmerhackl LB, Fretschner M, and Steinhausen M (1990) Cyclosporin reduces renal blood flow through vasoconstriction of arcuate arteries in the hydronephrotic rat model. Klin Wochenschr 68:166–174.

Zimmerman JJ, Ferron GM, Lim HK, and Parker V (1999) The effect of a high fat meal on the oral bioavailability of the immunosuppressant sirolimus (rapamycin). J Clin Pharmacol 39:1155–1161.

Zimmerman JJ, and Kahan BD (1997) Pharmacokinetics of sirolimus in stable renal transplant patients after oral administration. J Clin Pharmacol 37:405–415.

Zoja C, Furci L, Ghilardi F, Zilio P, Benigni A, and Remuzzi G (1986) Cyclosporin-induced endothelial cell injury. Lab Invest 55:455–462.

18

Cisplatin-Induced Nephrotoxicity

Ramon Bonegio and Wilfred Lieberthal

INTRODUCTION

Cis-diamminedichloroplatinum (cisplatin) is a potent chemotherapeutic agent that is widely used for the treatment of many malignancies. Cisplatin was first synthesized in 1844 by Peyrone who named the compound "Peyrone's chloride" (Peyrone, 1844). The elucidation of its chemical structure formed the basis for the Nobel prize winning work of Albert Werner (Werner, 1911). However, it was not until 1965 that the therapeutic potential of platinum-based agents became evident. Rosenberg and colleagues (Rosenberg et al., 1965) serendipitously discovered that exposure of *Escherichia coli* to an electric field inhibited bacterial proliferation, and found that this effect was due to the formation of a neutral, tetravalent platinum compound (diamminedichloroplatinum) created by electrolysis of the platinum-containing electrode placed into the ammonium chloride rich bacterial culture medium. Numerous studies subsequently demonstrated that it was predominantly the *cis* isomer of diamminedichloroplatinum (named cisplatin) that was responsible for inhibiting bacterial growth (Howle and Gale, 1970a, 1970b; Rosenberg et al., 1965, 1967, 1969). Cisplatin was shown to also inhibit the proliferation of mammalian cells (Harder and

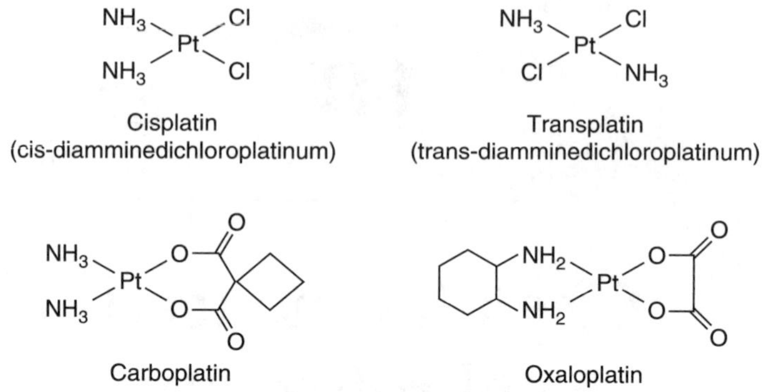

Figure 18.1 The structures of four organoplatinum compounds.

Rosenberg, 1970; Rosenberg and VanCamp, 1970; Rosenberg et al., 1969) and to slow the growth of metastatic germ cell (ovarian and testicular) tumors in humans (Ellerby et al., 1974; Hayat et al., 1979; Higby et al., 1974). These findings led to the development of cisplatin as a chemotherapeutic agent.

Cisplatin and its organoplatinum derivatives (Figure 18.1) have been used for more than a quarter century as part of a number of chemotherapeutic regimens developed for the treatment of many neoplasms. These include hematological and sarcomatous malignancies (Budd et al., 1993; Carli et al., 1987; Hill et al., 1975; Thigpen et al., 1986) and solid tumors involving the head and neck (Baur et al., 2002; Schoffski et al., 2000), lung (De Jager et al., 1980; Gatzemeier et al., 1992; Masuda, 2001; Scagliotti et al., 2002), esophagus (Ajani et al., 2002), colon (Giacchetti et al., 2000), uterus and cervix (Adachi et al., 1998; Ajani et al., 2001; Dazzi et al., 2000), testis (Bissett et al., 1990, Daugaard et al., 1988c; Fjeldborg et al., 1986; Groth et al., 1986; MacLeod et al., 1988; Osanto et al., 1992), ovary (Hannigan et al., 1993; Misset et al., 2001; Swenerton et al., 1992) and bladder (Bellmunt et al., 1997). Platinum-derived agents as a group are among the most commonly used chemotherapeutic agents, and new regimens containing these agents continue to be devised.

Cisplatin has a narrow therapeutic index. The neuro-, oto- and nephrotoxicity of cisplatin limit individual and cumulative doses and restrict its therapeutic potential. The direct renal tubular cell toxicity of cisplatin was recognized soon after therapeutic trials began (Safirstein et al., 1984; Gonzales-Vitale et al., 1977; Schabel, 1978).

Aggressive volume expansion with saline administration successfully attenuates the renal toxicity of cisplatin (see below). However, despite hydration with saline, patients treated with regimens containing cisplatin permanently lose 10 to 30% of their renal function (Bissett et al., 1990; Blachley and Hill, 1981; Daugaard et al., 1988a; Groth et al., 1986; Hansen et al., 1988; Hamilton et al., 1989; Hansen, 1992; Hartmann et al., 2000a, 2000b; Hayes et al., 1977; MacLeod et al., 1988; Osanto et al., 1992) (Table 18.1). In addition, acute or chronic renal dysfunction is one of the commonest adverse effects that necessitates a delay or change in chemotherapy. More than 10 analogs of cisplatin have been developed and clinically tested in an attempt to develop agents that reduce toxicity while retaining anti-neoplastic efficacy. Studies of these agents in humans have provided useful data that has improved the management of patients who require platinum-based chemotherapy (Sorensen et al., 1985; Goren et al., 1987; Gouyette et al., 1986; Doz, 2000; Cornelison and Reed, 1993; Cascinu et al., 2000; Boisdron-Celle et al., 2001; Larena et al., 2001; Mizumura et al., 2001; Garnuszek et al., 2002; Zhang et al., 2001).

This chapter discusses the mechanisms underlying cisplatin-induced cellular injury, which are also the basis of its antineoplastic activity. The renal handling of cisplatin and the mechanisms underlying cisplatin-associated nephrotoxicity are described, along with the many potential clinical manifestations of cisplatin nephrotoxicity. Finally, the strategies that have been developed to minimize adverse effects of organoplatinum compounds on the kidneys are examined.

CISPLATIN IS "ACTIVATED" UPON ENTRY INTO THE CELL

More than 90% of circulating cisplatin is bound to plasma proteins, and in this form it is not toxic. Furthermore, the high chloride concentration of extracellular fluid maintains the free cisplatin (not protein-bound) as a nonreactive neutral dichloride (Chu, 1994; Jordan and Carmo-Fonseca, 2000) (Figure 18.2). Cisplatin and its analogs become toxic only after they enter cells. The low concentration of chloride of intracellular fluid results in the displacement of the chloride ion of cisplatin by water, forming aquated cationic products (Figure 18.2). Aquated cisplatin is formed by replacement of the chloride ions by hydroxyl ions. Aquated cisplatin can interact with macromolecules within the cell that contain potential electron donors (such as oxygen, nitrogen, and sulfur atoms) to form cisplatin adducts (Jordan and Carmo-Fonseca, 2000;

Table 18.1 Long-Term Renal Effects of Cisplatin Therapy for Testicular Carcinoma

Study	*Regimen*	*Preventive measures*	*Chronic changes*
Fjeldborg et al., 1998 $n=22$	Cisplatin 20 mg/m^2 for 5 days + vinblastine and bleomycin	Saline and furosemide	GFR* decreased in 82% of patients GFR decreased by > 30% in 18% patients
Groth et al., 1986 $n=9$	Cisplatin 20 mg/m^2 for 5 days + vinblastine and bleomycin	Saline	GFR decreased by 30%
Daudaard et al., 1988 $n=30$	Cisplatin 20 mg/m^2 or cisplatin 40 mg/m^2 for 5 days + etoposide	Saline and 3% Mannitol	GFR decreased > 50% in 33% of patients Individual and cumulative dose both predictive Decline generally not reversible
Macleod et al., 1988 $n=22$	Cisplatin 100 mg/m^2 vinblastine and bleomycin	Saline and mannitol	GFR decreased in 85% by an average of 23% 55% still had chronic renal failure at 6 months Older patients at higher risk
Bissett et al., 1990 $n=74$	Cisplatin 400 mg/m^2 (average cumulative dose)	Saline	GFR decreased by an average of 30%
Osanto et al., 1992 $n=57$	Cisplatin 20 mg/m^2 for 5 days	Saline and mannitol	GFR decreased by an average of 15% No recovery in follow-up.

*GFR: glomerular filtration rate.

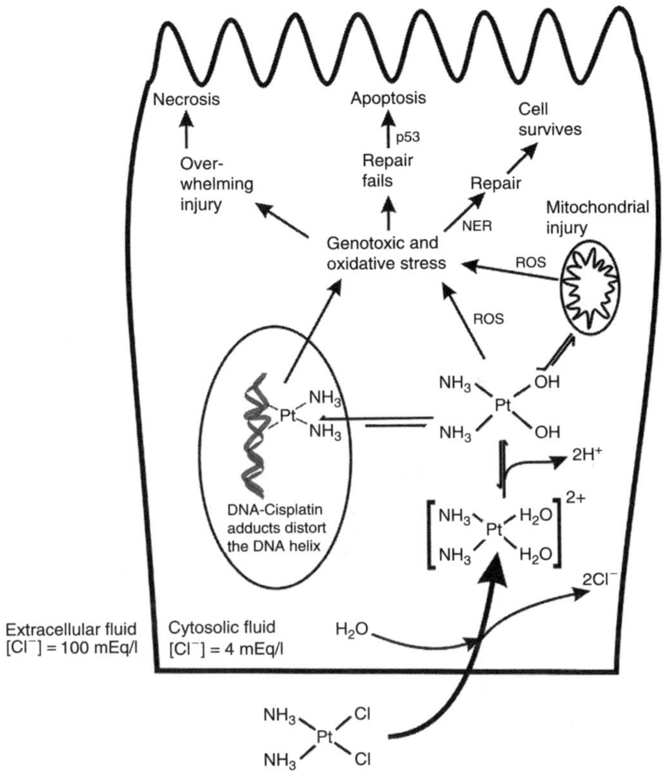

Figure 18.2 Mechanisms of acivation and cytotoxicity of cisplatin. Cisplatin becomes aquated after its entry into the hypochloric milieu of cytosolic fluid and induces injury by forming adducts with DNA and by inducing oxidative stress. Cells injured by cisplatin may survive if DNA injury can be repaired and if the cell is able to survive the oxidative stress. Irreparable DNA damage generally induces apoptosis, while overwhelming genotoxic and oxidative stress leads to necrosis. (NER) nucleotide excision repair complex, (ROS) reactive oxygen species.

Kartalou and Essigmann, 2001). The macromolecules with which cisplatin interacts include membrane phospholipid, DNA, RNA, and several of the R-groups of amino acids present in enzymes and structural proteins (Gonzalez et al., 2001; Johnson et al., 1980; Jordan and Carmo-Fonseca, 2000; Kartalou and Essigmann, 2001). Cisplatin adducts are highly toxic (see below).

The mechanisms involved in the cellular uptake of cisplatin remain uncertain. Available data are consistent with entry through transmembrane channels or via facilitated transport (Gately and Howell, 1993; Gonzalez et al., 2001). In addition, more-recent studies have implicated Ctr1, a high affinity copper transporting protein, in cisplatin uptake by cells. Cells lacking Ctr1 accumulate cisplatin relatively slowly and are resistant to cisplatin toxicity (Ishida et al., 2002; Lin et al., 2002). Copper, which downregulates the expression of Ctr1 expression, inhibits cellular uptake of cisplatin (Ishida et al., 2002). In addition to the Ctr1 transporter, the proximal tubular cationic transporter has been implicated in the disproportional accumulation of cisplatin in renal tubular cells (see below)(Endo et al., 2000; Groth et al., 1986; Grover et al., 2002: Safirstein et al., 1984).

MECHANISMS OF CISPLATIN-INDUCED CYTOTOXICITY: ADDITIVE ROLES OF DNA DAMAGE AND OXIDATIVE STRESS

Oxidative Stress

The antineoplastic and cytotoxic effects of cisplatin are mediated by the combined effects of cisplatin-adduct induced injury and oxidant stress. Cisplatin promotes the production of the hydroxyl radical. The hydroxyl radical is a potent free radical that injures cell membranes, proteins, and DNA, and also depletes intracellular antioxidant stores, such as glutathione (Levi et al., 1980; Masuda et al., 1994; Ries and Klastersky, 1986; Townsend et al., 2003) (Figure 18.3).

Hydroxyl radicals can originate from aquation intermediates of cisplatin and other organoplatinum molecules (Figure 18.2). In addition, cisplatin also increases hydroxyl radical formation indirectly by injuring mitochondria, an effect that occurs in many cells (including proximal tubular cells) (Brady et al., 1990; Chang et al., 2002) (Figure 18.3). Cisplatin-mediated injury to the mitochondrial membrane impairs oxidative phosphorylation thereby increasing the release of superoxide (Figure 18.3). Superoxide dismutase (SOD) converts superoxide to hydrogen peroxide, which can then be converted to the hydroxyl radicals via the iron requiring Fenton reaction (Figure 18.3). Cisplatin facilitates hydroxyl radical formation via the Fenton reaction by inducing the release of catalytically free iron from cytochrome P450 (Baliga et al., 1998b).

The oxidant stress induced by cisplatin can lead to cell dysfunction or death. Cisplatin-induced cell death usually occurs by apoptosis

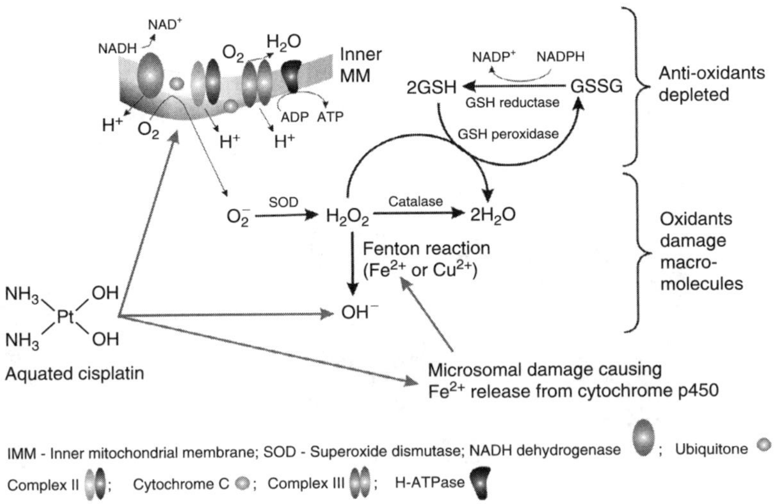

Figure 18.3 Mechanisms of cisplatin-induced oxidative stress. Oxidative stress may occur as the direct result of aquation of cisplatin, or indirectly as a result of damage to mitochondria. Impaired function of complexes I through IV of the respiratory chain generates superoxide. Superoxide is converted to hydrogen peroxide by a reaction catalyzed by superoxide dismutase. Hydrogen peroxide can then be converted to water by catalase. However, if free, ferrous iron is available, hydrogen peroxide can be converted to the highly reactive hydroxyl radical via the Fenton reaction. Cisplatin can promote hydroxyl radical formation by releasing catalytically free iron from cytochrome P450. Hydroxyl radical generation depletes intracellular antioxidant systems such as glutathione causing a vicious cycle. Oxidant stress leads to cell injury via peroxidation of plasma and membrane phopholipids, DNA, and proteins.

(Chang et al., 2002; Lieberthal et al., 1996) but in the presence of very high concentrations of cisplatin necrotic cell death can also occur (Chang et al., 2002; Lieberthal et al., 1996) (Figure 18.4). Most of the evidence suggesting the importance of oxidant stress in cisplatin-induced injury comes from numerous studies that demonstrate that antioxidants protect normal tissues from the effects of cisplatin and other organoplatinum compounds (Antunes et al., 2001; Babu et al., 1999; Baliga et al., 1998a; Bolaman et al., 2001; Camargo et al., 2001; Capizzi, 1999; DiPaola and Schuchter, 1999; Elsendoorn et al., 2001; Francescato et al., 2001; Glover et al., 1989; Glover et al., 1986; Gradishar et al., 2001; Hanada et al., 2000; Hara et al., 2001; Hartmann et al., 2000c;

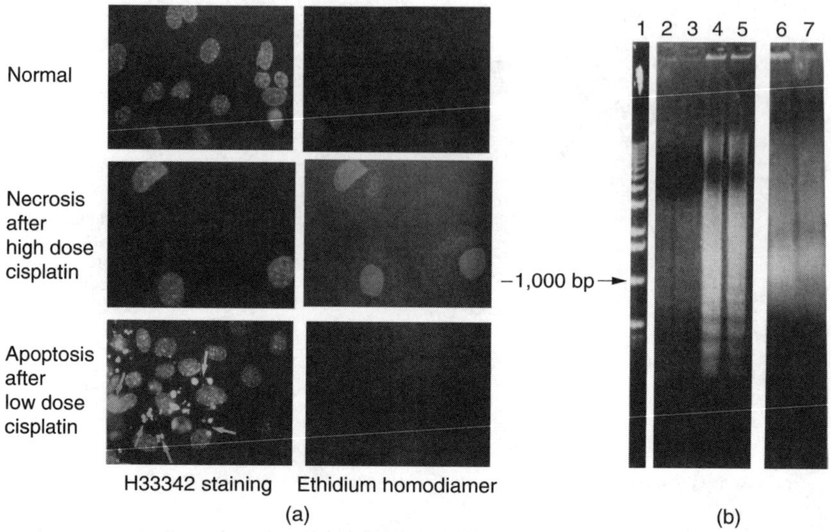

Figure 18.4 Cisplatin can induce apoptosis or necrosis of proximal tubular cells. (A) Fluorescence microscopy of primary cultured mouse proximal tubular (MPT) cells stained with Hoechst 33342 (H33342) and ethidium bromine homodimer (EH). H33342 enters viable as well as dead cells and intercalates with nuclear DNA. As a result, nuclei become fluorescent. By contrast, EH is only able to enter necrotic cells because the plasma membrane of the cell is severely injured. In normal (vehicle-treated) cells (upper panel), H33342 stained nuclei have normal morphology and the cells exclude EH. MPT cells exposed to a high dose cisplatin (800 μM) are necrotic, as evidenced by EH positive nuclei (middle panel). Many of the MPT cells subjected to low concentrations of cisplatin (50 μM) are apoptotic while some are still viable (bottom panel). The nuclei of apoptotic cells stained with H33342 are intensely fluorescent, condensed, and fragmented, while viable cells have nuclei with normal morphology. EH is excluded by both viable and apoptotic cells. (B) DNA extracted from MPT cells was subjected to agarose gel electrophosis. Lane 1: molecular size markers. Lanes 2 and 3: normal cells (exposed to vehicle), showing minimal fragmentation of DNA. Lanes 3 and 4: DNA from apoptotic cells showing fragmentation in a "ladder" pattern, typical of apoptosis. Lanes 6 and 7: necrotic MPT cells exposed to high dose cisplatin showing fragmentation of DNA in a "smear" pattern, characteristic of necrosis. (Taken from Lieberthal et al., 1996, with permission.)

Lieberthal et al., 1996; Plaxe et al., 1994; Sener et al., 2000; Sriswasdi et al., 2000; Walker and Gale, 1981). Interestingly, most studies have failed to show any effect of antioxidants on the antineoplastic efficacy of organoplatinum compounds.

DNA Damage ("Genotoxic Stress")

Cisplatin and its analogs (Figure 18.1) form DNA cross-links (adducts) by binding directly to purine bases (Bancroft et al., 1990; Stone et al., 1974). While platinum can also complex with RNA and proteins, these events require far higher intracellular concentrations of cisplatin and, as a result, represent less than 15% of the total cisplatin adducts formed at therapeutic doses (Akaboshi et al., 1992; Bancroft et al., 1990; Safirstein et al., 1986). A number of pieces of evidence suggest that the fate of a cell exposed to cisplatin is determined largely by the balance between the rates of formation and repair of cisplatin–DNA adducts. Cells with defective DNA repair systems are far more sensitive to cisplatin-induced damage than are normal cells (Dijt et al., 1988; Fornace and Seres, 1982; Gonzalez et al., 2001; Poll et al., 1982, 1985). Cisplatin adducts of DNA are planar and are thought to alter the three-dimensional structure of the double helix (Gelasco and Lippard, 1998; Huang et al., 1995a, 1995b; Kartalou and Essigmann, 2001; Paquet et al., 1996; Rice et al., 1988; Takahara et al., 1995).

The three-dimentional structure of the DNA adduct varies with the nature of the organoplatinum compound as well as the type of adduct (mono-, interstrand, or intrastrand). DNA adducts of different organoplatinum compounds are repaired with varying degrees of efficiency by the cell. These differences largely explain the variations in toxicity and efficacy between cisplatin and its analogs. For example, DNA adducts of transplatin (Figure 18.1), which is a relatively nontoxic organoplatinum compound with poor antineoplastic efficacy, are repaired more efficiently than DNA adducts induced by cisplatin and other organoplatinum compounds (see below).

CISPLATIN-INDUCED DNA DAMAGE

Impaired DNA Repair

Nucleotide Excision Repair Complex

Recognition and repair of DNA–platinum adducts is necessary for cell survival and normal cell function (Gonzalez et al., 2001). DNA–platinum adducts impair DNA replication and transcription by retarding the movement and accuracy of DNA and RNA polymerases. The nucleotide excision repair (NER) complex, which is responsible for repairing ultraviolet-induced pyrimidine cross-links, repairs platinum–DNA adducts (Brandsma et al., 1996; Mu et al., 1997; Zamble et al., 1996).

When RNA polymerase II (RNA Pol II), the complex that transcribes mRNA from DNA, encounters a platinum–DNA adduct it stalls at the site of DNA damage. Once stationary, the RNA pol II complex recruits the NER apparatus to the region of damaged DNA (Brandsma et al., 1996; Lindahl et al., 1997; Moggs et al., 1997; Zamble and Lippard, 1995; Zamble et al., 1996).

The NER complex repairs adducted bases by excising and replacing them using the opposite (normal) DNA strand as the template. Ribosomal DNA (rDNA) accumulates cisplatin-induced DNA adducts at a disproportionately high rate as compared with other forms of DNA because rDNA does not recruit the NER complex (Christians and Hanawalt, 1993; Fritz and Smerdon, 1995). The resultant dysfunction of rDNA is one of several mechanisms responsible for the inhibition of protein synthesis associated with cisplatin-induced injury.

Some chromosomally-associated proteins, such as transcription factors and high mobility group (HMG) proteins, have a high binding affinity for the DNA adducts caused by the organoplatinum compounds. (Jordan and Carmo-Fonseca, 2000). Chromosomally associated proteins such as the HMG proteins can directly compete with components of the NER for binding to cisplatin adducts, thereby impairing DNA repair (Gonzalez et al., 2001; Kartalou and Essigmann, 2001; Moggs et al., 1997).

Mismatch Repair Complex

The mismatch repair (MR) complex, which replaces mismatched bases in double-stranded DNA, also recognizes and attempts (unsuccessfully) to repair cisplatin-induced DNA damage. The MR complex competes with and inhibits NER complex mediated DNA repair. Since the MR complex is upregulated during the postreplicative period of the cell cycle, impaired function of the MR complex leads to the accumulation of more mutations in rapidly dividing cells than in quiescent cells (Comess et al., 1992; Hoffmann et al., 1995; Pillaire et al., 1995). As a result, regenerating tubular cells are highly susceptible to this form of DNA damage during the recovery phase of cisplatin-induced acute tubular necrosis.

Transplatin-induced adducts are repaired far more effectively than adducts induced by cisplatin and many other organoplatinum analogs. This is because the NER compelx is able to "recognize" adducts of transplatin more readily than other platinum-induced adducts.

Furthermore, chromosomally-associated proteins such as the HMG proteins bind to transplatin-induced DNA adducts with less affinity than to other organoplatinum adducts and therefore do not interfere with NER complex mediated repair of transplatin adducts as much as with other DNA adducts. The poor antineoplastic activity and low toxicity of transplatin can be explained by greater efficiency of repair of transplatin-induced DNA adducts as compared with adducts of other organoplatinum compounds (Hannemann and Baumann, 1990).

Cisplatin-Induced "Transcription Highjacking"

The term "transcription hijacking" refers to the consequences of the ability of certain transcription factors (such as UBF, TBP, LEF-1, SRY, and Ixr1) to bind to DNA adducts caused by organoplatinum compounds. This leads to the sequestration of these transcription factors from their usual promoter binding sites (Jordan and Carmo-Fonseca, 2000). Transcription hijacking represents one of a number of mechanisms responsible for the profound inhibition of protein synthesis associated with cisplatin-induced injury.

ROLE OF CYCLIN KINASE INHIBITORS IN AMELIORATING CISPLATIN-INDUCED CYTOTOXICITY

Growth arrest occurs in most cells when exposed to any form of injury or stress. Cell cycle arrest provides time for damaged DNA to be repaired before cell division proceeds. Cell cycle arrest after injury is mediated in part by the increased expression of cyclin dependent kinase inhibitors, such as $p21^{WAF1/CIP1/SDI1}$ (p21). Induction of p21 after cell injury delays the entry of injured cells into the S-phase of the cell cycle until the damaged DNA has been repaired.

In a series of elegant studies, Safirstein and his collaborators (Megyesi et al., 1996, 1998, 2001, 2002; Miyaji et al., 2001) provided convincing evidence that p21-induced cell cycle arrest and preproliferative DNA repair is important for functional recovery of the kidney in mice injured by cisplatin. Studies by the same group of investigators have demonstrated that when p21-deficient mice are subjected to cisplatin-induced acute renal failure (ARF), tubular cells can reenter the cell cycle prior to completion of DNA repair. As a result, apoptosis of tubular cells is increased and recovery of renal function delayed (Megyesi et al., 1996, 1998, 2001, 2002; Miyaji et al., 2001).

ROLE OF APOPTOSIS IN CISPLATIN-INDUCED TOXICITY

The cytotoxic effect of cisplatin was originally attributed to its ability to inhibit DNA synthesis. However, there is little correlation between inhibited DNA synthesis and cellular sensitivity to the drug (Jordan and Carmo-Fonseca, 2000; Pinto and Lippard, 1985; Sorenson et al., 1990). More-recent evidence has led to the realization that the antineoplastic and cytotoxic effects of cisplatin are predominantly due to its ability to trigger apoptosis (Lau, 1999; Lieberthal et al., 1996). Cell lines with defects in ability to repair DNA die by apoptosis in response to far lower concentrations of cisplatin than those required to inhibit DNA synthesis (Sorenson and Eastman, 1988). Apoptosis represents an essential mechanism for the elimination of cells with irreparable cisplatin-induced DNA damage before or soon after they enter the cell cycle. It is therefore not surprising that there are multiple and redundant pathways for mediating apoptosis of cells irreparably injured by cisplatin.

It has been shown that the predominant form of cisplatin-induced cell death is apoptosis rather than necrosis (Allday et al., 1995; Gonzalez et al., 2001; Lee et al., 2001; Lieberthal et al., 1996; Sorenson and Eastman, 1988; Sorenson et al., 1990). The first demonstration that high dose ($\approx 800\,\mu M$) cisplatin exposure induces necrosis of proximal tubular cells in culture while much lower concentrations ($< 200\,\mu M$) caused apoptosis of proximal tubular cells (Figure 18.4) was by the laboratory of this chapter's authors (Lieberthal et al., 1996). Antioxidants can ameliorate apoptotic cell death induced by relatively low concentrations of cisplatin but are ineffective in preventing necrosis induced by high concentrations of cisplatin (Lieberthal et al., 1996). These data highlight the importance of apoptotic cell loss at clinically relevant concentrations of cisplatin and support the notion that oxidant and genotoxic stress operate additively to initiate apoptosis in renal tissue (Figure 18.2 and Figure 18.3).

Two homologous proteins, p53 and p73, independently couple DNA damage to cell cycle arrest and apoptosis (Anthoney et al., 1996; Gallagher et al., 1997; Gong et al., 1999; Jost et al., 1997; Zamble et al., 1998). DNA damage stabilizes p53, a transcription factor that induces cell cycle arrest and activates the apoptotic pathway if DNA repair is ineffective (Anthoney et al., 1996; Gallagher et al., 1997). p53-independent activation of apoptosis in response to cisplatin has also been documented. Cells deficient in p53 are less sensitive, but not completely resistant, to the apoptosis-inducing effects of cisplatin (Agami et al., 1999; Yuan et al., 1999). Cisplatin-induced activation of stress-kinase cascades (Zanke et al., 1996) or telomeric loss (Ishibashi et al., 2002) also leads to apoptosis by inducing alterations in the relative expression of pro- and antiapoptotic Bcl-2 family proteins (Jones et al., 1998; Zanke et al., 1996) and by activating of the mitochondrial apoptotic cascade.

Cisplatin has been shown to induce apoptosis via a mitochondial-dependent biochemical pathway, which is also induced by a wide variety of other cell stresses including ischemia and oxidant stress (Bonegio and Lieberthal, 2002). Cisplatin promotes translocation of Bax, a proapoptotic protein, from the cytosol to the mitochondria. Bax inserts into and induces pore formation in the outer mitochondrial membrane leading to the release of cytochrome c into the cytosol (Lee et al., 2001; Park et al., 2002). In the cytosol, cytochrome c, in association with apoptosis activating factor, binds to and activates pro-caspase 9. Caspase 9 then activates the "executioner" caspase 3, which in turn activates other downstream caspases. The executioner caspases cleave a multitude of intracellular proteins and cause cell death associated with the characteristic morphologic features of apoptosis (Bonegio and Lieberthal, 2002; Lee et al., 2001). Overexpression of the antiapoptotic protein Bcl-2 has been shown to ameliorate cisplatin-induced apoptosis, presumably by binding to and neutralizing the proapoptotic effect of Bax (Lee et al., 2001).

CISPLATIN-INDUCED NEPHROTOXICITY

The Renal Handling of Cisplatin

Circulating cisplatin is predominantly (>90%) protein bound. The unbound drug, because it is small (≈300 Da) and uncharged, is freely filtered at the glomerulus and excreted by the kidney (Frick et al., 1979; Safirstein et al., 1984). In animals and humans, maximal red-cell concentrations of cisplatin are reached within 150 minutes of administration and then decline, with a terminal half-life of 36 to 47 days (Frick et al., 1979; Safirstein et al., 1984, 1986, 1987). Similarly, oxaliplatin is rapidly cleared from circulation by biotransformation and redistribution into red cells and tissues. In humans, only 15% of the administered dose remains in the systemic circulation 2 hours after administration. Like cisplatin, more than 90% of circulating oxaloplatin is irreversibly protein bound and this has a terminal half-life of less than 10 days (Ehrsson et al., 2002; Graham et al., 2000). In contrast, carboplatin is not bound to plasma proteins and not biotransformed. Carboplatin is rapidly eliminated with a postredistribution half-life of approximately 5.9 hours. Clearance of all three of these platinum compounds is predominantly effected by the kidney and correlates with glomerular filtration rate (GFR). Proximal tubular cell secretion of cisplatin also contributes to its overall clearance by the kidneys and, as a result, the renal clearance of cisplatin slightly exceeds GFR (Ries and Klastersky, 1986).

The proximal tubule accumulates cisplatin by peritubular uptake, with the highest concentrations reached within cells of the S_3 segment (pars recta) (Ries and Klastersky, 1986; Safirstein et al., 1984). The concentration of cisplatin in the S_3 segment exceeds by five to six times the levels found in any other cell within the kidney or any other organs (Leibbrandt et al., 1995; Safirstein et al., 1984). This explains why the S_3 segment is the part of the nephron most susceptible to cisplatin-induced toxicity (Hill et al., 1975; Lieberthal and Levine, 1996; Safirstein et al., 1984). Distal convoluted tubules and collecting ducts within the cortex and outer medulla are also injured by cisplatin (Gonzales-Vitale et al., 1977; Safirstein et al., 1984). However, concentrations in the inner medulla are similar to those in nonrenal tissues and nephron segments in this region are the least affected by cisplatin administration (Safirstein et al., 1984).

The mechanisms underlying renal tubular uptake of cisplatin are incompletely understood. Transport into proximal tubular cells has been shown to be saturable and can be partially inhibited by organic bases (such as tolazoline, mepiperphenidol, and thiamine). Transport is not affected by organic anions such as *p*-aminohippuric acid (PAH) or pyrazinoic acid(Kroning et al., 1999; Safirstein et al., 1986). In addition, Safirstein and colleagues (1984) have shown that uptake is energy requiring. Taken together these data suggest that selective cisplatin uptake by proximal tubular epithelium occurs via an active organic base transporter.

Mechanisms of Tubular Cell Toxicity

The adverse effects of cisplatin on tubular cell function are believed to occur via the same genotoxic and oxidant mechanisms that apply to cisplatin-induced injury to other cell types. Effects of cisplatin on epithelial cell function and survival have been implicated in all the renal complications associated with cisplatin-induced nephrotoxicity.

The mechanisms of cisplatin-induced ARF have been studied extensively in animals. ARF is generally induced in rats or mice by a single intraperitoneal injection with a dose ranging from 6 to 20 mg/kg (Anand and Bashey, 1993; Baliga et al., 1998a; Chopra et al., 1982, 1983; Clifton et al., 1982; Daugaard, 1990; Daugaard and Abildgaard, 1989; Daugaard et al., 1987, 1988b; Gordon and Gattone, 1986; Heyman et al., 2002; Jones et al., 1985; Lau, 1999; Lieberthal and Levine, 1996; Lieberthal et al., 1996; Megyesi et al., 1998; Ries and Klastersky, 1986; Rosen et al., 1994; Safirstein and Wiston, 1987; Safirstein et al., 1984, 1986, 1987). This relatively simple and reproducible model of reversible acute tubular necrosis can be induced by a single dose, as may occur in humans. In addition, the renal pathology is comparable with that in humans (Heyman et al., 2002; Hartmann et al., 1999), and predisposing factors to ARF in humans, such as volume depletion, are replicated in the animal model. Importantly, the tubular dysfunction seen in cisplatin-induced ARF in animals is comparable with that seen in humans (including glycosuria, hypomagnesemia, and hypokalemia) (Heyman et al., 2002).

While cisplatin has been shown to be directly nephrotoxic, the exact mechanisms by which cisplatin induces renal dysfunction remain controversial. Prerenal, intrarenal, and even postrenal mechanisms have been proposed to contribute to the acute and chronic renal failure induced by these drugs. A major factor contributing to nephrotoxicity is the accumulation of cisplatin and its analogs by tubular cells, especially the proximal tubular cells (Chopra et al., 1982; Safirstein et al., 1987). Cisplatin is more nephrotoxic than transplatin, carboplatin, and other organoplatinum compounds (Kroning et al., 1999; Leibbrandt et al., 1995) (Table 18.2). Cisplatin exposure in particular results in a noxious combination of severe and prolonged genotoxic stress induced by the formation of DNA adducts, and oxidant stress that results from cisplatin per se, mitochondrial dysfunction and prooxidant iron release from cytochrome enzymes (Masuda et al., 1994; Townsend and Hanigan, 2002; Townsend et al., 2003).

Cisplatin inhibits protein synthesis at the level of transcription and translation. Leibbrant and colleagues (1995) demonstrated that cisplatin, carboplatin, and CI-973 (another cisplatin analog) inhibit protein synthesis in cultured tubular epithelial cells by disrupting nucleolar

Table 18.2 Renal Effects of Non-Cisplatin Organoplatinum Compounds

Study	*Regimen*	*Preventive measures*	*Chronic changes*
Hannigan et al., 1993 $n = 288$	Cyclophosphamide with cisplatin 100 mg/m^2 or carboplatin 300 mg/m^2	Cisplatin: saline Carboplatin: none	Cisplatin: increase in serum creatinine in 58% of patients Carboplatin: increase in serum creatinine in 14% of patients
Swenerton et al., 1992 $n = 417$	Cyclophosphamide with cisplatin 75 mg/m^2 or carboplatin 300 mg/m^2	Cisplatin: saline Carboplatin: none	Cisplatin: increase in serum creatinine in 29% of patients Carboplatin: increase in serum creatinine in 9% of patients
Bellmunt et al., 1997 $n = 147$	MVA* with cisplatin 75 mg/m^2 or carboplatin 300mg/m^2	No data	No substantial chronic renal failure in either group
Misset et al., 2001 $n = 177$	Cyclophosphamide with cisplatin 100 mg/m^2 or oxaliplatin 130 mg/m^2	No data	No substantial chronic renal failure with oxaliplatin
Giachetti et al., 2001 $n = 200$	5-Fluorouracil with oxaliplatin or no oxaliplatin	None	No substantial chronic renal failure with oxaliplatin
Adachi et al., 1998 $n = 8$	Nedaplatin 100 mg/m^2	None	No renal toxicity reported

* MVA: mitomycin, vindesine, and adriamycin

structure and ribosomal function. Interestingly, despite equivalent accumulation of intracellular platinum in these and other studies, cisplatin proved to be more toxic than carboplatin, suggesting that differential uptake of the analogs is not the only reason for the differences in toxicity (Kroning et al., 1999).

The function of subcellular organelles is also affected. Mitochondria appear abnormal (Gordon and Gattone, 1986; Jones et al., 1985) and their O_2 consumption and accumulation of Rhodamine 123 (a measure of mitochondrial function) is abnormal (Leibbrandt et al., 1995). The resultant decrease in ATP generation and release of oxygen radicals delays the cells' return to normal function (Figure 18.3) (Brady et al., 1990; Chang et al., 2002; Kruidering et al., 1997; Kharbangar et al., 2000; Levi et al., 1980; Zhang and Lindup, 1993). Endocytosis and lysosomal function is also abnormal postplatinum and contributes to tubular proteinuria (Leibbrandt et al., 1995; Takano et al., 2002).

It has become recognized within the past decade that ischemic and toxic injury to the kidney induces an inflammatory response within the kidney that is an important contributor to the renal injury and dysfunction (Bonventre, 1998; Bonventre and Kelly, 1996; Burne-Taney and Rabb, 2003; Kelly et al., 1994, 1996, 1999; Rabb et al., 1994, 1995; Schrier, 2002; Sheridan and Bonventre, 2000, 2001; Thadhani et al., 1996). Exciting data have emerged to suggest that cytokine and chemokine production and an intrarenal inflammatory response contributes to cisplatin-induced ARF (Deng et al., 2001; Kelly et al., 1999; Ramesh and Reeves, 2002). Amelioration of cisplatin-induced ARF has been achieved in experimental animals by inhibiting tumor necrosis factor α (Ramesh and Reeves, 2002), leucocyte adhesion (with anti-CD54 antibody)(Kelly et al., 1999), or by the administration of the anti-inflammatory cytokine interleukin-10 (Deng et al., 2001).

Severely damaged tubular cells die after cisplatin exposure, contributing to renal dysfunction either by denuding areas of tubular epithelium or by causing intratubular casts that can produce obstruction (Kroning et al., 1999; Lieberthal et al., 1996, 2001). The presence of tubular cell necrosis, characterized by early loss of plasma membrane integrity, tubular cell swelling, and aberration of mitochondrial and other organelle structure, has clearly been demonstrated to follow cisplatin exposure, both *in vivo* and *in vitro* (Blachley and Hill, 1981; Chopra et al., 1982; Gonzales-Vitale et al., 1977; Kroning et al., 1999; Lieberthal et al., 1996, 2001; Safirstein and Wiston, 1987; Safirstein et al., 1984, 1986). Unlike animal models of toxic ARF, pathological evidence of frank necrosis is typically sparse in humans with toxic ARF (Schreiner and Maher, 1965) and so the extent to which necrosis contributes to the severity of acute renal failure in the clinical setting has been questioned.

RENAL SYNDROMES ASSOCIATED WITH CISPLATIN

Acute Renal Failure

ARF is the clinically most important renal complications of cisplatin treatment and is an important indication for delaying chemotherapy or necessitating a change to an alternative, and possibly less effective, regimen. GFR decreases in a dose-dependent manner after cisplatin treatment. GFR is influenced both by the size of the individual dose and by cumulative dose (Daugaard et al., 1988a). One of the earliest effects of cisplatin is intrarenal vasoconstriction, which is followed within hours by tubular injury and dysfunction. Polyuria is a prominent early clinical feature of cisplatin-induced tubular injury (Chopra et al., 1982; Clifton et al., 1982). Animal studies suggest that early on, prior to any change in the GFR, cisplatin causes polyuria due to a relative deficiency of vasopressin, caused by decreased release of vasopressin from the posterior pituitary (Clifton et al., 1982). Later, as renal dysfunction develops, medullary concentrating ability is impaired and a second phase of polyuria develops, which is refractory to vasopressin therapy (Safirstein et al., 1984). Glycosuria, enzymuria, and tubular proteinuria are common early markers of tubular damage. When ARF occurs it is typically nonoliguric and gradual in onset, developing over 3 to 5 days (Ries and Klastersky, 1986).

Recovery from ARF is often slow and incomplete. Several studies have documented persistent chronic renal dysfunction in long-term survivors of testicular carcinoma treated with cisplatin (Table 18.1). Chronic renal dysfunction is best predicted by the cumulative dose of cisplatin administered, and occurs most often when cisplatin is administered as part of a chemotherapeutic regimen for inducing bone marrow ablation (Merouani et al., 1997, 2002). Once ARF has been diagnosed, cisplatin therapy should be delayed until serum creatinine concentration decreases below 1.5 mg/dl. If the serum creatinine concentration continues to rise or remains above 3 mg/dl, cisplatin should be discontinued. Fortunately, cisplatin rarely causes progressive renal insufficiency.

Prevention of Cisplatin-Induced ARF

It was recognized early on that prerenal factors induced by volume depletion markedly exacerbate cisplatin-induced nephrotoxicity. ARF was a frequent problem. Prior to the routine use of saline-based hydration, the individual dose of cisplatin had to be limited to less than 50 mg/m^2. However, the administration of large volumes of normal or hypertonic saline before and during cisplatin treatment, with the aim of maintaining a urine output 100 to 150 ml/hour, has become standard practice (Bajorin et al., 1987) (Table 18.1). The administration of mannitol together with saline may provide added benefit, but the use of furosemide should be avoided (Al-Sarraf et al., 1982; Daugaard et al., 1988a). If mannitol is used, care must be taken to avoid volume depletion. Several phase II/III trials now recommend doses of cisplatin of up to 100 mg/m^2 (Glover et al., 1986; Hannigan et al., 1993; MacLeod et al., 1988; Misset et al., 2001). The use of hypertonic saline rather than isotonic saline has been recommended when the administration of high doses of cisplatin is contemplated (Daugaard et al., 1988a). While the mechanism underlying the beneficial effects of saline remain uncertain, probable factors include decreased tubular uptake of platinum, inhibition of the generation of reactive platinum compounds by the chloride load, and maintenance of rapid tubular flow (Ries and Klastersky, 1986).

In animals models, cisplatin-induced nephrotoxicity has been shown to be ameliorated with considerable success by a variety of antioxidants, including amifostine, sodium thiosulfate, sodium diethyldithiocarbamate, glutathione, selenium, melatonin, and iron chelators (Antunes et al., 2001; Babu et al., 1999; Baliga et al., 1998a; Bolaman et al., 2001; Camargo et al., 2001; Capizzi, 1999; DiPaola and Schuchter, 1999; Elsendoorn et al., 2001; Francescato et al., 2001; Glover et al., 1986, 1989; Gradishar et al., 2001; Hanada et al., 2000; Hara et al., 2001; Hartmann et al., 2000c; Lieberthal et al., 1996; Plaxe et al., 1994; Sener et al., 2000; Sriswasdi et al., 2000; Walker and Gale, 1981). Unfortunately, the efficacy of antioxidants in reducing cisplatin-induced renal injury has not yet been replicated in humans.

Since sodium thiosulfate ameliorates cisplatin-induced damage to both normal and malignant cells, it has been utilized clinically only in cases of accidental cisplatin overdose (Erdlenbruch et al., 2002) or in regimens that utilize intraperitoneal cisplatin or carboplatin. In these palliative regimens, the local effect of the platinum compound is considered the therapeutic target, and this is minimally affected by systemic administration of sodium thiosulfate. In this case, the absorbed organoplatinum, which mediates toxic rather than therapeutic effects, can be effectively detoxified with thiosulfate (Howell and Taetle, 1980).

Amifotine or W-2721, a thiophosphate originally designed by the military to protect troops from radiation exposure, is thought to detoxify cisplatin by donation of a protective thiol group (Capizzi, 1999). This agent apparently preferentially protects normal tissues from cisplatin-induced injury, without having much effect on its antineoplastic activity (Glover et al., 1986, 1989). Amifostine is the only agent currently approved by the U.S. Food and Drug Administration and recommended by the American Society of Clinical Oncology (Glover et al., 1986; Capizzi, 1999; Schuchter et al., 2002) for renal protection after cisplatin. However, continued concerns that amifostine may decrease the antineoplastic activity of cisplatin has limited general acceptance of this treatment (Gradishar et al., 2001).

Electrolyte Abnormalities

Magnesium is an important intracellular cation that is a cofactor for more than 300 enzymatic reactions. Whole-body magnesium is approximately equally divided between the intracellular fluid and that complexed in bone. Even though only about 1% of whole-body magnesium is extracellular, maintenance of normal extracellular levels is essential for nerve and muscle function (Dai et al., 2001; Dunn and Walser, 1966). Plasma magnesium exists in three forms: 33% is protein bound, 6% is complexed, and 61% exists as free, ionized magnesium. Only the free, metabolically active, magnesium is filtered at the glomerulus. Filtered magnesium is reabsorbed by the proximal tubule (5 to 15%), the thick ascending limb of the loop of Henle (70 to 80%), and the distal convoluted tubule (DCT) (5 to 10%) (Dai et al., 2001; Lajer and Daugaard, 1999).

Proximal tubular and DCT reabsorption is active, while in the loop of Henle magnesium is reabsorbed by a passive mechanism that is dependent on the negative transepithelial electrical gradient induced by the Na–K–2Cl transporter. Regulation of magnesium homeostasis depends largely on reabsorption by the DCT, which is influenced by many factors; including pH, potassium, phosphorus, calcium, peptide hormones (parathyroid hormone, calcitonin, glucagons, and vasopressin), α-adrenergic activity, prostaglandins (PGE_2), mineralocorticoids, and vitamin D (Dai et al., 2001).

Urinary magnesium wasting occurs in approximately 50% of patients treated with cisplatin, with hypomagnesemia developing in approximately only 20% (Daugaard et al., 1988a; Giaccone et al., 1985; Hartmann et al., 2000b, 2000c; Lajer and Daugaard, 1999; Lam and Adelstein, 1986). Hypomagnesenia is less common in patients receiving carboplatin, affecting 4 to 5% of those treated (English et al., 1999; Leyvraz et al., 1985). Symptoms generally develop when hypomagnesemia is severe (<0.5 m*M*) and respond well to oral supplementation (Bell et al., 1985). There is no linear correlation between blood magnesium levels and any clinical complication. This is not surprising given that the vast majority of total body magnesium is not in extracellular fluid. The incidence of hypomagnesemia depends primarily on the cumulative cisplatin dose rather than on the number of doses received. Several studies suggest that a cumulative dose of more than 300 mg/m^2 is usually required to produce magnesium wasting (Ariceta et al., 1997; Buckley et al., 1984).

Despite the high frequency of this complication, the mechanisms of cisplatin-induced magnesium depletion remain largely unexplained. Several factors are thought to contribute to cisplatin-induced magnesium wasting. First, saline infusion may decrease tubular reabsorption of magnesium by increasing sodium and water delivery to the loop of Henle and DCT. Second, distal tubular loss of magnesium from the loop of Henle may occur due to a disturbance in outer and inner medullary salt (Na–K–2Cl transporter) and urea (UT–A2 and UT–A4) transporters (Ecelbarger et al., 2001). Third, altered sensing of Ca/Mg and altered hormonal responses within the DCT may be induced by cisplatin-adducts within the DCT (Dai et al., 2001; Iida et al., 2000). Finally, there may be direct and perhaps selective toxicity to magnesium reabsorbing epithelial cells of the proximal and distal tubules (Dai et al., 2001; Mavichak et al.1985, 1988).

Hypocalcemia and hypokalemia are often associated with the magnesium wasting and are the result of complex and incompletely understood mechanisms. Hypocalcemia results, at least in part, from the combined effects of a relative decrease in parathyroid hormone secretion and peripheral resistance to parathyroid hormone action in bones and kidneys that is due to the hypomagnesemia (Dai et al., 2001). The mechanisms of the hypokalemia associated with magnesium wasting remain, elusive, although hypomagnesemia has been shown to induce a decrease in function of Na^+/K^+-ATPase pumps (Sheehan and Seelig, 1984; Wester and Dyckner, 1986). Hypocalcemia and hypokalemia are both refractory unless the magnesium deficiency has been corrected.

Magnesium wasting may become chronic and may persist in some patients for years after the cisplatin therapy (Ariceta et al., 1997; Schilsky et al., 1982; von der Weid et al., 1999). While it has been suggested that persistence of cisplatin-induced DNA adducts may contribute to the chronicity of hypermagnesuria, there is little direct evidence to support this notion. Long-term oral magnesium replacement is generally effective at maintaining magnesium levels in these patients (Mavichak et al., 1988).

Fanconi's syndrome

Severe proximal tubular dysfunction characterized by amino-aciduria, phosphaturia, glucosuria, and bicarbonaturia has been anecdotally associated with cisplatin therapy. However, it appears that cisplatin seldom causes Fanconi's syndrome in isolation, but rather appears to aggravate the proximal tubular damage caused by the concomitant administration of ifosfamide (Cachat et al., 1998; Rossi and Ehrich, 1993; Rossi et al., 1994a, 1994b; Suarez et al., 1991). Ifosfomide–cisplatin combinations are commonly used to treat childhood solid tumors, and 32 to 66% of these patients develop features of Fanconi's syndrome (Rossi et al., 1994a, 1994b; Suarez et al., 1991). Fortunately, renal function is usually unimpaired in these cases.

Hemolytic Uremic Syndrome

Cisplatin therapy has been associated hemolytic uremic syndrome (HUS) in several case reports and in three small trials (Angiola et al., 1990; Canpolat et al., 1994; Fisher et al., 1996; Gardner et al., 1989; Jackson et al., 1984; Kondo et al., 1998; Palmisano et al., 1998; Thurnher et al., 2001; van der Heijden et al., 1998; van der Meer et al., 1985; Watson et al., 1989; Weinblatt et al., 1987). The onset of the classic triad of renal dysfunction, thrombocytopenia, and microangiopathic hemolytic anemia is often delayed, and may occur several months after the drug exposure. Although rare, it is a serious complication of cisplatin therapy and is fatal in 73% of cases (Fisher et al., 1996). Anecdotal reports suggest efficacy of plasma exchange, high dose steroid therapy, and column exchange using a Staphylococcal protein A column (Fisher et al., 1996; Palmisano et al., 1998; Thurnher et al., 2001; Watson et al., 1989).

THE RELATIVE NEPHROTOXICITY OF CISPLATIN AND OTHER ORGANOPLATINUM COMPOUNDS

More than 10 platinum-based compounds have been developed. In two large clinical trials conducted in North America, patients were randomized to receive cisplatin- or carboplatin-based chemotherapy for ovarian cancers (Hannigan et al., 1993; Swenerton et al., 1992) (Table 18.2). The trials both documented equivalent treatment outcomes, with marked reductions (although not elimination) of renal complications in carboplatin treated patients. Importantly, these studies showed that hydration with saline is not necessary when carboplatin is used. This facilitates the use of carboplatin in an outpatient oncology setting. While oxaloplatin and nedaplatin have not been associated with nephrotoxicity in humans, the efficacy of these agents in treating different forms of cancer still needs to be elucidated. Cisplatin is still believed to be more efficacious than carboplatin and other analogs for the treatment of some malignancies and remains the organoplatinum compound of choice in these situations (Adachi et al., 1998; Giacchetti et al., 2000; Misset et al., 2001).

REFERENCES

Adachi S, Ogasawara T, Yamasaki N, Shibahara H, Tsuji Y, Takemura T, and Koyama K (1998) A pilot study of nedaplatin and etoposide for recurrent gynecological malignancies. *Oncol Rep* 5:881–884.

Agami R, Blandino G, Oren M, and Shaul Y (1999) Interaction of c-Abl and p73alpha and their collaboration to induce apoptosis. *Nature* 399:809–813.

Ajani JA, Baker J, Pisters PW, Ho L, Feig B, and Mansfield PF (2001) Irinotecan plus cisplatin in advanced gastric or gastroesophageal junction carcinoma. *Oncology(Huntingt)* 15:52–54.

Ajani JA, Baker J, Pisters PW, Ho L, Mansfield PF, Feig BW, and Charnsangavej C (2002) Irinotecan/cisplatin in advanced, treated gastric or gastroesophageal junction carcinoma. *Oncology (Huntingt)* 16:16–18.

Akaboshi M, Kawai K, Maki H, Akuta K, Ujeno Y, and Miyahara T (1992) The number of platinum atoms binding to DNA, RNA and protein molecules of HeLa cells treated with cisplatin at its mean lethal concentration. *Jpn J Cancer Res* 83:522–526.

Allday MJ, Inman GJ, Crawford DH, and Farrell P J (1995) DNA damage in human B cells can induce apoptosis, proceeding from G1/S when p53 is transactivation competent and G2/M when it is transactivation defective. *Embo J* 14:4994–5005.

Al-Sarraf M, Fletcher W, Oishi N, Pugh R, Hewlett JS, Balducci L, McCracken J, and Padilla F (1982) Cisplatin hydration with and without mannitol diuresis in refractory disseminated malignant melanoma: A southwest oncology group study. *Cancer Treat Rep* 66:31–35.

Anand AJ, and Bashey B (1993) Newer insights into cisplatin nephrotoxicity. *Ann Pharmacother* 27:1519–1525.

Angiola G, Bloss JD, DiSaia PJ, Warner AS, Manetta A, and Berman ML (1990) Hemolytic-uremic syndrome associated with neoadjuvant chemotherapy in the treatment of advanced cervical cancer. *Gynecol Oncol* 39:214–217.

Anthoney DA, McIlwrath AJ, Gallagher WM, Edlin AR, and Brown R (1996) Microsatellite instability, apoptosis, and loss of p53 function in drug- resistant tumor cells. *Cancer Res* 56:1374–1381.

Antunes LM, Darin JD, and Bianchi Nde L (2001) Effects of the antioxidants curcumin or selenium on cisplatin-induced nephrotoxicity and lipid peroxidation in rats. *Pharmacol Res* 43:145–150.

Ariceta G, Rodriguez-Soriano J, Vallo A, and Navajas A (1997) Acute and chronic effects of cisplatin therapy on renal magnesium homeostasis. *Med Pediatr Oncol* 28:35–40.

Babu E, Ebrahim AS, Chandramohan N, and Sakthisekaran D (1999) Rehabilitating role of glutathione ester on cisplatin induced nephrotoxicity. *Ren Fail* 21:209–217.

Bajorin D, Bosl GJ, and Fein R (1987) Phase I trial of escalating doses of cisplatin in hypertonic saline. *J Clin Oncol* 5:1589–1593.

Baliga R, Zhang Z, Baliga M, Ueda N, and Shah SV (1998a) In vitro and *in vivo* evidence suggesting a role for iron in cisplatin-induced nephrotoxicity. *Kidney Int* 53:394–401.

Baliga R, Zhang Z, Baliga M, Ueda N, and Shah SV (1998b) Role of cytochrome P-450 as a source of catalytic iron in cisplatin-induced nephrotoxicity. *Kidney Int* 54:1562–1569.

Bancroft DP, Lepre CA, and Lippard SJ (1990) 195Pt NMR kinetic and mechanistic studies of cis and trans-diamminedichloroplatimun(II) to DNA. *J Am Chem Soc* 112:6860–6871.

Baur M, Kienzer HR, Schweiger J, DeSantis M, Gerber E, Pont J, Hudec M, Schratter-Sehn AU, Wicke W, and Dittrich C (2002) Docetaxel/cisplatin as first-line chemotherapy in patients with head and neck carcinoma: A phase II trial. *Cancer* 94:2953–2958.

Bell DR, Woods RL, and Levi JA (1985) cis-Diamminedichloroplatinum-induced hypomagnesemia and renal magnesium wasting. *Eur J Cancer Clin Oncol* 21:287–290.

Bellmunt J, Ribas A, Eres N, Albanell J, Almanza C, Bermejo B, Sole LA, and Baselga J (1997) Carboplatin-based versus cisplatin-based chemotherapy in the treatment of surgically incurable advanced bladder carcinoma. *Cancer* 80:1966–1972.

Bissett D, Kunkeler L, Zwanenburg L, Paul J, Gray C, Swan IR, Kerr DJ, and Kaye SB (1990) Long-term sequelae of treatment for testicular germ cell tumours. *Br J Cancer* 62:655–659.

Blachley JD, and Hill JB (1981) Renal and electrolyte disturbances associated with cisplatin. *Ann Intern Med* 95:628–632.

Boisdron-Celle M, Lebouil A, Allain P, and Gamelin E (2001) [Pharmacokinetic properties of platinum derivatives]. *Bull Cancer* 88 Spec No. S14–9.

Bolaman Z, Koseoglu MH, Demir S, Atalay H, Akalin N, Hatip I, and Aslan D (2001) Effect of amifostine on lipid peroxidation caused by cisplatin in rat kidney. *J Chemother* 13:337–339.

Bonegio R, and Lieberthal W (2002) Role of apoptosis in the pathogenesis of acute renal failure. *Curr Opin Nephrol Hypertens* 11:301–308.

Bonventre J (1998) [Role of adhesion molecules in acute ischemic renal insufficiency]. *Nephrologie* 1957–1958.

Bonventre JV, and Kelly KJ (1996) Adhesion molecules and acute renal failure. *Adv Nephrol Necker Hosp* 25:159–176.

Brady HR, Kone BC, Stromski ME, Zeidel ML, Giebisch G, and Gullans SR (1990) Mitochondrial injury: An early event in cisplatin toxicity to renal proximal tubules. *Am J Physiol* 258:F1181–F1187.

Brandsma JA, de Ruijter M, Visse R, van Meerten D, van der Kaaden M, Moggs JG, and van de Putte P (1996) The *in vitro* more efficiently repaired cisplatin adduct cis-PtGG is *in vivo* a more mutagenic lesion than the relative slowly repaired cis-PtGCG adduct. *Mutat Res* 362: 29–40.

Buckley JE, Clark VL, Meyer TJ, and Pearlman NW (1984) Hypomagnesemia after cisplatin combination chemotherapy. *Arch Intern Med* 144:2347–2348.

Budd GT, Metch B, Weiss SA, Weick JK, Fabian C, Stephens RL, and Balcerzak SP (1993) Phase II trial of ifosfamide and cisplatin in the treatment of metastatic sarcomas: A Southwest Oncology Group study. *Cancer Chemother Pharmacol* 31:S213–S216.

Burne-Taney MJ, and Rabb H (2003) The role of adhesion molecules and T cells in ischemic renal injury. *Curr Opin Nephrol Hypertens* 12:85–90.

Cachat F, Nenadov-Beck M, and Guignard JP (1998) Occurrence of an acute Fanconi syndrome following cisplatin chemotherapy. *Med Pediatr Oncol* 31:40–41.

Camargo SM, Francescato HD, Lavrador MA, and Bianchi ML (2001) Oral administration of sodium selenite minimizes cisplatin toxicity on proximal tubules of rats. *Biol Trace Elem Res* 83:251–262.

Canpolat C, Pearson P, and Jaffe N (1994) Cisplatin-associated hemolytic uremic syndrome. *Cancer* 74:3059–3062.

Capizzi RL (1999) Amifostine reduces the incidence of cumulative nephrotoxicity from cisplatin: Laboratory and clinical aspects. *Semin Oncol* 26:72–81.

Carli M, Perilongo G, di Montezemolo LC, De Bernardi B, Ceci A, Paolucci G, Pianca C, Calculli G, Di Tullio MT, Grotto P, and et al (1987) Phase II trial of cisplatin and etoposide in children with advanced soft tissue sarcoma: A report from the Italian Cooperative Rhabdomyosarcoma Group. *Cancer Treat Rep* 71:525–527.

Cascinu S, Munao S, Mare M, Amadio P, Crucitta E, and Picone G (2000) [Tolerance profile of platinum compounds]. *Tumori* 86:S54–S55.

Chang B, Nishikawa M, Sato E, Utsumi K, and Inoue M (2002) L-Carnitine inhibits cisplatin-induced injury of the kidney and small intestine. *Arch Biochem Biophys* 405:55–64.

Chopra S, Kaufman J, and Flamenbaum W (1983) Cis-diamminedichloroplatinum (DDP) induced acute renal failure (ARF): Attempts at amelioration. *Clin Exp Dial Apheresis* 7:25–35.

Chopra S, Kaufman JS, Jones TW, Hong WK, Gehr MK, Hamburger RJ, Flamenbaum W, and Trump BF (1982) Cis-diamminedichlorplatinum-induced acute renal failure in the rat. *Kidney Int* 21:54–64.

Christians FC, and Hanawalt PC (1993) Lack of transcription-coupled repair in mammalian ribosomal RNA genes. *Biochemistry* 32:10512–10518.

Chu G (1994) Cellular responses to cisplatin The roles of DNA-binding proteins and DNA repair. *J Biol Chem* 269:787–790.

Clifton GG, Pearce C, O'Neill WM Jr and Wallin JD (1982) Early polyuria in the rat following single-dose cis-dichlorodiammineplatinum (II): Effects on plasma vasopressin concentration and posterior pituitary function. *J Lab Clin Med* 100:659–670.

Comess KM, Burstyn JN, Essigmann JM, and Lippard SJ (1992) Replication inhibition and translesion synthesis on templates containing site-specifically placed cis-diamminedichloroplatinum(II) DNA adducts. *Biochemistry* 31:3975–3990.

Cornelison TL, and Reed E (1993) Nephrotoxicity and hydration management for cisplatin, carboplatin, and ormaplatin. *Gynecol Oncol* 50:147–158.

Dai LJ, Ritchie G, Kerstan D, Kang HS, Cole DE, and Quamme GA (2001) Magnesium transport in the renal distal convoluted tubule. *Physiol Rev* 81:51–84.

Daugaard G (1990) Cisplatin nephrotoxicity: experimental and clinical studies. *Dan Med Bull* 37:1–12.

Daugaard G, and Abildgaard U (1989) Cisplatin nephrotoxicity A review. *Cancer Chemother Pharmacol* 25:1–9.

Daugaard G, Abildgaard U, Holstein-Rathlou NH, Bruunshuus I, Bucher D, and Leyssac PP (1988a) Renal tubular function in patients treated with high-dose cisplatin. *Clin Pharmacol Ther* 44:164–172.

Daugaard G, Abildgaard U, Larsen S, Holstein-Rathlou NH, Amtorp O, Olesen HP, and Leyssac PP (1987) Functional and histopathological changes in dog kidneys after administration of cisplatin. *Ren Physiol* 10:54–64.

Daugaard G, Holstein-Rathlou NH, and Leyssac PP (1988b) Effect of cisplatin on proximal convoluted and straight segments of the rat kidney. *J Pharmacol Exp Ther* 244:1081–1085.

Daugaard G, Rossing N, and Rorth M (1988c) Effects of cisplatin on different measures of glomerular function in the human kidney with special emphasis on high-dose. *Cancer Chemother Pharmacol* 21:163–167.

Dazzi C, Cariello A, Giannini M, Del Duca M, Giovanis P, Fiorentini G, Leoni M, Rosti G, Turci D, Tienghi A, Vertogen B, Zumaglini F, De Giorgi U, and Marangolo M (2000) A sequential chemo-radiotherapeutic treatment for patients with malignant gliomas: a phase II pilot study. *Anticancer Res* 20:515–518.

De Jager R, Longeval E, and Klastersky J (1980) High-dose cisplatin with fluid and mannitol-induced diuresis in advanced lung cancer: A phase II clinical trial of the EORTC Lung Cancer Working Party (Belgium). *Cancer Treat Rep* 64:1341–1346.

Deng J, Kohda Y, Chiao H, Wang Y, Hu X, Hewitt SM, Miyaji T, McLeroy P, Nibhanupudy B, Li S, and Star RA (2001) Interleukin-10 inhibits ischemic and cisplatin-induced acute renal injury. *Kidney Int* 60:2118–2128.

Dijt FJ, Fichtinger-Schepman AM, Berends F, and Reedijk J (1988) Formation and repair of cisplatin-induced adducts to DNA in cultured normal and repair-deficient human fibroblasts. *Cancer Res* 48:6058–6062.

DiPaola RS, and Schuchter L (1999) Neurologic protection by amifostine. *Semin Oncol* 26:82–88.

Doz F (2000) [Carboplatin in pediatrics]. *Bull Cancer* 87 Spec No. 25–29.

Dunn MJ, and Walser M (1966) Magnesium depletion in normal man. *Metabolism* 15:884–895.

Ecelbarger CA, Sands JM, Doran JJ, Cacini W, and Kishore BK (2001) Expression of salt and urea transporters in rat kidney during cisplatin- induced polyuria. *Kidney Int* 60:2274–2282.

Ehrsson H, Wallin I, and Yachnin J (2002) Pharmacokinetics of oxaliplatin in humans. *Med Oncol* 19:261–265.

Ellerby RA, Davis HL Jr, Ansfield FJ, and Ramirez G (1974) Phase I clinical trial of combined therapy with 5-FU (NSC 19893) and CIS-platinum (II) diaminedichloride (NSC 119875). *Cancer* 34:1005–1010.

Elsendoorn TJ, Weijl NI, Mithoe S, Zwinderman AH, Van Dam F, De Zwart FA, Tates AD, and Osanto S (2001) Chemotherapy-induced chromosomal damage in peripheral blood lymphocytes of cancer patients supplemented with antioxidants or placebo. *Mutat Res* 498:145–158.

Endo T, Kimura O, and Sakata M (2000) Carrier-mediated uptake of cisplatin by the OK renal epithelial cell line. *Toxicology* 146:187–195.

English MW, Skinner R, Pearson AD, Price L, Wyllie R, and Craft AW (1999) Dose-related nephrotoxicity of carboplatin in children. *Br J Cancer* 81:336–341.

Erdlenbruch B, Pekrun A, Schiffmann H, Witt O, and Lakomek M (2002) Topical topic: Accidental cisplatin overdose in a child: Reversal of acute renal failure with sodium thiosulfate. *Med Pediatr Oncol* 38:349–352.

Fisher DC, Sherrill GB, Hussein A, Rubin P, Vredenburgh JJ, Elkordy M, Ross M, Petros W and Peters WP (1996) Thrombotic microangiopathy as a complication of high-dose chemotherapy for breast cancer. *Bone Marrow Transplant* 18:193–198.

Fjeldborg P, Sorensen J, and Helkjaer PE (1986) The long-term effect of cisplatin on renal function. *Cancer* 58:2214–2217.

Fornace AJ Jr, and Seres DS (1982) Repair of trans-Pt(II) diamminedichloride DNA-protein crosslinks in normal and excision-deficient human cells. *Mutat Res* 94:277–284.

Francescato HD, Costa RS, Rodrigues Camargo SM, Zanetti MA, Lavrador MA, and Bianchi MD (2001) Effect of oral selenium administration on cisplatin-induced nephrotoxicity in rats. *Pharmacol Res* 43:77–82.

Frick GA, Ballentine R, Driever CW, and Kramer WG (1979) Renal excretion kinetics of high-dose cis-dichlorodiammineplatinum(II) administered with hydration and mannitol diuresis. *Cancer Treat Rep* 63:13–16.

Fritz LK, and Smerdon MJ (1995) Repair of UV damage in actively transcribed ribosomal genes. *Biochemistry* 34:13117–13124.

Gallagher WM, Cairney M, Schott B, Roninson IB, and Brown R (1997) Identification of p53 genetic suppressor elements which confer resistance to cisplatin. *Oncogene* 14:185–193.

Gardner G, Mesler D, and Gitelman HJ (1989) Hemolytic uremic syndrome following cisplatin, bleomycin, and vincristine chemotherapy: A report of a case and a review of the literature. *Ren Fail* 11:133–137.

Garnuszek P, Licianska I, Skierski JS, Koronkiewicz M, Mirowski M, Wiercioch R, and Mazurek AP (2002) Biological investigation of the platinum(II)-[*I]iodohistamine complexes of potential synergistic anti-cancer activity. *Nucl Med Biol* 29:169–175.

Gately DP, and Howell SB (1993) Cellular accumulation of the anticancer agent cisplatin: A review. *Br J Cancer* 67:1171–1176.

Gatzemeier U, Hossfeld DK, Neuhauss R, Reck M, Achterrath W, and Lenaz L (1992) Phase II, and III studies with carboplatin in small cell lung cancer. *Semin Oncol* 19:28–36.

Gelasco A, and Lippard SJ (1998) NMR solution structure of a DNA dodecamer duplex containing a cis-diammineplatinum(II) d(GpG) intrastrand cross-link, the major adduct of the anticancer drug cisplatin. *Biochemistry* 37:9230–9239.

Giacchetti S, Perpoint B, Zidani R, Le Bail N, Faggiuolo R, Focan C, Chollet P, Llory JF, Letourneau Y, Coudert B, Bertheaut-Cvitkovic F, Larregain-Fournier D, Le Rol A, Walter S, Adam R, Misset JL, and Levi F (2000) Phase III multicenter randomized trial of oxaliplatin added to chronomodulated fluorouracil-leucovorin as first-line treatment of metastatic colorectal cancer. *J Clin Oncol* 18:136–147.

Giaccone G, Donadio M, Ferrati P, Ciuffreda L, Bagatella M, Gaddi M, and Calciati A (1985) Disorders of serum electrolytes and renal function in patients treated with cis-platinum on an outpatient basis. *Eur J Cancer Clin Oncol* 21:433–437.

Glover D, Glick JH, Weiler C, Fox K, Turrisi A, and Kligerman MM (1986) Phase I/II trials of WR-2721 and cis-platinum. *Int J Radiat Oncol Biol Phys* 12:1509–1512.

Glover D, Grabelsky S, Fox K, Weiler C, Cannon L, and Glick J (1989) Clinical trials of WR-2721 and cis-platinum. *Int J Radiat Oncol Biol Phys* 16:1201–1204.

Gong JG, Costanzo A, Yang HQ, Melino G, Kaelin WG Jr, Levrero M, and Wang JY (1999) The tyrosine kinase c-Abl regulates p73 in apoptotic response to cisplatin-induced DNA damage. *Nature* 399:806–809.

Gonzales-Vitale JC, Hayes DM, Cvitkovic E, and Sternberg SS (1977) The renal pathology in clinical trials of cis-platinum (II) diamminedichloride. *Cancer* 39:1362–1371.

Gonzalez VM, Fuertes MA, Alonso C, and Perez JM (2001) Is cisplatin-induced cell death always produced by apoptosis? *Mol Pharmacol* 59:657–663.

Gordon JA, and Gattone VH 2nd (1986) Mitochondrial alterations in cisplatin-induced acute renal failure. *Am J Physiol* 250:F991-F998.

Goren MP, Forastiere AA, Wright RK, Horowitz ME, Dodge RK, Kamen BA, Viar MJ, and Pratt CB (1987) Carboplatin (CBDCA), iproplatin (CHIP), and high dose cisplatin in hypertonic saline evaluated for tubular nephrotoxicity. *Cancer Chemother Pharmacol* 19:57–60.

Gouyette A, Ducret JP, Caille P, Amiel JL, Rouesse J, Foka M, Carde P, Hayat M, and Sancho-Garnier H (1986) Preliminary phase I clinical study and pharmacokinetics of (1,2- diaminocyclohexane) (isocitrato) platinum (II) or PHIC. *Anticancer Res* 6:1127–1132.

Gradishar WJ, Stephenson P, Glover DJ, Neuberg DS, Moore MR, Windschitl HE, Piel I, and Abeloff MD (2001) A Phase II trial of cisplatin plus WR-2721 (amifostine) for metastatic breast carcinoma: An Eastern Cooperative Oncology Group Study (E8188). *Cancer* 92:2517–2522.

Graham MA, Lockwood GF, Greenslade D, Brienza S, Bayssas M, and Gamelin E (2000) Clinical pharmacokinetics of oxaliplatin: A critical review. *Clin Cancer Res* 6:1205–1218.

Groth S, Nielsen H, Sorensen JB, Christensen AB, Pedersen AG, and Rorth M (1986) Acute and long-term nephrotoxicity of cis-platinum in man. *Cancer Chemother Pharmacol* 17:191–196.

Grover B, Auberger C, Sarangarajan R, and Cacini W (2002) Functional impairment of renal organic cation transport in experimental diabetes. *Pharmacol Toxicol* 90:181–186.

Hamilton CR, Bliss JM, and Horwich A (1989) The late effects of cis-platinum on renal function. *Eur J Cancer Clin Oncol* 25:185–189.

Hanada K, Mukasa Y, Nomizo Y, and Ogata H (2000) Effect of buthionine sulphoximine, glutathione and methimazole on the renal disposition of cisplatin and on cisplatin-induced nephrotoxicity in rats: Pharmacokinetic-toxicodynamic analysis. *J Pharm Pharmacol* 52:1483–1490.

Hannemann J, and Baumann K (1990) Nephrotoxicity of cisplatin, carboplatin and transplatin A comparative *in vitro* study. *Arch Toxicol* 64:393–400.

Hannigan EV, Green S, Alberts DS, O'Toole R, and Surwit E (1993) Results of a Southwest Oncology Group phase III trial of carboplatin plus cyclophosphamide versus cisplatin plus cyclophosphamide in advanced ovarian cancer. *Oncology* 50 Suppl 2, 2–9.

Hansen SW (1992) Late-effects after treatment for germ-cell cancer with cisplatin, vinblastine, and bleomycin. *Dan Med Bull* 39:391–399.

Hansen SW, Groth S, Daugaard G, Rossing N, and Rorth M (1988) Long-term effects on renal function and blood pressure of treatment with cisplatin, vinblastine, and bleomycin in patients with germ cell cancer. *J Clin Oncol* 6:1728–1731.

Hara M, Yoshida M, Nishijima H, Yokosuka M, Iigo M, Ohtani-Kaneko R, Shimada A, Hasegawa T, Akama Y, and Hirata K (2001) Melatonin, a pineal secretory product with antioxidant properties, protects against cisplatin-induced nephrotoxicity in rats. *J Pineal Res* 30:129–138.

Harder HC, and Rosenberg B (1970) Inhibitory effects of anti-tumor platinum compounds on DNA, RNA, and protein syntheses in mammalian cells in virtro. *Int J Cancer* 6:207–216.

Hartmann JT, Fels LM, Franzke A, Knop S, Renn M, Maess B, Panagiotou P, Lampe H, Kanz L, Stolte H, and Bokemeyer C (2000a) Comparative study of the acute nephrotoxicity from standard dose cisplatin +/− ifosfamide and high-dose chemotherapy with carboplatin and ifosfamide. *Anticancer Res* 20:3767–3773.

Hartmann JT, Fels LM, Knop S, Stolt H, Kanz L, and Bokemeyer C (2000b) A randomized trial comparing the nephrotoxicity of cisplatin/ifosfamide- based combination chemotherapy with or without amifostine in patients with solid tumors. *Invest New Drugs* 18:281–289.

Hartmann JT, Knop S, Fels LM, van Vangerow A, Stolte H, Kanz L, and Bokemeyer C (2000c) The use of reduced doses of amifostine to ameliorate nephrotoxicity of cisplatin/ifosfamide-based chemotherapy in patients with solid tumors. *Anticancer Drugs* 11:1–6.

Hartmann JT, Kollmannsberger C, Kanz L, and Bokemeyer C (1999) Platinum organ toxicity and possible prevention in patients with testicular cancer. *Int J Cancer* 83:866–869.

Hayat M, Bayssas M, Brule G, Cappelaere P, Cattan A, Chauvergne J, Clavel B, Gouveia J, Guerrin J, Pommatau E, Muggia F, and Mathe G (1979) [Cis-diammino-dichloro-platinum therapy of cancers; phase II therapeutic trial]. *Nouv Presse Med* 8:1231–1234.

Hayes DM, Cvitkovic E, Golbey RB, Scheiner E, Helson L, and Krakoff I H (1977) High dose cis-platinum diammine dichloride: amelioration of renal toxicity by mannitol diuresis. *Cancer* 39:1372–1381.

Heyman SN, Lieberthal W, Rogiers P, and Bonventre JV (2002) Animal models of acute tubular necrosis. *Curr Opin Crit Care* 8:526–534.

Higby DJ, Wallace HJ Jr, Albert DJ, and Holland JF (1974) Diaminodichloroplatinum: A phase I study showing responses in testicular and other tumors. *Cancer* 33:1219–1225.

Hill JM, Loeb E, MacLellan A, Hill NO, Khan A, and King JJ (1975) Clinical studies of Platinum Coordination compounds in the treatment of various malignant diseases. *Cancer Chemother Rep* 59:647–659.

Hoffmann JS, Pillaire MJ, Maga G, Podust V, Hubscher U, and Villani G (1995) DNA polymerase beta bypasses *in vitro* a single d(GpG)-cisplatin adduct placed on codon 13 of the HRAS gene. *Proc Natl Acad Sci USA* 92:5356–5360.

Howell SB, and Taetle R (1980) Effect of sodium thiosulfate on cis-dichlorodiammine-platinum(II) toxicity and antitumor activity in L1210 leukemia. *Cancer Treat Rep* 64:611–616.

Howle JA, and Gale GR (1970a) Cis-dichlorodiammineplatinum (II) Persistent and selective inhibition of deoxyribonucleic acid synthesis *in vivo*. *Biochem Pharmacol* 19:2757–2762.

Howle JA, and Gale GR (1970b) cis-Dichlorodiammineplatinum(II): Cytological changes induced in Escherichia coli. *J Bacteriol* 103:258–259.

Huang H, Woo J, Alley SC, and Hopkins PB (1995a) DNA-DNA interstrand cross-linking by cis-diamminedichloroplatinum(II): N7(dG)-to-N7(dG) cross-linking at 5′-d(GC) in synthetic oligonucleotides. *Bioorg Med Chem* 3:659–669.

Huang H, Zhu L, Reid BR, Drobny GP, and Hopkins PB (1995b) Solution structure of a cisplatin-induced DNA interstrand cross-link. *Science* 270:1842–1845.

Iida T, Makino Y, Okamoto K, Yoshikawa N, Makino I, Nakamura T, and Tanaka H (2000) Functional modulation of the mineralocorticoid receptor by cis-diamminedi-chloroplatinum (II). *Kidney Int* 58:1450–1460.

Ishibashi T, Yano Y, and Oguma T (2002) A formula for predicting optimal dosage of nedaplatin based on renal function in adult cancer patients. *Cancer Chemother Pharmacol* 50:230–236.

Ishida S, Lee J, Thiele DJ, and Herskowitz I (2002) Uptake of the anticancer drug cisplatin mediated by the copper transporter Ctr1 in yeast and mammals. *Proc Natl Acad Sci USA* 99:14298–14302.

Jackson AM, Rose BD, Graff LG, Jacobs JB, Schwartz JH, Strauss GM, Yang JP, Rudnick MR, Elfenbein IB, and Narins RG (1984) Thrombotic microangiopathy and renal failure associated with antineoplastic chemotherapy. *Ann Intern Med* 101:41–44.

Johnson NP, Hoeschele JD, and Rahn RO (1980) Kinetic analysis of the *in vitro* binding of radioactive cis- and trans-dichlorodiammineplatinum(II) to DNA. *Chem Biol Interact* 30:151–169.

Jones NA, Turner J, McIlwrath AJ, Brown R, and Dive C (1998) Cisplatin- and paclitaxel-induced apoptosis of ovarian carcinoma cells and the relationship between bax and bak up-regulation and the functional status of p53. *Mol Pharmacol* 53:819–826.

Jones TW, Chopra S, Kaufman JS, Flamenbaum W, and Trump BF (1985) Cis-diamminedichloroplatinum (II)-induced acute renal failure in the rat Correlation of structural and functional alterations. *Lab Invest* 52:363–374.

Jordan P, and Carmo-Fonseca M (2000) Molecular mechanisms involved in cisplatin cytotoxicity. *Cell Mol Life Sci* 57:1229–1235.

Jost CA, Marin MC, and Kaelin WG Jr (1997) p73 is a simian [correction of human] p53-related protein that can induce apoptosis. *Nature* 389:191–194.

Kartalou M, and Essigmann JM (2001) Recognition of cisplatin adducts by cellular proteins. *Mutat Res* 478:1–21.

Kelly KJ, Meehan SM, Colvin RB, Williams WW, and Bonventre JV (1999) Protection from toxicant-mediated renal injury in the rat with anti-CD54 antibody. *Kidney Int* 56:922–931.

Kelly KJ, Williams WW Jr, Colvin RB, and Bonventre JV (1994) Antibody to intercellular adhesion molecule 1 protects the kidney against ischemic injury. *Proc Natl Acad Sci USA* 91:812–816.

Kelly KJ, Williams WW Jr, Colvin RB, Meehan SM, Springer TA, Gutierrez-Ramos JC, and Bonventre JV (1996) Intercellular adhesion molecule-1-deficient mice are protected against ischemic renal injury. *J Clin Invest* 97:1056–1063.

Kharbangar A, Khynriam D, and Prasad SB (2000) Effect of cisplatin on mitochondrial protein, glutathione, and succinate dehydrogenase in Dalton lymphoma-bearing mice. *Cell Biol Toxicol* 16:363–373.

Kondo M, Kojima S, Horibe K, Kato K, and Matsuyama T (1998) Hemolytic uremic syndrome after allogeneic or autologous hematopoietic stem cell transplantation for childhood malignancies. *Bone Marrow Transplant* 21:281–286.

Kroning R, Katz D, Lichtenstein AK, and Nagami GT (1999) Differential effects of cisplatin in proximal and distal renal tubule epithelial cell lines. *Br J Cancer* 79:293–299.

Kruidering M, Van de Water B, de Heer E, Mulder GJ, and Nagelkerke JF (1997) Cisplatin-induced nephrotoxicity in porcine proximal tubular cells: Mitochondrial dysfunction by inhibition of complexes I to IV of the respiratory chain. *J Pharmacol Exp Ther* 280:638–649.

Lajer H, and Daugaard G (1999) Cisplatin and hypomagnesemia. *Cancer Treat Rev* 25:47–58.

Lam M, and Adelstein DJ (1986) Hypomagnesemia and renal magnesium wasting in patients treated with cisplatin. *Am J Kidney Dis* 8:164–169.

Larena MG, Martinez-Diez MC, Monte MJ, Dominguez MF, Pascual MJ, and Marin JJ (2001) Liver organotropism and biotransformation of a novel platinum-ursodeoxycholate derivative, Bamet-UD2, with enhanced antitumour activity. *J Drug Target* 9:185–200.

Lau AH (1999) Apoptosis induced by cisplatin nephrotoxic injury. *Kidney Int* 56:1295–1298.

Lee RH, Song JM, Park MY, Kang SK, Kim YK, and Jung JS (2001) Cisplatin-induced apoptosis by translocation of endogenous Bax in mouse collecting duct cells. *Biochem Pharmacol* 62:1013–1023.

Leibbrandt ME, Wolfgang GH, Metz AL, Ozobia AA, and Haskins JR (1995) Critical subcellular targets of cisplatin and related platinum analogs in rat renal proximal tubule cells. *Kidney Int* 48:761–770.

Levi J, Jacobs C, Kalman SM, McTigue M, and Weiner MW (1980) Mechanism of cis-platinum nephrotoxicity: I Effects of sulfhydryl groups in rat kidneys. *J Pharmacol Exp Ther* 213:545–550.

Leyvraz S, Ohnuma T, Lassus M, and Holland JF (1985) Phase 1 study of carboplatin in patients with advanced cancer, intermittent intravenous bolus, and 24-hour infusion. *J Clin Oncol* 3:1385–1392.

Lieberthal W, Fuhro R, Andry CC, Rennke H, Abernathy VE, Koh JS, Valeri R, and Levine JS (2001) Rapamycin impairs recovery from acute renal failure: Role of cell-cycle arrest and apoptosis of tubular cells. *Am J Physiol Renal Physiol* 281:F693–F706.

Lieberthal W, and Levine JS (1996) Mechanisms of apoptosis and its potential role in renal tubular epithelial cell injury. *Am J Physiol* 271:F477–F488.

Lieberthal W, Triaca V, and Levine J (1996) Mechanisms of death induced by cisplatin in proximal tubular epithelial cells: Apoptosis vs necrosis. *Am J Physiol* 270:F700–F708.

Lin X, Okuda T, Holzer A, and Howell SB (2002) The copper transporter CTR1 regulates cisplatin uptake in Saccharomyces cerevisiae. *Mol Pharmacol* 62:1154–1159.

Lindahl T, Karran P, and Wood RD (1997) DNA excision repair pathways. *Curr Opin Genet Dev* 7:158–169.

MacLeod PM, Tyrell CJ, and Keeling DH (1988) The effect of cisplatin on renal function in patients with testicular tumours. *Clin Radiol* 39:190–192.

Masuda H, Tanaka T, and Takahama U (1994) Cisplatin generates superoxide anion by interaction with DNA in a cell-free system. *Biochem Biophys Res Commun* 203:1175–1180.

Masuda N (2001) Establishment of the standard regimen for non-small-cell lung cancer in Japan. *Oncology (Huntingt)* 15:13–18.

Mavichak V, Coppin CM, Wong NL, Dirks JH, Walker V, and Sutton RA (1988) Renal magnesium wasting and hypocalciuria in chronic cis-platinum nephropathy in man. *Clin Sci (Lond)* 75:203–207.

Mavichak V, Wong NL, Quamme GA, Magil AB, Sutton RA, and Dirks JH (1985) Studies on the pathogenesis of cisplatin-induced hypomagnesemia in rats. *Kidney Int* 28:914–921.

Megyesi J, Andrade L, Vieira JM Jr, Safirstein RL, and Price PM (2001) Positive effect of the induction of p21WAF1/CIP1 on the course of ischemic acute renal failure. *Kidney Int* 60:2164–2172.

Megyesi J, Andrade L, Vieira JM Jr, Safirstein RL, and Price PM (2002) Coordination of the cell cycle is an important determinant of the syndrome of acute renal failure. *Am J Physiol Renal Physiol* 283:F810–F816.

Megyesi J, Safirstein RL, and Price PM (1998) Induction of p21WAF1/CIP1/SDI1 in kidney tubule cells affects the course of cisplatin-induced acute renal failure. *J Clin Invest* 101:777–782.

Megyesi J, Udvarhelyi N, Safirstein RL, and Price PM (1996) The p53-independent activation of transcription of p21 WAF1/CIP1/SDI1 after acute renal failure. *Am J Physiol* 271:F1211–F1216.

Merouani A, Davidson SA, and Schrier RW (1997) Increased nephrotoxicity of combination taxol and cisplatin chemotherapy in gynecologic cancers as compared to cisplatin alone. *Am J Nephrol* 17:53–58.

Merouani A, Rozen R, Clermont MJ, and Genest J (2002) Renal function, homocysteine, and other plasma thiol concentrations during the postrenal transplant period. *Transplant Proc* 34:1159–1160.

Misset JL, Vennin P, Chollet PH, Pouillart P, Laplaige PH, Frobert JL, Castera D, Fabro M, Langlois D, Cortesi E, Lucas V, Gamelin E, Laadem A, and Otero J (2001) Multicenter phase II-III study of oxaliplatin plus cyclophosphamide vs cisplatin plus cyclophosphamide in chemonaive advanced ovarian cancer patients. *Ann Oncol* 12:1411–1415.

Miyaji T, Kato A, Yasuda H, Fujigaki Y, and Hishida A (2001) Role of the increase in p21 in cisplatin-induced acute renal failure in rats. *J Am Soc Nephrol* 12:900–908.

Mizumura Y, Matsumura Y, Hamaguchi T, Nishiyama N, Kataoka K, Kawaguchi T, Hrushesky WJ, Moriyasu F, and Kakizoe T (2001) Cisplatin-incorporated polymeric micelles eliminate nephrotoxicity, while maintaining antitumor activity. *Jpn J Cancer Res* 92:328–236.

Moggs JG, Szymkowski DE, Yamada M, Karran P, and Wood RD (1997) Differential human nucleotide excision repair of paired and mispaired cisplatin-DNA adducts. *Nucleic Acids Res* 25:480–491.

Mu D, Tursun M, Duckett DR, Drummond JT, Modrich P, and Sancar A (1997) Recognition and repair of compound DNA lesions (base damage and mismatch) by human mismatch repair and excision repair systems. *Mol Cell Biol* 17:760–769.

Osanto S, Bukman A, Van Hoek F, Sterk PJ, De Laat JA, and Hermans J (1992) Long-term effects of chemotherapy in patients with testicular cancer. *J Clin Oncol* 10:574–579.

Palmisano J, Agraharkar M, and Kaplan AA (1998) Successful treatment of cisplatin-induced hemolytic uremic syndrome with therapeutic plasma exchange. *Am J Kidney Dis* 32:314–317.

Paquet F, Perez C, Leng M, Lancelot G, and Malinge JM (1996) NMR solution structure of a DNA decamer containing an interstrand cross-link of the antitumor drug cis-diamminedichloroplatinum (II). *J Biomol Struct Dyn* 14:67–77.

Park MS, De Leon M, and Devarajan P (2002) Cisplatin induces apoptosis in LLC-PK1 cells via activation of mitochondrial pathways. *J Am Soc Nephrol* 13:858–865.

Peyrone M (1845) Über die Einwirkung des Ammoniaks auf platinchlorur. *Liebigs Ann Chem* 51:1–29.

Pillaire MJ, Hoffmann JS, Defais M, and Villani G (1995) Replication of DNA containing cisplatin lesions and its mutagenic consequences. *Biochimie* 77:803–807.

Pinto AL, and Lippard SJ (1985) Sequence-dependent termination of *in vitro* DNA synthesis by cis- and trans-diamminedichloroplatinum (II). *Proc Natl Acad Sci USA* 82:4616–4619.

Plaxe S, Freddo J, Kim S, Kirmani S, McClay E, Christen R, Braly P, and Howell S (1994) Phase I trial of cisplatin in combination with glutathione. *Gynecol Oncol* 55:82–86.

Poll EH, Arwert F, Joenje H, and Eriksson AW (1982) Cytogenetic toxicity of antitumor platinum compounds in Fanconi's anemia. *Hum Genet* 61:228–230.

Poll EH, Arwert F, Joenje H, and Wanamarta AH (1985) Differential sensitivity of Fanconi anaemia lymphocytes to the clastogenic action of cis-diamminedichloroplatinum (II) and trans-diamminedichloroplatinum (II). *Hum Genet* 71:206–210.

Rabb H, Mendiola CC, Dietz J, Saba SR, Issekutz TB, Abanilla F, Bonventre JV, and Ramirez G (1994) Role of CD11a and CD11b in ischemic acute renal failure in rats. *Am J Physiol* 267:F1052–F1058.

Rabb H, Mendiola CC, Saba SR, Dietz JR, Smith CW, Bonventre JV, and Ramirez G (1995) Antibodies to ICAM-1 protect kidneys in severe ischemic reperfusion injury. *Biochem Biophys Res Commun* 211:67–73.

Ramesh G, and Reeves WB (2002) TNF-alpha mediates chemokine and cytokine expression and renal injury in cisplatin nephrotoxicity. *J Clin Invest* 110:835–842.

Rice JA, Crothers DM, Pinto AL, and Lippard SJ (1988) The major adduct of the antitumor drug cis-diamminedichloroplatinum(II) with DNA bends the duplex by approximately equal to 40 degrees toward the major groove. *Proc Natl Acad Sci USA* 85:4158–4161.

Ries F, and Klastersky J (1986) Nephrotoxicity induced by cancer chemotherapy with special emphasis on cisplatin toxicity. *Am J Kidney Dis* 8:368–379.

Rosen S, Brezis M, and Stillman I (1994) The pathology of nephrotoxic injury: A reappraisal. *Miner Electrolyte Metab* 20:174–180.

Rosenberg B, and VanCamp L (1970) The successful regression of large solid sarcoma 180 tumors by platinum compounds. *Cancer Res* 30:1799–1802.

Rosenberg B, Van Camp L, Grimley EB, and Thomson A J (1967) The inhibition of growth or cell division in Escherichia coli by different ionic species of platinum(IV) complexes. *J Biol Chem* 242:1347–1352.

Rosenberg B, Van Camp L, and Krigas T (1965) Inhibition of cell division in Escherichia coli by electrolysis products from a platinum electrode. *Nature* 205:698–699.

Rosenberg B, VanCamp L, Trosko JE, and Mansour VH (1969) Platinum compounds: A new class of potent antitumour agents. *Nature* 222:385–386.

Rossi R, Danzebrink S, Hillebrand D, Linnenburger K, Ullrich K, and Jurgens H (1994a) Ifosfamide-induced subclinical nephrotoxicity and its potentiation by cisplatinum. *Med Pediatr Oncol* 22:27–32.

Rossi R, and Ehrich JH (1993) Partial and complete de Toni-Debre-Fanconi syndrome after ifosfamide chemotherapy of childhood malignancy. *Eur J Clin Pharmacol* 44:S43–S45.

Rossi R, Godde A, Kleinebrand A, Riepenhausen M, Boos J, Ritter J, and Jurgens H (1994b) Unilateral nephrectomy and cisplatin as risk factors of ifosfamide-induced nephrotoxicity: Analysis of 120 patients. *J Clin Oncol* 12:159–165.

Safirstein R, Miller P, and Guttenplan JB (1984) Uptake and metabolism of cisplatin by rat kidney. *Kidney Int* 25:753–758.

Safirstein R, Winston J, Goldstein M, Moel D, Dikman S, and Guttenplan J (1986) Cisplatin nephrotoxicity. *Am J Kidney Dis* 8:356–367.

Safirstein R, Winston J, Moel D, Dikman S, and Guttenplan J (1987) Cisplatin nephrotoxicity: Insights into mechanism. *Int J Androl* 10:325–346.

Safirstein R, and Wiston J (1987) Cisplatin nephrotoxicity. *J Uoeh* 9 (Suppl):216–222.

Scagliotti GV, De Marinis F, Rinaldi M, Crino L, Gridelli C, Ricci S, Matano E, Boni C, Marangolo M, Failla G, Altavilla G, Adamo V, Ceribelli A, Clerici M, Di Costanzo F, Frontini L, and Tonato M (2002) Phase III randomized trial comparing three platinum-based doublets in advanced non-small-cell lung cancer. *J Clin Oncol* 20:4285–4291.

Schabel FM Jr (Ed.) (1978) *Platinum complexes in cancer chemotherapy* Karger, Basel.

Schilsky RL, Barlock A, and Ozols RF (1982) Persistent hypomagnesemia following cisplatin chemotherapy for testicular cancer. *Cancer Treat Rep* 66:1767–1769.

Schoffski P, Wanders J, Verweij J, and Fumoleau P (2000) Docetaxel and cisplatin in head and neck cancer. *Ann Oncol* 11:1617.

Schreiner GE, and Maher JF (1965) Toxic Nephropathy. *Am J Med* 38:359–377.

Schrier RW (2002) Cancer therapy and renal injury. *J Clin Invest* 110:743–745.

Schuchter LM, Hensley ML, Meropol NJ, and Winer EP (2002) 2002 update of recommendations for the use of chemotherapy and radiotherapy protectants: Clinical practice guidelines of the American Society of Clinical Oncology. *J Clin Oncol* 20:2895–2903.

Sener G, Satiroglu H, Kabasakal L, Arbak S, Oner S, Ercan F, and Keyer-Uysa M (2000) The protective effect of melatonin on cisplatin nephrotoxicity. *Fundam Clin Pharmacol* 14:553–560.

Sheehan JP, and Seelig MS (1984) Interactions of magnesium and potassium in the pathogenesis of cardiovascular disease. *Magnesium* 3:301–314.

Sheridan AM, and Bonventre JV (2000) Cell biology and molecular mechanisms of injury in ischemic acute renal failure. *Curr Opin Nephrol Hypertens* 9:427–434.

Sheridan AM, and Bonventre JV (2001) Pathophysiology of ischemic acute renal failure. *Contrib Nephrol* 7–21.

Sorensen JB, Groth S, Hansen SW, Nissen MH, Rorth M, and Hansen HH (1985) Phase I study of the cisplatin analogue 1,1-diamminomethylcyclohexane sulfatoplatinum (TNO-6) (NSC 311056). *Cancer Chemother Pharmacol* 15:97–100.

Sorenson CM, Barry MA, and Eastman A (1990) Analysis of events associated with cell cycle arrest at G2 phase and cell death induced by cisplatin. *J Natl Cancer Inst* 82:749–755.

Sorenson CM, and Eastman A (1988) Influence of cis-diamminedichloroplatinum(II) on DNA synthesis and cell cycle progression in excision repair proficient and deficient Chinese hamster ovary cells. *Cancer Res* 48:6703–6707.

Sriswasdi C, Jootar S, and Giles FJ (2000) Amifostine and hematologic effects. *J Med Assoc Thai* 83:374–382.

Stone PJ, Kelman AD, and Sinex FM (1974) Specific binding of antitumour drug cis-Pt(NH3)2C12 to DNA rich in guanine and cytosine. *Nature* 251:736–737.

Suarez A, McDowell H, Niaudet P, Comoy E, and Flamant F (1991) Long-term follow-up of ifosfamide renal toxicity in children treated for malignant mesenchymal tumors: An International Society of Pediatric Oncology report. *J Clin Oncol* 9:2177–2182.

Swenerton K, Jeffrey J, Stuart G, Roy M, Krepart G, Carmichael J, Drouin P, Stanimir R, O'Connell G, and MacLean G (1992) Cisplatin-cyclophosphamide versus carboplatin-cyclophosphamide in advanced ovarian cancer: A randomized phase III study of the National Cancer Institute of Canada Clinical Trials Group. *J Clin Oncol* 10:718–726.

Takahara PM, Rosenzweig AC, Frederick CA, and Lippard SJ (1995) Crystal structure of double-stranded DNA containing the major adduct of the anticancer drug cisplatin. *Nature* 377:649–652.

Takano M, Nakanishi N, Kitahara Y, Sasaki Y, Murakami T, and Nagai J (2002) Cisplatin-induced inhibition of receptor-mediated endocytosis of protein in the kidney. *Kidney Int* 62:1707–1717.

Thadhani R, Pascual M, and Bonventre JV (1996) Acute renal failure. *N Engl J Med* 334:1448–1460.

Thigpen JT, Blessing JA, Orr JW Jr and DiSaia PJ (1986) Phase II trial of cisplatin in the treatment of patients with advanced or recurrent mixed mesodermal sarcomas of the uterus: A Gynecologic Oncology Group Study. *Cancer Treat Rep* 70:271–274.

Thurnher D, Kletzmayr J, Formanek M, Quint C, Czerny C, Burian M, and Kornek G (2001) Chemotherapy-related hemolytic-uremic syndrome following treatment of a carcinoma of the nasopharynx. Oncology 61:143–146.

Townsend DM, Deng M, Zhang L, Lapus MG, and Hanigan MH (2003) Metabolism of Cisplatin to a nephrotoxin in proximal tubule cells. *J Am Soc Nephrol* 14:1–10.

Townsend DM, and Hanigan MH (2002) Inhibition of gamma-glutamyl transpeptidase or cysteine S-conjugate beta-lyase activity blocks the nephrotoxicity of cisplatin in mice. *J Pharmacol Exp Ther* 300:142–148.

van der Heijden M, Ackland SP, and Deveridge S (1998) Haemolytic uraemic syndrome associated with bleomycin, epirubicin and cisplatin chemotherapy—a case report and review of the literature. *Acta Oncol* 37:107–109.

van der Meer J, de Vries EG, Vriesendorp R, Willemse PH, Donker AJ, and Aalders JG (1985) Hemolytic uremic syndrome in a patient on cis-platinum, vinblastine and bleomycin. *J Cancer Res Clin Oncol* 110:119–122.

von der Weid NX, Erni BM, Mamie C, Wagner HP, and Bianchetti MG (1999) Cisplatin therapy in childhood: renal follow up 3 years or more after treatment Swiss Pediatric Oncology Group. *Nephrol Dial Transplant* 14:1441–1444.

Walker EM Jr, and Gale GR (1981) Methods of reduction of cisplatin nephrotoxicity. *Ann Clin Lab Sci* 11:397–410.

Watson PR, Guthrie TH Jr and Caruana RJ (1989) Cisplatin-associated hemolytic-uremic syndrome Successful treatment with a staphylococcal protein A column. *Cancer* 64:1400–1403.

Weinblatt ME, Kahn E, Scimeca PG, and Kochen JA (1987) Hemolytic uremic syndrome associated with cisplatin therapy. *Am J Pediatr Hematol Oncol* 9:295–298.

Werner A (1911) New Ideas On Inorganic Chemistry (translation, by EP Hedley, of the second German edition, 1909), Longmans, Green & Co, London.

Wester PO, and Dyckner T (1986) Intracellular electrolytes in cardiac failure. *Acta Med Scand* Suppl 707:33–36.

Yuan ZM, Shioya H, Ishiko T, Sun X, Gu J, Huang YY, Lu H, Kharbanda S, Weichselbaum R, and Kufe D (1999) p73 is regulated by tyrosine kinase c-Abl in the apoptotic response to DNA damage. *Nature* 399:814–817.

Zamble DB, Jacks T, and Lippard SJ (1998) p53-Dependent and -independent responses to cisplatin in mouse testicular teratocarcinoma cells. *Proc Natl Acad Sci USA* 95:6163–6168.

Zamble DB, and Lippard SJ (1995) Cisplatin and DNA repair in cancer chemotherapy. *Trends Biochem Sci* 20:435–439.

Zamble DB, Mu D, Reardon JT, Sancar A, and Lippard SJ (1996) Repair of cisplatin—DNA adducts by the mammalian excision nuclease. *Biochemistry* 35:10004–10013.

Zanke BW, Boudreau K, Rubie E, Winnett E, Tibbles LA, Zon L, Kyriakis J, Liu FF, and Woodgett JR (1996) The stress-activated protein kinase pathway mediates cell death following injury induced by cis-platinum, UV irradiation or heat. *Curr Biol* 6:606–613.

Zhang JG, and Lindup WE (1993) Role of mitochondria in cisplatin-induced oxidative damage exhibited by rat renal cortical slices. *Biochem Pharmacol* 45:2215–2222.

Zhang, JS, Shen-Feng MA, Suenaga A, and Otagiri M (2001) Stability of a cisplatin-chondroitin sulfate A complex in plasma and kidney in terms of protein binding. *Biol Pharm Bull* 24:970–972.

19

The Pathogenesis and Prevention of Radiocontrast Medium Induced Renal Dysfunction

Nicholas Kaperonis, Mark Krause, and George L. Bakris

INTRODUCTION

With the advent of radiocontrast media (RCM) diagnostic radiology was advanced tremendously. However, despite ongoing refinement of these agents, the risk of contrast nephropathy (CN) remains present. Athough risk factors for CN have been elucidated, there have been few advances in prevention or treatment other than adequate prehydration and continued hydration after administration. In early reports, the third most common cause of hospital-acquired acute renal failure resulted from use of high-osmolar RCM for imaging studies (Shusterman et al., 1987). In more-recent analyses, this has become less common (Quader et al., 1998), however, it is not yet a rare occurrence. In spite of the current availability and use of low-osmolar RCM, acute renal dysfunction or failure continues to be observed in the hospital setting following radiographic studies (Lasser et al., 1997; Quader et al., 1998; Shusterman et al., 1987). Moreover, the number of reactions reported to the Food and Drug Administration that relate to the use of low-osmolar RCM are fourfold greater than for high-osmolar agents (Lasser et al., 1997), however there is an ever increasing number of radioimaging studies being performed in critically ill patients.

Since the 1960s, investigators have used various animal models to define the mechanisms that contribute to the development of RCM-induced renal dysfunction. Clinical studies have also attempted to outline and prevent this problem. This chapter presents an overview of the history, epidemiology, and pathophysiology of renal dysfunction following RCM administration. Preventive strategies derived from clinical studies are also discussed.

HISTORY

Radiocontrast agents were first used for *in vivo* angiographic studies in the early 1920s (Osborne et al., 1923). Agents used during this period include strontium bromide, thorium dioxide, and sodium bromide. These agents were associated with an increased incidence of malignancies and prolonged radioactivity (Osborne et al., 1923; Silpananta et al., 1983). Organic di-iodinated preparations were also used during the early 1920s (Sutton, 1987), and the first case report of renal dysfunction following their use was announced in 1931 (Pendergrass et al., 1942). Subsequently, these compounds were replaced in the mid-1950s with tri-iodinated compounds.

The presence of three atoms of iodine per molecule, as opposed to one or two, provided an ideal imaging substance. These new compounds were found to be less toxic, but more viscous, than the di-iodinated. The tri-iodinated RCM also lacked water solubility secondary to the high osmotic composition. The hyperosmolarity induced in serum following the administration of these compounds is thought to be a primary cause of acute decline in renal function (Alexander et al., 1978; Ansari and Baldwin, 1976; Barrett and Carlisle, 1993; Bartley et al., 1969; Byrd and Sherman, 1979; Diaz-Buxo et al., 1975; Krumlovsky et al., 1978).

Investigators in the early 1970s then developed the nonionic, monomeric RCM. These newer RCM had lower osmolality effects on serum than the older ones, although the osmolality of these agents is still double that of blood. There was an expectation, however, that the reduction in osmolality would result in reduced nephrotoxicity (Spataro, 1984). These newer agents are termed nonionic because an

Table 19.1 The Evolution of Clinically Used Radiocontrast Media

Agents	*Year used*
Sodium iodide	1918*, 1923
Strontium bromide	1923
Thorium dioxide	1923
Monoiodinated compounds: lopax and others	1929
Di-iodinated compounds: diodrast, skiodan, diodone and others	Early 1930s
Tri-iodinated compounds: sodium diatrizoate, diatrizoate meglumine, etc.	Mid 1950s
Low osmolar (nonionic) compounds: metrizamide, iohexol, iopamidol and others	Late 1970s, early 1980s

*Initially suggested for use as a radiocontrast agent

organic side chain has replaced the carboxyl group, and hence they do not ionize in solution. They also differ from the ionic RCM in the number of osmotic particles per iodine atom (roughly 50% less than ionic agents), which accounts for the low osmolarity (Alexander et al., 1978; Ansari and Baldwin, 1976; Barrett and Carlisle, 1993; Bartley et al., 1969; Byrd and Sherman, 1979; Diaz-Buxo et al., 1975; Krumlovsky et al., 1978; Pendergrass et al., 1942; Spataro, 1984; Sutton, 1987). In the 1980s, nonionic dimers were introduced (two nonionic tri-iodinated benzoic rings were attached), which have an osmolality similar to that of blood (isoosmolar agents) (Morcos and Thomsen, 2001).

Since the mid-1980s, attempts to develop nontoxic RCM have led to the development of gadolinium-diethyltriamine pentaacetic acid (Gd-DTPA) and carbon dioxide (Seeger et al., 1993; Spinosa et al., 1999). The historic evolution of RCM is summarized in Table 19.1.

PHARMACOLOGY AND PHYSIOLOGY

RCMs are organized into two groups: ionic and nonionic. The major difference between these groups is their divergent osmolalities, not iodine content or viscosity. Because the renal toxicity of an agent is more closely linked to its osmolality than to its ionic characteristics, lower osmolar agents are preferred (Benness, 1970; Bettmann, 1982; Burgener and Hamlin, 1981; Dean et al., 1978; Gaspari et al., 1997; Haustein et al.,

Table 19.2 Physical Properties of Example Radiocontrast Agents

	Osmolality (mOsm/l)	*Iodine (%)*	*Viscosity (37° mPas)*
Ionic			
Iothalamate meglumine (Conray 60)	1217	28	4.0
Sodium diatrizoate (Hypaque)+	1470	30	2.5
Meglumine-sodium diatrizoate (Renografin 76)	1690	37	9.1
Iothalamate-sodium (Conray 400)	1965	40	4.5
Nonionic			
Iotrol+	300	30	9.1
Metrizamide	450	28	5.0
Iopamidol+	570	28	3.8
Iohexol	620	28	4.8

Terms in brackets are trade name.
The range of molecular weights for radiocontrast medias mentioned above is 636 (diatrizoate) to 1626 (Iotrol).

1992; Morris and Fischer, 1986; Mudge, 1980, 1990; Rocco et al., 1996; Sage, 1983; Schiantarelli et al., 1973; Spataro et al., 1982; Talner, 1972; Ueda et al., 1998). RCM osmolalities are summarized in Table 19.2.

The RCM of each group that are used in medical practice are derivatives of 2,4,6-tri-iodinated benzoic acid (Figure 19.1).

The most widely used class of hyperosmolar RCM is the diatrizoate derivatives, which are water soluble and have a low pK_a due to their carboxyl group (Bettmann, 1982; Mudge, 1990). They exist as anions in biological systems and are confined to the extracellular space (Mudge, 1990). They do not bind significantly to serum proteins and are almost entirely cleared by the kidneys (99%), with a small component accounted for by gastrointestinal and hepatic clearance pathways (1%). The extrarenal removal of the diazoate derivatives becomes more pronounced in renal failure (Schiantarelli et al., 1973; Talner, 1972) and can account for up to 25 to 36% of the administered dose (Ackrill et al., 1976; Cattel et al., 1967). Despite this, the mean half-life of RCM is roughly doubled in dialysis patients when compared with persons with functional kidneys (Ackrill et al., 1976; Cattel et al., 1967; Millman and Christensen, 1974; Schindler et al., 2001; Waaler et al., 1990).

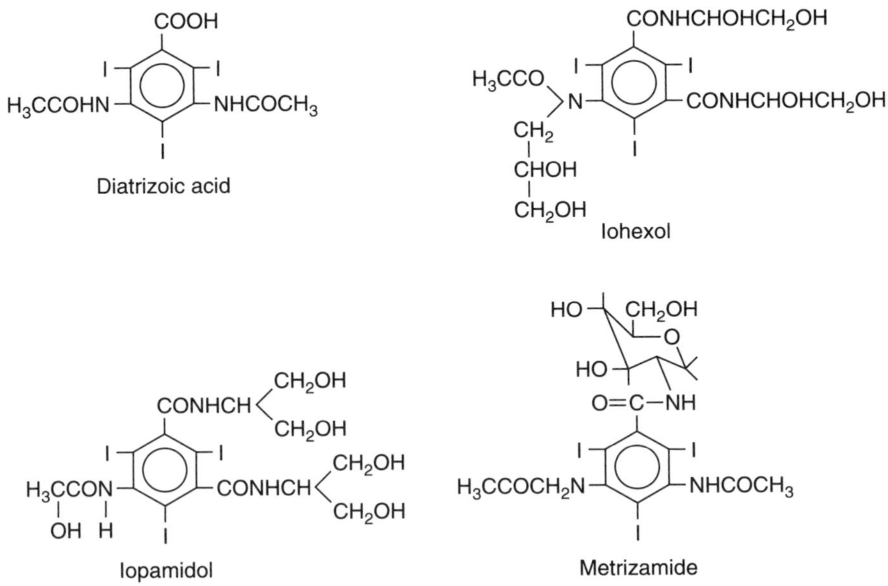

Figure 19.1 The chemical structures of hyperosmolar and low-osmolar RCMs commonly used in clinical practice.

Two types of RCM, tri-iodinated compounds and DTPA, are freely filtered by the glomerulus and neither secreted nor reabsorbed by the tubules (Burgener and Hamlin, 1981; Haustein et al., 1992). Consequently, these agents are similar to inulin, a marker of glomerular filtration rate (GFR). In persons with normal renal function, the plasma half-life of RCM is between 30 and 60 minutes (Burgener and Hamlin, 1981; Morris and Fischer, 1986), and the peak time for excretion of RCM is about 3 minutes following intravenous injection. Peak urine iodine concentrations occur approximately 1 hour after RCM administration (Mudge, 1980; Spataro et al., 1982; Ueda et al., 1998).

Although the initial concentration of RCM in the tubule is the same as in the plasma, the urinary concentration of RCM increases fivefold to tenfold as a result of proximal tubular sodium and water reabsorption. With meglumine derivatives, this increase is partially attenuated due to higher urine flow rate. Increased concentration of RCM may explain, in part, enhanced toxicity to the loop of Henle portion of the renal tubule.

Proximal tubular sodium reabsorption following RCM treatment is independent of hydration and is influenced by diuretics and osmotic load (Cattel et al., 1967; Talner, 1972). It is conceivable, therefore, that the concentration of RCM in the ultrafiltrate leaving the proximal tubule may be up to 50 or 100 times that entering the tubule (Benness, 1970; Cattel et al., 1967; Dean et al., 1978; Mudge, 1980; Burgener anad Hamlin, 1981; Morris and Fischer, 1986; Sage, 1983; Spataro et al., 1982; Talner, 1972; Ueda et al., 1998). Furthermore, in states of dehydration, the concentration of RCM can be further increased in the collecting duct system secondary to the increased levels of antidiuretic hormone.

In patients with end-stage renal disease, RCM is cleared by extrarenal pathways, as well as by dialysis, because they are not protein-bound and possess relatively low molecular weights. The clearance of these agents by hemodialysis, at blood flow rates between 172 and 250 ml/min, varies between 65 and 80% following a 4 hour treatment (Ackrill et al., 1976; Lehnert et al., 1998; Milman and Christeensen, 1974; Schindler et al., 2001; Waaler et al., 1990). The volume of distribution (Vd) of the RCM is limited due to their polar state in physiologic conditions. Reaching an equilibrium depends on the following factors: organ blood flow, capillary density and permeability, and interstitial diffusion distances (Burgener and Hamlin, 1981; Dean et al., 1978; Morris and Fischer, 1986; Sage, 1983). Following administration of RCM there is a rapid equilibration across capillary membranes. with the exception of the blood–brain barrier. However, RCM can enter the cerebrospinal fluid through fenestrae in the choroid plexus (Sage, 1983). During the first phase of distribution following RCM administration, there is a rapid fluid shift across the capillary membranes from the interstitial space due to the markedly increased intravascular osmolality (Sage, 1983; Schiantarelli et al., 1973; Morris and Fischer, 1986). It is during this phase that toxicity occurs.

MECHANISMS OF NEPHROTOXICITY

Overview

Nephrotoxicity has plagued the use of RCM since their inception. Through refinement, this side effect has become less frequent, but it still occurs at alarming rates. The etiology of RCM nephrotoxicity can be broadly grouped into two categories: hemodynamic effects (vasoconstriction) and tubular (cellular) injury. An overall schematic representation integrating these factors and the resulting consequences of RCM

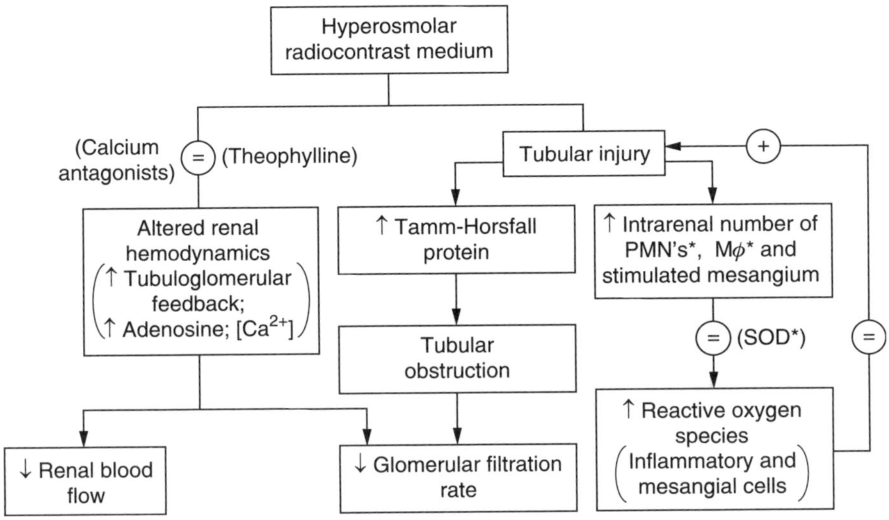

Figure 19.2 A proposed model of the mechanisms involved in RCM-induced renal dysfunction. (SOD) superoxide dismutase, (PMN) polymorphonuclear leukocytes, (MΦ) macrophages, (=) inhibits, (+) potentiates.

administration is summarized in Figure 19.2. Of these factors, the magnitude of RCM osmolality is central to both the acute hemodynamic and tubular changes following RCM administration (Caldicott et al., 1970; Harvey, 1960; Heyman et al., 1993; Katzberg et al., 1983; Morris et al., 1978; Norby and DiBona, 1975; Talner and Davidson, 1968). Nephrotoxicity from vasoconstriction has been postulated to be responsible for the majority of RCM nephrotoxicity (Bakris and Burnett, 1985; Byrd and Sherman, 1979; Caldicott et al., 1970; Harvey, 1960; Heyman et al., 1993; Katzberg et al., 1983; Larson et al., 1983; Morris et al., 1978; Norby and DiBona, 1975; Talner and Davidson, 1968).

It is well established that delivery of hyperosmolar solution to the renal vasculature results in a biphasic change in renal blood flow (RBF) (Arakawa et al., 1996; Arend et al., 1987; Bagnis et al., 1997; Bakris and Burnett, 1985; Bakris et al., 1990b, 1999; Deray et al., 1990; Drescher and Madsen, 1998; Drescher et al., 1998; Heyman et al., 1988, 1989; Katzberg et al., 1983; Larson et al., 1983; Margulies et al., 1990; Murphy et al., 1998; Schiantarelli et al., 1973; Vari et al., 1988) (Figure 19.3). After the administration of hyperosmolar solution there is an initial transient increase in RBF and GFR, followed by a prolonged decrease.

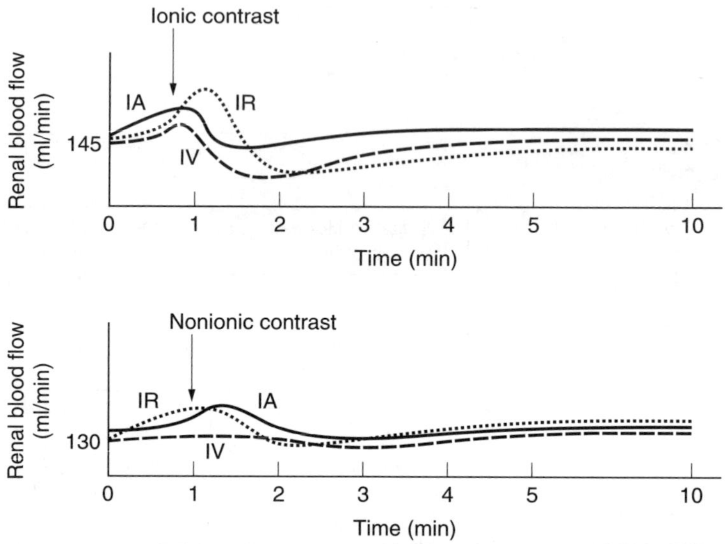

Figure 19.3 The renal vascular effects of a hyperosmolar ionic RCM (sodium diatrizoate) and nonionic RCM (iopamidol) in the normal dog. (IA) intra-arterial, (IR) intrarenal, (IV) intravenous. Note: results obtained from two separate dogs.

The decrease in RBF is approximately 30 to 40% of the baseline RBF. Importantly, since similar changes in RBF were reported when hypertonic saline was administered, the hyperosmolar quality of the solution appears to be responsible for the biphasic changed in RBF. Furthermore, the reduction in RBF has been shown to correlate positively with the osmolality of the infused solution (Gerber et al., 1979; Katzberg et al., 1977; Mudge et al., 1984). These observations have led investigators to hypothesize that the delivery of a high osmotic load, as associated with RCM administration, to the juxtaglomerular apparatus results in stimulation of a tubuloglomerular feedback processes. This, in turn, reduces GFR by altering the vascular tone of afferent arterioles.

Recovery from the vasoconstriction that follows hyperosmolar RCM injection in normal dogs is universal. However, in three volume-depleted dogs with a 1&$\frac{2}{3}$ nephrectomy there was no recovery from the reduction in RBF, even after 6 hours of observation (Figure 19.4). Such a prolonged effect on RBF supports the clinical observation that volume-depleted subjects with preexisting renal insufficiency are

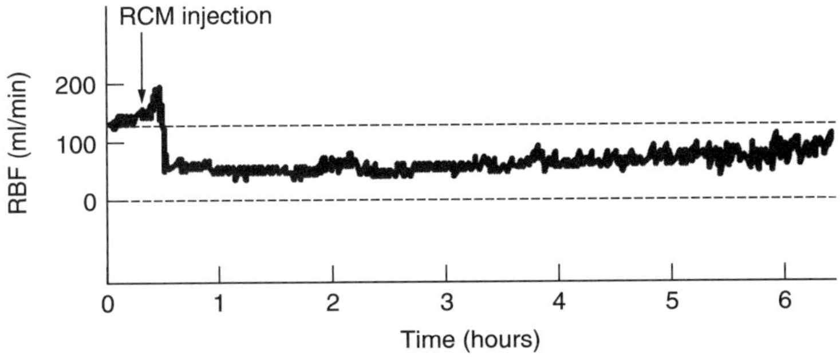

Figure 19.4 The renal blood flow response of a volume-depleted dog with a 1&2/3 nephrectomy following intrarenal injection of a hyperosmolar RCM, Renograffin. Note: renal blood flow never returns to baseline, even after 6 hours.

at higher risk for development of acute renal failure following RCM administration.

In spite of the importance of osmolarity in determining the renal hemodynamic response to RCM administration, there is variability in the vascular effects that is both dose-dependent and influenced by the route of administration (Burgener and Hamlin, 1981; Sage, 1983; Schiantarelli et al., 1973). Figure 19.3 demonstrates the time course and variability of renal vascular responsiveness between intrarenal, intra-arterial (thoracic aorta), and i.v. bolus injection of ionic and nonionic RCM in the dog. These observed renal vascular effects of RCM are also supported by other investigators (Caldicott et al., 1970; Morris et al., 1978; Norby and BiBona, 1975; Talner and Davidson, 1968). Similar observations were made in both femoral blood flow and RBF before and after the administration of both high- and low-osmolar RCM (Figure 19.5). From these studies, it is apparent that higher osmolar RCM have a higher risk of altered renal perfusion, and consequent renal dysfunction from hemodynamic insults.

The hyperosmolar RCM are also known to have direct toxic effects on renal tubule cells. Several *in vitro* studies clearly illustrate that hyperosmotic RCM are toxic within minutes to cultured proximal tubular cells (Bhandaru and Bakris, 1995; Humes et al., 1987; Messana et al., 1988a). Enhanced susceptibility to injury of mesangial cells has been demonstrated by comparing high-osmolar (diatrizoate) and low-osmolar (iohexol) RCM. Viability of mesangial cells was reduced to a greater degree under high-glucose conditions. The cytotoxic effects of two

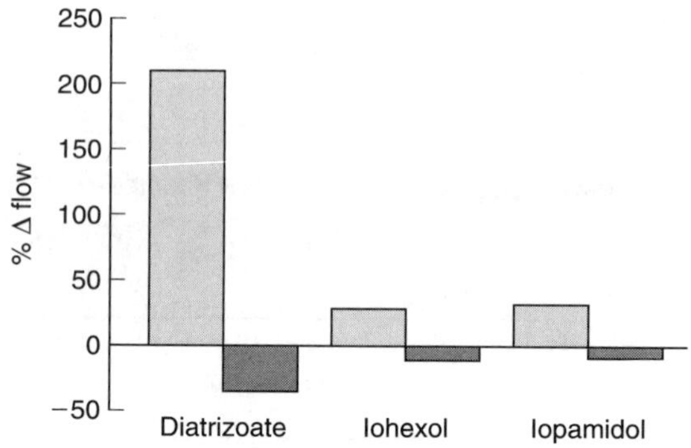

Figure 19.5 The percentage change in blood flow in either the renal or femoral arteries in a euvolemic dog following administration of either high- or low-osmolar radiocontrast agents. Femoral artery, rising right; renal artery, rising right. Note: results obtained from two separate dogs.

different RCM were similar if concentrations were of equivalent osmolality. Interestingly, exposure to mannitol did not influence cell viability unless the culture media contained very high glucose concentration (Wasaki et al., 2001). Other cellular studies, however, also report only minimal cytotoxicity in the presence of low-osmolar RCM (Bhandaru and Bakris, 1995).

In vitro evidence to support the clinical relevance of RCM-induced tubular injury comes from both animal and human studies. Several investigators have shown the presence of granular and hyaline casts in the urine within 2 hours following injection of a hyperosmolar RCM (Bhandaru and Bakris, 1995; Mudge et al., 1984; Wasaki et al., 2001). Moreover, renal biopsies from dogs given intrarenal injections of hyperosmolar RCM demonstrate the presence of polymorphonuclear leukocytes and macrophages in the interstitium of the kidney within 3 hours of hypertonic RCM injection (see below).

The following discussion focuses on specific factors implicated in both altered renal hemodynamics and tubular injury associated with RCM administration.

Hemodynamic Mechanisms of Nephrotoxicity

The Role of Calcium in RCM Vasoconstriction

Following the elucidation of a biphasic vasospastic response to high-osmolar RCM, investigators have attempted to either block this event or lessen its the impact on renal perfusion (Arakawa et al., 1996; Arend et al., 1987; Bakris and Burnett, 1985; Bakris et al., 1990b;Drescher and Madsen, 1998; Drescher et al., 1998; Deray et al., 1990; Heyman et al., 1989, 1993; Katzberg et al., 1983; Larson et al., 1983; Margulies et al., 1990; Murphy and Tublin, 1998; Vari et al., 1988, 1988). Medications that have been evaluated to intervene in the vasoconstriction that follows the administration of these agents include: renin–angiotensin antagonists, α-adrenergic antagonists, and vasodilators, such as hydralazine. In each case, these agents have failed to attenuate the vasoconstriction associated with RCM (Katzberg et al., 1983; Morris et al., 1978; Norby and DiBona, 1975; Talner and Davidson, 1968). However, in 1984, calcium channel blockers (CCBs) were reported to ameliorate the RCM-induced vasoconstriction (Bakris and Burnett, 1985). The renal protective effect of dihydropyridine CCBs is thought to depend, in part, on the inhibition of renal autoregulatory mechanisms mediated through the juxtaglomerular apparatus and tubuloglomerular feedback on afferent arteriolar "myogenic" responsiveness (Bhandaru and Bakris, 1995; Navar et al., 1986; Wasaki et al., 2001).

Normal rats are resistant to RCM-induced renal injury unless made susceptible (Vari et al., 1988). As nitric oxide (NO) and prostacycline are documented regulators of renal medulla perfusion, pretreatment with L-nitroarginine methyl ester (L-NAME) and indomethacin provides an alternative animal model, not requiring surgical intervention. In this model, injection of the high-osmolar RCM, diatrizoate (at doses of 6 or 8 ml/kg, 306 mg iodine/ml) resulted in significant, reversible increases in serum creatinine concentration, not observed with normal saline or smaller doses of diatrizoate. The effect of this RCM on renal function was greater than with an equal volume of low-osmolar nonionic monomer iopromide. The CCB, diltiazem, (in 2, 6, or 10 mg/kg doses), injected intraperitoneally 30 minutes prior to diatrizoate, reduced the magnitude of GFR decline in a dose-dependent manner (Wang et al., 2001).

CCBs have also been reported to have cytoprotective effects on renal cells; by suppressing influx of extracellular calcium after ischemic or toxic injuries, and by modulating mesangial traffic of macromolecules, they reduce free radical generation and renal hypertrophy (Baer and Navar, 1973; Bakris and Burnett, 1985; Esnault, 2002; Osborne et al., 1923). In addition, CCBs can inhibit the decrease of NO synthesis following RCM administration in humans (Esnault, 2002). Despite this evidence in animal models, the role of calcium channel blockade as a prophylactic agent against RCM in humans remains unproven.

The Role of Adenosine in RCM Vasoconstriction

Adenosine, a well-known vasodilator in the peripheral circulation, acts as a vasoconstrictor in the renal cortex (Hall et al., 1985; Osswald et al., 1978, 1995; Pflueger et al., 2000; Spielman and Thompson, 1982; Tagawa and Vander, 1970; Yao et al., 2001). Furthermore, hypertonic ultrafiltrates, including hyperosmolar RCM, have been shown to induce the release of adenosine from the macula densa cells of the distal tubule (Osswald et al., 1978). Several studies have demonstrated that RBF is altered similarly by both adenosine and hyperosmolar RCM (Arend et al., 1987; Bakris and Burnett, 1985; Deray et al., 1990; Drescher and Madsen, 1998; Hall et al., 1985; Spielman and Thompson, 1982; Tagawa and Vander, 1970). Taken together, the above data support a central role for adenosine in the mediation of the renal vasoconstrictor response following intrarenal hyperosmolar RCM injection. Adenosine has also been shown to be involved in renal autoregulation of tissue perfusion (Hall et al., 1985; Osswald et al., 1978). Inhibition of adenosine activity via an adenosine (A_1) receptor antagonist or its production via theophylline attenuates tubuloglomerular feedback and autoregulation mechanisms (Arend et al., 1987; Osswald et al., 1978, 1995; Tagawa and Vander, 1970).

The major source of adenosine is generation via 5′-nucleotidase, an enzyme activated by low levels of adenosine triphosphate (ATP) (which occurs during periods of hypoxia or ischemia). Adenosine-mediated vasoconstriction, unique for renal tissue, is induced by activation of A_1 receptors (predominant on afferent arteriole) and subsequent increase in intracellular calcium. RCM can directly activate A_1 receptors, while tubular osmotic load may increase transport mechanisms and ATP hydrolysis, leading to adenosine generation. Moreover, RCM reduces erythrocyte flexibility and activates adhesion molecules and leukocytes to endothelial cells. In diabetics, adenosine-induced vasoconstriction is 20- to 30-fold greater due to attenuated NO-dependent vasodilation, and possibly by upregulation of adenosine A_1 receptor density, as well as increased adenosine production by hyperfiltering diabetic kidney (Pflueger et al., 2000).

In an NO-depleted rat model in which L-NAME is given in drinking water for 8 weeks, GFR failed to drop in response to diatrizoate RCM in the presence of a selective adenosine A_1 receptor antagonist (KW-3902) administered intravenously 20 minutes prior to injection. Addition of mannitol had no effect on GFR changes following RCM in these rats (Yao et al., 2001). This suggests that the decrease in GFR was independent of osmotic load. KW-3902 induced a significant increase in urine volume and sodium excretion in addition to that observed by RCM, but it is unknown if this actually reflects a reduction in tubular damage.

The Role of Endothelin in RCM Vasoconstriction

Margulies and colleagues (1991) were first to demonstrate an increase in both plasma and urinary endothelin concentration following RCM administration in the dog. Unlike other autocoids, such as adenosine, endothelin release is not dependent on RCM tonicity (Harvey, 1960; Heyman et al., 1992a; Margulies et al., 1991; Pollock et al., 1997). Rather, both *in vivo* experiments in rats and *in vitro* experiments using endothelial cell culture have shown that neither iodine concentration nor osmotic content of the RCM alters endothelin release (Margulies et al., 1991). Therefore, the mechanism for increased endothelin concentrations following RCM in unclear. The lack of endothelin release with ioversol (Harvey, 1960) adds an additional layer of complexity. This exception is postulated to result from a lack of endothelial cell stimulation, but has not been investigated. Additional studies in rats have shown that endothelin receptor blockade with continuous infusion of BQ123 blocks the renal vasoconstrictor effects of both low- and high-osmolar RCM (Oldroyd et al., 1994; Pollock et al., 1997). Finally, a study using selective endothelin receptor blockade has also shown efficacy for preventing RCM induced decreases in renal function (Pollock et al., 1997).

These observations, taken together with previous studies, suggest that the vasoconstriction observed immediately after RCM administration is mediated by both adenosine and endothelin. These studies further suggest that there may be potentiation of their vasoconstrictor effects due to RCM-associated decreases in NO (Heyman et al., 1988, 1992a). Alternatively, there may be insufficient release of NO to compensate for medullary vasoconstriction in patients with preexisting renal dysfunction.

Mechanisms of Tubular Injury

The Role of Reactive Oxygen Species

Reactive oxygen species (ROS) include the superoxide anion, hydrogen peroxide, hydroxyl radical, and single oxygen. These molecules are released from renal cells in response to various stimuli, and act as paracrine and autocrine stimuli (Baud and Ardaillou, 1986; Baud et al., 1983; Cross et al., 1987; Messana et al., 1988b; Shah, 1989).

ROS are released by a variety of different cells, including polymorphonuclear leukocytes, macrophages, and glomerular mesangial cells (Baud and Ardaillou, 1986; Baud et al., 1983; Cross et al., 1987; Shah, 1989). This group of molecules has many functions, including antimicrobial activity. Their production can be inhibited by glucocorticoids and their effects reduced by specific scavengers, such as superoxide dismutase (SOD), glutathione, and dimethyl sulfoxide (Baud et al., 1983; Messana et al., 1988b; Scaduto et al., 1988; Shah, 1989). Renal biopsies performed within 3 hours of intrarenal RCM administration confirmed a large influx of polymorphonuclear leukocytes and macrophages in both the glomerular and tubular areas (Arakawa et al., 1996). It is hypothesized that these cells, in addition to mesangial cells, release ROS, which in turn contributes to the tubular injury initially induced by the hyperosmotic properties of RCM (Figure 19.2).

In vitro investigations utilizing electron spin resonance techniques demonstrated that exposure of human mesangial cells to an ionic RCM (diatrizoate sodium) produced an increase in ROS (Figure 19.6), including both superoxide and hydroxyl radical species. Both types of RCM increased the intracellular peroxide levels produced by mesangial cells, however D-α-tocopherol attenuated only the effect of the hyperosmolar RCM diatrizoate. This suggests that oxidative stress may contribute to injury under hyperosmolar conditions.

Measurement of direct tubular toxicity is difficult to assess in a clinical setting. Direct tubule toxicity is reflected by increased urinary excretion of lysosomal enzymes and low molecular weight proteins. It is therefore difficult to differentiate between direct toxicity and tubular injury caused by renal ischemia (Katholi et al., 1998). In normal, mildly volume-depleted dogs, the oxygen free radical scavenger SOD partially blocked the fall in GFR following hyperosmolar RCM treatment, but it had no effect on RBF (Arakawa et al., 1996). Yoshioka et al. (1992) found that the proximal tubular content of SOD was much lower in

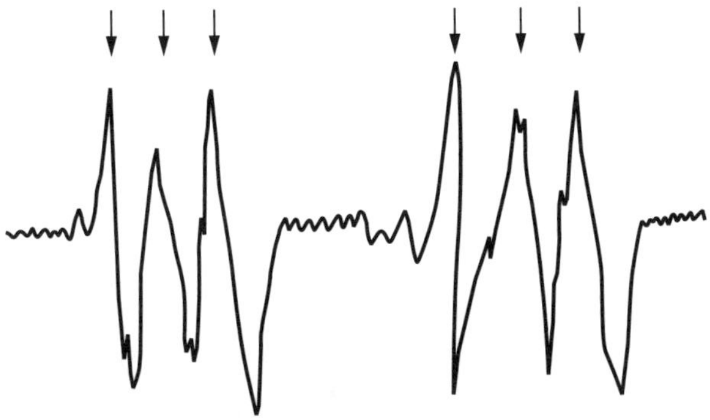

Figure 19.6 The electron spin resonance spectra following administration of a hyperosmolar, ionic RCM (sodium diatrizoate) administered to human mesangial cells and the resultant generation of reactive oxygen species measured by electron spin resonance. Arrows indicate generation of reactive oxygen species. Methodology adapted from Baud et al. (1983).

volume-depleted rats when compared with euvolemic animals. As expected, this group showed the greatest declines in GFR following ionic RCM administration.

It has also been argued that injection of RCM causes ischemia that decreases pO_2. In a separate study, Liss et al. (1997) measured oxygen tension in the rat kidney following intravenous injection of a high-osmolar RCM. They reported a small decrease in pO_2 in the renal cortex with a profound decrease (to nearly 45%) in the medulla.

Markers of Tubular Injury

Tamm-Horsfall Proteins

The Tamm-Horsfall protein is a large glycoprotein normally found only in the ascending limb of the loop of Henle extending into the very early portion of the distal tubule. Its presence in urine indicates tubular injury (Bakris et al., 1990a; Berdon et al., 1969; Hoyer and Seiler, 1987; Patel et al., 1964; Schwartz et al., 1970). The solubility of the Tamm-Horsfall protein in the urine depends on the pH, salt concentration, and concentration of proteins in the urine, as well as other factors (Hoyer and Seiler, 1987).

A number of studies demonstrate increases in urinary levels of Tamm-Horsfall protein following hyperosmolar RCM administration

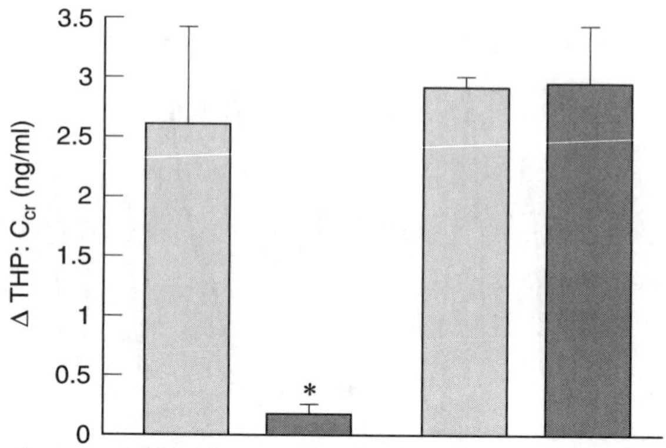

Figure 19.7 The effects of an intrarenal injection of the hyperosmolar RCM (sodium diatrizoate) on urinary Tamm-Horsfall protein. Data expressed as mean ± SEM and represent change in ratio of Tamm-Horsfall protein to creatinine clearance. *$p < 0.05$ compared with baseline. (Open bars) baseline, (dotted bars) superoxide dismutase, (light crosshatch bars) heat inactivated superoxide dismutase.

(Bakris et al., 1990a; Berdon et al., 1969; Nicot et al., 1984; Parvez et al., 1990; Patel et al., 1964; Schwartz et al., 1970) and multiple experiments were undertaken to evaluate the role of ROS (Bakris et al., 1990a). Studies conducted in volume-depleted dogs clearly demonstrate that SOD attenuates the increase in urinary Tamm-Horsfall protein, decreases urinary casts, and blunts declines in GFR following hyperosmolar RCM administration (Figure 19.7) (Bakris et al., 1990a). In addition, pilot studies in euvolemic dogs that received intrarenal injections of the nonionic, low-osmolar RCM, iopamidol, demonstrate a much smaller increase in urinary Tamm-Horsfall protein excretion as well as ROS generation when compared with hyperosmolar RCM (unpublished observations).

Recently, surface-enhanced laser desorption/ionization time of flight technology was used to determine urine protein perturbations after RCM administration. In patients with normal renal function, protein patterns returned to normal after 6 to 12 hours, but in patients with impaired renal function, no recovery was noted. Moreover, the renally impaired group displayed increases in peaks at 9.75 kD and 11.75 kD, representing heparin-binding epidermal growth factor like factor and $\beta 2$-microglobulin, respectively. Conversely, the peak at 66.4 kD, identified as albumin, was suppressed. These changes can be attributed to combined glomerular and tubular dysfunction after RCM injection (Hampel et al., 2001).

RENAL METABOLISM

Numerous studies have investigated the effects of high- and low-osmolar RCM on cellular enzyme activity, prostaglandin production, Na^+/K^+-ATPase activity, and renal oxygen metabolism (Brezis et al., 1989; DeRubertis and Craven, 1978; Hardiek et al., 2001; Kako et al., 1988; Kim and Akera, 1987; Lang and Lasser, 1975; Lear et al., 1990; Parvez et al., 1990; Talner et al., 1972; Workman et al., 1983). In general, these studies indicate that there is no reliable urinary enzymatic indicator of RCM injury, with the possible exception of the proximal tubular enzymes alanine aminopeptidase and γ-glutamyl transpeptidase (Parvez et al., 1990). These two enzymes were found in significantly higher concentrations in the urine of patients as early as 24 hours following either ionic or nonionic RCM, without any significant increase in serum creatinine concentration. Unfortunately, these studies do not provide follow-up data (>72 hours after RCM administration) to assess the clinical predictability of these enzyme markers on the development of renal dysfunction.

Although Na^+/K^+-ATPase activity is clearly inhibited by hyperosmolar RCM, the mechanism for this inhibition is unclear (Patel et al., 1964). Data from the laboratory of this chapter's authors and from three other separate investigators (Lang and Lasser, 1975; Talner et al., 1972; Workman et al., 1983) suggest that RCM-dependent ROS generation may play a role in Na^+/K^+-ATPase inhibition. ROS could cause either direct damage or inhibition of the Na^+/K^+-ATPase pump or inhibition of enzymatic activity secondary to tubular damage or necrosis.

In dog renal slices, both ionic (diatrizoate sodium) and nonionic (iohexol) RCM significantly reduced ouabain-sensitive Na^+/K^+-ATPase, but neither affected mitochondrial ATPase. Moreover, this reduction in Na^+/K^+-ATPase activity by RCM was attenuated in the presence of SOD. Decrease of Na^+/K^+-ATPase activity by RCM was also observed in *in vivo* dog studies. Natriuresis following RCM administration (diatrizoate sodium) was lower in dogs pretreated with SOD when compared with control aninmals (Bakris et al., 1990a).

Experimental studies of LLC-PK_1 cells (renal proximal tubule cell of porcine origin) showed that RCM resulted in shrinkage of the cells, but did not alter viability or induce apoptosis (Hardiek et al., 2001). Several additional effects were noted, including: cell proliferation (assessed by ^{3}H-thymidine incorporation) was reduced, mitochondrial dehydrogenase activity was inhibited (compatible with reversible alteration of mitochondrial function), and extracellular adenosine concentration, a marker of cellular stress, was increased. Interestingly, this study also found that the ionic properties of the RCM had a greater impact on altering mitochondrial function than the molecular structure or osmolality (Hardiek et al., 2001).

The effects of RCM, as well as other hypoxic factors, on the medullary thick ascending limb of the loop of Henle have also been evaluated (Brezis et al., 1989; Heyman et al., 1989; Lear et al., 1990; Vari et al., 1988). In unilaterally nephrectomized rats that were treated with indomethacin and administered a hyperosmolar RCM (iothalamate), a significant fall in creatinine clearance was noted after 24 hours (Vari et al., 1988). In addition, histological studies in this model demonstrate that the loop of Henle had the greatest amount of histologic injury. Consequently, these and other investigators have speculated that the loop of Henle is the most susceptible area of the nephron to hypoxic injury following ionic RCM injection. This is due to the low serum oxygen tension but high cellular oxygen demand (Brezis et al., 1989; Heyman et al., 1989; Vari et al., 1988).

Prostaglandins

Numerous studies have evaluated the modulatory role of prostaglandins of the E series (PGEs) on renal hemodynamics following RCM administration (Lear et al., 1990; Margulies et al., 1990; Osswald et al., 1995). PGE_2 administration has been shown to prevent hypoxic injury to the renal medulla by inhibiting oxygen consumption (Lear et al., 1990). However, prostaglandin inhibition in the rat has not been demonstrated to result in RCM-induced renal dysfunction, regardless of the osmotic content of the agent used (Workman et al., 1983). Prostaglandins do not, therefore, appear to play a major role in the pathogenesis of RCM nephropathy under normal conditions. Conversely, prostaglandin inhibition in rabbits with decreased renal mass clearly potentiates the vasoconstrictor response that follows RCM injection (Margulies et al., 1990).

It appears that prostaglandins are primarily important in pathophysiological states that are associated with an increased influence of endogenous vasoconstrictors, such as heart failure, diabetes, or renal insufficiency. Moreover, there are species differences in the prostaglandin response to vasoconstrictors. Therefore, studies evaluating the effects of various RCM on renal physiology need to be performed in pathophysiologically relevant models to provide clinically meaningful data. An appropriate model would be the volume-depleted dog with a 1&2/3 nephrectomy.

ANIMAL MODELS OF RCM NEPHROTOXICITY

Evaluations of the renal hemodynamic effects of RCM have largely been performed in normal dogs, rabbits, or rats. Therefore, any assertions from these studies to renal hemodynamic alterations in pathophysiological states (i.e., congestive heart failure, diabetic nephropathy, cirrhosis) are, at best, speculative.

Initially, only one clinically relevant animal model was developed for investigation of the renal effects of RCM, as well as possible prophylactic measures against these effects. Margulies and colleagues (1990) evaluated the influence of the RCM Vascoray in dogs with congestive heart failure, which was induced after 8 days of pacing the dogs at the ventricular rate of 250 beats/minute. RCM in the setting of heart failure resulted in a significant and more persistent decline in GFR than under normal circumstances, and this decline was attenuated when atrial natriuretic peptide (ANP) was infused. Unfortunately, a multicenter double-blinded placebo-controlled clinical trial, as well as other studies, failed to show a benefit from ANP for prevention of RCM nephropathy (Allgren et al., 1997; Kurnik et al., 1998).

The rat model of CN is probably not useful. Data from rat models have been misleading and have resulted in clinical trials that failed to show protection against RCM nephropathy, as is the case with furosemide, endothelin, and ANP (see section *Prophylactic Strategies for Prevention*, below). The rat is particularly refractory to renal hemodynamic changes following large doses of ionic RCM. Furthermore, some studies report that data derived from rat studies are difficult, if not impossible, to reproduce (Margulies et al., 1990). Dog models with 1&2/3 nephrectomy or congestive heart failure are closer to resembling human RCM nephropathy, however even these models lack a good correlation with human disease.

RISK FACTORS FOR HUMAN RCM NEPHROPATHY

Renal Insufficiency

By far the most relevant risk factor for predicting the development of RCM nephropathy is the presence of renal insufficiency, i.e., serum creatinine concentration ≥1.5 mg/dl. It is difficult to arrive at an actual incidence of RCM-associated renal dysfunction, largely due to varying definitions among the various studies. This incidence is extended between 0%, in a random population, and 92% in a high-risk population, i.e., those with diabetes or heart failure with preexising renal insufficiency (Alexander et al., 1978; Berns, 1989; Barrett et al., 1992; Berg et al., 1992; Byrd and Sherman, 1979; Carvallo et al., 1975; Cochran et al., 1983; Cramer et al., 1985; Davidson et al., 1989; D'Elia et al., 1982; Diaz-Buxo et al., 1975; Gale et al., 1984; Gomes et al., 1989; Harkonen and Kjellstrand, 1981; Katholi et al., 1993; Khoury et al., 1983; Kinnison et al., 1989; Krumlovsky et al., 1978; Mason et al., 1985; Parfrey et al., 1989; Pathria et al., 1987; Pelz et al., 1984; Pendergrass et al., 1942, 1955; Pillay et al., 1970; Quader et al., 1998; Roy et al., 1985; Rudnick et al., 1995; Schwab et al., 1989; Shafi et al., 1978; Shieh et al., 1982; Shusterman et al., 1987; Sunnegradh et al., 1990; Taliercio et al., 1986; Teruel et al., 1981; Waybill and Waybill, 2001; Weinrauch et al., 1977).

In a classic retrospective study by Byrd and Sherman (1979), subjects with either advanced age, prior renal insufficiency (serum creatinine concentration >1.6 mg/dl), dehydration, or hyperuricemia, had a greater than 50% risk for developing contrast nephrotoxicity. In cases of repeated administration of RCM within 24 hours or the presence of diabetes, the risk was increased by an additional 33% and 37%, respectively.

Since the mid-1980s, a number of well-designed prospective and retrospective studies have evaluated some of the risk factors in patients undergoing arteriographic studies (Barrett et al., 1992; Berg et al., 1992; Cramer et al., 1985; Harkonen and Kjellstrand, 1981; Katholi et al., 1993; Krumlovsky et al., 1978; Rudnick et al., 1995; Shafi et al., 1978; Shieh et al., 1982; Sunnegardh et al., 1990). Some studies have also compared the possible differences in renal effects following hyperosmolar RCM compared with low-osmolar RCM administration in patients with these and other risk factors (Davidson et al., 1989; Gale et al., 1984; Gomes et al., 1989; Katholi et al., 1993; Khoury et al., 1983; Kinnison et al., 1989; Krumlovsky et al., 1978; Parfrey et al., 1989;

Pathria et al., 1987; Roy and Robillard, 1985; Rudnick et al., 1995; Schwab et al., 1989).

A study by Parfrey et al. (1989) documented that the population at highest risk of developing CN following angiographic study was diabetic patients with preexisting renal insufficiency. The investigators found that, given equal hydration status, the presence of preexisting renal insufficiency alone resulted in a greater than fourfold higher risk for the development of acute renal failure when compared with the normal population. Interestingly, this was not true for diabetes alone.

A large multicenter trial, The Iohexol Cooperative Study, confirmed these observations (Rudnick et al., 1995). This study, of 1196 patients who underwent coronary angiography, demonstrated that preexisting evidence of renal insufficiency (serum creatinine concentration >1.6 mg/dl) was by far the most powerful risk factor for predicting renal failure following RCM injection. However, the presence of diabetes in subjects with renal insufficiency added to the risk profile. The risk of developing CN for patients with preexisting chronic renal insufficiency (CRI) was 21-fold higher compared with those with normal renal function (Waybill and Waybill, 2001). The risks due the to coexistence of diabetes in subgroups of patients were found to be as follows: diabetes and CRI, 19.7%, diabetes without CRI, 0.6%, CRI alone, 6% (Waybill and Waybill, 2001). Moreover, the study clearly demonstrated that the use of a low-osmolar RCM, instead of conventional RCM, reduced the incidence of renal impairment in those without diabetes but with renal insufficiency (12.2% CN after using low osmolar RCM versus 27% after using high-osmolar RCM).

Two additional studies, of participants undergoing coronary artery intervention, evaluated the impact of a low osmolar nonionic RCM on acute and long-term renal outcomes. In one study, patients with preexisting renal disease were studied (baseline creatinine concentration equal or greater than 1.8 mg/dl). CN was defined as an increase in serum creatinine concentration of at least 25%, and was reported to occur in 37% of the patients within 48 hours after the procedure. Independent predictors of renal function deterioration included left-ventricular ejection fraction, contrast volume, and blood transfusions. In-hospital and 1-year mortality for patients without CN were 4.9% and 19.4%, respectively, compared with 14.9% and 35.4% for those with CN (22.6% and 45.2% for patients who required dialysis treatment) (Gruberg et al., 2000).

In the second study, CN was defined as an increase in creatinine concentration greater than 0.5 mg/dl from baseline, which was observed in 3.3% of patients. The incidence of CN was associated with baseline creatinine concentration, acute myocardial infarction, shock, and RCM volume, together with age, diabetes, chronic heart failure, and peripheral vascular disease (Rihal et al., 2002). The risk among diabetics was higher only in the subgroup with baseline creatinine concentration of less than 2.0 mg/dl. In-hospital mortality was 22% among patients experiencing CN compared with only 1.4% in those without CN. Among survivors, the history of CN was associated with greater incidence of myocardial infarction and higher mortality (1-year survival, 12.1%, 5-year survival, 44.6%) compared with patients with no such a history (1-year mortality, 3.7%, 5-year mortality, 14.5%).

Miscellaneous Medical Conditions

A review of several prospective angiographic studies demonstrates that the probability of developing renal impairment following an RCM study is a function of the preexisting level of renal function. Moreover, if concomitant conditions, such as heart failure, diabetes, cirrhosis, or volume depletion, are present, the risk of renal dysfunction is substantially greater (Berns, 1989; Dudzinski et al., 1971; Murphy et al., 2000; Rudnick et al., 1995; Waybill and Waybill, 2001). For these groups, a reduced total volume of low-osmolar RCM is mandatory (Barrett and Carlisle, 1993).

It has been generally accepted clinical practice to dialyze a patient with end-stage renal disease immediately following an RCM-requiring procedure to eliminate the hyperosmolar RCM. However, about one-third of any given RCM is metabolized and eliminated prior to initiation of dialysis. Therefore, based on case reports and clinical experience, there is no need for dialysis following an RCM-requiring procedure unless the patient has heart failure or severe hypoalbuminemia, i.e., serum albumin < 2.0 g/l. Because low-osmolar RCM would be expected to cause less fluid shifts and fewer side effects, they are preferred for patients with end-stage renal disease. To support this concept further, in a study of patients with CRI (serum creatinine concentration > 2.3 mg/dl), prophylactic hemodialysis did not alter the rate of CN (Vogt et al., 2001). Dialysis sessions started 30 to 280 minutes after the procedure, used polysulfone membranes (blood flow 180 ± 42 ml/min) and lasted for 3.1 ± 0.7 hours. Thirteen patients in the dialysis group and nine in the nondialysis group developed CN. The incidence did not differ between subgroups stratified according to serum creatinine concentration or volume of RCM.

In conclusion, it seems that hemodialysis does not influence the development of CN. This is because renal injury occurs within 20 minutes after RCM injection.

COMPLICATIONS OF RCM PROCEDURES

Atheroembolic Disease

One of the less well known complications of RCM treatment is the development of atheroembolic-induced renal failure, which mainly occurs with the coexistence of smoking, peripheral vascular disease, age exceeding 50 years, and poorly controlled hypertension with subsequent progressive renal disease (Dudzinski et al., 1971; Kassirer, 1969; Meyrier et al., 1988; Smith et al., 1981; Vogt et al., 2001). Roughly two-thirds of the reported cases occur following a stimulus that triggers showers of microemboli from a damaged aorta (Kassirer, 1969; Meyrier et al., 1988; Smith et al., 1981).

Atheroembolic renal disease is attributed to occlusion of small arteries by fatty-type material from eroded plaques in the diseased aorta. These plaques are dislodged, in part due to the viscous nature of the RCM, and then travel through the renal artery to the arcuate and interlobular arterioles. These plaques occlude the vessels and result in ischemia to the glomerular tuft. Over time, this ischemia results in scarring and nephron loss.

Clinically, skin mottling (livedo reticularis) of the lower extremities, or "blue toes," is a common feature of this process (Coburn and Agre, 1993; Colt et al., 1988; Kassirer, 1969; Meyrier et al., 1988; Om et al., 1992; Smith et al., 1981). However, variable presentations, including abdominal pain, gastrointestinal bleeding, or pancreatitis, may also occur (Coburn and Agre, 1993; Colt et al., 1988; Kasinath et al., 1987; Kassirer, 1969; Meyrier et al., 1988; Om et al., 1992; Smith et al., 1981; Thurlbeck and Castleman, 1957). Laboratory indicators to the diagnosis include a transient eosinophilia, hypocomplementemia, and increased sedimentation rate (Kasinath et al., 1987; Kassirer, 1969; Meyrier et al., 1988).

The clinical distinction between RCM-induced nephrotoxicity and atheroembolic renal disease lies in the time course. RCM-induced nephrotoxicity is noted usually within hours, while atheroembolic renal disease has a time course of days to weeks after the procedure (Thurlbeck and Castleman, 1957). In order to minimize the risk of atheroemboli in high-risk subjects, some authors suggest a brachial rather than ileofemoral approach to cardiac catheterizations (Thurlbeck and Castleman, 1957). In this way the aorta, the repository for atheromatous plaques, is avoided.

PROPHYLACTIC STRATEGIES FOR PREVENTION

To achieve true prophylaxis against RCM-induced nephrotoxicity, the clinician must pretreat the patient prior to the injection of RCM. Numerous clinical studies have evaluated various therapeutic interventions shown to prevent declines in renal function in animal models following RCM. Pharmacological interventions investigated include:

- adequate hydration
- loop diuretics at various time periods throughout the contrast study
- osmotic diuretics, such as mannitol
- CCBs prior to RCM administration
- theophylline prior to a contrast procedure, to block adenosine
- nonselective and selective dopamine-1 receptor agonists
- acetylcysteine

Other interventions evaluated by clinical trials include infusions of ANP and nonspecific receptor blockade of endothelin, both of which failed to protect agonist CN (Allgren et al., 1997; Wang et al., 2000). A summary of prophylactic interventions shown to be effective for reducing the incidence of RCM-induced renal impairment is presented below. Only those procedures that have shown promise or have proven efficacy have been discussed in any depth.

Efficacious Prophylactic Approaches

Hydration

A decrease in effective circulating volume (volume depletion) is known to magnify the risk factors of RCM-induced renal dysfunction in patients (Byrd and Sherman, 1979; Louis et al., 1996; Mueller et al., 2002; Solomon et al., 1994; Stevens et al., 1999; Eisenberg et al., 1981). Therefore, pretreatment with normal saline should always be considered because it reduces the risk of CN. Interestingly, in most studies neither furosemide nor mannitol augmented protection offered by adequate hydration with saline alone. In fact, in two separate studies, patients with CRI who received hydration plus mannitol or furosemide were noted to have serum creatinine concentrations significantly increased as compared with patients who received hydration alone (Mueller et al., 2002; Solomon et al., 1994).

Recommendations include administration of 0.45% saline (1.0 to 1.5 ml/kg/min) starting 4 or 12 hours before and continuing for 8 to 12 or 12 to 24 hours after the procedure in patients at moderate or high risk, respectively. Although 0.45% sodium chloride is widely recommended, a recent prospective, randomized, controlled trial tested whether the use of isotonic or half-isotonic hydration would result in any difference to the incidence of CN (Mueller et al., 2002). The study included 1620 patients (20.7% with CRI, 15.7% with diabetes) who received low-osmolar nonionic RCM during coronary angioplasty. CN (increase in serum creatinine concentration of at least 0.5 mg/dl) developed in 0.7% of patients who received isotonic hydration and in 2.0% of those who received half-isotonic hydration ($p = 0.04$). The superiority of isotonic infusion was more pronounced in patients who underwent selective procedures (CN in 0.7% compared with 2.7%), but there was no difference between the two hydration schedules in the subgroup that underwent emergency angioplasty. Women, diabetics, and patients who received 250 ml or more of RCM benefited especially. Risk factors for CN were female gender and baseline serum creatinine concentration. In their discussion, Mueller and colleagues (2002) suggested that the superiority of isotonic saline was due to a more pronounced volume expansion and renin–angiotensin axis inhibition. However, as less than 2.0% of patients enrolled had severe CRI (creatinine concentration more than 1.6 mg/dl), additional studies are needed.

Adenosine Antagonists

Clinical studies have evaluated the effects of the adenosine antagonist theophylline for prophylactic use against RCM-induced-nephrotoxicity (Erley et al., 1994, 1999; Huber et al., 2001; Katholi et al., 1995; Kolonke et al., 1998). In one randomized, prospective double-blind study, subjects with GFRs below 75 ml/min were given either theophylline or placebo 45 minutes before injection with a low-osmolar RCM (Erley et al., 1994). While the number of participants was small ($n = 35$), the theophylline group had no significant reduction in renal hemodynamics compared with a 22% decrease in GFR after low-osmolar RCM alone. A second prospective randomized clinical study of 93 patients with coronary artery disease and serum creatinine concentration of <2.0 mg/dl corroborated the renal protective quality of theophylline (Katholi et al., 1995). Specifically, subjects were administered theophylline at a dose of 2.9 mg/kg every 12 hours for four doses prior to the RCM study. This regimen was found to prevent renal failure in a high-risk group that received low-osmolar RCM after adequate hydration.

More-recent clinical studies support the role of adenosine receptor blockade in the prevention of CN. Kolonke and colleagues (1998) studied 58 patients with renal insufficiency who were undergoing angiographic studies. Patients were randomized to 165 mg of theophylline or placebo before injection of a high-osmolar RCM. In the 24 hours following the procedure, those who did not receive theophylline showed significant reductions in GFR and increased $\beta 2$ microglobulin excretion.

In the placebo-controlled study by Erley et al. (1999), theophylline (270 mg in the morning, 540 mg in the evening orally started 2 days before and continued for 3 days after RCM) was administered together with hydration to patients with preexisting CRI (serum creatinine concentration >1.5 mg/dl). Serum creatinine concentration and its clearance level did not change significantly: serum creatinine concentration increased by ≥0.5 mg/dl in two patients who received theophylline (5.7%) and in one patient in the placebo group (3.4%). The only difference was a significant reduction of urine *N*-acetyl-β-glucosaminidase in the theophylline group. The authors concluded that theophylline did not afford an additional benefit (Erley et al., 1999). However, theophylline may be beneficial in patients for whom sufficient hydration may be impossible (e.g., congestive heart failure) (Erley et al., 1999).

Conversely, theophylline given intravenously at a dose of 200 mg/70 kg before the injection of low-osmolarity RCM in patients hospitalized in the intensive care unit (50% with CRI, 38% diabetics, use of nephrotoxic antibiotics) was associated with reduced incidence of CN (2%) (Huber et al., 2001). Specifically, serum creatinine concentration remained low 12, 24, and 48 hours after RCM exposure. Based on these studies, theophylline, at a dose <5 mg/kg could be started immediately prior injecting RCM, in addition to hydration for patients with CRI.

Low-Osmolality RCMs

Use of low-osmalority or isoosmolar RCM in combination with modern angiographic techniques (use of carbon dioxide as initial agent and iodinated RCM only when necessary, omission of left ventriculogram whenever possible, obstruction of femoral arteries during lumbar or renal arteriograms, and thus reduction of RCM dose, etc.) may lower the risk of development CN in patients with preexisting CRI (Aron et al., 1989; Gerber et al., 1982; Higgins et al., 1983; Lufft et al., 2002; Lund et al., 1984; Rieger et al., 2002; Sterner et al., 2001; Townsend et al., 2000).

Gadopendate dimeglumine, a gadolinium-based contrast medium, is widely used in magnetic resonance imaging with no adverse renal effects and is considered to be an alternative contrast agent for digital subtraction angiography (DSA) (Rieger et al., 2002; Townsend et al., 2000). Its use in patients with severe renal impairment (mean creatinine concentration 3.6 mg/dl), either alone or with carbon dioxide, was accompanied by no further renal deterioration in all patients but one (which eventually was diagnosed as a result of cholesterol embolism) (Townsend et al., 2000; Lufft et al., 2002; Rieger et al., 2002).

Previously, safety of gadolinium had been proved in a prospective, randomized, placebo-controlled study in patients with severe or moderate CRI (creatinine clearance 10 to 29 or 30 to 60 ml/min, respectively). With a dose of 0.2 mmol/kg, no increase in serum creatinine concentration greater than 0.5 mg/dl was observed during a 7 days follow-up period (Townsend et al., 2000).

A more-recent study compared spiral computed tomographic angiography CTA (intravenous injection of RCM) with DSA (direct RCM injection into the renal artery), with regard to CN during investigation for renal-artery stenosis in patients of whom one-third had preexisting renal impairment. Serum creatinine concentration increased and clearance decreased within 48 hours, but there were no significant differences between the CTA and DSA groups. Despite the greater dose of RCM in CTA, the incidence of CN was also similar and independently predicted by diabetes and preexisting renal dysfunction (Lufft et al., 2002).

RCM Dose Adjustment

It is important to monitor the dose of RCM in order to prevent renal toxicity. Patients with severe renal dysfunction should have an adjustment in the administered dose of RCM. Cigarroa and colleagues (1989) provided guidelines for dosing with RCM to reduce the risk of RCM-induced nephrotoxicity in patients with renal disease. These investigators demonstrated that appropriate dose reductions of RCM can prevent significant increases in RCM-induced renal dysfunction. The formula they suggest is:

$$\text{RCM limit} = \frac{5\,\text{ml RCM} \times \text{kg body weight(max.} = 300\text{ml)}}{\text{Serum creatinine concentration}} \qquad (19.1)$$

In this study of patients with serum creatinine concentrations greater than 1.8 mg/dl, CN developed in 2% of cases with use of limited RCM volume, but in 26% if the limit of quotation was exceeded (Cigarroa et al., 1989).

The type of RCM (hyperosmolar or low-osmolar) used is also of importance in preventing nephrotoxicity. A meta-analysis of 31 clinical trials clearly demonstrated that low-osmolar agents are the drugs of choice for any high-risk patient (Barrett and Carlisle, 1993). This type of RCM also offers benefits regarding cardiac effects: fewer arrhythymias, less hypotension, and less interaction with CCBs when compared with hyperosmolar agents (Gerber et al., 1982; Higgins et al., 1983; Shafi et al., 1978; Shieh et al., 1982).

Because of these meta-analyses and clinical trials, the question as to whether everyone should receive low-osmolar RCM has been posed. An early evaluation of the cost: enefit ratios for low-osmolar and hyperosmolar agents did not support the use of low-osmolar RCM in the population at high-risk for renal dysfunction (Fischer et al., 1986). However, it is clear from data submitted to the U.S. Food and Drug Administration that low-osmolar RCM is well tolerated in high-risk patients and is associated with a lower incidence of nephropathy and lower overall mortality (Lasser et al., 1997).

Other Recommendations

Before RCM is administered in diabetics, it is appropriate to discontinue non-steroidal anti-inflammatory drugs, dipyridamole as well as metformin, (associated with lactic acidosis in cases with renal impairment). It is also suggested that angiotensin converting enzyme inhibitors and diuretics be withheld 24 hours before the procedure.

Uncertain and Non-Efficacious Prophylactic Approaches

Selective Dopamine Receptor Stimulation

Some studies have produced data that suggest selective dopamine-1 (D1) receptor stimulation causes significant increase in GFR, RBF, and natriuresis (Chu and Cheng, 2001; Kini et al., 2002; Tumlin et al., 2002). An animal study evaluated the effect of fenoldopam, a selective D1 receptor agonist, on renal hemodynamics after RCM administration (Bakris et al., 1999). In the study, volume-depleted dogs treated with fenoldopam (0.01 μg/kg/min) 60 minutes after injection of RCM were found to have a blunted vasoconstriction response and no reduction in RBF and GFR. The action of fenoldopam was confirmed by using a D1 receptor antagonist (Schering 23390) in a separate experimental procedure.

Favorable outcomes for fenoldopam were initially found in uncontrolled (or using historical control) clinical studies and with a small number of patients. In two prospective studies with patients suffering from CRI, fenoldopam reduced the incidence of CN to 4.5% and 13%, compared with 18.8% and 38% in historical control groups. In both studies, hydration with sodium chloride (0.9%) was also used (Chu and Cheng, 2001).

Investigators have also tested whether fenoldopam would have any beneficial effect on preventing CN in patients with CRI (serum creatinine concentration equal or greater than 2.0 mg/dl) undergoing coronary intervention. The low-osmolar nonionic agent ioversol and hydration with sodium chloride (0.45%, before, during, and after the procedure) were used in all groups. Fenoldopam (0.03 to 0.1 μg/kg/min) was administered intravenously, starting 15 to 20 minutes before and continuing for 6 hours after the contrast agent. CN (increase in serum creatinine concentration more than 25% at 48 to 72 hours or absolute increase greater than 0.5 mg/dl) was observed in 3.8% of patients, with no significant difference between diabetics and nondiabetics.

Only age (greater than 75 years) was a strong independent predictor for CN. Other risk factors (diabetes, contrast volume, and baseline creatinine concentration) did not act in a similar way (Kini et al., 2002).

More-recently, a controlled multicenter study supported the role of fenoldopam in patients with preexisting CRI (serum creatinine concentration 2.0 to 5.0 mg/dl) after receiving nonionic low or isoosmotic RCM. One hour after the procedure, renal plasma flow (RPF) increased by 15.8% in the fenoldopam group and decreased by 33.2% in the control group. The incidence of CN (21% and 41%) did not reach significant level of difference between the two groups, however the peak of serum creatinine concentration at 72 hours (2.8 ± 0.35 and 3.6 ± 1.0 mg/dl) was significantly lower in the fenoldopam group. Data suggested that fenoldopam, unlike dopamine or endothelin blockade, increases the blood flow in both the cortex and the outer renal medulla (the outer renal medulla is thought to be the most susceptible region for contrast toxicity). Further analysis of this study (reduced RPF at 4 hours in fenoldopam group) suggests that higher concentrations and longer treatment periods may be necessary to produce a significant effect (Tumlin et al., 2002).

Acetylcysteine

There has been much interest regarding the role of acetylcysteine in the prevention of RCM nephropathy. Experimental evidence documented a role for ROS in contributing to RCM nephropathy, and acetylcysteine competes with superoxide to limit the production of peroxinitrite, while serving as a precursor of glutathione synthesis (to scavenge a variety of ROSs) (Bakris et al., 1990b; Safirstein et al., 2000). Moreover, acetylcysteine blocks the expression of vascular-cell adhesion molecule-1 and the activation of NF-κB in mesangial cells. Furthermore, acetylcysteine is considered to have vasodilatory properties, probably by increasing NO synthase activity or by potentiating the biologic effects of NO by reacting with it to form *S*-nitrosothiol, a more stable and potent vasodilator (Diaz-Sandoval et al., 2002; Safirstein et al., 2000).

The first study reporting favorable effects of acetylcysteine was a prospective study by Tepel et al. (2000). The study considered patients with CRI (mean serum creatinine concentration 2.4 mg/dl) who received low-osmolar iopromide for computed tomography imagine purposes. Acetylcysteine was administered at a dose of 600 mg twice daily on the day before and on the day of the procedure, together with hydration. The incidence of CN was significantly reduced in the treatment group (2% versus 21% in the control group). In fact, patients who received acetylcysteine were found to have serum creatinine concentration values decreased from their baseline 48 hours after RCM injection (Tepel et al., 2000).

Similar results were found in patients with CRI (serum creatinine concentration $\geq$1.4 mg/dl), who received low osmolar RCM during cardiac catheterization. CN was detected in only 8% in the acetylcysteine group compared with 45% in the placebo group. The incidence of CN was associated with use of RCM volume greater than 220 ml. Again, serum creatinine concentration decreased in the acetylcysteine group (Diaz-Sandoval et al., 2002).

The usefulness of acetylcysteine was also tested in a larger study of 183 patients with CRI who received iopromide for coronary and/or peripheral angiography and/or angioplasty. Lower percentage of patients with acetylcysteine (as the regimen by Tepel et al. [2000]) developed CN, 6.5% compared with 11% in the control group, but the difference was not significant. The amount of RCM was independently associated with CN, and only the subgroup with a low (<140 ml) RCM dose benefited by acetylcysteine (0% incidence of CN, compared with 8.5%) (Briguori et al., 2002).

Prostaglandins

In patients with renal impairment, PGE_1 (administered at a dose 20 ng/kg/min for 6 hours, starting prior to RCM) was associated with the lowest elevation in serum creatinine concentration 48 hours after a diagnostic procedure. This favorable effect was also observed in patients with diabetes, who displayed even greater declines in kidney function. Higher doses of PEG_1, however, manifested worsening kidney function; arterial pressure decreased resulting in hypotension and further limiting renal perfusion Koch et al., 2000).

Non-Efficacious Approaches

Unlike the previously discussed interventions, it is clear from numerous clinical studies that furosemide, mannitol, dopamine, endothelin blockade, calcium channel blockade, and atrial natriuretic peptide fail to offer any additional protection against RCM nephropathy over

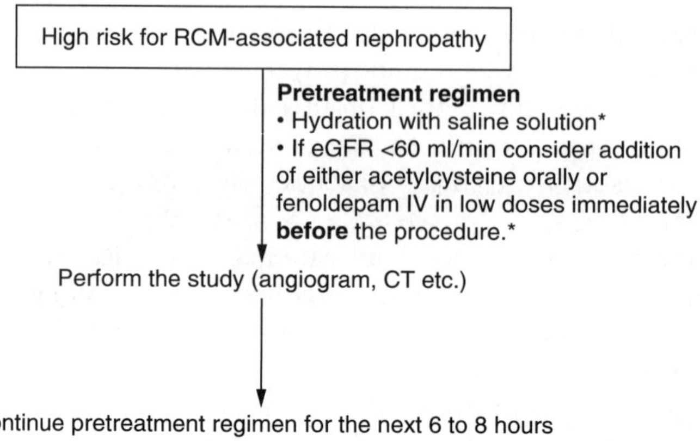

Figure 19.8 Proposed preventive strategy for prophylaxis against RCM nephropathy in high-risk patients. Hydration should be with either normal or half-normal saline for at least 4 to 6 hours because this is the minimum used in all studies performed to date. Some objective measure of hydration status show also is performed. The only adjunctive agents shown to be beneficial (in addition to hydration) in those with renal insufficiency and diabetes have been acetylcysteine or fenoldopam.

appropriate hydration and use of low-osmolar RCM (Abizaid et al., 1999; Carraro et al., 1996; Chertow et al., 1996; DeRubertis and Craven, 1978; Hans et al., 1998; Haylor and Morcos, 2001; Kassirer, 1969; Madsen et al., 1998; Meyrier et al., 1988; Murphy et al., 2000; Neumayer et al., 1998; Russo et al., 1990; Smith et al., 1981; Viklund-Spangberg et al., 1996; Vogt et al., 2001; Workman et al., 1983).

SUMMARY

Understanding of the mechanisms and pathophysiology of RCM-induced renal dysfunction has only minimally expanded since the mid-1980s. However, large clinical trials have now clearly demonstrated that patients with preexisting renal insufficiency and those who are volume depleted are at highest risk of RCM nephropathy. Moreover, decreased effective circulating volume markedly accentuates the small preexisting risk of RCM nephropathy in people who have diabetes or heart failure. The best evidence consistently supports adequate hydration as the best protection against nephropathy development, along with use of a low osmolar RCM and administration of the lowest dose possible for the procedure.

Other measures shown in human studies to be additionally efficacious for reducing the risk include use of either a selective dopamine-1 receptor agonist, adenosine receptor blockade with theophylline, or PGE1 infusion. Interventions with no evidence of protection include: mannitol, dopamine, furosemide and posttreatment dialysis. A proposed scheme for preventing RCM-induced nephrotoxicity is illustrated in Figure 19.8.

REFERENCES

Abizaid AS, Clark CE, Mintz GS, Dosa S, Popma JJ, Pichard AD, Satler LF, Harvey M, Kent KM, and Leon MD (1999) Effects of dopamine and aminophylline on contrast-induced acute renal failure after coronary angioplasty in patients with preexisting renal insufficiency. *Am J Cardiol* 83:260–263.

Ackrill P, McIntosh CS, Nimmon C, Baker LR, and Cattell WR (1976) A comparison of the clearance of urographic contrast medium (sodium diatrizzoate) by peritoneal and hemodialysis. *Clin Sci Mol Med* 50:69–74.

Alexander RD, Berkes SL, and Abuelo G (1978) Contrast media-induced oliguric renal fialure. *Arch Intern Med* 138:381–384.

Allgren RL, Marbury TC, Rahman SN, Weisberg LS, Fenves AZ, Lafayette RA, Sweet RM, Genter FC, Kurnik BR, Conger JD, and Sayegh MH (1997) Anaritide in acute tubular necrosis. Auriculin Anaritide Acute Renal Failure Study Group. *N Engl J Med* 336:828–834.

Ansari Z, and Baldwin DS (1976) Acute renal failure due to radiocontrast agents. *Nephron* 17:28–40.

Arakawa K, Suzuki H, Naitoh M, Matsumoto A, Hayashi K, Matsuda H, Ichihara A, Kubota E, and T Saruta (1996) Role of adenosine in the renal responses to contrast medium. *Kidney Int* 49:1199–1206.

Arend LJ, Bakris GL, Burnett JC Jr, Megerian C, Spielman WS (1987) Role of intrarenal adenosine in the renal hemodynamic response to contrast media. *J Lab Clin Med* 110:406–411.

Aron NB, Feinfeld DA, Peters AT, and Lynn RI (1989) Acute renal failure associated with ioxaglate, a low-osmolality radiocontrast agent. *Am J Kidney Dis* 13:189–193.

Baer PG, and Navar LG (1973) Renal vasodilatation and uncoupling of blood flow and filtration rate autoregulation. *Kidney Int* 4:12–21.

Bagnis C, Idee JM, Dubois M, Jacquiaud C, Maistre G, Jacobs C, and Deray G (1997) Role of endothelium derived nitric oxide endothelin balance in contrast medium-induced acute renal vasoconstriction in dogs. *Acad Radiol* 4:343–348.

Bakris GL, and Burnett JC Jr (1985) A role for calcium in radiocontrast-induced reduction in renal hemodynamics. *Kidney Int* 27:465–468.

Bakris GL, Gaber AO, and Jones JD (1990a) Oxygen free radical involvement in urinary Tamm-Horsfall protein excretion after intrarenal injection of contrast medium. *Radiology* 175:51–60.

Bakris GL, Lass N, Gaber AO, Jones JD, and Burnett JC Jr (1990b) Radiocontrast medium-induced declines in renal function a role for oxygen free radicals. *Am J Physiol* 258:F115–F120.

Bakris GL, Lass NA, and Glock D (1999) Reductions in renal hemodynamics in an experimental model of radiocontrast medium-induced renal dysfunction: A role for dopamine-1 receptors *Kidney Int* 56:206–210.

Barrett BJ, and Carlisle EJ (1993) Metaanalysis of the relative nephrotoxicity of high- and low-osmolality iodinated contrast media. *Radiology* 188:171–178.

Barrett BJ, Parfrey PS, Vavasour HM, McDonald J, Kent G, Hefferton D, O'Dea F, Stone E, Reddy R, and McManamon PJ (1992) Contrast nephropathy in patients with impaired renal function: High vs. low-osmolar media. *Kidney Int* 41:1274–1279.

Bartley O, Bengtsson U, and Cedarbon G (1969) Renal function before and after uroography and angiography with large doses of contrast media. *Acta Radiol* 8:9–12.

Baud L, and Ardaillou R (1986) Reactive oxygen species:production and role in the kidney. *Am J Physiol* 251:F765–F776.

Baud L, Hagege J, Sraer J, Rondeau E, Perez J, and Ardaillou R (1983) Reactive oxygen production by cultered rat glomerular mesangial cells during phagocytosis is associated with stimulation of lipoxygenase activity. *J Exp Med* 158:1836–1852.

Benness GT (1970) Urographic contrast agents. A comparison of sodium and methylglucamine salts. *Clin Radiol* 21:150–156.

Berdon WE, Schwartz RH, Becker J, and Baker DH (1969) Tamm-Horsfall proteinuria. Its relationship to prolonged nephrograms in infants and children and to renal failure following intravenous urography in adults with multiple myleoma. *Radiology* 92:714–722.

Berg KJ, Kolmannskog F, Lillevold PE, Nordal KP, Ressem L, Rootwelt K, and Svaland MG (1992) Iopentol in patients with chronic renal failure: Its effects on renal function and its use as glomerular filtration rate parameter. *Scand J Clin Lab Invest* 52:27–33.

Berns AS (1989) Nephrotoxicity of contrast media. *Kidney Int* 36:730–740.

Bettmann MA (1982) Angiographic contrast agents: Conventional and new media compared. *Am J Roentgenol* 139:787–794.

Bhandaru S, and Bakris GL (1995) Effects of high- and low-osmolar radiocontrast medium on mesangial cell growth and cell integrity (abstract). *J Invest Med* 43:103.

Brezis M, Rosen SN, and Epstein FH (1989) The pathophysiologic implications of medullary hypoxia. *Am J Kidney Dis* 13:253–258.

Briguori C, Manganelli F, Scarpato P, Elia PP, Golia B, Riviezzo G, Lepore S, Librera M, Villari B, Colombo A, and Ricciardelli B (2002) Acetylcysteine and contrast agent-associated nephrotoxicity. *J Am Coll Cardiol* 40:298–303.

Burgener FA, and Hamlin DJ (1981) Contrast enhancement in abdominal CT: bolus vs. infusion. *Am J Roentgenol* 137:351–358.

Byrd L, and Sherman RL (1979) Radiocontrast-induced acute renal failure: A clinical and pathophysiologic review. *Medicine (Baltimore)* 58:270–279.

Caldicott WJH, Hollenberg NK, and Abrams HL (1970) Characteristics of response of renal vascular bed to contrast media. Evidence for vasoconstriction induced by renin-angiotensin system. *Invest Radiol* 5:539–547.

Carraro M, Mancini W, Artero M, Stacul F, Grotto M, Cara M, and Faccini L (1996) Dose effect of nitrendipine on urinary enzymes and microproteins following nonionic radiocontrast administration. *Nephrol Dial Transplant* 11:444–448.

Carvallo A, Rakowski TA, Argy WP Jr and Schreiner GE (1975) Acute renal failure following drip infusion pyelography. *Am J Med* 65:38–45.

Cattell WR, Fry IK, Spencer AG, and Purkiss P (1967) Excretion urography, I. Factors determining the excretion of Hypaque. *Br J Radiol* 40:561–571.

Chertow GM, Sayegh MH, Allgren RL, and Lazarus JM (1996) Is the administration of dopamine associated with adverse or favorable outcomes in acute renal failure? Auriculin Anaritide Acute Renal Failure Study Group. *Am J Med* 101:49–53

Chu VL, and Cheng JW (2001) Fenoldopam in the prevention of contrast media-induced acute renal failure. *Ann Pharmacother* 35:1278–1282.

Cigarroa RG, Lange RA, Williams RH, and Hillis LD (1989) Dosing of contrast material to prevent contrast nephropathy in patients with renal disease. *Am J Med* 86:649–652.

Coburn JW, and Agre KL (1993) Atheroembolic Disease, in *Diseases of the Kidney*, 5th ed., Schrier RW, and Gottschalk CW (Eds), Little Brown, Boston, pp. 2119–2135.

Cochran ST, Wong WS, and Roe DJ (1983) Predicting angiography-induced acute renal function impairment: Clinical risk model. *Am J Roentgenol* 14:1027–1033.

Colt HG, Begg RJ, Saporito JJ, Cooper WM, and Shapiro AP (1988) Cholesterol emboli after cardiac catheterization. Eight cases and a review of the literature. *Medicine (Baltimore)* 67:389–400.

Cramer BC, Parfrey PS, Hutchinson TA, Baran D, Melanson DM, Ethier RE, and Seely JF (1985) Renal function following infusion of radiologic contrast material. A prospective controlled study. *Arch Intern Med* 145:87–89.

Cross CE, Halliwell B, Borish ET, Pryor WA, Ames BN, Saul RL, McCord JM, and Harman D (1987) Oxygen radicals and human disease. *Ann Intern Med* 107:526–545.

Davidson CJ, Hlatky M, Morris KG, Pieper K, Skelton TN, Schwab SJ, and Bashore TM (1989) Cardiovascular and renal toxicity of a nonionic radiographic contrast agent after cardiac catheterization. A prospective trial. *Ann Intern Med* 110:119–124.

Dean PB, Kivisaari L, and Kormano M (1978) The diagnostic potential of contrast enhancement pharmacokinetics. *Invest Radiol* 13:533–540.

D'Elia JA, Gleason RE, Alday M, Malarick C, Godley K, Warram J, Kaldany A, and Weinrauch LA (1982) Nephrotoxicity from angiographic contrast material. A prospective study. *Am J Med* 72:719–725.

Deray G, Martinez F, Cacoub P, Baumelou B, Baumelou A, and Jacobs C (1990) A role for adenosine, calcium and ischemia in radiocontrast-induced intrarenal vasoconstriction. *Am J Nephrol* 10:316–322.

DeRubertis FR, and Craven PA (1978) Effects of osmolality and oxygen availability on soluble cyclic AMP-dependent protein kinase activity of rat renal inner medulla. *J Clin Invest* 62:1210–1221.

Diaz-Buxo JA, Wagoner RD, Hattery RR, and Palumbo PJ (1975) Acute renal failure after excretory urography in diabetic patients. *Ann Intern Med* 83:155–158.

Diaz-Sandoval LJ, Kosowsky BD, and Losordo DW (2002). Acetylcysteine to prevent angiography-related renal tissue injury (the APART trial). *Am J Cardiol* 89:356–358.

Drescher P, and Madsen PO (1998) Receptor mediated mechanisms in contrast medium induced renal vasoconstriction. *Acad Radiol* 5(Suppl 1):S119–S122.

Drescher P, Kaes JM, and Madsen PO (1998) Prevention of contrast medium induced renal vasospasm by phosphodiesterase inhibition. *Invest Radiol* 33:56–862.

Dudzinski PJ, Petrone AF, Persoff M, and Callaghan EE (1971) Acute renal failure following high dose excretory urography in dehydrated patients. *J Urol* 106:619–628.

Eisenberg RI, Bank WO, and Hedgcock MW (1981) Renal failure after major angiography can be avoided with hydration. *Am J Roentgenol* 136:859–861.

Erley CM, Duda SH, Rehfuss D, Scholtes B, Bock J, Muller C, Osswald H, and Risler T (1999) Prevention of radiocontrast-media-induced nephropathy in patients with pre-existing renal insufficiency by hydration in combination with the adenosine antagonist theophylline. *Nephrol Dial Transplant* 14:1146–1149.

Erley CM, Duda SH, Schlepckow S, Koehler J, Huppert PE, Strohmaier WL, Bohle A, Risler T, and Osswald H (1994) Adenosine antagonist theophylline prevents reduction of glomerular filtration rate after contrast media application. *Kidney Int* 45:1425–1431.

Esnault VL (2002) Radiocontrast media-induced nephrotoxicity in patients with renal failure:rationale for a new double-blind, prospective, randomized trial testing calcium channel antagonists. *Nephrol Dial Tra*nsplant 17:1362–1364.

Fischer HW, Spataro RF, and Rosenberg RM (1986) Medical and economic considerations in using a new contrast medium. *Arch Intern Med* 146:1717–1721.

Gale ME, Robbins AH, Hamburger RJ, and Widrich WC (1984) Renal toxicity of contrast agents: Iopamidol, iothalamate, and diatrizoate. *Am J Roentgenol* 142:333–335.

Gaspari F, Perico N, and Remuzzi G (1997) Measurement of glomerular filtration rate *Kidney Int* (*Suppl*) 63:S151–S154.

Gerber JG, Branch RA, Nies AS, Hollifield JW, and Gerkens JF (1979) Influence of hypertonic saline on canine renal blood flow and renin release. *Am J Physiol* 237:F441–F446.

Gerber KH, Higgins CB, Yuh YS, and Koziol JA (1982) Regional myocardial hemodynamic and metabolic effects of ionic and nonionic contrast media in normal and ischemic states. *Circulation* 65:1307–1314.

Gomes AS, Lois JF, Baker JD, McGlade CT, Bunnell DH, and Hartzman S (1989) Acute renal dysfunction in high-risk patients after angiography comparison of ionic and nonionic contrast media. *Radiology* 170:65–68.

Gruberg L, Mintz GS, Mehran R, Dangas G, Lansky AJ, Kent KM, Pichard AD, Satler LF, and Leon MB (2000) The prognostic implications of further renal function deterioration within 48 h of interventional coronary procedures in patients with pre-existent chronic renal insufficiency. *J Am Coll Cardiol* 36(5):1542–1548.

Hall JE, Granger JP, and Hester RL (1985) Interactions between adenosine and angiotensin II in controlling glomerular filtration. *Am J Physiol* 248:F340–F346.

Hampel DJ, Sansome C, Sha M, Brodsky S, Lawson WE, and Goligorsky MS (2001) Toward proteomics in uroscopy: Urinary protein profiles after radiocontrast medium administration. *J Am Soc Nephrol* 12:1026–1035.

Hans SS, Hans BA, Dhillon R, Dmuchowski C, and Glover J (1998) Effect of dopamine on renal function after arteriography in patients with pre-existing renal insufficiency. *Am Surg* 64:432–436

Hardiek K, Katholi RE, Ramkumar V, and Deitrick C (2001) Proximal tubule cell response to radiographic contrast media. *Am J Physiol* 280:F61–F70.

Harkonen S, and Kjellstrand C (1981) Contrast nephropathy. *Am J Nephrol* 7:69–77.

Harvey RB (1960) Vascular resistance changes produced by hyperosmotic solutions. *Am J Physiol* 199:31–34.

Haustein J, Niendorf HP, Krestin G, Louton T, Schuhmann-Giampieri G, Clauss W, and Junge W (1992) Renal tolerance of gadolinium-DTPA dimeglumine in patients with chronic renal failure. *Invest Radiol* 27:153–156.

Haylor JL, and Morcos SK (2001) An oral ET(A)-selective endothelin receptor antagonist for contrast nephropathy? *Nephrol Dial Transplant* 16:1336–1337.

Heyman SN, Brezis M, Greenfeld Z, and Rosen S (1989) Protection role of furosemide and saline in radiocontrast-induced acute renal failure in the rat. *Am J Kidney Dis* 14:377–385.

Heyman SN, Brezis M, Reubinoff CA, Greenfeld Z, Lechene C, Epstein FH, and Rosen S (1988) Acute renal failure with selective medullary injury in the rat. *J Clin Invest* 82:401–412.

Heyman SN, Clark BA, Cantley L, Spokes K, Rosen S, Brezis M, and Epstein FH (1993) Effects of ioversol vs. iothalamate on endothelin release and radiocontrast nephropathy. *Invest Radiol* 28:313–318.

Heyman SN, Clark BA, Kaiser N, Spokes K, Rosen S, Brezis M, and Epstein FH (1992) Radiocontrast agents induce endothelin release *in vivo* and *in vitro*. *J Am Soc Nephrol* 3:58–65.

Higgins CB, Kuber M, and Slutsky RA (1983) Interaction between verapamil and contrast media in coronary arteriography: Comparison of standard ionic new nonionic media. *Circulation* 68:628–635.

Hoyer JR, and Seiler MW (1987) Pathophysiology of Tamm-Horsfall protein. *Kidney Int* 16:279–289.

Huber W, Jeschke B, Page M, Weiss W, Salmhofer H, Schweigart U, Ilgmann K, Reichenberger J, Neu B, and Classen M (2001) Reduced incidence of radiocontrast-induced nephropathy in ICU patients under theophylline prophylaxis: A prospective comparison to series of patients at similar risk. *Intensive Care Med* 27:1200–1209.

Humes HD, Hunt DA, and White MD (1987) Direct toxic effect of the radiocontrast agent diatrizoate on renal proximal tubule cells. *Am J Physiol* 252:F246–F255.

Kako K, Kato M, Matsuoka T, and Mustapha A (1988) Depression of membrane-bound Na^+-K^+-ATPase activity induced by free radicals and by ischemia of kidney. *Am J Physiol* 254:C330–C337.

Kasinath BS, Corwin HL, Bidani AK, Korbet SM, Schwartz MM, and Lewis EJ (1987) Eosinophilia in the diagnosis of atheroembolic renal disease. *Am J Nephrol* 7:173–177.

Kassirer JP (1969) Atheroembolic renal disease. *N Engl J Med* 280:812–818.

Katholi RE, Taylor GJ, Woods WT, Womack KA, Katholi CR, McCann WP, Moses HW, Dove JT, Mikell FL, and Woodruff RC (1993) Nephrotoxicity of nonionic low-osmolality vs. ionic high-osmolality contrast media: A prospective double-blind randomized comparison in human beings. *Radiology* 186:183–187.

Katholi RE, Taylor GJ, McCann WP, Woods Jr WT, Womack KA, McCoy CD, Katholi CR, Moses HW, Mishkel GJ, and Lucore CL (1995) Nephrotoxicity from contrast media: attenuation with theophylline. *Radiology* 195:17–22.

Katholi RE, Woods WT, Taylor GJ, Deitrick CL, Womack KA, Katholi CR, and McCann WP (1998) Oxygen free radicals and contrast nephropathy. *Am J Kidney Dis* 32:64–71.

Katzberg RW, Morris TW, Burgener FA, Kamm DE, and Fischer HW (1977) Renal renin and hemodynamic responses to selective renal artery catherization and angiography. *Invest Radiol* 12:381–388.

Katzberg RW, Schulman G, Meggs LG, Caldicott WJ, Damiano MM, and Hollenberg NK (1983) Mechanims of the renal response to contrast medium in dogs. Decrease in renal function due to hypertonicity. *Invest Radiol* 18:74–80.

Khoury GA, Hopper JC, Varghese Z, Farrington K, Dick R, Irving JD, Sweny P, Fernando ONand Moorhead JF (1983) Nephrotoxicity of ionic and nonionic contrast material in digital vascular imaging and selective renal arteriography. *Br J Radiol* 56:631–635.

Kim MS, and Akera T (1987) O_2 free radicals:cause of ischemia-reperfusion injury to cardiac Na^+-K^+-ATPase. *Am J Physiol* 252:H252–H257.

Kini AS, Mitre CA, Kamran M, Suleman J, Kim M, Duffy ME, Marmur JD, and Sharma SK (2002) Changing trends in incidence and predictors of radiographic contrast nephropathy after percutaneous coronary intervention with use of fenoldopam. *Am J Cardiol* 89:999–1002.

Kinnison ML, Powe NR, and Steinberg EP (1989) Results of randomized controlled trials of low- vs.high-osmolality contrast media. *Radiology* 170:381–389.

Koch JA, Plum J, Grabensee B, and Modder U (2000) Prostaglandin E1: A new agent for the prevention of renal dysfunction in high risk patients caused by radiocontrast media? PGE1 Study Group. *Nephrol Dial Transplant* 15:43–49.

Kolonke A, Wiecek A, and Kokot F (1998) The nonselective adenosine antagonist theophylline does prevent renal dysfunction induced by radiographic contrast agents. *J Nephrol* 11:151–156.

Krumlovsky FA, Simon N, Santhanam S, del Greco F, Roxe D, and Pomaranc MM (1978) Acute renal failure. Association with administration of radiographic contrast material. *JAMA* 239:125–127.

Kurnik BR, Allgren RL, Genter FC, Solomon RJ, Bates ER, and Weisberg LS (1998) Prospective study of atrial natriuretic peptide for prevention of radiocontrast nephropathy. *Am J Kidney Dis* 31:674–680.

Lang J, and Lasser EC (1975) Inhibition of adenosine triphosphate and carbonic anhydrase by contrast media. *Invest Radiol* 10:314–316.

Larson TS, Hudson K, Mertz JI, Romero JC, and Knox FG (1983) Renal vasoconstrictive response to contrast medium. The role of sodium balance and the renin-angiotensin system. *J Lab Clin Med* 101:385–391.

Lasser EC, Lyon SG, and Berry CC (1997) Reports of contrast media reactions:analysis of data from reports to the U.S. Food and Drug Administration. *Radiology* 203:605–610.

Lear S, Silva P, Kelley VE, and Epstein FH (1990) Prostaglandin E_2 inhibits oxygen consumption in rabbit medullary thick ascending limb. *Am J Physiol* 258:F1372–F1378.

Lehnert T, Keller E, Gondolf K, Schaffner T, Parenstadt H, and Schollmeyer P (1998) Effect of hemodialysis after contrast medium administration in patients with renal insufficiency. *Nephrol Dialy Transplant* 13:358–362.

Liss P, Nuggren A, Revsbech NP, and Ulfendall HR (1997) Measurements of oxygen tension in the rat kidney after contrast media using an oxygen microelectrode with a guard cathode. *Adv Exp Med Biol* 411:v569–576.

Louis BM, Hoch BS, Hernandez C, Namboodiri N, Neiderman G, Nissenbaum A, Foti FP, Magno A, Banayat G, Fata F, Manohar NL, and Lipner HI (1996) Protection from the nephrotoxicity of contrast dye. *Ren Fail* 18:639–646.

Lufft V, Hoogestraat-Lufft L, Fels LM, Egbeyong-Baiyee D, Tusch G, Galanski M, and Olbricht CJ (2002) Contrast media nephropathy: Intravenous CT angiography versus intraarterial digital subtraction angiography in renal artery stenosis: A prospective randomized trial. *Am J Kidney Dis* 40:236–242.

Lund G, Einzig S, Rysavy J, Borgwardt B, Salomonowitz E, Cragg A, and Amplatz K (1984) Role of ischemia in contrast-induced renal damage: An experimental study. *Circulation* 69:783–789.

Madsen JK, Jensen JW, Sandermann J, Johannesen N, Paaske WP, Egeblad M, and Pedersen EB (1998) Effect of nitrendipine on renal function and on hormonal parameters after intravascular iopromide. *Acta Radiol* 39:375–380

Margulies KB, McKinley LJ, Cavero PG, and Burnett JC Jr (1990) Induction and prevention of radiocontrast-induced nephropathy in dogs with heart failure. *Kidney Int* 38:1101–1108.

Margulies KB, Hildebrand FL, Heublein DM, and Burnett JC Jr (1991) Radiocontrast increases plasma and urinary endothelin. *J Am Soc Nephrol* 2:1041–1045.

Mason RA, Arbeit LA, and Giron F (1985) Renal dysfunction after arteriography. *JAMA* 253:1001–1004.

Messana JM, Cieslinski DA, Nguyen VD, and Humes HD (1988a) Comparison of the toxicity of the radiocontrast agents, iopanidol and diatrizoate to rabbit renal proximal tubule cells *in vitro*. *J Pharmacol Exp Ther* 244:1139–1144.

Messana JM, Cieslinski DA, O'Connor RP, and Humes HD (1988b) Glutathione protects against exogenous oxidant injury to rabbit renal proximal tubules. *Am J Physiol* 255:F874–F884.

Meyrier A, Buchet P, Simon P, Fernet M, Rainfray M, and Callard P (1988) Atheromatous renal disease. *Am J Med* 85:139–146.

Milman N, and Christensen E (1974) Elimination of diatrizoate by peritoneal dialysis in renal failure. *Acta Radiol Diagn (Stockholm)* 15:265–272.

Morcos SK, and Thomsen HS (2001) Adverse reactions to iodinated contrast media. *Eur Radiol* 11:1267–1275.

Morris TW, and Fischer HW (1986) The pharmacology of intravascular radiocontrast media. *Ann Rev Pharmacol Toxicol* 26:143–160.

Morris TW, Katzberg RW, and Fischer HW (1978) A comparison of the hemodynamic responses to metrizamide and meglumine sodium diatrizoate in canine renal angiography. *Invest Radiol* 13:74–78.

Mudge GH (1980) The maximal urinary concentration of diatrizoate. *Invest Radiol* 15:S67–S78.

Mudge GH (1990) Nephrotoxicity of urographic radiocontrast drugs. *Kidney Int* 18:540–552.

Mudge GH, Meier FA, and Ward KK (1984) in *Acute Renal Failure*, Solez K, and Whelton A (Eds), Marcel Dekker, New York, pp. 361–388.

Mueller C, Buerkle G, Buettner HJ, Petersen J, Perruchoud AP, Eriksson U, Marsch S, and Roskamm H (2002) Prevention of contrast media-associated nephropathy: Randomized comparison of 2 hydration regimens in 1620 patients undergoing coronary angioplasty. *Arch Intern Med* 162:329–336.

Murphy ME, Tublin ME, and Li S (1998) Influence of contrast media on the response of rate renal arteries to endothelin and nitric oxide: Influence of contrast media. *Invest Radiol* 33:356–365.

Murphy SW, Barrett BJ, and Parfrey PS (2000) Contrast nephropathy. *J Am Soc Nephrol* 11:177–182.

Navar LG, Champion WJ, and Thomas CE (1986) Effects of calcium channel blockade on renal vascular resistance responses to changes in perfusion pressure and angiotensin-converting enzyme inhibition in dogs. *Circ Res* 58:874–881.

Neumayer HH, Junge W, Kufner A, and Wenning A (1998) Prevention of radiocontrast media-induced nephrotoxicity by the calcium channel blocker nitrendipine: A prospective randomized clinical trial. *Nephrol Dial Transplant* 4:1030–1036.

Nicot GS, Merle LJ, Charmes JP, Valette JP, Nouaille YD, Lachatre GF, and Leroux-Robert C (1984) Transient glomerular proteinuria, enzymuria and nephrotoxic reaction induced by radiocontrast media. *JAMA* 252:2432–2434.

Norby LH, and DiBona GF (1975) The renal vascular effects of meglumine diatrizoate. *J Pharmacol Exp Ther* 193:932–940.

Oldroyd S, Slee SJ, Haylor J, Morcos SK, and Wilson C (1994) Role for endothelin in the renal responses to radiocontrast media in the rat. *Clin Sci* (*Lond*) 87:427–434.

Om A, Ellahham S, and DiSciasco G (1992) Cholesterol embolism:and under-diagnosed clinical entity. *Am Heart J* 124:1321–1326.

Osborne ED, Sutherland CG, and Scholl AJ (1923) Roentgenography or urinary tract during excretion of sodium iodide. *J Am Med Assoc* 80:368–372.

Osswald H, Gleiter C, and Muhlbauer B (1995) Therapeutic use of theophylline to antagonize renal effects of adenosine. *Clin Nephrol* 43 (Suppl 1):S33–S37.

Osswald H, Spielman WS, and Knox FG (1978) Mechanism of adenosine-mediated decreases in glomerular filtration rate in dogs. *Circ Res* 43:465–469.

Parfrey PS, Griffiths SM, Barrett BJ, Paul MD, Genge M, Withers J, Farid N, and McManamon PJ (1989) Contrast Material-induced renal failure in patients with diabetes mellitus, renal insufficiency or both. A prospective controlled study. *N Engl J Med* 320:143–149.

Parvez Z, Ramamurthy S, Patel NB, and Moncada R (1990) Enzyme markers of contrast media-induced renal failure. *Invest Radiol* 25 (Suppl 1):S133–S134.

Patel R, McKenzie JK, and McQueen EG, (1964) Tamm-Horsfall urinary microprotein and tubular obstruction by casts in acute renal failure. *Lancet* 1:41–46.

Pathria M, Somers S, and Gill G (1987) Pain during angiography: A randomized double-blind trial comparing ioxaglate and diatrizoate. *J Can Assoc Radiol* 38:32–34.

Pelz D, Fox AJ, and Vinuela F (1984) Clinical Trial of Iohexol vs. Conray 60 for cerebral angiography. *AJNR Am J Neuroradiol* 5:565–568.

Pendergrass EP, Chamberlin GW, and Godfrey EW (1942) A survey of deaths and unfavorable sequelae following the administration of contrast media. *Am J Radiol* 48:741–762.

Pendergrass EP, Hodges PJ, and Trondreau RJ (1955) Further consideration of deaths and unfavorable sequelae following administration of contrast media in urography in the United States. *Am J Roentgenol* 74:262–287.

Pflueger A, Larson TS, Nath KA, King BF, Gross JM, and Knox FG (2000) Role of adenosine in contrast media-induced acute renal failure in diabetes mellitus. *Mayo Clin Proc* 75:1275–1283.

Pillay VK, Robbins PC, Schwartz FD, and Kark RM (1970) Acute renal failure following intravenous urography in patients with long-standing diabetes mellitus and azotemia. *Radiology* 95:633–636.

Pollock DM, Polakowski JS, Wegner CD, and Opgenorth J (1997) Beneficial effect of ETA receptor blockade in a rat model of radiocontrast-induced nephropathy. *Ren Fail* 19:753–761.

Quader MA, Sawmiller C, and Sumeio BA (1998) Contrast induced nephropathy: Review of incidence and pathophysiology. *Am Vasc Surg* 12:612–620.

Rieger J, Sitter T, Toepfer M, Linsenmaier U, Pfeifer KJ, and Schiffl H (2002) Gadolinium as an alternative contrast agent for diagnostic and interventional angiographic procedures in patients with impaired renal function. *Nephrol Dial Transplant* 17:824–828.

Rihal CS, Textor SC, Grill DE, Berger PB, Ting HH, Best PJ, Singh M, Bell MR, Barsness GW, Mathew V, Garratt KN, and Holmes DR Jr (2002) Incidence and prognostic importance of acute renal failure after percutaneous coronary intervention. *Circulation* 105(19):2259–2264.

Rocco MV, Buckalew VM Jr, Moore LC, and Shihabi ZK (1996) Capillary electrophoresis for the determination of glomerular filtration rate using nonradioactive iohexol. *Am J Kidney Dis* 28:173–177

Roy P, Robillard P, L'Homme C, Li E, Eng FW, Diehl CA, and Shaw DD (1985) Iohexol: A new nonionic agent in adult peripheral arteriography. *J Can Assoc Radiol* 36:113–117.

Rudnick MR, Goldfarb S, Wexler L, Ludbrook PA, Murphy MJ, Halpern EF, Hill JA, Winniford M, Cohen MB, and VanFossen DB (1995) Nephrotoxicity of ionic and nonionic contrast media in 1196 patients: A randomized trial. The Iohexol Cooperative Study. *Kidney Int* 47:254–261.

Russo D, Testa A, Della Volpe L, and Sansone G (1990) Randomized prospective study on renal effects of two different contrast media in humans: protective role of a calcium channel blocker. *Nephron* 55:254–257.

Safirstein R, Andrade L, and Vieira JM (2000) Acetylcysteine and nephrotoxic effects of radiographic contrast agents—a new use for an old drug. *N Engl J Med* 343:210–212.

Sage MR (1983) Kinetics of water-soluble contrast media in the central nervous system. *Am J Roentgenol* 141:815–824.

Scaduto RC Jr, Gattone VH 2nd, Grotyohann LW, Wertz J, and Martin LF (1988) Effect of an altered glutathione content on renal ischemic injury. *Am J Physiol* 255:F911–F921.

Schiantarelli P, Peroni F, Tirone P, and Rosati G (1973) Effects of iodinated contrast media on erythrocytes. 1. Effects of canine erythrocytes on morphology. *Invest Radiol* 8:199–204.

Schindler R, Stahl C, Venz S, Ludat K, Krause W, and Frei U (2001) Removal of contrast media by different extracorporeal treatments. *Nephrol Dial Transplant* 16:1471–1474.

Schwab SJ, Hlatky MA, Pieper KS, Davidson CJ, Morris KG, Skelton TN, and Bashore TM (1989) Contrast nephrotoxicity: A randomized controlled trial of nonionic and ionic radiographic contrast agent. *N Engl J Med* 320:149–153.

Schwartz RH, Berdon WE, Wagner J, Becker J, and Baker DH (1970) Tamm-Horsfall urinary microprotein precipitation by urographic contrast agents: *In vitro* studies. *Am J Roentgenol Radium Ther Nucl Med* 108:698–701.

Seeger JM, Self S, Harward TR, Flynn TC, and Hawkins IF Jr (1993) Carbon dioxide gas as an arterial contrast agent. *Ann Surg* 217:688–698.

Shafi T, Chou SY, Porush JG, and Shapiro WB (1978) Infusion intravenous pyelography and renal function. Effects in patients with chronic renal insufficiency. *Arch Intern Med* 138:1218–1221.

Shah SV (1989) Role of reactive oxygen metabolites in experimental glomerular disease. *Kidney Int* 35:1093–1106.

Shieh SD, Hirsch SR, Boshell BR, Pino JA, Alexander LJ, Witten DM, and Friedman EA (1982) Low risk of contrast media-induced acute renal failure in nonazotemic type-2 diabetes mellitus. *Kidney Int* 21:739–743.

Shusterman N, Strom BL, Murray TG, Morrison G, West SL, and Maislin G (1987) Risk factors and outcome of hospital-acquired acute renal failure. Clinical epidemiologic study. *Am J Med* 83:65–71.

Silpananta P, Illescas FF, and Sheldon H (1983) Multiple malignant neoplasms 40 years after angiography with Thorotrast. *Can Med Assoc J* 128:289–292.

Smith MC, Chose MK, and Henry AR (1981) The clinical spectrum of renal cholesterol embolization. *Am J Med* 71:174–180.

Solomon R, Werner C, Mann D, D'Elia J, and Silva P (1994) Effects of saline, mannitol, and furosemide to prevent acute decreases in renal function induced by radiocontrast agents. *N Engl J Med* 331:1416–1420.

Spataro RF (1984) Newer contrast agents for urology. *Radiol Clin North Am* 22:365–380.

Spataro RF, Fischer HW, and Boglan L (1982) Urography with low-osmolality contrast media: Comparative urinary excretion of Iopamidol, Hexabrix, and Diatrizoaote. *Invest Radiol* 17:494–500.

Spielman WS, and Thompson CI (1995) A proposed role for adenosine in the regulation of renal hemodynamics and renin release. *Am J Physiol* 242:F423–F435.

Spinosa DJ, Matsumoto AH, Angle JF, Hagspiel KD, McGraw JK, and Ayers C (1999) Renal insufficiency: Usefulness of gadodiamide-enhanced renal angiography to supplement CO2-enhanced renal angiography for diagnosis and percutaneous treatment. *Radiology* 210:663–672.

Sterner G, Nyman U, Valdes T (2001) Low risk of contrast-medium-induced nephropathy with modern angiographic technique. *J Intern Med* 250:429–434.

Stevens MA, McCullough PA, Tobin KJ, Speck JP, Westveer DC, Guido-Allen DA, Timmis GC, and O'Neill WW (1999) A prospective randomized trial of prevention measures in patients at high risk for contrast nephropathy:results of the PRINCE study. *J Am Coll Cardiol* 33:403–411.

Sunnegardh O, Hietala SO, Wirell S, and Ekelund L (1990) Systemic pulmonary and renal hemodynamic effects of intravenously infused iopentol. *Acta Radiol* 31:395–399.

Sutton D (1987) Use of Radiocontrast Agents: A Century of Experience in *Textbook of Radiology and Imaging*, 4th ed., Sutton D (Ed.), Churchill-Livingstone, Edinburgh, pp. 692–699.

Tagawa H, and Vander AJ (1970) Effects of adenosine compounds on renal function and renin secretion in dogs. *Circ Res* 26:327–338.

Taliercio CP, Vlietstra RE, Fisher LD, and Burnett JC (1986) Risks for renal dysfunction with cardiac angiography. *Ann Intern Med* 104:501–504.

Talner LB (1972) Urographic contrast media in uremia. Physiology and pharmacology. *Radiol Clin North Am* 10:421–432.

Talner LB, and Davidson AJ (1968) Renal hemodynamic effects of contrast media. *Invest Radiol* 3:310–317.

Talner LB, Rushmer HN, and Coel MN (1972) The effect of renal artery injection of contrast material on urinary enzyme excretion. *Invest Radiol* 7:311–322.

Tepel M, van der Giet M, Schwarzfeld C, Laufer U, Liermann D, and Zidek W (2000) Prevention of radiographic-contrast-agent-induced reductions in renal function by acetylcysteine. *N Engl J Med* 343:180–184.

Teruel JL, Marcen R, Onaindia JM, Serrano A, Quereda C, and Ortuno J (1981) Renal function impairment caused by intravenous urography. A prospective study. *Arch Intern Med* 141:1271–1274.

Thurlbeck W, and Castleman B (1957) Atheromatous emboli to kidneys after aortic surgery. *N Engl J Med* 1957, 257:444–447.

Townsend RR, Cohen DL, Katholi R, Swan SK, Davies BE, Bensel K, Lambrecht L, and Parker J (2000) Safety of intravenous gadolinium (Gd-BOPTA) infusion in patients with renal insufficiency. *Am J Kidney Dis* 36:1207–1212.

Tumlin JA, Wang A, Murray PT, and Mathur VS (2002) Fenoldopam mesylate blocks reductions in renal plasma flow after radiocontrast dye infusion: A pilot trial in the prevention of contrast nephropathy. *Am Heart J* 143:894–903.

Ueda J, Nygren A, Sjoquist M, Jacobsson E, Ulfendahl HR, and Araki Y (1998) Iodine concentrations in the rat kidney measured by X-ray microanalysis. Comparison of concentrations and viscosities in the proximal tubules and renal pelvis after intravenous injections of contrast media. *Acta Radiol* 39:90–95.

Vari RC, Natarajan LA, Whitescarver SA, Jackson BA, and Ott CE (1988) Induction, prevention and mechanisms of contrast medium-induced acute renal failure. *Kidney Int* 33:699–707.

Viklund-Spangberg B, Berglund J, Nikonoff T, Nyberg P, Skan T, and Larsson R (1996) Does prophylactic treatment with felodipine, a calcium antagonist prevent low osmolar contrast-induced renal dysfunction in hydrated diabetic and nondiabetic patients with normal or moderately reduced renal function? *Scan J Urol Nephrol* 30:63–68.

Vogt B, Ferrari P, Schonholzer C, Marti HP, Mohaupt M, Wiederkehr M, Cereghetti C, Serra A, Huynh-Do U, Uehlinger D, and Frey FJ (2001) Prophylactic hemodialysis after radiocontrast media in patients with renal insufficiency is potentially harmful. *Am J Med* 111:692–8.

Waaler A, Svaland M, Fauchald P, Jakobsen JA, Kolmannskog F, and Berg KJ (1990) Elimination of iohexol, a low-osmolar nonionic contrast medium by hemodialysis in patients with chronic renal failure. *Nephron* 56:81–85.

Wang A, Holcslaw T, Bashore TM, Freed MI, Miller D, Rudnick MR, Szerlip H, Thames MD, Davidson CJ, Shusterman N, and Schwab SJ (2000) Exacerbation of radiocontrast nephrotoxicity by endothelin receptor antagonism. *Kidney Int* 57:1675–1680.

Wang YX, Jia YF, Chen KM, and Morcos SK (2001) Radiographic contrast media induced nephropathy: Experimental observations and the protective effect of calcium channel blockers. *Br J Radiol* 74:1103–1108.

Wasaki M, Sugimoto J, and Shirota K (2001) Glucose alters the susceptibility of mesangial cells to contrast media. *Invest Radiol* 36:355–362.

Waybill MM, and Waybill PN (2001) Contrast media-induced nephrotoxicity: Identification of patients at risk and algorithms for prevention. *J Vasc Interv Radiol* 12:3–9.

Weinrauch LA, Healy RW, Leland OS Jr, Goldstein HH, Kassissieh SD, Libertino JA, Takacs FJ, and D'Elia JA (1977) Coronary angiography and acute renal failure in diabetic azotemic nephropathy. *Ann Intern Med* 86:56–59.

Workman RJ, Shaff MI, Jackson RV, Diggs J, Frazer MG, and Briscoe C (1983) Relationship of renal hemodynamics and functional changes following intravascular contrast to the renin-angiotensin system and renal prostacycline in the dog. *Invest Radiol* 18:160–166.

Yao K, Heyne N, Erley CM, Risler T, and Osswald H (2001) The selective adenosine A1 receptor antagonist KW-3902 prevents radiocontrast media-induced nephropathy in rats with chronic nitric oxide deficiency. *Eur J Pharmacol* 414:99–104.

Yoshioka T, Fogo A, and Beckman JK (1992) Reduced activity of antioxidant enzymes underlies contrast media-induced renal injury in volume depletion. *Kidney Int* 41:1008–1015.

20

Analgesics and Nonsteroidal Anti-Inflammatory Drugs

Joan B. Tarloff

INTRODUCTION

Nonsteroidal anti-inflammatory drugs (NSAIDs), such as aspirin and ibuprofen, inhibit the cyclooxygenase activity of prostaglandin H synthase (PHS), the enzyme catalyzing the initial step in prostaglandin synthesis (formation of PGG_2 and PGH_2 from arachidonic acid) (Figure 20.1). Since prostaglandins are involved in inflammation, interrupting the synthesis of prostaglandins relieves the inflammation that accompanies many diseases, including rheumatoid arthritis, osteoarthritis, and systemic lupus erythematosus. Additionally, prostaglandins may be involved in body temperature elevation and pain production, accounting for the use of NSAIDs for fever reduction and analgesia. Overall, NSAID therapy is safe and effective, and the most significant adverse effects involve gastrointestinal irritation and ulceration.

Acute inhibition of prostaglandin synthesis in the kidney, while of minimal safety concern in normal healthy individuals, may lead to serious deterioration of renal function in patients with renal blood flow that is dependent on prostaglandins (e.g., congestive heart failure, nephrotic syndrome, cirrhosis, and salt depletion) (Harris, 2002). Additionally,

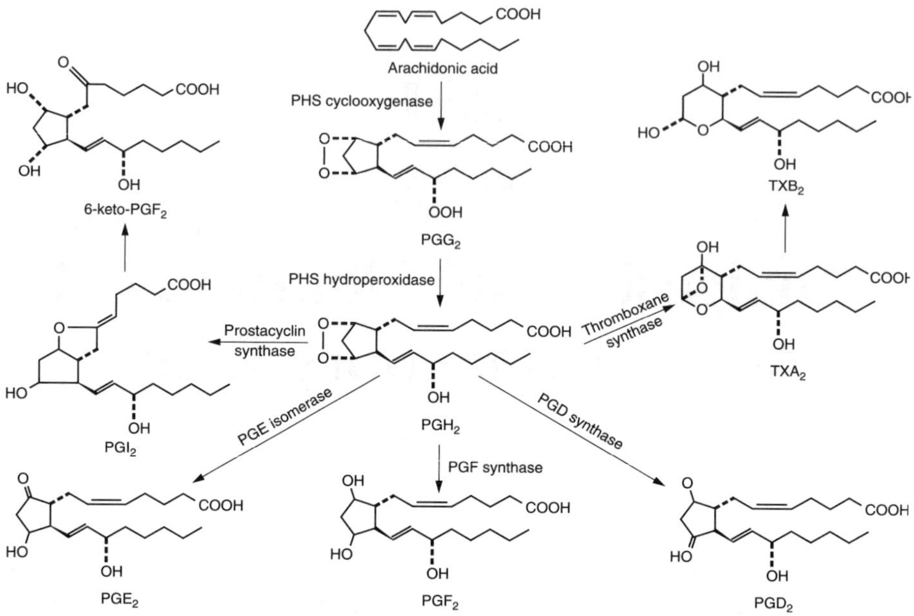

Figure 20.1 Formation of prostacyclins (PGG_2, PGH_2, and PGI_2), prostaglandins (PGE_2, $PGF_{2\alpha}$, and PGD_2) and thromboxanes (TXA_2, TXB_2) from arachidonic acid, catalyzed by prostaglandin H synthase (PHS). The fatty acid cyclooxygenase (COX) component of PHS is reversibly inhibited by NSAIDs and irreversibly inhibited by aspirin, accounting for the ability of these drugs to interrupt the synthesis of prostacyclines, prostaglandins, and thomboxanes.

NSAIDs exhibit other potentially serious renal toxicities, including acute tubular necrosis, interstitial nephritis with nephrotic syndrome, and papillary necrosis, that may or may not be related to inhibition of prostaglandin synthesis.

NSAIDs are among the most widely used drugs and are available both by prescription and over-the-counter. In the U.S., numerous NSAIDs are available to consumers, who receive more than 70 million prescriptions annually at a cost of more than $1 billion (Murray and Brater, 1993). The worldwide market for NSAIDs is more than $6 billion yearly. An estimated 2% of the general population use NSAIDs daily and one in seven Americans is likely to be treated with an NSAID for a chronic rheumatologic disorder (Clive and Stoff, 1984). DeMaria and Weir (2003) report that the incidence of renal side effects among NSAID users ranges from 1 to 5%. While this is a relatively low frequency, the large number of patients using NSAIDs suggests that renal complications may occur in a substantial number of people.

Traditionally, NSAIDs have been classified according to chemical categories (Table 20.1). The discovery of two distinct isoforms of PHS, cyclooxygenase-1 (COX-1) and cyclooxygenase-2 (COX-2), has led to the proposal that the beneficial effects of NSAID therapy are related to inhibition of COX-2 while the side effects of NSAID therapy are related to inhibition of COX-1 (DeWitt et al., 1993; Vane, 1994). Therefore, it is probably more informative to classify NSAIDs as selective COX-1 inhibitors, nonselective COX inhibitors, selective COX-2 inhibitors, and highly selective COX-2 inhibitors (Table 20.2) (Frolich, 1997; Vane et al., 1998).

PROSTAGLANDIN H SYNTHASE

The rate-limiting step in prostaglandin synthesis is the availability of free arachidonic acid. Arachidonic acid is esterified to membrane phospholipids and is deactylated under the influence of phospholipase A_2. Phospholipase A_2-mediated arachidonic acid release is stimulated by numerous agents, including antidiuretic hormone (ADH), bradykinin, angiotensin II, and norepinephrine. Released arachidonic acid is either reesterified back into membrane phospholipids or converted into biologically active eicosanoids (Palmer, 1995).

There are several pathways involved in the metabolism of arachidonic acid. In the pathway inhibited by NSAIDs, prostacyclins, prostaglandins, and thromboxanes are formed from arachidonic acid by the action of PHS (Figure 20.1). This membrane-bound enzyme contains both cyclooxygenase and peroxidase activities. Cyclooxygenase catalyzes the first step in the metabolism of arachidonic acid, converting this fatty acid to prostaglandin G_2 (PGG_2), a prostacyclin (Figure 20.1). COX-1 is expressed constitutively and is effectively inhibited by NSAIDs (Table 20.1). COX-2 is highly inducible and is also inhibited by some NSAIDs (Table 20.2).

The second enzyme activity associated with PHS is that of a peroxidase, specifically prostaglandin hydroperoxidase, which converts PGG_2 to a second prostacyclin, prostaglandin H_2 (PGH_2). There are no known specific inhibitors for PHS-catalyzed peroxidation (Eling and Curtis, 1992). PGH_2 is further metabolized by prostacyclin synthase (Figure 20.1) to generate prostaglandin I_2 (PGI_2). All three prostacyclins, PGG_2, PGH_2, and PGI_2, act as

Table 20.1 Comparison of PHS-1 IC_{50} Potency and Therapeutic Plasma Concentrations of some Nonsteroidal Anti-inflammatory Agents

	IC_{50} (μM)[a]			Pharmacokinetic profile	
Compound	*Murine COX-1*	*Human COX-1*	*Sheep COX-1*	*Dose (mg)*	*C_{max} (μM)[b]*
Salicylates					
Aspirin	1.67 ± 1.11[1]		83[4]	650	345
Sodium salicylate	219 ± 69[1]	>1000[3]		650	340
Aminophenols					
Acetaminophen[c]	17.9 ± 13.2[1]		>100	325	250
Enolic acids					
Phenylbutazone		16.0[3]	12.6[4]	100	27
Piroxicam	0.9–24[2]	17.7 ± 3.3[3]		10	3.6
Fenamic acids					
Mefenamic acid			2.1[4]	250	98.7
Meclofenamic acid	2.0–2.5[2]	1.5 ± 0.6[3]	0.19 ± 0.13[3]	50	14
Acetic acids					
Indomethacin	4.9–8.1[2]	13.5 ± 3.5[3]	0.67 ± 0.09[5]	25	8.3
Sulindac sodium	0.3–0.5[2]	1.3 ± 1.0[3]		150	50
Diclofenac	1.6 ± 0.6[1]	2.7 ± 1.0[3]		50	18.7
Propionic acids					
Ibuprofen	8.9–14[2]	4.0 ± 1.0[3]	1.5[4]	400	230
Naproxen	9.6 ± 4.5[1]	4.8 ± 1.8[3]	6.1[4]	250	130
Ketoprofen		31.5 ± 9.2[3]		50	25.5
Fluriprofen	0.46–0.50[2]	0.5 ± 0.1[3]	0.30 ± 0.03[4]	100	5.6

[a]Concentration *in vitro* required for 50% inhibition of oxygen consumption or PGE_2 synthesis.
[b]Calculated as dose/V_d. Volumes of distribution from Verbeeck et al. (1983) and C_{max} calculated assuming total body weight of 70 kg.
[c]IC_{30} because 50% inhibition was not achieved at concentrations up to 1 mg/ml.
[1]Mitchell et al., 1994.
[2]Meade et al., 1993.
[3]Laneuville et al., 1994.
[4]Flower, 1974.
[5]Kulmacz and Lands, 1985.

Table 20.2 Inhibition of COX-1 and COX-2 in Intact Cells

	COX-1	*COX-2*	*COX-2:COX-1*	*Reference*
Selective COX-1 inhibitors				
Aspirin	1.67	278	166	1
Acetaminophen	2.7 ± 2.0	20 ± 12	7.4	1
Nonselective COX inhibitors				
Indomethacin	0.028	1.68	60	1
Piroxicam	0.0024	0.60	250	2
Diclofenac	1.57	1.1	0.7	1
Ibuprofen	4.85	72.8	15	1
Naproxen	2.2 ± 0.9	1.3 ± 0.8	0.6	1
Selective COX-2 inhibitors				
Etodolac	34	3.4	0.1	2
Salicylate	254	725	2.8	1
Meloxicam	0.21	0.17	0.8	2
Nimesulide	9.2	0.52	0.06	2
Highly selective COX-2 inhibitors				
Celecoxib	15	0.04	0.003	3
Rofecoxib	>15	0.018	> 0.0012	4

[1]. IC_{50} value (μM). Bovine aortic endothelial cells were used to determine COX-1 activity and J774.2 macrophages were induced with lipopolysaccharide to express COX-2. Data from Mitchell et al.,1993.
[2]. Data from Vane et al., 1998.
[3]. Data from Penning et al., 1997.
[4]. Data from Ehrich et al., 1999.

vasodilators or vasoconstrictors, depending on the vascular bed. In the kidney, prostacyclins promote diuresis, natriuresis, and kaliuresis, suggesting that these substances promote vasodilation of renal blood vessels (Table 20.3). Prostacyclin production in the kidney may be monitored by measuring the urinary excretion of 6-keto-prostaglandin $F_{1\alpha}$ (6-keto-$PGF_{1\alpha}$), a stable product of PGI_2 metabolism (Figure 20.1).

PGH_2 may be converted either enzymatically or nonenzymatically to several prostaglandins, including prostaglandin E_2 (PGE_2), prostaglandin $F_{2\alpha}$ ($PGF_{2\alpha}$), and prostaglandin D_2 (PGD_2), illustrated in Figure 20.1. These prostaglandins, in particular PGE_2, perform numerous actions in the kidney, including: vasodilation of renal vessels, which causes an increase in renal blood flow (RBF) and glomerular filtration rate (GFR); stimulation of renin release; and inhibition of sodium reabsorption in the distal tubule, ADH-dependent water reabsorption in the collecting duct, and chloride reabsorption in the thick ascending limb of the loop of Henle (Table 20.3), resulting in diuresis, natriuresis, and kaliuresis.

PGH_2 may also be metabolized by thromboxane synthase (Figure 20.1) to thromboxane A_2 (TXA_2), which in turn spontaneously rearranges to yield thromboxane B_2 (TXB_2). Thromboxanes are potent vasoconstrictors and may antagonize the vasodilating effects of PGE_2 in the kidney (Table 20.3).

As illustrated in Figure 20.2, PHS-associated cyclooxygenase activities are present throughout the renal vasculature (afferent and efferent arterioles, glomerular and peritubular capillaries), in epithelial cells of the collecting tubule, and in interstitial cells adjacent to the thick ascending limb of the loop of Henle and in the medulla (Murray and Brater, 1993; Schlondorff, 1993). Immunoreactive COX-1 has been identified in epithelial cells of collecting ducts, smooth muscle and endothelial cells of renal blood vessels (arteries, arterioles, and veins), and in discrete interstitial cells. COX-1 staining

Table 20.3 Summary of Immunohistochemical Localization of COX-1 and COX-2 in Kidney

	COX-1				*COX-2*			
Site	*Dog*	*Rat*	*Monkey*	*Human*	*Dog*	*Rat*	*Monkey*	*Human*
Renal vasculature[a]	+	+	+	+	±	0	+	+
Glomerulus					0	0	+	+
TAL[b]					+↑	+↑	0	0
Macula densa					+↑	+↑	0	0
Collecting ducts	+++	+++	+++	++	0	0	0	0
Interstitium	0	+	+	+	+	+	0	0

[a]Thick ascending limb of the loop of Henle.
[b]Vasculature includes arteries, arterioles and veins.
0 = no staining; ± = variable staining; + = slight staining; ++ = moderate staining; +++ = marked staining.
↑ indicates change in intensity of staining following volume depletion.
Data from Khan et al., 1998.

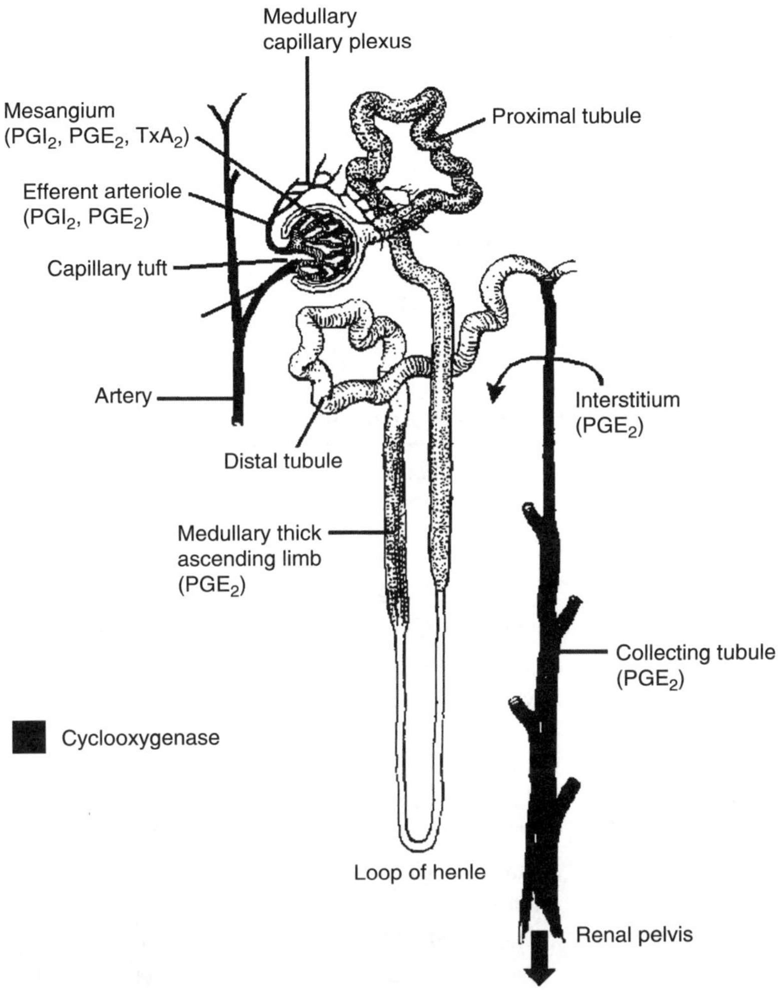

Figure 20.2 Sites of cyclooxygenase activity along the nephron. Prostaglandins are produced in the renal vasculature, collecting duct, and medullary and papillary interstitium. PGI_2 and PGE_2 act as vasodilators at the afferent arteriole and are particularly important in maintaining normal renal function in hypovolemic situations. PGE_2 produced by the collecting duct and interstitium functions in promoting sodium, potassium, and water excretion. Reproduced, with permission, from Murray and Brater (1993) *Annu Rev Pharmacol Toxicol* 32:435–436, by Annual Reviews, Inc.

is most intense in the papillary collecting ducts and less intense in cortical collecting ducts (Khan et al., 1998, 2002). Distribution of COX-1 mRNA (detected by *in situ* hybridization) and protein (detected by Western blotting) is similar in all species examined (Table 20.4) (Khan et al., 1998; Komhoff et al., 1997).

In contrast, there are marked species differences in COX-2 distribution in the kidney. In humans and monkeys, COX-2 is present in glomerular podocytes and interlobular arteries, arterioles, and veins in both the endothelial and smooth muscle cells (Khan et al., 1998;

Table 20.4 Effects of Arachidonic Acid Metabolites on the Kidney

	Vascular effects	*Tubular effects*
Prostacyclins		
$PGG_2/PGH_2/PGI_2$	Vasodilation ↑ renin release Maintain glomerular filtration rate Mesangial cell relaxation	Diuresis Natriuresis kaliuresis
Prostaglandins		
PGE_2	Vasodilation ↑ renal blood flow ↑ renin release Maintain glomerular filtration rate Mesangial cell relaxation	Diuresis, natriuresis, kaliuresis ↓ ADH-dependent water reabsorption ↓ Cl^- reabsorption in TAL[a] ↓ erythropoitein release
$PGF_{2\alpha}$	Vasodilation (humans)	Diuresis, natriuresis
PGD_2	↑ renin release	
PGA_2	Vasodilation ↑ renal blood flow	↑ erythropoietin release
Thromboxanes		
TXA_2/TXB_2	Vasoconstriction Decrease glomerular filtration rate Mesangial cell contraction	

[a]Thick ascending limb of the loop of Henle.

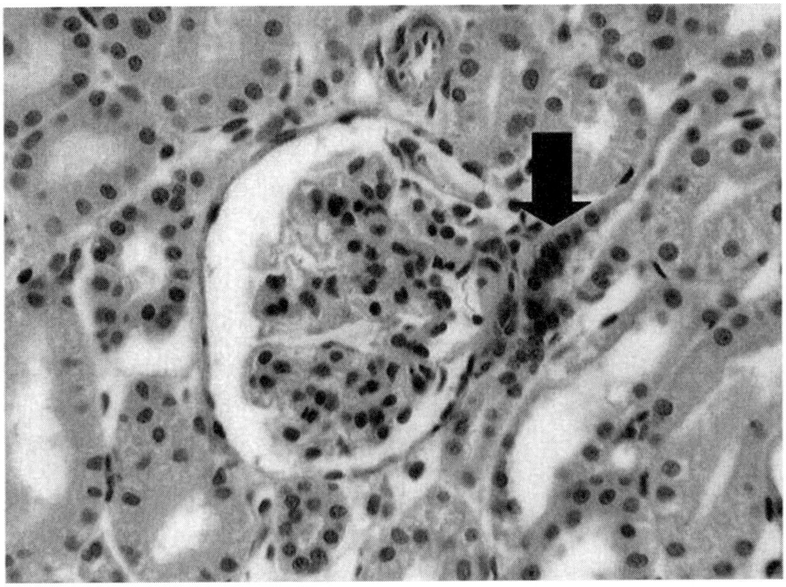

Figure 20.3 Localization of COX-2 in dog kidney. COX-2 immunoreactivity (dark staining) is localized to discrete cells of the macula densa, the portion of the nephron in contact with the glomerulus of origin. Magnification = 400×. Photomicrograph courtesy of Nasir Khan, G.D. Searle Co.

Komhoff et al., 1997). In rats and dogs, COX-2 is absent from the glomerular and vascular structures but is present in the thick ascending limb of the loop of Henle, in discrete cells in the macula densa portion of the tubule (Figure 20.3), and in medullary and papillary interstitial cells. The differential distribution of COX-2 in rats, dogs, monkeys, and humans is summarized in Table 20.4.

COX-2 immunoreactivity in the renal cortex is intensified when rats are placed on a salt-restricted diet, suggesting that COX-2 may be involved in autoregulation (Yang et al., 1998). A similar upregulation of COX-2 expression occurs on salt-restricted dogs (Khan et al., 1998). When rats are placed an a high-salt diet, COX-2 imunoreactivity in the cortex is suppressed while that in the medulla and papilla is intensified, suggesting that medullary and papillary COX-2 may participate in regulation of blood pressure (Yang et al., 1998).

PHYSIOLOGICAL ACTIONS OF PROSTAGLANDINS IN THE KIDNEY

Prostaglandins operate in conjunction with a variety of other mediators, including ADH, angiotensin II, norepinephrine, and kinins. However, under normal conditions, prostaglandins exert little or no important control over RBF (Palmer, 1995). Rather, prostaglandins are important in sustaining renal function under conditions where the vasoconstrictor system is activated, as by angiotensin II, norepinephrine, ADH, and endothelin (Schlondorff, 1993).

During activation of vasoconstrictor systems, such as in circulatory collapse or stress, unopposed vasoconstriction would lead to decreased RBF, GFR, and renal excretory capacity. To maintain normal renal function, prostaglandins are synthesized locally and attenuate these vasoconstrictor effects (Palmer, 1995; Schlondorff, 1993). Additionally, some of the vasoconstrictor hormones, such as angiotensin II and endothelin, stimulate local production of vasodilator prostaglandins, including PGE_2 and $PGF_{2\alpha}$, resulting in a negative feedback loop between vasoconstrictors and prostaglandins (Schlondorff, 1993). Interrupting prostaglandin synthesis by NSAID administration under conditions in which RBF is dependent on prostaglandins disturbs the normal balance between vasodilators and vasoconstrictors, allowing vasoconstriction to predominate (DeMaria and Weir, 2003; Palmer, 1995) (Figure 20.4).

Besides regulation of RBF, prostaglandins may modulate sodium and chloride transport in the thick ascending limb of the loop of Henle and ADH-mediated water transport in the collecting duct. Under conditions of volume depletion, angiotensin II, ADH, and other vasopressors tend to stimulate sodium, chloride, and water reabsorption. Renal prostaglandins blunt the sodium chloride and water retention associated with volume depletion. Inhibition of prostaglandin synthesis will, in low-volume conditions, allow unopposed sodium chloride and water retention (DeMaria and Weir, 2003).

In general, NSAID-induced adverse effects are relatively uncommon in the kidney. However, acute renal failure has been reported in patients with hypovolemia or reduced circulatory volume due to congestive heart failure, cirrhosis, or intensive diuretic therapy. Patients with nephrotic syndrome or preexisting renal impairment, including elderly patients and patients with diabetes, are also at heightened risk (Murray and Brater, 1993; Schlondorff, 1993). In all of

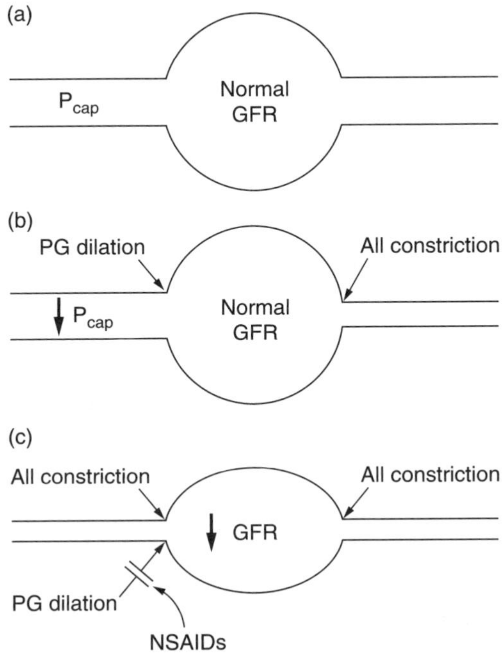

Figure 20.4 Balance between vasoconstrictors and vasodilators involved in renal blood flow and glomerular filtration. (A) Under normal conditions, the afferent and efferent arterioles maintain capillary hydrostatic pressure at about 60 mmHg, sufficient to cause filtration of about 20% of the plasma entering the glomerular capillaries. In low-volume states or under the influence of vasoconstrictors such as angiotensin II, ADH, or norepinephrine, capillary hydrostatic pressure is below 60 mmHg and GFR would be correspondingly decreased. Capillary hydrostatic pressure is elevated to about 60 mmHg due to the action of vasoconstrictors at the efferent arteriole. However, constriction at the efferent arteriole would decrease RBF, which would be undesirable. In these circumstances, prostaglandin production is important to cause vasodilation at the afferent arteriole (B), allowing maintenance of normal GFR. When NSAIDs are used in conditions where RBF and GFR are dependent on prostaglandin production, constriction of the afferent arteriole is unopposed (C). Consequently, RBF and FGR decline precipitously.

these conditions, RBF is at least partially dependent on prostaglandin production, and so NSAID administration would be expected to produce declines in RBF and GFR, possibly resulting in acute renal failure (Murray and Brater, 1993).

Specific COX-2 inhibitors were developed with the goal of achieving anti-inflammatory action with no or minimal gastrointestinal or renal side effects. However, current evidence suggests that COX-1 and COX-2 inhibitors may not differ with respect to renal effects. In particular, COX-2 is associated with synthesis of PGE_2, a prostaglandin important for sodium and water excretion. Therefore, both COX-1 and COX-2 inhibitors may produce sodium and water retention, albeit by different mechanisms (DeMaria and Weir, 2003).

ACUTE RENAL EFFECTS ASSOCIATED WITH NSAID THERAPY

Sodium and Water Retention

Under normal conditions, natriuresis is mediated by two prostaglandin-dependent effects: vasodilation resulting in increased RBF and reduced proximal tubular reabsorption of solutes and water (Ichikawa and Brenner, 1980), and direct inhibition of sodium reabsorption at the thick ascending limb of the loop of Henle (Kaojarern et al., 1983) (Table 20.3). Thus, NSAIDs would be expected to (and in fact do) cause a transient degree of sodium and water retention, as well as mild hypertension, even in clinically normal patients (Bennett et al., 1995). NSAID-induced antidiuresis and antinatriuresis occur in the absence of measurable changes in RBF or GFR (Kaojarern et al., 1983), suggesting that NSAIDs act, at least in part, by directly blocking the inhibitory effects of prostaglandins on tubular reabsorption of sodium. In individuals with preexisting renal dysfunction, clinically significant edema may develop during NSAID therapy because these patients are unable to excrete salt and water loads normally. In normal individuals, NSAID-induced fluid retention is minor, reversible on discontinuation of NSAID therapy, and easily managed in patients who require continued NSAID therapy (Whelton and Hamilton, 1991).

Transient changes in renal function have also been observed in laboratory animals following acute administration of NSAIDs. For example, sodium excretion was significantly reduced while urine volume was unaltered following acute administration of sodium meclofenamate to anesthetized, sodium-replete dogs (Blasingham and Nasjletti, 1980). In these normal dogs, NSAID administration had no effect on RBF or GFR. In laboratory animals as well as humans, salt and water retention following NSAID administration are transient and easily reversible when NSAID therapy is discontinued.

Hyperkalemia

Hyperkalemia is an unusual complication of NSAID therapy, although severe hyperkalemia has been reported in patients with mild renal insufficiency and who received indomethacin for gout (Clive and Stoff, 1984). Other predisposing factors for the development of NSAID-induced hyperkalemia include cardiac failure, diabetes, multiple myeloma, concurrent administration of potassium supplements or potassium-sparing diuretics, and administration of angiotensin converting enzyme inhibitors (Whelton and Hamilton, 1991) (Table 20.5). Prostaglandins stimulate renin release (Table 20.3) so that NSAID therapy would be expected to reduce renin release and angiotensin II formation (Schlondorff, 1993). Since angiotensin II is a prime stimulus for aldosterone secretion, NSAID therapy would be expected to secondarily decrease plasma aldosterone concentrations, as outlined in Figure 20.5. Aldosterone is the primary factor regulating plasma potassium concentration under normal circumstances. In the kidney, aldosterone stimulates potassium secretion, thereby promoting potassium excretion and maintaining potassium balance. With NSAID-induced hypoaldosteronism, potassium secretion would be decreased allowing the possibility for hyperkalemia to occur (Figure 20.5) (Clive and Stoff, 1984; Whelton and Hamilton, 1991).

NSAIDs (or at least indomethacin) may exert a direct effect on renal tubular cells to reduce potassium uptake (Clive and Stoff, 1984; Whelton and Hamilton, 1991). Investigators have described a

Table 20.5 Risk Factors for Development of NSAID-Induced Hyperkalemia

Mild renal insufficiency
Cardiac failure
Diabetes mellitus
Multiple myeloma
Concurrent administration of potassium supplements or potassium-sparing diuretics
Angiotensin converting enzyme inhibitors

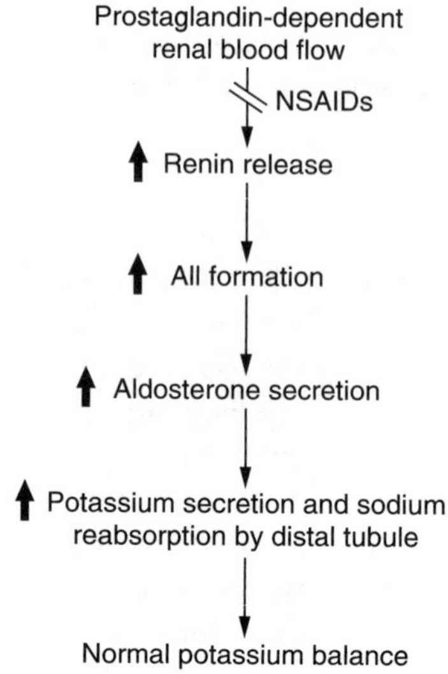

Figure 20.5 Mechanism of NSAID-induced hyperkalemia. Under normal circumstances, aldosterone secretion is stimulated by angiotensin II. Aldosterone promotes potassium secretion and sodium reabsorption in the distal tubule of the nephron. Angiotensin II formation is dependent on renin secretion to initiate the reactions that convert angiotensinogen to angiotensin II. Prostaglandins are among many stimuli for renin secretion and when prostaglandin synthesis is interrupted, as by NSAID therapy, renin secretion is decreased. Through the pathway outlined, inhibition of renin secretion may cause secondary hypoaldosteronism and hyperkalemia.

high-voltage K^+ conductance channel in the luminal membrane of collecting tubule cells that is regulated by intracellular calcium and arachidonic acid metabolites (Ling et al., 1992). These K^+ channels are proposed to open in response to cell swelling, allowing volume regulation partially driven by K^+ secretion (Ling et al., 1992). Schlondorff proposed that these channels may also contribute to K^+ secretion during ADH-dependent water conservation and suggested that both actions, ADH-dependent water reabsorption and K^+ secretion, may be mediated by PGE_2 (Schlondorff, 1993). Inhibition of PGE_2 synthesis by NSAIDs would hypothetically decrease the number of open K^+ channels, thereby decreasing K^+ secretion and contribute to NSAID-induced hyperkalemia (Schlondorff, 1993). Indeed, indomethacin delayed swelling-induced K^+ channel activation in rabbit cortical collecting tubule cells, and this delay was reversed when arachidonic acid was included in the incubation medium (Ling et al., 1992).

Acute Renal Failure

NSAID therapy may cause abrupt declines in RBF and GFR due to the vascular effects of PGE_2 withdrawal in patients with preexisting renal insufficiency (Murray and Brater, 1993) (Figure 20.6). Abrupt declines in renal function have been reported to occur with numerous NSAIDs. Any condition that causes a decrease in effective circulating blood volume makes RBF at least partially dependent on prostaglandin production. Such conditions include congestive heart failure, dehydration, hemorrhage, cirrhosis with ascites, and excessive diuretic therapy (Table 20.6). In all these conditions, the decrease in effective circulating blood volume acts as a stimulus to produce vasoconstrictors such as angiotensin II, ADH, norepinephrine, and endothelin. Angiotensin II, norepinephrine, and ADH also stimulate

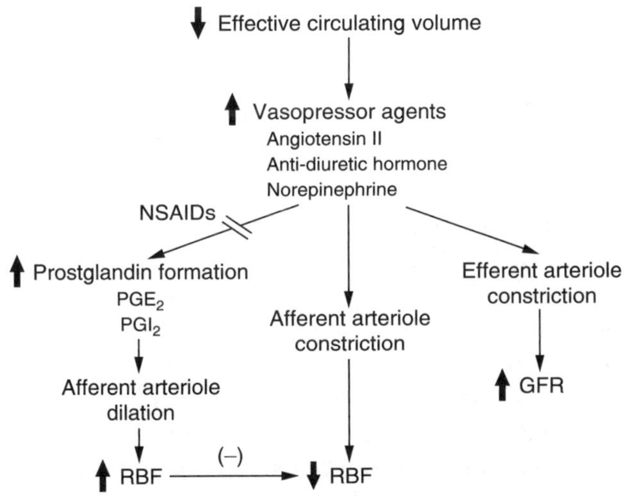

Figure 20.6 Mechanism of NSAID-induced decreases in RBF and GFR. Vasoconstrictor agents also stimulate prostaglandin production. The balance between vasoconstriction and vasodilation helps maintain RBF and GFR. When NSAIDs are used in these conditions, interruption of prostaglandin synthesis allows vasoconstriction to occur unopposed by vasodilation. When renal blood vessels are constricted, both RBF and GFR are reduced.

Table 20.6 Risk Factors for Development of NSAID-Associated Acute Renal Failure

Congestive heart failure
Hepatic cirrhosis
Nephrotic syndrome
Hemorrhage
Diuretic therapy
Hypoalbuminemia
Hepatic failure with ascites
Hypertension
Sepsis
Anesthesia
Diabetes mellitus
Volume depletion due to diuretics or extra-renal fluid loss

prostaglandin synthesis. In this setting, the vasodilator renal prostaglandins are necessary to counteract the vasoconstrictors and maintain RBF and GFR. When NSAID therapy is introduced, renal prostaglandin production is abruptly curtailed and RBF and GFR decline precipitously (Bennett et al., 1996) (Figure 20.6). This NSAID-induced acute renal failure is the most common renal side effect of NSAID therapy, occurring in 0.5 to 1% of patients taking NSAIDs on a chronic basis (Whelton and Hamilton, 1991). NSAID-induced acute renal failure is reversible and no structural damage is observed during or following the renal dysfunction.

In patients with renal insufficiency, but not in normal individuals, NSAIDs cause about 30% decreases in RBF (measured as *p*-aminohippurate clearance) and GFR (measured as inulin clearance) (Brater et al., 1985; Murray and Brater, 1993). In normal patients, therefore, NSAID administration does not alter renal hemodynamics, probably because RBF is not dependent on continuing prostaglandin production. In patients with renal insufficiency, however, in whom prostaglandin production partially counteracts the effects of vasoconstrictor agents such as angiotensin II and norepinephrine, NSAID administration causes declines in RBF and GFR (Schlondorff, 1993). In salt-deprived patients, aspirin administration produced 12 to 15% reductions in creatinine and inulin clearances (Muther et al., 1981), demonstrating that renal function may be depressed when effective circulatory volume is depleted, as by salt restriction. However, the hemodynamic and urinary changes caused by NSAID administration in both normal patients and patients with renal insufficiency are fully reversible within a few hours of NSAID administration (Brater et al., 1985; Murray and Brater, 1993). Furthermore, chronic therapy with NSAIDs for as long as 1 month caused no sustained decrement in renal function (Murray and Brater, 1993), suggesting again that the effects of NSAIDs on renal hemodynamics and glomerular filtration are fully reversible.

Similarly, in conscious, unstressed, sodium-replete dogs, administration of NSAIDs produced no significant changes in RBF or GFR, indicating that, under basal conditions, RBF is not dependent on prostaglandin synthesis (Blasingham and Nasjletti, 1980; Terragno et al., 1977). However, administration of NSAIDs to conscious or anesthetized sodium-deprived dogs (Blasingham et al., 1980), or the anesthetized surgically operated animals, produced significant decreases in RBF and increases in renal vascular resistance without changing GFR (Schlondorff, 1993). Similarly, in volume-depleted conscious dogs, indomethacin caused significant decreases in RBF in the absence of changes in GFR or renal solute excretion (Black et al., 1998). A selective COX-2 inhibitor, (3-(3,4-difluorophenyl)-4-(4-(methylsulfonyl)-phenyl-2-(*5H*)-furanone (similar to rofecoxib), had no effect on RBF in volume-depleted dogs (Black et al., 1998). The renal vasoconstrictor response in anesthetized, surgically operated animals is at least partially dependent on prostaglandin synthesis as the response to NSAID therapy was blunted or abolished by exogenous administration of PGE_1 (Chrysant, 1978). Additionally, NSAID-induced declines in RBF are at least partially dependent on activation of the renin-angiotensin system, as administration of saralasin blunted the effects of meclofenamate in sodium-depleted dogs (Blasingham and Nasjletti, 1980).

It is likely that metabolites of both COX-1 and COX-2 are involved in regulation of renal hemodynamics. Under normal conditions, COX-2 expression is fairly low in human kidneys (Harris, 2002). However, COX-2 expression in the renal cortex is increased under conditions of volume depletion or salt restriction (Harris, 2002). It has therefore been suggested that in normal patients, a COX-2-selective NSAID may be less likely to cause changes in RBF or GFR since COX-1 metabolites are available to counterbalance vasoconstriction

(Roig et al., 2002). However, in volume- or salt-depleted patients, COX-2 expression is likely to be elevated and RBF may become more dependent on COX-2 metabolites. In this scenario, either COX-1 or COX-2 selective NSAIDs may produce acute vasoconstriction and decrements in RBF and GFR (Roig et al., 2002).

Acute Tubular Necrosis

Oliguria and nonoliguric renal failure consistent with acute tubular necrosis has been reported following clinical overdoses of aspirin, acetaminophen, and other NSAIDs (indomethacin, phenylbutazone, fenoprofen, ibuprofen, and mefenamic acid, diclofenac) (Hickey et al., 2001; Nanra, 1983). Histological lesions ranging from minor tubular changes to frank tubular necrosis have been observed experimentally with acetaminophen, salicylates, phenylbutazone, ibuprofen, and mefenamic acid (Nanra, 1983). Histological damage with NSAIDs is largely confined to the proximal tubule of the nephron.

Salicylates

Sodium salicylate causes increases in blood urea nitrogen concentration and excretion of glucose, protein, blood, and tubular epithelial cells in urine of humans as well as experimental animals (Arnold et al., 1973; Kyle and Kocsis, 1985). The effects of sodium salicylate were transient and signs of regeneration were evident within 24 hours of administration to laboratory rats (Robinson et al., 1967). In rats, administration of 500 mg sodium salicylate/kg (i.p.) was associated with morphological changes, including dilation and vacuolization of proximal tubular epithelial cells within 6 hours of administration (Kyle and Kocsis, 1985). These morphological changes were transient and renal morphology in salicylate-treated rats was not different from control animals within 12 hours of administration (Kyle and Kocsis, 1985). Similarly, blood urea nitrogen concentrations were elevated from 2 hours through to 12 hours after salicylate administration, but returned to control values by 24 hours (Kyle and Kocsis, 1985).

Salicylate-induced proximal tubular damage was more severe and more persistent in 12-month old rats as compared with 3-month old rats (Kyle and Kocsis, 1985). For example, histological changes in 12-month old rats' kidneys included areas of focal proximal tubular necrosis and interstitial edema and these changes persisted through 12 hours following salicylate administration (Kyle and Kocsis, 1985). Regenerative changes were not observed in older rats until 24 hours after salicylate administration. Similarly, blood urea nitrogen concentrations increased to a greater extent and remained elevated for a longer period of time (through 24 hours) in 12-month old rats as compared with 3-month old rats receiving the same dosage of salicylate (Kyle and Kocsis, 1985). Although the precise mechanisms involved in salicylate nephrotoxicity remain undefined, covalent binding may play a role in the development of proximal tubular necrosis in rats. Specifically, radiolabel from salicylate became covalently bound to macromolecules in rat renal cortex but not medulla. In the cortex, greater than 50% of total covalently bound radioactivity in the renal cortex was associated with the mitochondrial fraction (Kyle and Kocsis, 1985). Therefore, salicylate-induced proximal tubular necrosis may involve bioactivation of salicylate to a reactive intermediate that undergoes covalent binding to mitochondrial proteins, thereby interrupting normal mitochondrial function (Mitchell et al., 1977).

Aspirin administration in rats produced mild proximal tubular damage, as indicated by enzymuria (GGT and *N*-acetyl-β-D-glucosaminidase), glucosuria, proteinuria, and elevations of blood urea nitrogen concentrations within 24 hours of aspirin administration to rats (Owen and Heywood, 1983). Massive proximal tubular necrosis was evident as early as 2 hours following oral administration of aspirin (900 mg/kg) to rats. Tubular necrosis was maximal at 12 hours and regeneration was observed 24 to 48 hours following aspirin administration (Arnold et al., 1973). Aspirin produced tubular necrosis in both males and females and cortical necrosis developed in some rats after dosages of aspirin as low as 100 mg/kg (Arnold et al., 1973).

Aspirin-induced tubular necrosis may be due to a reactive metabolite generated by cytochromes P450 (Mitchell et al., 1977). Alternately, since aspirin undergoes rapid deacetylation to form salicylate, the nephrotoxicity of aspirin may be due to salicylate rather than aspirin itself. Either aspirin or a metabolite underwent covalent binding with cellular macromolecules, significantly depleted renal glutathione content, and inhibited PHS activity in renal cortex and medulla (Caterson et al., 1978). Additionally, aspirin inhibited the hexose-monophosphate shunt metabolic pathway, which may contribute to further decreases in renal glutathione concentrations (Nanra, 1983). Aspirin uncoupled oxidative phosphorylation leading to decreases in intracellular ATP and increases in AMP in renal cortex but not medulla (Dawson, 1975; Quintanilla and Kessler, 1973). The lack of effect on renal medulla may relate to the fact that medullary metabolism is primarily glycolytic rather than aerobic (Silva, 1987) so that uncoupling of oxidative phosphorylation may have fewer dramatic effects on medullary metabolism.

Acetaminophen

Massive dosages of acetaminophen (500 mg/kg or greater) cause acute proximal tubular necrosis in humans, rats, and mice (Emeigh Hart et al., 1994; Prescott, 1982; Tarloff et al., 1989). In both rats and mice, acetaminophen-induced proximal tubular necrosis was accompanied by marked elevations of blood urea nitrogen concentrations and kidney weights within 24 hours of administration (Emeigh Hart et al., 1994; Tarloff et al., 1989). In addition, renal glutathione was transiently depleted and radiolabel from acetaminophen became covalently bound to renal proteins within a few hours of acetaminophen administration (Emeigh Hart et al., 1994; Newton et al., 1983).

Covalently bound material in both liver and kidneys of acetaminophen-treated mice reacted immunochemically with an antibody directly against the carbonyl group of acetaminophen (Emeigh Hart et al., 1994, 1991). Covalent binding of radiolabel to renal proteins is not a random event, but rather certain proteins are selective targets for acetaminophen. Specifically, proteins with molecular weights of 27, 33, and 56 to 58 kDa were arylated by acetaminophen (Emeigh Hart et al., 1994). The 56 to 58 kDa protein arylated by acetaminophen in kidney was also a target for acetaminophen in liver and has been identified as a 56 kDa selenium-binding protein in mouse liver (Bartolone et al., 1992). The proteins of 27 and 33 kDa arylated by acetaminophen in kidney have not been purified, sequenced, or identified as yet.

The mechanisms leading to acetaminophen-induced nephrotoxicity may be different in rats and mice. In mice, acetaminophen nephrotoxicity involves cytochrome P450-catalyzed oxidation of acetaminophen to *N*-acetyl-benzoquinoneimine, similar to the pathway involved in acetaminophen hepatotoxicity. As a reactive intermediate, *N*-acetyl-benzoquinoneimine may covalently bind to renal macromolecules or induce oxidative stress. In mice, acetaminophen nephrotoxicity was highly correlated with cytochrome P450 2E1 content (Hu et al., 1993). Male mice had a higher content of cytochrome P450 2E1 and were more susceptible to acetaminophen nephrotoxicity than were female mice (Hu et al., 1993).

Inhibition of cytochrome P450 activity with piperonyl butoxide protected mice against acetaminophen-induced nephrotoxicity, whereas inhibition of acetaminophen deacetylation with tri-*ortho*-tolyl phosphate, a carboxyesterase inhibitor, afforded no protection from toxicity (Emeigh Hart et al., 1991). These data suggest that in mice, acetaminophen nephrotoxicity involves a cytochrome P450-dependent pathway similar to the mechanism involved in acetaminophen-induced hepatotoxicity.

In rats, however, a different pathway, involving deacetylation of acetaminophen to yield *p*-aminophenol, seemed to be involved in acetaminophen nephrotoxicity (Newton et al., 1982). Pretreating rats with inhibitors and inducers of cytochrome P450 produced inconsistent effects on the severity of acetaminophen nephrotoxicity (McMurtry et al., 1978). Pretreating rats with bis-(*p*-nitrophenyl) phosphate, a carboxyesterase inhibitor, protected rats from acetaminophen-induced nephrotoxicity (Newton et al., 1985b), suggesting that deacetylation to *p*-aminophenol was an obligatory step.

Covalent binding of radiolabel derived from acetaminophen to renal macromolecules was greater when the radioactive tag was located on the aromatic ring rather than on the carbonyl moiety (Newton et al., 1985a). Using an antibody directed against the carbonyl group of acetaminophen, immunochemical binding was detectable in liver but not kidneys of acetaminophen-treated rats (Tarloff et al., 1996), supporting the hypothesis that a deacetylated metabolite of acetaminophen, probably *p*-aminophenol, is responsible for covalent binding and nephrotoxicity in rats.

Acute proximal tubular necrosis was produced in rats by *p*-aminophenol (Tarloff et al., 1989; Newton et al., 1982) and small amounts of *p*-aminophenol were excreted in urine from rats or hamsters treated with acetaminophen (Gemborys and Mudge, 1981; Newton et al., 1983). It is thus possible that acetaminophen-induced nephrotoxicity may involve deacetylation to *p*-aminophenol, which ultimately causes renal damage due to enzymatic or nonenzymatic oxidation (Boogaard et al., 1990; Calder et al., 1979; Josephy et al., 1983). However, both liver and kidney deacetylases catalyze formation of *p*-aminophenol from acetaminophen (Mugford and Tarloff, 1995; 1997) while *p*-aminophenol produces nephrotoxicity but not hepatotoxicity (Burnett et al., 1989; Newton et al., 1982). If autoxidation is responsible for the toxicity of *p*-aminophenol, then both liver and kidney should be damaged following administration of acetaminophen or *p*-aminophenol.

The precise mechanism by which *p*-aminophenol produces renal toxicity is not entirely clear. Radiolabel from *p*-aminophenol was preferentially incorporated into proteins in kidney as compared with liver (Crowe et al., 1977; Fowler et al.1993). Covalently bound material derived from radiolabeled *p*-aminophenol was distributed in mitochondria, microsomes, and cytosol and associated with protein, DNA, mitochondrial enzymes, and glucose-6-phosphatase (Crowe et al., 1977). In Wistar rats treated with *p*-aminophenol, glucose synthesis from lactate or pyruvate was inhibited in kidney but not liver within 30 to 60 minutes of treatment (Tange et al., 1977). Additionally, renal mitochondria obtained from Sprague-Dawley rats treated with *p*-aminophenol displayed decreased state 3 respiration and respiratory control ratio within 60 minutes of treatment (Crowe et al., 1977). Therefore, *p*-aminophenol may injure proximal tubular cells by interfering with mitochondrial function, intermediary metabolism, or a combination of events.

Other NSAIDs

Although acute tubular necrosis has been reported following overdoses of phenylbutazone, ibuprofen, and mefenamic acid (Nanra, 1983), nephrotoxicity due to these agents has not been investigated as thoroughly as with aspirin, salicylate, or acetaminophen. The time course, functional characteristics, and mechanisms leading to acute tubular necrosis with other NSAIDs are unclear, and the relevance to human consumption has not been established.

NSAIDs may interfere with mitochondrial respiration. When tested using *in vitro* preparations of kidney or liver mitochondria, several NSAIDs exhibited characteristics of classic uncouplers. For example, nimesulide and meloxicam stimulated mitochondrial state 3 (ATP-stimulated) and state 4 (substrate-supported) respiration, inhibited ATP synthesis, and collapsed the mitochondrial membrane potential (Moreno-Sanchez et al., 1999). In contrast, naproxen was ineffective as a mitochondrial toxicant (Moreno-Sanchez et al.1999).

In kidney mitochondria, acetylsalicylic acid, diclofenac, mefenamic acid, and piroxicam acted as uncouplers as well as inhibitors of oxidative phosphorylation (Mingatto et al., 1996). Several factors may make the kidney more susceptible than other organs, such as the liver, to damage from mitochondrial toxicants. The kidney, in particular the proximal tubule, is critically dependent on ATP for transport functions. Therefore, chemicals that interfere with ATP generation by mitochondria will have impact on proximal tubule reabsorption. In addition, many NSAIDs are organic anions and the proximal tubule contains transport proteins that allow organic anions to move from extracellular fluid into intracellular fluid in the first step of a secretory process. Several NSAIDs are efficiently transported *in vitro* by an organic anion transport protein (OAT1) (Apiwattanakul et al., 1999) and could theoretically achieve mitochondrial inhibitory concentrations *in vivo*. The role of mitochondria in NSAID-induced renal toxicity has not been fully explored.

CHRONIC RENAL EFFECTS ASSOCIATED WITH NSAID THERAPY

Interstitial Nephritis and Nephrotic Syndrome

Interstitial nephritis is a rare, idiosyncratic reaction to therapy with numerous NSAIDs (Harris, 2002). The incidence of interstitial nephritis is estimated as one of every 5,000 to 10,000 patients receiving NSAID therapy, and differs from acute ischemic renal insufficiency in onset, severity, and duration (Harris, 2002). It occurs more frequently with propionic acid NSAIDs (fenoprofen, naproxen, and ibuprofen) than with other classes (Ailabouni and Eknoyan, 1996). The elderly seem particularly susceptible to NSAID-induced interstitial nephritis (Ailabouni and Eknoyan, 1996).

Interstitial nephritis can occur within 1 week of NSAID administration but more often occurs several months to a year after the start of NSAID therapy (Murray and Brater, 1993). Whereas acute renal insufficiency causes minimal decrements in GFR, patients with interstitial nephritis may present with elevated serum creatinine concentrations (>6 mg/dl, compared with normal of 0.8 to 2.0 mg/dl) (Murray and Brater, 1993). In NSAID-induced acute renal insufficiency, withdrawal of the causative agent usually restores renal function to normal within a day or two. In contrast, NSAID-induced interstitial nephritis may persist for 1 to 3 months following discontinuation of NSAID therapy (Murray and Brater, 1993).

Clinical symptoms of NSAID-induced interstitial nephritis also include edema, oliguria, and proteinuria. Systemic symptoms of allergic interstitial nephritis, such as fever, drug rash, eosinophilia, and eosinophiluria, are absent with NSAID-induced interstitial nephritis (Whelton and Hamilton, 1991). Histological findings included diffuse interstitial edema with evidence of mild to moderate inflammation, and cellular infiltration of cytotoxic T cells and smaller numbers of B cells and eosinophils (Abt and Gordon, 1985; Bender et al., 1984). Glomerular membrane involvement, such as fusion of epithelial foot processes, is observed in NSAID-induced acute interstitial nephritis (Pirani et al., 1987). Microscopically, glomeruli resemble those seen in minimal-change nephrotic syndrome with effacement of foot processes of epithelial cells visible by electron microscopy but little or no abnormality visible by light microscopy (Kincaid-Smith, 1986).

Interstitial nephritis is not related to the dose of NSAID used and does not appear to be directly related to inhibition of prostaglandin synthesis *per se* (Tange et al., 1977). Rather, by inhibiting cyclooxygenase, NSAIDs may make more arachidonic acid available for lipoxygenase-catalyzed leukotriene formation (Whelton and Hamilton, 1991). Since leukotrienes are chemotactic agents, leukotrienes may participate in the development of interstitial nephritis and proteinuria by attracting T lymphocytes, promoting inflammation, and increasing vascular permeability (Bender et al., 1984). Fortunately, interstitial nephritis due to NSAID therapy is reversible by discontinuation of the drug.

Risk factors for NSAID-induced interstitial nephritis are not entirely clear. Preexisting renal insufficiency does not appear to predispose a patient to develop interstitial nephritis (Whelton and Hamilton, 1991). Old age may be a risk factor, but the increased incidence of interstitial nephritis in the elderly may be a reflection of the tendency for older patients to be the more likely candidates for chronic NSAID therapy.

Papillary Necrosis

Chronic administration of nearly all NSAIDs produces papillary necrosis in laboratory animals (Whelton and Hamilton, 1991). In some patients, analgesic use in the therapeutic range may cause renal impairment, such as an inability to excrete a sodium or water load, whereas cumulative ingestion of more than 3 kg of an analgesic agent has been associated with renal failure, such as papillary necrosis or oliguria (Schreiner et al., 1981). While the cumulative dose of 3 kg may seem quite high, patients with arthritis may easily consume 10 aspirin tablets per day (3.25 g) or about 1.2 kg aspirin per year.

Phenacetin was initially implicated as the causative agent of papillary necrosis and was removed from nonprescription analgesics as early as 1961 in Sweden and Denmark (Prescott, 1982). However, it is now apparent that papillary necrosis may be associated with a variety of analgesic agents in the absence of phenacetin ingestion (Murray and Brater, 1993). In humans, papillary necrosis has been reported following administration of aspirin (alone or in combination with other analgesics but not phenacetin), ibuprofen, fenoprofen, indomethacin, diclofenac, mefenamic acid, phenylbutazone, and piroxicam (Kincaid-Smith, 1986; Prescott, 1982).

The pathology of analgesic-associated papillary necrosis has been investigated by studying the lesion in laboratory animals. However, commonly used laboratory animals (dog, rat, and rabbit) are not reliably susceptible to salicylate-, phenacetin-, and NSAID-induced papillary necrosis so that these compounds must be given in large dosages of over long periods of time (Wiseman and Reinert, 1975). For example, papillary necrosis occurred in only 50% of rats given acetylsalicylic acid at a dose of 500 mg/kg/day for 60 weeks (Nanra and Kincaid-Smith, 1970). Renal papillary necrosis has been produced experimentally with phenacetin, aspirin, acetaminophen, antipyrine, aminopyrine, and a number of NSAIDs including phenylbutazone, indomethacin, mefenamic acid, flufenamic acid, meclofenamic acid, fenoprofen, and naproxen (Nanra, 1983; Wiseman and Reinert, 1975). When papillary necrosis is experimentally induced in laboratory animals (e.g., with 2-bromoethylamine), morphological changes similar to those observed in human kidneys are seen (Sabatini, 1988). Functional changes observed during the course of analgesic nephropathy in rats include impaired concentrating capacity, hypertension, reduced total body sodium, reduced renal vein plasma PGE_2, and increased renal vein plasma renin activity (Nanra 1983).

The mechanisms involved in papillary necrosis have received considerable attention, but remain unresolved. Duggin (1980) suggests that at least one of two factors must be present for the development of papillary necrosis. First, the tissue must have a metabolic or functional predisposition, such as susceptibility to ischemia from reduced RBF. It has been postulated that the mechanism of NSAID-induced papillary necrosis is the result of prostaglandin-mediated acute ischemic insult. Analgesic-associated nephropathy in rats occurs with a variety of structurally unrelated NSAIDs, indirectly supporting an ischemic mechanism (Murray and Brater, 1993). NSAIDs inhibit prostaglandin synthesis and with prolonged NSAID therapy, papillary ischemia followed by necrosis may occur (Murray and Brater, 1993).

The second requirement proposed by Duggin (1980) is that the causative agent must selectively concentrate within the renal papilla. Both aspirin and acetaminophen, agents associated with papillary necrosis, selectively concentrate within the papilla (Kincaid-Smith, 1986; Nanra, 1983). Although acetaminophen is reported to be an ineffective COX-1/COX-2 inhibitor outside of the central nervous system (Flower and Vane, 1972), acetaminophen effectively lowered PGE_2 and $PGF_{2\alpha}$ production in renal medullary slices from rat kidney (Zenser et al., 1978). The IC_{50} for acetaminophen-induced inhibition of prostaglandin synthesis was 100 to 200 μM and the effect of acetaminophen was reversible (Zenser et al., 1978). By comparison, 1 mM aspirin and 0.28 mM indomethacin reduced PGE_2 synthesis to the same extent as did 1 mM acetaminophen (Davis et al., 1981). Rats fed a combination of aspirin, phenacetin, and caffeine for 8 to 20 weeks had a 38.8% incidence of papillary necrosis when deprived of water for 16 hours/day whereas rats maintained in a constant diuretic state (by substituting 5% glucose for drinking water) had a 16.7% incidence of papillary necrosis (Nanra and Kincaid-Smith, 1970).

Biochemical mechanisms that may contribute to analgesic nephropathy include depletion of cellular glutathione by salicylates and metabolism of acetaminophen to a highly reactive intermediate that binds covalently to tissue proteins when glutathione concentrations are low (Kincaid-Smith, 1986; Mugford and Tarloff, 1995).

As previously mentioned, PHS, which is highly active in inner medulla and papilla, has a peroxidase activity that may participate in the formation of reactive intermediates (Davis et al., 1981). These reactive intermediates can acylate tissue nucleophiles, such as proteins, RNA, and DNA, to initiate cellular damage (Davis et al., 1981). It is possible that acetaminophen undergoes peroxidase-catalyzed cooxidation that may contribute to acetaminophen-induced papillary necrosis (Mohandas et al., 1981). However, acetaminophen inhibits PHS-associated cyclooxygenase activity with an IC_{50} of 0.1 m*M* acetaminophen for inhibition of PGE_2 formation by rabbit renal medullary microsomes (Mattammal et al., 1979). If acetaminophen selectively concentrates in the renal medulla and papilla, it is possible that regional concentrations could approach or exceed 0.1 mM, leading to inactivation of cyclooxygenase activity, and thereby actually protect the renal medulla and papilla from damage caused by reactive acetaminophen metabolites. In addition, the hypothesis of PHS-catalyzed bioactivation of aspirin is untenable because aspirin irreversibly inhibits PHS-associated cyclooxygenase activity (Table 20.1) (Mattammal et al., 1979), so that aspirin would be expected to inhibit its own PHS-associated cooxidation.

NSAIDs have the potential to cause direct cell injury. Salicylates depressed protein synthesis, probably due to inhibition of amino acyl tRNA synthetases, and salicylates depleted cells of ATP (Dawson, 1975). NSAIDs uncoupled oxidative phosphorylation and depleted ATP in rat liver mitochondria *in vitro*, and the rank order of potency for these effects was roughly: aspirin 1, ibuprofen 10, dinitrophenol 40 to 50, flufenamic acid 200 (Tokumitsu et al., 1977).

Papillary necrosis occurs with structurally dissimilar compounds that share the ability to inhibit prostaglandin synthesis, suggesting that redistribution of prostaglandin-dependent medullary blood flow may contribute to the development of papillary necrosis. With prolonged administration, NSAIDs may cause sustained papillary and medullary ischemia that ultimately may culminate in papillary necrosis if the causative agent is not discontinued.

Renal medullary interstitial cells express COX-2 and expression is upregulated under conditions of water deprivation. Hypertonicity in the renal medulla exposes interstitial cells to harsh conditions, and these cells respond by accumulating small organic solutes such as inositol, sorbital, and betaine (Moeckel et al., 2003). This accumulation is inhibited by COX-2-selective NSAIDs and stimulated by dehydration (Moeckel et al., 2003), suggesting that COX-2-derived prostaglandins may be involved in renal medullary interstitial cell function. Exposure of cultured mouse medullary interstitial cells (MICs) to various NSAIDs caused apoptotic cell death (Hao et al., 1999; Rocha et al., 2001). For example, 48 hour treatment of MICs with ibuprofen, sulindac sulfide, indomethacin, aspirin, SC-58236 (a COX-2-selective NSAID), and SC-58560 (a COX-1-selective NSAID) produced concentration-dependent cell death (Hao et al., 1999). In contrast, APAP and phenacetin had no effect on MIC viability (Hao et al., 1999). MICs are subject to changes in osmolarity and when abruptly switched from 300 to 500 mOsmol/l medium, more than 90% of MICs survive (Hao et al., 2000). The shift from isosmotic to hyperosmotic medium was accompanied by upregulation of COX-2 expression (Hao et al., 2000). When MICs were cultured with SC-58236, cell viability was dramatically decreased when cells were shifted from 300 to 500 mOsmol/l medium (Hao et al., 2000). Treatment of water-deprived rabbits with SC-58236 caused the appearance of patches of apoptotic cells in the renal medulla (Hao et al., 2000). These data suggest that renal medullary interstitial cells require COX-2-derived prostaglandins for survival under conditions of changing medullary osmolarity. Papillary necrosis may result from NSAID-induced cytotoxicity to medullary interstitial cells.

CONCLUSIONS

NSAIDs are indispensable therapeutic agents for a variety of disorders. In general, occasional use of NSAIDs for analgesia presents little or no risk for adverse renal effects. However, chronic use of NSAIDs, as for various types of arthritis, may predispose patients to potentially serious side effects. In particular, patients with preexisting risk factors, such as congestive heart failure, cirrhosis, or nephrotic syndrome, may be more vulnerable to serious renal toxicities associated with NSAIDs than the general population. In addition, the elderly may be more susceptible to NSAID toxicities than the general population. However, the potential benefits of NSAID therapy far outweigh the risks associated with NSAID use and these agents remain widely used in clinical medicine.

REFERENCES

Abt AB, and Gordon JA (1985) Drug-induced interstitial nephritis: Co-existence with glomerular disease. *Arch Intern Med* 145:1063–1067.

Ailabouni W, and Eknoyan G (1996) Nonsteroidal anti-inflammatory drugs and acute renal failure in the elderly. *Drugs Aging* 9:341–351.

Apiwattanakul N, Sekine T, Chairoungdua A, Kanai Y, Nakajima N, Sophasan S, and Endou H (1999) Transport properties of nonsteroidal anti-inflammatory drugs by organic anion transporter 1 expressed in *Xenopus laevis* oocytes. *Mol Pharmacol* 55:847–854.

Arnold L, Collins C, and Starmer GA (1973) The short-term effects of analgesics on the kidney with special reference to acetylsalicylic acid. *Pathology* 5:123–134.

Bartolone JB, Birge RB, Bulera SJ, Bruno MK, Nishanian EV, Cohen SD, and Khairallah EA (1992) Purification, antibody production, and partial amino acid sequence of the 58-kDa acetaminophen-binding liver proteins. *Toxicol Appl Pharmacol* 113:19–29.

Bender WL, Whelton A, Beschorner WE, Darwish MO, Hall-Craggs M, and Solez K (1984) Interstitial nephritis, proteinuria, and renal failure caused by nonsteroidal anti-inflammatory drugs: Immunological characterization of the inflammatory filtrate. *Am J Med* 76:1006–1012.

Bennett WM, Henrich WL, and Stoff JS (1996) The renal effects of nonsteroidal anti-inflammatory drugs: Summary and recommendations. *Am J Kidney Dis* 28:S56–S62.

Black SC, Brideau C, Cirino M, Belley M, Bosquet J, Chan C-C, and Rodger IW (1998) Differential effect of a selective cyclooxygenase-2 inhibitor versus indomethacin on renal blood flow in conscious volume-depleted dogs. *J Cardiovasc Pharmacol* 32:686–694.

Blasingham MC, and Nasjletti A (1980) Differential renal effects of cyclooxygenase inhibition in sodium-replete and sodium-deprived dog. *Am J Physio*l 239:F360–F365.

Blasingham MC, Shade RE, Share L, and Nasjletti A (1980) The effect of meclofenamate on renal blood flow in the unanesthetized dog: Relation to renal prostaglandins and sodium balance. *J Pharmacol Exp Ther* 214:1–4.

Boogaard PJ, Nagelkerke JF, and Mulder GJ (1990) Renal proximal tubular cells in suspension or in primary culture as *in vitro* models to study nephrotoxicity. *Chem-Biol* Interact 76:251–292.

Brater DC, Anderson SA, Brown-Cartwright D, Toto RD, Chen A, and Jacob GB (1985) Effect of etodolac in patients with moderate renal impairment compared with normal subjects. *Clin Pharmacol Ther* 38:674–679.

Burnett CM, Re TA, Rodriguez S, Loehr RF, and Dressler WE (1989) The toxicity of *p*-aminophenol in the Sprague-Dawley rat: Effects on growth, reproduction and foetal development. *Food Cosmet Toxic* 27:691–698.

Calder IC, Yong AC, Woods RA, Crowe CA, Ham KN, and Tange JD (1979) The nephrotoxicity of *p*-aminophenol. II. The effect of metabolic inhibitors and inducers. *Chem-Biol Interact* 27:245–254.

Caterson RJ, Duggin GG, Horvath J, Mohandas J, and Tiller D (1978) Aspirin, protein transacetylation and inhibition of prostaglandin synthetase in the kidney. *Br J Pharmacol* 64:353–358.

Chrysant SG (1978) Renal functional changes induced by prostaglandin E_1 and indomethacin in the anesthetized dog. *Arch Int Pharmacodyn* 234:156–163.

Clive DM, and Stoff JS (1984) Renal syndromes associated with nonsteroidal anti-inflammatory drugs. *N Engl J Med* 310:563–572.

Crowe CA, Calder IC, Madsen NP, Funder CC, Green CR, Ham KN, and Tange JD (1977) An experimental model of analgesic-induced renal damage – Some effects of *p*-aminophenol on rat kidney mitochondria. *Xenobiotica* 7:345–356.

Davis BB, Mattammal MB, and Zenser TV (1981) Renal metabolism of drugs and xenobiotics. *Nephron* 27:187–196.

Dawson AG (1975) Effects of acetylsalicylate on gluconeogenesis in isolated rat kidney tubules. *Biochem Pharmacol* 24:1407–1411.

DeMaria AN, and Weir MR (2003) Coxibs – Beyond the GI tract: Renal and cardiovascular issues. *J Pain Symptom Manage* 25:S41–S49.

DeWitt DL, Meade EA, and Smith WL (1993) PHS synthase isozyme selectivity: The potential for safer nonsteroidal antiinflammatory drugs. *Am J Med* 95:40S–44S.

Duggin GG (1980) Mechanisms in the development of analgesic nephropathy. *Kidney Int* 18:553–561.

Ehrich EW, Dallob A, DeLepeleire I, Van Hecken A, Riendeau D, Yuan W, Porras A, Wittreich J, Seibold JR, DeSchepper P, Mehlisch DR, and Gertz BJ (1999) Characterization of rofecoxib as a cyclooxygenase-2 isoform inhibitor and demonstration of analgesia in the dental pain model. *Clin Pharmacol Ther* 65:336–347.

Eling TE, and Curtis JF (1992) Xenobiotic metabolism by prostaglandin H synthase. *Pharmacol Ther* 53:261–273.

Emeigh Hart SG, BeierschmittWP, Wyand DS, Khairallah EA, and Cohen SD (1994) Acetaminophen nephrotoxicity in CD-1 mice. I. Evidence of a role for in situ activation in selective covalent binding and toxicity. *Toxicol Appl Pharmacol* 126:267–275.

Emeigh Hart SG, Beierschmitt WP, Bartolone JB, Wyand DS, Khairallah EA, and Cohen SD (1991) Evidence against deacetylation and for cytochrome P450-mediated activation in acetaminophen-induced nephrotoxicity in the CD-1 mouse. *Toxicol Appl Pharmacol* 107:1–15.

Flower RJ (1974) Drugs which inhibit prostaglandin biosynthesis. *Pharmacol Rev* 26:33–67.

Flower RJ, and Vane JR (1972) Inhibition of prostaglandin synthesis in brain explains the anti-pyretic activity of paracetamol (4-acetamidophenol). *Nature* 240:410–411.

Fowler LM, Foster JR, and Lock EA (1993) Effect of ascorbic acid, acivicin and probenecid on the nephrotoxicity of 4-aminophenol in the Fischer 344 rat. *Arch Toxicol* 67:613–621.

Frolich JC (1997) A classification of NSAIDs according to the relative inhibition of cyclooxygenase isoenzymes. *Trends Pharmacol Sci* 18:30–34.

Gemborys MW, and Mudge GH (1981) Formation and disposition of the minor metabolites of acetaminophen in the hamster. *Drug Metab Dispos* 9:340–351.

Hao C-M, Komhoff M, Guan Y, Redha R, and Breyer MD (1999) Selective targeting of cyclooxygenase-2 reveals its role in renal medullary interstitial cell survival. *Am J Physiol* 277:F352–F359.

Hao C-M, Yull F, Blackwell T, Komhoff M, Davis LS, and Breyer MD (2000) Dehydration activates an NF-κB-driven, COX2-dependent survival mechanism in renal medullary interstitial cells. *J Clin Invest* 106:973–982.

Harris RC (2002) Cyclooxygenase-2 inhibition and renal physiology. *Am J Cardiol* 89:10D–17D.

Hickey EJ, Rage RR, Reid VE, Gross SM, and Ray SD (2001) Diclofenac induced *in vivo* nephrotoxicity may involve oxidative stress-mediated massive genomic DNA fragmentation and apoptotic cell death. *Free Rad Biol Med* 31:139–152.

Hu JJ, Lee M-J, Vapiwala M, Reuhl K, Thomas PE, and Yang CS (1993) Sex-related differences in mouse renal metabolism and toxicity of acetaminophen. *Toxicol Appl Pharmacol* 122:16–26.

Ichikawa I, and Brenner BM (1980) Importance of efferent arteriolar vascular tone in regulation of proximal tubule fluid reabsorption and glomerulotubular balance in the rat. *J Clin Invest* 65:1192–1201.

Josephy PD, Eling TE, and Mason RP (1983) Oxidation of *p*-aminophenol catalyzed by horseradish peroxidase and prostaglandin synthase. *Mol Pharmacol* 23:461–466.

Kaojarern S, Chennavasin P, Anderson S, and Brater DC (1983) Nephron site of effect of nonsteroidal anti-inflammatory drugs on solute excretion in humans. *Am J Physiol* 244:F134–F139.

Khan KNM, Venturini CM, Bunch RT, Brassard JA, Koki AT, Morris DL, Trump BF, Maziasz TJ, and Alden CL (1998) Interspecies differences in renal localization of cyclooxygenase isoforms: Implications in nonsteroidal antiinflammatory drug-related nephrotoxicity. *Toxicol Pathol* 26:612–620.

Khan KNM, Paulson SK, Verburg KM, Lefkowith JB, and Maziasz TJ (2002) Pharmacology of cyclooxygernase-2 inhibition in the kidney. *Kidney Int* 61:1210–1219.

Kincaid-Smith P (1986) Effects of non-narcotic analgesics on the kidney. *Drugs* 32 (Suppl. 4):109–128.

Komhoff M, Grone HJ, Klein T, Seyberth HW, and Nusing RM (1997) Localization of cyclooxygenase-1 and -2 in adult and fetal human kidney: Implication for renal function. *Am J Physiol* 272:F460–F468.

Kulmacz RJ, and Lands WEM (1985) Stoichiometry and kinetics of the interactions of prostaglandin H synthase with anti-inflammatory agents. *J Biol Chem* 260:12572–12578.

Kyle ME, and Kocsis JJ (1985) The effect of age on salicylate-induced nephrotoxicity in male rats. *Toxicol Appl Pharmacol* 81:337–347.

Laneuville O, Breuer DK, Dewitt DL, Hla T, Funk CD, and Smith WL (1994) Differential inhibition of human prostaglandin endoperoxide H synthases-1 and -2 by nonsteroidal anti-inflammatory drugs. *J Pharmacol Exp Ther* 271:927–934.

Ling BN, Webster CL, and Eaton DC (1992) Eicosanoids modulate apical Ca^{2+}-dependent K^+ channels in cultured rabbit principal cells. *Am J Physiol* 263:F116–F126.

Mattammal MB, Zenser TV, Brown WW, Herman CA, and Davis BB (1979) Mechanism of inhibition of renal prostaglandin production by acetaminophen. *J Pharmacol Exp Ther* 210:405–409.

McMurtry RJ, Snodgrass WR, and Mitchell JR (1978) Renal necrosis, glutathione depletion, and covalent binding after acetaminophen. *Toxicol Appl Pharmacol* 46:87–100.

Meade EA, Smith WL, and DeWitt DL (1993) Differential inhibition of prostaglandin endoperoxide synthase (cyclooxygenase) isozymes by aspirin and other non-steroidal anti-inflammatory drugs. *J Biol Chem* 268:6610–6614.

Mingatto FE, Santos AC, Uyemura SA, Jordani MC, and Curti C (1996) *In vitro* interaction of nonsteroidal anti-inflammatory drugs on oxidative phosphorylation of rat kidney mitochondria: Respiration and ATP synthesis. *Arch Biochem Biophys* 334:303–308.

Mitchell JR, McMurtry RJ, Statham CN, and Nelson SD (1977) Molecular basis for several drug-induced nephropathies. *Am J Med* 62:518–526.

Mitchell JA, Akarasereenont P, Thiemermann C, Flower RJ, and Vane JR (1993) Selectivity of nonsteroidal antiinflammatory drugs as inhibitors of constitutive and inducible cyclooxygenase. *Proc Natl Acad Sci USA* 90:11693–11697.

Moeckel GW, Zhang L, Fogo AB, Hao C-M, and Breyer MD (2003) COX2-activity promotes organic osmolyte accumulation and adaptation of renal medullary interstitial cells to hypertonic stress. *J Biol Chem* 278:19352–19357.

Mohandas J, Duggin GG, Horvath JS, and Tiller DJ (1981) Metabolic oxidation of acetaminophen (paracetamol) mediated by cytochrome P-450 mixed-function oxidase and prostaglandin endoperoxide synthetase in rabbit kidney. *Toxicol Appl Pharmacol* 61 252–259.

Moreno-Sanchez R, Bravo C, Vasquez C, Ayala G, Silveira LH, and Martinez-Lavin M (1999) Inhibition and uncoupling of oxidative phosphorylation by nonsteroidal anti-inflammatory drugs. *Biochem Pharmacol* 57:743–752.

Mugford CA, and Tarloff JB (1995) Contribution of oxidation and deacetylation to the bioactivation of acetaminophen *in vitro* in liver and kidney from male and female Sprague-Dawley rats. *Drug Metab Dispos* 23:290–294.

Mugford CA, and Tarloff JB (1997) The contribution of oxidation and deacetylation to acetaminophen nephrotoxicity in female Sprague-Dawley rats. *Toxicol Lett* 93:15–22.

Murray MD, and Brater DC (1993) Renal toxicity of the nonsteroidal anti-inflammatory drugs. *Annu Rev Pharmacol Toxicol* 32:435–465.

Muther RS, Potter DM, and Bennett WM (1981) Aspirin-induced depression of glomerular filtration rate in normal humans: Role of sodium balance. *Ann Int Med* 94:317–321.

Nanra RS (1983) Renal effects of antipyretic analgesics. Am J Med 75:70–81.

Nanra RS, and Kincaid-Smith P (1970) Papillary necrosis in rats caused by aspirin and aspirin-containing mixtures. *Br Med J* 3:559–561.

Newton JF, Kuo C-H, Gemborys MW, Mudge GH, and Hook JB (1982) Nephrotoxicity of *p*-aminophenol, a metabolite of acetaminophen, in the Fischer 344 rat. *Toxicol Appl Pharmacol* 65:336–344.

Newton JF, Yoshimoto M, Bernstein J, Rush GF, and Hook JB (1983) Acetaminophen nephrotoxicity in the rat. I. Strain differences in nephrotoxicity and metabolism. *Toxicol Appl Pharmacol* 69:291–306.

Newton JF, Pasino DA, and Hook JB (1985a) Acetaminophen nephrotoxicity in the rat: Quantitation of renal metabolic activation *in vivo*. *Toxicol Appl Pharmacol* 78:39–46.

Newton JF, Kuo C-H, DeShone GM, Hoefle D, Bernstein J, and Hook JB (1985b) The role of *p*-aminophenol in acetaminophen-induced nephrotoxicity: Effect of bis(*p*-nitrophenyl)phosphate on acetaminophen and *p*-aminophenol nephrotoxicity and metabolism in Fischer 344 rats. *Toxicol Appl Pharmacol* 81:416–430.

Owen RA, and Heywood R (1983) Age-related susceptibility to aspirin-induced nephrotoxicity in female rats. *Toxicol Lett* 18:167–170.

Palmer BF (1995) Renal complications associated the use of nonsteroidal anti-cinflammatory agents. *J Invest Med* 43:516–533.

Penning TD, Talley JJ, Bertenshaw SR, Carter JS, Collins PW, Docter S, Graneto MJ, Lee LF, Malecha JW, Miyashiro JM, Rogers RS, Rogier DJ, Yu SS, Anderson GD, Burton EG, Cogburn JN, Gregory SA, Koboldt CM, Perkins WE, Seibert K, Veenhuizen AW, Zhang YY, and Isakson PC (1997) Synthesis and biological evaluation of the 1,5-diarylpyrazole class of cyclooxygenase-2 inhibitors: Identification of 4-[5-(4-methylphenyl)-3-(trifluoromethyl)-1H-pyrazole-1-yl]benzenesulfonamide (SC-58635, Celecoxib). *J Med Chem* 40:1347–1365.

Pirani CL, Valeri A, D'Agati V, and Appel GB (1987) Renal toxicity of nonsteroidal anti-inflammatory drugs. *Contr Nephrol* 55:159–175.

Prescott LF (1982) Analgesic nephropathy: A reassessment of the role of phenacetin and other analgesics. *Drugs* 23:75–149.

Quintanilla A, and Kessler RH (1973) Direct effect of salicylates on renal function in the dog. *J Clin Invest* 52:3143–3153.

Robinson MJ, Nichols EA, and Taitz L (1967) Nephrotoxic effect of acute sodium salicylate intoxication in the rat. *Arch Pathol* 84:224–226.

Rocha GM, Michea LF, Peters EM, Kirby M, Xu Y, Ferguson DR, and Burg MB (2001) Direct toxicity of nonsteroidal anti-inflammatory drugs for renal medullary cells. *Proc Natl Acad Sci USA* 98:5317–5322.

Roig F, Llinas MT, Lopez R, and Salazar FJ (2002) Role of cyclooxygenase-2 in the prolonged regulation of renal function. *Hypertension* 40:721–728.

Sabatini S (1988) Analgesic-induced papillary necrosis. *Semin Nephrol* 8:41–54.

Schlondorff D (1993) Renal complications of nonsteroidal anti-inflammatory drugs. *Kidney Int* 44:643–653.

Schreiner GE, McAnally JF, and Winchester JF (1981) Clinical analgesic nephropathy. *Arch Intern Med* 141:349–357.

Silva P (1987) Renal fuel utilization, energy requirements, and function. *Kidney Int* 32 (Suppl. 22):S9–S14.

Tange JD, Ross BD, and Ledingham JGG (1977) Effects of analgesics and related compounds on renal metabolism in rats. *Clin Sci Mol Med* 53:485–492.

Tarloff JB, Goldstein RS, Morgan DG, and Hook JB (1989) Acetaminophen and *p*-aminophenol nephrotoxicity in aging male Sprague-Dawley and Fischer 344 rats. *Fundam Appl Toxicol* 12:78–91.

Tarloff JB, Khairallah EA, Cohen SD, and Goldstein RS (1996) Sex- and age-dependent acetaminophen hepato- and nephrotoxicity in Sprague-Dawley rats: Role of tissue accumulation, nonprotein sulfhydryl depletion, and covalent binding. *Fundam Appl Toxicol* 30:13–22.

Terragno NA, Terragno DA, and McGiff JC (1977) Contribution of prostaglandins to the renal circulation in conscious, anesthetized, and laparotomized dogs. *Circ Res* 40:590–595.

Tokumitsu Y, Lee S, and Ui M (1977) In vitro effects of nonsteroidal anti-inflammatory drugs on oxidative phosphorylation in rat liver mitochondria. *Biochem Pharmacol* 26:2101–2106.

Vane JR, Bakhle YS, and Botting RM (1998) Cyclooxygenases 1 and 2. *Annu Rev Pharmacol Toxicol* 38:97–120.

Vane J (1994) Towards a better aspirin. *Nature* 367:215–216.

Verbeeck RK, Blackburn JL, and Loewen GR (1983) Clinical pharmacokinetics of non-steroidal anti-inflammatory drugs. *Clin Pharmacokinet* 8:297–331.

Whelton A, and Hamilton CW (1991) Nonsteroidal anti-inflammatory drugs: Effects on kidney function. *J Clin Pharmacol* 31:588–598.

Wiseman EH, and Reinert H (1975) Anti-inflammatory drugs and renal papillary necrosis. *Agents Actions* 5:322–325.

Yang T, Singh I, Pham H, Sun D, Smart A, Schnermann JB, and Briggs JP (1998) Regulation of cyclooxygenase expression in the kidney by dietary salt intake. *Am J Physiol* 274:F481–F489.

Zenser TV, Mattammal MB, Herman CA, Joshi S, and Davis BB (1978) Effect of acetaminophen on prostaglandin E2 and prostaglandin F2α synthesis in the renal inner medulla of rat. *Biochim Biophys Acta* 542:486–495.

21

Mycotoxins Affecting the Kidney

Evelyn O'Brien and Daniel R. Dietrich

INTRODUCTION

Although their existence and, indeed, some of their effects on human and animal health have been known for some time, the first mycotoxin was positively identified and chemically characterized in the early 1960s (Nesbit et al., 1962). Mycotoxins have been linked to one of the "ten plagues of Egypt" as described in both the Ipuwer papyrus and the Bible. In recent times, numerous additions have been made to the ever-growing list of mycotoxins and their analogues. This list currently encompasses more than 300 substances which have been isolated and chemically characterized from pure cultures (Steyn, 1995).

WHAT ARE MYCOTOXINS? A BRIEF OVERVIEW

Mycotoxins are secondary metabolites produced by a variety of mould and fungi species. While their actual function in moulds and fungi has not been definitively identified, one possibility is that they enable competition with other microorganisms for nutrients and space. For example, it has been suggested that production of ochratoxin A and citrinin by *Aspergillus* and *Penicillium* species may interfere with the uptake of iron in competing microorganisms (Stormer and Hoiby, 1996). A further suggestion is that the production of mycotoxins could contribute to the generation of favorable germination conditions for fungal spores. These theories remain to be clarified.

Of the 300 substances identified to date, approximately 20 may be found with disturbing regularity in foodstuffs and animal fodder (Steyn, 1995). The majority of these are produced by moulds of the *Aspergillus*, *Penicillium*, *Fusarium*, *Alternaria*, and *Claviceps* species and have been implicated in biological effects as diverse as mutagenicity, teratogenicity, neurotoxicity, immunotoxicity, to name but a few, in both animals and humans. In contrast, some mycotoxins have also been suggested to have beneficial activities, displaying antitumor, antimicrobial, and cytotoxic effects.

Many crops that are used for the production of both animal and human foodstuffs harbor fungi which, given the correct conditions of humidity and temperature, produce mycotoxins, either in the field or during suboptimal storage. Thus, mycotoxin contamination is a particular problem in moist, warm climates. Contaminated grain, which is harvested and milled, finds its way into both animal and human food supplies and, due to their heat stability, mycotoxins are not destroyed by industrial processing or domestic cooking. Whereas in former times dietary intake of mycotoxins depended on several factors, including culture, socioeconomic grouping, and local climatic conditions (Studer-Rohr et al., 2000), nowadays, due to the worldwide availability of practically every crop, the importance of geographical factors has diminished, making the complete avoidance of mycotoxin consumption practically impossible.

Structurally, mycotoxins form a diverse group of organic compounds of low molecular weight, which vary from simple compounds with a carbon chain length of four (e.g., moniliformin), to complex substances consisting of several ring structures (e.g., phomopsin) (Culvenor et al., 1989). It is likely that the myriad of biological effects reported for mycotoxins can probably be attributed to this structural diversity, at least in part (Figure 21.1).

THE KIDNEY AS A TARGET ORGAN

As a prerequisite for its role as an organ of excretion, reabsorption, and general homeostasis, the kidney has an extensive bloodflow, receiving

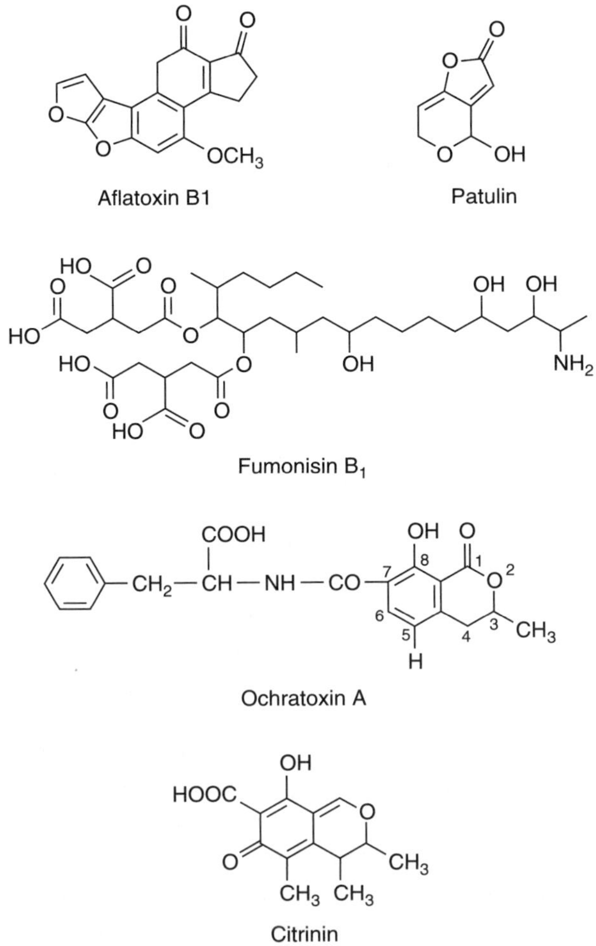

Figure 21.1 Molecular structures of the mycotoxins described in this chapter.

approximately 1.2 l/min and filtering on average 125 ml plasma/minute. The processes of reabsorption and secretion, particularly of organic acids and bases, may, however, lead to the accumulation of toxins within the tubules, making this vital organ more susceptible to toxic insults than other organs. The very nature of the kidney makes early clinical diagnosis of mycotoxin-induced nephrotoxicity difficult, if not impossible. Therefore, much research is now focusing on determining the toxic mode of action of these compounds both *in vivo* and *in vitro*, with particular emphasis on defining the species and sex differences in toxicity that are characteristic of much mycotoxin-mediated renal toxicity.

Not least among the factors crucial in the development of renal toxicity is the presence of multiple organic anion transporters (OATs). These transporters actively eliminate drugs and toxic compounds, and their metabolites, which may become hazardous upon accumulation. Most of these transporters are confined to the proximal tubule. These highly developed transport functions, coupled with its concentrating ability, render the proximal tubule as the region of the kidney most at risk from toxic insult. Numerous organic anion and organic cation transporter systems have been found and molecularly characterized with respect to affinity, kinetics, and inhibitor specificity. Some, such as the OAT1, are classical *p*-aminohippuric acid (PAH)/dicarbonate exchangers, which mediate high-affinity PAH uptake in a sodium-dependent manner. Others function as organic anion/glutathione antiports (Oatp1) or as peptide/H^+ symports (PEPT2).

The mechanism and actual physiological function of many transporters, however, remains to be determined. A complete review of organic anion transporter molecules is beyond the intention and scope of this manuscript, however several excellent reviews and papers are available (Koepsell et al., 1999; Russel et al., 2002; Uwai et al., 1998; van Aubel et al., 2000). These transporter systems have a particular relevance for the toxicity of mycotoxins, several of which have been demonstrated to be transported by renal organic anion transporters (Loe et al., 1997; Rappa et al., 1997; Tsuda et al., 1999). Indeed, the distribution of these transporters may play a role in the preferential toxicity of certain mycotoxins for specific organ systems.

NEPHROTOXIC MYCOTOXINS

Almost all mycotoxins examined to date possess at least some nephrotoxic potential. In many cases, this is dependent on species and mycotoxin concentration, and in some cases is secondary to the effects on other organ systems. Therefore, a complete and comprehensive review of all mycotoxins that have been demonstrated to cause renal damage is beyond the intention and scope of this text. The following sections will therefore concentrate on those known to have the kidney as one of their primary sites of action.

Aflatoxin B_1

After the discovery of their causal role in the deaths of thousands of turkeys, ducklings, and chicks in England in 1960, aflatoxins became the first mycotoxins to be extracted, identified, and chemically characterized (Asao et al., 1963; Nesbit et al., 1962). Up until the beginning of 2002, over 5,000 research papers on aflatoxins had been published, making them the most studied of all mycotoxins.

Aflatoxins are a family of substituted coumarins (Figure 21.1) containing a fused dihydrofuran moiety. They are produced mainly by *Aspergillus flavus* and *Aspergillus parasiticus*, but also by certain *Penicillium* and *Rhizopu* strains (Medicine, 2002; Searle, 1976). The family consists of four main members, aflatoxin B_1, G_1, G_2, and B_2. Aflatoxin B_1 (AFB_1) is the major toxin produced in culture and also the most toxic of the four – it is classified as a Group I hepatocarcinogen in humans by the International Agency for Research on Cancer (IARC) (IARC, 1993). Consequently, it is the best studied of the group.

As is the case for many, if not all, mycotoxins, large species differences in susceptibility are evident, and the route of administration is important. Of the common laboratory species, the guinea pig is one of the most sensitive to the effects of AFB_1, with an observed LD_{50} of 2 mg/kg body weight via oral gavage and intraperitoneal administration yields an LD_{50} of 1.4 mg/kg (Netke et al., 1997). More recently, exposure to feed containing 3 mg/kg AFB_1 has been shown to be sufficient to cause severe renal damage in chicks (Valdivia et al., 2001). An LD_{50} was, however, not determined in this study. Rats are slightly less sensitive, with an LD_{50} of 7 mg/kg body weight in feeding experiments (Rati et al., 1991), although this has been determined to be age-dependent, with younger F344 rats being less sensitive (LD_{50} >150 mg/kg body weight) (Croy, 1981). Large differences within species have been reported, with intraperitoneal LD_{50} values for various allogenic mouse strains ranging from 9 to 60 mg/kg body weight (Hayes and Campbel, 1986; Mako et al., 1971). In contrast, another study reported no mortality in four different inbred mouse strains (C57B1/6, CBA/J B10A, and Balb/c) despite exposure to 60 mg/kg body weight AFB_1 (Almeida et al., 1996). Similarly, non-inbred CD-1 Swiss mice display an LD_{50} greater than 150 mg/kg body weight (Croy, 1981). These species and strain differences appear to arise due to genetic variability in the expression levels of the cytochrome P450 mixed function oxidases, of which at least five are involved in the conversion of AFB_1 to its reactive metabolite, AFB_1-8,9-epoxide (Autrup et al., 1996; Eaton and Gallagher, 1994).

The relative insensitivity of mice to AFB_1-mediated hepatotoxicity toxicity has been attributed to a higher rate of transformation of AFB_1 to its demethylated derivative AFP_1 and to other water-soluble metabolites in mice than in other species, resulting in faster elimination of the mycotoxin and its metabolites (Almeida et al., 1996; Ramsdell and Eaton, 1990). These observations also correlate well with those of Wong and Hsieh (1980) who demonstrated that the mouse displays a high first-order elimination constant for AFB_1, whereas in the rat, the equilibrium transfer rate constant favors a relatively high concentration in plasma and hence a longer half-life. In contrast, Autrup and colleagues (1996), who noted a threefold higher concentration of adducts formed in murine kidney than in liver following AFB_1 exposure, have proposed that mice may in fact be more susceptible to its nephrotoxic effects than its hepatotoxic effects.

Until relatively recently, the bulk of research carried out has focused on the hepatotoxic potential of $AFB_{1,}$ which greatly overshadows the renal toxicity. Early observations of renal hypertrophy, tubular congestion, and epithelial degeneration, such as those of Newberne and coworkers (1964), were made for completeness in studies on hepatotoxicity rather than as detailed investigations of renal toxicity. Hayes (1980) reported that bovine renal tissue retained the highest concentration of AFB_1 and its metabolite M_1 following feeding with radioactively-tagged AFB_1. In contrast, in pigs, most of the administered AFB_1 was to be found in the liver following oral dosing, with only minor residues in renal tissue (Lüthy et al., 1980). Based on these observations and the findings of Wong and Hsieh (1980) that species susceptibility can be correlated to tissue distribution, the first study designed specifically to investigate the renal toxicity of AFB_1 in rats was carried out by Grosman and coworkers (1983). They described decreased glomerular filtration rate, decreased tubular glucose reabsorption, and decreased tubular transport of *p*-aminohippurate (PAH) in Wistar rats following a single intraperitoneal dose of AFB_1 (100 μg/kg body weight). These authors also observed increased urinary excretion of both sodium and potassium and a twofold increase in urinary γ-glutamyl transferase activity, which persisted for more than 48 hours after injection. These authors hypothesized that the nephrotoxic effects observed in rats following AFB_1 exposure were possibly due to effects on the glomerular basement membrane. This thesis is supported by the findings of Valdiva and coworkers (2001), who noted a thickening of the glomerular basement membrane following AFB_1 exposure in chicks.

Pathology

Morrissey and coworkers (1987) described characteristic histopathological changes in the kidneys of Sprague-Dawley rats exposed to 2 mg/kg body weight/day AFB_1 for 4 days, the most obvious of which was a dark red band between the medulla and the cortex, coupled with a relatively pale renal cortex. Closer examination revealed tubular necrosis, particularly in the inner cortex, with some pycnotic nuclei. Most nuclei were swollen and many displayed a clear center with the chromatin arranged around the periphery. This degeneration was observed to progress even following discontinuation of AFB_1 exposure, with the inner cortex becoming more necrotic, an increase in the number of pycnotic nuclei, and the presence of large quantities of nuclear debris in hematoxylin- and eosin-stained paraffin sections. These findings were supported by Rati et al. (1991) who observed nuclear enlargement in the tubular epithelium with proliferating anaplastic cells in the cortical region following subacute feeding experiments over 36 weeks with a 24 week washout phase. These authors also reported that one animal presented with massively necrotic and timorous kidneys. However, the tumor did not appear to arise as a consequence of the anaplastic cells in the cortical regions as none of the other similarly exposed animals presented with tumors up to 24 weeks following the exposure phase. Indeed, AFB_1 has not been associated with the development of renal tumors in any species tested to date. Guinea pigs respond to AFB_1 with minimal acute multifocal nephrosis of the renal tubules. This is only present following exposure to the LD_{50} dose (Netke et al., 1997).

All of the aforementioned renal effects occurred in combination with one or more severe hepatic lesions, including hepatoma and hepatocarcinoma (Rati et al., 1991), centrilobular necrosis, endothelial cell degeneration (Netke et al., 1997), and bile duct proliferation (Morrissey et al., 1987).

Mechanism of Action

It is generally accepted that both the hepatotoxicity and renal toxicity of AFB_1 are due to the generation of a reactive metabolite, namely AFB_1-8,9-epoxide (Almeida et al., 1996; Autrup et al., 1996; Busby and Wogan,

1984). This metabolite is produced rapidly by the action of at least five members of the mixed function oxidase family (Autrup et al., 1996; Eaton and Gallagher, 1994) and its concentration reaches a maximum six to twelve hours post dosing in F344 rats (Chou et al., 1993). AFB_1-8,9-epoxide reacts with DNA to yield the 8,9-dihydro-8-(N7-guanyl)-9-hydroxyaflatoxin B_1 adduct (AFB_1-N^7-Gua), which has been positively correlated with DNA strand breaks and hepatic tumor development in the rat (Bechtel, 1989). This adduct has also been shown to be positively correlated with the development of renal lesions in the mouse (Chou et al., 1997). In 1981, Croy and Wogan compared the generation of AFB_1-N^7-Gua in the livers and kidneys of both rats (F344) and mice (CD-1 Swiss). They reported a tenfold greater level of AFB_1 modification per nucleotide residue in rat liver than in rat kidney. In contrast, in the mouse, modification levels were threefold higher in the kidney than in the liver. This correlates well with both the relative species and organ sensitivities. Chou and colleagues (1993, 1997) have confirmed these results for the rat, but in contrast found the number of adducts in mouse liver and kidney to be similar. This difference may, however, be due to the use of a different mouse strain ($B6C3F_1$) in this study.

As a result of observations that the toxicity of AFB_1 correlates with the metabolic ability of a particular species or strain (Eaton and Gallagher, 1994), several groups have investigated the effects of reduced caloric intake on AFB_1-mediated toxicity. Indeed, a reduction of food intake results in reduced metabolic activity and hence reduced carcinogenic potential for a number of genotoxic substances (Shaddock et al., 1993). In the case of AFB_1, a similar phenomenon has been demonstrated with respect to hepatotoxicity and renal toxicity in rats (Chou et al., 1993) and in mice (Almeida et al., 1996; Chou et al., 1997), respectively. A 40% reduction in dietary caloric intake significantly reduces hepatic activation of AFB_1 in B63CF1 mice and this correlates with a reduction in the total AFB_1–DNA adduct formation in the mouse kidney (Chou et al., 1997). The authors suggest this to be due to a decrease in the activity of CYP2C11-dependent AFB_1 metabolizing enzyme or an increase in AFB_1-specific glutathione *S*-transferase activity (Chen et al., 1995). This theory is supported by the observation that glutathione reduces the ability of AFB_1-8,9-epoxide to bind to DNA and proteins, both *in vivo* and *in vitro*, and its depletion is associated with increased AFB_1-N^7-Gua adduct numbers and increased tumor rates in rats (Gopolan et al., 1993) and mice (Autrup et al., 1996).

Exposure to a single dose of AFB_1 has been demonstrated to reduce renal and hepatocellular proliferation in F344. In the liver, this reduction is then compensated by extensive and massive (140 to 250% of control levels) regenerative cell proliferation following termination of exposure (Chou et al., 1993). This proliferative response, which has been suggested to form the basis for the genesis of hepatic tumors, could be prevented by restriction of caloric intake. In the rat kidney, however, the rate of DNA synthesis was not increased above control levels and caloric restriction had no effect on the level of DNA synthesis. These results support the proposed organ specificity of AFB_1 toxicity in the rat.

In a study carried out by Grosman and colleagues (1983), AFB_1 was shown to impair the function of the organic acid transport system as measured by reduced PAH accumulation in renal slices. This was coupled with increased intracellular sodium and decreased intracellular potassium, suggesting a loss of normal control of membrane permeability; which could result in the observed renal effects. This is supported by a further study, which showed AFB_1 to dose-dependently reduce Na^+/P_i cotransport in opossum kidney (OK) cells, which were used as a model for proximal renal epithelial cells. This effect does not appear to be due to a generalized inhibition of sodium cotransport as sodium-dependent uptake of L-alanine was not affected, but rather a specific inhibitory effect on renal reabsorption of inorganic phosphate (Glahn et al., 1994). The authors suggest that this effect is also due to the conversion of AFB_1 to its active metabolite AFB_1-8,9-epoxide, which then binds to DNA and RNA. This alters cellular functions, such as hormone synthesis, the responsiveness of renal tissue to hormones such as insulin and parathyroid hormone, and the activity of protein kinases, adenylate cyclases, and cyclic nucleotide phosphodiesterases, leading to the observed renal toxicity.

In conclusion, the nephrotoxicity and hepatotoxicity of AFB_1 appear both to be mediated by the generation of the major active metabolite AFB_1-8,9-epoxide. However, whereas the downstream events leading to the generation of hepatic tumors have been relatively well characterized, much work remains to be done to deduce the order of events involved in its nephrotoxicity.

Citrinin

The organic anion citrinin is a benzopyran metabolite produced by toxic strains of *Penicillium* and *Aspergillus* species. As such, it can be coproduced with ochratoxin A and an isocoumarin ring is common to the structure of both (Figure 21.1). Both of these mycotoxins have been associated with the development of Balkan endemic nephropathy and urothelial tumors in humans. This is discussed in the section dealing with ochratoxin A and mycotoxin interactions (see below). Originally, citrinin was suggested for use as an antibiotic due to its marked antibacterial activity (Hetherington and Raistrick, 1931). However, animal tests demonstrated it to be severely nephrotoxic, with the detrimental effects far outweighing any potential benefits (Ambrose and DeEds, 1946; Blanpin, 1959). Citrinin has been demonstrated to be acutely toxic in several species, including rabbit, rat, mouse, and hamster, with LD_{50} values for intraperitoneal administration of 50, 64, 80, and 75 mg/kg body weight, respectively (Hanika and Carlton, 1983; Jordan and Carlton, 1977, 1978; Jordan et al., 1978b). Toxicity varies considerably with route of administration, with 134 mg/kg representing the oral LD_{50} in rabbits.

Some hepatotoxic effects have been reported for citrinin, however the lethal effects are largely due to a severe nephrotoxicity that is very similar in manifestation and progression in all species tested to date. It is characterized clinically by an increased excretion of dilute urine, which is thought to result from an impaired capacity for urine concentration. Other features include reduced glomerular filtration rate and renal blood flow resulting in increased blood urea nitrogen concentration, urinary lactic acid dehydrogenase, aspartate amino transferase, and isocitric dehydrogenase activities, as well as proteinuria, glucosuria, lowered urinary specific gravity, and the presence of necrotic cells in the urinary sediment (Jordan et al., 1978a; Kogika et al., 1996; Lockard et al., 1980).

These functional changes are associated with acute tubular necrosis. The location of the most severe tubular damage varies from species to species. In the mouse (Jordan and Carlton, 1977) and in hamsters (Jordan et al., 1978b) the distal portion of the nephron is mostly affected, whereas proximal segments are most at risk in the rat (Lockard et al., 1980), rabbit (Ambrose and DeEds, 1946), guinea pig (Thacker et al., 1977), and pig (Krogh et al., 1973). Despite this, the specific pathologies resulting from acute citrinin exposure are remarkably similar in all species. Initial pathological observations report the kidneys of acutely exposed mice to be swollen and pale, with stippling of the capsule cortex and outer medulla (Jordan and Carlton, 1977; Hanika and Carlton, 1983). Similar findings have been reported for other tested species (Hanika and Carlton, 1983; Jordan and Carlton, 1978; Jordan et al., 1978b).

Closer histopathological examinations of the kidneys of exposed animals of several species reveals a similar picture of citrinin-related damage, including necrosis and desquamation of renal epithelial cells of the proximal tubules, basement membrane thickening, tubule dilation, and proliferation of cells in the interstitium. In rats, the primary site of action appears to be the renal cortex and the outer part of the renal medulla, particularly the straight segments and the distal convoluted tubule (Jordan and Carlton, 1978), with the tubules in these areas displaying marked necrosis and deposition of protein casts. Almost identical observations have been made in hamsters (Jordan et al., 1978b).

Necrosis of individual cells or small groups of renal tubular cells has also been described in the straight segments and the distal convoluted tubule of the mouse (Jordan and Carlton, 1977) and in anesthetized dogs. In dogs, this is confined to the S_2 cells and is characterized further by a loss of brush border and apical vacuoles and a displacement of the organelles away from the luminal margin (Krejci et al., 1996). The protein casts observed in mice following citrinin exposure are, in contrast to the observations made in the rat, more commonly located to the tubules of the inner medullary area (Jordan and Carlton, 1977). This variation in the primary site of citrinin-mediated damage may be due to species differences in the function of various segments. Experiments carried out by Hanika and colleagues (Hanika and Carlton, 1983) demonstrated the renal injury caused by exposure to citrinin to be nonprogressive, and indeed reversible, at least in the rabbit. Renal tubule regeneration, starting in the convoluted segments and proceeding into the straight portions, could be observed as early as 3 days after exposure and recovery was almost complete by the seventh day. A similar regenerative response has also been made in one study using rats (Lockard et al., 1980). Hanika and colleagues also described the occurrence of a slight heterophilic inflammatory response in the kidneys of exposed rabbits. This has not been described for other species and the authors suggest that this may be responsible for the mild interstitial fibrosis evident in some animals following recovery.

A curious observation made by Jordan and Carlton (1977) is that although the number and severity of citrinin-induced renal lesions and pathology in mice can be positively correlated with the dose, multiple injections of a similarly toxic dose increases neither the number nor the severity of observed lesions nor does it increase mortality. These authors suggest two possible explanations. A citrinin-sensitive population of cells may be destroyed by the initial dose and the remaining cells are insensitive to subsequent doses of citrinin. Alternatively, the initial citrinin exposure may induce increased production of enzymes responsible for citrinin detoxification leading to a more rapid metabolism and excretion of subsequent doses.

Mechanism of Action

The mechanism of action of citrinin remains unclear. Direct effects on the kidney have been observed and the extrarenal vascular and parasympathomimetic (e.g., reduction in blood pressure) and local irritant actions of citrinin noted in *in vivo* experiments have also been suggested to indirectly affect renal function and ultrastructure (Ambrose and DeEds, 1946; Hanika and Carlton, 1983; Krejci et al., 1996). Indeed, some of the electrolyte disturbances that have been associated with renal effects may actually be due to dehydration as a result of emesis and diarrhea immediately following citrinin administration (Kitchen et al., 1977; Krejci et al., 1996).

Studies carried out by Berndt and Hayes (1982; Berndt, 1983) indicated the toxicity of citrinin to be related to its active tubular transport and accumulation in the kidney. Pretreatment of rats with the organic anion transporter blocker probenecid significantly reduces renal but not hepatic accumulation. In this study, probenecid was demonstrated not only to reduce citrinin accumulation in renal tissue, but also to ameliorate the effects of the toxin on rat mortality. These results correlated well with experiments into renal function, as measured by renal slice transport, carried out *in vitro* by the same authors, which demonstrated that accumulation of PAH into renal slices could be suppressed by citrinin and that this effect could be significantly reduced by preincubation with probenecid. These authors also described a reduced toxicity of citrinin in newborn rats and hypothesized this to be a result of either the lower level of renal transport (Tune, 1974) of the parent compound or to reduced renal or hepatic metabolism of citrinin to reactive metabolites in newborns.

In more recent studies, Chagas et al. (1992a, 1992b, 1995) have demonstrated citrinin to disrupt normal management of the mitochondrial membrane and hence normal membrane potential function in both liver and kidney mitochondria. The effects were primarily directed toward monovalent cation permeability, resulting in a disruption of the fluidity of the inner mitochondrial membrane. In these investigations, the kidney was clearly far more sensitive to these effects and this correlates well with the known organ specificity of citrinin toxicity.

Citrinin-mediated cytotoxicity has been demonstrated in an *in vitro* test system using Madin-Darby bovine kidney (MDBK) cell line and primary fetal bovine kidney (PFBK) cells (Yoneyama et al., 1986). Although these experiments were carried out using extremely high concentrations ($EC_{50} = 3.8 \times 10^{-4}$ M), and their relevance to the *in vivo* situation is hence questionable, the authors reported several findings that reflected those seen *in vivo*, including cellular swelling and loss of cell–cell contact. Moreover, PFBK cells were noted to require tenfold higher concentrations of citrinin to display similar levels of cytotoxicity to that observed in the continuous cell line, although primary cells would generally be expected to be more sensitive to toxic insult than continuous cell lines. The authors did not attempt to explain this difference, however there are two possible explanations. One possibility is that immature animals have an inherently lower expression level of transporter molecules, thus allowing reduced access of citrinin to the mitochondria. Alternatively, the preparation and culture procedure itself may reduce the number or activity of the membrane-bound transporters. Either way, the net result is a reduction in intracellular citrinin accumulation and an apparently reduced sensitivity. This supports the importance of the role of organic anion transporters in the renal toxicity mediated by citrinin and other mycotoxins.

The relative roles of the parent compound and reactive metabolites in citrinin-mediated toxicity remain to be definitively elucidated. Berndt and colleagues (Berndt and Hayes, 1982; Berndt, 1983) support the theory that citrinin itself is responsible as no renal metabolism of citrinin could be detected in their investigations. In contrast, the presence of the urinary metabolite dihydrocitrione has been demonstrated (Dunn et al., 1983), however, only 10 to 20% of citrinin is excreted as metabolites hence the possible role in renal toxicity is as yet, unknown.

In summary, the nephrotoxic actions of citrinin appear to result from active uptake into the kidney, probably of the parent compound, which then impairs the normal regulation of mitochondrial metabolism.

Ochratoxins

Like many other mycotoxins, ochratoxins belong to those that are produced by *Penicillium* and *Aspergillus* species (Scott et al., 1972). They may be produced in the field or, more commonly, as a result of

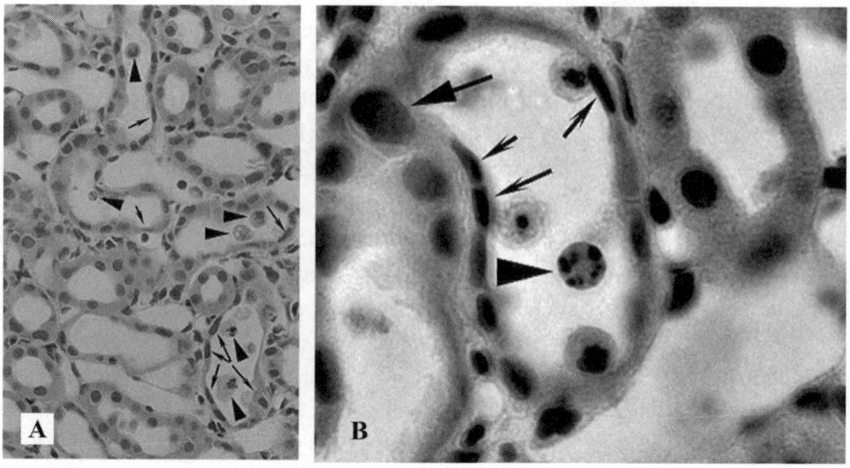

Figure 21.2 Hematoxylin-eosin stained kidney section of a male rat, treated with 1 mg/kg ochratoxin A for 7 days: (a) demonstrating a high number of necrotic (arrowheads) exfoliated or regenerative (small arrows) epithelial tubules cells in the inner part of the cortex (original magnification: ×400); (b) demonstrating exfoliated cells seemingly undergoing apoptotic necrosis (arrowheads). Regenerative epithelial cells (small arrows) as well as cells (large arrows) with giant nuclei can be observed within the same affected tubule (original magnification: ×800). (Reprinted from Rásonyi et al. (1999), with permission from Elsevier Science).

improper or suboptimal storage conditions of grain, coffee, dried fruits, etc., and are known to be common contaminants of human foodstuffs such as bread, cereals, beer, wine, etc. (Speijers and van Egmond, 1993; Studer-Rohr, 1995; Wolff et al., 2000). Three different ochratoxins, (A, B, and C) with differing toxicities have been isolated and characterized. Of these, ochratoxin A (OTA) is both the most commonly detected and the most toxic, followed by ochratoxin B (OTB) and C (OTC) (Li et al., 1997; van der Merwe et al., 1965). OTA is a weak organic acid and consists of a dihydroisocoumarin moiety linked to L-phenylalanine (Figure 21.1) and is classified under Group 2B (possibly carcinogenic to humans) by the IARC (IARC, 1993). A variety of other toxic effects, including teratogenesis (Fukui et al., 1992; Hood et al., 1976; Shreeve et al., 1977) and immunotoxicity (Harvey et al., 1992; Müller et al., 1999a, 1999b; Stoev et al., 2000), in several species have been attributed to OTA. It is, however, the renal toxicity, and in particular renal carcinogenic effects, that have commanded the most attention. OTB, differing from OTA only by the lack of a chloride at the 5-position on the isocoumarin-ring moiety (Figure 21.1), is significantly (10- to 20-fold) less toxic *in vitro* and *in vivo.*

OTA generally displays a relatively low acute toxicity, although large species differences are apparent in sensitivity (oral LD_{50} values range from approximately 20 and 46 to 58 mg/kg body weight in rats and mice, respectively, to 0.2 to 1 mg/kg body weight in pigs, cats, rabbits, and dogs). In contrast, it is the subchronic and chronic effects of OTA that are of greatest concern. OTA is considered causal for the nephropathies observed in several species of agricultural animals, particularly in pigs (Krogh et al., 1976; Stoev et al., 1998a), resulting in huge financial losses in agriculture and consequently in the food industry.

Functional deficits resulting from OTA exposure include increased urinary concentrations of glucose, proteins, leucine aminopeptidase, and γ-glutamyl transferase, coupled with a decrease in serum cholesterol and protein concentrations. Creatinine clearance rates are reduced and urinary specific gravity is markedly reduced. This decreased ability to produce concentrated urine is a direct consequence of impaired tubular function (Stoev et al., 1998b).

The renal pathological lesions caused by chronic OTA exposure, characterized by progressive tubular atrophy coupled with proliferation of fibroblastic connective tissue (Krogh et al., 1976) and progressing to activation and proliferation of vascular endothelial and adventitial cells (Stoev et al., 1998a), were first observed in pigs, and are described in the classical Danish model of mycotoxic nephropathy. This nephropathy results in reduced food intake and hence reduced weight gain by the animals. Furthermore, detection of OTA in meat and meat products leads, in certain countries (e.g., in Scandinavia), to the condemnation of this produce as unfit for consumption.

Chronic exposure to OTA is also thought to be involved in the etiology of two human kidney disease states, namely Balkan endemic nephropathy (BEN) and urothelial tumors. BEN is a chronic progressive kidney disease, first described for populations in the lowland regions of the river Danube. It is characterized by progressive tubulointerstitial nephropathy, leading to tubular atrophy, periglomerular fibrosis, and cortical cysts, inevitably leading to end-stage renal failure (Tatu et al., 1998). Urothelial tumors, i.e., the malignant tumors of the upper urinary tract, which often accompany BEN, are extremely aggressive in nature (Sostaric and Vukelic, 1991; Vukelic et al., 1991) and some studies have suggested a slightly higher disease incidence in females.

Although no direct link has been established, epidemiological data correlates a moderate increase in serum OTA levels with a higher incidence of nephropathy and urothelial tumors in humans. Studies carried out in several countries, where climatic conditions or suboptimal storage of grain and grain products promote OTA production by fungal species, have also indicated a link between dietary intake of OTA and the development of renal and urothelial tumors (Kuiper-Goodman, 1999; Maaroufi et al., 1995a, 1995b; Radic et al., 1997; Tatu et al., 1998).

The wealth of research that has been carried out into the effects and mode of action of OTA to date reflects its economic and social importance. Despite this, however, many questions remain unanswered.

Pathology

The pathology associated with OTA exposure varies between species. Initial examinations following feeding experiments in pigs revealed enlarged kidneys, which, when decapsulated, had a greyish appearance, indicative of fibrosis (Krogh et al., 1976; Elling, 1983). Closer examination revealed that the initial lesions occur in the proximal tubules and are characterized by desquamation and focal degeneration of the epithelial cells, coupled with focal peritubular fibrosis and thickening of the basement membrane. The severity of these lesions is dose- and time-dependent. Elling and colleagues also demonstrated the effects of OTA in young pigs to be more severe than in adult animals and that the renal damage caused by OTA exposure appears to be permanent as a return to contamination-free fodder did not reduce the incidence and severity of pathological changes (Elling, 1979, 1983). Renal tumors have so far not been observed in pigs. This may, however, be due to the relatively young age at which commercially raised pigs are slaughtered, thus not allowing for the relatively long latency and developmental periods typical for renal tumors.

Apart from a reduction in kidney size (in contrast to the increase seen in pigs), the pathomorphological effects noted in humans suffering from BEN are similar in nature to those noted in pigs. Kidneys also have a greyish appearance and are difficult to cut – giving the first indication of the underlying diffuse cortical fibrosis, which extends into the corticomedullary junction. In more advanced stages of the disease, the epithelium becomes severely degenerative and necrotic and hyperplastic arteriopathy is evident (Vukelic et al., 1991).

The similarities between the pathology observed for BEN and urothelial tumors in humans and that observed in pigs following chronic dietary intake of OTA have led to a number of laboratory studies that aimed to establish a definitive link between OTA exposure and renal disease. Chronic (two-year) studies in rats have demonstrated a clear causal relationship between OTA exposure and renal cortical tumor development, with 60% of male rats developing renal cell carcinoma (RCC), coupled with a distinct pathology of the pars recta (P_3) of the proximal tubule (Boorman, 1989; Boorman et al., 1992; Rásonyi et al., 1993). While no urothelial tumors or preneoplastic lesions of the transitional epithelium or the renal pelvis were reported, several prominent nonneoplastic lesions were observed in the renal cortex. These include degeneration of the renal tubular epithelium in the inner cortex and the outer stripe of the outer medulla, protein casts, karyomegalic nuclei, and renal cortical cysts, which were morphologically distinct from those present in aging rats. Hyperplastic lesions, which became apparent following nine months of exposure, were restricted to a single tubule. Longer exposure resulted in malignant renal cell adenomas and carcinomas which were often bilateral and/or multiple. In the same study, female rats were found to be much less susceptible to OTA-mediated toxicity, displaying a milder P_3 pathology and only a 6% tumor incidence. Even more pronounced sex differences have been noted in the carcinogenic response in mice where, despite being exposed to a 20-fold higher dose (4,800 mg/kg body weight/day for 2 years) than that employed in Boorman's study, only 28% of male mice presented with renal tumors, while female mice were totally refractive (Bendele et al., 1985).

Despite intensive research, the mechanism of OTA-induced carcinogenicity in rodents remains to be elucidated. In view of the decisive species differences and the lack of knowledge of the underlying mechanism, any human risk assessment is, at best, unreliable. The currently accepted virtually safe dose for human renal cancer risk of 0.2 ng/kg/day has been extrapolated from the rodent studies cited above. However, two important factors were not considered in this extrapolation. First, in rodents, OTA is primarily excreted via the biliary route, whereas in humans, the primary route of excretion is urinary (Appelgren and Arora, 1983; Fuchs and Hult, 1992; Fuchs et al., 1988), resulting in the delivery of higher concentrations of OTA to the human kidney. Second, and perhaps more critical, the half-life of OTA in humans (35.3 days) is approximately 14 times longer than that of the rat

(DFG, 1990; Li et al., 1997; Studer-Rohr et al., 2000). Hagelberg and Hult (Hagelberg et al., 1989) have proposed that this enormous species variation in half-life may be caused by differences in renal clearance rates due to different plasma protein binding characteristics. These authors also demonstrated that OTB possesses a far lower affinity for plasma proteins and is more rapidly eliminated than OTA in the species tested (fish, quail, mouse, rat, and monkey), observations that correspond well with the comparatively lower toxicity of OTB.

Mechanism of Action

It is currently unknown how OTA mediates its toxicity. The mechanistic background to the stark species and sex differences remains enigmatic. One possibility is that these differences are governed by specific renal handling of OTA; for example, through variations in the transporter and binding protein complements of renal cells from different species. Indeed, early experiments indicated OTA to be a substrate for the organic anion transport system (Sokol et al., 1988; Stein et al., 1985). Accumulation of OTA in rabbit renal basolateral membrane vesicles (BLMV) (Sokol et al., 1988) and OK cells (Gekle et al., 1994) was reported to occur solely via the PAH transport pathway. In contrast, Groves and colleagues, working with suspensions of isolated rabbit renal proximal tubules (Groves et al., 1998), concluded OTA accumulation to be a combination of passive diffusion and nonspecific binding and carrier-mediated processes. A more recent study has indicated OTA to be accumulated into mouse P_2 and P_3 renal cells stably transfected with the human organic anion transporters hOAT1 and hOAT3 (Jung et al., 2001). The authors describe saturable, dose- and time-dependent uptake of ^{3}H-OTA, which they assume to be localized to the basolateral membrane of the proximal tubule. The K_m values determined for the hOAT1 and hOAT3 in this study were two- to threefold higher than that determined by Groves et al. in their study with rabbit proximal tubule cells.

Heussner and colleagues (2002) describe the presence of at least one homogeneous binding component with low affinity but high capacity for OTA in renal cortical homogenates from pig, mouse, rat, and human, of both sexes. The binding of ^{3}H-OTA to these proteins could be competed by a range of substances known to have affinity for steroid receptors or for various organic anion transporters previously reported to be responsible for the transport of OTA (Tsuda et al., 1999). Heussner and colleagues reported a capacity ranking for specific OTA-binding of human > rat > pig ≥ mouse, which correlates with the toxicity ranking for experimental animals *in vivo* and, furthermore, suggests an even higher sensitivity for humans. Sex differences could, however, only be detected for the rat, with males having a higher binding capacity. Based on the pattern of protein binding competition, the authors suggested that the binding component does not belong to organic anion transporters previously described.

Similarly, a further study has demonstrated that primary renal epithelial cells originating from pigs or rats accumulate more OTA than their continuous cell line counterparts (O'Brien et al., 2001). Moreover, in this study, primary human renal epithelial cells were also shown to accumulate 10 to 15 times more OTA than the other cell types tested (LLC-PK_1, NRK-52E, NRK-49F, primary porcine kidney cells). Indeed, primary renal epithelial cells obtained from female donors were the most sensitive to the cytotoxic or antiproliferative effects of OTA in this study. A significant decrease in cell numbers could be detected following just 48 hours exposure to 1 nM OTA, a concentration that reflects normal serum levels of OTA in BEN areas. In agreement with the known toxicities of OTA and OTB, an approximately tenfold higher concentration of OTB was required to induce a comparable reduction in cell numbers. In this study, a maximal reduction in cell number was achieved following 48 hour exposure to 10 μM OTA. Approximately 15% of cells from all species tested survive and can reenter the cell cycle and proliferate following removal of the toxin, even following exposure to higher concentrations ($\leq$ 100 μM) and longer exposure times ($\leq$ 96 hours). The authors suggest the remaining cells to represent a subpopulation that are resistant to OTA-mediated toxicity and possibly apoptotic-defective (Dreger et al., 2000; O'Brien et al., 2001). Interestingly, although NRK-49F cells (a rat renal fibroblast cell line) accumulated comparable amounts of OTA to their epithelial (NRK-52E) counterparts, these cells were relatively insensitive to OTA-mediated cytotoxicity (O'Brien et al., 2001). Similar observations have been made for primary human renal fibroblasts (O'Brien, personal communication). The authors thus propose that BEN could be caused by a cytostatic or cytotoxic effect of OTA in epithelial cells, while fibroblasts are more refractive. This would lead to increased or maintained proliferation of fibroblasts coupled with reduced epithelial proliferation or epithelial cell death, resulting in the gradual and progressive replacement of healthy tissue with the fibrotic tissue characteristic for BEN.

The molecular mechanism of action remains controversial. OTA has been reported as both nonmutagenic (IARC, 1993; Kuiper-Goodman and Scott, 1989) and mutagenic (Obrecht-Pflumio et al., 1999) in a variety of microbial genotoxicity tests. The formation of DNA adducts by a reactive metabolite (Castegarno et al., 1998; Creppy et al., 1985; Pfohl-Leszkowicz et al., 1991, 1993), sister chromatid exchange (Föllmann et al., 1995), unscheduled DNA synthesis (Doerrenhaus et al., 2000), and the generation of reactive oxygen species have been proposed as candidate mechanisms. However, neither the oxidative changes nor the DNA adducts reported could be corroborated using HPLC-MS or LC-MS/MS (Gautier et al., 2001a; Zepnik et al., 2001). Furthermore, although the levels of DNA-adducts formed in male rats (determined by the ^{32}P postlabelling method) appeared to be higher than that in females, thus suggesting a correlation between DNA-adducts and the known sex differences in OTA-mediated renal carcinogenicity, no clear correlation could be determined between the level of DNA-adducts and the incidences of adenocarcinoma or karyomegaly reported (Castegarno et al., 1998). Indeed, while approximately 42% of female Dark Agouti (DA) rats were reported to present with DNA-adducts, some of which were at the same levels as the exposed male rats, none of the female DA rats presented with either karyomegaly or renal epithelial tumors, thus questioning the relevance of the adducts in the ^{32}P postlabeling method.

Obrecht-Pflumio and Dirheimer (2001) have reported the generation of DNA and deoxyguanosine-3′-monophosphate adducts using mouse microsomes. The incidence and number of adducts reported, using the ^{32}P postlabeling method, was much higher than that previously reported for the Lewis, Sprague-Dawley, or DA rat, which is in complete conflict with *in vivo* observations of renal tumor incidence. Furthermore, other researchers have reported poor metabolism of OTA via P450 peroxidases and little glutathione conjugate formation with liver or kidney microsomes or postmitochondrial supernatants from rat, mouse, or human tissue (Gautier et al., 2001a; Zepnik et al., 2001), making it unlikely that an OTA metabolite is responsible for the supposed DNA-adduct generation. Gautier and colleagues (Gautier et al., 2001a, 2001b) also reported an absence of OTA-induced DNA-adducts and have suggested that many of the ^{32}P-postlabeled adducts observed in previous studies with OTA may indeed be the result of cytotoxic effects of OTA or the ability of OTA to generate an oxidative stress response in rats or mice. This was corroborated by the absence of such adducts when animals are pretreated with superoxide dismutase, catalase, or other antioxidants.

Research into possible epigenetic mechanisms has so far concentrated on the demonstration of OTA-mediated inhibition of tRNA synthase (Creppy et al., 1979, 1983), on lipid peroxidation (Omar et al., 1990; Rahimtula et al., 1988) and on cytoskeletal changes (Heussner et al., 1998). Many of these studies however, used unrealistically high concentrations of OTA, which are irrelevant to the *in vivo* situation and are, in fact, close to the acute lethal doses in rats and mice. In contrast, several researchers have recently reported *in vitro* effects of OTA at nM (dietary-relevant) concentrations. OTA induced apoptosis in dedifferentiated MDCK-C7 cells (Gekle et al., 2000), immortalized human (IHKE) cells (Schwerdt et al., 1999) and human kidney epithelial (SB3) cells (Horvath et al., 2002). These observations, and some by other authors, were made in the absence of serum, in cells that had been previously transformed, or in cells that had been pretreated with hydroxyurea (Bondy et al., 1995; Doerrenhaus and Föllmann, 1997; Doerrenhaus et al., 2000; Dopp et al., 1999; Gekle et al., 1998; Maaroufi et al., 1999; Ueno et al., 1995), and as such may not be ideal model systems for the situation *in vivo*. In contrast to these reports, Seegers and colleagues (1994) reported that only a maximum of 5% of hamster kidney cells can be induced to undergo apoptosis *in vitro* following OTA exposure. These findings support the thesis proposed by Rásonyi and colleagues (1999), who described apoptotic cells in the lumen of affected tubules in rats following OTA exposure (Figure 21.2) and speculated that apoptosis may be a secondary or tertiary event in OTA toxicity, resulting from disruptions in cell–cell interactions or cell–basal lamina adhesion. This view is shared by other authors, who have found no evidence of the induction of apoptosis by OTA *in vitro* at dietary-relevant concentrations (Dreger et al., 2000; Heussner et al., 2000; O'Brien et al., 2001; Wolf et al., 2002). Cytotoxicity, however, could be demonstrated in primary cells exposed to nanomolar concentrations of OTA in serum-replete medium (O'Brien et al., 2001; Wolf et al., 2002). An interesting side-note is that following OTA exposure under these conditions, many floating cells can be seen in the culture flasks. If these are collected and reseeded, they will proliferate (Wolf and O'Brien, personal communication). It is possible that these cells could be resistant to OTA-mediated cytotoxicity. The occurrence of such cells *in vivo* could form the rudiments of the observed tumors via transformation or invasion into the transitional epithelium.

Several other possible mechanisms have been suggested, including an increase intracellular pH via a disruption in membrane anion conductance (Gekle et al., 1993), inhibition of mitochondrial transport (Meisner and Chan, 1974; Moore and Truelove, 1970), inhibition of mitochondrial respiration (Wei et al., 1985), and disruption of gap junction intercellular communication (Horvath et al., 2002). These theories could correlate to the previously described role of organic anion transporters. It remains to be seen if any or indeed all of these play a role in the toxicity of OTA.

Fumonisins

It can almost be said that the discovery of the fumonisins occurred as a byproduct of research into other mycotoxins produced by *Fusarium* species, namely the trichothecenes, which were responsible for several outbreaks of alimentary toxic aleukia in humans in Russia during World War II (Joffe, 1986). Alimentary toxic aleukia is caused by T-2 toxin and other, related trichothecenes. Of the various toxins produced by *Fusarium* species, the fumonisins are arguably the most significant having a variety of toxic effects in several species.

Fumonisins are a group of water-soluble bifuranocumarin mycotoxins (Figure 21.1) which, under suitable environmental or storage conditions, may be produced by several *Fusarium* species, in particular *Fusarium moniliforme* and *F. proliferatum* (Marasas et al., 1988). Several fumonisins have been identified to date, however of these fumonisin B_1 and B_2 (FB_1, FB_2) are the most abundant, making up 70% of the total concentration of fumonisins detected (Prozzi et al., 2000), and also the most toxic and hence the most investigated of the group (Sydenham et al., 1991). Commonly found as a contaminant of corn and in particular overwintered grain, it is thought that FB_1 is produced as a result of an endophytic relationship with grain (Bacon and Hinton, 1996), imparting an increased resistance to diseases or insects.

FB_1 has been determined to be neurotoxic, hepatotoxic, and toxic to the lung, and has also been associated with the development of esophageal cancer, particularly in areas where corn forms part of the staple diet (Gelderblom et al., 1988; Haschek et al., 1992; Norred, 1993). As in the case with many other mycotoxins, the concentrations required to cause these toxic syndromes vary both in effect induced and with the species tested.

One of the earliest diseases demonstrated to result from the feeding of animals with fumonisin-contaminated fodder is equine leucoencephalomalacia (Kellerman et al., 1990; Wilson et al., 1992). A similar syndrome has more recently been determined in rabbits (Bucci et al., 1996). Another disease of particular economic importance is porcine pulmonary edema, a fatal disease characterized by pulmonary edema and hydrothorax (Harrison et al., 1990). The economic relevance of fumonisins, coupled with their consistent presence in human foodstuffs (Sydenham et al., 1991), has resulted in fumonisins being one of the best-studied classes of mycotoxins. Although the kidney and the liver are primary organs for FB_1-mediated toxicity in several species, including mice (Howard et al., 2000b), sheep (Edrington et al., 1995), and rabbits (Gumprecht et al., 1995), display FB_1-mediated renal toxicity (Riley et al., 1994; Voss et al., 1993). The nephrotoxic aspects have only relatively recently been intensively investigated.

The feeding of FB_1-contaminated corn to rats has been reported to result in reduced body and absolute kidney weight, coupled with increased enzyme activities and elevated serum bilirubin levels (Norred et al., 1996). Gumprecht and colleagues (1995) demonstrated a number of specific renal effects in rabbits following intravenous administration of FB_1 (once-daily for 5 days). Serum creatinine and urea nitrogen concentrations were elevated and urinary glucose and protein concentrations were markedly increased following either single or multiple doses. Urine output was reduced. In contrast, Bondy and colleagues (1995) reported an increased production of dilute urine in Sprague-Dawley rats exposed to FB_1 under a similar dosage regimen (once-daily for 4 days) albeit at higher concentrations (7.5 to 10 mg/kg body weight) with intraperitoneal administration. These authors also reported an increased excretion of dilute urine coupled with elevated blood urea nitrogen concentrations and increased serum levels of several enzymes including alanine aminotransferase and serum alkaline phosphatases, well as chloride and potassium imbalances. These observations, and similar ones made by other researchers (Suzuki et al., 1995; Voss et al., 1998), imply a reduced concentrating ability, which is indicative of tubular damage.

Pathology

The clinical pathology effects associated with FB_1-exposure in rabbits have been shown to be coupled with a distinct pathology, which is

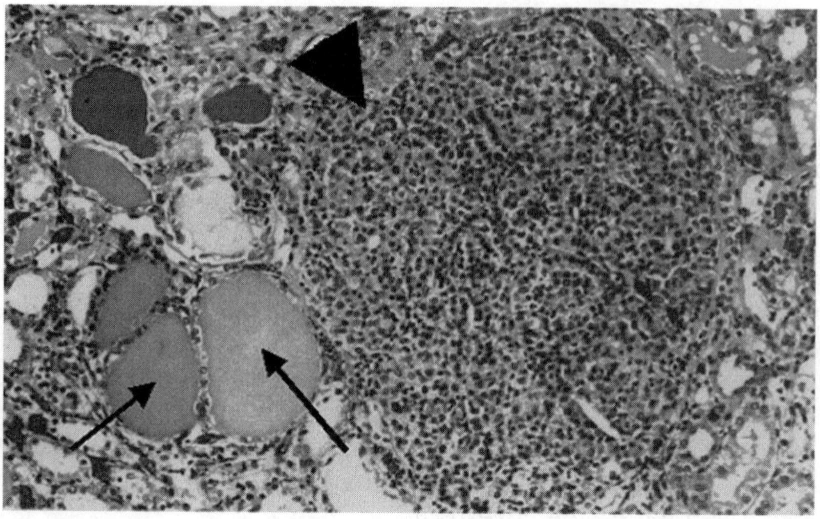

Figure 21.3 Hyaline casts (arrow) and a solid renal adenoma (arrowhead) in a male F344 rat treated with 150 mg FB_1/kg body weight for 2 years.

dependent on the number of doses applied as well as on the actual dose (Gumprecht et al., 1995). The primary lesion observed was necrosis of the proximal tubule, which was multifocal in the cortex and more extensive in the outer regions of the medulla. These authors also described the occurrence of mitotic figures and individual cell necrosis in the proximal tubular epithelium and vacuolization of tubular epithelial cells. The severity of the lesions was observed to increase, progressing into severe necrosis of the distal proximal tubule and denudation of the basement membrane, coupled with extensive hyaline casts (Figure 21.3), with repeated dosing. These authors also reported tubular regeneration.

Several authors have made similar observations in calves (Mathur et al., 2001), as well as in rats (Voss et al., 1998) and mice (Howard et al., 2000b), with mice being less sensitive to the effects of FB_1. Voss and colleagues (1998) described lesions in the outer medulla of the kidney of Sprague-Dawley rats, consisting of basophilic epithelial cells and the presence of condensed or pycnotic nuclei, indicating apoptosis. These apoptotic cells were sloughed off into the tubular lamina, and tubular cells displayed an altered morphology, appearing lower and cuboidal in shape. Mitotic figures were occasionally present and the

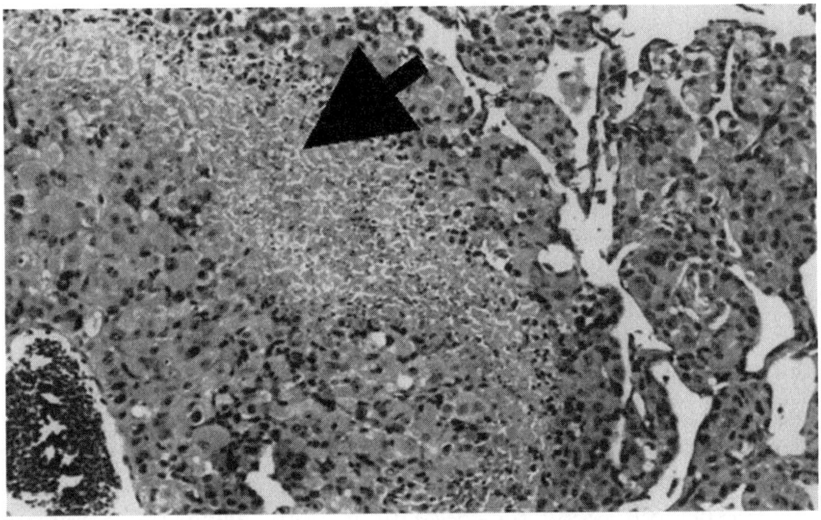

Figure 21.4 Invasive renal cell carcinoma presenting with areas of focal necrosis (arrow) in a male F344 rat treated with 150 mg FB_1/kg body weight for 2 years.

cells of the zona fasiculata showed cytoplasmic vacuolization. In contrast, apoptosis was not observed following exposure to the less toxic analogue FB_2.

Vacuolar degeneration of the tubular epithelium, detachment of epithelial cells and the presence of pycnotic nuclei have also been described by Prozzi and colleagues (2000) and by Hard and colleagues (2001) following chronic exposure to FB_1. Both of these groups also reported that the renal lesions were slightly more prominent in male than in female rats. In addition to the pathological changes previously described by other authors, Hard et al. (2001) observed the generation of a neoplastic response in the form of solitary foci of atypical tubule hyperplasia. These were generally located to the deep or midcortex and the corticomedullary in the kidneys of rats exposed to FB_1 in a two-year carcinogenicity study. These lesions progressed to yield renal tubule carcinomas (Figure 21.4), of a rare and highly malignant form, with marked cellular pleomorphism, locally invasive growth and a high rate of mitosis, as well as conventional solid (Figure 21.3) or papillary adenomas with higher doses of FB_1. Further publications from the same group report apoptotic cell death to predominate in short term exposure and conclude that these tumors observed in long term studies are a result of compensatory regenerative hyperplasia (Howard et al., 2000a, 2000b).

In all of the investigations discussed here, primary renal effects were accompanied by hepatic lesions. Based on these rodent studies, a provisional human tolerable daily intake for FB_1, FB_2, or FB_3, alone or in combination, of 2 μg/kg body weight has been established (Creppy, 2002).

Mechanism of Action

Considering the extremely short period since the discovery of fumonisins, the mechanism of FB_1-mediated toxicity is surprisingly well elucidated. As with many carcinogenic substances, the initial investigations were carried out into a possible genotoxic mechanism of action. The nonmutagenic nature of the fumonisins has been determined using a range of testing methods. The *Salmonella* mutagenicity test (Gelderblom and Snyman, 1991), *in vitro* DNA repair assays in primary rat hepatocytes (Gelderblom et al., 1992; Norred et al., 1992), the DNA repair assay in *Escherichia coli* (in the presence or absence of rat liver S9 fractions), and the micronuclei assay in primary rat hepatocytes have all proved negative.

The *in vivo* observations of apoptosis, necrosis, and regenerative processes gave rise to a number of *in vitro* investigations into the underlying mechanisms of FB_1 toxicity. FB_1 has been demonstrated to be antiproliferative (10 to 35 μM), cytotoxic (>35 μM), and to disrupt cell–cell contact in the LLC-PK_1 cell line (Yoo et al., 1992). These authors also demonstrated actively proliferating cells to be more susceptible to FB_1-mediated toxicity, which suggests a disruption of normal cell cycle control. FB_1 has been reported to cause rat hepatocytes to arrest in the G1 phase of the cell cycle and then to undergo apoptosis (Gelderblom et al., 1995). In contrast, other authors have demonstrated the induction of apoptosis without G1 arrest in cultured human keratinocytes (Tolleson et al., 1996). These differences may, however, be due to the use of differing cell types and dosage regimens, as both apoptosis and necrosis have been observed *in vivo*.

Further investigations have shown that, like so many other mycotoxins, the mechanism of fumonisin toxicity appears to be related to its structure. The fact that the backbone of fumonisin closely resembles that of sphinganine, sphingosine, and sphingoid bases, and that FB_1 inhibits the activity of ceramide synthase, which is a key enzyme in the generation of sphingomyelin and complex sphingolipids (Wang et al., 1991), gave the first clue to the nature of fumonisin toxicity. Interestingly, the *N*-acetylated forms of FB_1 (FA_1 and FA_2) have little or no cytotoxicity, whereas the aminophenol derivatives (AP_1 and AP_2), which share the backbone structure of FB_1, display a level of cytotoxicity similar to that of the parent compound (Yoo et al., 1992).

This disruption of sphingolipid metabolism leads to a depletion of complex sphingolipids and an accumulation of sphingoid bases. Indeed elevations of the ratio of free sphinganine to free sphingosine in serum, urine, kidneys, and liver can even be used as an indicator of exposure to FB_1 (Riley et al., 1993, 1994). Accumulation of sphingoid bases results in a cascade of events including inhibition of protein kinase C and Na^+/K^+-ATPase, release of intracellular calcium, promotion of retinoblastoma protein dephosphorylation and culminating in the induction of apoptosis (Merrill Jr. et al., 1995; Riley et al., 1996). All of these events have been observed *in vivo* or *in vitro* following FB_1 exposure (Riley et al., 1993, 1994; Mathur et al., 2001; Norred et al., 1998). Indeed, in the study carried out by Yoo and colleagues (1992), the cytotoxicity in LLC-PK_1 cell line exactly paralleled the inhibition of ceramide synthase. The regenerative response resulting from cell loss is a known risk factor for tumorigenesis, suggesting a tumor-promoting role for FB_1 (Howard et al., 2000a).

Sphingolipids play a crucial role in the regulation of cell–cell and cell–substrate contact, as well as in cellular growth and differentiation. Furthermore, many sphingolipids are uniquely expressed in the kidney (Shayman, 2000). Hard and colleagues (2001) have therefore proposed that apoptosis may not be the primary event leading to the observed pathology. These authors suggest that, instead, apoptosis occurs as a secondary event to anokisis – the loss of cell–substrate contact.

Two further theories have been proposed to explain the mechanism of action of FB_1: disruption of fatty acid metabolism and the induction of oxidative stress, and modulation of gene expression (Abado-Becongnee et al., 1998; Mobio et al., 2000). In contrast to the sphingolipid imbalance theory, which can account for all of the known effects of FB_1 either *in vivo* or *in vitro*, these alternative theories cannot.

Patulin

The cyclic γ-lactone mycotoxin, patulin (Figure 21.1) is also produced by members of the *Penicillium* and *Aspergillus* families, and is frequently found as a contaminant of a variety of foods and, in particular, processed apple products (e.g., apple juice and cider) (Le Bourhis, 1984; Thurm et al., 1979). The name comes from *Penicillium patulum*, where the toxin was first identified, and the substance is highly toxic to Gram-negative bacteria, certain fungi, and protozoans, and also to certain plants. Like citrinin, however, it has also been demonstrated to have toxic effects in several animal species.

Subacute and acute exposure leads to a massive increase in the incidence of fundic ulcers, associated with duodenal dilation and activation of the mesenteric and pancreatic duodenal lymph nodes in rats (Speijers et al., 1988). This is thought to be a result of a direct irritant action of patulin on the gastric mucosa. The presence of fibrosis in the submucosa and underlying muscle layer was also described in the same study. Due to the primarily gastric effects of patulin, relatively little research has been carried out into the renal effects. In contrast to many other mycotoxins, patulin has been determined to be noncarcinogenic in rats (Becci et al., 1981).

Initial investigations carried out into patulin-mediated toxicity revealed functional deficits of the kidney, including a dose-dependent occurrence of oligouria, decreased serum sodium levels (Becci et al., 1981; McKinley et al., 1982), and a decrease in creatinine clearance (Speijers et al., 1984). Becci and colleagues (1981) reported these effects to be of a transient nature, with urinary parameters returning to normal levels within two years. The effects of patulin have been reported to be independent of the route of administration although pulse dosing appears to increase the rate of mortality resulting from the gastric effects in gavage experiments when compared with administration via drinking water in subacute studies in rats (Speijers et al., 1984). Subsequent studies have aimed to identifying the pathological changes responsible for the functional effects observed in the kidney.

Speijers and colleagues (1988) observed an increase in relative and absolute kidney weights in rats following chronic exposure to patulin via drinking water. This was coupled with a slight increase in creatinine clearance in animals of the high-dose (approximately 27 mg/kg body weight/day) group. A slight reduction in urinary output and an increase in urinary protein and bilirubin concentration were also noted. Males appeared to be slightly more sensitive with respect to these effects.

Pathology

Curiously, despite the use of several different histological and staining techniques, no renal histopathological changes have been demonstrated. As a result of the lack of demonstrable pathology, research into the mechanism of action of patulin has been neglected. It seems unlikely that functional defects are not associated with at least some pathological indications. Future studies using more precise histopathological techniques could reveal these changes.

Mechanism of Action

Despite the lack of evidence of pathological changes resulting from patulin exposure, some research has been carried out into the mechanism of action of patulin using relevant *in vitro* model systems. Phillips and Hayes (1979) proposed that patulin interferes with transepithelial sodium transport in mouse brain and also directly inhibits Na^+/K^+-ATPase. These authors suggest that such an effect could explain the alterations in sodium and potassium levels observed *in vivo* (Becci et al., 1981; McKinley et al., 1982) and *in vitro* (Kreisberg and Wilson, 1988). This theory is supported by the work of Riley and colleagues (1990), who demonstrated that patulin altered the membrane function of the LLC-PK_1 cell line as manifested by a reduction in transepithelial transport of sodium. These and other authors (Hinton et al., 1989) have also demonstrated that patulin caused an increase in potassium efflux, relative to sodium, resulting in a transient hyperpolarization of the affected cells.

More detailed investigations by Riley and Showker (1991) described lipid peroxidation, abrupt calcium influx, extensive blebbling, depletion of nonprotein sulfhydryls, and a total release of cellular lactate dehydrogenase (LDH) in LLC-PK_1 and L6 cells. These observations, which occurred sequentially, were made following exposure to the relatively high concentration of 50 μM. Lower concentrations (5 to 10 μM) have been demonstrated to induce the same effects, albeit over a longer time span (Riley et al., 1990). The authors suggest that a progressive loss of integrity of the plasma membrane occurs rather than a specific interaction with Na^+/K^+-ATPase. The lipid peroxidation observed in this study could be prevented by coincubation with antioxidants, however this could not prevent the cells eventually dying as a result of patulin exposure. This is probably due to the continued ability of patulin to deplete nonprotein sulfhydryls, resulting in the observed cation leaks, and inhibition of ion pumps, changes in cellular volume and the occurrence of blebbing. Of particular relevance with respect to pathological potential is the ability of patulin to alter normal cellular calcium regulation, as described by Riley and Showker (1991), as alterations in calcium homeostasis have previously been demonstrated to be sufficient to cause DNA strand breakage (Cantoni et al., 1989).

The effects outlined above have been noted following exposure to very high concentrations of patulin, which are unlikely to be of relevance to the *in vivo* situation. It is disturbing that little or no research has been directed toward determining the pathological or toxic potential of patulin following chronic, low-level exposure. Indeed the long-term carcinogenic and the pathological potential of patulin have not been adequately investigated. Although Becci and colleagues (1981) found no evidence of tumorigenesis in rats, other species remain to be examined. The *in vitro* observations of Riley and colleagues (1990) that exposure to lower concentrations over a longer time period gives rise to similar effects to those observed following acute exposure to high concentrations also indicate an urgent need for further investigation.

Thus, patulin does appear to have at least the potential to cause pathological effects in the kidney, and this aspect of its toxicity should perhaps be more closely studied with state-of-the-art techniques in order to quantify and qualify possible latent risks that have long been neglected.

MYCOTOXIN INTERACTIONS

From the sections above, it is clear that many mycotoxins may be produced by a single mould. In particular, many of the *Aspergillus* and *Penicillium* species can produce several toxins simultaneously, depending on the environmental and substrate conditions. This begs the question of how do these mycotoxins interact with each other? Moreover, what effect do these interactions have on the nephrotoxic potential of these toxins? As most research has concentrated on the actions of pure toxins on the kidney, relatively little is known about potential additive or synergistic effects, which are arguably more relevant to the real-life situation. Some mycotoxins have, however, been shown to interact (Morrissey et al., 1987; Prozzi et al., 2000; Raju and Devegowda, 2000; Rati et al., 1991; Shinohara et al., 1976; Stoev et al., 2001) and exacerbate the toxicity resulting from individual administration. In view of the fact that under natural conditions single pure mycotoxins occur only very rarely, future investigations should aim to further elucidate whether additive or synergistic effects can occur so that relevant and accurate risk assessments may be carried out – as Oscar Wilde once said: "*The truth is rarely pure, and never simple*" (Wilde, 1895).

ACKNOWLEDGMENTS

Fumonisin pathology photographs (Figures 21.3 and 21.4) were kindly provided by Paul C. Howard, Ph.D., Director, NTP Centre for Phototoxicology, Division of Biochemical Toxicology, National Centre for Toxicological Research, U.S. Food and Drug Administration, Jefferson, AR, USA 72079.

REFERENCES

Abado-Becongnee K, Mobio TA, Ennamany R, Fleurat-Lessart F, Shier WT, Badria F, and Creppy EE (1998) Cytotoxicity of fumonisin B_1: Implication of lipid peroxidation and inhibition of protein and DNA synthesis. *Arch Toxicol* 72:233–236.

Almeida RMA, Correa B, Xavier JG, Mallozzi MAB, Gambale W, and Paula CR (1996) Acute effect of aflatoxin B1 on different inbred mouse strains II. *Mycopathologia* 133:23–29.

Ambrose AH, and DeEds F (1946) Some toxicological and pharmacological properties of citrinin. *J Pharmacol Exp Ther* 88:173.

Appelgren L-E, and Arora RG (1983) Distribution of 14C-labelled ochratoxinA in pregnant mice. *Food Chem Toxicol* 21:563–568.

Asao T, Buchi G, Abdel-Kader MM, Chang SB, Wick EL, and Wogan GN (1963) Aflatoxins B and G. *J Am Chem Soc* 85:1706.

Autrup H, Jorgensen CB, and Jensen O (1996) Aflatoxin B_1 induced *lac* I mutation in liver and kidney of transgenic mice C57BL/6N: Effect of phorone. *Mutagenesis* 11:69–73.

Bacon CW, and Hinton DM (1996) Symptomless endophytic colonization of maize by *Fusarium moniliforme. Can J Botany* 74:1195–1202.

Becci PJ, Hess FG, Johnson WD, Gallo MA, Babish JG, Dailey RE, and Parent RA (1981) Long-term carcinogenicity and toxicity studies of patulin in the rat. *J Appl Toxicol* 1:256–261.

Bechtel DH (1989) Molecular dosimetry of hepatic aflatoxin B1-DNA adducts; linear correlation with hepatic cancer risk. *Reg Toxicol Pharmacol* 10:74–81.

Bendele AM, Carlton WW, Krogh EB, and Lillehoj EB (1985) Ochratoxin A carcinogenesis in the (C57BL/&J X C*H)F_1 mouse. *J Natl Cancer Inst* 75:733–742.

Berndt WO (1983) Transport of citrinin by rat renal cortex. *Arch Toxicol* 54:35–40.

Berndt WO, and Hayes AW (1982) The effect of probenecid on citrinin-induced nephrotoxicity. *Toxicol Appl Pharmacol* 64:118–124.

Blanpin O (1959) La citrinine. Nouvelles donnes sur l'action pharmacodynamique de cet antibiotique. *Therapie* 14:677–658.

Bondy G, Suzuki C, Barker M, Armstrong C, Fernie S, Hierlihy L, Rowsell P, and Mueller R (1995) Toxicity of Fumonisin B_1 administered intraperitoneally to male Sprague-Dawley rats. *Fd Chem Toxicol* 33:653–665.

Boorman GA (1989) Toxicology and carcinogenesis studies of ochratoxin A in F344/N rats. *NTP Technical Report* NTP TR 358.

Boorman GA, McDonald MR, Imoto S, and Persing R (1992) Renal lesions induced by ochratoxin A exposure in the F344 rat. *Toxicol Pathol* 20:236–245.

Bucci TJ, Hanson DK, and LaBorde JB (1996) Leucoencephalomalacia and hemorrhage in the brain of rabbits gavaged with mycotoxin fumonisin B_1. *Natural Toxins* 4.

Busby WF, and Wogan GN (1984) Aflatoxins, in *Chemical Carcinogens 2nd Edition*, pp. 945–1136, American Chemical Society, Washington, D.C.

Cantoni O, Sestili P, Cattabeni F, Bellomo G, Pou S, and Cerutti P (1989) Calcium chelator quin 2 prevens hadrogen peroxide-induced DNA breakage and cytotoxicity. *Eur J Biochem* 182:209–212.

Castegarno M, Mohr U, Pfohl-Leszkowicz A, Esteve J, Steinmann J, Tillman T, Michelon J, and Bartsch H (1998) Sex- and strain-specific induction of renal tumours by ochratoxin A in rats correlates with DNA adduction. *Int J Cancer* 77:70–75.

Chagas GM, Campello AP, and Kluppel MLW (1992a) Mechanism of citrinin-induced dysfunction of mitochondria I. Effects on respiration, enzyme activities and membrane potential of renal cortical mitochondria. *J Appl Toxicol* 12:123–129.

Chagas GM, Oliveira MB, and Campello AP (1995) Mechanism of citrinin-induced dysfunction of mitochondria III. Effects on renal cortical and liver mitochondrial swelling. *J Appl Toxicol* 15:91–95.

Chagas GM, Oliveira MBM, Campello AP, and Kluppel MLW (1992b) Mechanism of citrinin-induces dysfunction of mitochondria II. Effect on respiration, enzyme activities and membrane potential of liver mitochondria. *Cell Biochem Funct* 10:209–216.

Chen W, Nichols J, Zhou Y, Chung KT, R.H. H, and Chou MW (1995) Effect of dietary restriction on glutathione S-transferase activity specific toward aflatoxin B1–8,9-epoxide. *Toxicol Lett* 78:235–243.

Chou MW, Chen W, Mikhailova MV, Nichols J, Weis C, Jackson CD, Hart RW, and Chung KT (1997) Dietary restriction modulated carcinogen-DNA adduct formation and the carcinogen-induced DNA strand breaks. *Toxicol Lett* 92:21–30.

Chou MW, Lu MH, Pegram RA, Gao P, Cao S, Kong J, and Hart RW (1993) Effect of caloric restriction on aflatoxin B_1-induced DNA synthesis, AFB_1-DNA binding and cell proliferation in Fischer 344 rats. *Mechanisms of Ageing and Development* 70:232–233.

Creppy EE (2002) Update, survey, regulation and toxic effects of mycotoxins in Europe. *Toxicol Lett* 127:19–28.

Creppy EE, Kane A, Dirheimer G, Lafarge-Frayssinet C, Mousset S, and Frayssinet C (1985) Genotoxicity of ochratoxin A in mice: DNA single-strand break evaluation in spleen, liver and kidney. *Toxicol Lett* 28:29–35.

Creppy EE, Lugnier AAJ, Beck G, Röschenthaler R, and Dirheimer G (1979) Action of ochratoxin A on cultured hepatoma cells-reversion of inhibition by phenylalanine. *FEBS Lett* 104:287–290.

Creppy EE, Størmer FC, Kern D, Röschenthaler R, and Dirheimer G (1983) Effect of ochratoxin A metabolites on yeast phenylalanyl-tRNA synthetase and on the growth and in vivo proteinsynthesis of hepatoma cells. *Chemico-Biol Interact* 47:239–247.

Croy RG (1981) Quantitative comparison of covalent aflatoxin-DNA adducts formed in rat and mouse livers and kidneys. *J Natl Cancer Inst* 66:761–768.

Croy RG, and Wogan GN (1981). Quantitiative comparison of covalent aflatoxin-DNA adducts formed in rat and mouse livers and kidneys. *J Natl Cancer Inst* 66:761–768.

Culvenor CCJ, Edgar JA, Mackay MF, Gorst-Allman CP, Marasas WFO, Steyn PS, Vleggaar R, and Wessels PL (1989) Structure elucidation and absolute configuration of phomopsin A, a heptapeptide mycotoxin produced by *Phomopsis leptostromiformis. Tetrahedron* 45:2351–2372.

DFG (1990) Ochratoxin A, Vorkommen und toxikologische Bewertung. *Deutsche Forschungsgemeinschaft, DFG.*

Doerrenhaus A, Flieger A, Golka K, Schulze H, Albrecht M, Degen GH, and Follman W (2000) Induction of unscheduled DNA synthesis in primary human urothelial cells by the mycotoxin ochratoxin A. *Toxicol Sci* 53:271–277.

Doerrenhaus A, and Föllmann W (1997) Effects of ochratoxin A on DNA repair in cultures of rat hepatocytes and porcine urinary bladder epithelial cells. *Arch Toxicol* 71:709–713.

Dopp E, Muller J, Hahnel C, and Schiffmann D (1999) Induction of genotoxic effects and modulation of the intracellular calcium level in syrian hamster embryo (SHE) fibroblasts caused by ochratoxin A. *Fd Chem Toxicol* 37:713–721.

Dreger S, O'Brien E, Satck M, and Dietrich D (2000) Antiproliferative effects and cell-cycle specific effects of ochratoxin A in LLC-PK1, NRK-52E, and porcine primary proximal kidney cells. *Toxicol Sci* 54:170.

Dunn BB, Stack ME, Park DL, Joshi A, Friedman L, and King RL (1983) Isolation and identification of dihydrocitrione, a urinary metabolite of citriinin in rats. *J Toxicol Environ Health* 12:283–289.

Eaton DL, and Gallagher EP (1994) Mechanisms of aflatoxin carcinogenesis. *Annu Rev Pharmacol Toxicol* 34:135–174.

Edrington TS, Kamps-Holtzappel CA, Harvey RB, Kubena LF, Elissalde MH, and Rottinghaus GE (1995) Acute hepatic and renal toxicity in lambs dosed with fumonisin-containing culture material. *J Animal Sci* 73:508–515.

Elling F (1979) Ochratoxin A-induced mycotoxic porcine nephropathy: Alterations in enzyme activity in tubular cells. *Acta Pathol Microbiol Scand Sect A* 87A:237–243.

Elling F (1983) Feeding Experiments with Ochratoxin A-contaminated Barley to Bacon Pigs. *Acta Agric Scand* 33:153–159.

Föllmann W, Hillebrand IE, Creppy EE, and Bolt HM (1995) Sister chromatid exchange frequency in cultured isolated porcine urinary bladder epithelial cells (PUBEC) treated with ochratoxin A, and alpha. *Arch Toxicol* 69:280–286.

Fuchs R, Appelgren L-E, and Hult K (1988) Distribution of 14C-OChratoxinA in the mouse monitored by whole-body autoradiography. *Pharmacol Toxicol* 63:355–360.

Fuchs R, and Hult K (1992) Ochratoxin A in blood and its pharmacokinetic properties. *Fd Chem Toxicol* 30:201–204.

Fukui Y, Hayasaka S, Itoh M, and Takeuchi Y (1992) Development of neurons and synapses in ochratoxin A-induced microcephalic mice: A quantitative assessment of somatosensory cortex. *Neurotoxicol Teratol* 14:191–196.

Gautier J, Richoz J, Welti DH, Markovic J, Gremaud E, Guengerich FP, and Turesky RJ (2001a) Metabolism of ochratoxin A: Absence of formation of genotoxic derivatives by human and rat enzymes. *Chem Res Toxicol* 14:34–45.

Gautier J-C, Holzhaeuser D, Markovic J, Gremaud E, Schilter B, and Turesky RJ (2001b) Oxidative Damage and Stress Response From Ochratoxin A Exosure in Rats. *Free Radic Biol Med* 30:1089–1098.

Gekle M, Gassner B, Freudinger R, Mildenberger S, Silbernagl S, Pfaller W, and Schramek H (1998) Characterization of an ochratoxin-A-dedifferentiated and cloned renal epithelial cell line. *Toxicol Appl Pharmacol* 152:282–291.

Gekle M, Oberleithner H, and Silbernagl S (1993) Ochratoxin A impairs "postproximal" nephron function *in vivo* and blocks plasma membrane anion conductance in Madin-Darby canine kidney cells *in vitro*. *Pflügers Arch Eur J Physiol* 425:401–408.

Gekle M, Schwerdt G, Freudinger R, Mildenberger S, Wilflingseder D, Pollack V, Dander M, and Schramek H (2000) Ochratoxin A induces JNK activation and apoptosis in MDCK-C7 cells at nanomolar concentrations. *J Pharmacol Exp Ther* 293:837–844.

Gekle M, Vogt R, Oberleithner H, and Silbernagel S (1994) The mycotoxin ochratoxin A deranges pH homeostasis in Madin-Darby canine kidney cells. *J Membr Biol* 139:183–190.

Gelderblom WCA, Kriek NPJ, Marasas WFO, and Thiel PG (1988) Fumonisins-Novel mycotoxins with cancer promoting activity produced by *Fusarium moniliforme*. *Appl Environ Microbiol* 54:1806–1811.

Gelderblom WCA, Semple E, and Farber E (1992) The cancer initiating potential of the fumonisin mycotoxins produced by *Fusarium moniliforme*. *Carcinogenesis* 13:433–437.

Gelderblom WCA, and Snyman SD (1991) Mutagenicity of potentially carcinogenic mycotoxins produced by Fusarium moniliforme. *Mycotoxin Research* 7:46–52.

Gelderblom WCA, Snyman SD, Van der Westhuizen L, and Marasas WFO (1995) Mitoinhibitory effect of fumonisin B_1 on rat hepatocytes in primary culture. *Carcinogenesis* 5:1047–1050.

Glahn RP, van Campen D, and Dousa TP (1994) Aflatoxin B_1 reduces Na^+-P_i co-transport in proximal renal epithelium: Studies in opossum kidney (OK) cells. *Toxicol* 92:91–100.

Gopolan P, Tsuji K, Lehmann K, Kimura M, Shinozuka H, Sato K, and Lotlikar PD (1993) Modulation of aflatoxin B1-induced glutathione S-transferase placental form hepatic foci by pretreatment of rats with phenobarbital and buthionine sulfoximine. *Carcinogenesis* 14:1469–1470.

Grosman ME, Elias MM, Comin EJ, and Rodriguez Garay EE (1983) Alterations in renal function induced by aflatoxin B_1 in the rat. *Toxicol Appl Pharmacol* 69:319–325.

Groves CE, Morales M, and Wright SH (1998) Peritubular transport of ochratoxin A in rabbit renal proximal tubules. *J Pharmacol Exp Ther* 284:943–948.

Gumprecht LA, Marcucci A, Weigel RM, Vesonder RF, Riley RT, Showker JL, Beasley VR, and Haschek WM (1995) Effects of intravenous fumonisin B_1 in rabbits: Nephrotoxicity and shingolipid alterations. *Natural Toxins* 3:395–404.

Hagelberg S, Hult K, and Fuchs R (1989) Toxicokinetics of ochratoxin A in several species and its plasma-binding properties. *J Appl Toxicol* 9:91–96.

Hanika C, and Carlton WW (1983) Citrinin mycotoxicosis in the rabbit. *Fd Chem Toxicol* 21:487–493.

Hard GC, Howard PC, Kovach RM, and Bucci TJ (2001) Rat kidney pathology induced by chronic exposure to fumonisin B_1 includes rare variants of renal tubule tumor. *Toxicol Pathol* 29:379–386.

Harrison LR, Colvin BM, Greene JT, Newman LE, and Cole JR (1990) Pulmonary edema and hydrothorax in swine produced by fumonisin B_1, a toxic metabolite of *Fusarium moniliforme. J Vet Diag Invest* 2:217–221.

Harvey RB, Elissalde MH, Kubena LF, Weaver EA, Corrier DE, and Clement BA (1992) Immunotoxicity of ochratoxin A to growing gilts. *Am J Vet Res* 53:1966–1970.

Haschek WM, Motelin G, Ness DK, Harlin KS, Hall WF, Vesonder RF, Peterson RE, and Beasley VR (1992) Characterization of fumonisin in orally and intravenously dosed swine. *Mycopathologia* 117:83–96.

Hayes AW (1980) Mycotoxins: A review of biological effects and their role in human diseases. *Clin Toxicol* 17:45–83.

Hayes JR, and Campbel TC (1986) Food additives and contaminants, in *Toxicology the basic science of poisons*, Casseret LJ (Ed.), McMillan, New York, pp. 771–800.

Hetherington AC, and Raistrick H (1931) Studies in the biochemistry of micro-organisms. Part XIV. On the production of and chemical constitution of of a new yellow coloring matter, citrinin, produced from glucose by Penicillium citrinum. *Phil Trans Royal Soc* 220:269–295.

Heussner A, Schwöbel F, and Dietrich DR (1998) Cytotoxicity of ochratoxin A, and B in vitro: Comparison of male and female rat primary renal cortex and distal cells and LLC-PK1 cells. *Toxicol Sci* 42:1416.

Heussner A, Stack M, Hochberg K, and Dietrich D (2000) Comparison of cytotoxic effects of ochratoxin A, and B on human, rat and porcine renal cells. *Toxicol Sci* 54:170.

Heussner AH, O'Brien E, and Dietrich DR (2002) Species- and sex-specific variations in binding of ochratoxin A by renal proteins *in vitro. Exp Toxicol Pathol* 54:151–159.

Hinton DM, Riley RT, Showker JL, and Rigsby W (1989) Patulin induced ion flux in renal cells and reversal by dithiothretol and glutathione: A scanning electron microscopy (SEM) X-ray microanalysis study. *J Biochem Toxicol* 4:47–54.

Hood RD, Naughton MJ, and Hays AW (1976) Prenatal effects of ochratoxin A in hamsters. *Teratol* 13:11–14.

Horvath A, Upham BL, Ganev V, and Trosko JE (2002) Determination of the epigenetic effects of ochratoxin in a human kidney and a rat liver epithelial cell line. *Toxicon* 40:273–282.

Howard PC, Eppley RM, Stack ME, Warbritton A, Voss KA, Lorentzen RJ, Kovach RM, and Bucci TJ (2000b) Carcinogenicity of fumonisin B_1 in a two-year bioassay with Fischer 344 rats and B6C3F_1 mice. *Mycopathologia* Supl. 99:45–54.

Howard PC, Warbritton A, Voss KA, Lorentzen RJ, Thurman JD, Kovach RM, and Bucci TJ (2000a) Compensatory regeneration as a mechanism for renal tubule carcinogenesis of fumonisin B_1 in the F344/N/Nctr BR rat. *Envion Health Perspect* 109:309–314.

IARC (1993) IARC monographs on the evaluation of carcinogenis risks to humans, Some naturally occurring substances, food items and constituents, heterocyclic aromatic amines and mycotoxins., in pp 56, 362.

Joffe AZ (1986) *Fusarium species, their biology and toxicology.* John Wiley & Sons, New York.

Jordan WH, and Carlton WW (1977) Citrinin mycotoxicosis in the mouse. *Fd Cosmet Toxicol* 15:29–34.

Jordan WH, and Carlton WW (1978) Citrinin mycotoxicosis in the rat. I Toxicology and pathology. *Fd Cosmet Toxicol* 16:431–449.

Jordan WH, Carlton WW, and Sansing GA (1978a) Citrinin mycotoxicosis in the rat. II Clinicopathological observations. *Fd Cosmet Toxicol* 16:441–447.

Jordan WH, Carlton WW, and Sansing GA (1978b) Citrinn mycotoxicosis in the Syrian hamster. *Fd Cosmet Toxicol* 16:355–363.

Jung KY, Takeda M, Kim DK, Tojo A, Narikawa S, Yoo BS, Hosoyamada M, Cha SH, Sekine T, and Endou H (2001) Characterization of ochratoxin A transport by human organic anion transporters. *Life Sciences* 69:2123–2135.

Kellerman TS, Marasas WFO, Thiel PG, Gelderblom WCA, Cawood M, and Coetzer AW (1990) Leucoencephalomalacia in two horses induced by oral dosing of fumonisin B_1. *Onderstepoort J Vet Res* 57:269–275.

Kitchen DN, Carlton WW, and Hinsman EJ (1977) Ochratoxin A, and citrinin-induced nephrosis in beagle dogs III. Terminal ultrastructural alterations. *Vet Pathol* 14:392–406.

Koepsell H, Gorbulev V, and Arndt P (1999) Molecular pharmacology of organic anion transporters in kidney. *J Membr Biol* 167:103–117.

Kogika MM, Hagiwara KM, and Mirandola RMS (1996) Experimental citrinin nephrotoxicosis in dogs. *Vet Human Toxicol* 35:136–140.

Kreisberg JI, and Wilson PD (1988) Renal cell culture. *J Electron Microsc Tech* 9:235–263.

Krejci ME, Bretz NS, and Koechel DA (1996) Citrinin produces acute adverse changes in renal function and ultrastructure in phenobarbital-anaesthetized dogs without concommitant reductions in $(\text{potassium})_{\text{plasma}}$. *Toxicol* 106:167–177.

Krogh P, Elling F, Gyrd-Hansen N, Hald B, Larsen AE, Lillehoj EB, Madsen A, Mortensen HP, and Ravnskov U (1976) Experimental porcine nephropathy: Changes of renal function and structure perorally induced by crystalline ochratoxin A. *Acta Pathol Microbiol Scand* [*A*] 84:429–434.

Krogh P, Hesselager E, and Friss P (1973) Occurrence of ochratoxin A, and citrinin in cereals associated with mycotoxic porcine nephropathy. *Acta Pathol Microbiol Scand* [*B*] 81:689–695.

Kuiper-Goodman T (1999) Approaches to the risk analysis of mycotoxins in the food supply, in *Third joint FAO/WHO/UNEP international conference on mycotoxins*, Tunis, Tunisia.

Kuiper-Goodman T, and Scott PM (1989) Risk assessment of the mycotoxin ochratoxin A. *Biomed Environ Sci* 2:179–248.

Le Bourhis B (1984) La patuline, un contaminant du jus de pomme. *Med et Nut* T.XX:23.

Li S, Marquardt RR, Frohlich AA, Vitti TG, and Crow G (1997) Pharmacokinetics of ochratoxin A, and its metabolites in rats. *Toxicol Appl Pharmacol* 145:82–90.

Lockard VG, Phillips RD, Wallace Hayes A, Berndt WO, and O'Neal RM (1980) Citrinin nephrotoxicity in rats: A light and electron microscopic study. *Exp Mol Pathol* 32:226–240.

Loe DW, Stweart RK, Massey TE, Deeley RG, and Cole SPC (1997) ATP-dependent transport of aflatoxin B1 and its glutathione conjugates by the product of the multidrug resistance protein (MRP). *Mol Pharmacol* 51:1034–1041.

Lüthy J, Sweifel U, and Schlatter C (1980) Metabolism and tissue distribution of ^{14}C-Aflatoxin B_1 in pigs. *Fd. Cosmet. Toxicol.* 18:253–256.

Maaroufi K, Achour A, Betbeder A-M, Hammami M, Ellouz F, Creppy EE, and Bacha H (1995a) Foodstuffs and human blood contamination by the mycotoxin ochratoxin A: correlation with chronic interstitial nephropathy in Tunisia. *Arch Toxicol* 69:552–558.

Maaroufi K, Achour A, Hammami M, el May M, Betheder AM, Ellouz F, Creppy EE, and Bacha H (1995b) Ochratoxin A in human blood in relation to nephropathy in Tunisia. *Hum Exp Toxicol* 14:609–614.

Maaroufi K, Zakhama A, Baudrimont I, Achour A, Abid S, Ellouz F, Dhouib S, Creppy EE, and Bacha H (1999) Karyomegaly of tubular cells as early stage marker of the nephrotoxicity induced by ochratoxin A in rats. *Hum Exp Toxicol* 18:410–415.

Mako M, Kuroda K, and Wogan GN (1971) Aflatoxin B_1 the kidney as site of action in the mouse. *Life Sciences* 10:495–501.

Marasas WFO, Kellerman TS, Gelderblom WCA, Coetzer JAW, Thiel PG, and Van der Lugt JJ (1988) Leukoencephalomalacia in horse induced by fumonisin B_1 isolated from Fusarium moniliforme. *Onderstepoort J Vet Res* 55:197–203.

Mathur S, Constable PD, Eppley RM, Waggonner AL, Tumbleson ME, and Haschek WM (2001) Fumonisin B_1 is hepatotoxix and nephrotoxic in milk-fed calves. *Toxicol Sci* 60:385 -396.

McKinley ER, Carlton WW, and Boon GD (1982) Patulin mycotoxicosis in the rat: Toxicology, pathology and clinical pathology. *Fd Chem Toxic* 20:289–300.

Medicine NLO (2002) Aflatoxins, in, Hazardous Substance Data Base.Toxnet (National Data Network).

Meisner H, and Chan S (1974) Ochratoxin A, in inhibitor of mitochondrial transport system. *Biochem* 13:2795–2800.

Merrill Jr. AH, Schmelz E-M, Wang E, Schroeder JJ, Dillehay DL, and Riley RT (1995) Role of dietary sphingolipids in and inhibitors of sphingoid metabolism in cancer and other diseases. *J Nutr* 125:1677S–1682S.

Mobio TA, Anane R, Baudrimont L, Carratu MR, Shier WT, Dano-Djedjc S, Ueno Y, and Creppy E-E- (2000) Epigenetic properties of fumonisin B_1: Cell cycle arrest and DNA base modification in C6 glioma cells. *Toxicol Appl Pharmacol* 164.

Moore JH, and Truelove B (1970) Ochratoxin A: Inhibition of mitochondrial respiration. *Science* 168:1102–1103.

Morrissey RE, Norred WP, and Hinton DM (1987) Combined effects of the mycotoxins aflatoxin B_1 and Cyclopiazonic acid on Sprague-Dawley rats. *Food Chem Toxicol* 25:837–842.

Müller G, Kielstein P, Berndt A, Heller M, and Köhler H (1999a) Studies on the influence of combined administration of ochratoxin A, fumonisin B_1, deoxynivalenol and T2 toxin on immune and defence reactions in weaner pigs. *Mycoses* 42:485–493.

Müller G, Kielstein P, Rosner H, Berndt A, Heller M, and Kohler H (1999b) Studies of the influence of ochratoxin A on immune and defence reactions in weaners. *Mycoses* 42:495–505.

Nesbit BF, O'Kelly J, Sargent K, and Sheridan A (1962) Toxic metabolites of Aspergillus flavus. *Nature* 195:1062–1063.

Netke SP, Roomi MW, Tsao C, and Niedzwiecki A (1997) Ascorbic acid protects guinea pigs from acute aflatoxin toxicity. *Toxicol Appl Pharmacol* 143:429–435.

Newberne PM, Carlton WA, and Wogan GN (1964) Hepatomas in rats and hepatorenal injury in ducklings fed peanut meal or *Aspergillus flavus* extract. *Pathol Vet* 1:105–132.

Norred WP (1993) Fumonisins-Mycotoxins produced by Fusarium moniliofrme. *J Toxicol Environ Health* 38:309–328.

Norred WP, Plattner RD, Vesonder RF, Bacon CW, and Voss KA (1992) Effects of selected secondary metabolites of Fusarium moniliforme on unscheduled synthesis of DNA by primary rat hepatocytes. *Food Chem Toxicol* 30:233–237.

Norred WP, Voss KA, Riley RT, Meredith FI, Bacon CW, and Merrill Jr. AH (1998) Mycotoxins and health hazards: Toxicological aspects and mechanism of action of fumonisins. *J Toxicol Sci* 23:160–164.

Norred WP, Voss KA, Riley RT, and Plattner RD (1996) Fumonisin toxicity and metabolism studies at the USDA, in *Fumonisins in food*, Jackson L (Ed.), Plenum Press, New York.

O'Brien E, Heussner AH, and Dietrich D (2001) Species-, sex- and cell type specific effects of ochratoxin A, and B. *Toxicol Sci* 63:256–264.

Obrecht-Pflumio S, and Dirheimer G (2001) Horseradish peroxidase mediates DNA, and deoxyguanosine 3′-monophosphate adduct formation in the presence of ochratoxin A. *Arch Toxicol* 75:583–590.

Obrecht-Pflumio S, Chassat T, Dirheimer G, and Marzin D (1999) Genotoxicity of ochratoxin A by Salmonella mutagenicity test after bioactivation by mouse kidney microsomes. *Mutat Res* 446:95–102.

Omar RF, Hasinoff BB, Mejilla F, and Rahimtula AD (1990) Mechanism of ochratoxin a stimulated lipid peroxidation. *Biochem Pharmacol* 40:1183–1191.

Pfohl-Leszkowicz A, Chakor K, Creppy EE, and Dirheimer G (1991) DNA adduct formation in mice treated with ochratoxin A, in *Mycotoxins, endemic nephropathy and urinary tract tumours*, Castegnaro M, Plestina R, Dirheimer G, Chernozemsky IN, and Bartsch H (Eds), International Agency for Research on Cancer, Lyon, pp. 245–253.

Pfohl-Leszkowicz A, Grosse Y, Kane A, Creppy EE, and Dirheimer G (1993) Differential DNA adduct formation and disappearance in three mouse tissues after treatment with the mycotoxin ochratoxin A. *Mutation Research* 289:265–273.

Phillips TD, and Hayes W (1979) Inhibition of electrogenic sodium transport across toad bladder by the mycotoxin patulin. *Toxicol* 13:17–24.

Prozzi CR, Correa B, Xavier JG, Direito GM, Orsi RB, and Matarazzo SV (2000) Effects of prolonged oral administratin of fumonisin B_1 and aflatoxin B_1 in rats. *Mycopathologia* 151:21–27.

Radic B, Fuchs R, Peraica M, and Lucic A (1997) Ochratoxin A in human sera in the areawith endemic nephropathy in Croatia. *Toxicol Lett* 91:105–109.

Rahimtula AD, Bereziat JC, Bussacchini GV, and Bartsch H (1988) Lipid peroxidation as a possible cause of ochratoxin A toxicity. *Biochem Pharmacol* 37:4469–4477.

Raju MVLN, and Devegowda G (2000) Influence of esterified-glucomannan on performance and organ morphology, serum biochemistry and haematology in briolers exposed to individual and combined mycotoxicosis (aflatoxin, ochratoxin and T-2 toxin. *Br Poult Sci* 41:640–650.

Ramsdell HS, and Eaton DL (1990) Mouse liver glutathione S-transferase isozyme activity toward aflatoxin B1–8, 9-epoxide and benzo(a)pyrene-7, 8-dihydro-9, 10-epoxide. *Toxicol Appl Pharmacol* 105:216–225.

Rappa G, Lorico A, Flavell RA, and Sartorelli AC (1997) Evidence that the multidrug resistance protein (MRP) functions as a co-transporter of glutathione and natural product toxins. *Cancer Res* 57:5232–5237.

Rásonyi T, Dietrich DR, Candrian R, Schlatter J, and Schlatter C (1993) The role of $\alpha 2\mu$-globulin in ochratoxin A induced kidney tumors. *The Toxicologist* 13:132.

Rásonyi T, Schlatter J, and Dietrich DR (1999) The role of a2u-globulin in ochratoxin A induced renal toxicity and tumors in F344 rats. *Toxicol Lett* 104:83–92.

Rati ER, Shantha T, and Ramesh HP (1991) Effect of long term feeding and withdrawal of aflatoxin B_1 and ochratoxin A on kidney cell transformation in albino rats. *Indian J Exp Biol* 29:813–817.

Riley RT, An NH, Showker JL, Yoo HS, Norred WP, Chamberlain WJ, Wang E, Merrill Jr. AH, Motelin G, Beasley VR, and Haschek WM (1993) Alteration of tissue and serum sphinganine to sphingosine ratio-an early biomarker of exposure to fumonisin containing feeds in pigs. *Toxicol Appl Pharmacol* 118:105–112.

Riley RT, Hinton DM, Chamberlain WJ, Bacon CW, Wang E, Merrill Jr. AH, and Voss KA (1994) Dietary fumonisin B_1 induces disruption of sphingolipid metabolism in Sprague-Dawley rats: A new mechanism of nephrotoxicity. *J Nutr* 124:594–603.

Riley RT, Hinton DM, Showker JL, Rigsby W, and Norred WP (1990) Chronology of patulin-induced alterations in membrane function of cultured renal cells, LLC-PK1. *Toxicol Appl Pharmacol* 102:128–141.

Riley RT, and Showker JL (1991) The mechanism of patulin's cytotoxicity and the antioxidant activity of indole tetramic acids. *Toxicol Appl Pharmacol* 109:108–126.

Riley RT, Wang E, Schroeder JJ, Smith ER, Plattner RD, Abbas H, Yoo HS, and Merrill Jr. AH (1996) Evidence for disruption of sphingolipid metabolism as a contributing factor in the toxicity and carcinogenicity of fumonisins. *Natural Toxins* 4:3–5.

Russel FGM, Masereeuw R, and van Aubel RAMH (2002) Molecular aspects of renal anionic drug transport. *Annu Rev Physiol* 64:563–594.

Schwerdt G, Freudinger R, Mildenberger S, Silbernagl S, and Gekle M (1999) The nephrotoxin ochratoxin A induces apoptosis in cultured human proximal tubule cells. *Cell Biol Toxicol* 15:405–415.

Scott PM, Van Walbeck W, Kennedy B, and Anyeti D (1972) Mycotoxins (ochratoxin A, citrinin and sterigmatocystin) and toxigenic fungi in grains and agricultural products. *J. Agric Food Chem* 20:1103–1109.

Searle CE (Ed.) (1976) *Chemical Carcinogens.* American Chemical Society, Washington, D.C..

Seegers JC (1994) A comparative study of ochratoxin A-induced apoptosis in hamster kidney and HeLa cells. *Toxicol Appl Pharmacol* 129:1–11.

Shaddock JG, Feuers RJ, Chou MW, Pegram RA, and Casciano DA (1993) Effects of aging and caloric restriction on the genotoxicity of four carcinogens in the *in vitro* rat hepatocyte/DMA assay. *Mutat Res* 295:19–30.

Shayman JA (2000) Sphingolipids. *Kidney Int* 58:11–26.

Shinohara Y, Arai M, Hirao K, Sugihara S, Nakanishi K, Tsunoda H, and Ito N (1976) Combination effects of citrinin and other chemicals on rat kidney tumorigenesis. *Gann* 67:147–155.

Shreeve BJ, Patterson SP, Pepin GA, Roberts BA, and Wrathall AE (1977) Effect of feeding ochratoxin to pigs during early pregnancy. *Br Vet J* 133:412–417.

Sokol PP, Ripich G, Holohan PD, and Ross CR (1988) Mechanism of ochratoxin A transport in kidney. *J Pharmacol Exp Ther* 246:460–465.

Sostaric B, and Vukelic M (1991) Characteristics of urinary tract tumours in the area of Balkan endemic nephropathy in Croatia, in *Mycotoxins, endemic nephropathy and urinary tract tumours*. Castegarno M, Plestina R, Dirheimer G, Chernozemsky IN, and Bartsch H (Eds), IARC Scientific Publications, Lyon, pp. 29–35.

Speijers GJA, Franken MAM, and van Leeuwen FXR (1988) Subacute toxicity study of patulin in the rat: Effects on the kidney and the gastro-intestinal tract. *Food Chem Toxicol* 26:23–30.

Speijers GJA, Kolkman R, Franken MAM, van Leeuwen FXR, and Danse LHJC (1984) Subactue toxiciteit van patuline in de rat., in, National Institute of Public Health and Environmental Hygiene.

Speijers GJA, and van Egmond HP (1993) Worldwide ochratoxin A levels in food and feeds, in *Human ochratoxicosis and its pathologies*, pp. 85–100, John Libbey Eurotext Ltd., Montrouge, France.

Stein AF, Phillips TD, Kubena LF, and Harvey RB (1985) Renal tubular secretion and reabsorption as factors in ochratoxicosis: Effects of probenecid on nephrotoxicity. *J Toxicol and Environ Health* 16:593–605.

Steyn PS (1995) Mycotoxins, general view, chemistry and structure. *Toxicol Lett* 82/83:843–851.

Stoev SD, Anguelov G, Ivanov I, and Pavlov D (2000) Influence of ochratoxin A, and an extract of artochoke on the vaccinal immunity and health in broiler chicks. *Exp Toxicol Pathol* 52:43–55.

Stoev SD, Hald B, and Mantle PG (1998a) Porcine nephropathy in Bulgaria: a progressive syndrome of complex or uncertain (mycotoxin) aetiology. *Vet Rec* 142:190–194.

Stoev SD, Stoeva JK, Anguelov G, Hald B, Creppy EE, and Radic B (1998b) Haematological, biochemical and toxicological investigations in spontaneous cases with different frequency of porcine nephropathy in Bulgaria. *Zentralbl Veterinarmed A* 45:229–236.

Stoev SD, Vitanov S, Anguelov G, Petkova-Bocharova T, and Creppy EE (2001) Experimental mycotoxic nephropathy in pigs provoked by a diet contaning ochratoxin A, and penicillic acid. *Vet Res Commun* 25:205–223.

Stormer FC, and Hoiby EA (1996) Citrinin, ocratoxin A, and iron. Possible implications for their biological function and induction of nephropathy. *Mycopathologia* 134:103–107.

Studer-Rohr I (1995) Ochratoxin A in Humans: Exposure, Kinetics and Risk Assessment, in p 100, Swiss Federal Institute of Technology Zürich, Zürich.

Studer-Rohr J, Schlatter J, and Dietrich DR (2000) Intraindividual variation in plasma levels and kinetic parameters of ochratoxin A in humans. *Arch Toxicol* 74:499–510.

Suzuki CAM, Hierlihy L, Barker M, Curran I, Mueller R, and Bondy G (1995) The effects of fumonisin B_1 on several markers of nephrotoxocity in rats. *Toxicol Appl Pharmacol* 133:207–214.

Sydenham EW, Shephard GS, Thiel PG, Marasas WFO, and Stockenstron S (1991) Fumonisin contamination of commercial corn-based foodstuffs. *J Agric Food Chem* 39:2014–2018.

Tatu CA, Orem WH, Finkelman RB, and Feder GL (1998) The etiology of Balkan endemic nephropathy: Still more questions than answers. *Environ Health Perspect* 106:689–699.

Thacker HL, Carlton WW, and Sansing GA (1977) Citrinin mycotoxicosis in the guinea pig. *Food Cosmet Toxicol* 15:553–562.

Thurm V, Paul P, and Koch C (1979) Zur hygienischer Vorkommen von Patulin in Obst und Gemüse. *Die Nahrung* 23:131.

Tolleson WH, Melchior Jr. WB, Morris SM, McGarrity LJ, Domon OE, Muskhelishvili L, James SJ, and Howard PC (1996) Apoptotic and antiproliferative effects of fumonisin B_1 in human keratinocytes, fibroblasts, esophageal epithelial cells and hepatoma cells. *Carcinogenesis* 17:239–249.

Tsuda M, Sekine T, Takeda M, Cha SH, Kanai Y, Kimura M, and Endou H (1999) Transport of ochratoxin A by renal multispecific organic anion transporter 1. *J Pharmacol Exp Ther* 289:1301–1305.

Tune BM (1974) Relationship between the transport and toxicity of cephalosporins in the kidney. *J Infect Dis* 132:189–194.

Ueno Y, Umemori K, Niimi E-C, Tanuma S-I, Nagata S, Sugamata M, Ihara T, Sekijima M, Kawai K-I, Ueno I, and Tashiro F (1995) Induction of apoptosis by T-2 toxin and other natural toxins in HL-60 human promyelotic leukemia cells. *Natural Toxins* 3:129–137.

Uwai Y, Okuda M, Takami K, Hashimoto Y, and Inui KI (1998) Fuctional characterization of the rat multispecific organic anion transporter Oat1 mediating basolateral uptake of anionic drugs in the kidney. *FEBS Lett* 438:321–324.

Valdivia AG, Martinez A, Damian FJ, Quezada T, Ortiz R, Martinez C, Llamas J, Rodriguez ML, Yamamoto L, Jaramillo F, Loarca-Pina MG, and Reyes JL (2001) Efficacy of N-acetylcysteine to reduce the effects of aflatoxin B_1 intoxication in broiler chickens. *Poult Sci* 80:727–734.

van Aubel RAMH, Masereeuw R, and Russel FGM (2000) Molecular pharmacology of renal organic anion transporters. *Am J Physiol Renal Physiol* 279:F 216–F232.

van der Merwe KJ, Steyn PS, Fourie L, Scott DB, and Theron JJ (1965) Ochratoxin A, a toxic metabolite produced by Aspergillus ochraceus Wilh. *Nature* 205:1112–1113.

Voss KA, Chamberlain WJ, Bacon CW, and Norred WP (1993) A preliminary investigation on renal and hepatic toxicity in rats fed purified fumonisin B_1. *Natural Toxins* 1:222–228.

Voss KA, Plattner RD, Riley RT, Meredith FI, and Norred WP (1998) *In vivo* effects of fumonisin B_1-producing and fumonisin B_1-nonproducing *Fusarium moniliforme* isolates are similar: FumonisinsB_2 and B_3 cause hepato- and nephrotoxicity in rats. *Mycopathologia* 141:45–58.

Vukelic M, Sostaric B, and Fuchs R (1991) Some Pathomorphological features of Balkan endemic nephropathy in Croatia, in *Mycotoxins, endemic nephropathy and urinary tract tumours.* Castegnaro M, Plestina R, Dirheimer G, Chernozemsky IN, and Bartsch H (Eds), IARC Scientific Publications, Lyon, pp. 37–42.

Wang E, Ross PF, Wilson TM, Riley RT, and Merrill AH (1991) Increases in serum sphingosine and sphinganine and decreases in complex sphingolipids in ponies given feed containing fumonisin. *J Nutr* 122:1706–1716.

Wei Y-H, Lu C-Y, Lin T-N, and Wei R-D (1985) Effect of ochratoxin A on rat liver mitochondrial respiration and oxidative phosphorylation. *Toxicol* 36:119–130.

Wilde O (1895) The Importance of Being Earnest, in, London.

Wilson TM, Ross PF, Owens DL, Rice LG, Green SA, Jenkins SJ, and Nelson HA (1992) Experimental reproduction of ELEM-A study to determine the minimum toxic dose in ponies. *Mycopathologia* 117:115–120.

Wolf P, O'Brien E, Heussner AH, Stack ME, Thiel R, and Dietrich DR (2002) Sex- and age- specific effects of ochratoxin A in primary human kidney cells (HKC). *Toxicol Sci* 66:400.

Wolff J, Bresch H, Cholmakow-Bodechtel C, Engel G, Erhardt S, Gareis M, Majerus P, Rosner H, and Scheuer R (2000) Belastung des Verbrauchers und der Lebensmittel mit Ochratoxin A, in p 243, forschungsverbund Produkt- und Ernährungsforschung des Bundesministeriums für Ernährung, Landwirtschaft und Forsten.

Wong ZA, and Hsieh DPH (1980) The comparative metabolism and toxicokinetics of aflatoxin B1 in the monkey, rat and mouse. *Toxicol Appl Pharmacol* 55:115–125.

Yoneyama M, Sharma RP, and Kleinschuster SJ (1986) Cytotoxicity of citrinin in cultured kideny epithelial cell systems. *Ecotoxicol Environ Saf* 11:100–111.

Yoo HS, Norred WP, Wang E, Merrill Jr. AH, and Riley RT (1992) Fumonisin inhibition of *de novo* sphingolipid biosynthesis and cytotoxicity are correlated in LLC-PK1 cells. *Toxicol Appl Pharmacol* 113:9–15.

Zepnik H, Pahler A, Schauer U, and Dekant W (2001) Ochratoxin A-Induced Tumor Formation: Is There a Role of Reactive Ochratoxin A Metabolites? *Toxicol Sci* 59:59–67.

22

Nephrotoxicology of Metals

Rudolfs K. Zalups and Gary L. Diamond

INTRODUCTION

There are numerous metals that can exert nephrotoxic effects in humans and other mammals, so it would be an insurmountable task to provide useful information pertaining to the nephrotoxicology of all of them within the confines of a single chapter of a text dealing with the general subject of renal toxicology. This chapter, therefore, focuses on four of the more environmentally and occupationally relevant heavy metals that are nephrotoxic: cadmium, mercury, lead, and uranium. It is hoped that the discussions of these four metals will make it clear to the reader how complex and varied the mechanisms are that induce injurious effects in the kidneys.

NEPHROTOXICOLOGY OF CADMIUM

Renal Handling of Cadmium

In humans exposed to cadmium (Cd) via oral or pulmonary routes, the kidney is the primary organ affected adversely by Cd (ATSDR, 1999). Risk assessment data indicate that when the renal concentration of Cd begins to exceed 50 μg/g kidney (wet weight) there is a significant risk for the induction of renal tubular injury and impaired renal function. Additionally, there are findings that indicate that ingestion of approximately 30 μg of Cd per day may result in mild forms of renal dysfunction in about 1% of the adult population, depending on individual variations in absorption and sensitivity to the toxic effects of Cd. Despite all of the studies implicating the kidney as a target organ affected adversely by Cd, very little is known about the mechanisms involved in the renal handling and toxicity of Cd.

Unlike other nephrotoxic metals, such as mercury (Hg) and uranium, which exert their toxic effects mainly in the straight portions of the proximal tubule, Cd appears to adversely affect the proximal convoluted tubule. One should have a good understanding of the mechanisms involved in the uptake and transport of Cd in the target epithelial cells in order to gain a more precise understanding of the mechanisms by which Cd induces its toxic effects.

Some investigators have suggested that under conditions generated by chronic exposure to Cd, complexes of Cd–metallothionein (MT) (formed in hepatocytes in response to the uptake of Cd) are released from necrotic hepatocytes and are delivered (via systemic circulation) to the kidneys. It appears that the complexes are taken up in the kidney and induce proximal tubular injury and death (Zalups and Ahmad, 2003). Although there is evidence indicating that Cd–MT is delivered to the kidneys and that it induces proximal tubular injury and death (Zalups and Ahmad, 2003), it is clearly not the only species of Cd delivered to, or taken up by, the kidneys. This is particularly evident after an acute pulmonary or parenteral exposure to a nonhepatotoxic dose of a Cd-salt, such as $CdCl_2$. Based on the chemical properties of Cd, there is little doubt that thiol-containing molecules, such as albumin, glutathione (GSH) and cysteine (Cys), form conjugates with Cd, and that these molecules probably serve as molecular shuttles that deliver Cd to the epithelial cells that take up and transport Cd within the kidneys. Support for this hypothesis comes from the *in vivo* data of Zalups (2000a), which show that the net renal uptake and accumulation of Cd increases significantly when Cd is administered systemically as a conjugate of Cys or GSH.

Intrarenal Disposition and Accumulation of Cadmium

When relatively low doses of Cd are injected intravenously into rats as $CdCl_2$, only a very small percentage of each dose becomes localized in the combined renal mass, while between 50% and 60% of the dose localizes in the liver (Zalups, 2000a). Approximately 1.5 to 1.8% of the administered dose of Cd accumulates in the total renal mass during the first hour after exposure (Zalups, 2000a; Zalups and Barfuss, 2002a). During the remainder of the first 24 hours, the renal burden of Cd increases only slightly, to about 2% of the dose (Zalups, 1997a, 2000a; Zalups and Barfuss, 2002). The majority of the renal burden of Cd is present in segments of nephron present in the renal cortex and outer stripe of the outer medulla.

Dorian et al. (1992) provided light microscopic, autoradiographic findings demonstrating the intrarenal distribution and disposition of ^{109}Cd in mice treated intravenously with $^{109}CdCl_2$. They showed that Cd was distributed evenly throughout the lengths of the proximal tubules in the renal cortex and outer medulla. At a lower dose (0.1 mg Cd/kg), Cd apparently localized in both proximal and distal segments of the nephron. After administration of a large dose (3 mg Cd/kg), Cd became localized mainly in proximal tubular segments. Additionally, the concentration of Cd in the epithelial cells of both pars convoluta and pars recta segments appeared to be similar, as well as being distributed evenly throughout the cytoplasm.

In this same study, Dorian et al. (1992) also evaluated the renal disposition of ^{109}Cd in mice treated intravenously with ^{109}Cd–MT. Their findings ostensibly showed that ^{109}Cd became localized mainly in the epithelial cells lining the S_1 and S_2 segments of the proximal tubule situated in the renal cortex. The relative concentration of ^{109}Cd in the basal and apical portions of the affected cells was similar when a nonnephrotoxic dose of Cd–MT was injected. By contrast, ^{109}Cd was distributed mainly in the apical portions of the proximal tubular cells when a nephrotoxic dose was used. The findings from this study also showed that the renal burden of Cd increased rapidly until approximately 85% of the administered dose of Cd was present in the kidneys. Subsequently, the renal content of Cd remained constant for up to seven days after treatment. The data from this study support the hypothesis that the nephropathy induced by Cd–MT may be due (at least in part) to the preferential uptake of Cd–MT by the epithelial cells lining the S_1 and S_2 (convoluted) segments of the proximal tubule (presumably by an endocytotic mechanism).

Filterability of Cadmium at the Renal Glomerulus

In order for Cd to be taken up from the luminal compartment of the nephron, it has to first pass through the glomerular filter or be secreted by one or more of the different tubular epithelial cells. Therefore, it is of great importance to determine, and gain a thorough understanding of, the molecular species of Cd that are filtered at the glomerulus following the different types of exposure.

The only study in which data on the filterability of Cd (at the renal glomerulus) have been obtained is the micropuncture study of Felley-Bosco and Diezi (1989). These investigators showed that only about 20% of the Cd in plasma, when infused as Cd–acetate, is filterable at the glomerulus, while 100% of the Cd, in the form of Cd–DTPA (diethylene triaminepenta acetate), is freely filterable. The findings obtained with Cd–acetate likely represent the glomerular filtration of Cd when it is administered systemically in the form of inorganic salts, while the findings obtained with Cd–DTPA likely reflect the glomerular handling of Cd–MT or Cd *S*-conjugates of Cys or GSH. It was also shown that much of the Cd ions that were ultrafiltered into the lumen of superficial nephrons were taken up mainly by the pars convoluta of proximal tubules. No transepithelial movement of Cd-DTPA was detected, probably because of the polar nature of DTPA.

Luminal Handling of Cadmium along the Proximal Tubule

By using the isolated perfused tubule technique, Robinson and colleagues (1993) discovered that all three segments (i.e., S_1, S_2, and S_3 segments) of the rabbit proximal tubule absorbed Cd avidly when Cd ions were perfused through the lumen in the form of $CdCl_2$. S_1 segments of the proximal tubule accumulated Cd more rapidly and developed a more severe form of tubular injury (at concentrations greater than 500 μM) than the other two segments. These findings are consistent with the histopathological documentation that the pars convoluta of the proximal tubule is the most sensitive segment of the nephron to the toxic effects of Cd. In addition, under the conditions studied, saturation of absorptive transport could not be demonstrated in S_1 segments, but was demonstrable in S_2 and S_3 segments. Their findings also showed that only about 10% of the Cd taken up from the lumen was transported across the basolateral membrane, indicating that 90% of the absorbed Cd was retained in the proximal tubular epithelial cells.

Using *in vivo* tubular microperfusion, Felley-Bosco and Diezi (1987) studied the luminal handling of various forms of Cd. They microperfused into the lumen of superficial proximal and distal segments of the rat nephron. In experiments where ^{109}Cd was microinjected into superficial proximal convoluted tubules as $^{109}CdCl_2$, the fractional absorption of Cd was approximately 70%, which is consistent with the high rates of luminal disappearance flux (J_D) of Cd detected in the isolated perfused study of Robinson et al. (1993). By contrast, almost 90% of the Cd microinjected (as $^{109}CdCl_2$) into superficial distal convoluted tubules was recovered in the pelvic urine from the ipsilateral kidney, indicating that most of the absorption of Cd ions occurs in proximal tubular segments.

Endo and colleagues (Endo, 2002; Endo et al., 1998a, 1998b, 1998c, 1999) suggested that uptake of Cd by a highly transformed line of immortalized porcine proximal tubular (LLC-PK_1) cells (exposed to $CdCl_2$) occurs through a mechanism that is both temperature- and pH-dependent, and that can be inhibited by $ZnCl_2$ or $CuCl_2$. They have also suggested that Cd is taken up at the apical plasma membrane, in part by an inorganic anion-exchanger. Moreover, following efflux experiments in the presence tetraethylammonium (TEA), these investigators suggested that Cd is secreted from within LLC-PK_1 cells, across the apical plasma membrane, by a proton driven antiport (H^+/Cd^{2+}) of the organic cation transport system. Similar findings have apparently been obtained using luminal membrane vesicles isolated from the kidneys and small intestine of rats. Figure 22.1 provides a diagrammatic summary of putative mechanisms involved in the proximal tubular uptake of Cd.

Cd–MT is a form of Cd that is generated in the liver and kidneys (as well as other organs) *in vivo* after the exposure to salts of Cd. During chronic exposure to Cd, hepatocytes can become intoxicated, and once they undergo necrosis or apoptosis, Cd–MT is released into system circulation (Dudley et al., 1985; Tanaka et al., 1975). This pool of Cd–MT has been implicated in the nephropathy induced by chronic exposure to Cd. Nomiyama et al. (1998) provided data showing a relationship between increased plasma levels of Cd–MT and the induction of hepatic injury in rabbits treated chronically with a daily subcutaneous dose of Cd. They also noted that this relationship correlated with the levels of induced renal dysfunction. It should be kept in mind, however, that circulating Cd–MT arising from intestinal sources (Cherian, 1979; Cherian et al., 1978; Sugawara and Sugawara, 1991) may also contribute to the proximal tubular nephropathy associated with oral exposure to Cd. Based on its small size, Cd–MT is filtered readily at the glomerulus. Experimental evidence indicates that once this complex enters into the lumen of the proximal tubule, some of the Cd is absorbed by the

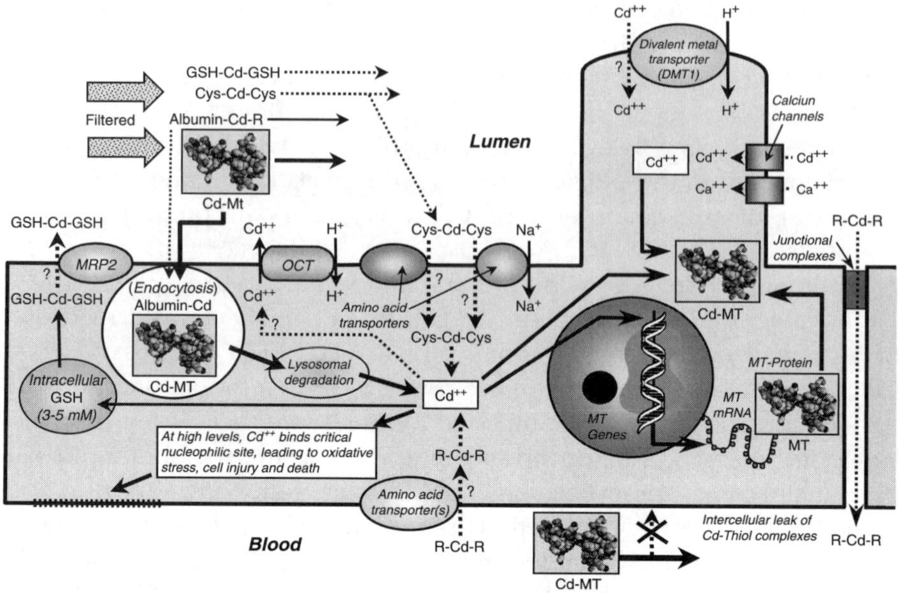

Figure 22.1 Diagrammatic representation of mechanisms involved in the luminal and basolateral handling of cadmium in proximal tubular epithelial cells. In this scheme, cysteine (Cys), glutathione (GSH) and albumin *S*-conjugates of Cd, as well as complexes of Cd bound to metallothionein (MT), are shown to be present in the lumen of the proximal tubule. This scheme shows that a significant fraction of filtered Cd that is bound to proteins (such as albumin or MT) is transported into the intracellular compartment during the endocytosis of the proteins to which the Cd ions are bound. There is a strong association between the *in vivo* administration of Cd–MT complexes and the induction of cellular necrosis in proximal tubular cells. Cys *S*-conjugates of Cd may act as molecular homologs of amino acids. It is possible that these conjugates enter into the proximal tubular epithelial cells via amino acid transporters located on luminal or basolateral plasma membranes. Current evidence indicates that when Cd is coadministered to rats with GSH or Cys there is a significant increase in the renal basolateral uptake and accumulation of Cd, perhaps as a result of these carrier systems. Cd may also be taken up at the luminal membrane of proximal tubular epithelial cells by calcium (Ca) channels. Although it is improbable that Cd exists as an unbound cation in the proximal tubular lumen, Cd ions may be delivered to the binding-sites of Ca-channels in the form of thiol- or protein-conjugates. It is also possible that Cd may "leak" through leaky junctional complexes into the basolateral compartment. This may particularly be the case when the epithelial cells begin to become intoxicated by Cd. It is highly unlikely that complexes of Cd–MT are taken up at the basolateral membrane because proximal tubular epithelial cells lack the

proximal tubular epithelial cells, presumably as the result of the endocytosis of Cd–MT (Felley-Bosco and Diezi, 1987; Nomiyama and Foulkes, 1977).

It is interesting, however, that findings from earlier studies by Zalups et al. (1995) and Liu et al. (1994) showed that the rates of luminal uptake of metals in the form of metal–MT complexes range from low to negligible in *in vitro* preparations of proximal tubules or their epithelial cells. Moreover, Felley-Bosco and Diezi (1987) demonstrated that when ^{109}Cd–MT is microinjected into the lumen of superficial proximal tubules, the fractional uptake of Cd is very low (8 to 17 per cent).

endocytotic machinery to accommodate such transport. The scheme presented also shows that a pool of Cd accumulates within the proximal tubular epithelial cells from the different sites of entry and that this pool can interact with various components within the cells. The pool of Cd that accumulates during and after periods of exposure provides a source of Cd ions that can induce the transcription of the genes for MT-1 and MT-2 by mechanisms that have not yet been fully defined. Numerous lines of evidence indicate that expression of renal MT genes, and especially in proximal tubular epithelial cells, increases markedly after exposure to Cd. If the intracellular pool of exchangeable Cd increases beyond what the protective elements inside the proximal tubular epithelial can handle, oxidative stress is induced. This can in turn alter mitochondrial respiratory activity and lead to lipid peroxidation in the plasma membrane and other perturbations in cellular metabolism. All of these effects can, and do, lead to the induction of cell death by either necrosis or apoptosis, which results in the release of Cd from within the necrotic or apoptotic proximal tubular epithelial cells into the tubular lumen. One intracellular molecule that protects cells from oxidative injury and binds Cd is GSH. This molecule is the most abundant nonprotein thiol in the body, and is present in proximal tubular epithelial cells at concentrations of between 3 and 5 mM. Some of the Cd that forms *S*-conjugates with GSH may be transported into the proximal tubular lumen by the ATP-binding-cassette protein MRP2, which has been shown to be expressed in the luminal plasma membrane of proximal tubular epithelial cells and has been shown to transport GSH *S*-conjugates of various molecules. Pathways depicted by solid black arrows are supported either directly or indirectly by experimental evidence. The thicker solid black lines denote pathways for which there is substantive evidentiary support. By contrast, the dashed lines represent pathways of secondary importance or pathways that are based primarily on indirect findings or speculation. Question marks associated with a pathway denote uncertainty for the occurrence of that pathway *in vivo*.

These findings suggest that Cd–MT is not absorbed very efficiently along the proximal tubule *in vivo*. Furthermore, Ishido et al. (1999) have shown that Cd–MT induces apoptosis in the kidneys of rats *in vivo*, but not in cultured proximal tubular (LLC-PK_1) cells, which adds to the body of findings indicating that the mechanisms involved in the proximal tubular uptake of Cd–MT *in vivo* have not yet been characterized fully. Liu et al. (1998) suggested that the acute administration of Cd–MT *in vivo* might not be an ideal model system for study of the nephropathy induced by chronic exposure to Cd salts.

Zalups (2000a) demonstrated that when Cd is coadministered with Cys or GSH, the uptake of Cd at both luminal and basolateral membranes increases significantly. Therefore, Cys and GSH *S*-conjugates of Cd may serve as transportable substrates and may provide an efficient means of delivering Cd to luminal and basolateral transporters of Cd situated along the nephron. Felley-Bosco and Diezi (1987) also demonstrated that the fractional absorption of Cd increases (to 82%) along the proximal tubule when Cd and Cys are comicroperfused into the lumen of the early proximal convoluted tubules of the rat kidney. These findings indicate that there is at least one mechanism, other than endocytosis, that is involved in the luminal absorption of Cd along the nephron.

Because both Cd and Hg are group IIB metals and share some of the same chemical properties, one must consider the possibility that Cd and Hg may enter into proximal tubular epithelial cells by one or more of the same mechanisms. A number of experimental findings from the laboratories of Zalups and Barfuss (Zalups, 1995; Zalups and Barfuss, 1998b; Cannon et al., 2000, 2001) indicate that specific amino acid transporters serve as the primary mechanisms responsible for transporting mercuric conjugates of Cys (i.e., Cys–Hg–Cys) across the luminal plasma membrane into proximal tubular epithelial cells. They have postulated that the absorption of the mercuric conjugate 2-amino-3-(2-amino-2-carboxyethylsulfanyl mercuric sulfanyl)-propionic acid (Cys–*S*–Hg–*S*–Cys) occurs via a mechanism involving molecular "mimicry" or homology, where Cys–*S*–Hg–*S*–Cys acts as a molecular homolog or "mimic" of the amino acid cystine (Cys–Cys) and Cys at the site of one or more amino acid carrier proteins (such as systems $B^{0,+}$, ASC, or $b^{0,+}$) involved in the luminal absorption of these amino acids. Based on these studies, it seems possible that Cys *S*-conjugates of Cd may also be transported by one or more of these same transporters.

Absorptive Transport of Cadmium in Distal Segments of the Nephron

Most studies on the renal disposition and handling of Cd have focused on the uptake of Cd by proximal segments of the nephron. However, several sets of investigators have provided both *in vivo* and *in vitro* lines of evidence implicating distal portions of the nephron in the renal handling of Cd (Dorian et al., 1992; Felley-Bosco and Diezi, 1987; Ferguson et al., 2001; Friedman and Gesek, 1994; Olivi et al., 2001). Potential transport mechanisms that have been implicated in the luminal uptake of Cd in distal segments of the nephron include calcium (Ca) channels and the divalent metal transporter DMT1. Both of these transporters are present in the epithelial cells of distal portions of the nephron and collecting duct.

Retention of Cadmium within Proximal Tubular Epithelial Cells

One would predict that Cd ions, once they have gained entry into the cytosolic compartment of proximal tubular epithelial cells, would tend to bind to intracellular protein thiols, largely because of their abundance in the intracellular milieu. Findings obtained from LLC-PK_1 cells exposed to Cd indicate that Cd ions taken up across the luminal membrane were bound primarily to cellular proteins shortly after exposure, as might have been predicted (Felley-Bosco and Diezi, 1991). Within three hours, the intracellular burden of Cd was shown to be associated with a low molecular weight pool of molecules. It is likely that MT was one of the primary ligands in that pool; significant induction of MT was demonstrated in the LLC-PK_1 cells within three to six hours of exposure.

In some *in vitro* preparations (such as in isolated perfused tubules; Robinson et al., 1993) it has been difficult to get a precise assessment of the intracellular distribution of Cd in proximal tubular cells. This is especially the case when acid precipitation methods are used to isolate and separate cellular proteins, because a significant amount of Cd can be liberated from cellular proteins and added to an acid-soluble compartment during the extraction and isolation of proteins. Such events make it almost impossible to gain an accurate representation of the intracellular localization of Cd in renal or other epithelial cells.

Secretion of Cadmium into the Tubular Lumen

As mentioned above, when LLC-PK_1 cells are exposed to $CdCl_2$, they apparently have the ability to secrete Cd at the apical plasma membrane by a proton-driven H^+/Cd^{2+} exchanger (Endo et al., 1998a, 1998b, 1998c, 1999; Endo, 2002). It is not clear, however, whether such a mechanism is functional in proximal tubular epithelial cells *in vivo*, especially when considering that the urinary excretion of Cd is extremely low after acute exposures to Cd salts.

If there is some level of secretion of Cd by any of the proximal tubular epithelial cells *in vivo*, the activity of multidrug resistance protein 2 (MRP2) may be involved. It would seem that proximal tubular epithelial cells possess the potential to secrete GSH *S*-conjugates of Cd into the lumen; epithelial cells contain high intracellular concentrations of GSH (Zalups and Lash, 1990; Zalups et al., 1999a, 1999b, 1999c), and MRP2, which is localized in the luminal membrane of proximal tubular epithelial cells, has been shown to transport various GSH *S*-conjugates (Schaub et al., 1997, 1999). Further studies are clearly needed to address this hypothesis.

Basolateral Uptake of Cadmium along the Nephron

One of the first investigators to suggest that there may be a basolateral mechanism in the renal uptake of Cd was Foulkes (1974). His suggestion was based on indirect findings from single-pass experiments in the rabbit. Subsequently, Diamond and colleagues (1986) provided *in vitro* evidence for a basolateral mechanism using the isolated perfused rat kidney. Specifically, they showed that occlusion of the ureter did not decrease significantly the net accumulation Cd in isolated kidneys perfused *in vitro* with 1 μM $CdCl_2$ in a protein-free buffer.

In a more recent study, Liu et al. (1994) demonstrated that when LLC-PK_1 cells were exposed to $CdCl_2$ or Cd–MT at their basolateral membrane, there was a significant level of association of Cd with these cells. A number of other studies have also examined basolateral handling of Cd *in vitro* using LLC-PK_1 cells grown on a permeable membrane insert (Bruggeman et al., 1992; Kimura et al., 1996; Prozialeck, 2000; Prozialeck and Lamar, 1993; Prozialeck et al., 1993). The findings from some of these studies indicate that the level of association of Cd is greater in these cells following basolateral exposure to Cd than luminal exposure to Cd. Regrettably, some of these findings are clouded by the occurrence of cellular intoxication induced by the Cd (Prozialeck et al., 1993; Prozialeck and Lamar, 1993). Although these findings tend to indicate that Cd can bind to the basolateral surface of, and be taken up into, the cultured cells, it is not clear whether the findings reflect the manner by which Cd is taken up at the basolateral membrane *in vivo*.

Zalups (2000a) provided strong *in vivo* evidence indicating that basolateral uptake of Cd does occur in the kidneys. This evidence shows that a significant level of renal uptake and accumulation of Cd occurred in rats treated with mannitol and bilateral ureteral ligation to reduce glomerular filtration rate (GFR) to negligible levels. In fact, the renal burden of Cd in these animals was as much as 70 to 80% of that in control rats. Based on these findings, one is lead to believe that between 70 to 80% of the acute renal uptake and accumulation of Cd that occurs in the kidneys of rats, following exposure to $CdCl_2$, is probably due to one or more basolateral mechanisms. Zalups (2000a) also demonstrated that when a low dose of $CdCl_2$ was coadministered with Cys or GSH, both luminal and basolateral uptake of Cd increased by at least 50%. These findings indicate that Cys or GSH *S*-conjugates of Cd are probably transportable substrates at the basolateral membrane and that they provide a more efficient means of delivering Cd to the sites of the transporters that take up Cd (Figure 22.1). Although it appears that basolateral uptake of Cd occurs in both the renal cortex and outer stripe of the outer medulla, it is unclear which specific segments of the nephron are involved in this process. However, since the three segments of the proximal tubule are present in these two zones of the kidney, they would appear to be logical choices for initial investigations.

In a series of recent studies, Zalups and colleagues provided an extensive amount of both *in vivo* and *in vitro* evidence implicating the classical *p*-aminohippurate-sensitive organic anion transport system, more specifically the organic anion exchanger (OAT1), in the basolateral uptake of mercuric conjugates of Cys, *N*-acetylcysteine (NAC) and GSH (Aslamkhan et al., 2003; Zalups; 1995, 1998a, 1998b; Zalups and Barfuss,1995a, 1995b, 1998a, 1998b, 2002b; Zalups and Minor, 1995). Due to the fact that Cd and Hg have a strong affinity for SH-groups, and that the basolateral uptake of both metals apparently increases when they are bonded to Cys or GSH, it is quite reasonable to postulate that the organic anion transport system may be involved in the basolateral uptake of Cys or GSH *S*-conjugates of Cd along the three segments of the proximal tubule. However, there is also the distinct possibility that one or more completely different mechanisms not involving the organic anion transporters are involved in the basolateral uptake of Cd.

Elimination and Excretion of Cadmium

Cd that is absorbed enterically is eliminated very slowly by urinary and fecal excretory mechanisms. It has been calculated that the half-life of absorbed Cd in mice, rats, rabbits, and monkeys ranges from several months to several years (Kjellstorm and Nordberg, 1985). The urinary and fecal excretion of absorbed Cd in animals have been observed as being approximately equal (Kjellstrom and Nordberg, 1978), although fecal excretion predominates after parenteral exposure to inorganic salts of Cd and urinary excretion of Cd predominates after parenteral exposure to Cd–MT. Findings from animal experiments indicate that the urinary excretion of Cd increases when Cd-induced renal tubular damage occurs (Friberg, 1984). The most likely explanation for this response is that Cd is added to the luminal compartment of nephrons from necrotic or apoptotic proximal tubular epithelial cells (Zalups et al., 1992a).

Urinary Excretion of Cadmium

Excretion of Cd in the urine is greatly dependent on the type of exposure. Following acute parenteral exposure to inorganic forms of Cd, the urinary elimination of Cd is negligible. Zalups (1997a) demonstrated in normal, uninephrectomized and 75% nephrectomized rats that less than 1% of an intravenous dose (8.9 μmol Cd/kg) was excreted by the end of the first week after exposure. Although the cumulative urinary excretion of Cd was extremely low in all three groups of rats, the 75% nephrectomized rats excreted significantly more Cd than the uninephrectomized or normal control rats during each day of the seven-day study.

When data on the renal accumulation and urinary excretion of Cd are assessed together, following a low parenteral dose of Cd, they indicate the following.

- The filtered load of Cd, over the initial hours and days subsequent to exposure, is very low, accounting for no more than 1 to 2% of the dose.
- Very little of the filtered load escapes absorption along the nephron.
- There is very little to no net secretion of Cd along the nephron.

Following exposure to Cd–MT (or chronic exposure to Cd-salts), the Cd–MT in blood filters readily into the luminal compartment of the nephron and the urinary excretion of Cd increases (relative to the level of Cd excreted in the urine following an acute exposure to $CdCl_2$). By having Cd in a filterable form, more Cd gets delivered to the luminal surface of proximal tubular epithelial cells, where Cd–MT can be potentially absorbed by one or more endocytotic processes. Interestingly, Zalups et al. (1995) demonstrated that the luminal uptake of Hg–MT was very low to negligible when the complex was perfused through the lumen of S_1, S_2, or S_3 segments of the rabbit proximal tubule. Zalups et al. (1993) also showed that both the renal accumulation of Hg and the urinary excretion of Hg increased greatly during the initial 72 hours in rats injected intravenously with a single nonnephrotoxic dose of Hg–MT (relative to that in rats treated with the same dose of Hg in the form of $HgCl_2$). By the end of the first three days after treatment, the cumulative urinary excretion of Hg was almost ninefold greater in rats treated with Hg–MT than in rats treated with $HgCl_2$. Additionally, Felley-Bosco and Diezi (1987) also demonstrated that the fractional uptake of Cd was very low when ^{109}Cd–MT was microinjected into the lumen of superficial proximal tubules. All of these findings indicate that Cd–MT and other divalent cationic forms of MT are not absorbed very efficiently along the proximal tubule *in vivo*. Thus, it appears that when divalent metal complexes of MT are delivered to the kidneys, the complexes are filtered readily into the tubular lumen and there is a significant level of tubular absorption and accumulation of the metal. However, the absorptive uptake is not very efficient, allowing a significant amount of the filtered load to be excreted in the urine.

Ottenwalder and Simon (1987) demonstrated a fourfold increase in the urinary excretion of Cd in rats given NAC (up to 100 mg/kg daily on six consecutive days, i.p.) after treatment with $CdCl_2$. It is not clear from this study how treatment with NAC promoted the urinary excretion of Cd, but it is likely that the Cd that entered into the luminal compartment was in the form of a nonabsorbable, highly polar, conjugate of NAC, which would promote the urinary excretion of Cd. Evidence supporting this hypothesis comes from two recent studies, in which the renal tubular absorption of Hg was very low when the Hg was in the form of a mercuric conjugate of NAC (Zalups, 1998b; Zalups and Barfuss, 1998a).

It has also been reported that repeated daily oral administration of certain chelating agents, such as diethylenetriaminepentaacetic, ethylenediaminetetraacetic acid (EDTA), and 2,3-dimercaptosuccinic acid (DMPS), after exposure to Cd caused the urinary excretion of Cd to increase (Klaassen et al., 1984). Despite these findings, there is considerable uncertainty about the putative palliative effects of widely used dithiol chelating agents, such as DMPS or dimercaptosuccinic acid (DMSA), on the renal burden and urinary excretion of Cd. Preliminary data from Zalups and colleagues indicate that treatment with either DMPS or DMSA does not reduce significantly the renal burden of Cd in rats or mice treated with $CdCl_2$. Only further research will be able to determine if there are indeed beneficial effects of treating humans exposed to Cd with these dithiol chelating agents.

Urinary excretion of Cd is also dependent on the structural and functional integrity of the proximal tubular epithelium. When tubular pathology is induced by Cd (or other nephrotoxicants), the urinary excretion of Cd increases, largely due to decreased absorption of filtered species of Cd plus the release of Cd from necrotic or apoptotic tubular epithelial cells. This point is exemplified in the study by Zalups et al. (1992a), in which the cumulative urinary excretion of Cd (factored by renal mass) was greater in uninephrectomized (NPX) rats than corresponding control rats 24 hours after the two groups received one of three doses of Cd–MT (factored by kg body weight). At each of three doses of Cd–MT studied, the NPX rats developed a more severe form of the nephropathy induced by Cd–MT (as determined histopathologically and by the urinary excretion of selective plasma solutes and cellular enzymes) than corresponding control rats with two kidneys. These findings clearly indicate that as renal injury is induced or made more severe, there is a corresponding increase in the urinary excretion of Cd.

Markers of Renal Injury Induced by Cadmium

Proteinuria and decreases in GFR have been documented in humans exposed to Cd chronically (Friberg, 1950; Wang and Foulkes, 1984). The proteinuria appears to be characterized by the urinary excretion of low molecular weight proteins, such as β_2-microglobulin, lysozyme, ribonuclease, immunoglobulin light chains, and retinol binding protein (Piscator, 1966). The urinary excretion of β_2-microglobulin or retinol binding protein has been used widely as an indicator of renal tubular dysfunction (Piscator, 1984; Roels et al., 1981; Smith et al., 1998). Retinol binding protein is more stable than β_2-microglobulin (Bernard and Lauwerys, 1981), although the urinary excretion of this protein is not a specific indicator of renal tubular injury induced by Cd.

Urinary excretion of MT has been reported to correlate with the levels of Cd in the kidneys, liver, and urine (Shaikh and Smith, 1980). Shaikh et al. (1989) demonstrated that in rats that had been exposed to Cd orally for up to two years the urinary excretion of Cd correlated with the urinary excretion of MT. It is presently unclear as to what this correlation means in terms of the nephropathy induced by Cd.

Evidence for high molecular weight proteins, such as albumin, being excreted as part of the nephropathy induced by Cd has been documented in humans and experimental animals (Bernard et al., 1979; Elinder et al., 1985b; Mason et al., 1988; Roels et al., 1989; Thun et al., 1989; Zalups et al., 1992a). In more severe cases of the nephropathy, enzymuria and a Fanconi-like syndrome may occur, which is characterized by decreased tubular absorption and increased urinary excretion of filtered solutes such as amino acids, glucose, Ca, copper (Cu), and inorganic phosphate (P_i) (Elinder et al., 1985a, 1985b; Falck et al., 1983; Gompertz et al., 1983). Increased rates of urinary excretion of *N*-acetyl-β-D-glucosaminidase (Chia et al., 1989; Kawada et al., 1990), γ-glutamyltransferase (γ-GT), alkaline phosphatase, and alanine aminopeptidase (Mueller et al., 1989) have been documented with the nephropathy induced by Cd. Nephrolithiasis has also been documented in some individuals manifesting signs of the nephropathy induced by Cd (Kazantzis 1979).

NEPHROTOXICOLOGY OF MERCURY

To fully understand the biology and toxicology of mercury (Hg) in humans and other mammals, one must have a thorough comprehension of the bonding interactions that occur between mercuric ions and the broad range of molecular ligands present in the various compartments of the body. Although mercuric ions bind to numerous nucleophilic groups on molecules, they have a particularly strong propensity to bond to reduced sulfur atoms, especially those on endogenous thiol-containing molecules, such as GSH, Cys, homocysteine (hCys), NAC, MT, and albumin. The affinity constant for the bonding of mercuric ions to reduced sulfhydryl groups or thiolate anions is approximately 10^{15} to 10^{20}. By contrast, the affinity constants for mercuric ions bonding to oxygen- or nitrogen-containing ligands (such as carbonyl or amino groups) are about ten orders of magnitude lower. Thus, it is likely that the biological and toxicological effects of inorganic or organic Hg in the kidneys can be attributed, at least in part, to the interactions that occur between these forms of Hg and critical sulfhydryl-containing domains on molecules present in intracellular or extracellular compartments.

Besides seeking a better understanding of the chemical properties of Hg-containing compounds and the bonding reactions that occur within and around target epithelial cells, other factors must be considered to define more precisely the biochemical and molecular mechanisms of action of Hg-containing compounds in the kidney. Particular attention may need to be paid to the potential role of "molecular mimicry" and the species of Hg involved in the renal (proximal) tubular uptake and transport of mercuric ions. Several lines of evidence have tended to support the role of molecular mimicry in the epithelial uptake and transport of Hg (Zalups, 2000b).

Susceptibility to the injurious effects of mercuric ions is modified greatly by a number of intracellular and extracellular factors. The role of some of these factors in providing protection from the intoxicating effects of mercuric ions has served, and continues to serve, as the theoretical basis for most of the currently employed therapeutic strategies used to treat individuals who have been exposed to Hg.

Intrarenal Distribution and Tubular Localization of Mercury

The kidneys are the primary organs that accumulate mercuric ions after exposure to any of the different forms of Hg. In fact, the renal uptake and accumulation of Hg *in vivo* is very rapid, which is likely to be an important contributing factor responsible for the rapid induction of renal tubular injury. As much as 50% of low doses of inorganic Hg is present in the kidneys of rats within a few hours after exposure (Zalups, 1993a). Significant amounts of Hg also accumulate in the kidneys after exposure to organic forms of Hg (Clarkson, 1970a, 1970b; Friberg, 1957; Magos and Butler, 1976; Magos et al., 1981, 1985; McNeil et al., 1988; Norseth and Prickett et al., 1950; Zalups et al., 1992b), although the level of accumulation is much less than that which occurs after exposure to inorganic or elemental forms of Hg.

Within the kidneys, mercuric ions become localized primarily in the cortex and outer stripe of the outer medulla (Bergstrand et al., 1959; Berlin, 1963; Berlin and Ullberg, 1963a, 1963b; Friberg et al., 1957; Taugner et al., 1966; Zalups, 1991a, 1991b, 1991c, 1993a, 1993b; Zalups and Barfuss, 1990; Zalups and Cherian, 1992a, 1992b; Zalups and Lash, 1990). Histochemical and autoradiographic data from mice and rats

(Hultman and Enestrom, 1986, 1992; Hultman et al., 1985; Magos et al., 1985; Rodier et al., 1988; Taugner et al., 1966; Zalups, 1991a), and tubular microdissection data from rats and rabbits (Zalups, 1991b; Zalups and Barfuss, 1990) indicate that this localization is due primarily to the uptake and retention of mercuric ions along convoluted and straight segments of the proximal tubule. It should be made clear, however, that although the three segments of the proximal tubule appear to be the primary sites where mercuric ions are taken up and accumulated, there are currently insufficient data to exclude the possibility that other segments of the nephron and/or collecting duct may also, to a minor extent, transport and accumulate various forms of Hg. Based on a preponderance of dispositional findings, the tubular sites where the toxic effects of Hg are expressed correlate well with the tubular sites in which mercuric ions are taken up and accumulated.

Mechanisms Involved in the Transport of Mercury along the Proximal Tubule

An overwhelming body of evidence indicates that mercuric ions enter proximal tubular epithelial cells by mechanisms localized at both the luminal (Zalups, 1995, 1997b, 1998b, 1998c; Zalups and Barfuss, 1993a, 1998a, 1998b; Zalups and Lash, 1997a; Zalups and Minor, 1995; Zalups et al., 1991, 1998) and basolateral (Zalups, 1995, 1997, 1998a, 1998b, 1998c; Zalups and Barfuss, 1993a, 1995a, 1998b; Zalups and Lash, 1997a; Zalups and Minor, 1995) membranes. Experimental findings coming from both whole-animal and isolated perfused tubule studies (Cannon et al., 2000, 2001) indicate that the luminal mechanism is largely dependent on the actions of γ-GT and cysteinylglycinase, and appears to involve the transport of mercuric *S*-conjugates of Cys, mainly in the form Cys–*S*–Hg–*S*–Cys. Data from isolated perfused tubule experiments indicate that the mercuric conjugates of Cys are transported into proximal tubular epithelial cells by one or more amino acid transport systems (Cannon et al., 2000, 2001). It has been hypothesized by Zalups and colleagues that the luminal absorption of Cys–*S*–Hg–*S*–Cys by proximal tubular epithelial cells occurs through a mechanism where Cys–*S*–Hg–*S*–Cys acts as a molecular homolog or "mimic" of the amino acid cystine at one or more of the transporters that take up this amino acid. Figure 22.2 provides a diagrammatic summary of a number of mechanisms involved in the uptake of mercuric ions along proximal tubular segments of the nephron.

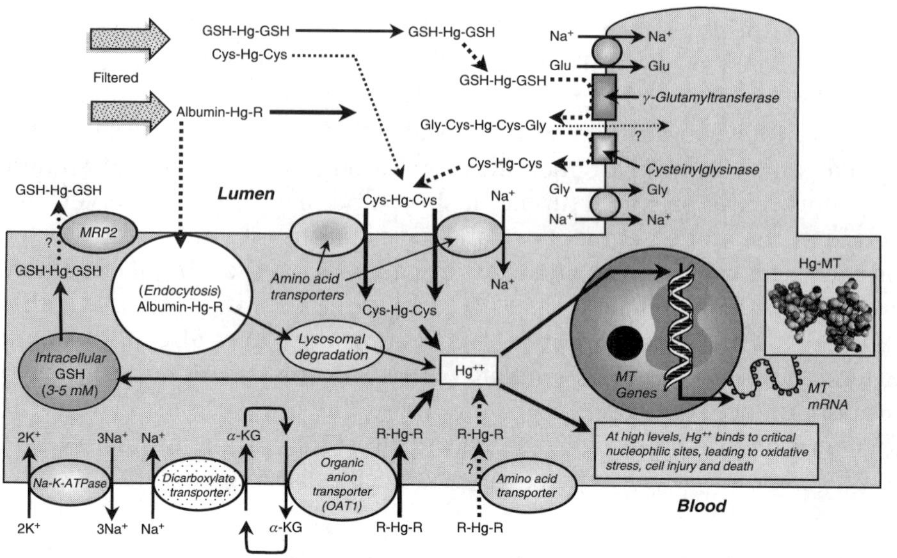

Figure 22.2 Diagrammatic representation of mechanisms involved in the luminal and basolateral handling of mercury (Hg) in proximal tubular epithelial cells. In this scheme, cysteine (Cys), glutathione (GSH) and albumin *S*-conjugates are shown to be present in the lumen of the proximal tubule. Although it is not known currently precisely which chemical species of Hg are filtered at the glomerulus, these conjugates have been implicated in the absorptive luminal transport of Hg. This scheme shows that a fraction of filtered Hg that is bound to albumin may be transported into the intracellular compartment during the endocytosis of this protein. There is insufficient evidence to assert that this occurs *in vivo*. Experimental data indicate that if GSH *S*-conjugates of Hg (in the form GSH–Hg–GSH) are filtered into the proximal tubular lumen, they are degraded rapidly in the tubular lumen to cysteinylglycine S-conjugates of Hg (Cys–Gly–Hg–Cys–Gly) by γ-glutamyltransferase, and then to Cys S-conjugates of Hg (Cys–Hg–Cys) by cysteinylglycinase. The resulting Cys *S*-conjugates of Hg enter the proximal tubular epithelial cells via sodium-dependent and sodium-independent amino acid transport systems in the luminal plasma membrane. Because the mercuric conjugates of Cys are structurally similar to the amino acid cystine, Zalups and colleagues have hypothesized that the mercuric conjugate Cys–*S*–Hg–*S*–Cys serves as a molecular homolog or mimic of cystine at the site of one or more of the amino acid transport systems involved in the absorptive transport of this amino acid. On the basis of a strong body of experimental evidence, the organic anion transport systems are thought to be involved in the basolateral uptake of mercuric conjugates. The manner in which these systems operate is that intracellular generation of α-ketoglutarate (as a result of normal metabolic processes) creates a chemical gradient facilitating the movement of this dicarboxylate out of the cell.

When intracellular concentrations of α-ketoglutarate are high enough, it exits proximal tubular cells at the basolateral membrane by exchanging with organic anions at one of the organic anion exchangers. Attempts are made to reclaim the α-ketoglutarate by the sodium-dependent dicarboxylate symport. This transporter is driven by the sodium gradient generated by the Na^+/K^+-ATPase localized in the basolateral membrane. According to the scheme presented, inorganic mercury enters proximal tubular epithelial cells (presumably as a conjugate of GSH or Cys) via the organic anion transporters, OAT1 and OAT3, in exchange for intracellular α-ketoglutarate. The most likely species of inorganic Hg taken up at the basolateral membrane by the organic anion exchanger include mercuric conjugates of GSH (GSH–Hg–GSH), Cys (Cys–Hg–Cys) and other small molecules possessing a negative charge ($^-$R–Hg–R$^-$), such as *N*-acetylcysteine (NAC). Support for this notion comes from the fact that the basolateral uptake of mercuric ions can be inhibited by *p*-aminohippurate (PAH), probenecid and dicarboxylic acids, which are competitive substrates at the organic anion exchangers and/or the sodium-dependent dicarboxylate transporters. The scheme presented also shows that a pool of Hg accumulates within the proximal tubular epithelial cells from the different sites of entry, and that this pool can interact with various components within the cells. The pool of Hg that accumulates, during and after periods of exposure, provides a source of mercuric ions that can induce the transcription of the genes for MT-1 and MT-2, by mechanisms that have not yet been fully defined. Numerous lines of evidence indicate that expression of renal MT genes, and especially in proximal tubular epithelial cells, increases markedly after exposure to Hg. If the intracellular pool of exchangeable Hg increases beyond what the protective elements inside the proximal tubular epithelium can handle, oxidative stress is induced, which in turn can alter mitochondrial respiratory activity and lead to lipid peroxidation in the plasma membrane and other perturbations in cellular metabolism. All of these effects can, and do, lead to the induction of cell death by either necrosis or apoptosis, which results in the release of Hg from within the necrotic or apoptotic proximal tubular epithelial cells into the tubular lumen. One intracellular molecule that protects cells from oxidative injury and binds Hg is GSH. This molecule is the most abundant nonprotein thiol in the body, and is present in proximal tubular epithelial cells at concentrations of between 3 and 5 mM. Some of the Hg that forms *S*-conjugates with GSH may be transported into the proximal tubular lumen by the ATP-binding-cassette protein MRP2, which has been shown to be expressed in the luminal plasma membrane of proximal tubular epithelial cells and has been shown to transport GSH *S*-conjugates of various molecules. Pathways depicted by solid black arrows are supported either directly or indirectly by experimental evidence. The thicker solid black lines denote pathways for which there is substantive evidentiary support. By contrast, the dashed lines represent pathways of secondary importance or pathways that are based primarily on indirect findings or speculation. Question marks associated with a pathway denote uncertainty for the occurrence of that pathway *in vivo*.

Both *in vivo* and *in vitro* data indicate that the primary mechanism involved in the basolateral uptake of Hg involves the uptake of mercuric thiol *S*-conjugates by the classic, *p*-aminohippurate (PAH)-sensitive, organic anion transport systems. Most recently, data obtained from Madin-Darby canine kidney (MDCK) cells that had been genetically manipulated to overexpress the human form of the organic anion/dicarboxylate exchanger (hOAT1), indicate that hOAT1 can transport inorganic mercury into the cytosolic compartment when the cells were exposed to mercuric conjugates of NAC or Cys (Aslamkhan et al., 2003). Moreover, the cytotoxic effects of inorganic Hg were manifested at lower concentrations of mercuric conjugates of NAC or Cys, and during shorter periods of exposure, in the MDCK cells that overexpressed hOAT1 than in corresponding control MDCK cells. Inhibitors of OAT1, such as PAH, probenecid, and glutarate, inhibited the uptake of inorganic Hg and decreased the associated severity of the cytotoxic effects of inorganic only in the transfected MDCK cells. These molecular data indicate that it is likely OAT1 plays an important role in the basolateral uptake of inorganic Hg and the expression of the toxic effects of Hg in proximal tubular epithelial cells. Additionally, these data also support the hypothesis that mercuric conjugates of certain low molecular weight thiol-containing molecules are transportable substrates of OAT1. Finally, molecular data from *Xenopus* oocytes injected with mRNA for OAT3 indicate that OAT3, which is also present in proximal tubular epithelial cells, can transport some of the same mercuric conjugates transported by OAT1 (Aslamkhan et al., 2003).

Preliminary data from the laboratory of Zalups and colleagues provide insights into at least one of the mechanisms that participate in the luminal uptake of Cys–*S*–Hg–*S*–Cys, which is formed in the luminal compartment of the proximal tubule by the enzymatic degradation of mercuric conjugates of GSH. These data indicate that system $b^{0,+}$, which is a dimeric complex made up of the sodium-independent amino acid transporter $b^{0,+}$ and the anchoring protein rBAT, plays a key role in transporting Cys–*S*-Hg-*S*-Cys (and perhaps other mercuric conjugates) from the luminal compartment into the cytosolic compartment of epithelial cells lining the proximal tubule. More importantly, some of these data indicate that it is likely Cys–*S*-Hg-*S*-Cys serves as a molecular mimic or homolog of the amino acid cystine at the binding site of $b^{0,+}$. It should be made clear, however, that one of the primary absorptive transporters of cystine and basic amino acids along the proximal tubule is system $b^{0,+}$. Further studies are currently underway to define the role of molecular mimicry in the uptake of mercuric ions by proximal tubular epithelial cells.

Mercuric Ions as Nephrotoxicants

All forms of Hg can induce acute or chronic pathological changes (or both) in the kidneys (Cuppage and Tate, 1967; Fowler, 1972; Ganote et al., 1974; Gritzka and Trump, 1968; Klein et al., 1973; Magos and Clarkson, 1977; McDowell et al., 1976; Zalme et al., 1976; Zalups, 1991b; Zalups and Barfuss, 1996b; Zalups and Diamond 1987a; Zalups and Lash, 1990, 1994; Zalups et al., 1988, 1991b). Inorganic forms of Hg are by far much more acutely nephrotoxic than organic forms of Hg, which generally require multiple exposures to relatively large doses to induce renal pathology (Chang et al., 1973; Magos et al., 1985; McNeil et al., 1988). Under most conditions, renal tubular injury induced by inorganic Hg is generally expressed fully within the initial 24 hours following exposure (Zalups, 2000b; Zalups and Diamond 1987b; Zalups et al., 1988). In rats, the oral LD_{50} for mercuric chloride was reported to be in the range of 25.9 to 77.7 mg/kg (Kostial et al., 1978). A lower range of doses of inorganic Hg (10 to 42 mg/kg) has been estimated to be fatal in humans (Gleason et al., 1957).

Interestingly, the dose–response relationships for the toxicity of inorganic Hg are extremely steep under most *in vivo* and *in vitro* conditions (Zalups, 2000b). A threshold effect is generally observed, in that no cellular necrosis (death) occurs up until a certain dose or concentration. Above that dose or concentration, however, cellular death progresses rapidly, and in some systems an all-or-none response is observed. One possible explanation for this effect is that endogenous thiol-containing ligands, such as GSH and MT, bind Hg and act as a buffer to prevent injurious events from occurring. Above a certain dose or concentration of Hg, the buffer becomes depleted and mercuric or mercurous ions begin to bind more readily to critical nucleophilic groups in the cell; thereby impairing the normal homeostatic functions of the tubular epithelial cells.

Site of Renal Tubular Injury Induced by Mercury

It has been well-documented that the nephrotoxic effects of Hg are expressed predominantly along the proximal nephron, although glomerular injury (mainly of an immunological nature) has been documented in some species and strains of mice and rats exposed to inorganic Hg (Zalups, 2000b). Histopathological evaluation has determined that the pars recta (straight segment) of the proximal tubule, especially at the junction of the cortex and outer medulla, is the region of the nephron that is most vulnerable to the toxic effects of both inorganic and organic forms of Hg (Cuppage and Tate, 1967; Cuppage et al., 1972; Fowler, 1972; Ganote et al., 1974; Gritzka and Trump, 1968; Klein et al., 1973; McDowell et al., 1976; Rodin and Crowson, 1962; Verity and Brown, 1970; Zalme et al., 1976; Zalups, 1991b; Zalups and Barfuss, 1996b; Zalups and Diamond 1987a,b; Zalups and Lash, 1990; Zalups et al., 1988, 1991b).

The toxic effects of Hg can be elicited very rapidly in the kidneys. Certain degenerative changes have been detected in portions of the proximal tubule of rats as early as one hour after exposure to a very high (100 mg $HgCl_2$/kg) dose of inorganic Hg (Rodin and Crowson, 1962). At lower doses (1 to 5 mg $HgCl_2$/kg), significant pathological changes are generally not detected with light microscopy until approximately six to eight hours after exposure (Ganote et al., 1974; Rodin and Crowson, 1962). However, at the electron microscopic level, cellular pathology can be detected in the epithelial cells lining the pars recta in the kidneys of rats in as little as three hours following a subcutaneous dose of 4 mg/kg of inorganic Hg (Gritzka and Trump, 1968). Some of the pathological features that can be detected include swelling of the mitochondrial matrix with loss of matrix granules, dilation of cisternae of rough endoplasmic reticulum (ER), loss of ribosomes from the rough ER, dispersion of ribosomes, increase in number and size of the cisternae of the smooth ER, and single membrane-limited inclusion bodies. By the end of the initial 12 hours after exposure to nephrotoxic doses of inorganic Hg, cellular necrosis along the pars recta of the proximal tubule is prominent at both light and electron microscopic levels (Gritzka and Trump, 1968; Rodin and Crowson, 1962; Zalups, 2000b).

Convoluted portions of proximal tubules, and sometimes distal segments of the nephron, become involved when the nephropathy is very severe (Gritzka and Trump, 1968; Rodin and Crowson, 1962). The involvement of segments of the nephron distal to the proximal tubule probably represents secondary effects elicited by the severe damage to the pars recta of proximal tubules. However, due to a paucity of data on the toxic effects of mercuric compounds on segments of the nephron distal to the proximal tubule, the cause of pathological changes in the distal nephron remains speculative.

If the exposure to a nephrotoxic dose of inorganic Hg is not fatal, the proximal tubular epithelium usually regenerates completely during the initial two weeks after induction of tubular pathology. For example, complete re-lining of the proximal tubular epithelium has been demonstrated in rats as early as four days after receiving a 1.5 mg/kg i.v. dose of mercuric chloride (Cuppage et al., 1972).

It is interesting that in contrast to the effects of Hg *in vivo*, all three segments (S_1, S_2, and S_3) of the proximal tubule (of the rabbit) become intoxicated with either inorganic Hg or methylmercury when the Hg-containing compounds are perfused through the lumen of these segments *in vitro* (Barfuss et al., 1990; Zalups and Barfuss, 1993a; Zalups et al., 1991). The differences between the *in vivo* and *in vitro* findings are somewhat perplexing since all segments of the proximal tubule accumulate Hg under both experimental conditions. Another interesting difference between the *in vivo* and *in vitro* situation is that, *in vitro*, organic Hg (specifically methylmercury) is more toxic to proximal tubular epithelial cells than inorganic Hg. This has been demonstrated in primary cultures of proximal tubular epithelial cells (Aleo et al., 1987, 1992) and in isolated perfused segments of the proximal tubule (Zalups and Barfuss, 1993a).

Markers of Renal Cellular Injury and Impaired Renal Function

A number of methods have been used to detect renal tubular injury induced by Hg. One noninvasive method that has been employed frequently is measuring the urinary excretion of a number of cellular enzymes (Buchet et al., 1980; Ellis et al., 1973; Gotelli et al., 1985; Kirschbaum, 1979; Planas-Bohne, 1977; Price, 1982; Stonard et al., 1983; Stroo and Hook, 1977; Zalups and Diamond, 1987b). The rationale for using the urinary excretion of cellular enzymes as an indicator of renal tubular injury is based on the close association between renal cellular necrosis and enzymuria. After renal epithelial cells undergo cellular necrosis or apoptosis, most, if not all, of the contents of the dead epithelial cells, including numerous cellular enzymes, are released into the tubular lumen and are excreted in the urine. The usefulness of any particular cellular enzyme as a marker of renal cellular injury or necrosis depends on the stability of the enzyme in urine, whether the enzyme or the activity of the enzyme is greatly influenced by the toxicant that is being studied, and the subcellular localization of the enzyme relative to the subcellular site of injury.

During the early stages of the nephropathy induced by Hg, prior to tubular necrosis, cells along the proximal tubule undergo a number of degenerative changes and begin to lose some of their luminal (brush-border) membrane (Zalme et al., 1976). Evidence from several studies indicates that the urinary excretion of the brush-border enzymes, alkaline phosphatase and γ-GT, increases during the nephropathy induced by Hg-containing compounds (Gotelli et al., 1985; Price, 1982; Zalups et al., 1988, Zalups, 1991). When tubular injury becomes severe and necrosis of tubular epithelial cells is apparent, the urinary excretion of a number of intracellular enzymes, such as lactate dehydrogenase, aspartate aminotransferase, alanine aminotransferase and *N*-acetyl-β-D-glucosaminidase (NAG), increases (ATSDR, 1999; Planas-Bohne, 1977; WHO, 1991; Zalups and Diamond, 1987b; Zalups et al., 1988, Zalups, 1991).

After a significant number of proximal tubules have become functionally compromised by the toxic effects of Hg, the capacity for the reabsorption of filtered plasma solutes and water is diminished greatly. As a consequence of this diminished absorptive capacity, there is increased urinary excretion of both water and a number of plasma solutes, such as glucose, amino acids, albumin, and other plasma proteins (Diamond, 1988; Price, 1982; Zalups and Diamond, 1987b; Zalups et al., 1988). In a recent study of workers exposed to Hg vapor, it was demonstrated that increased urinary excretion of Tamm-Horsfall glycoprotein and tubular antigens and decreased urinary excretion of prostaglandin E2 and F2α and thromboxane B2 can also be used as indices of renal pathology induced by Hg (Cardenas et al., 1993).

Moreover, the urinary excretion of Hg (factored by the total renal mass) correlates very closely with the level of injury in pars recta segments of proximal tubules during the acute nephropathy induced by low toxic doses of inorganic Hg (Zalups and Diamond, 1987b; Zalups et al., 1988). Overall, it appears that as the level of renal injury increases in the kidneys, there is a corresponding increase in the urinary excretion of Hg. Other nephrotoxic agents also decrease the retention of Hg in the kidney and increase the excretion of Hg in the urine (Clarkson and Magos, 1967; Magos and Stoychev, 1969; Trojanowska et al., 1971), presumably by causing the release of Hg from, and decreased luminal uptake by, proximal tubular epithelial cells undergoing cellular death. Although the urinary excretion of Hg appears to correlate well with the level of acute renal injury induced by mercuric chloride, there does not appear to be a close correlation between the severity of renal injury and the renal concentration or content of Hg (Zalups and Diamond, 1987b; Zalups et al., 1988).

When renal tubular injury becomes severe during the acute nephropathy induced by Hg, overall GFR decreases (Barenberg et al., 1968; McDowell et al., 1976; Zalups, 1991). This can be documented readily by measuring the concentration of creatinine in the plasma. The mechanisms responsible for the decreased GFR are not known at present, but are likely to be complex and involve several factors. Besides causing decreases in GFR, Hg causes the fractional excretion of sodium and potassium to increase (McDowell et al., 1976). These functional changes likely reflect a significant decrease in the total number of functioning nephrons, inasmuch as similar changes occur in rats and mice when their total renal mass is reduced significantly (Zalups 1989; Zalups and Henderson, 1992; Zalups et al., 1985). As part of the severe nephropathy induced by Hg, blood urea nitrogen (BUN) also increases as plasma creatinine increases as GFR decreases. Thus, measurements of plasma creatinine and BUN may be used as indicators of impaired renal function induced by Hg (McDowell et al., 1976). However, it is preferable to use the clearance of creatinine or inulin over measuring BUN as an index of renal function. BUN can be elevated by more nonrenal causes than creatinine, and therefore is not as sensitive an indicator of renal function. After exposure to high doses of Hg, an oliguric or anuric acute renal failure ensues. The factors that lead to acute renal failure are complex, involving multiple systems. Clearly, further research is needed to better understand the mechanisms involved in the induction of acute renal failure following exposure to Hg.

Influence of Intracellular Thiols on the Renal Accumulation and Toxicity of Mercury

GSH and MT are abundant intracellular thiols that can greatly influence the renal accumulation of Hg and, ultimately, susceptibility to Hg-induced renal cellular injury. Johnson (1982), Baggett and Berndt (1986), Berndt et al. (1985), Zalups and Lash (1997b), and Zalups et al. (1999a, 1999b, 1999c) demonstrated that depletion of intracellular GSH or nonprotein thiols is accompanied by decreases in the renal accumulation of inorganic Hg in animals treated with mercuric chloride. In the studies by Berndt and colleagues (Baggett and Berndt, 1986; Berndt et al., 1985), depletion of intracellular GSH appeared to increase the severity of renal injury induced by mercuric chloride. Zalups and Lash (1990) also found a close correlation between intrarenal concentrations of GSH and the accumulation of inorganic Hg. Recently, Zalups and Lash (1997b) and Zalups et al. (1999a, 1999b, 1999c) demonstrated in rats that acute depletion of GSH in the kidneys and liver by treatment with diethyl maleate caused a significant decrease in the renal uptake and accumulation of Hg during the initial hour after the administration of a low, nontoxic, dose of mercuric chloride. Interestingly, while the renal accumulation of Hg decreased following treatment with diethyl maleate, the net hepatic accumulation of Hg increased. Thus, depletion of renal and hepatic GSH has mixed effects on the disposition of Hg.

Zalups and Lash (1997b) and Zalups et al. (1999a) also showed in rats that acute depletion of renal GSH with buthionine sulfoximine does not affect the early aspects of the accumulation of inorganic Hg in the kidneys. By contrast, Zalups et al. (1999b, 1999c) demonstrated that pretreatment with buthionine sulfoximine did cause significant decreases in the net renal content of Hg 24 hours after treatment with inorganic Hg (Zalups et al., 1999b, 1999c). These findings indicate that there are significant temporal factors with respect to the effects of buthionine sulfoximine on the renal disposition of Hg.

It has also been shown that when intraluminal degradation of GSH is prevented (in the proximal tubule) by irreversible chemical inhibition of γ-GT, the urinary excretion of GSH and inorganic Hg increases and the net renal accumulation of inorganic Hg decreases in rats and mice (Berndt et al., 1985; Tanaka et al., 1990; Zalups, 1995). Tanaka et al. (1990) also demonstrated in mice that pretreatment with 1,2-dichloro-4-nitrobenzene, which depletes the hepatic content of GSH (prior to injection of mercuric chloride), causes a marked reduction in the renal accumulation of Hg and a significant decrease in the severity of renal cellular injury induced by inorganic Hg. Overall, these findings tend to suggest that hepatically synthesized GSH and the activity of γ-GT are involved in the renal uptake of Hg. Additional findings from a set of recent studies, in which bile flow was either diverted or prevented from entering the small intestine of rats, demonstrate that some aspect of hepatic function is linked to a component of the renal uptake and accumulation of Hg (Zalups, 1998a; Zalups and Barfuss, 1996a).

Acute biliary ligation causes significant increases in the renal and hepatic contents of GSH in rats (Zalups et al., 1999c).

Zalups et al. (1999c) suggested that the observed increased in the renal concentration of GSH induced by biliary ligation is due to a hepatic mechanism. They postulate that as the concentration of GSH in the biliary canaliculi increase (after biliary ligation), the transport of GSH out of the hepatocytes is redirected down a concentration gradient into the sinusoidal blood. They also postulate that as GSH is continually added to the blood, plasma concentrations of this thiol increase, which in turn provides more GSH to be taken up at the luminal and basolateral membranes of proximal tubular epithelial cells in the kidneys. Interestingly, biliary ligation has been shown to cause the net accumulation of Hg in the liver to increase and the net accumulation of Hg in the kidneys to decrease in rats during the initial 24 hours after intravenous injection of a nontoxic dose of inorganic Hg. What makes these findings interesting is that the renal accumulation of Hg decreased despite an increase in the renal cellular content of GSH, which is contrary to what one might expect. It has been postulated that decreased renal accumulation of Hg in animals that have undergone biliary ligation is not due to the content of GSH in the kidney, but rather to the content of GSH in the liver, where the accumulation of Hg increases. Such findings also confirm that some aspects of hepatic function contribute to renal disposition of Hg.

Induction of MT in the kidneys by various pretreatments (mainly with metals) has been demonstrated to be associated with increased intrarenal accumulation of Hg and decreased severity of the nephropathy induced by either organic or inorganic Hg (Fukino et al., 1984, 1986; Zalups and Cherian, 1992a, 1992b). Recent findings from MT-null mice (in which MT-1 and MT-2 genes have been genetically "knocked-out") also confirm a beneficial role of MT in the renal handling and toxicity of inorganic Hg. These findings indicate that the severity of renal injury induced by inorganic Hg is greater in the MT-null mice than corresponding control mice. Obviously, there is a complex interplay between protein and nonprotein thiols in the renal disposition of Hg, and a significant amount of new research is warranted to better understand the role of nonprotein thiols in the renal disposition and toxicity of Hg.

Mercury-Induced Renal Autoimmunity

Besides the well-characterized tubular nephropathy that is induced by the different forms of Hg, there is strong evidence that Hg can also induce glomerulonephritis under certain conditions in some strains of experimental animals. Evidence from studies with rabbits (Roman-Franco et al., 1978), inbred Brown-Norway rats (Druet et al., 1978), and a cross between Brown-Norway and Lewis rats, indicate that multiple exposures to inorganic Hg can lead to the production of antibodies against the glomerular basement membrane, which results in an immunologically mediated, membranous glomerular nephritis (Bigazzi, 1988,1992). This glomerular nephropathy is characterized by the binding of antibodies to the glomerular basement membrane, followed by the deposition of immune complexes in the glomerulus (Druet et al., 1978; Roman-Franco et al., 1978; Sapin et al., 1977). There is also evidence from studies using several strains of both mice and rats that repeated exposures to inorganic Hg can lead to the deposition of immune complexes in the mesangium and glomerular basal lamina, which leads to an immune-complex glomerulonephritis (Bigazzi, 1988; Enestrom and Hultman, 1984; Hultman and Enestrom, 1992). Whether Hg can induce an autoimmune glomerulonephritis in humans is not clear at present. It should be pointed out that the majority of the cases of glomerulonephritis (of an immunological origin) in humans is classified as idiopathic. Thus, until research proves otherwise, it remains possible that some forms of glomerulonephritis could be induced by exposure to Hg or other environmental or occupational toxicants.

It is likely that the autoimmunity induced by Hg reflects some complex effects of mercuric ions on cell-signaling and gene-expression events in immune cells, such as in monocytes and lymphocytes. For example, Koropatnick and Zalups (1997, 1999) demonstrated that exposure of human monocytes to low, nontoxic, doses of the inorganic Hg causes a rapid suppression of activation signaling events that are normally induced in these cells by lipopolysaccharide or phorbol ester.

Treatment for Mercury-Induced Nephropathy

Currently, the only effective treatment for the nephropathy induced by Hg is chelation therapy, employing one of the effective, watersoluble, dithiol-chelating agents, namely 2,3-Dimercaptopropane-1-sulfonate (DMPS) or 2,3-mesoDimercaptosuccinic acid (DMSA) (Aposhian, 1983; Aposhian and Aposhian, 1990). Both of these agents can reduce the renal burden of Hg very rapidly. Zalups et al. (1992b) have demonstrated that the renal burden of Hg can be reduced by as much as 85% in

24 hours by the administration of two intravenous doses of DMPS. DMPS can also serve as a rescue agent against the toxic effects of inorganic Hg when it is administered shortly after exposure to mercuric chloride (Zalups et al., 1991). DMPS is slightly more effective as a chelator of mercuric ions in the kidneys than DMSA. This probably relates to the efficiency and rate of transport of DMPS or DMSA by proximal tubular epithelial cells. Strong evidence indicates that at least DMPS is transported into proximal tubular cells by the PAH-sensitive organic anion transport system localized on the basolateral membrane (Klotzbach and Diamond, 1988; Stewart and Diamond, 1987, 1988). It is fortuitous that the primary site where DMPS is presumed to be transported along the nephron corresponds to the same site along the nephron where inorganic Hg accumulates avidly.

By studying the effects of DMPS in isolated perfused pars recta segments of the proximal tubule of the rabbit, Zalups and colleagues (1998) obtained data that defined some of the mechanisms of action of DMPS. Their findings show clearly that the therapeutic efficacy of DMPS in the nephropathy induced by inorganic Hg is in part linked to its transport at the basolateral membrane by the organic anion transport system. Moreover, it is also linked to the ability of DMPS to extract accumulated inorganic Hg from within proximal tubular epithelial cells and then to deliver preferentially this Hg in the form of a non-reabsorbable complex into the tubular lumen to promote the urinary excretion and elimination of Hg. Another therapeutic effect of DMPS is the prevention of significant proximal tubular uptake of inorganic Hg, from either the luminal fluid or blood, subsequent to the formation of mercuric conjugates of DMPS.

NEPHROTOXICOLOGY OF LEAD

The nephrotoxic effects of lead (Pb) are characterized by a proximal tubular nephropathy, glomerular sclerosis and interstitial fibrosis (Goyer, 1989; Loghman-Adham, 1997). Functional deficits in humans that have been associated with excessive Pb exposure include enzymuria, low and high molecular weight proteinuria, impaired transport of organic anions and glucose, and depressed GFR. A few studies have revealed histopathological features of renal injury in humans, including intra-nuclear inclusion bodies and cellular necrosis in the proximal tubule and interstitial fibrosis (Cramer et al., 1974; Wedeen et al., 1975, 1979; Biagini et al., 1977).

The above findings are consistent with observations made in animal models. In rats, proximal tubular injury involves the convoluted and straight portions of the tubule (Aviv et al., 1980; Dieter et al., 1993; Khalil-Manesh et al., 1992a, 1992b; Vyskocil et al., 1989), with greater severity, at least initially, in the straight (S_3) segment (Fowler et al., 1980; Murakami et al., 1983). Typical histological features include, in the acute phase, the formation of intranuclear inclusion bodies in proximal tubular cells (see below for further discussion), abnormal morphology (e.g., swelling and budding) of proximal tubular mitochondria (Fowler et al., 1980; Goyer and Krall, 1969), karyomegaly and cytomegaly, and cellular necrosis at sufficiently high dosage. These changes appear to progress, in the chronic phase of toxicity and with sufficient dosage, to tubular atrophy and interstitial fibrosis (Goyer, 1971; Khalil-Manesh et al., 1992a, 1992b). Glomerular sclerosis has also been reported (Khalil-Manesh et al., 1992a). Adenocarcinomas of the kidney have been observed in long-term studies in rodents in which animals also developed proximal tubular nephropathy (Moore and Meredith, 1979; Goyer, 1993).

Mechanisms Involved in the Nephrotoxicology of Lead

Structural and functional abnormalities in renal proximal tubular mitochondria are consistent features of Pb-induced nephropathy (Fowler et al., 1980; Goyer, 1968; Goyer and Krall, 1969). Mitochondria isolated from intoxicated rats contain Pb, principally associated with the intramembrane space or bound to the inner and outer membranes, and show abnormal respiratory function, including decreased respiratory control ratio during pyruvate/malate- or succinate-mediated respiration (Fowler et al., 1980; Oskarsson and Fowler, 1985). Pb inhibits the uptake of Ca into isolated renal mitochondria and may enter mitochondria as a substrate for a Ca transporter (Kapoor et al., 1985). This would be consistent with evidence that Pb can interact with Ca binding proteins and, thereby, affect Ca-mediated or Ca-regulated events in a variety of tissues (Fullmer et al., 1985; Goldstein and Ar, 1983; Goldstein, 1993; Habermann et al., 1983; Platt and Busselberg, 1994; Pounds, 1984; Richdardt et al., 1986; Rosen and Pounds, 1989; Simons and Pocock, 1987; Sun and Suszkiw, 1995; Tomsig and Suszkiw, 1995; Watts et al., 1995). Impairments of oxidative metabolism could conceivably contribute to transport deficits and cellular degeneration; however, the exact role this plays in the nephropathy induced by Pb has not been elucidated.

Exposure to Pb also appears to produce an oxidative stress of unknown, and possibly multi-pathway, origin (Dagggett et al., 1998; Ding et al., 2001; Hermes-Lima et al., 1991; Lawton and Donaldson, 1991; Monteiro et al., 1991; Nakagawa, 1991; Sugawara et al., 1991; Sandhir et al., 1994). Secondary responses to Pb-induced oxidative stress include induction of nitric oxide (NO) synthase, GSH *S*-transferase, and transketolase in the kidney (Daggett et al., 1998; Moser et al., 1995; Witzmann et al., 1999; Wright et al., 1998; Vizari et al., 2001). Depletion of NO has been implicated as a contributor to Pb-induced hypertension in the rat (Carmignani et al., 2000; Gonick et al., 1997; Vizari et al., 1997, 1999a, 1999b) and, thereby, may contribute to impairments in glomerular filtration and, possibly, in the production of glomerular lesions. However, a direct role for this mechanism in Pb-induced proximal tubular injury has not been elucidated. Both Pb and L-N-(G)-nitro arginine methyl ester (L-NAME), an inhibitor of NO synthetase, increased the release of NAG from isolated rat kidneys perfused with an albumin-free perfusate (Dehpour et al., 1999). The addition of L-arginine decreased the effect of Pb on NAG release. This observation is consistent with an oxidative stress mechanism possibly contributing to Pb-induced enzymuria and increased urinary excretion of NAG.

Experimental studies in laboratory animals have shown that Pb can depress GFR and renal blood flow (Aviv et al., 1980; Khalil-Manesh et al., 1992a, 1992b). In rats, depressed GFR appears to be preceded by a period of increased filtration (Khalil-Manesh et al., 1992a; O'Flaherty et al., 1986). The mechanism by which Pb alters GFR is unknown and, as important for risk assessment, its mechanistic connection to Pb-induced hypertension has not been fully elucidated. Glomerular sclerosis or proximal tubular injury and impairment could directly affect renin release (Boscolo and Carmignani, 1988) or renal insufficiency could secondarily contribute to hypertension.

Transport of Lead in the Kidneys

Measurements of the renal clearance of radioisotopic Pb in dogs have shown that Pb can be absorbed from and secreted into the lumen of the nephron depending on the acid–base status of the animal (Vander et al., 1977; Victery et al., 1979a). Net absorption was observed during acidosis induced by infusion of hydrogen chloride and infusion of sodium chloride, whereas net secretion was observed during alkalosis induced by infusion of sodium bicarbonate. Further evidence for bidirectional transfer of Pb was provided by observations of the effects of ureteral ligation on Pb uptake by the kidney. Glomerular filtration ceases after ligation of the ureter during mannitol-induced diuresis (Malvin and Wilde, 1973). Therefore, Pb uptake observed in ligated kidneys (stop-flow) can be assumed to have resulted primarily from transfer of Pb to the kidney from the plasma (basolateral) side of the tubule, rather than from the glomerular filtrate or luminal side of the tubule. The difference between uptake observed during free-flow (ureter-open) and stop-flow conditions provides an estimate of the luminal uptake component. Uptake of Pb under stop-flow conditions in the dog was shown to be approximately 50% of the uptake that occurs under free-flow conditions, suggesting that approximately half of the Pb-uptake in the kidney may occur at the basolateral side of the tubular epithelium (Victery et al., 1979b). During alkalosis induced by infusion of sodium bicarbonate, the magnitude of the basolateral component was twice that observed during normal conditions (sodium chloride infusion) and was estimated to contribute approximately 30% to total Pb uptake. The estimated luminal uptake component was four times the filtered load of Pb that was observed under free-flow conditions, suggesting that a major component of luminal uptake may represent Pb that was secreted into the tubular lumen. These observations are consistent with the clearance measurements, which showed net tubular secretion of Pb in alkalosis and net reabsorption under normal conditions (Vander et al., 1977; Victery et al., 1979a).

The mechanisms of secretory and absorptive transfer of Pb along the nephron have not yet been characterized. Studies conducted in preparations of mammalian small intestine support the existence of saturable and non-saturable pathways of Pb-transfer and suggest that Pb can interact with transport mechanisms for Ca and iron (Fe) (Diamond, 2000). Although these observations may be applicable to the kidney, empirical evidence for specific transport mechanisms in the nephron is lacking. Victery et al. (1979b) examined the kinetics of urinary excretion of Pb, sodium, and inulin during the transition from stop-flow to free-flow conditions in the dog. The concentration-profiles of urine excreted immediately after removal of the ureteral ligature reflect the profile achieved along the lumen of the nephron during stop-flow conditions in combination with any further processing of the tubular fluid (e.g., absorption or secretion) that occurs after release of the ureteral ligature, the latter is ideally minimized by intense free-flow diuresis (Malvin and Wilde, 1973). Stop-flow analysis indicated that

Pb was absorbed from the tubular lumen at a site distal to the proximal tubule (i.e., co-localized with the minimum sodium concentration) and in the proximal tubule (i.e., co-localized with the plateau of the sodium concentration). Alkalosis induced by infusion of sodium bicarbonate enhanced the proximal absorption of Pb.

Uptake of Pb in rabbit renal cortical slices appears to be at least partially dependent on metabolic energy, in that it is depressed under anoxic conditions and by metabolic inhibitors (e.g., dinitrophenol, sodium cyanide, sodium iodoacetate), but not by the Na^+/K^+-ATPase inhibitor ouabain (Vander et al., 1979). Uptake of Pb, but not PAH, was decreased when slices were incubated with $SnCl_4$ and in the presence of 20 g/l bovine serum albumin. Cys or citrate increased the uptake of Pb by approximately 20% (Vander and Johnson, 1981).

Victery et al. (1984) studied interactions between Pb and isolated apical membrane vesicles of rat renal proximal tubules. Incubation of the isolated membranes with Pb resulted in saturable binding of Pb to the membranes that was insensitive to transient electrochemical gradients of Na^+, K^+, H^+, or amino acid. Binding was blocked by Sn^{2+}, Sn^{4+}, La^{3+}, Fe^{2+}, Fe^{3+}, or Cu^{2+}, but not by Mg^{2+}, Zn^{2+}, Cd^{2+}, Hg^{2+}, or Ca^{2+}. The extensive binding to the apical membrane vesicles may have hindered detection of carrier-mediated transport of Pb in this preparation, if such mechanisms exist. The same can be said for studies of Ca-transport in isolated apical membrane-vesicles (Gmaj et al., 1979). Thus, to the extent that Pb-transport has been studied in isolated renal apical membrane vesicles, the characteristics of Pb-binding seem to be different from those of Ca.

The possibility of Pb crossing cell membranes through cation channels may be relevant to Pb transport across the apical membrane along the nephron. Pb has been shown to enter cells through voltage-gated L-type Ca^{2+} channels in bovine adrenal medullary cells (Legare et al., 1998; Simons and Pocock, 1987; Tomsig and Suszkiw, 1991) and through store-operated Ca^{2+} channels in pituitary GH3, glial C3, human embyronic kidney, and bovine brain capillary endothelial cells (Kerper and Hinkle, 1997a, 1997b). Evidence that such mechanisms play an important role in the transport of Pb in epithelia is lacking, although there is growing evidence suggesting an important role of cation channels in apical membrane transport of Ca^{2+} in the kidney (Friedman and Gesek, 1995).

Transfer of Pb across the plasma membrane of erythrocytes appears to be coupled to the activity of a 4,4′-diisothiocyanostilbene-2,2′-disulfonic acid (DIDS)-sensitive anion exchanger (Bannon et al., 2000; Simons, 1986a, 1986b). However, involvement of an anion exchanger in Pb transport has not been revealed in kidney. Uptake of Pb in MDCK cells is not DIDS-sensitive, although these cells express a functional anion exchange mechanism (Bannon et al., 2000).

Intracellular Disposition of Lead

A characteristic histologic feature of Pb nephrotoxicity is the formation of intranuclear inclusion bodies in the renal proximal tubule (Choie and Richter, 1972; Goyer et al., 1970a, 1970b). Inclusion bodies contain Pb complexed with protein (Moore et al., 1973). Appearance of nuclear inclusion bodies is associated with a shift in compartmentalization of Pb from the cytosol to the nuclear fraction (Oskarsson and Fowler, 1985). Sequestration of Pb in nuclear inclusion bodies can achieve Pb concentrations that are 100-times higher (μg Pb/mg protein) than that in kidney cytosol (Goyer et al., 1970a, 1970b; Horn, 1970); thus, the bodies can have a profound effect on the intracellular disposition of Pb in the kidney. The sequestration of Pb in intranuclear inclusion bodies may limit or prevent toxic interactions with other molecular targets of Pb. In rats exposed to nephrotoxic doses of Pb acetate, few intranuclear inclusion bodies occurred in the S_3 segment of the proximal tubule, where acute injury was most severe, whereas, intranuclear inclusion bodies were more numerous in the S_2 segment, where the injury was less severe (Murakami et al., 1983).

The exact identity of the Pb–protein complex in inclusion bodies remains unknown as is the mechanism of formation of the inclusion body itself. Although proteins that appear to be unique to Pb-induced inclusion bodies have been isolated, their role in the Pb sequestration has not been elucidated (Shelton and Egle, 1982). Cytosolic proteins may serve as carriers of Pb or intermediary ligands for uptake of Pb into the nucleus. Two cytosolic proteins, which appear to be cleavage products of β2-microgobulin (Fowler and DuVal, 1991), have been isolated from rat kidney cytosol that have high affinity binding sites for Pb (K_d = 13 and 40 nM, respectively) and that can mediate uptake of Pb into isolated nuclei (Mistry et al., 1985, 1986). These proteins can also participate in ligand exchange reactions with other cytosolic binding sites, including δ-aminolevulinic dehydratase, which binds to and is inhibited by Pb (Goering and Fowler, 1984, 1985). Other high-affinity Pb binding proteins

(K_d approximately 14 nM) have been isolated in human kidney, two of which have been identified as a 5 kD peptide, thymosin $\beta 4$ and a 9 kD peptide, acyl-CoA binding protein (Smith et al., 1998). Pb also binds to MT, but does not appear to be a significant inducer of the protein in comparison with the inducers of Cd and zinc (Zn) (Eaton et al., 1980; Waalkes and Klassen, 1985). *In vivo*, only a small fraction of the Pb in the kidney is bound to MT, and it appears to have a binding affinity that is less than Cd^{2+}, but higher than Zn^{2+} (Ulmer and Vallee, 1969). Thus, Pb will more readily displace Zn from MT than Cd (Goering and Fowler, 1987; Nielson et al., 1985; Waalkes et al., 1984). The precise role of cytosolic Pb binding proteins in inclusion body formation has not been determined, although it has been hypothesized that aggregations of $\beta 2$-microgobulin may contribute to the Pb–protein complex observed in nuclear inclusion bodies (Fowler and DuVal, 1991).

NEPHROTOXICOLOGY OF URANIUM

The pathophysiology of acute nephrotoxicity of uranium (U) has been examined extensively, and reviewed by Diamond (1989). Interest in the toxicology of U originated with the Manhattan Project and the atomic energy program, which required the processing and handling of large amounts of hexavalent U compounds (U^{6+}) and which presented significant occupational health-challenges (Hodge, 1973). As new techniques for studying renal morphology and function have developed, research on the nephrotoxicology of U has continued with a focus on using hexavalent U compounds, particularly uranyl (UO_2^{2+}) compounds (e.g., uranyl acetate, uranyl fluoride, uranyl nitrate) to explore mechanisms of acute renal failure (Flamenbaum, 1973). Recent reports have provided a fairly detailed description of the histopathology and functional disturbances that occur in response to an acutely nephrotoxic dose of U. Only a few studies of long-term exposures have been reported, which have revealed a nephrotoxic response to UO_2^{2+} that is consistent with the acute studies, in terms of the general characteristics of renal lesions (Gilman et al., 1998a, 1998b, 1998c; McDonald-Taylor et al., 1997). The nephropathy induced by UO_2^{2+} continues to serve as a useful model for studying the pathophysiology and pharmacology of acute renal failure (Appenroth et al., 2001; Katayama et al., 1999; Lee et al., 2000; Sano et al., 2000; Sun et al., 2000; Van Crugten et al., 2000).

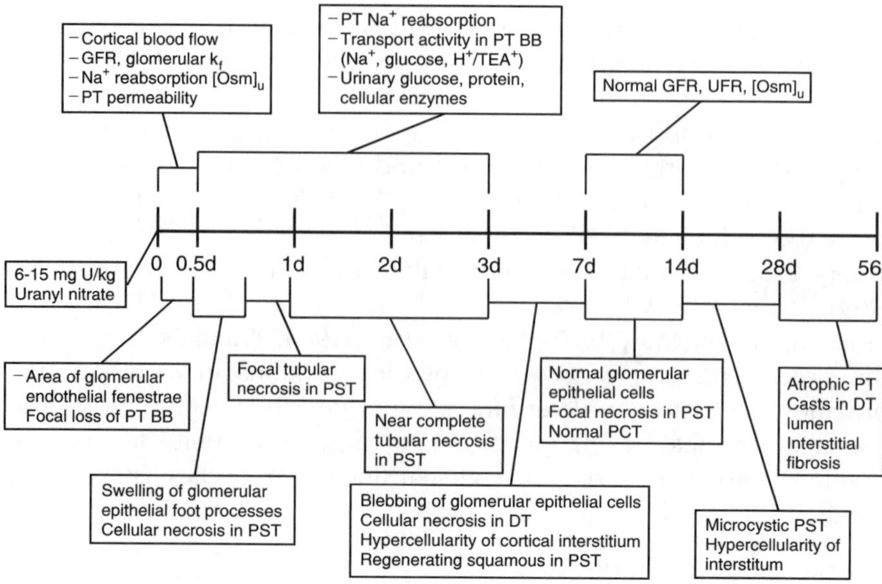

Figure 22.3 Time course for the functional and morphologic changes after parenteral UO_2^{2+} is administered to the rat. Abbreviations: BB, brush border; DT, distal tubule; GFR, glomerular filtration rate; K_f, glomerular ultrafiltration coefficient; Osm, osmolarity; PCT, proximal convoluted tubule; PST, proximal straight tubule; PT, proximal tubule; TEA^+, tetraethyl ammonium; UFR, urine flow rate. (Avasthi et al., 1980; Blantz, 1977; Flamenbaum et al., 1974; Haley, 1982; Haley et al., 1982; Hori et al., 1985; Olbricht et al., 1977)

Functional and morphologic changes that follow a single parenteral dose of uranyl nitrate (6 to 15 mg U/kg) administered to rats are summarized in Figure 22.3. A qualitatively similar sequence of changes has been established for dogs and rabbits exposed to uranyl acetate (Kobayashi et al., 1984; Sudo et al., 1977) and, in less detail, for dogs and rats exposed to uranyl fluoride (Diamond et al., 1989; Leach et al., 1973; Morrow et al., 1981). Principal features of this sequence can be divided into glomerular effects (including perfusion and filtration defects and morphological abnormalities), and tubular effects (including transport and permeability defects, and necrosis). Lesions progress in severity for the first five days, culminating in renal failure in some animals during the first week after exposure (Blantz, 1977; Sudo et al., 1977). The mechanisms for the glomerular and tubular lesions have been explored in animal models, but have yet to be completely elucidated.

Effects of Uranium on Glomerular Filtration

Decreased GFR is a consistently observed outcome of parenterally administered UO_2^{2+} in rats, rabbits, and dogs (Flamenbaum et al., 1972, 1974; Nomiyama and Foulkes, 1968). Single nephron filtration rate, assessed by micropuncture, is depressed within six hours after a parenteral of dose of 6 to 9 mg U/kg in the rat (Blantz, 1977; Flamenbaum, 1974) and within 48 hours after a dose of 3 mg U/kg in the dog (Stein et al., 1975). Whole-animal GFR, assessed by clearance techniques, also decreases in response to an acute dose of UO_2^{2+} (e.g., Appenroth et al., 2001; Morrow et al., 1981; Nomiyama and Foulkes, 1968; Zalups et al., 1988). However, interpretation of clearance measurements is confounded in U-treated animals by the possibility of toxicity-induced changes in permeability along the nephron, which could result in an underestimation of GFR from clearance measurements (Blantz, 1977; Nomiyama and Foulkes, 1968).

The mechanism by which U decreases GFR is not completely understood. A consistently reported feature of U-induced acute renal failure in dogs, rabbits, and rats is a decrease in renal outer cortical blood flow and glomerular perfusion (Flamenbaum et al., 1972, 1974; Nomiyama and Foulkes, 1968; Stein et al., 1975; Sudo et al., 1977). Decreased glomerular perfusion could lower GFR by decreasing glomerular capillary hydraulic pressure or by shifting the point of filtration pressure equilibrium towards the afferent end of the glomerular capillary. This decreases the length of the capillary that is available for filtration. In at least one study, GFR was observed to decrease in the absence of significant changes in renal plasma flow in rats treated with uranyl nitrate (Blantz, 1977). Therefore, decreased renal plasma flow does not appear to be required for decreasing GFR, and other factors must be involved.

A rise in tubular hydraulic pressure as a result of impaired tubular solute and fluid absorption or tubular blockage may also contribute to decreased GFR. Although elevated tubular hydraulic pressure has been reported to occur in rats exposed to uranyl nitrate (Mason et al., 1977) and in isolated perfused dog kidneys exposed *in vitro* to uranyl nitrate (Nizet, 1981), it is not a consistent finding (Flamenbaum et al., 1974; Stein et al., 1975). Thus, it cannot be the sole mechanism responsible for decreasing GFR.

A partial quantitative assessment of Starling forces in the glomerulus is possible in the Munich-Wistar rat, which possesses a subcapsular population of glomeruli that can be micropunctured for measurement of glomerular hydrostatic and oncotic pressures (Brenner et al., 1971). In this animal, a 50% decrease in single nephron filtration rate was observed during the first six hours after injection of uranyl nitrate (15 mg U/kg bw), concurrent with significant elevation of net glomerular filtration pressure, and in the absence of a change in renal plasma flow (Blantz, 1977). This suggests that U may alter the glomerular ultrafiltration coefficient.

Induction of Proteinuria by Uranium

Proteinuria is a consistent feature of the acute nephropathy induced by UO_2^{2+} (Bentley et al., 1985; Blantz, 1977; Diamond et al., 1989; Domingo et al., 1987; Gilman et al., 1998c; Leach et al., 1984; Zalups et al., 1988). Parenteral doses of 0.5 to 1.0 mg U/kg in dogs, rats, and guinea pigs results in a transient proteinuria, consisting mainly in the urinary excretion of proteins having a molecular weight of 60 to 70 kDa, including serum albumin (Flamenbaum et al., 1974; Leach et al., 1984; Morrow et al., 1981). While this suggests a possible glomerular origin to the proteinuria, other mechanisms are likely to be involved. In rabbits, the renal clearance of dextrans having molecular weights of 70 to 90 kDa was not affected by 0.2 mg U/kg administered as uranyl acetate (Foulkes, 1971). This suggests that mechanisms other than a change in the size-selectivity of the glomerular filter contribute to proteinuria. Enhanced excretion of low molecular weight proteins has been observed in rats after parenteral doses of UO_2^{2+} (0.05 to 0.25 mg U/kg bw), suggesting the possibility of impaired tubular absorption of filtered proteins (Bentley et al., 1985).

Histopathology of Glomerular Injury Induced by Uranium

Abnormalities in the ultrastructure of the glomerulus occur in UO_2^{2+}-treated animals concurrent with decreases in GFR. In rats, the diameter and density of the glomerular endothelial fenestrae decrease seven hours after a parenteral dose of 9 to 15 mg U/kg as uranyl nitrate (Avasthi et al., 1980). Such changes may give rise to a decreased ultrafiltration coefficient through a decrease in surface area available for filtration. Endothelial changes are followed by distortions of the glomerular visceral epithelium (Avasti et al., 1980; Haley, 1982), suggesting that the endothelium, rather than the epithelium, may be the primary target of

U in the renal corpuscle of the rat. In rabbits, concurrent changes in glomerular endothelial and epithelial cells have been reported (Kobayashi et al., 1984). Epithelial changes observed in rabbits include spreading and flattening of the visceral epithelial cells (podocytes), and loss of podocyte processes, leading to decreased filtration area.

Effects of Uranium on the Nephron and Collecting Duct

Uranyl compounds induce cellular necrosis that originates, at the lowest effective dose levels, in the pars recta of the proximal tubule. In rats, morphological changes in the pars recta segments of the proximal tubule are evident within six hours after a 6 mg U/kg dose of UO_2^{2+}, including focal loss of periodic acid-Schiff (PAS) positive staining and loss of brush border from the proximal tubular epithelial cells (Flamenbaum et al., 1974; Haley, 1982). Cellular injury progresses in severity for the first five days, culminating in near complete necrosis of the pars recta segment of proximal tubules, with the pars convoluta segment remaining relatively free from injury (Haley, 1982; Sun et al., 2000). Distortion, focal necrosis, and casts in the thick ascending limb of the loop of Henle, distal convoluted tubule, and cortical collecting duct follow proximal tubular injury induced by uranyl compounds (Diamond et al., 1989; Haley, 1982). Regeneration of epithelium lining proximal and distal portions of the nephron begins within one week after dosing, during the time of peak severity of injury (Diamond et al., 1989; Sano et al., 2000; Sun et al., 2000; Zalups et al., 1988). After a parenteral dose of uranyl fluoride (0.7 to 13 mg U/kg), restoration of the normal morphology of the tubule is largely completed within 35 days after dosing (Diamond et al., 1989). However, atrophic tubules, intraluminal casts, and interstitial fibrosis have been reported 8 and 20 weeks after administering acutely toxic doses of uranyl nitrate (approximately 5 to 6 mg U/kg) to rats (Appenroth et al., 2001; Haley et al., 1982) and 90 days after cessation of exposure of rabbits to uranyl nitrate (41 mg U/kg per day) in drinking water for 90 days (Gilman et al., 1998a). Thus, there appear to be chronic sequelae to tubular injury induced by U.

Abnormalities in tubular function become apparent within two hours after parenteral doses of UO_2^{2+} (6 to 15 mg U/kg) are administered. Severity of pathophysiological changes progresses during the first 1 to 5 days (Figure 22.3). Some of these changes include polyuria, decreased urine osmolality, deceased net tubular absorption of sodium and glucose (Blantz, 1977; Flamenbaum et al., 1974; Zalups et al., 1988), decreased tubular transport maxima for glucose, PAH, and amino acids (Nizet, 1981; Nomiyama and Foulkes, 1968) and enzymuria (Bentley et al., 1985; Morrow et al., 1981; Zalups et al., 1988). In rats, glucosuria has been observed at doses between 0.02 and 0.05 mg U/kg (Leach et al., 1984), with multiple abnormalities in tubular function apparent after dose levels exceeding 0.1 mg U/kg (Diamond et al., 1989; Zalups et al., 1988). Enzymuria and proteinuria have been observed in rabbits administered approximately 1 mg U/kg of uranyl nitrate in drinking water (Gilman et al., 1998c).

Abnormal solute permeability of the tubular epithelium may also occur after a large acute dose of UO_2^{2+} (Nomiyama and Foulkes, 1968). Increased permeability of the proximal tubule to mannitol and inulin has been reported to occur in rats (Blantz, 1977; Olbricht et al., 1977) and dogs (Stein et al., 1975) administered uranyl nitrate (15 or 6 mg U/kg, respectively). As noted above, abnormal solute permeability along the nephron has implications for the use of clearance techniques for assessing whole-kidney GFR in U-poisoned animals. The effect is most likely dose-related for a given species and may only be an important factor at relatively high-doses. Tubular permeability has been reported to be unaltered in rats 2 days after a 6 mg U/kg dose of uranyl nitrate (Flamenbaum et al., 1974) and, in rabbits, after a 0.2 mg U/kg dose of uranyl acetate (Foulkes, 1971).

The mechanism by which UO_2^{2+} selectively damages the terminal segment of the proximal tubule is not understood. Damage to the proximal tubule may result from selective uptake of U in this region of the tubule. Autoradiographs of animals that were administered ^{232}U are consistent with this view. In rats, when U is administered as ^{232}U-uranyl nitrate, it concentrates in the outer stripe of the outer medulla and along the medullary rays, penetrating the inner cortex. This is consistent with a location of the pars recta of the proximal tubule (Tannenbaum et al., 1951). However, more direct evidence for transport of U in specific nephron segments has not been established.

REFERENCES

Aleo MD, Taub ML, and Kostyniak PJ (1992) Primary cultures of rabbit renal proximal tubule cells. III. Comparative cytotoxicity of inorganic and organic mercury. *Toxicol Appl Pharmacol* 112:310–317.

Aleo MD, Taub ML, Olson JR, Nickerson PA, and Kostyniak PJ (1987) Cellular uptake and response of primary cultures of rabbit renal proximal tubule cells exposed to mercuric chloride and methylmercury chloride, in *In vitro Toxicology: Approaches to Validation*, Goldberg AM (Ed.), Mary Ann Liebert, Inc., New York, pp. 211–226.

Aposhian HV (1983) DMSA and DMPS-Water soluble antidotes for heavy metal poisoning. *Annu Rev Pharmacol Toxicol* 23:193–215.

Aposhian HV, and Aposhian MM (1990) meso-2,3-Dimercaptosuccinic acid: Chemical, pharmacological and toxicological properties of an orally effective metal chelating agent. *Annu Rev Pharmacol Toxicol* 30:279–306.

Appenroth D, Lupp A, Kriegsmann J, Sawall S, Splinther J, Sommer M, Stein G, and Fleck C (2001) Temporary Warm Ischaemia, 5/6 Nephrectomy and Single Uranyl Nitrate Administration—Comparison of Three Models Intended to Cause Renal Fibrosis in Rats. *Exp Toxicol Pathol* 53:316–324.

Aslamkhan AG, Han Y-H, Zalups RK, and Pritchard JB. (2003) Human renal organic anion transporter 1 (hOAT1) dependent uptake and toxicity of mercuric SH-conjugates in MDCK cells. *Mol Pharmaco*l 63:590–596.

ATSDR (1999) *Toxicological Profile for Mercury*. United States: U.S. Department of Health and Human Services, Public Health Service, Agency for Toxic Substance and Disease Registry, TP-93/10.

Avasthi PS, Evan AP, and Hay D (1980) Glomerular endothelial cells in uranyl nitrate-induced acute renal failure in rats. *J Clin Invest* 65:121–127.

Aviv A, John E, Bernstein J, Goldsmith DI, and Spitzer A (1980) Lead intoxication during development: Its late effect on kidney function and blood pressure. *Kidney Int* 17:430–437.

Baggett J McC, and Berndt WO (1986) The effect of depletion of nonprotein sulfhydryls by diethyl maleate plus buthionine sulfoximine on renal uptake of mercury in the rat. *Toxicol Appl Pharmacol* 83:556–562.

Bannon DI, Olivi L, and Bressler J (2000) The role of anion exchange in the uptake of Pb by human erythrocytes and Madin-Darby Canine Kidney Cells. *Toxicology* 147:101–107.

Barenberg RL, Solomon S, Papper S, and Anderson R (1968) Clearance and micropuncture study of renal function in mercuric chloride treated rats. *J Lab Clin Med* 72:473–484.

Barfuss DW, Robinson MK, and Zalups RK (1990) Inorganic mercury transport in the proximal tubule of the rabbit. *J AM Soc Nephrol* 1:910–917.

Bentley KW, Stockwell DR, Britt KA, and Kerr CB (1985) Transient proteinuria and aminoaciduria in rodents following uranium intoxication. *Bull Environ Contam Toxicol* 34:407–416.

Bergstrand A, Friberg L, Mendel L, and Odeblad E (1959) The localization of subcutaneously administered radioactive mercury in the rat kidney. *J Ultrastruct Res* 3:238–239.

Bernard A, Buchet JP, Roels H, Masson P, and Lauwerys R (1979) Renal excretion of proteins and enzymes in workers exposed to cadmium. *Eur J Clin Invest* 9:11–22.

Bernard AM, and Lauwerys RR (1981) Retinol binding protein in urine: A more practical index than urinary 2-microglobulin for the routine screening of renal tubular function. *Clin Chem* 27: 1781–1782.

Berlin M (1963). Accumulation and retention of mercury in the mouse. III. An autoradiographic comparison of methylmercuric dicyandiamide with inorganic mercury. *Arch Environ Health* 6:610–616.

Berlin M, and Ullberg S (1963a) Accumulation and retention of mercury in the mouse. I. An autoradiographic study after a single intravenous injection of mercuric chloride. *Arch Environ Health* 6:582–601.

Berlin M, and Ullberg S (1963b) Accumulation and retention of mercury in the mouse. II. An autoradiographic comparison of phenylmercuric acetate with inorganic mercury. *Arch Environ Health* 6:602–609.

Berndt WO, Baggett J McC, Blacker A, and Houser M (1985) Renal glutathione and mercury uptake by kidney. *Fund Appl Toxicol* 5:832–839.

Biagini G, Caudarella R, and Vangelista A (1977) Renal morphological and functional modification in chronic lead poisoning. *Clin Chem Toxicol Metals* 15:123–126.

Bigazzi PE (1988) Autoimmunity induced by chemicals. *Clin Toxicol* 26:125–156.

Bigazzi PE (1992) Lessons from animal models: The scope of mercury-induced autoimmunity. *Clin Immunol Immunopathol* 65:81–84.

Blantz RC (1977) The mechanism of acute renal failure after uranyl nitrate. *J Clin Invest* 55:621–635.

Boscolo P, and Carmignani M (1988) Neurohumoral blood pressure regulation in lead exposure. *Environ Health Perspect* 78:101–106.

Brenner BM, Troy JL, and Daugharty TM (1971) The dynamics of glomerular ultrafiltration in the rat. *J Clin Invest* 50:1776–1789.

Bruggeman IM, Temmink JHM, and Van Bladeren PJ (1992) Effect of glutathione on apical and basolateral uptake and toxicity of $CdCl_2$ in kidney cells (LLC-PK_1). *Toxicol in vitro* 6:195–200.

Buchet JP, Roels H, Bernard A, and Lauwerys R (1980) Assessment of renal function of workers exposed to inorganic lead, cadmium or mercury vapor. *J Occup Med* 22:741–750.

Cannon VT, Barfuss DW, and Zalups RK (2000). Molecular homology ("mimicry") and the mechanisms involved in the luminal uptake of inorganic mercury in the proximal tubule of the rabbit. *J Am Soc Nephrol* 11:394–402.

Cannon VT, Zalups RK, and Barfuss DW (2001) Amino acid transporters involved in the luminal transport of mercuric conjugates of cysteine in the rabbit proximal tubule.*J Pharmacol Exp Ther* 298:780–789.

Cardenas A, Roels H, Bernard AM, Barbon R, Buchet JP, Lauwreys RR, Rosello J, Hotter G, Mutti A, and Franchini I, et al. (1993) Markers of early renal changes induced by industrial pollutants. I. Application to workers exposed to mercury vapour. *Br J Ind Med* 50:17–27.

Carmignani M, Volpe AR, Boscolo P, Qiao N, Di Gioacchino M, Grilli A, and Felaco M (2000) Catecholamine and nitric oxide systems as targets of chronic lead exposure in inducing selective functional impairment. *Life Sci* 68:401–415.

Chang LW, Ware RA, and Desnoyers PA (1973) A histochemical study on some enzymes changes in the kidney, liver and brain after chronic mercury intoxication in the rat. *Food Cosmet Toxicol* 11:283–286.

Cherian MG (1979) Metabolism of orally administered cadmium-metallothionein in mice. Environ Health Perspect 28:127–130.

Cherian MG, Goyer RA, and Valberg LS (1978) Gastrointestinal absorption and organ distribution of oral cadmium chloride and cadmium-metallothionein in mice. *J Toxicol Environ Health* 4:861–868.

Chia KS, Ong C, Ong HY, and Endo G (1989) Renal tubular function of workers exposed to low levels of cadmium. *Br J Ind Med* 46:165–170.

Choie DD, and Richter GW (1972) Lead poisoning: Rapid formation of intranuclear inclusions. Science 177:1194–1195.

Clarkson TW, and Magos L (1967) The effect of sodium maleate on the renal disposition and excretion of mercury. *Br J Pharmacol* 31:560–567.

Cramer K, Goyer RA, Jagenburg R, and Wilson MH (1974) Renal ultrastructure, renal function, and parameters of lead toxicity in workers with different periods of lead exposure. *Br J Ind Med* 31:113–127.

Cuppage PE, and Tate A (1967) Repair of the nephron following injury with mercuric chloride. *Am J Pathol* 51:405–429.

Cuppage PE, Chiga M, and Tate A (1972) Cell cycle studies in the regenerating rat nephron following injury with mercuric chloride. *Lab Invest* 26:122–126.

Daggett DA, Oberley TD, Nelson SA, Wright LS, Kornguth SE, and Siegel FL (1998) Effects of lead on rat kidney and liver: GST expression and oxidative stress. *Toxicology* 128:191–206.

Dehpour AR, Essalat M, Ala S, Ghazi-Khansari M, and Ghafourifar P (1999) Increase by NO synthase inhibitor of lead-induced release of *N*-acetyl-β-D-glucosaminidase from perfused kidney. *Toxicology* 132:119–125.

Diamond GL (1988) Biological monitoring of urine for exposure to toxic metals, in *Biological Monitoring of Toxic Metals*, Clarkson TW, Friberg L, Nordberg GF, and Sager PR (Eds), Plenum Publishing Corp., New York, pp. 515–529.

Diamond GL (1989) Biological consequences of exposure to soluble forms of natural uranium. Radiat Protect Dosim 26:23–33.

Diamond GL (2000) Transport of metals in the gastrointestinal system and kidneys, in *The Molecular Biology and Toxicology of Metals*, Zalups RK, and Korpatnick DJ (Eds), Taylor & Francis, London, pp. 300–344.

Diamond GL, Cohen JJ, and Weinstein SL (1986) Renal handling of cadmium in perfused rat kidney and effects on renal function and tissue composition. *Am J Physiol* 251:F784–F794.

Diamond GL, Morrow PE, Panner BJ, Gelein RM, and Baggs RB (1989) Reversible uranyl fluoride nephrotoxicity in the Long Evans rat. *Fund Appl Toxicol* 13:65–78.

Dieter MP, Matthews HB, Jeffcoat RA, and Moseman RF (1993) Comparison of lead bioavailability in F344 rats: Lead acetate, lead oxide, lead sulfide, or lead ore concentrate from Skagway, Alaska. *J Toxicol Environ Health* 39:79–93.

Ding Y, Gonick HC, Vaziri ND, Liang K, and Wei L (2001) Lead-induced hypertension. Increased hydroxyl radical production. *Am J Hypertens* 14:169–173.

Domingo JL, Llobet JM, Tomas JM, and Corbella J (1987) Acute toxicity of uranium in rats and mice. *Bull Environ Contam Toxicol* 39:168–174.

Dorian C, Gattone VH 2nd, and Klaassen CD (1992) Renal cadmium deposition and injury as a result of accumulation of cadmium-metallothionein (CdMT) by the proximal convoluted tubules-A light microscopy autoradiography study with ^{109}CdMT. *Toxicol Appl Pharmacol* 114:173–181.

Dudley RE, Gammal LM, and Klaassen CD (1985) Cadmium-induced hepatic and renal injury in chronically exposed rats: Likely role of hepatic cadmium-metallothionein in nephrotoxicity. *Toxicol Appl Pharmacol* 77:414–426.

Druet P, Druet E, Potdevin F, and Sapin C (1978) Immune type glomerulonephritis induced by $HgCl_2$ in the Brown-Norway rat. *Ann Immunol (Institute Pasteur)* 129C:777–792.

Eaton DL, Stacey NH, Wong K-L, and Klaassen CD (1980) Dose response effects of various metal ions on rat liver metallothionein, glutathione, heme oxygenase, and cytochrome P-450. *Toxicol Appl Pharmacol* 55:393–402.

Elinder C-G, Edling C, Lindberg E, Kagedal B, and Vesterberg O (1985a). β{-2}-microglobulinuria among workers previously exposed to cadmium: follow-up and dose-response analyses. *Am J Ind Med* 8:553–564.

Elinder CG, Edling C, Lindberg E, Kagedal B, and Vesterberg O (1985b). Assessment of renal function in workers previously exposed to cadmium. *Br J Ind Med* 42:754–760.

Ellis BG, Price RG, and Topham JC (1973) The effect of tubular damage by mercuric chloride on kidney function and some urinary enzymes in the dog. *Chem-Biol Interact* 7:101–113.

Endo T (2002) Transport of cadmium across the apical membrane of epithelial cells lines. *Comp Biochem Physiol C Toxicol Pharmacol* 131:223–239.

Endo T, Kimura O, and Sakata M (1999) Further analysis of cadmium uptake from apical membrane of LLC-PK_1 cells via inorganic anion exchanger. *Pharmacol Toxicol* 84:187–192.

Endo T, Kimura O, and Sakata M (1998a) Bidirectional transport of cadmium across apical membrane of renal epithelial cell lines via H^+-antiporter and inorganic anion exchanger. *Toxicology* 131:183–192.

Endo T, Kimura O, and Sakata M (1998d). pH-dependent transport of cadmium in rat renal brush border membrane vesicles: cadmium efflux via H^+-antiport. *Toxicol Lett* 99:99–107.

Endo T, Kimura O, and Sakata M (1998c). Cadmium uptake from apical membrane of LLC-PK_1 cells via inorganic anion exchanger. *Pharmacol Toxicol* 82:230–235.

Endo T, Kimura O, Hatakeyama M, Takada M, and Sakata M (1998d) Effects of zinc and copper on cadmium uptake by brush border membrane vesicles. *Toxicol Lett* 91:111–120.

Enestrom S, and Hultman P (1984) Immune-mediated glomerulonephritis induced by mercuric chloride in mice. *Experientia* 40:1234–1240.

Falck FY, Fine LJ, Smith RG, McClatchey KD, Annesley T, England B, and Schork AM (1983) Occupational cadmium exposure and Renal status. *Am J Ind Med* 4:541–549.

Felley-Bosco E, and Diezi J (1987). Fate of cadmium in rat renal tubules: A microinjection study. *Toxicol Appl Pharmacol* 91:204–211.

Felley-Bosco E, and Diezi J (1989) Fate of cadmium in rat renal tubules: A micropuncture study. *Toxicol Appl Pharmacol* 98:243–251.

Felley-Bosco E, and Diezi J (1991) Cadmium uptake and induction of metallothionein synthesis in a renal epithelial cell line (LLC-PK_1). *Arch Toxicol* 65:160–163.

Ferguson CJ, Wareing M, Ward DT, Green R, Smith CP, and Riccardi D (2001) Cellular localization of divalent metal transporter DMT-1 in rat kidney. *Am J Physiol* 280:F803–F814.

Flamenbaum W (1973) Pathophysiology of acute renal failure. *Arch Intern Med* 131:911–928.

Flamenbaum W, Huddleston ML, McNeil JS, and Hamburger RJ (1974) Uranyl nitrate-induced acute renal failure in the rat: Micropuncture and hemodynamic studies. *Kidney Int* 6:408–418.

Flamenbaum W, McNeil JS, Kotchen TA, and Saladino AJ (1972) Experimental acute renal failure induced by uranyl nitrate in the dog. *Circ Res* 31:682–697.

Foulkes EC (1971) Glomerular filtration and renal plasma flow in uranium poisoned rabbits. *Toxicol Appl Pharmacol* 20:380–385.

Foulkes EC (1974) Excretion and retention of cadmium, zinc, and mercury by rabbit kidney. *Am J Physiol* 227:1356–1360.

Fox MR, Jacobs RM, Jones AO, Fry BE Jr, and Stone CL (1980) Effects of vitamin C and iron and cadmium metabolism. *Ann NY Acad Sci* 355:249–261.

Fowler BA (1972) The morphological effects of dieldrin and methyl mercuric chloride on pars recta segments of the rat kidney proximal tubules. *Am J Pathol* 69:163–178.

Fowler BA, and DuVal G (1991) Effects of lead on the kidney: Roles of high-affinity lead-binding proteins. *Environ Health Perspect* 91:77–80.

Fowler BA, Kimmel CA, Woods JS, McConnell EE, and Grant LD (1980) Chronic low-level lead toxicity in the rat. *Toxicol Appl Pharmacol* 56:59–77.

Friberg L (1950) Health hazards in the manufacture of alkaline accumulators with special reference to chronic cadmium poisoning. From the industrial hygiene department of the Swedish Public Health Institute, Stockholm, and the Department for Occupational Medicine at the Medical Clinic of Karolinska Sjukhuset, Stockholm.

Friberg L (1959) Studies on the metabolism of mercuric chloride and methyl mercury dicyandiamide. Experiments on rats given subcutaneous injections with radioactive mercury (^{203}Hg). *AMA Arch Industr Health* 20:42–49.

Friberg L (1984) Cadmium and the kidney. *Environ Health Perspect* 54:1–11.

Friberg L, Odeblad E, and Forssman S (1957) Distribution of 2 mercury compounds in rabbits after a single subcutaneous injection. *AMA Arch Industr Health* 16:163–168.

Friedman HL (1957) Relationship between chemical structure and biological activity in mercurial compounds. *Ann NY Acad Sci* 65:461–470.

Friedman PA, and Gesek FA (1994). Cadmium uptake by kidney distal convoluted tubule cells. *Toxicol Appl Pharmacol* 128:257–263.

Friedman PA, and Gesek FA (1995) Cellular calcium transport in renal epithelia: Measurement, mechanisms, and regulation. *Physiol Rev* 75:429–471.

Fukino H, Hirai M, Hsueh YM, and Yamane Y (1984) Effect of zinc pretreatment on mercuric chloride-induced lipid peroxidation in the rat kidney. *Toxicol Appl Pharmacol* 73:395–401.

Fukino H, Hirai M, HsuehYM, Moriyasu S, and Yamane Y (1986) Mechanism of protection by zinc against mercuric chloride toxicity in rats: Effects of zinc and mercury on glutathione metabolism. *J Toxicol Environ Health* 19:75–89.

Fullmer CS, Edelstein S, and Wasserman RH (1985) Lead-binding properties of intestinal calcium binding proteins. *J Biol Chem* 260:6816–6819.

Ganote CE, Reimer KA, and Jennings RB (1974) Acute mercuric chloride nephrotoxicity: An electron microscopic and metabolic study. *Lab Invest* 31:633–647.

Gilman AP, Moss MA, Villeneuve DC, Secours VE, Yagminas AP, Tracy BL, Quinn JM, and Valli VE (1998a) Uranyl nitrate: 91-Day exposure and recovery studies in the male New Zealand white rabbit. *Toxicol Sci* 41:138–151.

Gilman AP, Villeneuve DC, Secours VE, Yagminas AP, Tracy BL, Quinn JM, Valli VE, Willes RJ, and Moss MA (1998b) Uranyl nitrate: 28-Day and 91-day toxicity studies in the Sprague-Dawley rat. *Toxicol Sci* 41:117–128.

Gilman AP, Villeneuve DC, Secours VE, Yagminas AP, Tracy BL, Quinn JM, Valli VE, and Moss MA (1998c) Uranyl nitrate: 91-Day toxicity studies in the New Zealand white rabbit. *Toxicol Sci* 41:129–137.

Gleason MN, Gosselin RE, and Hodge DC (1957) *Clinical Toxicology of Commercial Products.*Williams and Wilkins Co., Baltimore, pp. 154.

Gmaj P, Murer H, and Kinne R (1979) Calcium ion transport across plasma membranes isolated from rat kidney cortex. *Biochem J* 178:549–557.

Goering PL, and Fowler BA (1987) Metal constitution of metallothionein influences inhibition of δ-aminolaevulinic acid dehydratase (porphobilinogen synthase) by lead. *Biochem J* 245:339–345.

Goering PL, and Fowler BA (1985) Mechanisms of renal lead-binding protein protection against lead-inhibition of delta-aminolevulinic acid dehydratase. *J Pharmacol Exp Ther* 234:365–371.

Goering PL, and Fowler BA (1984) Regulation of lead inhibition of δ-aminolevulinic acid dehydratase by high affinity renal lead binding protein. *J Pharmacol Exp Ther* 231:66–71.

Goldstein GW (1993) Evidence that lead acts as a calcium substitute in second messenger metabolism. *Neurotoxicology* 14:97–102.

Goldstein GW, and Ar D (1983) Lead activates calmodulin sensitive processes. *Life Sci* 33:1001–1006.

Gompertz D, Chettle DR, Fletcher JG, Mason H, Perkins J, Scott MC, Smith NJ, Topping MD, and Blindt M (1983) Renal dysfunction in cadmium smelters: relation to *in vivo* liver and kidney cadmium concentrations. *Lancet* 1:1185–1187.

Gonick HC, Ding Y, Bondy SC, Ni Z, and Vaziri ND (1997) Lead-induced hypertension: Interplay of nitric oxide and reactive oxygen species. *Hypertension* 30:1487–1492.

Gotelli CA, Astolfi E, Cox C, Cernichiari E, and Clarkson TW (1985) Early biochemical effects of organic mercury fungicide on infants: "Dose makes the poison." *Science* 227:638–640.

Goyer RA (1993) Lead toxicity: Current concerns. *Environ Health Perspect* 100:177–187.

Goyer RA (1989) Mechanisms of lead and cadmium nephrotoxicity. *Toxicol Lett* 46:153–162.

Goyer RA (1971) Lead toxicity: A problem in environmental pathology. *Am J Pathol* 64: 167–179.

Goyer RA (1968) The renal tubule in lead poisoning. I. Mitochondrial swelling and aminoaciduria. *Lab Invest* 19:71–77.

Goyer RA, and Krall R (1969) Ultrastructural transformation in mitochondria isolated from kidneys of normal and lead-intoxicated rats. *J Cell Biol* 41:393–400.

Goyer RA, Leonard DL, Moore JF, Rhyne B, and Krigman MR (1970a) Lead dosage and the role of the intranuclear inclusion body. *Arch Environ Health* 20:705–711.

Goyer RA, May P, Cates M, and Krigman MR (1970b) Lead and protein content of isolated intranuclear inclusion bodies from kidneys of lead-poisoned rats. *Lab Invest* 22:245–251.

Gritzka TL, and Trump BF (1968) Renal tubular lesions caused by mercuric chloride: Electron microscopic observations: Degeneration of the pars recta. *Am J Pathol* 52:1225–1277.

Habermann HC, Crowell K, and Janicki P (1983) Lead and other metals can substitute for Ca^{2+} in calmodulin. *Arch Toxicol* 54:61–70.

Haley DP (1982) Morphologic changes in uranyl nitrate-induced acute renal failure in saline- and water-drinking rats. *Lab Invest* 46:196–208.

Haley DP, Bulger RE, and Dobyan DC (1982) The long-term effect of uranyl nitrate on the structure and function of the rat kidney. *Virchows Arch (Cell Pathol)* 4: 181–192.

Hermes-Lima M, Periera B, and Bechara EJH (1991) Are free radicals involved in lead poisoning?" *Xenobiotica* 21:1085–1090.

Hodge HC (1973) A history of uranium poisoning, in *Uranium, Plutonium, Transplutonic Elements*, Hodge HC, Stannard JN, and Hursh JB (Eds), Springer-Verlag, New York, pp. 5–68.

Horn J (1970) Isolation and examination of inclusion bodies of the rat kidney after chronic lead poisoning. *Virchows Arch (Zellpathol)* 6:313.

Hultman P, and Enestrom S (1986) Localization of mercury in the kidney during experimental acute tubular necrosis studied by the cytochemical silver amplification method. *Br J Exp Pathol* 67:493–503.

Hultman P, and Enestrom S (1992) Dose-response studies in murine mercury-induced autoimmunity and immune-complex disease. *Toxicol Appl Pharmacol* 113:199–208.

Hultman P, Enestrom S, and von Schenck H (1985) Renal handling of inorganic mercury in mice. The early excretion phase following a single intravenous injection of mercuric chloride studied by the silver amplification method. *Virchows Arch (Cell Pathol)* 49:209–224.

Ishido M, Tohyama C, and Suzuki T (1999) Cadmium-bound metallothionein induces apoptosis in rat kidneys, but not in cultured kidney LLC-PK_1 cells. *Life Sci* 64:797–804.

Johnson DR (1982) Role of renal cortical sulfhydryl groups in development of mercury-induced renal toxicity. *J Toxicol Environ Health* 9:119–126.

Kapoor SC, Van Rossum GDV, O'Neill KJ, and Mercolrella I (1985) Uptake of inorganic lead *in vitro* by isolated mitochondria and tissue slices of rat renal cortex. *Biochem Pharmacol* 34:1439–1448.

Katayama H, Yasuhara M, and Hori R (1999) Effect of acute renal failure on the disposition of cefoperazone. *J Pharm Pharmacol* 51:361–366.

Kawada T, Tohyama C, and Suzuki S (1990) Significance of the excretion of urinary indicator proteins for a low level of occupational exposure to cadmium. *Int Arch Occup Environ Health* 62:95–100.

Kazantzis G (1979) Renal tubular dysfunction and abnormalitites of calcium metabolism in cadmium workers. *Environ Health Perspect* 28:155–159.

Kerper LE, and Hinkle PM (1997a) Cellular uptake of lead is activated by depletion of intracellular calcium stores. *J Biol Chem* 272:8346–8352.

Kerper LE, and Hinkle PM (1997b) Lead uptake in brain capillary endothelial cells: Activation by calcium store depletion. *Toxicol Appl Pharmacol* 146:127–133.

Khalil-Manesh F, Gonick HC, Cohen AH, Alinovi R, Bergmaschi E, Mutti A, and Rosen VJ (1992a) Experimental model of lead nephropathy. I. Continuous high dose lead administration. *Kidney Int* 41:1192–1203.

Khalil-Manesh F, Gonick HC, Cohen A, Bergamaschi E, and Mutti A (1992b) Experimental model of lead nephropathy. II. Effect of removal from lead exposure and chelation treatment with dimercaptosuccinic acid (DMSA). *Environ Res* 58:35–54.

Kimura O, Endo T, and Sakata M (1996) Comparison of cadmium uptakes from apical and basolateral membranes of LLC-PK_1 cells. *Toxicol Appl Pharmacol* 137:301–306.

Kirschbaum BB (1979) Alanine aminopeptidase excretion after mercuric chloride renal failure. *Biochem Med* 21:220–225.

Kjellstrom T, and Nordberg GF (1978) A kinetic model of cadmium metabolism in the human being. *Environ Res* 16:248–269.

Kjellstrom T, and Nordberg GF (1985) Kinetic model of cadmium metabolism, in *Cadmium and Health: A Toxicological and Epidemiological Appraisal. Vol. I. Exposure, Dose and Metabolism.* Friberg L, Elinder CG, Kjellstorm T, et al. (Eds), CRC Press, Boca Raton, FL, pp. 179–197.

Klaassen CD, Waalkes MP, and Cantilena LR Jr (1984) Alteration of tissue disposition of cadmium by chelating agents. *Environ Health Perspect* 54:233–242.

Klein R, Herman SP, Bullock BC, and Talley FA (1973) Methyl mercury intoxication in rat kidneys. *Arch Pathol* 96:83–90.

Klotzbach JM, and Diamond GL (1988) Complexing activity and excretion of 2,3-dimercapto-1-propane sulfonate in the rat kidney. *Am J Physiol* 254:F871–F878.

Kobayashi S, Nagase M, Honda N, and Hishida A (1984) Glomerular alterations in uranyl acetate-induced acute renal failure in rabbits. *Kidney Int* 26:808–815.

Koropatnick DJ, and Zalups RK (1997) Effect of non-toxic mercury, zinc or cadmium pretreatment on the capacity of human monocytes to under lipopolysaccharide-induced activation. *Br J Pharmacol* 120:797–806.

Koropatnick DJ, and Zalups RK (2000) Effect of toxic and essential metals on cellular responsiveness to cell signals, in *Molecular Biology and Toxicology of Metals.* Zalups RK, and Koropatnick DJ (Eds), Taylor & Francis, London, pp. 551–576.

Kostial K, Kello D, and Jugo S (1978) Influence of age on metal metabolism and toxicity. *Environ Health Perspect* 25:81–86.

Lawton LJ, and Donaldson WE (1991) Lead-induced tissue fatty acid alterations and lipid peroxidation. *Biol Trace Element Res* 28:83–97.

Leach LJ, Yuile CL, Hodge HC, Sylvester GE, and Wilson HB (1973) A five-year inhalation study with natural uranium dioxide dust. II. Postexposure retention and biologic effects in the monkey, dog and rat. *Health Phys* 25:239–258.

Leach LJ, Gelein RM, Panner BJ, Yuile CL, Cox CC, Balys MM, and Rolchigo PM (1984) The acute toxicity of the hydrolysis products of uranium hexafluoride when inhaled by the rat and guinea pig. U.S. Department of Energy Report K/SUB/81-9039/3.

Lee YH, Park KH, and Ku YS (2000) Pharmacokinetic changes of cyclosporine after intravenous and oral administration to rats with uranyl nitrate-induced acute renal failure. *Int J Pharm* 194:221–227.

Legare ME, Barhoumi R, Hebert E, Bratton GR, Burghardt RC, Tiffany-Castiglioni E (1998) Analysis of Pb^{2+} entry into cultured astroglia. *Toxicol Sci* 46:90–100.

Liu J, Liu Y, and Klaassen CD (1994) Nephrotoxicity of $CdCl_2$ and Cd-metallothionein in cultured rat kidney proximal tubules and LLC-PK_1 cells. *Toxicol Appl Pharmacol* 128:264–270.

Liu J, Habeebu SS, Liu Y, and Klaassen CD (1998) Acute CdMT injection is not a good model to study chronic Cd nephropathy: Comparison of chronic $CdCl_2$ and CdMT exposure with acute CdMT injection in rats. *Toxicol Appl Pharmacol* 153:48–58.

Loghman-Adham M (1997) Renal effects of environmental and occupational lead exposure. *Environ Health Perspect* 105:928–939.

Magos L, and Butler WH (1976) The kinetics of methylmercury administered repeatedly to rats. *Arch Toxicol* 35:25–39.

Magos L, and Clarkson TW (1977) Renal injury and urinary excretion, in *Handbook of Physiology, Section 9, Renal Physiology,* Lee DHK (Ed.), American Physiological Society, Bethesda, pp. 503–512.

Magos L, and Stoychev T (1969) Combined effect of sodium maleate and some thiol compounds on mercury excretion and redistribution in rats. *Br J Pharmacol* 35:121–126.

Magos L, Peristianis GG, Clarkson TW, Brown A, Preston S, and Snowden RT (1981) Comparative study of the sensitivity of male and female rats to methylmercury. *Arch Toxico*l 48:11–20.

Magos L, Brown AW, Sparrow S, Bailey E, Snowden RT, and Skipp WR (1985) The comparative toxicology of ethyl- and methyl-mercury. *Arch Toxicol* 57:260–267.

Malvin RL, and Wilde MS (1973) Stop-flow technique, in *Handbook of Physiology, Section 8, Renal Physiology,* Vol. 1., Orloff J, and Berliner RW (Eds), American Physiology Society, Washington, pp. 119–128.

Mason J, Olbricht C, Takabatake T, and Thurau K (1977) The early phase of experimental acute renal failure. I. Intratubular pressure and obstruction. Pflugers Arch 370:155–163.

Mason HJ, Davison AG, Wright AL, Guthrie CJG, Fayers PM, Venables KM, Smith NJ, Chettle DR, Franklin DM, Scott MC, Holden H, Gompertz D, and Newman-Taylor AJ (1988) Relations between liver cadmium, cumulative exposure, and renal function in cadmium alloy workers. *Br J Ind Med* 45:793–802.

McDonald-Taylor CK, Singh A, and Gilman A (1997) Uranyl nitrate-induced proximal tubule alterations in rabbits: A quantitative analysis. *Toxicol Pathol* 25:381–389.

McDowell EM, Nagle RB, Zalme RC, McNeil JS, Flamenbaum W, and Trump BF (1976) Studies on the pathophysiology of acute renal failure. I. Correlation of ultrastructure and function in the proximal tubule of the rat following administration of mercuric chloride. *Virchows Arch (Cell Pathol)* 22:173–196.

McNeil SI, Bhatnagar MK, and Turner CJ (1988) Combined toxicity of ethanol and methylmercury in rat. *Toxicology* 53:345–363.

Mistry P, Lucier GW, and Fowler BA (1985) High affinity lead binding proteins from rat kidney cytosol mediate cell-free nuclear translocation of lead. *J Pharmacol Exp Ther* 232:462–469.

Mistry P, Mastri C, and Fowler BA (1986) Influence of metal ions on renal cytosolic lead-binding proteins and nuclear uptake of lead in the kidney. *Biochem Pharmacol* 35:711–713.

Monteiro HP, Bechara EJH, and Absalla DSP (1991) Free radicals involvement in neurological porphyrias and lead poisoning. *Mol Cell Biochem* 103:73–83.

Moore MR, and Meredith PA (1979) The carcinogenicity of lead. *Arch Toxicol* 42:87–94.

Moore JF, Goyer RA, and Wilson M (1973) Lead-induced inclusion bodies: Solubility, amino acid content, and relationship to residual acidic nuclear proteins. *Lab Invest* 29:488–494.

Morrow PE, Leach LJ, Smith FA, Gelein RM, Scott JM, Beiter HD, Amato FJ, Picano JJ, Yuile CL, and Consler TG (1981) *Metabolic fate and evaluation of injury in rats and dogs following exposure to the hydrolysis products of uranium hexafluoride.* U.S. Nuclear Regulatory Commission Report NUREG/CR-2268.

Moser R, Oberly TD, Daggett DA, Friedman AL, Johnson JA, and Siegel FL (1995) Effects of lead on developing rat kidney. I. Glutathione S-transferase isozymes. *Toxicol Appl Pharmacol* 131:85–93.

Mueller PW, Smith SJ, Steinberg KK, and Thun MJ (1989) Chronic renal tubular effects in relation to urine cadmium levels. *Nephron* 52:45–54.

Murakami M, Kawamura R, Nishii S, and Katsunuma H (1983) Early appearance and localization of intranuclear inclusions in the segments of renal proximal tubules of rats following ingestion of lead. *Br J Exp Pathol* 64:144–154.

Nakagawa K (1991) Decreased glutathione S-transferase activity in mice livers by acute treatment with lead, independent of alteration in glutathione content. *Toxicol Lett* 56:13–17.

Nielson KB, Atkin CL, and Winge DR (1985) Distinct metal-binding configurations in metallothionein. *J Biol Chem* 260:5342–5350.

Nizet A (1981) Influence of uranyl nitrate upon tubular reabsorption and glomerular filtration in blood perfused isolated dog kidneys. *Pflugers Arch* 391:296–300.

Nomiyama K, and Foulkes EC (1968) Some effects of uranyl acetate on proximal tubular function in rabbit kidney. *Toxicol Appl Pharmacol* 13:89–98.

Nomiyama K, and Foulkes EC (1977) Reabsorption of filtered cadmium metallothionein in the rabbit kidney. *Proc Soc Exp Biol Med* 156:97–99.

Nomiyama K, Nomiyama H, Kameda N (1998) Plasma cadmium-metallothionein, a biological exposure index for cadmium-induced renal dysfunction, based on the mechanism of its action. *Toxicology* 129:157–168.

Norseth T, Clarkson TW (1970a) Studies on the biotransformation of ^{203}Hg-labeled methylmercury chloride in rats. *Arch Environ Health* 21:717–727.

Norseth T, Clarkson TW (1970b) Biotransformation of methylmercury salts in the rat studied by specific determination of inorganic mercury. *Biochem Pharmacol* 19:2775–2783.

O'Flaherty EJ, Adams WD, Hammond PB, and Taylor E (1986) Resistance of the rat to development of lead-induced renal functional deficits. *J Toxicol Environ Health* 18:61–75.

Olbricht C, Mason J, Takabatake T, Hohibrugger G, and Thurau K (1977) The early phase of experimental acute renal failure. II. Tubular leakage and the reliability of glomerular markers. *Pflugers Arch* 372:1–258.

Olivi L, Sisk J, and Bressler J (2001) Involvement of DMT1 in uptake of Cd in MDCK cells: role of protein kinase C. *Am J Physiol* 281:C793–C800.

Oskarsson A, and Fowler BA (1985) Effects of lead inclusion bodies on subcellular distribution of lead in rat kidney: The relationship to mitochondrial function. *Exp Mol Pathol* 43:397–408.

Ottenwalder H, and Simon P (1987). Differential effect of *N*-acetylcysteine on excretion of the metals Hg, Cd, Pb and Au. *Arch Toxicol* 60:401–402.

Piscator M (1966a) Proteinuria in chronic cadmium poisoning. III. Electrophoretic and immunoelectrophoretic studies on urinary proteins from cadmium workers, with special reference to low molecular weight proteins. *Arch Environ Health* 12:335–344.

Piscator M (1966b) Gel filtration and ion exchange chromatography of urinary proteins from cadmium workers. *Arch Environ Health* 12:345–359.

Piscator M (1984) Long-term observations on tubular and glomerular function in cadmium-exposed persons. *Environ Health Perspect* 54: 175–179.

Planas-Bohne F (1977) The effect of mercuric chloride on the excretion of two urinary enzymes in the rat. *Arch Toxicol* 37:219–225.

Platt B, and Busselberg D (1994) Combined actions of Pb^{2+}, Zn^{2+}, and Al^{3+} on voltage-activated calcium channel currents. *Cell Mol Neurobiol* 14:831–840.

Pounds JG (1984) Effect of lead intoxication on calcium homeostatsis and calcium-mediated cell function: A review. *Neurotoxicology* 5:295–332.

Price RG (1982) Urinary enzymes, nephrotoxicity and renal disease. *Toxicology* 23:99–134.

Prickett CS, Laug EP, and Kunze FM (1950) Distribution of mercury in rats following oral and intravenous administration of mercuric acetate and phenylmercuric acetate. *Proc Soc Exp Biol Med* 73:585–588.

Prozialeck WC, and Lamar PC (1993) Surface binding and uptake of cadmium (Cd^{2+}) by LLC-PK_1 cells on permeable membrane supports. *Arch Toxicol* 67:113–119.

Prozialeck WC, Wellington DR, and Lamar PC (1993) Comparison of the cytotoxic effects of cadmium chloride and cadmium-metallothionein in LLC-PK_1 cells. *Life Sci* 53:337–342.

Prozialeck WC (2000). Evidence that E-cadherin may be a target for cadmium toxicity in epithelial cells. *Toxicol Appl Pharmacol* 164:231–249.

Richdardt G, Federolf G, and Haberman E (1986) Affinity of heavy metal ions to intracellular Ca^{2+}-binding proteins. *Biochem Pharmacol* 35:1331–1335.

Robinson MK, Barfuss DW, and Zalups RK (1993) Cadmium transport and toxicity in isolated perfused segments of the renal proximal tubule. *Toxicol Appl Pharmacol* 121:103–111.

Rodier PM, Kates B, and Simons R (1988) Mercury localization in mouse kidney over time: Autoradiography versus silver staining. *Toxicol Appl Pharmacol* 257:235–245.

Rodin AE, and Crowson CN (1962) Mercury nephrotoxicity in the rat. I. Factors influencing the localization of tubular lesions. *Am J Pathol* 41:297–313.

Roels HA, Lauwerys RR, Buchet J-P, Bernard A, Chettle DR, Harvey TC, and Al-Hadda IK (1981) In vivo measurement of liver and kidney cadmium in workers exposed to this metal: its significance with respect to cadmium in blood and urine. *Environ Res* 26:217–240.

Roels HA, Lauwerys RR, Buchet JP, Bernard AM, Vos A, and Oversteyns M (1989) Health significance of cadmium induced renal dysfunction: a five year follow up. *Br J Ind Med* 46:755–764.

Roman-Franco AA, Turiello M, Albini B, Ossi E, Milgrom F, and Andres GA (1978) Anti-basement membrane antibodies and antigen-antibody complexes in rabbits injected with mercuric chloride. *Clin Immunol Immunopathol* 9:464–481.

Rosen JF, and Pounds JG (1989) Quantitative interactions between Pb^{2+} and Ca^{2+} homeostasis in cultured osteoclastic bone cells. *Toxicol Appl Pharmacol* 98:503–543.

Sandhir R, Julka D, and Gill KD (1994) Lipoperoxidative damage on lead exposure in rat brain and its implications on membrane bound enzymes. *Pharmacol Toxicol* 74:66–71.

Sano K, Fujigaki Y, Miyaji T, Ikegaya N, Ohishi K, Yonemura K, and Hishida A (2000) Role of apoptosis in uranyl acetate-induced acute renal failure and acquired resistance to uranyl acetate. *Kidney Int* 57:1560–1570.

Sapin C, Dreut E, and Dreut P (1977) Induction of anti-glomerular basement membrane antibodies in the brown Norway rat by mercuric chloride. *Clin Exp Immunol* 28:173–179.

Schaub TP, Kartenbeck J, Konig J, Vogel O, Witzgall R, Kriz W, Keppler D (1997) Expression of the conjugate export pump encoded by the mrp2 gene in the apical membrane of kidney proximal tubules. *J Am Soc Nephrol* 8:1213–1221.

Schaub TP, Kartenbeck J, Konig J, Spring H, Dorsam J, Staehler G, Storkel S, Thon WF, and Keppler D (1999) Expression of the MRP2 gene-encoded conjugate export pump in human kidney proximal tubules and in renal cell carcinoma. *J Am Soc Nephrol* 10:1159–1169.

Shaikh ZA, and Smith JC (1980) Metabolism of orally ingested cadmium in humans. In Mechanisms of Toxicity and Hazard Evaluation, in *Cadmium and health: A Toxicological and epidemiological appraisal.* Elsevier/North-Holland Biomedical Press, New York, NY, pp. 247–255.

Shaikh ZA, Harnett KM, Perlin SA, and Huang PC (1989) Chronic cadmium intake results in dose-related excretion of metallothionein in urine. *Experientia* 45: 146–148.

Shelton KR, and Egle PM (1982) The proteins of lead-induced intranuclear inclusion bodies. *J Biol Chem* 257:1180211807.

Simons TJB (1986a) Passive transport and binding of lead by human red blood cells. *J Physiol* 378:267–286.

Simons TJB (1986b) The role of anion transport in the passive movement of lead across the human red cell membrane. *J Physiol* 378:287–312.

Simons TJB, and Pocock G (1987) Lead enters bovine adrenal medullary cells through calcium channels. *J Neurochem* 48:383–389.

Smith DR, Kahng MW, Quintanilla-Vega B, and Fowler BA (1998) High-affinity renal lead-binding proteins in environmentally-exposed humans. *Toxicol Appl Pharmacol* 115:39–52.

Stein JH, Gottschall J, Osgood W, and Ferris TF (1975) Pathophysiology of a nephrotoxic model of acute renal failure. *Kidney Int* 6:27–41.

Steward JR, and Diamond GL (1987) Renal tubular secretion of the alkanesulfonate 2,3-dimercaptopropane-1-sulfonate. *Am J Physiol* 252:F800–F810.

Steward JR, and Diamond GL (1988) In vivo renal tubular secretion and metabolism of the disulfide of 2,3-dimercaptopropane-1-sulfonate. *Drug Metab Dispos* 16: 189–195.

Stonard MD, Chater BV, Duffield DB, Nevitt AL, O'Sullivan JJ, and Steele GT (1983) An evaluation of renal function in workers occupationally exposed to mercury vapor. *Int Arch Occup Environ Health* 52:177–189.

Stroo WE, and Hook JB (1977) Enzymes of renal origin in urine as indicators of nephrotoxicity. *Toxicol Appl Pharmacol* 39:423–434.

Sudo M, Honda N, Hishida A, and Nagase M (1977) Renal hemodynamics in uranyl acetate-induced acute renal failure of rabbits. *Kidney Int* 11:35–43.

Sugawara N, and Sugawara C (1991) Gastrointestinal absorption of Cd-metallothionein and cadmium chloride in mice. *Arch Toxicol* 65:689–692.

Sugawara E, Nakamura K, Miyake T, Fukumura A, and Seki Y (1991) Lipid peroxidation and concentration of glutathione in erythrocytes from workers exposed to lead. *Br J Ind Med* 48:239–242.

Sun LR, and Suszkiw JB (1995) Extracellular inhibition and intracellular enhancement of Ca^{2+} currents by Pb^{2+} in bovine adrenal chromaffin cells. *J Neurophysiol* 74:574–581.

Sun DF, Fujigaki Y, Fujimoto T, Yonemura K, and Hishida A (2000) Possible involvement of myofibroblasts in cellular recovery of uranyl acetate-induced acute renal failure in rats. *Am J Pathol* 157:1321–1335.

Tanaka K, Sueda K, Onosaka S, and Okahara K (1975) Fate of ^{109}Cd-labeled metallothionein in rats. *Toxicol Appl Pharmacol* 33:258–266.

Tanaka T, Naganuma A, and Imura N (1990) Role of γ-glutamyltranspeptidase in renal uptake and toxicity of inorganic mercury in mice. *Toxicology* 60:187–198.

Tannenbaum A, Silverstone H, and Koziol J (1951) Tracer studies of the distribution and excretion of uranium in mice, rats and dogs, in *Toxicology of Uranium Compounds*, Tannenbaum A (Ed.), McGraw Hill, New York, pp. 128–181.

Taugner R, Winkel K, and Iravani J (1966) Zur Lokalization der Sublimatanreicherung in der Rattenneire. *Virchows Arch Pathol Anat Physiol* 340:369–383.

Thun MJ, Osorio AM, Schober S, Hannon WH, Lewis B, and Halperin W (1989) Nephropathy in cadmium workers: Assessment of risk from airborne occupational exposure to cadmium. *Br J Ind Med* 46:689–697.

Tomsig JL, and Suszkiw JB (1991) Permeation of Pb^{2+} through calcium channels: Fura-2 measurements of voltage- and dihydropyridine-sensitive Pb^{2+} entry in isolated bovine chromaffin cells. *Biochem Biophys Acta* 1069:197–200.

Tomsig JL, and Suszkiw JB (1995) Multisite interactions between Pb^{2+} and protein kinase C and its role in norepinephrine release from bovine adrenal chromaffin cells. *J Neurochem* 64:2667–2673.

Trojanowska B, Piotrowski JK, and Szendzikowski S (1971) The influence of thioacetamide on the excretion of mercury in rats. *Toxicol Appl Pharmacol* 18:374–386.

Ulmer DD, and Vallee BL (1969) Effects of lead on biochemical systems, in *Trace Substances in Environmental Health*, Vol. 12, Hemphill DD (Ed.), University of Missouri Press, Columbia, pp. 7–27.

Van Crugten JT, Somogyi AA, and Nation RL (2000) Effect of uranyl nitrate-induced renal failure on morphine disposition and antinociceptive response in rats. *Clin Exp Pharmacol Physiol* 27:74–79.

Vander AJ, Taylor DL, Kalitis K, Mouw DR, and Victery W (1977) Renal handling of lead in dogs: Clearance studies. *Am J Physiol* 233(6):F532–F538.

Vander AJ, and Johnson B (1981) Accumulation of lead by renal slices in the presence of organic anions. *Proc Soc Exp Biol Med* 166:583–586.

Verity MA, and Brown WJ (1970) Hg^{2+}-induced kidney necrosis: Subcellular localization and structure-linked lysosomal enzyme changes. *Am J Pathol* 61:57–74.

Victery W, Miller CR, and Fowler BA (1984) Lead accumulation by rat renal brush-border membrane vesicles. *J Pharmacol Exp* Ther 231:589–596.

Victery W, Vander AJ, and Mouw DR (1979a) Effect of acid-base status on renal excretion and accumulation of lead in dogs and rats. *Am J Physiol* 237:F398-F407.

Victery W, Vander AJ, and Mouw DR (1979b) Renal handling of lead in dogs: Stop-flow analysis. *Am J Physiol* 237: F408-F414.

Vizari ND, Ding Y, and Ni Z (2001) Compensatory up-regulation of nitric oxide synthase isoforms in lead-induced hypertension; reversal by a superoxide dismutase-mimetic drug. *J Pharmacol Exp Ther* 298:679–685.

Vizari ND, Ding Y, and Ni Z (1999a) Nitric oxide synthase expression in the course of lead-induced hypertension. *Hypertension* 34:558–562.

Vizari ND, Liang K, and Ding Y (1999b) Increased nitric oxide inactivation by reactive oxygen species in lead-induced hypertension. *Kidney Int* 56:1492–1498.

Vizari ND, Ding Y, Ni Z, and Gonick HC (1997) Altered nitric oxide metabolism and increased oxygen free radical activity in lead-induced hypertension: Effect of lazaroid therapy. *Kidney Int* 52:1042–1046.

Vyskocil A, Pancl J, Tusl M, Ettlerova E, Semecky V, Kasparova L, Lauwerys R, and Bernard A (1989) Dose-related proximal tubular dysfunction in male rats chronically exposed to lead. *J Appl Toxicol* 9:395–400.

Waalkes MP, Harvey MJ, and Klaassen CD (1984) Relative *in vitro* affinity of hepatic metallothionein for metals. *Toxicol Lett* 20:33–39.

Waalkes MP, and Klaassen CD (1985) Concentration of metallothionein in major organs of rats after administration of various metals. *Fund Appl Toxicol* 5:473–477.

Wang XP, and Foulkes EC (1984) Specificity of acute effects of Cd on renal function. *Toxicology* 30:243–247.

Watts SW, Chai S, and Webb RC (1995) Lead acetate-induced contraction in rabbit mesenteric artery: Interaction with calcium and protein kinase C. *Toxicology* 99:55–65.

Wedeen RP, Mallik DK, and Batuman V (1979) Detection and treatment of occupational lead nephropathy. *Arch Intern Med* 139:53–57.

Wedeen RP, Maesaka JK, Weiner B, Lipat GA, Lyons MM, Vitale LF, and Joselow MM (1975) Occupational lead nephropathy. *Am J Med* 59:630–641.

WHO (1991) *Environmental Health Criteria 118: Inorganic Mercury*. World Health Organization, Geneva.

Witzmann FA, Fultz CD, Grant RA, Wright LS, Kornguth SE, and Siegel FL (1999) Regional protein alterations in rat kidneys induced by lead exposure. *Electrophoresis* 20:943–951.

Wright LS, Kornguth SE, Oberley TD, and Siegel FL (1998) Effects of lead on glutathione S-transferase expression in rat kidney. A dose-response study. *Toxicol Sci* 46:254–259.

Zalme RC, McDowell FM, Nagle RB, McNeil JS, Flamenbaum W, and Trump BF (1976) Studies on the pathophysiology of acute renal failure. II. A histochemical study of the proximal tubule of the rat following administration of mercuric chloride. *Virchows Arch [Cell Pathol]* 22:197–216.

Zalups RK (1989) Effect of dietary K^+ and 75% nephrectomy on the morphology of principal cells in CCDs. *Am J Physiol* 257:F387–F396.

Zalups RK (1991a) Autometallographic localization of inorganic mercury in the kidneys of rats: Effect of unilateral nephrectomy and compensatory renal growth. *Exp Mol Pathol* 54:10–21.

Zalups RK (1991b) Method for studying the *in vivo* accumulation of inorganic mercury in segments of the nephron in the kidneys of rats treated with mercuric chloride. *J Pharmacol Meth* 26:89–104.

Zalups RK (1991c) Renal accumulation and intrarenal distribution of inorganic mercury in the rabbit: Effect of unilateral nephrectomy and dose of mercuric chloride. *J Toxicol Environ Health* 33:213–228.

Zalups RK (1993a) Early aspects of the intrarenal distribution of mercury after the intravenous administration of mercuric chloride. *Toxicology* 79: 215–228.

Zalups RK (1993b) Influence of 2,3-dimercaptopropane-1-sulfonate (DMPS) and meso-2,3-dimercaptosuccinic acid (DMSA) on the renal disposition of mercury in normal and uninephrectomized rats exposed to inorganic mercury. *J Pharmacol Exp* Ther 267:791–800.

Zalups RK (1995) Organic anion transport and action of γ-glutamyltranspeptidase in kidney linked mechanistically to renal tubular uptake of inorganic mercury. *Toxicol Appl Pharmacol* 132:289–298.

Zalups RK (1997a) Influence of different degrees of reduced renal mass on the renal and hepatic disposition of administered cadmium. *J Toxicol Environ Health* 51:245–264.

Zalups RK (1997b) Enhanced renal outer medullary uptake of mercury associated with uninephrectomy: Implication of a luminal mechanism. *J Toxicol Environ Health* 50:173–194.

Zalups RK (1998a) Intestinal handling of mercury in the rat: Implication of intestinal secretion of inorganic mercury following biliary ligation or cannulation. *J Toxicol Environ Health* 53:615–636.

Zalups RK (1998b) Basolateral uptake of inorganic mercury in the kidney. *Toxicol Appl Pharmacol* 150:1–8.

Zalups RK (1998c) Basolateral uptake of mercuric conjugates of *N*-acetylcysteine and cysteine in the kidney involves the organic anion transport system. *J Toxicol Environ Health* 54:101–117.

Zalups RK (2000a) Evidence for basolateral uptake of cadmium in the kidneys of rats. Toxicol Appl Pharmacol 164:15–23.

Zalups RK (2000b) Molecular interactions with mercury in the kidney. *Pharmacol Rev* 52:113–143.

Zalups RK, and Ahmad S (2003) Molecular handling of cadmium in transporting epithelia. *Toxicol Appl Pharmacol* 186:163–188.

Zalups RK, and Barfuss DW (1990) Accumulation of inorganic mercury along the renal proximal tubule of the rabbit. *Toxicol Appl Pharmacol* 106:245–253.

Zalups RK, and Barfuss DW (1993a) Transport and toxicity of methylmercury along the proximal tubule of the rabbit. *Toxicol Appl Pharmacol* 121:176–185.

Zalups RK, and Barfuss DW (1995a) Pretreatment with p-aminohippurate inhibits the renal uptake and accumulation of injected inorganic mercury in the rat. *Toxicology* 103:23–35.

Zalups RK, and Barfuss DW (1995b) Accumulation and handling of inorganic mercury in the kidney after coadministration with glutathione. *J Toxicol Environ Health* 44:385–399.

Zalups RK, and Barfuss DW (1996a) Diversion or prevention of biliary outflow from the liver diminishes the renal uptake of injected inorganic mercury. *Drug Metab Dispos* 24:480–486.

Zalups RK, and Barfuss DW (1996b) Nephrotoxicity of inorganic mercury co-administered with L-cysteine. *Toxicology* 109:15–29.

Zalups RK, and Barfuss DW (1998a) Small aliphatic dicarboxylic acids inhibit renal uptake of administered mercury. *Toxicol Appl Pharmacol* 148:183–193.

Zalups RK, and Barfuss DW (1998b) Participation of mercuric conjugates of cysteine, homocysteine, and *N*-acetylcysteine in mechanisms involved in the renal tubular uptake of inorganic mercury. *J Am Soc Nephrol* 9:551–561.

Zalups RK, and Barfuss DW (2002a) Simultaneous coexposure to inorganic mercury and cadmium: A study of the renal and hepatic disposition of mercury and cadmium. *J Toxicol Environ Health* 65:101–120.

Zalups RK, and Barfuss DW (2002b) Renal organic anion transport system: A mechanism for the basolateral uptake of mercury-thiol conjugates along the *pars recta* of the proximal tubule. *Toxicol Appl Pharmacol* 182: 234–243.

Zalups RK, and Cherian MG (1992a) Renal metallothionein metabolism after a reduction of renal mass. I. Effect of unilateral nephrectomy and compensatory renal growth on basal and metal-induced renal metallothionein metabolism. *Toxicology* 71:83–102.

Zalups RK, and Cherian MG (1992b) Renal metallothionein metabolism after a reduction of renal mass. II. Effect of zinc pretreatment on the renal toxicity and intrarenal accumulation of inorganic mercury. *Toxicology* 71:103–117.

Zalups RK, and Diamond GL (1987a) Intrarenal distribution of mercury in the rat: Effect of administered dose of mercuric chloride. *Bull Environ Contam Toxicol* 38:67–72.

Zalups RK, and Diamond GL (1987b) Mercuric chloride-induced nephrotoxicity in the rat following unilateral nephrectomy and compensatory renal growth. *Virchows Arch B (Cell Pathol)* 53:336–346.

Zalups RK, and Henderson DA (1992) Cellular morphology in outer medullary collecting duct: Effect of 75% nephrectomy and K^+ depletion. *Am J Physiol* 263:F1119–F1127.

Zalups RK, and Lash LH (1990) Effects of uninephrectomy and mercuric chloride on renal glutathione homeostasis. *J Pharmacol Exp Ther* 254:962–970.

Zalups RK, and Lash LH (1994) Advances in understanding the renal transport and toxicity of mercury. *J Toxicol Environ Health* 42:1–44.

Zalups RK, and Lash LH (1997a) Binding of mercury in renal brush-border and basolateral membrane-vesicles: Implication of a cysteine conjugate of mercury involved in the luminal uptake of inorganic mercury. *Biochem Pharmacol* 53:1889–1900.

Zalups RK, and Lash LH (1997b) Depletion of glutathione in the kidney and the renal disposition of administered inorganic mercury. *Drug Metab Dispos* 25:516–523.

Zalups RK, and Minor KH (1995) Luminal and basolateral mechanisms involved in the renal tubular uptake of inorganic mercury. *J Toxicol Environ Health* 46:73–100.

Zalups RK, Gelein RM, and Cherian MG (1992a) Shifts in the dose-effect relationship for the nephropathy induced by cadmium-metallothionein in rats after a reduction of renal mass. *J Pharmacol Exp Ther* 262:1256–1266.

Zalups RK, Barfuss DW, and Kostyniak PJ (1992b) Altered intrarenal accumulation of mercury in uninephrectomized rats treated with methylmercury chloride. *Toxicol Appl Pharmacol* 115:174–182.

Zalups RK, Barfuss DW, and Lash LH (1999a). Disposition of inorganic mercury following biliary obstruction and chemically-induced glutathione depletion: Dispositional changes one hour after the intravenous admimistration of mercuric chloride. *Toxicol Appl Pharmacol* 154:135–144.

Zalups RK, Barfuss DW, and Lash LH (1999b) Effects of biliary ligation and modulation of GSH-status on the renal and hepatic disposition of inorganic mercury in rats. *Toxicologist* 48:330–331.

Zalups RK, Barfuss DW, and Lash LH (1999c) Relationships between alterations in glutathione metabolism and the disposition of inorganic mercury in rats: Effects of biliary ligation and chemically induced modulation of glutathione status. *Chem-Biol Interact* 123:171–195.

Zalups RK, Cherian MG, and Barfuss DW (1995) Lack of luminal or basolateral uptake and transepithelial transport of mercury in isolated perfused proximal tubules exposed to mercury-metallothionein. *J Toxicol Environ Health* 44:101–113.

Zalups RK, Cox C, and Diamond GL (1988) Histological and urinalysis assessment of nephrotoxicity induced by mercuric chloride in normal and uninephrectomized rats, in *Biological Monitoring of Toxic Metals,* Clarkson TW, Friberg L, Nordberg GF, and Sager PR (Eds), Plenum Publishing Corporation, New York, pp. 531–545.

Zalups RK, Gelein RM, and Cernichiari E (1991) DMPS as a rescue agent for the nephropathy induced by mercuric chloride. *J Pharmacol Exp Ther* 256:1–10.

Zalups RK, Cherian MG, and Barfuss DW (1993) Mercury-metallothionein and the renal accumulation and handling of mercury. *Toxicology* 83:61–78.

Zalups RK, Gelein RM, Morrow PE, and Diamond GL (1988) Nephrotoxicity of uranyl fluoride in uninephrectomized and sham-operated rats. *Toxicol Appl Pharmacol* 94:11–22.

Zalups RK, Parks L, Cannon VT, and Barfuss DW (1998) Mechanisms of action of 2,3-dimercaptopropane-1-sulfonate and the transport, disposition, and toxicity of inorganic mercury in isolated perfused segments of rabbit proximal tubules. *Mol Pharmacol* 54:353–363.

Zalups RK, Stanton BA, Wade JB, and Giebisch G (1985) Structural adaptation in initial collecting tubule following reduction in renal mass. *Kidney Int* 27:636–642.

23

Chemical-Induced Nephrotoxicity Mediated by Glutathione *S*-Conjugate Formation

Wolfgang Dekant

INTRODUCTION

Renal transport and xenobiotic metabolism play an important role in the detoxication and excretion of potentially toxic xenobiotics. However, recent experimental evidence has demonstrated that renal xenobiotic metabolism and renal transport processes also play important roles in the nephrotoxicity of xenobiotics and xenobiotic metabolites. The high blood flow to the kidney combined with its ability to concentrate solutes may expose the kidney to high concentrations of xenobiotics and xenobiotic metabolites present in the systemic circulation. In addition xenobiotic metabolites may be targeted to the kidney by the presence of efficient renal transport systems (Anders, 1988, 1991; Lock, 1989, 1993; Monks and Lau, 1992, 1998; Monks et al., 1990; Lash, 1994). Glutathione *S*-conjugate formation from xenobiotics represents a novel pathway of biotransformation resulting in nephrotoxicity, and at least three types of toxic glutathione *S*-conjugates have been identified.

Toxic glutathione *S*-conjugates are synthesized from 1,2-dihaloalkanes, several polyhalogenated alkenes, hydroquinones, and 4-aminophenol. The *S*-conjugates formed from 1,2-dihaloalkanes are electrophilic episulfonium ions and glutathione conjugation of hydroquinones targets redox-cycling molecules to the kidney. Glutathione *S*-conjugates formed from polyhalogenated alkenes require processing to the cysteine *S*-conjugates and bioactivation by renal cysteine conjugate β-lyase. The intermediate formation of these *S*-conjugates has been linked with the renal toxicity observed with the parent compounds and represents a mechanism to explain the renal toxicity of certain haloalkanes, haloalkenes, hydroquinones, and aminophenols.

ROLE OF PHASE II BIOTRANSFORMATION IN TARGET ORGAN TOXICITY

The organ-specific toxicity of many xenobiotics present in the systemic circulation is often due to biotransformation reactions by both phase I and phase II enzymes (Anders, 1985). Phase I reactions are mainly catalyzed by cytochrome P450 enzymes, which often exhibit their highest activity in the liver (Guengerich, 2001; Guengerich and Liebler, 1985). Due to the short-lived intermediates formed by many cytochrome P450 catalyzed oxidations of xenobiotics, extrahepatic toxicity due to cytochrome P450 dependent activation reactions may be due to the distribution of specific P450 enzymes or availability of the substrate in high concentrations in the target organ.

A variety of phase II enzyme catalyzed bioactivation reactions have been identified (Anders and Dekant, 1994). For example, the bladder and colon carcinogenicity of certain aromatic amines is due to a complex sequence of phase I and phase II bioactivation reactions ultimately resulting in the formation of covalently binding metabolites in the target cells. Further examples of phase II enzyme mediated bioactivation reactions have been elucidated for glucuronide and sulfate conjugation, for acetylation, and for glutathione conjugation. The conjugates initially produced may be stable in the tissue where formed; however, when transported to other organs for excretion or further processing, they may yield reactive metabolites due to changes in pH, oxygen tension, or further enzymatic transformation. Due to the much wider variety of phase II reactions, the role of active transport mechanisms in the distribution and excretion of phase II metabolites, and the often observed interplay between different phase II enzymes during the biotransformation of a xenobiotic, phase II dependent bioactivation reactions have often been associated with extrahepatic toxicity. For some relevant examples, these reactions form the basis for the organ-specific toxicity or carcinogenicity of xenobiotics (Anders and Dekant, 1994; Dekant and Neumann, 1992; Miller and Miller, 1966).

Glutathione conjugation represents an important phase-II reaction and is usually associated with detoxication. Glutathione (γ-glutamylcysteinylglycine) is a major low molecular weight peptide in mammalian cells. Due to the nucleophilicity of the sulfur atom and the antioxidant properties of glutathione, this tripeptide is an important factor in the detoxication of xenobiotics and reactive oxygen species (Boyland and Chasseaud, 1969). However, glutathione *S*-conjugate formation may also result in toxicity and represent a bioactivation reaction. Formation of toxic glutathione *S*-conjugates has been demonstrated with some halogenated alkanes, halogenated alkenes, and hydroquinones. The formation of toxic glutathione conjugates with direct electrophilicity (formed from vic-dihaloalkanes) or requiring further bioactiviation (formed from several polyhalogenated alkenes) is an accepted mechanism to explain tissue- and cell-specific toxicity of these compounds to the kidney (Anders and Dekant, 1998; Dekant and Vamvakas, 1996; Monks et al., 1990; Lash, 1994).

BIOSYNTHESIS OF NEPHROTOXIC GLUTATHIONE *S*-CONJUGATES

Biosynthesis of nephrotoxic glutathione *S*-conjugates has been demonstrated to occur with haloalkanes, haloalkenes, hydroquinones, and 4-aminophenol.

Halogenated Alkanes

The vicinal dihaloalkanes 1,2-dibromoethane, 1,2-dichloroethane, and 1-bromo-2-chloroethane are toxic and carcinogenic (Spencer et al., 1951; Weisburger, 1977). Their biotransformation involves cytochrome P450 dependent oxidation and glutathione *S*-conjugate formation (Guengerich et al., 1980). The cytochrome P450 dependent oxidation results in formation of haloacetaldehydes; these reactive aldehydes are thought to be responsible for the covalent binding of metabolites of 1,2-dibromoethane and 1,2-dichloroethane to proteins (Guengerich et al., 1980). The second pathway of 1,2-dihaloalkane bioactivation involves formation of toxic glutathione *S*-conjugates. *S*-(2-Bromoethyl) glutathione is a biliary metabolite excreted after administration of 1,2-dibromoethane to rats; 1,2-dibromoethane is metabolized to *S*-(2-bromoethyl)glutathione by glutathione *S*-transferases (Cmarik et al., 1990; Humphreys et al., 1990; Inskeep and Guengerich, 1984; Kim and Guengerich, 1989, 1990; Ozawa and Tsukioka, 1990). 1,2-Dichloroethane and 1-bromo-2-chloroethane are also metabolized to *S*-(2-chloroethyl)glutathione in rat liver and are excreted with bile (Jean and Reed, 1992).

The nematocide 1,2-dibromo-3-chloropropane has been widely used as a soil fumigant. The toxicity of 1,2-dibromo-3-chloropropane is characterized by necrosis of the renal proximal tubules, testicular atrophy, and occasional liver damage (Torkelson et al., 1961). 1,2-Dibromo-3-chloropropane is metabolized to many polar metabolites, which are largely derived from glutathione conjugates (Jones et al., 1979a). Results of mechanistic studies on 1,2-dibromo-3-chloropropane biotransformation indicate that *S*-(3-chloro-2-bromopropyl)glutathione (Figure 23.1) is an intermediate in 1,2-dibromo-3-chloropropane metabolism (Pearson et al., 1990); it spontaneously cyclizes to an episulfonium ion and its hydrolysis accounts for a major part of *S*-(3-chloro-2-hydroxypropyl) glutathione and *S*-(2,3-hydroxypropyl)-glutathione excreted in bile. This episulphonium ion reacts with DNA and two isomers of *S*-[1-(hydroxymethyl)-2-(N^7-guanyl)ethyl]glutathione and *S*-[bis(N^7-guanyl)methyl]glutathione have been identified as major 1,2-dibromo-3-chloropropane-derived adducts in DNA (Humphreys et al., 1991).

Formation of toxic glutathione conjugates is most likely responsible for the renal and testicular toxicity of 1,2-dibromo-3-chloropropane. Based on the absence of significant isotope effects with perdeutero-1,2-dibromo-3-chloropropane, the breaking of a carbon–hydrogen bond (assumed to occur by a cytochrome P450 catalyzed oxidation) is not the rate-limiting step in 1,2-dibromo-3-chloropropane-induced renal and testicular toxicity. Moreover, renal and testicular necrosis and the ability of 1,2-dibromo-3-chloropropane to induce DNA-damage in these organs are independent of cytochrome P450 but require glutathione (Holme et al., 1991; Lag et al., 1989a, 1989b). These observations further support the hypothesis that the toxicity of 1,2-dibromo-3-chloropropane

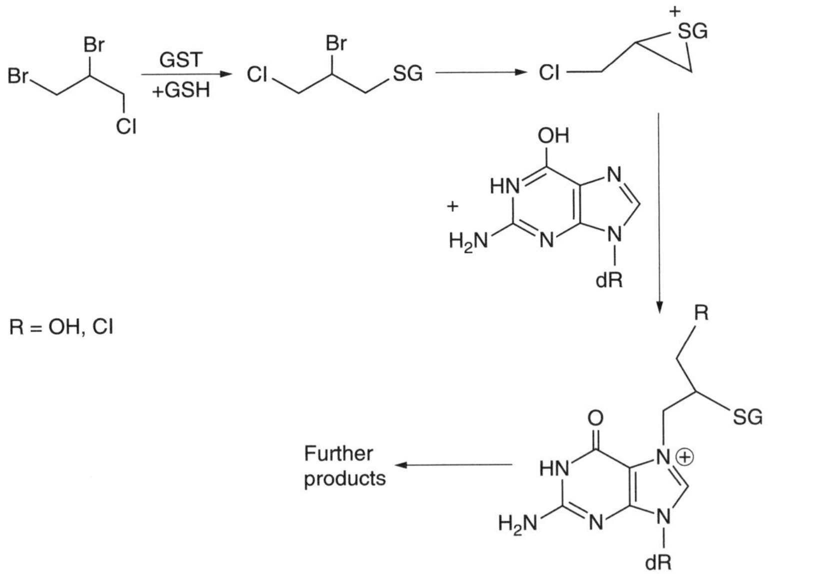

Figure 23.1 Bioactivation of 1,2-dibromo-3-chloropropane by glutathione conjugation and formation of DNA-adducts (modified from Humphreys et al., 1991).

requires a glutathione-dependent pathway involving formation of a reactive episulphonium ion. Metabolic episulphonium formation by glutathione conjugation may also occur during the biotransformation of 1,2,3-trichloropropane (Mahmood et al., 1991; Weber and Sipes 1990, 1992; Winter et al., 1992).

The flame retardant tris(2,3-dibromopropyl)phosphate is nephrotoxic in animals (Elliot et al., 1982; Osterberg et al., 1977; Söderlund et al., 1984) and a renal carcinogen (IARC, 1990). Tris(2,3-dibromopropyl)phosphate is metabolized to bis(2,3-dibromopropyl)phosphate, 2,3-dibromopropanol, 2-bromoacrolein, and several polar metabolites (Lynn et al., 1982; Söderlund et al., 1984). In rats, radioactivity from ^{14}C-tris(2,3-dibromopropyl)phosphate bound covalently to proteins, with binding to kidney proteins five times greater than binding to liver proteins.

Formation of reactive glutathione conjugates seems to play an important role in the toxicity of tris(2,3-dibromopropyl)phosphate. Incubation of tris(2,3-dibromopropyl)phosphate with cytosolic enzymes, ^{35}*S*-glutathione and DNA resulted in ^{35}*S*-glutathione binding to DNA

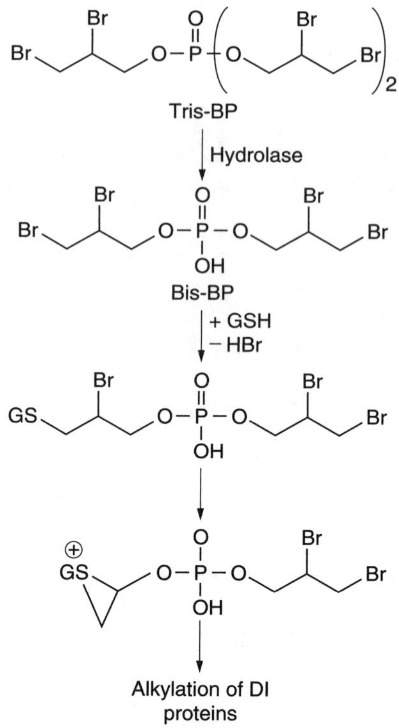

Figure 23.2 Bioactivation of tris(2,3-dibromopropyl)phosphate by hydrolysis, glutathione conjugation and episulfonium ion formation.

(Inskeep and Guengerich, 1984) (Figure 23.2) and the mutagenicity of tris(2,3-dibromopropyl)phosphate was markedly increased in *Salmonella typhimurium* expressing human glutathione *S*-transferases (Simula et al., 1993). The structures of the glutathione conjugates identified in rats given tris(2,3-dibromopropyl)phosphate also support a role of episulphonium ion intermediate formation, probably by glutathione *S*-transferases, in tris(2,3-dibromopropyl)phosphate toxicity (Pearson et al., 1993) (Figure 23.2).

Halogenated Alkenes

Hexachlorobutadiene, perfluoropropene, chlorotrifluoroethene, compound A (2-(fluoromethoxy)-1,1,3,3,3-pentafluoro-1-propene, a degradation product of the inhalation anesthetic sevoflurane), and dichloroethyne are nephrotoxic and induce proximal tubular damage in rodents (Ishmael et al., 1982; Jin et al., 1995; Kharasch et al., 1997; Potter et al., 1981; Reichert et al., 1975). In addition, the solvents trichloroethene (NCI, 1986b) and tetrachloroethene, and also dichloroethyne (NCI, 1986a) and hexachlorobutadiene (Kociba et al., 1977), induced tumors of the proximal tubules in rats after administration of high doses.

Glutathione-dependent pathways have been implicated in the renal toxicity of these compounds. Nephrotoxic haloalkenes are metabolized to glutathione *S*-conjugates by microsomal and cytosolic glutathione *S*-transferases. The biotransformation of hexachlorobutadiene (Dekant et al., 1988a, 1988b; Wolf et al., 1984), 1,1,2-trichloro-3,3,3-trifluoropropene (Vamvakas et al., 1989b), trichloroethene (Dekant et al., 1986a, 1990; Green et al., 1997; Lash et al., 1995, 1998b, 1999b), and tetrachloroethene (Dekant et al., 1987b, 1998; Lash et al., 1998a) yields exclusively *S*-(haloalkenyl)glutathione *S*-conjugates (Figure 23.3).

Identical glutathione *S*-conjugates have also been observed in bile obtained in isolated rat livers perfused with the haloalkenes and in the bile of rats given hexachlorobutadiene (Nash et al., 1984), 1,1,2-trichloro-3,3,3-trifluoropropene (Vamvakas et al., 1989b), trichloroethene (Dekant et al., 1990), or tetrachloroethene (Vamvakas et al., 1989a). The corresponding mercapturic acids are urinary metabolites of the respective haloalkenes (Dekant et al., 1986a, 1986b, 1990; Reichert and Schuetz 1986; Vamvakas et al., 1989b). Hexachlorobutadiene seems to be metabolized *in vivo* exclusively by glutathione conjugate formation (Wallin et al., 1988). In contrast, both trichloroethene and tetrachloroethene are mainly metabolized by cytochrome P450 (Dekant et al., 1984), both in humans and in rodents (Birner et al., 1996, 1997; Lash et al., 1998a, 1998b, 1999a, 1999b; Völkel et al., 1998).

In contrast to chloroalkenes, perfluoropropene and compound. A undergo both addition and addition-elimination reactions with glutathione (Figure 23.4). The enzymatic reaction of glutathione with perfluoropropene yields both *S*-(1,1,2,3,3,3-hexafluoropropyl) glutathione and *S*-(1,2,3,3,3-pentafluoro-propenyl)glutathione as products (Koob and Dekant, 1990). Compound A is transformed to four glutathione *S*-conjugates, two diasteromers of *S*-[2-(fluoromethoxy)-1,3,3, 3-tetrafluoro-1-propenyl]glutathione and to *S*-[1,1-difluoro-2-(fluoromethoxy)-2-(trifluoromethyl)ethyl]glutathione and (Z)-*S*-[1-fluoro-2-(fluoromethoxy)-2-(trifluoromethyl)-vinyl]glutathione (Jin et al., 1996). Other fluoroalkenes are metabolized by glutathione *S*-transferases to

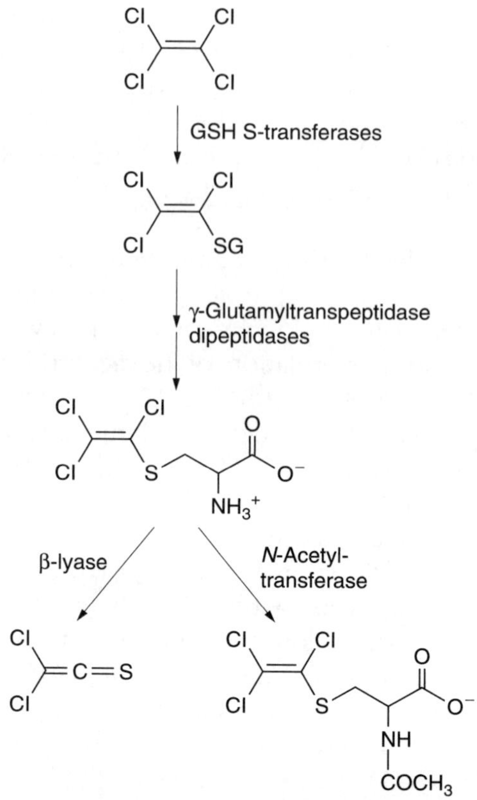

Figure 23.3 Biosynthesis of glutathione conjugates from perchloroethene as an example for polychlorinated alkenes, renal processing, and bioactivation of cysteine *S*-conjugates by cysteine conjugate *β*-lyase.

give *S*-(fluoroalkyl)glutathione conjugates. For example, chlorotrifluoroethene and tetrafluoroethene are metabolized to *S*-(1-chloro-1,1,2-trifluoroethyl)glutathione (Dohn and Anders, 1982a) and *S*-(1,1,2, 2-tetrafluoro-ethyl)glutathione (Odum and Green, 1984), respectively. The nephrotoxic chloro-fluoroalkene 1,1-dichloro-2,2-difluoroethene is metabolized to *N*-acetyl-*S*-(1,1-dichloro-2,2-difluoroethyl)-L-cysteine in rats (Commandeur et al., 1987).

The highly nephrotoxic dichloroethyne (Reichert et al., 1975) is converted by addition of glutathione to *S*-(1,2-dichlorovinyl)glutathione; in rats, *N*-acetyl-*S*-(1,2-dichlorovinyl)-L-cysteine is a major urinary metabolite of dichloroacetylene (Kanhai et al., 1989, 1991).

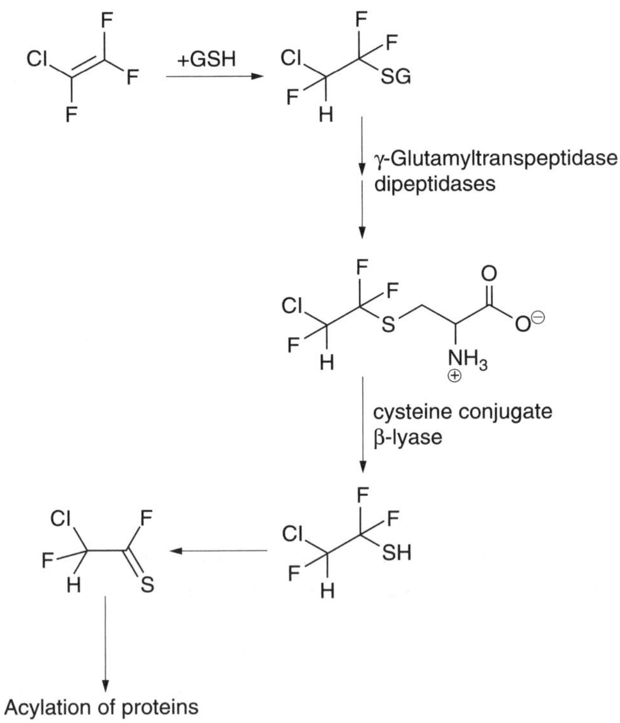

Figure 23.4 Biosynthesis of glutathione conjugates from chlorotrifluoroethene as an example for polyfluorinated alkenes, renal processing, and bioactivation of cysteine *S*-conjugates by cysteine conjugate *β*-lyase.

Hydroquinones and Aminophenol

Bromohydroquinone is a major toxic metabolite of bromobenzene and is easily converted to bromoquinone (Lau et al., 1984). Glutathione-dependent reactions have been implicated in bromohydroquinone induced nephrotoxicity since bromoquinone readily reacts with glutathione (Monks et al., 1985) to give several glutathione *S*-conjugates present in bile of rats treated with bromohydroquinone (Lau and Monks, 1990; Monks et al., 1985).

In bile of rats given the nephrotoxicant and renal carcinogen hydroquinone, the major biliary *S*-conjugate identified was 2-glutathion-*S*-yl-hydroquinone (Figure 23.5), and additional products were 2,5-diglutathion-*S*-yl-hydroquinone, 2,6-diglutathion-*S*-yl-hydroquinone and

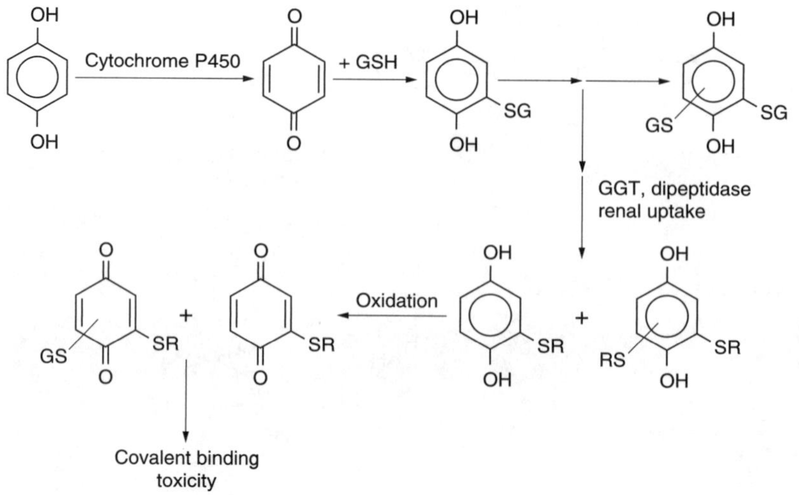

Figure 23.5 Suggested pathways of biotransformation of hydroquinone in the rat: biosynthesis of toxic glutathione *S*-conjugates and renal accumulation.

2,3,5-triglutathion-*S*-yl-hydroquinone (Hill et al., 1993; Kleiner et al., 1992). The food antioxidants tert-butyl-4-hydroxyanisole and tert-butyl-hydroquinone have been shown to promote kidney and bladder carcinogenesis in the rat. tert-Butyl-hydroquinone is also metabolized to glutathione *S*-conjugates *in vivo* (Peters et al., 1996a, 1996b) and 2-tert-butyl-5-glutathion-*S*-yl-hydroquinone, 2-tert-butyl-6-glutathion-S-yl-hydroquinone, and 2-tert-butyl-3,6-bisglutathion-*S*-yl-hydroquinone were identified as biliary metabolites of tert-butyl-hydroquinone in rats.

4-Aminophenol is nephrotoxic and causes necrosis of the pars recta of the proximal tubules in rats (Gartland et al., 1989a, 1989b). 4-Aminophenol is oxidized to benzoquinone imine, which reacts with glutathione to give the nephrotoxic glutathione *S*-conjugates 1-amino-3-(glutathione-*S*-yl)-4-hydroxybenzene and 1-amino-2-(glutathione-*S*-yl)-4-hydroxybenzene (Klos et al., 1992).

METABOLISM OF *S*-CONJUGATES AND UPTAKE BY THE KIDNEY

The biosynthesis of toxic glutathione *S*-conjugates seems to occur mainly in the liver. With more-reactive substrates for the glutathione *S*-transferases, glutathione conjugation may also occur in the kidney. For induction of nephrotoxicity, the *S*-conjugates formed in the liver must be translocated to the kidney (Inoue et al., 1984c; Okajima et al., 1983). The role of interorgan cooperation in the disposition of *S*-conjugates is of specific relevance for *S*-conjugates formed from haloalkenes, which require further bioactivation by cysteine conjugate *β*-lyase in the kidney.

Efflux of glutathione *S*-conjugates from hepatocytes occurs mainly across the canalicular membrane by active transport mechanisms into bile (Vore, 1993, 1994) since the molecular weight of most glutathione *S*-conjugates biosynthesized in the liver is above the threshold for biliary transport in rats, and the presence of polar groups in the molecules prevents passive diffusion through membranes (Awasthi 1990; Awasthi et al., 1989; Inoue et al., 1984a, 1984b). Biliary excretion is further supported by the observation that biliary cannulation protects male rats from the nephrotoxicity of hexachlorobutadiene (Nash et al., 1984) and 4-aminophenol, indicating that glutathione *S*-conjugate formation and biliary excretion are the first steps in the disposition of nephrotoxic *S*-conjugates. The glutathione *S*-conjugates formed are transported out of cells for further processing by *γ*-glutamyltranspeptidase and dipeptidases that catalyze the sequential removal of the glutamy and glycyl moieties. These proteins are not equally distributed between cells, but are localized predominantly to the apical surfaces of epithelial tissues.

Mercapturic acid formation and delivery to the kidney is generally considered to be an interorgan process, with the liver serving as the major site of glutathione *S*-conjugate formation, and the kidney as the primary site of processing to cysteine *S*-conjugates. Cysteine *S*-conjugates may be transported back to the liver for *N*-acetylation and may then be delivered to the kidney for excretion. The role of the organs involved in the disposition of *S*-conjugates is difficult to evaluate due to the complexity of the pathways and the lack of adequate experimental systems. The present model of interorgan cooperation is largely based on the distribution of the enzymes involved in mercapturic acid formation, focusing on the high activity of *γ*-glutamyltranspeptidase in rat kidney. *γ*-Glutamyl transpeptidase is the only enzyme known with the ability to initiate the breakdown of glutathione *S*-conjugates, thus the following pathways are suggested by the available experimental evidence.

After secretion into bile, glutathione *S*-conjugates derived from halogenated alkenes are either transported intact to the small intestine or degraded to the corresponding cysteine *S*-conjugates by the sequential action of γ-glutamyltransferase and dipeptidases, which are present in the luminal membrane of the bile duct epithelium and in the bile canalicular membrane of hepatocytes (Inoue et al., 1983; Meister, 1988). *S*-conjugates translocated to the small intestine may either be excreted with the feces, or reabsorbed from the gut and, after passage through the liver, translocate to the kidney or reenter enterohepatic circulation (Dekant et al., 1988a). Glutathione *S*-conjugates excreted in the bile may also be reabsorbed by the liver from bile after breakdown to cysteinyl glycine or cysteine *S*-conjugates.

Glutathione *S*-conjugates absorbed by hepatocytes at the luminal surface will again be excreted with bile and thus undergo enterohepatic circulation. Due to their molecular weight, cysteine *S*-conjugates taken up by the liver may be excreted into the blood circulation, may be *N*-acetylated by hepatic *N*-acetyltransferases, and may enter the systemic circulation as mercapturic acids (Duffel and Jakoby, 1982; Green and Elce, 1975).

Glutathione *S*-conjugates leaving the liver via the sinusoidal membrane and glutathione *S*-conjugates biosynthesized in organs other than the liver may be delivered intact to the kidney. The high blood flow to the kidneys and their capacity to process and activate glutathione *S*-conjugates are regarded as major determinants of *S*-conjugate toxicity (Monks and Lau, 1987, 1989; Monks et al., 1990; Rush et al., 1984). Once in the kidney, *S*-conjugates may reach their target in the straight portion of the proximal tubular cells by glomerular filtration or by transport across the basolateral membrane, or both. The kidney is rich in γ-glutamyltransferase activity, dipeptidase activity, and aminoacylase activity; thus all *S*-conjugates delivered to the kidney may be efficiently converted to cysteine *S*-conjugates in the proximal tubular cells (Anderson et al., 1990; Guder and Ross, 1984; Guder and Wirthensohn, 1985; Hughey et al., 1978).

Glutathione *S*-conjugates present in blood may be filtered at the glomerulus or may enter the peritubular circulation. Studies with *S*-(pentachlorobutadienyl)glutathione in the isolated perfused rat kidney demonstrate that this glutathione *S*-conjugate is preferentially removed by nonfiltering mechanisms (Schrenk et al., 1988).

The filtered glutathione *S*-conjugates may be metabolized by γ-glutamyltransferase and dipeptidases or in the renal brush-border membrane (Figure 23.3, Figure 23.4, and Figure 23.5) (Hughey et al., 1978; Jones et al., 1979b; Moldeus et al., 1978; Tsao and Curthoys, 1980). The glutathione *S*-conjugates in the peritubular capillaries may either be metabolized by γ-glutamyltransferase present in the renal vasculature or basolateral membrane, but transport of intact glutathione into renal epithelial cells has also been observed (Lash and Jones, 1984; Rankin and Curthoys, 1982).

The probenecid-sensitive organic anion transporter present on the basolateral side of the proximal tubular cells seems to play the most important role in the accumulation of *S*-conjugates in proximal tubular cells and in the organ-selective toxicity. Probenecid is a selective inhibitor of the organic anion transporter. Haloalkene-derived mercapturates have high affinity for the organic anion transporter, but glutathione and cysteine *S*-conjugates with lipophilic substituents on sulfur are also substrates (Pombrio et al., 2001; Ullrich and Rumrich 1988; Ullrich et al., 1988, 1989a, 1989b). The inhibitory effects of probenecid on the toxicity of several haloalkene cysteine *S*-conjugates and mercapturic acids support a central role for the renal organic anion transporter in the renal accumulation of *S*-conjugates.

It has been demonstrated that mercapturic acid formation may also occur intrahepatically and that these mercapturic acids may be delivered to the kidney by the systemic circulation. In the kidney, mercapturic acids formed from halogenated alkenes require deacetylation by acylases to generate the substrates for cysteine conjugate β-lyase. Intrahepatic mercapturic acid formation may be relevant in species with higher γ-glutamyltranspeptidase activity in the liver, such as the guinea pig (Hinchman and Ballatori, 1994; Hinchman et al., 1998). The role of species differences in mercapturic acid biosynthesis and disposition in the toxicity of haloalkenes and haloalkanes is not well defined.

NEPHROTOXICITY OF *S*-CONJUGATES

The *S*-conjugates formed are toxic metabolites of the parent haloalkanes, haloalkenes, hydroquinones, and aminophenols. These metabolites seem to be accumulated in the kidney. For a better characterization of the mechanisms of nephrotoxicity induced by the formed *S*-conjugates, many studies have been performed to investigate the nephrotoxicity of synthetic *S*-conjugates (Anders and Dekant 1998; Lock, 1988, 1989).

The sulfur half mustards biosynthesized from 1,2-dihaloalkanes are strong alkylating agents that react with nucleic acids (Jean and Reed, 1989). Their electrophilicity is attributable to neigh-boring group assistance in nucleophilic displacement (Dohn and Casida, 1987). Despite their high reactivity with water, proteins, and nucleic acids,

S-(2-chloroethyl)glutathione and *S*-(2-chloroethyl)-DL-cysteine administered to rats were found to be selectively nephrotoxic (Kramer et al., 1987).

The first report on the toxicity of cysteine *S*-conjugates appeared when *S*-(1,2-dichlorovinyl)-L-cysteine present in trichloroethene-extracted soybean meal caused aplastic anemia in cattle (McKinney et al., 1959).

In contrast to the hematopoetic toxicity of *S*-(1,2-dichlorovinyl)-L-cysteine observed in cattle, this compound and structurally related *S*-conjugates were nephrotoxic in mice, rats, guinea pigs, and dogs, causing necrosis of the renal tubular epithelium and a perturbation of kidney function (Jaffe et al., 1984; Koechel et al., 1991; Terracini and Parker, 1965). Several other haloalkenyl cysteine *S*-conjugates caused identical toxicity to the proximal tubules after administration to rodents (Dohn et al., 1985; Green and Odum, 1985; Iyer et al., 1997; Kharasch et al., 1997; Odum and Green, 1984). The corresponding glutathione *S*-conjugates were also nephrotoxic in rats, causing renal damage identical to that seen after administration of the cysteine *S*-conjugates (Dohn et al., 1985; Elfarra et al., 1986; Iyer and Anders, 1997; Nash et al., 1984).

The role of γ-glutamyltranspeptidase and cysteine conjugate β-lyase-catalyzed bioactivation in the renal toxicity of haloalkene-derived *S*-conjugates has been confirmed based on the modulation of activity of γ-glutamyltranspeptidase and pyridoxal phosphate dependent enzymes (Elfarra et al., 1986; Lash and Anders, 1986). The results suggest the following pathways for bioactivation of haloalkene-derived glutathione *S*-conjugates (Figure 23.3 and Figure 23.4).

β-Lyase-catalyzed β-elimination reactions of cysteine *S*-conjugates derived from chloroalkenes give unstable thiolates, pyruvate, and ammonia as products (Dohn and Anders, 1982b). The thiolates rearrange to thioacylating intermediates that react with tissue nucleophiles, forming covalently bound adducts (Chen et al., 1990; Darnerud et al., 1988, 1989). For example, N^{ε}-(dichloroacetyl)-L-lysine was identified as a modified amino acid residue in rats exposed to tetrachloroethene and *S*-(1,2,2-trichlorovinyl)-L-cysteine (Birner et al., 1994; Pähler et al., 1999; Völkel et al., 1999). A β-lyase-catalyzed β-elimination reaction with 1,1-difluoroalkene-derived cysteine *S*-conjugates as the substrates gives 1,1-difluoro-2,2-dihaloalkylthiolates as initial products, which lose fluoride to give electrophilic thioacyl fluorides (Dekant et al., 1987a; Commandeur et al., 1989). The halogenated thioacetyl fluorides formed by the β-lyase-catalyzed biotransformation of 1,1-difluoroalkene-derived cysteine *S*-conjugates also react with tissue nucleophiles (Bruschi et al.,

1993; Hargus and Anders 1991; Harris et al., 1992; Hayden et al., 1991a, 1991b, 1992). Covalent adduct formation is associated with cysteine *S*-conjugate-induced cytotoxicity (Chen et al., 1990).

Glutathione *S*-conjugates biosynthesized from several quinones and aminophenols are also nephrotoxic in rats (Hill et al., 1992; Lau et al., 1988b, 2001; Monks and Lau 1998; Redegeld et al., 1991). The tert-butylhydroquinone metabolites (2-tert-butyl-5-(glutathion-*S*-yl)hydroquinone, 2-tert-butyl-6-(glutathion-*S*-yl)hydroquinone, and 2-tert-butyl-3,6-bis-(glutathion-*S*-yl)hydroquinone) are nephrotoxic in rats, and 2-tert-butyl-3,6-bis-(glutathion-*S*-yl)hydroquinone was the most potent of these glutathione *S*-conjugates. In addition to being nephrotoxic, 2-tert-butyl-3,6-bis-(glutathion-*S*-yl)hydroquinone was toxic to the bladder (Peters et al., 1996b). Other quinone-derived *S*-conjugates are also neprotoxic in rodents. For example, administration of 2,5-dichloro-3-(glutathion-*S*-yl)-1,4-benzoquinone and 2,5,6-trichloro-3-(glutathion-*S*-yl)-1,4-benzoquinone to rats caused dose-dependent renal proximal tubular necrosis (Hill et al., 1992; Lau et al., 1995; Monks et al., 1991). Also, 2,3,5-(tris-glutathion-*S*-yl)hydroquinone, a biliary metabolite of hydroquinone, caused tubular necrosis in the S_3 segment of the proximal tubule when administered to rats, and 2-bromo-(diglutathion-*S*-yl) hydroquinone was a potent nephrotoxicant causing severe histological alterations to renal proximal tubules. In addition, the corresponding mercapturic acids are selective nephrotoxicants after administration to rats.

4-Amino-3-*S*-glutathionylphenol, a metabolite of the renal toxicant 4-aminophenol, also produced necrosis of the proximal tubular epithelium and altered renal excretory function in rats. The lesion was specific to the pars recta region of nephrons and was very similar to that produced by 4-aminophenol. Both direct and indirect evidence for the involvement of glutathione conjugates as a transport form for 4-aminophenol metabolites has been obtained (Fowler et al., 1991, 1994; Gartland et al., 1990).

The glutathione *S*-conjugates formed from hydroquinones and aminophenol (Kleiner et al., 1992) are accumulated by the kidney in a γ-glutamyltranspeptidase-dependent pathway. Their toxicity is diminished or increased by inhibition of γ-glutamyltranspeptidase but not by inhibition of β-lyase (Monks and Lau, 1990; Monks et al., 1988). These experiments suggest that β-lyase-mediated cleavage does not play a role in the toxicity of these hydroquinone *S*-conjugates. The glutathione conjugates of hydroquinones and quinones may serve as transport forms to γ-glutamyltranspeptidase-rich organs and may accumulate there. In the rat, γ-glutamyltranspeptidase is almost exclusively present in the kidney. Moreover, substitution of bromohydroquinones with cysteine alters the redox potential and may make these hydroquinone moieties more prone to oxidation to toxic quinones. The reduction of nephrotoxicity of bromohydroquinone glutathione *S*-conjugates by simultaneous administration of ascorbic acid suggests that processes involving oxidation to a reactive quinone and, presumably, peroxidative mechanisms are involved in the nephrotoxicity of these compounds (Lau et al., 1988a, 1988b; Lau et al., 1990; Monks and Lau, 1990).

CONCLUSIONS

The results summarized here indicate that, besides playing an important role in detoxication, glutathione conjugation may play a role in directing toxic compounds to the kidney. Kidney-specific toxicity of xenobiotics activated by glutathione conjugate formation in many cases is due to the capability of the kidney to accumulate intermediates formed by processing of glutathione *S*-conjugates and to bioactivate them to toxic metabolites.

REFERENCES

Anders MW (Ed.) (1985) *Bioactivation of Foreign Compounds*. Academic Press, Orlando.

Anders MW (1988) Bioactivation mechanisms and hepatocellular damage, in *The Liver: Biology and Pathology*, 2nd ed, Arias IM, Jakoby WB, Popper H, Schachter D, and Shafritz DA (Eds), Raven Press, Ltd., New York, pp. 389–400.

Anders MW (1991) Glutathione-dependent bioactivation of xenobiotics. *FASEB J* 4:87–92.

Anders MW, and Dekant W (Eds) (1994) *Conjugation-Dependent Carcinogenicity and Toxicity of Foreign Compounds*. Academic Press, San Diego.

Anders MW, and Dekant W (1998) Glutathione-dependent bioactivation of haloalkenes. *Annu Rev Pharmacol Toxicol* 38:501–37.

Anderson ME, Naganuma A, and Meister A (1990) Protection against cisplatin toxicity by administration of glutathione ester. *FASEB J* 4:3251–3255.

Awasthi YC (1990) The interrelationship between *p*-glycoprotein and glutathione *S*-conjugate transporter(s). *TIBS* 15:376–377.

Awasthi YC, Singh SV, Ahmad H, Wronski LW, Srivastava SK, and Labelle EF (1989) ATP dependent primary active transport of xenobiotic-glutathione conjugates by human erxthrocyte membrane. *Cell Biochem* 91:131–136.

Birner G, Bernauer U, Werner M, and Dekant W (1997) Biotransformation, excretion and nephrotoxicity of haloalkene-derived cysteine *S*-conjugates. *Arch Toxicol* 72:1–8.

Birner G, Richling C, Henschler D, Anders MW, and Dekant W (1994) Metabolism of tetrachloroethene in rats: Identification of N^{ε}-(dichloroacetyl)-L-lysine and N^{ε}-(trichloroacetyl)-L-lysine as protein adducts. *Chem Res Toxicol* 7:724–732.

Birner G, Rutkowska A, and Dekant W (1996) *N*-Acetyl-*S*-(1,2,2-trichlorovinyl)-L-cysteine and 2,2,2-trichloroethanol: Two novel metabolites of tetrachloroethene in humans after occupational exposure. *Drug Metab Dispos* 24:41–48.

Boyland E, and Chasseaud LF (1969) Role of glutathione and glutathione *S*-transferases in mercapturic acid biosynthesis. *Adv Enzymol* 32:173–177.

Bruschi SA, West K, Crabb JW, Gupta RS, and Stevens JL (1993) Mitochondrial HSP60 (P1protein) and a HSP70-like protein (mortalin) are major targets for modification during *S*-(1,1,2,2-tetrafluoroethyl)-L-cysteine-induced nephrotoxicity. *J Biol Chem* 268:23157–23161.

Chen Q, Jones TW, Brown PC, and Stewens JL (1990) The mechanism of cysteine conjugate cytotoxicity in renal epithelial cells. *J Biol Chem* 265:21603–21611.

Cmarik JL, Inskeep PB, Meredith MJ, Meyer DJ, Ketterer B, and Guengerich FP (1990) Selectivity of rat and human glutathione *S*-transferases in activation of ethylene dibromide by glutathione conjugation and DNA binding and induction of unscheduled DNA synthesis in human hepatocytes. *Cancer Res* 50:2747–2752.

Commandeur JNM, De Kanter FJJ, and Vermeulen NPE (1989) Bioactivation of the cysteine-*S*-conjugate and mercapturic acid of tetrafluoroethylene to acylating reactive intermediates in the rat: Dependence of activation and deactivation activities on acetyl coenzyme A availability. *Mol Pharmacol* 36:654–663.

Commandeur JNM, Oostendorp RAJ, Schoofs PR, Xu B, and Vermeulen NPE (1987) Nephrotoxicity and hepatotoxicity of 1,1-dichloro-2,2-difluoroethylene in the rat. *Biochem Pharmacol* 36:4229–4237.

Darnerud PO, Brandt I, Feil VJ, and Bakke JE (1988) *S*-(1,2-Dichloro-(14C)vinyl)-L-cysteine (DCVC) in the mouse kidney: Correlation between tissue-binding and toxicity. *Toxicol Appl Pharmacol* 95:423–434.

Darnerud PO, Brandt I, Feil VJ, and Bakke JE (1989) Dichlorovinyl cysteine (DCVC) in the mouse kidney: Tissue-binding and toxicity after glutathione depletion and probenecid treatment. *Arch Toxicol* 63:345–350.

Dekant W, Birner G, Werner M, and Parker J (1998) Glutathione conjugation of perchloroethene in subcellular fractions from rodent and human liver and kidney. *Chem-Biol Interact* 116:31–43.

Dekant W, Koob M, and Henschler D (1990) Metabolism of trichloroethene – *in vivo* and *in vitro* evidence for activation by glutathione conjugation. *Chem-Biol Interact* 73:89–101.

Dekant W, Lash LH, and Anders MW (1987a) Bioactivation mechanism of the cytotoxic and nephrotoxic *S*-conjugate *S*-(2-chloro-1,1,2-trifluoroethyl)-L-cysteine. *Proc Natl Acad Sci USA* 84:7443–7447.

Dekant W, Martens G, Vamvakas S, Metzler M, and Henschler D (1987b) Bioactivation of tetrachloroethylene – role of glutathione *S*-transferase-catalyzed conjugation versus cytochrome P-450-dependent phospholipid alkylation. *Drug Metab Dispos* 15:702–709.

Dekant W, Metzler M, and Henschler D (1984) Novel metabolites of trichloroethylene through dechlorination reactions in rats, mice and humans. *Biochem Pharmacol* 33:2021–2027.

Dekant W, Metzler M, and Henschler D (1986a) Identification of *S*-1,2,2-trichlorovinyl-*N*-acetylcysteine as a urinary metabolite of tetrachloroethylene: Bioactivation through glutathione conjugation as a possible explanation of its nephrocarcinogenicity. *J Biochem Toxicol* 1:57–72.

Dekant W, Metzler M, and Henschler D (1986b) Identification of *S*-1,2-dichlorovinyl-*N*-acetyl-cysteine as a urinary metabolite of trichloroethylene: A possible explanation for its nephrocarcinogenicity in male rats. *Biochem Pharmacol* 35:2455–2458.

Dekant W, and Neumann H-G, (Eds) (1992) *Tissue specific toxicity: Biochemical mechanisms.* Academic Press, London.

Dekant W, Schrenk D, Vamvakas S, and Henschler D (1988a) Metabolism of hexachloro-1,3-butadiene in mice: *In vivo* and *in vitro* evidence for activation by glutathione conjugation. *Xenobiotica* 18:803–816.

Dekant W, and Vamvakas S (1996) Biotransformation and membrane transport in nephrotoxicity. *Crit Rev Toxicol* 26:309–334.

Dekant W, Vamvakas S, Henschler D, and Anders MW (1988b) Enzymatic conjugation of hexachloro-1,3-butadiene with glutathione: Formation of 1-(glutathion-*S*-yl)-1,2,3,4,4-pentachlorobuta-1,3-diene and 1,4-bis(glutathion-*S*-yl)-1,2,3,4-tetrachlorobuta-1,3-diene. *Drug Metab Dispos* 16:701–706.

Dohn DR, and Anders MW (1982a) The enzymatic reaction of chlorotrifluoroethylene with glutathione. *Biochem Biophys Res Commun* 109:1339–1345.

Dohn DR, and Anders MW (1982b) A simple assay for cysteine conjugate β-lyase activity with *S*-(2-benzothiazolyl)cysteine as the substrate. *Anal Biochem* 120:379–386.

Dohn DR, and Casida JE (1987) Thiiranium ion intermediates in the formation and reactions of *S*-(2-haloethyl)-L-cysteines. *Bioorganic Chemistry* 15:115–124.

Dohn DR, Leininger JR, Lash LH, Quebbemann AJ, and Anders MW (1985) Nephrotoxicity of *S*-(2-chloro-1,1,2-trifluoroethyl)glutathione and *S*-(2-chloro-1,1,2-trifluoroethyl)-L-cysteine, the glutathione and cysteine conjugates of chlorotrifluoroethene. *J Pharmacol Exp Ther* 235:851–857.

Duffel MW, and Jakoby WB (1982) Cysteine *S*-conjugate *N*-acetyltransferase from rat kidney microsomes. *Mol Pharmacol* 21:444–448.

Elfarra AA, Jakobson I, and Anders MW (1986) Mechanism of *S*-(1,2-dichlorovinyl) glutathione-induced nephrotoxicity. *Biochem Pharmacol* 35:283–288.

Elliot WC, Lynn RK, Hougton DC, Kennish JM, and Bennett WM (1982) Nephrotoxicity of the flame retardant tris(2,3-dibromopropyl)phosphate, and its metabolites. *Toxicol Appl Pharmacol* 63:179–182.

Fowler LM, Foster JR, and Lock EA (1994) Nephrotoxicity of 4-amino-3-*S*-glutathionylphenol and its modulation by metabolism or transport inhibitors. *Arch Toxicol* 68:15–23.

Fowler LM, Moore RB, Foster JR, and Lock EA (1991) Nephrotoxicity of 4-aminophenol glutathione conjugate. *Hum Exper Toxicol* 10:451–459.

Gartland KPR, Bonner FW, and Nicholson JK (1989a) Investigations into the biochemical effects of region-specific nephrotoxins. *Mol Pharmacol* 35:242–250.

Gartland KPR, Bonner FW, Timbrell JA, and Nicholson JK (1989b) Biochemical characterisation of *para*-aminophenol-induced nephrotoxic lesions in the F334 rat. *Arch Toxicol* 63:97–106.

Gartland KPR, Eason CT, Bonner FW, and Nicholson JK (1990) Effects of biliary cannulation and buthionine sulphoximine pretreatment on the nephrotoxicity of para-aminophenol in the Fisher 334 rat. *Arch Toxicol* 64:14–25.

Green RM, and Elce JS (1975) Acetylation of S-substituted cysteines by rat liver and kidney microsomal *N*-acetyltransferase. *Biochem J* 147:283–289.

Green T, Dow J, Ellis MK, Foster JR, and Odum J (1997) The role of glutathione conjugation in the development of kidney tumours in rats exposed to trichloroethylene. *Chem-Biol Interact* 105:99–117.

Green T, and Odum J (1985) Structure/activity studies of the nephrotoxic and muta genic action of cysteine conjugates of chloro- and fluoroalkenes. *Chem-Biol Interact* 54:15–31.

Guder WG, and Ross BD (1984) Enzyme distribution along the nephron. *Kidney Int* 26:101–111.

Guder WG, and Wirthensohn G (1985) Enzyme distribution and unique biochemical pathways in specific cells along the nephron, in *Renal Heterogeneity and Target Cell toxicity*, Bach PH, and Lock EA (Eds), Wiley and Sons, Chinchester, pp. 195–198.

Guengerich FP (2001) Common and uncommon cytochrome P450 reactions related to metabolism and chemical toxicity. *Chem Res Toxicol* 14:611–650.

Guengerich FP, Crawford WMJ, Domoradzki JY, Macdonald TL, and Watanabe PG (1980) In vitro activation of 1,2-dichloroethane by microsomal and cytosolic enzymes. *Toxicol Appl Pharmacol* 55:303–317.

Guengerich FP, and Liebler DC (1985) Enzymatic activation of chemicals to toxic metabolites. *CRC Crit Rev Toxicol* 14:259–307.

Hargus SJ, and Anders MW (1991) Immunochemical detection of covalently modified kidney proteins in *S*-(1,1,2,2-tetrafluoroethyl)-L-cysteine-treated rats. *Biochem Pharmacol* 42:R17–R20.

Harris JW, Dekant W, and Anders MW (1992) *In vivo* detection and characterization of protein adducts resulting from bioactivation of haloethene cysteine *S*-conjugates by ^{19}F NMR: Chlorotrifluoroethene and tetrafluoroethene. *Chem Res Toxicol* 5:34–41.

Hayden PJ, Ichimura T, McCann DJ, Pohl LR, and Stevens JL (1991a). Detection of cysteine conjugate metabolite adduct formation with specific mitochondrial proteins using antibodies raised against halothane metabolite adducts. *J Biol Chem* 266:18415–18418.

Hayden PJ, Welsh CJ, Yang Y, Schaefer WH, Ward AJI, and Stevens JL (1992) Formation of mitochondrial phospholipid adducts by nephrotoxic cysteine conjugate metabolites. *Chem Res Toxicol* 5:231–237.

Hayden PJ, Yang Y, Ward AJI, Dulik DM, McCann DJ, and Stevens JL (1991b) Formation of difluorothionoacetyl-protein adducts by *S*-(1,1,2,2-tetrafluoroethyl)-L-cysteine metabolites: Nucleophilic catalysis of stable adduct formation by histidine and tyrosine. *Biochemistry* 30:5935–5943.

Hill BA, Kleiner HE, Ryan EA, Dulik DM, Monks TJ, and Lau SS (1993) Identification of multi-*S*-substituted conjugates of hydroquinones by HPLC-coulometric electrode array analysis and mass spectroscopy. *Chem Res Toxicol* 6:459–469.

Hill BA, Monks TJ, and Lau SS (1992) The effects of 2,3,5-(triglutathion-*S*-yl)hydroquinone on renal mitochondrial respiratory function *in vivo* and *in vitro*: Possible role in cytotoxicity. *Toxicol Appl Pharmacol* 117:165–171.

Hinchman CA, and Ballatori N (1994) Glutathione conjugation and conversion to mercapturic acids can occur as an intrahepatic process. *J Toxicol Environ Health* 41:387–409.

Hinchman CA, Rebbeor JF, and Ballatori N (1998) Efficient hepatic uptake and concentrative biliary excretion of a mercapturic acid. *Am J Physiol* 275:G612–G619.

Holme JA, Hongslo JK, Bjorge C, and Nelson SD (1991) Comparative cytotoxicity effects of acetaminophen (N-acetyl-p-aminophenol), a non-hepatotoxic regiosomer acetyl-m-aminophenol and their postulated reactive hydroquinone and quinone metabolites in monolayer cultures of mouse hepatocytes. *Biochem Pharmacol* 42:1137–1142.

Hughey RP, Rankin BB, Elce JS, and Curthoys NP (1978) Specificity of a particulate rat renal peptidase and its localization along with other enzymesof mercapturic acid synthesis. *Arch Biochem Biophys* 186:211–217.

Humphreys WG, Kim DH, Cmarik JL, Shimada T, and Guengerich FP (1990). Comparison of the DNA-alkylating properties and mutagenic responses of a series of *S*-(2-haloethyl)-substituted cysteine and glutathione derivatives. *Biochemistry* 29:10342–10350.

Humphreys WG, Kim DH, and Guengerich FP (1991) Isolation and characterization of N^7-guanyl adducts derived from 1,2-dibromo-3-chloropropane. *Chem Res Toxicol* 4:445–453.

IARC (1990) *Some Flame Retardants and Textile Chemicals, and Exposure in the Textile Manufacturing Industry.* IARC-Monographs, International Agency for Research on Cancer, Lyon.

Inoue M, Akerboom TP, Sies H, Kinne R, Thao T, and Arias IM (1984a) Biliary transport of glutathione *S*-conjugate by rat liver canalicular membrane vesicles. *J Biol Chem* 259:4998–5002.

Inoue M, Kinne R, Tran T, and Arias IM (1984b) Glutathione transport across hepatocyte plasma membranes. Analysis using isolated rat-liver sinusoidal-membrane vesicles. *Eur J Biochem* 138:491–495.

Inoue M, Kinne R, Tran T, Biempica L, and Arias IM (1983) Rat liver canalicular membrane vesicles – isolation and topological characterization. *J Biol Chem* 258:5183–5188.

Inoue M, Okajima K, and Morino Y (1984c) Hepato-renal cooperation in biotransformation, membrane transport, and elimination of cysteine *S*-conjugates of xenobiotics. *J Biochem Tokyo* 95:247–254.

Inskeep PB, and Guengerich FP (1984) Glutathione-mediated binding of dibromoalkanes to DNA: Specificity of rat glutathione-*S*-transferases and dibromoalkane structure. *Carcinogenesis* 5:805–808.

Ishmael J, Pratt I, and Lock EA (1982) Necrosis of the pars recta (S3 segment) of the rat kidney produced by hexachloro-1:3-butadiene. *J Pathol* 138:99–113.

Iyer RA, and Anders MW (1997) Cysteine conjugate β-lyase-dependent biotransformation of the cysteine *S*-conjugates of sevoflurane degradation product 2-(fluoromethoxy)-1,1,3,3,3-pentafluoro-1-propene (compound A). *Chem Res Toxicol* 10:811–819.

Iyer RA, Baggs RB, and Anders MW (1997) Nephrotoxicity of the glutathione and cysteine S-conjugates of the sevoflurane degradation product 2-(fluoromethoxy)-1,1,3,3,3- pentafluoro-1-propene (Compound A) in male Fischer 344 rats. *J Pharmacol Exp Ther* 283:1544–1551.

Jaffe DR, Gandolfi AJ, and Nagle RB (1984) Chronic toxicity of *S*-(trans-1,2-dichlorovinyl)-L-cysteine in mice. *J Appl Toxicol* 4:315–319.

Jean PA, and Reed DJ (1989) In vitro dipeptide, nucleoside, and glutathione alkylation by S-(2-chloroethyl)glutathione and S-(2-chloroethyl)-L-cysteine. *Chem Res Toxicol* 2:455–460.

Jean PA, and Reed DJ (1992) Utilization of glutathione during 1,2-dihaloethane metabolism in rat hepatocytes. *Chem Res Toxicol* 5:386–391.

Jin L, Davis MR, Kharasch ED, Doss GA, and Baillie TA (1996) Identification in rat bile of glutathione conjugates of fluoromethyl 2,2-difluoro-1-(trifluoromethyl)vinyl ether, a nephrotoxic degradate of the anesthetic agent sevofluorane. *Chem Res Toxicol* 9:555–561.

Jin LX, Baillie TA, Davis MR, and Kharasch ED (1995) Nephrotoxicity of sevoflurane compound A [fluoromethyl-2,2-difluoro-1-(trifluoromethyl)vinyl ether] in rats: Evidence for glutathione and cysteine conjugate formation and the role of renal cysteine conjugate beta-lyase. *Biochem Biophys Res Commun* 210:498–506.

Jones AR, Fakhouri G, and Gadiel P (1979a) The metabolism of the soil fumigant 1,2-dibromo-3-chloropropane. *Experienta* 35:1432–1434.

Jones DP, Moldeus P, Stead AH, Ormstad K, Joernvall H, and Orrenius S (1979b) Metabolism of glutathione and a glutathione conjugate by isolated kidney cells. *J Biol Chem* 254:2787–2792.

Kanhai W, Dekant W, and Henschler D (1989) Metabolism of the nephrotoxin dichloroacetylene by glutathione conjugation. *Chem Res Toxicol* 2:51–56.

Kanhai W, Koob M, Dekant W, and Henschler D (1991) Metabolism of ^{14}C-dichloroethyne in rats. *Xenobiotica* 21:905–916.

Kharasch ED, Thorning D, Garton K, Hankins KC, and Kilty CG (1997) Role of renal cysteine *S*-conjugate β-lyase in the mechanism of Compound A neprotoxicity in rats. *Anesthesiol* 86:160–171.

Kim D-H, and Guengerich FP (1989) Excretion of the mercapturic acid S-[2-(N7-guanyl)ethyl]-N-acetylcysteine in urine following administration of ethylene dibromide to rats. *Cancer Res* 499:5843–5847.

Kim D-H, and Guengerich FP (1990) Formation of the DNA adduct S-[2-(N7-guanyl)ethyl]glutathione from ethylene dibromide: Effects of modulation of glutathione and glutathione S-transferase levels and lack of a role for sulfation. *Carcinogenesis* 11:419–424.

Kleiner HE, Hill BA, Monks TJ, and Lau SS (1992) *In vivo* and *in vitro* formation of several S-conjugates of hydroquinone. *Toxicologist* 12:1350–1350.

Klos C, Koob M, Kramer C, and Dekant W (1992) p-Aminophenol nephrotoxicity: Biosynthesis of toxic glutathione conjugates. *Toxicol Appl Pharmacol* 115:98–106.

Kociba RJ, Keyes DG, Jersey GC, Ballard JJ, Dittenber DA, Quast JF, Wade LE, Humiston CG, and Schwetz BA (1977) Results of a two year chronic toxicity study with hexachlorobutadiene in rats. *Am Ind Hyg Assoc J* 38:589–602.

Koechel DA, Krejci ME, and Ridgewell RE (1991) The accute effects of *S*-(1,2-dichlorovinyl)-L-cysteine and related chemicals on renal function and ultrastructure in the pentobarbital-anesthetized dog: Structure-activity relationships, biotransformation, and unique site-specific nephrotoxicity. *Fundam Appl Toxicol* 17:17–33.

Koob M, and Dekant W (1990) Metabolism of hexafluoropropene – evidende for bioactivation by glutathione conjugate formation in the kidney. *Drug Metab Dispos* 18:911–916.

Kramer RA, Foureman G, Greene KE, and Reed DJ (1987) Nephrotoxicity of *S*-(2-chloroethyl)glutathione in the Fischer rat: Evidence for γ-glutamyltranspeptidase-independent uptake by the kidney. *J Pharmacol Exp Therap* 242:741–748.

Lag M, Omichinski JG, Soderlund EJ, Brunborg G, Holme JA, Dahl JE, Nelson SD, and Dybing E (1989a) Role of P-450 activity and glutathione levels in 1,2-dibromo-3-chloropropane tissue distribution, renal necrosis and *in vivo* DNA damage. *Toxicology* 56:273–288.

Lag M, Soderlund EJ, Brunborg G, Dahl JE., Holme JA, Omichinski JG, Nelson SD, and Dybing E (1989b) Species differences in testicular necrosis and DNA damage, distribution and metabolism of 1,2-dibromo-3-chloropropane (DBCP). *Toxicology* 58:133–144.

Lash LH (1994) Role of renal metabolism in risk to toxic chemicals. *Environ Health Perspect* 102:75–79.

Lash LH, and Anders MW (1986) Cytotoxicity of *S*-(1,2-dichlorovinyl)glutathione and *S*-(1,2-dichlorovinyl)-L-cysteine in isolated rat kidney cells. *J Biol Chem* 261:13076–13081.

Lash LH, and Jones DP (1984) Renal glutathione transport. Characteristics of the sodium-dependent system in the basal-lateral membrane. *J Biol Chem* 259:14508–14514.

Lash LH, Lipscomb JC, Putt DA, and Parker JC (1999a) Glutathione conjugation of trichloroethylene in human liver and kidney: Kinetics and individual variation. *Drug Metab Dispos* 27:351–359.

Lash LH, Putt DA, Brashear WT, Abbas R, Parker JC, and Fisher JW (1999b) Identification of S-(1,2-dichlorovinyl)glutathione in the blood of human volunteers exposed to trichloroethylene. *J Toxicol Environ Health* 56:1–21.

Lash LH, Qian W, Putt DA, Desai K, Elfarra AA, Sicuri AR, and Parker JC (1998a) Glutathione conjugation of perchloroethylene in rats and mice *in vitro*: Sex-, species-, and tissue-dependent differences. *Toxicol Appl Pharmacol* 150:49–57.

Lash LH, Qian W, Putt DA, Jacobs K, Elfarra AA, Krause RJ, and Parker JC (1998b) Glutathione conjugation of trichloroethylene in rats and mice: Sex-, species-, and tissue-dependent differences. *Drug Metab Dispos* 26:12–19.

Lash LH, Xu YP, Elfarra AA, Duescher RJ, and Parker JC (1995) Glutathione-dependent metabolism of trichloroethylene in isolated liver and kidney cells of rats and its role in mitochondrial and cellular toxicity. *Drug Metab Dispos* 23:846–853.

Lau SS, Hill BA, Highet RJ, and Monks TJ (1988a) Sequential oxidation and glutathione addition to 1,4-benzoquinone: Correlation of toxicity with increased glutathione substitution. *Mol Pharmacol* 34:829–836.

Lau SS, Jones TW, Highet RJ, Hill B, and Monks TJ (1990) Differences in the localization and extent of the renal proximal tubular necrosis caused by mercapturic acid and glutathione conjugates of 1,4-naphthoquinone and menadione. *Toxicol Appl Pharmacol* 104:334–350.

Lau SS, Kleiner HE, and Monks TJ (1995) Metabolism as a determinant of species susceptibility to 2,3,5-(triglutathion-*S*-yl)hydroquinone-mediated nephrotoxicity – The role of *N*-acetylation and *N*-deacetylation. *Drug Metab Dispos* 23:1136–1142.

Lau SS, McMenamin MG, and Monks TJ (1988b) Differential uptake of isomeric 2-bromohydroquinone-glutathione conjugates into kidney slices. *Biochem Biophys Res Commun* 152:223–230.

Lau SS, and Monks TJ (1990) The *in vivo* disposition of 2-bromo-[14C]hydroquinone and the effect of γ-glutamyl transpeptidase inhibition. *Toxicol Appl Pharmacol* 103:121–132.

Lau SS, Monks TJ, Everitt JI, Kleymenova E, and Walker CL (2001) Carcinogenicity of a nephrotoxic metabolite of the "nongenotoxic" carcinogen hydroquinone. *Chem Res Toxicol* 14:25–33.

Lau SS, Monks TJ, and Gillette JR (1984) Identification of 2-bromohydroquinone as a metabolite of bromobenzene and o-bromophenol: Implications for bromobenzene-induced nephrotoxicity. *J Pharmacol Exp Ther* 230:360–366.

Lock EA (1988) Studies on the mechanism of nephrotoxicity and nephrocarcinogenicity of halogenated alkenes. *CRC Crit Rev Toxicol* 19:23–42.

Lock EA (1989) Mechanism of nephrotoxic action due to organohalogenated compounds. *Toxicol Letters* 46:93–106.

Lock EA (1993) Responses of the kidney to toxic compounds, in *General and Applied Ttoxicology*, Vol. 1 and 2, Ballantyne B, Marrs T, and Turner P (Eds), Macmillan Press, New York, pp. 507–536.

Lynn RK, Garvie-Gould C, Wong K, and Kennish JM (1982) Metabolism, distribution, and excretion of the flame retardant tris(2,3-dibromopropyl)phosphate (Tris-BP) in the rat: Identification of mutagenic and nephrotoxic metabolites. *Toxicol Appl Pharmacol* 63:105–119.

Mahmood NA, Overstreet D, and Burka LT (1991) Comparative disposition and metabolism of 1,2,3-trichloropropane in rats and mice. *Drug Metab Dispos* 19:411–418.

McKinney LL, Picken JCJ, Weakley FB, Eldridge AC, Campbell RE, Cowan JC, and Biester HE (1959) Possible toxic factor of trichloroethylene-extracted soybean oil meal. *J Am Chem Soc* 81:909–915.

Meister A (1988) Glutathione metabolism and its selective modification. *J Biol Chem* 263:17205–17208.

Miller EC, and Miller JA (1966) Mechanisms of chemical carcinogenesis: Nature of proximate carcinogens and interactions with macromolecules. *Pharmacol Rev* 18:805–838.

Moldeus P, Jones DP, Ormstad K, and Orrenius S (1978) Formation and metabolism of a glutathione-S-conjugate in isolated rat liver and kidney cells. *Biochem Biophys Res Commun* 83:195–200.

Monks TJ, Anders MW, Dekant W, Stevens JL, Lau SS, and van Bladeren PJ (1990) Glutathione conjugate mediated toxicities. *Toxicol Appl Pharmacol* 106:1–19.

Monks TJ, Highet RJ, and Lau SS (1988) 2-Bromo-(diglutathion-*S*-yl)hydroquinone nephrotoxicity: Physiological, biochemical, and electrochemical determinants. *Mol Pharmacol* 34:492–500.

Monks TJ, Jones TW, Hill BA, and Lau SS (1991) Nephrotoxicity of 2-bromo-(cystein-S-yl)hydroquinone and 2-bromo-(*N*-acetyl-L-cystein-*S*-yl)hydroquinone thioethers. *Toxicol Appl Pharmacol* 111:279–298.

Monks TJ, and Lau SS (1987) Commentary: Renal transport processes and glutathione conjugate-mediated nephrotoxicity. *Drug Metab Disposs* 15:437–441.

Monks TJ, and Lau SS (1989) Sulphur conjugate-mediated toxicity. *Rev Biochem Toxicol* 10:41–90.

Monks TJ, and Lau SS (1990) Glutathione, γ-glutamyl transpeptidase, and the mercapturic acid pathway as modulators of 2-bromohydroquinone oxidation. *Toxicol Appl Pharmacol* 103:557–563.

Monks TJ, and Lau SS (1992) Toxicology of quinone-thioethers. *Crit Rev Toxicol* 22:243–270.

Monks TJ, and Lau SS (1998) The pharmacology and toxicology of polyphenolic-glutathione conjugates. *Annu Rev Pharmacol Toxicol* 38:229–255.

Monks TJ, Lau SS, Highet RJ, and Gillette JR (1985) Glutathione conjugates of 2-bromohydroquinone are nephrotoxic. *Drug Metab Dispos* 13:553–559.

Nash JA, King LJ, Lock EA, and Green T (1984) The metabolism and disposition of hexachloro-1:3-butadiene in the rat and its relevance to nephrotoxicity. *Toxicol Appl Pharmacol* 73:124–137.

NCI (1986a) Carcinogenesis bioassay of tetrachloroethylene. *National Toxicology Program Technical Report* TR 232.

NCI (1986b) Carcinogenesis bioassay of trichloroethylene. *National Toxicology Program Technical Report* TR 311.

Odum J, and Green T (1984) The metabolism and nephrotoxicity of tetrafluoroethylene in the rat. *Toxicol Appl Pharmacol* 76:306–318.

Okajima K, Inoue M, Itoh K, Horiuchi S, and Morino Y (1983) Interorgan cooperation in enzymic processing, in *Glutathione*: *Storage, Transport and Turnover in Mammals*, Sakamoto Y, Higashi T, and Tateishi N (Eds), Japan Scientific Societies Press, Tokyo, pp. 129–144.

Osterberg RE, Bierbower GW, and Hehir RM (1977) Renal and testicular damage following dermal application of the flame retardant tris(2,3-dibromopropyl)phosphate. *J Toxicol Environ Health* 3:979–987.

Ozawa H, and Tsukioka T (1990) Gas chromatographic separation and determination of chloroacetic acids in water by a difluoroanilide derivatisation method. *Analyst* 115:1343–1347.

Pähler A, Völkel W, and Dekant W (1999) Quantitation of *N* -(dichloroacetyl-L-lysine in proteins after perchloroethene exposure by gas chromatography – mass spectrometry using chemical ionisation and negative ion detection following immunoaffinity chromatography. *J Chromatogr* 847:25–34.

Pearson PG, Omichinski JG, Holme JA, McClanahan RH, Brunborg G., Soderlund EJ, Dybing E, and Nelson SD (1993) Metabolic activation of tris(2,3-dibromopropyl)-phosphate to reactive intermediates. II. Covalent binding, reactive metabolite formation, and differential metabolite-specific DNA damage *in vivo*. *Toxicol Appl Pharmacol* 118:186–195.

Pearson PG, Soderlund EJ, Dybing E, and Nelson SD (1990) Metabolic activation of 1,2-dibromo-3-chloropropane: Evidence for the formation of reactive episulfonium ion intermediates. *Biochemistry* 29:4971–4977.

Peters MM, Lau SS, Dulik D, Murphy D, van Ommen B, van Bladeren PJ, and Monks TJ (1996a) Metabolism of tert-butylhydroquinone to S-substituted conjugates in the male Fischer 344 rat. *Chem Res Toxicol* 9:133–9.

Peters MMCG, Rivera MI, Jones TW, Monks TJ, and Lau SS (1996b) Glutathione conjugates of tert-butyl-hydroquinone, a metabolite of the urinary tract tumor promoter 3-tert-butyl-hydroxyanisole, are toxic to kidney and bladder. *Cancer Res* 56:1006–1011.

Pombrio JM, Giangreco A, Li L, Wempe MF, Anders MW, Sweet DH, Pritchard JB, and Ballatori N (2001) Mercapturic acids (*N*-acetylcysteine *S*-conjugates) as endogenous substrates for the renal organic anion transporter-1. *Mol Pharmacol* 60:1091–1099.

Potter CL, Gandolfi AJ, Nagle R, and Clayton JW (1981) Effects of inhaled chlorotrifluoroethylene and hexafluoropropene on the rat kidney. *Toxicol Appl Pharmacol* 59:431–440.

Rankin BB, and Curthoys NP (1982) Evidence for the renal paratubular transport of glutathione. *FEBS Lett* 147:193–196.

Redegeld FAM, Hofman GA, Loo PGF, Koster AS, and Noordhoek J (1991) Nephrotoxicity of glutathione conjugate of menadione (2-methyl-1,4-naphthoquinone) in the isolated perfused rat kidney. Role of metabolism by γ-glutamyltranspeptidase and probenecid-sensitive transport. *J Pharmacol Exp Ther* 256:665–669.

Reichert D, Ewald D, and Henschler D (1975) Generation and inhalation toxicity of dichloroacetylene. *Fd Cosmet Toxicol* 13:511–515.

Reichert D, and Schuetz S (1986) Mercapturic acid formation is an activation and intermediary step in the metabolism of hexachlorobutadiene. *Biochem Pharmacol* 35:1271–1275.

Rush GF, Smith JH, Newton JF, and Hook JB (1984) Chemically induced nephrotoxicity: Role of metabolic activation. *Crit Rev Toxicol* 13:99–160.

Schrenk D, Dekant W, and Henschler D (1988) Metabolism and excretion of S-conjugates derived from hexachlorobutadiene in the isolated perfused rat kidney. *Mol Pharmacol* 34:407–412.

Simula TP, Glancey MJ, Söderlund EJ, Dybing E, and Wolf CR (1993) Increased mutagenicity of 1,2-dibromo-3-chloropropane and tris(2,3-dibromopropyl)phosphate in *Salmonella* TA100 expressing human glutathione *S*-transferases. *Carcinogenesis* 14:2303–2307.

Söderlund EJ, Gordon WP, Nelson SD, Omichinski JG, and Dybing E (1984) Metabolism *in vitro* of tris(2,3-dibromopropyl)-phosphate: Oxidative debromination and bis(2,3-dibromopropyl)phosphate formation as correlates of mutagenicity and covalent protein binding. *Biochem Pharmacol* 33:4017–4023.

Spencer HC, Rowe VK, Adams EM, McCollister DD, and Irish DD (1951) Vapor toxicity of ethylene dichloride determined by experiments on laboratory animals. *Arch Ind Hyg Occup Med* 4:482–493.

Terracini B, and Parker VH (1965) A pathological study on the toxicity of *S*-dichlorovinyl-L-cysteine. *Fd Cosmet Toxicol* 3:67–74.

Torkelson TR, Sadek SE, Rowe VK, Kodama IK, Anderson HH, Loquvam GS, and Hine CH (1961) Toxicological investigation of 1,2-dibromo-3-chloropropane. *Toxicol Appl Pharmacol* 3:545–553.

Tsao B, and Curthoys NP (1980) The absolute asymetry of orientation of γ -glutamyltranspeptidase and aminopeptidase on the external surface of the rat renal brush border membrane. *J Biol Chem* 255:7708–7711.

Ullrich KJ, and Rumrich G (1988) Contraluminal transport systems in the proximal renal tubule involved in secretion of organic anions. *Am J Physiol 254*:453–462.

Ullrich KJ, Rumrich G, and Kloess S (1988) Contraluminal *para*-aminohippurate (PAH) transport in the proximal tubule of the rat kidney. *Pfluegers Arch* 413:134–146.

Ullrich KJ, Rumrich G, and Klöss S (1989a) Contraluminal organic anion and cation transport in the proximal renal tubule: V. Interaction with sulfamoyl- and phenoxy diuretics, and with β-lactam antibiotics. *Kidney Int* 36:78–88.

Ullrich KJ, Rumrich G, Wieland T, and Dekant W (1989b) Contraluminal para-aminohippurate (PAH) transport in the proximal tubule of the rat kidney. *Europ J Physiol* 415:342–350.

Vamvakas S, Herkenhoff M, Dekant W, and Henschler D (1989a) Mutagenicity of tetrachloroethylene in the Ames-test – metabolic activation by conjugation with glutathione. *J Biochem Toxicol* 4:21–27.

Vamvakas S, Kremling E, and Dekant W (1989b) Metabolic activation of the nephrotoxic haloalkene 1,1,2-trichloro-3,3,3-trifluoro-1-propene by glutathione conjugation. *Biochem Pharmacol* 38:2297–2304.

Völkel W, Friedewald M, Lederer E, Pähler A, Parker J, and Dekant W (1998) Biotransformation of perchloroethene: Dose-dependent excretion of trichloroacetic acid, dichloroacetic acid and *N*-acetyl-*S*-(trichlorovinyl)-L-cysteine in rats and humans after inhalation. *Toxicol Appl Pharmacol* 153:20–27.

Völkel W, Pähler A, and Dekant W (1999) Gas chromatography-negative ion chemical ionisation mass spectrometry as a powerfull tool for the detction of mercapturic acids, DNA, and protein adducts as biomarkers of exposure to halogenated olefins. *J Chromatogr* 847:35–46.

Vore M (1993) Canalicular transport: Discovery of ATP-dependent mechanisms. *Toxicol Appl Pharmacol* 118:2–7.

Vore M (1994) Phase III elimination: Another two-edged sword. *Environ Health Perspect* 102:422–423.

Wallin A, Gerdes RG, Morgenstern R, Jones TW, and Ormstad K (1988) Features of microsomal and cytosolic glutathione conjugation of hexachlorobutadiene in rat liver. *Chem-Biol Interactions* 68:1–11.

Weber GL, and Sipes IG (1990) Covalent interactions of 1,2,3-trichloropropane with hepatic macromolecules: Studies in the male F-344 rat. *Toxicol Appl Pharmacol* 104:395–402.

Weber GL, and Sipes IG (1992) In vitro metabolism and bioactivation of 1,2,3-trichloropropane. *Toxicol Appl Pharmacol* 113:152–158.

Weisburger EK (1977) Carcinogenicity studies on halogenated hydrocarbons. *Environ Health Perspect* 21:7–16.

Winter SM, Weber GL, Gooley PR, Mackenzie NE, and Sipes IG (1992). Identification and comparison of the urinary metabolites of [1,2,3–13C3] acrylic acid and [1,2,3–13C3]propionic acid in the rat by homonuclear 13C nuclear magnetic resonance spectroscopy. *Drug Metab Dispos* 20:665–672.

Wolf CR, Berry PN, Nash JA, Green T, and Lock EA (1984) Role of microsomal and cytosolic glutathione *S*-transferases in the conjugation of hexachloro-1:3-butadiene and its possible relevance to toxicity. *J Pharmacol Exp Ther* 228:202–208.

IV

Risk and Safety Assessment

24

The Role of Epidemiology in Human Nephrotoxicity

Margaret R.E. McCredie and John H. Stewart

INTRODUCTION

The epidemiology of nephrotoxicity is as diverse as its pathophysiology. The circumstances under which exposure occurs, whether a toxicant acts more or less directly on the kidney or activates extrarenal immune or metabolic events that harm the kidney, the interval between exposure and appearance of disease, and the clinical and laboratory manifestations of renal damage, all vary greatly. With stricter controls on the environment, especially the workplace, several noxious agents that once were important are no longer significant causes of renal disease. However, new renal toxicants are coming into prominence, the majority being pharmaceuticals.

This chapter first describes relevant epidemiological methods, defining their terminology, and then outlines the authoritative evidence bearing on some of the principal nephrotoxic exposures and diseases.

EPIDEMIOLOGICAL METHODS

Epidemiology is the study of the distribution and determinants of disease in human populations. As one cannot test suspected toxicants in humans, only epidemiological studies provide definitive conclusions on the role of toxicants in the etiology of human kidney disease. Even so, any such conclusions are based on evidence that, by its very nature, is often circumstantial. For this reason, methodology for gathering and interpreting evidence must adhere to the most stringent guidelines, including the use of statistics.

Markers of Disease

Markers of disease that are frequently used in epidemiology, e.g., serum creatinine concentration, creatinine clearance, the excretion of various urinary proteins, and renal morbidity or mortality, detect renal injury with varying degrees of sensitivity (see Box 24.1). Usually they are specific for kidney disease, but the most common (serum creatinine concentration, creatinine clearance, and renal morbidity or mortality) do not identify the site of injury within the kidney (see Chapter 3). This results in loss of statistical power and allows confounding (see *Measures of Exposure*, below) to occur when a risk factor for another type of renal pathology is associated in some way with exposure to the toxicant under investigation. Hence, it is desirable to use more specific, and generally more expensive and invasive, markers of disease, such as histopathology, imaging, or special laboratory tests. The use of such clinical markers of disease may be justified only in a limited group of subjects, restricting the size and composition of the group to be studied, thereby perhaps compromising the generalizability of the findings.

Population-Based Disease Registration

For population-based epidemiology, which avoids the likelihood of selection bias that is inherent in most clinical series that employ specific

Box 24.1 Attributes of a Diagnostic or Screening Test

	Disease	
Test	*Present*	*Absent*
Positive	*a* (True positive)	*b* (False positive)
Negative	*c* (False negative)	*d* (True negative)

Sensitivity $= a/(a+c)$
Specificity $= d/(b+d)$
Positive predictive value $= a/(a+b)$.

markers of disease, it is necessary to rely on population-based disease registers, either of mortality (death certification) or of incidence. In nephrology, the only morbidity registers generally available are those recording treated end-stage renal disease. All such registers record basic demographic information (age, sex, place of residence), date of death or diagnosis, and, with less accuracy, cause of death or diagnosis of disease.

Registers of End-Stage Renal Disease

Since the 1960s, patients with total and irreversible renal failure, termed end-stage renal disease (ESRD), have been treated by maintenance dialysis or transplantation. The organizational and funding problems that had to be overcome in order to offer these treatments to all suitable patients fostered the establishment of population-based registries in North America, Europe, Japan, and Australasia, with the aim of recording all patients entering dialysis or renal transplantation programs. The most complete register, although small, is ANZDATA, which covers Australia and New Zealand since 1971. ANZDATA does not only record virtually 100% of patients entering ESRD programs, but also has a higher proportion of cases for whom the diagnosis of the disease causing renal failure has been verified (Table 24.1) than do the larger registers.

For epidemiological purposes, the onset of ESRD is taken to be the date of first dialysis or kidney transplantation (if without prior dialysis), usually when the glomerular filtration rate has fallen to 5 to 10 ml/min (i.e., when approximately 90 to 95% of renal function has been lost). There has been no reliable record kept of patients dying with advanced renal failure if not enrolled in an ESRD program; these are chiefly the elderly and those with significant co-morbidity, but will include a disproportionate number of patients who are culturally, socially, economically, or geographically disadvantaged. During the 30 or so years that ESRD programs have been in operation, barriers to treatment have been relaxed. This has resulted in a gradual increase in crude or age-standardized incidence, other things being equal, due to greater acceptance of patients who are old, with significant co-morbidity (especially diabetes), or from indigenous or minority peoples. All of these groups have a high incidence of renal failure (Table 24.1).

The underlying kidney disease can be identified with some certainty in over 90% of cases, based on clinical findings, modern imaging,

Table 24.1 Number of Cases and Age-Standardized Rates of End-Stage Renal Disease in Australia and New Zealand According to Cause[a]

	1982–1991			*1992–2001*		
Cause	*n*	*%*	*Rate[b] (per million)*	*n*	*%*	*Rate[b] (per million)*
Environmental						
Lead	53	0.6	0.24	30	0.2	0.11
Other	7	0.1	0.03	2	0	0.01
Medication, self administered						
Analgesics[c]	1113	12.3	5.22	1022	5.7	3.54
Other	2	0	0.01	7	0	0.03
Medication, prescribed						
NSAIDs	0	0	0	7	0	0.02
Lithium	2	0	0.01	60	0.3	0.23
Cyclosporin	1	0	0	46	0.3	0.18
Radiation	11	0.1	0.05	22	0.1	0.08
Other	7	0.1	0.03	24	0.1	0.08
Other						
Interstitial nephritis[d]	119	1.3	0.56	250	1.4	0.91
Unknown diagnosis	561	6.2	2.52	1094	6.1	3.71
All causes[e]	9055	100	51.24	18,081	100	80.04

[a]When a toxic etiology is given as either the primary or a secondary cause.
[b]Directly age-standardized to the 'world' population (dos Santos Silva, 1999).
[c]Of all patients with this diagnosis, 24% were aged 65 years and over in 1982–91 and 60% in 1992–2001.
[d]Includes cases with a pathological diagnosis of either acute or chronic interstitial nephritis, chiefly of unknown etiology as cases are placed preferentially in other diagnostic categories when the etiology is known or presumed.
[e]Of all patients, 17% were aged 65 years and over in 1982–91 and 34% in 1992–2001.

Source: Australian and New Zealand Dialysis and Transplant Registry (ANZDATA).

and renal biopsy, including immunofluorescence microscopy (Table 24.1). The diagnosis is usually expressed in pathologic terms; often the etiology can be inferred from the pathology, but in certain categories of renal disease, especially interstitial nephritis and, to a lesser extent, glomerulonephritis, the etiology is unknown or can be inferred only from collateral evidence. Investigations to establish the underlying renal disease are more likely to be undertaken in young than in elderly patients. The possibility that clinicians have overlooked a nephrotoxic etiology is especially likely in registers in which the diagnosis is frequently recorded as unknown or is nonspecific (e.g., chronic interstitial nephritis) (Dieperink, 1989). Conversely, chronic renal failure may be ascribed to a toxicant on anecdotal rather than conclusive epidemiological, histopathologic, or imaging evidence.

ANZDATA records suggest that between 1992 and 2001, about 7% of all ESRD in Australia and New Zealand was caused by recognized nephrotoxicants, predominantly antipyretic-analgesics (which were heavily abused in several Australian states prior to 1979) (Table 24.1). Only about 1% of ESRD was caused by other recognized nephrotoxicants, chiefly prescribed drugs. Some of the 1% of cases with interstitial nephritis without specified etiology and the 6% with unknown diagnosis were possibly toxic in origin. Not separately identified in Table 24.1 were cases of glomerulonephritis that were possibly caused by direct or indirect effects of nephrotoxicants, such as drugs, silicon, or organic solvents (see below). Finally, no account has been taken in Table 24.1 of the contributory role of nephrotoxicants (e.g., lead, solvents, silicon, acetaminophen, NSAIDs) to the progression of chronic renal failure due primarily to renal disease that itself is not toxic in origin (see below).

Measures of Disease Frequency

Commonly used measures of "incidence" are given in Box 24.2; comparable measures of "mortality" are calculated in the same way. "Prevalence" records the amount of disease present in a population at a particular time.

Epidemiology often involves comparison of rates for a particular disease between two different populations, or for the same population at different times (dos Santos Silva, 1999). For such comparison, "crude rates" can be misleading because of variations in the age structure of the populations when, as with most diseases, the incidence varies greatly at different ages. Age-standardization, using either the "direct" or the "indirect" method, allows for changing or different age structure. A "directly age-standardized rate" is a theoretical rate that would have occurred if the observed age-specific rates applied in a reference or standard population. The alternative, "indirect age-standardization,"

Box 24.2 Measures of Disease Frequency

Measure	*Definition*
Number	New cases in defined period, indicates public health burden
Population at risk	Generally approximated by mid-year population obtained from census
Incidence rate	Frequency of new cases occurring in a defined population of disease-free individuals in a specified period of time
Crude incidence rate	Total number of new cases divided by population at risk
Age-specific incidence rate	Number of new cases in particular age group divided by population at risk in that age group (usually determined separately for males and females)
Age-standardized incidence rate	Enables comparison of rates in populations with different age structures (necessary when incidence varies greatly throughout life) (see text)
Cumulative rate	For ages 0–74 years approximates lifetime risk of developing the disease; the sum of the annual age-specific rates, often expressed as a percentage

compares observed and expected numbers of cases; the expected number of cases is calculated by applying a standard (reference) set of age-specific rates to the population of interest. The observed to expected ratio is generally expressed as a percentage by multiplying by 100, giving the "standardized incidence ratio" or "standardized mortality ratio" (SIR or SMR) (dos Santos Silva, 1999).

Measures of Exposure

Sources of exposures have been categorized as part of the general environment (e.g., air, soil, water), local environment (e.g., home, workplace, recreational sites), or personal environment (e.g., recreational and medicinal drugs, diet) (Armstrong et al., 1994). In analytical epidemiology, measurements are made of exposures that are known or suspected to be linked with disease. However, the suspected agent may be only a contributing, rather than the sole, etiologic factor; there may be a long latent period between exposure and appearance of disease; there may be no measurable indicator of exposure, especially if not recent; or the measurement may be imprecise. In the absence of measurable markers of exposure, reliance must be placed on medical, occupational, or other records, or on the memory of the subjects of the study (the latter being especially prone to error). Sources of error in the measurement of exposures include: inaccurate job-exposure matrices, incomplete occupational histories, unrepresentative samples in respect of place or time (e.g., of an environmental pollutant), or whether the exposed individual was unaware of the exposure.

Each exposure will leave a trace, although this may be transient rather than permanent (Armstrong et al., 1994). The agent itself, its metabolic products, or biologic effects resulting from contact with the agent, may be measured in body fluid or tissue samples (e.g., blood lead, urinary cadmium, tissue levels of trace metals by x-ray fluorescence, urinary *N*-acetyl-*p*-aminophenol, antibodies to coronavirus, DNA adducts with ochratoxin A or aristolochic acid, or etiologically specific histopatholgic appearances). Few nephrotoxicants cause a renal lesion that is sufficiently characteristic in itself for identification of the etiology with certainty. However, a specific pathology for which an association with a suspected toxicant has been validated (e.g., renal papillary necrosis in an analgesic habitué, nephrogenic diabetes insipidus in a psychiatric patient, or chronic interstitial nephritis in a resident of a village where Balkan endemic nephropathy is prevalent) provides stronger epidemiological evidence than when a nonspecific marker of disease (such as raised serum creatinine concentration or chronic renal failure/ESRD of any type) is employed.

Relating Exposure to Disease

When evaluating a cause-and-effect hypothesis it is necessary to decide whether an association found between an exposure and a disease is valid (Angell, 1990). Alternative explanations include chance (which is quantifiable statistically), "bias" and "confounding." Bias results from systematic error in the way that subjects (cases or controls) were selected, or the information about the exposure or the disease was obtained. Confounding, in which a third factor is linked to both the exposure and the disease (e.g., septicemia in aminoglycoside treated patients who subsequently develop acute renal failure; earlier phenacetin consumption when investigating the role of acetaminophen in patients with analgesic nephropathy), may produce a spurious result.

A valid exposure–disease relationship having been established, criteria are applied to judge causality. These include: proof that exposure preceded disease, the strength of the association, that risk increased with exposure (dose–response relationship), the specificity of the association, consistency with the findings of other independent investigations, and biologic plausibility.

Aspects of personal behavior or lifestyle, environmental exposures, or inherited characteristics that are shown to be associated with an increased risk for disease, are designated "risk factors," a term which can be used in the absence of experimental proof of causality.

Study Design

Essentially, analytical epidemiology collects information about both disease and exposure in individuals in such a way as to prove or disprove causal relationships. In cross-sectional, case-control, and cohort studies, the investigator examines the relationship between disease and exposure but does not intervene. In a clinical trial, the investigator observes the effect of an intervention.

Cross-Sectional Studies

Cross-sectional studies, the least rigorous of the alternatives, are often employed to investigate occupational hazards. The frequency and level of exposure and of disease in the subject population are measured *at a particular point in time* to quantify disease prevalence according to type and amount of exposure. The study design should include gathering information on possible confounders. Such studies cannot prove causal relationships without collateral evidence that exposure preceded disease. Moreover, as cross-sectional studies are based on prevalent rather than incident cases, patients with a short duration of disease (because of rapid recovery or death) are less likely to be included. Cross-sectional studies may lack power to implicate rare conditions or rare exposures.

Case-Control Studies

In these studies, persons selected because they have a specific disease (cases) are compared with like persons who do not have the disease (controls), with regard to retrospectively obtained (usually by means of a questionnaire or by reviewing medical or occupational records) estimates of exposure (to one or more risk factors suspected of being causal) and potential confounders. Preferably, incident cases (rather than prevalent cases, which may introduce survival bias) are ascertained through a disease register (which should be population-based) and controls ascertained through a population register (such as an electoral roll). Selection bias may occur when either cases or controls are identified from lists of patients in hospitals or clinics, when volunteers are used or inducements offered, or when more than a small proportion of subjects fails to participate.

Case-control studies are considered inferior to the cohort design because of the possibility of bias, particularly in selection of controls and differential recall of past exposures by cases and controls; the latter can be reduced by the use of an objective measure of exposure (e.g., a biologic marker), especially if the observer is "blind" to disease status. Case-control studies are much less costly than cohort studies, and are more flexible, for example when studying several exposures or measuring confounding factors. As case-control studies inevitably entail some error (due to bias or misclassification) they are not suitable for investigating rare exposures, unless that exposure is responsible for a large proportion of the cases.

The results of several large case-control studies designed to determine whether volatile solvents, silicon, antipyretic-analgesics other than those containing phenacetin, or the newer NSAIDs, cause chronic renal failure, have been inconclusive. This has been due to the use of nonspecific markers of disease, not taking into account potential confounders, or attempting to investigate relatively infrequent or low-level exposures that account for no more than a small proportion of the cases.

Cohort Studies

Cohort (or follow-up) studies are more rigorous than cross-sectional or case-control studies. They start with a defined population for which exposure information is collected; the group is then followed forward in time to ascertain which persons develop the disease. Examples include cohorts of workers for whom employment records, measures of the work environment, or biological markers are used to assess exposure to occupational hazards. Subsequent mortality or, less often, morbidity from the disease in question is then determined. When only a few of the exposed are likely to develop the disease, even large cohorts may lack the power to show an exposure–disease relationship (Churchill et al., 1983). Cohort studies are expensive and time-consuming; moreover, only prospectively identified risk factors can be investigated. Sources of error include: the possibility of a "healthy worker effect" (dos Santos Silva, 1999; Nuyts et al., 1993), changes in exposure status and diagnostic criteria over time that may alter the classification of individuals, absence of information about alternative risk factors or confounders, bias in the detection or ascertainment of disease introduced by knowledge of exposure status (avoided by "blinding" the observer to exposure status) or when a significant proportion of the cohort is lost to follow-up, and the accuracy of the marker of disease (often death certification). The principal reason for the high epidemiological standing of cohort studies is that exposure is measured before disease onset and so is unlikely to be biased. Also, the cohort can be selected so as to investigate the effect of a rare exposure.

Clinical Trials

In a clinical trial, subjects are allocated to one of at least two alternative treatments (which may include a placebo) and followed to determine specific outcomes, such as efficacy or toxicity. Confounding is minimized in a sufficiently large study by truly random allocation of subjects to treatment or control groups. Bias is addressed by random allocation, employing objective markers of outcome, and concealing treatment type from both patient and observer ("double blinding").

The randomized, double-blind, clinical trial, the closest that clinical medicine comes to the laboratory experiment, is invaluable for clarifying associations and judging the relative benefits of interventions. Although selection bias, information bias, and confounding are minimized, it may be difficult to ensure compliance and avoid contamination (by the employment of treatments not covered by the trial design) when the trial is large or prolonged. Clinical trials are designed to evaluate both the effect of the intervention on the disease and its toxicity, but they lack power to identify infrequent side effects. The generalizability of the results of a clinical trial depends upon how representative the trial subjects were of the general run of patients with the disease being treated.

Measures of Exposure–Disease Association

Traditional (population health) epidemiological research is designed to elucidate exposure–disease relationships in representative groups. This involves the use of relative measures of association (Box 24.3). In contrast, clinical epidemiology focuses on individuals, and generally requires absolute measures of association.

Statistical Measures: Relative Risk, Odds Ratio, and Attributable Risk

The "relative risk" (RR) of disease associated with an exposure can be calculated in a cohort study by comparing the proportion that develops the disease (the incidence rate) in the exposed group with that in the non-exposed group (dos Santos Silva, 1999). If RR differs from unity then the exposure is associated with the risk of disease; positive if $RR > 1$, negative if $RR < 1$ (indicating that the factor is "protective").

Case-control studies primarily calculate rates of exposure rather than rates of disease. Nevertheless, a case-control study can provide a close estimate of the relative risk associated with an exposure, the "odds ratio" (OR) (see Box 24.3), provided that the disease is comparatively rare and the controls are recruited from the population by a rigorous sampling scheme (i.e., not influenced by the likelihood of being exposed).

The RR and OR are generally reported together with a 95% confidence interval (CI), the range within which the true population risk would lie (with 95% probability). If the 95% CI excludes unity, this is interpreted to mean that the estimate is statistically significant. When several estimates have been calculated from one set of data, a more stringent criterion (e.g., 99% or 99.9% CI) is appropriate.

By combining information on the distribution of exposures with estimates of relative risk, it is possible to calculate the proportion of cases in the population explained by the exposure: the "population

Box 24.3 Measures of Disease–Exposure Relationship

In a cohort study:

Relative risk (RR) of disease associated with an exposure

$$= \frac{\text{rate of disease in exposed group}}{\text{rate of disease in unexpected group}}$$

In a case-control study:

Odds ratio (OR), an estimate of the relative risk,

$$= ad/bc$$

where:

	Number of	
	Cases	*Controls*
Exposed	*a*	*b*
Non-exposed	*c*	*d*

Population attributable risk (PAR), the proportion of cases occurring in the population that can be explained by the exposure, is given by:

$$= \frac{p(r-1)}{pr + (1-p)}$$

where: p is proportion of exposed persons in the population (given by $b/(b+d)$) and r is the odds ratio.

attributable risk" (see Box 24.3). Unlike relative measures of association, the attributable risk is affected substantially by the underlying incidence or prevalence of the disease resulting from all other causes, and so may not be the same in different populations, even when exposures are identical.

EPIDEMIOLOGICAL EVIDENCE IMPLICATING INDIVIDUAL NEPHROTOXICANTS

Perhaps the most logical classification of nephrotoxicants, from the epidemiological perspective, is according to the type of exposure.

Environmental (often occupational) and medicinal (self-administered or prescribed) are the major categories used in this chapter. The epidemiological issues will differ according to the relationship between exposure and disease: whether immediate or delayed; deterministic (as in most chemically-induced toxicity) or stochastic (as in carcinogenicity); direct (when the toxicant or one of its metabolites itself causes the renal damage) or indirect (when the injury is caused by an immunologic, e.g., glomerulonephritis or acute interstitial nephritis, or metabolic, e.g., tumor lysis, sequences of events initiated by the toxicant). The classification preferred by biologists and clinicians, namely according to site and type of kidney pathology, is less useful for epidemiology.

When the exposure is to a single substance, the pathogenesis deterministic, and the interval between exposure and disease short (as is the case with most prescribed nephrotoxicants), epidemiology has little role in proving etiology, but it may contribute to defining safe levels of exposure or evaluating protective strategies. Conversely, epidemiology, often sophisticated, may be necessary to prove a suspected cause–effect relationship if the latent interval is long or the onset of disease insidious, if the exposure–disease relationship is indirect or stochastic, or if the suspected nephrotoxic exposure comprises more than one component.

In the following sections, eight nephrotoxicants are used to illustrate epidemiological issues relating to identification and prevention of nephrotoxicty. Neither carcinogenicity (see Table 24.2) nor the nephrotoxic consequences of infection have been included.

Environmental Nephrotoxicants – Metals

Lead

That heavy exposure to lead early in life can cause later chronic renal failure was established by astute clinical research in the Australian state of Queensland (Henderson 1954, 1955; Nye, 1953). The observation that lead content of bone was twice as high in Queenslanders who died of chronic renal failure with presumed lead nephropathy as in three control groups (Henderson and Inglis, 1957) prompted the introduction of the lead mobilization test (in which urinary lead excretion is measured following injection of EDTA) as a marker of distant exposure to lead (Emmerson, 1963). This test has since been used in cross-sectional studies to associate lead with renal impairment in occupationally exposed persons (Weeden et al., 1975, 1979), patients with gout

Table 24.2 Implication of Chemical and Physical Carcinogens in Cancers of the Kidney and Urinary Tract

Risk factor	*Odds ratio or risk ratio*	*Best evidence for cause–effect relationship*
Renal parenchyma[a]		
Tobacco	≈2	Consistent population-based case-control studies; dose-response (McLaughlin et al., 1995)
Thorotrast	?	Case reports
Urothelium		
Tobacco	3–6[b,c] 2–3[d]	Consistent population-based case-control studies; dose-response (McCredie and Stewart, 1992; McCredie et al., 1983; McLaughlin et al., 1983; Piper et al., 1986)
Phenacetin-containing analgesics	5–12[b] 3–6[d]	Consistent population-based case-control studies; dose-response (McCredie et al., 1983, 1993; Piper et al., 1986)
Balkan nephropathy	29–57[c] 4–12[d]	Cross-sectional studies (Cukuranovic et al., 1991; Nikolov et al., 1996)
Ochratoxin A		DNA adducts in Balkan nephropathy-related tumors (Nikolov et al., 1996)
Thorotrast	?	Case reports
Aristolochic acid[c]	?	Cross-sectional studies; dose response (Nortier et al., 2000); DNA adducts (Schmeiser et al., 1996)

[a]Epidemiology has failed to substantiate the suspected role for phenacetin-containing analgesics, asbestos and cadmium.
[b]For renal pelvis.
[c]For renal pelvis and ureter.
[d]For bladder.

(Batuman et al., 1981), or with supposedly essential hypertension (Batuman et al., 1983). X-ray fluorescence is now preferred for measuring bone lead (Hu et al., 1990).

Environmental lead exposures, mainly from air, soil, food, and water, are falling (Pirkle et al., 1998); the epidemiological challenge now is to ascertain whether current levels are a cause of significant kidney disease in the general population. Two cross-sectional studies have shown a correlation between blood lead (which reflects recent exposure) and reduced creatinine clearance (Staessen et al., 1992) or raised serum creatinine (Kim et al., 1996), and a third found similar results in men but not women (Staessen et al., 1990). A fourth study was negative (Pocock et al., 1984). Neither positive study was representative of the general population. One deliberately drew half the participants from two cadmium-polluted areas, and had an unacceptably low (39%) response rate from two of their four population samples (Staessen et al., 1992), while in the other, which recruited volunteers, selection bias is evident from its racial composition (Kim et al., 1996). There is at most a small effect of low-level lead exposure on renal function.

However, in a carefully conducted observational study of moderate chronic renal failure not caused primarily by lead exposure, kidney function worsened more quickly in patients with a high-normal blood lead than in those with low-normal levels. Patients from the high-normal blood lead group who were then randomly allocated to lead-chelating therapy with EDTA preserved their remaining renal function a little better than those not so treated (Lin et al., 2001). These observations, yet to be confirmed independently, suggest that lead burden may contribute to progression of renal failure of any primary etiology.

Cadmium

Cadmium has infrequently been blamed for clinically overt kidney pathology, except for "itai-itai" disease, in which severe osteomalacia, osteoporosis, and hypercalciuria resulted from proximal tubular dysfunction of the Fanconi type in a heavily polluted region of Japan (Kido and Nordberg, 1998). However, the probability that cadmium causes insidious and irreversible chronic renal damage in those living or working in contaminated areas has generated large cross-sectional studies in Belgium (CadmiBel) and Sweden. Unlike lead, from which the nervous system is more at risk than the kidney, nephrotoxicity and associated bone disease constitutes the earliest and most serious consequence of long-term exposure to cadmium (Nordberg, 1992).

The most valuable outcomes of the CadmiBel project have been to define the sensitivity of markers of disease and the validity of markers of exposure.

Markers of Exposure

Blood levels of cadmium chiefly reflected recent (i.e., in the previous few months) cadmium uptake (Lauwerys et al., 1979), except when exposure had ceased (Jarup et al., 1997). Conversely, urinary excretion of cadmium or metallothionein (a carrier protein that preferentially binds cadmium) reflected cumulated body burden, except when there was overt renal injury, in which case urinary cadmium and metallothionein excretion rose disproportionately due to inability of the damaged kidney to conserve the metal–protein complex (Chang et al., 1980; Roels et al., 1981). Similarly, the concentration of cadmium in the renal cortex *in vivo*, measured by neutron activation analysis, rose with duration of exposure (and therefore with body burden) to a maximum of 200 to 300 mg/kg, then declined with the onset of renal dysfunction (Roels et al., 1983). The finding that urinary cadmium and metallothionein so closely reflect cadmium uptake by the kidney validates their use as markers of exposure in epidemiological studies, and for monitoring workers at risk prior to the onset of overt renal damage.

Markers of Disease

Markers of early renal damage were investigated in a cross-sectional study of Belgian workers, half of whom had been heavily exposed to cadmium, predominantly by inhalation (Roels et al., 1993). Subjects exposed to other nephrotoxicants or with kidney disease of other cause were excluded. Urinary 6-keto-$PGF_{1\alpha}$ and sialic acid rose above the threshold level (the 90th percentile in the reference group) when urinary cadmium exceeded 2 μmol cadmium/mol creatinine. Some proximal tubular enzymes or antigens, albumin and transferrin appeared at about 4 μmol cadmium/mol creatinine, and other proximal tubular enzymes or antigens, markers of tubular proteinuria, and glycosaminoglycans appeared at about 10 μmol cadmium/mol creatinine. Tubular proteinuria, but not raised excretion of the other biochemical markers, had been shown to be associated with mild irreversible and progressive renal impairment (Roels et al., 1989, 1991). Accordingly, 10 μmol cadmium/mol creatinine was recommended as the threshold urinary concentration above which workers must be removed from exposure.

A parallel study of persons exposed in the domestic environment, and therefore taking in cadmium principally by ingestion, indicated that the threshold urinary cadmium concentration for significantly abnormal renal function was less than half of that found in the cadmium workers (Bernard et al., 1992). However, a follow-up of the rural participants in the CadmiBel study (i.e., exposed chiefly domestically) found no association between decreased renal function and increased body burden of cadmium (Hotz et al., 1999).

Swedish research also found that dose-related proximal tubular damage, with associated reduction of glomerular filtration rate, persisted long after exposure ceased (Jarup et al., 1995, 1997), but indicated a lower threshold (urinary cadmium 1 μmol/mol creatinine) at which significant kidney damage occurred (Jarup et al., 2000). Moreover, the incidence of ESRD rose with level of exposure in a Swedish population working in, or living within 2 km of, a cadmium battery plant (Hellstrom et al., 2001). Although suggestive, this evidence falls short of proof in view of the nonspecific marker of disease (renal failure of any type) and because of the possibility of confounding inherent in the study design. Conversely, a supposedly negative Belgian case-control study of patients with chronic renal failure lacked statistical power (Nuyts et al., 1995b).

The evidence is inconclusive but suggests that cadmium rarely causes significant renal disease unless exposure is both heavy and prolonged.

Environmental Exposures Possibly Causing Glomerulonephritis

Apart from glomerular syndromes of metabolic or vascular origin, the pathogenesis of glomerular injury nearly always involves an immune process. All such diseases are grouped as glomerulonephritis. While there is good understanding of the relationship of the immunologic derangement to the histologic and immunopathologic appearances (which are used for diagnostic classification), rarely is the initiating etiology identified, be it a specific antigen or an agent that alters either the activity of the immune system or the susceptibility to immune surveillance. It is known that at least some cases of glomerulonephritis are caused by external exposures, for example to drugs (NSAIDs, gold, penicillamine, heroin) or infective agents (Streptococcus, hepatitis B, HIV). However, it has been suspected increasingly in recent decades that occupational and other environmental toxicants may also be implicated. Apart from mercury, for which there is persuasive anecdotal and clinical evidence, most suspicion falls on organic solvents (chiefly hydrocarbons) and silicon compounds. For neither is there any conclusive clinical, biologic, or experimental proof, and epidemiological investigations have been hampered by not using objective markers of exposure or specific markers of disease.

Volatile Hydrocarbons and Organic Solvents

The difficulty of conducting methodologically sound epidemiological studies is illustrated by the efforts, published since 1975, to prove an association between glomerulonephritis and exposure, chiefly occupational, to volatile hydrocarbons or organic solvents. Prior to 1980, six case-control studies had been published, five showing a positive association. All but possibly one (Ravnskov et al., 1979) of those studies were deemed to be unsatisfactory (Churchill et al., 1983).

Since then, there have been a further 10 studies, two using prevalent cases of ESRD due to glomerulonephritis, hypertension, or interstitial nephritis (Steenland et al., 1990), or chronic renal failure of any primary cause (Nuyts et al., 1995b). Both found positive associations, but for different categories of solvent. One study implicated "solvents used as cleaning agents or degreasers" (i.e., halogenated hydrocarbons) (OR = 2.50, 95% CI = 1.56 to 3.95) (Steenland et al., 1990), while the other implicated "oxygenated hydrocarbons" (i.e., paint removers, varnishes, glues) (OR = 5.3, 95% CI = 1.8 to 6.2) (Nuyts et al., 1995b). In the latter study, the association was no stronger for glomerulonephritis than for other renal diseases. In both studies, the participation rate of the population-based controls was unacceptably low (50 to 61%); in neither was there a convincing relationship between dose and response.

Two case-control studies enrolled patients from consecutive hospital series of biopsy-proven glomerulonephritis, excluding cases associated with systemic disease. In one, patients with proliferative glomerulonephritis (other than mesangial IgA or mesangio-capillary glomerulonephritis) had a mean exposure, chiefly occupational, to solvents that was more than four times that of matched hospital controls ($p < .01$) (Bell et al., 1985). The other, which included all types of primary glomerulonephritis, showed a rising ratio of observed to expected cases when exposure was ranked according to occupation as recorded at a census ($p = .001$) (Ravnskov et al., 1983). In this study, there was a difference in the ages of the cases and controls, and the controls (but not the cases) excluded persons not working for at least 1 hour/week.

The remaining six case-control studies must be interpreted with caution because of flawed methodology, i.e.: unblinded interviewers ($n=3$), non-expert and unblinded allocation of exposure status ($n=2$), use of hospital controls ($n=4$) or a method of selecting population-based controls open to bias ($n=1$), unacceptably low ($n=2$) or unstated ($n=1$) participation rates by controls, the inclusion of retrospectively identified (and therefore prevalent rather than incident) cases ($n=4$), and not all subjects accounted for in the analysis ($n=1$). Four studies showed a positive association for glomerulonephritis (Harrison et al., 1986; Porro et al., 1992; Yaqoob et al., 1992) or for nephropathy in Type I diabetes (Yaqoob et al., 1994). One study was negative (Harrington et al., 1989), and one found an association only in male patients and after *post hoc* subdivision of cases by severity of disease and exposure (Stengel et al., 1995).

Superficially, the epidemiological evidence appears to implicate occupational volatile hydrocarbon and organic solvent exposure, but there is no consistency between studies in respect of the type of hydrocarbon compounds incriminated or the type of glomerulonephritis resulting, nor has glomerulonephritis, when investigated, been more frequently involved than other primary renal diseases.

Silicon

The epidemiological evidence indicting compounds of silicon falls into four classes. Three cross-sectional studies (Boujemaa et al., 1994; Hotz et al., 1995; Ng et al., 1992) attested that workers with current heavy exposure or with silicosis had increased urinary albumin, *N*-acetyl-β-D-glucosaminidase and either α-1-microglobulin or retinol binding protein, indicative of proximal tubular dysfunction rather than glomerulopathy. A fourth showed doubling of the risk for raised serum creatinine in patients with silicosis (Rosenman et al., 2000).

Of four cohort studies, one that enrolled 16,661 workers in 17 manmade mineral fiber plants in the U.S. found a standardized mortality ratio of 146 ($p<.01$) for deaths (observed $n=56$) from "nephritis and nephrosis" (International Classification of Diseases, 8th Revision (ICD-8) 580 to 584, which includes glomerulonephritis, interstitial nephritis, and renal sclerosis) in the period 1946 to 1985 (Marsh et al., 1990). This study is convincing because of the negligible rate of loss to follow-up (2.4%) or missing death certificates (2.2%), and the consistency of the raised SMR in each of the three periods of follow-up (1946 to 1977, 1978 to 1982, and 1983 to 1985), but it does not identify the types of nephritis involved. The other three cohort studies showed a threefold to fourfold excess risk for ESRD due to: glomerular or interstitial nephritis in gold miners (Calvert et al., 1997), any type of renal disease in ceramic workers (Rapiti et al., 1999), and from any type of renal disease (and specifically glomerulonephritis) in industrial sand workers (Steenland et al., 2001). The latter study showed a convincing dose–response relationship.

Two large case-control studies incriminated occupational exposures (assessed from questionnaires) in 325 men with ESRD due to glomerulonephritis, interstitial nephritis, or nephrosclerosis (Steenland et al., 1990) and in 272 men and women with chronic renal failure of any etiology (Nuyts et al., 1995b). Both found significantly raised odds ratios for any occupational exposure to silicon compounds (1.67 and 2.51, respectively), but each to a different subgroup (brick and foundry = 1.92, grain dust = 2.96). In the study in which the case group was subdivided according to type of renal disease, there was no greater likelihood of patients with glomerulonephritis having been exposed to silicon than of those with any other renal disease (Nuyts et al., 1995b).

Three case-control studies specifically sought evidence for occupational exposure to silicon in hospital-based series of patients with pauci-immune vasculitic glomerulonephritis defined in different ways (Gregorini et al., 1993; Nuyts et al., 1995a; Stratta et al., 2001). While all yielded significantly positive results, numbers were small, selection of cases may have been biased by inclusion of the index cases, and in each the control group was not ideal (hospital patients for two, population-based in the other but with only a 50% participation rate).

Therefore, while silicon-containing compounds do appear to cause chronic kidney disease or to contribute to its progression, their implication in the etiology of glomerulonephritis in general, or of pauci-immune renal vasculitis in particular, is as yet based on suggestive rather than conclusive evidence.

Environmental Exposures Where the Nephrotoxicant is Unidentified

Balkan Endemic Nephropathy

Given its striking geographic distribution, high incidence in affected communities, and the possibility of early detection employing a sensitive and specific marker of disease (Hrabar et al., 1991), it would be expected that epidemiology would have identified the etiology of Balkan endemic nephropathy. Although several promising avenues have been explored, none has been proved to account for this disease, and most now seem unlikely (Table 24.3). Any credible etiologic hypothesis must explain the high association of urothelial cancer with Balkan endemic nephropathy.

Apart from a single cohort study that showed β_2-microglobulinuria predicted subsequent nephropathy (Hrabar et al., 1991), the epidemiological studies have been cross-sectional or observational, and often with small sample sizes or not designed to allow differentiation of the effect of the risk factor under investigation from that of other, co-occurring, environmental factors. Some merely show that the geographic distribution of the putative risk factor matches that of the disease (Feder et al., 1991; Puntaric et al., 2001)

The best epidemiological evidence on this subject supports, but falls short of proving, a role for ochratoxin A, a common mutagenic mycotoxin believed to cause porcine nephropathy, in the etiology of Balkan endemic nephropathy and the associated urinary tract cancers. Cross-sectional studies have shown that blood or urine levels of ochratoxin A were higher in patients than in control subjects living in non-endemic areas, with intermediate levels in relatives of patients, other persons living in "endemic" villages, or persons living in nearby "non-endemic" villages (Nikolov et al., 1996; Petkova-Bocharova and Castegnaro, 1991; Radic et al., 1997; Stoev et al., 1998). They also have been generally higher in the Balkan endemic regions, especially Bulgaria, than elsewhere in Europe (Hald, 1991). The finding that ochratoxin A–DNA adducts were present in endemic nephropathy-related urotheliomas, but not in similar tumors from persons without endemic nephropathy, nor in non-cancerous tissues, is further evidence incriminating this mycotoxin (Nikolov et al., 1996). However, the studies published to date have rarely investigated, and therefore not ruled out, a role for other mycotoxins that co-occur with ochratoxin A.

Self-Administered Medicinal Nephrotoxicants

Antipyretic-Analgesics

From the perspective of nephrotoxicity, a distinction must be drawn between the older antipyretic-analgesics (salicylates, phenacetin and

Table 24.3 Epidemiological Evidence for the Various Etiologies Proposed for Balkan Endemic Nephropathy

Risk factor	*Epidemiological evidence in support*	*Biological plausibility*	
		Chronic interstitial nephritis	*Urothelioma*
Heavy metals/trace elements	Weakly supported by geographical distribution (Long et al., 2001)	Yes	Possible
Viruses, e.g., papilloma, coronavirus, Hantavirus, West Nile	Limited cross-sectional study of prevalence of antibodies to papilloma virus and coronavirus, not validated (Nastac et al., 1984; Uzelac-Keserovic et al., 1999); isolation of coronavirus from affected tissue (Uzelac-Keserovic et al., 1999)	Yes	Weak
Ochratoxin A	Cross-sectional studies showing higher exposure in endemic villages or in patients (Petkova-Bocharova and Castegnaro, 1991; Radic et al., 1997); DNA adducts in related tumors (Nikolov et al., 1996)	Yes	Yes
Other mycotoxins	Cross-sectional studies indicating exposure to citrinin (Petkova-Bocharova and Castegnaro, 1991; Vrabcheva et al., 2000)	Yes	Possible
Selenium deficiency	Geographical distribution (Maksimovic, 1991; Long et al., 2001)	No	Possible
Pliocene lignite products in water	Geographical distribution (Feder et al., 1991; Tatu et al., 1998)	No	Probable
Genetic predisposition	Conflicting case reports, family studies, genetic epidemiology (Ceovic et al., 1985; Toncheva et al., 1998)	Possible	Yes
Aristolochic acid	Limited, unconfirmed evidence of exposure (Ivic, 1970)	Yes	Yes

acetaminophen/paracetamol, pyrazolones such as phenazone/antipyrin) which have little or no pharmacological activity or acute toxicity in the kidney, apart from the rare instance of tubular necrosis due to acetaminophen in high dosage (Björck et al., 1988), and the newer nonsteroidal anti-inflammatory drugs (NSAIDs), whose renal toxicity, often acute, is largely due to potent inhibition of renal cyclooxygenase.

Phenacetin-Containing Compound Analgesics

"Classical" analgesic nephropathy is an insidious, chronic disease, only proven to have occurred after prolonged (averaging about 20 years) heavy (averaging about five tablets or powders every day) consumption of analgesic-antipyretic preparations that contain phenacetin together with either aspirin or a pyrazolone, and also caffeine (usually), codeine, or another habituating agent; "2+" analgesics, as defined by Elseviers and De Broe (1995). Because of its characteristic renal lesion (papillary necrosis, often calcified, associated with chronic interstitial nephritis that relatively spares the columns of Bertin), analgesic nephropathy can be diagnosed with a high degree of specificity by contrast pyelography when kidney function is good (McCredie et al., 1982) or by noncontrast CT scans, which can be read irrespective of kidney function (Elseviers et al., 1995).

Epidemiological proof that heavy consumption of 2+ analgesic preparations containing phenacetin caused analgesic nephropathy was provided from two sources. A case-control study assessed exposure by means of a questionnaire and used clinic controls, but the marker of disease (presence of renal papillary necrosis) was relatively specific (McCredie et al., 1982). The OR was 18 (95% CI = 9 to 35) with a convincing dose-response. A cohort study used a specific marker of exposure (*N*-acetyl-*p*-aminophenol in randomly collected urine) but relied upon nonspecific markers of disease. The markers of disease that were used were: abnormal renal function (RR = 8.1; 95% CI = 2.8 to 23), mortality from all causes (RR = 2.1; 95% CI = 1.5 to 3.1), and mortality from unspecified urologic and renal disease (RR = 12.5; 95% CI = 3.2 to 48, with a convincing dose–response relationship) (Dubach et al., 1983, 1991). Salicylate consumption, which also had been monitored by regular urinary testing in this cohort, conferred no increase in risk.

Invariably, when phenacetin-containing 2+ analgesics have been banned, the age-specific incidence of ESRD due to analgesic nephropathy has fallen, immediately in younger persons, and after a delay in older patients (Gault and Barrett, 1998; Michielsen and De Schepper, 2001; Stewart et al., 1994). In younger patients, the immediate benefit may to some extent have been due to increased public awareness of the dangers of analgesic abuse, better medical care of non-end-stage analgesic nephropathy, or to the earlier removal of phenacetin from some popular brands of 2+ analgesics. The delay in older persons was more apparent than real, additional new cases being accounted for, in part, by the gradual relaxation of restrictions upon entry into dialysis programs, and in part to autonomous progression of renal disease despite exposure having ceased.

Analgesics Other Than Phenacetin

The epidemiological evidence considered above was obtained in populations where analgesic abuse was common, intake huge, and nearly all heavy consumers had at some time taken phenacetin-containing 2+ analgesics. Whether analgesics other than phenacetin, alone or in combination, cause analgesic nephropathy or chronic renal failure can be determined only in populations not heavily exposed to phenacetin at any time, or by incorporating design controls that ensure that any contribution of phenacetin is taken into account in the analysis. Moreover, rarely have analgesic preparations that do not contain phenacetin been taken in the quantities, or over such a period of time, as was phenacetin in former times. It also must be kept in mind that case-control studies are subject to recall bias, and are not suitable for investigating rare or low-level exposures unless these are responsible for a high proportion of cases, while cohort studies have not always recorded all relevant prior exposures.

Of six case-control studies that have examined the role of the various commonly taken antipyretic-analgesics, retrospectively in hospital- or population-based series of patients with chronic or end-stage renal failure, one was negative (Murray et al., 1983), and two made no adjustment for phenacetin consumption (Pommer et al., 1989; Steenland et al., 1990). Two others, which included cases with ESRD irrespective of primary renal diagnosis and adjusted for phenacetin intake, found significantly raised ORs of about 2. One implicated salicylates, with or without pyrazolones, but there was no dose–response relationship and hospital controls were used (Morlans et al., 1990), and the other implicated acetaminophen, but only for the highest levels of intake, which were "more often than one pill per day" or "more than 5000 pills in total" (Perneger et al., 1994). In neither was there specificity for any type of kidney disease. In the American study (Perneger et al., 1994), the population controls differed from the patient group markedly in respect of race and sex, and there was no attempt to distinguish analgesic consumption before from consumption after onset of renal disease.

The sixth case-control study, which enrolled all patients with newly diagnosed chronic renal failure, except when a systemic or familial condition was responsible, was conducted in a region of the U.S. from which analgesic abuse and analgesic nephropathy have been reported (Sandler et al., 1989). It exonerated aspirin and implicated phenacetin-containing mixtures (with a dose–response relationship), but suggested that only the highest level of intake (daily use for at least 1 year) of acetaminophen might be harmful. (Even in this group, consumption was low by Australian or European standards.) The results (OR = 3.21; 95% CI = 1.05 to 9.80, when adjusted for intake of other analgesics) did not show a dose–response relationship, and were no longer significant when adjusted for person interviewed (case/control vs. proxy) as well as all other factors (OR = 2.76; 95% CI = 0.82 to 9.31). For both phenacetin and acetaminophen, the ORs were highest when the diagnosis was interstitial nephritis or "renal insufficiency."

This study alone cannot be used as proof that acetaminophen causes chronic renal failure. Reservations include: information having been obtained in proxy interviews for 55% of cases but only 10% of controls, absence of a specific marker of disease, the large proportion of the original case group that was excluded after review of the medical record or who were too ill, had died, could not be contacted, or refused to participate, and the population-based control group that was less poor and better educated than the case group.

In 1995, participants in the Physicians' Health Study completed a self-administered questionnaire about analgesic consumption during the 14 years since the cohort was assembled, and provided a blood sample between 1995 and 1997 (in this respect it was a cross-sectional study in survivors). No association of any type of analgesic consumption was found with raised serum creatinine or calculated creatinine clearance. However, there were few subjects in the highest category of exposure, for which the threshold was low (only one tablet per day taken for 7 years) (Rexrode et al., 2001).

A cohort of heavy analgesic abusers (averaging 5 to 6 tablets or powders per day for about 20 years), chiefly patients identified by doctors or pharmacists, followed for 7 years in Belgium yielded an RR for reduced creatinine clearance of 6.1 (95% CI = 1.4 to 26) compared with population-based controls (Elseviers and De Broe, 1995). All 18 analgesic abusers who developed renal impairment had been taking 2+ analgesics. Although only three were taking phenacetin when enrolled, all but one had taken preparations that had formerly contained phenacetin (Michielsen and De Schepper, 1996).

In summary, the epidemiological evidence exonerates salicylates alone, condemns phenacetin when taken in huge quantities with either a salicylate or a pyrazolone, but has been inconclusive about other antipyretic-analgesics, taken singly or in combination.

Prescribed (Iatrogenic) Nephrotoxicants

No drug with significant toxicity will be prescribed unless there is a clear margin of safety or else its therapeutic efficacy, and lack of a nontoxic alternative, warrants the risk. Nephrotoxicants frequently encountered in medical practice fall into the latter category, offering tangible benefits that justify what usually is asymptomatic and fully reversible impairment of renal function. When, as often is the case, the injury is acute, the cause–effect relationship generally is obvious without formal epidemiological proof, but epidemiology can estimate 'safe' levels of exposure or validate measures for control.

Nonsteroidal Anti-Inflammatory Drugs

NSAIDs are drugs that inhibit cyclooxygenase (prostaglandin H synthetase) throughout the body; they do not include phenacetin or acetaminophen, whose site of therapeutic action is in the central nervous system. In the family of NSAIDs there is a difference between the newer agents, which potently inhibit cyclooxygenase within the kidney, and NSAIDs that were introduced more than 60 years ago, chiefly aspirin, which only weakly inhibit renal cyclooxygenases (Dunn and Zambraski, 1980). It is clear on clinical grounds that the newer NSAIDs cause various acute renal syndromes, but evaluation of their role in chronic kidney disease has been clouded by the interchangeable use of the terms "analgesic" and "NSAID," and the failure to distinguish NSAID-induced chronic kidney disease from "classical" analgesic nephropathy (Nanra, 1993). A third distinct nephropathy may be the chronic interstitial nephritis reported with 5-aminosalicylic acid given for inflammatory bowel disease (De Broe et al., 1998).

The three most substantial case series associating NSAIDs with chronic interstitial nephritis (diagnosed by biopsy) or papillary necrosis (demonstrated by imaging) are sufficiently impressive to implicate the drugs "on the balance of the evidence." However, in none of the studies were the diagnostic criteria specific or applied by an observer blind to the exposure status, nor was there an objective measure of exposure (Nanra, 1993; Schwarz et al., 2000; Segasothy et al., 1994).

Four of the epidemiological studies that explored the risk of chronic renal failure or ESRD attributable to antipyretic-analgesics also assessed the risk from NSAIDs. Two yielded results that were inconsistent, the ORs being raised in some, but not all, categories of patient or exposure (Perneger et al., 1994; Sandler et al., 1991), while the other two were negative (Nuyts et al., 1995b; Rexrode et al., 2001).

It is undeniable that NSAIDs cause chronic renal disease, but unlike the phenacetin-containing 2+ antipyretic-analgesics, they are insignificant as a cause of ESRD (Table 24.1).

Lithium

Although there are clinical and experimental grounds for believing that lithium causes nephrogenic diabetes insipidus (Bendz et al., 1996), and a number of patients are recorded with lithium-related ESRD (Table 24.1, and Presne et al., 2002), there is no epidemiological proof that this agent results in chronic renal disease.

Nephrogenic diabetes insipidus can only be distinguished reliably from psychogenic polydipsia by measuring urinary concentration in a subject given vasopressin after being fluid deprived under observation for at least 12 hours, a test with obvious practical difficulties for epidemiological research. Reduced glomerular filtration rate, nearly always of mild degree, was present in 15% (on average) of subjects in a number of case series; this was deemed likely to be an overestimate due to selection bias and the inclusion of an unknown proportion of cases of chronic renal failure of other etiology (Boton et al., 1987). Chronic tubulointerstitial nephritis with characteristic tubular lesions has been proposed as a specific marker of disease (Batlle and Dourhout-Mees,

1998), but whether primary glomerular abnormality can be caused by lithium is unproven (Markowitz et al., 2000).

Therapeutic drug monitoring is advocated to reduce nephrotoxicity, but has not been evaluated epidemiologically.

SUMMARY

Both success and failure are recorded against attempts to solve questions about nephrotoxicity by using an epidemiological approach. Resourceful detective work backed by intelligent use of markers of exposure have proved that accumulated lead stored for years in the skeleton caused ESRD in Queensland, incorrect formulation of a herbal slimming regimen resulted in progressive interstitial nephritis and urothelial cancer in Belgium (Schmeiser et al., 1996; Vanherweghem et al., 1993), and phenacetin, not salicylate, was the principal cause of analgesic nephropathy.

Massive epidemiological resources have been deployed to determine whether other antipyretic-analgesics, particularly acetaminophen, are significant causes of chronic kidney disease, but these have returned "not guilty" verdicts on salicylates, and "not proven" on acetaminophen, pyrazolones, and the newer NSAIDs. Whether this represents success or failure of the epidemiological approach is yet to be agreed.

Epidemiology has indicted organic solvents and silicon as possible causes of glomerulonephritis, but the failure to identify a specific pathology resulting from these exposures represents a serious barrier to proof of their role.

Monitoring of cadmium exposure, minimizing the nephrotoxic impact of radiographic contrast (see Chapter 19), and establishing the therapeutic range for safe and effective cyclosporin dosage (see Chapter 17) all are indebted to epidemiology, which, however, has been unable to prove that lithium causes chronic renal failure, let alone provide a means of using this drug safely. With respect to aminoglycosides, the epidemiological approach has yielded little, the currently preferred strategy for preventing toxicity being based entirely on deductive reasoning using biologic knowledge, retrospectively endorsed by clinical trials (see Chapter 16). Moreover, epidemiology has failed to identify the toxicant responsible for Balkan endemic nephropathy despite 40 years of research.

Evidence-based medicine has properly assumed an authoritative role in medical decisions. When the evidence is sound, comprehensive, and applicable to the problem in question, it must reign supreme. Often, as illustrated in this chapter, it has been necessary to make do with information that is, at least in some respects, of less than satisfactory quality.

ACKNOWLEDGMENTS

The authors thank the Australia and New Zealand Dialysis and Transplant Registry (anzdata@anzdata.org.au) who provided the numbers of incident cases of end-stage renal failure, by primary renal disease, that were used in Table 24.1.

REFERENCES

Angell M (1990) The interpretation of epidemiologic studies. *N Engl J Med* 323:823–825.

Apostolov K, and Spasic P (1975) Evidence of a viral aetiology in endemic (Balkan) nephropathy. *Lancet* 2:1271–1273.

Armstrong BK, White E, and Saracci R (1994) Principles of Exposure Measurement in Epidemiology. Oxford University Press New York.

Batlle D, and Dourhout-Mees EJ (1998) Lithium and the kidney, in *Clinical Nephrotoxins. Renal Injury from Drugs and Chemicals*, De Broe ME, Porta GA, Bennett WM, and Verpooten GA (Eds), Kluwer Academic Publications, Dordrecht, pp. 383–395.

Batuman V, Maesaka JK, Haddad B, Tepper E, Landy E, and Wedeen RP (1983) The role of lead in gout nephropathy. *N Engl J Med* 309:17–21.

Batuman V, Landy E, Maesaka JK, and Wedeen RP (1981) Contribution of lead to hypertension with renal impairment. *N Engl J Med* 304:520–523.

Bell GM, Gordon ACH, Lee P, Doig A, MacDonald MK, Thomson D, Anderton JL, and Robson JS (1985) Proliferative glomerulonephritis and exposure to organic solvents. *Nephron* 40:161–165.

Bendz H, Sjödin I, and Aurell M (1996) Renal function on and off lithium in patients treated with lithium for 15 years or more. A controlled, prospective lithium-withdrawal study. *Nephrol Dial Transplant* 11:457–460.

Bernard AM, Roels H, Buchet JP, Cardenas A, and Lauwerys R (1992) Cadmium and health: The Belgian experience, in *Cadmium in the Human Environment: Toxicity and Carcinogenicity*, Nordberg GF, Herber RFM, and Alessio L (Eds), IARC Scientific Publications No 118. International Agency for Research on Cancer, Lyon, pp. 15–33.

Björck S, Svalander CT, and Aurell M (1988) Acute renal failure after analgesic drugs including paracetamol (acetaminophen). *Nephron* 49:45–53.

Boton R, Gaviria M, and Batlle DC (1987) Prevalence, pathogenesis, and treatment of renal dysfunction associated with chronic lithium therapy. *Am J Kidney Dis* 10:329–345.

Boujemaa W, Lauwerys R, Bernard A (1994) Early indicators of renal dysfunction in silicotic workers. *Scand J Work Environ Health* 20:188–191.

Calvert GM, Steenland K, and Palu S (1997) End-stage renal disease among silica-exposed goldminers. A new method for assessing incidence among epidemiologic cohorts. *JAMA* 277:1219–1223.

Ceovic S, Hrabar A, and Radonic M (1985) An etiological approach to Balkan endemic nephropathy based on the investigation of two genetically different populations. *Nephron* 40:175–179.

Chang CC, Lauwerys R, Bernard A, Roels H, Buchet JP, and Garvey JS (1980) Metallothionein in cadmium exposed workers. *Environ Res* 23:422–428.

Churchill DN, Fine A, and Gault MH (1983) Association between hydrocarbon exposure and glomerulonephritis. An appraisal of the evidence. *Nephron* 33:169–172.

Cukuranovic R, Ignjatovic M, and Stefanovic V (1991) Urinary tract tumors and Balkan nephropathy in the South Morava River basin. *Kidney Int* 40 (Suppl 34): 80–84.

De Broe ME, Stolear J-C, Nouwen EJ, and Elseviers MM (1998) 5-Aminosalicylic acid and chronic interstitial nephritis, in *Clinical Nephrotoxins. Renal Injury from Drugs and Chemicals*, De Broe ME, Porta GA, Bennett WM, and Verpooten GA (Eds), Kluwer Acadmic Publications, Dordrecht, pp. 217–222.

Dieperink HH (1989) Identification of groups at risk for renal diseases (including nephrotoxicity). *Toxicol Lett* 46:257–268.

dos Santos Silva I (1999) *Cancer Epidemiology: Principles and Methods.* International Agency for Research on Cancer, Lyon.

Dubach UC, Rosner B, and Pfister E (1983) Epidemiologic study of abuse of analgesics containing phenacetin. Renal morbidity and mortality (1968–1979). *N Engl J Med* 308:357–362.

Dubach UC, Rosner B, and Stürmer T (1991) An epidemiologic study of abuse of analgesic drugs. Effects of phenacetin and salicylate on mortality and cardiovascular morbidity (1968 to 1987). *N Engl J Med* 324:155–160.

Dunn MJ, and Zambraski EJ (1980) Renal effects of drugs that inhibit prostaglandin synthesis. *Kidney Int* 18:609–622.

Elseviers MM, and De Broe ME (1995) A long-term prospective controlled study of analgesic abuse in Belgium. *Kidney Int* 48:1912–1919.

Elseviers MM, De Schepper A, Corthouts R, Bosmans J-L, Cosyn L, Lins RL, Lornoy W, Matthys E, Roose R, VanCaesbroeck D, Waller I, Horackova M, Schwarz A, Svrcek P, Bonvichi D, Franek E, Morlans M, and De Broe ME (1995) High diagnostic performance on CT scan for analgesic nephropathy in patients with incipient to severe renal failure. *Kidney Int* 48:1316–1323.

Emmerson BT (1963) Chronic lead nephropathy: The diagnostic use of calcium EDTA, and the association with gout. *Australas Ann Med* 12:310–324.

Feder GL, Radovanovic Z, and Finkleman RB (1991) Relationship between weathered coal deposits and the etiology of Balkan endemic nephropathy. *Kidney Int* 40 (Suppl 34): 9–11.

Gault MH, and Barrett BJ (1998) Analgesic nephropathy. *Am J Kidney Dis* 32:351–360.

Gregorini G, Ferioli A, Donato F, Tira P, Morassi L, Tardanico R, Lancini L, and Maiorca R (1993) Association between silica exposure and necrotizing crescentic glomerulonephritis with p-ANCA, and anti-MPO antibodies: A hospital-based case-control study. *Adv Exp Med Biol* 336:435–440.

Hald B (1991) Ochratoxin A in human blood in European countries, in *Mycotoxins, Endemic Nephropathy and Urinary Tract Tumours*, Castegnaro M, Pleština R, Dirheimer G, Chernozemsky IN, and Bartsch H (Eds), IARC Scientific Publications No 115. International Agency for Research on Cancer, Lyon, pp. 159–164.

Harrington JM, Whitby H, Gray CN, Reid FJ, Aw TC, and Waterhouse JA (1989) Renal disease and occupational exposure to organic solvents: A case referent approach. *Br J Ind Med* 46:643–650.

Harrison DJ, Thomson D, and MacDonald MK (1986) Membranous glomerulonephritis. *J Clin Pathol* 39:167–171.

Hellstrom L, Elinder CC, Dahlberg B, Lundberg M, Jarup L, Persson B, and Axelson O (2001) Cadmium exposure and end-stage renal disease. *Am J Kidney Dis* 38:1001–1008.

Henderson DA (1954) A follow-up of cases of plumbism in children. *Australas Ann Med* 3:219–224.

Henderson DA (1955) Chronic nephritis in Queensland. *Australas Ann Med* 4:163–177.

Henderson DA, and Inglis JA (1957) The lead content of bone in chronic Bright's disease. *Australas Ann Med* 6:145–154.

Hotz P, Buchet JP, Bernard A, Lison D, and Lauwerys R (1999) Renal effects of low-level cadmium exposure: 5-year follow-up of a sub-cohort from the Cadmibel study. *Lancet* 354:1508–1513.

Hotz P, Gonzalez-Lorenzo J, Siles E, Trujillano G, Lauwerys R, and Bernard A (1995) Subclinical signs of kidney dysfunction following short exposure to silicon in the absence of silicosis. *Nephron* 70:438–442.

Hrabar A, Aleraj B, Čeović S, Čvoriščec D, Vacca C, and Hall PW III (1991) A fifteen year cohort based evaluation of β_2-microglobulin as an early sign of Balkan endemic nephropathy. *Kidney Int* 40 (Suppl 34):41–43.

Hu H, Milder FL, and Burger DE (1990) X-ray fluorescence measurements of lead burden in subjects with low-level community exposure. *Arch Environ Health* 45:335–341.

Ivic M (1970) The problem of endemic nephropathy. *Acta Fac Med Naissensis* 1:29–38.

Jarup L, Persson B, and Elinder CG (1995) Decreased glomerular filtration rate in solderers exposed to cadmium. *Occup Environ Med* 52:818–822.

Jarup L, Persson B, and Elinder CG (1997) Blood cadmium as an indicator of dose in a long-term follow-up of workers previously exposed to cadmium. *Scand J Work Environ Health* 23:31–36.

Jarup L, Hellstrom L, Alfven T, Carlsson MD, Grubb A, Persson B, Pettersson C, Spang G, Schutz A, and Elinder CG (2000) Low level exposure to cadmium and early kidney damage: The OSCAR study. *Occup Environ Med* 57:668–672.

Kido T, and Nordberg G (1998) Cadmium-induced renal effects in the general environment, in *Clinical Nephrotoxins. Renal Injury from Drugs and Chemicals*, De Broe ME, Porta GA, Bennett WM, and Verpooten GA (Eds), Kluwer Academic Publications, Dordrecht, pp. 345–361.

Kim R, Rotnitzky A, Sparrow D, Weiss ST, Wager C, and Hu H (1996) A longitudinal study of low-level lead exposure and impairment of renal function. *JAMA* 275:1177–1181.

Lauwerys RR, Roels HA, Buchet JP, Bernard A, and Stanescu D (1979) Investigation on the lung and kidney function in workers exposed to cadmium. *Environ Health Perspect* 28:137–145.

Lin J-L, Tan D-T, Hsu K-H, and Yu C-C (2001) Environmental lead exposure and progressive renal insufficiency. *Arch Intern Med* 161:264–271.

Long DT, Icopini G, Ganev V, Petropoulos E, Havezov I, Voice T, Chou K, Spassov A, and Stein A (2001) Geochemistry of Bulgarian soils in villages affected and not affected by Balkan endemic nephropathy: A pilot study. *Int J Occup Med Environ Health* 14:193–196.

Maksimovic ZJ (1991) Selenium deficiency and Balkan endemic nephropathy. *Kidney Int* 40 (Suppl 34):12–14.

Markowitz GS, Rodhakrisnan J, Kambham N, Valeri N, Hines WH, and D'Agati VD (2000) Lithium nephtotoxicity: A progressive combined glomerular and tubulointerstitial nephropathy. *J Am Soc Nephrol* 11:1439–1448.

Marsh GM, Enterline PE, Stone RA, and Henderson VL (1990) Mortality among a cohort of US man-made fiber workers: 1985 follow-up. *J Occup Med* 32: 594–604.

McCredie M, and Stewart JH (1992) Risk factors for kidney cancer in New South Wales – I. Cigarette smoking. *Eur J Cancer* 28A:2050–2054.

McCredie M, Stewart JH, and Day NE (1993) Different roles for phenacetin and paracetamol in cancer of the kidney and renal pelvis. *Int J Cancer* 53:245–249.

McCredie M, Stewart JH, Ford JM, and MacLennan RA (1983) Phenacetin-containing analgesics and cancer of the bladder and renal pelvis in women. *Br J Urol* 55:220–224.

McCredie M, Stewart JH, and Mahony JF (1982) Is phenacetin responsible for analgesic nephropathy in New South Wales? *Clin Nephrol* 17:134–140.

McLaughlin JK, Lindblad P, Mellemgaard A, McCredie M, Mandel JS, Schlehofer B, Pommer W, and Adami H-O (1995) International renal cell cancer study. I. Tobacco use. *Int J Cancer* 60:194–198.

McLaughlin JK, Blot WJ, Mandel JS, Schuman LM, Mehl ES, and Fraumeni JF Jr (1983) Etiology of cancer of the renal pelvis. *J Natl Cancer Inst* 71:287–291.

Michielsen P, and De Schepper P (1996) Combination analgesic involvement in the pathogenesis of analgesic nephropathy (letter). *Am J Kidney Dis* 28: 959–960.

Michielsen P, and De Schepper P (2001) Trends of analgesic nephropathy in two high-endemic regions with different legislation. *J Am Soc Nephrol* 12:550–556.

Morlans M, Laporte J-R, Vidal X, Cabeza D, and Stolley PD (1990) End-stage renal disease and non-narcotic analgesics: A case-control study. *Br J Clin Pharmacol* 30:717–723.

Murray TG, Stolley PD, Anthony JC, Schinnar R, Helper-Smith E, and Jeffreys JL (1983) Epidemiologic study of regular analgesic use and end-stage renal disease. *Arch Intern Med* 143:1687–1693.

Nanra RS (1993) Analgesic nephropathy in the 1990s – an Australian perspective. *Kidney Int* 44 (Suppl 42):86–92.

Nastac E, Stoian M, Hozoc M, Iosipenco M, and Melencu M (1984) Further data on the prevalence and serum antibodies to papova viruses (BK, and SV40) in subjects from the Romanian area with Balkan endemic nephropathy. *Virologie* 55:65–67.

Ng TP, Ng YL, Lee HS, Chia KS, and Ong HY (1992) A study of silica nephrotoxicity in exposed silicotic and non-silicotic workers. *Br J Ind Med* 49:35–37.

Nikolov IG, Petkova-Bocharova D, Castegnaro M, Pfohl-Leskowicz A, Gill C, Day N, and Chernozemsky IN (1996) Molecular and epidemiological approaches to the etiology of urinary tract tumors in an area with Balkan endemic nephropathy. *J Environ Pathol Toxicol Oncol* 15:201–207.

Nordberg GF (1992) Application of the 'critical effect' and 'critical concentration' concept to human risk assessment for cadmium, in *Cadmium in the Human Environment: Toxicity and Carcinogenicity*, Nordberg GF, Herber RFM, and Alessio L (Eds), IARC Scientific Publications No 118. International Agency for Research on Cancer, Lyon, pp. 3–14.

Nortier JL, Martinez M-CM, Schmeiser HH, Arlt VM, Bieler CA, Petein M, Depierreux MF, De Pauw L, Abramowicz D, Vereerstraeten P, and Vanherweghem J-L (2000) Urothelial carcinoma associated with the use of a Chinese herb (*Aristolochia fangchi*). *N Engl J Med* 342:1686–1692.

Nuyts GD, Elseviers MM, and De Broe ME (1993) Healthy worker effect in a cross-sectional study of lead workers. *J Occup Med* 35:387–391.

Nuyts GD, Van Vlem E, De Vos A, Daelemans RA, Rorive G, Elseviers MM, Schurgers M, Segaert M, D'Haese PC, and De Broe ME (1995a) Wegener granulomatosis is associated to exposure to silicon compounds: A case-control study. *Nephrol Dial Transplant* 10:1162–1165.

Nuyts GD, Van Vlem E, Thys J, De Leersnijder D, D'Haese PC, Elseviers MM, and De Broe ME (1995b) New occupational risk factors for chronic renal failure. *Lancet* 346:7–11.

Nye LJJ (1953) An investigation of extraordinary incidence of chronic nephritis in young people in Queensland. *Australas Ann Med* 2:145–159.

Perneger TV, Whelton PK, and Klag MJ (1994) Risk of kidney failure associated with the use of acetaminophen, aspirin, and nonsteroidal antiinflammatory drugs. *N Engl J Med* 331:1675–1679.

Petkova-Bocharova T, and Castegnaro M (1991) Ochratoxin A in human blood in relation to Balkan endemic nephropathy and renal tumours in Bulgaria, in *Mycotoxins, Endemic Nephropathy and Urinary Tract Tumours*, Castegnaro M, Pleština R, Dirheimer G, Chernozemsky IN, and Bartsch H (Eds), IARC Scientific Publications No 115. International Agency for Research on Cancer, Lyon, 1991, pp. 135–138.

Piper JM, Matanoski GM, and Tonascia J (1986) Bladder cancer in young women. *Am J Epidemiol* 123:1033–1042.

Pirkle JL, Kaufmann RB, Brody DJ, Hickman T, Gunter EW, and Paschal DC (1998) Exposure of the U.S. population to lead, 1991–1994. *Environ Health Perspect* 106:745–750.

Pocock SJ, Shaper AG, Ashby D, Delves T, and Whitehead TP (1984) Blood lead concentrations, blood pressure, and renal function. *Br Med J* 289:872–874.

Pommer W, Bronder E, Greiser E, Helmert U, Jesdinsky HJ, Klimpel A, Borner K, and Molzahn M (1989) Regular analgesic intake and the risk of end-stage renal failure. *Am J Nephrol* 9:403–412.

Porro A, Lomonte C, Coratelli P, Passavanti G, Ferri GM, and Assennato G (1992) Chronic glomerulonephritis and exposure to solvents: A case-referent study. *Br J Ind Med* 49:738–742.

Presne C, Fakouri F, Kenouch S, Stengel B, Kreis W, and Grunfeld JP (2002) [Progressive renal failure caused by lithium nephropathy.] *Presse Medicale* 31:828–833.

Puntaric D, Bosnir J, Smit Z, Skes I, and Baklaic Z (2001) Ochratoxin A in corn and wheat: Geographical association with endemic nephropathy. *Croat Med J* 42:175–180.

Radic B, Fuchs R, Peraica M, and Lucic A (1997) Ochratoxin A in human sera in the area with endemic nephropathy in Croatia. *Toxicol Lett* 91:105–109.

Rapiti E, Sperati A, Micelli M, Forastiere F, Di Lallo D, Cavariani F, Goldsmith D, and Perucci C (1999) End stage renal disease among ceramic workers exposed to silica. *Occup Environ Med* 56:559–561.

Ravnskov U, Forsberg B, and Skerfving S (1979) Glomerulonephritis and exposure to organic solvents. A case-control study. *Acta Med Scand* 205:575–579.

Ravnskov U, Lundström S, and Nordén Å (1983) Hydrocarbon exposure and glomerulonephritis: Evidence from patients' occupations. *Lancet* 2:1214–1216.

Rexrode KM, Buring JE, Glynn RJ, Stampfer MJ, Youngman LD, and Gaziano JM (2001) Analgesic use and renal function in men. *JAMA* 286:315–321.

Roels HA, Bernard AM, Cárdenas A, Buchet JP, Lauwerys RR, Hotter G, Ramis I, Mutti A, Franchini I, Bundschuh I, Stolte H, De Broe ME, Nuyts BD, Taylor SA, and Price RG (1993) Markers of early renal changes induced by industrial pollutants. III. Application to workers exposed to cadmium. *Br J Ind Med* 50:37–48.

Roels HA, Lauwerys RR, Bernard AM, Buchet JP, Vos A, and Oversteyns M (1991) Assessment of the filtration reserve capacity of the kidney in workers exposed to cadmium. *Br J Ind Med* 48:365–374.

Roels HA, Lauwerys RR, Buchet JP, Bernard A, Chettle DR, Harvey TC, and Al-Haddad IK (1981) *In vivo* measurement of liver and kidney cadmium in workers exposed to this metal. *Environ Res* 26:217–240.

Roels HA, Lauwerys RR, Buchet JP, Bernard AM, Vos A, and Oversteyns M (1989) Health significance of cadmium induced renal dysfunction: A five year follow up. *Br J Ind Med* 46:755–760.

Roels HA, Lauwerys RR, and Dardenne AN (1983) The critical level of cadmium in human renal cortex: A reevaluation. *Toxicol Lett* 15:357–360.

Rosenman K, Moore-Fuller M, and Reilly M (2000) Kidney disease and silicosis. *Nephron* 85:14–19.

Sandler DP, Smith JC, Weinberg CR, Buckalew VM Jr, Dennis VW, Blythe WB, and Burgess WP (1989) Analgesic use and chronic renal disease. *N Engl J Med* 320:1238–1243.

Sandler DP, Burr FR, and Weinberg CR (1991) Nonsteroidal anti-inflammatory drugs and the risk for chronic renal disease. *Ann Intern Med* 115:165–172.

Schmeiser HH, Bieler CA, Wiessler M, van Ypersele de Strihou C, and Cosyns JP (1996) Detection of DNA adducts formed by aristolochic acid in renal tissue from patients with Chinese herbs nephropathy. *Cancer Res* 56:2025–2028.

Schwarz A, Krause PH, Kunzendorf U, Keller F, and Distler A (2000) The outcome of acute interstitial nephritis: Risk factors to the transition from acute to chronic interstitial nephritis. *Clin Nephrol* 54:179–190.

Segasothy M, Samad SA, Zulfigar A, and Bennett WM (1994) Chronic renal disease and papillary necrosis associated with the long-term use of nonsteroidal anti-inflammatory drugs as the sole or predominant analgesic. *Am J Kidney Dis* 24:17–24.

Staessen JA, Lauwerys RR, Buchet J-P, Bulpitt CJ, Rondia D, Vanrenterghem Y, Amery A, and the Cadmibel Study Group (1992) Impairment of renal function with increasing blood lead concentrations in the general population. *N Engl J Med* 327:151–156.

Staessen JA, Yeoman WB, Fletcher AE, Markowe HL, Marmot MG, Rose G, Semmence A, Shipley MJ, and Bulpitt CJ (1990) Blood lead concentrations, blood pressure, and renal function. *Br J Ind Med* 47:442–447.

Steenland NK, Thun MJ, Ferguson CW, and Port FK (1990) Occupational and other exposures associated with male end-stage renal disease: A case/control study. *Am J Public Health* 80:153–157.

Steenland NK, Sanderson W, and Calvert GM (2001) Kidney disease and arthritis in a cohort study of workers exposed to silica. *Epidemiology* 12:405–412.

Stengel B, Cénée S, Limasset J-C, Protois J-C, Marcelli A, Brochard P, and Hémon D (1995) Organic solvent exposure may increase the risk of glomerular nephropathies with chronic renal failure. *Int J Epidemiol* 24:427–434.

Stewart JH, McCredie M, Disney APS, and Mathew TH (1994) Trends in end-stage renal failure in Australia, 1972–1991. *Nephrol Dial Transplant* 9:1377–1382.

Stoev SD (1998) The role of ochratoxin A as a possible cause of Balkan endemic nephropathy and its risk evaluation. *Vet Hum Toxicol* 40:352–360.

Stratta P, Messuerotti A, Coravese C, Coen M, Luccoli L, Bussolati B, Malavenda P, Cacciabue M, Bugiani M, Bo M, Ventura M, Camussi G, and Fubini B (2001) The role of metals in autoimmune vasculitis: Epidemiological and pathogenic study. *Sci Total Environ* 270:179–190.

Tatu CA, Orem WH, Finkelman RB, and Feder GL (1998) The etiology of Balkan endemic nephropathy: Still more questions than answers. *Environ Health Perspect* 106:689–700.

Toncheva D, Dimitrov T, and Stojanova S (1998) Etiology of Balkan endemic nephropathy: A multifactorial disease? *Eur J Epidemiol* 14:389–394.

Uzelac-Keserovic B, Spasic P, Bojanic N, Dimitrijevic J, Lako B, Lepsanovic Z, Kuljic-Kapulica N, Vasic D, and Apostolov K (1999) Isolation of a coronavirus from kidney biopsies of endemic Balkan nephropathy patients. *Nephron* 81:141–145.

Vanherweghem JL, Depierreux M, Tielemans C, Abramowicz D, Dratwa M, Jadoul M, Richard C, Vandervelde D, Verbeelen D, Vanhaelen-Fastre R, and Vanhaelen M (1993) Rapidly progressive interstitial renal fibrosis in young women: Association with slimming regimen including Chinese herbs. *Lancet* 341:387–391.

Vrabcheva T, Usleber E, Dietrich R, and Martlbauer E (2000) Co-occurrence of ochratoxin A and citrinin in cereals from Bulgarian villages with a history of Balkan endemic nephropathy. *J Agric Food Chem* 48:2483–2488.

Weeden RP, Maesaka JK, Weiner B, Lipat GA, Lyons MM, Vitale LF, and Joselon NM (1975) Occupational lead nephropathy. *Am J Med* 59:630–641.

Weeden RP, Mallik DK, and Batuman V (1979) Detection and treatment of occupational lead nephropathy. *Arch Intern Med* 139:53–57.

Yaqoob M, Bell GM, Percy DF, and Finn R (1992) Primary glomerulonephritis and hydrocarbon exposure: A case-control study and literature review. *Q J Med* 83:409–418.

Yaqoob M, Patrick AW, McClelland P, Stevenson A, Mason H, Percy DF, White DC, and Bell GM (1994) Occupational hydrocarbon exposure in diabetic nephropathy. *Diabet Med* 11:789–793.

25

Age, Sex, and Species Differences in Nephrotoxic Response

Rani S. Sellers and Kanwar Nasir M. Khan

OVERVIEW

The cellular and anatomical complexity of the kidney is essential to its function. Alterations in renal cellular function may increase the risk of toxic damage to the kidney, and also to other organ systems. Notable age, sex, and species related differences in renal physiology and anatomy affect susceptibility to exogenous insults. Nephrotoxicity can be a common finding in laboratory animals and humans following exposure to xenobiotics. Laboratory animals, including rats, dogs, and monkeys, are often used in the risk assessment of new therapeutics and chemicals or to understand etiopathogenesis of potential renal toxicities in humans. In conventional nonclinical safety assessment studies, laboratory animals are exposed to xenobiotics during all stages of their lifespan, including fetal, perinatal, juvenile, and adult. Therefore, refinement of risk assessment requires not only specific knowledge of the molecular, biochemical, and structural effects of drugs and chemicals, but also of how an animal's age, sex, species, breed, and strain contribute to the toxic effects of xenobiotics. This chapter emphasizes the role of age, gender, and species in renal physiology and nephrotoxicity, the mechanisms by which changes occur, and their relationship with risk assessment in the human.

DEVELOPMENTAL CONSIDERATIONS IN NEPHROTOXIC RESPONSES

Comparative Renal Morphology

Basic familiarity with the complex structure of the mammalian kidney and interspecies differences in renal anatomy can provide a basis for comprehending the multitude of functional characteristics and nephrotoxic responses. The kidneys of mammals can be classified as unipyramidal (unipapillary) or multipyramidal (multipapillary). The medulla can be further characterized into simple, in which vascular bundles have only descending and ascending vasa recta, and complex, in which the descending and ascending vasa recta surround the short loops of Henle and tend to fuse, forming larger complex vascular structures. Animals with complex medullas have greater urinary concentrating capacity than those with simple medullas. Similary, there are species-related differences in the number of nephrons with short loops of Henle versus long loops of Henle. The latter have greater urinary concentrating capacity. Differences in urinary concentrating capacity may result in different susceptibilities to nephrotoxicants. The relative corticomedullary width, number of nephrons per kidney, and the organization of medulla can vary between different species (Table 25.1), which may also contribute to differential susceptibility to nephrotoxicants (Bovee, 1984; Clapp and Tisher, 1989; Khan and Alden, 2002).

In humans, nephron formation completes prior to birth, with up to 60% of the nephrons developing in the third trimester (Manalich et al., 2000). The time of completion of nephrogenesis varies in different mammalian species. In rats, nephrogenesis ends by eighth postnatal day (Clapp and Tisher, 1989; Guron and Friberg, 2000). In dogs, nephrogenesis is not complete until 3 weeks after birth, after which renal growth occurs by a 235% increase in tubular volume and a 33% increase in glomerular volume (Bovee, 1984). Changes in glomerular and tubular growth in the dog are essentially complete by approximately 2 months of age as reflected by assessments of glomerular, cortical, and medullary sizes (Bovee, 1984).

Table 25.1 Comparative Renal Structure and Function

Renal Characteristic	*Rat*	*Dog*	*Monkey*	*Human*
Body weight (kg)	0.24	9.1	3.8	70
Kidney weight (g)	0.75	31	9	157
Renal organization	Unipapillary	Unipapillary	Unipapillary	Multipapillary
Medullary structure	Complex	Simple	Simple	Simple
Nephrons per g body weight	128	45	49	16
Number of nephrons (thousands)	30	340–490	103–123	300–1100
Glomeruli radius (μm)	61	90	83	100
Proximal tubule length (mm)	12	20	—[a]	16
Tubule radius (μm)	29	33	—[a]	36
Percentage long loops	28	100	—[a]	14
Relative medullary thickness	5.8	4.3	—[a]	3.0
Maximum Osmolality (mOsmol/kg)	2610	2610	1900	1400
Inulin clearance (ml/min/kg)	6.0	4.3	1.9	2.0
Maximal GFR (ml/min/m^2)	35	104	—[a]	75

[a]Data not available.

Genetic and environmental factors can affect the number of nephrons in a kidney (Lelievre-Pegorier and Merlet-Benichou, 2000; Manalich et al., 2000; Merlet-Benichou, 1999). Glomerular number and size can differ according to race; for example, kidneys from African-American donors have larger and fewer glomeruli than their Caucasian counterparts, independent of donor age (Abdi et al., 1998). Nephron number may vary between rat strains (Battista et al., 2002). Similarly, environmental factors, such as restriction in fetal growth, can affect the glomerular number per kidney (Lelievre-Pegorier and Merlet-Benichou, 2000).

Developmental Biology of the Kidney Relevant to Renal Function

Changes in the *in utero* microenvironment can have a significant effect on renal development in the fetus (Lelievre-Pegorier and

Merlet-Benichou, 2000; Manalich et al., 2000; Marchand and Langley-Evans, 2001; Merlet-Benichou, 1999). Intrauterine fetal growth restriction and low birth weight have been associated with smaller kidneys and fewer nephrons in both humans and rats. In human full-term pregnancies with low birth weight caused by delayed fetal growth, there is a reduction in renal size and a 20% decrease in the number of nephrons relative to normal infants; similar findings have been shown in rats (Battista et al., 2002; Manalich et al., 2000; Marchard and Langley-Evans, 2001). Fetal growth restriction is usually associated with the development of essential hypertension, chronic renal failure, and renal disease (Clark and Bertram, 1999), which have been attributed, in part, to decreased numbers of nephrons (Langley-Evans et al., 1998; Lelievre-Pegorier and Merlet-Benichou, 2000; Manalich et al., 2000; Merlet-Benichou, 1999).

Reductions in nephron number are thought to cause glomerular hypertension, promoting progressive renal degeneration and reduced resistance to renal damage later in life (Langley-Evans et al., 1998; Manalich et al., 2000). Further evidence that decreased nephron number contributes to hypertension is in the spontaneously hypertensive rats (SHR), which have 20% fewer nephrons than their normotensive counterparts (Wistar Kyoto rats) (Battista et al., 2002).

Low birth weight has also been associated with increased progression and poor prognosis in IgA glomerulonephritis in children (Zidar et al., 1998). Renal alterations in low birth weight babies may be related to *in utero* reductions in protein and vitamin A, exposure to hyperglycemia, or various drug effects (Lelievre-Pegorier and Merlet-Benichou, 2000). Mild to moderate protein restriction in rats during pregnancy (i.e., less than 12% total dietary protein as compared with 18% in a normal diet) results in offspring with lower birth weight, smaller kidneys with fewer nephrons, and increased blood pressure. Additionally, these offspring exhibit lower renal blood flow, higher plasma and pulmonary angiotensin converting enzyme (ACE) activity, and lower plasma renin activity (Langley-Evans et al., 1998). These observations suggest that there are significant alterations in the renin-angiotensin-aldosterone system (RAS) in growth and protein restricted fetuses.

RAS is an important pathway for regulating vascular tone and extracellular fluid volume and composition, but also appears important for normal nephrogenesis (Guron and Friberg, 2000). Immunohistochemical detection of renin in the rat kidney has demonstrated it to be more widespread in the fetus and neonate than in the adult

(Lumbers, 1995). Many of the renin expressing cells in the fetal kidney are around the vasculature and shift from the interlobar and arcuate arteries in the fetus to the afferent arterioles in the adult (Gomez et al., 1989; Reddi et al., 1998). Growth restricted stillborn human fetuses, unlike their normal counterparts, fail to shift renin expression from the deep cortex to the superficial cortex with increased gestational age, despite normal histological maturation (Kingdom et al., 1999).

ACE inhibitors given to humans during pregnancy have been associated with fetal renal tubular dysgenesis, and reductions in RAS during fetal renal development in rats results in decreased numbers of nephrons (developmental nephrotoxicity) (Guron and Friberg, 2000; Jose et al., 1994). Mice deficient in angiotensin II (ATII) type-1 (AT_1) receptor-mediated effects during nephrogenesis have numerous developmental disturbances. These include papillary atrophy, thickening of the intrarenal arterioles, tubular atrophy with interstitial expansion, and impaired urinary concentrating ability. In addition, targeted inactivation of angiotensinogen, ACE or AT_1 receptor demonstrates similar changes (Guron and Friberg, 2000). Deficiency of ATII type-2 receptor in mice is also associated with abnormalities in renal development (Guron and Friberg, 2000). Although causality has not been shown, these finding may be consistent with the observed renal developmental alterations found in growth-restricted animals and subsequent hypertension (Kingdom et al., 1999).

Similar to RAS, disruptions in cycloxygenase enzyme (COX) activity in the kidney may contribute to developmental nephrotoxicity. Two isoforms of COX have been identified, COX-1 and COX-2, which generate prostanoids from arachadonic acid (Needleman and Isakson, 1997). Analogous to age-related changes in renal distribution of renin, the expression of renal cortical COX-2 changes with age in humans, rats, and dogs (Khan et al., 2002). In comparison with adults, the fetal kidneys of humans, rats, and dogs exhibit high levels of COX-2 and associated prostaglandins (PGs) in the macula densa (Khan et al., 2001). In rats, COX-2 is not present in the embryonic kidney but its highest expression is noted between gestation day 16 and postnatal day 14 (Stanfield et al., 2000), after which it progressively declines (Stanfield et al., 2000). This period coincides with the development and maturation of the kidney in this species. A similar pattern of COX-2 expression has been reported in the immature human and canine kidneys (Khan et al., 1999, 2001).

This temporal expression suggests that PGs are involved in renal differentiation and growth and *in vitro* and *in vivo* data further support this hypothesis. Metanephric cell cultures demonstrate that *in vitro* treatment with the cyclooxygenase metabolite PGE_1 is necessary for both growth and differentiation of these cells (Avner et al., 1985). *In vivo* data suggesting the importance of PGs in renal maturation have been obtained for COX-2-knockout mice. These mice have normal appearing kidneys at birth (Dinchuk et al., 1995; Morham et al., 1995). However, by 6 weeks of age, the kidneys of these mice appear small and pale and are comprised of small immature glomeruli and small tubules consistent with hypoplasia. By 8 weeks of age, the majority of the animals develop renal failure (Dinchuk et al., 1995; Morham et al., 1995).

Similar histological changes have been reported in the kidneys of human and nonhuman primate fetuses exposed to conventional nonsteroidal anti-inflammatory drugs (NSAIDs) *in utero* (nonspecific COX-1 and COX-2 inhibitors, such as indomethacin and ibuprofen) during the third trimester of pregnancy (Dudley and Hardie, 1985; van der Heijden et al., 1994). In many of the reported cases of infants born to mothers treated with NSAIDs, oligohydraminos was observed, indicating decreased fetal urine output suggestive of fetal renal malfunction (Dudley and Hardie, 1985; Novy, 1978).

Genetic polymorphisms may contribute to renal changes occuring in late adulthood. Studies examining the polymorphisms in genes encoding components of the RAS indicate that certain polymorphisms are associated with an increased risk of end-stage renal disease (ESRD). Patients with the ACE DD allele progressed more rapidly to ESRD than those with the DI or II alleles (Lovati et al., 2001). Polymorphisms in the angiotensin gene, in particular the M235T polymorphism, in normal and diabetic patients correlate with the development and progression of chronic renal failure (Gumprecht et al., 2000). Diabetic patients with this polymorphism demonstrate increased susceptibility and progression to ESRD (Gumprecht et al., 2000).

CONSIDERATION OF AGE-, SEX-, AND SPECIES-RELATED FACTORS IN THE NEPHROTOXIC RESPONSE

There are notable age, sex, and species differences in renal function, including glomerular filtration rate (GFR) and tubular transport efficiency that could affect nephrotoxic responses. The important factors altering renal function and their associated mechanisms are discussed below.

Table 25.2 Important Factors in the Consideration of Nephrotoxic Responses

Variable	*Contributing factors*
Age	■ Alterations in renal mass, nephron number, and function
	■ Changes in tissue glutathione concentrations
	■ Altered responsiveness to vasoactive mediators and hormones
	■ Altered renal cytokine profile
	■ Concomitant diseases (e.g., diabetes mellitus, congestive heart failure, hypertension)
	■ Enzyme expression (e.g., cyclooxygenases)
Sex	■ Hormonal status
	■ Variations xenobiotic metabolism by glutathione conjugation
	■ Variations in xenobiotic metabolism due to P450 polymorphisms
	■ Specific protein production (e.g., α2-microglobulin)
	■ Spontaneous syndromes (e.g., chronic progressive nephropathy)
	■ Differences in glomerular filtration rate
	■ Variations in responsiveness to vasoactive mediators
Species	■ Renal anatomy
	■ P450 polymorphisms
	■ Renal enzyme presence and distribution (e.g., cysteine conjugate β-lyase, cyclooxygenase)
	■ Specific protein production (e.g., α2-microglobulin)
	■ Spontaneous syndromes (e.g., chronic progressive nephropathy)

The interspecies differences in renal structure and function are summarized in Table 25.1. Age-related differences in xenobiotic metabolism and the nephrotoxic response are discussed in detail in a later section, and important factors in the consideration of nephrotoxic responses are detailed in Table 25.2.

Age-Related Changes in Glomerular Function in Different Species

Age-related variations in GFR can have a significant impact on renal susceptibility to nephrotoxicity. Before 34 weeks of gestational age, the functional demands on the human fetal kidney are minimal, but they increase dramatically in the last part of the third trimester and with birth (Aperia et al., 1983; Arant, 1987). While the adult kidney receives

20 to 25% of the cardiac output, the human fetal kidney receives only 4%, which increases to 10% by the end of the first postnatal week (Jose et al., 1994; Lindeman, 1992; Lumbers, 1995). Renal function in neonates remains different from that of adults, and GFR and renal blood flow (RBF) corrected for body size is not comparable with adult values until the child is 12 months old. GFR gradually increases until adolescence, where it is maintained at adult levels (Arant, 1987).

After the age of 40, there is a progressive decrease in renal function (Beck, 1998; Epstein, 1996; Lindeman, 1990; Lumbers, 1995), which is primarily the result of progressive loss of renal mass and nephron number. Although not all individuals are affected similarly, creatinine clearance decreases, on average, by 1 ml/min/year after the age of 40 (Lindeman, 1990). This change may not be reflected by increases in serum creatinine concentration, due to generalized decreases in muscle mass in the elderly (Lindeman, 1992). Age-related decreases in GFR are more rapid in men than women and may be attributed to both physiological and physical alterations in the vasculature (e.g., glomerular damage) (Baylis and Schmidt, 1996).

Similar to humans, GFR gradually decreases with age in the rat. By 18 months, GFR is decreased to approximately 68% of its rate in young adults. These changes are manifested in decreased urine output, decreased sodium excretion, increased blood urea nitrogen concentration and proteinuria. Proteinuria along with urinary loss of albumin is evident as early as 6 months of age (Montgomery et al., 1990; Owen and Heywood, 1986).

In dogs, there is a gradual increase in GFR in the first three postnatal weeks. This increase in GFR is primarily the result of continuing glomerular differentiation in the outer cortical region. With progressive nephron loss in older dogs, there are decreases in GFR similar to those in humans (Bovee, 1984).

Decreases in RBF and GFR in the aged kidney make it more susceptible to hypoxia and ischemic damage. Age-related renal damage contributes significantly to abnormalities in GFR; age-associated decreases in renal responsiveness to certain cytokines and peptide hormones, such as renin, angiotensin, atrial natriuretic peptide (ANP), PGs, and nitric oxide (NO), can contribute significantly to changes in renal function and subsequent risk of nephrotoxicity. The potential involvement of those cytokines and peptide hormones are further discussed below.

Renin-Angiotensin-Aldosterone System

As mentioned previously, RAS is an important pathway for regulating vascular tone and extracellular fluid volume and composition (Guron et al., 2000). Alterations in RAS contribute to renal dysfunction in aging, and include decreased synthesis and release of renin, decreased renal responsiveness to renin, and altered renal responsiveness to angiotensin. In humans there is a 40 to 60% decrease in plasma renin activity (PRA) in the elderly, with defects in both synthesis and release, despite normal renin mRNA expression (Baylis and Corman, 1998; Clark, 2000; Lubran, 1995). Rats, too, have age-related decreases in PRA, which are associated with decreases in renal renin content. In addition to reductions in PRA, there is blunting of the renin response in aged humans after sodium depletion. The PRA response in old Sprague-Dawley rats in stressed conditions is blunted, but not in unstressed conditions (Baylis and Corman, 1998). Causes for decreased PRA and blunted renin responses have not been elucidated, but potential contributing factors include age-related increases in serum ANP (which inhibits renin release) or reductions in renal nerve activity (which promotes renin release) (Baylis and Corman, 1998).

The RAS has an important role in controlling GFR. ATII causes vasoconstriction of the efferent arteriole, increasing glomerular pressure. ATII regulation and renal responsiveness are altered with age (Clark, 2000). Notably, the aged kidney has increases in local ATII levels (Baylis and Schmmidt, 1996), which may contribute to afferent arteriolar vasoconstriction causing reductions in GFR. Despite increased ATII levels within the kidney, the aged kidney appears to also have an exaggerated response to ATII inhibition, such as with ACE inhibitors (Clark, 2000). In addition to altering GFR, increased activity of ATII in the kidney is also likely to promote physical damage to the kidney by promoting fibrosis and sclerosis of glomeruli and the tubular interstitium through cell proliferation and the activation of other factors associated with renal fibrosis, such as transforming growth factor β (TGFβ), platelet-derived growth factor (PDGF), osteopontin, and type IV collagen (Baylis and Corman, 1998; Guron et al., 2000).

ACE inhibitors have been associated with ischemic damage to the kidney, and decreased GFR in the elderly increases ACE inhibitor associated renal damage through decreased drug clearance (Knox and Martof, 1995). The higher plasma concentrations of the ACE inhibitor combined with increased sensitivity of the aged kidney to ACE inhibition results in hypotension and renal hypoperfusion. Species differences in susceptibility to ACE inhibitor associated renal damage have been demonstrated with quinoxins; rats are two to three times more susceptible to nephrotoxicity than dogs (Matzke and Frye, 1997; McGuire et al., 1996).

Vasoactive Mediators

Because control of RBF serves as a protective mechanism for the kidney, physiologic alterations in vascular responses decrease the renal capacity to modify GFR and increase susceptibility to ischemic damage (Clark, 2000). Renal vasoconstriction, a common cause of decreased GFR in the aged kidney, is likely a result of both altered vascular responses to mediators and altered expression of mediators (Baylis and Schmidt, 1996; Clark, 2000). The aged kidney has decreased vascular responses to vasorelaxants such as PGs (mediated by COX enzymes), ANP, pyrogen, acetylcholine, and amino acids, as well as decreased expression of nitric oxide synthase (NOS) (Bovee et al., 1984; Clark, 2000; Khan et al., 2001; Noris and Remuzzi, 1999).

Prostaglandins generated by COX act in a paracrine fashion to regulate blood flow through the glomerulus and medulla. Age and species differences in COX expression, especially COX-2 in the macula densa and renal medulla, are evident in human, rat, dog, and monkey (Khan and Alden, 2002; Khan et al., 1999, 2001; Stanfield et al., 2000). These distinctions may contribute to differences in nephrotoxicity associated with NSAIDs (Figure 25.1). COX-2 protein expression in the macula densa is high in the fetus and progressively decreases in the adulthood (Khan and Alden, 2002; Khan et al., 1999, 2001; Stanfield et al., 2000). Similarly, there are age-related decreases in vascular ratios of the vasodilatory PG (PGI_2) to the vasoconstrictive PG thromboxane (Clark, 2000). These observations are corroborated by the finding of decreased urinary excretion of the vasodilator PGE_2 in both humans and rats (Clark, 2000). In addition, COX-2 expression in the macula densa of kidneys with severe renal disease is markedly increased, and may be an effort on the part of the kidney to maintain renal function.

Decreases in renal prostaglandins with NSAIDs result in reduced afferent blood flow and GFR with subsequent renal hypoxia and ischemic damage (Matzke and Frye, 1997; Ruiz and Lowenthal, 1997). The aged kidney often has deficits in GFR, RBF, and PG concentrations, which account for the age-associated increased susceptibility to

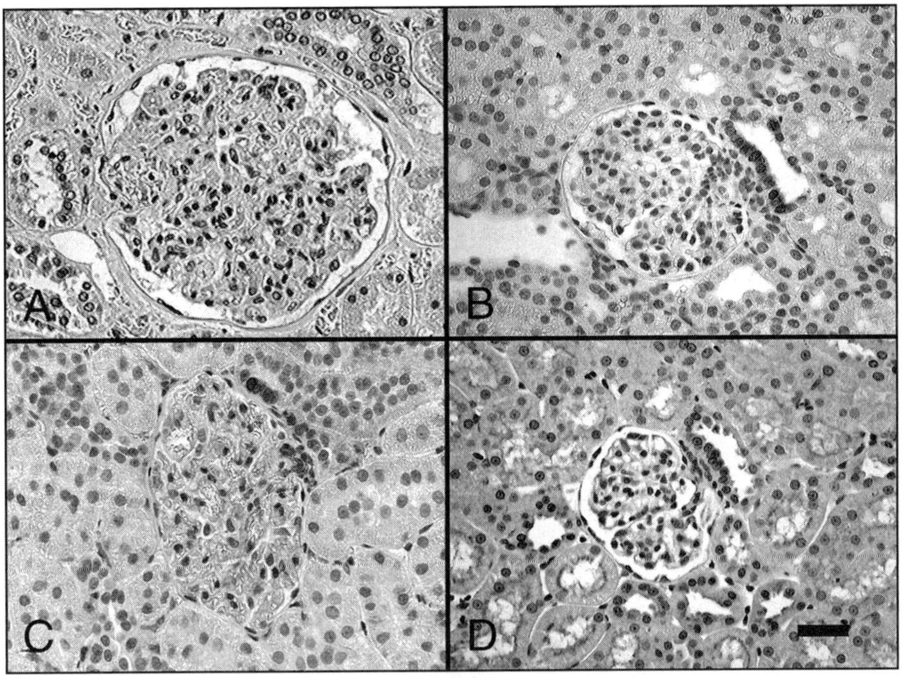

Figure 25.1 Species differences in COX-2 expression in the macula densa of normal adult animals. Kidney histological sections stained immunohistochemically for COX-2 in: (A) human, (B) rat, (C) dog, (D) monkey. Bar = 30 μm.

NSAID-associated nephrotoxicity. Renal papillary necrosis (RPN) is a common manifestation of NSAID toxicity in animals, but not in humans. This may be related to a relatively high dependence of the renal papilla on COX-2 to maintain papillary blood flow in animals, especially rats and dogs, as is suggested by high COX-2 expression in the renal papillary interstitial cells of these animals (Brix, 2002). Specific examples of NSAIDs that have a high risk of RPN in rodents and dogs are naproxen sodium, indomethacin, mefenamic acid, and ibuprofen (Brix, 2002). This damage may be compounded by concomitant use of ACE inhibitors (Adhiyaman et al., 2001). Dehydration as well as underlying heart or renal disease increases the risk of NSAID-associated nephrotoxicity (Ruiz and Lowenthal, 1997).

GFR is also regulated by ANP, which promotes vasodilation in the renal arteries, suppresses renin release, and promotes natriuresis (Beck, 1998). ANP release is enhanced by increased sodium or volume load to promote increased GFR and renal sodium excretion. The elderly are reported to have high basal plasma concentrations of ANP, which is likely to be due to both alterations in ANP release and excretion as compared with the young. However, renal responsiveness to the vasodilatory and natriuretic effects of ANP may be blunted (Beck, 1998; Epstein, 1996). Defects in ANP release and responsiveness have little effect on renal sodium handling in rats. In the rat, injections of ANP have not altered renal hemodynamics but do promote marked natriuresis (Baylis and Corman, 1998).

NO, is a potent vasodilator, generated by two constitutive forms of NOS, neuronal NOS (nNOS) and endothelial NOS (eNOS), and one inducible form of NOS (iNOS) (Calderone, 2003). NO regulates tubular reabsorption, glomerular filtration, and renal renin secretion (Noris and Remuzzi, 1999; Wang, 2002). The majority of nNOS is localized to the macula densa, and appears to have a key role in mediating renal hemodynamics (Ollerstam and Persson, 2002; Wang, 2002). NO is reduced in the aged rat kidney as well as in experimental models of renal reduction, and this reduction has been postulated to play a role in the increased vascular tone of the efferent arteriole in the aged human kidney (Clark, 2000; Noris and Remuzzi, 1999). Unlike for rats, however, there are few conclusive studies in humans on the role of NO on alterations in renal hemodynamics (Ollerstam and Persson, 2002; Wang, 2002).

Age-Related Changes in Tubular Function in Different Species

Changes in tubular function, similar to GFR, occur with age and are important in age-dependent differences in susceptibility to xenobiotic mediated nephrotoxicities (e.g., aminoglycosides and vancomycin). Fetuses and neonates have inefficient urine concentrating ability, both as a result of short proximal tubules and inefficient tubular transport (Arant, 1987; Jose et al., 1994). The human fetal kidney begins to make urine at approximately 9 weeks of gestation with progressive development of glomerular and tubular functions. At 28 weeks of gestation, the fractional excretion of sodium is as high as 5 to 6%, and is at its highest in the first 10 days postpartum (Jose et al., 1984). Mitochondrial enzymes essential for normal tubular transport continue to mature up to the postnatal day 15, as does the kidney tubular basement membrane (Fleck and Braunlich, 1995). This inefficient concentrating capacity is essential for normal growth and development of the fetus.

The earliest vital function by the kidney is to maintain the amniotic fluid volume, which is essential for normal development of the lung, limb, and gut (Lumbers, 1995). The placenta, rather than the kidney, maintains fluid and electrolyte homeostasis in the fetus (Jose et al., 1994; Lumbers, 1995). By 1 month of age, tubular function increase significantly, and fractional sodium excretion is reduced to less than 0.4% (Jose et al., 1994; Lumbers, 1995). The human kidney reaches adult capacity by 12 to 24 months of age and rat kidney by 6 weeks of age (Fleck and Braunlich, 1995; Jose et al., 1994). While the neonate has decreased renal tubular transport capacity as compared with adults, it is generally of no clinical significance (Aperia et al., 1983; Arant, 1987).

Tubular transport efficiency is greatly reduced in the aged kidney as a result of progressive nephron loss, decreased mitochondrial function, and reduced numbers of proximal tubular ion transporters (Beck, 1998; Epstein, 1996; Lindeman, 1990; Lindeman and Goldman, 1986; Lubran, 1995). The aged kidney has functional reductions in sodium-hydrogen exchange and sodium-dependent phosphate transport, potassium transport, and lower renal tubular concentrations of sodium-potassium ATPase and other enzymes (Beck, 1998; Clark, 2000). Loss of mitochondrial function reduces the efficiency of tubular transport, resulting in abnormalities in ion balance (Nicholls, 2002). Age-related reductions in tubular function such as decreased concentrating ability, ion balance, and acid–base balance parallel decreases in GFR, with decreases in function increasing after the age of 50 (Epstein, 1996; Lindeman, 1990; Lubran, 1995). Under normal circumstances, the aged kidney can maintain its homeostatic capacity. While most elderly can maintain sodium, potassium, acid, and water homeostasis under normal conditions, the response of the aged kidney to environmental factors, disease states, or treatment-related stress is often inadequate. In the face of dehydration, the aged kidney does not effectively alter urine flow or urine osmolality to prevent renal water loss; the renal defect is compounded by a decreased sense of thirst in the elderly (Beck, 1998).

Decreased urinary concentrating capacity by the aged kidney is in part due to loss of solute from the medullary gradient from defects in tubular transport as well as abnormalities in renal responsiveness to the effects of antidiuretic hormone (ADH) (Epstein, 1996; Lindeman, 1990). Secretion of ADH by the hypophyseal-pituitary-axis in response to sodium chloride infusion has been demonstrated to be higher in aged humans (Beck, 1998; Epstein, 1996), and the osmoreceptor in aged humans is more sensitive than that in the young. A cause for the decreased renal responsiveness to the effects of ADH has yet to be elucidated (Epstein, 1996; Lindeman, 1990). Dehydration may increase the risk of nephrotoxicity for certain drugs, such as NSAIDs, fungicides, and aminoglycoside antibiotics (Adelman et al., 1987a, 1987b; Appenroth and Braunlich, 1984; Appenroth and Winnefeld, 1993; Bhatt-Mehta et al., 1999; Braunlich and Otto, 1984; Fanos and Cataldi, 1999; Jose et al., 1994).

Rats also develop age-related changes in normal tubular transport function. This is characterized by progressive decreases in urinary concentrating ability and decreased chloride, calcium, and potassium retention. Defects in tubular transport with increasing age are demonstrated by decreases in *p*-aminohippurate uptake by proximal tubular cells. Between the age of 12 months and 27 months, there is a 40% decrease in active renal tubular transport in rats. Changes in tubular transport function are also evident as 5 to 10% decreases in urine osmolality in rats as early as 6 months of age (Baylis and Corman, 1998; Wabner and Chen, 1985).

Similar to rats and humans, age-related differences in tubular function are noted in dogs. Renal tubular function in dogs does not reach adult levels until after 1 to 2 months of age, as evidenced by reduced renal amino acid reabsorption in puppies as compared with adult dogs (Bovee et al., 1984). With age, dogs develop reductions in urinary concentrating capacity.

Age-related changes in tubular function as well as hydration status can affect susceptibility to xenobiotic-induced nephrotoxicity. Because the neonatal renal tubular function for the transport of organic ions is low, neonates reduce the accumulation of nephrotoxic substances into the proximal tubular epithelial cell (Ali, 1995; Fanos and Cataldi, 1999). Increased renal volume to body volume ratio in neonates is also believed to reduce the nephrotoxic susceptibility of neonates to xenobiotics (Ali, 1995; Bhatt-Mehta et al., 1999; Fanos and Cataldi, 1999). Therefore, neonates are generally less likely to have xenobiotic-induced nephrotoxicity. Examples of specific xenobiotics, in which adults have a higher risk of nephrotoxicity than neonates, include aminoglycosides, vancomycin, cadmium, cisplatin, acetaminophen, and others (Table 25.3) (Fleck and Braunlich, 1995).

Age-related increases in nephrotoxicity from aminoglycosides is considered to be related to reductions in GFR (resulting in higher plasma

Table 25.3 Risk of Xenobiotic-Associated Nephrotoxicity Associated with Increased Age

Xenobiotic	*Examples*	*Species*	*Risk factors*
ACE inhibitors	Enalapril	Human, dog	Dehydration
	Captopril		Decreased GFR
Aminoglycosides	Gentamicin	Human, rat, dog	Dehydration
	Netilmicin		Decreased GFR
	Tobramicin		Tubular deficits
	Amikacin		
NSAIDs	Ibuprofen	Human, rat	Dehydration
	Indomethacin		Decreased GFR
	Naproxin sodium		ACE inhibitors
			Heart disease
Antifungals	Amphopotericin B	Human	Dehydration

concentrations of the drug), decreases in nephron number (higher individual nephron exposure to the drug), and decreased responsiveness of the aged kidney to regeneration and repair (Ali, 1995; Paterson et al., 1998). In addition, gentamicin has been associated with changes in prostaglandin expression as reflected by increased urinary thromboxane A_2 (TXA_2), a vasoconstrictive prostaglandin, and decreased urinary PGE : TXA_2 ratios (Ali, 1995). Gentamicin-associated nephrotoxicity risk is exacerbated by dehydration, which is common in the elderly population, and concomitant use of other nephrotoxic drugs such as ACE inhibitors (Adhiyaman et al., 2001; Ali, 1995; Paterson et al., 1998).

MAJOR MECHANISMS OF AGE-ASSOCIATED RENAL DAMAGE IN DIFFERENT SPECIES

Age-related renal damage contributes significantly to abnormalities in the GFR and tubular function. This results in different susceptibility of the aged population to nephrotoxic response as compared with neonates, juveniles, and young adults. Age-related changes in renal function are primarily the result of physical damage to the kidney, with decreased nephron number and subsequent decreases in concentrating ability, secretion, and acidification capacity (Beck, 1998; Clark, 2000; Lindeman, 1990). Glomerular damage and loss result in abnormal glomerular filtration, cortical shunting of blood, and nephron loss, which decrease

GFR, RBF, and tubular functions (Baylis and Corman, 1998). This decrease in GFR, tubular function, and renal vascular responsiveness to hormones and cytokines, as discussed previously, increases the risk of renal damage from xenobiotics.

Oxidative Damage

Accumulated parenchymal injury from reactive oxygen species plays an important role in age-related nephron loss (Lindeman, 1990; Linton et al., 2001; Nicholls, 2002). Tissues with high metabolic rates, such as the kidney, tend to accumulate more oxidized proteins than metabolically inactive organs (Linton et al., 2001), suggesting a higher generation of free radicals and reactive oxygen species. Free radicals and reactive oxygen species (e.g., hydroxyl radical, superoxide anion, NO, and hydrogen peroxide) are generated from normal metabolic or inflammatory processes and xenobiotic metabolism, and contribute to age-related renal alterations (Linton et al., 2001). These reactive species interact with cellular macromolecules causing the formation of cross-linked and aggregated protein, and nucleic acid adducts or strand breaks. Damage to proteins and nucleic acids leads to cell dysfunction, cell death, or, potentially, neoplasia. Reactive oxygen species also promote the expression of matrix metalloproteinases and collagenases, which result in alterations to the glomerular basement membrane and matrix (Yoo et al., 2002). Oxidative damage to mitochondria, particularly in cells with high respiratory requirements such as the kidney, inhibits normal respiration, and therefore the function of the cell. Mitochondrial oxidative stress has also been demonstrated to promote aging (Linton et al., 2001; Nicholls, 2002). Mitochondrial dysfunction may lead to the generation of incompletely oxidized materials, resulting in the generation of oxygen free radicals and oxidative stress (Epstein, 1996; Nicholls, 2002).

Two important components of the cellular mechanism protecting against the damage from reactive molecules are glutathione peroxidase and glutathione (γ-glutamyl-L-cysteinylglycine) (GSH). Glutathione peroxidase eliminates peroxidase-generated free radicals through the oxidation of GSH, the primary source of nonprotein sulfhydryl groups in the cytoplasm (Parkinson, 1996). It has been demonstrated that with increasing age, concentrations of antioxidants, glutathione peroxidase, and GSH are decreased in many cell types, including renal tubular epithelial cells. These molecules serve to scavenge free radicals and reactive metabolites and detoxify them, thereby protecting the cell from electrophilic and peroxidative damage (Parkinson, 1996).

GSH depletion increases the risk of nephrotoxicity (Appenroth and Winnefeld, 1993; Briguori et al., 2002; Hirata et al., 1990; Inselmann et al., 1994; Mizutani et al., 1990; Nakagawa et al., 1995). Because the antioxidant glutathione peroxidase and GSH concentrations are decreased in the aged kidney, it becomes more vulnerable to electrophilic xenobiotic metabolites. Xenobiotics that have increased toxicity as a result of GSH depletion include acetaminophen, cisplatin, cyclosporin A, paraquat, and arsenite (Fleck and Braunlish, 1995; Hirata et al., 1990; Inselmann et al., 1994; Nakagawa et al., 1995; Tarloff et al., 1996). Increased renal GSH after treatment with *N*-acetylcysteine protects against cisplatin-, acetaminophen-, cyclosporin-, and contrast agent induced nephrotoxicity in humans (Briguori et al., 2002; Eguia and Materson, 1997; Kelly, 1998; Nisar and Feinfeld, 2002; Tariq et al., 1999). Female rats, in general, tend to be more vulnerable to oxidative damage due to lower nonprotein sulfhydryl (NPSH) content in the kidneys as compared with male rats (Hong et al., 2001; Nyarko et al., 1997; Rankin et al., 2001).

Advanced Glycosylation Products

Protein glycation is a normal nonoxidative protein modification that yields advanced glycosylation end products (AGEs). However, with age and with high-protein diets, there is accumulation of AGEs in the kidney (Linton et al., 2001), which promotes extracellular matrix synthesis by the glomerular mesangial cells (Baylis and Corman, 1998; Lindeman, 1990). It has been suggested that the accumulation of AGEs within the glomeruli is an important cause of renal lesions in aged humans and rats. Rats given aminoguanidine, an AGE inhibitor (also an inhibition of iNOS), have decreased accumulation of AGEs in the kidney and reduced glomerular damage and nephron loss (Baylis and Corman, 1998; Lindeman, 1990).

Hyperfiltration and Hyperperfusion

In order to maintain RBF and GFR after nephron damage and loss, blood flow is redistributed to the remaining intact nephrons. The redistribution of blood flow from damaged to healthy glomeruli results in increased individual glomerular blood flow with subsequent glomerular hyperfiltration and hyperperfusion (Clark, 2000; Lindeman, 1992; Lindeman and Goldman, 1986). Data from animal models suggest that hyperfiltration alters the integrity of the glomerular capillary membrane, resulting in urinary protein loss, increased mesangial protein deposits, basement membrane changes, and progressive loss of renal function through glomerular sclerosis (Baylis and Corman, 1998; Clark, 2000). In addition to nephron loss, glomerular hyperfiltration and subsequent progressive glomerulosclerosis have been attributed to high dietary protein intake (Lindeman, 1992), although this hypothesis is controversial, and does not appear to have a deleterious effect on the dog kidney (Bovee, 1991).

Evidence that glomerular hypertension contributes to glomerular damage and subsequent aging has been demonstrated with the use of ACE inhibitors. The progression of renal disease can be slowed down with the use of ACE inhibitors in humans and rats (Lewis et al., 1993). Long-term follow-up studies in unilateral nephrectomy patients indicate that hyperfiltration in a normal kidney does not adversely affect renal function (Clark, 2000; Lindeman, 1992). It is therefore likely that hyperfiltration-associated progression of renal disease is enhanced by concomitant age-related alterations in renal responsiveness to vasoactive compounds (e.g., PGs) and renal cytokine expression (e.g., TGFβ and fibroblast growth factor) (Clark, 2000).

Transforming Growth Factor β

Overexpression of growth factors such as TGFβ and PDGF in the aging kidney is likely to contribute to the development of renal fibrosis and glomerulosclerosis (Clark, 2000). The causes of increased TGFβ expression in the kidney probably include high dietary protein and increased local ATII activity (Clark, 2000). TGFβ promotes matrix synthesis and cell proliferation in fibroblasts and other cells, as well as increasing collagenases and matrix metalloproteases, which degrade matrix and therefore contribute to the progressive age-related changes in the human kidney (Clark, 2000). Older rats also overexpress TGFβ ?in the renal interstitium, which probably contributes to the age-related increase in interstitial fibrosis. TGFβ ?expression in the rat kidney results in reduced age-related interstitial fibrosis and glomerulosclerosis, and can be reduced by ATII inhibition or decreased dietary protein (Clark, 2000; Montgomery and Seely, 1990).

SEX-RELATED DIFFERENCES IN THE NEPHROTOXIC RESPONSE

Sex-related differences in hormonal status can affect renal function and thus responses to xenobiotics. Specific examples of sex-related differences in metabolism of xenobiotics due to differences in GSH conjugation and cytochrome P450 polymorphisms will be discussed in detail elsewhere in this section, and important factors in the consideration of nephrotoxic responses are detailed in Table 25.2.

Hormonal differences contribute to variations in the response of the renal vasculature and tubular function. For example, there are notable differences between age-matched men and women in the basal GFR (10% lower in women), development of ESRD (faster progression and earlier onset in men), and susceptibility to nephrotoxicants such as aminoglycosides (men more susceptible than women). Chronic progressive nephropathy (CPN) and α2-microglobulin nephropathy are two sex-related renal manifestations in rats that can affect the nephrotoxic response to xenobiotics. CPN occurs earlier and more substantially in male rats than female rats, and progresses with age (Figure 25.2), and is characterized by proteinuria, with subsequent alterations in urinary concentrating capacity and GFR, as well as increased renal size, weight, and cortical scarring (Montgomery and Seely, 1990).

The incidence of CPN in rats is very low in the first 20 weeks of life, with only 4 to 5% of males and females having CPN lesions. This increases to 55% and 19% in males and females, respectively, between 49 and 62 weeks of age and can reach up to 50% in females and 100% in males after 90 weeks of age (Burek et a., 1988). Sex hormones appear to be important in the development of this disease; castration reduces the onset and progression of the renal lesions, while testosterone, thyroxin, and adrenal hormones promote the progression (Montomery et al., 1990). This spontaneous chronic renal disease is more severe in male Fischer 344 (F344) and Sprague-Dawley (SD) rats (Couser and Stilmant, 1975; Montgomery and Seely, 1990; Neugarten et al., 1999; Owen and Heywood, 1986).

Histologically, the earliest changes appear to be thickening of the basement membrane of the proximal tubules and thickened and enlarged glomerular tufts (Baylis and Corman, 1998; Owen and Heywood, 1986). These changes progress to extensive loss of tubules and glomeruli, marked interstitial fibrosis (attributed to deposition of thrombospondin and fibronectin), interstitial mononuclear cell infiltrates,

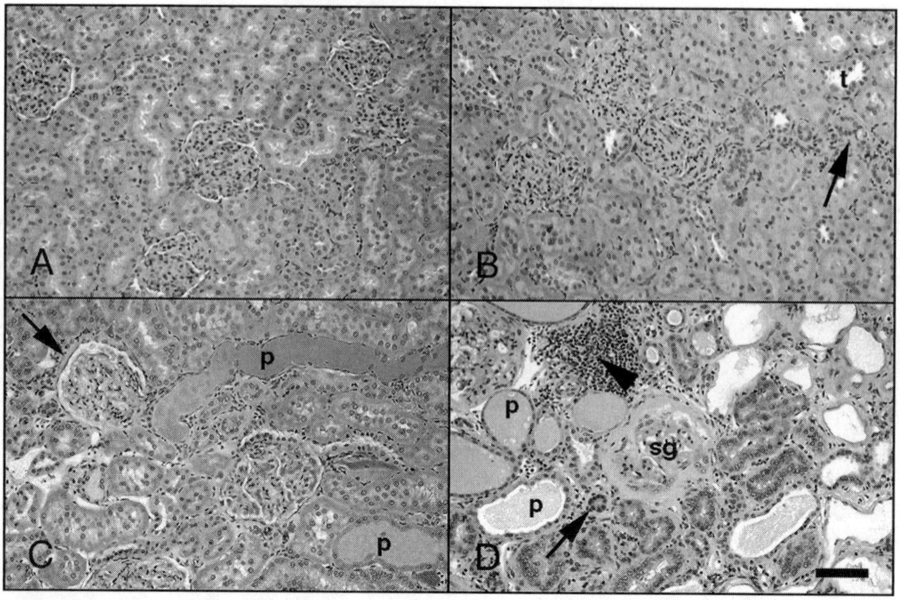

Figure 25.2 Chronic progressive nephropathy in rats. Kidney histological sections stained with hematoxylin and eosin. (A) At 1.5 months, the kidney is histologically normal. (B) At 6 months of age, there are scattered basophilic tubules (arrow) and mild tubular dilation (t). (C) By 12 months, tubules are dilated by proteinaceous fluid (p) and basement membrane thickening is evident around Bowman's capsule (arrow). (D) By 18 months, additional changes include lymphocytic and histiocytic interstitial inflammatory infiltrates (arrowhead), tubular atrophy (arrow), glomerular sclerosis (sg), thickened tubular basement membranes, and an expanded interstitium. Bar = 60 μm.

and widespread mineralization, culminating in ESRD. Interestingly, Wistar, Long-Evans, and brown Norway strains are not as affected by CPN as SD or F344 rats, and Osborne-Mendel (OM) and Buffalo strains of rats are relatively resistant to CPN (Montgomery and Seely, 1990; Owen and Heywood, 1986; Solleveld and Boorman, 1986).

Many xenobiotics, such as gentamicin, hydroquinone, heavy metals (e.g., cadmium, mercury), and *d*-limonene, have been associated with increased incidence and severity of CPN in rats (Hard et al., 1997; Khan and Alden, 2002). The mechanism of increased nephrotoxicity associated with CPN is uncertain, but reductions in nephron number in rats with CPN may increase the single nephron exposure to nephrotoxic xenobiotics (Seely et al., 2002). However, not all nephrotoxic xenobiotics result in increased CPN, suggesting diversity in the mechanisms by which these drugs alter CPN (Seely et al., 2002).

Xenobiotic-induced α2-microglobulin nephropathy syndrome is exclusively manifested in male rats. The xenobiotic or its metabolites bind reversibly and specifically with α2-microglobulin in the kidney. The associated nephropathy is characterized by the accumulation of hyaline droplets in the cytoplasm of the proximal tubular epithelial cells. These hyaline droplets correlate ultrastructurally with large crystalloid phagolysosomes from protein reabsorption, mostly of α2-microglobulin. Only male rats synthesize large amounts of α2-microglobulin in their livers, which are excreted through the kidney.

This poorly hydrolyzable protein accumulates in phagolysosomes after being resorbed by the proximal tubules. Xenobiotics that increase the production of α2-microglobulin cause an accumulation of the protein in proximal tubular epithelial cells. This accumulation may overload the cell, resultin in apoptosis and necrosis (Montgomery and Seely, 1990). Chronically, it may cause hyperplasia of the pelvic urothelium, as well as marked intratubular mineral deposits in the inner zone of the renal medulla and increased incidence of CPN and tubular tumors (Khan and Alden, 2002). Examples of xenobiotics associated with this nephropathy are jet fuels (JP-4, JP-TS, JP-7, RJ-5, JP-10), dimethyl methylphosphonate, hexachloroethane, isophorone, *d*-limonene, lindane, paradichlorobenzene, pentachloroethane, and unleaded gasoline (Khan and Alden, 2002). α2-microglobulin nephropathy does not occur in female rats or in other species with these compounds.

SPECIFIC EXAMPLES OF AGE, SEX, AND SPECIES RELATED DIFFERENCES IN XENOBIOTIC METABOLISM AND THE NEPHROTOXIC RESPONSE

Nephrotoxicity occurs most commonly in the highly metabolically active proximal convoluted tubules of the kidney, and less frequently involves the distal tubules and collecting ducts. Nephrotoxicity frequently results from the accumulation of toxic xenobiotics or their metabolites in the kidney or the intrarenal biotransformation of xenobiotics to toxic metabolites (Khan and Alden, 2002). The majority of xenobiotic metabolism occurs in the liver, and to a lesser extent in the kidney, either through GSH conjugation or through cytochrome P450 (P450)-mediated oxidation (Anders et al., 1988; Dekant, 1996). The severity of nephrotoxicity induced by xenobiotics is not uniform between genders, species, and ages. Variations in metabolism, either in GSH conjugation or P450-mediated oxidation, and hormonal melieu have been demonstrated to cause variations in susceptibility to xenobiotic-associated nephrotoxicity (Anders et al., 1988; Baliga et al., 1999; Dekant, 1996; Monks et al., 1990). The majority of examples in this section are derived from rat studies, with effects in other species noted where applicable. Important factors in the consideration of nephrotoxic responses are detailed in Table 25.2.

Sex Related Differences in Glutathione Conjugation of Xenobiotics Influencing Toxicity

GSH conjugation of xenobiotics in the liver, and to lesser extent kidney, predominately generates inactive metabolites that are secreted into bile (Parkinson, 1996). However, in some cases, conjugation of xenobiotics to GSH results in a bioactive *S*-conjugated metabolite that, through a multistep bioactivation process, may cause nephrotoxicity. A GSH *S*-conjugated xenobiotic or metabolite may be metabolized by γ-glutamyl transferase (GGT) at the level of the bile canaliculi or renal proximal tubules resulting in a cysteine-conjugate, which is taken up by proximal tubular cells through a basolaterally oriented electrogenic, sodium-coupled and probenecid-sensitive transporter. This cysteine-conjugate may be directly nephrotoxic, or, more commonly, bioactivated within the renal proximal tubular cytoplasm to a reactive electrophilic molecule (Anders et al., 1988; Baliga et al., 1999; Dekant, 1996, 2001; Monks et al., 1990). These electrophilic metabolites interact with cellular macromolecules and nucleic acids, causing cellular dysfunction and cell death. The primary GSH conjugated nephrotoxic xenobiotics that have been extensively examined are the halogenated alkenes, halogenated alkanes, and polyphenolic compounds (Table 25.4) (Monks et al., 1990).

Nephrotoxicity of Glutathione-Metabolized Xenobiotics in Males

The majority of GSH-activated nephrotoxicants that have a more severe effect in male as compared with female rats are haloalkenes and polyphenolic compounds. In both cases, the mechanism of this gender

Table 25.4 Gender Difference in Xenobiotic-Induced Nephrotoxicity: Glutathione Conjugation

Compound	*Cytochrome P450*	*Gender*	*Toxic intermediate*	*Mechanism*
Halogenated alkenes	Trichloroethylene Percholoroethyelene Hexachlorobutadiene	Male > female	*S*-(1,2-dichlorovinyl)cysteine *S*-(1,2-dichlorovinyl)glutatione *S*-(-trichlorovinyl)cysteine	*S*-conjugation
Halogenated alkanes	Vicinal dihaloalkanes Ethylene dibromide Ethylene dichloride	Female > male	Half-sulfur mustard	Uncertain
Phenols and quinones	Hydroquinone Ethoxyquin	Male > female	Benzoquinone	*S*-conjugation

difference in nephrotoxicity is attributed to higher hepatic GSH conjugation efficiency, as characterized by greater concentrations of GSH-conjugated compounds in male rats as compared with female rats (Lash et al., 1998a).

In the case of haloalkenes, such as trichloroethylene (Tri), an industrial chemical, perchloroetheylene (Per), which is a dry-cleaning solvent, and hexachlorabutadiene (HCBD), the liver generated *S*-conjugate is hydrolyzed by GGT at the level of the bile canaliculi and proximal tubular epithelium to form cysteine conjugates (Dekant, 1996, 2001). The cysteine conjugates are absorbed by the proximal tubular epithelial cells and bioactivated into cytotoxic reactive thiols by renal pyridoxal phosphate dependent enzymes, such as cysteine conjugate β-lyase (Birner et al., 1995, 1998; Bruning and Bolt, 2000; Lash et al., 1998a, 1998b). In the case of Per, GSH conjugation occurs in both the liver and kidney. With all three compounds, there is a greater incidence of nephrotoxicity in males (Birner et al., 1995, 1998; Bruning and Bolt, 2000; Lash et al., 1998a, 1998b).

Metabolites associated with nephrotoxicity due to Tri exposure have also been demonstrated in humans, with males having 3.4-fold greater metabolite concentration in their blood than females (Bruning and Bolt, 2000; Lash et al., 1999). Therefore, the risk of nephrotoxicity for human males after exposure to Tri may be greater than that for human females. Humans also excrete the mercapturic acid of Per metabolites, indicating that they too conjugate Per to GSH, suggesting that it may be nephrotoxic in humans (Volkel et al., 1998). However, humans have 10 to 20% less renal cysteine conjugate β-lyase activity than rats, and therefore may not form sufficient quantities of the reactive thiol to induce nephrotoxicity (Lash et al., 1998a). The metabolism of HCBD by P450 may also be important in the gender difference in nephrotoxicity for this compound, and is discussed later (Birner et al., 1998).

In the case of polyphenolic compounds, GSH *S*-conjugation of oxidized metabolites results in the generation of highly reactive electrophiles (Dekant, 1996; Lau et al., 2001; Monks et al., 1990). Examples of polyphenolic compounds that are associated with more severe nephrotoxicity in male than female rats are hydroquinone (HQ) and ethoxyquin (EQ) (Monks and Lau, 1998). The cytotoxicity of these species is in their ability to generate reactive oxygen species, and their early targets are the plasma membrane and nucleus. While polyphenolic-GSH conjugates are toxic, metabolites, such as thioethers, are often more reactive than the parent polyphenol, and can cause oxidative stress as well as form covalent adducts (Corley et al., 2000; Lau et al., 2001; Monks et al., 1990).

Nephrotoxicity of Glutathione Metabolized Xenobiotics in Females

Examples of enhanced nephrotoxicity in female rats as compared with male rats as the result of GSH metabolism of xenobiotics are less common. The haloalkane ethylene dichloride has been demonstrated to cause nephrotoxicity in female F344 rats, but not in male or female SD or OM rats (Morgan et al., 1990). The general mechanism of nephrotoxicity of haloalkanes is through the generation of a directly toxic episulfonium ion from GSH-conjugated xenobiotics. Episulfonium ions are highly electrophilic and interact with nucleic acids, forming DNA and RNA adducts. However, the mechanism of this sex difference has not been elucidated (Dekant, 1996; Monks et al., 1990).

Sex and Species Polymorphisms in Tubular Cytochrome P450 Metabolism Influencing Xenobiotic Toxicity

P450 proteins are comprised of a superfamily of hemethiolate proteins that oxidize compounds (Baliga et al., 1999). There are multiple Cytochrome P family members, but Cytochrome P1, Cytochrome P2, and Cytochrome P3 are the most important in human xenobiotic metabolism. While these enzymes are predominately found in the liver, many extra-hepatic tissues, including kidney, have notable Cytochrome P activity. Cytochrome Ps have complex sex-, tissue-, age-, and species-specific expression, which probably contribute to sex and species difference in xenobiotic-associated nephrotoxicity (Baliga et al., 1999; Mugford and Kedderis, 1998). While there are many examples of gender-enhanced or gender-specific differences in Cytochrome P450 activity in rats and humans, only a subset have been associated with xenobiotic-induced nephrotoxicity (Table 25.5) (Chen et al., 1999; de Wildt et al., 1999; Kalsotra et al., 2002; Pfohl-Leszkowicz et al., 1998). However, their role in gender differences in xenobiotic metabolism and toxicity may be important.

The majority of studies on sex-dependent differences in xenobiotic metabolism have used the rat, which has more prominent differences in sex-dependent xenobiotic metabolism as compared with the human. Because of extensive inbreeding, interanimal variation in the rat is minimal, as compared with humans. It is therefore important to note that

Table 25.5 Gender Difference in Xenobiotic-Induced Nephrotoxicity: P450

Compound	*Cytochrome P450*	*Gender*	*Toxic intermediate*	*Mechanism*
Acetaminophen	CYP3A4	Female > male	*N*-acetyl-*p*-benzoquinoneimine *p*-aminophenol	Estrogen-related
1,2-dichloropropane	CYP2E1	Male > female		Testosterone induced
HCBD	CYP3A	Male > female	*N*-acetyl-*S*-(pentachlorobutadienyl)-L-cysteine sulfoxide	Metabolite only in males
NDPS	—[a]	Female > male	*N*-(3,5-dichlorophenyl) 2-hydroxysuccinimide *N*-(3,5-dichlorophenyl)-2-hydroxylsucciniamic acid	Possibly increased ability of females to form sulfate conjugates from alcohols
Gentamicin	—[a]	Female > male	Reactive oxygen species	Estrogen induced

[a]Uncertain which P450.

human and rat Cytochrome P450 enzymes are divergent, and toxicity in rats can only be compared to humans if metabolism is through the male-predominate Cytochrome P1A or Cytochrome P2E. However, animal models used to predict sex-dependent differences in xenobiotic toxicities do not translate well from rats to humans or other animals. For example, Cytochrome P2C, which is highly expressed in rats and is a sex-specific isoform of Cytochrome P450, is not expressed in humans (Mugford and Kedderis, 1998).

Nephrotoxicity of Cytochrome P450 Metabolized Xenobiotics in Males

Examples of P450 enzymes that have higher levels in male rats than female rats and which may contribute to a higher incidence of nephrotoxicity in males include Cytochrome P3A and Cytochrome P2E1. Sex-related differences in nephrotoxicity in rats have been directly related to these enzymes in the cases of HCBD (Cytochrome P3A), 1,2-dichloropropane (DCP) (Cytochrome P2E1), and acetaminophen (Cytochrome P2E1) (Birner et al., 1995, 1998; Hoivik et al., 1995;Tarloff et al., 1996).

The Cytochrome P3A subfamily is the most abundant family in the liver in humans and has at least three isoforms: Cytochrome P3A4, 3A5, and 3A7 (de Wildt et al., 1999; Monks and Lau, 1998). The Cytochrome P3A subfamily accounts for the majority of xenobiotic metabolism in the human liver, and Cytochrome P3A4 is important in the metabolism of many drugs (de Wildt et al., 1999). HCBD, a potent nephrotoxicant in the rat, is also differentially metabolized by P450 in males and females. Metabolites of HCBD in male rats may be oxidized by members of the Cytochrome P3A family, which are male specific, to a cysteine sulfoxide that is only evident in male rats. This sulfoxide is a Michael acceptor (an electrophilic metabolite) and interacts with cellular proteins and nucleic acids, resulting in cytotoxicity. Because this metabolite has been demonstrated only in male rat, this may be the primary mechanism of gender-related nephrotoxicity, rather than a GSH-mediated mechanism (Birner et al., 1995, 1998; Werner et al., 1995).

Cytochrome P2E1 bioactivates low molecular weight toxicants and is highly inducible by ethanol (Chen et al., 1999). In rodents, Cytochrome P2E1 has been implicated in sex differences in xenobiotic-induced nephrotoxicity associated with P450 metabolism (Chen et al., 1999). In mice, Cytochrome P2E1 activity is only evident in males (Speerschneider and Dekant, 1995). The basis of the sex differences involves testosterone and Cytochrome P2E1; Cytochrome P2E1 is induced by testosterone while inhibition of testosterone is associated with decreased hepatic

Cytochrome P2E1 activity (Chen et al., 1999). Studies of DCP in Wistar rats have demonstrated that nephrotoxicity in males can be prevented by castration (Odinecs et al., 1995). In addition, pretreatment of female and castrated male rats with testosterone increases DCP nephrotoxicity. Therefore, the mechanism for gender differences in DCP nephrotoxicity probably lies in the effect of testosterone on the DCP oxidizing cytochrome Cytochrome P2E1 (Odinecs et al., 1995).

Cytochrome P2E1 has also been implicated in the nephrotoxicity of acetaminophen (APAP) in male, but not female, CD-1 mice. Testosterone treatment of CD-1 female mice increases renal Cytochrome P2E1, as well as acetaminophen-induced nephrotoxicity. Further evidence that Cytochrome P2E1 was involved in APAP-induced nephrotoxicity was obtained from *in vitro* studies of microsomes from mice pretreated with testosterone. Renal, but not hepatic, microsomes from mice treated with testosterone demonstrated increased APAP activation. These data indicate that renal biotransformation of APAP to toxic electrophiles is important in APAP-induced nephrotoxicity in CD-1 mice (Hoivik et al., 1995). These findings are in contrast to those for young rats, in which APAP causes more severe nephrotoxicity in females as compared with males (Tarloff et al., 1996).

Nephrotoxicity of Cytochrome P450 Metabolized Xenobiotics in Females

Examples of Cytochrome P450 enzymes found at higher levels in females than males and which may contribute to a higher incidence of nephrotoxicity in females, as compared with males, have not been extensively examined. It appears that estrogen is important in the regulation of these differences, however few specific Cytochrome P450 enzymes associated with nephrotoxicity have been identified. An example of a Cytochrome P450 that is estrogen dependent and more highly expressed in female rats is Cytochrome P4F (Kalsotra et al., 2002). While no specific xenobiotic-associated nephrotoxicity has been correlated with this cytochrome, its role in gender differences in xenobiotic metabolism and toxicity may be important. In human females, Cytochrome P3A4 levels are higher than in males; however, sex differences in nephrotoxicity attributed to this cytochrome may actually be the result of lower P-glycoprotein activity in females (Harris et al., 1995; Meibohm et al., 2002). Examples of Cytochrome P450 metabolized xenobiotics in which there is a higher incidence of nephrotoxicity in female rats as compared with males are *N*-(3,5-dichlorophenyl)succinimide (NDPS), gentamicin, and acetaminophen.

NDPS is a fungicide used in agriculture. Female rats, when given NDPS intraperitoneally, are twice as sensitive as male rats to nephrotoxicity, and renal damage is differentially located. Proximal tubular damage in females is localized to the S_3 segment of the proximal tubule, whereas in males, it is localized to the S_1 and S_2 segments. The mechanism of toxicity appears to be Cytochrome P450-based oxidation of the succinimide ring in NDPS, which results in several nephrotoxic metabolites (Birner et al., 1995; Nyarko et al., 1997; Rankin et al., 2001). An exact mechanism for the gender difference in rats has not been elucidated, but has been suggested to be a greater ability of females to form the nephrotoxic sulfate conjugates from alcohols (Hong et al., 2001; Nyarko et al., 1997; Rankin et al., 2001).

There is evidence that both female rats and female humans are more susceptible to gentamicin-induced nephrotoxicity. The mechanism of gentamicin-induced nephrotoxicity is uncertain, but altered mitochondrial respiration resulting in the generation of incompletely reduced oxygen species (reactive oxygen species) has been implicated (Ali, 1995; Baliga et al., 1999). Studies in Wistar rats show that rats with normal or elevated estrogen levels had renal functional impairment after gentamicin treatment. Testosterone supplementation partially protected male and female rats from gentamicin-induced nephrotoxicity. Estradiol administration promoted renal deficits in males and females treated with gentamicin (Carraro-Eduardo et al., 1993). The exact mechanisms involved in gender differences in nephrotoxicity have not been elucidated, but hormonally mediated differences in the P450 system appear to underlie these differences (Ali, 1995).

Unlike mice, APAP causes nephrotoxicity in young (3-month old) female rats but not male rats. This sex-difference is eliminated by 18 months of age, at which time males and females are equally susceptible to the nephrotoxic effects of APAP. Interestingly, there is also an age-related vulnerability to nephrotoxicity: 18-month old rats develop nephrotoxicity at 40% lower doses than 3-month old rats. A mechanism for the sex-based differences in nephrotoxicity is not completely clear. However, it is evident that females have three times more covalently bound APAP in the kidney than do males, as well as greater reductions in nonprotein sulfhydryl groups than their male counterparts after APAP treatment. These data suggest that there is greater bioactivation of APAP in the kidney of young female rats and that both oxidative and deacetylation of APAP are important in the development of nephrotoxicity (Tarloff et al., 1996).

RENAL CARCINOGENESIS AND THE NEPHROTOXIC RESPONSE

Approximately 2% of all noncutaneous malignancies in humans arise in the kidney, the majority of which are adenocarcinomas arising from the proximal tubules (Hard and Khan, 1998). These renal carcinomas occur after 40 years of age, with peak incidence in the sixth and seventh decades of life. Males are affected approximately two to three times as frequently as females. Renal adenomas are much more frequent, with incidence as high as 20% of autopsied specimens (Khan and Alden, 2002). In the aged dog, renal adenomas and carcinomas are less common than in humans, comprising approximately 1% of all neoplasms (Maxie and Prescott, 2002). Aged rats develop renal adenomas and carcinomas more frequently than other common laboratory animal species (Hard, 1990). One study demonstrated the overall incidence of spontaneous renal tumors at 2 years of age is 0.08% and 0.28% in SD and F344 rats, respectively (Chandra et al., 1993). Other studies have shown that the ranked incidence of spontaneous kidney tumors is higher in male as compared with female F344 rats (1.34% and 0.55%, respectively) and B6C3F1 mice (0.46% and 0.25%, respectively) (Wolf, 2002). The male preponderance of spontaneous renal tumors in rodents is similar to that in male humans (Khan and Alden, 2002).

While rodents are commonly used to study xenobiotic-induced renal carcinogenesis, interpretation of results relative to relevance in humans may be difficult. Renal tubular neoplasms occurring as a result of xenobiotic exposure cannot be morphologically differentiated from spontaneously occurring tubular tumors. Weak tumor responses with incidence less than 8.5% in a group size of 50 at a 99% confidence level may not be recognized in cancer bioassays (Khan and Alden, 2002).

In addition, rat-specific lesions, such as CPN and the production of α2-microglobulin nephropathy in male rats, may be associated with increased susceptibility to xenobiotic-induced renal carcinogenesis as compared with humans. Indeed, significantly increased incidence and severity of CPN on a group basis has been reported to correlate with a weak tumor response, due to increases in the replication rate on a nephron basis. However, severity of CPN does not necessarily correlate with the development of renal tubular carcinomas (Seely et al., 2002).

Toxic renal tumorigens also tend to increase the incidence or severity of CPN, with increased number of nephrons showing high mitotic activity (Hard and Khan, 1998). Because of the importance of growth stimulation in tumorigenesis, this individual nephron replication rate increase may have significance when considered in association with the age-associated chromosome alterations in rats (Bucher et al., 1990a, 1990b; Hard et al., 1997; Khan and Alden, 2002). CPN is a rat-specific disease, and so CPN-associated neoplasia in rat carcinogenicity studies using chemicals and therapeutics has little relevance for predicting human renal neoplasia with these agents. Examples of xenobiotics associated with CPN-related renal carcinogenicity have been previously discussed. Males are more susceptible to xenobiotic-induced CPN, in both the incidence and severity. Similarly, males are more prone to develop CPN-associated renal neoplasia.

Specific subchronic lesions of the α2-microglobulin nephropathy syndrome, as discussed previously, also appears to have a potential role in renal tumorigenesis. Examples of chemicals exhibiting the induced specific subchronic lesions and tumor response after being administered for at least 1 year with sacrifice at 2 years of age include: jet fuels (JP-4, JP-TS, JP-7, RJ-5, JP-10), dimethyl methylphosphonate, hexachloroethane, isophorone, *d*-limonene, lindane, paradichlorobenzene, pentachloroethane, and unleaded gasoline (Khan and Alden, 2002).

Because toxic injury is accompanied by increased apoptosis, replication rate increase, and hyperplastic changes in tubular epithelium, nephrotoxicity in rats has been associated with increased risk of renal tumorigenesis (Hard and Khan, 1998). Sex differences in xenobiotic induced nephrotoxicity may be the result of differences in xenobiotic metabolism and subsequent nephrotoxicity as described previously. The relevance of this rodent response for higher species, including humans, has not been demonstrated. In part, this may be because the high level of xenobiotic administration required to induce precursor lesions is unlikely to be encountered outside the laboratory setting.

Several additional kinds of renal neoplasms, which rarely occur spontaneously, are associated with xenobiotic exposure in the rat. The highly malignant renal mesenchymal tumor, composed of fibroblastic spindle cells, smooth muscle cells, vascular cells, and, occasionally, striated muscle and cartilage, occurs in young rats and is associated with genotoxic insult at an immature age (Frank et al., 1992). The nephroblastoma is a tumor of the young rat, rarely occurring spontaneously, but inducible with specific *in utero* xenobiotic insult (Hard, 1985).

Even though toxicity *per se* does not enable a confident prediction for a renal neoplastic response, it certainly prevents a confidential prediction that no renal neoplasia will ensue in lifetime studies. Again, renal tumorigens are consistently reported to cause tubular toxicity in short-term studies. The absence of genotoxic and renal toxicity (in subchronic studies) enables a confident prediction that the test will not induce renal neoplasia (Bucher et al., 1990a, 1990b; Hard et al., 1997).

CONCLUSIONS

There are notable age-, sex-, and species-related differences in renal anatomy and function, as well as xenobiotic metabolism, that could contribute to susceptibility to xenobiotic-induced nephrotoxicity. This chapter has detailed and summarized the important factors that need to be considered for understanding the nephrotoxic response. Risk assessment for new chemicals and therapeutics requires understanding of not only the mechanistic effects of xenobiotics, but also how the animal's age, sex, species, breed, or strain may contribute to the development of renal lesions, and whether this change translates to humans.

REFERENCES

Abdi R, Slakey D, Kittur D, and Racusen LC (1998) Heterogeneity of glomerular size in normal donor kidneys: Impact of race. *Am J Kidney Dis* 32:43–46.

Adelman RD, Wirth F, and Rubio T (1987a) A controlled study of the nephrotoxicity of mezlocillin and gentamicin plus ampicillin in the neonate. *J Pediatr* 111:888–893.

Adelman RD, Wirth F, and Rubio T (1987b) A controlled study of the nephrotoxicity of mezlocillin and amikacin in the neonate. *Am J Dis Child* 141:1175–1178.

Adhiyaman V, Asghar M, Oke A, White AD, and Shah IU (2001) Nephrotoxicity in the elderly due to co-prescription of angiotensin converting enzyme inhibitors and nonsteroidal anti-inflammatory drugs. *J R Soc Med* 94:512–514.

Ali BH (1995) Gentamicin nephrotoxicity in humans and animals: Some recent research. *Gen Pharmacol* 26:1477–1487.

Anders MW, Lash L, Dekant W, Elfarra AA, and Dohn DR (1988) Biosynthesis and biotransformation of GSH S-conjugates to toxic metabolites. *Crit Rev Toxicol* 18:311–341.

Aperia A, Broberger O, Broberger U, Herin P, and Zetterstrom R (1983) Glomerular tubular balance in preterm and fullterm infants. *Acta Paediatr Scand Suppl* 305:70–76.

Appenroth D, and Braunlich H (1984) Age differences in cisplatinum nephrotoxicity. *Toxicology* 32:343–353.

Appenroth D, and Winnefeld K (1993) Role of glutathione for cisplatin nephrotoxicity in young and adult rats. *Ren Fail* 15:135–139.

Arant BS Jr (1987) Postnatal development of renal function during the first year of life. *Pediatr Nephrol* 1:308–313.

Avner ED, Sweeney WE, Jr., Piesco NP, and Ellis D (1985) Growth factor requirements of organogenesis in serum-free metanephric organ culture. *in vitro Cell Dev Biol* 21:297–304.

Baliga R, Ueda N, Walker PD, and Shah SV (1999) Oxidant mechanisms in toxic acute renal failure. *Drug Metab Rev* 31:971–997.

Battista MC, Oligny LL, St Louis J, and Brochu M (2002) Intrauterine growth restriction in rats is associated with hypertension and renal dysfunction in adulthood. *Am J Physiol Endocrinol Metab* 283:E124–E131.

Baylis C, and Corman B (1998) The aging kidney: Insights from experimental studies. *J Am Soc Nephrol* 9:699–709.

Baylis C, and Schmidt R (1996) The aging glomerulus. *Semin Nephrol* 16:265–276.

Beck LH (1998) Changes in renal function with aging. *Clin Geriatr Med* 14:199–209.

Bhatt-Mehta V, Schumacher RE, Faix RG, Leady M, and Brenner T (1999) Lack of vancomycin-associated nephrotoxicity in newborn infants: A case-control study. *Pediatrics* 103:e48.

Birner G, Werner M, Ott MM, and Dekant W (1995) Sex differences in hexachlorobutadiene biotransformation and nephrotoxicity. *Toxicol Appl Pharmacol* 132:203–212.

Birner G, Werner M, Rosner E, Mehler C, and Dekant W (1998) Biotransformation, excretion, and nephrotoxicity of the hexachlorobutadiene metabolite (E)-*N*-acetyl-S-(1,2,3,4,4-pentachlorobutadienyl)-L-cysteine sulfoxide. *Chem Res Toxicol* 11:750–757.

Bovee KC (1984) *Canine Nephrology*. Harwal Publishing Company, Philadelphia.

Bovee KC (1991) Influence of dietary protein on renal function in dogs. *J Nutr* 121:S128–S139.

Bovee KC, Jezyk PF, and Segal SC (1984) Postnatal development of renal tubular amino acid reabsorption in canine pups. *Am J Vet Res* 45:830–832.

Braunlich H, and Otto R (1984) Age-related differences in nephrotoxicity of cadmium. *Exp Pathol* 26:235–239.

Briguori C, Manganelli F, Scarpato P, Elia PP, Golia B, Riviezzo G, Lepore S, Librera M, Villari B, Colombo A, and Ricciardelli B (2002) Acetylcysteine and contrast agent-associated nephrotoxicity. *J Am Coll Cardiol* 40:298–303.

Brix AE (2002) Renal papillary necrosis. *Toxicol Pathol* 30:672–674.

Bruning T, and Bolt HM (2000) Renal toxicity and carcinogenicity of trichloroethylene: Key results, mechanisms, and controversies. *Crit Rev Toxicol* 30:253–285.

Bucher JR, Huff J, Haseman JK, Eustis SL, Elwell MR, Davis WE Jr. and Meierhenry EF (1990a) Toxicology and carcinogenicity studies of diuretics in F344 rats and B6C3F1 mice. 1. Hydrochlorothiazide. *J Appl Toxicol* 10:359–367.

Bucher JR, Huff J, Haseman JK, Eustis SL, Davis WE Jr and Meierhenry EF (1990b) Toxicology and carcinogenicity studies of diuretics in F344 rats and B6C3F1 mice. 2. Furosemide. *J Appl Toxicol* 10:369–378.

Burek JC, Duprat P, Owen R, Peter C, and Van Zweiten M (1988) Spontaneous renal disease in laboratory animals. *Int Rev Exp Path* 30:319.

Calderone C (2003) The therapeutic effect of natriuretic peptides in heart failure; differential regulation of endothelial and inducible nitric oxide synthases. *Heart Failure Reviews* 8:55–70.

Carraro-Eduardo JC, Oliveira AV, Carrapatoso ME, and Ornellas JF (1993) Effect of sex hormones on gentamicin-induced nephrotoxicity in rats. *Braz J Med Biol Res* 26:653–662.

Chandra M, Riley MG, and Johnson DE (1993) Spontaneous renal neoplasms in rats. *J Appl Toxicol* 13:109–116.

Chen GF, Ronis MJ, Ingelman-Sundberg M, and Badger TM (1999) Hormonal regulation of microsomal cytochrome P4502E1 and P450 reductase in rat liver and kidney. *Xenobiotica* 29:437–451.

Clapp WL, Tisher CC (1989) Gross Anatomy and Development of the Kidney in *Renal Pathology with Clinical and Functional Correaltions* (Tisher CC, and Brenner BM, Eds), pp. 67–91. J.B. Lipincott Company, Philadelphia.

Clark AT, and Bertram JF (1999) Molecular regulation of nephron endowment. *Am J Physiol* 276:F485–F497.

Clark B (2000) Biology of renal aging in humans. *Adv Ren Replace Ther* 7:11–21.

Corley RA, English JC, Hill TS, Fiorica LA, and Morgott DA (2000) Development of a physiologically based pharmacokinetic model for hydroquinone. *Toxicol Appl Pharmacol* 2000:163–174.

Couser WG, and Stilmant MM (1975) Mesangial lesions and focal glomerular sclerosis in the aging rat. *Lab Invest* 33:491–501.

de Wildt SN, Kearns GL, Leeder JS, and van den Anker JN (1999) Cytochrome P450 3A: Ontogeny and drug disposition. *Clin Pharmacokinet* 37:485–505.

Dekant W (1996) Biotransformation and renal processing of nephrotoxic agents. *Arch Toxicol Suppl* 18:163–172.

Dekant W (2001) Chemical-induced nephrotoxicity mediated by glutathione *S*-conjugate formation. *Toxicol Lett* 124:21–36.

Dinchuk JE, Car BD, Focht RJ, Johnston JJ, Jaffee BD, Covington MB, Contel NR, Eng VM, Collins RJ, and Czerniak PM (1995) Renal abnormalities and an altered inflammatory response in mice lacking cyclooxygenase II. *Nature* 378:406–409.

Dudley DK, and Hardie MJ (1985) Fetal and neonatal effects of indomethacin used as a tocolytic agent. *Am J Obstet Gynecol* 151:181–184.

Eguia L, and Materson BJ (1997) Acetaminophen-related acute renal failure without fulminant liver failure. *Pharmacotherapy* 17:363–370.

Epstein M (1996) Aging and the kidney. *J Am Soc Nephrol* 7:1106–1122.

Fanos V, and Cataldi L (1999) Antibacterial-induced nephrotoxicity in the newborn. *Drug Saf* 20:245–267.

Fleck C, and Braunlich H (1995) Renal handling of drugs and amino acids after impairment of kidney or liver function—influences of maturity and protective treatment. *Pharmacol Ther* 67:53–77.

Frank AA, Hiedel JR, Thompson DJ, Carlton WW, and Beckwith JB (1992) Renal transplacental carcinogenicity of 3,3-dimethyl-1-phenyltriazene in rats: Relationship of renal mesenchymal tumor to congenital mesoblastic nephroma and intralobar nephrogenic rests. *Toxicol Pathol* 20:313–322.

Gomez RA, Lynch KR, Sturgill BC, Elwood JP, Chevalier RL, Carey RM, and Peach MJ (1989) Distribution of renin mRNA, and its protein in the developing kidney. *Am J Physiol* 257:F850–F858.

Gumprecht J, Zychma MJ, Grzeszczak W, and Zukowska-Szczechowska E (2000) Angiotensin I-converting enzyme gene insertion/deletion and angiotensinogen M235T polymorphisms: Risk of chronic renal failure. End-Stage Renal Disease Study Group. *Kidney Int* 58:513–519.

Guron G, and Friberg P (2000) An intact renin-angiotensin system is a prerequisite for normal renal development. *J Hypertens* 18:123–137.

Hard GC (1985) Differential renal tumor response to N-ethylnitrosourea and dimethylnitrosamine in the Nb rat: Basis for a new rodent model of nephroblastoma. *Carcinogenesis* 6:1551–1558.

Hard GC (1990) Tumours of the Kidney, Renal Pelvis and Ureter in *Pathology of Tumours in Laboratory Animals* (Turusov V, and Mohr U, Eds), pp. 301–344, International Agency for Research on Cancer, Lyon.

Hard GC, and Khan KN (1998) Mechanisms of chemically induced renal carcinogenesis in the laboratory rodent. *Toxicol Pathol* 26:104–112.

Hard GC, Whysner J, English JC, Zang E, and Williams GM (1997) Relationship of hydroquinone-associated rat renal tumors with spontaneous chronic progressive nephropathy. *Toxicol Pathol* 25:132–143.

Harris RZ, Benet LZ, and Schwartz JB (1995) Gender effects in pharmacokinetics and pharmacodynamics. *Drugs* 50:222–239.

Hirata M, Tanaka A, Hisanaga A, and Ishinishi N (1990) Effects of glutathione depletion on the acute nephrotoxic potential of arsenite and on arsenic metabolism in hamsters. *Toxicol Appl Pharmacol* 106:469–481.

Hoivik DJ, Manautou JE, Tveit A, Hart SG, Khairallah EA, and Cohen SD (1995) Gender-related differences in susceptibility to acetaminophen-induced protein arylation and nephrotoxicity in the CD-1 mouse. *Toxicol Appl Pharmacol* 130:257–271.

Hong SK, Anestis DK, Valentovic MA, Ball JG, Brown PI, and Rankin GO (2001) Gender differences in the potentiation of *N*-(3,5-dichlorophenyl)succinimide metabolite nephrotoxicity by phenobarbital. *J Toxicol Environ Health A* 64:241–256.

Inselmann G, Lawerenz HU, Nellessen U, and Heidemann HT (1994) Enhancement of cyclosporin A induced hepato- and nephrotoxicity by glutathione depletion. *Eur J Clin Invest* 24:355–359.

Jose PA, Fildes RD, Gomez RA, Chevalier RL, and Robillard JE (1994) Neonatal renal function and physiology. *Curr Opin Pediatr* 6:172–177.

Kalsotra A, Anakk S, Boehme CL, and Strobel HW (2002) Sexual dimorphism and tissue specificity in the expression of Cytochrome P4F forms in Sprague-Dawley rats. *Drug Metab Dispos* 30:1022–1028.

Kelly GS (1998) Clinical applications of N-acetylcysteine. *Altern Med Rev* 3:114–127.

Khan KNM, Alden CL (2002) Kidney, in *Handbook of Toxicologic Pathology* (Haschek WM, Rousseaux CG, Wallig MA, Eds), pp. 255–330. Academic Press, New York.

Khan KN, Paulson SK, Verburg KM, Lefkowith JB, and Maziasz TJ (2002) Pharmacology of cyclooxygenase-2 inhibition in the kidney. *Kidney Int* 61:1210–1219.

Khan KNM, Silverman LR, and Baron DA (1999) Cyclooxygenase-2 (cox-2) expression in the developing canine kidney. *Vet Pathol* 36:486.

Khan KN, Stanfield KM, Dannenberg A, Seshan SV, Baergen RN, Baron DA, and Soslow RA (2001) Cyclooxygenase-2 expression in the developing human kidney. *Pediatr Dev Pathol* 4:461–466.

Kingdom JC, Hayes M, McQueen J, Howatson AG, Lindop GB (1999) Intrauterine growth restriction is associated with persistent juxtamedullary expression of renin in the fetal kidney. *Kidney Int* 55:424–429.

Knox DM, and Martof MT (1995) Effects of drug therapy on renal function of healthy older adults. *J Gerontol Nurs* 21:35–40.

Langley-Evans SC, Gardner DS, and Welham SJ (1998) Intrauterine programming of cardiovascular disease by maternal nutritional status. *Nutrition* 14:39–47.

Lash LH, Putt DA, Brashear WT, Abbas R, Parker JC, and Fisher JW (1999) Identification of S-(1,2-dichlorovinyl)glutathione in the blood of human volunteers exposed to trichloroethylene. *J Toxicol Environ Health A* 56:1–21.

Lash LH, Qian W, Putt DA, Desai K, Elfarra AA, Sicuri AR, and Parker JC (1998a) Glutathione conjugation of perchloroethylene in rats and mice *in vitro*: Sex-, species-, and tissue-dependent differences. *Toxicol Appl Pharmacol* 150:49–57.

Lash LH, Qian W, Putt DA, Jacobs K, Elfarra AA, Krause RJ, and Parker JC (1998b) Glutathione conjugation of trichloroethylene in rats and mice: Sex-, species-, and tissue-dependent differences. *Drug Metab Dispos* 26:12–19.

Lau SS, Monks TJ, Everitt JI, Kleymenova E, and Walker CL (2001) Carcinogenicity of a nephrotoxic metabolite of the "nongenotoxic" carcinogen hydroquinone. *Chem Res Toxicol* 14:25–33.

Lelievre-Pegorier M, and Merlet-Benichou C (2000) The number of nephrons in the mammalian kidney: Environmental influences play a determining role. *Exp Nephrol* 8:63–65.

Lewis EJ, Hunsicker LG, Bain RP, and Rohde RD (1993) The effect of angiotensin-converting-enzyme inhibition on diabetic nephropathy. The Collaborative Study Group. *N Engl J Med* 329:1456–1462.

Lindeman RD (1990) Overview: Renal physiology and pathophysiology of aging. *Am J Kidney Dis* 16:275–282.

Lindeman RD (1992) Changes in renal function with aging. Implications for treatment. *Drugs Aging* 2:423–431.

Lindeman RD, and Goldman R (1986) Anatomic and physiologic age changes in the kidney. *Exp Gerontol* 21:379–406.

Linton S, Davies MJ, and Dean RT (2001) Protein oxidation and ageing. *Exp Gerontol* 36:1503–1518.

Lovati E, Richard A, Frey BM, Frey FJ, and Ferrari P (2001) Genetic polymorphisms of the renin-angiotensin-aldosterone system in end-stage renal disease. *Kidney Int* 60:46–54.

Lubran MM (1995) Renal function in the elderly. *Ann Clin Lab Sci* 25:122–133.

Lumbers ER (1995) Development of renal function in the fetus: A review. *Reprod Fertil Dev* 7:415–426.

Manalich R, Reyes L, Herrera M, Melendi C, and Fundora I (2000) Relationship between weight at birth and the number and size of renal glomeruli in humans: A histomorphometric study. *Kidney Int* 58:770–773.

Marchand MC, and Langley-Evans SC (2001) Intrauterine programming of nephron number: The fetal flaw revisited. *J Nephrol* 14:327–331.

Matzke GR, and Frye RF (1997) Drug administration in patients with renal insufficiency. Minimising renal and extrarenal toxicity. *Drug Saf* 16:205–231.

Maxie MG, and Prescott JF (2002) The Urinary System in *Pathology of Domestic Animals* (Jubb KVF, Kennedy PC, and Palmer N, Eds), pp. 447–538, Academic Press, Inc., San Diego.

McGuire EJ, Anderson JA, Gough AW, Herman JR, Pegg DG, Theiss JC, and de la Iglesia FA (1996) Preclinical toxicology studies with the angiotensin-converting enzyme inhibitor quinapril hydrochloride (Accupril). *J Toxicol Sci* 21:207–214.

Meibohm B, Beierle I, and Derendorf H (2002) How important are gender differences in pharmacokinetics? *Clin Pharmacokinet* 41:329–342.

Merlet-Benichou C (1999) Influence of fetal environment on kidney development. *Int J Dev Biol* 43:453–456.

Mizutani T, Ito K, Nomura H, and Nakanishi K (1990) Nephrotoxicity of thiabendazole in mice depleted of glutathione by treatment with DL-buthionine sulphoximine. *Food Chem Toxicol* 28:169–177.

Monks TJ, and Lau SS (1998) The pharmacology and toxicology of polyphenolic-glutathione conjugates. *Annu Rev Pharmacol Toxicol* 38:229–255.

Monks TJ, Anders MW, Dekant W, Stevens JL, Lau SS, and van Bladeren PJ (1990) Glutathione conjugate mediated toxicities. *Toxicol Appl Pharmacol* 106:1–19.

Montgomery CA, and Seely JC (1990) Kidney in *Pathology of the Fischer Rat* (Boorman GA, Eustis SL, Elwell MR, Mongomery CA, and MacKenzie WF, Eds), pp. 127–153, Academic Press, Inc., San Diego.

Morgan DL, Bucher JR, Elwell MR, Lilja HS, and Murthy AS (1990) Comparative toxicity of ethylene dichloride in F344/N, Sprague-Dawley and Osborne-Mendel rats. *Food Chem Toxicol* 28:839–845.

Morham SG, Langenbach R, Loftin CD, Tiano HF, Vouloumanos N, Jennette JC, Mahler JF, Kluckman KD, Ledford A, and Lee CA (1995) Prostaglandin synthase 2 gene disruption causes severe renal pathology in the mouse. *Cell* 83:473–482.

Mugford CA, and Kedderis GL (1998) Sex-dependent metabolism of xenobiotics. *Drug Metab Rev* 30:441–498.

Nakagawa I, Suzuki M, Imura N, and Naganuma A (1995) Enhancement of paraquat toxicity by glutathione depletion in mice *in vivo* and *in vitro*. *J Toxicol Sci* 20:557–564.

Needleman P, and Isakson PC (1997) The discovery and function of COX-2. *J Rheumatol* 24 Suppl 49:6–8.

Neugarten J, Gallo G, Silbiger S, and Kasiske B (1999) Glomerulosclerosis in aging humans is not influenced by gender. *Am J Kidney Dis* 34:884–888.

Nicholls D (2002) Mitochondrial function and dysfunction in the cell: Its relevance to aging and aging-related disease. *Int J Biochem Cell Biol* 34:1372–1381.

Nisar S, and Feinfeld DA (2002) N-acetylcysteine as salvage therapy in cisplatin nephrotoxicity. *Ren Fail* 24:529–533.

Noris M, and Remuzzi G (1999) Physiology and pathophysiology of nitric oxide in chronic renal disease. *Proc Assoc Am Physicians* 111:602–610.

Novy MJ (1978) Effects of indomethacin on labor, fetal oxygenation, and fetal development in rhesus monkeys. *Adv Prostaglandin Thromboxane Res* 4:285–300.

Nyarko AK, Kellner-Weibel GL, and Harvison PJ (1997) Cytochrome P450-mediated metabolism and nephrotoxicity of N-(3,5-dichlorophenyl)succinimide in Fischer 344 rats. *Fundam Appl Toxicol* 37:117–124.

Odinecs A, Maso S, Nicoletto G, Secondin L, and Trevisan A (1995) Mechanism of sex-related differences in nephrotoxicity of 1.2-dichloropropane in rats. *Ren Fail* 17:517–524.

Ollerstam A, and Persson AE (2002) Macula densa neuronal nitric oxide synthase. *Cardiovasc Res* 56:189.

Owen RA, and Heywood R (1986) Age-related variations in renal structure and function in Sprague- Dawley rats. *Toxicol Pathol* 14:158–167.

Parkinson A (1996) Biotransformation of Xeniobiotics in *Casarett & Doull's Toxicology: The Basic Science of Poisons* (Klaasen CD, Ed.), pp. 113–186, McGraw-Hill, New York.

Paterson DL, Robson JM, and Wagener MM (1998) Risk factors for toxicity in elderly patients given aminoglycosides once daily. *J Gen Intern Med* 13:735–739.

Pfohl-Leszkowicz A, Pinelli E, Bartsch H, Mohr U, and Castegnaro M (1998) Sex- and strain-specific expression of cytochrome P450s in ochratoxin A-induced genotoxicity and carcinogenicity in rats. *Mol Carcinog* 23:76–85.

Rankin GO, Hong SK, Anestis DK, Lash LH, and Miles SL (2001) In vitro nephrotoxicity induced by N-(3,5-dichlorophenyl)succinimide (NDPS) metabolites in isolated renal cortical cells from male and female Fischer 344 rats: Evidence for a nephrotoxic sulfate conjugate metabolite. *Toxicology* 163:73–82.

Reddi V, Zaglul A, Pentz ES, and Gomez RA (1998) Renin-expressing cells are associated with branching of the developing kidney vasculature. *J Am Soc Nephrol* 9:63–71.

Ruiz JG, and Lowenthal DT (1997) NSAIDS, and nephrotoxicity in the elderly. *Geriatr Nephrol Urol* 7:51–57.

Seely JC, Haseman JK, Nyska A, Wolf DC, Everitt JI, and Hailey JR (2002) The effect of chronic progressive nephropathy on the incidence of renal tubule cell neoplasms in control male F344 rats. *Toxicol Pathol* 30:681–686.

Solleveld HA, and Boorman GA (1986) Spontaneous renal lesions in five rat strains. *Toxicol Pathol* 14:168–174.

Speerschneider P, and Dekant W (1995) Renal tumorigenicity of 1,1-dichloroethene in mice: The role of male-specific expression of cytochrome P450 2E1 in the renal bioactivation of 1,1-dichloroethene. *Toxicol Appl Pharmacol* 130:48–56.

Stanfield KS, Kelce WR, Lisowski A, English ML, Baron DA, and Khan KNM (2000) Expression of cyclooxygenase-2 in the embryonic and fetal tissues during organogenesis and late pregnancy. *Vet Pathol* 37:558.

Tariq M, Morais C, Sobki S, Al Sulaiman M, and Al Khader A (1999) N-acetylcysteine attenuates cyclosporin-induced nephrotoxicity in rats. *Nephrol Dial Transplant* 14:923–929.

Tarloff JB, Khairallah EA, Cohen SD, and Goldstein RS (1996) Sex- and age-dependent acetaminophen hepato- and nephrotoxicity in Sprague-Dawley rats: Role of tissue accumulation, nonprotein sulfhydryl depletion, and covalent binding. *Fundam Appl Toxicol* 30:13–22.

van der Heijden BJ, Carlus C, Narcy F, Bavoux F, Delezoide AL, and Gubler MC (1994) Persistent anuria, neonatal death, and renal microcystic lesions after prenatal exposure to indomethacin. *Am J Obstet Gynecol* 171:617–623.

Volkel W, Friedewald M, Lederer E, Pahler A, Parker J, and Dekant W (1998) Biotransformation of perchloroethene: Dose-dependent excretion of trichloroacetic acid, dichloroacetic acid, and *N*-acetyl-*S*-(trichlorovinyl)-L-cysteine in rats and humans after inhalation. *Toxicol Appl Pharmacol* 153:20–27.

Wabner CL, and Chen TS (1985) Aging changes in the renal clearance in rats. *Pharmacologist* 27:169.

Wang T (2002) Role of iNOS, and eNOS in modulating proximal tubule transport and acid– base balance. *Am J Physiol Renal Physiol* 283:F658–F662.

Werner M, Birner G, and Dekant W (1995) The role of cytochrome P4503A1/2 in the sex-specific sulfoxidation of the hexachlorobutadiene metabolite, *N*-acetyl-*S*-(pentachlorobutadienyl)-L-cysteine in rats. *Drug Metab Dispos* 23:861–868.

Wolf JC (2002) Characteristics of the spectrum of proliferative lesions observed in the kidney and urinary bladder of Fischer 344 Rats and $B6C3F_1$ mice. *Toxicol Pathol* 30:657–662.

Yoo HG, Shin BA, Park JS, Lee KH, Chay KO, Yang SY, Ahn BW, and Jung YD (2002) IL-1beta induces MMP-9 via reactive oxygen species and NF-kappaB in murine macrophage RAW 264.7 cells. *Biochem Biophys Res Commun* 298:251–256.

Zhang M-Z, Wang J-L, Cheng J-F, Harris RC, and McKanna JA (1997) Cyclooxygenase-2 in rat nephron development. *Am J Physiol* 273:F994–F1002.

Zidar N, Cavic MA, Kenda RB, Koselj M, and Ferluga D (1998) Effect of intrauterine growth retardation on the clinical course and prognosis of IgA glomerulonephritis in children. *Nephron* 79:28–32.

26

Risk Assessment of Nephrotoxic Metals

Gary L. Diamond

GENERAL PRINCIPLES

Toxicological and epidemiologic investigations seek to describe the pharmacodynamics and pharmacokinetics of toxic agents. Risk assessment is a process for utilizing this information to predict the likelihood that an adverse effect will result from human contact with toxic chemicals. The outcome of this process, as it relates to nephrotoxic metals, is a mathematical description of the relationship between human exposure to a metal and the probability of impaired kidney function.

A general model of the risk assessment process is depicted in Figure 26.1. In this model, the various chemical and biological processes for which there are links between exposure and impaired function are mathematically represented in a series of submodels. Exposure models link concentrations of metals in specific environmental media with rates of contact of humans with those media to arrive at estimates of chemical intakes, the "external dose" (e.g., μg ingested or inhaled per kg body weight per day; μg/kg/day). Pharmacokinetics models translate estimates of external dose into estimates of the dose that is absorbed, the "internal dose." Depending on the objective and the metal of interest, the internal dose may represent a rate of absorption (e.g., μg absorbed/

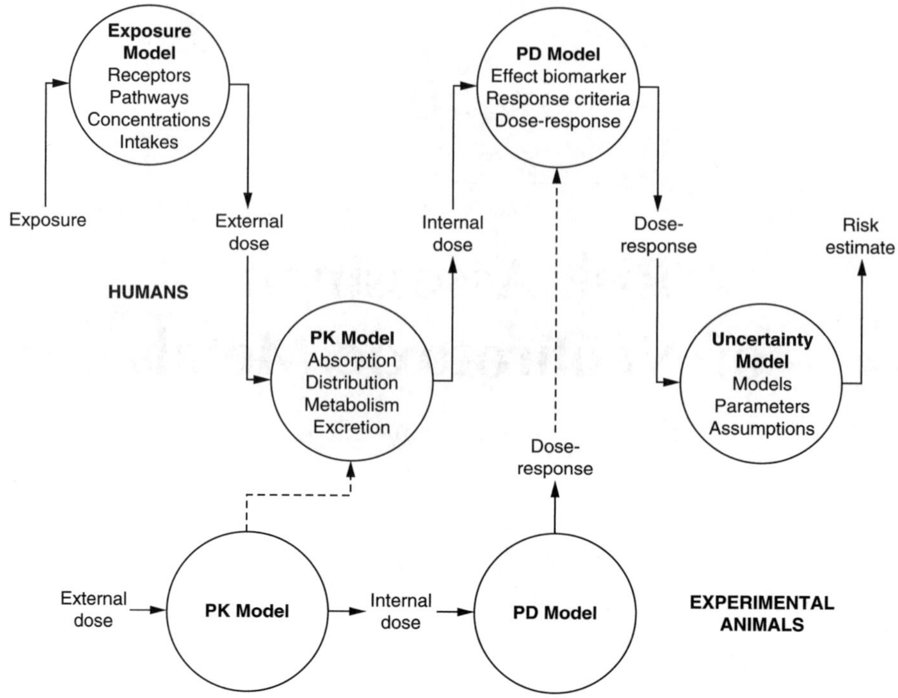

Figure 26.1 General model of the risk assessment process. Dotted lines connect models based on experimental animals with human risk models. (PD) pharmacodynamics, (PK) pharmacokinetics.

kg/day), a cumulative absorbed dose (e.g., mg), or may be a dose to a specific tissue or organelle (e.g., μg/g kidney).

For nephrotoxic metals that exert toxicity through direct interactions with biochemical receptors in the kidney (e.g., cadmium, uranium), kidney burden or concentration of the metal in the kidney may be the most appropriate metric of internal dose, if such data are available for modeling. This approach may not be appropriate for a metal that exerts nephrotoxicity by interacting with receptors outside of the kidney (e.g., mercury-induced autoimmunity). Other commonly used metrics of internal dose include the concentration of metal in the systemic circulation or its rate of excretion in urine.

Pharmacodynamics models relate the internal (or external) dose to the probability of an adverse response to yield a "dose–response relationship" (probability). The dose–response relationship is established from measurements of changes in one or more "effect biomarkers."

Table 26.1 Effect Biomarkers Used in Human Epidemiological Studies of Metal Nephrotoxicity

Endpoint	*Effect*	*Biomarker*
Glomerular filtration	Glomerular injury	Creatinine clearance
	Cardiovascular disturbance	Serum creatinine concentration
		Blood urea nitrogen
Enzymuria	Tubular cell injury	Alanine aminopeptidase
		Alkaline phosphatase
		Aspartate aminotransferase
		γ-Glutamyltranserase
		Lactate dehydrogenase
		N-acetyl-β-glucosaminidase
Proteinuria	Glomerular injury	Total protein
	Impaired reabsorption	Albumin
		Transferrin
LMW proteinuria	Impaired reabsorption	α_2-Microglobulin
		α_1-Microglobulin
		Retinol binding protein
Urinary LMW solutes	Impaired reabsorption	Calcium
		Glucose
		α-Aminonitrogen
		Phosphate
Urinary eicosanoids	Vascular injury or increased vascular resistance	Prostaglandin E_2
		6-keto-prostaglandin $F_{1\alpha}$
		Prostaglandin $F_{2\alpha}$
		Thromboxane 2

LMW = low molecular weight.
See Cardenas et al. (1993a, 1993b); Clarkson et al. (1988); Fels et al. (1998); Pesce and First (1979); Pless-Mulloli et al. (1998); and Roels et al. (1993) for relevant reviews and examples of applications.

Examples of effect biomarkers that are commonly used in human epidemiological studies of metal nephrotoxicity are provided in Table 26.1. Each of these is a continuous variable whose probability distribution is expected to change in some detectable way in response to increasing dose, relative to the distribution in a reference population that is assumed to have had no (or minimal) exposure.

A common practice is to convert the continuous variables to a quantal form for modeling dose–response relationships. One approach to this conversion is to establish a "response criterion," a boundary between values of the continuous variable that are indicative of an adverse effect (i.e., response) and values considered to be nonadverse (i.e., non-response). In typical applications of this approach, the response criterion represents a parameter of the distribution of the continuous variable in a reference population, for example, a specific percentile value of the reference population. The value selected for the response criterion can have a substantial impact on the resulting dose–response relationship because it establishes the expected background response rate and affects the slope of the dose–response relationship (Kodell et al., 1995). A variety of alternative approaches to the modeling of continuous variables have been described (Chen and Gaylor, 1992; Gaylor and Slikker, 1990; Kodell and West, 1993). An uncertainty model is needed in order to take into account the simplifications inherent to each of the models and the limitations in the information available to parameterize the models and to establish plausible values for model parameters.

While the ultimate objective of the risk assessment process is to estimate risks in humans, very often the toxicological or epidemiological experience for a given metal will not adequately support, by itself, risk estimations. In these cases, toxicology studies conducted in other mammalian species can provide information about dose–response relationships that can be extrapolated, with greater uncertainty, to humans.

Not all of the components of the general model shown in Figure 26.1 will be needed in every risk assessment; however, the extent to which each are understood will determine the certainty in the ultimate risk estimate and the extent to which risk estimates can be made for a variety of different exposure scenarios. In the sections that follow, cadmium and lead serve to illustrate practical applications of the model shown in Figure 26.1.

CADMIUM

Cadmium risk assessment has been substantially advanced by the three major developments: the discovery that low molecular weight (LMW) proteinuria is a prominent sign of cadmium nephrotoxicity (Friberg, 1950), the development of urinalysis techniques to assess LMW proteinuria in humans (Bergagard and Bearn, 1968), and the development of models of the pharmacokinetics of cadmium in humans that allow estimates to be made of the dose to the kidney (Kjellström and Nordberg, 1978). These advances have made it possible to express dose–response relationships derived from disparate epidemiologic studies in terms of common dose and response metrics. When this is done, risk estimates based on individual studies that might otherwise be incomparable can be integrated into a combined pharmacokinetics and pharmacodynamics (PK/PD) model for estimating cadmium risks.

LMW Proteinuria as a Risk Indicator

LMW proteinuria in humans was first described as a syndrome in alkaline battery workers who had been exposed to cadmium (Friberg, 1950), and has since been observed in a variety of forms of renal disease, including Fanconi's syndrome, Wilson's disease, familial progressive renal tubulopathy (Furuse et al., 1992), insulin-dependent diabetes mellitus (Ginevri et al., 1993; Kordonouri et al., 1992), burn-related acute renal failure (Schiavon et al., 1988), idiopathic membranous nephropathy (Reichert et al., 1995), and Balkan endemic nephropathy (Ceovic et al., 1985, 1991). Therefore, while LMW proteinuria appears to be a prominent feature of cadmium-induced nephrotoxicity, potential confounding from other causes of LMW proteinuria need to be considered in epidemiologic studies in which it is used to assess cadmium toxicity.

The most commonly used biomarkers of LMW proteinuria are urinary β_2-microglobulin ($\beta_2\mu$G) and retinal binding (RBP). $\beta_2\mu$G is labile in acidic urine and, for this reason, the pH of fresh urine samples must be adjusted to a pH greater than 5.5 to ensure preservation of the protein in the bladder urine or urine sample (Bernard and Lauwerys, 1981). Procedures to alkalinize the urine *in situ* (e.g., by having subjects ingest sodium bicarbonate) are rarely performed in large-scale epidemiology studies. If the pH of bladder urine is less than 5.5, some loss of $\beta_2\mu$G can be expected, even if the urine sample is alkalinized. More recently, α_1-microglobulin ($\alpha_1\mu$G) has been explored as an alternative to $\beta_2\mu$G as an acid-stable marker of LMW proteinuria (Cai et al., 2001; Järup et al., 2000; Pless-Mulloli et al., 1998).

LMW proteinuria, in general, is considered a "preclinical" biomarker of proximal tubular dysfunction (Lapsley et al., 1991; Pesce and First, 1979). A transient LMW proteinuria, detected as an increased urinary excretion of RBP and $\beta_2\mu$G, precedes and is indicative of the subsequent development of albuminuria associated with diabetes-related nephropathy in children (Ginevri et al., 1993) and of renal failure in nephrosis (Reichert et al., 1995).

The origin of the LMW proteinuria in cadmium toxicity appears to be an impairment of the reabsorption of LMW proteins in the renal proximal tubule. This results in an increase in the excretion of filtered LMW proteins and, providing that the rate of glomerular filtration of LMW protein remains relatively constant, an increase in the rate of excretion of LMW proteins in urine (Aoshima et al., 1995; Cai et al., 1992; Thun et al., 1989). The constraint of constancy of filtration has two implications. First, in people who have low glomerular filtration rates related to kidney disease, cadmium-induced LMW proteinuria may be more difficult to detect from measurement of the rate of urinary excretion of LMW proteins (Bernard et al., 1988; Mutti, 1989). Second, cadmium-induced decrease in GFR, if it were to occur, would introduce a bias that would tend to result in an underestimate of the severity (but not necessarily estimates of the prevalence) of LMW proteinuria associated with any given level of exposure to cadmium. The potential magnitude of this source of bias has not been addressed in most of the major epidemiological studies of cadmium toxicity that form the basis for the dose–response analyses presented below (Table 26.2). Decrements in glomerular filtration rate, as indicated by changes in serum creatinine concentrations, have been reported in workers whose mean urinary cadmium concentrations were 6, 9, 11, or 16 μg/g creatinine: Mueller et al. (1992), Thun et al. (1989), Bernard et al. (1979), and Falck et al. (1983), respectively. This range, 9–16 μg/g creatinine, encompasses the high end of the range in which LMW proteinuria has been observed (Table 26.2).

LMW proteinuria associated with exposures to cadmium can be highly persistent, if not irreversible, and an indicator of more-severe decrements in renal function (Hellström et al., 2001; Järup and Elinder, 1994; Järup et al., 1997; Roels et al., 1997). However, with relatively low exposures, progression to more serious outcomes is not always observed (Hotz et al., 1999). Additional evidence for the toxicological significance of LMW proteinuria derives from mortality studies that have revealed an association between elevated urinary excretion of $\beta_2\mu$G and increased mortality in cadmium exposed populations (Arisawa et al., 2001; Iwata et al., 1991a, 1991b; Nakagawa et al., 1993; Nishijo et al., 1994). However, these studies do not directly establish causality between increased mortality and cadmium exposure. They support the concept that increased urinary $\beta_2\mu$G may be associated with or indicative of more-severe physiological impairments that contribute to mortality and, as such, provide further support for its use as a biomarker of an adverse effect of cadmium for use in risk assessment.

Comparison of Modeled Risk Estimates from Epidemiological Studies

Table 26.2 summarizes the outcomes of various studies from which dose–response functions for cadmium-induced renal tubular proteinuria can be derived. Included are studies of populations having primarily occupational exposures, populations exposed to dietary sources resulting from environmental contamination of watersheds, and general populations without occupational exposure for whom exposure was mainly from the diet. The cadmium-dose metric varied among studies, and included urinary cadmium, cumulative cadmium intake, kidney cadmium burden, or liver cadmium concentration.

Some studies reported more than one response variable. However, for consistency, the dose–response analysis can be confined to urinary excretion of $\beta_2\mu$G. Where both $\beta_2\mu$G and retinol binding protein (RBP) were evaluated in the same study, the two response variables yielded similar outcomes (Buchet et al., 1990; Roels et al., 1993). A response was defined as urinary excretion of $\beta_2\mu$G in an individual that exceeded a specific value representing an upper percentile of a reference population (i.e., the response criterion). Response criteria varied among the studies and ranged from approximately 200 to 1100 μg $\beta_2\mu$G/g creatinine (84th to 97.5th percentile of reference populations).

The extent to which multiple explanatory variables other than dose were included in the various regression models reported varied between studies. In general, age and gender were important factors in most of the reported models and, for this reason, renal cortex cadmium concentrations and associated risks are modeled for females and males separately and focused on the age range 50 to 60 years (in which renal cortex cadmium concentration is predicted to reach the highest levels) (Figure 26.2). Table 26.2 presents the estimates of the dose (in units reported in each study) associated with a probability of 0.1 of a response. These values are not adjusted for background response rates which can only be roughly estimated from the percentile of the response criterion used in each study. However, in general, a 0.1 probability of response corresponds to an added risk, above background, of less than 0.1.

Table 26.2 Selected Studies of Dose–Response Relationship for Cadmium-Induced Renal Tubular Proteinuria

Study	*Population*	*n*	*Dose metric*	*Effect biomarker*	*Response criterion*[a]	*Dose–response model*	ED_{10}[b]
Buchet et al., 1990	General population, no occupational exposure (Belgium)	1699, FM	Urine Cd (μg/24 hr)	Urine β2μG Urine RBP	283 μg/24 hr (95%) 338 μg/24 hr (95%)	Logistic: UCd, age, gender[c]	3 (6), FM
Suwazono et al., 2000	General population, no occupational exposure (Japan)	1648, F 1105, M	Urine Cd (μg/g Cr)	Urine β2μG	400 μg/g Cr, F (84%)[d] 507 μg/g Cr, M (84%)[d]	Logistic: UCd, age, smoking	1.3 (2.4), F 0.05 (1.5), M
Yamanaka et al., 1998	General population, no occupational exposure (Japan)	743, F 558, M	Urine Cd (μg/g Cr)	Urine β2μG	403 μg/g Cr, F (84%)[d] 492 μg/g Cr, M (84%)[d]	Logistic: UCd, age	0.4 (1.0), F 0.1 (1.2), M
Järup et al., 2000	Residents living near battery plant (Sweden)	544, F 479, M	Urine Cd (μg/g Cr)	Urine α1μG	6.1 mg/g Cr, F (95%) 4.6 mg/g Cr, M (95%)	Logistic: UCd, age	1 (2), FM
Nogawa et al., 1989	Residents of contaminated river basin (Japan)	972, F 878, M	Cumulative intake (g)	Urine β2μG	1,000 μg/g Cr (97.5%)	Linear: intake	2.2 (2.4), F 2.2 (2.5), M
Ishizaki et al., 1989	Residents of contaminated river basin (Japan)	1754, F 1424, M	Urine Cd (μg/g Cr)	Urine β2μG	1,059 μg/g Cr, F (97.5%) 1,129 μg/g Cr, M (97.5%)	Probit: UCd	5.8 (6.6), F 4.9 (5.6), M
Kido and Nogawa 1993	Residents of contaminated river basin in Japan	972, F 878, M	Cumulative intake (g)	Urine β2μG	1,000 μg/g Cr (97.5%)	Logistic: intake, age	3.8 (4.3), F 4.4 (3.8), M
Hochi et al., 1995	Residents of contaminated river basin (Japan)	972, F 878, M	Cumulative intake (g)	Urine β2μG	1,000 μg/g Cr (97.5%)	Linear: intake, age, gender	3.1 (3.9), F 5.0 (5.8), M

Hayano et al., 1996	Residents of contaminated river basin (Japan)	1754, F 1428, M	Urine Cd (μg/g Cr)	Urine $\beta 2\mu$G	1,000 μg/g Cr (97.5%)	Logistic: UCd, age	13 (15), F 8.9 (10), M
Cai et al., 1998	Residents of contaminated river basin (China)	112, F 219, M	Cumulative intake (g)	Urine $\beta 2\mu$G	Age-specific 300–2500 μg/g Cr (90%)	Polynomial: intake	1.7 (2.9), FM
Wu et al., 2001	Residents of contaminated river basin (China)	125, F 122, M	Urine Cd (μg/g Cr)	Urine $\beta 2\mu$G	900 μg/g Cr, F (95%) 800 μg/g Cr, M (95%)	Linear: UCd	2.0 (3.5), FM
Ellis et al., 1984	Cadmium production workers (U.S.)	82, M	Kidney Cd (mg)	Urine $\beta 2\mu$G	200 μg/g Cr (≈95%)	Logistic: KCd	23 (25), M
Järup and Elinder, 1994	Battery workers (Sweden)	94, F 300, M	Urine Cd (μg/g Cr)	Urine $\beta 2\mu$G	190 μg/g Cr (95%)	Probit: UCd, age- stratified	4 (5), FM <60 yr 1.5 (2) >60 yr
Roels et al., 1981	Cadmium production workers (Belgium)	309, M	Liver Cd (μg/g)	Urine $\beta 2\mu$G	200 μg/g Cr (≈95%)	Logistic: LCd[e]	39 (42), M
Roels et al., 1993	Cadmium/zinc smelter workers (Belgium)	80, M	Urine Cd (μg/g Cr)	Urine $\beta 2\mu$G Urine RBP	279 μg/l (95%) 184 μg/l (95%)	Logistic: UCd	10 (12), M

$\alpha 1\mu$G = α1-microglobulin; $\beta 2\mu$G = β2-microglobulin; Cd = cadmium; Cr = creatinine; F = females; FM = females and males combined; KCd = kidney cadmium; LCd = liver cadmium; M = males; n = number of subjects; RBP = retinal binding protein; UCd = urine cadmium

[a]Values in parentheses are the percentile of a reference population.

[b]Effective dose; dose associated with a 10% risk of renal tubular proteinuria (p = .1), numbers in parentheses are added risk above background (p–p_0), where background risk was estimated based on the reported response criterion (e.g., p_0 = .05 if the response criterion was the 95th percentile of a reference population).

[c]Adjustments were made for other significant covariates of urinary $\beta 2\mu$G excretion (see reference).

[d]Two standard deviations above mean of reference population (approximately 84th percentile of a normal distribution).

[e]Regression function derived by authors from data presented in Table 4 of Roels et al. (1981) (Odds = LCd 0.086–5.56)

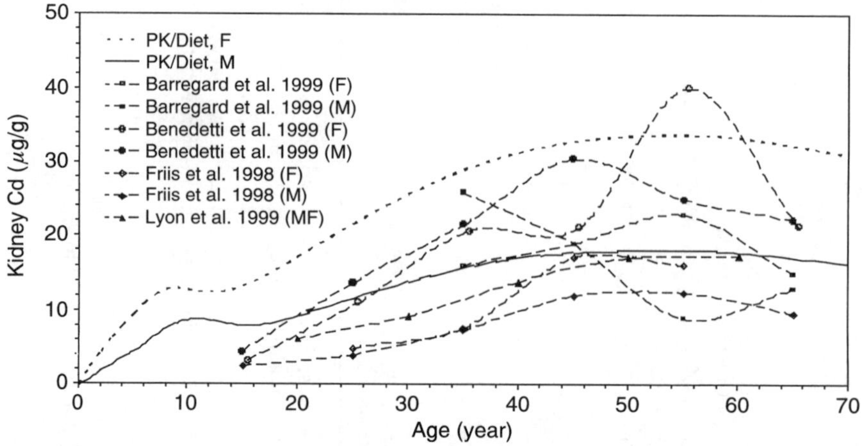

Figure 26.2 Comparison of kidney cadmium concentrations predicted by the pharmacokinetics (PK) model with observed concentrations in humans. The graph shows model predictions assuming estimated age-specific dietary cadmium intakes for U.S. female (PK/Diet, F) or male (PK/Diet, M) nonsmokers without occupational exposure (Choudhury et al., 2001). Other plots show reported means of observations made in general populations, based on measurements made postmortem or from biopsy samples in Canada (Benedetti et al., 1999, Figure 1; *n* = 289), U.K. (Lyon et al., 1999, Figure 4; *n* = 301), and Sweden (Barregård et al., 1999; *n* = 34, and Friis et al., 1998; *n* = 128). Values for the Canadian study are concentrations of cadmium in whole kidney; all other observed values are concentrations in renal cortex (which tend to be 1.2 to 1.5 times higher than whole kidney). (Data points digitized from cited figures and lines connecting data points were smoothed by the author.)

The pharmacokinetics model developed by Kjellström and Nordberg (1978), with modifications for computing growth and urine creatinine levels (Choudhury et al., 2001), provides a means for converting dose units for direct comparisons of the dose–response relationships. Comparisons of model predictions with observations indicate that the model simulates the observed age-pattern of renal cortex concentrations, reflecting a slow accumulation of cadmium in the kidney over the lifetime and peak levels at ages 40 to 60 years (Figure 26.2). The model also predicts the observed relationship between urinary and renal cortex concentrations for observations less than 2 μg Cd/g creatinine and less than 70 μg Cd/g cortex, although it overpredicts renal cortex cadmium concentrations at urinary cadmium concentrations exceeding approximately 2 μg/g creatinine (Diamond et al., 2003). This error may reflect an increase in the clearance of the cadmium from plasma or kidney to urine, related to the onset of toxicity (Ellis, 1985; Ellis et al., 1984; Roels et al., 1981), which is not represented accurately in the model. The model may thus overpredict renal cortex concentrations, and underpredict urinary cadmium levels in people who have experienced sufficiently high exposures to produce renal impairment. The analysis presented below suggests that accurate prediction of renal cortex concentrations of less than 200 μg Cd/g cortex would be sufficient for most risk assessment applications where the major concern would be to establish whether or not risks exceed 0.05 to 0.1, corresponding to renal cortex concentrations less than 200 μg Cd/g cortex.

A realistic simulation of absorption of ingested cadmium is also crucial for translating oral intakes into renal cortex concentrations and associated risk. Estimates of cadmium absorption in humans range from 2.2 to 9.0% (Flanagan et al., 1978; McLellan et al., 1978; Newton et al., 1984; Rahola et al., 1972; Shaikh and Smith, 1980); with the mean $\pm$ SE of the estimates being 4.7 ± 1.0 (Diamond and Goodrum, 1998). However, the distribution of the bioavailability estimates for individual subjects is skewed, in part because iron status was a covariable in these studies. This is most readily illustrated from the estimates from Flanagan et al. (1978) which, when stratified by serum ferritin levels, show that absorption among subjects who were iron-deficient (0 to 20 μg/l serum ferritin) was approximately fourfold higher than in iron-replete subjects. Shaikh and Smith (1980) also reported higher bioavailability in two subjects who had low serum ferritin levels (i.e., less than 10 μg/l).

More-recent studies have provided a plausible mechanistic basis for interactions between iron nutritional status and cadmium absorption. Cadmium and iron appear to share a common transporter, divalent metal ion transporter (DMT1), in the rat small intestine (Park et al., 2002). On this basis, coupled with the observation that U.S. females have a prevalence of marginal iron deficiency (serum ferritin less than 20 μg/l) that is approximately 10-times greater than males (29% and 2.4%, respectively), values of 5% and 10% were assumed in the PK model for the absorption fraction for U.S. males and females, respectively (Choudhury et al., 2001).

The Kjellström and Nordberg (1978) model, with the above assumptions regarding absorption, reliably predicts the higher urinary cadmium excretion observed in U.S. nonsmoking females, compared with males, who have had no occupational exposures to cadmium and whose cadmium exposures can be assumed to have been primarily from

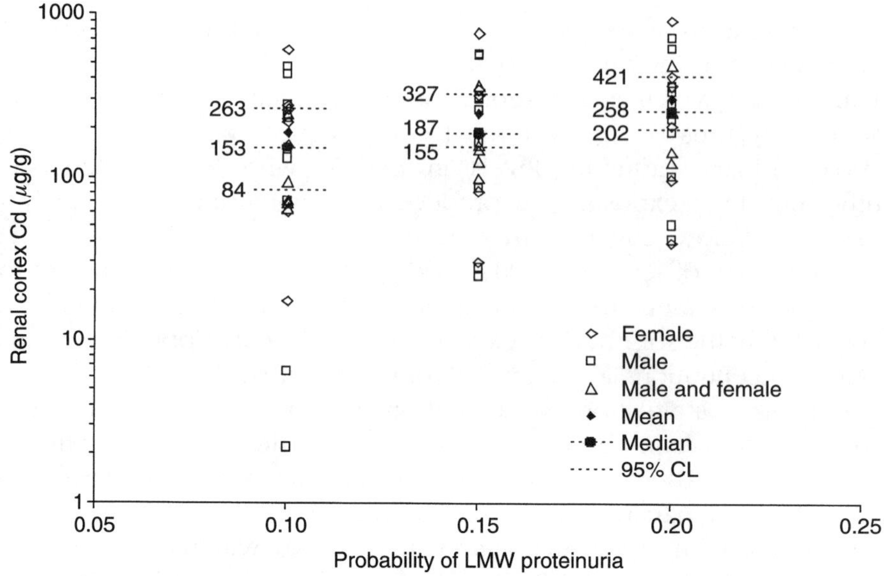

Figure 26.3 Renal cortex concentration associated with specified risks of LMW proteinuria. Each point represents an estimate from an individual study (see Table 26.2). Doses reported in each study were converted to an equivalent renal cortex concentration (μg/g), simulated using the Kjellström and Nordberg (1978) model.

dietary sources (Choudhury et al., 2001). This suggests that the model provides a reasonably accurate overall simulation of the relationships between cadmium intake, absorption, and body burden, for the U.S. population.

Figure 26.3 presents the distribution of the estimates of renal cortex cadmium concentrations associated with .1, .15, or .2 probabilities (RC_{10}, RC_{15}, and RC_{20}, respectively) after conversion of the cadmium dose units used in each study presented in Table 26.2 into units of renal cortex concentration. This range of probabilities is shown for comparison purposes because it is within the prediction limits of all of the models presented in Table 26.2. Most of the models could not predict probabilities below .1 because of convergence with the response criterion (e.g., most response criteria established a background response prevalence of 5 to 10%). Predicting probabilities greater than .2 is irrelevant to most risk assessment applications, where the objective is usually to limit risks to levels below .2. The distributions of RC values are skewed, with the means exceeding the medians. Values for RC_{10},

Table 26.3 Predictions of External and Internal Dose–Response Relationships for Cadmium-Induced LMW Proteinuria in Humans

Risk	*Renal cortex Cd (μg/g)*	*Cd intake (μg/kg/day)*	
		Females	*Males*
All estimates ($n=22$)			
.10	153 (190)[a]	2.0	4.3
	(84–263)	(1.0–3.6)	(2.2–7.7)
.15	187 (250)[a]	2.5	5.4
	(155–327)	(2.0–4.6)	(4.4–9.7)
.20	258 (308)[a]	3.6	7.6
	(202–421)	(2.7–6.0)	(5.8–12.6)
Estimates from occupational and environmental contamination studies ($n=17$)			
.10	231 (236)[a]	3.2	6.7
	(73–241)	(0.8–3.3)	(1.8–7.0)
Estimates from general population studies ($n=5$)			
.10	31[a] (17)[b]	0.4 (0.2)	0.7 (0.4)
	(2–69)[c]	(0.02–0.8)	(0.05–1.7)

Values are medians and with 95% confidence limits in parenthesis, except where noted otherwise.
[a]Arithmetic mean.
[b]Median.
[c]Range.

RC_{15}, and RC_{20} can be translated into equivalent chronic cadmium intakes using the Kjellström and Nordberg (1978) model; these are presented in Table 26.3. The median value for RC_{10} was 153 μg/g cortex (95% CI = 84 to 263) and the corresponding values for cadmium intake were 2.0 μg/kg/day (95% CI = 1.0 to 5.0) in females and 4.3 μg/kg/day (95% CI = 2.2 to 7.7) in males. A lower cadmium intake in females is predicted to result in a similar renal cortex concentration as in males because females were assumed, in the PK model, to absorb a larger fraction (10% and 5%, respectively) of an ingested cadmium dose (Choudhury et al., 2001).

The inclusion of all of these studies in the derivation of central-tendency risk estimates introduces a weight of evidence that would not be realized if a only single study were used. However, this approach may also introduce errors of simplification by not adequately adjusting for differences in the study designs, such as the types and levels of exposure, response criteria, power, and error (e.g., measurement error, adequacy of adjusting for covariables and confounders) in each study. Ideally, a complete meta-analysis of the primary data from these studies would be needed to develop a composite dose–response function that reflects the variability and uncertainty inherent to each data set. The advantage of the simpler approach depicted in Figure 26.3 is that it can be implemented without access to the primary data and can be readily updated with additional parameter estimates as they become available. Combining estimates of dose–response parameters for males and females was justified based on the observation that estimates of the RC_{10} (and RC_{15} and RC_{20}) were not significantly different for studies of female and male populations. This outcome is expected if the concentration of cadmium in the kidney is the dominant predictor of risk of LMW proteinuria. On the other hand, RC_{10} and RC_{15} values for general population exposures were significantly less than that for occupational exposures and exposures from highly contaminated environments (e.g., contaminated river basins). The observation that the RC estimates showed positive skew, rather then being symmetrically distributed around a central tendency, also suggests the possibility that the data set may represent a mixed distribution.

One variable that would be expected to influence the dose–response parameter estimates is the response criterion for urinary $\beta 2\mu G$, which ranged from the 84th to the 97.5th percentile (Table 26.2). A relatively weak but statistically significant relationship ($R^2 = 0.31$) between the two variables was observed, with the RC_{10} increasing with increasing percentile of the response criterion, which may have contributed to some of the skew in the estimates. More importantly, two of three studies of studies of general population exposures (Suwazano et al., 2000; Yamanaka et al., 1998) used the lowest percentile for the response criterion (84%) and also yielded the four lowest values for $RC_{10.}$ (2 to 61 μg Cd/g cortex). This outcome is not surprising because use of the lower percentile as the response criterion would tend to result in classifying more subjects as responders than would occur if a higher percentile were adopted. Exclusion of the studies of general population exposures from the analysis results in a much more symmetric and narrower distribution of RC_{10} values, and higher values for the RC_{10} (231 μg Cd/g cortex, 95% CL = 73 to 241).

Given the uncertainties in the numerous uncontrolled factors, which might contribute to variability in the dose–response parameters across studies, use of a lower confidence limit on the median of the parameter estimates for the full data set should account for some of the uncertainty in combining the estimates. This approach is recommended for risk estimates, and is analogous to the benchmark-dose approach (in which the lower confidence on the mean dose associated with a .10 probability of response is estimated for individual studies) (Crump, 1984; Gaylor et al., 1999).

LEAD

Although the literature on lead-induced nephrotoxicity in humans is at least as abundant as that for cadmium, the lack of a relatively specific biomarker of lead-induced nephropathy (e.g., LMW proteinuria), together with uncertainties related to the use of blood-lead concentration as an internal dose metric, has made quantitative risk assessment for lead far less certain that for cadmium. Nevertheless, the large number of studies of lead nephropathy in humans and, in particular, the results of more-recent large-scale epidemiological studies provide a basis for establishing blood-lead concentration ranges of concern. Developments in modeling of pharmacokinetics of lead have provided a means to translate blood-lead concentrations into estimates of corresponding lead intakes.

Clinical and Epidemiologic Basis for Risk Estimates

Lead nephrotoxicity in humans and in experimental animals is characterized by proximal tubular nephropathy, glomerular sclerosis, and interstitial fibrosis (Goyer, 1989; Loghman-Adham, 1997). Figure 26.4 summarizes the results of studies in which various indicators of renal functional impairment or injury have been examined in people exposed to lead to varying degrees (Table 26.4). Most of these studies are of adults whose exposures were of occupational origin; however, a few environmental and mixed exposures are represented and a few studies of children are also included (Bernard et al., 1995; Fels et al., 1998; Verberk et al., 1996).

The studies plotted in Figure 26.4 are sorted by the central tendency for blood-lead concentration reported in each study; details about the

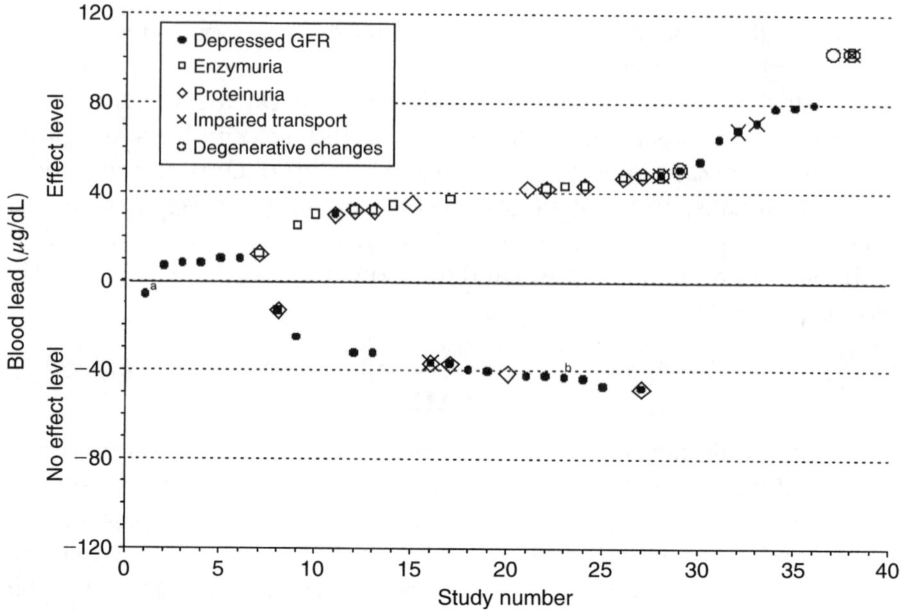

Figure 26.4 Indicators of renal functional impairment observed at various blood-lead concentrations in humans. Degenerative changes include cellular necrosis in proximal tubule, glomerular sclerosis, and interstitial fibrosis. [a]Significant increase in serum creatinine concentration (Hu, 1991). [b]Significant increase in creatinine clearance (Roels et al., 1994). (GFR) glomerular filtration rate.

subjects and exposures are provided in Table 26.4. Endpoints of kidney status captured in this data set include various measures of glomerular and tubular dysfunction. Data on changes in glomerular filtration rate represent measurements of either creatinine clearance or serum creatinine concentration. Measurements of enzymuria, mainly urinary *N*-acetyl-β-D-glucosaminidase (NAG), are also represented. Increased excretion of NAG, but with no increased excretion of other proximal tubule enzymes (e.g., alanine aminopeptidase, alkaline phosphatase, γ-glutamyltransferase), has been found in lead-exposed workers (Pergande et al., 1994). Data points indicating proteinuria refer to total urinary protein, urinary albumin, or urinary LMW protein (e.g., $\beta_2\mu G$ or RBP). Indices of impaired transport include clearance or transport maxima for organic anions (e.g., *p*-aminohippurate, urate) or glucose (Biagini et al., 1977; Hong et al., 1980; Wedeen et al., 1975).

A few studies have provided histopathological confirmation of proximal tubular injury (Biagini et al., 1977; Wedeen et al., 1975, 1979).

Figure 26.4 illustrates a few general trends regarding the relationship between blood-lead concentration and qualitative aspects of the kidney response. A cluster of observations of decrease in glomerular filtration rate appear at the low end of the blood-lead concentration range (less than 20 μg/dl); the significance of these studies is discussed in greater detail below. Outcomes for the various renal toxicity endpoints are mixed over the blood-lead concentration range of 20 to 50 μg/dl. Enzymuria or proteinuria were detected in most studies in which these endpoints were evaluated, whereas indications of depressed glomerular filtration rate were, with only one exception, not observed over this blood-lead concentration range. At blood-lead concentrations greater than 50 μg/dl, functional deficits, including enzymuria, proteinuria, impaired transport, and depressed glomerular filtration rate, dominate the observations.

The overall dose–effect pattern suggests an increasing severity of nephrotoxicity associated with increasing blood-lead concentration, with effects on glomerular filtration evident at blood-lead concentrations below 20 μg/dl, enzymuria and proteinuria becoming evident above 30 μg/dl, and severe deficits in function and pathological changes occurring in association with blood-lead concentrations exceeding 50 μg/dl.

Blood-Lead Concentration as an Internal Dose Metric for Nephrotoxicity Risk

The above findings provide some quantitative support for reliance on blood-lead concentration as an internal dose metric for assessing risks of lead-induced nephropathy. However, extensive use of blood lead as a dose metric mainly reflects the greater feasibility of incorporating blood-lead measurements into clinical or epidemiological studies compared with other potential dose indicators, such as lead in kidney, plasma, urine, or bone (Skerfving, 1988). Noninvasive methods for measuring lead levels in kidney have not been developed for practical use and the relationship between kidney lead levels and kidney toxicity has not been adequately defined to support risk estimates.

Lead in blood is extensively bound (>99%) to proteins in red blood cells (Bergdahl et al., 1998; DeSilva, 1981; Ong and Lee, 1980) and therefore the blood-lead concentration may not be a good reflection of the amount of lead in plasma that is available to the kidney or other tissues that are targets for lead toxicity. The concentration of lead in plasma is extremely difficult to measure accurately because levels in plasma are near the limits of quantitation for most analytical techniques (e.g., approximately 0.4 μg/l at blood-lead concentration of 100 μg/l (Bergdahl and Skerfving, 1997; Bergdahl et al., 1997) and because hemolysis that occurs with typical analytical practices can contribute substantial measurement error (Cavelleri et al., 1978; Bergdahl et al., 1998; Smith et al., 1998). Advances in inductively-coupled plasma mass spectrometry (ICP-MS) has made the technique sufficiently sensitive for measurement of lead in plasma (Schütz et al., 1996).

An additional concern is that the blood-lead concentration can change relatively rapidly (e.g., over the course of several weeks) in response to changes in exposure; thus blood-lead concentration can be influenced by short-term variability in exposure that may have only minor effects on lead body burden and the long-term cumulative dose to the kidney or other tissues. In adults, more than 90% of the lead body burden resides in bone (Barry and Mossman, 1970; Gross et al., 1975; Schroeder and Tipton, 1968). Changes in nutrition or bone metabolism that affect rates of bone deposition (e.g., growth) or resorption (e.g., pregnancy, lactation, menopausal osteoporosis) can alter rates of deposition or release of bone lead (Gulson et al., 1997, 1999; Leggett, 1993; O'Flaherty, 1991, 1992, 1993; Tsaih et al., 2001); these factors can introduce variability to the lead dose to the kidney and other tissues that may not be reflected in measurements of blood-lead concentration. Time-integrated measurements of blood-lead concentration may provide a means for accounting for some of these factors (Roels et al., 1995).

Recent developments in non-invasive x-ray fluorescence techniques for measuring bone lead levels have enabled the application of bone lead measurements to the epidemiology of lead exposure and toxicity (Hu et al., 1990, 1991; Todd et al., 1992). Tibial bone lead appears to be a better predictor of hypertension than blood-lead concentration (Hu et al., 1996; Korrick et al., 1999), and may be better than blood-lead concentration as a surrogate for plasma lead (Tsiah et al., 1999). Bone-lead levels may also be superior to blood-lead levels for predicting other outcomes in lead-exposed populations, including low birth weight (González-Cossío et al., 1997) and declines in blood hemoglobin and hematocrit (Hu et al., 1994).

Table 26.4 Selected Studies of Lead-Induced Nephrotoxicity in Humans

Number	*Study*	*Exposure type*	*Number of subjects*	*Age (year)*	*Exposure duration (year)*	*Blood-lead concentration (μg/dl)*	*Biomarker evaluated*
1	Hu, 1991	Environmental	22	55	NA	6	CCr
2	Lin et al., 2001	Unknown	55	57	NA	7	CCr
3	Staessen et al., 1992	Environmental	1981	48	NA	8	CCr, SCr
4	Payton et al., 1994	Environmental	744	64	NA	8	CCr
5	Kim et al., 1996	Unknown	459	57	NA	10	SCr
6	Staessen et al., 1990	Environmental	531	48	NA	10	SCr
7	Bernard et al., 1995	Environmental	154	13	NA	12	UNAG, URBP
8	Fels et al., 1998	Environmental	62	10	NA	13	SCr, UE, UP, ULMWP
9	Sonmez et al., 2002	Occupational	13	32	0.14	25	UNAG
10	Chia et al., 1994	Occupational	128	28	3	30	UNAG
11	Chia et al., 1995	Occupational	137	28	>0.5	30	SCr, $S\beta_2\mu G$
12	Mortada et al., 2001	Occupational	43	33	10	32	SCr, UNAG, UAlb
13	Gerhardsson et al., 1992	Occupational	100	37–68	14–32	32	CCr, CCr, $U\beta_2\mu G$, UNAG
14	Verberk et al., 1996	Environmental	151	4.6	NA	34	UNAG
15	Factor-Litvak et al., 1999	Environmental	394	6	6	35	UP
16	Omae et al., 1990	Occupational	165	18–57	0.1–26	37	CCr, CUA, $U\beta_2\mu G$, $C\beta_2\mu G$
17	Cardozo dos Santos et al., 1994	Occupational	166	33	4.5	37	SCr, UNAG, UAlb, UP
18	Wedeen et al., 1975	Occupational	4	36	5–8	40	GFR, RPF, TMPAH, HP
19	Hsiao et al., 2001	Occupational	30	38	13	40	SCr

20	Huang et al., 1988	Occupational	40	30	5	41	U$\beta_2\mu$G, UP
21	Fels et al., 1994	Occupational	81	30	7	42	UP
22	Pergande et al., 1994	Occupational	82	30	7	42	SCr, UP, UE
23	Roels et al., 1994	Occupational	76	44	6–36	43	CCr, UNAG
24	Kumar et al., 1995	Occupational	22	32.5	NA	43	CCr, U$\beta_2\mu$G, UNAG
25	de Kort et al., 1987	Occupational	53	42	12	47	SCr, BUN
26	Verschoor et al., 1987	Occupational	155	30–51	<2–>10	47	UNAG, URPB
27	Cardenas et al., 1993b	Occupational	41	39	14	48	SCr, UP, U$\beta_2\mu$G, UNAG, UTBX, UPG
28	Wedeen et al., 1975	Occupational	1	40	5	48	GFR, TMPAH, HP
29	Wedeen et al., 1979	Occupational	15	41	14	51	GFR, HP
30	Ehrlich et al., 1998	Occupational	382	41	12	54	SCr, SUA
31	Pinto et al., 1987	Occupational	52	38	NA	64	SCr
32	Hong et al., 1980	Occupational	6	35	7	68	GFR, TMG
33	Wedeen et al., 1975	Occupational	3	28	3–5	72	GFR, RPF, TMPAH, HP
34	Baker et al., 1979	Occupational	160	29–62	4–31	77	GFR, BUN
35	Lilis et al., 1968	Occupational	102	32–61	>10	79	GFR, SCr
36	Lilis et al., 1980	Occupational	449	NA	12	80	SCr, BUN
37	Cramer et al., 1974	Occupational	7	45	9	103	GFR, HP
38	Biagini et al., 1977	Occupational	11	44	12	103	GFR, CPAH, HP

Study numbers refer to Figure 26.4. Blood-lead concentrations are reported central tendencies.
BUN = blood urea nitrogen; CCr = creatinine clearance; C$\beta_2\mu$G = $\beta_2\mu$G; CPAH = *p*-aminohippurate clearance; CUA = uric acid clearance; GFR = glomerular filtration rate; HP = histopathology; RPF = renal plasma flow; S$\beta_2\mu$G = serum $\beta_2\mu$G; SCr = serum creatinine; SUA = serum uric acid; TMG = transport maximum for glucose; TMPAH = transport maximum for *p*-aminohippurate; UAlb = urine albumin; U$\beta_2\mu$G = urinary $\beta_2\mu$G; UE = urine enzymes; ULMWP = urine low molecular weight proteins; UNAG = urine *N*-acetyl-β-D-glucosaminidase; UP = urine protein; UPG = urine prostaglandins; URBP = urine retinol binding protein; UTBX = urine thromboxane.

Dose–response Relationships for Decrements in Glomerular Filtration Rate

Experimental studies in laboratory animals have shown that exposures to lead that result in blood-lead concentrations exceeding 50 μg/dl can depress glomerular filtration rate and renal blood flow (Aviv et al., 1980; Khalil-Mensch et al., 1992a, 1992b). In rats, depressed glomerular filtration rate appears to be preceded with a period of increased filtration (Kahlil-Manesh et al., 1992a; O'Flaherty et al., 1986). The mechanism by which lead alters glomerular filtration rate is unknown and, as important for risk assessment, its mechanistic connection to lead-induced hypertension has not been fully elucidated. Glomerular sclerosis or proximal tubule injury and impairment could directly affect renin release and renal insufficiency could secondarily contribute to hypertension.

In humans, reduced glomerular filtration rate (i.e., assessed by creatinine clearance or serum creatinine concentration) has been observed in association with exposures resulting in blood-lead concentrations exceeding 50 μg/dl; however, at lower blood levels, study outcomes have been mixed (Figure 26.4, Table 26.4). The results of the larger epidemiological studies have shown a significant effect of age on the relationship between glomerular filtration rate (assessed from creatinine clearance of serum creatinine concentration) and blood-lead concentration (Kim et al., 1996; Payton et al.1994; Staessen et al., 1990, 1992). Furthermore, hypertension can be both a confounder in studies of associations between lead exposure and creatinine clearance (Perneger et al., 1993) as well as a covariable with lead exposure (Harlan et al., 1985; Hu et al., 1994; Pirkle et al., 1985; Pocock et al., 1984, 1988; Weiss et al., 1986). These factors may explain some of the variable outcomes of smaller studies in which the age and hypertension effects were not fully taken into account.

When age and other covariables that might contribute to glomerular disease are factored into the dose–response analysis, decreased glomerular filtration rate has been consistently observed in populations that have average blood-lead concentrations less than 20 μg/dl (Table 26.5). In the Kim et al. (1996) study, a significant relationship between serum creatinine and blood-lead concentrations was evident in subjects who had blood-lead concentrations below 10 μg/dl (serum creatinine increased 0.14 mg/dl per tenfold increase in blood-lead concentration). Assuming a glomerular filtration rate of approximately 90 to 100 ml/min in the studies reported in Table 26.5, a change in

Table 26.5 Summary of Dose–Response Relationships for Effects of Lead Exposure on Biomarkers of Glomerular Filtration Rate

Study	*Exposure*	*n*	*Blood-lead concentration, mean and range (μg/dL)*	*Endpoint*	*Change in endpoint (per tenfold increase in blood lead)*
Payton et al., 1994	Mixed[a]	744, M	8.1 (4–26)	CCr (ml/min)	−10[b]
Staessen et al., 1992	Environmental	1016, F 965, M	7.5 (1.7–65)	CCr (ml/min)	−10, F[c] −13, M
Kim et al., 1996	Mixed[a]	459, M	9.9 (0.2–54)	SCr (mg/dl)	0.08[d] 0.14[e]
Staessen et al., 1990	Environmental	133, F 398, M	12 (6–35)	SCr (mg/dl)	0.07 M[f]

M = males; F = females; CCr = creatinine clearance; SCr = serum creatinine concentration; *n* = number of subjects.
[a]U.S. Veterans Administration Normative Aging Study.
[b]Partial regression coefficient: −0.040 ln ml/min creatinine clearance per ln μmol/l blood lead concentration.
[c]Partial regression coefficient, −9.51 ml/min creatinine clearance per log μmol/l blood-lead concentration.
[d]Partial regression coefficient, 2.89 μmol/l serum creatinine per ln μmol/l blood-lead concentration.
[e]In subjects with blood-lead concentrations less than 10 μg/dl, the partial regression coefficient was 5.29 μmol/l serum creatinine per ln μmol/l blood-lead concentration.
[f]Reported 0.6 increase in serum creatinine (μmol/l) per 25% increase in blood-lead concentration (μmol/l, log-transformed) in males (two subjects with serum creatinine concentrations exceeding 180 μmol/l excluded; regression coefficient not reported for females).

creatinine clearance of 10 to 14 ml/min would represent a 9 to 16% change in glomerular filtration rate per tenfold increase in blood-lead concentration. Estimating the change in glomerular filtration rate from the incremental changes in serum creatinine concentration reported in Table 26.5 is far less certain because decrements in glomerular filtration do not necessarily give rise to proportional increases in serum creatinine concentrations. A 50% decrement in glomerular filtration rate can occur without a measurable change in serum creatinine excretion (Brady et al., 2000). Nevertheless, the changes reported in Table 26.5

(0.07 to 0.14 mg/dl) would represent a 6 to 16% increase, assuming a mean serum creatinine concentration of 0.9 to 1.2 mg/dl. This suggests at least a similar, and possibly a substantially larger, decrement in glomerular filtration rate.

The confounding and covariable effects of hypertension are also relevant to the interpretation of the regression coefficients reported in these studies. Given the evidence for an association between lead exposure and hypertension, and that decrements in glomerular filtration rate can be a contributor to hypertension, it is possible that the reported hypertension-adjusted regression coefficients may underestimate the actual slope of the blood-lead concentration relationship with serum creatinine concentration or creatinine clearance.

The observations derived from the studies presented in Table 26.5 that suggest a relationship between blood-lead concentrations and decrements in glomerular filtration rate are consistent with those of a smaller prospective clinical study. In that study, progression of renal insufficiency was related to higher lead body burden among patients whose blood-lead concentrations were less than 15 μg/dl (Lin et al., 2001). Mean blood-lead concentrations in a high lead body burden group (ethylenediamine tetraacetic acid provocation test yielded >600 μg excreted/72 hours) were 6.6 μg/dl (range of 1.0 to 15) compared with 3.9 μg/dl (range of 1 to 7.9) in a low lead body burden group.

The above observations suggest that significant decrements in glomerular filtration rate may occur in association with blood-lead concentrations below 20 μg/dl and, possibly, below 10 μg/dl (Kim et al., 1996). This range is used as the basis for estimates of lead intakes that would place individuals at risk for renal functional deficits.

Pharmacokinetics Modeling of External Dosage Associated with Decrements in Glomerular Filtration Rate

Several pharmacokinetics models that simulate blood-lead levels and tissue distribution of absorbed lead in adult humans have been described (Bert et al., 1989; Leggett, 1993; O'Flaherty, 1991, 1993, 1995). The International Commission on Radiological Protection (ICRP) model (Leggett, 1993; Pounds and Leggett, 1998) has been used by the ICRP, U.S. Department of Energy, and U.S. Environmental Protection Agency to develop cancer-risk coefficients for internal radiation exposures to lead and other alkaline-earth elements that have biokinetics similar to those

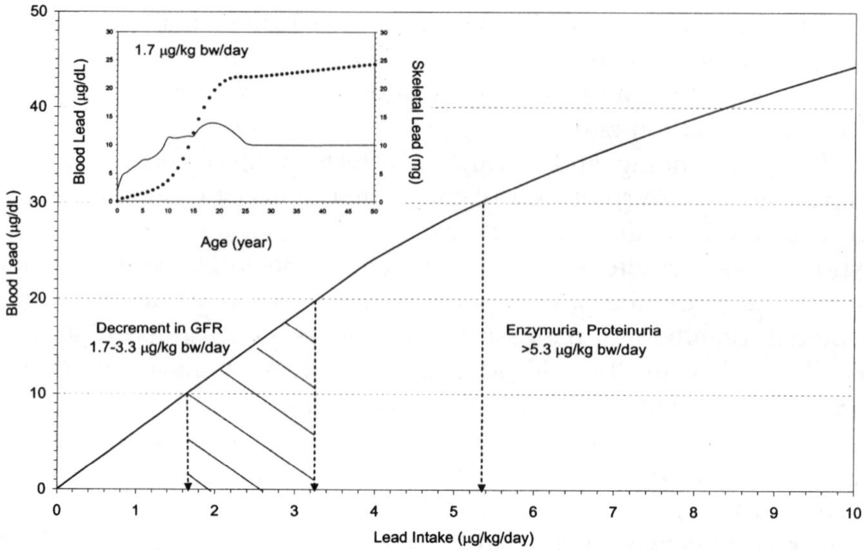

Figure 26.5 Lead intakes predicted to result in renal functional impairment in humans. The main panel shows the relationship between chronic lead intake (µg/kg bodyweight/day) and steady state blood-lead concentration predicted by the ICRP lead model (solid line). The inset panel shows the age pattern for blood-lead concentration and bone-lead burden that is predicted to result from an intake of 1.7 µg/kg/day, beginning at birth.

of calcium (ICRP, 1993; U.S. EPA, 2000). The model simulates age-dependent kinetics of uptake of lead from the gastrointestinal and respiratory tracts, distribution of absorbed lead to blood, bone, liver, kidney and other soft tissues, and excretion of lead.

Figure 26.5 shows the relationship between the chronic lead ingestion rate and steady state blood-lead concentration predicted by the ICRP model. For blood-lead concentrations below 25 µg/dl, the relationship is linear; however, a distinct curvature is evident as the blood-lead concentration increases above this level; this is thought to result from capacity limited uptake of lead into red blood cells (Bergdahl et al., 1998; DeSilva, 1981; Manton and Cook, 1984; Ong and Lee, 1980).

The model predicts that a constant daily intake (per kg bodyweight), beginning in infancy, would result in a steady state blood-lead concentration after approximately 25 years of continuous exposure. Slightly higher peak blood-lead concentrations would occur during the ages 15 to 20 years, which corresponds to the period of rapid adolescent growth and skeletal accrual of lead (Leggett, 1993; O'Flaherty, 1991, 1993). Steady state blood-lead concentrations in the range 10 to 20 μg/dl, the estimated range associated with lead-related decrements in glomerular filtration rate (see Table 26.5) would be achieved with chronic daily oral intakes of 1.7 to 3.3 μg/kg bodyweight/day (Figure 26.5). A steady state blood-lead concentration of 30 μg/dl, the approximate lower end of the range associated with enzymuria, proteinuria, and more severe renal functional deficits, would be achieved with chronic daily oral intakes of 5.3 μg/kg bodyweight/day (Figure 26.5).

CONCLUDING REMARKS

Risk assessment of metal-induced nephrotoxicity has advanced considerably in recent decades, to the point where it is possible to foresee the development of risk models that integrate information on exposure, pharmacokinetics, pharmacodynamics, and uncertainty for a variety of metals. The examples described here for cadmium and lead demonstrate, at least conceptually, that such integrated approaches are feasible. Improved risk models for cadmium, lead, and other nephrotoxic metals will undoubtedly result from continued advances in analytical techniques, discovery of new and more specific biomarkers of exposure, internal dose, and nephrotoxicity, advances in our understanding of variability of human responses to toxic metals, and continued accrual of knowledge about the mechanisms by which metals exert their effects on the kidney.

REFERENCES

Aoshima K, Kawanishi Y, Fan J, Cai Y, Katoh T, Teranishi H, and Kasuya M (1995) Cross-sectional assessments of renal function in inhabitants of a cadmium-polluted area. *Ann Clin Lab Med* 25:493–501.

Arisawa K, Nakano A, Saito H, Liu X-J, Yokoo M, Soda M, Koba T, Takahashi T, and Kinoshita K (2001) Mortality and cancer incidence among a population previously exposed to environmental cadmium. *Int Arch Occup Environ Health* 74:255–262.

Aviv A, John E, Bernstein J, Goldsmith DI, and Spitzer A (1980) Lead intoxication during development: Its late effect on kidney function and blood pressure. *Kidney Int* 17:430–437.

Baker EL Jr, Landrigan PJ, Barbour AG, Cox DH, Folland DS, Ligo RN, and Throckmorton J (1979) Occupational lead poisoning in the United States: Clinical and biochemical findings related to blood lead levels. *Br J Ind Med* 36:314–322.

Barregård L, Svalander C, Schütz A, Westberg G, Sällsten G, Blohmé., Mölne J, Attman P-O, and Haglind P (1999) Cadmium, mercury, and lead in kidney cortex of the general swedish population: A study of biopsies from living kidney donors. *Environ Health Perspect* 107:867–871.

Barry PSI, and Mossman DB (1970) Lead concentration in human tissues. *Br J Ind Med* 27:339–351.

Benedetti J-L, Samuel O, Dewailly É, Gingras S, and Lefebvre MA (1999) Levels of cadmium in kidney and liver tissues among a Canadian population (province of Quebec). *J Toxicol Environ Health* 56 (Part A):145–163.

Bergagard I, and Bearn AG (1968) Isolation and properties of a low molecular weight β2-globulin in human biological fluids. *J Biol Chem* 243:4095–4103.

Bergdahl IA, and Skerfving S (1997) Partition of circulating lead between plasma and red cells does not seem to be different for internal and external sources of lead. Letter to the Editor. *Am J Ind Med* 32:317–318.

Bergdahl IA, Sheveleva M, Schutz A, Artamonova VG, and Skerfving S (1998) Plasma and blood lead in humans: Capacity-limited binding to δ-aminolevulinic acid dehydratase and other lead-binding components. *Toxicol Sci* 46:247–253.

Bergdahl IA, Schutz A, Gerhardsson L, Jenson A, and Skerving S (1997) Lead concentrations in human plasma, urine and whole blood. *Scand J Work Environ Health* 23:359–363.

Bernard A, Vyskocyl A, Mahieu P, and Lauwerys R (1988) Effect of renal insufficiency on the concentration of free retinol-binding protein in urine and serum. *Clinica Chimica Acta* 171:85–94.

Bernard A, Buchet JP, Roels H, Masson P, and Lauwerys R (1979) Renal excretion of proteins and enzymes in workers exposed to cadmium. *Eur J Clin Invest* 9:11–22.

Bernard AM, and Lauwerys RR (1981) Retinol binding protein in urine: A more practical index than urinary β2-microglobulin for the routine screening of renal tubular function. *Clin Chem* 27:1781–1782.

Bernard AM, Vyskocil A, Roels H, Kriz J, Kodl M, and Lauwerys R (1995) Renal effects in children living in the vicinity of a lead smelter. *Environ Res* 68:91–95.

Bert JL, Van Dusen LJ, and Grace JR (1989) A generalized model for the prediction of lead body burdens. *Environ Res* 48:117–127.

Biagini G, Caudarella R, and Vangelista A (1977) Renal morphological and functional modification in chronic lead poisoning. *Clin Chem Chem Toxicol Metals* 123–126.

Brady HR, Brenner BM, Clarkson MR, and Lieberthal W (2000) Acute Renal Failure, in *The Kidney*, Brenner BM (Ed.), W.B. Saunders Co., pp. 1202.

Buchet J-P, Lauwerys R, Roels H, Bernard A, Bruaux P, Clayes F, Ducoffre G, De Plaen P, Staessen J, Amery A, Lijnin P, Thijs L, Rondia D, Sartor F, Saint Remy A, and Nick L (1990) Renal effects of cadmium body burden of the general population. *Lancet* 336:699–702.

Cai Y, Heranishi H, Aoshima K, Katoh T, Arai Y, and Kasuya M (2001) Development of the fluorometric ELISA method for determination of α1-microglobulinuria in a cadmium-polluted area in Japan. *Int Arch Occup Environ Health* 74:514–518.

Cai S, Yue I, Jin T, and Nordberg G (1998) Renal dysfunction from cadmium contamination of irrigation water: Dose-response analysis in a Chinese population. *Bull WHO* 76:153–159.

Cai S, Wang J, Xue J, Zhu X, Wang J, and Wang Y (1992) A judgment of attribution of increase in urine β2-microglobulin after cadmium exposure. *Biomed Environ Sci* 5:130–135.

Cardenas A, Roels H, Bernard AM, Barbon R, Buchet JP, Lauwerys RR, Rosello J, Ramis I, Mutti A, Franchini I, Fels LM, Stolte H, De Broe ME, Nuyts GD, Taylor SA, and Price RG (1993a) Markers of early renal changes induced by industrial pollutants. I. Application to workers exposed to mercury vapour. *Br J Ind Med* 50:17–27.

Cardenas A, Roels H, Bernard AM, Barbon R, Buchet JP, Lauwerys RR, Rosello J, Ramis I, Mutti A, Franchini I, Fels LM, Stolte H, De Broe ME, Nuyts GD, Taylor SA, and Price RG (1993b) Markers of early renal changes induced by industrial pollutants. II. Application to workers exposed to lead. *Br J Ind Med* 50:28–36.

Cardozo dos Santos A, Colacciopo S, Dal Bo CMR, and Guinaim dos Santos NA (1994) Occupational exposure to lead, kidney function tests, and blood pressure. *Am J Ind Med* 26:635–643.

Cavalleri A, Minoia C, Pozzoli L, and Baruffini A (1978) Determination of plasma lead levels in normal subjects and lead-exposed workers. *Br J Ind Med* 35:21–26.

Ceovic S, Plestina R, Miletic-Medved M, Stavljenic A, Mitar J, and Vukelic M (1991) Epidemiological aspects of Balkan endemic nephropathy in a typical focus in Yugoslavia. *IARC Sci Publ* 115:5–10.

Ceovic S, Hrabar A, and Radonic M (1985) An etiological approach to Balkan endemic nephropathy based on the investigation of two genetically different populations. *Nephron* 40:175–179.

Chen JJ, and Gaylor DW (1992) Dose-response modeling of quantitative response date for risk assessment. *Comm Statist Theory Meth* 21:2367–2381.

Chia KS, Jeyaratnam J, Tan C, Ong HY, Ong CN, and Lee E (1995) Glomerular function of lead-exposed workers. *Toxicol Lett* 77:319–328.

Chia KS, Mutti A, Tan C, Ong HY, Jeyaratnam J, Ong CN, and Lee E (1994) Urinary N-acetyl-β-D-glucosaminidase activity in workers exposed to inorganic lead. *Occup Environ Med* 51:125–129.

Choudhury H, Harvey T, Thayer WC, Lockwood TF, Stiteler WM, Goodrum PE, Hassett JM, and Diamond GL (2001) Urinary cadmium elimination as a biomarker of exposure for evaluating a cadmium dietary exposure-biokinetics model. *J Toxicol Environ Health* 63 (Part A):101–130.

Clarkson TW, Hursh JB, Sager PR, and Syversen TLM (1988) Mercury. In *Biological Monitoring of Toxic Metals*, Plenum Press, New York and London, pp. 199–246.

Cramer K, Goyer RA, Jagenburg R, and Wilson MH (1974) Renal ultrastructure, renal function, and parameters of lead toxicity in workers with different periods of lead exposure. *Br J Ind Med* 31:113–127.

Crump KS (1984) A new method for determining allowable daily intakes. *Fundam Appl Toxicol* 4:854–871.

de Kort WLAM, Verschoor MA, Wibowo AAE, and van Hemmen JJ (1987) Occupational exposure to lead and blood pressure: A study in 105 workers. *Am J Ind Med* 11:145–156.

DeSilva PE (1981) Determination of lead in plasma and studies on its relationship to lead in erythrocytes. *Br J Ind Med* 38:209–217.

Diamond GL, Thayer WC, and Choudhury H (2003) Pharmacokinetics/pharmacodynamics (PK/PD) modeling of risks of kidney toxicity from exposure to cadmium: Estimates of dietary risks in the U.S. population. *J Toxicol Environ Health* 66 (Part A):1–24.

Diamond GL, and Goodrum PH (1998) Gastrointestinal absorption of metals. *Drug Chem Toxicol* 20:345–368.

Ehrlich R, Robins T, Jordaan E, Miller S, Mbuli S, Selby P, Wynchank S, Cantrell A, De Broe M, D'Haese P, Todd A, and Landrigan P (1998) Lead absorption and renal dysfunction in a South African battery factory. *Occup Environ Med* 55:453–460.

Ellis KJ (1985) Dose-response analysis of heavy metal toxicants in man: Direct *in vivo* assessment of body burden. *Trace Subst Environ Health* 19:149–159.

Ellis KJ, Yuen K, Yasumura S, and Cohn SH (1984) Dose-response analysis of cadmium in man: Body burden vs kidney dysfunction. *Environ Res* 33:216–226.

Factor-Litvak P, Wasserman G, Kline JK, and Graziano J (1999) The Yugoslavia prospective study of environmental lead exposure. *Environ Health Perspect* 107: 9–15.

Falck FY, Fine LJ, Smith RG, McClatchey KD, Annesley T, England B, and Schork AM (1983) Occupational cadmium exposure and renal status. *Am J Ind Med* 4:541–549.

Fels LM, Wunsch M, Baranowski J, Norska-Borowka I, Price RG, Taylor SA, Patel S, De Broe M, Elsevier MM, Lauwerys R, Roels H, Bernard A, Mutti A, Gelpi E, Rosello J, and Stolte H (1998) Adverse effects of chronic low level lead exposure on kidney function risk group study in children. *Nephrol Dial Transplant* 13:2248–2256.

Fels LM, Herbort C, Pergande M, Jung K, Hotter G, Rosello J, Gelpi E, Mutti A, De Broe M, and Stolte H (1994) Nephron target sites in chronic exposure to lead. *Nephrol Dial Transplant* 9:1740–1746.

Flanagan PR, McLellan JS, Haist J, Cherian G, Chamberlain MJ, and Valberg LS (1978) Increased dietary cadmium absorption in mice and human subjects with iron deficiency. *Gastroenterology* 7:841–846.

Friberg L (1950) *Health hazards in the manufacture of alkaline accumulators with special reference to chronic cadmium poisoning.* From the Industrial Hygiene Department of the Swedish Public Health Institute, Stockholm, and the Department for Occupational Medicine at the Medical Clinic of Karolinska Sjukhuset, Stockholm.

Friis L, Petersson L, and Edling C (1998) Reduced cadmium levels in human kidney cortex in Sweden. *Environ Health Perspect* 106:175–178.

Furuse A, Futagoishi Y, Karashima S, Hattori S, and Matsuda I (1992) Familial progressive renal tubulopathy. *Clin Nephrol* 37:192–197.

Gaylor DW, and Slikker W Jr (1990) Risk assessment for neurotoxic effects. *Neurotoxicol* 11:211–218.

Gaylor DW, Kodell RL, Chen JJ, and Krewski D (1999) A unified approach to risk assessment for cancer and noncancer endpoints based on benchmark doses and uncertainty/safety factors. *Reg Toxicol Pharmacol* 29:151–157.

Gerhardsson L, Chettle DR, Englyst V, Nordberg GF, Nyhlin H, Scott MC, Todd AC, and Vesterberg O (1992) Kidney effects in long term exposed lead smelter workers. *Br J Ind Med* 49:186–192.

Ginevri F, Piccotti E, Alinovi R, DeToni T, Biagini C, Chiggeri GM, and Gusmano R (1993) Reversible tubular proteinuria precedes microalbuminuria and correlates with the metabolic status in diabetic children. *Pediatr Nephrol* 7:23–26.

González-Cossío T, Peterson KE, Sanín L-H, Fishbein E, Palazuelso E, Aro A, Hernández-Avila M, and Hu H (1997) Decrease in birth weight in relation to maternal bone-lead burden. *Pediatrics* 100:856–862.

Goyer RA (1989) Mechanisms of lead and cadmium nephrotoxicity. *Toxicol Lett* 46:153–162.

Gross SB, Pfitzer EA, Yeager DW, and Kehoe RA (1975) Lead in human tissues. *Toxicol Appl Pharmacol* 32:638–651.

Gulson BL, Pounds JG, Mushak PA, Thomas BJ, Gray B, and Korsch MJ (1999) Estimation of cumulative lead releases (lead flux) from the maternal skeleton during pregnancy and lactation. *J Lab Clin Med* 134:631–640.

Gulson BL, Jameson CW, Mahaffey KR, Mizon KJ, Korsch MJ, and Vimpani G (1997) Pregnancy increases mobilization of lead from maternal skeleton. *J Lab Clin Med* 130:51–62.

Harlan WR, Landis JR, Schmouder RL, Goldstein NG, and Harlan LC (1985) Blood lead and blood pressure. Relationship in the adolescent and adult U.S. population. *JAMA* 253:530–534.

Hayano M, Nogawa K, Kido T, Kobayashi E, Honda R, and Turitani I (1996) Dose–response relationship between urinary cadmium concentration and β2-microglobulinuria using logistic regression analysis. *Arch Environ Health* 51:162–167.

Hellström L, Elinder C-G, Dahlberg B, Lundberg M, Järup L, Persson B, and Axelson O (2001) Cadmium exposure and end-stage renal disease. *Am J Kidney Dis* 38:1001–1008.

Hochi Y, Kido T, Nogawa K, Kito H, and Shaikh Z (1995) Dose–response relationship between total cadmium intake and prevalence of renal dysfunction using general linear models. *J Appl Toxicol* 15:109–116.

Hong CD, Hanenson IB, Lerner S, Hammond PB, Pesce AJ, and Pollak VE (1980) Occupational exposure to lead: Effects on renal function. *Kidney Int* 18:489–494.

Hotz P, Buchet JP, Bernard A, Lison D, and Lauwerys R (1999) Renal effects of low-level environmental cadmium exposure: 5-year follow-up of a subcohort from the Cadmibel study. *Lancet* 354:1508–1513.

Hsiao C-Y, Wu H-DI, Lai J-S, and Kuo H-W (2001) A longitudinal study of the effects of long-term exposure to lead among lead battery factory workers in Taiwan (1989–1999). *Sci Total Environ* 279:151–158.

Hu H (1991) A 50-year follow-up of childhood plumbism. Hypertension, renal function, and hemoglobin levels among survivors. *Am J Dis Child* 145:681–687.

Hu H, Aro A, Payton M, Korrick S, Sparrow D, Weiss ST, and Rotnitsky A (1996) The relationship of bone and blood lead to hypertension. The normative aging study. *JAMA* 275:1171–1176.

Hu H, Watanabe H, Payton M, Korrick S, and Rotnitzky A (1994) The relationship between bone lead and hemoglobin. *JAMA* 272:1512–1517.

Hu H, Milder FL, and Burger DE (1991) The use of K X-ray fluorescence for measuring lead burden in epidemiological studies: High and low lead burdens and measurement uncertainty. *Environ Health Perspect* 94:107–110.

Hu H, Milder FL, and Burger DE (1990) X-ray fluorescence measurements of lead burden in subjects with low-level community lead exposure. *Arch Environ Health* 45:335–341.

Huang J, He F, Wu Y, and Zhang S (1988) Observations on renal function in workers exposed to lead. *Sci Total Environ* 71:535–537.

ICRP (1993) *Age-specific Biokinetics Models for the Alkaline Earth Elements.* ICRP Publication No. 67, International Commission on Radiological Protection, Pergamon Press, NY, pp. 95–120.

Ishizaki M, Kido T, Honda R, Tsuritaini I, Yamada Y, Nakagawa H, and Nogawa K (1989) Dose–response relationship between urinary cadmium and β2-microglobulin in Japanese environmentally cadmium exposed population. *Toxicology* 58:121–131.

Iwata K, Saito H, Moriyama M, and Nakano A (1991a) Association between renal tubular dysfunction and mortality among residents in a cadmium-polluted area, Nagasaki, Japan. *Tohoku J Exp Med* 164:93–102.

Iwata K, Saito H, and Nakano A (1991b) Association between cadmium-induced renal dysfunction and mortality: Further evidence. *Tohoku J Exp Med* 164:319–330.

Järup L, and Elinder C-G (1994) Dose-response relations between urinary cadmium and tubular proteinuria in cadmium-exposed workers. *Am J Ind Med* 26:759–769.

Järup L, Hellström L, Alfvén T, Carlsson MD, Grubb A, Persson B, Pettersson C, Spang G, Schutz A, and Elinder CG (2000) Low level exposure to cadmium and early kidney damage: The OSCAR study. *Occup Environ Med* 57:668–672.

Järup L, Persson B, and Elinder C-G (1997) Blood cadmium as an indicator of dose in a long-term follow-up of workers previously exposed to cadmium. *Scand J Work Environ Health* 23:31–36.

Khalil-Manesh F, Gonick HC, Cohen AH, Alinovi R, Bergmaschi E, Mutti A, and Rosen VJ (1992a) Experimental model of lead nephropathy. I. Continuous high dose lead administration. *Kidney Int* 41:1192–1203.

Kahlil-Manesh F, Gonick HC, Cohen A, Bergamaschi E, and Mutti A (1992b) Experimental model of lead nephropathy. II. Effect of removal from lead exposure and chelation treatment with dimercaptosuccinic acid (DMSA). *Environ Res* 58:35–54.

Kido T, and Nogawa K (1993) Dose–response relationship between total cadmium intake and β2-microglobulinuria using logistic regression analysis. *Toxicol Lett* 69:113–120.

Kim R, Rotnitzky A, Sparrow D, Weiss ST, Wager C, and Hu H (1996) A longitudinal study of low-level lead exposure and impairment of renal function. *JAMA* 275:1177–1181.

Kjellström T, and Nordberg GF (1978) A kinetic model of cadmium metabolism in the human being. *Environ Res* 16:248–269.

Kodell RL, and West RW (1993) Upper confidence limits on excess risk for quantitative responses. *Risk Anal* 13:177–182.

Kodell RL, Chen JJ, and Gaylor DW (1995) Neurotoxicity modeling for risk assessment. *Regul Toxicol Pharmacol* 22:24–29.

Kordonouri O, Jorres A, Muller C, Enders I, Gahl GM, and Weber B (1992) Quantitative assessment of urinary protein and enzyme excretion—A diagnostic programme for the detection of renal involvement in Type I diabetes mellitus. *Scand J Clin Lab Invest* 52:781–790.

Korrick S, Hunter D, Rotnitzky A, Hu H, and Speizer F (1999) Lead and hypertension in a sample of middle-aged women. *Am J Pub Health* 89:330–335.

Kumar BD, and Krishnaswamy K (1995) Detection of occupational lead nephropathy using early renal markers. *Clin Toxicol* 33:331–335.

Lapsley M, Sansom PA, Marlow CT, Flynn FV, and Norden AG (1991) Beta 2-glycoprotein-1 (apolipoprotein H) excretion in chronic renal tubular disorders: Comparison with other protein markers of tubular malfunction. *J Clin Pathol* 44:812–816.

Leggett RW (1993) An age-specific kinetic model of lead metabolism in humans. *Environ Health Perspect* 101:598–615.

Lilis R, Fischbein A, Valciukas JA, Blumberg W, and Selikoff IJ (1980) Kidney function and lead: Relationships in several occupational groups with different levels of exposure. *Am J Ind Med* 1:405–412.

Lilis R, Gavrilescu N, Nestorescu B, Dumitriu C, and Roventa A (1968) Nephropathy in chronic lead poisoning. *Br J Ind Med* 25:196–202.

Lin J-L, Tan D-T, Hsu K-H, Yu C-C (2001) Environmental lead exposure and progressive renal insufficiency. *Arch Intern Med* 161:264–271.

Loghman-Adham M (1997) Renal effects of environmental and occupational lead exposure. *Environ Health Perspect* 105:928–939.

Lyon TDB, Aughey E, Scott R, and Fell GS (1999) Cadmium concentrations in human kidney in the UK: 1978–1993. *J Environ Monitor* 1:227–231.

Manton WI, and Cook JD (1984) High accuracy (stable isotope dilution) measurements of lead in serum and cerebrospinal fluid. *Br J Ind Med* 41:313–319.

McLellan JS, Flanagan PR, Chamberlain MJ, and Valberg LS (1978) "Measurement of dietary cadmium absorption in humans. *J Toxicol Environ Health* 4:131–138.

Mortada WI, Sobh MA, El-Defrawy MM, and Farahat SE (2001) Study of lead exposure from automobile exhaust as a risk for nephrotoxicity among traffic policeman. *Am J Nephrol* 21:274–279.

Mueller PW, Paschal DC, Hammel RR, Klincewicz SL, MacNeil ML, Spierto B, and Steinberg KK (1992) Chronic renal effects in three studies of men and women occupationally exposed to cadmium. *Arch Environ Contam Toxicol* 23:125–136.

Mutti A (1989) Detection of renal diseases in humans: Developing markers and methods. *Toxicol Lett* 46:177–191.

Nakagawa H, Nishijo M, Morikawa Y, Tabata M, Senma M, Kitagawa Y, Kawano S, Ishizaki M, Sugita N, and Nishi M (1993) Urinary beta 2-microglobulin concentration and mortality in a cadmium-polluted area. *Arch Environ Health* 48:428–435.

Newton D, Johnson P, Lally AE, Pentreath RJ, and Swift DJ (1984) The uptake by man of cadmium ingested in crab meat. *Human Toxicol* 3:23–28.

Nishijo M, Nakagawa H, Morikawa Y, Tabata M, Senma Y, Kitagawa S, Kawano M, Ishizaki N, Sugita M, Nishi T, Kido T, and Nagawa K (1994) Prognostic factors of renal dysfunction induced by environmental cadmium pollution. *Environ Res* 64:112–121.

Nogawa K, Honda R, Kido T, Tsuritani I, Yamada Y, Ishizaki M, and Yamaya H (1989) A dose-response analysis of cadmium in the general environment with special reference to total cadmium intake limit. *Environ Res* 48:7–16.

O'Flaherty EJ (1991) Physiologically based models for bone-seeking elements. III. Human skeletal and bone growths. *Toxicol Appl Pharmacol* 111:332–341.

O'Flaherty EJ (1992) Modeling bone mineral metabolism, with special reference to calcium and lead. *Neurotoxicol* 13:789–798.

O'Flaherty EJ (1993) Physiologically based models for bone seeking elements: IV. Kinetics of lead disposition in humans. *Toxicol Appl Pharmacol* 118:16–29.

O'Flaherty EJ (1995) Physiologically based models for bone-seeking elements. V. Lead absorption and disposition in childhood. *Toxicol Appl Pharmacol* 131:297–308.

O'Flaherty EJ, Adams WD, Hammond PB, and Taylor E (1986) Resistance of the rat to development of lead-induced renal functional deficits. *J Toxicol Environ Health* 18:61–75.

Omae K, Sakurai H, Higashi T, Muto T, Ichikawa M, and Sasaki N (1990) No adverse effects of lead on renal functions in lead-exposed workers. *Ind Health* 28:77–93.

Ong CN, and Lee WR (1980) High affinity of lead for fetal haemoglobin. *Br J Ind Med* 37:292–298.

Park JD, Cherrington NJ, and Klaassen CD (2002) Intestinal absorption of cadmium is associated with divalent metal transporter 1 in rats. *Toxicol Sci* 68:288–294.

Payton M, Hu H, Sparrow D, and Weiss ST (1994) Low-level lead exposure and renal function in the normative aging study. *Am J Epidemiol* 140:821–829.

Pergande M, Jung K, Precht S, Fels LM, Herbort C, and Stolte H (1994) Changed excretion of urinary proteins and enzymes by chronic exposure to lead. *Nephrol Dial Transplant* 9:613–618.

Perneger TW, Nieto FJ, Whelton PK, Klag MJ, Comstock GW, and Szklo M (1993) A prospective study of blood pressure and serum creatinine: Results from the 'clue' study and the ARIC study. *JAMA* 269:488–493.

Pesce AJ, and First MJ (1979) *Proteinuria: An Integrated Review.* Marcel Dekker: New York, NY, pp. 175–201.

Pinto de Almeida AR, Carvalho FM, Spinola AG, and Rocha H (1987) Renal dysfunction in Brazilian lead workers. *Am J Nephrol* 7:455–458.

Pirkle JL, Schwartz J, Landis JR, and Harlan WR (1985) The relationship between blood lead levels and blood pressure and its cardiovascular risk implications. *Am J Epidemiol* 121:246–258.

Pless-Mulloli T, Boettcher M, Steiner M, and Berger J (1998) α1-Microglobulin: Epidemiological indicator for tubular dysfunction induced by cadmium. *Occup Environ Med* 55:440–445.

Pocock SJ, Shaper AG, Ashby D, Delves HT, and Clayton BE (1988) The relationship between blood lead, blood pressure, stroke, and heart attacks in middle-aged British men. *Environ Health Perspect* 78:23–30.

Pocock SJ, Shaper AG, Ashby D, Delves T, and Whitehead TP (1984) Blood lead concentration, blood pressure, and renal function. *Br Med J* 289:872–874.

Pounds JG, and Leggett RW (1998) The ICRP age-specific biokinetic model for lead: Validations, empirical comparisons, and explorations. *Environ Health Perspect* 106:1505–1522.

Rahola T, Aaran R-K, and Miettinen JK (1972) Half-time studies of mercury and cadmium by wholebody counting, in *Assessment of Radioactive Contamination in Man,* IAEA-SM-150/13, International Atomic Energy Agency, Unipublisher, New York, pp. 553–562.

Reichert LJ, Koene RA, and Wetzels JF (1995) Urinary excretion of beta 2-microglobulin predicts renal outcome in patients with idiopathic membranous nephropathy. *J Am Soc Nephrol* 6:1666–1669.

Roels HA, Van Assche FJ, Oversteyns M, De Groof M, Lauwerys RR, and Lison D (1997) Reversibility of microproteinuria in cadmium workers with incipient tubular dysfunction after reduction of exposure. *Am J Ind Med* 31:645–652.

Roels H, Konings J, Green S, Bradley D, Chettle D, and Lauwerys R (1995) Time-integrated blood lead concentration is a valid surrogate for estimating the cumulative lead dose assessed by tibial lead measurement. *Environ Res* 69:75–82.

Roels H, Lauwerys R, Konings J, Buchet J-P, Bernard A, Green S, Bradley D, Morgan W, and Chettle D (1994) Renal function and hyperfiltration capacity in lead smelter workers with high bone lead. *Occup Environ Med* 51:505–512.

Roels H, Bernard AM, Cardenas A, Buchet J-P, Lauwerys RR, Hotter G, Ramis I, Mutti A, Franchini I, Bundschuh I, Stolte H, De Broe MD, Nuyts GD, Taylor SA, and Price RG (1993) Markers of early renal changes induced by industrial pollutants. III. Application to workers exposed to cadmium. *Br J Ind Med* 50:37–48.

Roels HA, Lauwerys RR, Buchet J-P, Bernard A, Chettle DR, Harvey TC, and Al-Hadda IK (1981) *In vivo* measurement of liver and kidney cadmium in workers exposed to this metal: Its significance with respect to cadmium in blood and urine. *Environ Res* 26:217–240.

Schiavon M, Di Landro D, Baldo M, De Silvestro G, and Chiarelli A (1988) A study of renal damage in seriously burned patients. *Burns Incl Therm Inj* 14:107–112.

Schroeder HA, and Tipton IH (1968) The human body burden of lead. *Arch Environ Health* 17:965–978.

Schütz A, Bergdahl IA, Ekholm A, and Skerfving S (1996) Measurement by ICP-MS of lead in plasma and whole blood of lead workers and controls. *Occup Environ Med* 53:736–740.

Shaikh ZA, and Smith JC (1980) Metabolism of orally ingested cadmium in humans, in *Mechanism of Toxicity and Hazard Evaluation*, Holmstedt B, Lauwerys R, Mercier M, and Roberfroid M (Eds), Elsevier/North-Holland Biomedical Press, pp. 569–574.

Skerfving S (1988) Biological monitoring of exposure to inorganic lead, in *Biological Monitoring of Toxic Metals*, Clarkson TW, Friberg L, Nordberg GF, and Sager PR (Eds), Plenum Press, pp. 169–197.

Smith DR, Ilustre RP, and Osterloh JD (1998) Methodological considerations for the accurate determination of lead in human plasma and serum. *Am J Ind Med* 33:430–438.

Sonmez F, Donmez O, Sonmez HM, Keskinoglu A, Kabasakal C, and Mir S (2002) Lead exposure and urinary N-acetyl β-D-glucosaminidase activity in adolescent workers in auto repair workshops. *J Adolesc Health* 30:213–216.

Staessen JA, Lauwerys RR, Buchet J-P, Bulpitt CJ. Rondia D, Vanrenterghem Y, and Amery A. (1992) Cadmibel study group. Impairment of renal function with increasing blood lead concentrations in the general population. *N Engl J Med* 327:151–156.

Staessen J, Yeoman WB, Fletcher AE, Markowe HLJ, Marmot MG, Rose G, Semmence A, Shipley MJ, and Bulpitt CJ (1990) Blood lead concentration, renal function, and blood pressure in London civil servents. *Br J Ind Med* 47:442–447.

Suwazono Y, Kobayashi E, Okubo Y, Nogawa K, Kido T, and Nakagawa H (2000) Renal effects of cadmium exposure in cadmium nonpolluted areas in Japan. *Environ Res* 84:44–55.

Thun MJ, Osorio AM, Schober S, Hannon WH, Lewis B, and Halperin W (1989) Nephropathy in cadmium workers: Assessment of risk from airborne occupational exposure to cadmium. *Br J Ind Med* 46:689–697.

Todd AC, McNeill FE, Palethorpe JE, Peach DE, Chettle DR, Tobin MJ, Strosko SJ, and Rosen JC (1992) In vivo X-ray fluorescence of lead in bone using K X-ray excitation with ^{109}Cd sources: Radiation dosimetry studies. *Enviro Res* 57:117–132.

Tsaih S-W, Korrick S, Schwartz J, Lee M-LT, Amarasiriwardena C, Aro A, Sparrow D, and Hu H (2001) Influence of bone resorption on the mobilization of lead from bone among middle-aged and elderly men: The normative aging study. *Environ Health Perspect* 109:995–999.

Tsaih S-W, Schartz J, Lee M-LT, Amarasiriwardena C, Aro A, Sparrow D, and Hu H (1999) The independent contribution of bone and erythrocyte lead to urinary lead among middle-aged and elderly men: The normative aging study. *Environ Health Perspect* 107:391–396.

U.S. EPA (2000) Cancer Risk Coefficients for Environmental Exposure to Radionuclides. *Federal Guidance Report No. 13*, EPA 402-C-99-001.

Verberk MM, Willems TEP, Verplanke AJW, and DeWolff FA (1996) Environmental lead and renal effects in children. *Arch Environ Health* 51:83–87.

Verschoor M, Wibowo A, Herber R, van Hemmen J, and Zielhuis R. (1987) Influence of occupational low-level lead exposure on renal parameters. *Am J Ind Med* 12:341–351.

Wedeen RP, Mallik DK, and Batuman V (1979) Detection and treatment of occupational lead nephropathy. *Arch Intern Med* 139:53–57.

Wedeen RP, Maesaka JK, Weiner B, Lipat GA, Lyons MM, Vitale LF, and Joselow MM (1975) Occupational Lead Nephropathy. *Am J Med* 59:630–641.

Weiss ST, Munoz A, Stein A, Sparrow D, and Speizer FE (1986) The Relationship of Blood Lead to Blood Pressure in a Longitudinal Study of Working Men. *Am J Epidemiol* 123:800–808.

Wu X, Jin T, Wang Z, Ye T, Kong Q, and Nordberg G (2001) Urinary calcium as a biomarker of renal dysfunction in a general population exposed to cadmium. *J Occup Environ Med* 43:898–904.

Yamanaka O, Kobayashi E, Nogawa K, Suwazono Y, Sakurada I, and Kido T (1998) Association between renal effects and cadmium exposure in cadmium-nonpolluted area in Japan. *Environ Res* 77:1–8.

27

Risk Assessment for Selected Therapeutics

Kanwar Nasir M. Khan and Rani S. Sellers

OVERVIEW

The kidney is a frequent target organ of toxicity in laboratory animals and humans following exposure to chemicals or therapeutics. This nephrotoxicity typically depends on selective concentration of the toxic moiety at the target cell or subcellular organelle. The kidney receives more blood flow per gram of tissue than other organs and functions to concentrate and excrete solutes (including xenobiotics) and metabolize xenobiotics. Because of this, the kidney is at a risk of greater exposures to compounds, increasing the potential for cellular damage. In addition, kidney has the capacity both to dissociate toxicants from protein-bound states and to transform solutes, by changes in pH, to active and reactive forms. Renal metabolism of xenobiotics to reactive electrophilic intermediates can cause renal injury following covalent or peroxidative reactions with the target cell macromolecules. Physiological changes in basal renal functions may also be altered by xenobiotics or their metabolites without producing any morphologic changes.

Nephrotoxicants can be categorized into several groups according to intrinsic structural or functional characteristics of the therapeutics. This chapter discusses examples of risk assessment of therapeutics that can produce nephrotoxicity either through the accumulation of materials within the urineferous tubules (xemilofiban and orbofiban) or through direct effects (cyclooxygenase inhibition) on renal function.

TOXICITY ASSESSMENT OF XEMILOFIBAN AND ORBOFIBAN

The fibans are a group of compounds whose pharmacological activity is to block the glycoprotein IIb-IIIa receptor on platelets. This receptor ordinarily binds fibrinogen, which is the final step of platelet aggregation, allowing platelets to aggregate *in vivo* and *in vitro*. The orally available fibans, such as xemilofiban and orbofiban, reversibly block the receptor in a concentration-dependent manner, leading to a titratable inhibition of platelet aggregation. These compounds are ester prodrugs (Figure 27.1), which are absorbed after oral administration and are deesterified to their pharmacologically active acids (Levin et al., 1999). These two drugs have been shown to be pharmacologically active in dogs and humans, affecting platelet aggregation while pharmacological activity was limited in rats.

During the research and development of xemilofiban and orbofiban, conventional repeated-dose preclinical safety studies were conducted in rats and dogs. The longest studies in rats and dogs were up to 6 and 12 months in duration, respectively. Additionally, a variety of mechanistic studies, including studies in animals of different age groups, were conducted to understand the potential pathogenesis of nephrotoxicity and the risk of development of lesions in humans. Renal toxicological assessments in animals included: histology, histochemistry, cytology of

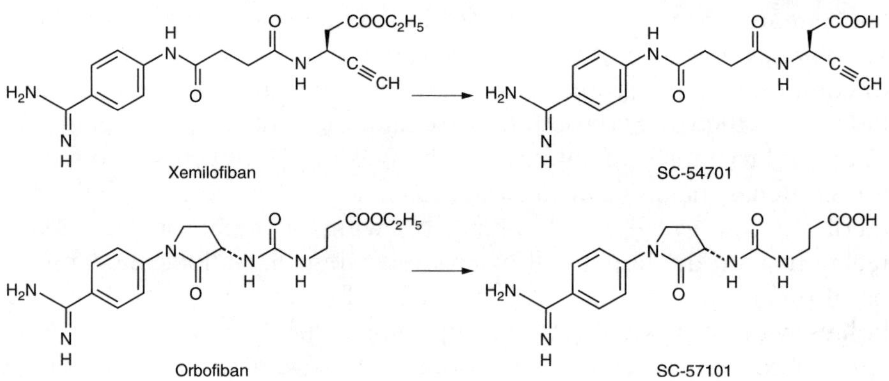

Figure 27.1 The structures of xemilofiban and its acid metabolite (SC-54701), and orbofiban (SC-57099B) and its acid metabolite (SC-57101).

urine sediments, urine analyses (including urine chemistry), and renal function tests. These data are discussed below.

Morphologic Changes Associated with Xemilofiban and Orbofiban in the Kidney

No renal toxicity was seen in dogs with either compound. Severe dose- and duration-dependent renal toxicity was present in rats with both compounds. Nephrotoxicity in rats was observed as early as 4 days after treatment. Grossly, renal lesions were generally bilateral and were characterized by multifocal (0.5 to 3 mm diameter) white foci, which on cut surface extended through the cortex, medulla, and papilla (Figure 27.2). The kidneys with severe lesions were slightly enlarged and had an irregular contour, with dilated renal pelvis sometimes containing small collections of pale white calculi (Figure 27.2). Microscopic evaluations of the calculi and urine sediment showed the presence of bundles of birefringent needle-shaped crystals (Figure 27.3).

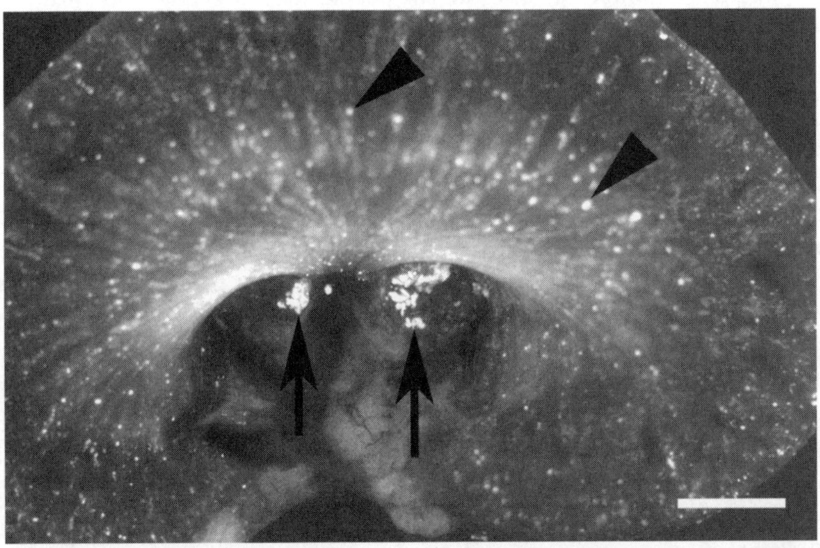

Figure 27.2 A longitudinal section of a rat kidney treated with xemilofiban (350 mg / kg/day) for two weeks. Note grayish white foci (arrowheads) in cortex, medulla, and papilla, representing obstructed tubules and collecting ducts. Renal pelvis contains collections of numerous small calculi in the renal pelvis (arrows). Bar = 1.5 mm.

Figure 27.3 Urinary sediment from a rat treated with xemilofiban (350 mg / kg / day). Note large number of birefringent needle-shaped crystals (a) with pointed ends (b). Magnification: a = 4×; b = 40×.

Histologically, renal lesions were comprised of a combination of suppurative to pyogranulomatous inflammation, tubular epithelial degeneration, necrosis, and regeneration (basophilic tubules), and intratubular amphophilic radiating casts associated with marked tubular dilatation (Figure 27.4, Figure 27.5, Figure 27.6). The earliest inflammation associated with tubular dilation and tubular casts were infiltrates of neutrophils (Figure 27.5). Some animals died shortly after renal insult, as evidenced by a lack of histological evidence of inflammation. These animals frequently had markedly dilated tubules, with or without radiating casts (Figure 27.4). Ethanol-fixed and Diff-Quick stained kidney sections demonstrated that the radiating casts seen in formalin-fixed hematoxylin and eosin stained sections were birefringent crystals, which were presumably dissolved following formalin fixation and routine processing. Morphologic differences in the crystals were evident: the xemilofiban crystals were sharp spikes radiating out from a central core, whereas the orbofiban crystals were more blunt.

Kidneys from some animals also contained numerous microgranulomas, which were scattered throughout the renal parenchyma. These

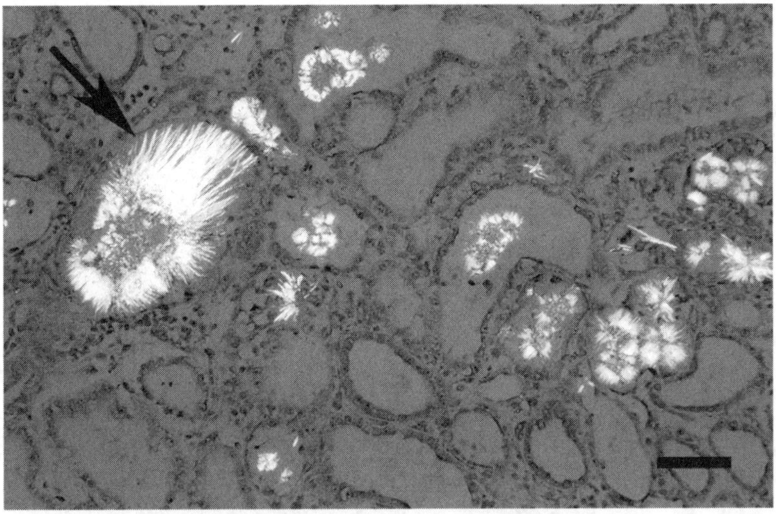

Figure 27.4 Intratubular birefringent crystals resulting in tubular dilatation in a rat kidney. Diff-Quick stain. Bar = 100 μm.

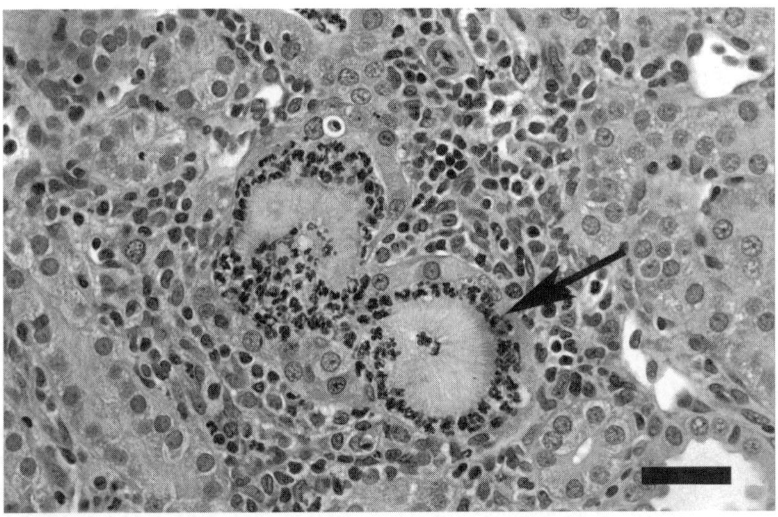

Figure 27.5 Intratubular amphophilic cast surrounded by neutrophilic infiltrate in a rat kidney treated with xemilofiban (350 mg/kg/day). The casts likely represent proteins trapped in the interstices of a crystal that dissolved during the tissue fixation, processing, or staining procedures. Hematoxylin and eosin stain, Bar = 30 μm.

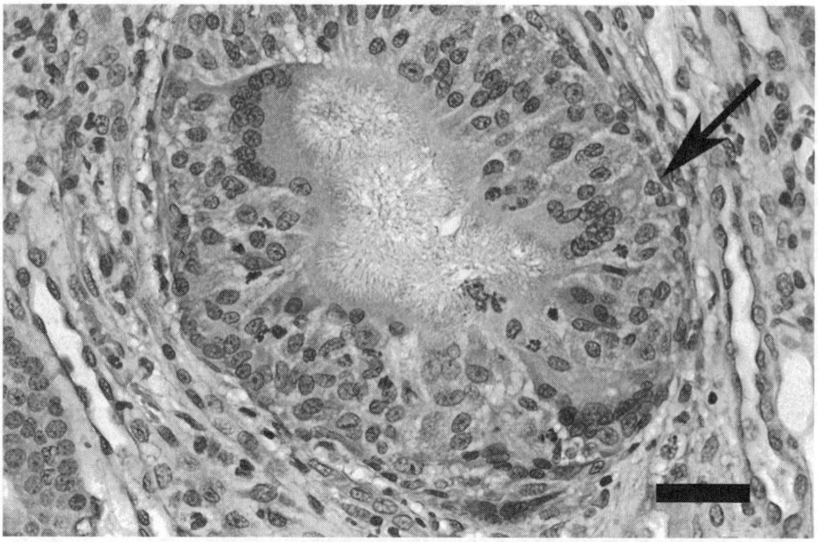

Figure 27.6 Focal granulomatous inflammation surrounding tubular casts in the renal cortex of a rat treated with xemilofiban (350 mg / kg / day). Hematoxylin and eosin stain, Bar = 30 μm.

micrograngulomas were comprised of predominantly epithelioid macrophages and lesser numbers of multinucleated giant cells, lymphocytes and neutrophils (Figure 27.6). Granulomas also contained birefringent needle-shaped crystals indicating a foreign-body reaction to the crystals. Mild to moderate nephrocalcinosis was evident in chronic cases. This change was probably mineralization secondary to derangement in calcium metabolism due to renal disease.

The gross and microscopic appearance of the kidney lesions with both fiban compounds varied somewhat in terms of the amount of the precipitates, the location of precipitates within the kidney, and the length of time the animals survived after tubular obstruction by precipitates. Xemilofiban-related crystalline precipitates were found with about equal frequency in all parts of the kidney (cortex, medulla, and papilla) whereas orbofiban precipitates were located predominately in the papilla. The propensity for orbofiban to precipitate in the papilla often led to complete obstruction of urine outflow; these animals died more quickly after the precipitates formed and thus tended to have relatively little inflammation. Xemilofiban was often associated with a greater amount of kidney inflammation. The randomness of the xemilofiban lesion locations, and perhaps the timing of the precipitation events, apparently permitted the animals to live longer after precipitation.

Serum and urinary assessments showed elevations in blood urea nitrogen and serum creatinine concentrations, increased urinary protein, and increased number of white blood cells and degenerated renal cells in the urine sediment. These changes correlated with the severity of renal toxicity.

Based on morphologic presentation, the renal toxicity was attributed to obstruction of tubules by casts and crystals, resulting in tubular damage with secondary inflammatory reaction. Additional studies were conducted to characterize the chemical nature and source of tubular casts, the dosages needed to produce nephrotoxicity, and the contribution of age-related lesions in the kidney to increase the susceptibility of rats to nephrotoxicity.

Identification of Xemilofiban and Orbofiban Crystals in the Kidney

As seen with xemilofiban and orbofiban, urinary calculi or crystals have been observed with other compounds (e.g., sulfa drugs), and have been produced experimentally in rats using calcium oxalate, calcium phosphate (de Bruijn et al., 1994; Khan and Glenton, 1995), xanthopterin (Garber et al., 1995), and steroidal natriuretic hormones (Bricker et al., 1993). Urinary calculi or crystals (uroliths) represent the aggregations of precipitated compounds, urinary solutes, urinary proteins, and proteinaceous debris. The most important determinant in urolithiasis is an increased urinary concentration of the crystals' constituents, such that they exceed their solubility in renal tubular fluids and urine (supersaturation).

It was considered that the tubular radiating casts in fiban studies might represent host-derived proteins (e.g., uromucoid) adsorbed on to the birefringent crystals or host proteins which coprecipitated with fiban compounds. Infrared spectroscopy, Raman spectroscopy, time-of-flight secondary ion mass spectroscopy (TOF-SIMS) and liquid chromatography mass spectroscopy (LCMS) were used to identify the crystals *in situ*.

TOF-SIMS is a surface analytical technique that uses the impact of a primary ion beam to dislodge and eject intact molecular or fragment ions from the surface into the gas phase for time-of-flight mass analysis. TOF-SIMS combines the analytical power of mass spectrometry with the imaging ability of microprobe techniques.

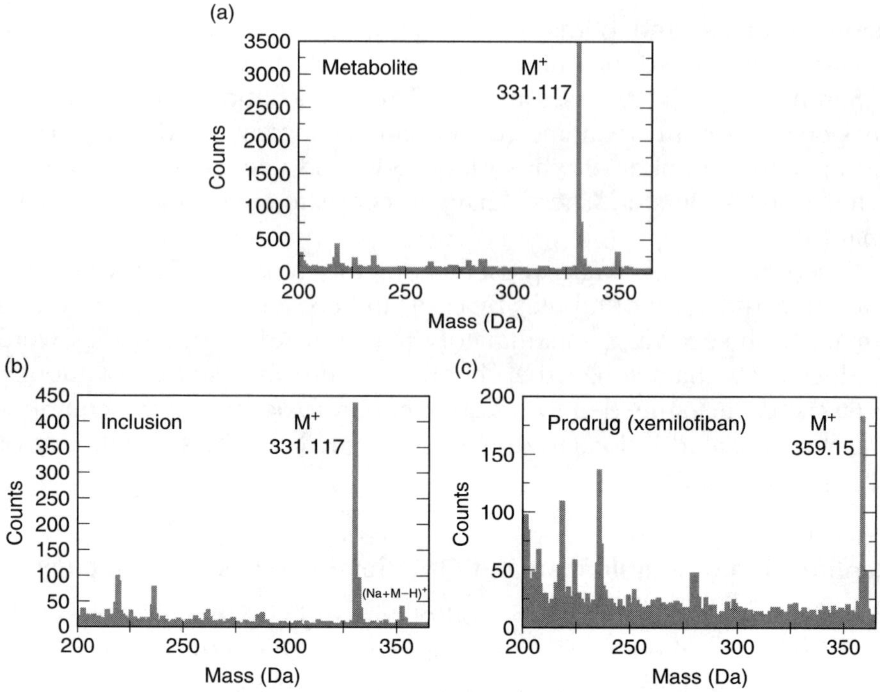

Figure 27.7 TOF-SIMS spectra: (a) xemilofiban, (b) xemilofiban's acid metabolite, (c) a renal crystal. The spectra demonstrate that the crystal is the acid metabolite.

Rat kidney sections containing the crystals caused by xemilofiban were evaluated using TOF-SIMS. Figure 27.7 shows TOF-SIMS spectra of purified specimens of xemilofiban, its acid metabolite (SC-54701), and the crystal within the rat kidney. These evaluations showed that the spectra of SC-54701 and the renal crystal were identical, registering with the same masses as SC-54701 and its fragments.

LCMS is the direct combination of liquid chromatography with mass spectrometry through an interface that provides ionized samples to the mass spectrometer. The material collected directly from the kidney of rats evaluated with LCMS confirmed that the crystalline material was SC-54701.

Infrared and Raman spectroscopy are techniques widely used to determine molecular structure and to identify compounds. Frozen kidney sections containing crystals formed after orbofiban administration were

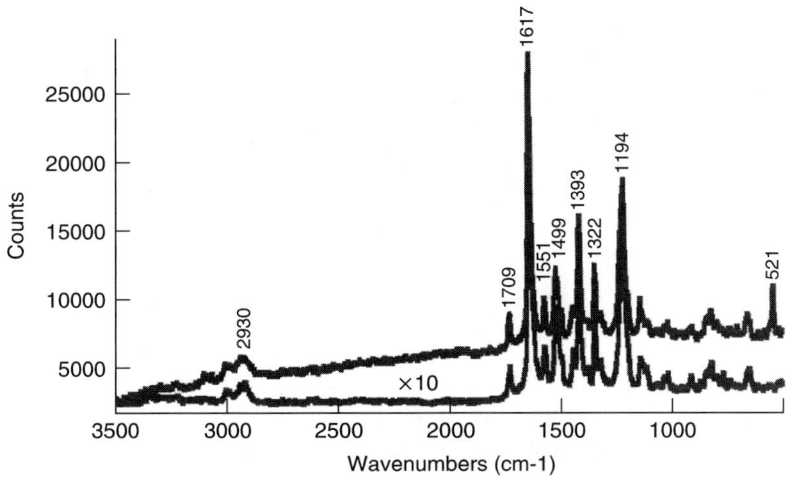

Figure 27.8 Raman spectra of the acid metabolite of orbofiban and a renal crystal show that the two are the same. Top line shows spectra of crystal in rat tissue and the bottom line show spectra of acid metabolite.

evaluated by this method. The spectra of crystalline method were virtually identical to those of SC-57101 (Figure 27.8).

Collectively, results of the state-of-the-art analytical methods have confirmed that the intratubular material was in fact local accumulation of active metabolites of xemilofiban and orbofiban, which contributed to the nephrotoxicity.

Effect of Age on Susceptibility to Xemilofiban-Related Nephrotoxicity

To determine whether the toxic effects of xemilofiban were a manifestation of the duration of exposure or physiological changes in the animals as they grew and aged, an experiment was conducted in which xemilofiban was administered to rats of various ages (6 weeks, 6 months, 12 months, and 18 months) for two weeks. Control untreated animals were available for each age group. A dosage was chosen that had been seen as nontoxic in young rats for short durations of administration (2 weeks) but which were nephrotoxic in old rats after chronic administration (26 weeks).

In this study, the plasma levels of both xemilofiban and SC-54701 showed a clear trend toward increased plasma exposures with age (Table 27.1). Similarly, there was a clear trend toward an increase in

Table 27.1 Incidence of Xemilofiban-Related Nephrotoxicity and Plasma Concentrations of Active Metabolite (SC-54701) in Male Sprague Dawley Rats of Various Ages

Group	*Age (months)*	*Body weight (mg)*	*Incidence*[a] *(%)*	C_{max} *(μg/ml)*[b] *Day 1*	*Day 15*
2	1.5	193	0	6.49	5.01
4	6	649	40	11.2	8.35
6	12	772	90	c	c
8	18	839	80	39.7	18.8

[a]Incidence of xemilofiban-related nephrotoxicity.
[b]Xemilofiban dosed at 500 mg/kg/day on Day 1 and at 350 mg/kg/day on Day 15.
[c]Because of high mortality and the presence of outliers, data from this group were excluded for comparison across groups.

Table 27.2 Severity Grades of Xemilofiban-Related Renal Changes in Male Sprague Dawley Rats of Various Ages

	Number of animals affected			
Age	6 weeks	6 months	12 months	18 months
Severity grades				
Minimal	0	0	1	2
Mild	0	1	2	0
Moderate	0	0	3	3
Marked	0	3	2	3
Severe	0	0	1	0

Total number of animals per group was 10.

the incidence of compound-related mortality and the incidence and severity of renal lesions with increasing age of the rat. Because of high xemilofiban-related mortality in older rats, the dose of xemilofiban was reduced to 350 mg/kg in all groups on Day 3. After 2 weeks, nephrotoxicity was observed in all 6-, 12-, and 18-month old rats, with increasing severity with age, but not in all 6-week old rats (Table 27.2).

As shown in the literature (Corman and Owen, 1992), an age-related increase in the incidence and severity of chronic progressive nephropathy (CPN) was observed in this study. The incidence ranged from

Table 27.3 Creatinine Clearance, Urinary Protein Excretion and Urine Output in Male Sprague Dawley Rats of Various Ages Given Xemilofiban

Age	*Group*	*Creatinine clearance [a] ml/min)*	*Urine protein mg/22 hours)*	*Urine output[a] ml/22 hours)*
6 weeks				
	1. Control	0.383 ± 0.066	7.3 ± 1.91	9.26 ± 3.20
	2. Xemilofiban	0.323 ± 0.131	6.2 ± 2.6	9.23 ± 4.67
6 months				
	3. Control	0.367 ± 0.069	9.4 ± 2.45	3.66 ± 1.40
	4. Xemilofiban	0.333 ± 0.131	10.9 ± 4.9	4.18 ± 1.31
12 months				
	5. Control	0.331 ± 0.074	18.5 ± 8.9	3.44 ± 1.20
	6. Xemilofiban	0.184 ± 0.167	32.8 ± 47.79	2.01 ± 0.90
18 months				
	7. Control	0.260 ± 0.095	103.4 ± 122.43	2.49 ± 1.15
	8. Xemilofiban	0.249 ± 0.083	95.5 ± 101.73	3.51 ± 1.19

[a]Creatinine clearance and urine output are calculated per 100 gram body weights. Values represent means ± standard deviations.

0% in 6-week old rats to 100% in 18-month old rats. When data were analyzed on an individual animal basis, there was no correlation between the severity of xemilofiban-related nephrotoxicity and concomitant occurrence of CPN, suggesting that this chronic rodent-specific condition was unlikely to influence the age-related increase in plasma concentrations of the drug or the associated nephrotoxicity. The changes in creatinine clearance (18-month old rats) and urinary protein excretion in older rats were considered secondary to age-related renal lesions of CPN. A further slight decrease in mean creatinine clearance in 12-month old rats given xemilofiban was considered secondary to xemilofiban-related nephrotoxicity (Table 27.3).

The factors that appeared to be most influential on the increased susceptibility of the older rats to xemilofiban toxicity were body weight and urine volume (Table 27.3). Although body weights increased with age, the urine volume remained remarkably constant. The body weight adjusted urine output was approximately 2.5, 2.7, and 3.7 times lower respectively, in 6-, 12- and 18-month old rats compared with 6-week old rats. Other studies have produced similar findings. Since animals are dosed typically on a mg/kg basis in toxicity studies, the older (and thus larger) animals were administered larger total amounts of the fibans. Both drugs are almost entirely excreted by the kidneys. Thus, as animals grow, the increasingly larger total amount of the administered drug must be excreted in the same volume of urine. This led to supersaturation and precipitation of the drug in the urine. A low urine volume in metabolically normal subjects is also expected to favor supersaturation of solutes, leading to crystal or calculus formation (Cotran et al., 1994).

An age-related increase in the susceptibility of rats to nephrotoxicity also has been observed with other drugs, such as salicylate (Kyle and Kocsis, 1985), acetaminophen (Beierschmitt et al., 1986), cephaloridine (Goldstein et al., 1986), gentamicin (McMartin and Engel, 1982), and quinidine (Agarwal and Rao, 1994). Similar to fibans, the plasma concentrations of cephaloridine (Goldstein et al., 1986) and acetaminophen (Beierschmitt et al., 1986) have been reported to be elevated in older rats, suggesting an age-dependent effect on pharmacokinetics. These differences in pharmacokinetics in older rats may result in part from age-related changes in body composition (decreased lean body mass, increased body fat), resulting in alterations in volume of distribution. Fibans and their active metabolites are zwitterionic in nature, containing a tertiary amine and a carboxylic acid and thus will distribute primarily into lean body mass. Such poorly lipophilic compounds when given on a body weight basis to older rats (which are often obese) most likely result in marked elevations in plasma levels as a result of a decreased volume of distribution. Volume of distribution in older rats does not increase in proportion to body weight.

Zwitterions, such as probenicid and *p*-aminohippurate, are secreted by active tubular transport mechanisms located in the proximal tubule of the kidney (Besseghir and Roch-Ramel, 1987). Tubular secretion of these xenobiotics may result in selective accumulation of the secreted compound in the lumen of the nephron, and may therefore reach concentrations many times higher than the concentration in the plasma and in the remainder of the body. The increased luminal concentration associated with the active transport may result in direct renal tubular epithelial toxicity or result in luminal precipitation or crystalization of the compound leading to obstructive nephropathy. The histopathologic nature and distribution of the fiban-related lesions indicate that nephrotoxicity was the result of increased luminal precipitation and crystallization of those compounds. The prevention of renal accumulation and, therefore, renal toxicity, could be accomplished by inhibiting the active transport of the xenobiotics into the renal tubules. For example, probenicid can prevent renal accumulation and toxicity of cephaloridine, cisplatin, and citrinin (Berndt and Hayes, 1982; Ross and Gale, 1979; Tune and Fravert, 1980).

Clinical Risk Assessment of Nephrotoxicity Associated with Xemilofiban and Orbofiban

Investigative mechanistic evaluations have indicated that the acid metabolites of xemilofiban and orbofiban are precipitated in rat kidney tubules when the urine became supersaturated with large administered doses of these compounds. Development of age-related chronic renal disease (i.e., CPN) was considered less likely to contribute to the susceptibility of older rats to toxicity of xemilofiban or orbofiban. No obstructive nephropathy was seen in dogs in studies up to 52 weeks. Antiplatelet effects and associated hemorrhages prevented use of high dosages in this species. Thorough renal assessments were completed in humans in clinical studies conducted at therapeutic exposures. There was no evidence of renal toxicity in humans, indicating that nephrotoxicity of fibans was specific to the rats and was a high-dose phenomenon.

Other Therapeutic Agents Associated with Obstructive Nephropathy

Obstructive nephropathy is a relatively common condition in animals and human. It is classified according to the degree, duration, and site of the urinary tract obstruction. Obstruction can occur anywhere from the level of renal tubules to the urethral meatus. The causes of obstructive nephropathy may be intra or extra renal and may be congenital or acquired. Among acquired intrarenal causes, accumulation of drugs and chemicals in the urinary tract is an important condition, especially in susceptible populations. Table 27.4 summarizes some important examples of drugs that have been associated with obstructive nephropathy.

Intrarenal obstruction occurs secondary to accumulation of poorly soluble materials within the tubules, such as following high dosages of xemilofiban, naproxen, methotrexate, acyclovir, and triamterene. Nonsteroidal anti-inflammatory drugs (NSAIDs) and analgesics can lead to ureteral obstruction as a result of papillary necrosis. Obstructive nephropathy with older sulfonamides was related to their poor solubility; however, newer sulfonamides are more soluble in acid urine and the incidence of obstructive nephropathy due to sulfonamide crystals

Table 27.4 Examples of Drugs and Chemicals Associated with Obstructive Nephropathy

Drugs/chemicals	*Therapeutic class*	*Location of accumulation*
Alkylating agents	Anticancer	Tubules (uric acid crystals)
Sulfonamides		Tubules
Acyclovir	Antiviral	Tubules
Fibans (orbofiban and xemilofiban)	Antithrombotics	Tubules
Triamterene		Tubules
Methotrexate		Tubules
Radiocontrast media	Imaging	Tubules
NSAIDs	Anti-inflammatory and analgesics	Papilla
Levodopa (α-adrenergic properties)		Urinary bladder
Anticholinergic drugs		Urinary bladder
Methysergide		Extrarenal

has declined remarkably. Sulfadiazine is still used in AIDS patients for the treatment of toxoplasmosis, which has led to the resurgence of crystal-associated obstructive nephropathy and acute renal failure (Molina et al., 1991). The crystals that occur through sulfadiazine use are formed by its primary metabolite, acetylsulfadiazine, and resemble shocks of wheat (Simon et al., 1990).

Uric acid crystal deposition can occur with acid urine, usually after cancer chemotherapy with alkylating agents, and the risk of its development is related directly to the plasma uric acid concentrations (Conger, 1981; Curhan and Zeidel, 1998).

Acyclovir, when given intravenously in high doses (500 mg/m^2), can produce obstructive nephropathy (Cronin and Henrich, 1998; Sawyer et al., 1988). Acyclovir is excreted via glomerular filtration and tubular secretion and has low solubility in the urine leading to intratubular precipitation and obstruction. Needle-shaped crystals can be visualized under polarized light upon urinalysis. Histologically, marked interstitial inflammation is seen adjacent to the obstructed tubules (Sawyer et al., 1988). Dehydration predisposes to obstructive nephropathy with acyclovir, which can be prevented by vigorous hydration before and during the therapy.

There are various examples of drugs associated with extrarenal obstructive nephropathy. For example, methysergide-induced retroperitoneal fibrosis can lead to extrarenal urinary obstruction by physical compression. Functional obstruction can be caused by anticholinergic agents, which decrease bladder contractility, or by levodopa, because of its α-adrenergic-mediated increase in bladder outlet resistance (Murdock et al., 1975; Novicki and Willscher, 1979).

Effect of Obstructive Nephropathy on Renal Function

In the adult human kidney, approximately two liters of urine flow through the renal papilla daily. Any obstruction to this unidirectional flow can lead to build up of urine, and pressure affecting renal functions. Urinary symptoms may or may not occur along with changes in urine output. Mild episodes of polyuria (high urine output) may alternate with periods of oliguria (minimal urine output) or occasionally anuria (no urine output). A complete obstruction of short duration results in profound alterations in renal hemodynamics and glomerular filtration, with minimal anatomic changes. Within 24 hours after unilateral urinary obstruction, glomerular filtration rate (GFR) and renal blood flow (RBF) are reduced to ≈25% of normal and in bilateral obstruction to ≈40 to 0% (Dal Canton et al., 1977, 1980; Klahr, 2000). In the postobstructive period, the degrees of change in RBF and GFR are dependent on the duration of obstruction and species involved (Curhan and Zeidel, 1998; Harris and Yarger, 1974; Yarger and Griffith, 1974). Within 24 hours of release of unilateral obstruction, GFR is reduced to 25% of normal in rats and less than 50% of normal in dogs; whereas RBF is markedly reduced in both species. After release of bilateral obstruction, RBF is less impaired than in unilateral circumstances, but GFR is markedly impaired. The reduction in GFR is considered secondary to poor perfusion of the majority of glomeruli.

Several abnormalities in tubular function may occur in obstructive nephropathy. These include decreased reabsorption of solutes and water, inability to concentrate the urine, and impaired excretion of hydrogen and potassium. Changes in renal enzymes associated with urinary obstruction include increases in urinary glucose-6-phosphate dehydrogenase and phosphogluconate dehydrogenase and decreases in urinary alkaline phosphatase, Na^+/K^+-ATPase, glucose-6-phosphate, succinate dehydrogenase, NADH/NADPH dehydrogenase, and glycerol-3-phosphate dehydrogenase. Renal interstitial fibrosis is a common finding in patients with long-term obstructive uropathy. Several factors (macrophages, growth factors, hypoxia, and cytokines) are involved in the pathogenesis of interstitial fibrosis. It has been shown that angiotensin converting enzyme inhibitors ameliorate the interstitial fibrosis in animals with obstructive uropathy.

TOXICITY ASSESSMENT OF THE CYCLOOXYGENASE-2 INHIBITORS

Cyclooxygenase (COX) catalyzes the committed step in prostaglandin (PG) biosynthesis. Mammalian cells contain two related but unique isozymes, COX-1 and COX-2. COX-1 is expressed constitutively and is involved in the production of PGs that modulate normal physiologic functions in several organ systems, including the kidneys, gastrointestinal tract, and platelets. COX-2 expression is inducible by bacterial endotoxins, cytokines, and growth factors and is involved in the production of PGs that modulate physiological events in development, cell growth, and inflammation (Masferrer et al., 1994; Needleman and Isakson, 1997).

Role of Prostaglandins in the Kidney

In general, NSAIDs are nonselective inhibitors of both COX isoforms and their use has been associated with significant renal toxicities in both animals and humans. In contrast to picomolar circulating concentrations of PGs, urinary PG levels are in the nanomolar range, indicating high intrarenal synthesis (Breyer et al., 1998).

Local PG synthesis in the kidney supports important roles in the regulation of renal function and homeostasis (Table 27.5) and NSAIDs can exert their effects at multiple points along the nephron. Therefore, these effects likely reflect pharmacological inhibition of renal COX-1 or COX-2, or both. Currently, the relationships between specific COX isoforms and renal function are not well understood (Schneider et al., 1999). However, the available immunoreactivity data suggest that simple relationships are not likely, and that COX isoforms may support a region-specific, and even redundant, function in the kidney (Khan et al., 1998a, 1998b; Stanfield et al., 2000; Vio et al., 1997; Whelton, 1999).

Table 27.5 Expression of COX-2 in the Kidney

Location	*COX-1*	*COX-2*	*Prostaglandins*	*Possible functions*
Glomerulus	+	+	PGE_2, PGI_2	Podocyte contractility and maintenance of GFR
Thick ascending limb of loop of Henle	–	+[a]	PGE_2	Enhance excretion of sodium and chloride
Macula densa	–	+[a]	PGE_2	Regulation of renin release; mediates tubuloglomerular feedback
Collecting ducts	+	+[b]	PGE_2	Enhance excretion of sodium, chloride, and water
Papillary interstitial cells	+	+	PGE_2, TXA_2	Enhance vasodilatation and natriuresis
Renal vasculature	+	+	PGE_2, PGI_2, TXA_2	Regulation of regional blood flow, antagonism of angiotensin II induced vasoconstriction

TXA_2 = thromboxane A_2.
[a]Not seen in normal human and non-human primates but seen in elderly humans.
[b]Observed in mouse only.
Sources: Khan et al., 1998b; Whelton, 1999; Clive and Stoff, 1984; Delmas, 1995; Murray and Brater, 1993; Komhoff et al., 1997, 2000b; Nantel et al., 1999; Stanfield et al., 2001; Harris et al., 1994; Ferguson et al., 1999.

The most frequent renal effects of NSAIDs in humans are functional, and include interference (generally mild and transient) with fluid and electrolyte homeostasis. The less frequent, but biologically significant, renal effects of NSAIDs include acute renal failure, interstitial nephritis, renal papillary necrosis, nephrotic syndrome, interference with hypertensive and diuretic therapy, allergic type interstitial nephritis with peripheral eosinophilia, eosinophiluria in humans, and outer cortical atrophy with interstitial fibrosis in human infants. In contrast to a wide range of human renal effects associated with COX inhibition, only a few have been seen in animals, including antinatriuresis (sodium retention), decrements in RBF, renal papillary necrosis, and subcapsular cortical atrophy (Bennett et al., 1996; Khan et al., 1997, 1999, 1998b, 2002; Whelton and Hamilton, 1991).

Recently, selective COX-2 inhibitors were introduced into the clinical medicine, with an emphasis on improved safety linked to the potential sparing of COX-1. This has helped understanding of some of the renal effects that could be linked to COX-1 or COX-2 inhibition. Celecoxib was the first selective COX-2 inhibitor approved for clinical use. This chapter's authors have summarized published preclinical and clinical data with celecoxib and have compared the results with similar information from other selective COX-2 inhibitors and nonselective NSAIDs where available. Dogs are extremely sensitive to the gastrointestinal toxicity of COX-1 inhibition; therefore, long-term toxicity studies with nonselective NSAIDs in this species were limited. However, the dog was the main nonrodent species in the toxicity assessments for all recent selective COX-2 inhibitors.

Renal Effect of COX-2 Inhibition in Animals

The majority of selective COX-2 inhibitors and nonselective NSAIDs are for chronic use, therefore preclinical safety studies with these drugs were conducted for up to 6 months in rats and up to 12 months in dogs or monkeys. In celecoxib repeated-dose toxicity studies, rats were dosed for up to 6 months at dosages ranging from 20 to 600 mg/kg/day, and dogs for 12 months at dosages ranging from 25 to 250 mg/kg/day. The plasma exposures of celecoxib in these studies were up to 20-fold above those associated with the maximum clinical dosages. Renal effects typically seen with nonselective NSAIDs are discussed below, along with celecoxib data.

Antinatriuresis

Studies included in the preclinical safety database for celecoxib reported transient sodium retention (up to 60% compared with the concurrent control group), termed antinatriuresis, in rats up through 6 weeks of treatment (Khan et al., 2002). No antinatriuretic effects were seen in the 26-week rat study or the 52-week dog study. Antinatriuresis is also a consequence of use of other selective COX-2 inhibitors and NSAIDs, both in animals and in humans (Whelton and Hamilton, 1991). For example, dogs treated with naproxen sodium (5 to 30 mg/kg/day) develop marked (60 to 93%) reductions in sodium excretion within a few days of treatment (Table 27.6).

Natriuresis, or sodium excretion, is stimulated by increases in renal interstitial volume and is PG-dependent (Haas and Knox, 1996). COX-1 is highly expressed in the collecting ducts of both laboratory animals and humans, an area of the nephron that is active in the regulation of sodium

Table 27.6 Renal Effects of Naproxen Sodium in Dogs after Daily Treatment for 14 or 28 Days

	Day 14		*Day 28*	
	Control	*Naproxen*	*Control*	*Naproxen*
Urine output (ml/4.5 hours)	159 ± 69.5	65 ± 35.9	153 ± 71	37 ± 9.5
Inulin clearance (ml/min/kg)	4.36 ± 0.527	4.28 ± 0.89	5.79 ± 2.69	2.47 ± 1.22
RBF (ml/min/kg)	10.12 ± 3.63	9.70 ± 2.8	10.40 ± 2.9	5.6 ± 3.4
RPF (ml/min/kg)	6.47 ± 2.5	6.13 ± 1.9	6.47 ± 1.8	4.51 ± 2.9
Sodium excretion (mmol/4.5 hrs)	29.5 ± 13.71	11.8 ± 8.4	33.9 ± 15.12	2.4 ± 2.38
Potassium excretion (mmol/4.5 hrs)	11.3 ± 5.17	9.3 ± 3.28	9.1 ± 1.7	10.8 ± 6.7
Chloride excretion (mmol/4.5 hrs)	19.9 ± 11.21	8.3 ± 7.9	22.8 ± 14.3	1.5 ± 0.0
Urinary 6-keto-$PGF_{1\alpha}$ (pg/ml)	—	—	454.0 ± 234.0	21.4 ± 4.54

Naproxen sodium was given at 5 mg/kg/day for first 2 weeks followed by 30 mg/kg/day for another 2 weeks. Values represent means ± standard deviations.

excretion (Currie and Needleman, 1984). In addition, COX-2 is colocalized with Na^+/K^+-ATPase in the thick ascending limb of the rat nephron, further suggesting a role in sodium regulation in that species (Vio et al., 1997). Antinatriuretic effects with celecoxib occurred at systemic exposures within the index of COX-2 selectivity (Khan et al., 2002). PGE_2 concentrations in renal tissue, but not urine, were concomitantly decreased at these exposures, suggesting that the antinatriuretic effect might be related partly to the inhibition of COX-2-dependent PGs. The lack of effect of celecoxib on basal urine PGE_2 concentrations may represent sparing of a COX-1-dependent PG pool. Gross and colleagues (1999) have shown a sparing of sodium excretion with COX-2-selective inhibitors during forced diuresis.

Other studies suggest that COX-2 is upregulated in response to increased salt load and volume overload. Yang and colleagues (1998) characterized the COX-2 expression in the kidneys of rats subjected to either high or low salt dietary intake for 7 days. COX-2 mRNA was increased up to 4.5-fold in the inner medulla by high salt diet, whereas no change in COX-2 mRNA was observed with the low salt diet. In contrast, cortical COX-2 mRNA was altered by both high and low salt dietary regimens: a 2.9-fold decrease was observed with the high salt diet, whereas a 3.3-fold increase was observed with the low salt diet. COX-1 mRNA was not affected in any region of the kidney with either diet. The expression of COX-2 was localized primarily in the macula densa and cortical thick ascending limb, and medullary interstitial cells. The authors suggest that the increased expression of COX-2 in the medullary interstitium may promote excretion of excess sodium in response to increased pressure (i.e., stretch) and volume overload in the medullary interstitium.

The upregulation of COX-2 expression observed in the cortex under conditions of low salt intake was noted by Yang and his colleagues (1998) to be consistent with a role for COX-2 in the preservation of cortical circulation in the volume-contracted state. Clearly, this study's results illustrate the potential for multiplicity of PG function and regulation in various regions of the kidney. The reciprocal decrease in COX-2 expression in the cortex provides evidence that COX-2 is regulated differentially in specific regions of the kidney in response to changes in functional demands on the kidney.

Changes in Renal Blood Flow

There was no evidence of changes in renal function, including RBF and GFR, in repeated-dose rat and dog studies of celecoxib. In contrast, naproxen sodium (5 to 30 mg/kg/day) when given to dogs resulted in marked alterations in renal functions, including decrements in renal blood flow (Table 27.6), indicating that inhibition of COX-1 or simultaneous inhibition of both COX isoforms may be needed to produce these effects. In addition to decreases in RBF, inulin clearance, urine output, and sodium and chloride excretion were decreased in naproxen treated dogs. These alterations corresponded with marked reductions in urinary levels of 6-keto-$PGF_{1\alpha}$, a marker for renal prostaglandin production.

A few published studies tried to elucidate the roles of COX isoforms in RBF in normal animals. For example, Brooks and colleagues (1998) studied the effects of indomethacin, a potent nonselective NSAID, 6-MNA (the active metabolite of nabumetone), and celecoxib on RBF in normal conscious dogs following intravenous administration. Only celecoxib was reported to decrease RBF, which was attributed to its COX-2-selective action. The data for this study, however, raise some questions. Failure to control for vehicle effect (for the organic solvent used for intravascular delivery of celecoxib) and the unexplained lack of effect with the positive control (indomethacin) limit the conclusions regarding pharmacological inhibition of COX.

In contrast, SC-046, a COX-2-selective inhibitor closely related to celecoxib, given orally to normal conscious dogs produced no diminution of renal blood flow at doses that produced plasma concentrations adequate for maximal and selective inhibition of COX-2 (Freeman and Venturini, 1999). These results were corroborated by a study using another COX-2-selective agent, nimesulide (Rodriguez et al., 2000). COX-2 selective doses of this agent, confirmed by the absence of an effect on platelet aggregation, were found to produce no effect on mean arterial blood pressure, glomerular filtration, or RBF in normal dogs. Decreases in sodium excretion, urine volume, and lithium excretion were observed with nimesulide, which were considered as tubular effects. Collectively, these two studies suggest that COX-2 does not govern basal RBF in laboratory animals.

Atrophy of Subcapsular Cortex

No renal cortical lesions were seen in rats or dogs given celecoxib. Dogs given NSAIDs, such as naproxen, at doses that are tolerated over long periods develop a progressive tubular atrophy and interstitial fibrosis that is primarily limited to the outer renal cortex (Silverman and Khan, 1999). These typically mild lesions are associated with modest elevations in blood urea nitrogen and decreases in RBF, GFR, and urinary PGs (Table 27.6) without remarkable changes in serum creatinine concentration, and are consistent with a mild, progressive ischemic etiology. The restriction of these changes to the outer cortex differentiates them from both acute ischemic insult, characterized by acute tubular necrosis throughout the cortex, and wedge-shaped focal ischemic cortical necrosis resulting from obstruction of the arcuate artery. Outer cortical lesions resembling those observed in dogs have not been observed in adult humans following treatment with NSAIDs. This may reflect interspecies differences in renal anatomy or the expression and distribution of COX isoforms, especially COX-2 (Khan et al., 1998b); however, no such morphologic changes were seen in preclinical studies with celecoxib.

Similar outer cortical lesions are seen in COX-2 knockout mice (in which the gene for COX-2 has been deleted) and neonatal rat kidneys exposed to selective COX-2 inhibitors, and in human fetal kidneys exposed to nonselective NSAIDs during the third trimester (Khan et al., 2002; Matson et al., 1981; van der Heijden et al., 1994). The functional significance of COX-2 in renal development in rodents is evident from studies in COX-2 knockout mice (Dinchuk et al., 1995; Morham et al., 1995). These animals are born with apparently normal kidneys expected for the neonatal mice, but during early postnatal periods they develop marked morphologic alterations characterized by immature glomeruli, tubular atrophy, basophilic tubular epithelium, proteinuria, and interstitial fibrosis. Eventually, most of these mice die of renal failure. Komhoff and colleagues (2000b) have also demonstrated that inhibition of COX-2 in the neonatal rat can impair glomerulonephrogenesis and produce renal cortical damage. However, it should be noted that in contrast with humans, where the kidney is fully developed at birth, renal maturation in rats and dogs continues for another 2 to 4 weeks after birth; therefore, extrapolation of the rodent data to human neonates is questionable.

Renal Papillary Necrosis

Celecoxib did not produce renal papillary necrosis (RPN) in preclinical repeated-dose studies in rats or dogs (Khan et al., 1997, 2002). Similarly, RPN was not a major manifestation for other selective COX-2 inhibitors. In contrast, NSAIDs readily produce RPN in dogs and rats (Figure 27.9 and Figure 27.10); these species appear sensitive to renal papillary injury for a variety of reasons related to papillary structure and tortuosity of renal blood flow (Alden and Khan, 2002). This form of nephrotoxicity can be seen in these species even after acute exposures with subtherapeutic doses of NSAIDs (Table 27.7), including indomethacin, naproxen, nabumetone, and meloxicam (Bach and Hardy, 1985; Khan et al., 1998a). In contrast, celecoxib, when given chronically to dogs for up to 12 months and at dosages as high 16-fold above the therapeutic efficacy, did not produce RPN. RPN with nonselective NSAIDs occurred at dosages greater than those that produced gastrointestinal toxicity, indicating that kidney is relatively less susceptible to COX inhibition than the gastrointestinal tract.

RPN caused by NSAID use in humans is rare, and usually is associated with the simultaneous presence of other risk factors, e.g., age,

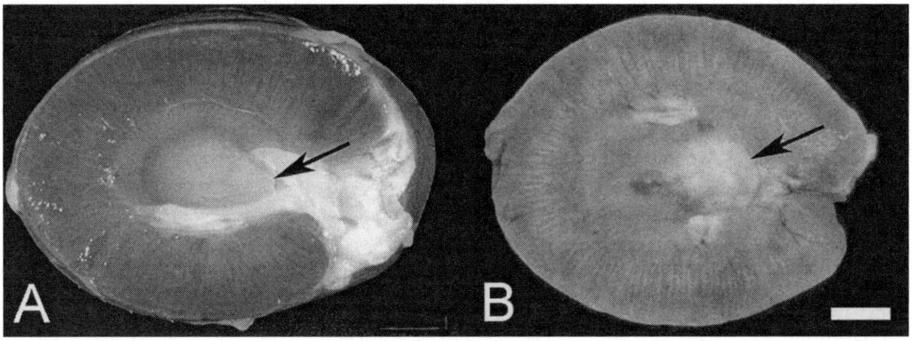

Figure 27.9 Cross sections of kidneys from dogs given (A) naproxen or (B) meloxicam. Note pale white discoloration (arrows) extending from tip of papilla into the inner medulla, representing moderate to marked renal papillary necrosis. Bar = 5 mm.

Figure 27.10 Microscopic section of kidney from a rat treated with indomethacin. Note marked renal papillary necrosis involving approximately two-thirds of the papilla. Hematoxylin and eosin stain. Bar = 300 μm.

Table 27.7 Incidence of Renal Papillary Necrosis in Dogs Given NSAIDs vs. Celecoxib

	Meloxicam			*Nabumetone*		*Naproxen*		*Celecoxib*
Dose (mg/kg/day)	0.1	1.0	3.0	300	600	10	50	35
Therapeutic exposures (fold)*	1	3	7	< 1	< 1	< 1	7	16
RPN incidence (%)	0	50	50	25	50	0	50	0
Gastrointestinal toxicity (%)	0	50	100	75	100	100	100	

Note: There were 4, 4, 6, and 8 dogs in meloxicam, nabumetone, naproxen, and celecoxib treatment groups, respectively.
*Therapeutic exposures based on efficacy in adjuvant arthritis rat model.

atherosclerotic heart disease, congestive heart failure, concomitant use of diuretics, dehydration, or analgesic (acetaminophen and phenacetin) abuse (Bennett et al., 1996; Clive and Stoff, 1984; Delmas, 1995; Murray and Brater, 1993; Sabatini, 1988).

The mechanisms of renal papillary necrosis are poorly understood despite the appearance of this toxicity across a broad range of chemicals (Bach and Hardy, 1985). It is speculated that the papillae of rats and dogs are susceptible to chemical insult due to sluggish regional blood flow, which predisposes to ischemia and accumulation of toxic substances. Therefore, local inhibition of papillary PG synthesis and disruption of regional blood flow may play a role in NSAID-induced renal papillary necrosis. NSAID-induced renal papillary necrosis is initially manifested as morphological changes of loss of extracellular matrix and degeneration of the interstitial cells of the papilla; thus, these cells are likely to be the early targets of NSAIDs (Burrell et al., 1991).

Since interstitial cells of the renal papilla in rats and dogs show approximately equal expression of both COX-1 and COX-2, it might be difficult to correlate the papillary injury seen with nonselective NSAIDs with inhibition of a specific COX isoform. However, RPN was not observed in preclinical toxicology studies with celecoxib (Khan et al., 1997) while it was readily produced by nonselective NSAIDs (meloxicam and nabumetone). After 52 weeks of dosing, total celecoxib concentrations measured in the papilla of rats were approximately 1000-fold the IC_{50} for COX-2 (40 n*M*) and threefold the IC_{50} for COX-1 (15 *M*). Estimates of free concentrations of this highly bound drug would support the conclusion that papillary COX-2 was maximally inhibited while any inhibition of COX-1 was submaximal. Therefore, it may be inferred that COX-2 inhibition alone is not a primary target in NSAID-induced interstitial cell death.

While RPN seen with NSAIDs is a rare clinical event, the failure to produce this toxicity with COX-2-selective inhibitors in sentinel species suggests that these drugs may offer some increment in renal safety over nonselective NSAIDs. Assuming that complete ablation of papillary COX activity is required for renal papillary necrosis, there may be little risk for renal papillary necrosis with COX-2-selective drugs since COX-1 is the principal isoform in the human papilla (Khan et al., 1998b; Komhoff et al., 1997).

Clinical Studies with Selective COX-2 Inhibitors

Because both COX-1 and COX-2 probably play some role in renal function, clinical differentiation of COX-2-specific inhibitors from conventional nonselective NSAIDs may not be as clear in the kidney (Khan et al., 2002). This possibility is underscored by studies with celecoxib and rofecoxib in healthy elderly subjects (Catella-Lawson et al., 1999; helton, 1999). Mild and transient sodium retention that reproduced the effect seen with naproxen and indomethacin (the comparator drugs in these studies) was observed with both COX-2-specific drugs. Both studies suggest that sodium excretion in humans can be regulated by COX-2. However, neither COX-2-specific drug affected glomerular filtration, in contrast with the decreases seen with both nonselective NSAIDs. It can therefore be inferred that COX-1 governs basal glomerular filtration in healthy individuals (i.e., those without the risk factors previously discussed). COX-2-specific inhibitors thus produce some mild disturbance of sodium excretion that is characteristic of conventional NSAIDs, but unlike NSAIDs, these drugs spare glomerular filtration. The latter represents a clinical advantage for COX-2-specific agents in the treatment of elderly patients in which glomerular filtration is already deteriorating.

These conclusions were confirmed for celecoxib by the results of a 6-month safety study in arthritis patients (Silverstein et al., 2000). This study, termed the Celecoxib Long-term Arthritis Safety Study (CLASS), was conducted to compare the incidence of ulcer complications in patients receiving either celecoxib or conventional NSAIDs. The study design included 4000 patients receiving celecoxib at a dose of 400 mg bid (twice the maximum arthritis dose), and 2000 patients who received standard therapeutic doses of either diclofenac (75 mg bid) or ibuprofen (800 mg tid). Although the primary purpose was to collect data on endpoints related to gastrointestinal safety, the study broadly included evaluations of cardiovascular, hemostasis, hepatic, and renal function.

The incidence of any renal adverse event in celecoxib-treated patients was similar to patients receiving diclofenac, and less than the incidence in patients receiving ibuprofen. The incidence of withdrawals due to adverse renal events was small and similar across all treatment groups. Low incidences of emergent hypertension, aggravated hypertension, and edema (generalized and peripheral) were seen with celecoxib and diclofenac. Patients receiving ibuprofen, however, exhibited significantly higher incidences of emergent hypertension, generalized edema, and peripheral edema as compared with the celecoxib group. A higher percentage of patients in both of the nonselective NSAID groups exhibited laboratory values that were suggestive of diminished RBF (i.e., blood urea nitrogen >40 mg/dl and/or creatinine >2 mg/dl) as compared with the celecoxib group. Collectively, these results show that celecoxib has some advantage in renal safety over nonselective NSAIDs, based on its COX-1-sparing action.

Taken together, the available animal and human data suggest that selective COX-2 inhibitors have superior renal safety profiles compared with nonselective NSAIDs. Extensive renal evaluations of celecoxib in animals suggest that this drug may affect natriuresis, but is devoid of many renal effects seen with other nonselective NSAIDs in animals.

REFERENCES

Agarwal AK, and Rao SS (1994) Susceptibility of ageing kidney to quinidine. *Pharmacol Res* 29:145–154.

Alden CL, and Khan KNM (2002) Kidney, in *Handbook of Toxicologic Pathology*, Haschek WM, Rousseaux CG, and Wallig MA (Eds), Academic Press, New York, pp. 255–330.

Bach PH, and Hardy TL (1985) Relevance of animal models to analgesic-associated renal papillary necrosis in humans. *Kidney Int* 28:605–613.

Beierschmitt WP, Keenan KP, and Weiner M (1986) Age-related increased susceptibility of male Fischer 344 rats to acetaminophen nephrotoxicity. *Life Sci* 39:2335–2342.

Bennett WM, Henrich WL, and Stoff JS (1996) The renal effects of nonsteroidal anti-inflammatory drugs: Summary and recommendations. *Am J Kidney Dis* 28:S56–S62.

Berndt WO, and Hayes AW (1982) The effect of probenecid on citrinin-induced nephrotoxicity. *Toxicol Appl Pharmacol* 64:118–124.

Besseghir K, and Roch-Ramel F (1987) Renal excretion of drugs and other xenobiotics. *Ren Physiol* 10:221–241.

Breyer MD, Zhang Y, Guan YF, Hao CM, Hebert RL, and Breyer RM (1998) Regulation of renal function by prostaglandin E receptors. *Kidney Int* 54, Suppl. 67:S-88–S-94.

Bricker NS, Zea L, Shapiro M, Sanclemente E, and Shankel S (1993) Biologic and physical characteristics of the non-peptidic, non-digitalis-like natriuretic hormone. *Kidney Int* 44:937–947.

Brooks DP, DePalma PD, Pullen M, Elliott JD, Ohlstein EH, and Nambi P (1998) SB 234551, A novel endothelin – A receptor antagonist, unmasks endothelin-induced renal vasodilatation in the dog. *J Cardiovasc Pharmacol* 31 Suppl 1:S339–S341.

Burrell JH, Yong JL, and MacDonald GJ (1991) Irreversible damage to the medullary interstitium in experimental analgesic nephropathy in F344 rats. *J Pathol* 164:329–338.

Catella-Lawson F, McAdam B, Morrsion BW, Kapoor S, Kujubu D, Antes L, Lasseter K, Quan H, Gertz B, and Fitzgerald G (1999) Effects of specific inhibition of cyclooxygenase-2 on sodium balance, hemodynamics, and vasoactive eicosanoids. *J Pharmacol Exp Ther* 289:735–741.

Clive DM, and Stoff JS (1984) Renal syndromes associated with nonsteroidal antiinflammatory drugs. *N Engl J Med* 310:563–572.

Conger JD (1981) Acute acid nephropathy. Semin Nephrol 1:69–74.

Corman BJ, and Owen RA (1992) Normal development, growth and aging of the kidney, in *Pathobiology of Aging Rat,* Mohr U, Dungworth DL, and Capen CC (Eds), ILSI Press, Washington, D.C., pp. 195–209.

Cotran RS, Kumar V, and Robbins SL (1994) The Kidney, in *Robbins Pathologic Basis of Disease*, Cotran RS, Kumar V, and Robbins SL (Eds), WB Saunders Co., Philadelphia, pp. 927–990.

Cronin RE, and Henrich WL (1998) Toxic Nephropathy, in *Brenner Rector's The Kidney* Brenner BM, and Rector FC (Eds), WB Saunders Company. Philadelphia, pp. 1680–1711.

Curhan GC, and Zeidel ML (1998) Urinary Tract Obstruction, in *Brenner and Rector's The Kidney,* Brenner BM, and Rector FC (Eds), WB Saunders Company, Philadelphia, pp. 1936–1958.

Currie MG, and Needleman P (1984) Renal arachidonic acid metabolism. *Annu Rev Physiol* 46:327–341.

Dal Canton A, Corradi A, Stanziale R, Maruccio G, and Migone L (1980) Glomerular hemodynamics before and after release of 24-hour bilateral ureteral obstruction. *Kidney Int* 17:491–496.

Dal Canton A, Stanziale R, Corradi A, Andreucci VE, and Migone L (1977) Effects of acute ureteral obstruction on glomerular hemodynamics in rat kidney. *Kidney Int* 12:403–411.

de Bruijn WC, Boeve ER, van Run PR, van Miert PP, Romijn JC, Verkoelen CF, Cao LC, and Schroder FH (1994) Etiology of experimental calcium oxalate monohydrate nephrolithiasis in rats. *Scanning Microsc* 8:541–549.

Delmas PD (1995) Non-steroidal anti-inflammatory drugs and renal function. *Br J Rheumatol* 34 Suppl 1:25–28.

Dinchuk JE, Car BD, Focht RJ, Johnston JJ, Jaffee BD, Covington MB, Contel NR, Eng VM, Collins RJ, and Czerniak PM (1995) Renal abnormalities and an altered inflammatory response in mice lacking cyclooxygenase II. *Nature* 378:406–409.

Ferguson S, Herbert RL, and Laneuville O (1999) NS-398 upregulates constitutive cyclooxygenase-2 expressionin the M-1 cortical collecting duct cell line. *A Am Soc Nephrol* 10:2261–2271.

Freeman RH, and Venturini CM (1999) Renal function in normal conscious dogs during cyclooxygenase-2 blockade. *FASEB J* 13:A722.

Garber SL, Salmassi J, Arruda JA, and Dunea G (1995) Xanthopterin-induced renal dysfunction: A reversible model of crystal nephropathy. *Nephron* 69:71–78.

Goldstein RS, Pasino DA, and Hook JB (1986) Cephaloridine nephrotoxicity in aging male Fischer-344 rats. *Toxicology* 38:43–53.

Gross JM, Dwyer JE, and Knox FG (1999) Natriuretic response to increased pressure is preserved with COX-2 inhibitors. *Hypertension* 34:1163–1167.

Haas JA, and Knox FG (1996) Effect of meclofenamate or ketoconazole on the natriuretic response to increased pressure. *J Lab Clin Med* 128:202–207.

Harris RC, McKanna JA, Akai Y, Jacobson HR, Dubois RN, and Breyer MD (1994) Cyclooxygenase-2 is associated with the macula densa of rat kidney and increases with salt restriction. *J Clin Invest* 94:2504–2510.

Harris RH, and Yarger WE (1974) Renal function after release of unilateral ureteral obstruction in rats. *Am J Physiol* 227:806–815.

Khan KN, Alden CL, Gleissner SE, Gessford MK, and Maziasz TJ (1998a) Effect of papillotoxic agents on expression of cyclooxygenase isoforms in the rat kidney. *Toxicol Pathol* 26:137–142.

Khan KN, Paulson S, Seibert K, and Maziasz T (1997) Gastrointestinal and renal safety of a specific COX-2 inhibitor versus NSAIDs in dogs. *Vet Pathol* 34:509.

Khan KN, Paulson SK, Verburg KM, Lefkowith JB, and Maziasz TJ (2002) Pharmacology of cyclooxygenase-2 inhibition in the kidney. *Kidney Int* 61:1210–1219.

Khan KN, Venturini CM, Bunch RT, Brassard JA, Koki AT, Morris DL, Trump BF, Maziasz TJ, and Alden CL (1998b) Interspecies differences in renal localization of cyclooxygenase isoforms: Implications in nonsteroidal antiinflammatory drug-related nephrotoxicity. *Toxicol Pathol* 26:612–620.

Khan KNM, Silverman LR, and Baron DA (1999) Cyclooxygenase-2 (COX-2) expression in the developing dog kidney. *Vet Pathol* 35:486.

Khan SR, and Glenton PA (1995) Deposition of calcium phosphate and calcium oxalate crystals in the kidneys. *J Urol* 153:811–817.

Klahr S (2000) Obstructive nephropathy. *Intern Med* 39:355–361.

Komhoff M, Grone HJ, Klein T, Seyberth HW, and Nusing RM (1997) Localization of cyclooxygenase-1 and -2 in adult and fetal human kidney: Implication for renal function. *Am J Physiol* 272:F460–F468.

Komhoff M, Jeck ND, Seyberth HW, Grone HJ, Nusing RM, and Breyer MD (2000a) Cyclooxygenase-2 expression is associated with the renal macula densa of patients with Bartter-like syndrome. *Kidney Int* 58:2420–2424.

Komhoff M, Wang JL, Cheng HF, Langenbach R, McKanna JA, Harris RC, and Breyer MD (2000b) Cyclooxygenase-2-selective inhibitors impair glomerulogenesis and renal cortical development. *Kidney Int* 57:414–422.

Kyle ME, and Kocsis JJ (1985) The effect of age on salicylate-induced nephrotoxicity in male rats. *Toxicol Appl Pharmacol* 81:337–347.

Levin S, Friedman RM, Cortez E, Hribar J, Nicholas M, Schlessinger S, Fouant M, and Khan N (1999) Lesions and identification of crystalline precipitates of glycoprotein IIb-IIIa antagonists in the rat kidney. *Toxicol Pathol* 27:38–43.

Masferrer JL, Zweifel BS, Manning PT, Hauser SD, Leahy KM, Smith WG, Isakson PC, and Seibert K (1994) Selective inhibition of inducible cyclooxygenase 2 *in vivo* is antiinflammatory and nonulcerogenic. *Proc Natl Acad Sci USA* 91:3228–3232.

Matson JR, Stokes JB, and Robillard JE (1981) Effects of inhibition of prostaglandin synthesis on fetal renal function. *Kidney Int* 20:621–627.

McMartin DN, and Engel SG (1982) Effect of aging on gentamicin nephrotoxicity and pharmacokinetics in rats. *Res Commun Chem Pathol Pharmacol* 38:193–207.

Molina JM, Belenfant X, Doco-Lecompte T, Idatte JM, and Modai J (1991) Sulfadiazine-induced crystalluria in AIDS patients with toxoplasma encephalitis. *AIDS* 5:587–589.

Morham SG, Langenbach R, Loftin CD, Tiano HF, Vouloumanos N, Jennette JC, Mahler JF, Kluckman KD, Ledford A, and Lee CA (1995) Prostaglandin synthase 2 gene disruption causes severe renal pathology in the mouse. *Cell* 83:473–482.

Murdock MI, Olsson CA, Sax DS, and Krane RJ (1975) Effects of levodopa on the bladder outlet. *J Urol* 113:803–805.

Murray MD, and Brater DC (1993) Renal toxicity of the nonsteroidal anti-inflammatory drugs. *Annu Rev Pharmacol Toxicol* 33:435–465.

Nantel F, Meadows E, Denis D, Connolly B, Metters KM, and Giaid A (1999) Immunolocalization of cyclooxygenase-2 in the macula densa of human elderly. *FEBS Lett* 457:475–477.

Needleman P, and Isakson PC (1997) The discovery and function of COX-2. *J Rheumatol* 24 Suppl 49:6–8.

Novicki DE, and Willscher MK (1979) Case profile: Anticholinergic-induced hydronephrosis. *Urology* 13:324–325.

Rodriguez F, Llinas MT, Gonzalez JD, Rivera J, and Salazar FJ (2000) Renal changes induced by a cyclooxygenase-2 inhibitor during normal and low sodium intake. *Hypertension* 36:276–281.

Ross DA, and Gale GR (1979) Reduction of the renal toxicity of cis-dichlorodiammine-platinum(II) by probenecid. *Cancer Treat Rep* 63:781–787.

Sabatini S (1988) Analgesic-induced papillary necrosis. *Semin Nephrol* 8:41–54.

Sawyer MH, Webb DE, Balow JE, and Straus SE (1988) Acyclovir-induced renal failure. Clinical course and histology. *Am J Med* 84:1067–1071.

Schneider A, Harendza S, Zahner G, Jocks T, Wenzel U, Wolf G, Thaiss F, Helmchen U, and Stahl RA (1999) Cyclooxygenase metabolites mediate glomerular monocyte chemoattractant protein-1 formation and monocyte recruitment in experimental glomerulonephritis. *Kidney Int* 5:430–441.

Silverman LR, and Khan KN (1999) Nonsteroidal anti-inflammatory drug-induced renal papillary necrosis in a dog. *Toxicol Pathol* 27:244–245.

Silverstein FE, Faich G, Goldstein JL, Simon LS, Pincus T, Whelton A, Makuch R, Eisen G, Agrawal NM, Stenson WF, Burr AM, Zhao WW, Kent JD, Lefkowith JB, Verburg KM, and Geis GS (2000) Gastrointestinal toxicity with celecoxib vs nonsteroidal anti-inflammatory drugs for osteoarthritis and rheumatoid arthritis: The CLASS study: A randomized controlled trial. Celecoxib Long-term Arthritis Safety Study. *JAMA* 284:1247–1255.

Simon DI, Brosius FC III, and Rothstein DM (1990) Sulfadiazine crystalluria revisited. The treatment of Toxoplasma encephalitis in patients with acquired immunodeficiency syndrome. *Arch Intern Med* 150:2379–2384.

Stanfield KS, Kelce WR, Lisowski A, English ML, Baron DA, and Khan KNM (2000) Expression of cyclooxygenase-2 in the embryonic and fetal tissues during organogenesis and late pregnancy. *Vet Pathol* 37:558.

Stanfield KS, Khan KNM, and Gralinski MR (2001) Localization of cyclooxygenase isozymes in cardiovascular tissues of dogs treated with naproxen. *Vet Immuno* 80:309–314.

Tune BM, and Fravert D (1980) Mechanisms of cephalosporin nephrotoxicity: A comparison of cephaloridine and cephaloglycin. *Kidney Int* 18:591–600.

van der Heijden BJ, Carlus C, Narcy F, Bavoux F, Delezoide AL, and Gubler MC (1994) Persistent anuria, neonatal death, and renal microcystic lesions after prenatal exposure to indomethacin. *Am J Obstet Gynecol* 171:617–623.

Vio CP, Cespedes C, Gallardo P, and Masferrer JL (1997) Renal identification of cyclooxygenase-2 in a subset of thick ascending limb cells. *Hypertension* 30:687–692.

Whelton A (1999) Renal safety and tolerability of celecoxib, a novel cyclooxygenase-2 inhibitor. *Am J Therapeutics* 7:159–175.

Whelton A, and Hamilton CW (1991) Nonsteroidal anti-inflammatory drugs: Effects on kidney function. *J Clin Pharmacol* 31:588–598.

Yang T, Singh I, Pham H, Sun D, Smart A, Schnermann JB, and Briggs JP (1998) Regulation of cyclooxygenase expression in the kidney by dietary salt intake. *Am J Physiol* 274:F481–F489.

Yarger WE, and Griffith LD (1974) Intrarenal hemodynamics following chronic unilateral ureteral obstruction in the dog. *Am J Physiol* 227:816–826.

Index

A

B

C

D

E

H

I

J

K

L

M

P

Q

R

S

T

U

V